DIGITAL COMMUNICATION

Third Edition

DIGITAL COMMUNICATION

Third Edition

John R. Barry
Georgia Institute of Technology

Edward A. Lee
University of California at Berkeley

David G. Messerschmitt
University of California at Berkeley

Volume I

SPRINGER SCIENCE+BUSINESS MEDIA, LLC

Library of Congress Cataloging-in-Publication Data

Barry, John R., 1963-
Digital Communication/ John R. Barry, Edward A. Lee, and David G. Messerschmitt. --3rd ed.
p.cm.
Rev. ed. of: Digital Communication / Edward A. Lee, David G. Messerschmitt. 2nd ed. c 1994.
Includes bibliographical references and index.
ISBN 978-0-7923-7548-7 ISBN 978-1-4615-0227-2 (eBook)
DOI 10.1007/978-1-4615-0227-2
1. Digital Communications. I. Lee, Edward A., 1957- II. Messerschmitt, David G. III. Lee, Edward A., 1957- Digital Communication. IV. Title.

TK5103.7.L44 2003
621.382—dc22

2003054667

Printed on acid-free paper.

To Annis, Reid, and Neil
Rhonda, Helen, and Katalina
Dody and Laura

CONTENTS

Preface

This book concerns digital communication. Specifically, we treat the transport of bit streams from one geographical location to another over various physical media, such as wire pairs, coaxial cable, optical fiber, and radio. We also treat multiple-access channels, where there are potentially multiple transmitters and receivers sharing a common medium.

Ten years have elapsed since the Second Edition, and there have been remarkable advances in wireless communication, including cellular telephony and wireless local-area networks. This Third Edition expands treatment of communication theories underlying wireless, and especially advanced techniques involving multiple antennas, which turn the traditional single-input single-output channel into a multiple-input multiple-output (MIMO) channel. This is more than a trivial advance, as it stimulates many advanced techniques such as adaptive antennas and coding techniques that take advantage of space as well as time. This is reflected in the addition of two new chapters, one on the theory of MIMO channels, and the other on diversity techniques for mitigating fading. The field of error-control coding has similarly undergone tremendous changes in the past decade, brought on by the invention of turbo codes in 1993 and the subsequent rediscovery of Gallager's low-density parity-check codes. Our treatment of error-control coding has been rewritten to reflect the current state of the art. Other materials have been reorganized and reworked, and three chapters from the previous edition have been moved to the book's Web site to make room. For this third edition we have added a third author, John Barry, who carried the major burden of these revisions.

The general approach of this book is to extract the common principles underlying a range of media and applications and present them in a unified framework. It is relevant to the design of a variety of systems, including voice and video digital cellular telephone, digital CATV distribution, wireless LANs, digital subscriber loop, metallic ethernet, voiceband data modems, and satellite communication systems.

This book is intended for *designers* and *would-be designers* of digital communication systems. To limit the length we have been selective in topics covered and in the depth of coverage. For example, the coverage of advanced information, coding, and detection theory is limited to those aspects directly relevant to the design of digital communication systems. This emphasis on topics important to designers results in more detailed treatment of some topics than is traditional in academic textbooks, for example in our coverage of synchronization (timing and carrier recovery).

This book is suitable as a first-year graduate textbook, and should also be of interest to many industry professionals. We have attempted to make the book attractive to both audiences through the inclusion of many practical examples and a practical flavor in the choice of topics. In addition, we have increased the readability by relegating many of the more detailed derivations to appendices and exercise solutions, both of which are included in the book.

The book has a Web site at http://www.ece.gatech.edu/~barry/digital/, where the reader may find the chapters "Physical Media and Channels," "Spectrum Control," and "Echo Cancellation" from the Second Edition, other supplementary materials, useful links, a problem solutions manual, and errata.

For this third edition, we owe a debt of gratitude to Abdallah Said Al-Ahmari, Anuj Batra, Richard T. Causey, Elizabeth Chesnutt, Arumugam Kannan, Piya Kovintavewat, Renato da Rocha Lopes, Aravind R. Nayak, Faith Nilo, Joon Hyun Sung, Badri Varadarajan, Deric Waters, and to several anonymous reviewers. In addition we gratefully acknowledge the many people who helped shape the first two editions. Any remaining errors are the full responsibility of the authors.

We hope the result is a readable and useful book, and always appreciate comments and suggestions from the readers.

John R. Barry
Edward A. Lee
David G. Messerschmitt

Atlanta, Georgia
Berkeley, California
June 10, 2003

Changes from the Second Edition

The Third Edition differs from the Second Edition in three significant ways. First, chapters 6 through 9 from the Second Edition have been reorganized and streamlined to highlight pulse-amplitude modulation, becoming the new chapters 5 through 7. Second, new material on recent advances in wireless communications, error-control coding, and multiuser communications has been added. As a result, two new chapters have been added. Finally, the three chapters "Physical Media and Channels," "Spectrum Control," and "Echo Cancellation" from the Second Edition have been moved to the book Web site.

Here is a chapter-by-chapter summary of the major changes for this Third Edition:

Chapter 2. The upconverter, downconverter, and complex envelope are defined in mathematical terms for later use. More details of linear waveform spaces, including orthonormal bases and the Gram-Schmidt procedure, are added.

Chapter 5. This chapter focuses exclusively on pulse-amplitude modulation (PAM), both passband and baseband. It consolidates material that was previously spread across four different chapters. The new organization retains the intuitive flow that characterized the Second Edition, with initial emphasis on the deterministic features of PAM systems. The minimum-distance receiver is proposed, first for an isolated pulse and then for dispersive channels with intersymbol interference, which leads to such concepts as the correlator, matched filter, whitened-matched filter and the Viterbi algorithm. The statistics of the noise do not enter into the picture until the very end, where we analyze the performance of various PAM techniques with minimum-distance detection in the presence of white Gaussian noise. Separating PAM like this offers two key advantages. First, PAM and its derivatives (like multicarrier modulation) are a preferred choice for many emerging applications, making PAM a worthy topic of study in its own right. A practicing engineer may wish to study PAM only, and the new organization makes this easy to do. Second, focusing on PAM allows us to introduce important concepts (such as minimum-distance detection and power-bandwidth trade-offs) in a highly focused and motivated setting. In a sense, we use PAM as a case study, which facilitates the generalization to arbitrary M-ary modulation schemes that follows.

Chapter 6. This chapter moves beyond PAM to examine advanced modulation techniques such as orthogonal modulation and orthogonal pulse-amplitude modulation, which includes code-division multiple access and multicarrier modulation (such as OFDM) as special cases. Also covered are advanced topics such as modulation with memory, the relationship between bandwidth and dimensionality, and the normalized SNR as a means for comparing modulation to Shannon capacity.

Chapter 7. This chapter adopts a probabilistic approach to the detection problem, thus expanding our scope beyond white-Gaussian noise. The Viterbi algorithm is formulated in probabilistic terms, and it is related to the forward-backward (or BCJR) algorithm for a posteriori-probability detection, which plays a key role in turbo decoding.

Chapter 10. This new chapter examines the fundamentals of multiple-input multiple-output communications, with particular emphasis on the detection problem, also known as the multiuser detection problem.

Chapter 11. This new chapter describes diversity techniques for mitigating multipath fading, which exploit antenna arrays at either the transmitter or the receiver (or both). We examine beamforming and combining techniques as well as space-time codes and spatial multiplexing.

Chapter 12. New material on low-density parity-check codes has been added, including Tanner graphs, message-passing decoding, and density evolution. We also describe parallel-concatenated and serial-concatenated turbo codes and turbo-decoding algorithms, with repeat-accumulate codes and turbo equalization treated as special cases.

Notes to an Instructor

This book can be used as a textbook for advanced undergraduates, or for a first course in digital communication for graduate students. We presume a working knowledge of transforms, linear systems, and random processes, and review these topics in chapters 2 and 3 at a depth suitable only for establishing notation. This treatment also serves to delimit the background assumed in the remainder of the book, or with more advanced students can be omitted or made optional. We include a more detailed treatment of basic topics important to digital communication but which may not be familiar to a first-year graduate student, including signal space (chapter 2), Markov chains and their analysis (chapter 3), Poisson processes and shot noise (chapter 3), the basic boundaries of communication from information theory (chapter 4), and maximum-likelihood detection and the Viterbi algorithm (chapter 7). These treatments are self-contained and assume only the basic background mentioned earlier. These basic topics can be covered at the beginning of the course, or delayed until the first time they are used. Our own preference is the latter, since the immediate application of the techniques serves as useful reinforcement.

The core of book is the treatment modulation, detection and equalization (chapters 5 through 9), MIMO channels (chapters 10 and 11), coding (chapters 12 and 13), and synchronization (chapters 14 through 16). These topics are covered in considerable depth. After completing a course based on this book, students should be highly motivated to take advanced courses in information theory, algebraic coding, detection and estimation theory, and communication networks, and will have a prior appreciation of the utility of these topics.

There is sufficient material for two semesters, although it can easily be used for a single-semester course by selectively covering topics. At Georgia Tech we use this book for a one-semester graduate course that has as prerequisites undergraduate courses in systems and transforms and probability and random processes. We do not presume any prior exposure to signal space, Markov chains, or the Poisson process. In this course we rely on the students to review Chapters 1 through 4, and we cover Chapters 5 through 8 and Chapter 10 and 11 in lecture. Chapters 9, 12 and 13 are skipped because adaptive filtering and error-control coding techniques are covered in other courses.

1

Introduction

But let your communication be, Yea, yea; Nay, nay:
for whatever is more than these, cometh of evil.

— The Gospel According to St. Matthew (5:37)

Digital transmission of information has sufficiently overwhelming advantages that it increasingly dominates communication systems, and certainly all new designs. In computer-to-computer communication, the information to be transported is inherently digital. But information that at its source is inherently continuous time (or continuous space) and continuous amplitude, like voice, music, pictures, and video, can be represented, not exactly but accurately, by a collection of bits. Why does it make sense to do so?

- The encoding of analog signals in digital form has benefited from advances in compression algorithms, which reduce dramatically the bit rate while maintaining high perceptual accuracy.
- Signal processing and coding techniques have dramatically increased the bit rate that can be supported by physical channels like wire pairs or radio.
- Integrated circuits make complex signal processing and coding functions cost effective.

- Optical fiber has reduced the cost of transmitting high bit rates over long distances.

As a result of the happy convergence of these elements, virtually all communication is either digital or in the process of being converted to digital. An essential element is the transmission of higher and higher bit rates over a given physical transmission medium, the subject of this book.

1.1. Applications of Digital Communication

Digital communication is used for signals that are inherently analog and continuous-time, such as speech and images, and signals that are inherently digital, such as text files. The demands placed on a communication system are, however, different. Modern communication networks support a mixture of services, and hence must take into account the demands of all.

1.1.1. Continuous-Time Signals

It is common in modern practice to convert analog continuous-time and continuous-amplitude signals such as voice, music, pictures, and video to digital form for communication. The first step is *digitization*, and the second step is *compression*. The digitization step is shown in Fig. 1-1. The continuous-time signal is first sampled at a rate f_s Hz (or samples / sec), which must be greater than twice the highest frequency component in the signal (see Section 2.3). Often this sampling operation is preceded by a lowpass filter, not shown, which ensures that the signal is properly bandlimited. Each sample is then converted to an n-bit binary word by an analog-to-digital converter (A/D). The output is a bit stream with a bit rate of nf_s bits per second (often written b/s or bps). Before actual transmission or storage, the digitized signal is usually compressed by various algorithms that remove redundancy or perceptually insignificant information (familiar examples include MP3 for music, JPEG for pictures, and MPEG for video). After reception, the compression can be reversed and the resulting digitized representation converted back to continuous time and amplitude using a digital-to-analog converter (D/A) and a reconstruction low-pass filter.

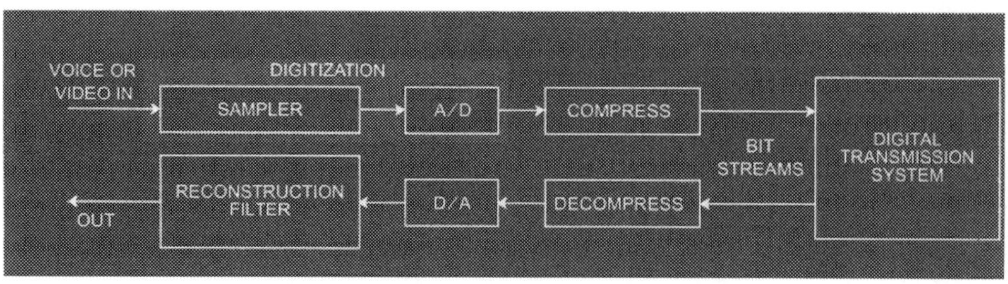

Fig. 1-1. A continuous-time signal sent across a digital transmission system.

From the viewpoint of the signals or data being transmitted over a digital network, the digital transmission links in that network can be abstracted and adequately characterized by four parameters:

- the *bit rate*,
- the *propagation* and *processing delay*,
- the *probability of error*, which indicates how likely the bits arriving at the destination are to differ from the transmitted bits, and
- the *timing jitter* in the arriving bit stream (variation from a uniform rate of arrival).

Both bit errors and timing jitter can cause some degradation in the quality of a recovered continuous-time signal, and excessive delay can cause subjective impairment in interactive applications like telephony, video conferencing, or interactive data retrieval, so the designer of the digital communication system must control these impairments.

Another application of digital communication techniques is storage systems using magnetic or optical media. In this case the objective is not transmission "from here to there" but rather "from now to then." These media have unique impairments, different from those in transmission media, but many of the same basic techniques apply.

1.1.2. Data

The Internet has greatly increased the importance of data transmitted from one computer to another, and uses a general technique called packet switching to accommodate a variety of different information sources and media such as word processing files, email messages, and Web pages.

1.2. Digital Networks

A digital communication link is able to transmit data from one geographical location to another, but this rarely suffices by itself. It is impractical to provision communication links from every location to every other location (except over limited geographic areas using wireless). This problem is addressed by building a digital network that includes not only communication links but also switches that route data from one edge terminal to another. A network supports many simultaneous connections between many pairs of terminals.

Some networks are specialized to particular applications and media, like telephony and video distribution, and others are specialized to data. Increasingly, data networks like the Internet have been adapted to the conversion and transmission of continuous-time signals (this is often called continuous-media *streaming*).

Telecommunication networks are often classified according to geographical extent. *Local-area networks* are designed to communicate at very high bit rates over a small geographical area (on the order of one km in extent) using wireless, wire pair, or optical fiber. *Metropolitan-area networks* cover a larger area (approximately 50 km) using optical fiber. *Wide-area networks* like the Internet encompass the world using a optical fiber (for high density uses) and satellite (for areas with light traffic) to cover large distances, including over or under the oceans.

Often different types of network applications share common digital transmission facilities. Beyond this, there is a trend toward not only common transmission facilities, but also a single network supporting all these types of applications. This so-called *integrated network* can support computer-to-computer communications, telephony, video distribution and video on demand, and other applications.

While some earlier networks (like the telephone network) use circuit switching, data networks and integrated networks use packet switching. In *circuit switching*, the network connects a constant-rate bit stream to the destination for a relatively long period of time (of the order of minutes or longer). (This first arose in the context of voice networks, where the circuit is formed for the duration of one telephone call.) In *packet switching*, the data is encapsulated into packets, which are collections of bits to which a header (with destination address and other information meaningful to the network itself) is appended. Each packet can be routed through the network by the switches based on the header information. One advantage of packet switching is that the bit stream between source and destination can have a variable bit rate — this is accomplished by generating only the needed packets. This leads to greater efficiency for sources of data that have a large ratio of peak to average bit rate, and for connections that are very short-lived.

1.3. Digital vs. Analog Communications

Given that the world got along with mostly analog transmission for many decades (and it is still used in some applications like broadcast radio), it is useful to review why digital communication has become so prevalent. There are some important advantages.

Interface Abstraction

Digital communication links are relatively simple to characterize from a user perspective. This is in contrast to analog communication, where there are many more and more complex ways in which a transmission can be degraded. For example, the signal-to-noise ratio may be poor, or the signal may suffer second or third order intermodulation distortion, or crosstalk from one user to another, or the system may clip the input signal. A digital communication system has only four parameters important to users mentioned earlier. The impairments of the physical medium (which may be quite severe) are largely hidden from the user, as are implementation details. This property we call an *interface abstraction*. The power of this abstraction was perhaps first appreciated by Claude Shannon in his classic 1948 paper which started the field of information theory (Chapter 4). He showed that theoretically there is nothing lost by defining the interface between the signal to be transmitted and the transmission system to be a bit stream, regardless of whether the signal is analog or digital.

The Regenerative Effect

Consider the challenge of transporting a bit stream over a long distance. The degradation of an uninterrupted physical medium, such as a cable or optical fiber, may be unacceptable. The digital solution is to place *regenerative repeaters* at periodic intervals in the physical medium, as shown in Fig. 1-2. Each of these repeaters includes a receiver, which detects the bit stream (just as at the eventual destination), and a re-transmitter similar to the one at the origination. In the absence of transmission bit errors, the bit stream that is regenerated in this repeater is identical to the one that was originally transmitted. Any effects due to noise and distortion on the physical medium have been completely removed by the regenerative repeater (except insofar as they cause occasional bit errors). Contrast this to analog transmission, where the equivalent repeaters consist basically of amplifiers, and the noise and distortion cannot be removed and accumulates as the distance increases.

In practice, the digital communication system will introduce some bit errors in the detection of the bit stream. When the probability of error at each repeater is low, the total probability of error for m repeaters is approximately m times the probability of error for a single repeater. Assuming a maximum number of repeaters, we can derive from an overall probability of error objective and reference this to a bit-error requirement for a single repeater. We can then meet this latter objective by adjusting the design of the repeater, the design of the signal, and the spacing between repeaters (closer spacing will result in a lower error rate for a given physical medium of transmission).

The same regenerative effect applies to the storage of signals. Consider for example the high quality possible with digital audio (like CD, DVD Audio, or MP3). Part of the reason for this is that each time the audio is copied onto a new medium (for example from a CD to a computer hard disk to a portable device), its digital representation is regenerated, and any degradations peculiar to the original or intermediate physical media are removed.

Economics of Implementation

We have seen that digital communication has some powerful technical advantages. What are the relative economics of digital and analog communication? Digital communication receivers are internally logically much more complex than analog. In the technologies available before the integrated circuit (vacuum tubes and transistors), this complexity would have been prohibitively expensive. However, decades of Moore's law have improved the cost-performance characteristics of semiconductor technology to the point that remarkable amounts

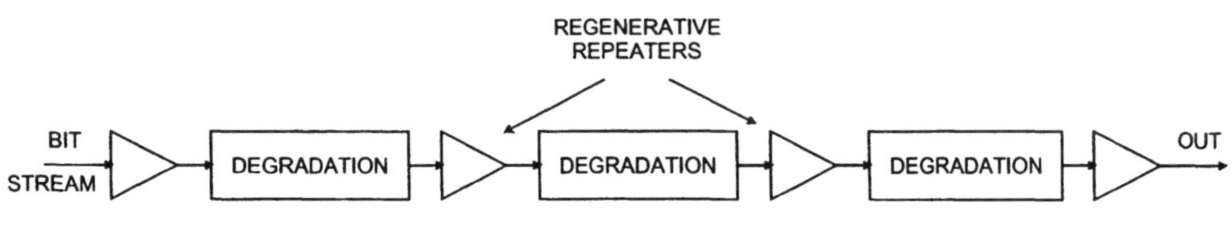

Fig. 1-2. A chain of regenerative repeaters reduces the effect of cascaded degradations by regenerating the digital signal at intermediate points in the transmission.

of complexity and remarkably high processing rates can be achieved at modest cost, even subject to the limitations of battery powered portable devices. Further, the semiconductor technologies favor a digital implementation, as digital circuits are easier to design and manufacture and test. Software implementations (also intrinsically digital) have compelling advantages where they are economical, because they allow a variety of functions from a single piece of hardware, are relatively easy to design, and can be modified and upgraded even in the field.

Communication systems incorporate a physical channel, which is intrinsically analog. Thus, all communication systems require analog as well as digital circuitry. At the high bit-rate end, fiber optics transmitters and receivers are largely analog and logically relatively simple. At the low bit-rate end, and particularly for wireless channels, the transmitters and receivers are logically quite complex, and the trend is toward reducing the analog circuitry to the bare minimum (basically the radio-frequency amplifiers, modulators, and the digitization function).

1.4. Plan of the Book

This book concentrates on the techniques that are used to design a digital communication system starting with any of the common physical media. Our concern is thus with how to get a bit stream from one location to another, and not so much with how this bit stream is used. In the particular context of a digital network, this aspect of the system is called the *physical layer*. We also address the problems of multiple bit streams sharing a common medium, called *multiple access*.

In Chapters 2–4 some basics required for the understanding of later material are covered. Many readers will have a prior background in many of these basics, in which case only a superficial reading to pick up notation is required. We have also covered some basic topics with which the reader may not be so familiar. These include spectral factorization of rational transfer functions (Chapter 2), signal space (Chapter 2), Markov chains and Poisson processes (Chapter 3), and information-theoretic bounds (Chapter 4).

Chapters 5–13 cover the theory of modulation, detection, and coding that is necessary to understand how a single bit stream is transported over a physical medium. The need for this theory arises because all physical media are analog and continuous-time in nature. It is ironic that much of the design of a digital communication system is inevitably related to the analog and continuous-time nature of the medium, even though this is not evident at the abstracted interface to the user.

The design of a digital communication system or network raises many difficult *synchronization* issues that are covered in Chapters 14–16. Often a large part of the effort in the design of a digital communication system involves phase-locked loops, timing, and carrier recovery.

A complete telecommunication network requires that many bit streams originating with many different users be transported simultaneously while sharing facilities and media. This leads to the important topic of ***multiple access*** of more than one user to a single physical medium for transmission. This is covered in Chapter 17.

1.5. Further Reading

There are a number of excellent books on digital communication. While these books have a somewhat different emphasis from this one, they provide very useful supplementary material. The books by Roden [1], Benedetto, Biglieri, and Castellani [2], and Gitlin, Hayes, and Weinstein [3], cover similar material to this one, perhaps with a bit less practical emphasis. The books by Blahut [4], and Bingham [5] are valued for their practical orientation. Two texts provide additional detail on topics in this book: the recent book by Proakis [6] is an excellent treatise on applied information theory and advanced topics such as coding, spread spectrum, and multipath channels; the book by Viterbi and Omura [7] gives a detailed treatment of source and channel coding as applied to digital communication, as does Biglieri, Divsalar, McLane, and Simon [8]. The recent book by Wilson provides excellent coverage of digital modulation and coding [9]. An excellent treatment of the statistical communication theory as applied to digital communication is given by Schwartz [10]. On the topics of modulation, equalization, and coding the book by Lucky, Salz, and Weldon is somewhat dated but still recommended reading [11]. The same applies to the book by Wozencraft and Jacobs, which emphasizes principles of detection [12]. Books by Keiser and Strange [13] and Bellamy [14] give broad coverage of digital transmission at a descriptive level. Practical details of digital transmission can be found in a book published by AT&T Technologies [15], in the book by Bylanski and Ingram [16], and in the book by Cattermole [17]. A couple of books expand on our brief description of digital switching, including McDonald [18] and Pearce [19]. For the information theory background that gives a solid theoretical foundation for digital communication, the books by Gallager [20], Cover and Thomas [21], Blahut [22], and McEliece [23] are recommended. Schwartz [24] and Bertsekas and Gallager [25] are recommended comprehensive texts on computer networks. There are also many elementary texts that cover both digital and analog communication, as well as the basic systems, transforms, and random process theory. Simulation techniques for communication systems are covered comprehensively in Jeruchim, Balaban, Shanmugan [26].

Problems

Problem 1-1. For an A/D converter, define a signal-to-error ratio as the signal power divided by the quantization error power, expressed in dB. A uniform quantizer, which has equally-spaced thresholds, has two parameters: the number of bits n and the step size Δ.

(a) If we were to increase n by one, to $n + 1$, for the same input signal, what would be the appropriate change to Δ?

(b) Without doing a detailed analysis, what do you suppose would be the effect on signal-to-error ratio of increasing from n to $n + 1$ bits/sample?

(c) What effect will this same change have on the bit rate?

(d) Using the prior results, what is the *form* of the relationship between signal-to-error ratio and the bit rate? (You may have unknown constants in your equation.)

Problem 1-2. An analog signal is transmitted using the system of Fig. 1-1. Discuss qualitatively the effects of bit errors on the recovered analog signal.

Problem 1-3. Discuss qualitatively the sources of delay that you would expect in a digital system like the one shown in Fig. 1-1.

Problem 1-4. Suppose you have a source of data that outputs a bit stream with a bit rate that varies with time, but also has a peak or maximum bit rate. Describe qualitatively how you might transmit this bit stream over a link that provides a constant bit rate.

References

1. M. S. Roden, *Digital And Data Communication Systems*, Prentice- Hall, Englewood Cliffs, N.J., 1982.
2. S. Benedetto, E. Biglieri, and V. Castellani, *Digital Transmission Theory*, Prentice-Hall, Inc., Englewood Cliffs, NJ, 1987.
3. R. D. Gitlin, J. F. Hayes, and S. B. Weinstein, *Data Communications Principles*, Plenum Press, New York and London, 1992.
4. R. E. Blahut, *Digital Transmission of Information*, Addison- Wesley, New York, 1990.
5. J. A. C. Bingham, *The Theory and Practice of Modem Design*, John Wiley & Sons, New York, 1988.
6. J. G. Proakis, *Digital Communications*, Fourth Edition, McGraw-Hill, New York, 2001.
7. A. J. Viterbi and J. K. Omura, *Principles of Digital Communication and Coding*, McGraw-Hill, 1979.

8. E. Biglieri, D. Divsalar, P. J. McLane, and M. K. Simon, *Introduction to Trellis-Coded Modulation with Applications*, Macmillan, New York, 1991.

9. S. G. Wilson, *Digital Modulation and Coding*, Prentice Hall, Upper Saddle River, NJ, 1996.

10. M. Schwartz, *Information Transmission, Modulation, and Noise*, McGraw-Hill, New York, 1980.

11. R. W. Lucky, J. Salz, and E. J. Weldon, Jr., *Principles of Data Communication*, McGraw-Hill Book Co., New York, 1968.

12. J. M. Wozencraft and I. M. Jacobs, *Principles of Communication Engineering*, Wiley, New York, 1965.

13. B. E. Keiser and E. Strange, *Digital Telephony and Network Integration*, Van Nostrand Reinhold, New York, 1985.

14. J. Bellamy, *Digital Telephony*, John Wiley, New York, 1982.

15. Members of Technical Staff, Bell Laboratories, *Transmission Systems for Communications*, Western Electric Co., Winston- Salem N.C., 1970.

16. P. Bylanski and D. G. W. Ingram, *Digital Transmission Systems*, Peter Peregrinus Ltd., Stevenage England, 1976.

17. K. Cattermole, *Principles of Pulse-Code Modulation*, Iliffe Books Ltd., London England, 1969.

18. J. C. McDonald, *Fundamentals of Digital Switching*, Plenum Press, New York, 1983.

19. J. G. Pearce, *Telecommunications Switching*, Plenum, New York, 1981.

20. R. Gallager, *Information Theory and Reliable Communication*, John Wiley and Sons, Inc., New York, 1968.

21. T. M. Cover, and J. A. Thomas, *Elements of Information Theory*, Wiley, New York, 1991.

22. R. E. Blahut, *Principles and Practice of Information Theory*, Addison-Wesley, New York, 1987.

23. R. J. McEliece, *The Theory of Information and Coding*, Addison Wesley Pub. Co., 1977.

24. M. Schwartz, *Telecommunication Networks: Protocols, Modeling, and Analysis*, Addison-Wesley, Reading, Mass., 1987.

25. D. Bertsekas and R. Gallager, *Data Networks*, Prentice-Hall, Englewood Cliffs, N.J., 1987.

26. M. C. Jeruchim, P. Balaban, and K. S. Shanmugan, *Simulation of Communication Systems*, Plenum Press, New York, 1992.

2

Deterministic Signal Processing

In this chapter we review some basic concepts in order to establish the notation used in the remainder of the book. In addition, we cover in more detail several specific topics that some readers may not be familiar with, including complex signals and systems, the convergence of bilateral Z-transforms, and signal-space geometry. The latter allows simple geometric interpretation of many signal processing operations, and demonstrates relationships among many seemingly disparate topics.

2.1. Signals

A *continuous-time signal* is a function $x(t)$ of the real-valued variable t, usually denoting time. A *discrete-time signal* is a sequence $\{x_k\}$, where k usually indexes a discrete progression in time. Throughout this book we will see systems containing both continuous-time and discrete-time signals. Often a discrete-time signal results from *sampling* a continuous-time signal; this is written $x_k = x(kT)$, where T is the *sampling interval*, and $1/T$ is the sampling frequency. The sampling operation can be represented as

$$x_k = x(kT) = \int_{-\infty}^{\infty} x(\tau)\delta(\tau - kT)\, d\tau\,, \tag{2.1}$$

where $\delta(\tau)$ is the *Dirac delta function* or *continuous-time impulse*. The discrete-time signal x_k has a continuous-time *pulse-amplitude modulation (PAM)* representation

$$\hat{x}(t) = \sum_{k=-\infty}^{\infty} x_k\, \delta(t - kT)\,, \tag{2.2}$$

in terms of impulses.

A continuous-time signal can be constructed from a discrete-time signal as represented symbolically in Fig. 2-1. A discrete-time input to a continuous-time system implies first the generation of the continuous-time impulse train in (2.2), and then its application to a continuous-time filter $G(f)$, yielding

$$y(t) = \sum_{k=-\infty}^{\infty} x_k g(t - kT)\,. \tag{2.3}$$

In some applications $g(t)$ is said to be an *interpolation* filter, but more often in this book it is said to be a *pulse-shaping* filter.

2.1.1. Complex-Valued Signals

In digital communication systems, complex-valued signals are often a convenient mathematical representation for a pair of real-valued signals. A complex-valued signal consists of a *real* signal and an *imaginary* signal, which may be visualized as two voltages induced across two resistors, or as two sequences of numbers.

Example 2-1.

A complex-valued signal we encounter frequently is the complex exponential,

$$x_k = e^{-j\omega k} = \cos(\omega k) - j\sin(\omega k)\,,$$
$$x(t) = e^{-j2\pi ft} = \cos(2\pi ft) - j\sin(2\pi ft)\,. \tag{2.4}$$

We consistently use j to represent $\sqrt{-1}$.

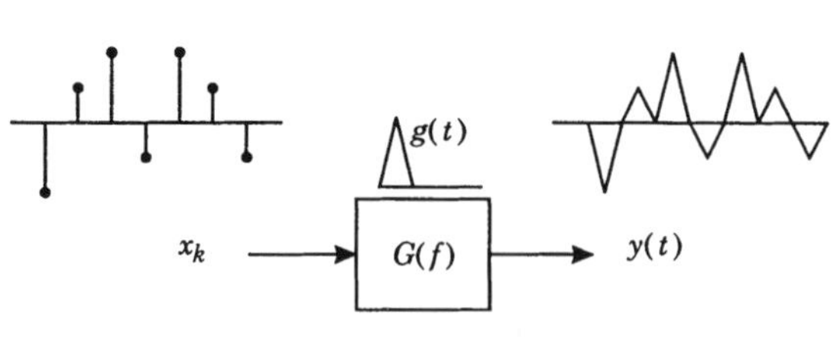

Fig. 2-1. Construction of a continuous-time signal from a discrete-time signal. When we show a discrete-time input to a continuous-time system, we imply first the generation of the impulse train in (2.2). An example is shown above the system.

Complex-valued signals are processed just as real-valued signals are, except that the rules of complex arithmetic are followed.

Exercise 2-1.

Draw diagrams specifying the addition and multiplication of two complex-valued continuous-time signals in terms of real-valued additions and multiplications.

(The reader is reminded that the solution to this and all exercises can be found at the back of the book.)

The real part of the signal $x(t)$ is written $\mathrm{Re}\{x(t)\}$ and the imaginary part $\mathrm{Im}\{x(t)\}$. Furthermore, we write the complex conjugate of a signal $x(t)$ as $x^*(t)$, and the squared modulus as $|x(t)|^2 = x(t)x^*(t)$. We don't use any special notation to distinguish real-valued from complex-valued signals because it will generally be clear from the context.

2.1.2. Energy and Power

The energy of a signal $x(t)$ or $\{x_k\}$ is defined to be

$$\int_{-\infty}^{\infty} |x(t)|^2 dt, \qquad \sum_{k=-\infty}^{\infty} |x_k|^2. \tag{2.5}$$

The average power is

$$\lim_{\tau \to \infty} \frac{1}{2\tau} \int_{-\tau}^{\tau} |x(t)|^2 dt, \qquad \lim_{K \to \infty} \frac{1}{2K+1} \sum_{k=-K}^{K} |x_k|^2. \tag{2.6}$$

2.2. LTI Systems and Fourier Transforms

The Fourier transform is valuable in the analysis of modulation systems and linear time-invariant systems. For the convenience of the reader, the properties of both discrete and continuous-time Fourier transforms are summarized in Appendix 2-A. In this section we establish notation and review a few basic facts.

2.2.1. Linear Time-Invariant (LTI) Systems

If a system is *linear* and *time invariant* (LTI), then it is characterized by its impulse response h_k (for a discrete-time system) or $h(t)$ (for a continuous-time system). The output of the LTI system can be expressed in terms of the input and impulse response as a convolution; for the discrete-time case,

$$y_k = x_k * h_k = \sum_{m=-\infty}^{\infty} x_m h_{k-m}, \tag{2.7}$$

and the continuous-time case,

$$y(t) = x(t) * h(t) = \int_{-\infty}^{\infty} x(\tau)h(t-\tau)\,d\tau\,. \tag{2.8}$$

An LTI system is *real* if its impulse response is real-valued, and *complex* if its impulse response is complex-valued.

Exercise 2-2.

Show that if a complex system has a real-valued input it can be implemented using two real systems and sketch the configuration. Show that the same is true if a real system has a complex-valued input, and again sketch the configuration.

Exercise 2-3.

The notion of linearity extends to complex LTI systems. Demonstrate that if the four real systems required to implement a complex system are linear, then the resulting complex system is linear. It follows immediately that real-valued LTI systems are linear with respect to complex-valued inputs.

2.2.2. The Fourier Transform

The Fourier transform pair for a continuous-time signal $x(t)$ is

$$X(f) = \int_{-\infty}^{\infty} x(t)e^{-j2\pi ft}\,dt\,, \qquad x(t) = \int_{-\infty}^{\infty} X(f)e^{j2\pi ft}\,df\,, \tag{2.9}$$

while the discrete-time Fourier transform (DTFT) pair for x_k is

$$X(e^{j\theta}) = \sum_{k=-\infty}^{\infty} x_k e^{-jk\theta}\,, \quad x_k = \frac{1}{2\pi}\int_{-\pi}^{\pi} X(e^{j\theta})e^{jk\theta}\,d\theta\;. \tag{2.10}$$

The notation $X(e^{j\theta})$ deserves some explanation. $X(e^{j\theta})$ is the Z-transform $X(z)$, defined as

$$X(z) = \sum_{k=-\infty}^{\infty} x_k z^{-k}\,, \tag{2.11}$$

evaluated at $z = e^{j\theta}$. Furthermore, the argument of the function, $e^{j\theta}$, is periodic in θ, emphasizing that the DTFT itself is periodic in θ with period 2π.

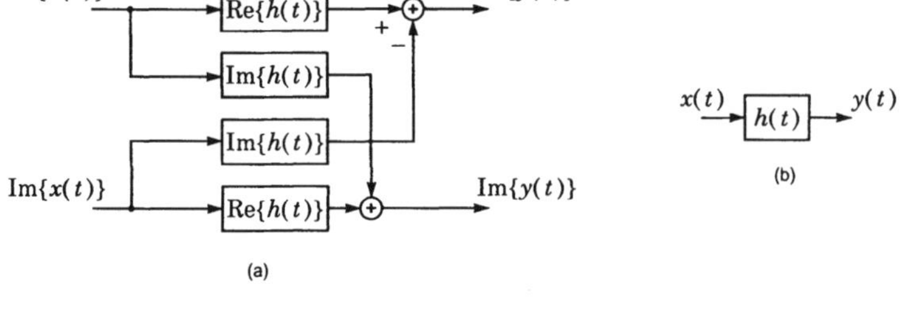

Fig. 2-2. A complex-valued LTI system with a complex-valued input and output.

If $h(t)$ is the impulse response of a continuous-time system, then the Laplace transform $H(s)$ is called the *transfer function*, and the Fourier transform $H(f)$ is called the *frequency response*. Correspondingly, for a discrete-time impulse response h_k, the transfer function is $H(z)$ and the frequency response is $H(e^{j\theta})$. Discrete-time and continuous-time systems will often be distinguished only by the form of the argument of their transfer function or frequency response.

Exercise 2-4.
Starting with the convolution, show that the Fourier transform of the output of an LTI system is

$$Y(f) = H(f)X(f) , \qquad Y(e^{j\theta}) = H(e^{j\theta})X(e^{j\theta}) , \tag{2.12}$$

for the continuous-time and discrete-time cases, respectively, where $X(f)$ and $X(e^{j\theta})$ are the Fourier transforms of the input signals.

The magnitude of the frequency response $|H(f)|$ or $|H(e^{j\theta})|$ is called the *magnitude response*. The argument of the frequency response $\angle(H(f))$ or $\angle(H(e^{j\theta}))$ is called the *phase response*. The reason for these terms is explored in Problem 2-2.

A fundamental result allows us to analyze any system with a combination of continuous-time and discrete-time signals.

Exercise 2-5.
Given the definition (2.2) of a continuous-time PAM signal $\hat{x}(t)$ derived from a discrete-time signal x_k, show that for all f

$$\hat{X}(f) = X(e^{j2\pi fT}) . \tag{2.13}$$

In words, the *continuous-time* Fourier transform of a PAM representation of a discrete-time signal is equal to the DTFT of the discrete-time signal evaluated at $\theta = 2\pi fT$.

2.3. The Nyquist Sampling Theorem

Suppose that we sample a continuous-time signal $x(t)$ to get

$$x_k = x(kT) . \tag{2.14}$$

From (2.2) we obtain

$$\hat{x}(t) = x(t)\sum_{k=-\infty}^{\infty} \delta(t - kT) . \tag{2.15}$$

Multiplication in the time-domain corresponds to convolution in the frequency domain, so

$$\hat{X}(f) = X(f) * \frac{1}{T}\sum_{m=-\infty}^{\infty} \delta(f - m/T)$$

$$= \frac{1}{T}\int_{-\infty}^{\infty} X(\nu)\sum_{m=-\infty}^{\infty} \delta(f - \nu - m/T)\, d\nu$$

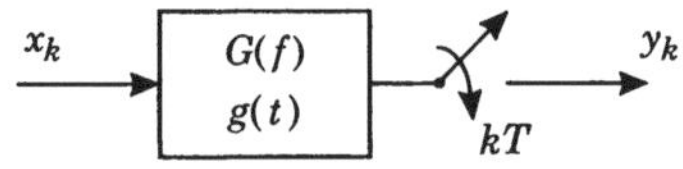

Fig. 2-3. A discrete-time system using a continuous-time filter.

$$= \frac{1}{T}\sum_{m=-\infty}^{\infty} X(f - m/T) . \tag{2.16}$$

Combining this with (2.13) we get the very important relation

$$X(e^{j2\pi fT}) = \frac{1}{T}\sum_{m=-\infty}^{\infty} X(f - m/T) . \tag{2.17}$$

This fundamental *sampling theorem* relates the signals $x(t)$ and x_k in the frequency domain. Systems with both discrete and continuous-time signals can now be handled easily.

Exercise 2-6.

Use (2.17) to show that the frequency response of a completely discrete-time system equivalent to that in Fig. 2-3 is

$$G(e^{j2\pi fT}) = \frac{1}{T}\sum_{m=-\infty}^{\infty} G(f - m/T) . \tag{2.18}$$

Notice that in (2.17) a component of $X(f)$ at any $f = f_0$ is indistinguishable in the sampled version from a component at $f_0 + m/T$ for any integer m. This phenomenon is called *aliasing*.

Example 2-2.

Given a signal $x(t)$ with Fourier transform $X(f)$ shown in Fig. 2-4(a), the Fourier transform of the sampled signal $\hat{X}(f) = X(e^{j2\pi fT})$ is shown in Fig. 2-4(b). The overlap evident in Fig. 2-4(b), called *aliasing distortion*, makes it very difficult to recover $x(t)$ from its samples.

Exercise 2-7. (Nyquist sampling theorem.)

Show that, from (2.17), a continuous-time signal can be reconstructed from its samples if it is sampled at a rate at least twice its highest frequency component. More precisely, if a signal $x(t)$ with Fourier transform $X(f)$ is sampled at frequency $1/T$ Hz, then $x(t)$ can be reconstructed from the samples if $X(f) = 0$ for all $|f| > 1/(2T)$.

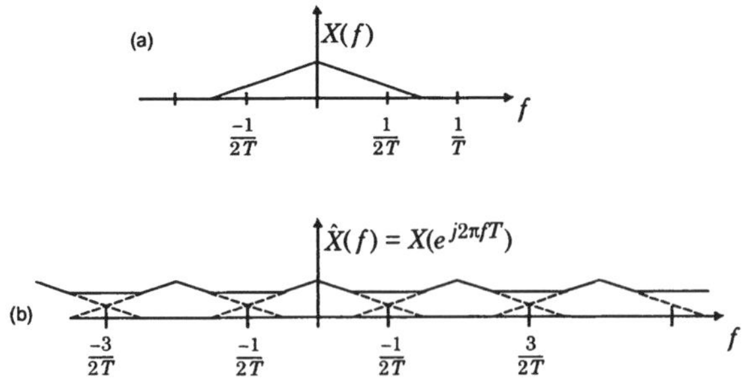

Fig. 2-4. The Fourier transform of a continuous-time signal (a) and its sampled version (b), where the sample rate is $1/T$.

The sampling theorem gives a sufficient but not necessary condition for reconstructing a signal from its samples. In the absence of aliasing distortion, a lowpass signal can be reconstructed from its samples using an ideal low-pass filter with cutoff frequency $1/(2T)$,

$$x(t) = \hat{x}(t) * \left[\frac{\sin(\pi t/T)}{\pi t/T}\right] = \sum_{m=-\infty}^{\infty} x_m \left[\frac{\sin[\pi(t-mT)/T]}{\pi(t-mT)/T}\right]. \tag{2.19}$$

2.4. Downconversion and Complex Envelopes

Real-valued signals may be represented in terms of a complex signal called a *complex envelope*, which is defined in terms of a building block called a *downconverter*. These ideas play a central role in the analysis and implementation of many communication systems. In this section we define the downconverter in terms of a phase splitter and a Hilbert transformer, and then summarize how to convert back and forth between a signal and its complex envelope.

2.4.1. Downconversion, Phase Splitters, and the Hilbert Transform

A real-valued signal whose Fourier transform is concentrated at high frequencies (away from d.c.) is said to be *passband*, while a signal whose Fourier transform is concentrated at low frequencies (near d.c.) is said to be *baseband*. The first step a receiver often takes is to convert a real-valued passband signal into an equivalent low-frequency baseband signal that is complex valued, a process known as *downconversion*. A *downconverter* is a system that performs downconversion in a mathematically precise manner, as shown in Fig. 2-5(a). Let $s(t)$ denote the downconverter input; by assumption, it is always real valued. The downconverter first applies $s(t)$ to a phase splitter, and then multiplies by a complex exponential. These two steps are detailed below.

The *phase splitter* of Fig. 2-5(a) is a filter with impulse response $\phi(t)$ and transfer function $\Phi(f) = u(f)$, or:

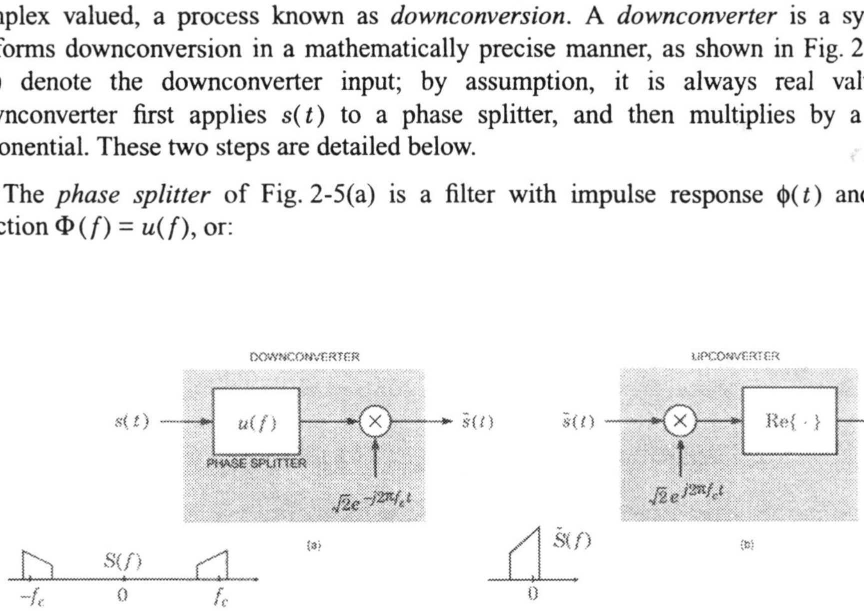

Fig. 2-5. A downconverter (a) consists of a phase splitter, which passes positive frequencies and eliminates negative frequencies, followed by a complex exponential, which scaled the spectrum by $\sqrt{2}$ and shifts it to the left by f_c. The downconverter of (a) is reversed by the upconverter of (b).

$$\Phi(f) = \begin{cases} 1, & f > 0 \\ 0, & f < 0 \end{cases}. \tag{2.20}$$

This filter is most often applied to passband signals with no d.c. content, in which case its d.c. gain is irrelevant. For those rare instances where it matters, however, we define the phase splitter d.c. gain to be $\Phi(0) = 1/2$. The filter passes positive frequencies and rejects negative frequencies. Clearly, since $\Phi(f)$ does not display complex-conjugate symmetry, the impulse response $\phi(t)$ is complex valued. Regardless of the input to a phase splitter, the output will not have any negative frequency components. A signal with only nonnegative frequency components is said to be *analytic*. Such a signal must be complex-valued in the time domain.

It is often convenient to decompose the phase splitter into its even and odd parts, yielding:

$$\Phi(f) = \tfrac{1}{2}(1 + jH(f)), \tag{2.21}$$

where the filter

$$H(f) = -j\,\text{sign}(f) \tag{2.22}$$

is called a *Hilbert transformer*. The *Hilbert transform* of a real signal $s(t)$ is denoted by $\hat{s}(t)$, and is defined as the output of a Hilbert transformer when $s(t)$ is the input. Because $H(f)$ has complex-conjugate symmetry, the impulse response of a Hilbert transformer is real valued, and hence the Hilbert transform $\hat{s}(t)$ will also be real valued. A Hilbert transformer does not modify the magnitude spectrum of the input, but does give a $-\pi$ phase shift at all frequencies. From (2.21) it is evident that the time-domain output of a phase splitter can be written as:

$$\frac{1}{2}\Big(s(t) + j\hat{s}(t)\Big). \tag{2.23}$$

Thus, the real part of the phase-splitter output is half of the input, and the imaginary part is half of the Hilbert transform of the input.

Referring again to Fig. 2-5(a), the downconverter multiplies the phase splitter output by the complex exponential $\sqrt{2}e^{-j2\pi f_c t}$, where the *carrier frequency* f_c is a free parameter that must be specified in order to fully characterize a downconverter.

2.4.2. The Complex Envelope

The *complex envelope* $\tilde{s}(t)$ of a real signal $s(t)$ *with respect to the carrier* f_c is defined simply as the output of a downconverter with carrier f_c. The relationship between $s(t)$ and $\tilde{s}(t)$ is thus defined by the downconverter of Fig. 2-5(a). Equivalently, given (2.23), the complex envelope may be defined mathematically by:

$$\tilde{s}(t) = \frac{1}{\sqrt{2}}\Big(s(t) + j\hat{s}(t)\Big)e^{-j2\pi f_c t}. \tag{2.24}$$

Taking the Fourier transform yields the equivalent frequency-domain relationship:

$$\tilde{S}(f) = \sqrt{2}\, u(f+f_c)S(f+f_c)\ . \tag{2.25}$$

Exercise 2-8.
Show that energy of a complex envelope $\tilde{s}(t)$ is precisely equal to that of $s(t)$. This equal-energy property is important when we deal with noise and signal-to-noise ratios.

A crucial feature of the downconverter is that it is reversible. The complex envelope uniquely defines a signal, regardless of both its spectral content and the choice for the carrier frequency parameter. Specifically, any real signal $s(t)$ may be recovered from its complex envelope through the equation

$$s(t) = \sqrt{2}\,\mathrm{Re}\{\, \tilde{s}(t)e^{j2\pi f_c t}\}\ . \tag{2.26}$$

This equation defines the *complex-envelope representation* of a real signal $s(t)$ with respect to f_c. It is easily verified by substituting (2.24) into (2.26). Any real-valued signal can be written in this form. Not surprisingly, (2.26) defines what is called an *upconverter*, because it reverses the downconversion process. A block diagram of an upconverter is shown in Fig. 2-5(b).

While complex envelopes may be defined for any real signal, they are most useful for *passband* signals $s(t)$ with a spectrum concentrated in the vicinity of $f = f_c$. Since $s(t)$ is real-valued, its spectrum contains energy near $-f_c$ as well as f_c. The phase splitter discards the negative frequency components, and multiplying the resulting analytic signal by $\sqrt{2}\, e^{-j2\pi f_c t}$ shifts the remaining positive frequency components to near d.c., yielding the complex envelope $\tilde{s}(t)$. The benefit of downconversion in this case stems from the fact that $\tilde{s}(t)$ is a baseband or low-pass signal, meaning that its energy is concentrated near d.c. This simplifies subsequent receiver processing, be it analog or digital.

The discontinuity of the phase splitter and Hilbert transformer at d.c. would make the downconverter difficult to implement in practice. Fortunately, however, the implementation is much easier when the input is a passband signal, because then the transfer function in the vicinity of d.c. does not matter.

Two alternative representations of a real signal are derivatives of the complex envelope representation. The real and imaginary parts of the complex envelope $\tilde{s}(t)$ are referred to as the *in-phase* and *quadrature* components of $s(t)$ with respect to f_c, respectively, and are denoted $s_I(t) = \mathrm{Re}\{\, \tilde{s}(t)\}$ and $s_Q(t) = \mathrm{Im}\{\, \tilde{s}(t)\}$, respectively. Substituting $\tilde{s}(t) = s_I(t) + js_Q(t)$ into (2.26) defines the *in-phase and quadrature representation* of a real signal $s(t)$ with respect to f_c:

$$\begin{aligned} s(t) &= \sqrt{2}\,\mathrm{Re}\Big\{\big(s_I(t) + js_Q(t)\big)\, e^{j2\pi f_c t}\Big\} \\ &= \sqrt{2}s_I(t)\cos(2\pi f_c t) - \sqrt{2}\, s_Q(t)\sin(2\pi f_c t)\ . \end{aligned} \tag{2.27}$$

Alternatively, define the *envelope* $e(t) = \sqrt{2}\,|\,\tilde{s}(t)\,|$ and *phase* $\theta(t) = \angle\tilde{s}(t)$, so that we may write $\sqrt{2}\tilde{s}(t) = e(t)e^{j\theta(t)}$. Substituting these definitions into (2.26), we arrive at the *envelope and phase representation* of a real signal:

$$s(t) = \mathrm{Re}\Big\{\big(e(t)e^{j\theta(t)}\big)\, e^{j2\pi f_c t}\Big\}$$

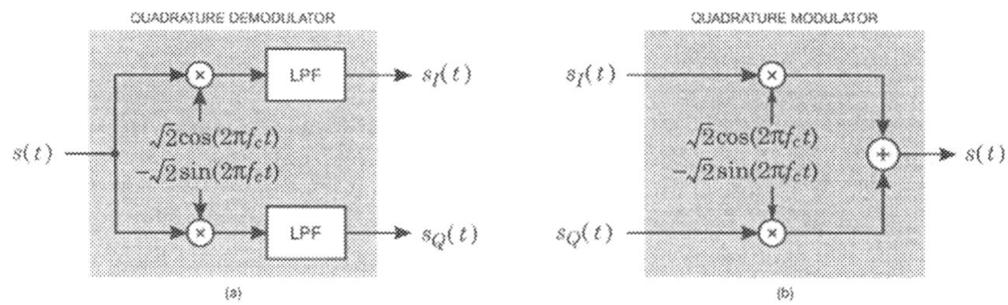

Fig. 2-6. A quadrature (a) demodulator and (b) modulator. These are usually interchangeable with the downconverter and upconverter of Fig. 2-5. The two low-pass filters (LPF) are identical and ideal with a gain of unity and a cutoff frequency equal to the carrier frequency f_c.

$$= e(t)\cos\left(2\pi f_c t + \theta(t)\right) . \tag{2.28}$$

Example 2-3.

The complex envelope of $\sqrt{2}\cos(200\pi t)$ with respect to a 100 Hz carrier frequency is 1, which implies that its in-phase and quadrature components are the constants 1 and 0, respectively. The answers change when we change the carrier frequency. For example, the complex envelope of $\sqrt{2}\cos(200\pi t)$ with respect to a 99 Hz carrier frequency is $e^{j2\pi t}$. In this case, its in-phase and quadrature components are $\cos(2\pi t)$ and $\sin(2\pi t)$, respectively. Observe that the real envelope $e(t) = \sqrt{2}\,|\,\tilde{s}(t)\,| = \sqrt{2}$ is the same for both carrier frequencies, as one might expect from an intuitive interpretation of the term *envelope*.

Exercise 2-9.

Show that the envelope $e(t)$ of any real signal $s(t)$ is independent of the carrier frequency f_c.

The downconverter of Fig. 2-5(a) first filters, then shifts the spectrum. In many circumstances the order can be reversed, and the downconverter of Fig. 2-5(a) may be substituted by the *quadrature demodulator* shown in Fig. 2-6(a). The input is first multiplied by a pair of sinusoids which differ in phase by $\pi/2$, and the results are then filtered by identical ideal low-pass filters with unity gain and bandwidth equal to the carrier frequency f_c. The only time this substitution is invalid is when the bandwidth of the input exceeds twice the carrier frequency (Problem 2-12), a condition rarely met in practice. Thus, a phase splitter is often not necessary to perform downconversion. The *quadrature modulator* of Fig. 2-6(b) is a direct implementation of the in-phase and quadrature representation (2.27), and hence it always reverses the downconverter of Fig. 2-5(a). It also reverses the quadrature demodulator of Fig. 2-6(a) when the downconverter input is bandlimited to twice the carrier frequency.

The impact of filtering a real signal can be modeled by filtering its complex envelope. Suppose a real signal $s(t)$ is filtered by a real LTI system with impulse response $h(t)$, producing the real output $r(t) = s(t) * h(t)$. By definition, the complex envelope of the output has Fourier transform $\tilde{R}(f) = \sqrt{2}\,u(f + f_c)R(f + f_c)$, which using $R(f) = S(f)H(f)$ reduces to:

$$\tilde{R}(f) = \sqrt{2}\,u(f + f_c)S(f + f_c)H(f + f_c) \tag{2.29}$$

$$= \tilde{S}(f)H(f+f_c) \tag{2.30}$$

$$= \frac{1}{\sqrt{2}}\tilde{S}(f)\sqrt{2}\,u(f+f_c)H(f+f_c) \tag{2.31}$$

$$= \frac{1}{\sqrt{2}}\tilde{S}(f)\tilde{H}(f)\,. \tag{2.32}$$

From (2.30) we see that the complex envelope of the filter input is modified according to the frequency response of the passband channel near the carrier frequency, which makes intuitive sense. Converting (2.32) back to the time domain, we see that the filter input and output complex envelopes are related by $\tilde{r}(t) = \tilde{s}(t) * \tilde{h}(t)/\sqrt{2}$. Thus, to convert a real system to an equivalent complex system, simply replace all signals by their complex envelopes, and replace all filter impulse responses by their complex envelopes divided by $\sqrt{2}$.

2.5. Z Transforms and Rational Transfer Functions

The Z transform, which is closely related to the DTFT, is particularly useful in the study of rational transfer functions. The Z transform of a sequence $\{h_k\}$ is defined as

$$H(z) = \sum_{k=-\infty}^{\infty} h_k z^{-k}\,, \tag{2.33}$$

where z is a complex variable. As pointed out before, the DTFT is the Z transform evaluated at $z = e^{j\theta}$, or on the unit circle in the z plane, as shown in Fig. 2-7. This justifies the notation $H(e^{j\theta})$ for the DTFT. When $\{h_k\}$ is the impulse response of a discrete-time LTI system, then $H(z)$ is called the transfer function of the system. The transfer function on the unit circle is called the frequency response.

2.5.1. One-Sided Sequences

A *causal* sequence $\{h_k\}$ has $h_k = 0$ for $k < 0$. An *anticausal* sequence has $h_k = 0$ for $k > 0$. A *right-sided* sequence is one for which, for some K, $h_k = 0$ for $k < K$. A *left-sided* sequence correspondingly has $h_k = 0$ for $k > K$ for some K. When h_k is the impulse response of an LTI

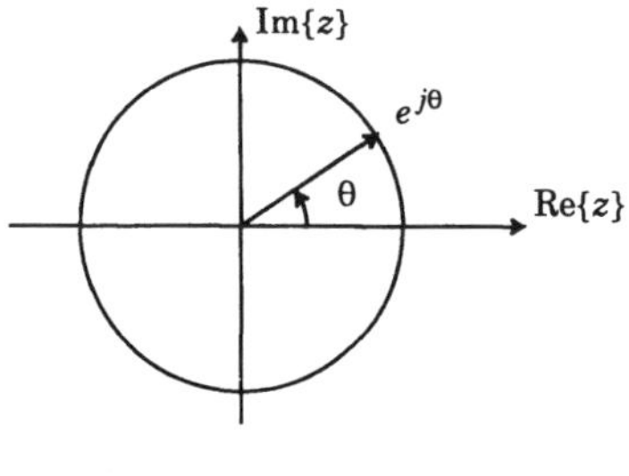

Fig. 2-7. The Fourier transform of a discrete-time signal is the Z transform evaluated on the unit circle.

system, that *system* is causal (anticausal) if the impulse response is right-sided (left-sided) for $K = 0$. While physically realizable real-time LTI systems are causal, we will frequently find it useful to model systems as non-causal.

Example 2-4.

Assume a communication channel has the impulse response shown in Fig. 2-8(a). We can think of this channel as having a *flat propagation delay* of M samples in addition to the non-causal response $\{h_k\}$ as shown in Fig. 2-8(b). Often the flat delay will not be an essential feature of the channel, in which case we ignore it.

Example 2-5.

Suppose we come up with a non-causal filter $H(z)$ in a theoretical development. This need not concern us too much, since such a filter can be approximated by a causal filter $G(z)$ together with an additional flat delay z^{-M}. This extra flat delay which did not arise in the theoretical development will often not harm the system.

A particularly important class of causal or anticausal sequences have a unity-valued sample at time zero ($h_0 = 1$). These sequences are said to be *monic*. A similar terminology is used for polynomials, which are monic if the constant term is unity. It is easy to see from (2.33) that $H(z)$ is the Z transform of a causal and monic sequence if and only if $H(\infty) = 1$. Similarly, it is anticausal and monic if and only if $H(0) = 1$. When a sequence $\{h_k\}$ is monic, then $H(z)$, as a polynomial in z or z^{-1}, is also monic.

The *region of convergence* (ROC) of the Z transform is the region of the z plane where the series in (2.33) is absolutely summable,

$$\sum_{k=-\infty}^{\infty} |h_k z^{-k}| < \infty . \tag{2.34}$$

Note that for any $z \in$ ROC, $|H(z)| < \infty$, because the triangle inequality implies that

$$|H(z)| \leq \sum_{k=-\infty}^{\infty} |h_k z^{-k}| < \infty . \tag{2.35}$$

Since the Fourier transform is the Z transform evaluated on the unit circle, for signals with a Fourier transform the ROC includes the unit circle.

A *bounded-input bounded-output (BIBO) stable* system has the property that any bounded input sequence with $|x_k| < L$ produces a bounded output sequence with $|y_k| < K$. We will often use the term "stable" to denote "BIBO stable."

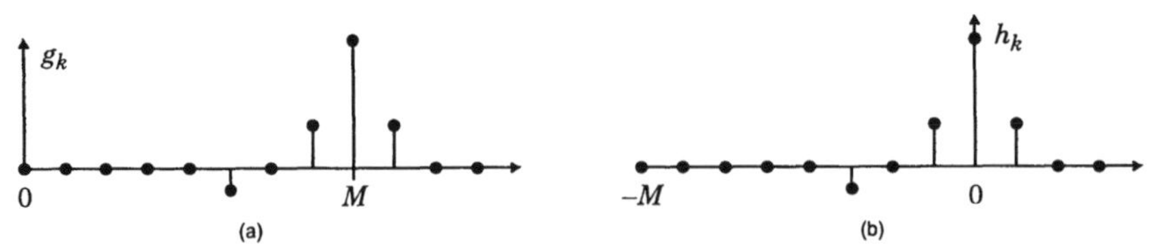

Fig. 2-8. Illustration of the usefulness of a non-causal channel model. (a) Actual channel impulse response. (b) A non-causal version of the impulse response where the flat propagation delay is ignored.

Exercise 2-10.

Show that a system with impulse response $\{h_k\}$ is BIBO stable if and only if

$$S = \sum_{k=-\infty}^{\infty} |h_k| < \infty . \tag{2.36}$$

A consequence of this is that a system is BIBO stable if and only if the ROC includes the unit circle. To see this, note that (2.34) can be rewritten as

$$\sum_{k=-\infty}^{\infty} |h_k|\,|z|^{-k} < \infty . \tag{2.37}$$

On the unit circle, $|z| = 1$, so this sum equals S in (2.36). By analogy with BIBO stable systems, a *sequence* $\{h_k\}$ (not necessarily an impulse response) is said to be stable if it is absolutely summable, as in (2.36).

It is evident from (2.37) that the ROC depends only on $|z|$; that is, it is of the form of an *annulus* or doughnut-shaped region. For a causal sequence, the ROC will be of the form $|z| > R$ for some constant R. In words, the ROC will be the region outside a circle of radius R. If the sequence is also stable, then $R < 1$, as shown in Fig. 2-9(a). To see this, note that for a causal sequence, the summation in (2.37) becomes

$$\sum_{k=0}^{\infty} |h_k|\,|z|^{-k} < \infty . \tag{2.38}$$

All the terms in the summation are positive powers of $|z|^{-1}$, and hence get smaller as $|z|$ gets larger. Thus, if absolute convergence occurs for some $|z_1| > R$, it will occur for all z such that $|z| \ge |z_1|$.

If the sequence is right-sided but not causal, (2.37) becomes

$$\sum_{k=K}^{\infty} |h_k|\,|z|^{-k} < \infty , \tag{2.39}$$

for some $K < 0$. The positive powers of $|z|$ do not converge at $z = \infty$, but do converge at all other z. Thus, the ROC cannot include $|z| = \infty$, and should be written $R < |z| < \infty$. Similar results apply to left-sided sequences.

Exercise 2-11.

(a) Show that the ROC of a left-sided stable sequence is of the form $0 < |z| < R$ for $R > 1$.

(b) Show that a left-sided sequence is anticausal if and only if its ROC includes the origin, $0 \le |z| < R$, as shown in Fig. 2-9(b).

To summarize, a right-sided sequence has an ROC consisting of the region outside a circle. That region includes $|z| = \infty$ if and only if the sequence is causal. A left-sided sequence has an ROC consisting of the inside of a circle. That region includes $z = 0$ if and only if the sequence is anticausal. In all cases, the ROC includes the unit circle if and only if the sequence is stable.

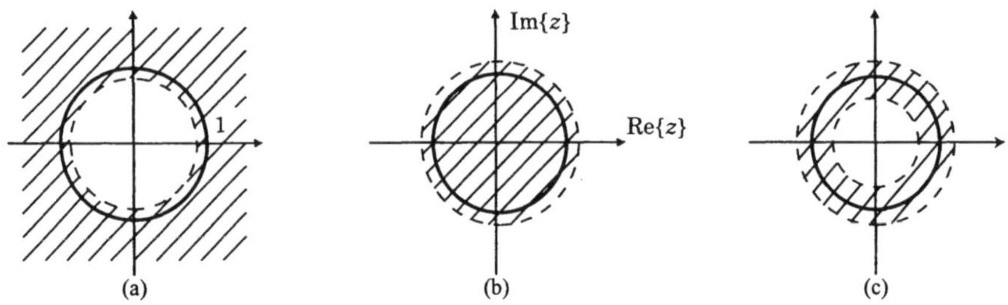

Fig. 2-9. The ROC of the Z transform of a stable sequence must include the unit circle. Three cases of stable sequences are illustrated: (a) A right-sided, (b) left-sided, and (c) two-sided sequence. The ROC includes $|z| = \infty$ in (a) if the sequence is causal. It includes $z = 0$ in (b) if the sequence is anticausal.

2.5.2. Rational Transfer Functions

A *rational transfer function* can be written in any of the forms

$$H(z) = z^r \cdot \frac{\sum_{k=0}^{M} b_k z^{-k}}{\sum_{k=0}^{N} a_k z^{-k}} = A \cdot z^r \cdot \frac{\prod_{k=1}^{M}(1 - c_k z^{-1})}{\prod_{k=1}^{N}(1 - d_k z^{-1})} = A \cdot z^m \cdot \frac{\prod_{k=1}^{M}(z - c_k)}{\prod_{k=1}^{N}(z - d_k)}, \quad (2.40)$$

where $A = b_0 / a_0$ and $m = N - M + r$. Notice that in the middle form, the numerator and denominator polynomials are both monic. The ratio of two such monic polynomials is also monic (as can be verified by carrying out the long division).

The system has M *zeros* (roots of the numerator) at $\{c_1, \ldots, c_M\}$, and N *poles* (roots of the denominator) at $\{d_1, \ldots, d_N\}$. The factor z^m represents merely an advance or delay in the impulse response. If $m > 0$ this factor introduces m zeros at the origin and m poles at $|z| = \infty$ (conversely for $m < 0$). If h_k is real valued, then $H(z)$ in (2.40) has real-valued coefficients, and the zeros and poles are always either real valued or come in complex-conjugate pairs (Problem 2-25).

Including poles and zeros at $z = 0$ and $|z| = \infty$, every rational transfer function has the same number of poles and zeros. This will be illustrated by two examples.

Example 2-6.

The causal FIR transfer function $H(z) = 1 - 0.5z^{-1}$ has one zero at $z = 1/2$ and one pole at $z = 0$. The only possible ROC is $|z| > 0$, which is a degenerate case of Fig. 2-9(a).

Example 2-7.

The anticausal FIR transfer function $H(z) = 1 - 0.5z$ has one zero at $z = 2$ and one pole at $|z| = \infty$. The only possible ROC is $|z| < \infty$, which is a degenerate case of Fig. 2-9(b).

The ROC cannot include any of the poles, since $H(z)$ is unbounded there. Moreover, for rational transfer functions, the ROC is bordered by poles. Referring to Fig. 2-9, for a causal and stable $H(z)$, all poles must be inside the unit circle. For an anticausal and stable $H(z)$, all poles must be outside the unit circle. No stable $H(z)$ can have poles on the unit circle, although it can certainly have zeros on the unit circle.

Exercise 2-12.

LTI systems that can actually be implemented with computational hardware can be represented by linear constant-coefficient difference equations with zero initial conditions. Show that the system represented by

$$y_k = \frac{1}{a_0}\left(\sum_{l=0}^{M} b_l x_{k-l} - \sum_{l=1}^{N} a_l y_{k-l}\right) \tag{2.41}$$

has transfer function given by (2.40) with $r = 0$.

When the denominator in (2.40) is unity ($N = 0$), the system has a *finite impulse response (FIR)*, otherwise it has an *infinite impulse response (IIR)*. FIR systems are always stable, and are often a good approximation to physical systems. They can have poles only at $z = 0$ and $|z| = \infty$, and the ROC therefore includes the entire z plane with the possible exception of $z = 0$ and $|z| = \infty$. If an FIR system is causal, it has no poles at $|z| = \infty$. If it is anticausal, it has no poles at $z = 0$.

Example 2-8.

Physical channels, such as a coaxial cable, usually do not have, strictly speaking, a rational transfer function. However, they can be adequately approximated by a rational transfer function. Often the simplest approximation is FIR, obtained by simply truncating the actual impulse response for sufficiently large M. Alternatively, it may be possible to approximate the response with fewer parameters using an IIR transfer function.

2.5.3. Reflected Transfer Functions

Given a transfer function $H(z)$, we define the *reflected* transfer function to be $H^*(1/z^*)$. It has impulse response h^*_{-k}, as is easy to verify. (A filter with such an impulse response will be called a *matched filter* in Chapter 5.) For a rational transfer function in the second form of (2.40), the reflected transfer function can be written

$$H^*(1/z^*) = A^* \cdot z^{-r} \cdot \frac{\prod_{k=1}^{M}(1 - c_k^* z)}{\prod_{k=1}^{N}(1 - d_k^* z)} . \tag{2.42}$$

The zeros c_k of $H(z)$ become zeros at the conjugate-reciprocal locations $1/c_k^*$ in $H^*(1/z^*)$. The poles are similarly reflected through the unit circle, which explains the term *reflected*. If $H(z)$ is stable and causal (all poles are inside the unit circle) then $H^*(1/z^*)$ is stable and anticausal (all poles are outside the unit circle). Zeros of $H(z)$ on the unit circle have corresponding zeros of $H^*(1/z^*)$ at identical locations on the unit circle.

To see the relationship between the frequency response of $H(z)$ and $H^*(1/z^*)$, just evaluate them at $z = e^{j\theta}$, getting $H(e^{j\theta})$ and $H^*(e^{j\theta})$. Hence the frequency response of a reflected transfer function is simply the complex-conjugate of the original frequency response. That is, any transfer function and its reflected transfer function have the same magnitude frequency response, and their phase responses are the negative of one another.

2.5.4. Allpass Transfer Functions

An *allpass* transfer function is any transfer function where the magnitude frequency response is unity for all θ,

$$|A(e^{j\theta})| = 1 . \tag{2.43}$$

This can be written as

$$A(e^{j\theta})A^*(e^{j\theta}) = 1 . \tag{2.44}$$

Applying the inverse DTFT, we get

$$a_k * a^*_{-k} = \delta_k , \tag{2.45}$$

where a_k is the impulse response of $A(e^{j\theta})$. Taking Z transforms we see that

$$A(z)A^*(1/z^*) = 1 . \tag{2.46}$$

For rational Z transforms, (2.46) implies that every pole of $A(z)$ is cancelled by a zero of $A^*(1/z^*)$. Therefore, if $A(z)$ has a pole at $z = b$, then $A^*(1/z^*)$ has a zero at $z = b$. The latter implies that $A(z)$ has a zero at $z = 1/b^*$. Therefore, any zero or pole must be accompanied by a matching pole or zero at the *conjugate-reciprocal* location.

Example 2-9.

A first-order allpass transfer function is given by

$$A(z) = \frac{z^{-1} - b^*}{1 - bz^{-1}} . \tag{2.47}$$

The pole-zero plot is shown in Fig. 2-10. Note that the pole and zero form a conjugate-reciprocal pair. For the value of c shown in the figure, the impulse response will be complex valued.

Observe that b and $1/b^*$ have the same angle in the Z plane, but their magnitudes are the reciprocal of one another, as shown in Fig. 2-10.

Example 2-10.

Since $H(z)$ and $H^*(1/z^*)$ have the same magnitude response, the ratio $H(z)/H^*(1/z^*)$ defines an allpass system, regardless of $H(z)$.

Even stronger, *every* rational allpass transfer function can be written as the ratio of one FIR transfer function to its reflection, up to a constant and a delay. Specifically, consider a transfer function of the form

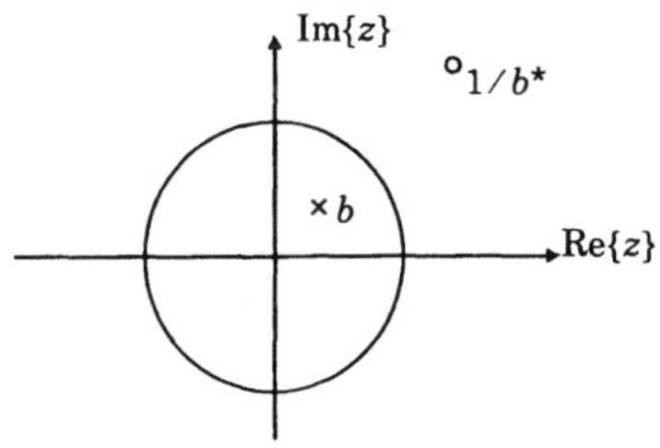

Fig. 2-10. The pole-zero plot for a first-order allpass filter.

$$A(z) = e^{j\alpha} z^M \cdot \frac{z^{-N} + b_1 z^{-N+1} + \ldots + b_N}{1 + b_1^* z^{-1} + \ldots + b_N^* z^{-N}} \; . \tag{2.48}$$

This can be rewritten as

$$A(z) = e^{j\alpha} \frac{B(z)}{z^{N-M} B^*(1/z^*)} , \qquad B(z) = 1 + b_1 z + \ldots + b_N z^N \; . \tag{2.49}$$

Such a transfer function is allpass,

$$|A(e^{j\theta})| = \left| \frac{B(e^{j\theta})}{e^{j(N-M)\theta} B^*(e^{j\theta})} \right| = 1 \; . \tag{2.50}$$

Note that if $N \geq M$, the term $z^{(N-M)}$ in the denominator contributes $N-M$ poles at $z = 0$ and $N-M$ zeros at $|z| = \infty$, the conjugate-reciprocal of 0. If $N < M$, then this term puts zeros $z = 0$ and poles at $|z| = \infty$. Thus, poles and zeros at zero and infinity also come in conjugate-reciprocal pairs. Since poles and zeros must come in conjugate-reciprocal pairs, any rational allpass transfer function can be written in the form of (2.49).

2.5.5. Minimum and Maximum-Phase Transfer Functions

A rational transfer function is said to be *strictly minimum-phase* when all its poles and zeros are inside the unit circle. See Problem 2-28 for the intuition behind the term "minimum phase." A rational transfer function is *strictly maximum-phase* when all poles and zeros are outside the unit circle. Neither strictly minimum-phase nor strictly maximum-phase transfer functions are allowed to have zeros on the unit circle itself. If they have zeros on the unit circle, then we call them *loosely minimum-phase* or *loosely maximum-phase*.

A minimum-phase transfer function is both *causal* and *stable*. To see why, observe that there can be no poles at $z = \infty$, which is clearly outside the unit circle, and hence the transfer function $H(z) = \sum_k h_k z^{-k}$ cannot contain any positive powers of z. This implies that the sequence h_k is causal, a special case of right-sided, which implies that the ROC is outside the outermost pole. Finally, since the outermost pole is still inside the unit circle, the ROC must include the unit circle, which implies stability.

The importance of minimum-phase filters stems largely from the property that, of all transfer functions with a given magnitude response, only a minimum-phase transfer function can be inverted by a causal and stable (and hence physically realizable) filter. This is because zeros of the transfer function become poles of the inverse, and for a strictly minimum-phase transfer function, all such poles will end up inside the unit circle. (A loosely minimum-phase transfer function does not necessarily have a BIBO stable inverse, but it is still useful because it may display a mild form of instability.) If $H(z)$ is minimum-phase and monic, then its reflection $H^*(1/z^*)$ is maximum-phase and monic, and *vice versa*.

A *minimum-phase sequence* is defined as the impulse response of a minimum-phase transfer function. Such sequences have useful properties. As noted above, a minimum-phase sequence is both causal and stable. Furthermore, a minimum-phase sequence must be nonzero at time zero, to prevent a zero at $z = \infty$. Another property heavily exploited by linear prediction (Chapter 3) and decision-feedback equalization (Chapter 8) is that, of all sequences with a given magnitude response, a minimum-phase sequence has its energy maximally concentrated in the vicinity of time zero (see Problem 2-28).

Any causal and stable rational transfer function can be factored as

$$H(z) = A(z)H_{\text{min}}(z)H_{\text{zero}}(z) , \tag{2.51}$$

where $A(z)$ is causal, stable, and allpass, $H_{\text{min}}(z)$ is strictly minimum phase, and $H_{\text{zero}}(z)$ is FIR with only zeros on the unit circle and corresponding poles at $z = 0$. The construction of this factorization, illustrated in Fig. 2-11, is straightforward. All zeros of $H(z)$ on the unit circle are assigned to $H_{\text{zero}}(z)$. To ensure that $H_{\text{zero}}(z)$ is causal, it is given a pole at $z = 0$ for each zero on the unit circle. Such poles may be canceled by zeros at $z = 0$ in $H_{\text{min}}(z)$, if necessary. All remaining poles in $H(z)$, which must lie inside the unit circle, are assigned to $H_{\text{min}}(z)$. All zeros that lie inside the unit circle are also assigned to $H_{\text{min}}(z)$. Each zero outside the unit circle is assigned to $A(z)$. To make sure $A(z)$ is allpass, it is assigned a pole at the conjugate-reciprocal location of each such zero. That pole will be inside the unit circle, ensuring that $A(z)$ is causal and stable. To cancel the effect of that pole, a zero at the same location is assigned to $H_{\text{min}}(z)$. When all is done, $H_{\text{min}}(z)$ should have an equal number of poles and zeros, all inside the unit circle.

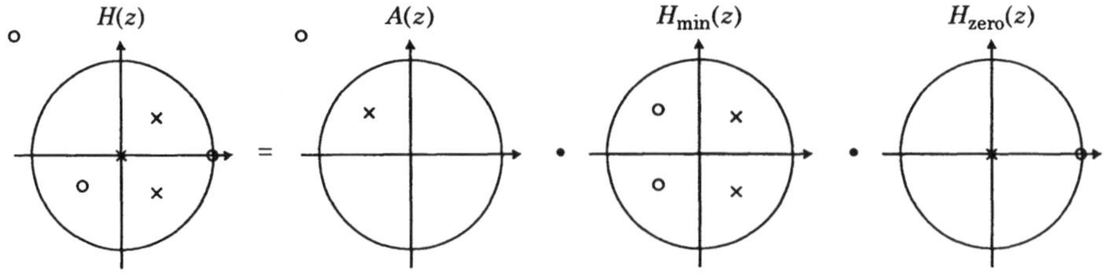

Fig. 2-11. Factorization of a causal and stable rational transfer function into an allpass transfer function, a strictly minimum-phase transfer function, and a causal transfer function with only zeros on the unit circle. Notice that the number of poles at $z = 0$ is chosen so that there are no poles at $z = \infty$, ensuring that each transfer function is causal.

We can develop another useful factorization of a stable (not necessarily causal) rational transfer function by dividing poles and zeros into four classes: those inside, on, and outside the unit circle, plus some zeros at the origin. The factorization is

$$H(z) = B \cdot z^L \cdot H_{\min}(z) \cdot H_{\max}(z) \cdot H_{\text{zero}}(z) \tag{2.52}$$

where $H_{\min}(z)$ is a strictly minimum-phase transfer function containing all the poles and zeros inside the unit circle, except possibly for some zeros at the origin, while $H_{\max}(z)$ is a strictly maximum-phase transfer function containing all the poles and zeros outside the unit circle. $H_{\text{zero}}(z)$ is an FIR transfer function containing all the zeros on the unit circle (stability rules out poles on the unit circle) with corresponding poles at the origin. In addition, we choose constants B and L so that $H_{\min}(z)$ and $H_{\text{zero}}(z)$ are causal and monic ($H_{\min}(\infty) = H_{\text{zero}}(\infty) = 1$), and $H_{\max}(z)$ is anticausal and monic ($H_{\max}(0) = 1$).

With these constraints, the factorization is unique. In particular, the terms can always be written in the form

$$H_{\min}(z) = \frac{\prod_{k=1}^{M}(1 - c_k z^{-1})}{\prod_{k=1}^{N}(1 - d_k z^{-1})}, \quad |c_k| < 1, \ \ |d_k| < 1,$$
$$H_{\text{zero}}(z) = \prod_{k=1}^{K}(1 - e_k z^{-1}), \quad |e_k| = 1,$$
$$H_{\max}(z) = \frac{\prod_{k=1}^{I}(1 - f_k z)}{\prod_{k=1}^{J}(1 - g_k z)}, \quad |f_k| < 1, \ \ |g_k| < 1. \tag{2.53}$$

An example will serve to illustrate how we turn a general rational transfer function into this canonical form.

Example 2-11.

Given the rational transfer function

$$H(z) = \frac{(1 - 0.5z^{-1})(1 - z^{-1})}{1 - 1.25z^{-1}}, \tag{2.54}$$

we can write

$$H(z) = -0.8z(1 - 0.5z^{-1})\frac{1}{1 - 0.8z}(1 - z^{-1}). \tag{2.55}$$

We can identify $B = -0.8$, $L = 1$, and

$$H_{\min}(z) = 1 - 0.5z^{-1}, \quad H_{\max}(z) = \frac{1}{1 - 0.8z}, \quad H_{\text{zero}}(z) = 1 - z^{-1}. \tag{2.56}$$

Given a transfer function in the second or third form of (2.40), the factorization in (2.52) is simple to obtain.

2.5.6. Transfer Functions that are Real and Nonnegative

In the study of random processes, transfer functions $S(z)$ that are real and nonnegative on the unit circle, satisfying $S(e^{j\theta}) \geq 0$ for all θ, arise frequently. (We will see in Chapter 3 that such transfer functions are valid power spectral densities.) In this subsection, we study the *deterministic* properties of such transfer functions.

Arithmetic and Geometric Means

We will frequently encounter the arithmetic and geometric means of a real nonnegative function $S(e^{j\theta})$. The *arithmetic mean* of $S(e^{j\theta})$ is defined as

$$\langle S \rangle_A = \frac{1}{2\pi}\int_{-\pi}^{\pi} S(e^{j\theta})\, d\theta\,, \tag{2.57}$$

which has the interpretation as the average value of $S(e^{j\theta})$. Similarly, the *geometric mean* is defined as

$$\langle S \rangle_G = \exp\left\{\frac{1}{2\pi}\int_{-\pi}^{\pi} \log S(e^{j\theta})\, d\theta\right\}. \tag{2.58}$$

The base of the exponential and logarithm must agree but otherwise do not affect the value of the geometric mean. The geometric mean is a generalization of the more familiar two-variable mean $\sqrt{ab}$ or three-variable mean $(abc)^{1/3}$ to a continuum of variables; it can be derived by taking N uniformly spaced samples of $S(e^{j\theta})$ over the interval $[-\pi, \pi)$, raising the product of these N samples to $1/N$, then letting N go to infinity.

The integral form of the geometric mean does not always behave as expected. For example, although the geometric mean of a *finite* set of numbers will be zero whenever one of the numbers is zero, the geometric mean of a *rational* function $S(e^{j\theta})$ will never be zero, even if there are several values of θ for which $S(e^{j\theta}) = 0$. (This fact is a consequence of Theorem 2-1 below.)

Example 2-12.
The geometric mean of $S(e^{j\theta}) = 2 - 2\sin(\theta)$ is unity, despite the fact that this function is zero at $\theta = \pi/2$. The arithmetic mean is two.

The arithmetic and geometric means have several useful properties. Jensen's inequality implies that

$$\langle S \rangle_G \leq \langle S \rangle_A\,, \tag{2.59}$$

with equality if and only if $S(e^{j\theta})$ is constant for all θ. In this case, both means reduce to the constant itself,

$$\langle a \rangle_A = \langle a \rangle_G = a\,. \tag{2.60}$$

Similarly,

$$\langle a \cdot S \rangle_A = a \cdot \langle S \rangle_A, \qquad \langle a \cdot S \rangle_G = a \cdot \langle S \rangle_G\,. \tag{2.61}$$

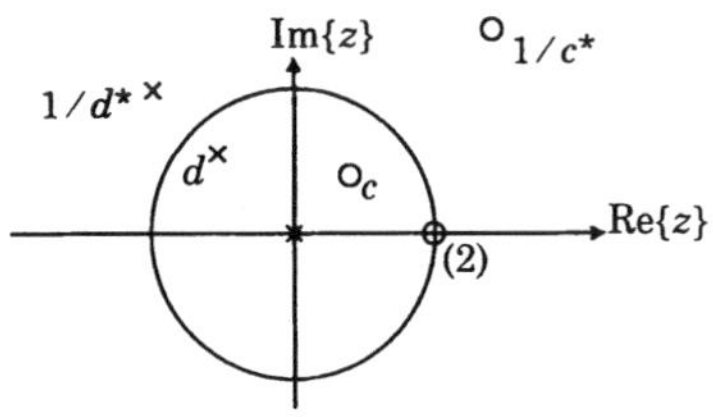

Fig. 2-12. An example of a pole zero plot for an $S(z)$ that is real valued on the unit circle.

Let $S(e^{j\theta})$, $P(e^{j\theta})$, and $Q(e^{j\theta})$ all be nonnegative real. Then the arithmetic mean also obeys the distributive law,

$$\langle a \cdot S + b \cdot P \rangle_A = a \cdot \langle S \rangle_A + b \cdot \langle P \rangle_A \ , \tag{2.62}$$

but the geometric mean does not. Conversely, the geometric mean obeys the multiplicative law

$$\langle \frac{SP}{Q} \rangle_G = \frac{\langle S \rangle_G \langle P \rangle_G}{\langle Q \rangle_G} , \tag{2.63}$$

which the arithmetic mean does not.

Spectral Factorization

Example 2-13.

The product of any transfer function and its reflection,

$$S(z) = H(z)H^*(1/z^*), \tag{2.64}$$

is always real and nonnegative, regardless of $H(z)$, since:

$$S(e^{j\theta}) = H(z)H^*(1/z^*)\Big|_{z = e^{j\theta}} = |H(e^{j\theta})|^2 \ . \tag{2.65}$$

Furthermore, observe that:

$$S(z) = S^*(1/z^*) . \tag{2.66}$$

Obviously, the right-hand side in this equation is zero (or infinity) whenever the left-hand side is zero (or infinity). This implies that, a zero (or pole) of $S(z)$ at z_0 must be accompanied by a zero (or pole) at $1/z_0^*$. Hence, the zeros (and poles) come in conjugate-reciprocal pairs.

Example 2-14.

An example pole-zero plot for a transfer function of the form (2.64) is shown in Fig. 2-12. The zero at $z = c$ has a matching zero at $z = 1/c^*$. The pole at $z = d$ also has a matching pole. The double zero at $z = 1$ illustrates another implication of (2.64); any zero on the unit circle must be double. The pole at $z = 0$ has a matching pole at $|z| = \infty$. Although the latter pole is not explicitly shown, it is implied because only three poles are shown, compared to four zeros. Note that the impulse response s_k is not real for this example, but nonetheless the frequency response $S(e^{j\theta})$ is.

The conjugate-reciprocal symmetry of (2.66) is not the same as that found for allpass filters. Rather than having zeros inside the unit circle matched by zeros outside, and poles inside matched by poles outside, as shown in Fig. 2-12, an allpass filter has poles matching zeros, and zeros matching poles. While allpass filters can be causal, only a trivial (constant) non-negative real transfer function in the form of (2.64) can be causal.

The above examples show that multiplying a rational transfer function by its reflection produces a rational transfer function that is real and nonnegative. In fact, *every* such transfer function can be written in this way, and the decomposition is unique when one of the factors is constrained to be monic and loosely minimum phase.

Theorem 2-1. (Spectral Factorization Theorem.) Any rational $S(z)$ that is real and nonnegative on the unit circle can be factored *uniquely* as

$$S(z) = \gamma^2 M(z) M^*(1/z^*) , \tag{2.67}$$

where $M(z)$ is a *monic* and *loosely minimum-phase* rational transfer function,

$$M(z) = \frac{\prod_{k=1}^{M}(1 - c_k z^{-1})}{\prod_{k=1}^{N}(1 - d_k z^{-1})} , \quad |c_k| \le 1, \ |d_k| < 1 , \tag{2.68}$$

and where $\gamma^2 = \langle S \rangle_G$ is the geometric mean of $S(e^{j\theta})$,

$$\gamma^2 = \exp\left\{ \frac{1}{2\pi} \int_{-\pi}^{\pi} \log S(e^{j\theta})\, d\theta \right\} . \tag{2.69}$$

Equation (2.67) is the *monic minimum-phase spectral factorization* of $S(z)$. It is obtained from $S(z)$ by accumulating within $M(z)$ all the poles and zeros of $S(z)$ within the unit circle, plus half of each of the zeros of $S(z)$ on the unit circle. This factorization is illustrated in Fig. 2-13, where we have, from left to right, the original $S(z)$, the loosely minimum-phase term (poles and zeros inside or on the unit circle), and its reflected transfer function. It is a remarkable fact that the constant γ^2, which is real and positive and was ostensibly chosen to allow $M(z)$ to be monic, is equal to the geometric mean of $S(e^{j\theta})$.

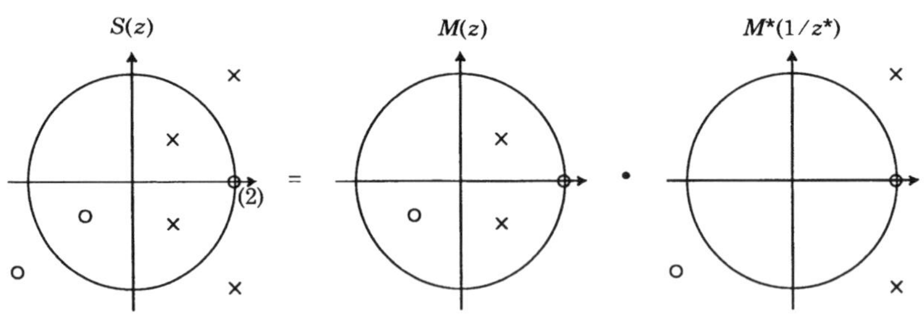

Fig. 2-13. Spectral factorization of a transfer function $S(z)$, which is non-negative real on the unit circle. The zero at $z = 1$ has multiplicity two (in general its multiplicity could be any even integer). These two zeros on the unit circle are split between $M(z)$ and $M^*(1/z^*)$.

The spectral factorization theorem is proved in Appendix 2-B for rational spectra. Nevertheless, the theorem applies to nonrational spectra as well, provided that the geometric mean in (2.69) is nonzero and finite, which implies that $S(z)$ satisfies the Paley-Wiener condition [8].

2.6. Signals as Vectors

It is possible to abstractly represent the signals in a digital communication system as vectors in a *linear space* or *vector space*, much like the familiar three-dimensional vectors in our physical world. This representation does not allow us to solve any problems that cannot be solved by other methods, but it gives valuable intuition.

2.6.1. Linear Spaces and Subspaces

A *linear space* or *vector space* is a set of elements called *vectors* together with two operators, addition of vectors and multiplication by a scalar.

Example 2-15.
Ordinary *Euclidean space* is the most familiar example of a linear space. In n-dimensional Euclidean space, a vector is specified by its n coordinates,

$$\mathbf{X} = [x_1, x_2, \ldots, x_n]^T, \tag{2.70}$$

where the superscript $(\cdot)^T$ denotes a transpose. There are rules for adding two vectors (sum the individual components) and multiplying a vector by a scalar (multiply each of the components by that scalar).

Example 2-16.
A space of some importance in this book is the *Euclidean space of complex-valued vectors*. Vectors in this space are identical to (2.70) except that the components x_k of the vector are complex-valued. Ordinary (real) Euclidean space is a special case, in which the imaginary parts of the vectors are zero.

More abstract linear spaces may be defined, as long as the addition and scalar multiplication operations satisfy certain constraints. The addition rule must produce a new vector $\mathbf{X} + \mathbf{Y}$ that must be in the linear space. Addition must obey familiar rules of arithmetic, such as the commutative and associative laws,

$$\mathbf{X} + \mathbf{Y} = \mathbf{Y} + \mathbf{X}, \qquad \mathbf{X} + (\mathbf{Y} + \mathbf{Z}) = (\mathbf{X} + \mathbf{Y}) + \mathbf{Z}. \tag{2.71}$$

The direct sum of two vectors has the interpretation illustrated in Fig. 2-14(a) for the two-dimensional Euclidean space. A linear space must include a zero vector $\mathbf{0}$, and every vector must have an *additive inverse*, denoted $-\mathbf{X}$, such that

$$\mathbf{0} + \mathbf{X} = \mathbf{X}, \qquad \mathbf{X} + (-\mathbf{X}) = \mathbf{0}. \tag{2.72}$$

Multiplication by a scalar α produces a new vector $\alpha \cdot \mathbf{X}$ that must be in the vector space. Multiplications must obey the associative law,

$$\alpha\cdot(\beta\cdot\mathbf{X}) = (\alpha\beta)\cdot\mathbf{X} \tag{2.73}$$

and also follow the rules

$$1\cdot\mathbf{X} = \mathbf{X}, \qquad 0\cdot\mathbf{X} = \mathbf{0}. \tag{2.74}$$

The geometric interpretation of multiplying a vector by a scalar is shown in Fig. 2-14(b). Finally, addition and multiplication must obey the distributive laws,

$$\alpha\cdot(\mathbf{X}+\mathbf{Y}) = \alpha\cdot\mathbf{X} + \alpha\cdot\mathbf{Y}, \qquad (\alpha + \beta)\cdot\mathbf{X} = \alpha\cdot\mathbf{X} + \beta\cdot\mathbf{X}. \tag{2.75}$$

Real linear spaces are defined in terms of real-valued scalars, while *complex* linear spaces have complex-valued scalars. We will encounter both types.

Euclidean space as defined earlier meets all of these requirements, and is therefore a linear space. There are less obvious examples.

Example 2-17.
The set of all complex-valued discrete-time signals of the form $\{y_k\}$ that have finite energy, so that,

$$\Sigma_k |y_k|^2 < \infty, \tag{2.76}$$

is a linear space; it can be viewed as a generalization of the n-dimensional Euclidean space of (2.70) to the case in which the number of components is infinite. Scalar multiplication and vector addition are the same as for Euclidean spaces.

Example 2-18.
The set of complex-valued continuous-time signals $y(t)$ that have finite energy,

$$\int_{-\infty}^{\infty} |y(t)|^2 dt < \infty, \tag{2.77}$$

is a linear space. We can think of this space as a strange Euclidean space with a continuum of coordinates. The definition of multiplication of a signal vector by a scalar and the addition of two signal vectors are the natural and obvious, and the definition of a zero vector is the zero-valued signal.

Fig. 2-14. Elementary operations in a two-dimensional linear space.
(a) Sum of two vectors. (b) Multiplication of a vector by a scalar.

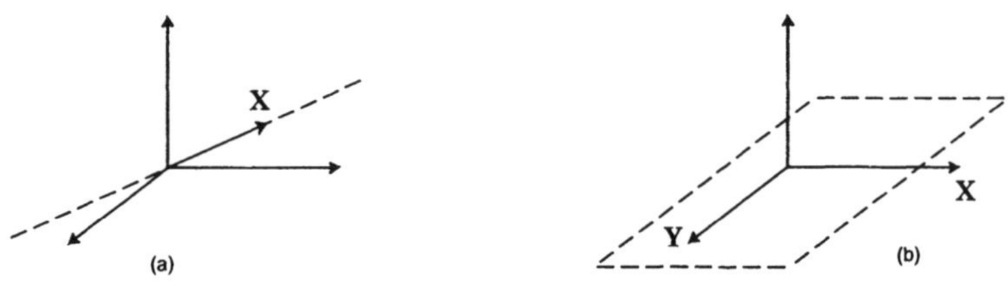

Fig. 2-15. Subspaces in three-dimensional Euclidean space.

A *subspace* of a linear space is a *subset* of the linear space that is itself a linear space. Roughly speaking this means that the sum of any two vectors in the subspace must also be in the subspace, and the product of any vector in the subspace by any scalar must also be in the subspace.

Example 2-19. ---

An example of a subspace in three-dimensional Euclidean space is either a line or a plane in the space, where in either case the vector **0** must be in the subspace.

Example 2-20. ---

A more general subspace is the set of vectors obtained by forming all possible weighted linear combinations of n vectors $\mathbf{X}_1, \ldots, \mathbf{X}_n$. The subspace so formed is said to be *spanned* by the set of n vectors. This is illustrated in Fig. 2-15 for three-dimensional Euclidean space. In Fig. 2-15(a), the subspace spanned by $\mathbf{X}$ is the dashed line, which is infinite in length and co-linear with the vector $\mathbf{X}$. Any vector on this line can be obtained by multiplying $\mathbf{X}$ by the appropriate scalar. In Fig. 2-15(b), the subspace spanned by $\mathbf{X}$ and $\mathbf{Y}$ is the plane of infinite extent (depicted by the dashed lines) that is determined by the two vectors. Any vector in this plane can be formed as a linear combination of the two vectors multiplied by appropriate scalars.

The most famous subspace encountered in digital communications is called *signal space*, and is defined in the context of a digital transmitter that transmits one of M possible signals $\{s_1(t), s_2(t), \ldots s_M(t)\}$. In this setting, the *signal space* $\mathcal{S}$ is defined as the span of the M signals:

$$\mathcal{S} = \text{span}\{s_1(t), s_2(t), \ldots s_M(t)\} . \tag{2.78}$$

In other words, $\mathcal{S}$ is the set of all signals that can be expressed as a linear combination of $\{s_1(t), s_2(t), \ldots s_M(t)\}$.

Example 2-21. ---

Given a binary signal set with $s_1(t)$ and $s_2(t)$ shown in Fig. 2-16(a), the signal space $\mathcal{S}$ contains an uncountably infinite number of signals, four examples of which are shown in Fig. 2-16(b). Because adding any two signals in $\mathcal{S}$ produces another signal in $\mathcal{S}$, it is itself a linear space.

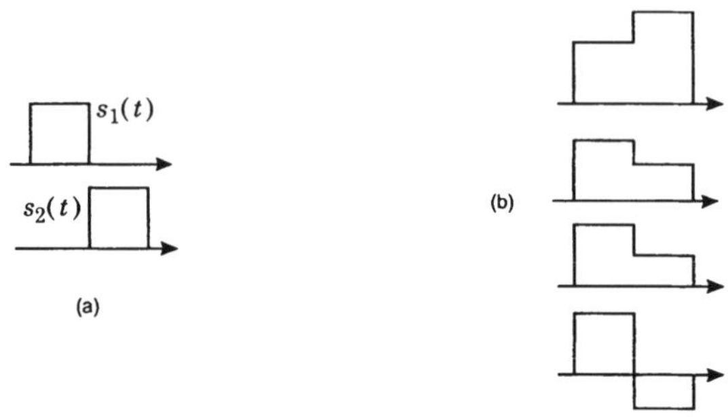

Fig. 2-16. (a) Two signals $s_1(t)$, $s_2(t)$; (b) four examples of signals that are in $\mathcal{S} = \text{span}\{s_1(t), s_2(t)\}$.

2.6.2. Inner Products

The definition of a linear space does not capture the most important properties of Euclidean space; namely, its *geometric* structure. This structure includes such concepts as the length of a vector, the distance between two vectors, and the angle between two vectors. All these properties of Euclidean space can be deduced from the definition of an *inner product* of two vectors, defined for n-dimensional Euclidean space by

$$\langle \mathbf{X}, \mathbf{Y} \rangle = \sum_{i=1}^{n} x_i \, y_i^* \\ = \mathbf{Y}^* \mathbf{X}, \tag{2.79}$$

where y_i^* is the complex conjugate of the scalar y_i, and $\mathbf{Y}^*$ is the *conjugate transpose* of the vector $\mathbf{Y}$. The squared length of the vector is denoted $\|\mathbf{X}\|^2$, and is defined by $\|\mathbf{X}\|^2 = \langle \mathbf{X}, \mathbf{X} \rangle = \mathbf{X}^*\mathbf{X} = \sum_{i=1}^{n} |x_i|^2$.

The inner product as applied to Euclidean space can be generalized to the other linear spaces of interest. The important consequence is that the geometric concepts familiar in Euclidean space can be applied to these spaces as well. Let $\mathbf{X}$ and $\mathbf{Y}$ be elements of a linear space; they might represent vectors (Example 2-15), or discrete-time sequences (Example 2-15), or continuous-time waveforms (Example 2-15). Then the inner product $\langle \mathbf{X}, \mathbf{Y} \rangle$ is a mapping from two elements of the linear space to a scalar (real or complex, depending on the scalar field associated with the linear space), and it must obey *the rules*

$$\langle \mathbf{X} + \mathbf{Y}, \mathbf{Z} \rangle = \langle \mathbf{X}, \mathbf{Z} \rangle + \langle \mathbf{Y}, \mathbf{Z} \rangle \\ \langle \alpha \cdot \mathbf{X}, \mathbf{Y} \rangle = \alpha \langle \mathbf{X}, \mathbf{Y} \rangle, \qquad \langle \mathbf{X}, \mathbf{Y} \rangle = \langle \mathbf{Y}, \mathbf{X} \rangle^* \\ \langle \mathbf{X}, \mathbf{X} \rangle > 0, \quad \text{for } \mathbf{X} \neq \mathbf{0}. \tag{2.80}$$

We again adopt the shorthand notation $\|\mathbf{X}\|^2 = \langle \mathbf{X}, \mathbf{X} \rangle$, where $\|\mathbf{X}\|$ is called the *norm* of $\mathbf{X}$.

These rules are all obeyed by the familiar Euclidean space inner product of (2.79), as can be easily verified. For the other linear spaces of interest, analogous definitions of the inner product satisfying the rules can be made. In particular, for discrete-time signals, the inner product and norm are defined by

$$\langle x_k, y_k \rangle = \sum_{k=-\infty}^{\infty} x_k y_k^* \,, \qquad \| x_k \|^2 = \sum_{k=-\infty}^{\infty} |x_k|^2 \,, \tag{2.81}$$

which is a natural extension of the finite-dimensional case. For continuous-time signals, the summation becomes an integral, and the inner product is defined as

$$\langle x(t), y(t) \rangle = \int_{-\infty}^{\infty} x(t) y^*(t) dt \,, \qquad \| x(t) \|^2 = \int_{-\infty}^{\infty} |x(t)|^2 \, dt \,. \tag{2.82}$$

Exercise 2-13.

Verify that the definitions of inner product of (2.81) and (2.82) satisfy the properties of (2.80).

All valid inner products satisfy the *Schwarz inequality*, which bounds their magnitudes:

Exercise 2-14.

(The Schwarz Inequality.) Show that for two elements $\mathbf{X}$ and $\mathbf{Y}$ of an inner product space,

$$|\langle \mathbf{X}, \mathbf{Y} \rangle| \le \|\mathbf{X}\| \cdot \|\mathbf{Y}\| \,, \tag{2.83}$$

with equality if and only if $\mathbf{X} = K \cdot \mathbf{Y}$ for some scalar K.

In the context of continuous-time signals, the inner product (2.82) is also called the *correlation* of $x(t)$ with $y(t)$, and its computation arises frequently. Intuitively, the correlation measures how closely the two signals resemble one another. At one extreme, when $x(t)$ and $y(t)$ are identical, the correlation reduces to the energy $\int |x(t)|^2 dt$. At the other extreme, the two signals are said to be *orthogonal* when the correlation is zero, $\langle x(t), y(t) \rangle = 0$, which is sometimes written as $x(t) \perp y(t)$. In general, two element signals or vectors $\mathbf{X}$ and $\mathbf{Y}$ are said to be orthogonal whenever $\langle \mathbf{X}, \mathbf{Y} \rangle = 0$, in which case the shorthand $\mathbf{X} \perp \mathbf{Y}$ is used.

The geometric properties are so important that the special name *inner-product space* is given to a linear space on which an inner product is defined. Thus, both Example 2-17 and Example 2-18 defined earlier are inner-product spaces. If the inner product space has the additional property of *completeness*, then it is defined to be a *Hilbert space*. Intuitively the notion of completeness means that there are no "missing" vectors that are arbitrarily close to vectors in the space but are not themselves in the space. Since the spaces used in this book are all complete and hence formally Hilbert spaces, we will not dwell on this property further.

2.6.3. Projection onto a Subspace

Let $\mathcal{H}$ be a Hilbert space, and let $\mathcal{S}$ be a subspace of $\mathcal{H}$. We are often interested in finding an element of $\mathcal{S}$ that best matches a given element of $\mathcal{H}$. We define the *projection on* $\mathcal{S}$ of any $\mathbf{X} \in \mathcal{H}$ as the unique element $\hat{\mathbf{X}} \in \mathcal{S}$ that is "closest" to $\mathbf{X}$, satisfying

$$\|\mathbf{X}-\hat{\mathbf{X}}\| = \min_{\mathbf{Y}\in\mathcal{S}} \|\mathbf{X}-\mathbf{Y}\| . \tag{2.84}$$

The *projection theorem*, which is proven in [1], states that $\hat{\mathbf{X}} \in \mathcal{S}$ is the projection if and only if the projection *error* $\mathbf{X}-\hat{\mathbf{X}}$ is orthogonal to the subspace onto which $\mathbf{X}$ is being projected.

Theorem 2-2. (Projection Theorem.) Let $\mathcal{H}$ be a Hilbert space, and let $\mathcal{S}$ be a subspace of $\mathcal{H}$. Then the *projection* $\hat{\mathbf{X}} \in \mathcal{S}$ of $\mathbf{X}$ onto $\mathcal{S}$ satisfies (2.84) if and only if

$$(\mathbf{X}-\hat{\mathbf{X}}) \perp \mathbf{Y}, \qquad \text{for all } \mathbf{Y} \in \mathcal{S}. \tag{2.85}$$

It is instructive to show how this orthogonality condition implies (2.84). Because $\mathbf{X} \in \mathcal{H}$ and $\mathbf{Y} \in \mathcal{S}$, we can decompose the difference $\mathbf{X}-\mathbf{Y}$ as the sum of $\mathbf{X}-\hat{\mathbf{X}} \perp \mathcal{S}$ and $\hat{\mathbf{X}}-\mathbf{Y} \in \mathcal{S}$; this decomposition is illustrated below:

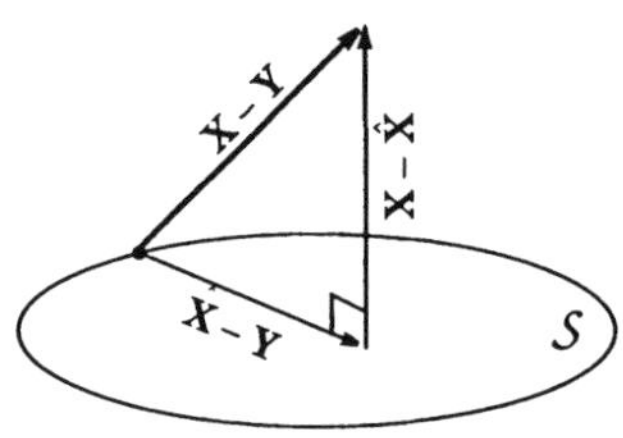

As a consequence, we have

$$\begin{aligned}\|\mathbf{X}-\mathbf{Y}\|^2 &= \|\mathbf{X}-\hat{\mathbf{X}}+\hat{\mathbf{X}}-\mathbf{Y}\|^2 \\ &= \|\mathbf{X}-\hat{\mathbf{X}}\|^2 + \|\hat{\mathbf{X}}-\mathbf{Y}\|^2 + 2\mathrm{Re}\{\langle \mathbf{X}-\hat{\mathbf{X}}, \hat{\mathbf{X}}-\mathbf{Y}\rangle\} . \end{aligned} \tag{2.86}$$

The projection theorem guarantees that $\hat{\mathbf{X}}-\mathbf{Y} \in \mathcal{S}$ is orthogonal to the projection error, so that the last term is zero, yielding:

$$\|\mathbf{X}-\mathbf{Y}\|^2 = \|\mathbf{X}-\hat{\mathbf{X}}\|^2 + \|\hat{\mathbf{X}}-\mathbf{Y}\|^2 . \tag{2.87}$$

This is a *Pythagorean theorem*: the left-hand side is the squared hypotenuse of a right-angle triangle, and the right-hand side adds the squared height to the squared base. The squared height is independent of $\mathbf{Y} \in \mathcal{S}$. Clearly, therefore, the sum is minimized by choosing $\mathbf{Y} = \hat{\mathbf{X}}$, which collapses the base to zero, thus implying (2.84).

Example 2-22.

A projection is illustrated in Fig. 2-17 for three-dimensional Euclidean space, where the subspace $\mathcal{S}$ is the plane formed by the x-axis and y-axis and $\mathbf{X}$ is an arbitrary vector. The projection is the result of dropping a perpendicular line from $\mathbf{X}$ down to the plane (this is the dashed line). The resulting vector $(\mathbf{X}-\hat{\mathbf{X}})$ is the vector shown parallel to the dashed line. It is orthogonal to the plane $\mathcal{S}$, and hence to every vector in $\mathcal{S}$.

Example 2-23.

Let $\mathcal{S} = \mathrm{span}\{s_1(t), s_2(t)\}$ be the linear subspace spanned by two rectangular waveforms of Example 2-21. Then the projection $\hat{r}(t)$ of a truncated sinusoid $r(t)$ onto $\mathcal{S}$ is sketched in Fig. 2-18.

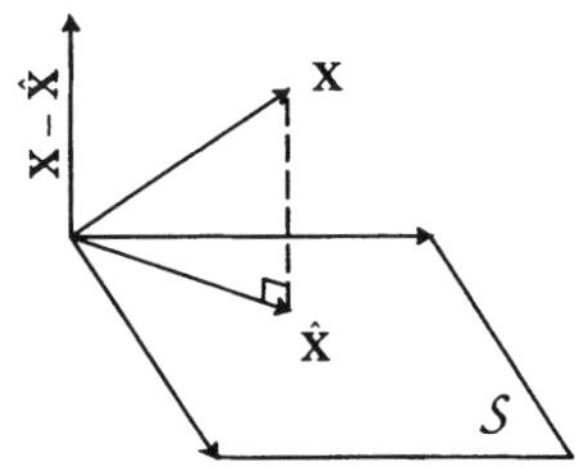

Fig. 2-17. The vector in $\mathcal{S}$ closest to **X** is evidently $\hat{\mathbf{X}}$; any other vector in $\mathcal{S}$ is farther from **X**.

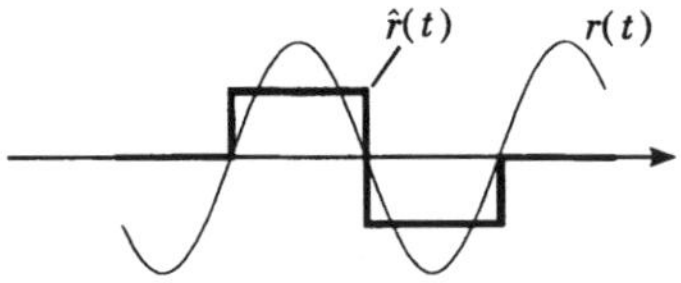

Fig. 2-18. The projection of a sinusoid onto the subspace spanned by two rectangular pulses.

2.6.4. Projection onto the Signal Space

When a digital transmitter transmits one of M signals $\{s_1(t), \ldots s_M(t)\}$, it is common for the receiver to project the received signal $r(t)$ onto the signal space spanned by the M signals. (The motivation for this projection will be explained in later chapters.) In concept this projection can be found by searching in a brute-force manner the subspace for the signal closest to the $r(t)$, but in practice it is much more efficient to project $r(t)$ onto a collection of one-dimensional subspaces and combine the results, as described below.

A set of functions $\{\phi_1(t), \phi_2(t), \ldots\}$ is said to be *orthonormal* when each signal has unit energy, and distinct signals are orthogonal, so that:

$$\int_{-\infty}^{\infty}\phi_i(t)\phi_k^*(t)dt = \delta_{ik} = \begin{cases} 1, & i = k \\ 0, & i \neq k \end{cases}. \tag{2.88}$$

An *orthonormal basis* for the signal space $\mathcal{S} = \text{span}\{s_1(t), \ldots s_M(t)\}$ is a minimal set of N orthonormal functions $\{\phi_1(t), \ldots \phi_N(t)\}$ with the property that every element $s(t) \in \mathcal{S}$ can be expressed as a linear combination of basis functions:

$$s(t) = \sum_{i=1}^{N} s_i\phi_i(t). \tag{2.89}$$

Any signal $r(t)$ can now be expressed as a sum $r(t) = \hat{r}(t) + e(t)$, where $\hat{r}(t) = \sum_{i=1}^{N} r_i\phi_i(t)$ is the projection of $r(t)$ onto $\mathcal{S}$, and $e(t) = r(t) - \hat{r}(t)$ is the projection error. The j-th expansion coefficient r_j of the projection is found by taking the inner product of $r(t)$ with the j-th basis function:

$$\begin{aligned}\int_{-\infty}^{\infty} r(t)\phi_j^*(t)\,dt &= \langle\, r(t), \phi_j(t)\,\rangle \\ &= \langle\, \hat{r}(t) + e(t), \phi_j(t)\,\rangle \\ &= \langle\, \hat{r}(t), \phi_j(t)\,\rangle + \langle\, e(t), \phi_j(t)\,\rangle \\ &= \langle\, \textstyle\sum_{i=1}^{N} r_i\phi_i(t), \phi_j(t)\,\rangle + 0 \\ &= r_j\,. \end{aligned} \tag{2.90}$$

The second equality follows from the fact that the projection error $e(t)$ is orthogonal to everything in $\mathcal{S}$, including $\phi_j(t)$.

Since the k-th signal $s_k(t)$ is an element of $\mathcal{S}$, it too can be expanded in terms of a basis:

$$s_k(t) = \sum_{i=1}^{N} s_{k,i}\phi_i(t)\,. \tag{2.91}$$

This expansion is most useful when the number of basis functions is smaller than the size of the original signal set, or $N < M$.

Example 2-24.

A brute force implementation of an M-ary transmitter might consist of a bank of M waveform generators, one for each waveform, with a switch connecting the selected waveform to the transmission medium. This is very inefficient when M is large (and M as high as 256 is not uncommon). In contrast, a transmitter that generates $s_i(t)$ using the expansion (2.91) would need only N waveform generators.

Naturally, there are two pertinent questions: How small can N be? And what are the corresponding N basis functions? The answers to both questions can be found using the Gram-Schmidt orthonormalization procedure.

The Gram-Schmidt Orthonormalization Procedure

The Gram-Schmidt orthonormalization procedure is a systematic method for converting a set of M waveforms $\{s_1(t), \ldots s_M(t)\}$ into an orthonormal basis $\{\phi_1(t), \ldots \phi_N(t)\}$ of size N for their span, with N as small as possible. Let the first basis function be:

$$\phi_1(t) = \frac{s_1(t)}{\|s_1(t)\|}, \tag{2.92}$$

where the denominator is the square root of the energy of $s_1(t)$. (The signals may have to be reordered to ensure that this energy is nonzero.) This defines a one-dimensional subspace $\mathcal{S}_1 = \text{span}\{\phi_1(t)\}$. If we let $\hat{s}_2(t)$ denote the projection of $s_2(t)$ onto this subspace, then the projection error will be orthogonal to $\mathcal{S}_1$; in particular, $s_2(t) - \hat{s}_2(t) \perp \phi_1(t)$. Thus, if we normalize the projection error to have unit energy, we produce a second basis function:

$$\phi_2(t) = \frac{s_2(t) - \hat{s}_2(t)}{\|s_2(t) - \hat{s}_2(t)\|}\,. \tag{2.93}$$

Now $\phi_1(t)$ and $\phi_2(t)$ are unit-energy and orthogonal. But what if the denominator in (2.93) is zero? This would imply that already $s_2(t) \in \mathcal{S}_1$, and so the projection error is zero. In this case we do not use (2.93), but instead we move $s_2(t)$ to the end of the list and *renumber* the signals so that $s_2(t)$ becomes the new $s_M(t)$, $s_3(t)$ becomes the new $s_2(t)$, $s_4(t)$ becomes the new $s_3(t)$, etc.

The Gram-Schmidt procedure repeats the above process M times. Let $\hat{s}_k(t)$ denote the projection of $s_k(t)$ onto the span of $\{\phi_1(t), \ldots \phi_{k-1}(t)\}$ for $k \in \{2, \ldots, M\}$. If the projection error $s_k(t) - \hat{s}_k(t)$ is zero, renumber the signals by moving $s_k(t)$ to the end of the list. Otherwise, the k-th basis function can be found by normalizing the k-th projection error:

$$\phi_k(t) = \frac{s_k(t) - \hat{s}_k(t)}{\| s_k(t) - \hat{s}_k(t) \|} . \tag{2.94}$$

The Gram-Schmidt orthonormalization procedure is defined by the recursion (2.94), together with the initialization (2.92) and the renumbering strategy. If all $M - 1$ of the projection errors are nonzero, then the signals are linearly independent and the final number of basis functions N will be the same as the number of signals, $N = M$. Each zero projection error would decrease the number of basis functions by one, yielding $N < M$. The *dimension* of $\mathcal{S}$ is the number N of basis functions. The basis functions themselves are not unique; reordering the signals $\{s_i(t)\}$ will generally lead to a different basis. However, the dimension N is fixed and independent of the basis.

Example 2-25.

Recall the signal space $\mathcal{S}$ spanned by the two rectangular pulses of Fig. 2-16(a). There are an infinite number of possible bases for this space. Three examples are shown below:

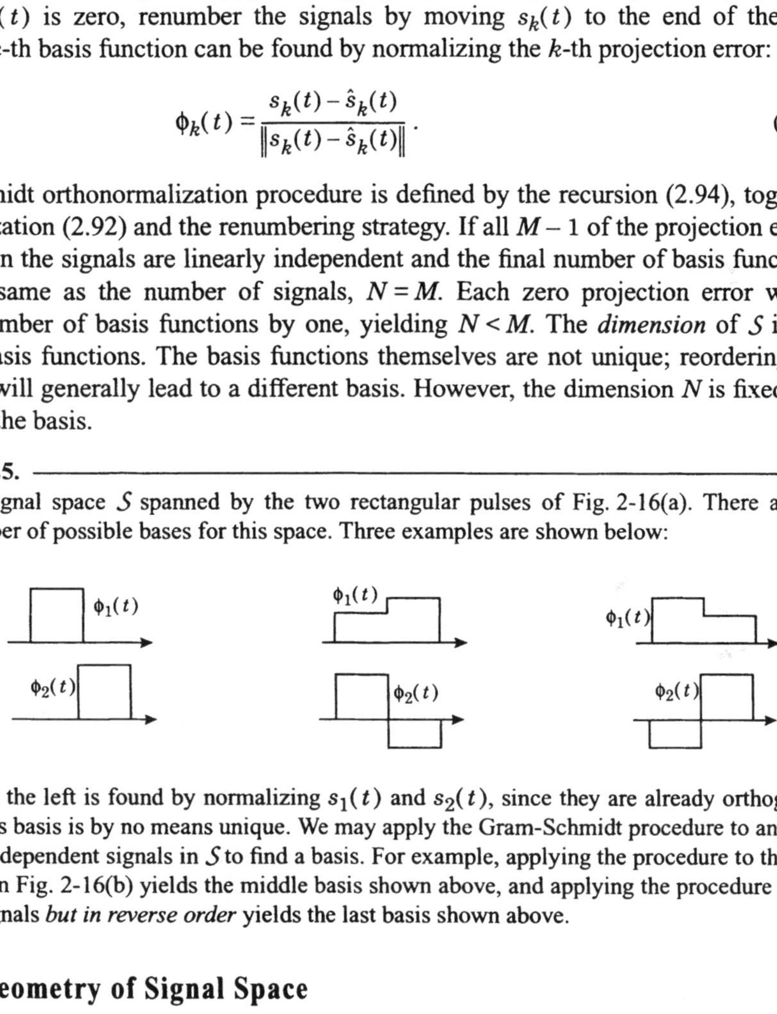

The basis on the left is found by normalizing $s_1(t)$ and $s_2(t)$, since they are already orthogonal. However, this basis is by no means unique. We may apply the Gram-Schmidt procedure to any pair of linearly independent signals in $\mathcal{S}$ to find a basis. For example, applying the procedure to the first two signals in Fig. 2-16(b) yields the middle basis shown above, and applying the procedure to the same two signals *but in reverse order* yields the last basis shown above.

2.6.5. The Geometry of Signal Space

Once we find an orthonormal basis $\{\phi_1(t), \ldots \phi_N(t)\}$ for the signal space $\mathcal{S}$ spanned by $\{s_1(t), \ldots s_M(t)\}$, the k-th signal $s_k(t)$ may be expressed as a linear combination of basis functions according to (2.91), where the coefficients are given by:

$$s_{k,i} = \int_{-\infty}^{\infty} s_k(t)\phi_i^*(t)dt\,. \tag{2.95}$$

Hence, in the context of a basis $\{\phi_1(t), \ldots \phi_N(t)\}$, the k-th signal is uniquely specified by the N expansion coefficients. Thus, we may associate with each signal $s_k(t) \in \mathcal{S}$ the vector

$$\mathbf{s}_k = [s_{k,1}, s_{k,2}, \ldots s_{k,N}]^T, \tag{2.96}$$

which uniquely specifies the signal.

A remarkable consequence is that operations on $s_1(t) \ldots s_M(t)$ can be interpreted as like operations on the signal vectors $\mathbf{s}_1 \ldots \mathbf{s}_M$. In particular, observe that:

$$\begin{aligned}
\langle s_j(t), s_k(t) \rangle &= \int_{-\infty}^{\infty} s_j(t)s_k^*(t)dt \\
&= \int_{-\infty}^{\infty}\left(\sum\nolimits_{i=1}^{N} s_{j,i}\phi_i(t)\right)\left(\sum\nolimits_{m=1}^{N} s_{k,m}^*\phi_m^*(t)\right)dt \\
&= \sum\nolimits_{i=1}^{N} s_{j,i}\sum\nolimits_{m=1}^{N} s_{k,m}^*\int_{-\infty}^{\infty}\phi_i(t)\phi_m^*(t)\,dt \\
&= \sum\nolimits_{i=1}^{N} s_{j,i}\sum\nolimits_{m=1}^{N} s_{k,m}^*\delta_{im} \\
&= \sum\nolimits_{i=1}^{N} s_{j,i}s_{k,i}^* \\
&= \mathbf{s}_k^*\mathbf{s}_j \\
&= \langle \mathbf{s}_j, \mathbf{s}_k \rangle.
\end{aligned} \tag{2.97}$$

Hence, the inner product between two signals in $\mathcal{S}$ is identical to the inner product of the corresponding coefficient vectors in N-dimensional complex Euclidean space. Furthermore, as a special case of (2.97) when $j = k$, we have:

$$\| s_k(t) \|^2 = \int_{-\infty}^{\infty} | s_k(t) |^2 dt = \sum\nolimits_{i=1}^{N} | s_{k,i} |^2 = \| \mathbf{s}_k \|^2\,. \tag{2.98}$$

In words, the energy of the signal is the squared length of its corresponding vector. Finally, if we apply this result to the difference between two signals, we find that:

$$\| s_j(t) - s_k(t) \|^2 = \int_{-\infty}^{\infty} | s_j(t) - s_k(t) |^2 dt = \| \mathbf{s}_j - \mathbf{s}_k \|^2\,. \tag{2.99}$$

Hence, the energy in the error between two signals is equal to the squared distance between the two corresponding vectors.

Observe that (2.98) is a form of *Parseval's relationship*: the energy of a signal in one domain (the time domain) is equal to its energy in another. For this reason, (2.97) may be viewed as a *generalized* Parseval's relationship: the inner product in one domain is equal to the inner product in another. See Problem 2-4 and Problem 2-5.

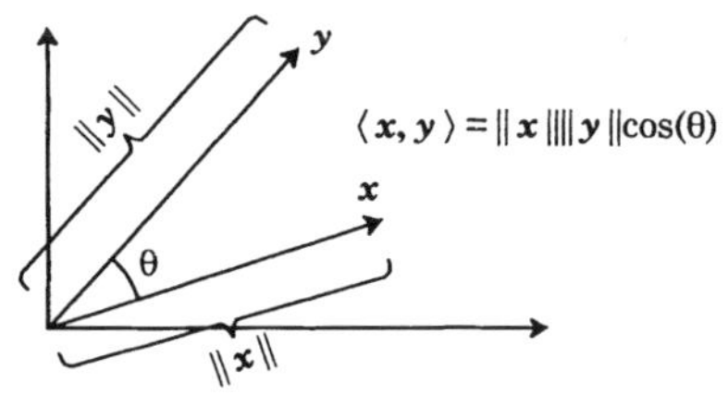

Fig. 2-19. Geometrical interpretation of inner product, illustrated in the plane spanned by x and y.

The implication of the equivalence (2.97) is that, by equating signals with their vector counterparts in Euclidean space, we can now adopt geometric intuition when thinking about continuous-time signals. The power of this tool stems from our considerable experience and intuition regarding Euclidean space. For example, we may think of signals as being close to one another when the distance between their vectors is small, i.e., when the energy of the error is small. Or we may think of two signals as being orthogonal, or at right angles, when their correlation is zero.

When the signals are real, the signal vectors $\{\mathbf{s}_1, \ldots \mathbf{s}_M\}$ live in *real* Euclidean space of N dimensions. In this case, the inner product defines the *angle* between two signals. Consider the correlation between two signals $x(t)$ and $y(t)$; as illustrated in Fig. 2-19, the correlation is equal to the product of the length of the first vector, the length of the second vector, and the cosine of the angle between the vectors. The angle θ between two signals $x(t)$ and $y(t)$ is the angle between the two signal vectors x and y. Note that $\|x\|\cos(\theta)$ is the length of the component of x in the direction of y. Hence we get a particularly useful interpretation of correlation: $\int_{-\infty}^{\infty} x(t)y(t)\,dt / (\int_{-\infty}^{\infty} |y(t)|^2 dt)^{1/2} = \langle x(t), y(t) \rangle / \| y(t) \|$ is the length of the component of $x(t)$ in the direction of $y(t)$, and $\langle y(t), x(t) \rangle / \| x(t) \|$ is the length of the component of $y(t)$ in the direction of $x(t)$. Here, the length of a signal is the square root of its energy.

Example 2-26.

Consider $s_1(t)$ and $s_2(t)$ as defined in Fig. 2-16(a), and define $s_3(t)$ through $s_6(t)$ as the four signals in Fig. 2-16(b) from top to bottom. Given a basis for the span $\mathcal{S}$ of these signals, each can be represented as a two-dimensional vector of expansion coefficients. For example, in terms of the first basis of Example 2-25, the vectors $\mathbf{s}_1$ through $\mathbf{s}_6$ corresponding to $s_1(t)$ through $s_6(t)$ are shown in Fig. 2-20(a). Fig. 2-20(b) shows how the same signals map to vectors when the second basis of Example 2-25 is used. Finally, Fig. 2-20(c) corresponds to the last basis of Example 2-25. Observe that the picture in (b) is a reflected and rotated version of that in (a), while the picture in (c) is a rotated version (by about 34°) of that in (a). As a result, the geometric relationships between signals is invariant to the choice of the basis: the distance between $\mathbf{s}_2$ and $\mathbf{s}_3$ (for example, or the angle between $\mathbf{s}_4$ and $\mathbf{s}_5$) is the same regardless of which basis is chosen. This invariance is a general result that follows from the generalized Parseval's relationship of (2.97): since the left-hand side of (2.97) is clearly independent of the basis, the right-hand side must be as well.

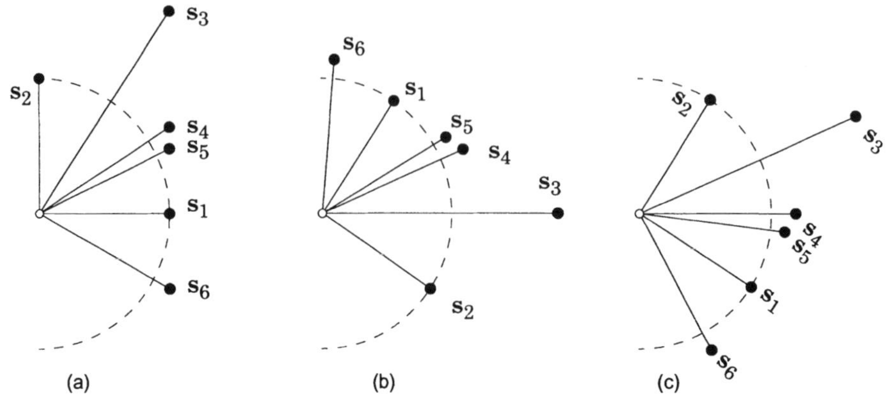

Fig. 2-20. A geometric representation of the six waveforms of Fig. 2-16. Regardless of whether basis (a), basis (b), or basis (c) is used, the geometric relationships between the different signal vectors remain the same.

Further Reading

Many textbooks cover the topics of this chapter in a more introductory and complete fashion than we do here. McGillem and Cooper [2], Oppenheim and Willsky [3], and Ziemer, Tranter, and Fannin [4] are useful for techniques applicable to both continuous and discrete-time systems. For discrete-time techniques only, the texts by Oppenheim and Schafer [5] and Jackson [6] are recommended. For continuous-time systems, with some discussion of discrete-time systems, we recommend Schwarz and Friedland [7]. To explore the Fourier transform in more mathematical depth, we recommend Papoulis [8] and Bracewell [9].

Appendix 2-A. Properties of the Fourier Transform

The properties of both discrete and continuous-time Fourier transforms are summarized in this appendix. We define the even part $f_e(x)$ and odd part $f_o(x)$ of a function $f(x)$ to be

$$f_e(x) = \frac{1}{2}[f(x) + f^*(-x)],$$
$$f_o(x) = \frac{1}{2}[f(x) - f^*(-x)], \tag{2.100}$$

so for example, $$X_e(e^{j\theta}) = \frac{1}{2}[X(e^{j\theta}) + X_e^*(e^{-j\theta})].$$

We define the rectangular function as follows,

$$\operatorname{rect}(t, T) = \begin{cases} 1; & |t| \le T \\ 0, & |t| > T \end{cases}, \tag{2.101}$$

and the unit step function for continuous-time and discrete-time signals as

$$u(x) = \begin{cases} 1; & x \ge 0 \\ 0, & x < 0 \end{cases}, \qquad u_k = \begin{cases} 1; & k \ge 0 \\ 0, & k < 0 \end{cases} \tag{2.102}$$

Table 2-1. Fourier Transform Symmetries.

Continuous time			Discrete time		
$x(t)$	$\leftrightarrow$	$X(f)$	x_k	$\leftrightarrow$	$X(e^{j\theta})$
$x(-t)$	$\leftrightarrow$	$X(-f)$	x_{-k}	$\leftrightarrow$	$X(e^{-j\theta})$
$x^*(t)$	$\leftrightarrow$	$X^*(-f)$	x^*_k	$\leftrightarrow$	$X^*(e^{-j\theta})$
$x^*(-t)$	$\leftrightarrow$	$X^*(f)$	x^*_{-k}	$\leftrightarrow$	$X^*(e^{j\theta})$
$\mathrm{Re}\{x(t)\}$	$\leftrightarrow$	$X_e(f)$	$\mathrm{Re}\{x_k\}$	$\leftrightarrow$	$X_e(e^{j\theta})$
$j\mathrm{Im}\{x(t)\}$	$\leftrightarrow$	$X_o(f)$	$j\mathrm{Im}\{x_k\}$	$\leftrightarrow$	$X_o(e^{j\theta})$
$x_e(t)$	$\leftrightarrow$	$\mathrm{Re}\{X(f)\}$	$x_{e,k}$	$\leftrightarrow$	$\mathrm{Re}\{X(e^{j\theta})\}$
$x_o(t)$	$\leftrightarrow$	$j\mathrm{Im}\{X(f)\}$	$x_{o,k}$	$\leftrightarrow$	$j\mathrm{Im}\{X(e^{j\theta})\}$

Table 2-2. Fourier Transform Properties.

Continuous time			Discrete time		
$ax(t)+by(t)$	$\leftrightarrow$	$aX(f)+bY(f)$	ax_k+by_k	$\leftrightarrow$	$aX(e^{j\theta})+bY(e^{j\theta})$
$x(t)*y(t)$	$\leftrightarrow$	$X(f)Y(f)$	$x_k * y_k$	$\leftrightarrow$	$X(e^{j\theta})Y(e^{j\theta})$
$x(t)y(t)$	$\leftrightarrow$	$X(f)*Y(f)$	$x_k y_k$	$\leftrightarrow$	$\frac{1}{2\pi}\int_{-\pi}^{\pi} X(e^{j\alpha})Y(e^{j(\theta-\alpha)})d\alpha$
$x(at)$	$\leftrightarrow$	$\frac{1}{\lvert a\rvert}X\left(\frac{f}{a}\right)$			
$x(t-\tau)$	$\leftrightarrow$	$e^{-j2\pi f\tau}X(f)$	x_{k-N}	$\leftrightarrow$	$e^{-jN\theta}X(e^{j\theta})$
$e^{j2\pi f_0 t}x(t)$	$\leftrightarrow$	$X(f-f_0)$	$e^{j\theta_0 k}x_k$	$\leftrightarrow$	$X(e^{j(\theta-\theta_0)})$
$\cos(2\pi f_0 t)x(t)$	$\leftrightarrow$	$\frac{1}{2}X(f-f_0)+\frac{1}{2}X(f+f_0)$	$\cos(\theta_0 k)x_k$	$\leftrightarrow$	$\frac{1}{2}X(e^{j(\theta-\theta_0)})+\frac{1}{2}X(e^{j(\theta+\theta_0)})$
$\frac{d^m}{dt^m}x(t)$	$\leftrightarrow$	$(j2\pi f)^m X(f)$			
$(-j2\pi t)^m x(t)$	$\leftrightarrow$	$\frac{d^m}{df^m}X(f)$			
$\int_{-\infty}^{t} x(t)\,d\tau$	$\leftrightarrow$	$\frac{1}{j2\pi f}X(f)+\frac{1}{2}\delta(f)\int_{-\infty}^{\infty}x(t)\,d\tau$			

Table 2-3. Fourier Transform Pairs[1]

$e^{j2\pi f_0 t}$	$\leftrightarrow$	$\delta(f - f_0)$	$e^{j\theta_0 k}$	$\leftrightarrow$	$2\pi\delta(\theta - \theta_0)$
$\delta(t - t_0)$	$\leftrightarrow$	$e^{-j2\pi f t_0}$	δ_{k-d}	$\leftrightarrow$	$e^{-jd\theta}$
$\cos(2\pi f_0 t)$	$\leftrightarrow$	$\frac{1}{2}\delta(f - f_0) + \frac{1}{2}\delta(f + f_0)$	$\cos(\theta_0 k)$	$\leftrightarrow$	$\pi\delta(\theta - \theta_0) + \pi\delta(\theta + \theta_0)$
$\sin(2\pi f_0 t)$	$\leftrightarrow$	$\frac{1}{2j}\delta(f - f_0) - \frac{1}{2j}\delta(f + f_0)$	$\sin(\theta_0 k)$	$\leftrightarrow$	$\frac{\pi}{j}\delta(\theta - \theta_0) - \frac{\pi}{j}\delta(\theta + \theta_0)$
$\frac{\sin \pi W t}{\pi W t}$	$\leftrightarrow$	$\frac{1}{W}\,\mathrm{rect}(f, \frac{W}{2})$	$\frac{\sin W k T}{W k T}$	$\leftrightarrow$	$\frac{\pi}{WT}\,\mathrm{rect}(\theta, W)$
$e^{-at}u(t)$	$\leftrightarrow$	$\frac{1}{j2\pi f + a}$; $\mathrm{Re}\{a\} > 0$	$r^{-k}u_k$	$\leftrightarrow$	$\frac{1}{1 - r^{-1}e^{-j\theta}}$
$\sum_{m=-\infty}^{\infty} \delta(t - mT)$	$\leftrightarrow$	$\frac{1}{T}\sum_{k=-\infty}^{\infty} \delta\left(f - \frac{k}{T}\right)$	$\sum_{m=-\infty}^{\infty} \delta_{k-mN}$	$\leftrightarrow$	$\frac{2\pi}{N}\sum_{k=-\infty}^{\infty} \delta\left(\theta - \frac{2\pi k}{N}\right)$
$\mathrm{rect}(t, T)$	$\leftrightarrow$	$2T\frac{\sin \pi f T}{\pi f T}$			
$\frac{1}{\pi t}$	$\leftrightarrow$	$-j\,\mathrm{sign}(f)$			

1. The discrete-time Fourier transform expressions $X(e^{j\theta})$ in the right column are valid only in the range $-\pi \le \theta < \pi$. To extend this range, the given expression should be repeated periodically.

Appendix 2-B. Spectral Factorization

The derivation of the spectral factorization theorem (2.67) is straightforward, relying almost entirely on two facts: If $S(z)$ is real and nonnegative on the unit circle, then

(i) its poles and zeros must come in conjugate-reciprocal pairs; and
(ii) any zeros on the unit circle must be of even order.

To prove (i), observe that if $S(e^{j\theta})$ is real (but not necessarily nonnegative) then $S(e^{j\theta}) = S^*(e^{j\theta})$ for all θ. Taking the inverse DTFT yields $s_k = s_{-k}^*$, and taking the Z transform yields $S(z) = S^*(1/z^*)$. Since the right-hand side is zero whenever the left-hand side is zero, it follows that a zero (or pole) of $S(z)$ at z_0 must be accompanied by a zero (or pole) at $1/z_0^*$. Hence, the zeros (and poles) come in conjugate-reciprocal pairs.

The result (i) exploits only the real nature of $S(e^{j\theta})$, and hence it does not rule out the possibility that $S(e^{j\theta})$ might change sign at one or more values of θ. When $S(e^{j\theta})$ cannot change sign, we find that the zeros of $S(z)$ are further constrained such that *any zeros on the unit circle must be of even order.* This is fact (ii).

The proof of fact (ii) is by contradiction; any zeros of odd order would imply that $S(e^{j\theta})$ changes sign. Specifically, suppose $z_0 = e^{j\theta_0}$ is a zero of order k, so that $S(z)$ can be factored as

$$S(z) = (z - e^{j\theta_0})^k \tilde{S}(z), \tag{2.103}$$

where $\tilde{S}(z)$ is what remains after all zeros at z_0 have been factored out, so that $\tilde{S}(z_0) \neq 0$. To compare the sign of $S(e^{j\theta})$ just before and just after θ_0, consider the ratio:

$$\begin{aligned}\frac{S(e^{j(\theta_0+\varepsilon)})}{S(e^{j(\theta_0-\varepsilon)})} &= \left(\frac{e^{j(\theta_0+\varepsilon)} - e^{j\theta_0}}{e^{j(\theta_0-\varepsilon)} - e^{j\theta_0}}\right)^k \frac{\tilde{S}(e^{j(\theta_0+\varepsilon)})}{\tilde{S}(e^{j(\theta_0-\varepsilon)})} \\ &= \left(\frac{e^{j\varepsilon} - 1}{e^{-j\varepsilon} - 1}\right)^k \frac{\tilde{S}(e^{j(\theta_0+\varepsilon)})}{\tilde{S}(e^{j(\theta_0-\varepsilon)})} \\ &= (-1)^k e^{jk\varepsilon} \frac{\tilde{S}(e^{j(\theta_0+\varepsilon)})}{\tilde{S}(e^{j(\theta_0-\varepsilon)})} \ .\end{aligned} \tag{2.104}$$

This ratio will be negative when $S(e^{j\theta})$ has a different sign at $\theta_0 - \varepsilon$ than at $\theta_0 + \varepsilon$, and will be positive when the sign is the same. In the limit as $\varepsilon \to 0$, the sign of the ratio determines whether or not $S(e^{j\theta})$ changes sign as θ passes through θ_0. But

$$\begin{aligned}\lim_{\varepsilon \to 0} \frac{S(e^{j(\theta_0+\varepsilon)})}{S(e^{j(\theta_0-\varepsilon)})} &= (-1)^k \lim_{\varepsilon \to 0} e^{jk\varepsilon} \frac{\tilde{S}(e^{j(\theta_0+\varepsilon)})}{\tilde{S}(e^{j(\theta_0-\varepsilon)})} \\ &= (-1)^k,\end{aligned} \tag{2.105}$$

where $\tilde{S}(e^{j(\theta_0+\varepsilon)})/\tilde{S}(e^{j(\theta_0-\varepsilon)}) \to 1$ as $\varepsilon \to 0$ follows from the continuity of rational functions. It is thus clear that if $S(e^{j\theta})$ does not change sign at θ_0 then the order k of the zero must be even.

Having proven (i) and (ii), the spectral factorization theorem now follows by construction. First, assign all poles and zeros that are inside the unit circle to $M(z)$. Second, assign half of each of the zeros that are on the unit circle to $M(z)$. Together, (i) and (ii) imply that $M^*(1/z^*)$ will automatically account for all remaining poles and zeros. Finally, choose the constant γ^2 so that $M(z)$ is monic.

Before we derive γ^2, we must establish the following fact:

$$\langle\, |1 - be^{-j\theta}|^2 \,\rangle_G = 1\,, \quad \text{for any } |b| \leq 1. \tag{2.106}$$

In other words, the geometric mean of $|M(e^{j\theta})|^2$ is unity for any first-order filter $M(z)$ that is monic and loosely minimum phase. To see why, write b in polar form, $b = ae^{j\phi}$ with $0 < a \leq 1$. Then the logarithm of the geometric mean reduces to

$$\frac{1}{2\pi}\int_{-\pi}^{\pi}\log|1-ae^{j\phi}e^{-j\theta}|^2\,d\theta = \frac{1}{\pi}\int_0^{\pi}\log(1+a^2-2a\cos(\theta))\,d\theta\ ,\quad 0<a\leq 1\ . \tag{2.107}$$

Note that the angle ϕ of the zero does not affect the integral. Integral (2.107) can be found in standard integral tables, which show that it evaluates to zero, thus establishing (2.106).

To derive γ^2, replace z by $e^{j\theta}$ in (2.68) to get

$$S(e^{j\theta}) = \gamma^2\frac{\prod_{k=1}^{M}\left|1-c_ke^{-j\theta}\right|^2}{\prod_{k=1}^{N}\left|1-d_ke^{-j\theta}\right|^2},\quad |c_k|\leq 1,\quad |d_k|<1\ . \tag{2.108}$$

Taking the geometric mean of both sides, and using (2.63) and (2.106), yields

$$\begin{aligned}\langle S\rangle_G = \langle\gamma^2\rangle_G\frac{\prod_{k=1}^{M}\langle\left|1-c_ke^{-j\theta}\right|^2\rangle_G}{\prod_{k=1}^{N}\langle\left|1-d_ke^{-j\theta}\right|^2\rangle_G}\\ = \gamma^2\ .\end{aligned} \tag{2.109}$$

This establishes (2.69).

Problems

Problem 2-1. A system with a complex-valued input and output can be described in terms of systems with real-valued inputs and outputs, as shown in Fig. 2-2. Show that if the impulse response of the system is real-valued, then there is no crosstalk (or cross-coupling) between the real and imaginary parts, whereas if the impulse response is complex-valued then there is crosstalk.

Problem 2-2.

(a) Show that $e^{j2\pi ft}$ is an *eigenfunction* of a continuous-time LTI system with impulse response $h(t)$, meaning that the response to this input is the same complex exponential multiplied by a complex constant called the *eigenvalue*.

(b) Show that $e^{j2\pi fkT}$ is an eigenfunction of a discrete-time LTI system with impulse response h_k.

(c) Show that for a fixed f the eigenvalue in (b) is the Fourier transform $H(e^{j2\pi fT})$ of the discrete-time impulse response h_k. Specifically, show that when the input is $e^{j2\pi fkT}$, the output can be written

$$y_k = H(e^{j2\pi fT})e^{j2\pi fkT}\ . \tag{2.110}$$

Hence that magnitude response $|H(e^{j2\pi fT})|$ gives the gain of the system at each frequency, and the phase response $\angle(H(e^{j2\pi fT}))$ gives the phase change.

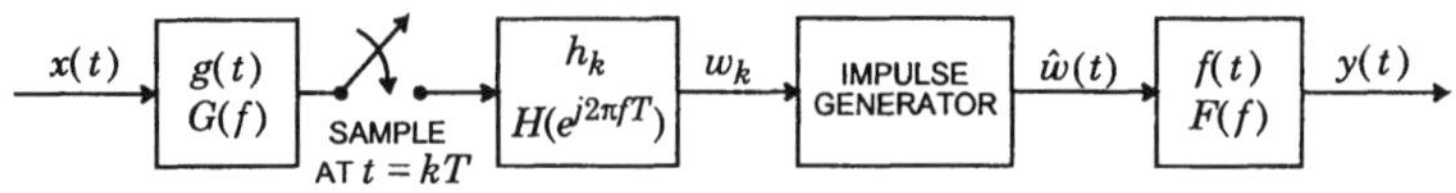

Fig. 2-21. The mixed continuous and discrete-time system of Problem 2-3.

Problem 2-3. Consider the mixed discrete and continuous-time system in Fig. 2-21. The impulse generator converts w_k to its PAM representation $\hat{w}(t)$, according to (2.2). The rate is the same as that of the sampler, namely $1/T$.

(a) Find the Fourier transform of $y(t)$.

(b) Is the system linear? Justify.

(c) Find conditions on $G(f)$, $H(e^{j2\pi fT})$, and/or $F(f)$ such that the system is time invariant.

Problem 2-4. Derive *Parseval's relationships* for the energy of a signal:

$$\int_{-\infty}^{\infty} |x(t)|^2\, dt = \int_{-\infty}^{\infty} |X(f)|^2\, df\,, \qquad \sum_{k=-\infty}^{\infty} |x_k|^2 = \frac{1}{2\pi}\int_{-\pi}^{\pi} |X(e^{j\theta})|^2\, d\theta\ . \tag{2.111}$$

Problem 2-5. Derive the *generalized Parseval's relationships*:

$$\int_{-\infty}^{\infty} x(t)y^*(t)\, dt = \int_{-\infty}^{\infty} X(f)Y^*(f)\, df\,, \qquad \sum_{k=-\infty}^{\infty} x_k y_k^* = \frac{1}{2\pi}\int_{-\pi}^{\pi} X(e^{j\theta})Y^*(e^{j\theta})\, d\theta\ . \tag{2.112}$$

Problem 2-6. Given that a discrete-time signal x_k is obtained from a continuous-time signal $x(t)$ by sampling, can you relate the energy of the discrete-time signal to the energy of the continuous-time signal? What if the continuous-time signal is known to be properly bandlimited?

Problem 2-7. Given a discrete-time system with impulse response $h_k = \delta_k + \delta_{k-1}$, what is its transfer function and frequency response? If the input is $h_k = \cos(\theta_0 k)$ what is the output? Show that the system has a phase response that is piecewise linear in frequency.

Problem 2-8. Show that the phase response $\phi(f) = \angle(H(f))$ of a real system is antisymmetric.

Problem 2-9. What is the impulse response of a real system that produces a constant phase shift of ϕ and unity gain at all frequencies? Such a system is called a *phase shifter.*

Problem 2-10. Find the Fourier transform of

$$x(t) = \sum_{m=-\infty}^{\infty} \frac{1}{j(t-mT)+a}\ .$$

Problem 2-11. Show that the output of an LTI system cannot contain frequencies not present in the input.

Problem 2-12. Show that the quadrature demodulator outputs $s_I(t)$ and $s_Q(t)$ of Fig. 2-6(a) combine to give the complex envelope $\tilde{s}(t) = s_I(t) + js_Q(t)$ if and only if the real input is bandlimited to twice the carrier frequency, so that the Fourier transform $S(f)$ satisfies $S(f) = 0$ for $f > f_c$.

Problem 2-13. Consider a *Hilbert transformer*, which is a linear filter with impulse response and transfer function given as

$$h(t) = \frac{1}{\pi t}, \qquad H(f) = -j\text{sign}(f) = \begin{cases} j; & f < 0 \\ -j; & f \geq 0 \end{cases}. \tag{2.113}$$

Show that if $x(t) = \cos(2\pi f_0 t)$ is the input, then $y(t) = \sin(2\pi f_0 t)$ is the output. Show further that if $x(t) = \sin(2\pi f_0 t)$ is the input, then $y(t) = -\cos(2\pi f_0 t)$ is the output. Any sinusoidal input experiences a 90 degree phase change.

Problem 2-14. Suppose a complex signal $z(t) = \text{Re}\{z(t)\} + j\text{Im}\{z(t)\}$ is analytic, so that its Fourier transform is zero for negative frequencies.

(a) Show that $\text{Im}\{z(t)\}$ can be obtained from $\text{Re}\{z(t)\}$ by filtering $\text{Re}\{z(t)\}$ with the Hilbert transformer of Problem 2-13. In other words,

$$\text{Im}\{z(t)\} = \frac{1}{\pi t} * \text{Re}\{z(t)\}. \tag{2.114}$$

(b) Show that $-\text{Re}\{z(t)\}$ can be obtained from $\text{Im}\{z(t)\}$ using the same filter.

Problem 2-15. Show that if $h(t) = 0$ for $t > 0$ (i.e. $h(t)$ is *anticausal*), then the real and imaginary parts are related by a Hilbert transform in the frequency domain

$$\text{Im}\{H(f)\} = \frac{1}{\pi f} * \text{Re}\{H(f)\}. \tag{2.115}$$

Problem 2-16. Consider a discrete-time signal $z_k = \text{Re}\{z_k\} + j\text{Im}\{z_k\}$ satisfying

$$Z(e^{j2\pi fT}) = 0 \qquad \text{for} \quad f \in [-1/(2T), 0)\,,$$

where T is the sampling interval. An example is shown in the following figure:

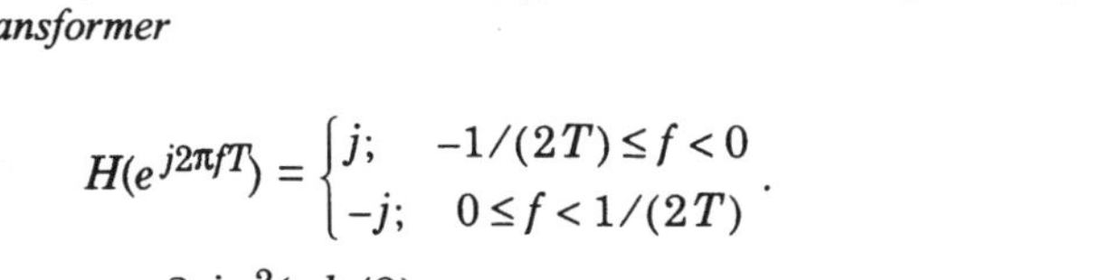

By analogy, such signals are called *discrete-time analytic signals*, although the term "analytic" does not mathematically apply to sequences. Show that $\text{Im}\{z_k\}$ can be obtained by filtering $\text{Re}\{z_k\}$ with the *discrete-time Hilbert transformer*

$$H(e^{j2\pi fT}) = \begin{cases} j; & -1/(2T) \leq f < 0 \\ -j; & 0 \leq f < 1/(2T) \end{cases}. \tag{2.116}$$

The impulse response is $h_k = \dfrac{2\sin^2(\pi k/2)}{\pi k}$, $k \neq 0$, with $h_0 = 0$, as shown in Fig. 2-22.

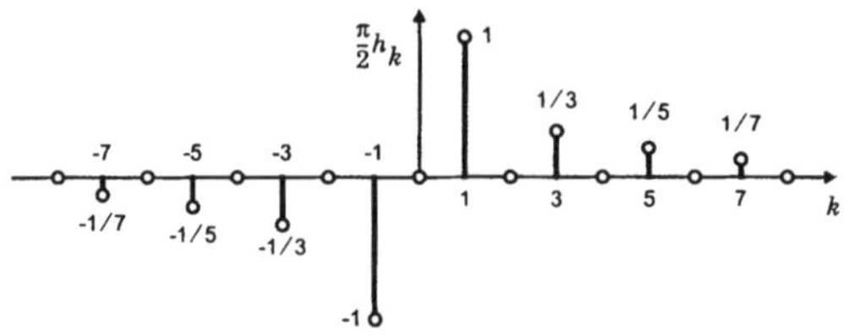

Fig. 2-22. The impulse response of the discrete-time Hilbert transformer.

Problem 2-17. Assume you are given a discrete-time signal x_k with sample interval T and the discrete-time Fourier transform as shown in the following figure:

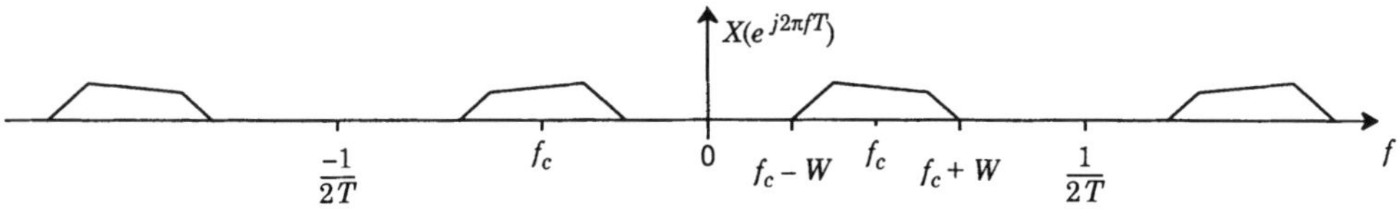

Suppose you are told it is the real part of a discrete-time analytic signal z_k.

(a) Show that $\text{Im}\{z_k\}$ is the result of filtering x_k with the *real* filter h_k with frequency response given in the following figure:

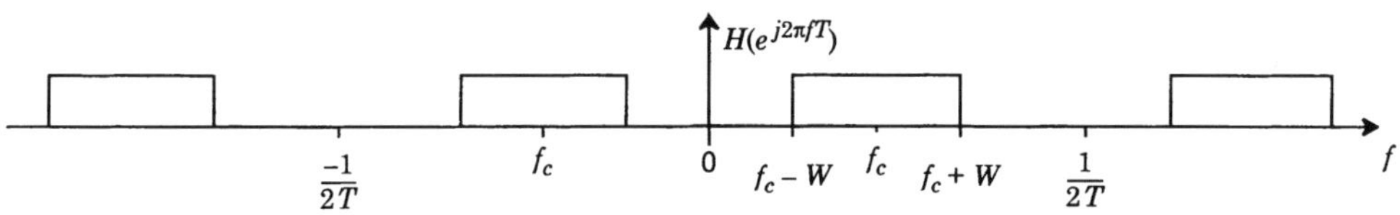

(b) Assume you are given a practical FIR low pass filter g_k approximating the ideal low pass filter transfer function shown below:

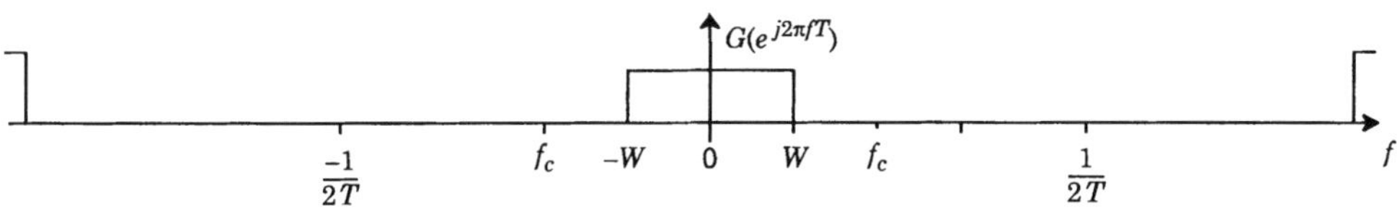

Design a practical filter h_k approximating the filter in part (a).

Problem 2-18.

(a) Using (2.7), show that for any complex number z, the sequence z^k is an eigenfunction of a discrete-time LTI system. That is, the response to this signal is

$$y_k = H(z)z^k . \tag{2.117}$$

(b) How is this related to the frequency response result discussed in Problem 2-2, part (c)?

Problem 2-19. Repeat Problem 2-18 using only the definition of a discrete-time LTI system and not using the convolution sum. (*Hint:* Note that $z^{k+m} = z^k z^m$.)

Problem 2-20. Calculate the Z transform of $x_k = a^k u_k$, where u_k is the unit step of (2.102).

Problem 2-21. Show that for any complex number z, z^t is an eigenfunction of any continuous-time LTI system. Also show that for any z there exists an s such that $e^{st} = z^t$. Relate the eigenvalue of the system for a fixed z to the *Laplace transform*

$$H_L(s) = \int_{-\infty}^{\infty} e^{-st} h(t) dt \,. \tag{2.118}$$

The Fourier transform is the Laplace transform evaluated at $s = j2\pi f$, or $H(f) = H_L(j2\pi f)$.

Problem 2-22.

(a) Show that the signals $x_k = a^k u_k$ and $y_k = -a^k u_{-k-1}$ have the same Z transform.

(b) What are the ROC for the two cases?

(c) Under what conditions are the two signals stable? Relate this to the ROC.

Problem 2-23. Let $X(z) = \frac{z}{z-a}$, and find the time domain signals for both possible ROC. Do this directly without using the results of Problem 2-22.

Problem 2-24. Given $X(z) = \dfrac{z^2}{z^2 - (a+b)z + ab}$ where $|a| < 1 < |b|$, find the corresponding time-domain signal for the following two cases:

(a) The time domain signal is known to be causal.

(b) The time domain signal is known to be neither causal nor anticausal.

(c) Comment on whether the signal is stable in each case, and state your reasons.

Problem 2-25. Show that when the transfer function $H(z)$ given in (2.40) has real-valued coefficients, the zeros and poles are always either real valued or come in complex-conjugate pairs.

Problem 2-26. Given a transfer function in the middle form of (2.40) with $r = 0$, $A = 1$, zeros at $\frac{3}{2} e^{\pm j\pi/4}$ and $\pm j$, and poles at $\frac{1}{2} e^{\pm j\pi/8}$, Find all the terms in the factorization (2.52). Write them in terms of polynomials with *real-valued* coefficients.

Problem 2-27. Give an example of a transfer function $S(z)$ that is real and nonnegative on the unit circle, and whose arithmetic and geometric means on the unit circle are 3 and 1, respectively. Sketch the pole-zero plot for this $S(z)$.

Problem 2-28.

(a) Let h_k be a causal strictly minimum-phase sequence with a rational Z transform, and let g_k be another causal sequence obtained by taking a zero of $H(z)$ at c and replacing it with a zero at $1/c^*$. Show that $|H(e^{j\theta})| = |G(e^{j\theta})|$. *Hint:* Find a transfer function $A(z)$ that when multiplied by $H(z)$ yields $G(z)$.

(b) Show that

$$\sum_{k=0}^{N} |h_k|^2 \geq \sum_{k=0}^{N} |g_k|^2 \,, \tag{2.119}$$

for all $N \geq 0$. *Hint:* Define $F(z) = H(z)/(1 - cz^{-1})$ and write g_k and h_k in terms of f_k.

(c) Show that for any two rational transfer functions $H(z)$ and $G(z)$ such that $H(z)$ is minimum phase and $|H(e^{j\theta})| = |G(e^{j\theta})|$, (2.119) is true for all $N \geq 0$. Thus, among all sequences with the same magnitude response, minimum-phase sequences are maximally concentrated near $k = 0$ in the mean-square sense. From Parseval's formula plus the unit magnitude of an allpass filter, clearly both sides of (2.119) approach one another as $N \to \infty$.

Problem 2-29. Pass a causal input signal x_k through a first-order stable causal allpass filter such as that in Example 2-9 to yield a causal output signal y_k. Show that for any $N \geq 0$

$$\sum_{k=0}^{N} |x_k|^2 \geq \sum_{k=0}^{N} |y_k|^2, \tag{2.120}$$

and hence the allpass filter is dispersive in the sense that it reduces the signal energy in the first N samples while keeping the total signal energy the same (since it has unit magnitude frequency response). *Hint:* Consider a solution method similar to Problem 2-28.

Problem 2-30. Use (2.52) and Example 2-10 to derive the factorization in (2.51).

Problem 2-31. A filter with impulse response $g(t) = h^*(-t)$ is said to be matched to $h(t)$. What is the frequency response $G(f)$ of the matched filter?

Problem 2-32. Given three signals $s_1(t)$, $s_2(t)$, and $s_5(t)$:

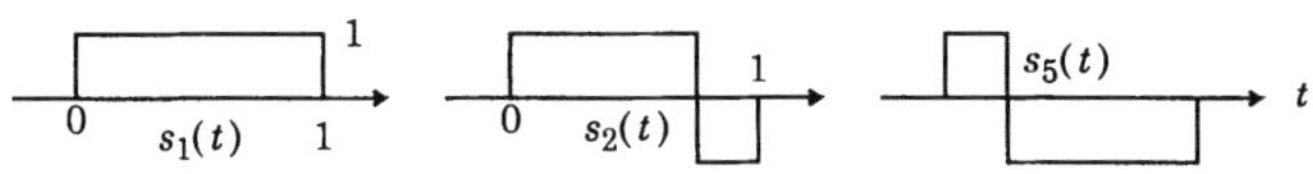

(a) Find the norm of $s_1(t)$ and $s_2(t)$ and the inner product of these two signals in a linear space. What is the angle between the two signals?

(b) Find the norm of the signal $s_1(t) + s_2(t)$.

(c) Find a signal $s_3(t)$ that is orthogonal to both $s_1(t)$ and $s_2(t)$.

(d) Find a signal $s_4(t)$ that is in the subspace spanned by $s_1(t)$ and $s_2(t)$ and is orthogonal to $s_1(t)$.

(e) Find the signal in the subspace spanned by $s_1(t)$ and $s_2(t)$ that is closest to $s_5(t)$.

Problem 2-33. Consider the set B of all finite-energy continuous-time signals that are bandlimited to $|f| \leq W$ Hz.

(a) Show that this set of signals B is a subspace of the set of all finite-energy continuous-time signals that have no bandwidth constraint.

(b) Characterize the subspace consisting of all signals orthogonal to every signal in B.

(c) Find the projection of the signal $\mathbf{S}_1$ in Problem 2-32 on B for $W = 1$ Hz.

Problem 2-34. Two subspaces $\mathcal{S}_1$ and $\mathcal{S}_2$ of a Hilbert space are said to be orthogonal if every vector in $\mathcal{S}_1$ is orthogonal to every vector in $\mathcal{S}_2$. The sum of the two subspaces $\mathcal{S}_1 \oplus \mathcal{S}_2$ is the subspace consisting of vectors that can be expressed as the sum of a vector in $\mathcal{S}_1$ and a vector in $\mathcal{S}_2$. Given two orthogonal subspaces $\mathcal{S}_1$ and $\mathcal{S}_2$ of a Hilbert space $\mathcal{H}$ and an arbitrary vector $\mathbf{X} \in \mathcal{H}$, show that the projection $\hat{\mathbf{X}}$ of $\mathbf{X}$ onto $\mathcal{S}_1 \oplus \mathcal{S}_2$ can be expressed uniquely as

$$\hat{\mathbf{X}} = \hat{\mathbf{X}}_1 + \hat{\mathbf{X}}_2, \tag{2.121}$$

where $\hat{\mathbf{X}}_1$ is the projection of $\mathbf{X}$ onto $\mathcal{S}_1$, and $\hat{\mathbf{X}}_2$ is the projection of $\mathbf{X}$ onto $\mathcal{S}_2$.

Problem 2-35. Given a transmitted pulse $h(t)$ it is useful to define an *autocorrelation function*

$$\rho_h(k) = \int_{-\infty}^{\infty} h(t)h^*(t - kT)\, dt \ . \tag{2.122}$$

Show that

$$|\rho_h(k)| \le \rho_h(0), \tag{2.123}$$

or in words, the autocorrelation function of a pulse can never be larger than the energy of the pulse.

References

1. A. W. Naylor and G. R. Sell, *Linear Operator Theory in Engineering and Science*, Holt, Rinehart and Winston, Inc., New York, 1971.
2. C. D. McGillem and G. R. Cooper, *Continuous and Discrete Signal and System Analysis*, Holt, Rinehart, and Winston, 1984.
3. A. V. Oppenheim, A. S. Willsky, and I. T. Young, *Signals and Systems*, Prentice Hal, 1983.
4. R. E. Ziemer, W. H. Tranter, and D. R. Fannin, *Signals and Systems: Continuous and Discrete*, Macmillan Publishing Co., NY, 1983.
5. A. V. Oppenheim and R. W. Schafer, *Digital Signal Processing*, Prentice-Hall, Inc., 1989.
6. L. Jackson, *Digital Filters and Signal Processing*, Kluwer Academic Publishers, Boston, MA, 1985.
7. R. J. Schwarz and B. Friedland, *Linear Systems*, McGraw-Hill Book Co., 1965.
8. A. Papoulis, *The Fourier Integral and its Applications*, McGraw- Hill Book Co., New York, 1962.
9. R. N. Bracewell, *The Fourier Transform and its Applications*, McGraw-Hill Book Co., New York, 1965.

3

Stochastic Signal Processing

Although modulation and demodulation are deterministic, the information to be transmitted, as well as the noise encountered in the physical transmission medium, is random or stochastic. These phenomena cannot be predicted in advance, but they have certain predictable characteristics which can be summarized in a random process model. The design of a digital communication system heavily exploits these characteristics.

In this chapter we review the notation that will be used for random variables and processes, and cover several topics in detail that may be new to some readers and are particularly important in the sequel. These include Chernoff bounding techniques, Bayes' rule, and mixtures of discrete-time and continuous-time random processes. Markov chains are discussed in Section 3.3, and will be used in a diverse set of applications in Chapters 7, 8, 13, and 17. Section 3.4, on Poisson processes, uses the Markov chain results to describe Poisson processes and shot noise, which are important to the understanding of optical fiber systems.

3.1. Random Variables

Before reviewing the theory of the stochastic process, we review some theory and notation associated with random variables. In digital communication it is common to encounter combinations of discrete and continuous-valued random variables, so this will be emphasized.

We denote a *random variable* by a capital letter, such as X, and an *outcome* of the random variable by a lower-case letter, such as x. The random variable is a real or complex-valued function defined on the *sample space* Ω of all possible outcomes. An *event* E is a set of possible outcomes and is assigned a probability, written $\Pr[E]$, where $0 \le \Pr[E] \le 1$. Since an event is a set, we can define the union of two events, $E_1 \cup E_2$ or the intersection of events $E_1 \cap E_2$. The basic formula

$$\Pr[E_1 \cup E_2] = \Pr[E_1] + \Pr[E_2] - \Pr[E_1 \cap E_2] \tag{3.1}$$

leads to the very useful *union bound*,

$$\Pr[E_1 \cup E_2] \le \Pr[E_1] + \Pr[E_2] \ . \tag{3.2}$$

The *cumulative distribution function (c.d.f.)* of a real valued random variable X is the probability of the event $X \le x$,

$$F_X(x) = \Pr[X \le x] \ . \tag{3.3}$$

Where there can be no confusion, we often omit the subscript, writing the c.d.f. as $F(x)$. For a complex-valued random variable Y, the c.d.f. is

$$F_Y(y) = \Pr[\mathrm{Re}\{Y\} \le \mathrm{Re}\{y\}, \mathrm{Im}\{Y\} \le \mathrm{Im}\{y\}] \ . \tag{3.4}$$

Here and elsewhere, we adopt the shorthand $\Pr[A, B]$ for $\Pr[A \cap B]$. For a continuous real-valued random variable, the *probability density function (p.d.f.)* $f_X(x)$ is defined such that for any interval $I \subseteq \Re$,

$$\Pr[X \in I] = \int_I f_X(x)\, dx \ . \tag{3.5}$$

For a complex-valued random variable, I is a region in the complex plane. For a real-valued random variable X,

$$f_X(x) = \frac{d}{dx} F_X(x) \ , \tag{3.6}$$

where the derivative exists. We will often use the generalized derivative, so that when the c.d.f. includes a step function the corresponding p.d.f. has a Dirac delta function.

Example 3-1.

For the c.d.f. shown below,

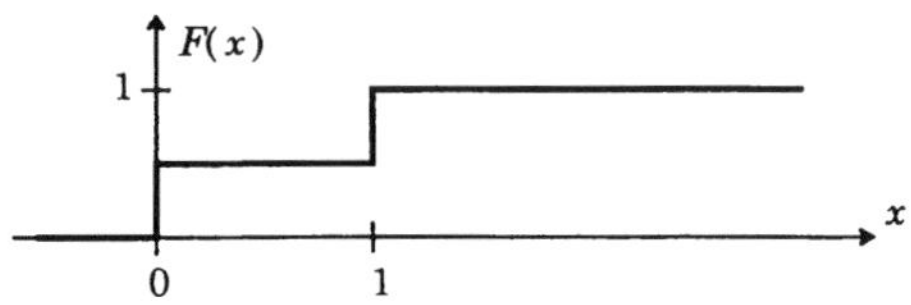

the p.d.f. consists exclusively of Dirac delta functions,

$$f(x) = \tfrac{1}{2}\delta(x) + \tfrac{1}{2}\delta(x-1)\,. \tag{3.7}$$

Such a density is characteristic of a discrete random variable.

For a discrete-valued random variable X, we will denote the probability of an outcome $x \in \Omega$ as

$$p_X(x) = \Pr[X = x], \tag{3.8}$$

where we will again omit the subscript where there can be no confusion. The p.d.f. can be written as

$$f_X(x) = \sum_{y \in \Omega_x} p_X(y)\delta(x-y)\,. \tag{3.9}$$

The *expected value* or *mean* of X is defined as

$$E[X] = \int_{-\infty}^{\infty} x f_X(x)\,dx\,, \qquad \text{or} \qquad E[X] = \sum_{y \in \Omega_x} x \cdot p_X(x)\,, \tag{3.10}$$

for continuous-valued and discrete-valued random variables, respectively. For a complex-valued random variable Y, we integrate over the complex plane,

$$E[Y] = \int_{-\infty}^{\infty}\int_{-\infty}^{\infty} (x + jz) f_Y(x + jz)\,dx\,dz\,. \tag{3.11}$$

The *fundamental theorem of expectation* states that if $g(\cdot)$ is any function defined on the sample space of X, then

$$E[g(X)] = \int_{-\infty}^{\infty} g(x) f_X(x)\,dx\,. \tag{3.12}$$

Especially important expectations are the mean μ and variance σ_X^2, defined as

$$\mu = E[X], \qquad \sigma_X^2 = E[(X - E[X])^2] = E[X^2] - E[X]^2\,. \tag{3.13}$$

For complex-valued random variables, the variance is defined similarly as

$$\sigma_X^2 = E[\,|X - E[X]|^2] = E[\,|X|^2] - |E[X]|^2\,. \tag{3.14}$$

The *joint c.d.f.* of two real-valued random variables X and Y is

$$F_{X,Y}(x, y) = \Pr[X \le x, Y \le y] = \int_{-\infty}^{x}\int_{-\infty}^{y} f_{X,Y}(\alpha, \beta)\, d\alpha\, d\beta\,, \tag{3.15}$$

where $f_{X,Y}(x, y)$ is the *joint p.d.f.* The joint p.d.f. can be written in terms of the joint c.d.f. as

$$f(x, y) = \frac{\partial^2}{\partial x \partial y} F(x, y)\,, \tag{3.16}$$

where we have omitted the subscripts as before. The *marginal density* $f_X(x)$ of a random variable X can be found from the joint p.d.f. using

$$f_X(x) = \int_{-\infty}^{\infty} f_{X,Y}(x, y)\, dy\,. \tag{3.17}$$

The random variables X and Y are *independent* or *statistically independent* if for all intervals I and J,

$$\Pr[X \in I \cap Y \in J] = \Pr[X \in I]\Pr[Y \in J]\,, \tag{3.18}$$

which is equivalent to

$$f_{X,Y}(x, y) = f_X(x) f_Y(y) \qquad \text{or} \qquad F_{X,Y}(x, y) = F_X(x) F_Y(y)\,. \tag{3.19}$$

Independence implies that the *correlation* reduces to the product of the means, or

$$E[XY] = E[X]E[Y]\,. \tag{3.20}$$

When (3.20) is satisfied, the random variables are said to be *uncorrelated.* Two random variables can be uncorrelated and yet not be independent.

3.1.1. Moment Generating Function and Chernoff Bound

The *characteristic function* of X is defined as

$$\Phi_X(s) = E[e^{sX}] = \int_{-\infty}^{\infty} e^{sx} f_X(x)\, dx\,, \tag{3.21}$$

for a complex variable s. This is the Laplace transform of $f_X(x)$ evaluated at $-s$. When s is real-valued, which will suffice for applications in this book, (3.21) is called the *moment generating function.*

Exercise 3-1.
If X and Y are independent and $Z = X + Y$, show that

$$\Phi_Z(s) = \Phi_X(s)\Phi_Y(s)\,. \tag{3.22}$$

Exercise 3-2.

Show that

$$E[X] = \frac{\partial}{\partial s}\Phi_X(s)|_{s=0}, \qquad \text{and} \qquad E[X^2] = \frac{\partial^2}{\partial s^2}\Phi_X(s)|_{s=0}. \tag{3.23}$$

The *Chernoff bound*, based on the moment generating function, is very useful for bounding the tail probability for a random variable where an exact evaluation is intractable.

Exercise 3-3.

(a) Show that the probability of event $X > x$ is bounded by

$$1 - F_X(x) = \Pr[X > x] \le e^{-sx}\Phi_X(s) \tag{3.24}$$

for any real-valued $s \ge 0$. This establishes that the tail of the p.d.f. decreases at least exponentially for any distribution for which the moment generating function exists. (*Hint:* Write the probability as the integral against a step function, and bound the step function by an exponential.)

(b) Find the similar bound

$$F_X(x) \le e^{sx}\Phi_X(-s) \tag{3.25}$$

for $s \ge 0$.

(c) Show that the s that minimizes the bound (makes it tightest) in (a) and (b) must satisfy

$$x\Phi_X(s) = \frac{\partial}{\partial s}\Phi_X(s), \qquad -x\Phi_X(-s) = \frac{\partial}{\partial s}\Phi_X(-s), \tag{3.26}$$

respectively.

3.1.2. Conditional Probabilities and Bayes' Rule

The *conditional probability* that a continuous-valued random variable X is in the interval I given that Y is in the interval J is defined for all J such that $\Pr[Y \in J] \ne 0$ to be

$$\Pr[X \in I \mid Y \in J] = \frac{\Pr[X \in I \cap Y \in J]}{\Pr[Y \in J]}, \tag{3.27}$$

where $\Pr[Y \in J]$ is called a *marginal probability* because it does not consider the possible effects of X on Y. For complex or vector-valued random variables, I and J are regions or volumes, rather than intervals. If X and Y are independent, then $\Pr[X \in I \mid Y \in J] = \Pr[X \in I]$. The joint probability can be written in terms of the conditional probabilities,

$$\Pr[X \in I \cap Y \in J] = \Pr[X \in I \mid Y \in J]\Pr[Y \in J]. \tag{3.28}$$

Equivalently,

$$f_{X|Y}(x \mid y) = \frac{f_{X,Y}(x, y)}{f_y(y)}, \tag{3.29}$$

where $f_Y(y)$ is called a *marginal density*. The *conditional density* $f_{X|Y}(x|y)$ is well defined only for y such that $f_Y(y) \neq 0$. Since $f_{X,Y}(x, y) = f_{Y,X}(y, x)$, (3.29) implies that

$$f_{X|Y}(x|y)f_Y(y) = f_{Y|X}(y|x)f_X(x), \tag{3.30}$$

which is a form of *Bayes' rule*.

It is common in digital communication systems to encounter both discrete-valued and continuous-valued random variables in the same system. In this case, (3.30) has Dirac delta functions.

Exercise 3-4.

Suppose that Y is discrete-valued and X is continuous-valued. Show that by integrating (3.30) over small intervals about y, we get the mixed form of Bayes' rule,

$$f_{X|Y}(x|y)p_Y(y) = p_{Y|X}(y|x)f_X(x). \tag{3.31}$$

This involves both probabilities and probability density functions. It has no delta functions as long as X is continuous-valued. If X is also discrete-valued, show that then

$$p_{X|Y}(x|y)p_Y(y) = p_{Y|X}(y|x)p_X(x), \tag{3.32}$$

which has only discrete probabilities.

For discrete-valued distributions, the marginal probability can be written in terms of the conditional probabilities as

$$p_Y(y) = \sum_{x \in \Omega} p_{Y|X}(y|x)p_X(x) = \sum_{x \in \Omega} p_{Y,X}(y, x), \tag{3.33}$$

where Ω is the countable sample space for X. This relation shows us how to obtain the marginal probabilities of a random variable given only joint probabilities, or given only conditional probabilities and the marginal probabilities of the other random variable. Using this relation, we can write the conditional probability of X given Y in terms of the conditional probability of Y given X and the marginal probability of X

$$p_{X|Y}(x|y) = \frac{p_{Y|X}(y|x)p_X(x)}{\sum_{x \in \Omega} p_{Y|X}(y|x)p_X(x)}. \tag{3.34}$$

This relation is known as *Bayes' theorem*. The analogous Bayes' theorem for continuous-valued random variables is

$$f_{X|Y}(x|y) = \frac{f_{Y|X}(y|x)f_X(x)}{\int_{x \in \Omega} f_{Y|X}(y|x)f_X(x)dx}. \tag{3.35}$$

3.1.3. Gaussian Random Variables and the Central Limit Theorem

A *Gaussian* or *normal* random variable has the p.d.f.

$$f_X(x) = \frac{1}{\sqrt{2\pi\sigma^2}} e^{-(x-\mu)^2/(2\sigma^2)}, \tag{3.36}$$

where σ^2 is the variance and μ is the mean. As a shorthand notation, we use $X \sim \mathcal{N}(\mu, \sigma^2)$ to specify X as a random variable with the above p.d.f. The c.d.f. can be expressed only as an integral,

$$F_X(x) = \frac{1}{\sqrt{2\pi\sigma^2}} \int_{-\infty}^{x} e^{-(t-\mu)^2/(2\sigma^2)}\, dt, \tag{3.37}$$

for which there is no closed-form expression. The *standard Gaussian* random variable is a zero-mean Gaussian random variable U with unit variance $\sigma^2 = 1$, *i.e.*, $U \sim \mathcal{N}(0, 1)$. The *complementary distribution function* of this standard Gaussian is denoted by the special notation $Q(x)$,

$$Q(x) = \Pr[U > x] = 1 - F_U(x) = \frac{1}{\sqrt{2\pi}} \int_{x}^{\infty} e^{-t^2/2} dt. \tag{3.38}$$

$Q(x)$, therefore, is the integral of the tail of the Gaussian density. It is plotted in Fig. 3-1 using a log scale for probability. The function is related to the well-tabulated *error function* $\text{erf}(x)$ and the *complementary error function* $\text{erfc}(x)$ by

$$Q(x) = \tfrac{1}{2}\text{erfc}(x/\sqrt{2}) = \tfrac{1}{2}\Big(1 - \text{erf}(x/\sqrt{2})\Big). \tag{3.39}$$

Exercise 3-5.

Show that for a Gaussian random variable X with mean μ and variance σ^2,

$$\Pr[X > x] = Q\left(\frac{x-\mu}{\sigma}\right). \tag{3.40}$$

Although $Q(\cdot)$ can only be tabulated or numerically determined, a useful bound follows from the Chernoff bound of Exercise 3-3.

Exercise 3-6.

(a) Show that the moment generating function of a Gaussian random variable with mean μ and variance σ^2 is

$$\log\Phi_X(s) = \mu s + \sigma^2 s^2/2. \tag{3.41}$$

(b) Show from the Chernoff bound (Exercise 3-3) that

$$1 - F_X(x) \le e^{-(x-\mu)^2/(2\sigma^2)}, \tag{3.42}$$

and thus that

$$Q(x) \leq e^{-x^2/2}. \tag{3.43}$$

Tighter bounds are derived in Problem 3-3 and plotted in Fig. 3-1.

Use of the Gaussian distribution for modeling noise phenomena can be justified on physical grounds by the *central limit theorem*. It states, roughly, that the Gaussian distribution is a good model for the cumulative effect of a large number of independent random variables, regardless of the nature of their individual distributions. More precisely, let $\{Y_1, \ldots, Y_N\}$ denote a set of N statistically independent zero-mean random variables, each with the same p.d.f. $f(y)$ and finite variance σ^2. That is, the random variables are *independent and identically distributed* (i.i.d.). Define a random variable Z that is a normalized sum of the $\{Y_i\}$,

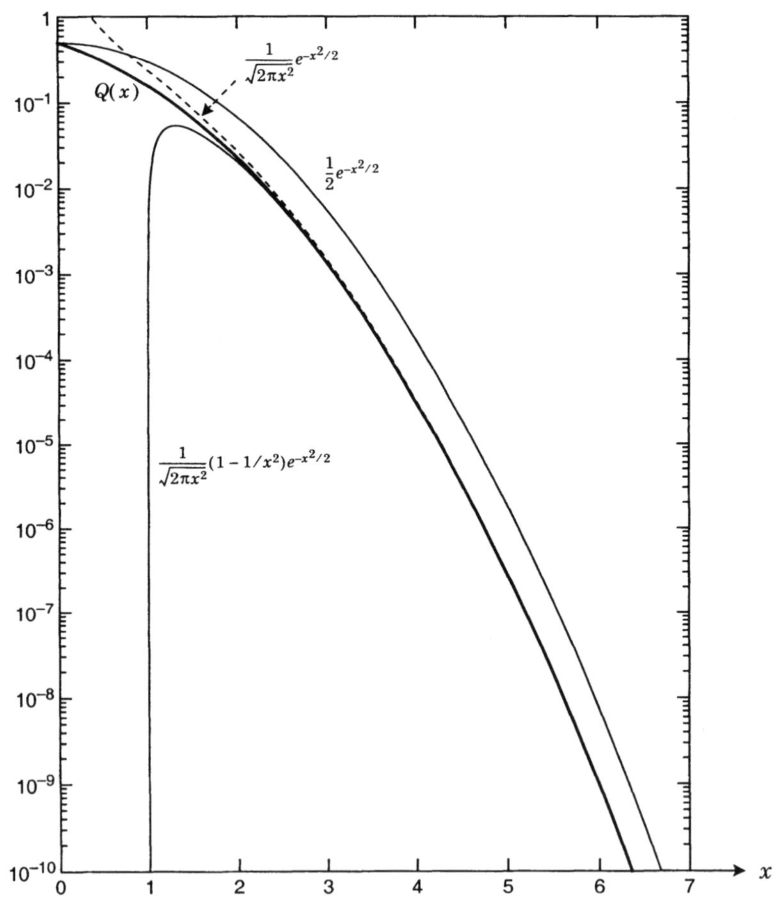

Fig. 3-1. The probability $Q(x)$ that a zero-mean, unit-variance Gaussian random variable X (the standard Gaussian) exceeds x, plotted on a log scale.

$$Z = \frac{1}{\sqrt{N}} \sum_{i=1}^{N} Y_i \ . \tag{3.44}$$

Then the distribution function of Z approaches Gaussian, $(1 - Q(z/\sigma))$, as $N \to \infty$. If each random variable Y_i represents some individual physical phenomenon, and Z is the cumulative effect of these phenomena, then as N gets large, the distribution of Z becomes Gaussian, regardless of the distribution of each Y_i.

In view of this theorem, it is hardly surprising that the sum of independent Gaussian random variables is Gaussian.

Exercise 3-7.

For an arbitrary linear combination of N zero mean independent Gaussian random variables X_i, each with variance σ^2,

$$Z = a_1 X_1 + \ldots + a_N X_N \ , \tag{3.45}$$

use the moment generating function to show that Z is itself zero-mean Gaussian with variance

$$\sigma_Z^2 = (a_1^2 + \ldots + a_N^2)\sigma^2 \ . \tag{3.46}$$

Two zero-mean Gaussian random variables with variance σ^2 are *jointly Gaussian* if and only if their joint p.d.f. can be expressed as

$$f_{X,Y}(x, y) = \frac{1}{2\pi\sigma^2\sqrt{1-\rho^2}} e^{-\frac{x^2 - 2\rho xy + y^2}{2\sigma^2(1-\rho^2)}} \ , \tag{3.47}$$

for some constant ρ. The parameter ρ is called the *correlation coefficient*, because

$$\rho = \frac{E[XY]}{\sigma^2} \ . \tag{3.48}$$

Note that $-1 \le \rho \le 1$, and if X and Y are uncorrelated then $\rho = 0$.

Exercise 3-8.

Show that two jointly Gaussian random variables are statistically independent if and only if they are uncorrelated.

This definition can be extended to $N > 2$ jointly Gaussian random variables. If a random vector $\mathbf{X}$ has components that are jointly zero-mean independent Gaussian random variables with the same variance σ^2, then the joint p.d.f. is

$$f_{\mathbf{X}}(\boldsymbol{x}) = \frac{1}{(2\pi\sigma^2)^{N/2}} e^{-\|\boldsymbol{x}\|^2/(2\sigma^2)} \ , \tag{3.49}$$

where N is the number of components in the vector and $\|\boldsymbol{x}\|$ is the Euclidean norm of the vector. When $\mathbf{X}$ is complex-valued with independent real and imaginary parts, (3.49) still holds. Any linear combination of jointly Gaussian random variables is Gaussian (as we saw in Exercise 3-7 for independent zero-mean Gaussian random variables).

This can be further generalized. A vector **X** with N jointly Gaussian real-valued random variables has p.d.f.

$$f_{\mathbf{X}}(\boldsymbol{x}) = \frac{1}{(2\pi)^{N/2}|\mathbf{C}|^{1/2}} e^{-\frac{1}{2}(\boldsymbol{x}-\boldsymbol{m})^T \mathbf{C}^{-1}(\boldsymbol{x}-\boldsymbol{m})}, \tag{3.50}$$

where

$$\mathbf{C} = \mathrm{E}[(\mathbf{X}-\boldsymbol{m})(\mathbf{X}-\boldsymbol{m})^T] \tag{3.51}$$

is the *covariance matrix*, $|\mathbf{C}|$ is its determinant, and $\boldsymbol{m} = \mathrm{E}[\mathbf{X}]$ is the vector mean. In the special case that the vector mean is zero and elements of the random vector are independent with equal variances, **C** becomes diagonal and (3.50) reduces to (3.49). An important observation from (3.50) is that the p.d.f. of a Gaussian random vector is completely specified by the vector mean and the pairwise covariances contained in the covariance matrix. Consequently, these two sets of parameters completely specify all the statistical properties of a Gaussian random vector.

Circularly Symmetric Gaussian Random Variables

Let X and Y be real-valued zero-mean Gaussian random variables, and let us create a complex-valued random variable according to:

$$Z = (X + jY)\,. \tag{3.52}$$

Consider the mean of the square of Z:

$$\mathrm{E}[Z^2] = \mathrm{E}[X^2] - \mathrm{E}[Y^2] + 2j\mathrm{E}[XY]\,. \tag{3.53}$$

Observe that X and Y are *identically distributed* (have the same variance) and *independent* ($\mathrm{E}[XY] = 0$) if and only if the following condition holds:

$$\mathrm{E}[Z^2] = 0. \tag{3.54}$$

A complex-valued Gaussian random variable Z is *circularly symmetric* if $\mathrm{E}[Z^2] = 0$ [14][15]. Based on the above observation, this is equivalent to saying that the real and imaginary parts are independent and identically distributed. The source of the terminology is that the probability density function of $Z = X + jY$ is circularly symmetric, or

$$p_{X,\,Y}(x, y) = \frac{1}{2\pi\sigma^2} e^{-(x^2+y^2)/(2\sigma^2)}\,, \tag{3.55}$$

where σ^2 is the variance of the real and imaginary parts. In other words, Z and $e^{j\theta}Z$ have the same distribution for any angle θ. We use $Z \sim \mathcal{CN}(0, 2\sigma^2)$ as shorthand for identifying Z as having the above pdf, so that the real and imaginary parts of Z are *i.i.d.* $\sim \mathcal{N}(0, \sigma^2)$.

3.1.4. Geometric Interpretation

Random variables can be interpreted geometrically using the linear-space approach of Section 2.6. In particular, consider the set of all complex-valued random variables X with bounded second moments, $\mathrm{E}[|X|^2] < \infty$.

Exercise 3-9.

Make reasonable definitions for the operations of addition of random variables, multiplication by a scalar, the vector element, and the additive inverse. Show that the set of such vectors form a linear space (Section 2.6.1).

An inner product on this space can be defined as

$$\langle X, Y \rangle = \mathrm{E}[XY^*] \ . \tag{3.56}$$

Exercise 3-10.

Show that (3.56) is a legitimate inner product (Section 2.6.2).

This geometric interpretation pays dividends in understanding the results of linear prediction theory (Section 3.2.3).

3.2. Random Processes

A discrete-time random process $\{X_k\}$ is a sequence of random variables indexed by integers k, while a continuous-time random process $X(t)$ is indexed by a real variable t. We write an *outcome* of $\{X_k\}$ or $\{X(t)\}$ as the lower case deterministic signal $\{x_k\}$ or $\{x(t)\}$. When there can be no confusion between a *signal* and a *sample of the signal*, we omit the braces $\{ \cdot \}$. Each random sample X_k or $X(t)$ may be complex, vector-valued, or real-valued.

Example 3-2.

A real-valued random process $X(t)$ is a *Gaussian random process* if its samples $\{X(t_1), \ldots, X(t_N)\}$ are jointly Gaussian random variables for any N and for any $\{t_1, \ldots, t_N\}$.

The first and second moments of the random process are the *mean*

$$m_k = E[X_k] \ , \qquad m(t) = E[X(t)] \tag{3.57}$$

and the *autocorrelation*

$$R_{XX}(k, i) = E[X_k X_i^*] \ , \qquad R_{XX}(t_1, t_2) = E[X(t_1)X^*(t_2)] \ . \tag{3.58}$$

where X^* is the complex conjugate of X.

Example 3-3.

Consider a real-valued, zero-mean Gaussian random process. A random vector $\mathbf{X}$ can be constructed from some arbitrary set of samples. For such a vector, the covariance matrix of (3.51) can be obtained from the autocorrelation function (3.58). Consequently, the joint p.d.f. (3.50) of any set of samples can be obtained from the autocorrelation function. Thus, the statistical properties of a zero-mean real-valued Gaussian random process are completely specified by its autocorrelation function.

A random process is *strict-sense stationary* if the p.d.f. for any sample is independent of the time index of the sample, and the joint p.d.f. of any set of samples depends only on the time differences between samples, and not on the absolute time of any sample. It is *wide-sense stationary (WSS)* if its mean is independent of the time index, and its autocorrelation depends only on the time difference between samples, and not on the absolute time. In other words, m_k or $m(t)$ must be constant and $R_{XX}(k, i)$ or $R_{XX}(t_1, t_2)$ must be a function only of the difference $k - i$ or $t_1 - t_2$. Strict sense stationarity implies wide-sense stationarity, but not the reverse, unless the process is Gaussian.

Example 3-4.
A real-valued WSS Gaussian random process is also strict-sense stationary. The autocorrelation function and mean of such a process can be used to construct the covariance matrix (3.51) for any set of samples. Since the process is WSS, the entries in the matrix will be independent of the absolute time index of the samples, and will depend instead only on the time differences between samples. Consequently, the joint p.d.f. (3.50) of any set of samples will depend only on these time differences. Hence the process is strict-sense stationary.

For a WSS random process the autocorrelation function can be written in terms of the time difference between samples, $m = k - i$ or $\tau = t_1 - t_2$, yielding the simpler notation

$$R_{XX}(m) = \mathrm{E}[X_{k+m}X_k^*] , \qquad R_{XX}(\tau) = \mathrm{E}[X(t+\tau)X(t)^*] . \tag{3.59}$$

When there is no ambiguity we sometimes use only a single subscript, writing $R_X(\cdot)$ instead. Observe that $R_X(0)$ is the second moment of the samples

$$R_X(0) = \mathrm{E}[|X_k|^2] , \qquad R_X(0) = \mathrm{E}[|X(t)|^2] , \tag{3.60}$$

and can be interpreted as the *power* of a random process. For a WSS random process, the *power spectral density* or *power spectrum* is the Fourier transform of the autocorrelation function,

$$S_X(e^{j\theta}) = \sum_{m=-\infty}^{\infty} R_X(m)e^{-jm\theta} , \qquad S_X(f) = \int_{-\infty}^{\infty} R_X(\tau)e^{-j2\pi f\tau}\, d\tau . \tag{3.61}$$

The power therefore is the integral of the power spectrum,

$$R_X(0) = \frac{1}{2\pi}\int_{-\pi}^{\pi} S_X(e^{j\theta})\, d\theta , \qquad R_X(0) = \int_{-\infty}^{\infty} S_X(f)\, df . \tag{3.62}$$

The power spectrum is real-valued since the autocorrelation function is conjugate symmetric, $R_X(-m) = R_X^*(m)$ or $R_X(-\tau) = R_X^*(\tau)$. It is also non-negative (see Problem 3-9). Furthermore, if X_k or $X(t)$ is real-valued, then the power spectrum is symmetric about zero frequency. We can also write the power spectrum as a Z transform or a Laplace transform,

$$S_X(z) = \sum_{m=-\infty}^{\infty} R_X(m)z^{-m} , \qquad S_X(s) = \int_{-\infty}^{\infty} R_X(\tau)e^{-s\tau}\, d\tau . \tag{3.63}$$

Evaluating $S_X(z)$ on the unit circle or $S_X(s)$ on the imaginary axis yields (3.61).

Example 3-5.
Consider a zero-mean random process $\{X_k\}$ where the samples X_k are all independent and identically distributed (i.i.d.) zero-mean random variables with variance σ_x^2. In this case $R_X(k) = \sigma_x^2\delta_k$ and the power spectrum is a constant, $S_X(e^{j\theta}) = \sigma_x^2$, independent of the frequency, with power $R_X(0) = \sigma_x^2$.

Any zero-mean process with a constant power spectrum is said to be a *white random process*. This may or may not imply that the samples of the random process are independent, although for the important Gaussian case they are.

Example 3-6.
As in Example 3-5, consider a continuous-time random process $\{X(t)\}$ with the autocorrelation function

$$R_X(\tau) = \frac{N_0}{2}\delta(\tau). \tag{3.64}$$

The power spectrum of this process is a constant, $S_X(f) = N_0/2$, so $\{X(t)\}$ is white. The power $R_X(0)$ of this continuous-time white process is infinite. So we immediately run into mathematical difficulties for the continuous-time case that we did not encounter in the discrete-time case.

Although the continuous-time white random process of Example 3-6 leads to the non-physical condition that the power is infinite (or undefined), it is an extremely important model. It would appear from the fact that $R_X(\tau) = 0$ for all $\tau \neq 0$ that any two distinct samples of a continuous-time white random process are uncorrelated, but, unfortunately, this makes no mathematical or physical sense. Sampling a continuous-time white random process is an ill-defined concept. Roughly speaking, a continuous-time white random process varies so quickly that it is not possible to determine its characteristics at any instant in time.

In spite of these mathematical difficulties, the continuous-time white random processes is useful as a model for noise which has an approximately constant power spectrum over a bandwidth larger than the bandwidth of the system we are considering. In such a system we will always band-limit the noise to eliminate any out-of-band component. In this event, it makes no difference if we start with a white noise or a more accurate model; the result will be very nearly the same. But using the white noise model results in significantly simpler algebraic manipulation. In this book we will often use the white noise model, and take care to always band-limit this noise process prior to other operations such as sampling. After band-limiting, we obtain a well-behaved process with finite power.

Example 3-7.
Thermal or *Johnson* noise in electrical resistors has a power spectrum that is flat to more than 10^{12} Hz, a bandwidth much greater than most systems of interest (see [1]). Thus, we can safely use white noise as a model for this thermal noise without compromising accuracy. The noise in the model at frequencies greater than 10^{12} Hz will always be filtered out at the input to our system anyway. By contrast, in optical systems, thermal noise is generally insignificant at optical frequencies. Thermal noise is modeled as a Gaussian random process, from the central limit theorem, since it is comprised of the superposition of many independent events (thermal fluctuations of individual electrons).

3.2.1. Cross-Correlation and Complex Processes

Given two random processes $X(t)$ and $Y(t)$, we can define a *cross-correlation* function,

$$R_{XY}(t_1, t_2) = E[X(t_1)Y^*(t_2)] \ . \tag{3.65}$$

If $X(t)$ and $Y(t)$ are each wide-sense stationary, then they are *jointly wide-sense stationary* if $R_{XY}(t_1, t_2)$ is a function only of $t_1 - t_2$.

A complex-valued random process $X(t)$ is defined as

$$X(t) = \text{Re}\{X(t)\} + j\text{Im}\{X(t)\} \ , \tag{3.66}$$

where $\text{Re}\{X(t)\}$ and $\text{Im}\{X(t)\}$ are real-valued random processes. The second order statistics of such a process consist of the two autocorrelation functions of the real and imaginary parts, as well as their cross-correlation functions. Complex Gaussian random processes are very important in digital communication systems; they have some special properties that are considered in detail in Section 3.2.7.

3.2.2. Filtered Random Processes

A particular outcome x_k or $x(t)$ of a random process is a signal, and therefore may be filtered or otherwise processed. We can also talk about filtering the random process X_k or $X(t)$ itself, rather than an outcome. Then we get a new random process with a sample space that is obtained by applying every element of the sample space of the original random process to the input of the filter.

Example 3-8.
A filtered Gaussian random process is a Gaussian random process. Intuitively, this is true because filtering is linear, and any linear combination of jointly Gaussian random variables is a Gaussian random variable.

Consider the two continuous-time LTI systems shown in Fig. 3-2 with WSS continuous-time random process inputs.

Exercise 3-11.
Show that the output of the filter $h(t)$ is WSS, and that its autocorrelation function and power spectrum are given by

$$R_W(\tau) = h(\tau) * h^*(-\tau) * R_X(\tau) \ , \qquad S_W(f) = S_X(f)\,|H(f)|^2 \ , \tag{3.67}$$

$$S_W(s) = S_X(s)H(s)H^*(-s^*) \ . \tag{3.68}$$

Fig. 3-2. Two linear systems with WSS random process inputs.

Exercise 3-12.

Show that if a WSS discrete-time random process X_k is filtered by a filter that has impulse response h_k, and the result is W_k, then W_k is WSS and

$$R_W(m) = h_m * h^*_{-m} * R_X(m) , \qquad S_W(e^{j\theta}) = S_X(e^{j\theta}) \, | H(e^{j\theta}) |^2 , \tag{3.69}$$

$$S_W(z) = S_X(z) H(z) H^*(1/z^*) . \tag{3.70}$$

The *cross-spectral density* of two jointly WSS random processes at the filter inputs in Fig. 3-2 is defined as the Fourier transform of the cross-correlation function,

$$S_{XY}(f) = \int_{-\infty}^{\infty} R_{XY}(\tau) e^{-j2\pi f\tau} \, d\tau , \qquad R_{XY}(\tau) = E[X(t+\tau)Y^*(t)] . \tag{3.71}$$

Exercise 3-13.

Show that the cross-power spectrum of the outputs in Fig. 3-2 is

$$S_{WU}(f) = H(f) G^*(f) S_{XY}(f) . \tag{3.72}$$

3.2.3. The Innovations Process

Given a wide-sense stationary random process $\{X_k\}$ with power spectrum $S_X(e^{j\theta})$, a natural *innovations representation* of that random process follows from the monic minimum-phase spectral factorization of $S_X(z)$ (see (2.67) in Section 2.5.6). In particular, since $S_X(z)$ is real-valued and non-negative on the unit circle, it can be decomposed as

$$S_X(z) = \gamma_x^2 M_x(z) M_x^*(1/z^*) , \tag{3.73}$$

where $M_x(z)$ is a monic loosely minimum-phase causal filter, and γ_x^2 is a constant to be interpreted shortly. If $S_X(z)$ has no zeros on the unit circle ($S_X(e^{j\theta}) > 0$ for all θ), then $M_x(z)$ is strictly minimum phase. In this case, its inverse filter $M_x^{-1}(z)$ is stable, and is also a monic minimum-phase causal filter. If we filter the process X_k with the filter $M_x^{-1}(z)$, as shown in Fig. 3-3, then from (3.70), the output I_k is a white random process with power spectrum $S_I(z) = \gamma_x^2$. The random process $\{I_k\}$ is called the *innovations process*. Its power is γ_x^2.

The innovations process and the filter $M_x(z)$ can be used to generate the random process X_k, as shown in Fig. 3-3. This helps to explain the terminology. Since I_k is white, each new sample is uncorrelated with previous samples. Thus each new sample brings new information (an "innovation") about the random process X_k. Viewed another way, the *whitening filter* $M^{-1}(z)$ removes redundant information from X_k by removing correlated components in the

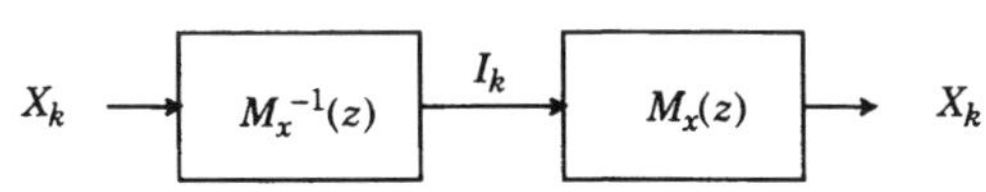

Fig. 3-3. Generation of the innovations I_k from X_k, and the recovery of X_k from its innovations.

samples. What is left has only uncorrelated samples. Thus we can think of X_k as having two components; the innovation is the new or "random" part, while the remainder is a linear combination of past innovations.

3.2.4. Linear Prediction

A *linear predictor* forms an estimate of the current sample of a discrete-time random process by exploiting the correlation between present and past samples. Specifically, if X_k is the random sequence to be predicted, a linear predictor generates an estimate $\hat{X}_k$ of X_k at time k by forming a linear combination of the past:

$$\hat{X}_k = \sum_{n=1}^{\infty} p_n X_{k-n} . \tag{3.74}$$

This estimate can be viewed as the output of an LTI *prediction filter* whose transfer function $P(z)$ is causal,

$$P(z) = \sum_{n=1}^{\infty} p_n z^{-n} . \tag{3.75}$$

A strictly causal prediction filter ensures that only past samples are used in constructing the prediction. A well-designed linear predictor would choose $P(z)$ so that $\hat{X}_k$ matches X_k as closely as possible. A convenient measure of closeness is the mean-square error (MSE), defined by MSE = $E[|\hat{X}_k - X_k|^2]$. The prediction filter that minimizes MSE is closely linked to the spectral factorization theorem.

Theorem 3-1. Given a random process X_k with PSD $S_X(z) = \gamma_x^2 M_x(z) M_x^*(1/z^*)$ factored according to the spectral factorization theorem of Section 2.5.6, the optimal linear prediction filter (minimizing MSE) is

$$P(z) = 1 - \frac{1}{M_x(z)} . \tag{3.76}$$

The resulting minimal MSE is γ_x^2.

This result is easy to prove. The *prediction error* $E_k = X_k - \hat{X}_k$ may be generated by passing the random sequence X_k through a *prediction-error filter* with transfer function

$$D(z) = 1 - P(z) . \tag{3.77}$$

To be a legitimate prediction-error filter, $D(z)$ must be monic, causal, and stable. The PSD of the prediction error can now be written as

$$\begin{aligned} S_E(z) &= S_X(z) D(z) D^*(1/z^*) \\ &= \gamma_x^2 M_x(z) M_x^*(1/z^*) D(z) D^*(1/z^*) \\ &= \gamma_x^2 B(z) B^*(1/z^*) , \end{aligned} \tag{3.78}$$

where we have introduced $B(z) = M_x(z)D(z)$. Since both $M_x(z)$ and $D(z)$ are monic and causal ($M_x(\infty) = D(\infty) = 1$), it follows that $B(z)$ must also be monic and causal ($B(\infty) = 1$), with impulse response b_k satisfying $b_k = 0$ for $k < 0$ and $b_0 = 1$. The goal is to minimize:

$$\begin{aligned}
\text{MSE} = \mathrm{E}[\,|E_k|^2] &= R_E(0) \\
&= \frac{1}{2\pi}\int_{-\pi}^{\pi} S_E(e^{j\theta})\, d\theta \\
&= \frac{1}{2\pi}\int_{-\pi}^{\pi} \gamma_x^2\, |B(e^{j\theta})|^2 \\
&= \gamma_x^2 \sum_{k=-\infty}^{\infty} |b_k|^2 \\
&= \gamma_x^2\left(1 + \sum_{k=1}^{\infty} |b_k|^2\right),
\end{aligned} \tag{3.79}$$

where the fourth line follows from Parseval's relationship, and the last line from the monic and causal property of b_k. Clearly, the MSE is minimized by choosing $b_1 = b_2 = b_3 = \ldots = 0$, or equivalently by choosing $B(z) = 1$, yielding a minimum MSE of γ_x^2. But since $B(z) = M_x(z)D(z)$, it follows that the optimal prediction-error filter is $D(z) = M_x^{-1}(z)$. Finally, solving $D(z) = 1 - P(z)$ for $P(z)$ yields (3.76).

The prediction error E_k for the optimal predictor is precisely the innovation I_k, and the resulting prediction error variance is γ_x^2, which is the geometric mean $\langle S_X \rangle_G$ of the PSD of X_k. In contrast, the variance of X_k (before linear prediction) is the *arithmetic* mean of its PDF, $\langle S_X \rangle_A$. From the arithmetic-geometric inequality, the variance after linear prediction is always less than the variance before, unless X_k was a white random process to begin with. Linear prediction reduces variance by the ratio $\langle S_X \rangle_G / \langle S_X \rangle_A$.

Example 3-9. ——————————————————————S

uppose the PSD of X_k is $S_X(e^{j\theta}) = 25 - 24\cos(\theta)$, which has arithmetic mean 25 and geometric mean 16. Therefore, optimal linear prediction would reduce the variance of X_k by a factor of $16/25$.

The optimal prediction-error filter $D(z) = M_x^{-1}(z)$ is said to be a *whitening filter* because its output is white. The whitening property of the optimal prediction-error filter is not surprising, given that the predictor is exploiting the correlation of input samples. Intuitively, if the prediction error were not white, there would still be correlation left to further exploit. However, this intuitive explanation is incomplete. The optimal prediction-error filter does more than just whiten its output. Indeed, there are many monic and causal whitening filters, but only one that minimizes MSE. This is explored in the following exercise.

Exercise 3-14.

Let $A(z)$ be a causal allpass filter with $a_0 \neq 0$, and let its impulse response length be greater than one (thus ruling out trivial filters of the form $A(z) = e^{j\alpha}$ for constant α). Then $A(z)/a_0$ is monic and causal, and its magnitude response is a constant $1/|a_0|$. Show that $|a_0| < 1$.

We can thus multiply the optimal prediction-error filter by the monic and causal filter $A(z)/a_0$, thus producing another valid (monic and causal) prediction-error filter that is also a whitener. However, because of Exercise 3-14, the resulting MSE $\gamma_x^2/|a_0|^2$ would be strictly greater than the minimal MSE γ_x^2.

This exercise is instructive, because it shows that any nontrivial causal and monic filter with a flat frequency response will amplify its inputs. The optimal prediction error filter $M_x^{-1}(z)$ thus has two key properties: it is a *whitening filter*, resulting in a white prediction error, and it is *minimum phase*. The whitening-filter property of the prediction-error filter (if not the minimum-phase property) can also be demonstrated by orthogonality arguments (see Problem 3-5), and has a simple geometric interpretation (see Problem 3-6).

3.2.5. Sampling a Random Process

A finite power continuous-time random process $X(t)$ can be sampled, yielding a discrete-time random process $Y_k = X(kT)$. Since we will be performing this sampling operation often in digital communication systems, it is important to relate the statistics of the continuous-time random processes with those of the discrete-time random process obtained by sampling it. Assuming $X(t)$ is WSS,

$$R_{YY}(k, i) = E[X(kT)X^*(iT)] = R_X(mT) , \tag{3.80}$$

where $m = k - i$, so the sampled process is WSS with autocorrelation equal to a sampled version of the autocorrelation $R_X(\tau)$ of the original continuous-time signal. From (2.17), the power spectrum of the continuous-time random process and its sampled discrete-time process are related by

$$S_Y(e^{j2\pi fT}) = \frac{1}{T}\sum_{m=-\infty}^{\infty} S_X(f - \tfrac{m}{T}) . \tag{3.81}$$

As in the deterministic case, aliasing distortion results when the bandwidth is greater than ½ the sampling rate, where bandwidth in this case is defined in terms of the power spectrum.

Using the techniques discussed so far, we should have no difficulty considering systems that mix discrete and continuous-time random processes as well as deterministic signals. However, there are some subtleties. Consider a discrete-time random process X_k filtered by a continuous-time filter with impulse response $h(t)$ in the sense defined in Section 2.1. The output can be written

$$Y(t) = \sum_{k=-\infty}^{\infty} X_k h(t - kT). \tag{3.82}$$

This is *pulse-amplitude modulation* (PAM), described in detail in Chapter 5.

Example 3-10.

The transmission of a discrete-time sequence of data symbols X_m over a continuous-time channel often takes the form of the random process in (3.82). Suppose that $h(t)$ is as shown in Fig. 3-4(a) and that X_k is a random sequence with i.i.d. samples taking values ±1 with equal probability. A possible outcome is shown in Fig. 3-4(b). The first important observation is that the process $Y(t)$ is not wide-sense stationary because $E[Y(t+\tau)Y(t)]$ is not independent of t. For example,

$$E[Y(T/4)Y(0)] = E[X_0^2] = 1 \neq E[Y(T)Y(3T/4)] = 0.$$

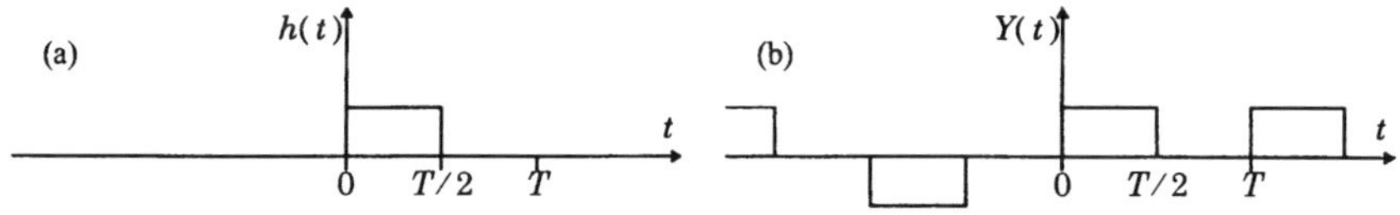

Fig. 3-4. (a) An example of a pulse shape for transmitting bits. (b) An example of a waveform using this pulse shape.

This process is actually *cyclostationary*, a weaker form of stationarity. Since this process is not wide-sense stationary, its power spectrum is not defined.

The fact that $Y(t)$ in (3.82) is not wide-sense stationary is a major inconvenience. A common gimmick changes our random process into a wide-sense stationary process. Define the random variable Θ, called a *random phase epoch*, that is uniformly distributed on $[0, T)$ and independent of $\{X_k\}$. Then define the new random process

$$Z(t) = Y(t - \Theta) = \sum_{k=-\infty}^{\infty} X_k h(t - kT - \Theta). \tag{3.83}$$

This process has a *random phase* which is constant over time but chosen randomly at the beginning of time. Physically, this new process reflects our uncertainty about the phase of the signal; the origin in the time axis is of course arbitrary. This redefined process is wide-sense stationary, as shown in Appendix 3-A, with power spectrum

$$S_Z(f) = \frac{1}{T} |H(f)|^2 S_X(e^{j2\pi fT}) . \tag{3.84}$$

Note the dependence on the power spectrum of the discrete-time process and the magnitude-squared spectrum of the pulse $h(t)$.

Example 3-11.

Consider transmission of a random sequence of uncorrelated random variables X_k with equally probable values ± 1 using a pulse shape $h(t)$. The sequence X_k is white and the variance is unity, so the power spectrum of the data sequence is

$$S_X(e^{j2\pi fT}) = 1, \tag{3.85}$$

and the power spectrum of the random phase transmitted signal is

$$S_Z(f) = \frac{1}{T} |H(f)|^2. \tag{3.86}$$

With a white data sequence, the power spectrum has the shape of the magnitude squared of the Fourier transform of the pulse.

3.2.6. Reconstruction of Sampled Signal

It might appear that (3.81) establishes the conditions under which a random process can be recovered from its samples, just as (2.17) does for deterministic signals. However, this appearance is deceiving because two random processes can have the same power spectrum and not be "equal" in any sense. The power spectrum is merely a second-order statistic, not a full characterization of the process. By a derivation similar to that in Appendix 3-A, we can investigate the recovery of the original continuous-time random process from its samples. A method of sampling and recovering a random process analogous to the deterministic case is shown in Fig. 3-5. We first filter the random process using an antialiasing filter $F(f)$, then sample, and finally recover using recovery filter $H(f)$ to yield the random process $Y(t)$. To make $Y(t)$ WSS we must again introduce a random phase. The way to tell whether the system recovers the input random process is not to calculate the output power spectrum, but rather to investigate the error signal between input and output. In particular, define

$$E(t) = X(t-\Theta) - Y(t-\Theta)\,, \tag{3.87}$$

where Θ is uniformly distributed over $[0, T)$. We would conclude that the recovery is exact (in a mean-square sense) if

$$\mathrm{E}[\,|E(t)|^2] = 0\,. \tag{3.88}$$

This is not the same as showing that $E(t) = 0$, which cannot be shown using second order statistics only. However, (3.88) is just as good for engineering purposes. The conditions under which (3.88) is valid can be inferred from the following exercise, which can be solved using similar techniques to those used in Appendix 3-A.

Exercise 3-15.

Show that the power spectrum of $E(t)$ is

$$S_E(f) = \frac{1}{T^2}|H(f)|^2 \sum_{m \neq 0} S_X\left(f - \frac{m}{T}\right)\left|F\left(f - \frac{m}{T}\right)\right|^2 + \left|1 - \frac{1}{T}H(f)F(f)\right|^2 S_X(f). \tag{3.89}$$

Examining (3.89), the first term is aliasing distortion resulting from a signal at the output of the antialiasing filter, if it is not sufficiently bandlimited. In particular, if $H(f) = 0$ and $F(f) = 0$ for $|f| \geq 1/(2T)$ then this term is identically zero. The second term is in-band distortion due to an improper reconstruction filter $H(f)$ and also distortion due to band-limiting of the input prior to sampling. For an ideal reconstruction filter,

$$H(f)F(f) = T, \quad |f| \leq 1/(2T)\,, \tag{3.90}$$

in which case the error signal has power spectrum

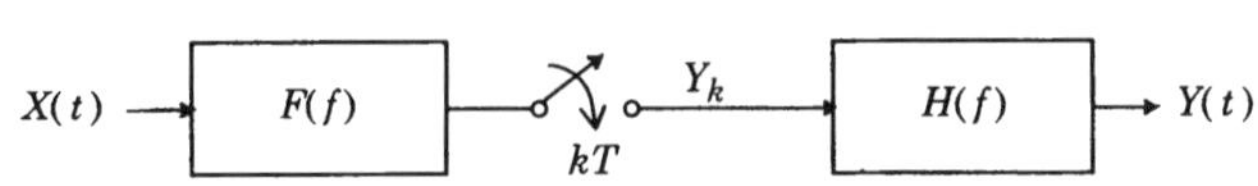

Fig. 3-5. Sampling and recovery of a random process using antialiasing filter $F(f)$ and recovery filter $H(f)$.

$$S_E(f) = \begin{cases} 0, & |f| < 1/(2T) \\ S_X(f), & |f| \geq 1/(2T) \end{cases}, \tag{3.91}$$

and the total error power is

$$\mathrm{E}[\,|E(t)|^2] = 2\int_{1/(2T)}^{\infty} S_X(f)\, df. \tag{3.92}$$

The fact that the reconstruction error is just the error in initially band-limiting $X(t)$ is not surprising, and corresponds to the deterministic signal case.

The results of this subsection are important not only for their implications to the recovery of sampled random processes, but also in the techniques used. We will find the need for similar techniques in the optimization problems of Chapter 7.

3.2.7. Complex-Valued Gaussian Processes

At the beginning of Section 3.2, the real-valued Gaussian random process was defined as a process for which any arbitrary set of samples is jointly Gaussian. A *complex-valued Gaussian random process* consists of two jointly Gaussian real-valued processes, a real part and an imaginary part. By jointly Gaussian, we mean that any arbitrary set of samples of the real and imaginary parts is a jointly Gaussian set of random variables. Such processes are important in digital communications in modeling the complex envelope of noise. Complex-valued processes have some special properties that distinguish them from real-valued processes.

Let $Z(t)$ be a zero-mean complex-valued Gaussian process. Since $Z(t)$ is complex-valued, it consists of two real-valued processes,

$$R(t) = \mathrm{Re}\{Z(t)\}\,, \qquad I(t) = \mathrm{Im}\{Z(t)\}\ . \tag{3.93}$$

By assumption both $R(t)$ and $I(t)$ are zero-mean Gaussian processes. To fully characterize the statistics of $Z(t)$, we must specify the *joint* statistics of $R(t)$ and $I(t)$. Because they are Gaussian and zero-mean, $R(t)$ and $I(t)$ are fully characterized by their second order statistics,

$$R_R(\tau) = \mathrm{E}[R(t+\tau)R(t)]\,, \quad R_I(\tau) = \mathrm{E}[I(t+\tau)I(t)]\,, \quad R_{RI}(\tau) = \mathrm{E}[R(t+\tau)I(t)]\,. \tag{3.94}$$

The complex-valued process $Z(t)$ is strictly stationary if $R(t)$ and $I(t)$ are jointly wide-sense stationary, and hence jointly strictly stationary (since they are Gaussian). $R(t)$ and $I(t)$ are jointly wide-sense stationary if the correlation functions $\mathrm{E}[R(t+\tau)R(t)]$, $\mathrm{E}[I(t+\tau)I(t)]$, and $\mathrm{E}[R(t+\tau)I(t)]$ are not functions of t, as indicated in (3.94).

According to the definition, the *complex* process $Z(t)$ is wide-sense stationary if the autocorrelation function

$$R_Z(\tau) = \mathrm{E}[Z(t+\tau)Z^*(t)]\,, \tag{3.95}$$

is not a function of t. This is not the same as saying that the real and imaginary parts are jointly wide-sense stationary. Clearly, $R_Z(\tau)$ could not by itself contain information equivalent to $R_R(\tau)$, $R_I(\tau)$, and $R_{RI}(\tau)$, since (3.94) constitutes *three* functions of τ, while $R_Z(\tau)$ specifies only *two* functions of τ, its real and imaginary parts. Thus, we require more than $R_Z(\tau)$ to fully specify the statistics of $Z(t)$. In addition to $R_Z(\tau)$, it suffices to know the *complementary autocorrelation function*, defined as

$$\tilde{R}_Z(\tau) = \mathrm{E}[Z(t+\tau)Z(t)] \ . \tag{3.96}$$

Again, since $\tilde{R}_Z(\tau)$ can be expressed in terms of $R_R(\tau)$, $R_I(\tau)$, and $R_{RI}(\tau)$, it must not be a function of t if $Z(t)$ is to be strict-sense stationary.

Using the relations $2R(t) = Z(t) + Z^*(t)$ and $2jI(t) = Z(t) - Z^*(t)$, it is easy to show that

$$2R_R(\tau) = \mathrm{Re}\{R_Z(\tau)\} + \mathrm{Re}\{\tilde{R}_Z(\tau)\}\ , \qquad 2R_I(\tau) = \mathrm{Re}\{R_Z(\tau)\} - \mathrm{Re}\{\tilde{R}_Z(\tau)\}\ ,$$
$$2R_{RI}(\tau) = \mathrm{Im}\{\tilde{R}_Z(\tau)\} - \mathrm{Im}\{R_Z(\tau)\} \ . \tag{3.97}$$

For the special case of a real-valued $Z(t)$, $R_I(\tau) = R_{RI}(\tau) = 0$, and $R_Z(\tau) = \tilde{R}_Z(\tau)$. Given *both* the autocorrelation and complementary autocorrelation functions, (3.97) allows us to determine the full complement of joint statistics of $R(t)$ and $I(t)$. Conversely, neither the autocorrelation nor the complementary autocorrelation function is sufficient by itself to fully specify the statistics of $Z(t)$.

$Z(t)$ is wide-sense stationary if it has an autocorrelation function $\mathrm{E}[Z(t+\tau)Z^*(t)] = R_Z(\tau)$ that is independent of t. In that case, its power spectrum $S_Z(f)$ is the Fourier transform of $R_Z(\tau)$. However, in the case of complex Gaussian processes, wide-sense stationarity does *not* imply strict sense stationarity, because even for a wide-sense stationary process $\mathrm{E}[Z(t+\tau)Z(t)]$ may be a function of t. However, (3.97) implies that if a Gaussian process is wide-sense stationary, and in addition $\mathrm{E}[Z(t+\tau)Z(t)]$ is not a function of t, then the real and imaginary parts are jointly wide-sense stationary, and $Z(t)$ is strictly stationary.

Based on these considerations, there are two important differences between real-valued and complex-valued Gaussian processes:

- A complex-valued zero-mean Gaussian process is fully specified by both the autocorrelation and complementary autocorrelation functions, but not by either one alone. In particular, it is not fully characterized by its power spectrum. In contrast, a real-valued zero-mean Gaussian process requires only the autocorrelation function, and is thus fully characterized by its power spectrum.
- A wide-sense stationary complex-valued zero-mean Gaussian process is not necessarily strictly stationary, but it is strictly stationary if *both* the autocorrelation and complementary autocorrelation functions are not functions of t. In contrast, a real-valued zero-mean Gaussian process is strictly stationary if and only if it is wide-sense stationary.

Although there are significant differences between real-valued and complex-valued Gaussian processes, there is an important special case, considered next, where the two have similar properties.

3.2.8. Circularly Symmetric Gaussian Processes

In Section 3.1.3 we saw that a zero-mean Gaussian random *variable* is circularly symmetric if and only if $E[Z^2] = 0$. We now generalize to random *processes*. A complex-valued zero-mean Gaussian process is circularly symmetric if

$$E[Z(t+\tau)Z(t)] = 0, \qquad \text{for all } t \text{ and } \tau. \tag{3.98}$$

Such processes have a number of important simplifying properties, and further, most Gaussian processes encountered in digital communication are circularly symmetric. Note that a nonzero real-valued process *cannot* be circularly symmetric since for such a process $R_Z(\tau) = \tilde{R}_Z(\tau)$.

First of all, a circularly symmetric Gaussian process is strictly stationary if and only if it is wide-sense stationary, since then the real and imaginary parts are jointly wide-sense stationary. For a wide-sense stationary circularly symmetric Gaussian process, (3.97) simplifies to

$$2R_R(\tau) = \mathrm{Re}\{R_Z(\tau)\}, \qquad 2R_I(\tau) = \mathrm{Re}\{R_Z(\tau)\}, \qquad 2R_{RI}(\tau) = -\mathrm{Im}\{R_Z(\tau)\}. \tag{3.99}$$

Based on (3.99), circularly symmetric Gaussian processes have several nice properties:

- The real and imaginary parts individually have identical statistics, by virtue of having the same autocorrelation function $\frac{1}{2}\mathrm{Re}\{R_Z(\tau)\}$.
- Since $R_Z(0)$ must be real valued, $\mathrm{Im}\{R_Z(0)\} = R_{RI}(0) = 0$. This implies that for any given time t, $R(t)$ and $I(t)$ are uncorrelated and hence statistically independent, although they are not necessarily uncorrelated nor independent when sampled at different times.
- Circularly symmetric processes with a *real-valued* autocorrelation function $R_Z(\tau)$ have a real and imaginary part that are independent at *all* times, since $R_{RI}(\tau) = 0$. (Note that a real-valued $R_Z(\tau)$ does not imply that the process is real-valued. In fact, a circularly symmetric $Z(t)$ with real-valued $R_Z(\tau)$ *cannot* be a real-valued process!) $R_Z(\tau)$ is real-valued when the power spectrum of the process (which is *always* real-valued) has even symmetry about $f = 0$.

Circular symmetry is preserved by linear time-invariant filtering. That is, if we apply a circularly symmetric Gaussian process $Z(t)$ to a linear time-invariant system with impulse response $h(t)$, then the output is given by a convolution

$$V(t) = \int_{-\infty}^{\infty} h(\tau)Z(t-\tau)\,d\tau, \tag{3.100}$$

and

$$E[V(t+\tau)V(t)] = \int_{-\infty}^{\infty}\int_{-\infty}^{\infty} h(u)h(v)E[Z(t+\tau-u)Z(t-v)]dudv = 0, \tag{3.101}$$

since the integrand is identically zero. More generally, the circularly symmetric property is preserved by time-varying linear systems, such as modulators.

Example 3-12.

Assume that $V(t)$ is a circularly symmetric stationary zero-mean Gaussian random process, and define $Z(t) = e^{j2\pi f_c t}V(t)$. $Z(t)$ is also stationary and circularly symmetric, since

$$R_Z(\tau) = \mathrm{E}[Z(t+\tau)Z^*(t)] = e^{j2\pi f_c\tau}R_V(\tau)\,, \tag{3.102}$$

$$\tilde{R}_Z(\tau) = e^{j2\pi f_c(2t+\tau)}\mathrm{E}[V(t+\tau)V(t)] = 0\,. \tag{3.103}$$

Further, the power spectrum of $Z(t)$ is $S_V(f-f_c)$. $\mathrm{Re}\{Z(t+\tau)\}$ and $\mathrm{Im}\{Z(t)\}$ are thus independent for all τ if and only if $S_V(f)$ is symmetric about f_c; that is, $S_V(f_c+\Delta f) = S_V(f_c-\Delta f)$ for all Δf. If $V(t)$ is a narrowband process, this implies that $V(t)$ has only positive-frequency components in its power spectrum.

A complex-valued Gaussian process obtained from a real-valued Gaussian process by modulating a complex exponential is neither stationary nor circularly symmetric, as illustrated by the following example.

Example 3-13.

Let $N(t)$ be a real-valued zero-mean Gaussian random process. Thus, $N(t)$ cannot be circularly symmetric, and, not surprisingly, neither is $Z(t) = e^{j2\pi f_c t}N(t)$. In particular,

$$R_Z(\tau) = \mathrm{E}[Z(t+\tau)Z^*(t)] = e^{j2\pi f_c\tau}R_N(\tau)\,, \tag{3.104}$$

$$\mathrm{E}[Z(t+\tau)Z(t)] = e^{j2\pi f_c(2t+\tau)}R_N(\tau)\,. \tag{3.105}$$

$Z(t)$ is wide-sense stationary, but it is neither strictly stationary nor circularly symmetric. That $Z(t)$ is non-stationary is not surprising, since at certain times (the zero-crossings of the carrier) the real part of $Z(t)$ is identically zero, and similarly for the imaginary part. That $Z(t)$ is wide-sense stationary in spite of not being strictly stationary is perhaps surprising to those accustomed to real-valued Gaussian processes.

Discrete-Time Gaussian Processes

All the properties we have described carry over to discrete-time zero-mean Gaussian random processes. In particular, such a complex-valued process Z_k is fully characterized by $R_Z(m)$ and $\tilde{R}_Z(m)$. By definition, it is circularly symmetric if

$$\mathrm{E}[Z_{k+m}Z_k] = 0\,, \qquad \text{for all } m \text{ and } k\,. \tag{3.106}$$

If $Z_k = Z(kT)$ is obtained by sampling a circularly symmetric continuous-time process, it will itself be circularly symmetric. If Z_k is circularly symmetric, its real and imaginary parts have the same variance, and are independent at a given time t. Further, the real and imaginary parts are statistically independent for all time if and only if the autocorrelation $R_Z(k)$ is real-valued. As in continuous time, circular symmetry is preserved by linear time-invariant filtering, and more generally by linear operations.

White Gaussian Processes

An important subclass of zero-mean complex Gaussian processes are *white*. Such processes have an autocorrelation function

$$R_Z(\tau) = N_0\delta(\tau)\,, \qquad R_Z(k) = N_0\delta_k\,, \tag{3.107}$$

for continuous and discrete time respectively. The convention is that $\sigma^2 = N_0/2$ is the variance of the real part or the imaginary part, so that $2\sigma^2 = N_0$ is the variance of the complex process.

For real-valued processes, in continuous time the white property implies that $Z(t + \tau)$ and $Z(t)$ are uncorrelated and hence independent for all $\tau \neq 0$, and for discrete time, $Z(k + m)$ and $Z(k)$ are uncorrelated and hence independent for all $m \neq 0$.

A white complex-valued Gaussian process is not necessarily strict-sense stationary. However, if the process is both *white* and *circularly symmetric*, then the following properties hold:

- The real and imaginary parts of the process are identically distributed, and are each white real-valued Gaussian processes.
- The real and imaginary parts are independent of one another, since the autocorrelation function is real-valued.

Thus, circularly symmetric zero-mean white complex Gaussian processes are maximally random, in the sense that (a) the samples of the process are mutually independent and (b) the real and imaginary parts are independent.

Another important observation is that any Gaussian process obtained by a time-invariant linear filtering of a circularly symmetric white Gaussian process is itself circularly symmetric, although in general it will not be white (unless the filter is allpass).

The Complex Envelope of White Gaussian Noise

In this section we characterize the statistics of the *complex envelope* of white-Gaussian noise. In other words, we characterize the output of a downconverter when the input is white and Gaussian. We will see that the complex envelope is a circularly symmetric Gaussian random process that is *almost* white. The results have important implications in later chapters.

Consider a real-valued white Gaussian random process $N(t)$ with power spectrum $N_0/2$ at the input to a downconverter, as shown below:

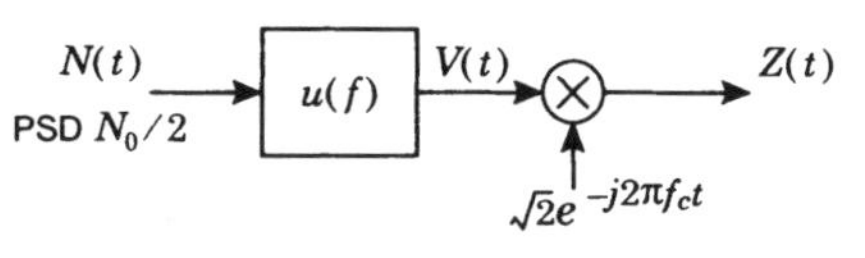

The output $V(t)$ of the phase splitter is clearly complex-valued, since the phase-splitter frequency response is not Hermitian symmetric, and $V(t)$ is also a Gaussian random process, since filtering preserves Gaussianity. But is $V(t)$ circularly symmetric?

The answer is yes, because $V(t)$ can be decomposed according to (2.23), namely:

$$V(t) = \tfrac{1}{2}\big(N(t) + j\hat{N}(t)\big) , \tag{3.108}$$

where $\hat{N}(t)$ is the Hilbert transform of $N(t)$, namely $\hat{N}(t) = N(t) * h(t)$ where $h(t) = \frac{1}{\pi t}$ is the Hilbert transform impulse response, with frequency response $H(f) = -\mathrm{sign}(f)$. Therefore:

$$\mathrm{E}[V(t+\tau)V(t)] = \tfrac{1}{4}\mathrm{E}\Big[\big(N(t+\tau) + j\hat{N}(t+\tau)\big)\Big]\Big[\big(N(t) + j\hat{N}(t)\big)\Big]$$

$$= \frac{1}{4}\Big(R_N(\tau) - R_{\hat{N}}(\tau) \Big) + j\frac{1}{4}\Big(R_{\hat{N}N}(\tau) - R_{\hat{N}N}(-\tau) \Big) . \quad (3.109)$$

Since the Hilbert transformer has unity magnitude response ($|H(f)|^2 = 1$), both $\hat{N}(t)$ and $N(t)$ have the same power spectrum. Therefore, the real part above is zero. Furthermore, from Exercise 3-13, the cross-spectrum between the Hilbert transform output and input is:

$$S_{\hat{N}N}(f) = H(f) S_{NN}(f) , \quad (3.110)$$

which implies that the crosscorrelation is $R_{\hat{N}N}(\tau) = (N_0/2)h(\tau) = N_0/(2\pi\tau)$. In particular $R_{\hat{N}N}(\tau)$ is an odd function, so that the imaginary part of (3.109) is also zero. Since the phase-splitter output $V(t)$ is complex Gaussian and (3.109) is zero, we conclude that $V(t)$ is circularly symmetric. Clearly, the power spectrum of $V(t)$ is $S_V(f) = (N_0/2)u(f)$.

We have already seen (Example 3-13) that circular symmetry is invariant to modulation by a complex exponential. Therefore, since the phase-splitter output $V(t)$ is circularly symmetric, so too is the downconverter output $Z(t) = \sqrt{2}e^{-j2\pi f_c t}V(t)$. Furthermore, the resulting power spectrum is:

$$S_Z(f) = 2S_V(f + f_c) = N_0 u(f + f_c) . \quad (3.111)$$

The bottom line is that the complex envelope of real white Gaussian noise is complex circularly symmetric Gaussian noise with a power spectrum given by (3.111). In some circumstances the power spectrum of the complex envelope beyond the carrier frequency is irrelevant. This would be the case, for example, when the complex envelope is applied to a filter that is strictly bandlimited to $|f| < f_c$. In such circumstances we may model the complex envelope of white noise as being white.

3.3. Markov Chains

A *discrete-time Markov process* $\{\Psi_k\}$ is a random process that satisfies

$$p(\psi_{k+1} | \psi_k, \psi_{k-1}, \ldots) = p(\psi_{k+1} | \psi_k) . \quad (3.112)$$

In words, the future sample Ψ_{k+1} is independent of past samples $\Psi_{k-1}, \Psi_{k-2}, \ldots$ if the present sample $\Psi_k = \psi_k$ is known. The particular case of a Markov process where the samples take on values from a discrete and countable set Ω is called a *Markov chain.* In this section, we will often take Ω to be a set of integers. Markov chains are a useful model of a finite state machine with a random input, where the samples of the random input are statistically independent of one another. Since any digital circuit with internal memory (flip flops, registers, or RAMs) is a finite state machine, most digital communication systems contain finite state machines. Markov chains are useful signal generation models for digital communication systems with intersymbol interference or convolutional coding (Chapters 5, 12, and 13). Markov chain theory is also useful in the analysis of error propagation in decision-feedback equalizers (Chapter 8) and in the calculation of the power spectrum of line codes. The following treatment uses Z-transform techniques familiar to the readers of this book. Sections 3.3.2

through 3.3.4, as well as Appendix 3-B, can be skipped on a first reading, since the techniques are not used until Chapter 8.

3.3.1. State Transition Diagrams

Consider a random process Ψ_k (real, complex, or vector valued) whose sample outcomes are members of a finite or countably infinite set Ω of values. The random process Ψ_k is a Markov chain if (3.112) is satisfied. The next sample Ψ_{k+1} of a Markov chain is independent of the past samples Ψ_{k-1}, Ψ_{k-2}, ... given the present sample Ψ_k. Furthermore, *all* future samples of the Markov chain are independent of the past given knowledge of the present, as shown in the following exercise.

Exercise 3-16.

Let Ψ_k be a Markov chain and show that for any $n > 0$,

$$p(\psi_{k+n} \mid \psi_k, \psi_{k-1}, \ldots) = p(\psi_{k+n} \mid \psi_k). \tag{3.113}$$

Since knowledge of the current sample Ψ_k makes the past samples irrelevant, Ψ_k is all we need to predict the future behavior of the Markov chain. For this reason, Ψ_k is said to be the *state* of the Markov chain at time k, and Ω is the set of all possible states.

Example 3-14.

A *shift register process* is shown in Fig. 3-6. If X_k is a random sequence that is independent of its distant past X_{k-M-1}, X_{k-M-2}, ... and if we define the vector

$$\Psi_k = [X_{k-1}, \ldots, X_{k-M}], \tag{3.114}$$

where M is the memory of the system, then the Markov property (3.112) is satisfied. This follows since Ψ_{k+1} is a function of X_{k+1} and Ψ_k only. Hence $\{\Psi_k\}$ is a vector-valued discrete-time Markov process. If the inputs X_k are discrete-valued, then it is also a Markov chain.

A Markov chain can be described graphically by a *state transition diagram*. This graph displays each state of the Markov chain as a node, and also displays the input and output or some other relevant properties for the transitions between states.

Example 3-15.

The *parity* of a bit stream X_k is defined to be the accumulated modulo-two summation of the bits, and is computed by the circuit in Fig. 3-7(a). This circuit is also known as a binary *accumulator*. It is sufficient for the input bits to be independent for the random process $\Psi_k = Y_{k-1}$ to be Markov. It is easily seen from the diagram that Ψ_{k+1} depends only on the current state Ψ_k and the current input X_k. Ψ_k has a finite sample space $\Omega = \{0, 1\}$, so the parity checker can be represented by the

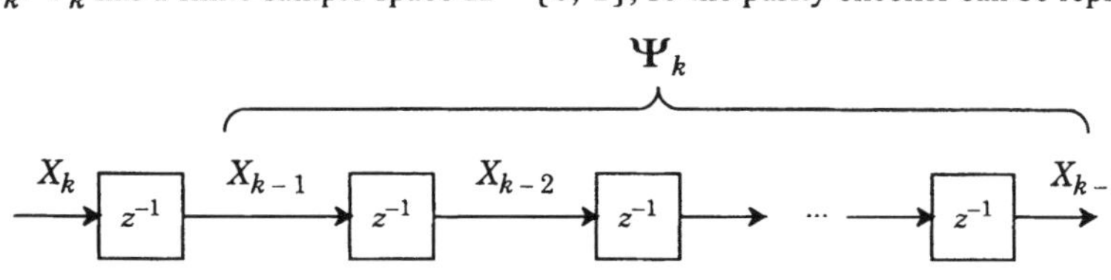

Fig. 3-6. A shift register process with independent inputs X_k is a Markov chain with state Ψ_k.

state transition diagram Fig. 3-7(b), where the arcs are labeled with the input that stimulates the state transition and the output resulting from the transition. The arcs of such a state diagram can alternatively be labeled with the transition probabilities, if the transition probabilities are independent of time.

A Markov chain Ψ_k is called *homogeneous* if the conditional probability $p(\psi_k \mid \psi_{k-1})$ is not a function of k. Homogeneity is therefore a kind of stationarity or time invariance. A homogeneous Markov chain can be characterized by its *state transition probabilities*, which we write with the shorthand

$$p(j \mid i) = p_{\Psi_{k+1}|\Psi_k}(j|i), \tag{3.115}$$

for $i, j \in \Omega$.

Example 3-16.
If in the previous example the incoming bits are not only independent but also identically distributed, then the Markov chain is homogeneous. If furthermore the incoming bits are equally likely to be one and zero, then the state transition probabilities are all 0.5.

It is often convenient to define a random process that is some real-valued function of the state trajectory of a Markov chain,

$$X_k = f(\Psi_k) \ . \tag{3.116}$$

This is encountered in the modeling of modulation with memory such as continuous-phase modulation. The transmitted power spectrum is an important property of such techniques, and thus we need to calculate the power spectrum of (3.116). This problem is considered in Appendix 3-B.

3.3.2. Transient Response of a Markov Chain

For a homogeneous Markov chain, we can find a relation for the evolution of the state probabilities with time. Using (3.33) we write

$$p_{k+1}(j) = \sum_{i \in \Omega} p(j \mid i) p_k(i) \tag{3.117}$$

for all $j \in \Omega$, where we have defined a notation for the probability of being in state i at time k,

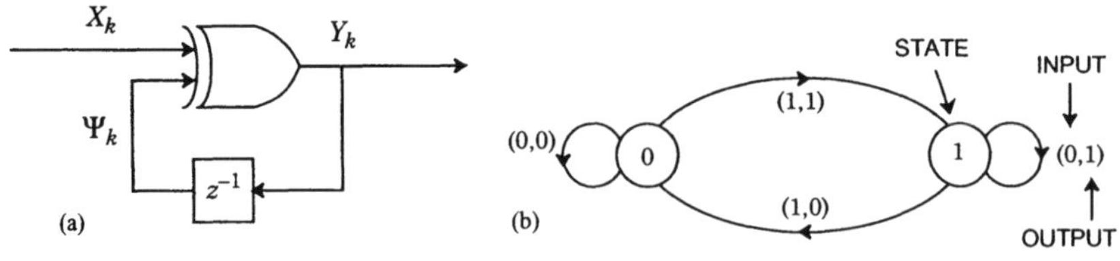

Fig. 3-7. (a) A circuit that computes the parity of the bit stream X_n. (b) The state transition diagram of the corresponding Markov chain.

$$p_k(i) = \Pr[\Psi_k = i]. \tag{3.118}$$

The new notation emphasizes that $p_k(i)$ is a discrete-time sequence. In applications we often want to determine the probability of being in a certain state j at a certain time k given a set of probabilities for being in those states at initial time $k = 0$. We can accomplish this by analyzing (3.117), a system of *time-invariant* difference equations, using Z-transform techniques. If we define $p_k(j) = 0$ for $k < 0$, then the Z-transform of the state probability for state j is

$$P_j(z) = \sum_{k=0}^{\infty} p_k(j)z^{-k}. \tag{3.119}$$

Exercise 3-17.

Take the Z-transform of both sides of (3.117) to show that

$$P_j(z) = p_0(j) + \sum_{i \in \Omega} p(j \mid i)z^{-1}P_i(z). \tag{3.120}$$

If there are N states, (3.120) gives us N equations with N unknowns $P_j(z)$. These equations can be solved and the inverse Z-transform calculated to determine the state probability $p_k(i)$.

Example 3-17.

Continuing Example 3-15, the parity check circuit, suppose that the initial state is equally likely to be either zero or one, so

$$p_0(0) = p_0(1) = 1/2. \tag{3.121}$$

Suppose further that the incoming bits X_k are equally likely to be zero or one, so the transition probabilities are all $1/2$. Then (3.120) becomes

$$P_0(z) = \tfrac{1}{2} + \tfrac{1}{2}z^{-1}P_0(z) + \tfrac{1}{2}z^{-1}P_1(z),$$
$$P_1(z) = \tfrac{1}{2} + \tfrac{1}{2}z^{-1}P_1(z) + \tfrac{1}{2}z^{-1}P_0(z). \tag{3.122}$$

Solving this set of two simultaneous equations, the Z-transforms of the state probabilities are equal,

$$P_0(z) = P_1(z) = \frac{1/2}{1 - z^{-1}}. \tag{3.123}$$

Using the Z-transform pair in (2.15) we can invert the Z-transform to get

$$p_k(0) = p_k(1) = \tfrac{1}{2}u_k, \tag{3.124}$$

where u_k is the unit step function. The chain is therefore equally likely to be in either state at any point in time beginning at $k = 0$. A Markov chain in which the state probabilities are independent of time is called *stationary*.

3.3.3. Signal Flow Graph Analysis

Translation of a state diagram into a set of equations to be solved is often made easier using signal flow graphs. A signal flow graph is a graphical representation of a linear equation, and in particular can represent the system of equations given by (3.120). Its value lies in the fact that the state diagram can be directly translated into a topologically equivalent signal flow graph representing the equations. In fact, the experienced can write down the signal flow graph directly without ever generating a state diagram. The idea of a graph representing linear equations is illustrated by the following simple example.

Example 3-18.
The equation $w = au + bx$ can be represented by the signal flow graph shown in Fig. 3-8(a). The nodes of the graph represent the variables u, w, and x, while the two arcs represent the multiplication of the variables by constants, and also the addition. The signal flow graph in Fig. 3-8(b) represents the recursive equation $x = au + bw + cx$.

In general, a node in a signal flow graph represents a variable that is equal to the sum of the incoming arcs. A weight on an arc is a multiplicative factor. Our interest is in signal flow graphs in which the variables are all Z-transforms.

Example 3-19.
The signal flow graph in Fig. 3-8(c) represents a dynamical system described by the equations $X(z) = z^{-1}Y(z) + W(z)$ and $Y(z) = z^{-1}W(z)$.

From the last example, it is clear that the equations (3.120) can be represented using a signal flow graph for any given Markov chain, as shown in Fig. 3-9. Shown are just two of the states, i and j. Each of the states is represented by two nodes of the graph, one for the Z-transform of the state probability sequence, $P_i(z)$, and the other for the initial probability of that state $p_0(i)$ (the latter is not a variable in the equations, but a constant). In many cases the initial probability is zero so the corresponding node can be omitted.

Example 3-20.
Returning to the parity check example of Example 3-15, the equations (3.122) are represented by the signal flow graph in Fig. 3-10. Note that this one figure takes the place of the state diagram of Fig. 3-7 and the set of equations of (3.122).

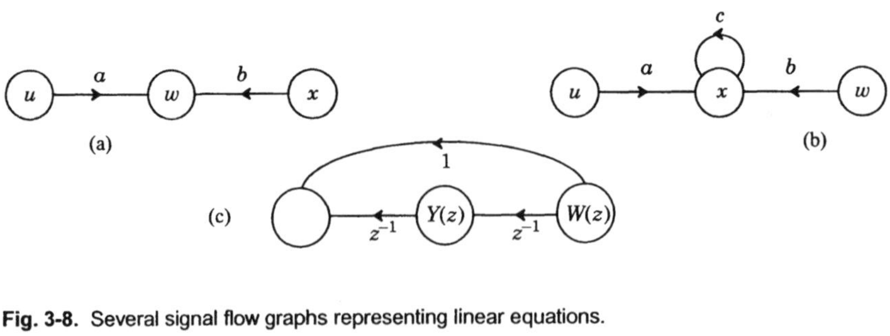

Fig. 3-8. Several signal flow graphs representing linear equations.

In retrospect, the signal flow graph is intuitive. Each state transition has a delay operator z^{-1} corresponding to the time it takes for that transition to occur, as well as the probability of that transition. The arcs from the initial state probabilities have no such delay since the initialization is instantaneous, and we can think of that transition as occurring only once at $k = 0$. For Markov chains that start in a particular state, there will only be one such node corresponding to the starting state.

Once we have a signal flow graph, we can easily write down the set of equations and then solve them for the Z-transform of the state probabilities. For some problems, a shortcut known as *Mason's gain formula* allows us to solve these equations directly by inspection of the signal flow graph [2][3][4][5].

3.3.4. First Passage Problem

When Markov chains are used to model the behavior of framing recovery circuits or error propagation (Chapter 8), we would like to calculate the *average first passage time* for an *absorption state* of the chain. An absorption state is defined as a state with an entry but no exit,

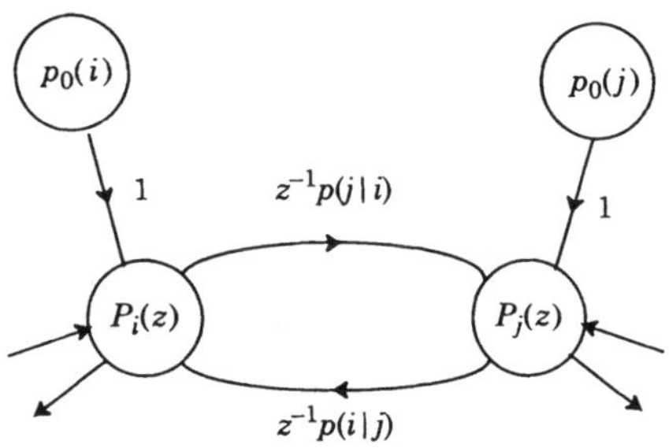

Fig. 3-9. A signal flow graph representation of the Markov chain dynamical equations (3.120).

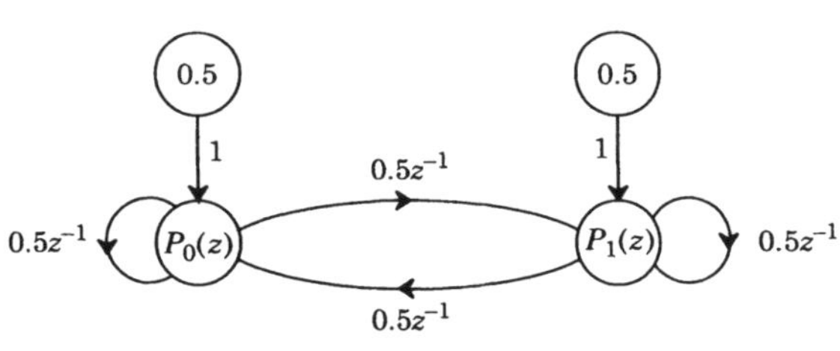

Fig. 3-10. A signal flow graph representation of the system of state probabilities for the parity examples.

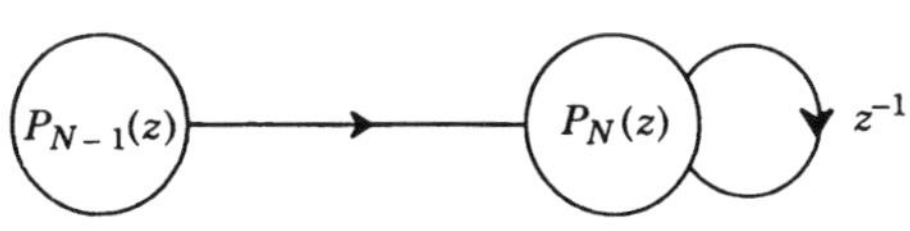

Fig. 3-11. A part of a signal flow graph for a Markov chain in which state N is an absorption state, with only one entry from the outside, namely from state $N-1$.

so that the steady-state probability of that state is unity. This is illustrated in Fig. 3-11 for the case where the absorption state is N. An absorption state must have a self-loop with gain z^{-1} indicating that the chain stays in that state forever. The figure also assumes that there is only one way to get to the absorption state, from state $N-1$, although that is not necessary for the following analysis.

What we are often interested in is the *first passage time to state N*, which is defined as the time-index of the first time we enter that state. Define the probability of entering state N at time k as $q_k(N)$. Then we have that

$$p_k(N) = p_{k-1}(N) + q_k(N) \ , \tag{3.125}$$

or in words, the probability of being in state N at time k is equal to the probability of being in that state at time $k-1$ plus the probability of first entry into that state at time k. This relation follows from the fact that there are only two mutually exclusive ways to be in state N at time k — either we were there before or else we entered the state at time k. From (3.125) we can relate the first passage probability to the state probability that has already been calculated. Assuming that $p_0(N) = 0$, taking the Z transform of (3.125) we get

$$Q_N(z) = (1 - z^{-1})P_N(z) \ . \tag{3.126}$$

Since $P_N(z)$ is an absorption state, it turns out that it will always have a factor of $(1 - z^{-1})$ in the denominator which will be canceled, resulting in a $Q_N(z)$ which is simpler than the $P_N(z)$ that we started with.

If we define the average or expected time for first entry into state N as f_N, then it turns out that we can find this time without the need to take the inverse Z-transform of $Q_N(z)$.

Exercise 3-18.

Show that the mean first passage time is

$$f_N = -\frac{\partial}{\partial z} Q_N(z)\Big|_{z=1} . \tag{3.127}$$

Example 3-21.

If we toss a fair coin, what is the average number of tosses until we have seen two heads in a row? The signal flow graph for this example is shown below:

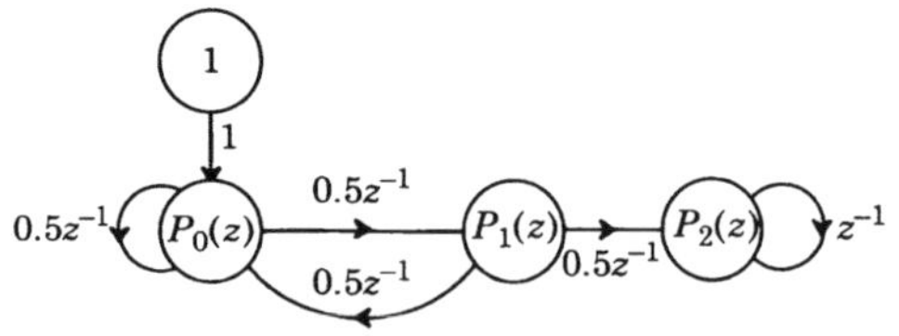

The numbering of states is the number of heads in a row. We assume that we start with zero heads in a row. At each toss the number of heads in a row increases by one with probability ½, or goes back to zero with probability ½ (that is, we get a tail). We define state two (two heads in a row) as an absorption state so that we can calculate the first passage time. Solving the linear equations, we get

$$P_2(z) = \frac{z}{4z^3 - 6z^2 + z + 1} = \frac{z}{(z-1)(4z^2 - 2z - 1)} \ . \tag{3.128}$$

Finally,

$$f_N = -\frac{\partial}{\partial z}\left(\frac{1}{4z^4 - 2z - 1}\right)\Bigg|_{z=1} = 6 \ . \tag{3.129}$$

3.4. The Poisson Process and Queueing

There was a time when no random processes could challenge the Gaussian process for the attention of communication theorists. However, the *Poisson process*, and its generalization, the *birth and death process* can reasonably claim to hold that distinction. The question often arises in communications as to the distribution for the times of discrete events, such as the arrivals of messages at a digital communication multiplex, or the arrivals of photons in a light beam at an optical detector in an optical communication system. The Poisson process models the most random such distribution, and is an excellent model for many of these situations.

To proceed, we need to define the notion of *random points in time*, where a point in time might denote the arrival of a message from a random source or a photon at a photodetector. Defining some notation, let the time of the k-th arrival be denoted by t_k, where of course $t_k \geq t_j$ for $k > j$. Further, define a continuous-time random process $N(t)$ that equals the number of arrivals from some starting time t_0 to the current time t. We call $N(t)$ a *counting process* since it counts the accumulated number of random points in time. Thus, $N(t)$ assumes only non-negative integer values, has initial condition $N(t_0) = 0$, and at each random point in time t_k, $N(t)$ increases by one. Such a counting process is pictured in Fig. 3-12(a), where the arrival times and the value of the counting process are pictured for one typical outcome.

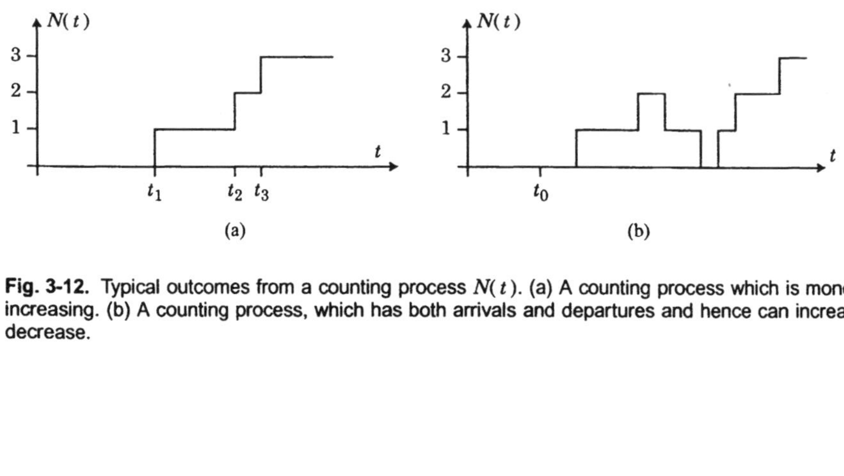

Fig. 3-12. Typical outcomes from a counting process $N(t)$. (a) A counting process which is monotone increasing. (b) A counting process, which has both arrivals and departures and hence can increase or decrease.

In some situations there are only arrivals, so that a counting process of the type pictured in Fig. 3-12(a) is the appropriate model. In other situations, there are departures as well as arrivals. A typical situation is the *queue* pictured in Fig. 3-13. We can define a counting process $N(t)$ to be the difference between the accumulated number of arrivals and the accumulated number of departures.

Example 3-22.

Consider a computer communication system that stores arriving messages in a buffer before retransmitting them to some other location. $N(t)$ gives a current count of the number of messages in the system at time t. A typical outcome of such a process is pictured in Fig. 3-12(b), where it should be noted that the process can never go below zero (since nothing can depart if there is nothing in the buffer).

In many instances of practical importance, the count $N(t)$ at time t is all we need to know to predict the future evolution of the system after time t. The manner in which system reached $N(t)$ is irrelevant in terms of predicting the future. For this case, the counting process denotes the *state of the system* in the same sense as Markov chains in the last section. In particular, we say that the system is in state j at time t if $N(t) = j$. This is similar to a Markov chain with one important distinction — a Markov chain can only change states at discrete points in time, whereas we now allow the state to change at any continuous point in time. Like Markov chains, a sample of the counting process $N(t_0)$ is a discrete-valued random variable. Just as for Markov chains (3.118), we define a probability of being in state j at time t as

$$q_j(t) = \Pr[N(t) = j] = P_{N(t)}(j). \tag{3.130}$$

This notation emphasizes that this probability is a continuous-time function. The only real distinction between (3.118) and (3.130) is that the latter is defined for continuous-time and the former for discrete-time.

In the following subsections, we analyze a counting process under the specific conditions appropriate for optical communication (Section 3.4.3) and statistical multiplexing (Section 3.4.2).

3.4.1. Birth and Death Process

The cases of interest to us are subsumed by a general process called a *birth and death process*, which is a mathematician's macabre terminology for a counting process with both arrivals and departures. This analysis is given in this section.

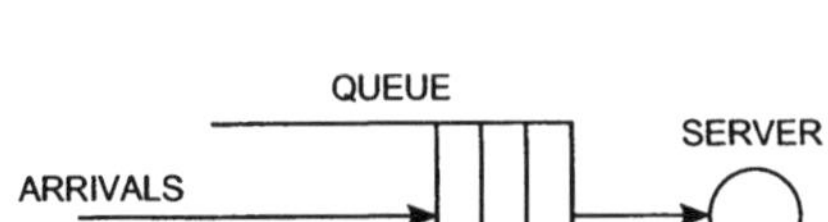

Fig. 3-13. A queueing system, which models among other things the status of a buffer in a communication system.

We have to somehow model the evolution of the system from one state to another. The approach for the Markov chain in (3.115) is inappropriate, since the probability of transition between any two states at any point in time t is most likely zero! While we cannot characterize the *probability* of transition, what we can characterize is the *rate* of transition between two states. Suppose for two particular states, the rate of transitions between one state and the other is a constant R. What we mean by this is that in a time δt we can expect an average $R\delta t$ transitions. If δt is very small, then $R\delta t$ is a number much smaller than unity, and the probability of more than one transition in time δt is vanishingly small. Under these conditions, we can think of $R\delta t$ as the probability of one transition in time δt, and $(1 - R\delta t)$ as the probability of no transition.

This logic leads us to a transition diagram and associated set of differential equations. The transition diagram in Fig. 3-14 associates a node with each state, and within that node we put the probability of being in that state at time t, which we denote $q_j(t)$. Each transition in the diagram is labeled with the rate at which that transition occurs, where the rates in the general case are allowed to be time-varying (non-homogeneous). Each rate is labeled with a subscript indicating the state in which it originates, where $\lambda(t)$ is the rate for transitions corresponding to births or arrivals and $\mu(t)$ corresponds to deaths or departures. Reiterating, the interpretation of these rates is as follows: for a very small time interval δt, the probability of a particular transition is equal to the rate times the time interval.

The set of differential equations which describe the evolution of the birth and death process are

$$\frac{dq_j(t)}{dt} = \lambda_{j-1}(t)q_{j-1}(t) + \mu_{j+1}(t)q_{j+1}(t) - [\lambda_j(t) + \mu_j(t)]q_j(t), \; j \geq 0 , \tag{3.131}$$

with initialization $q_{-1}(t) = 0$. These equations can be derived rigorously from fundamental principles [6], but for our purposes they are evident from intuitive considerations. The equations say that the rate of increase of a probability with time for state j is equal to the rate at which transitions into that state from states $j-1$ and $j+1$ are occurring (times the current probability of those states) minus the rate at which transitions out of state j are occurring (times the current probability of state j).

We must also specify an initial condition, which for our purposes specifies that the process starts in state zero (no arrivals) at time t_0,

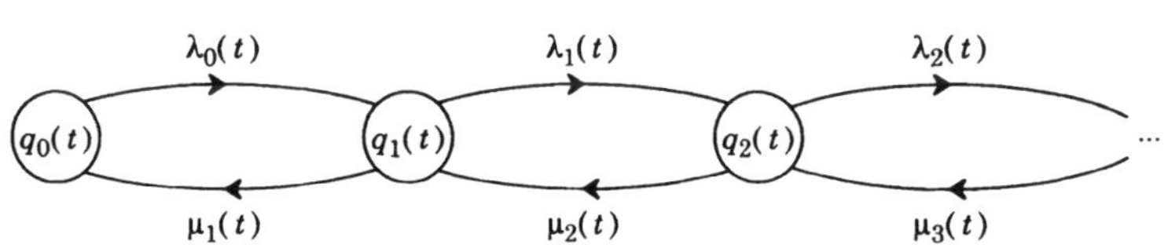

Fig. 3-14. State transition diagram for a birth and death process.

$$q_j(t_0) = \begin{cases} 1, & j = 0 \\ 0, & j > 0 \end{cases}. \tag{3.132}$$

The first order differential equations can be solved for many special cases.

Example 3-23.

Consider the important case of a *pure birth process* in which $\mu_j(t) = 0$. Also assume the birth rates are all the same and a constant with time, $\lambda_j(t) = \lambda$. The transition diagram for this model is shown in Fig. 3-15. This corresponds to the important case where the arrival rate does not depend on the state of the system, the usual case in the problems that we will encounter. Then (3.131) becomes

$$\frac{dq_j(t)}{dt} - \lambda q_j(t) = \lambda q_{j-1}(t). \tag{3.133}$$

which is a simple first order differential equation with constant coefficients. Assume that the initial condition is

$$q_0(0) = 1 \tag{3.134}$$

implying that the initial count at $t = 0$ is 0. We can solve this using very similar techniques to our solution of the Markov chain, but use the Laplace transform in place of the Z-transform. In analogy to (3.119), defining the Laplace transform of the state probability,

$$Q_j(s) = \int_0^{\infty} q_j(t) e^{-st}\, dt . \tag{3.135}$$

Taking the Laplace transform of both sides of (3.133),

$$sQ_j(s) - q_j(0) + \lambda Q_j(s) = \lambda Q_{j-1}(s) \tag{3.136}$$

Using (3.132), with $t_0 = 0$, this becomes

$$Q_0(s) = \frac{1}{s+\lambda}, \qquad Q_j(s) = \frac{\lambda}{s+\lambda} Q_{j-1}(s), \quad j > 0 . \tag{3.137}$$

This set of iterative equations for the state probability Laplace transform is easily solved by iteration,

$$Q_j(s) = \frac{\lambda^j}{(s+\lambda)^{j+1}}, \tag{3.138}$$

and taking the inverse Laplace transform, we find that for $t \geq 0$ and $j \geq 0$,

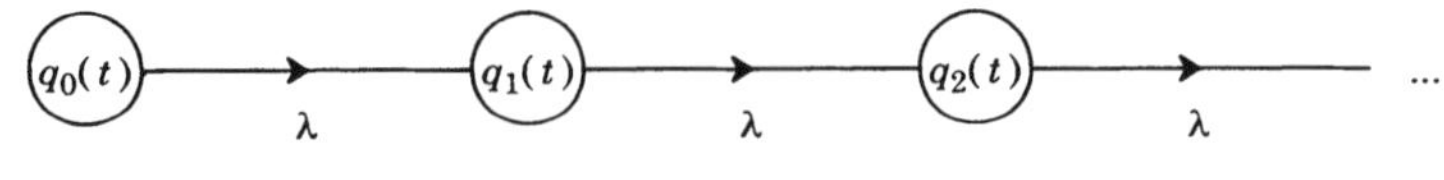

Fig. 3-15. State transition diagram for a constant-rate pure birth process.

$$\Pr[N(t) = j] = q_j(t) = \frac{(\lambda t)^j}{j!} e^{-\lambda t} . \tag{3.139}$$

This is the well-known *Poisson distribution* with parameter λt. For this reason, the pure birth process $N(t)$ we have just analyzed is called a *Poisson process* with constant rate. We will generalize this to a variable rate in the next subsection.

The Poisson distribution is an important one in the theory of birth and death processes, so we summarize its properties in the following exercise.

Exercise 3-19.

Consider a Poisson distribution with parameter a,

$$p_N(k) = e^{-a} \frac{a^k}{k!} . \tag{3.140}$$

(a) Show that the mean and variance of this distribution are

$$\mathrm{E}[N] = a, \quad \mathrm{Var}[N] = a . \tag{3.141}$$

(*Hint:* Form a power series for e^a and differentiate it twice.)

(b) Show that the moment generating function is given by

$$log\Phi_N(s) = a(e^s - 1) . \tag{3.142}$$

The last example can be generalized with respect to the initial condition.

Exercise 3-20.

Show that if $N(t_0) = k$ (there have been k counts up to time t_0), then

$$q_j(t) = \frac{(\lambda(t - t_0))^{j-k}}{(j-k)!} e^{-\lambda(t - t_0)}, \quad j \geq k, \ t \geq t_0 . \tag{3.143}$$

This result implies that the number of counts starting at $t = t_0$ is a Poisson distribution with parameter $\lambda(t - t_0)$, which is the expected number of arrivals since the start time. Furthermore, index $j - k$ of the Poisson distribution is the number of counts since the start time. The important conclusion is that the number of arrivals in the interval starting at $t = t_0$ has a distribution which does not depend in any way on what happened prior to t_0. This is roughly the definition of a *Markov process*, and a Poisson counting process is in fact a Markov process. For such a process, the number of arrivals in the interval $[t_0, t]$ is statistically independent of the number of arrivals in any other nonoverlapping interval of time. It is in this sense that the Poisson process is the most random among all monotone non-decreasing counting processes.

Exercise 3-21. *(Pure death process.)* For $\lambda_j(t) = 0$, consider the case where departures from the system are proportional to the state index, $\mu_j(t) = j\mu$. This is an appropriate model for a system in which the departure or death rate is proportional to the size of the population, as in a human population. Further, assume that the initial state at $t = 0$ is n. Draw the state transition diagram and show that the state probabilities obey a binomial distribution,

$$\Pr[N(t) = j] = q_j(t) = \binom{n}{j} p_j(t)(1 - p(t))^{n-j}, \qquad p(t) = e^{-\mu t}. \tag{3.144}$$

Now we give an example of a problem in which both births and deaths occur. This is an example of a *queueing problem*, and it is appropriate at this point to define some terminology used in queueing, particularly as it relates to digital communication. A queue is a *buffer* or *memory* which stores messages. There is some mechanism which clears messages from the queue, which is usually the transmission of the message to another location. This mechanism is called the *server* to the queue. Assume a server can process only one message at a time, so that if more than one message is being processed (there are multiple communication channels for transmission of messages), then there are an equivalent number of servers. Typically the buffer contains space for a maximum number of messages to wait for service, and the number of messages that can be waiting at any time is called the *number of waiting positions*. The *state of the system*, which naturally tracks a counting process, is the number of messages waiting for service plus the number of messages currently being served. Messages arrive at the queue (births) at random times, and they depart from the queue (deaths) due to the completion of service.

Exercise 3-22.

(Queue with one server and no waiting positions.) Assume that a queue has constant arrival rate λ, a single server which clears a message being served at rate μ, and no waiting positions. If a message arrives while the server is busy then since there are no waiting positions that arrival is lost and leaves the system permanently. Draw the state transition diagram for the system and show that the probability that the server is not busy is

$$q_0(t) = \frac{\mu}{\lambda + \mu} + \left(q_0(0) - \frac{\mu}{\mu + \lambda}\right)e^{-(\mu + \lambda)t}. \tag{3.145}$$

The differential equation approach we have described is capable of describing the transient response of a system starting with any initial condition. Often, however, it is sufficient to know what the state probabilities are in the steady state. There is no such steady state distribution for a Poisson process, since the state grows without bound. However, for queueing systems where the service rate is always guaranteed to be higher than the arrival rate, and where all the rates are independent of time, there will be a steady state distribution. This distribution can be obtained by letting $t \to \infty$ in the transient solution we have obtained, or can be obtained much more simply by setting the time derivatives in the differential equations to zero and solving for the resulting probabilities.

Example 3-24.

Continuing Exercise 3-22, letting $t \to \infty$ in (3.145), the steady state probability is

$$q_0(\infty) = \frac{\mu}{\mu + \lambda}. \tag{3.146}$$

We can get this same result without solving the differential equation by setting the derivative in (3.131) to zero.

In the following two sections we will specialize the general birth and death process to two situations of particular interest to us.

3.4.2. M/M/1 Queue

Consider the following queueing model which characterizes a single server queue with the most mathematically tractable assumptions. This model is actually a combination of the pure birth process of Example 3-23 and a pure death process (Exercise 3-21). Assume arrivals occur at a constant rate λ independent of the number of waiting positions occupied, there are an infinite number of waiting positions so that no arrival ever encounters a full buffer, arrivals wait indefinitely for service, and there is a single server with service rate μ. The departure rate is independent of the number of messages waiting in the queue, as long as there is at least one. The state transition diagram for this queueing model is shown in Fig. 3-16.

As in most queueing problems, we are content to know the steady state distribution of states. This distribution will only exist if the service rate μ is greater than the arrival rate λ, because otherwise the buffer size will grow to infinity. Making that assumption, the differential equations governing the queue are

$$\frac{dq_j(t)}{dt} = \lambda q_{j-1}(t) + \mu q_{j+1}(t) - (\lambda + \mu) q_j(t), \quad j > 0,$$

$$\frac{dq_0(t)}{dt} = \mu q_1(t) - \lambda q_0(t) \tag{3.147}$$

with initial condition (assuming there are no positions occupied at time $t = 0$),

$$q_j(0) = \delta_j. \tag{3.148}$$

We could attempt to solve this system of differential equations, but since we are content with the steady state solution, set the derivatives to zero,

$$0 = \lambda q_{j-1} + \mu q_{j+1} - (\lambda + \mu) q_j , \; j > 0 ,$$

$$0 = \mu q_1 - \lambda q_0 , \tag{3.149}$$

where we have also taken the liberty of suppressing the time dependence since we are looking only at the steady state. These equations are easily solved.

Exercise 3-23.
Show that the solution to (3.149) is

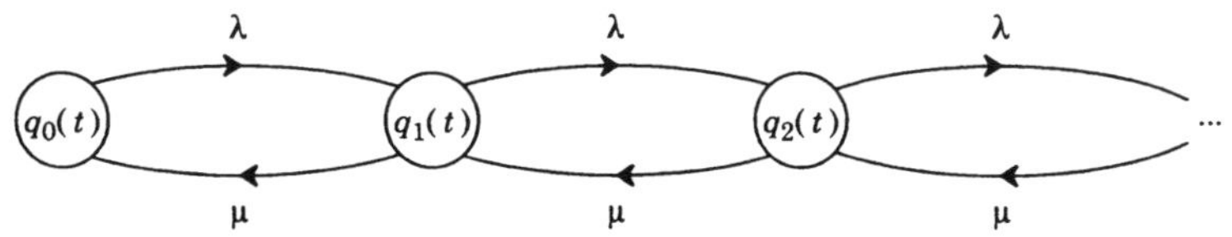

Fig. 3-16. The state transition diagram for the single server queue with an infinite number of waiting positions.

$$q_j = \rho^j(1-\rho) \tag{3.150}$$

where ρ is called the *offered load*,

$$\rho = \lambda/\mu \tag{3.151}$$

and is less than unity by assumption. Note from (3.150) that the probability that the single server is busy is $1 - q_0 = \rho$, which is obvious since the server has more "capacity" than the arrivals require by a factor of μ/λ. Thus, ρ is also called the *server utilization*.

In many queueing problems the most critical parameter is the delay that a new arrival experiences before being served. This is also called the *queueing delay*, and represents a significant impairment in communication systems that utilize a buffer delay discipline to increase the capacity of a communication link (Chapter 17). A related parameter is the *waiting time*, which is defined to be the queueing delay plus the service time. The calculation of the delay is a little more complicated than what we have done heretofore, so we will simply state the results [6]. The mean delay is given by

$$D = \frac{\rho}{\mu(1-\rho)}\,. \tag{3.152}$$

Note that as the offered load or server utilization approaches unity, the mean delay grows without bound; conversely, as the utilization approaches zero, the lightly loaded queue, the delay approaches zero. The mean queueing delay is equal to the average service time $1/\mu$ for a utilization of $\rho = \frac{1}{2}$.

3.4.3. Poisson Process With Time-Varying Rate

In optical communication systems, the counting process which gives the accumulated number of arrival times for photons is a Poisson process. The Poisson process is a pure birth process where the arrival rate is independent of the state of the system, and we have already been exposed to it in Example 3-23 for a constant arrival rate. In optical communication, the arrival rate is actually signal dependent, so in this section we discuss that case.

The Poisson process with time-varying rate is the pure birth process in which the incoming rate $\lambda(t)$ is independent of the state of the system. Thus, the system is governed by a first-order differential equation with time-varying coefficients,

$$\frac{dq_j(t)}{dt} + \lambda(t)q_j(t) = \lambda(t)q_{j-1}(t)\,, \quad q_{-1}(t) = 0\,, \tag{3.153}$$

and we assume the system starts at time t_0 in state $j = 0$. Because of the time-varying coefficients, the Laplace transform is of no help, and we must resort to solving the differential equation directly. This is straightforward (since it is a first order equation), but tedious, so the solution is relegated to Appendix 3-C. Define

$$\Lambda(t) = \int_{t_0}^{t} \lambda(u)\, du\ , \tag{3.154}$$

which has the interpretation as the average total number of arrivals in the interval $[t_0, t]$. Then the probability of n arrivals in the interval $[t_0, t]$ is governed by a Poisson distribution with parameter $\Lambda(t)$,

$$q_n(t) = \frac{\Lambda^n(t)}{n!} e^{-\Lambda(t)} . \tag{3.155}$$

This reduces to the solution given in Example 3-23 for the constant rate case.

As a reminder, (3.155) specifies the number $N(t)$ of arrivals during the time interval $[t_0, t]$. This random number of arrivals is Poisson distributed with parameter $\Lambda(t)$, and hence has mean and variance

$$E[N(t)] = \Lambda(t) , \qquad \sigma^2_{N(t)} = \Lambda(t) . \tag{3.156}$$

As in the constant rate case, it can be shown that the number of arrivals in any two non-overlapping intervals are statistically independent.

3.4.4. Shot Noise

In optical communication, a waveform is generated in the photodetector by generating impulses at times corresponding to random arrival times of photons and then filtering these impulses. This is known as a *filtered Poisson process*, or a *shot noise process*.

If a Poisson process is characterized by a set of arrival times t_k for the k-th arrival, and given a filter with impulse response $h(t)$, then a shot noise process is a continuous-time random process $X(t)$ with outcome

$$x(t) = \sum_k h(t - t_k) . \tag{3.157}$$

An outcome of this random process is illustrated in Fig. 3-17 for a particular impulse response. In this figure, it is assumed that qualitatively the duration of the impulse response is short

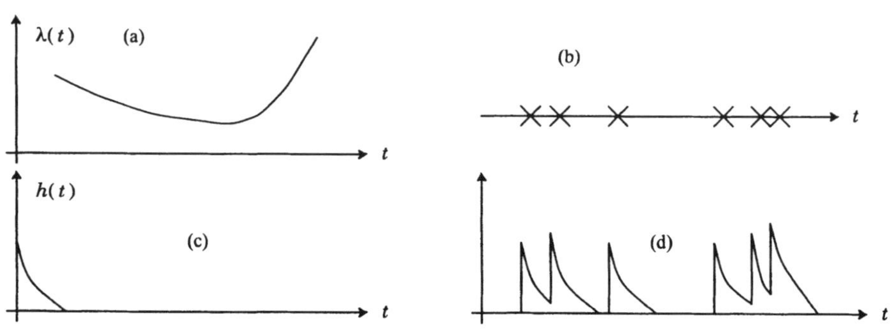

Fig. 3-17. Illustration of an outcome of a shot noise process. (a) The average arrival rate vs. time. (b) The random actual times of arrival, where arrivals occur at the average rate given in (a). (c) The impulse response of the filter. (d) The corresponding outcome.

relative to the average time between arrivals. If the impulse response were long, this would have an averaging effect resulting in a much smoother outcome.

It is shown in Appendix 3-D that the moment generating function of the shot noise process at time t is

$$\log \Phi_{X(t)}(s) = \lambda(t) * (e^{sh(t)} - 1) . \tag{3.158}$$

The mean and variance of shot noise are easily derived from (3.158).

Exercise 3-24.

Show that the mean value of shot noise is the convolution of the filter impulse response with the arrival rate,

$$m_X(t) = \mathrm{E}[X(t)] = \lambda(t) * h(t) \tag{3.159}$$

and that the variance is the convolution of the square of the filter impulse response with the arrival rate,

$$\sigma_X^2(t) = E[X^2(t)] - m_X^2(t) = \lambda(t) * h^2(t) . \tag{3.160}$$

These relations are known as *Campbell's theorem.*

3.4.5. High-Intensity Shot Noise

When the intensity of shot noise is high, the statistics become that of a Gaussian random process. The intuition behind this is that $X(t)$ is the sum of a large number of independent events, and hence approaches a Gaussian by the central limit theorem. To demonstrate this more rigorously, we will show that the moment generating function of shot noise approaches a Gaussian moment generating function in the limit of high intensity.

In order to avoid an infinitely large power of shot noise, as the intensity grows we need to scale the size of the impulse response $h(t)$ also. Therefore, let us use a scaling constant β, which we will allow to grow to infinity, and let

$$\lambda(t) = \beta\lambda_0(t) , \qquad h(t) = \frac{1}{\sqrt{\beta}} h_0(t) . \tag{3.161}$$

With this scaling, we get from Campbell's theorem that

$$m_X(t) = \sqrt{\beta}\,\lambda_0(t) * h_0(t) , \qquad \sigma_X^2(t) = \lambda_0(t) * h_0^2(t) . \tag{3.162}$$

Hence, as the scaling factor β grows, the variance of the process stays constant and the mean value grows without bound. We cannot help this, because as the intensity grows the variance becomes a smaller fraction of the mean. In this sense high-intensity shot noise approaches a deterministic signal $m_X(t)$ as the intensity grows.

Only two terms in the moment generating function are important as the scaling constant β grows.

Exercise 3-25.

Show that for large β the only significant terms in the moment generating function of (3.158) are

$$\log\Phi_{X(t)}(s) \approx \sqrt{\beta}\lambda_0(t) * h_0(t) + \tfrac{1}{2}s^2\lambda_0(t) * h_0{}^2(t) . \tag{3.163}$$

Comparing this with the Gaussian moment generating function of (3.41), we see that high intensity shot noise is approximately Gaussian with mean and variance given by (3.162).

3.4.6. Random-Multiplier Shot Noise

In optical communication systems, it is sometimes appropriate to introduce a random multiplier into the shot noise process, *viz.*

$$X(t) = \sum_k G_k h(t - t_k) , \tag{3.164}$$

where G_k is a sequence of mutually statistically independent identically distributed random variables which are also statistically independent of the arrival times t_j for all j.

Exercise 3-26.

Use Campbell's theorem and the assumptions to show that the mean-value of (3.164) is

$$m_X(t) = E[G]\lambda(t) * h(t) , \tag{3.165}$$

and the variance is

$$\sigma_X{}^2(t) = E[G^2]\lambda(t) * h^2(t) , \tag{3.166}$$

where $E[G]$ and $E[G^2]$ are the mean-value and second moment of the random multiplier G_k for all k.

3.5. Further Reading

For a general introduction to random variables and processes, Papoulis [7], Stark and Woods [8], and Ross [9] are recommended. Papoulis has more of an engineering perspective. Both books have comprehensive treatments of Markov chains and Poisson and shot noise processes. An excellent introduction to Poisson processes can be found in Ross [10]. There are a number of books that give comprehensive treatment to the application of Poisson and birth and death processes to queueing models, such as Cooper [6], Hayes [11], and Kleinrock [12].

Appendix 3-A. Power Spectrum of A Cyclostationary Process

In this appendix we determine the power spectrum of the PAM random process with a random phase epoch (3.83). Calculating the autocorrelation function of (3.83),

$$\mathrm{E}[Z(t+\tau)Z^*(t)] = \mathrm{E}\Big[\sum_{m=-\infty}^{\infty}\sum_{n=-\infty}^{\infty} X_m X_n^* h(t+\tau-mT-\Theta)h^*(t-nT-\Theta)\Big]. \quad (3.167)$$

Assuming we can interchange expectation and summation, we use the fact that Θ is independent of X_k to get

$$\mathrm{E}[Z(t+\tau)Z^*(t)] = \sum_{m=-\infty}^{\infty}\sum_{n=-\infty}^{\infty} \mathrm{E}[X_m X_n^*]\mathrm{E}[h(t+\tau-mT-\Theta)h^*(t-nT-\Theta)]. \quad (3.168)$$

The first expected value is simply the autocorrelation function $R_X(m-n)$. The second expected value can be computed using the definition of expectation and the p.d.f. of the uniform random variable Θ

$$\mathrm{E}[h(t+\tau-mT-\Theta)h^*(t-nT-\Theta)] = \int_0^T \frac{1}{T} h(t+\tau-mT-\theta)h^*(t-nT-\theta)\,d\theta. \quad (3.169)$$

Changing variables, letting $i = m - n$, using (3.169), and exchanging summations, we get

$$\mathrm{E}[Z(t+\tau)Z^*(t)] = \frac{1}{T}\sum_{i=-\infty}^{\infty} R_X(i)\sum_{n=-\infty}^{\infty}\int_{t-nT}^{t-nT+T} h(\alpha+\tau-iT)h^*(\alpha)\,d\alpha. \quad (3.170)$$

The second summation is the sum of integrals with adjoining limits, so it can be replaced with a single infinite integral

$$E[Z(t+\tau)Z^*(t)] = \frac{1}{T}\sum_{i=-\infty}^{\infty} R_X(i)\int_{-\infty}^{\infty} h(\alpha+\tau-iT)h^*(\alpha)\,d\alpha, \quad (3.171)$$

which is independent of t, so the process $Z(t)$ is wide sense stationary. To get the power spectrum, we take the Fourier transform with τ as the time index

$$S_Z(f) = \frac{1}{T}\sum_{i=-\infty}^{\infty} R_X(i)\Big[\int_{-\infty}^{\infty} h^*(\alpha)e^{j2\pi f(\alpha-iT)}d\alpha\Big]. \quad (3.172)$$

The expression in brackets is $e^{-j2\pi fiT}H^*(f)$, getting

$$S_Z(f) = \frac{1}{T}H(f)H^*(f)\sum_{i=-\infty}^{\infty} R_X(i)e^{-j2\pi fiT}. \quad (3.173)$$

The summation is simply the discrete-time Fourier transform $S_X(e^{j2\pi fT})$ of the autocorrelation function. The final result is

$$S_Z(f) = \frac{1}{T}\,|H(f)|^2 S_X(e^{j2\pi fT}). \quad (3.174)$$

Appendix 3-B. Power Spectrum of A Markov Chain

In this appendix we solve the problem of finding the power spectrum of the random process (3.116). The power spectrum only exists if the random process is wide sense stationary. Strictly speaking, this requires that the Markov chain be running over all time, although we can interpret the results as indicative of the power spectrum for a chain that was initialized but has been running long enough to be in the steady-state. We approach this by assuming that the initial probability of each state is the same as its steady-state probability, so that the state probability is in fact constant with time (a *stationary* Markov chain).

We first determine the autocorrelation function of (3.116),

$$R_X(n) = \mathrm{E}[f(\Psi_k)f(\Psi_{k+n})] \; , \tag{3.175}$$

assuming $f(\cdot)$ is a real-valued function. Assuming wide-sense stationarity, we can take $k = 0$ and this can be written

$$R_X(n) = \sum_{i \in \Omega} \sum_{j \in \Omega} f(i)f(j)p_{0,n}(i, j). \tag{3.176}$$

where by Bayes' rule

$$p_{0,n}(i, j) = p_{n\,|\,0}(j\,|\,i)p_0(i) \tag{3.177}$$

is the joint probability of being in state i at time 0 and state j at time n. Assuming we have already calculated the steady-state state probabilities $p(i)$ for the chain, by the stationarity assumption we can write

$$p(i) = p_0(i). \tag{3.178}$$

One way to think of this is as forcing the initial state probability to equal the steady-state probability, thus suppressing any transient solution. Finally, we must carefully note the d.c. component of the random process, since it contributes a delta-function to the power spectrum that can easily be lost if we are not careful. Specifically, the d.c. component is

$$\mu_X = \sum_{i \in \Omega} f(i)p(i). \tag{3.179}$$

The power spectrum is simply the Z transform of the autocorrelation function (see (3.61)). Rather than calculate the Z transform $S_X(z)$ directly, let us first concentrate on the quantity

$$S_X^+(z) = \sum_{n=0}^{\infty} R_X(n)z^{-n} \tag{3.180}$$

that includes only the positive index terms in the summation making up the Z transform. From (3.176), (3.177), and (3.178), this can be written as

$$S_X^+(z) = \sum_{i \in \Omega} \sum_{j \in \Omega} f(i)f(j)p(i) \cdot P_{j\,|\,i}(z) \tag{3.181}$$

where

$$P_{j\,|\,i}(z) = \sum_{n=0}^{\infty} p_{n\,|\,0}(j\,|\,i)z^{-n} . \tag{3.182}$$

This latter quantity can be interpreted as the Z-transform of $p_{n\,|\,0}(j\,|\,i)$, which is in turn the probability of being in state j at time n given that we started (with probability one) in state i at time 0. This quantity is easy to calculate using the techniques we have previously displayed, since it is simply the Z-transform of a transient solution starting with probability one in a particular state. The signal flow graph for this solution is shown in Fig. 3-18, where only the states i and j are shown. This signal flow graph must be solved for $P_{j\,|\,i}(z)$ for all (i, j) for which $f(i)f(j)$ is non-zero in (3.181).

Example 3-25.
Again returning to the parity check circuit of Example 3-15, let us compute $S_X^+(z)$. In this case $f(i) = i$, so that the random process $X_k = f(\Psi_k) = \Psi_k$ assumes the values 0 and 1. For that case, we only need evaluate one term in (3.176), corresponding to $i = j = 1$, and all the others are zero. This term is shown by the signal flow graph in Fig. 3-18. Solving this flow graph, we get

$$P_{1\,|\,1}(z) = \frac{1 - 0.5z^{-1}}{1 - z^{-1}} , \tag{3.183}$$

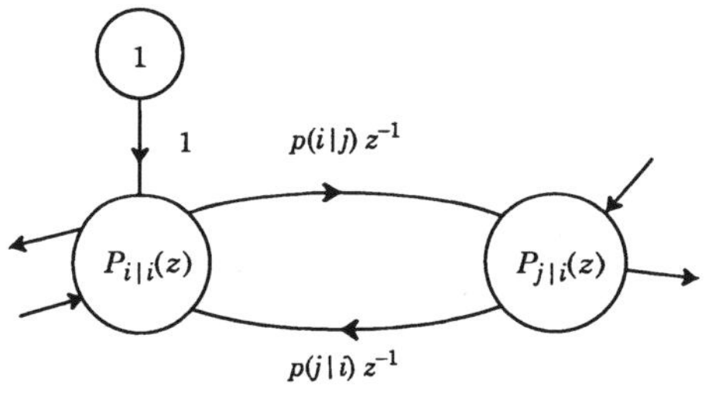

Fig. 3-18. Signal flow graph representation of equations that must be solved to find $P_{j\,|\,i}(z)$.

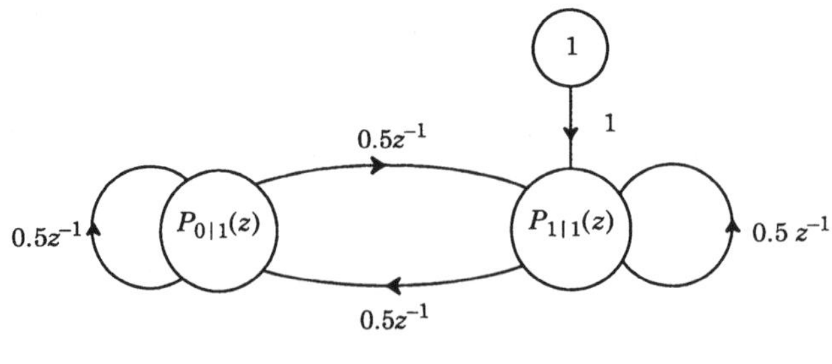

Fig. 3-19. Signal flow graph for the parity check circuit.

and

$$S_X^+(z) = 0.5\frac{1 - 0.5z^{-1}}{1 - z^{-1}}, \tag{3.184}$$

since there is only one term in the sum and $p(0) = p(1) = ½$. Inverting the Z transform, we find that

$$R_X(n) = \begin{cases} 1/2, & n = 0 \\ 1/4, & n > 0 \end{cases}. \tag{3.185}$$

This result says that the power of the process is ½, which is obvious, and that the process has a d.c. component of ½ since the autocorrelation function approaches ¼ for large n, which is also obvious.

We have determined the one-sided terms in the power spectrum, and we must generate the two-sided spectrum $S_X(z)$. However, before doing this, we must first remove any d.c. component, since that d.c. component can be represented by the one-sided transform but is problematic in the two-sided transform. This is simple, since we only need to replace $S_X^+(z)$ by

$$S_X^+(z) - \frac{\mu_X^2}{1 - z^{-1}} \tag{3.186}$$

to remove this d.c. component. Alternatively we could have defined a new random process with the d.c. component removed, although that method is often harder.

Example 3-26.
For the parity check circuit of Example 3-15, the d.c. component is $\mu_X = ½$, and subtracting the appropriate term from (3.184),

$$S_X^+(z) - \frac{\mu_X^2}{1 - z^{-1}} = \frac{1}{4}. \tag{3.187}$$

Note that for this process this result would have been much more difficult to obtain if we had defined a d.c. free random process, since then we would have to evaluate all four terms in (3.181) rather than just one.

We must now turn the one-sided version of the power spectrum into a two-sided version. The Z transform of the autocorrelation function can be written

$$S_X(z) = \sum_{m=0}^{\infty} R_X(m)z^{-m} = \sum_{m=0}^{\infty} R_X(m)z^{-m} + \sum_{m=0}^{\infty} R_X(m)z^{m} - R_X(0), \tag{3.188}$$

where we have used the symmetry of the autocorrelation function. Noting that $R_X(0) = S_X^+(\infty)$, we get finally

$$S_X(z) = S_X^+(z) + S_X^+(z^{-1}) - S_X^+(\infty) . \tag{3.189}$$

Example 3-27.

To finish with the parity check example of Example 3-15,

$$S_X(z) = \frac{1}{4} + \frac{1}{4} - \frac{1}{4} = \frac{1}{4}, \tag{3.190}$$

and the process is white with power $\frac{1}{4}$. However, recall that this power spectrum does not include the d.c. term, so that in fact

$$S_X(e^{j2\pi fT}) = \frac{1}{4} + \frac{1}{4}\delta(f) . \tag{3.191}$$

The area of the delta function has been chosen to be $\frac{1}{4}$, the power of the d.c. component.

Appendix 3-C. Derivation of a Poisson Process

In this appendix we show that the Poisson distribution for the accumulated number of arrivals as given by (3.155) is valid. To begin with, we need the solution to a first-order differential equation, which is given in the following exercise [13].

Exercise 3-27.

Consider the following first order differential equation,

$$\dot{x}(t) + a(t)x(t) = b(t) . \tag{3.192}$$

(a) Let $\dot{A}(t) = a(t)$ and show that

$$\frac{d}{dt}[e^{A(t)}x(t)] = b(t)e^{A(t)} . \tag{3.193}$$

(b) Integrate both sides of (3.193) to obtain the solution for $x(t)$

$$x(t) = x(t_0)\, e^{\Lambda(t)} + e^{-\Lambda(t)}\int_{t_0}^{t} b(u)e^{\Lambda(u)}\, du\,, \qquad \Lambda(t) = \int_{t_0}^{t} a(v)\, dv . \tag{3.194}$$

Returning to the Poisson process, identify

$$a(t) = \lambda(t)\,, \qquad b(t) = \lambda(t)q_{j-1}(t). \tag{3.195}$$

Therefore, given the definition of (3.154) for $\Lambda(t)$,

$$q_j(t) = q_j(t_0)e^{-\Lambda(t)} + e^{-\Lambda(t)}\int_{t_0}^{t} \lambda(u)q_{j-1}(u)e^{-\Lambda(u)}\, du. \tag{3.196}$$

The solution follows immediately for $j = 0$ using the initial condition of (3.153),

$$q_0(t) = e^{-\Lambda(t)}\,, \tag{3.197}$$

and the rest is easy!

Exercise 3-28.
Verify the validity of (3.155) by induction on (3.196).

Appendix 3-D. Moment Generating Function of Shot Noise

In this appendix we derive the moment generating function of a shot noise process $X(t)$ corresponding to impulse response $h(t)$. A sample function of such a process is given by (3.157).

To find the moment generating function, divide the time axis into small intervals of length δt, where the k-th interval is $[(k - \frac{1}{2}) \cdot \delta t, (k + \frac{1}{2}) \cdot \delta t]$. Group all the arrivals in the k-th interval together into a single impulse of height N_k located at time $k \cdot \delta t$, where N_k is the number of arrivals in the k-th interval. Thus, the shot noise of (3.157) becomes approximately

$$X(t) = \sum_{k=-\infty}^{\infty} N_k h(t - k \cdot \delta t) \tag{3.198}$$

where this equation becomes increasingly accurate as $\delta t \to 0$.

Since the intervals are non-overlapping, the N_k are independent Poisson random variables with parameter $\lambda(k \cdot \delta t) \cdot \delta t$, the average number of arrivals in the interval. The moment generating function of N_k is therefore

$$\log\Phi_{N_k}(s) = \lambda(k \cdot \delta t) \cdot \delta t\,(e^s - 1)\,, \tag{3.199}$$

and the moment generating function of (3.198) is

$$\Phi_{X(t)}(s) = \mathrm{E}[\exp\{s\sum_{k=-\infty}^{\infty} N_k h(t - k \cdot \delta t)\}] = \prod_{k=-\infty}^{\infty} \mathrm{E}[\exp\{sN_k h(t - k \cdot \delta t)\}]$$

$$= \prod_{k=-\infty}^{\infty} \Phi_{N_k}(s\,h(t - k \cdot \delta t))\,. \tag{3.200}$$

Taking the logarithm of the moment generating function, and substituting from (3.199),

$$\log\Phi_{X(t)}(s) = \sum_{k=-\infty}^{\infty} \lambda(k \cdot \delta t)(e^{sh(t - k\cdot\delta t)} - 1) \cdot \delta t\,, \tag{3.201}$$

and as $\delta t \to 0$ this approaches the integral

$$\log\Phi_{X(t)}(s) = \int_{-\infty}^{\infty} \lambda(\tau)(e^{sh(t-\tau)} - 1)\,d\tau\,, \tag{3.202}$$

which we recognize as the convolution of (3.158).

Problems

Problem 3-1. Use the moment generating function of (3.41) to show that the mean of the Gaussian distribution is μ and the variance σ^2.

Problem 3-2. Show that the marginal p.d.f.s of X and Y in (3.47) are those of a zero-mean Gaussian random variable with variance σ^2.

Problem 3-3. Show that for $y > 0$

$$\frac{1}{y\sqrt{2\pi}}e^{-y^2/2}\left(1-\frac{1}{y^2}\right) < Q(y) < \frac{1}{y\sqrt{2\pi}}e^{-y^2/2}. \tag{3.203}$$

These bounds are plotted in Fig. 3-1. *Hint:* Write the definition of $Q(\cdot)$ from (3.38) and integrate by parts.

Problem 3-4. Let X and Y be two complex-valued random variables.

(a) Form an estimate of X as $\hat{X} = a \cdot Y$ for some complex number a. Find the a that minimizes the mean-square error $E[\,|\hat{X} - X|^2]$.

(b) Reformulate the problem of (a) in terms of linear space and inner products.

(c) Re-solve the problem of (a) using the projection theorem of Section 2.6.3.

Problem 3-5. Let E_k be a prediction error generated by filter $E(z)$ such that

$$E[E_{k+m}X_k^*] = R_{EX}(m) = 0\,, \qquad m > 0\,, \tag{3.204}$$

and let E'_k be the output generated by any other causal and monic filter.

(a) Show that

$$\mathrm{E}[\,|E'_k\,|]^2 = \mathrm{E}[\,|E'_k - E_k\,|^2] + \mathrm{E}[\,|E_k\,|^2], \tag{3.205}$$

thus establishing that the output MSE is minimized when $E'_k = E_k$.

(b) Show that it follows from the orthogonality property of (3.204) that $R_E(m) = 0$ for all $m \neq 0$, and hence the optimal prediction error must be white.

Problem 3-6.

(a) Restate the results of Problem 3-5 in geometric terms, using the interpretation of Section 3.1.4.

(b) Re-derive the results of Problem 3-5 using the projection theorem of Section 2.6.3.

Problem 3-7. Given a WSS random process $X(t)$ with power $R_X(0)$, show that the sampled random process $Y_k = X(kT)$ has the same power,

$$E[\,|Y_k|^2] = R_Y(0) = R_X(0) \ . \tag{3.206}$$

Problem 3-8. Given a sequence of i.i.d. random variables A_k which take on values ± 1 with equal probability, find an expression for $E[A_p A_q A_r A_s]$.

Problem 3-9. Consider a random process $X(t)$ filtered by an ideal bandpass filter with frequency response

$$H(f) = \begin{cases} 1, & f_a < |f| < f_b \\ 0, & \text{otherwise} \end{cases} .$$

Let $Y(t)$ be the output of the filter. Show that

$$R_Y(0) = \int_{f_a}^{f_b} S_X(f)\, df \, .$$

Use this to show that $S_X(f) \ge 0$ for all f.

Problem 3-10. Extending Exercise 2-6 to random signals, assume the input to the possibly complex-valued LTI system shown in Fig. 2-3 is a wide sense stationary complex-valued discrete-time random process with power spectral density $S_X(e^{j2\pi fT}) = N_0$. Show that the autocorrelation of the output is

$$R_Y(k) = N_0\, f(kT) * f^*(-kT) = N_0 \sum_m f(mT) f^*((m-k)T) \ . \tag{3.207}$$

Problem 3-11. Show that the cross-correlation function has symmetry

$$R_{XY}(-\tau) = R^*_{YX}(\tau). \tag{3.208}$$

Is the cross-spectral density of two random processes necessarily real-valued?

Problem 3-12. Where a Markov chain has unique steady-state probabilities $p_k(i) = p(i)$, they can be found from the condition that the state probabilities will not change with one time increment. Assume $\Omega = \{0, \ldots, M\}$, define the matrix of state transition probabilities $\mathbf{P}$ to contain $p(j \mid i)$ in its $(i, j)^{th}$ entry, and define the vector $\pi = [p(0), \ldots, p(M)]$ to contain the steady-state probabilities, if they exist. Show that the steady-state probabilities can be obtained by solving the system of equations $\pi = \pi \boldsymbol{P}$ with the constraint

$$\sum_{i=0}^{M} p(i) = 1 \ . \tag{3.209}$$

Problem 3-13. Assume you toss a coin that is not fair, where p is the probability of a tail and $q = 1 - p$ is the probability of a head.

(a) Draw a signal flow graph representation for a Markov chain representing the number of heads tossed in a row. Define N as an absorption state, since in part (c) we will be interested in the first passage time to state N.

(b) Show that

$$P_N(z) = \frac{a^N z(z-q)}{(z-1)(z^{N+1} - z^N + pq^N)} . \tag{3.210}$$

(c) Show that the first passage time to N heads in a row is

$$f_N = \frac{1-q^N}{pq^N} . \tag{3.211}$$

(d) Interpret this equation for $p \approx 1$ and N large.

Problem 3-14. Show that for a Markov chain Ψ_k,

$$p(\psi_0,\psi_1,\ldots,\psi_n) = p(\psi_n \mid \psi_{n-1})p(\psi_{n-1} \mid \psi_{n-2}) \cdots p(\psi_1 \mid \psi_0)p(\psi_0).$$

In words, show that the joint probability of the states at times zero through n is the product of the initial state probability $p(\psi_0)$ and the transition probabilities $p(\psi_k \mid \psi_{k-1})$.

Problem 3-15. Show that for a Markov chain Ψ_k

$$p(\psi_n \mid \psi_{n+1},\psi_{n+2}, \ldots, \psi_{n+m}) = p(\psi_n \mid \psi_{n+1}). \tag{3.212}$$

In words, show that a Markov chain is also Markov when time is reversed.

Problem 3-16. Show that for the Markov chain Ψ_k, the future is independent of the past if the present is known. In other words, for any $n > r > s$,

$$p(\psi_n,\psi_s \mid \psi_r) = p(\psi_n \mid \psi_r)p(\psi_s \mid \psi_r).$$

Problem 3-17. Consider the parity checker example in Fig. 3-7. Suppose that the initial state is zero, $p_0(0) = 1$. Sketch the signal flow graph describing the state probabilities. Compute $p_k(0)$ and $p_k(1)$ as a function of k. Sketch these functions. Is the Markov chain stationary?

Problem 3-18. Consider tossing a fair coin. We are interested in the probability that at the kth toss we have seen at least two heads in a row. Define the random process Ψ_k to have value two if there have been two heads in a row, to have value one if not and the last toss was heads, and to have value zero otherwise.

(a) Show that the random process Ψ_k is Markov and sketch the state diagram of the Markov chain.

(b) Sketch the signal flow graph describing the state probabilities. Assume that the coin is fair.

(c) Solve for the probability that at the kth toss we have seen at least two heads in a row. You may leave the solution in the Z domain.

Problem 3-19. Using the results of Exercise 3-3, show that the Chernoff bounds on the distribution function for a Poisson random variable N with parameter a are

$$1 - F_N(n) \le \left(\frac{a}{n}\right)^n e^{n-a} , \quad a < n, \qquad F_N(n) \le \left(\frac{a}{n}\right)^n e^{n-a} , \quad a > n . \tag{3.213}$$

Problem 3-20. Find the mean and variance at time t_0 of a Poisson process $N(t)$ with constant rate λ.

Problem 3-21. Show that if $t_1 < t_2$ then

$$P_{N(t_1),\,N(t_2)}(k, k+n) = \frac{\lambda^{k+n}(t_2 - t_1)^n t_1^k}{n!k!} e^{-\lambda t_2}.$$

Problem 3-22. Consider a pure birth process in which the birth rate is proportional to the state ($\lambda_j(t) = j\lambda$), as might model the growth of a biological population. Assume the initial condition is $q_1(0) = 1$, that is we start with a population of one. Find $Q_j(s)$ for all j.

Problem 3-23. Shot noise can be generated from a Poisson process by linear filters as shown in Fig. 3-20. Assume without further justification that expectation and differentiation can be interchanged; that is, the mean value of $dN(t)/dt$ is $d\mathrm{E}[N(t)]/dt$.

(a) For $N(t)$ a Poisson process, show that the mean value of $dN(t)/dt$ is $\lambda(t)$.

(b) Similarly show that the mean value of $X(t)$ is given by (3.159).

(c) For a random process $N(t)$, show that the derivative of this process $\dot{N}(t)$ has autocorrelation

$$R_{\dot{N}\dot{N}}(t_1, t_2) = \frac{\partial^2 R_{NN}(t_1, t_2)}{\partial t_1 \partial t_2}. \tag{3.214}$$

(d) Consider a linear time-invariant system with input $W(t)$ and output $X(t)$, where $W(t)$ has autocorrelation function $R_{WW}(t_1, t_2)$. Show that

$$R_{WX}(t_1, t_2) = R_{WW}(t_1, t_2) * h(t_2), \qquad R_{XX}(t_1, t_2) = R_{WX}(t_1, t_2) * h(t_1). \tag{3.215}$$

Problem 3-24. For the Poisson process $N(t)$ in Fig. 3-20, consider two times $0 < t_1 < t_2$, and note the statistical independence of $(N(t_1) - N(0))$ and $(N(t_2) - N(t_1))$. Using this fact, and assuming $N(0) = 0$, show that

$$R_{NN}(t_1, t_2) = \Lambda(t_1)[1 + \Lambda(t_2)], \quad t_1 \le t_2, \tag{3.216}$$

where $\Lambda(t)$ is defined in (3.154). Exchange the role of t_1 and t_2 to show that

$$R_{NN}(t_1, t_2) = \Lambda(t_2)[1 + \Lambda(t_1)], \quad t_1 \ge t_2. \tag{3.217}$$

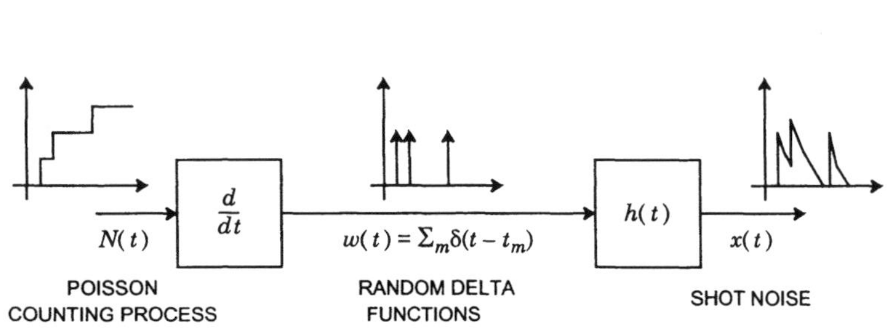

Fig. 3-20. The generation of the shot noise from a Poisson counting process.

Problem 3-25. Using the results of Problem 3-23 and Problem 3-24, show that the autocorrelation of shot noise is

$$R_X(t_1, t_2) = [\lambda(t_1) * h(t_1)][\lambda(t_2) * h(t_2)] + [\lambda(t_2)h(t_1 - t_2)] * h(t_2), \tag{3.218}$$

and evaluating at $t_1 = t_2 = t$,

$$R_X(t, t) = [\lambda(t) * h(t)]^2 + \lambda(t) * h^2(t), \tag{3.219}$$

thereby establishing Campbell's theorem (3.160) by a different method.

Problem 3-26. For the constant rate case ($\lambda(t) = \lambda$), the shot noise process is wide-sense stationary. Find the autocorrelation and power spectrum.

Problem 3-27. Let a Poisson process have rate

$$\lambda(t) = \begin{cases} 0, & t < 0 \\ \lambda_0, & t \geq 0 \end{cases}.$$

Show that a shot noise with this rate has mean value proportional to the step function of the system.

Problem 3-28. Consider a shot noise with rate function

$$\lambda(t) = \lambda_0 + \lambda_1 \cos(2\pi f_1 t).$$

Find the mean value of this shot noise.

Problem 3-29. Show that the power spectrum of the output of the parity checker of Fig. 3-7 when the input bits are not equally probable is

$$S_X(z) = \frac{p(1-p)}{(1-(1-2p)z^{-1})(1-(1-2p)z)}, \tag{3.220}$$

where p is the probability of a one-bit.

References

1. R. E. Ziemer and W. H. Tranter, *Principles of Communications: Systems Modulation and Noise*, Houghton Mifflin Co., Boston (1985).
2. S. J. Mason, "Feedback Theory — Some Properties of Signal Flow Graphs," *Proc. IEEE*, vol. 41, Sep. 1953.
3. S. J. Mason, "Feedback Theory — Further Properties of Signal Flow Graphs," Proc. *IRE*, vol. 44, no. 7, p. 920, July 1956.
4. B. C. Kuo., *Automatic Control Systems*, Prentice-Hall, Englewood Cliffs, N.J. (1962).

5. C. L. Phillips and R. D. Harbor, *Feedback Control Systems*, Prentice-Hall, Englewood Cliffs, N.J. (1988).

6. R. B. Cooper, *Introduction to Queueing Theory*, MacMillan, New York (1972).

7. A. Papoulis, *Probability, Random Variables, and Stochastic Processes*, McGraw-Hill, New York (1991).

8. H. Stark, and J. W. Woods, *Probability, Random Processes, and Estimation Theory for Engineers*, Prentice-Hall, Englewood Cliffs, NJ (1986).

9. S. M. Ross, *Stochastic Processes*, John Wiley & Sons, New York (1983).

10. S. M. Ross, *Introduction to Probability Models*, 2nd Ed., Academic Press, New York (1980).

11. J. F. Hayes, *Modeling and Analysis of Computer Communication Networks*, Plenum Press, New York (1984).

12. L. Kleinrock, *Queueing Systems. Volume I: Theory*, John Wiley & Sons, New York (1975).

13. E. A. Coddington, An *Introduction to Ordinary Differential Equations*, Prentice Hall, Englewood Cliffs, N.J. (1961).

14. D. B Williams and D. H. Johnson, "On Resolving 2M-1 Narrow-Band Signals with an M Sensor Uniform Linear Array," *IEEE Trans. on Signal Processing*, p. 707 (March 1992).

15. N. R. Goodman, "Statistical Analysis based on a Certain Multivariate Complex Gaussian Distribution (An Introduction)," *The Annals of Mathematical Statistics*, Vol. 34, (1) pp. 152-177 (March 1963).

4

Limits of Communication

In the late 1940's, Claude Shannon of Bell Laboratories developed a mathematical theory of information that profoundly altered our basic thinking about communication, and stimulated considerable intellectual activity, both practical and theoretical. This theory, among other things, gives us some fundamental boundaries within which communication can take place. Often we can gain considerable insight by comparing the performance of a digital communication system design with these limits.

Information theory provides profound insights into the situation pictured in Fig. 4-1, in which a *source* is communicating over a *channel* to a *sink*. The source and channel are both modeled statistically. The objective is to provide the source information to the sink with the

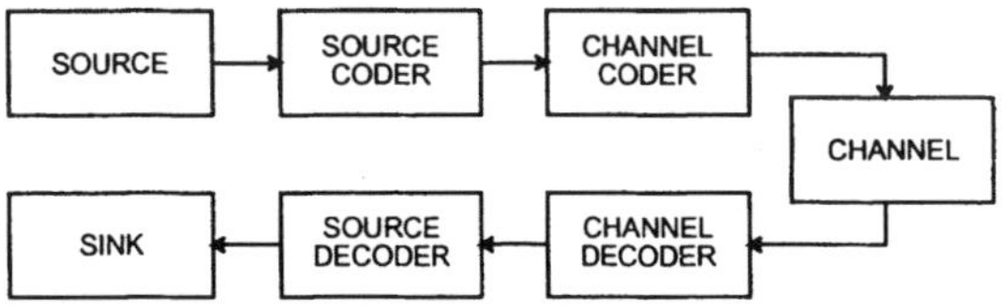

Fig. 4-1. A general picture of a source communicating over a channel using source and channel coding.

greatest fidelity. To that end, Shannon introduced the general idea of *coding*. The objective of *source coding* is to minimize the bit rate required for representation of the source at the output of a source coder, subject to a constraint on fidelity. Shannon showed that the interface between the source coder and channel coder can be, without loss of generality, a bit stream, regardless of the nature of the source and channel. The objective of *channel coding* is to maximize the information rate that the channel can convey sufficiently reliably (where reliability is normally measured as a bit error probability). Our primary focus in this book will be on the channel and the associated channel coder, although understanding source coding will also be helpful.

Given the statistics of a source, modeled as a discrete-time random process, the minimum number of bits per unit time required to represent it at the output of the source coder with some specified distortion can be determined. The *source coding theorem* is the key result of this *rate distortion theory* (see for example [1]). This theory offers considerable insight into the bit rates required for digital communication of an analog signal (Chapter 1).

Example 4-1.

We limit our attention here to the simple special case of a *discrete-time discrete-valued* random process $\{X_k\}$ with independent and identically distributed (i.i.d.) samples. Because the process is discrete-valued, it is possible to encode the signal as a bit stream with *perfect fidelity*. In fact, the minimum average number of bits required to represent each sample without distortion is equal to the *entropy* of X, defined to be

$$H(X) = E[-\log_2 p_X(X)] = -\sum_{x \in \Omega} p_X(x) \log_2 p_X(x) , \tag{4.1}$$

where Ω is the alphabet (sample space) of X. This result is developed in Section 4.1.

Since the entropy determines the number of bits required to represent a sample at the output of the source coder, it is said to determine the amount of *information* in the sample, measured in bits. This concept is explained in Section 4.1.

A second concept due to Shannon is the *capacity* of a noisy communication channel, defined as the maximum bit rate that can be transmitted over that channel with a vanishingly small error rate. The various forms of the *channel coding theorem* specify the capacity. The fact that an error rate approaching zero can be achieved was very surprising at the time, and it motivated the practical forms of channel coding to be discussed in Chapters 12 and 13.

Example 4-2.

Consider transmitting a random process $\{X_k\}$, with similar characteristics to Example 4-1, over a noisy discrete-time memoryless channel, defined as one for which the current output Y_k is dependent on only the current input X_k. Because the channel is memoryless, the samples Y_k are also independent and identically distributed. The capacity of this channel can be obtained from the *mutual information* between the input random variable X and the output random variable Y,

$$I(X, Y) = H(X) - H(X \mid Y) , \tag{4.2}$$

where $H(X \mid Y)$ is the *conditional entropy*. The channel capacity equals the mutual information maximized over all possible probability distributions for the input X. This result is developed in Section 4.2.

The result of Example 4-2 can also be used to determine the channel capacity of a bandlimited continuous-time channel using the Nyquist sampling theorem, as will be discussed in Section 4.3.

4.1. Just Enough Information About Entropy

Intuitively, *observing the outcome* of a random variable gives us *information*. Rare events carry more information than common events.

Example 4-3.

You learn very little if I tell you that the sun rose this morning, but you learn considerably more if I tell you that San Francisco was destroyed by an earthquake this morning. The reason the latter observation carries more information is that it has a lower prior probability.

In 1928 Hartley proposed a logarithmic measure of information that reflects this intuition. Consider a random variable X with sample space $\Omega = \{a_1, a_2, \ldots, a_K\}$. The *self-information* in an outcome a_m is defined to be

$$h(a_m) = -\log_2 p_X(a_m) \ . \tag{4.3}$$

The self-information of a rare event is greater than the self-information of a common event, conforming with intuition. Furthermore, the self-information is non-negative. But why the logarithm? One intuitive justification arises from considering two independent random variables X and Y, where $\Omega = \{b_1, b_2, \ldots, b_N\}$. The information in the joint events a_m and b_n intuitively should be the sum of the information in each. The self information defined in (4.3) has this property,

$$\begin{aligned} h(a_m, b_n) &= -\log_2 p_{X,Y}(a_m, b_n) = -\log_2 p_X(a_m) - \log_2 p_Y(b_n) \\ &= h(a_m) + h(b_n) \ . \end{aligned} \tag{4.4}$$

The *average information* $H(X)$ in X, defined in (4.1), is also called the *entropy* of X because of its formal similarity to thermodynamic entropy. Equivalent interpretations of $H(X)$ are

- the average information obtained by observing an outcome,
- the average uncertainty about X before it is observed, and
- the average uncertainty removed by observing X.

Because of the base-two logarithm in (4.1), information is measured in *bits*.

Example 4-4.

Consider a binary random variable X with alphabet $\Omega = \{0, 1\}$. Suppose that $q = p_X(1)$, so

$$H(X) = -q\log_2 q - (1-q)\log_2(1-q) \ . \tag{4.5}$$

This is plotted as a function of q in Fig. 4-2. Notice that the entropy peaks at 1 bit when $q = 1/2$ and goes to zero when $q = 0$ or $q = 1$. This agrees with our intuition that there is no information in certain events.

Although the intuitive justification given so far may seem adequate, the key to the interpretation of entropy as an information measure lies in the *asymptotic equipartition theorem*, which is further justified in Appendix 4-A. Define the random vector $\mathbf{X} = (X_1, \ldots, X_n)$ where X_i are independent trials of a discrete random variable X with entropy $H(X)$. Define the vector $\boldsymbol{x}$ to be an outcome of the random vector $\mathbf{X}$. The theorem says that asymptotically as $n \to \infty$, there is a set of "typical" outcomes S for which

$$p_{\mathbf{X}}(\boldsymbol{x}) \approx 2^{-nH(X)}, \quad \boldsymbol{x} \in S, \tag{4.6}$$

and the total probability that the outcome is in S is very close to unity. Since the "typical" outcomes all have approximately the same probability, there must be approximately $2^{nH(X)}$ outcomes in S. This approximation becomes more accurate as n gets large.

We can now conceptually design a source coder as follows. This source coder will assign to each outcome $\boldsymbol{x}$ a binary word, called the *code*. If n is large, we can assign binary words only to the "typical" outcomes, and ignore the "nontypical" ones. If we use $nH(X)$-bit code words, we can encode each of the $2^{nH(X)}$ typical outcomes with a unique binary word, for an average of $H(X)$ bits per component of the vector $\boldsymbol{x}$. Since each outcome of the component random variable X requires on average $H(X)$ bits, $H(X)$ is the average information obtained from the observation. It is important to note, however, that this argument applies only if we encode a large number of components collectively, and not each component separately. The statement that $H(X)$ is the average number of bits required to encode a component X applies only to an average of n components, not to an individual component.

We will now state (but not prove) the *source coding theorem* for discrete-amplitude discrete-time sources. If a source can be modeled as repeated independent trials of a random variable X at r trials per second, we define the *rate* of the source to be $R = rH(X)$. The source can be encoded by a source coder into a bit stream with bit rate less than $R + \varepsilon$ for any $\varepsilon > 0$.

Constructing practical codes that come close to R is difficult, but constructing good suboptimal codes is often easy.

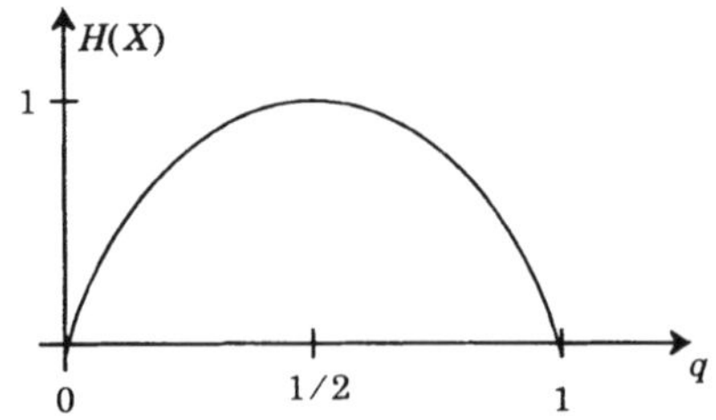

Fig. 4-2. The entropy of a binary random variable as a function of the probability $q = p_X(1)$.

Example 4-5.

For the source of Example 4-4, if $q = \frac{1}{2}$ then $H(X) = 1$. This implies that to encode repeated outcomes of X we need one bit per outcome, on average. In this case, this is also adequate for each sample, not just on average, since the source is binary. A source coder that achieves rate R just transmits outcomes of X unaltered.

Example 4-6.

When $q = 0.1$ in Example 4-4,

$$H(X) = -0.1 \cdot \log_2(0.1) - 0.9 \cdot \log_2(0.9) \approx 0.47 , \tag{4.7}$$

implying that less than half a bit per outcome is required, on average. This is not so intuitive; however, there are coding schemes in which the average number of bits per outcome will be lower than unity but greater than 0.47. One simple coding scheme takes a *pair* of outcomes and assigns them bits according to the following table.

outcomes	bits
0, 0	0
0, 1	10
1, 0	110
1, 1	111

A bit stream formed by repeated trials can be easily decoded. The average number of bits produced by this coder is 0.645 bits per trial. But note that the pair of trials (1, 1) requires three bits, or 1.5 bits per trial. This emphasizes that the entropy is an average quantity.

Example 4-7.

Consider a particularly unfair coin that *always* comes up heads. Then

$$H(X) = 0, \tag{4.8}$$

using the identity $0 \log_2 0 = 0$. This says that no bits are required to specify the outcome, which is valid.

Exercise 4-1.

It is clear from the definition of entropy that $H(X) \geq 0$. Use the inequality $\log(x) \leq x - 1$ to show that

$$H(X) \leq \log_2 K , \tag{4.9}$$

where K is the size of the alphabet of X, with equality if and only if the outcomes of X are equally likely.

The conclusion of Exercise 4-1 is that $\log_2 K$ bits *always* suffices to specify the outcomes, as is obvious since $2^{\log_2 K} = K$ possible outcomes can be encoded by a straightforward assignment, at least when K is a power of two. The less obvious conclusion is that the maximum number of bits, $\log_2 K$, is *required* only when the outcomes are equally likely.

4.2. Capacity of Discrete-Time Channels

The concept of entropy and information can be extended to channels, yielding considerable information about their fundamental limits. This section considers discrete-time channels, deferring continuous-time channels to Section 4.3. We consider three different types of discrete-time channels: discrete-valued inputs and outputs, discrete-valued inputs and continuous-valued outputs, and continuous-valued inputs and outputs.

4.2.1. Discrete-Valued Inputs and Outputs

Consider a *discrete-time* channel with input random process $\{X_k\}$ and output $\{Y_k\}$. We consider here only *memoryless channels* for which the current output Y_k is independent of all inputs except X_k. Such a channel is fully characterized by the conditional probabilities $p_{Y|X}(y \mid x)$ for all $x \in \Omega_X$ and $y \in \Omega_Y$.

Example 4-8.

Consider a channel with input and output alphabet $\Omega_X = \Omega_Y = \{0, 1\}$ such that $p_{Y|X}(0 \mid 1) = P_{Y|X}(1 \mid 0) = p$. This *binary symmetric channel (BSC)* offers a useful model of a channel that introduces independent random errors with probability p. The transition probabilities may be illustrated by a diagram:

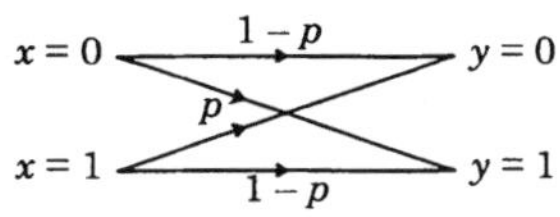

If the input samples are independent, the information per sample at the input is $H(X)$ and the information per n samples is $nH(X)$. The question is how much of this information gets through the channel. We can answer this question by finding the uncertainty in X after observing the output of the channel Y. Suppose that y is an outcome of Y. Then the uncertainty in X given the event $Y = y$ is

$$H(X \mid y) = E[-\log_2 p_{X|Y}(X \mid y)] = -\sum_{x \in \Omega_X} p_{X|Y}(x \mid y) \log_2 p_{X|Y}(x \mid y) . \tag{4.10}$$

To find the average uncertainty in X after observing Y, we must average this over the distribution of Y, yielding a quantity called the *conditional entropy,*

$$H(X \mid Y) = \sum_{y \in \Omega_Y} H(X \mid y) p_Y(y) = -\sum_{y \in \Omega_Y} \sum_{x \in \Omega_X} p_{X,Y}(x, y) \log_2 p_{X|Y}(x \mid y) . \tag{4.11}$$

This conditional entropy, on a channel such as the BSC, is a measure of the average uncertainty about the input of the channel after observing the output.

The uncertainty about X must be larger before observing Y than after; the difference is a measure of the information passed through the channel on average. Thus we define

$$I(X, Y) = H(X) - H(X \mid Y) \tag{4.12}$$

as the *average mutual information* (as in (4.2)). In other words, $I(X, Y)$ is interpreted as the uncertainty about X that is removed by observing Y, or the information about X in Y.

Exercise 4-2.

(a) Show that $I(X,Y)$ can be written directly in terms of the transition probabilities (channel) and the input distribution (input) as

$$I(X,Y) = \sum_{x \in \Omega_X} p_X(x) \sum_{y \in \Omega_Y} p_{Y|X}(y\,|\,x)\log_2\left[\frac{p_{Y|X}(y|x)}{\sum_{x \in \Omega_X} p_X(x)p_{Y|X}(y|x)}\right]. \tag{4.13}$$

(b) Show that (4.12) can be written alternatively as

$$I(X, Y) = H(Y) - H(Y\,|\,X) = I(Y, X)\;. \tag{4.14}$$

Thus, the information about X in Y is the same as the information about Y in X.

The transition probabilities are fixed by the channel. The input probabilities are under our control through the design of the channel coder. The mutual information (information conveyed through the channel) is a function of both transition and input probabilities. It makes intuitive sense that we would want to choose the input probabilities so as to maximize this mutual information. The *channel capacity per symbol* is defined as the maximum information conveyed over all possible input probability distributions,

$$C_s = \max_{p_X(x)} I(X, Y)\;. \tag{4.15}$$

This capacity is in bits/symbol, where a symbol is one sample of X. If the channel is used s times per second, then the channel capacity in bits per second is

$$C = sC_s\;. \tag{4.16}$$

Exercise 4-3.

For the BSC of Example 4-8, let the probability of the two inputs be q and $1 - q$.

(a) Show that the mutual information is

$$I(X,Y) = H(Y) + p\log_2 p + (1 - p)\log_2(1 - p)\;. \tag{4.17}$$

(b) By maximizing over q, show that the channel capacity per symbol is

$$C_s = 1 + p\log_2 p + (1 - p)\log_2(1 - p)\;. \tag{4.18}$$

The capacity is zero if $p = ½$, since then the channel inputs and outputs are independent, and is unity when $p = 0$ or $p = 1$, since then the channel is binary and noiseless.

Using the channel capacity theorem and the source coding theorem, we will now state (but not prove) a general *channel capacity theorem*. Given a source with rate $R = rH(X)$ bits/second, and a channel with capacity $C = sC_s$ bits/sec, then if $R < C$ there exists a combination of source and channel coders such that the source can be communicated over the channel with fidelity arbitrarily close to perfect. If the source is a bit stream, the channel coder can achieve *arbitrarily low probability of error* if the bit rate is below the channel capacity. In practice, achieving vanishingly small error probability requires arbitrarily large computational complexity and processing delay. Nevertheless, the channel capacity result is very useful as an ideal against which to compare practical modulation and coding systems.

4.2.2. Discrete Inputs and Continuous Outputs

Another useful channel model is a discrete-time channel with a discrete-valued input and a continuous-valued output.

Example 4-9.
In an *additive noise channel*, the output is

$$Y = X + N \tag{4.19}$$

where X is a discrete random input to the channel and N is a continuous noise variable. This model arises often in this book in the situation where a discrete *data symbol* taking on a finite number of possible values is transmitted over a channel with additive Gaussian noise (*i.e.* N is Gaussian).

This model is useful because most communications media have continuous-valued outputs, due to thermal noise, whereas digital signals are discrete-valued.

The previous definitions of entropy carry over to continuous-valued random variables, if we are careful about replacing summations with integrals. For example, the entropy of a continuous-valued random variable Y is defined as

$$H(Y) = \mathrm{E}[-\log_2 f_Y(Y)] = -\int_{\Omega_Y} f_Y(y) \log_2 f_Y(y)\, dy\,. \tag{4.20}$$

Just as with discrete-valued random variables, it is possible to bound the entropy of a continuous-valued random variable.

Exercise 4-4.
Show that if Y has zero mean and variance σ^2, then

$$0 \le H(Y) \le \tfrac{1}{2}\log_2(2\pi e \sigma^2)\,, \tag{4.21}$$

with equality if and only if Y is Gaussian. *Hint:* Show that

$$H(Y) \le -\int_{-\infty}^{\infty} f_Y(y) \log_2 g(y)\, dy \tag{4.22}$$

for any probability density function $g(y)$, using the inequality $\log(x) \le x - 1$. Then substitute a Gaussian p.d.f. for $g(y)$.

It is important to note that we have constrained the variance of the random variable in this exercise. A different constraint would lead to a different bound; or, no constraint could lead to *unbounded* entropy.

The conditional entropy is a little trickier because it involves both discrete and continuous-valued random variables. Following the second expression in (4.11), we can define

$$H(Y|X) = \sum_{x \in \Omega_X} p_X(x) \int_{\Omega_Y} f_{Y|X}(y|x) \log_2 f_{Y|X}(y|x)\, dy. \tag{4.23}$$

Exercise 4-5.

Consider the additive Gaussian noise of Example 4-9. Show that $H(Y|X) = H(N)$. This result is intuitive, since after observing the outcome of X, the uncertainty in Y is precisely the entropy of the noise.

The mutual information and capacity are defined as before, in (4.12) and (4.15).

Exercise 4-6.

Following (4.13), the mutual information can be written in terms of the channel transition probability $f_{Y|X}(y|x)$ and the probability distribution of the input $p_X(x)$,

$$I(X, Y) = \sum_{x \in \Omega_X} p_X(x) \int_{\Omega_Y} f_{Y|X}(y|x) \log_2 \left[\frac{f_{Y|X}(y|x)}{\sum_{x \in \Omega_X} p_X(x) f_{Y|X}(y|x)} \right] dy. \tag{4.24}$$

Derive this from (4.20) and (4.11).

The channel capacity for the continuous-output channel depends on the values in the discrete input Ω_X. For example, on an additive noise channel, we would expect the capacity of a channel with inputs ± 100 to be larger than the capacity with inputs ± 1 when the noise is the same. The set Ω_X of channel inputs is called the input *alphabet*.

Example 4-10.

Some common channel alphabets that we will encounter in Chapter 5 are shown in Fig. 4-3. The M-PAM alphabets are real-valued, containing M equally spaced points centered at the origin. The remaining alphabets are complex-valued, as appropriate for complex-valued discrete-time channels. The noise in this case is assumed to be complex white Gaussian noise, where the real and imaginary parts have the same power but are independent of one another and of the channel input.

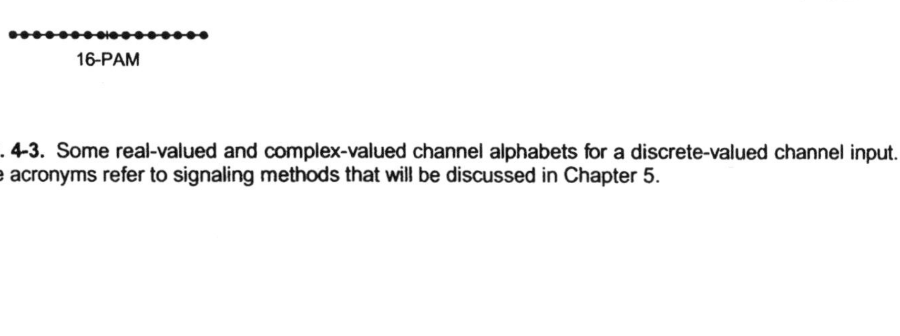

Fig. 4-3. Some real-valued and complex-valued channel alphabets for a discrete-valued channel input. The acronyms refer to signaling methods that will be discussed in Chapter 5.

One approach to calculating channel capacity would be not to constrain the alphabet at all; this is done in Section 4.2.3. Another approach is to choose an input alphabet, getting the discrete-input channel model of this subsection, and then determine the capacity by maximizing the mutual information over the probabilities of the inputs using (4.24). Going one step further, we can assume a particular distribution for the input alphabet, and then find the information $I(X, Y)$ conveyed by the channel. In a classic paper that is credited with establishing the practical importance of *trellis coding* (Chapter 13), Ungerboeck makes this calculation assuming that the input symbols in the alphabet are equally likely and that the channel adds independent Gaussian noise [2]. He computes the information conveyed by the channel as a function of the *signal-to-noise ratio* (SNR) for the input alphabets in Fig. 4-3. The results are shown in Fig. 4-4.

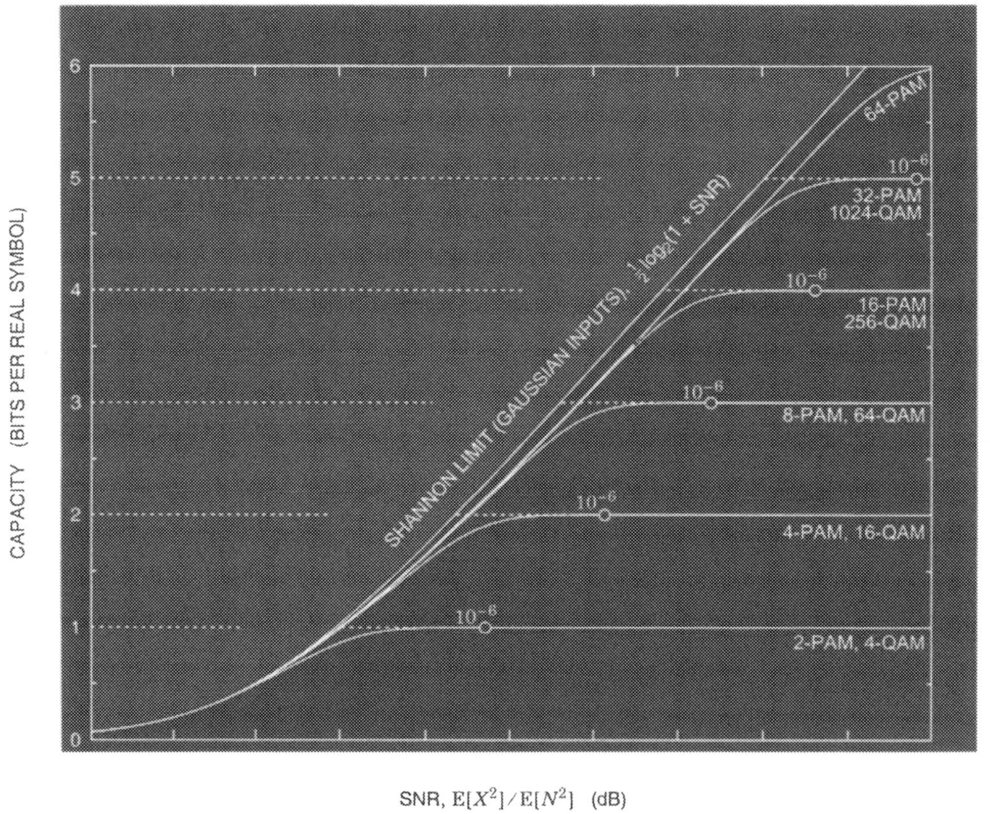

Fig. 4-4. Bounds on the information conveyed by a real-valued discrete-time channel with additive white Gaussian noise as a function of SNR for input alphabets of the type defined in Fig. 4-3. It is assumed that the symbols in the alphabet are equally likely. Also shown is the channel capacity without any constraint on the channel inputs. The points labeled 10^{-6} indicate the SNR at which a probability of error of 10^{-6} is achieved with direct techniques (no coding). The significance of these points will be discussed further in Chapter 13. The SNR is defined by the ratio of the symbol variance to the noise variance, whether the symbols be real or complex. (After Ungerboeck [2].)

Example 4-11. ————————————————————

Consider the curve corresponding to 4-PAM. As the signal to noise ratio increases, the information conveyed approaches two bits per symbol. This is intuitive because if the noise is small, nearly two bits per symbol can be sent with an alphabet of four symbols with low probability of error. For each input alphabet Ω_X with size $|\Omega_X|$, the information conveyed asymptotically approaches $\log_2 |\Omega_X|$ as the signal to noise ratio increases. While a capacity of two bits per symbol is not achievable with 4-PAM, it is achievable with 8-PAM for an SNR as low as 13 dB. Furthermore, using 16-PAM to transmit two bits per symbol does not gain much noise immunity. This suggests that there is very little lost if we use 8-PAM to transmit two bits per symbol. This observation is exploited in Chapter 13, where we discuss trellis coding.

4.2.3. Continuous-Valued Inputs and Outputs

The question arises as to what is lost by choosing a specific discrete alphabet at the channel input. We can answer this question by determining the capacity with a continuous-valued input, which is an infinite alphabet. For the additive Gaussian channel considered in Example 4-9, for any given SNR, we lose very little in capacity by choosing a discrete input alphabet, as long as the alphabet is sufficiently large (the higher the SNR, the larger the required alphabet). This result is important in that it justifies many of the digital communication techniques used in practice (Chapter 5).

Let X be a continuous-valued random variable. The entropy of Y is still given by (4.20), but the summation over x in the conditional entropy (4.23) must be replaced by an integral,

$$H(Y|X) = \int_{\Omega_X} f_X(x) \int_{\Omega_Y} f_{Y|X}(y|x) \log_2 f_{Y|X}(y|x)\, dy\,. \tag{4.25}$$

We obtain the channel capacity by maximizing $I(X,Y)$ over $f_X(x)$.

Scalar Additive Gaussian Noise Channel

Assume an additive Gaussian noise channel, $Y = X + N$ where N is an independent zero-mean Gaussian random variable with variance σ^2. What is the capacity under the constraint that the variance of X is σ_X^2? The result of Exercise 4-5 is trivially extended to get $H(Y|X) = H(N)$, which is not a function of the input distribution, so the channel capacity is obtained by maximizing $H(Y)$. The variance of Y is constrained to be $\sigma_X^2 + \sigma^2$, so from (4.21),

$$H(Y) \le \tfrac{1}{2}\log_2[2\pi e(\sigma_X^2 + \sigma^2)]\,, \tag{4.26}$$

with equality if and only if Y is Gaussian. Fortunately, Y is Gaussian if X is Gaussian, so the bound can in fact be achieved. Therefore channel capacity is achieved with a Gaussian input, and from (4.14),

$$C_s = \tfrac{1}{2}\log_2[2\pi e(\sigma_X^2 + \sigma^2)] - \tfrac{1}{2}\log_2(2\pi e\sigma^2) = \tfrac{1}{2}\log_2(1 + \sigma_X^2/\sigma^2)\,, \tag{4.27}$$

in bits per symbol. This channel capacity is plotted in both Fig. 4-4, where $\text{SNR} = \sigma_X^2/\sigma^2$. Note that this capacity is very similar to the capacity for any particular discrete alphabet at low SNR, and diverges significantly at large SNR.

The conclusion is that for the Gaussian channel and any particular SNR, there is a sufficiently large discrete input alphabet that has a capacity close to the continuous-input capacity. (At very high SNR there is an asymptotic penalty of 1.53 dB with uniform discrete inputs relative to the unconstrained capacity; this penalty is examined in Chapter 13.) This result gives a solid theoretical underpinning to the practical use of discrete input alphabets, which are also very convenient for implementation (Chapter 5).

Capacity of Vector Additive Gaussian Noise Channel

These results for the additive Gaussian channel are easily extended to a vector channel model. This extension will prove to be critically important in Chapter 7, where we consider continuous-time bandlimited Gaussian channels. We will show there that, for a given finite time interval, such a channel can be reduced to a vector Gaussian channel. Consider a channel modeled by

$$\mathbf{Y} = \mathbf{X} + \mathbf{N} \tag{4.28}$$

where **X**, **Y**, and **N** are N-dimensional vectors, **X** and **N** are independent, and the components of **N** are independent Gaussian random variables each with variance σ^2. It is easily shown, as a generalization of Exercise 4-5, that

$$I(\mathbf{X},\mathbf{Y}) = H(\mathbf{Y}) - H(\mathbf{Y}\,|\,\mathbf{X}) = H(\mathbf{Y}) - H(\mathbf{N}) \tag{4.29}$$

and that

$$H(\mathbf{N}) = \frac{N}{2}\log_2(2\pi e\sigma^2)\,. \tag{4.30}$$

The entropy of a random vector is the same as that of a scalar random variable, (4.1) or (4.20), except that the sample space has vector-valued members. The noise entropy is proportional to the dimension N because each component of the noise contributes the same entropy as in the scalar case. All that remains, then, is to find the maximum of $H(\mathbf{Y})$ over all input distributions $f_{\mathbf{X}}(\boldsymbol{x})$.

Exercise 4-7.

(a) Generalize (4.22) to show that

$$H(\mathbf{Y}) \le -\int_{\Omega_Y} f_{\mathbf{Y}}(\boldsymbol{y})\log_2 g(\boldsymbol{y})\, d\boldsymbol{y}\,, \tag{4.31}$$

for any probability density function $g(\boldsymbol{y})$.

(b) Substitute a vector Gaussian density with independent components with mean zero and variance $(\sigma^2 + \sigma_{x,n}^2)$ for the n-th component to obtain

$$H(\mathbf{Y}) \le \tfrac{1}{2}\sum_{n=1}^{N}\log_2[2\pi e(\sigma^2 + \sigma_{x,n}^2)]\,, \tag{4.32}$$

and thus show that

$$I(\mathbf{X}, \mathbf{Y}) \le \tfrac{1}{2} \sum_{n=1}^{N} \log_2\left(1 + \frac{\sigma_{x,n}^2}{\sigma^2}\right), \tag{4.33}$$

with equality if **Y** is Gaussian with independent zero-mean components. Fortunately, this upper bound can be achieved if the input vector **X** is chosen to have independent Gaussian components, each with mean zero and with variance $\sigma_{x,n}^2$ for the n-th component.

(c) Using the inequality $\log(x) \le (x - 1)$, show that if the variance of **X** is constrained to some σ_x^2,

$$\sigma_x^2 = \sum_{n=1}^{N} \sigma_{x,n}^2, \tag{4.34}$$

then

$$I(\mathbf{X}, \mathbf{Y}) \le \frac{N}{2} \log_2\left(1 + \frac{\sigma_x^2}{N\sigma^2}\right), \tag{4.35}$$

with equality if and only if all the components of **X** have equal variance.

The conclusion is that the capacity of the vector Gaussian channel with input variance constrained to $E[\|\mathbf{X}\|^2] = \sigma_x^2$ is given by

$$C = \frac{N}{2} \log_2\left(1 + \frac{\sigma_x^2}{N\sigma^2}\right), \tag{4.36}$$

and the input distribution that achieves capacity is a zero-mean Gaussian vector with independent components, each with variance σ_x^2/N. The interpretation of this result is that the capacity is N, the number of degrees of freedom, times $0.5 \cdot \log_2(1 + SNR)$, where the signal to noise ratio $SNR = \sigma_x^2/N\sigma^2$ is the total input signal power divided by the total noise power. Equivalently, the SNR can be interpreted as the signal power per dimension divided by the noise power dimension.

4.3. Further Reading

Abramson [3] gives a short elementary introduction to information theory, particularly the channel coding theorem. Gallager [4] has long been a standard advanced text and includes an extensive discussion of continuous-time channels. McEliece [5] provides a readable introduction with qualitative sections devoted to describing the more advanced work in the field. An excellent text is by Cover and Thomas [6]. Also recommended is the text by Blahut [7]. A collection of key historical papers, edited by Slepian [8] provides an easy way to access the most important historical papers, including twelve by Shannon. "A Mathematical Theory of Communication" and "Communication in the Presence of Noise," two of Shannon's best known papers, are highly recommended reading, for their lucidity, relevance, and historical value. Especially interesting, and mandatory reading for anyone with an interest in the subject,

Shannon gives an axiomatic justification of entropy as a measure of information. He simply assumes three properties that a reasonable measure of information should have, and derives entropy as the only measure that has these properties [9][10]. Viterbi and Omura [11] provide an encyclopedic coverage of information theory, with an emphasis throughout on convolutional codes. Finally, Wolfowitz [12] gives a variety of generalizations of the channel coding theorem.

Appendix 4-A. Asymptotic Equipartition Theorem

In this appendix we give a non-rigorous derivation of the asymptotic equipartition theorem that gives a great deal of insight. Define a random process Y_k in which each sample is an independent trial of the random variable Y with alphabet $\Omega = \{b_1, \ldots, b_K\}$. Let there be n trials, and define n_i to be the number of outcomes equal to b_i. The relative-frequency interpretation of probabilities tells us that if n is large, then with high probability,

$$\frac{n_i}{n} \approx p_Y(b_i) \,. \tag{4.37}$$

(A rigorous development depends mainly on defining precisely what we mean by "high probability." One approach is to show that given any $\varepsilon > 0$, the probability that $[p_Y(b_i) - \varepsilon] < n_i/n < [p_Y(b_i) + \varepsilon]$ approaches unity as n gets large.) Suppose that we are interested in the product of the n observations. We can write the product as

$$y_1 \cdots y_n = (b_1)^{n_1} \cdots (b_K)^{n_K} = \left[(b_1)^{n_1/n} \cdots (b_K)^{n_K/n} \right]^n$$

$$= \left[2^{\frac{n_1}{n}\log_2 b_1} \cdots 2^{\frac{n_K}{n}\log_2 b_K} \right]^n = \left[2^{\sum_{i=1}^{K} \frac{n_i}{n}\log_2 b_i} \right]^n . \tag{4.38}$$

Then using (4.37),

$$y_1 \cdots y_n \approx \left[2^{\sum_{i=1}^{K} p_Y(b_i)\log_2 b_i} \right]^n = \left[2^{E[\log_2 Y]} \right]^n , \tag{4.39}$$

with high probability. A rigorous proof is left to Problem 4-16. Since (4.39) is true for any discrete-valued random variable Y, it is certainly true for a random variable

$$Y = f(X) \,, \tag{4.40}$$

where f is any function defined on the alphabet Ω_X of X. Define $f(x) = p_X(x) = \Pr[X = x]$ for all $x \in \Omega_X$, certainly a legitimate function defined on the alphabet of X. Then (4.39) implies that for large n

$$y_1 \cdots y_n = f(x_1) \cdots f(x_n) = \prod_{i=1}^{n} f(x_i) = \prod_{i=1}^{n} p_X(x_i)$$

$$\approx \left[2^{\mathrm{E}[\log_2 p_X(X)]} \right]^n = 2^{-nH(X)} \tag{4.41}$$

with high probability. Since the X_i are independent,

$$p_{\mathbf{X}}(\boldsymbol{x}) = \prod_{i=1}^{n} p_X(x_i) , \tag{4.42}$$

so with high probability (4.6) holds.

Problems

Problem 4-1. Consider an unfair coin that produces heads with probability $1/4$. What is the entropy of the coin flip outcome? Suppose the coin is flipped once per second. What is the rate in this source? Devise a coder to encode successive coin flips outcomes so that the average number of bits per flip is less than one. How does your coder compare with the rate of the source?

Problem 4-2. Consider a random variable X with alphabet $\Omega_X = \{a_1, a_2, a_3, a_4\}$ and probabilities

$$p_X(a_1) = 1/2, \qquad p_X(a_2) = 1/4, \qquad p_X(a_3) = 1/8, \qquad p_X(a_4) = 1/8 . \tag{4.43}$$

Find the entropy of the random variable. Suppose independent trials of the random variable occur at rate $r = 100$ trials/second. What is the rate of the source? Devise a coder that exactly achieves the rate of the source.

Problem 4-3. The well known *Jensen's inequality* from probability theory implies that

$$\mathrm{E}[\log_2 X] \le \log_2 \mathrm{E}[X].$$

Use this to prove the *p-q inequality*: Given p_i and q_i, both strictly positive and defined for $i \in \{1, 2, \ldots, M\}$ such that

$$\sum_{i=1}^{M} p_i = 1 \qquad \text{and} \qquad \sum_{i=1}^{M} q_i = \alpha > 0$$

(so p_i could be a probability distribution) then

$$-\sum_{i=1}^{M} p_i \log_2 p_i \le -\sum_{i=1}^{M} p_i \log_2 q_i + \log_2 \alpha ,$$

with equality if and only if $q_i = \alpha p_i$ for all i.

Problem 4-4. For a discrete-valued random variable X, use the p-q inequality of Problem 4-3 to give another derivation of the results in Exercise 4-1.

Problem 4-5. Let $\mathbf{X}$ denote a vector of n i.i.d. random variables each taking the value zero or one. Show that

$$H(\mathbf{X}) \le n \tag{4.44}$$

with equality if and only if the two outcomes have equal probability.

Problem 4-6. Consider the following discrete memoryless channel, where all transition probabilities are $1/3$:

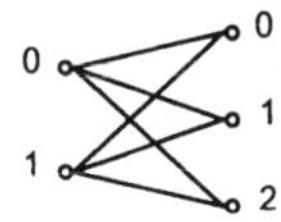

(a) Find the two conditional entropies and the mutual information in terms of the input and output entropies.

(b) Find the channel capacity.

Problem 4-7. Repeat Problem 4-6 for the following channel:

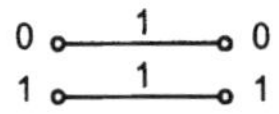

Problem 4-8. Repeat Problem 4-6 for the so-called *binary erasure* channel, shown below:

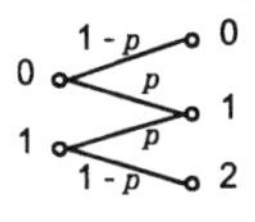

(Answer to (b): $C_s = 1 - p$.)

Problem 4-9.

(a) Show that when p_i and q_i are probability distributions,

$$-\sum_i p_i \log_2 p_i \le -\sum_i p_i \log_2 q_i \, . \tag{4.45}$$

(b) Use (4.45) to establish the result of Exercise 4-1.

Problem 4-10. Consider a cascade of L BSC's each with the same transition probability, where the output of each BSC is connected to the input of the next.

(a) Show that the resulting overall channel is a BSC.

(b) Find the error probability of the overall channel as a function of L.

(c) What happens as $L \to \infty$?

Problem 4-11. Consider a distribution $\{p_i,\ 1 \le i \le K\}$, where $p_1 > p_2$. Further define a second distribution $\{q_i,\ 1 \le i \le K\}$, where $q_1 = p_1 - \delta$ and $q_2 = p_2 + \delta$ and $q_i = p_i,\ i > 2$, where $\delta > 0$. Show that the second distribution has larger entropy. *Hint:* Use the results of Problem 4-9.

Problem 4-12. Consider a continuous-valued random variable X uniformly distributed on the interval $[-a, a]$:

(a) What is its entropy?

(b) How does its entropy compare to that of a Gaussian distribution with the same variance?

Problem 4-13. Use the p-q inequality of Problem 4-3 to show the following.

(a) For any two discrete-valued random variables X and Y, $I(X, Y) \geq 0$.

(b) $H(X) \geq H(X \mid Y)$

(c) $H(X) + H(Y) \geq H(X,Y)$

(d) When are these inequalities equalities?

Problem 4-14. Show that by replacing the summations in (4.24) with integrals, the mutual information of two continuous-valued random variables can be written

$$I(X, Y) = \int_{\Omega_X} \int_{\Omega_Y} f_{X,Y}(x, y) \log_2 \frac{f_{X,Y}(x, y)}{f_X(x) f_Y(y)} \, dy \, dx. \tag{4.46}$$

Problem 4-15. Investigate the capacity of the vector Gaussian channel of (4.36) as the number of degrees of freedom N increases. Interpret the result.

Problem 4-16. Consider a random process $\{X_k\}$, where the components are independent observations of a random variable X. The law of large numbers for sums of random variables states that for any $\varepsilon > 0$,

$$\Pr\left[E[X] - \varepsilon < \frac{1}{n}(X_1 + \ldots + X_n) < E[X] + \varepsilon\right] \rightarrow 1 \quad \text{as} \quad n \rightarrow \infty. \tag{4.47}$$

Use this to prove (4.39).

Problem 4-17. Consider the following betting game. You bet \$100 and toss a die. If a six comes up, you win \$500, otherwise you win \$20. Next you bet your \$500 or \$20 and the game is repeated with the same rate of return. In other words, on the n-th iteration, you bet M_n dollars (your previous winnings) and win $5M_n$ if a six comes up and $M_n / 5$ otherwise. What is the expected value of the money you have after n flips? Show that with high probability, M_n goes to zero for large n. Would you play this game?

Problem 4-18. Consider an analog continuous-time communication circuit with cascaded amplifiers. Suppose that the amplifiers have random gain, each independently taken from the same distribution. If the number of amplifiers is large, which is a better estimate of the gain of the system, (a) the expected value of the product of the gains of the amplifiers, or (b) the expected value of the sum of the gains expressed in dB?

References

1. T. Berger, *Rate Distortion Theory*, Prentice-Hall, Englewood Cliffs, NJ, 1971.
2. G. Ungerboeck, "Channel Coding with Multilevel/Phase Signals," *IEEE Trans. on Information Theory*, vol. IT-28, No. 1, Jan. 1982.
3. N. Abramson, *Information Theory and Coding*, McGraw-Hill Book Co., New York, 1963.
4. R. Gallager, *Information Theory and Reliable Communication*, John Wiley and Sons, Inc., New York, 1968.
5. R. J. McEliece, *The Theory of Information and Coding*, Addison Wesley Pub. Co., 1977.
6. T. M. Cover and J. A. Thomas, *Elements of Information Theory*, Wiley, New York, 1991.
7. R. E. Blahut, *Principles and Practice of Information Theory*, Addison-Wesley, New York, 1987.
8. D. Slepian (editor), *Key Papers in the Development of Information Theory*, IEEE Press, New York, 1974.
9. C. E. Shannon, "A Mathematical Theory of Communication," *Bell System Technical Journal*, Oct. 1948.
10. C. E. Shannon and W. Weaver, *The Mathematical Theory of Communication*, University of Illinois Press, Urbana, Illinois, 1963.
11. A. J. Viterbi and J. K. Omura, P*rinciples of Digital Communication and Coding*, McGraw-Hill, 1979.
12. J. Wolfowitz, *The Coding Theorems of Information Theory*, 3d ed., Springer-Verlag, Berlin, 1978.

5

Pulse-Amplitude Modulation

An information-bearing signal must conform to the limitations of its channel. While the bit streams we wish to transmit are inherently discrete-time, all physical media are continuous-time in nature. Hence, we need to represent the bit stream as a continuous-time signal for transmission, a process called *modulation*.

Rather than examine the modulation process in full generality, this chapter specializes to a special class of modulation techniques called *pulse-amplitude modulation (PAM)*. There are two reasons for this restriction. First, PAM is widely used in a variety of applications and is an extremely important technique in its own right. Second, the simplicity of PAM facilitates our development of the basic principles of receiver design. The main results of this chapter will be generalized to arbitrary modulation schemes in Chapter 6.

We start with *baseband PAM*, in which a sequence of time-translates of a basic pulse is amplitude-modulated by a sequence of data symbols. Baseband PAM is commonly used for metallic media, such as wire pairs, where the signal spectrum is allowed to extend down to zero frequency (d.c.). We then extend PAM to *passband transmission* by introducing a sinusoidal carrier signal. Passband PAM is commonly used on media with highly constrained bandwidth, such as radio. It uses two sinusoidal carriers of the same frequency (with a ninety degree phase difference) which are modulated by the real and imaginary parts of a complex-valued baseband signal. Special cases of passband PAM are the commonly used *phase-shift keying* (PSK), *amplitude and phase modulation* (AM-PM) and *quadrature amplitude modulation* (QAM). By treating these techniques as special cases of passband PAM we avoid the alphabet soup that pervades most comprehensive treatments of digital communications, where every minor variation is given a new acronym and treated as a separate topic.

We then shift our focus to the receiver, where we study the problem of receiver design. In particular, this chapter proposes the *minimum-distance strategy* for receiver design that has many advantages: it is easy to describe and intuitively reasonable; it leads to many important receiver structures (such as correlators, matched filters, whitened-matched filters, folded spectra, and even the Viterbi algorithm); and it turns out to be optimal under certain constraints on the statistics of the noise. The precise conditions under which the minimum-distance criterion is optimal will be established later, in Chapter 7. For now, we just consider minimizing distance for its own sake. We do not consider the noise statistics until the last section, when we analyze the error-probability performance of PAM with minimum-distance receivers when the noise happens to be white and Gaussian.

The choice of modulation scheme depends on the characteristics of the medium. Many channels can be approximated as either *baseband* or *passband*; a baseband channel has the frequency response of a low-pass filter, while a passband channel has the frequency response of a bandpass filter, as sketched below:

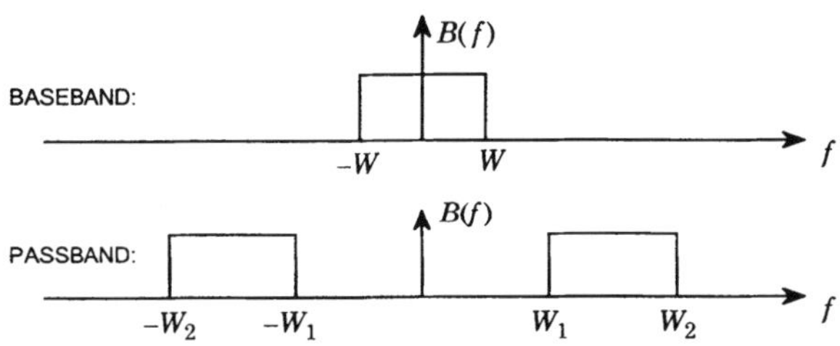

A baseband channel calls for a form of PAM called *baseband PAM*, as described in Section 5.1. A passband channel calls for *passband PAM*, which is described in Section 5.2.

5.1. Baseband PAM

A baseband PAM transmitter sends information by modulating the amplitudes of a series of pulses, so that the transmitted signal is:

$$s(t) = \sum_{m=-\infty}^{\infty} a_k g(t - kT) , \tag{5.1}$$

where $1/T$ is the symbol rate, where $g(t)$ is the *pulse shape*, and where the set of amplitudes $\{a_k\}$ are referred to as *symbols*. This signal can be interpreted as a sequence of possibly overlapping pulses with the amplitude of the k-th pulse determined by the k-th symbol. Such signals are termed *pulse-amplitude modulated* (PAM) signals, regardless of the pulse shape. PAM and its generalization to passband are by far the most common signaling methods in digital communications. There is a confusing array of techniques (e.g., QAM, PSK, BPSK, PRK, QPSK, DPSK, and AM-PM) which are all special cases of passband PAM, perhaps with some special coding.

Example 5-1.

A PAM transmitter is usually represented schematically as shown in Fig. 5-1, where a sequence of source bits is mapped to a sequence of symbols $\{a_k\}$, which in turn drives a *transmit filter* with impulse response $g(t)$. Also shown is a sketch of the resulting transmitted signal for the special case of a rectangular pulse shape. In practice, however, the large bandwidth of the rectangular pulse makes it ill-suited to bandlimited channels.

As shown in Fig. 5-1, an incoming bit stream is converted to the modulating symbol stream by a *mapper*. In practice, the symbols are restricted to a finite *alphabet* $\mathcal{A}$, so that $a_k \in \mathcal{A}$. It is particularly convenient when the size of this alphabet is a power of two, $|\mathcal{A}| = 2^b$ for some integer b, since then each symbol can be uniquely associated with a block of b source bits.

Example 5-2.

The simplest mapper translates the bits into symbols with the same values, so the alphabet is $\{0, 1\}$. A slightly more complicated mapper might use alphabet $\{-1, 1\}$ so that the symbols have zero mean if the bits are equally likely to be "0" and "1". A more complicated mapper might map pairs of bits from the set $\{00, 01, 10, 11\}$ to one of four levels from the alphabet $\{-3, -1, 1, 3\}$. All of these mappers are used in practice.

Since the mapper may map multiple bits into a single data symbol, we must make a distinction between the *symbol rate* and the *bit rate*. The symbol rate is also called the *baud rate*, after the French telegraph engineer Baudot.

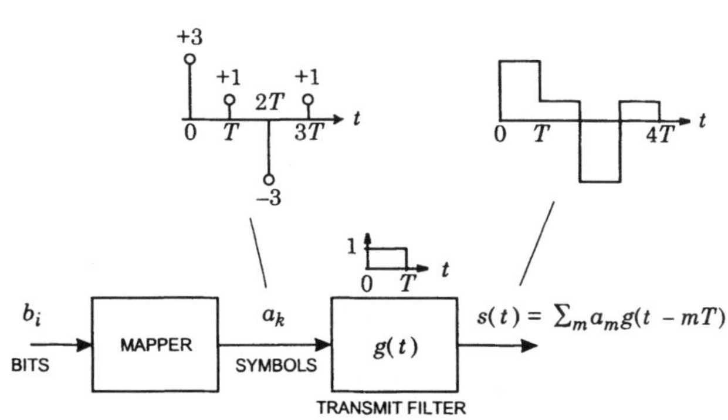

Fig. 5-1. A PAM transmitter with a rectangular transmit pulse shape. The symbol rate is $1/T$ symbols per second. A sample symbol sequence and corresponding continuous-time signal are also shown.

Example 5-3.

If the mapper maps two bits into a symbol with an alphabet size of four, the symbol rate is half the bit rate.

In the examples thus far, there is a one-to-one mapping between blocks of input bits and the alphabet. A mapper may also increase the alphabet size, usually in order to introduce *redundancy*. In such cases, the mapper is no longer memoryless but contains memory, and is more generally referred to as a *coder*. For example, the coder might convert an input bit into a symbol from an alphabet of size three. Alternatively, the coder could convert an input bit into a sequence of two or more symbols, in which case the symbol rate would be higher than the bit rate. These possibilities are discussed in Chapters 12 and 13, where it is shown that redundancy can be used to reduce errors. For the purposes of this chapter, we will assume that the mapper does not introduce redundancy. Specifically, we will usually assume that the symbols coming from the mapper are independent and identically distributed, forming a white discrete-time random process.

5.1.1. Nyquist Pulse Shapes

The job of a receiver is to recover the transmitted symbols from a continuous-time PAM signal that has been distorted by a noisy channel, but let us consider for a moment the *noiseless* PAM signal of (5.1). The noiseless case is sufficient to explore the relationship between bandwidth and symbol rate, which is a primary objective of this section. To recover the symbols $\{a_k\}$ from $s(t)$, one might try sampling $s(t)$ at multiples of the symbol period, so that the k-th sample is given by:

$$\begin{aligned} s(kT) &= \sum_{m=-\infty}^{\infty} a_m g(kT - mT) \\ &= a_k * g(kT) , \end{aligned} \tag{5.2}$$

which can be interpreted as a discrete-time convolution of the symbol sequence with a sampled version of the pulse shape. Decomposing the convolution sum into two parts yields:

$$s(kT) = g(0)a_k + \sum_{m \neq k} a_m g(kT - mT) , \tag{5.3}$$

where the first term is desired, and the second term represents interference from neighboring symbols, or *intersymbol interference (ISI)*. Under what conditions will the k-th sample reproduce the k-th symbol exactly, so that $s(kT) = a_k$? In other words, when is there no ISI? Clearly, only when the second term in (5.3) is zero, so that the sampled pulse shape reduces to a delta function:

$$g(kT) = \delta_k . \tag{5.4}$$

Equivalently, taking the Fourier transform of each side of (5.4) and using (2.17) we have

$$\frac{1}{T}\sum_{m=-\infty}^{\infty} G\left(f - \frac{m}{T}\right) = 1 . \tag{5.5}$$

This is called the *Nyquist criterion.* A pulse satisfying (5.5) (and thus also (5.4)) is said to be a *Nyquist pulse.* It does not induce ISI when sampled properly.

The Nyquist criterion is the key that ties symbol rate to bandwidth. Specifically, the Nyquist criterion implies the existence of a *minimum bandwidth* for transmitting at a certain symbol rate with no ISI. Alternatively, given a certain bandwidth, there is a maximum symbol rate for avoiding ISI.

Example 5-4.

The following plot shows $\frac{1}{T}\Sigma_m G(f-\frac{m}{T})$ for a particular pulse shape $g(t)$ whose bandwidth W is less than $1/(2T)$:

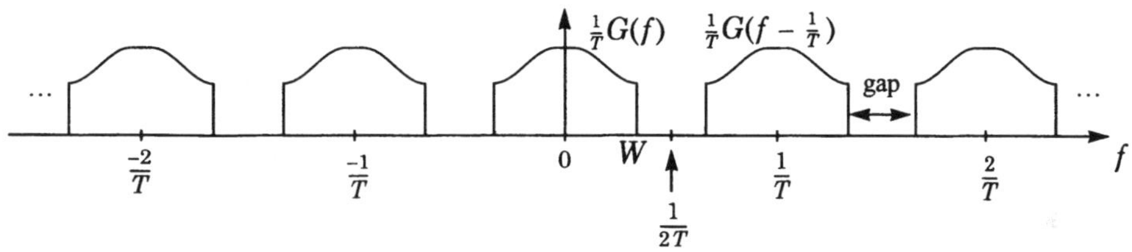

The effect of sampling is to place an *image* of $G(f)$ at each multiple of the sampling rate. Regardless of the shape of $G(f)$, it is clear that there will always be a *gap* between images whenever the pulse shape bandwidth is less than *half* the symbol rate. Such gaps prevent the images from adding to a constant, as required by the Nyquist criterion (5.5).

From the previous example it is clear that the minimum bandwidth required to avoid ISI is half the symbol rate, $1/(2T)$. But not just any pulse with this bandwidth will do. A bandwidth of $1/(2T)$ eliminates the gap between aliases, but to ensure that the aliases add to a constant, each alias must itself have a rectangular shape, so that $G(f)$ is as sketched below:

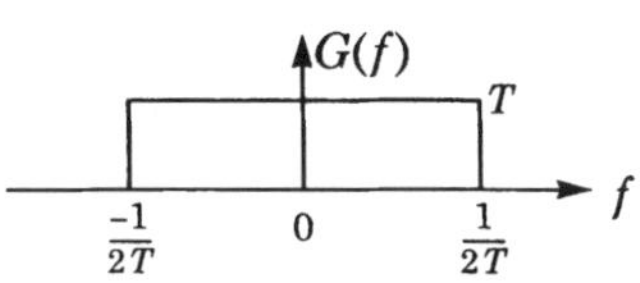

Thus, taking the inverse Fourier transform, we conclude that the minimum-bandwidth pulse satisfying (5.5) is the sinc function:

$$g(t)=\frac{\sin(\pi t/T)}{\pi t/T}\ , \tag{5.6}$$

as sketched below:

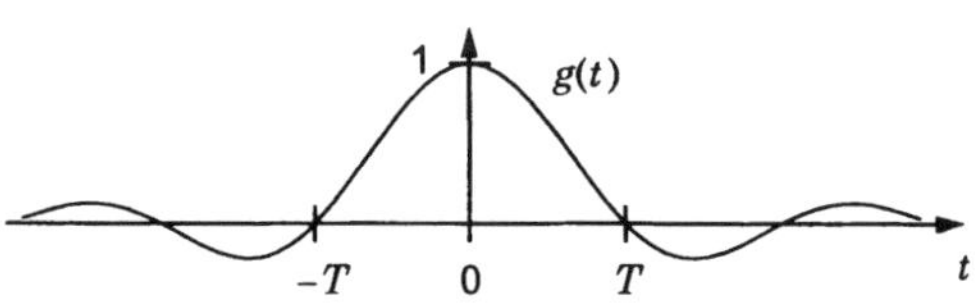

Observe that the pulse has zero crossings at all multiples of T except at $t=0$, where $g(0)=1$.

Example 5-5.

Consider two successive symbols with values $a_0 = 1$ and $a_1 = 2$ and pulse shape (5.6). The contribution of these two symbols to the signal (5.1) is shown below:

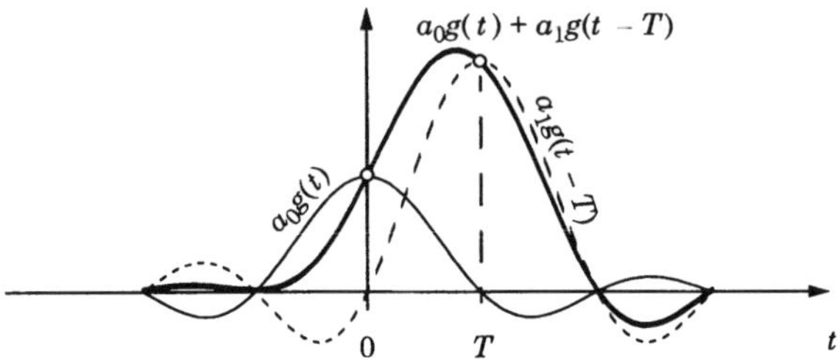

Sampling at times 0 and T recovers a_0 and a_1, respectively. Even though the neighboring pulses overlap, they do not interfere with one another at the proper sampling times; i.e., there is no ISI.

The Nyquist criterion also implies a maximum symbol rate for a given bandwidth. Specifically, if we are constrained to frequencies $|f| < W$, the maximum symbol rate that can be achieved with zero ISI is $1/T = 2W$.

Minimum bandwidth is desirable, but the ideal bandlimited pulse is impractical. The bandwidth W of a practical pulse is larger than its minimum value by a factor of $1 + \alpha$:

$$W = \frac{1+\alpha}{2T}, \tag{5.7}$$

where α is called the *excess-bandwidth parameter*. Usually excess bandwidth is expressed as a percentage; for example, 100% excess bandwidth corresponds to $\alpha = 1$ and a bandwidth of $1/T$, or twice the minimum. Practical systems usually have an excess bandwidth in the range of 10% to 100%. Increasing the excess bandwidth simplifies implementation (simpler filtering and timing recovery), but of course requires more channel bandwidth.

The zero-excess-bandwidth pulse is unique — the ideal bandlimited pulse. With non-zero excess bandwidth, the pulse shape is no longer unique. Commonly used pulses with nonzero excess bandwidth that satisfy the Nyquist criterion are the *raised-cosine pulses*, given by

$$g(t) = \left(\frac{\sin(\pi t/T)}{\pi t/T}\right)\left(\frac{\cos(\alpha\pi t/T)}{1-(2\alpha t/T)^2}\right), \tag{5.8}$$

which have Fourier transforms

$$G(f) = \begin{cases} T, & |f| \le \frac{1-\alpha}{2T} \\ T\cos^2\left[\frac{\pi T}{2\alpha}\left(|f| - \frac{1-\alpha}{2T}\right)\right], & \frac{1-\alpha}{2T} < |f| \le \frac{1+\alpha}{2T} \\ 0, & \frac{1+\alpha}{2T} < |f| \end{cases}. \tag{5.9}$$

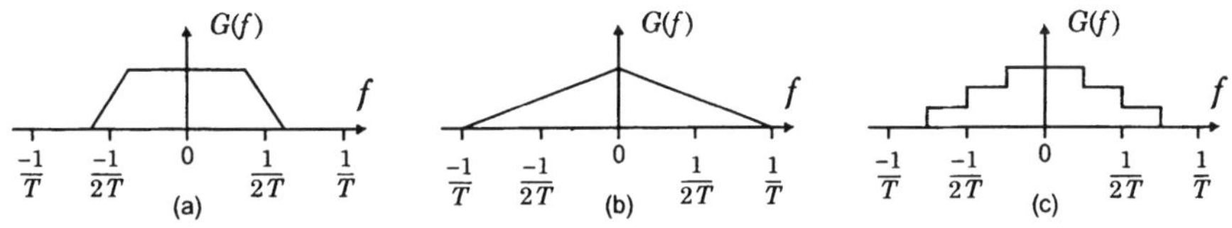

Fig. 5-3. The Fourier transform of some pulses that satisfy the Nyquist criterion.

These pulses and their Fourier transforms are plotted in Fig. 5-2 for a few values of α. For $\alpha = 0$, the pulse is identical to the ideally bandlimited pulse (5.6). For other values of α, the energy rolls off more gradually with increasing frequency, so α is also called the *roll-off* factor. The shape of the roll-off is that of a cosine raised above the abscissa, which explains the name. The pulse for $\alpha = 0$ is the pulse with the smallest bandwidth that has zero crossings at multiples of T; larger values of α require excess bandwidth varying from 0% to 100% as α varies from 0 to 1. In the time domain, the tails of the pulses are infinite in extent. However, as α increases, the size of the tails diminishes. For this reason, these pulses can be practically approximated using FIR filters by truncating the pulse at some multiple of T.

There are an infinite number of pulses that satisfy the Nyquist criterion and hence have zero crossings at nonzero multiples of T. Some examples are shown in Fig. 5-3.

5.1.2. The Impact of Filtering on PAM

To make our above discussion on Nyquist pulses more realistic, we must consider the impact of a channel. Many important channels of interest are adequately modeled as a linear time-invariant filter with impulse response $b(t)$ and additive noise $n(t)$, as shown below:

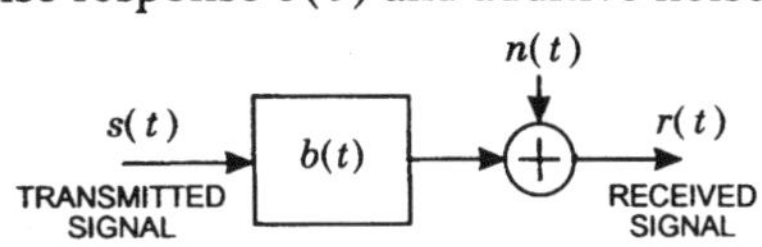

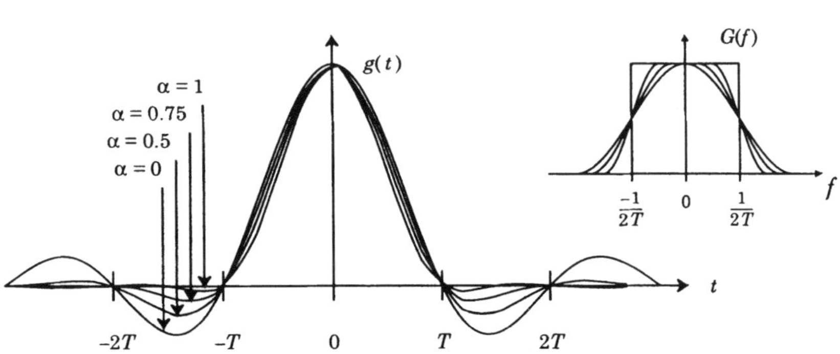

Fig. 5-2. A family of pulses with zero crossings at multiples of T, for four values of α, the roll-off factor. The Fourier transform of the pulses is also shown. Note the raised-cosine shape, and the excess bandwidth that increases with α from 0% to 100%.

(Major exceptions to the linearity assumption are the microwave radio and magnetic recording channels. A major exception to the additive noise model is the signal shot noise encountered in optical fiber channels. These channels therefore require special techniques, to be described separately.)

If we take the PAM signal of (5.4) as the transmitted signal, and apply it to a linear channel with impulse response $b(t)$ and additive noise $n(t)$, the received signal will be:

$$r(t) = \int_{-\infty}^{\infty} b(\tau) \sum_{m=-\infty}^{\infty} a_m g(t - mT - \tau) d\tau + n(t) . \tag{5.10}$$

This can be rewritten as

$$r(t) = \sum_{m=-\infty}^{\infty} a_m h(t - mT) + n(t), \tag{5.11}$$

where $h(t) = g(t) * b(t)$ is the convolution of $g(t)$ with $b(t)$,

$$h(t) = \int_{-\infty}^{\infty} g(\tau) b(t - \tau) d\tau \tag{5.12}$$

and is called the *received pulse*. The key point here is that when the transmitted signal is PAM, so too is the received signal, but with a different pulse shape and with added noise.

The design of a good receiver, one that minimizes the probability of error, is important and complicated enough to dominate the design effort as well as this book. For now we will simply point out that, given a received signal of the form (5.11), a typical receiver front end consists of a *receive filter* $f(t)$ followed by a sampler, as illustrated in Fig. 5-4.

The receive filter can perform several functions, such as compensating for the distortion of the channel and diminishing the effect of additive noise. Roughly speaking, the receive filter conditions the received signal before sampling. For example, if the bandwidth of the additive noise is wider than that of the transmitted signal, the receive filter can reject the out-of-band noise. Furthermore, the receive filter might be chosen to avoid ISI after sampling. (The optimal design of the receive filter is examined in Section 5.4.) In Chapter 9, adaptive techniques are described, in which the receive filter transfer function is adaptively adjusted to learn an unknown channel or to track changing channel responses. Adaptive filtering allows a modem to be designed without knowledge of the specific channel over which it will operate, since an adaptive receiver can learn the channel characteristics after it is deployed in the field.

The output of the receive filter — and thus the input to the sampler — will be:

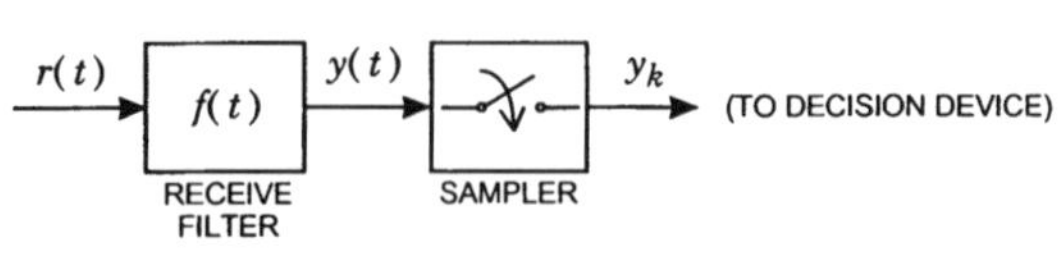

Fig. 5-4. The front-end of a typical baseband PAM receiver consists of a receive filter and a sampler.

$$y(t) = \sum_{m=-\infty}^{\infty} a_m p(t - mT) + n'(t)\ , \tag{5.13}$$

where $p(t) = g(t) * b(t) * f(t)$ is called the *overall pulse shape*, representing the cascade of the transmit pulse shape, the channel impulse response, and the receive filter. Alternatively, since the received pulse is $h(t) = g(t) * b(t)$, the overall pulse shape can be thought of as the output of the receive filter when the received pulse $h(t)$ is the input. The noise $n'(t)$ is a filtered version of the received noise $n(t)$. From (5.13) we see that the receive filter output is another PAM signal, but one whose pulse shape has been modified to $p(t) = g(t) * b(t) * f(t)$, and with added noise.

To avoid ISI using the receiver of Fig. 5-4, the *overall* pulse shape $p(t) = g(t) * b(t) * f(t)$ must be Nyquist, not the transmit pulse shape $g(t)$. In other words, according to (5.4) and (5.5), we must have $p(kT) = \delta_k$, or equivalently $\sum_m P(f - \frac{m}{T}) = T$. When this condition is satisfied, the k-th sample $y(kT)$ reduces to a_k plus noise, with no interference from $\{a_l \neq k\}$. Since $P(f) = G(f)B(f)F(f)$, a bandwidth limitation on the channel necessarily leads to the same bandwidth limitation on the overall pulse shape. Thus, it is the *channel* bandwidth W that determines the maximum symbol rate, namely $1/T = 2W$.

Example 5-6.
Consider a channel bandlimited to $|f| \leq 1500$ Hz. The absolute maximum PAM symbol rate is 3000 symbols per second. If we use a pulse with 100% excess bandwidth, then the maximum symbol rate is 1500 symbols per second.

5.1.3. ISI and Eye Diagrams

In a baseband PAM system, we are free to design the transmit and receiver filters ($g(t)$ and $f(t)$), but not the channel $b(t)$. The impulse responses $g(t)$ and $f(t)$ can be chosen to force the ISI to zero, so that the overall pulse shape satisfies the Nyquist criterion. One difficulty with such a strategy is that the channel is rarely known at the time the filters are designed. Furthermore, even when the channel is known, the filters required to exactly satisfy the Nyquist criterion may be difficult or expensive to realize. In practice, the overall pulse shape is rarely Nyquist.

With suboptimal filtering, it is useful to quantify the degradation of the signal. A useful graphical illustration of the degradation is the *eye diagram*, so called because its shape is similar to that of the human eye. An eye diagram is easily generated using an oscilloscope to observe the output of the receive filter, where the symbol timing serves as the trigger. Such displays have historically served as a quick check of the performance of a modem in the field. The eye diagram is also a useful design tool during the analytical and simulation design phase of the system.

An eye diagram consists of many overlaid traces of small sections of a signal, as shown in Fig. 5-5. If the data symbols are random and independent, it summarizes visually all possible intersymbol interference waveforms. It summarizes several features of the signal, as shown in Fig. 5-6. In the presence of intersymbol interference, when the pulse shape does not satisfy the

Nyquist criterion, the eye diagram will tend to close vertically. For error-free transmission in the absence of noise, the eye must maintain some vertical opening, since otherwise there are intersymbol interference waveforms that will cause errors.

When there is incomplete vertical closure, the intersymbol interference will reduce the size of the additive noise required to cause errors. Hence, the wider the vertical opening, the greater the noise immunity. The ideal sampling instant is at the point of maximum (vertical) eye opening, but this can never be achieved precisely by a practical timing recovery circuit. Thus the horizontal eye opening is also practically important, since the smaller this opening the greater the sensitivity to errors in timing phase (the instant at which the signal is sampled).

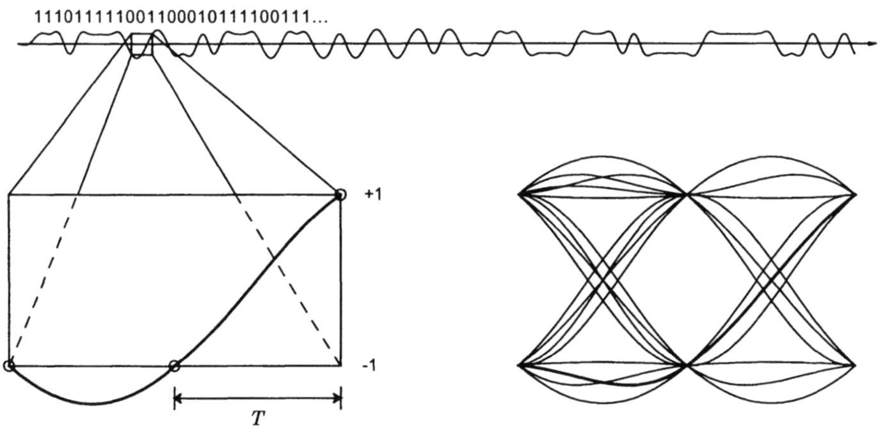

Fig. 5-5. A binary PAM signal made with 50% excess-bandwidth raised-cosine pulses. A segment of length $2T$ is shown in detail in (a). The small circles indicate the sample points where symbols are unperturbed by neighboring symbols. In (b), an eye diagram is made by overlaying sections of length $2T$. The component from part (a) is shown darkened. This display is typical of an oscilloscope display of a signal, where the oscilloscope is triggered at the symbol rate.

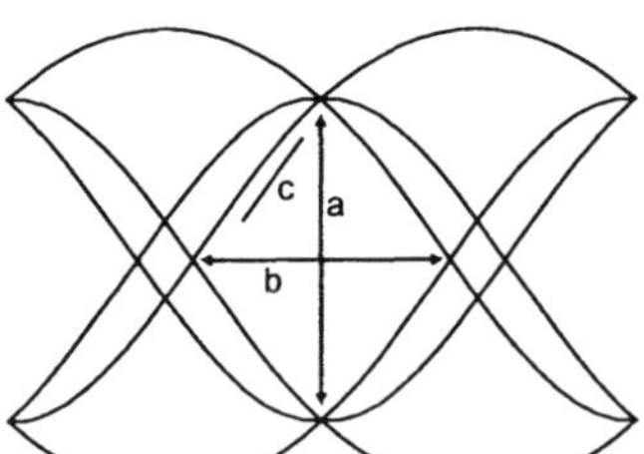

Fig. 5-6. A summary of the salient features of an eye diagram. The vertical eye opening (a) indicates the immunity to noise. The horizontal eye opening (b) indicates the immunity to errors in the timing phase. The slope of the inside eye lid (c) indicates the sensitivity to jitter in the timing phase. The ZF criterion is satisfied if all traces pass through the two symbol values in the center of the eye.

Fig. 5-7. Eye diagrams for (a) 25% and (b) 100% excess bandwidth raised-cosine pulses. Note that the pulse with larger tails and less bandwidth (25%) has a smaller eye opening.

Fig. 5-8. An eye diagram for a baseband PAM signal made with 25% excess-bandwidth raised-cosine pulses and an alphabet with four equally spaced symbols.

The shape of the eye is determined by the pulse shape. In particular, the vertical eye opening is determined by the size of the pulse at multiples of T, and the horizontal eye opening is determined by the size of the tails of the pulse $p(t)$. In Fig. 5-7 are shown two eye diagrams for 25% and 100% excess-bandwidth raised-cosine pulses. It is important to note the beneficial effect of increasing the excess bandwidth in terms of horizontal eye opening. However, more bandwidth might allow more noise to be admitted by the receiver front end, if we do not carefully design the receive filter. Thus, there is a basic system tradeoff between excess bandwidth, noise immunity, and the complexity of the timing recovery circuitry. In particular, without special coding (such as partial response line coding) it is futile to try to achieve zero excess bandwidth because the horizontal eye opening is zero (see Problem 5-5).

An eye diagram for a four-level PAM signal is shown in Fig. 5-8.

5.1.4. Bit Rate and Spectral Efficiency

If the symbols are chosen independently and uniformly from an alphabet $\mathcal{A}$ of size $|\mathcal{A}|$, so that the mapper does not introduce redundancy, then each symbol conveys $\log_2|\mathcal{A}|$ bits of information.

Example 5-7.
With the 4-PAM alphabet $\mathcal{A} = \{-3, -1, 1, 3\}$, each symbol conveys two source bits.

Furthermore, since $1/T$ symbols are transmitted per second, the *bit rate* of a PAM signal is:

$$R_b = \frac{\log_2|\mathcal{A}|}{T} \quad \text{b/s} . \tag{5.14}$$

The bit rate can be increased by increasing the size of the alphabet, or by increasing the symbol rate. As we have seen, the symbol rate is bounded by the bandwidth constraints of the channel; if we wish to avoid interference between symbols, then symbols may be transmitted at a rate no greater than twice the bandwidth of the channel. Furthermore, the size of the alphabet is constrained by the allowable transmitted power and by the severity of the additive noise on the channel. The combination of these two constraints — on symbol rate and alphabet size — limits the available bit rate for a given channel.

The ratio of the information bit rate R_b to the required bandwidth W is called the *spectral efficiency* [1]:

$$\nu = \frac{\text{bit rate}}{\text{bandwidth}} = \frac{R_b}{W} . \tag{5.15}$$

Spectral efficiency has the units of bits/sec-Hz, or loosely, bits/sec/Hz.

Example 5-8.
In point-to-point microwave radio, the spectrum is a public resource and it is important to use it efficiently for maximum public benefit. Therefore, the regulatory agencies have placed minimum requirements on the bit rates achieved as well as the bandwidth allowed. For example, 500 MHz of bandwidth is allocated in the United States for digital radio at 4 GHz divided into channels with a spacing of 20 MHz. Each channel is required to carry 90 Mb/s, for a spectral efficiency of 4.5 bits/sec-Hz [2]. Higher efficiencies would of course be desirable.

Example 5-9.
A typical voiceband data modem operates at a bit rate of 28.8 Kb/s. If the bandwidth is 3.2 KHz, then the spectral efficiency is 9 bits/sec-Hz. Lest the reader conclude that the voiceband modem designers must be smarter, it should also be noted that the digital radio operates at bit rates about 5000 times faster, making the implementation problems somewhat more severe. Also, the SNRs for the two systems are different.

For the special case of baseband PAM, spectral efficiency simplifies. The previous section showed that the symbol rate $1/T$ of a baseband PAM signal is related to the channel bandwidth W by the relationship $W = (1 + \alpha)/(2T)$, where α is the excess-bandwidth parameter. Thus, combining this with (5.14) and (5.15), we conclude that the spectral efficiency of baseband PAM is:

$$\nu = \frac{R_b}{W} = \frac{\log_2|\mathcal{A}|/T}{(1+\alpha)/(2T)} = \frac{2\log_2|\mathcal{A}|}{1+\alpha} . \tag{5.16}$$

In particular, the *maximal* spectral efficiency of baseband PAM (corresponding to zero excess bandwidth) is:

$$\nu_{\max} = 2\log_2|\mathcal{A}| . \tag{5.17}$$

Alternatively, solving (5.16) for the alphabet size yields:

$$|\mathcal{A}| = 2^{(1+\alpha)\nu/2} . \tag{5.18}$$

Example 5-10.
To achieve a bit rate of 28.8 kb/s over an ideal low-pass channel with bandwidth 3.2 KHz using baseband PAM with 11.1% ($\alpha = 1/9$) excess bandwidth would require an alphabet size of 32.

5.2. Passband PAM

Many practical communication channels are passband in nature, meaning that their frequency response is that of a bandpass filter, as sketched in Fig. 5-9. Such channels do not support transmission of baseband signals. Most physical transmission media are incapable of transmitting frequencies at d.c. and near d.c., whereas baseband PAM signals as discussed in the last section usually contain d.c. and low-frequency components.

Example 5-11.
Telephone channels, designed for voice, carry signals in the frequency range of about 300-3300 Hz with relatively little distortion. Radio channels are restricted to specified frequency bands by government regulatory bodies, such as the Federal Communications Commission (FCC) in the U.S., and constrain these channels to a bandwidth which is small relative to the center frequency.

The next section shows how PAM can be adapted to accommodate a passband channel.

5.2.1. Three Representations of Passband PAM

This section describes several strategies for communicating across a passband channel. Our starting point will be a suboptimal strategy known as *pulse-amplitude-modulation double-sideband (PAM-DSB)*; although this strategy is inefficient and not recommended, it is nevertheless a useful stepping stone to more efficient strategies.

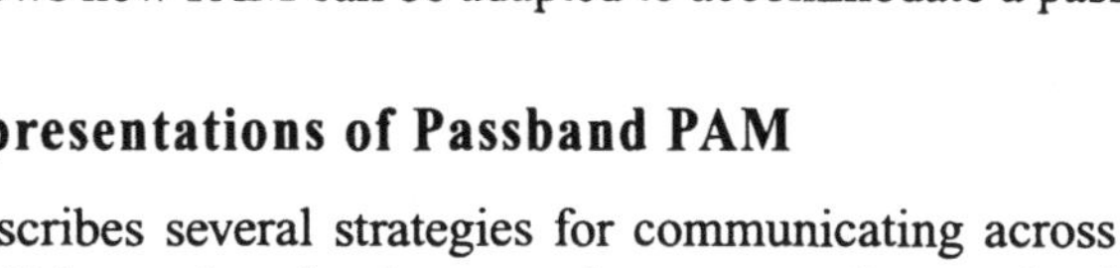

Fig. 5-9. A passband channel with bandwidth B calls for passband PAM.

The passband channel of Fig. 5-9 has bandwidth is B. The basic idea of PAM-DSB is to start with a real-valued baseband PAM signal with bandwidth $B/2$, and to *modulate* it by multiplying it by a sinusoid with frequency equal to the channel center frequency f_c. Specifically, a PAM-DSB signal takes the form:

$$s(t) = \sqrt{2}\cos(2\pi f_c t)\Sigma_k a_k g(t - kT) . \tag{5.19}$$

The above signal will pass undistorted through the channel of Fig. 5-9 when the pulse shape $g(t)$ is low-pass with bandwidth $B/2$. To avoid ISI, the maximal symbol rate is twice the pulse shape bandwidth, or $1/T = B$. The maximal spectral efficiency of PAM-DSB with real alphabet $\mathcal{A}$ is thus $\log_2 |\mathcal{A}|$, which is half that of baseband PAM.

There are two ways to modify PAM-DSB so as to make it more efficient. The first is to recognize that the upper and lower sidebands of $s(t)$ are redundant; because the underlying baseband PAM signal is real, its Fourier transform displays Hermitian symmetry, meaning that its negative frequencies are uniquely determined by its positive frequencies. We can thus double the spectral efficiency by adopting the *single-sideband (SSB)* strategy of transmitting only one of the sidebands, say the upper sideband. PAM-SSB can be implemented by passing the baseband PAM signal through the phase splitter of (2.20), which rejects the negative frequencies, before modulating by the sinusoid. One disadvantage of PAM-SSB is the difficulty in realizing the sharp discontinuity of the phase splitter near zero frequency.

A more common way to modify PAM-DSB so as to improve efficiency is to recognize from (5.19) that a PAM-DSB carries information only in the in-phase component; the *quadrature* component of the PAM-DSB signal is zero. The quadrature component represents an extra resource that the PAM-DSB strategy ignores. (Recall from Section 2.4 that the in-phase and quadrature components of a passband signal can be separated at a receiver without interfering with one another.) Thus, we can double the spectral efficiency of PAM-DSB by transmitting a second baseband PAM signal in quadrature, leading to:

$$s(t) = \sqrt{2}\cos(2\pi f_c t)\Sigma_k a_k^I g(t - kT) - \sqrt{2}\sin(2\pi f_c t)\Sigma_k a_k^Q g(t - kT) . \tag{5.20}$$

This defines *passband PAM*. The bandwidth of this signal is no greater than that of PAM-DSB, yet it conveys twice as much information. We have assumed that both of the baseband PAM signals use the same pulse shape, and that the symbols modulating the in-phase and quadrate components are denoted $\{a_k^I\}$ and $\{a_k^Q\}$, respectively. When the two sequences $\{a_k^I\}$ and $\{a_k^Q\}$ are chosen independently from the same real alphabet, (5.20) is said to be *quadrature-amplitude modulation (QAM)*. In the general case, when $\{a_k^I\}$ and $\{a_k^Q\}$ may be chosen jointly, we refer to (5.20) as passband PAM. A block diagram of a passband PAM transmitter is shown in Fig. 5-10.

We can equivalently represent (5.20) in terms of its complex envelope:

$$s(t) = \sqrt{2}\mathrm{Re}\{\tilde{s}(t)e^{j2\pi f_c t}\}, \tag{5.21}$$

where the complex envelope of a passband PAM signal is:

$$\tilde{s}(t) = \Sigma_k a_k g(t - kT) , \tag{5.22}$$

where we have introduced a *complex symbol*:

$$a_k = a_k^I + j a_k^Q \ . \tag{5.23}$$

The complex model for passband PAM will prove to be extremely useful. Observe from (5.22) that the complex envelope of a passband PAM signal looks exactly like the real-valued baseband PAM signal of the previous section; the only difference is that the symbol alphabet is now complex. This leads to the following succinct definition: a passband PAM signal is a signal whose complex envelope is the baseband PAM signal (5.22) with complex symbols and a real pulse shape.

A realization of a passband PAM transmitter based on the complex view of (5.21) is shown in Fig. 5-11. While equivalent to Fig. 5-10, an implementation based on Fig. 5-11 would be inefficient because Fig. 5-11 suggests that the imaginary part of $\tilde{s}(t)e^{j2\pi f_c t}$ is computed, and then thrown away, while in Fig. 5-10 the imaginary part is not computed. Nevertheless, in the remainder of this book we will tend to use the complex-valued notation of Fig. 5-11 because it is much more compact, and because it is easy to recognize situations where the computation of the imaginary part of a signal can be avoided.

Although both passband PAM and PAM-SSB double the spectral efficiency of PAM-DSB, they go about it in different ways. The SSB strategy fixes the bit rate but cuts the bandwidth in half. On the other hand, the passband-PAM strategy doubles the bit rate while keeping the

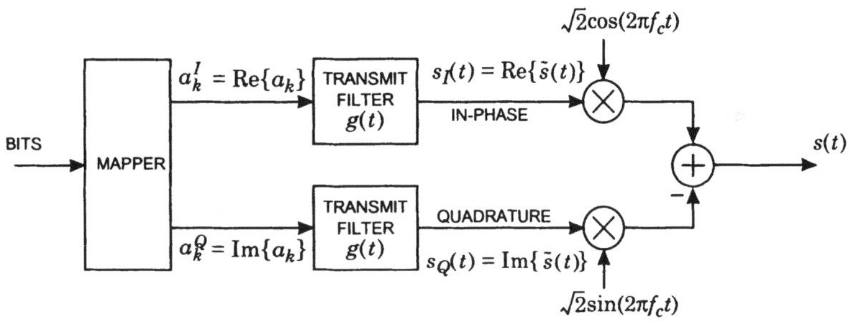

Fig. 5-10. A passband PAM transmitter. The in-phase and quadrature components of the transmitted signal are each real-valued *baseband* PAM signals.

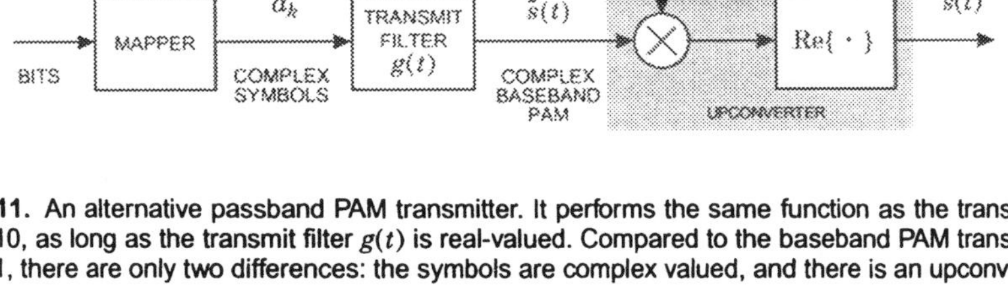

Fig. 5-11. An alternative passband PAM transmitter. It performs the same function as the transmitter in Fig. 5-10, as long as the transmit filter $g(t)$ is real-valued. Compared to the baseband PAM transmitter of Fig. 5-1, there are only two differences: the symbols are complex valued, and there is an upconverter.

bandwidth fixed. Although the spectral efficiencies are identical, passband PAM is generally preferred because it is easier to implement. For this reason, we limit the remainder of our discussion to passband PAM.

We have described two ways to represent a passband PAM: in terms of its in-phase and quadrature components (Fig. 5-10), and in terms of its complex-envelope (Fig. 5-11). A third representation of passband PAM follows by representing the data symbols a_m in terms of their magnitude and angle (polar coordinates),

$$a_m = c_m e^{j\theta_m}, \tag{5.24}$$

so that

$$\begin{aligned} s(t) &= \sqrt{2}\,\mathrm{Re}\left\{ \sum_{m=-\infty}^{\infty} c_m e^{j(2\pi f_c t + \theta_m)} g(t - mT) \right\} \\ &= \sqrt{2} \sum_{m=-\infty}^{\infty} c_m \cos(2\pi f_c t + \theta_m) g(t - mT). \end{aligned} \tag{5.25}$$

Each pulse $g(t - mT)$ is multiplied by a carrier, where the amplitude and phase of the carrier is determined by the amplitude and angle of a_m. This is sometimes called AM/PM, for amplitude modulation and phase modulation. It suggests that *phase-shift keying (PSK)*, in which data is conveyed only on the phase of the carrier, is a special case of passband PAM. This is in fact true, and will be explored further in Section 5.2.2.

5.2.2. Constellations

The *alphabet* is the set $\mathcal{A}$ of symbols that are available for transmission. A baseband signal has a real-valued alphabet that is simply a set of real numbers, for example $\mathcal{A} = \{-3, -1, +1, +3\}$. A passband PAM signal has an alphabet that is a set of complex numbers, for example $\mathcal{A} = \{-1, -j, +1, +j\}$. Both of these example alphabets have size $|\mathcal{A}| = 4$; each symbol can represent $\log_2 |\mathcal{A}| = 2$ bits. A complex-valued alphabet is best described by plotting the alphabet as a set of points in a complex plane. Such a plot is called a *signal constellation.* There is a one-to-one correspondence between the points in the constellation and the signal alphabet. Two popular constellations are illustrated in the following examples.

Example 5-12.

The 4-PSK alphabet is $\mathcal{A} = \{-b, -jb, +b, +jb\}$, and its constellation is shown in Fig. 5-12(a). It consists of four symbols of magnitude b, each with a different phase. Hence the symbols may be written

$$a_m = b e^{j\phi_m} \tag{5.26}$$

and the transmitted signal may be written (from (5.25))

$$s(t) = \sqrt{2}\, b \sum_{m=-\infty}^{\infty} \cos(2\pi f_c t + \phi_m) g(t - mT), \tag{5.27}$$

where $\phi_m \in \{0,\ \pi/2,\ \pi,\ 3\pi/2\}$. The information is carried on the phase of the carrier, while the amplitude of the carrier is constant, which explains the term *phase-shift keying* (PSK). The 4-PSK constellation is also called *quadrature phase-shift keying* (QPSK).

Example 5-13.

The 16-QAM constellation shown in Fig. 5-12(b) has 12 possible phases and three amplitudes. Because of the rectangular nature of the constellation, the rectangular coordinate system is preferable to the polar coordinates that are natural for PSK.

In a baseband PAM system, the real-valued alphabet can also be plotted as a one-dimensional constellation, although this is perhaps less informative.

Example 5-14.

The binary signal constellation below corresponds to a binary baseband PAM system:

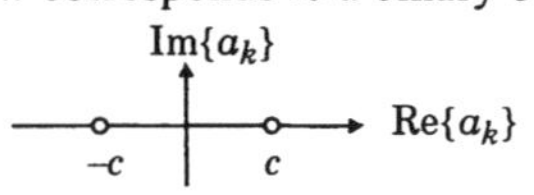

This is called a *binary antipodal* signal constellation. It can be used for passband as well as baseband signaling, in which case it is sometimes called *binary phase-shift keying* (BPSK).

Since baseband PAM is a special case of passband PAM, we will concentrate on passband PAM for the remainder of this chapter.

The constants b and c in Fig. 5-12 are not arbitrary. We now show that they directly control the energy or power of the transmitted signal. Assume that all symbols in the alphabet are equally likely, and assume that the pulse shape is normalized to have unit energy. Then the expected energy E of a single passband PAM pulse that is transmitted *in isolation*, so that $s(t) = \sqrt{2}\mathrm{Re}\{\tilde{s}(t)e^{j2\pi f_c t}\}$ with $\tilde{s}(t) = ag(t)$, is:

$$E = \mathrm{E}[\int_{-\infty}^{\infty} s^2(t)dt] = \mathrm{E}[\int_{-\infty}^{\infty} |\tilde{s}(t)|^2 dt]$$

$$= \mathrm{E}[|a|^2]\int_{-\infty}^{\infty} g(t)^2 dt$$

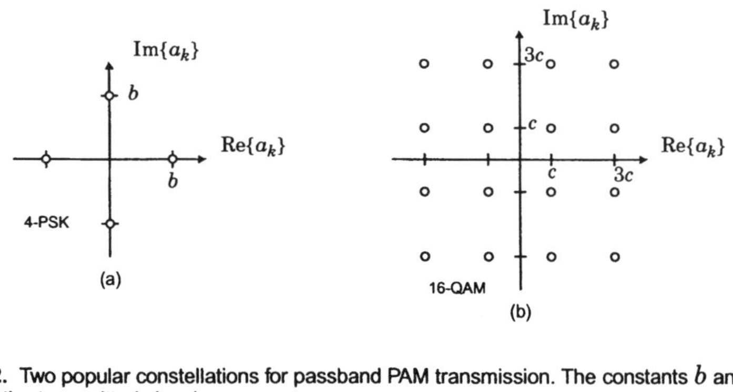

Fig. 5-12. Two popular constellations for passband PAM transmission. The constants b and c affect the power of the transmitted signal.

$$= \frac{1}{|\mathcal{A}|} \sum_{a \in \mathcal{A}} |a|^2. \qquad (5.28)$$

Thus, we can equate the energy of a single pulse with the energy of the alphabet.

There is also a simple relationship between the alphabet energy and the transmitted *power* of an infinite sequence of passband PAM pulses when the symbol sequence is white, so that the power spectrum is a constant, $S_a(e^{j\theta}) = E$. This assumption is usually valid. From Appendix 3-A, the power spectrum of the passband PAM complex envelope (5.22) is then

$$S_{\tilde{s}}(f) = \frac{E}{T} |G(f)|^2, \qquad (5.29)$$

where $G(f)$ is the Fourier transform of the pulse shape. The power of the complex-valued baseband PAM signal $\tilde{s}(t)$, which is also the power of the passband signal after upconversion, can be found by integrating the power spectrum of (5.29), yielding:

$$P = E / T, \qquad (5.30)$$

where we have exploited unit-energy assumption for $g(t)$. Intuitively this makes sense, since power is energy per unit of time. Thus, a power constraint of P on the transmitted signal translates to an energy constraint of $E = PT$ on the symbol alphabet.

Example 5-15.

On the telephone channel, the average transmitted power is constrained by regulation to be comparable to the power of voice signals. Many long distance facilities, particularly of the analog variety, are carefully designed under assumptions on the average power of each voiceband channel. On radio channels the average power is often constrained by regulation to avoid interference with other radio services. In addition, nonlinearities in the RF circuitry become more severe as the signal power gets larger. In wire-pair channels, the signal power is constrained so as to limit crosstalk interference with other cable pairs.

Alphabet Design

The distance between points in the constellation determines the likelihood that one point will be confused with another. (This will be explored with rigor in later chapters.) Furthermore, two points are more likely to be confused for one another if they are closer together than if they are farther apart. Hence the *minimum distance* between points in the constellation, denoted d_{min}, is a key parameter of the constellation. Two constellations can be considered to have approximately the same noise immunity if the minimum distance d_{min} is the same. But to make d_{min} the same, constellations with more points (such as 16-QAM vs. 4-PSK) require more transmit power. Hence there is either a power or an error-probability penalty associated with using larger constellations.

Intuitively, the objective of signal constellation design is to maximize the distance between symbols while not exceeding the power constraint. Optimal constellations are often difficult to derive, and modems that use them may be unnecessarily costly. In this section, we describe some popular constellations that have close to optimal performance. We assume an average power constraint, but the results are easily extended to a peak power constraint.

Since the performance of a constellation depends only on the distances among symbols, we expect the performance of a constellation to be invariant under translation. Hence we should translate a constellation so that its power is minimized. We now show that its power will be minimized if it has zero mean. In other words, given a set of symbols $\{a_i\}$, we wish to translate them with a complex number m such that the power

$$E[\,|a - m|^2] = \sum_{i=1}^{M} p_a(a_i)\,|a_i - m|^2 \tag{5.31}$$

of the translated symbol set $\{a_i - m\}$ is minimized. Note that (5.31) is precisely the expression for the moment of inertia of M point masses, where the mass of the ith point is $p_a(a_i)$ and its position is $(a_i - m)$ [3]. This is easily shown to be minimized if m is the taken to be the centroid (center of gravity) of the untranslated point masses; in other words, translate the system so that the centroid is at the origin. Thus the best choice for a translation is

$$m = E[a]. \tag{5.32}$$

To prove this, note that for any other translation n,

$$\begin{aligned} E[\,|a - n|^2] &= E[\,|(a - m) + (m - n)|^2] \\ &= E[\,|a - m|^2] + 2(m - n)(E[a] - m) + |m - n|^2 \\ &= E[\,|a - m|^2] + |m - n|^2 \end{aligned} \tag{5.33}$$

where the last step follows from (5.32). The mean energy under the translation n is larger than the mean energy under the translation m by the amount $|m - n|^2$. From this it is clear that alphabets with zero mean always perform better than translated alphabets with non-zero mean under an average power constraint.

Aside from ensuring zero mean (which is easy), the problem of optimal design of the constellation is complicated. A group of theoretical papers in the early 1960's [4][5][6][7] developed some basic design techniques. We will concentrate on constellations that are used in practice.

Some *quadrature-amplitude modulation* (QAM) constellations are shown in Fig. 5-13. The constellations are classified according to the number of bits b per symbol that they can convey. The number of points in the constellation is therefore $M = 2^b$. We have restricted b to

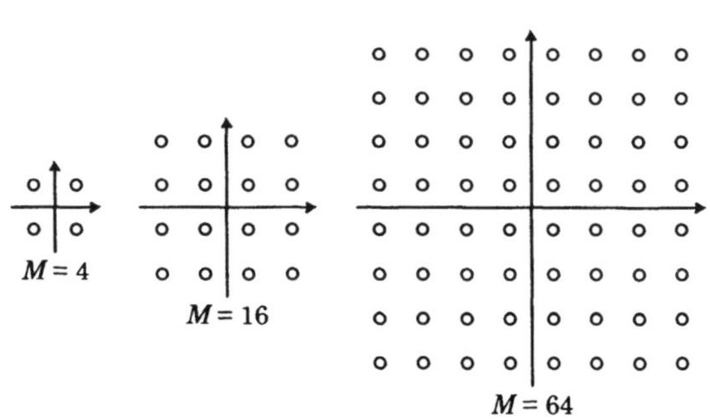

Fig. 5-13. Some QAM constellations.

be even so that half the bits are represented by the value along the imaginary axis and half by the value along the real axis. There are two significant practical advantages to these types of constellations. First, the in-phase and quadrature signals are independent $(b / 2)$ level PAM signals, so the design of the mapper is simple. Second, because of the regular rectangular pattern, noisy symbols may be transformed into symbol decisions by independently applying a thresholds along the real and imaginary axes, simplifying implementation of the decision device. There is a price for this convenience. Rectangular constellations are not the most efficient in power for a given minimum distance between symbols.

Example 5-16.

Mathematically, the alphabet for M-ary QAM takes on a simple form when M can be written as $M = L^2$ for some L that is a power of two. In this case, the real and imaginary parts a_I and a_Q of the symbol are both chosen from the real L-ary PAM alphabet $\{\pm c, \pm 3c, \ldots \pm(L-1)c\}$, for some scaling constant c. The m-th element of this alphabet can be written as $c(-L+1+2m)$ for $m \in \{0, \ldots, L-1\}$. The average energy of the M-QAM alphabet is:

$$
\begin{aligned}
E = \mathrm{E}[\,|a|^2] &= \mathrm{E}[a_I^2\,] + \mathrm{E}[a_Q^2\,] = 2\mathrm{E}[a_I^2\,] \\
&= 2\sum_{m=0}^{L-1} c^2(-L+1+2m)^2 \\
&= \tfrac{2}{3}c^2(L^2-1) \;=\; \tfrac{2}{3}c^2(M-1)\,.
\end{aligned} \tag{5.34}
$$

Thus, in terms of the average symbol energy E, the constant c is given by $c = \sqrt{\frac{3E/2}{M-1}}$. For example, we have $c = \sqrt{E/2}$ for 4-QAM, and $c = \sqrt{E/10}$ for 16-QAM.

Constellations for an odd number b of bits per symbol are also possible (and practical), as shown in Fig. 5-14. The $b = 1$ constellation is the familiar binary antipodal signal constellation. Recall that since the symbols are real-valued, this signal can be transmitted in baseband, if a baseband channel is available. Because of their shape, the $b = 5$ and $b = 7$ ($M =$ 32 and $M = 128$) constellations are called *cross constellations*. These constellations require slightly more complicated mappers than the QAM constellations and are often used in

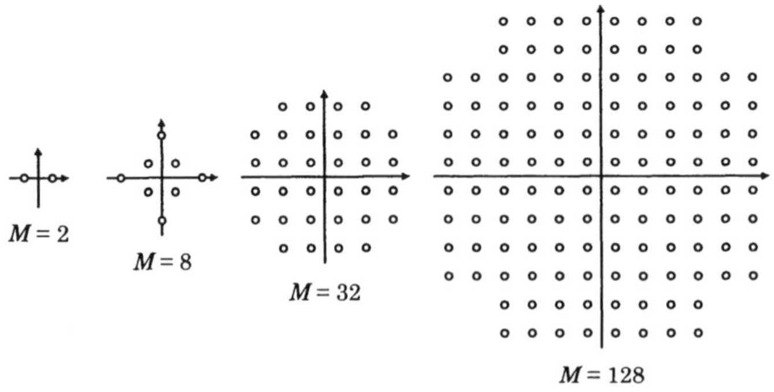

Fig. 5-14. Cross constellations.

combination with *trellis coding* (see Chapter 13 and Problem 5-11). Both the QAM and the cross constellations are called *rectangular constellations* because the symbols are on a rectangular lattice.

Another family of constellations with simple geometry is based on phase-shift keying, sometimes combined with amplitude modulation, shown in Fig. 5-15. The 2-PSK signal is identical to the binary antipodal signal, and is sometimes called *binary phase-shift keying* (BPSK). The 4-PSK signal, which we saw in Example 5-12, is sometimes called QPSK or 4-QAM. In general, the M elements of the M-PSK alphabet can be written as:

$$a = \sqrt{E}\, e^{j2\pi m/M}, \quad \text{for } m \in \{0, \ldots, M-1\} . \tag{5.35}$$

Traditionally, pure PSK (without any amplitude modulation) has been used because the signal has a constant envelope which makes it robust to amplifier nonlinearities. Furthermore, because all the information about the signal is borne by the phase of the carrier, the receiver can immediately apply a hard limiter without erasing the locations of the zero crossings, which indicate the phase. Thus, the hard-limited signal can then be processed inexpensively with digital logic without requiring A/D conversion.

A performance improvement can be achieved with hexagonal constellations, some examples of which are shown in Fig. 5-16. The symbols lie on the vertices of equilateral triangles. The term *hexagonal* refers to the shape of the decision regions, shown for $M = 16$ in Fig. 5-16. For large M, hexagonal constellations minimize the extent of a constellation for a given distance between points (try penny packing), and were suggested very early [8]. However, the improvement over rectangular constellations is slight, and the mapper and decision device are significantly more complicated [9][10].

A systematic method for constructing optimal constellations is given in [9], but because of the limited benefit and serious complication of the detector these more elaborate constellations have not found widespread use. Constellations where the probability of the symbols is not uniform have also been proposed, but practical difficulties often dominate. Some performance gains can also be accomplished using constellations of higher dimension, such as four or eight. Essentially, several symbols are selected by the mapper simultaneously and transmitted sequentially [11] or simultaneously on different carriers. (We will see multidimensional constellations again in Section 13.1).

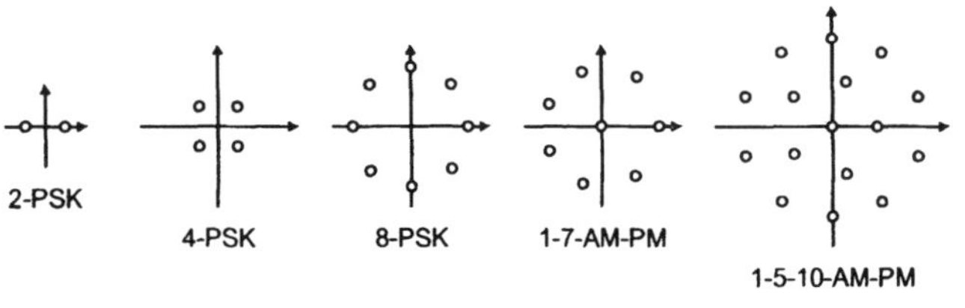

Fig. 5-15. Constellations using phase-shift keying and amplitude modulation.

5.2.3. Spectral Efficiency

The spectral efficiency of passband PAM is easily quantified. The key equations of (5.14) and (5.15) don't change; the bit rate is still $\log_2|\mathcal{A}|$ times the symbol rate, and the spectral efficiency is still the ratio of the bit rate to the bandwidth. What changes for the passband case is the relationship between symbol rate and bandwidth. With a baseband channel of bandwidth W, the maximal symbol rate using baseband PAM is $2W$. With a *passband* channel with bandwidth W, however, the maximal symbol rate using passband PAM is only W. This is because the bandwidth of a passband PAM signal is equal to the twice the bandwidth of the pulse shape. (Recall from Section 2.4 that the upconversion process doubles the bandwidth). It follows that the spectral efficiency of passband PAM with excess bandwidth α is half that of (5.16):

$$\nu = \frac{R_b}{W} = \frac{\log_2|\mathcal{A}|}{1+\alpha} \,. \tag{5.36}$$

Lest the reader infer that passband PAM has lower spectral efficiency than baseband PAM, keep in mind that the complex alphabet in (5.36) is often much larger than the real alphabet in (5.16). In fact, if we use QAM and transmit L levels on each of the two quadrature carriers, the spectral efficiency is

$$\nu = \log_2 L^2 = 2 \cdot \log_2 L \quad \text{bits/sec–Hz}\,, \tag{5.37}$$

the same as for the baseband PAM with L levels. So the efficiency of baseband and passband PAM are effectively identical, other considerations being equal.

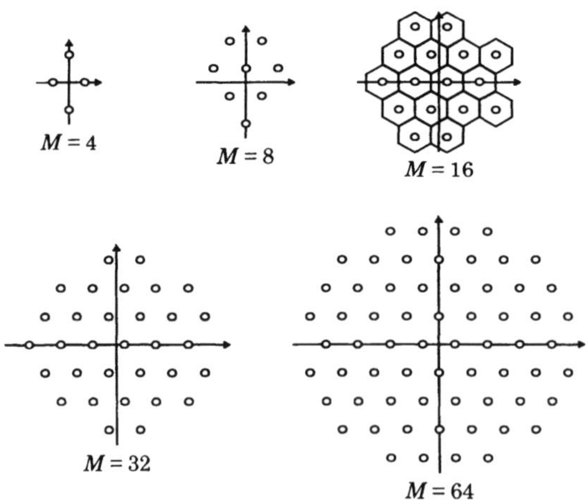

Fig. 5-16. Optimal hexagonal constellations. For the $M = 16$ constellation we have shown the hexagonal decision regions. The outer decision regions are approximated as hexagonal for uniformity.

Example 5-17.

Consider the following baseband channel with bandwidth W:

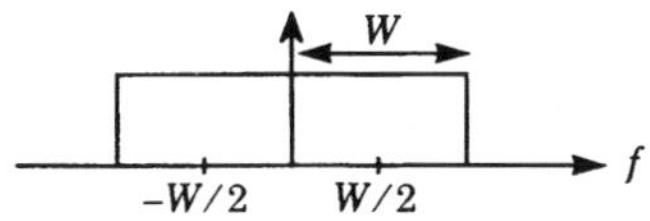

From (5.17), baseband PAM with symbol rate $2W$ and a real 4-PAM alphabet $\mathcal{A}_4 = \{\pm 1, \pm 3\}$ would achieve a spectral efficiency of $2\log_2 |\mathcal{A}_4| = 4$ b/s/Hz. However, an unconventional but conceptually legitimate view of this channel is as a *passband* channel with center frequency $W/2$ and bandwidth W. We could thus use passband PAM with symbol rate W, with carrier frequency $W/2$, and with a complex 16-QAM alphabet $\mathcal{A}_4 + j\mathcal{A}_4 = \{\pm 1 \pm j, \pm 1 \pm 3j, \pm 3 \pm j, \pm 3 \pm 3j\}$. Both 4-PAM with symbol rate $2W$ and 16-QAM with symbol rate W will achieve the same bit rate and both will require the same bandwidth. Thus, they have identical spectral efficiency, namely, 4 b/s/Hz.

Example 5-18. To achieve 4.5 bits/sec-Hz in a digital radio system, (5.36) implies an alphabet size of at least $M = 23$, but considering the need for some excess bandwidth, and the convenience implied if M is power of two, M will be larger in practice. Let B be the spacing between carriers in a frequency-division-multiplexed digital radio system. Then the nominal bandwidth available on each carrier is B and a zero excess bandwidth system would have a symbol rate of $1/T = B$. In fact, the FCC transmission mask can be met for a digital radio system with $1/T = (3/4) \cdot B$ and raised-cosine shaping with $\alpha = 0.5$ [1]. The signal bandwidth is therefore $3/2 \cdot 1/T = 9/8 \cdot B$ or 12.5% larger than the available bandwidth. This is acceptable, since the resulting interference with the adjacent carrier is small (the band edges of the raised-cosine pulse are small enough). The resulting spectral efficiency is

$$\nu = \frac{\log_2 M}{BT} = \frac{3}{4} \log_2 M , \tag{5.38}$$

and 4.5 bits/sec-Hz can be achieved with $M = 64$. Thus, the number of points in the constellation is more than twice as great as with zero excess bandwidth. This is the price paid for practical filtering characteristics and tolerance for timing errors (Chapter 16).

5.3. The One-Shot Minimum-Distance Receiver

Up to now this chapter has examined the design of a PAM transmitter. We now shift our attention to the receiver.

Our ultimate goal is to describe the receiver that optimally accounts for both the presence of ISI and the presence of noise. However, in this section we temporarily eliminate ISI from consideration by assuming that only one pulse is transmitted. This is called the *isolated pulse* or the *one-shot* case. In this context, we will develop a systematic approach to designing a PAM receiver based on a *minimum-distance* philosophy. This philosophy will lead to an important structure known as a *sampled matched filter*.

5.3.1. The Minimum-Distance Criterion

Suppose the transmitter sends only a single symbol, so that the receiver observation is:

$$r(t) = ah(t) + n(t) , \tag{5.39}$$

where $a \in \mathcal{A}$ is the transmitted symbol, $h(t)$ is the received pulse shape, and $n(t)$ is the noise. This model applies to baseband PAM, for which the symbol, received pulse, and noise are real-valued. This model also applies to passband PAM when we interpret $r(t)$ as the complex envelope of the received signal. In this case, the symbol, the received pulse, and the noise are all complex-valued. Because baseband PAM can be viewed as a special case of passband PAM, we will focus on the more general passband case in the following.

The receiver design problem is to infer from $r(t)$ which of the symbols was transmitted. The minimum-distance design strategy chooses the alphabet symbol that best representations the received waveform in a minimum-distance sense. This will be illustrated by example.

Example 5-19. Consider the case of binary antipodal signaling with a zero-excess bandwidth pulse and an alphabet of $\{\pm 1\}$, so that noiseless received signal is either $h(t)$ or $-h(t)$, as shown in Fig. 5-17. Suppose the receiver observes the noisy waveform $r(t)$, also shown in the figure. What decision should the receiver make? Intuitively, it is easy for us to see that $r(t)$ is "closer" to $-h(t)$ than $h(t)$, so we would make the decision $\hat{a} = -1$ by inspection. But a receiver requires a systematic technique for automating this intuition. A simple and intuitive approach is to form the difference between the received signal $r(t)$ and each of the candidate signals $\pm h(t)$. As shown in Fig. 5-18 for the example of Fig. 5-17, when we form the difference between the received signal

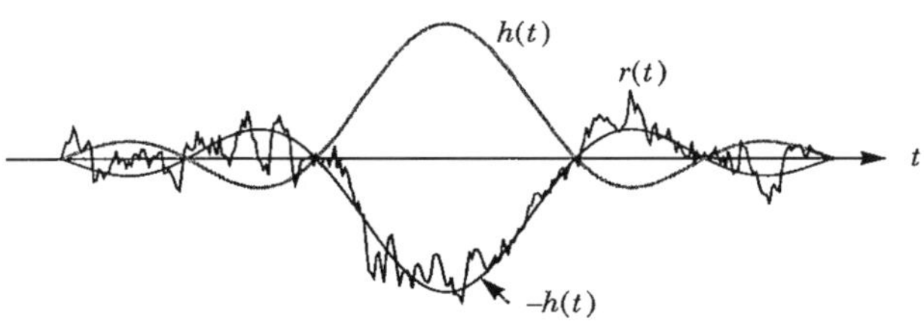

Fig. 5-17. A pair of binary antipodal signals $\pm h(t)$ and a received signal $r(t)$.

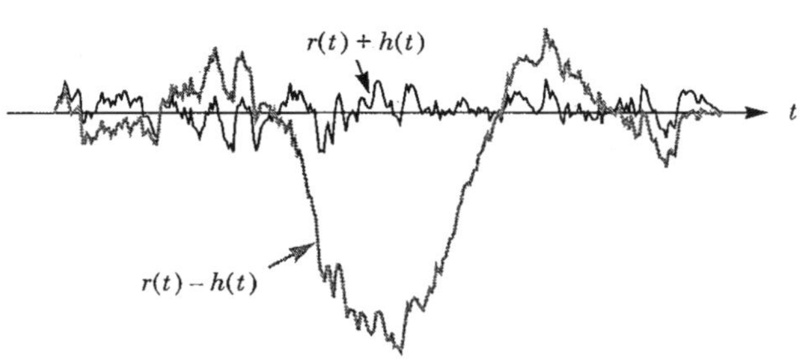

Fig. 5-18. The difference between the received signal and each of the known signals for the signal set and received signal in Fig. 5-17.

$r(t)$ and $-h(t)$, it is much smaller than if we form the difference between $r(t)$ and $h(t)$. Usually this difference signal will be much "smaller" for the actual signal than for the wrong signals. One easy way to quantify the term "smaller" is to calculate the energy of the difference signal. A receiver following this strategy would calculate

$$\int_{-\infty}^{\infty} |r(t)-h(t)|^2\,dt \quad \text{and} \quad \int_{-\infty}^{\infty} |r(t)+h(t)|^2\,dt\,, \tag{5.40}$$

and compare them. The minimum-distance receiver decides $\hat{a}=1$ if the first energy is smaller, and it decides $\hat{a}=-1$ if the second energy is smaller.

Generalizing this binary example to arbitrary alphabets, we arrive at the following. The *minimum-distance receiver* chooses the symbol that best represents the observation in a minimum-distance sense, namely:

$$\hat{a} = \underset{a \in \mathcal{A}}{\arg\min} \int_{-\infty}^{\infty} |r(t)-ah(t)|^2\,dt\,. \tag{5.41}$$

The minimum-*distance* terminology stems from Section 2.6, where signals were interpreted as vectors in a vector space. In that context, we saw in (2.98) that the energy in the error between two signals is equal to the squared distance between their corresponding vectors.

While we do not explicitly consider the effects of noise in this section, the minimum-distance criterion is primarily motivated by noise. Later in this chapter (Section 5.5), we will calculate the receiver probability of error for additive Gaussian noise on the channel, when the receiver is designed according to the principles described in this section. In Chapter 7, we will show that for additive white Gaussian noise, the minimum-distance criterion of this chapter is in fact the "optimal" receiver structure, according to definitions of optimality defined there.

As defined so far, the minimum-distance receiver must calculate the $M = |\mathcal{A}|$ integrals of (5.41), one for each element of the alphabet. We now derive a more efficient implementation that requires only a single integral. From (5.41), the minimum-distance decision is the candidate symbol a that minimizes the following cost function:

$$J = \int_{-\infty}^{\infty} |r(t)-ah(t)|^2\,dt$$

$$= \int_{-\infty}^{\infty} |r(t)|^2\,dt - 2\mathrm{Re}\left\{a^*\int_{-\infty}^{\infty} r(t)h^*(t)\,dt\right\} + |a|^2\int_{-\infty}^{\infty} |h(t)|^2\,dt \tag{5.42}$$

$$= E_r - 2\mathrm{Re}\{a^*y\} + |a|^2 E_h\,, \tag{5.43}$$

where we have introduced:

$$E_r = \int_{-\infty}^{\infty} |r(t)|^2\,dt\,, \qquad E_h = \int_{-\infty}^{\infty} |h(t)|^2\,dt\,, \tag{5.44}$$

the energies of $r(t)$ and of $h(t)$, respectively, and where we have also introduced:

$$y = \int_{-\infty}^{\infty} r(t)h^*(t)dt\,. \tag{5.45}$$

We may interpret y as the *correlation* or inner product between the observation and the received pulse, $y = \langle r(t), h(t) \rangle$.

Observe that the first term E_r in the cost function (5.43) is independent of the candidate symbol a, and hence the receiver can ignore it. Furthermore, of the remaining two terms, only the first depends on the observation waveform $r(t)$, and it does so only through the correlation y. It follows that y is a *sufficient statistic* for determining the minimum-distance decision; it collapses everything important about the observation waveform into a single scalar quantity. Rather than perform the M integrals of (5.41), the minimum-distance receiver need only perform the single integral (5.45), and then minimize the last two terms in (5.43).

There are two ways to implement the correlation integral of (5.45). A direct implementation would multiply the observation waveform by a replica of the conjugated received pulse, and integrate the result, as shown below:

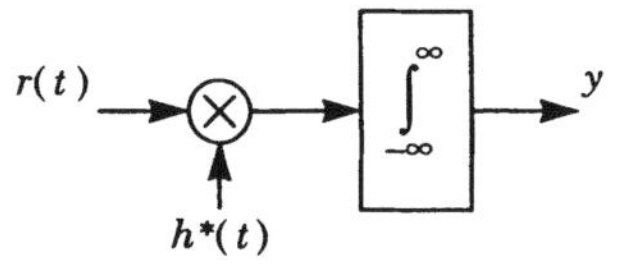

This is called a *correlator*. An alternative implementation is to apply $r(t)$ to a filter with impulse response $h^*(-t)$, and to sample the filter output at time zero, as shown below:

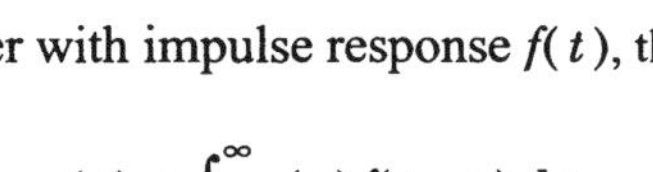

A filter with impulse response $h^*(-t)$ is said to be *matched* to $h(t)$. The above structure is thus called a *sampled matched filter (MF)*. Mathematically, the correlator and MF approaches are equivalent and interchangeable, in the sense that they produce identical outputs.

Example 5-20.

If $r(t)$ is applied to a filter with impulse response $f(t)$, the output is:

$$y(t) = \int_{-\infty}^{\infty} r(\tau)f(t-\tau)d\tau\,. \tag{5.46}$$

Sampling at time zero yields:

$$y(0) = \int_{-\infty}^{\infty} r(\tau)f(-\tau)d\tau\,. \tag{5.47}$$

If the impulse response is $f(t) = h^*(-t)$, the above integral reduces to the desired correlation of (5.45).

In practice the MF approach is often preferred over the correlator because it offers some implementation advantages, such as the ability to compensate for synchronization errors by adjusting the timing of the sampler. In contrast, to function properly, the correlator requires that the two inputs $r(t)$ and $h^*(t)$ be synchronized ahead of time.

We now return to the minimum-distance cost function of (5.43), and simplify it further. Factoring out the term E_h, it can we rewritten as:

$$J = E_h\left(\frac{E_r}{E_h} - 2\mathrm{Re}\left\{a^*\frac{y}{E_h}\right\} + |a|^2\right)$$

$$= E_h\left|\frac{y}{E_h} - a\right|^2 - \frac{|y|^2}{E_h} + E_r. \tag{5.48}$$

Because only the first term depends on the candidate symbol a, the minimum-distance receiver reduces to:

$$\hat{a} = \arg\min_{a \in \mathcal{A}} \left|\frac{y}{E_h} - a\right|^2, \tag{5.49}$$

where y is the correlation of (5.45). The minimum distance decision is the symbol $a \in \mathcal{A}$ closest to the normalized correlation $z = y/E_h$. In other words, the decision can be found by "rounding" or quantizing z to the nearest alphabet symbol. The decision device that performs this operation is called a *slicer*, and is drawn schematically as shown below:

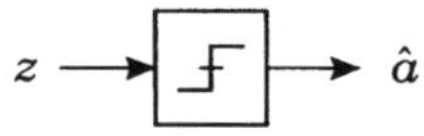

Often this can be implemented by applying a series of *decision thresholds* to the input signal.

Example 5-21.
If the data symbols are drawn from the alphabet $\{-1, 0, 1\}$, the minimum-distance slicer would apply decision thresholds at $-\frac{1}{2}$ and $\frac{1}{2}$, as shown below:

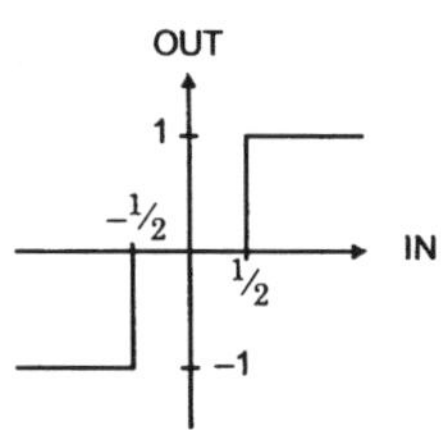

Similarly, a slicer for the 16-QAM alphabet $\{\pm1 \pm j, \pm1 \pm 3j, \pm3 \pm j, \pm3 \pm 3j\}$ would independently apply thresholds at $\{0, \pm2\}$ to the real and imaginary parts of the input.

We can summarize the main results of this section with the help of Fig. 5-19, which shows the block diagram of the minimum-distance receiver for an isolated pulse of passband PAM. First, if applicable, the complex envelope is extracted using a downconverter. (For baseband

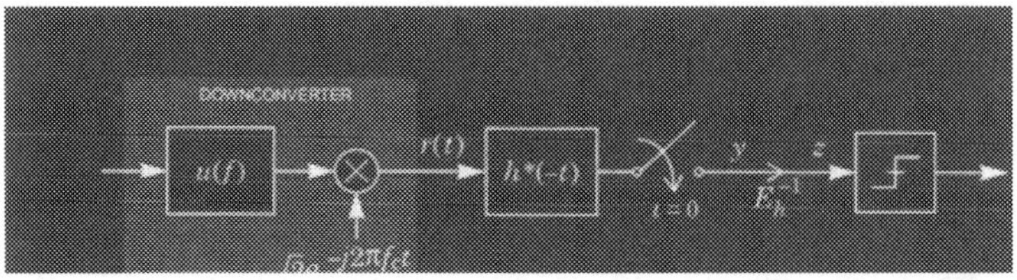

Fig. 5-19. The minimum-distance receiver for an isolated pulse of passband PAM. The receiver consists of a downconverter, a matched filter, a sampler, and a slicer. (In practice, the normalizing gain of E_h^{-1} can be absorbed into either the matched filter or the slicer.) The minimum-distance receiver for baseband PAM has exactly the same form, except there is no downconverter.

PAM, the received signal is already baseband PAM, so that the downconverter is not needed.) Second, the complex envelope is correlated with the known received pulse using a sampled matched filter, producing (5.45). Finally, the correlation is normalized and quantized using a slicer, according to (5.49). The same structure applies to an isolated pulse of baseband PAM as well, with one minor modification: there is no downconverter in the baseband case.

5.3.2. Properties of the Matched Filter

The previous section derived the MF as a simple means for implementing the minimum-distance receiver. The MF has another useful interpretation as the receive filter that maximizes SNR. Specifically, suppose we generalize the receiver in Fig. 5-19 by replacing the MF by a more general receiver filter with impulse response $f(t)$, leaving the rest (the sampler and slicer) unchanged. Because we are specializing to the isolated pulse case, the receive filter does not have to be concerned about ISI. From (5.47), the sampler output is:

$$y = \int_{-\infty}^{\infty} r(t) f(-t)\, dt\,. \tag{5.50}$$

Substituting (5.39) for $r(t)$ yields:

$$y = a\int_{-\infty}^{\infty} h(t) f(-t)\, dt + \int_{-\infty}^{\infty} n(t) f(-t)\, dt$$

$$= S + N\,. \tag{5.51}$$

The first term (S) is the signal and the second term (N) is the noise. If $n(t)$ is assumed to be white with PSD N_0, the energy of the noise term is easily shown to be

$$\mathrm{E}[\,|N|^2] = N_0 E_f\,, \qquad \text{where } E_f = \int_{-\infty}^{\infty} |f(-t)|^2 dt\,. \tag{5.52}$$

We may interpret E_f as either the energy in $f(t)$ or the energy in $f(-t)$, since they are the same. In this case, the *signal-to-noise ratio* is defined as:

$$SNR = \frac{\mathrm{E}[|S|^2]}{\mathrm{E}[|N|^2]}$$

$$= \frac{\left|\int h(t) f(-t) dt\right|^2}{E_f} \cdot \frac{E_a}{N_0}, \tag{5.53}$$

where $E_a = \mathrm{E}[\,|a|^2]$ is the symbol energy. We can uniquely choose the receive filter $f(t)$ to maximize (5.53). The integral form of the *Schwarz inequality* tells us that, for any two complex integrable functions $h(t)$ and $g(t)$ with energies E_h and E_g, respectively, their correlation must satisfy:

$$\left|\int_{-\infty}^{\infty} h(t) g^*(t) dt\right|^2 \le E_h E_g, \tag{5.54}$$

with equality if and only if $g(t) = Kh(t)$ for some constant K [12]. This is actually the same as the Schwarz inequality of Section 2.6, but using the L_2 inner product. Dividing both sides of (5.54) by E_g, and taking $g(t) = f^*(-t)$, the Schwarz inequality implies that:

$$SNR \le E_a E_h / N_0 . \tag{5.55}$$

This relationship is known as the *matched-filter bound* on the SNR. Furthermore, the Schwarz inequality tells us that equality is reached if and only if $g(t) = Kh(t)$, or equivalently $f(t) = Kh^*(-t)$. In the sequel, we choose $K = 1$, since any other choice affects the signal and noise terms equally. This proves that the MF maximizes SNR.

The matched filter $h^*(-t)$ has frequency response $H^*(f)$. The matched filter performs perfect *phase equalization*, since the transfer function of the overall pulse at the output of the matched filter, $|H(f)|^2$, is real-valued.

Additional motivation for the matched filter comes from the geometry of signal space (Section 2.6). The sampled matched filter takes the inner product of the received signal with the known pulse, or equivalently, it calculates the component of the received signal in the direction of the known pulse. Components in other directions must be due to the noise, and thus it makes sense that the minimum-distance can discard them without penalty.

If $h(t)$ is causal, as will usually be the case, then the matched filter is anticausal. To implement it in practice, $h(t)$ is assumed to be finite in length,

$$h(t) = 0 \quad \text{for} \quad t \ge t_{\max}, \tag{5.56}$$

for some constant $t_{\max}$, and the causal matched filter

$$f'(t) = h^*(t_{\max} - t) \tag{5.57}$$

is implemented. The output is then sampled at $t_{\max}$ instead of at time zero. A similar assumption is required to compute the integral in the correlation receiver in finite time.

5.3.3. Matched Filter and ISI

It should be emphasized that the preceding optimization ignored the effect of ISI by assuming an isolated pulse. When a *sequence* of pulses is transmitted, using a matched filter as a receive filter will generally introduce ISI. There are exceptions, however.

Exercise 5-1.

Show that if the received pulse is time-limited to one symbol interval, so that $h(t) = 0$ for $t \notin [0, T)$ then the overall pulse shape $p(t) = h(t) * h(-t)$ at the output of the matched filter is time-limited to two symbol intervals, $-T \le t \le T$, and furthermore goes to zero at $t = -T$ and $t = T$. Thus, the overall pulse shape satisfies the Nyquist criterion, and there is no ISI.

More generally, we can say that if the overall pulse shape at the output of the matched filter obeys the Nyquist criterion, then the matched filter is the optimal receive filter for both the isolated-pulse case and the sequence-of-pulses case, in the sense that it maximizes the SNR. For a received pulse $h(t)$, the overall pulse shape at the output of the matched filter has Fourier transform $|H(f)|^2$. The Nyquist criterion thus becomes, at the output of the matched filter,

$$S_h(f) = \frac{1}{T}\sum_{m=-\infty}^{\infty}\left|H\left(f - \frac{m}{T}\right)\right|^2 = 1\,. \tag{5.58}$$

The quantity $S_h(f)$ is called the *folded spectrum* of the received pulse. It will play a key role in Chapters 7 and 8, where we consider ISI in detail. Equation (5.58) depends only on the magnitude $|H(f)|$, as illustrated by the following example.

Example 5-22.

The raised cosine pulses given in (5.8) have a Fourier transform (5.9) that is real-valued and non-negative for all f. Therefore, a simple way to satisfy (5.58) is to use a pulse $h(t)$ and receive filter $f(t)$ with Fourier transforms equal to the square root of the raised cosine,

$$H(f) = F(f) = P(f)^{1/2}\,, \tag{5.59}$$

where $P(f)$ is given by (5.9). The corresponding time domain pulse shapes are [13],

$$h(t) = f(t) = \frac{4\alpha}{\pi\sqrt{T}} \cdot \frac{\cos((1+\alpha)\pi t/T) + T\sin((1-\alpha)\pi t/T)/(4\alpha t)}{1-(4\alpha t/T)^2}\,. \tag{5.60}$$

Convolving such a pulse with itself will yield the raised cosine pulse of (5.8), so using such a pulse and receive filter results in no ISI at the receive filter output. Such pulses are called *square-root raised cosine* pulses. They are particularly easy to implement if the channel is assumed to be flat, so that the transmit pulse $g(t)$ is the same as the received pulse $h(t)$.

5.3.4. Passband PAM Receivers

The passband PAM minimum-distance receiver of Fig. 5-19 first downconverts, then filters, and then samples. In practice, as described in this section, the order in which these functions are performed is sometimes rearranged in order to simplify the implementation. To aid our discussion we start with the general receiver structure shown in Fig. 5-20(a), which first downconverts a real passband signal $\hat{r}(t)$, yielding its complex envelope $r(t)$; then filters by a complex-valued baseband receive filter $f(t)$, yielding $y(t)$; and finally samples at the baud rate, yielding $y_k = y(kT)$. The receive filter $f(t)$ may be a matched filter, for example.

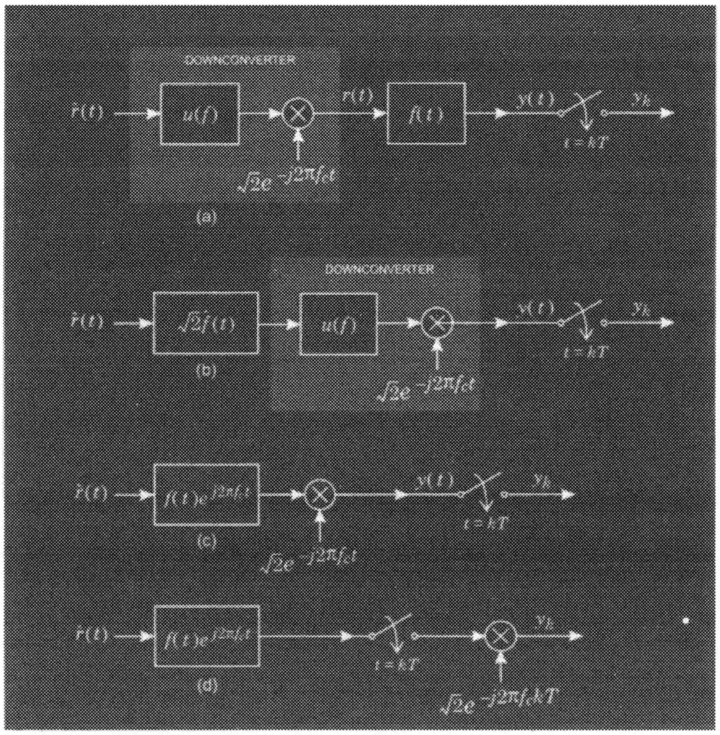

Fig. 5-20. Four equivalent ways to implement a passband PAM receiver front end. The usual approach (a), using a downconverter, a baseband receive filter, and a baud-rate sampler. An equivalent structure (b) in which the baseband receive filter is implemented in passband, before the downconverter. An equivalent structure (c) in which the phase splitter and passband filter are combined to form a analytic passband filter. An equivalent structure (d) in which the order of the sampling and mixing is reversed, so that the mixer can be implemented in discrete time.

A fundamental result from Chapter 2, namely (2.32), tells us that the convolution of two complex envelopes is a factor of $\sqrt{2}$ larger than the complex envelope of the convolution of the corresponding passband signals. In terms of Fig. 5-20 this means that we can calculate the receive filter output by first filtering $\hat{r}(t)$ by a real passband filter with impulse response $\sqrt{2}\hat{f}(t) = 2\text{Re}\{f(t)e^{j2\pi f_c t}\}$, and then downconverting. Therefore, the system of Fig. 5-20(a) is equivalent to that of Fig. 5-20(b), where a real passband filter with impulse response $\sqrt{2}\hat{f}(t)$ is applied before the downconverter. This is increasingly done in practice.

We can simplify the receiver further by recognizing that the Fourier transform of $\sqrt{2}\hat{f}(t)$ is $F(f - f_c) + F^*(-f - f_c)$. This filter can be combined with the phase splitter filter $u(f)$ in the downconverter to yield a single filter. As long as the bandwidth of the original receive filter is less than the carrier frequency, which is almost always true in practice, the frequency response of the combined filter will be $F(f - f_c)$, and the impulse response will be $f(t)e^{j2\pi f_c t}$. This leads to the equivalent receiver structure of Fig. 5-20(c).

The combined filter $f(t)e^{j2\pi f_c t}$ is still a passband filter, but it passes only positive frequency components, and not negative frequency components. Since the impulse response of this filter, $f(t)e^{j2\pi f_c t}$, is an analytic signal, this filter is called an *analytic passband filter*. The impulse response $f(t)e^{j2\pi f_c t}$ is always complex-valued, so the filter will require two real filters for implementation, as shown below:

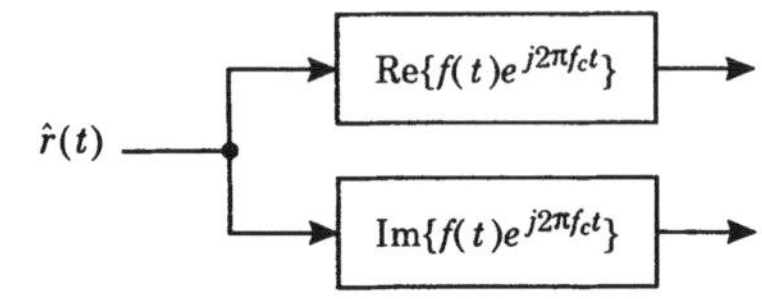

Finally, Fig. 5-20(d) shows a further simplification that arises from reversing the order of the filter and downconverter. Since the symbol-rate sampler immediately follows the mixer in Fig. 5-20(c), the sampler and mixer can change places. The mixer can then be performed in the discrete-time domain. This has important practical consequences, because it is common to coordinate the choice of symbol rate and carrier frequency at the transmitter so that the quantity $f_c T$ assumes a convenient value. For example, if $f_c T = 1/N$, then the values of $e^{j2\pi f_c kT}$ can be easily generated from a lookup table with N entries. This is often much simpler to implement than generating $e^{j2\pi f_c t}$ and performing the multiplication in continuous time.

5.3.5. More Elaborate PAM Receivers: A Preview

The receiver that we have derived in this section consists of a downconverter, a filter, and a slicer. However, passband PAM receivers can be much more elaborate, as we will see in subsequent chapters. In order to motivate those chapters, we give here a qualitative description of a typical passband PAM receiver in Fig. 5-21. It is a practical receiver, although there are

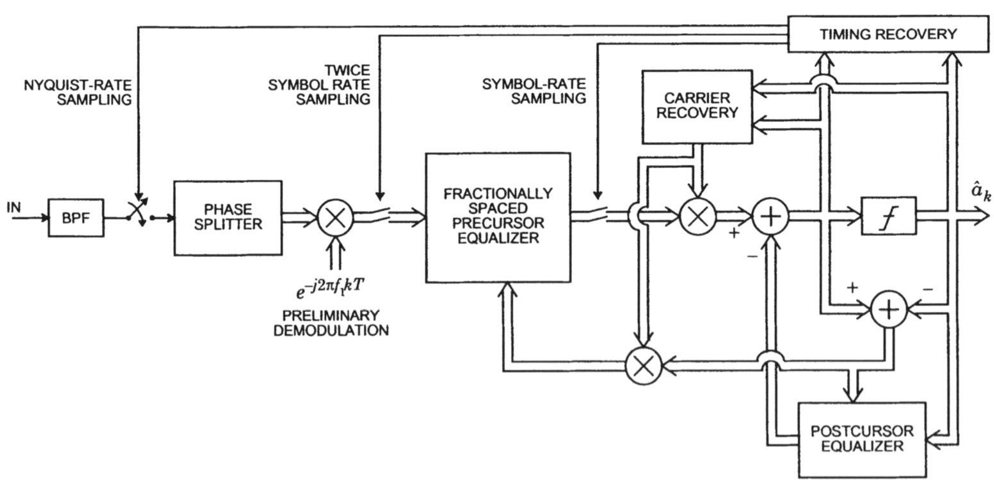

Fig. 5-21. Block diagram of a typical passband PAM receiver. Specific parts of this receiver will be discussed in detail in subsequent chapters.

many variations. The parts of the receiver that are already familiar are the bandpass filter on the front end, the phase splitter, the demodulator, and the slicer. In fact, the front end consisting of a BPF followed by a phase splitter is much like the structure shown in Fig. 5-20(b).

In the next section we will learn how to deal with intersymbol interference in an optimal way. The complexity can be prohibitive, however. Careful compromises lead to structures that look more promising and are made fully practical in Chapter 8. One such structure is a *decision-feedback equalizer* that consists of a *fractionally spaced precursor equalizer* (often also called a "forward equalizer") and a *postcursor equalizer* (often called a "feedback equalizer"), as shown in Fig. 5-21. The fractionally-spaced precursor equalizer is a filter that performs the function of the matched filter, and also equalizes the *precursor* portion of the ISI, which is defined as the interference from future data symbols. The postcursor equalizer then removes the *postcursor* portion of the ISI, defined as the interference from past data symbols. In Chapter 9 we show how the parameters of these filters can be *adapted* automatically so that characteristics of the channel do not have to be precisely known by the designer of the receiver.

Timing recovery is required to derive a symbol-rate clock from the PAM waveform itself, as shown in Fig. 5-21 and explained in Chapter 16. There are many different timing recovery schemes available; the one shown here is *decision-directed*, which means that it uses the receiver decisions to update the phase and frequency of the clock. It is also shown producing three different sampling rates, all related by rational multiples. The Nyquist-rate sampling at the front end is required if the phase splitter is implemented in discrete time. Of course it need not be, and in fact can be combined with the bandpass filter at the front end, in which case this sampling operation will not be required. The second sampling rate is at twice the symbol rate; this explains the terminology "fractionally-spaced" for the subsequent equalizer. The final sampling operation is at the symbol rate, since the slicer requires samples only at the symbol rate.

There are also connections from the output of the slicer (the decisions) to the two equalizers. These connections are required for adaptation of the equalizers, and imply that adaptation is also decision-directed.

Also shown in Fig. 5-21 is the *carrier recovery*, which will be explained in Chapter 15. Until Chapter 15 we will consistently assume that the precise carrier frequency and phase are available at the receiver (except for a brief discussion of incoherent detection in Chapter 6), but in practice this is not true. After the phase splitter in Fig. 5-21, a preliminary demodulation is done using a carrier with frequency f_1. This carrier frequency is not expected to match the transmitter carrier frequency precisely, so phase errors result from the demodulation. These phase errors are corrected by further demodulation, shown as a complex multiplication following the fractionally-spaced precursor equalizer. The reason for this two-step demodulation is that the carrier recovery is decision-directed, like the timing recovery. A loop is formed that includes the slicer, the carrier recovery, and a complex multiplier, as shown in Fig. 5-21. It will become clear in Chapter 15 that the performance of this structure is considerably improved if there is no additional filtering inside the loop (the postcursor equalizer is harmless in this configuration). Consequently the final demodulation should be done as close to the slicer as possible. The preliminary demodulation, however, is required in order to bring the signal down close to baseband so that the receiver does not have to operate on the high frequency signal. Sometimes this first demodulation can be performed simply by sampling the signal below the Nyquist rate, without using the complex multiplier shown in Fig. 5-21.

Some possible variations on the receiver shown in Fig. 5-21 include the use of error correcting codes (Chapter 12) or trellis codes (Chapter 13), the use of a Viterbi detector instead of the slicer and equalizers, or the omission of the postcursor equalizer (Chapters 8 and 9). It is also practical to design passband signals that are not PAM signals, for example FSK (below) or *continuous-phase modulation*, in which case the receivers are significantly different. Baseband receivers can also be more elaborate than those discussed here, using for example adaptive equalization (Chapters 8 and 9).

5.4. Minimum-Distance Sequence Detection

The previous section considered receiver design for an isolated pulse. In this section we extend the minimum-distance concept to the ISI case, when the transmitter sends a *sequence* of symbols $\{a_0, a_1 \dots a_{L-1}\}$ of length L. If each symbol is drawn from an alphabet of size $M = |\mathcal{A}|$, there are M^L possible symbol sequences in all.The received signal is then

$$r(t) = \sum_{k=0}^{L-1} a_k h(t - kT) + n(t) , \tag{5.61}$$

where $h(t)$ is the received pulse shape, and we make no assumptions about $h(t)$ being time-limited or satisfying the Nyquist criterion. However, we do assume that $h(t)$ has finite energy E_h and that $n(t)$ is unknown noise with finite energy.

5.4.1. The Minimum-Distance Sequence Detector and the Folded Spectrum

A straightforward extension of the minimum-distance philosophy of Section 5.3 leads to the following. The minimum-distance *sequence detector* for the PAM model of (5.61) chooses the symbol sequence $\{a_0, \dots, a_{L-1}\}$ that best represents the observation waveform in a minimum-distance sense, namely:

$$\{\hat{a}_k\} = \arg \min_{\{a_k\} \in \mathcal{A}^L} \int_{-\infty}^{\infty} |r(t) - \sum_{k=0}^{L-1} a_k h(t - kT)|^2 \, dt . \tag{5.62}$$

In other words, of the M^L *candidate* signals of the form $\sum_{k=0}^{L-1} a_k h(t - kT)$, the receiver chooses that which is closest to the actual observation $r(t)$.

The sequence detector described above is a significant departure from the one-shot receiver. Specifically, the sequence detector makes a decision about *the entire sequence* all at once, rather than making individual symbol decisions one by one. Conceptually, the sequence detector must wait until all symbols have been transmitted and received before making any decisions. Among other things, this implies that there is thus an inherent decoding delay that grows with the block length L.

The above sequence detector is conceptually very simple: calculate M^L integrals, and find the smallest. Implementation, on the other hand, is a challenge, especially when L is large. The objective of this section is to make implementation manageable.

The cost function of (5.62) can be written as:

$$J = \int_{-\infty}^{\infty} | r(t) - \sum_{k=0}^{L-1} a_k h(t - kT) |^2 \, dt$$

$$= E_r - 2\mathrm{Re}\left\{ \sum_{k=0}^{L-1} a_k^* \int_{-\infty}^{\infty} r(t) h^*(t - kT) \, dt \right\}$$

$$+ \sum_{k=0}^{L-1} \sum_{j=0}^{L-1} a_k a_j^* \int_{-\infty}^{\infty} h(t - kT) h^*(t - jT) dt$$

$$= E_r - 2\mathrm{Re}\left\{ \sum_{k=0}^{L-1} a_k^* y_k \right\} + \sum_{k=0}^{L-1} \sum_{j=0}^{L-1} a_k a_j^* \rho_h(j - k) , \tag{5.63}$$

where we have used E_r from (5.44) as the energy in $r(t)$, and where we have made the following two definitions:

$$y_k = \int_{-\infty}^{\infty} r(t) h^*(t - kT) \, dt , \tag{5.64}$$

$$\rho_h(k) = \int_{-\infty}^{\infty} h(t) h^*(t - kT) \, dt . \tag{5.65}$$

We recognize y_k as the correlation of $r(t)$ with a delayed version of the received pulse, namely $y_k = \langle r(t), h(t - kT) \rangle$. Therefore, the k-th correlation y_k can be calculated by sampling the output of a matched filter at time kT, as shown below:

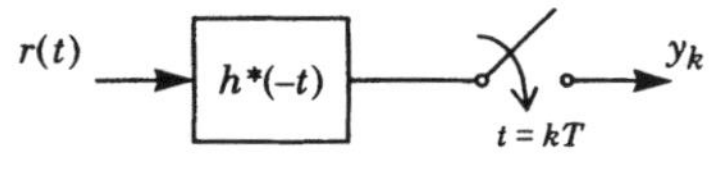

At this point we can draw an important conclusion. The first term (E_r) in the cost function of (5.63) is independent of the candidate sequence $\{a_k\}$, and hence the minimum-distance receiver can ignore it and optimize only the remaining terms of (5.63). Furthermore, these remaining terms depend on the received waveform $r(t)$ only through y_k. Therefore, the sequence of correlations $\{y_k\}$ represents a *sufficient statistic* for solving the minimum-distance sequence detection problem. Everything important about the received waveform is captured by this sequence of correlations. Since the matched filter produces these correlations, we conclude that the sampled matched filter provides sufficient statistics. This generalizes the result of Section 5.3.1, which showed that the matched filter was optimal for the special case of an isolated pulse. Here, we see that it is also optimal for the general case of a sequence of pulses.

The matched filter attenuates those frequencies that the channel attenuates, and it amplifies those frequencies that the channel amplifies. In this sense, the matched filter *accentuates* the channel distortion.

Example 5-23.

Suppose the Fourier transform $H(f)$ of the received pulse is as sketched below:

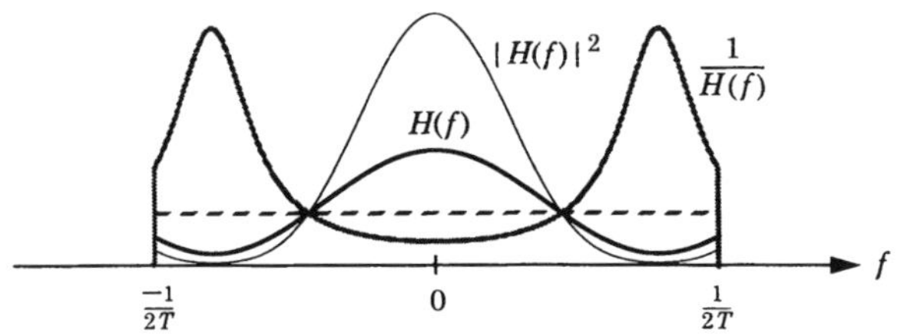

Also shown above is $|H(f)|^2$, which is the Fourier transform of the overall pulse shape after a matched filter. We see that $|H(f)|^2$ fluctuates over a range wider than $H(f)$. In a sense, the matched filter has made the ISI more severe. This is the price paid for optimally accounting for the noise. Also shown above is a sketch of the channel inverse $1/H(f)$, which is the transfer function of a zero-forcing linear *equalizer*; it forces the overall pulse to be the ideal rectangular Nyquist pulse indicated with a dotted line. Comparing the two, we see that the matched filter is very different from a linear equalizer.

The previous example resolves an apparent conflict in the design of the receive filter. Intuitively we would like the receive filter to do two things: reject as much noise as possible, and shape the pulse so as to satisfy the Nyquist criterion and thus avoid ISI. But we cannot do both simultaneously. The minimum-distance criterion chooses the receive filter to minimize noise, not avoid ISI. The ISI is then compensated for afterwards, not with linear filtering, but through nonlinear signal processing that minimizes the cost function J of (5.63).

The quantity $\rho_h(k)$ of (5.65) is called the *sampled autocorrelation* of the received pulse. We recognize it as the correlation between a delayed and non-delayed version of the received pulse, namely $\rho_h(k) = \langle h(t), h(t-kT) \rangle$. It plays a key role in receiver design, and so we will take a moment to explore some of its properties. First, at time zero, $\rho_h(k)$ reduces to $\rho_h(0) = E_h$, the energy in the received pulse. Second, it displays Hermitian symmetry, so that $\rho_h(-k) = \rho_h^*(k)$. Third, the Schwarz inequality of (5.54) implies that the magnitude of the sampled autocorrelation achieves a maximum value at time $k = 0$, so that $|\rho_h(k)| < \rho_h(0)$ for all nonzero k. Fourth, it is instructive to view the sampled autocorrelation as the output of a sampled matched filter when the received pulse is the input, as shown below:

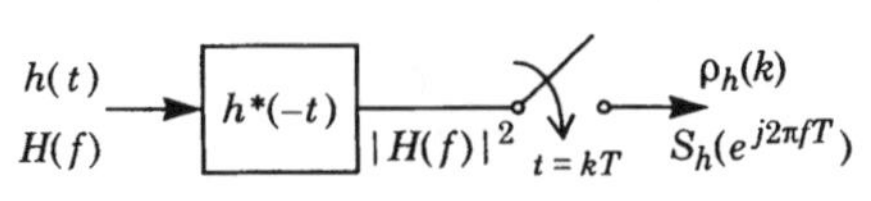

The Fourier transform of $\rho_h(k)$, denoted $S_h(e^{j2\pi fT})$, is easy to derive using the above figure. In particular, because the filter input has Fourier transform $H(f)$, and because the matched filter has Fourier transform $H^*(f)$, the filter output has Fourier transform $|H(f)|^2$. According to (2.17), sampling the filter output aliases the Fourier transform, so that the discrete-time Fourier transform of the sampled autocorrelation $\rho_h(k)$ is given by:

$$S_h(e^{j2\pi fT}) = \frac{1}{T}\sum_{m=-\infty}^{\infty} |H(f - m/T)|^2 . \tag{5.66}$$

Because this represents an aliased version of the matched filter output, $S_h(e^{j2\pi fT})$ is called the *folded spectrum*. We see that it is everywhere real and nonnegative, and thus it would be a valid power spectrum for a random process.

The overall pulse after a matched filter satisfies the Nyquist criterion only if its Fourier transform — given by $|H(f)|^2$ — aliases to a constant. In light of (5.66), this will be true only if the folded spectrum is a constant. Specifically, only if $S_h(e^{j2\pi fT}) = E_h$, so that the sampled autocorrelation reduces to $\rho_h(k) = E_h\delta_k$.

The Nyquist criterion stipulates that the bandwidth of a PAM signal be at least half the symbol rate. Thus, the sampling theorem would dictate a sampling rate *greater* than the symbol rate. On the other hand, the minimum-distance receiver requires only baud-rate sampling. Therefore, it introduces aliasing. We will see in Chapter 9 that it is common to choose a sampling rate higher than the symbol rate in practice, to address practical concerns.

5.4.2. Minimizing Distance in Discrete-Time

A basic result in Section 2.5, (2.67), provides a factorization of a rational transfer function that is non-negative real on the unit circle. This spectral factorization will now prove useful in deriving a basic minimum-distance receiver structure for ISI.

A direct calculation of the minimum-distance cost J of (5.63) appears to be impractical from two perspectives:

- Calculating the cost J for just one candidate symbol sequence $\{a_0, \ldots, a_{L-1}\}$ requires on the order of L^2 additions and multiplications. In practical implementations, we can only provide a constant processing rate, or a total computational resource that is proportional to L. Thus, the calculation of (5.63) is impractical for large L.
- Even worse, the total number of symbol sequences $\{a_0, \ldots, a_{L-1}\}$ for which (5.63) must be calculated is M^L, which grows exponentially with L. Again, this is impractical to implement.

In this section we will prove that the *continuous-time* minimum-distance receiver can equivalently be implemented using a *discrete-time* minimum-distance receiver. This receiver structure solves the first problem, since the computational overhead for one symbol sequence $\{a_0, \ldots, a_{L-1}\}$ is proportional to L rather than L^2. A solution to the second problem (the Viterbi algorithm) will be described in Section 5.4.4.

The fundamental enabling result is a spectral factorization of the folded spectrum. Since the folded spectrum $S_h(e^{j\theta})$ is non-negative real-valued, it can be factored in the form (see (2.67)),

$$S_h(z) = \gamma^2 M(z) M^*(1/z^*) , \tag{5.67}$$

where $M(z)$ is a monic loosely minimum-phase transfer function. Expressed in the time domain, this spectral factorization is

$$\rho_h(k) = \gamma^2 m_k * m^*_{-k} . \tag{5.68}$$

Consider Fig. 5-22, which appends to the output of the sampled matched filter a "precursor equalizer" whose transfer function is:

$$\frac{1}{\gamma^2 M^*(1/z^*)} . \tag{5.69}$$

(Generally this will be a stable filter, but there are instances where filters with isolated poles on the unit circle are allowable, so we do not rule out this case.) Let $\{z_k\}$ denote the output of this filter, which is related to the sampled matched filter output $\{y_k\}$ by

$$y_k = \gamma^2 z_k * m^*_{-k} = \gamma^2 \sum_{l=k}^{\infty} z_l m^*_{l-k} . \tag{5.70}$$

It turns out that the continuous-time minimum distance criterion of (5.63) is a simple function of z_k. This result can be demonstrated by completing the square. In particular, using the inner product notation $\langle x_k, y_k \rangle$ of (2.81) as shorthand for $\Sigma_k x_k y_k^*$, and $\langle x_k, x_k \rangle = \| x_k \|^2$, (5.63) can be written as:

$$\begin{aligned} J &= E_r - 2\mathrm{Re}\{\langle y_k, a_k \rangle\} + \sum_{k=0}^{L-1} \sum_{j=0}^{L-1} a_k a_j^* \rho_h(j-k) \\ &= E_r - 2\gamma^2 \mathrm{Re}\{\langle z_k, a_k * m_k \rangle\} + \gamma^2 \| a_k * m_k \|^2 \\ &= E_r + \gamma^2 \| z_k - a_k * m_k \|^2 - \gamma^2 \| z_k \|^2 , \end{aligned} \tag{5.71}$$

where we made use of (5.68) and (5.70). Observe the similarity of this expression with that of (5.48) for the isolated pulse case.

Since only the second term above depends on a_k, the continuous-time minimum distance criterion of (5.63) reduces to the following equivalent *discrete-time* minimum-distance criterion:

$$\{ \hat{a}_k \} = \arg \min_{\{a_k\} \in \mathcal{A}^L} \sum_{k=0}^{\infty} | z_k - \sum_{l=0}^{L-1} a_l m_{k-l} |^2 . \tag{5.72}$$

We can think of this receiver structure as a generalized slicer; the isolated data symbol a_k is replaced by a filtered sequence of data symbols $a_k * m_k$. Instead of comparing each data symbol independently to a single sample, we compare a filtered sequence of data symbols to a

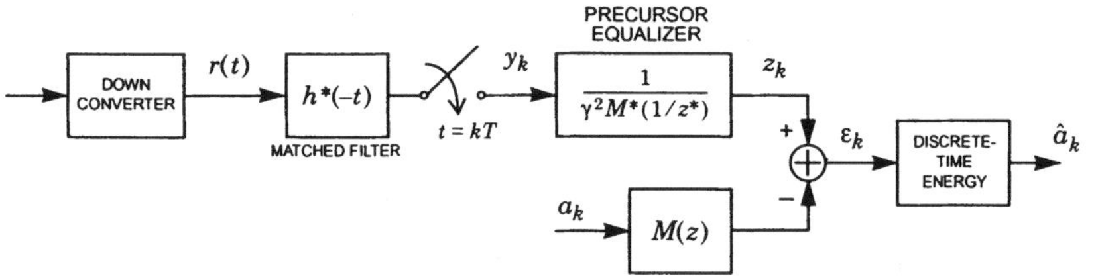

Fig. 5-22. A minimum-distance sequence detector for PAM with ISI, in which the continuous-time minimum distance criterion is transformed into a discrete-time minimum distance criterion.

sequence of discrete-time samples z_k, repeating this comparison for all allowed data-symbol sequences.

Observe that the range of the summation index in (5.72) is $0 \le k \le \infty$, even though the sequence of data symbols is finite. This is because the ISI may be infinite in extent, and thus the entire received signal sequence is considered by the minimum-distance criterion. If $M(z)$ were FIR with memory μ, the index range would change to $0 \le k \le L + \mu - 1$.

For one symbol sequence $\{a_0, \ldots, a_{L-1}\}$, and when $M(z)$ is FIR, (5.72) requires a computational load proportional to L. Thus, this form of the minimum-distance criterion reduces the computation for each candidate symbol sequence from the order of L^2 to L. If $M(z)$ is not FIR, then it can usually be accurately approximated by an FIR response. In addition to the practical implications of (5.72), it is also of great theoretical interest, because it demonstrates the equivalence of two minimum-distance criteria, (5.63) in continuous time and (5.72) in discrete time.

The receiver structure corresponding to criterion (5.72) is shown in Fig. 5-22. It measures the energy in the error between two discrete-time signals. One signal, $\{z_k\}$, is a filtered version of the sampled matched filter output. The symbol-rate discrete-time filter in this upper path has transfer function given by (5.69), and it is called a *precursor equalizer*. The reason for the terminology "equalizer" is that, as we will see, this filter eliminates a portion (although not all) of the ISI at its input. That is, it inverts (but only partially inverts) the equivalent response of the channel and the sampled matched filter. It is called a "precursor equalizer" because it eliminates the "anticausal" or "precursor" response of the channel and sampled matched filter. This terminology will be explained further in Chapter 8.

The second signal used in the discrete-time energy calculation is a filtered version of the candidate data-symbol sequence. The filter in this path is an equivalent discrete-time model for the response of the transmit filter, channel, matched filter, and precursor equalizer to the input data symbols, as we will see. The distance between precursor equalizer output and the filtered version of the candidate data-symbol sequences is calculated for all possible sequences of L data symbols. The sequence that minimizes that discrete-time distance is chosen. The distance must be recalculated many times, once for each allowable sequence of L data symbols. In the presence of ISI ($M(z) \neq 1$) it *never* reduces to a series of symbol-by-symbol decisions. The distance should be calculated only for feasible sequences of data symbols, reflecting any redundancy built into the coder at the transmitter.

Since $M(z)$ is minimum-phase, $M^*(1/z^*)$ is maximum-phase, and so is $1/M^*(1/z^*)$. Thus, to be stable this filter must be anticausal, which is impractical, although if it is FIR or can be approximated as FIR, it can be implemented as a causal filter in combination with a delay.

Example 5-24.
The lowest-order rational all-pole folded spectrum is

$$S_h(z) = \frac{\gamma^2}{(1 - cz^{-1})(1 - c^*z)} , \qquad |c| < 1 . \tag{5.73}$$

For this case, $M(z) = 1/(1 - cz^{-1})$, and the resulting receiver structure is shown in Fig. 5-23. The precursor equalizer $(1 - c^*z) / \gamma^2$ is an FIR filter. Although it is anticausal, it is very easily implemented as a causal FIR filter plus a single-sample delay. The candidate data symbols are filtered by a single-pole IIR filter with impulse response $c^k u_k$ (where u_k is the unit-step function).

Example 5-25.
The lowest-order rational all-zero folded spectrum is

$$S_h(z) = \gamma^2(1 - cz^{-1})(1 - c^*z) , \qquad |c| < 1 . \tag{5.74}$$

For this case, $M(z) = 1 - cz^{-1}$, and the resulting receiver structure is shown in Fig. 5-24. The precursor equalizer $1/(1 - c^*z)$ is an anticausal IIR filter with impulse response $(c^*)^{-k}u_{-k}$, and because it is IIR it cannot be implemented directly. Rather, it can only be approximated by an anticausal FIR filter. The candidate data sequence is filtered by the impulse response $\delta_k - c\delta_{k-1}$.

We will now further motivate the reasons for the equivalence of (5.72) and (5.63). When the input signal is given by (5.61), the output of the sampled matched filter is

$$y_k = \sum_{m=0}^{L-1} a_m \rho_h(k - m) + n_k = a_k * \rho_h(k) + n_k , \tag{5.75}$$

for some unknown noise n_k. The signal portion of this output is represented by an equivalent discrete-time filter with impulse response $\rho_h(k)$ (transfer function $S_h(z)$) and input a_k.

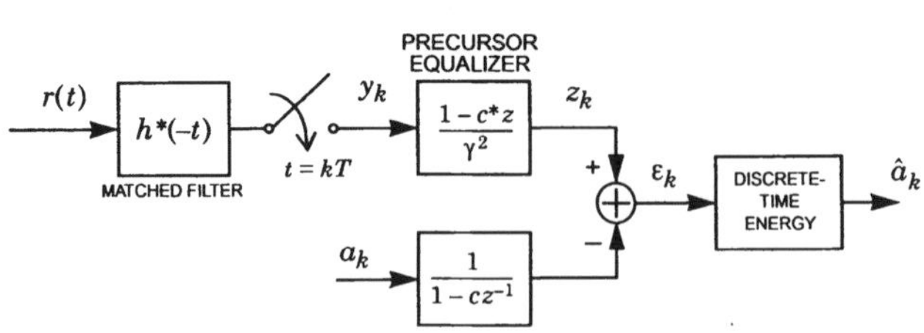

Fig. 5-23. The minimum-distance receiver for a first-order all-pole folded spectrum.

Consider what happens if we put the sampled matched filter output y_k through a precursor equalizer with transfer function $1/(\gamma^2 M^*(1/z^*))$ as shown in Fig. 5-22. The overall transfer function to data symbols is

$$\frac{S_h(z)}{\gamma^2 M^*(1/z^*)} = M(z) , \tag{5.76}$$

and hence the output will be

$$z_k = \sum_{l=0}^{L-1} a_l m_{k-l} + n'_k , \tag{5.77}$$

where n'_k is the filtered noise. This output signal has the desirable property — heavily exploited in Chapters 8 and 9 — that the resulting overall discrete-time channel is causal. The precursor equalizer has thus removed the anticausal portion of the ISI; that is, the equivalent response up to the precursor equalizer input has a two-sided impulse response, and the precursor equalizer output turns this into a causal impulse response. In Fig. 5-22, the actual sequence z_k, corrupted by noise, is compared to what it would be for a candidate sequence of data symbols, namely $z_k * m_k$, using a Euclidean distance measure.

Special Case: Orthogonal Pulses

It is instructive to examine how the minimum-distance sequence detector simplifies for the special case of no ISI. Suppose successive received pulses are orthogonal, or equivalently the pulse shape at the output of the matched filter satisfies the Nyquist criterion. This implies that the folded spectrum is a constant, namely $S_h(e^{j2\pi fT}) = E_h$, which admits a trivial spectral factorization (5.67) with $\gamma^2 = E_h$ and $M(z) = 1$. The precursor filter of (5.69) reduces to a constant gain of E_h^{-1}. In this case, the minimum-distance sequence detector of (5.72) reduces to:

$$\{\hat{a}_k\} = \arg \min_{\{a_k\} \in \mathcal{A}^L} \sum_{k=0}^{L-1} | z_k - a_k |^2 . \tag{5.78}$$

This makes intuitive sense, since the output of the sampled matched filter is $y_k = E_h a_k + n_k$, where n_k is unknown noise. That is, the signal component at the sampled matched filter output is free of ISI. This receiver structure is illustrated in Fig. 5-25(a). The difference between the sampled matched filter output and the normalized sequence of data symbols is formed, and the

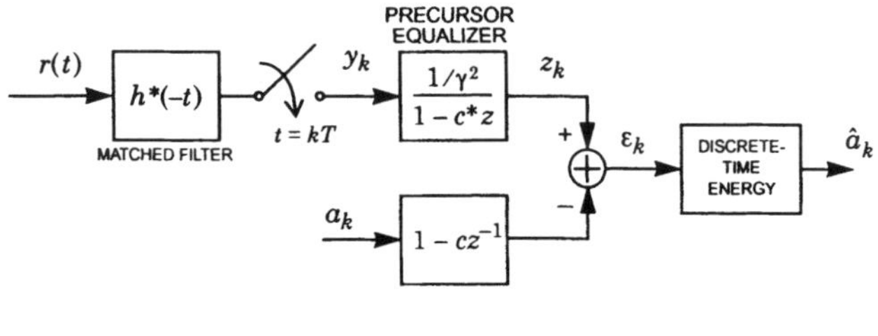

Fig. 5-24. The minimum-distance receiver for a first-order all-zero folded spectrum.

energy in discrete-time is calculated. This is repeated for all possible sequences of data symbols, and the one that minimizes the error energy is chosen.

When the data symbols are chosen independently of one another, (5.78) can be minimized symbol-by-symbol: since all the terms in the sum are non-negative, the sum is minimized by minimizing each term, where each term is precisely the criterion for a minimum-distance slicer design. Thus, in this case, the simplified structure of Fig. 5-25(b) can be used. This is exactly the same structure of Fig. 5-19 that was derived for the special case of an isolated pulse, except that here the matched filter output is sampled every symbol period instead of only once at time zero. Thus, with independent symbols and no ISI, the minimum-distance sequence detector reduces to a sequence of independent one-shot minimum-distance detectors.

The following example illustrates when the more general structure of Fig. 5-25(a) must be used.

Example 5-26.

A simple technique for ensuring no d.c. content in a baseband PAM waveform is a coding technique called *alternate-mark inversion (AMI)*. Briefly, the data symbol a_k is chosen from an alphabet of size three, $\{\pm 1, 0\}$ in order to communicate one bit of information. A zero bit is transmitted as $a_k = 0$, and a one bit is transmitted by $| a_k | = 1$, where the sign of a_k is chosen to be the opposite sign from the last non-zero a_k. Since the non-zero a_k alternate in sign, there is no d.c. content to the sequence of data symbols. If we were to use the simplified structure of Fig. 5-25(b), the three-level slicer could detect sequences of data symbols that violate the known constraints on the data symbols. For example, it might detect two positive or two negative data symbols in a row. The correct procedure is to use the generalized structure of Fig. 5-25(a), and calculate the energy for valid sequences of data symbols only. In particular, for sequences of L user bits at the input to the transmitter, there are 2^L valid sequences of data symbols, and not 3^L as might be suggested by the alphabet of size three.

This example illustrates a case where the sequence of data symbols embodies *redundancy*; that is, there are restrictions on the sequence of data symbols imposed by the coder that make the

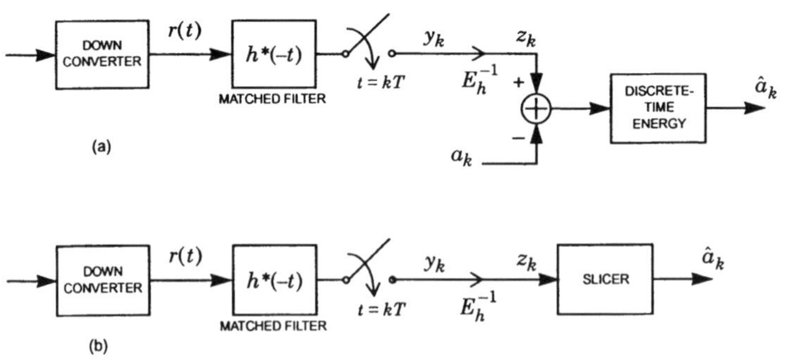

Fig. 5-25. The minimum-distance receiver design for PAM when the overall pulse at the output of the MF satisfies the Nyquist criterion. (a) The receiver that applies to any set of data-symbol sequences, and (b) the special case where data symbols are chosen independently.

number of sequences less than M^L for an alphabet of size M. There are many other examples of this, for example for the purpose of controlling the transmitted signal power spectrum and for combating noise on the channel (Chapter 12). In each of these cases, the redundancy can be accounted for in the minimum-distance receiver of Fig. 5-25(a) by considering only valid sequences of data symbols. This has the desirable side effect of restricting the number of sequences for which the energy must be recalculated.

Non-Uniqueness of the Discrete-Time Criterion

We will now show that the minimum-distance receiver structure that substitutes a discrete-time energy for a continuous-time energy is not unique. The discrete-time channel model in (5.77) (ignoring for a moment the noise) is minimum-phase, because of the minimum-phase spectral factorization. However, the spectral factorization of $S_h(z)$ need not be minimum-phase. There are many front-end discrete-time filters that are equivalent in the sense that they result in the same detected sequence of data symbols $\{a_0, \ldots, a_{L-1}\}$.

This is illustrated in Fig. 5-26, where the error sequence $\{\varepsilon_k\}$ defined in Fig. 5-22 is filtered by an arbitrary rational allpass filter before the energy is calculated. Since the allpass filter has a unit magnitude response, it does not change the energy, i.e.

$$\sum_{m=0}^{\infty} |\varepsilon_m|^2 = \sum_{m=0}^{\infty} |\varepsilon_m'|^2 , \tag{5.79}$$

as long as we assume that $\varepsilon_m = 0$ for $m < 0$. Hence, the allpass filter will not change the sequence of data symbols chosen by the receiver. Now, we can move the allpass filter through the summation in Fig. 5-22, replacing the filter $1/\gamma^2 M^*(1/z^*)$ by $A(z)/\gamma^2 M^*(1/z^*)$ and replacing $M(z)$ by $A(z)M(z)$. This replacement has no effect on the data-symbol sequence chosen by the receiver. If this allpass filter has all poles inside the unit circle (and hence all zeros outside), then it is a stable causal filter, and does not destroy the causality of the channel model in (5.77) either. Effectively, this changes the discrete-time channel model from minimum-phase to non-minimum-phase. While this change would not appear to be harmful, it is shown in Problem 2-28 that a causal minimum-phase sequence has the property that, among all sequences with the same Fourier transform magnitude, it is *maximally concentrated* near zero delay. Thus, in this sense the impulse response of the minimum-phase channel model has minimum intersymbol interference, among all impulse responses with the same Fourier transform magnitude. Stating this another way, Problem 2-29 shows that the impulse response of the filter $M(z)$ is more concentrated near the origin than the impulse response of the modified channel model $A(z)M(z)$, and thus has less ISI. While this property does not affect the minimum-distance receiver, in the sense that it chooses the same data-symbol sequence, it

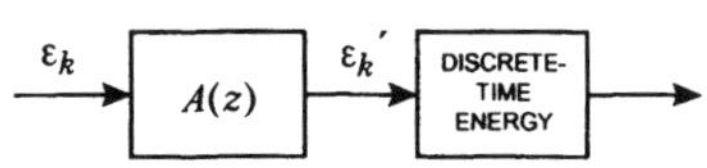

Fig. 5-26. Introducing a rational allpass filter $A(z)$ before the energy is calculated in Fig. 5-22.

is very important for another practically important receiver structure based on a similar decomposition (the decision-feedback precursor equalizer of Chapter 8).

5.4.3. The Whitened-Matched Filter

The main result of the previous section was the discrete-time minimum-distance criterion of (5.72). In this section we provide a simple geometric interpretation based on signal space. In the process, we introduce an important concept known as the whitened-matched filter.

Define the *signal space* $\mathcal{S}$ as the span of all translates $\{h(t-kT)\}$ of the received pulse. Also, define a new signal $\phi(t)$ according to its Fourier transform, when it exists:

$$\Phi(f) = \frac{H(f)}{\gamma M(e^{j2\pi fT})}, \tag{5.80}$$

where γ and $M(z)$ are defined by the folded spectrum factorization $S_h(z) = \gamma^2 M(z)M^*(1/z^*)$.

Fact:

The set of translates $\{\phi(t-kT)\}$ forms an *orthonormal basis* for the signal space $\mathcal{S}$.

Proof:

To demonstrate orthonormality, the following integral must reduce to $\delta_{m,n}$:

$$\int_{-\infty}^{\infty} \phi(t-mT)\phi^*(t-nT)\,dt = \int_{-\infty}^{\infty} \phi(t)\phi^*(t-kT)\,dt\,, \tag{5.81}$$

where we substituted $k = n - m$. Taking the discrete-time Fourier transform of (5.81) yields:

$$\frac{1}{T}\sum_{k=-\infty}^{\infty} |\Phi(f-k/T)|^2 = \frac{1}{T}\sum_{k=-\infty}^{\infty}\left|\frac{H(f-k/T)}{\gamma M(e^{j2\pi(f-k/T)T})}\right|^2$$

$$= \frac{1}{\gamma^2|M(e^{j2\pi fT})|^2}\,\frac{1}{T}\sum_{k=-\infty}^{\infty}|H(f-k/T)|^2 = 1\,. \tag{5.82}$$

Reverting back to the time domain, we see that (5.81) reduces to $\delta_{m,n}$, so that the set $\{\phi(t-kT)\}$ is orthonormal.

To make $\{\phi(t-kT)\}$ a basis we must also show that it *spans*, so that every element $s(t)$ of $\mathcal{S}$ can be written as a linear combination of $\{\phi(t-kT)\}$. If $s(t) \in \mathcal{S}$, there exists a sequence $\{a_k\}$ such that $s(t) = \sum_k a_k h(t-kT)$. Since (5.80) implies that $H(f) = \gamma M(e^{j2\pi fT})\Phi(f)$, it follows that $s(t)$ is the output of the following cascade of filters:

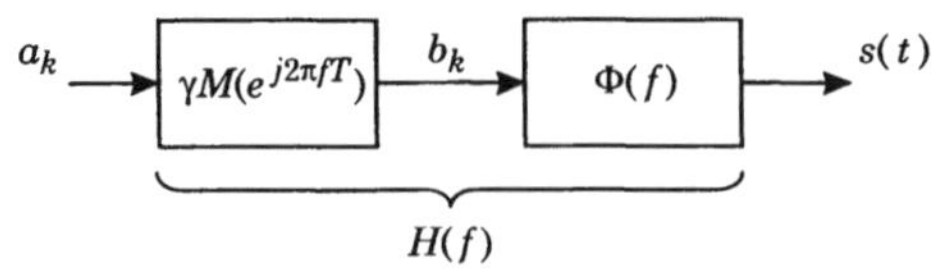

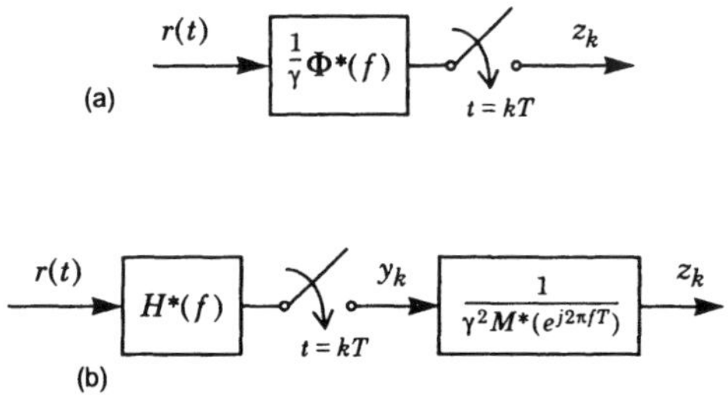

Fig. 5-27. Two equivalent implementations of the whitened-matched filter: (a) with the precursor equalizer embedded inside the receive filter; (b) with a separate discrete-time precursor equalizer after the sampler. We can convert (b) to (a) by moving the precursor equalizer to the other side of the sampler, and then absorbing it into the definition of the receive filter. Either way, the above structures project their input onto the signal space spanned by $\{h(t-kT)\}$.

From this picture it is clear that $s(t) = \sum_k b_k \phi(t-kT)$ where $b_k = \gamma\, a_k * m_k$. This proves that every element $s(t)$ of the signal space can be expressed as a linear combination of $\{\phi(t-kT)\}$, which concludes our proof that $\{\phi(t-kT)\}$ is an orthonormal basis. Indirectly, we have also shown that the projection of $s(t) = \sum_k a_k h(t-kT)$ onto $\phi(t-kT)$ is $b_k = \gamma\, a_k * m_k$.

Having identified a basis for $\mathcal{S}$, let us revisit the minimum-distance sequence detector of (5.62). Let $s(t) = \sum_k a_k h(t-kT)$ denote the candidate PAM signal, and let $\hat{r}(t)$ denote the projection of the received waveform $r(t)$ onto $\mathcal{S}$. Since $\hat{r}(t) \in \mathcal{S}$, we must be able to expand it in terms of the basis; that is, there must exist a sequence of coefficients $\{z_k\}$ such that:

$$\hat{r}(t) = \gamma \sum_k z_k \phi(t-kT)\,. \tag{5.83}$$

(The motivation for the constant factor γ will become clear shortly.) In particular, the k-th coefficient z_k can be found by correlating $r(t)$ with the k-th basis function:

$$z_k = \frac{1}{\gamma}\langle\, r(t),\, \phi(t-kT)\,\rangle\,. \tag{5.84}$$

This correlation can be implemented by sampling at time kT the output of a receive filter with transfer function $\Phi^*(f)/\gamma$ and impulse response $\phi^*(-t)/\gamma$, as illustrated in Fig. 5-27(a). The receive filter is thus matched to $\phi(t)/\gamma$. For reasons that will become clear shortly, the receive filter of Fig. 5-27(a) is called a *whitened-matched filter*.

Because of (5.80), we can equivalently implement the whitened-matched filter of Fig. 5-27(a) using the system of Fig. 5-27(b), which consists of a conventional sampled matched filter followed by a discrete-time filter with frequency response $1/(\gamma^2 M^*(e^{j2\pi fT}))$. In effect, we have factored out the term $1/(\gamma^2 M^*(e^{j2\pi fT}))$ from the receive filter and then moved it to the other side of the sampler. Because of its equivalence to Fig. 5-27(a), the system of Fig. 5-27(b) — including the sampler — is also known (loosely) as the whitened-matched filter. We immediately recognize the structure of Fig. 5-27(b) as the front-end of the minimum-distance receiver of Fig. 5-22, where it was motivated in an ad hoc fashion. We now have a useful *geometric* interpretation of the whitened-matched filter structure of Fig. 5-27(b) (and thus of Fig. 5-22): it *projects* the received waveform onto the signal space.

The minimum-distance sequence detector can equivalently minimize distance after the whitened-matched filter, as was argued in the discussion leading up to (5.72). Based on our geometric interpretation of the whitened-matched filter, we now present an alternative derivation of (5.72), starting with the continuous-time cost function of (5.62):

$$J = \int_{-\infty}^{\infty} |r(t) - s(t)|^2 dt$$

$$= \int_{-\infty}^{\infty} |r(t) - \hat{r}(t) + \hat{r}(t) - s(t)|^2 dt$$

$$= \int_{-\infty}^{\infty} |r(t) - \hat{r}(t)|^2 dt + \int_{-\infty}^{\infty} |\hat{r}(t) - s(t)|^2 dt \tag{5.85}$$

$$= E_e + \sum_{k=-\infty}^{\infty} |\gamma z_k - \gamma a_k * m_k|^2 \tag{5.86}$$

$$= E_e + \gamma^2 \sum_{k=-\infty}^{\infty} |z_k - a_k * m_k|^2 , \tag{5.87}$$

where E_e is the energy in the projection error $r(t) - \hat{r}(t)$. The equality in (5.85) follows from the fact that the projection error $r(t) - \hat{r}(t)$ is orthogonal to all of $\mathcal{S}$, including $\hat{r}(t) - s(t)$. The equality in (5.86) is based on Parseval's relationship of (2.99). Since E_e is independent of the candidate symbol sequence $\{a_k\}$, the minimum-distance receiver can equivalently minimize the second term in (5.87). In other words, the minimum-distance sequence detector reduces to (5.72).

The whitened-matched filter derives its name from two properties: like the conventional matched filter, it provides sufficient statistics for the minimum-distance receiver, and yet unlike the matched filter, it transforms white noise into white noise. Specifically, if the continuous-time noise $n(t)$ at the input to the whitened-matched filter is white with power spectrum N_0, the discrete-time noise after the whitened-matched filter will also be white, with power spectrum N_0/γ^2. In contrast, the noise after a matched filter will have power spectrum proportional to the folded spectrum, namely $N_0 S_h(z)$, and thus will not generally be white.

5.4.4. The Viterbi Algorithm

The minimum-distance sequence detection requires that (5.72) be calculated M^L times, one for each possible sequence of L symbols from an M-ary alphabet. The computation is therefore exponential in time (because time is proportional to L). In this section we describe a dynamic programming algorithm known as the *Viterbi algorithm*, originally proposed by A. Viterbi in 1967 [14], which achieves a computational load that is linear in time, which corresponds to a fixed computational *rate*. The Viterbi algorithm is used in many applications beyond sequence detection, such as in the decoding of convolutional codes, but to simplify its presentation we restrict our attention to the ISI channel.

Finite-State Machine Signal Generator

The relationship between the output of the whitened-matched filter $\{z_k\}$ and the transmitted symbols $\{a_k\}$ is captured by the whitened-matched filter model of Fig. 5-28(a); the transmitted symbols are filtered by $M(z)$, and noise is added. An expanded view of this model is shown in Fig. 5-28(b) for the special case when $M(z)$ is FIR with memory μ, namely $M(z) = 1 + m_1 z^{-1} + \ldots + m_\mu z^{-\mu}$. In theory, $M(z)$ is FIR if and only if $S_h(z)$ is an all-zero filter, or equivalently if and only if a finite set of translates $h(t - kT)$ are non-orthogonal. In practice, even an IIR filter can be well-approximated by an FIR filter with large memory; therefore, from now on we assume that μ is finite.

With our assumption of finite memory, the signal s_k in Fig. 5-28 (before the noise) is generated by a *finite-state machine (FSM)*. Let us define the *state* at time k by the vector of μ previous inputs:

$$\psi_k = [a_{k-1}, a_{k-2}, \ldots a_{k-\mu}] . \tag{5.88}$$

Since each element of this vector is drawn from an alphabet $\mathcal{A}$, we have $\psi_k \in \mathcal{A}^\mu$, so that the number of permissible states is $|\mathcal{A}|^\mu$. Beyond the fact that there are only a finite number of states, the FSM is characterized by two additional properties:

- its output s_k depends only on the current state ψ_k and current input a_k;
- the next state ψ_{k+1} depends only on the current state ψ_k and current input a_k.

We will find it convenient to characterize the FSM output as a function of a state *transition*. Specifically, since ψ_{k+1} uniquely determines the input a_k, we can represent the FSM output as a function of a state transition:

$$s_k = g(\psi_k, \psi_{k+1}) , \tag{5.89}$$

where $g(\cdot, \cdot)$ is a memoryless function, namely:

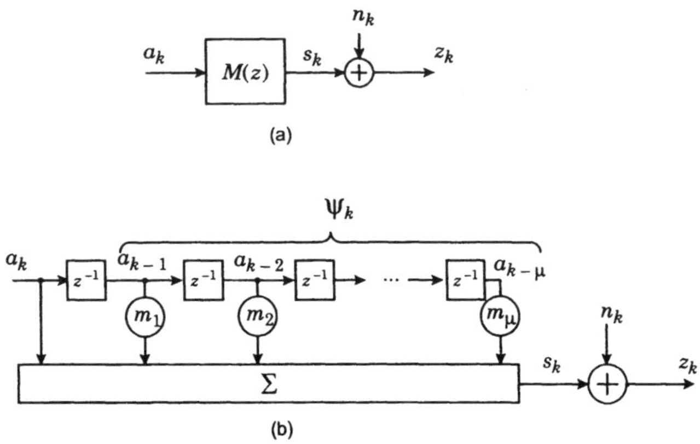

Fig. 5-28. The whitened-matched filter channel model for the case when $M(z)$ is FIR.

$$s_k = \sum_{l=0}^{\mu} m_l a_{k-l} . \tag{5.90}$$

Example 5-27.

A simple example that we will carry along for illustrative purposes is the binary OOK alphabet $\mathcal{A} = \{0, 1\}$ and the following ISI model

$$m_k = \delta_k + \tfrac{1}{2}\delta_{k-1} , \tag{5.91}$$

as shown in Fig. 5-29(a). The memory is $\mu = 1$, and the state is $\psi_k = a_{k-1}$, the previous input. Since the alphabet is binary, there are only two possible states. The state transition diagram for this FSM is shown in Fig. 5-29(b), where the arcs are labeled with the input/output pair (a_k, s_k).

One of the characteristics of the output s_k produced by the FSM signal generator is redundancy. In Example 5-27, the redundancy occurs because s_k takes on four possible levels $\{0, 0.5, 1, 1.5\}$, even though only one bit of information is carried. The ISI which introduces this redundancy is assumed to be an undesired property of the channel, although there are situations where it is introduced deliberately — for example with *partial response*. Similarly, the error-control coding techniques of Chapter 12 introduce redundancy to help mitigate the effects of noise. Although the form and function of an error-control coder is very different from the ISI channel, the principles of sequence detection developed in this section apply equally to both cases. In the remainder of this section we will discuss the technique of sequence detection, and leave detailed discussion of its applications to later chapters.

The Trellis Diagram

The state transition diagram in Example 5-27 is a traditional representation of a FSM. D. Forney suggested in 1967 a valuable alternative representation called a *trellis diagram* [15], which shows the possible progression of states over time.

To this point we have assumed that the transmitter sends only L symbols $\{a_0, \ldots, a_{L-1}\}$, which is equivalent to sending an infinite sequence of symbols $\{a_k: -\infty < k < \infty\}$ with the restrictions that $a_k = 0$ for $k < 0$ and $a_k = 0$ for $k \geq L$. This assumption leads to a minor complication when zero is not a part of the input alphabet, as is often the case. One approach would be to define an augmented alphabet as $\mathcal{A}' = \mathcal{A} \cup \{0\}$, but this introduces complications of its own, such as a time-varying state space: the first few state vectors $\{\psi_0, \psi_1, \ldots\}$ would be drawn from $\mathcal{A}'^{\mu}$, whereas later states would be drawn from the strict subset $\mathcal{A}^{\mu}$. We opt for a

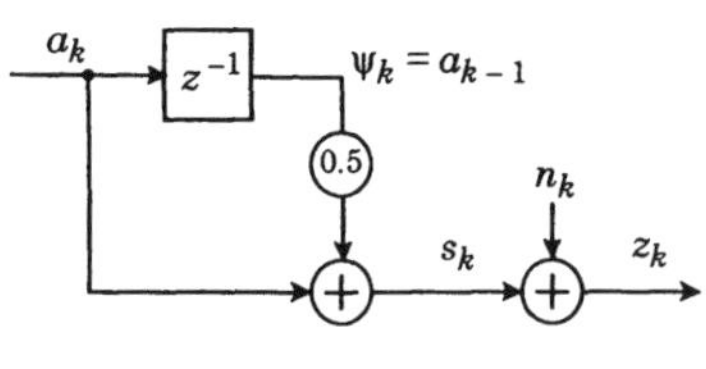

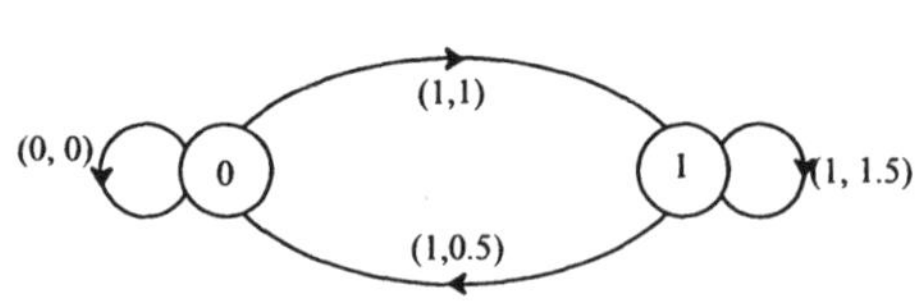

Fig. 5-29. The ISI signal generator (a) of Example 5-27. The state transition diagram (b) assuming binary input symbols. The arcs are labeled with the input bit/signal output pair (a_k, s_k).

simpler solution. Specifically, in what follows we assume that the transmitter sends an *idle symbol* $a^0 \in \mathcal{A}$ before the first data symbol and after the last data symbol, so that $a_k = a^0$ for $k < 0$ and $a_k = a^0$ for $k \geq L$. For example, given a 4-PAM alphabet $\{\pm1, \pm3\}$, the idle symbol might be $a^0 = -3$. Of course, $a^0 = 0$ is the natural choice when $0 \in \mathcal{A}$.

Let the $Q = |\mathcal{A}|^{\mu}$ permissible states be labeled by the integers $\{0, 1, 2, \ldots Q-1\}$. Furthermore, assume that the zero label is reserved for the *all-idle state* $[a^0, a^0, \ldots a^0]$. We will thus write $\psi_k = 0$ as shorthand for $\psi_k = [a^0, a^0, \ldots a^0]$, and similarly for other states. Because $a_k = a^0$ for $k < 0$, it is clear from (5.88) that the state will satisfy $\psi_k = 0$ for all $k \leq 0$. Similarly, because $a_k = a^0$ for $k \geq L$, (5.88) implies that $\psi_k = 0$ for all $k \geq L + \mu$.

The trellis diagram is a essentially a plot of all possible state progressions versus time. Because the state is known to be zero for all $k \leq 0$ and for all $k \geq L + \mu$, the time axis need only range from $k = 0$ to $L + \mu$.

Example 5-28.

Recall Example 5-27, for which the alphabet was binary and the memory was $\mu = 1$. The trellis diagram for this example is shown in Fig. 5-30(a). The starting and ending conditions are $\psi_0 = 0$ and $\psi_{L+1} = 0$. Each small circle is a *node* of the trellis, and corresponds to the FSM being in a particular state at a particular time. Each arc in the diagram is called a *branch*, and corresponds to a particular state transition at a particular time. Thus, the single node at the left indicates that the FSM begins in state $\psi_0 = 0$ at time $k = 0$. The next state can be either state $\psi_1 = 0$ or state $\psi_1 = 1$, depending on the value of a_0, so transitions to both are shown. At time $k = 1$, the FSM for this example may branch (transition) from any node (state) to any other node (state), until it reaches the terminal node of the trellis in state $\psi_{L+1} = 0$. Each branch in the trellis corresponds to one state transition that is triggered by a particular input a_k and produces the output s_k, and thus there is a one-to-one correspondence at time k between a branch, the state transition, and both the input and output of the FSM. One *stage* of the trellis is shown in Fig. 5-30(b) with the input and output pairs (a_k, s_k) labeled for each transition.

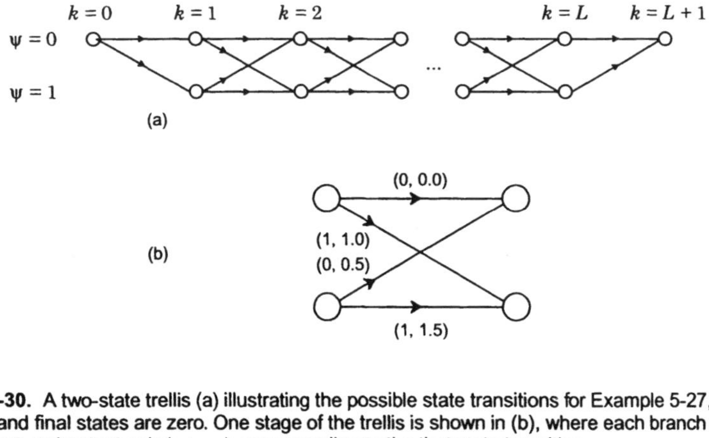

Fig. 5-30. A two-state trellis (a) illustrating the possible state transitions for Example 5-27, assuming the initial and final states are zero. One stage of the trellis is shown in (b), where each branch is labeled with the input and output pair (a_k, a_k) corresponding to the that state transition.

Example 5-29.

Suppose a sequence of $L = 5$ symbols drawn from the 4-PSK alphabet $\{-1, -j, 1, j\}$ are filtered by $M(z) = 1 - z^{-2}$. The trellis diagram for this example is shown in Fig. 5-31, assuming the idle symbol is $a^0 = -1$. This figure illustrates important features of a general trellis. Written above each stage of the trellis is the symbol that causes the corresponding transition. For example, the state is initially zero at time zero, and the value of the zero-th symbol a_0 determines which of the four states is reached at time one. Written below each stage is the corresponding received observation. There are $L + \mu$ stages in the trellis, the first L being due to the data symbols, while the remaining μ are due to the idle symbols. There are $|\mathcal{A}| = 4$ branches emanating from all nodes in the first L stages, one for each possible input symbol, while there is only one branch emanating from each node in the last μ stages. All states are not reachable until time μ (in this case, $\mu = 2$), since that is how long it takes for the memory elements of the tapped-delay line to fill up. The next $L - \mu$ stages (in this case three) of the trellis are identical. At time L, the filter input reverts to the idle symbol, and hence succeeding transitions are restricted in such a way that state zero is reached at time $L + \mu$. The branches are labeled by the unique output associated with the corresponding state transition. The inputs are not labeled but are uniquely specified by associating the ordered alphabet $\{-1, -j, 1, j\}$ with the 4 branches emanating from each node from top to bottom. For example,

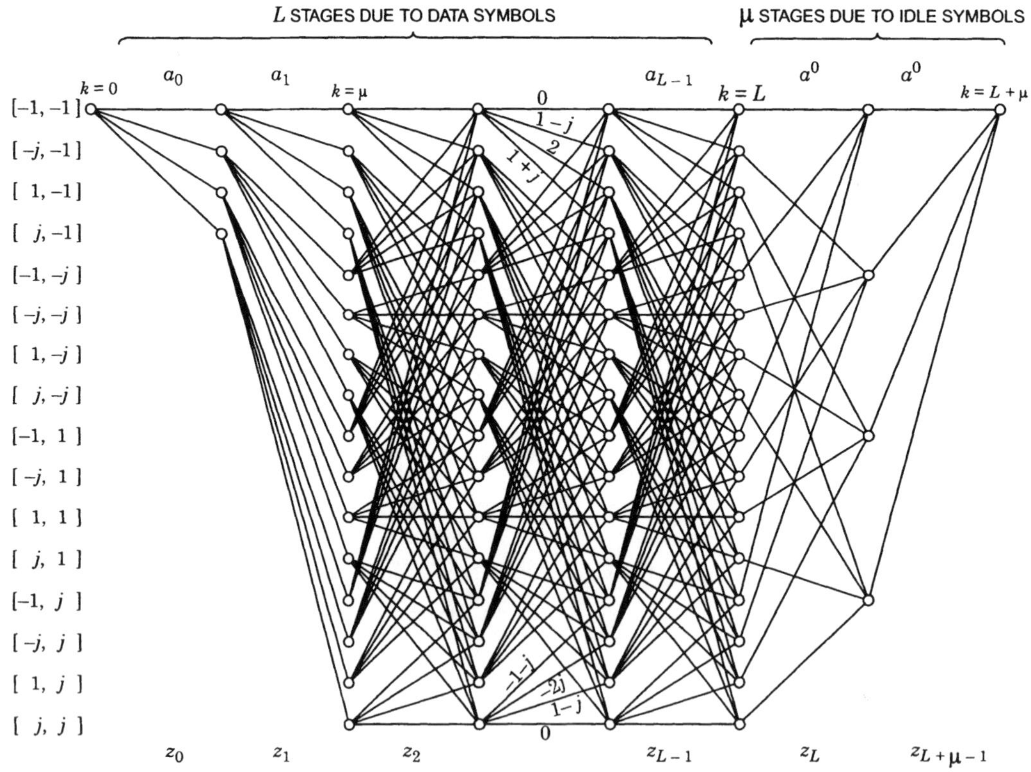

Fig. 5-31. A 16-state trellis for a sequence of 5 symbols drawn from the 4-PSK alphabet when the channel has memory 2.

when the state is [−1, −1] and the input is $-j$, the output is $1 - j$, as indicated in the second branch emanating from state zero.

A sequence of branches through the trellis diagram from the beginning to terminal nodes is called a *path*. Every possible path corresponds to a unique input sequence $\{a_0, \ldots a_{L-1}\}$. Thus, we may rephrase the goal of the minimum-distance sequence detector: based on the observation of s_k corrupted by noise, detect the minimum-distance *path* through the trellis diagram.

Recall from (5.72) that the minimum-distance sequence detector chooses the symbol sequence $\{a_k\}$ that minimizes the following cost function:

$$J' = \sum_{k=0}^{L+\mu-1} |z_k - s_k|^2, \tag{5.92}$$

where s_k depends on $\{a_k\}$ according to $s_k = \sum_{l=0}^{L-1} a_l m_{k-l}$. We can relate this cost function to the trellis diagram. Recall that there is a signal s_k associated with each branch of the trellis at each stage k of the trellis. For each stage k there is also an observation z_k. After observing z_k, we can assign to each branch of the trellis a numerical value called the *branch metric* that is low if z_k is close to s_k and high otherwise, namely

$$\text{branch metric} = |z_k - s_k|^2 \ . \tag{5.93}$$

Then for each path through the trellis, we can calculate the *path metric*, which is the sum of the branch metrics. The preferred path will be the one with the lowest path metric.

The minimum-distance sequence detector first calculates the branch metrics for every branch in the trellis diagram. It then calculates the path metric for every path in the trellis diagram, and chooses the path for which this path metric is minimum. The detected input sequence is then the sequence corresponding to this path. This straightforward approach of exhaustively calculating the path metric for each and every path through the trellis will clearly fail in practice because the number of paths grows exponentially with L. Usually L will be very large, corresponding to the entire time that communication takes place (usually minutes, hours, or even decades!). The Viterbi algorithm is a computationally efficient algorithm that exploits the special structure of the trellis to achieve a complexity that grows only linearly with L, or in other words, requires a constant computation rate (per unit time).

Consider one node in the trellis diagram, and all paths through the trellis that pass through this node.

Example 5-30.
The particular case of two incoming branches and two outgoing branches is shown below:

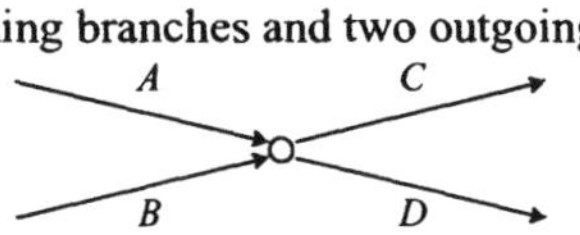

The incoming branches are labeled A and B, and the outgoing branches are labeled C and D. There are a large number of paths passing through this node (increasing exponentially with L), but all these paths follow one of just four routes through this node, AC, AD, BC, and BD.

The path metric for a particular path through the node is the sum of the *partial path metrics* for the portion of the path to the left and the portion to the right of the node. Among all possible partial paths to the left, the detector will always prefer the one with the smallest partial path metric, called the *survivor path* for that node. We can immediately remove from consideration all partial paths to the left other than the survivor path, because any other partial path to the left has by definition a larger partial path metric, and if it replaced the survivor path in any overall path, the path metric would be larger. This is the basis of the Viterbi algorithm, which allows us to reject many possible paths at each stage of the trellis.

The Viterbi algorithm finds the path with the minimum path metric by sequentially moving through the trellis and at each node retaining only the survivor paths. At each stage of the trellis we do not know which node the optimal path passes through, so we must retain one survivor path for each node. When we reach the terminal node of the trellis, we find the optimal path, which is the single survivor path for that node. The algorithm thus determines, at each time increment k, the survivor path for each of the Q nodes. The trick, then, is finding these Q survivor paths based on the information developed up to time $k-1$. This is pictured below for the case where there are two incoming branches to a given node at time k:

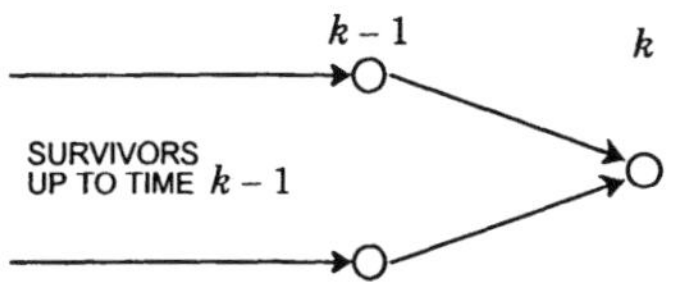

The only incoming paths to a node at time k that are candidates to be survivors are those consisting of survivors at time $k-1$ followed by branches to time k. (The number of such candidates is equal to the number of incoming branches to that node.) We therefore determine the partial path metrics for each of those candidate paths by summing the partial path metric of the survivor at time $k-1$ and the metric of the branch to time k. The survivor path at time k for a given node is the candidate path terminating on that node with the smallest partial path metric. We must store, for each node at time k, the survivor path and the associated partial path metric, for the algorithm to proceed to time $k+1$. We will illustrate the Viterbi algorithm with an example.

Example 5-31.
The trellis shown in Fig. 5-32 is marked with branch metrics corresponding to the transmission of $L = 3$ symbols and observing $\{0.2, 0.6, 0.9, 0.1\}$ for Example 5-27. The path metrics $|z_k - s_k|^2$ are labeled in Fig. 5-32. A simple instantaneous slicer (not a sequence detector) would decide that the transmitted bits were $\{0, 1, 1\}$, but the minimum-distance sequence detector takes into account knowledge of the ISI and selects $\{0, 1, 0\}$. An iterative procedure for making this decision is illustrated in Fig. 5-32. The survivor paths at each node and the partial path metric of each surviving path are shown.

The computational complexity of the Viterbi algorithm is the same at each time increment, except for end effects at the originating and terminating nodes, and hence the total computational complexity is proportional to the length of time L.

One practical problem remains. The algorithm does not determine the optimal path until the terminal node of the trellis; that is, it does not reach a conclusion on the entire sequence until the end of the sequence. Further, while the computation at each step is the same, the memory required to store the survivor paths grows linearly with time. In digital communications systems, sequences may be very long, and we cannot afford the resulting long delay in making decisions nor the very large memory that would be required. It is helpful if at some iteration k, all the survivor paths up to iteration $k - d$ coincide, for some d.

Example 5-32.

In Example 5-31, when $k \geq 2$, all the survivor paths coincide from $k = 0$ to $k = 1$. The minimum-distance detector decision for the first state transition can be made when $k = 2$. It is not necessary to wait until the terminal node of the trellis.

When all the survivor paths at some time k coincide up to some time $k - d$, we say that the partial paths have *merged* at depth d, and we can make a decision on all the inputs or states up to time $k - d$. Unfortunately, we cannot depend on the good fortune of a merge, as it is possible

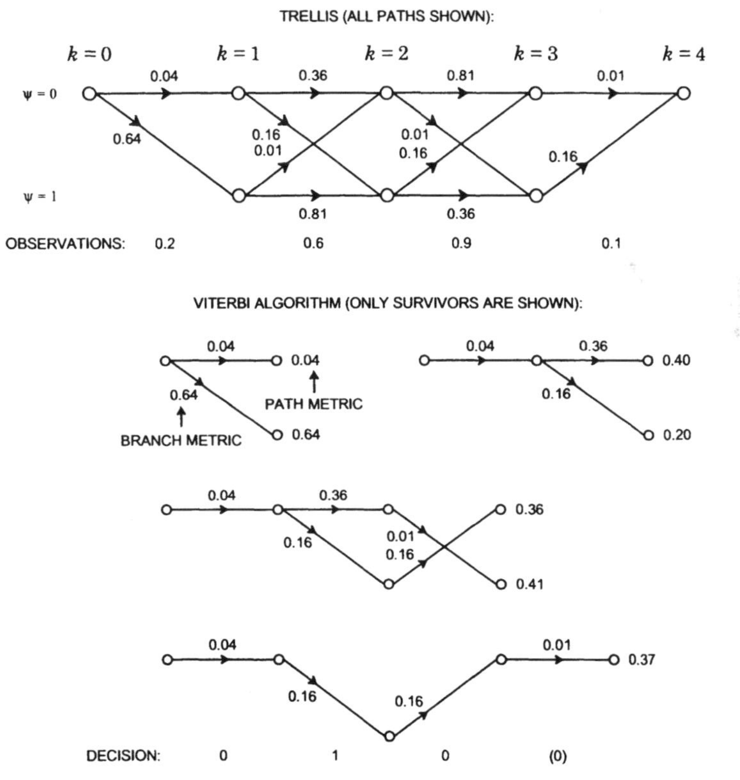

Fig. 5-32. A two-state trellis with the branch metrics of the transitions marked and the Viterbi algorithm illustrated. The Viterbi algorithm iteratively finds the path with the minimum path metric without ever considering more than two paths at once.

for no merges to occur. It is usual therefore to make a modification to the algorithm by forcing a decision at time k on all transitions prior to time $k - d$, for some *truncation depth* d. The usual approach is to compare all the partial path metrics for the N partial paths at time k, and note which one is the smallest. The decision on the input or state transition at time $k - d$ is then the transition at $k - d$ of this survivor path. Since the decision has been made, there is no need to store the survivor paths beyond a depth of d transitions in that path. If d is chosen to be large enough, this modification will have negligible impact on the probability of detecting the correct sequence.

5.5. Performance Analysis in AWGN

A communications system must make efficient use of two precious resources: bandwidth and signal power. There is a fundamental trade-off between these two quantities. Some modulation schemes make very efficient use of signal power, but require a wide signal bandwidth, while other schemes are very bandwidth efficient but require high signal power. The overall aim of this section is to quantify this tradeoff for the special case of PAM. In order to accomplish this we will need to make some assumptions about the characteristics of the noise, so that we may tie signal power to the reliability of the communication system. We will assume that the noise is additive, white, Gaussian, and independent of the transmitted signal; this assumption is valid in a wide range of applications (but not all). With this assumption we will determine the probability that the minimum-distance receiver makes an error, expressed as a function of the signal power and noise power spectral density. Ultimately this analysis will lead to a quantification of the power-bandwidth tradeoff for PAM schemes.

5.5.1. Probability of Symbol Error

In this section we will evaluate the performance of an isolated pulse of passband PAM in the presence of additive white Gaussian noise. The results will also apply to the general case when a *sequence* of pulses is transmitted, provided that the transmitted symbols are independent and the folded spectrum is a constant (which implies that the received pulse after the matched filter satisfies the Nyquist criterion), since in that case the minimum distance sequence detector reduces to a sequence of one-shot minimum-distance detectors. We thus consider the system shown in Fig. 5-33, which consists of a one-shot passband transmitter, a channel that adds white Gaussian noise, and a one-shot minimum-distance receiver. As demonstrated in the previous section, the receiver observation after downconversion is:

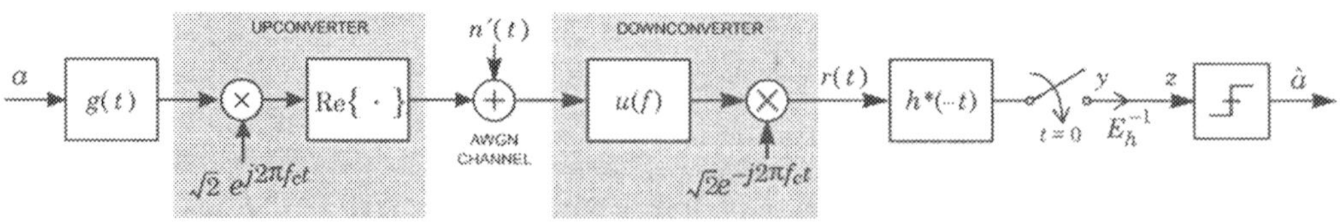

Fig. 5-33. A one-shot passband PAM transmitter, a channel that adds Gaussian noise with PSD $N_0/2$, and a one-shot minimum-distance receiver.

$$r(t) = ah(t) + n(t)\,, \tag{5.94}$$

where $a \in \mathcal{A}$ is the transmitted symbol, $h(t)$ is the received pulse shape, and $n(t)$ is the downconverted noise.

So far this chapter has paid little attention to the details of the noise; instead, we have treated it as some unknown but deterministic quantity. In this section, to achieve our goal of performance analysis, we need a full statistical characterization of the noise. We will assume that the channel noise before down conversion, call it $n'(t)$, is real-valued, white and Gaussian with PSD $N_0/2$. Before we proceed we need to explore the statistics of its complex envelope $n(t)$.

The statistics of the noise complex envelope can be derived with the aid of the following picture, which shows white Gaussian noise $n'(t)$ with PSD $N_0/2$ as the downconverter input:

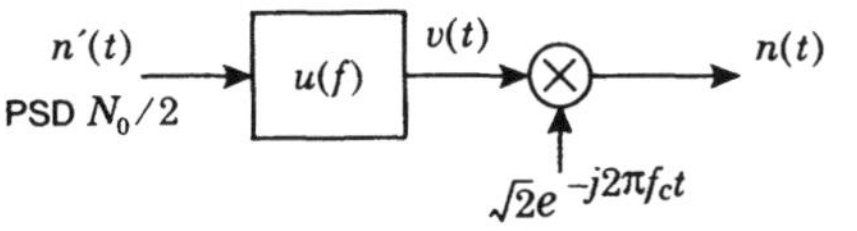

Because linear filtering preserves Gaussianity, the output $v(t)$ of the above filter is a Gaussian random process with PSD $S_v(f) = (N_0/2)u(f)$. Furthermore, as demonstrated in Section 3.2.8, the phase-splitter output is circularly symmetric. Therefore, the PSD of the noise complex envelope (after the mixer) is circularly symmetric and Gaussian with power spectrum $S_n(f) = N_0 u(f + f_c)$. The PSD is thus a constant for all frequencies greater than $-f_c$, and zero for all frequencies less than $-f_c$. Therefore, strictly speaking, the complex envelope of white noise is no longer white. However, a practical receiver will always filter white noise. In particular, the minimum-distance receiver starts out with a matched filter. In this case, as long as the bandwidth of the MF is less than the carrier frequency f_c — which will happen whenever the bandwidth of the transmit pulse is less than the carrier frequency, or equivalently when the passband spectrum does not overlap d.c. — the MF output will be circularly symmetric Gaussian noise with power spectrum $N_0 |H(f)|^2$. The very same PSD would arise if we assumed that the complex envelope of white noise is white.

According to Section 5.3, the one-shot minimum-distance receiver correlates $r(t)$ with the received pulse using a MF, and then normalizes by the received pulse energy, yielding:

$$\begin{aligned} z &= E_h^{-1}\int_{-\infty}^{\infty} r(t)h^*(t)\,dt \\ &= a + n\,, \end{aligned} \tag{5.95}$$

where

$$n = E_h^{-1}\int_{-\infty}^{\infty} n(t)h^*(t)\,dt\,. \tag{5.96}$$

Fact:

When $n(t)$ is the complex envelope of real white Gaussian noise with PSD $N_0/2$, the real and imaginary parts of n in (5.96) are i.i.d. zero-mean Gaussian random variables with variance $N_0/(2E_h)$. In other words, $n \sim \mathcal{CN}(0, N_0/E_h)$.

Proof:

We have already demonstrated that the output of the MF is a circularly symmetric Gaussian process with PSD $N_0 | H(f) |^2$. Sampling a complex circularly symmetric Gaussian random process leads to a circularly symmetric Gaussian random variable. The power of this random variable can be found by integrating the PSD, yielding:

$$\int_{-\infty}^{\infty} N_0 | H(f) |^2 df = N_0 E_h . \tag{5.97}$$

Finally, dividing by E_h changes the power from $N_0 E_h$ to N_0 / E_h, so that $\mathrm{E}[| n |^2] = N_0 / E_h$.

Although (5.95) was derived for the case of passband PAM, for which the alphabet is generally complex-valued, the same model also applies to baseband PAM, for which the alphabet is necessarily real-valued. We might choose to modify (5.94) slightly for the baseband case by forcing the imaginary part of the noise to zero, but even this minor modification is not really necessary. Even if we leave the imaginary part of the noise where it is, the minimum-distance slicer will ignore any imaginary components in z, and so it ultimately has no impact on performance.

The Signal-to-Noise Ratio

Let E denote the average *received* signal energy for a single isolated pulse of passband PAM, ignoring the noise. From (5.94), and exploiting the fact that the energy of a passband signal and its complex envelope are the same, we have $E = E_a E_h$, where $E_a = \mathrm{E}[| a |^2]$ denotes the energy of the alphabet, and where E_h denotes the energy of the received pulse shape. When the transmit pulse has unit energy, E_a also represents the average transmitted energy per pulse. In this context, the *signal-to-noise ratio (SNR)* for white noise is defined by the ratio of the received signal energy to twice the two-sided noise power-spectral density, or:

$$SNR = \frac{E}{N_0} . \tag{5.98}$$

One way to derive this is to start with (5.95), and to think of the SNR as the ratio of the energy of the signal term to the energy in the noise term:

$$\mathrm{SNR} = \frac{E[|a|^2]}{E[|n|^2]} . \tag{5.99}$$

Since the numerator is E_a and the denominator is N_0 / E_h, this reduces to (5.98).

The minimum distance receiver will apply z to a slicer, and choose the symbol $a \in \mathcal{A}$ that minimizes $| z - a |^2$. A *symbol error* occurs when the slicer chooses a symbol different from the actual transmitted symbol. Other definitions of error, such as bit error or block error, will be considered later. The main objective of this section is to determine the probability that the one-shot minimum-distance receiver makes a decision error when there is additive white Gaussian noise.

Derivations of Error Probability

In the following series of examples we will evaluate the probability of error for the most popular forms of PSK and QAM schemes.

Example 5-33.

The best binary alphabet — subject to an average energy constraint — is the so-called binary antipodal alphabet $\mathcal{A} = \{\pm\sqrt{E_a}\}$. For passband PAM this is also known as the binary phase-shift keying (BPSK) alphabet. In this case, the minimum-distance decision is determined by the sign of z. If the negative symbol is transmitted, the pdf for z will be as sketched below:

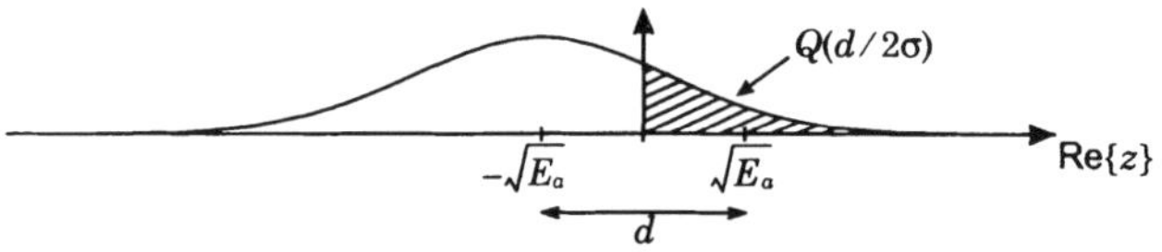

Specifically, the conditional pdf for z will be $\mathcal{N}(-\sqrt{E_a}, \sigma^2)$, where we have introduced

$$\sigma^2 = \frac{N_0/2}{E_h} \tag{5.100}$$

as the noise variance *per dimension* (real and complex) after the sampled and normalized MF. (Keep in mind that the complex noise has double the variance, since $\mathrm{E}[|n|^2] = 2\sigma^2$.) Let $d = 2\sqrt{E_a}$ denote the *distance* between the two alphabet symbols. From the above picture it is clear that the negative symbol will be erroneously detected as a positive symbol if and only if the noise exceeds half the distance between the two transmitted symbols. The probability of this event is $Q(d/2\sigma)$, the area under the tail of the conditional pdf. as depicted by the shaded region above. Conversely, when the positive symbol is transmitted, an error will occur whenever the noise is less than $-d/2$, which also occurs with probability $Q(d/2\sigma)$. The error probability is thus symmetric: if either of the symbols is transmitted, the probability of the other symbol being chosen by the slicer is $Q(d/2\sigma)$. The overall error probability for BPSK is thus:

$$\Pr[\text{error}] = Q(d/2\sigma) = Q(\sqrt{2E/N_0}), \tag{5.101}$$

where to arrive at the second equality we used $d = 2\sqrt{E_a}$, $E = E_a E_h$, and (5.100).

When $M > 2$ it is not always possible to find an exact expression for the error probability. There are exceptions, however.

Example 5-34.

Consider the 4-PAM alphabet $\mathcal{A} = \{\pm c, \pm 3c\}$, as shown in the following figure:

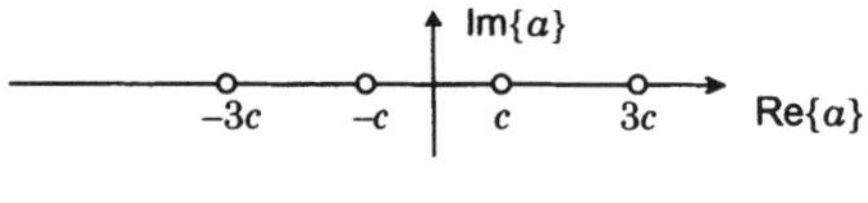

where the constant c is related to the transmitted energy by $c = \sqrt{E_a/5}$. If the transmitted symbol is

$-c$, then the p.d.f. of z is shown in the following figure:

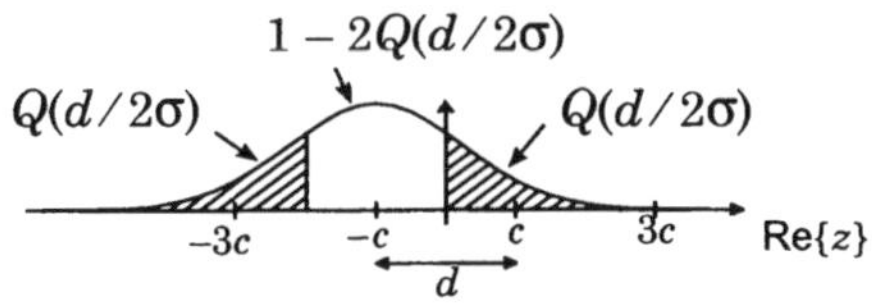

The probability that the received sample z is closer to a symbol other than $-c$ is equal to the sum of the areas of the shaded regions. The shaded regions each have area $Q(d/2\sigma)$, so

$$\Pr[\text{symbol error} \mid a = \pm c] = 2Q(d/2\sigma) . \tag{5.102}$$

On the other hand,

$$\Pr[\text{symbol error} \mid a = \pm 3c] = Q(d/2\sigma) , \tag{5.103}$$

so if the symbols are equally likely then

$$\Pr[\text{symbol error}] = \tfrac{3}{2} Q(d/2\sigma) = \tfrac{3}{2} Q\Big(\sqrt{\tfrac{2}{5}E\ /N_0}\Big) . \tag{5.104}$$

The coefficient $^3\!/_2$ is the average number of nearest neighbors. The last equality follows from the fact that $d/2 = c = \sqrt{E_a/5}$ and (5.100).

Example 5-35.

Consider the 4-QAM alphabet $\mathcal{A} = \{\pm c \pm jc\}$, as sketched in Fig. 5-32, where $c = \sqrt{E_a/2}$. By symmetry the probability of error is independent of which symbol is transmitted, so let us assume that the lower-left symbol is transmitted, $a = -c - jc$. The decision region for this symbol is the shaded region in the figure. Rather than calculating the probability of error directly, in this case it is more convenient to first calculate the probability of being *correct*:

$$\begin{aligned}\Pr[\text{correct}] &= \Pr[\text{correct} \mid a = -c - jc] \\ &= \Pr[\text{Re}\{z\} < c,\ \text{Im}\{z\} < c] \\ &= \Pr[\text{Re}\{z\} < c]\Pr[\text{Im}\{z\} < c] \\ &= (1 - Q(d/2\sigma))^2 \\ &= 1 - 2Q(d/2\sigma) + Q^2(d/2\sigma) ,\end{aligned} \tag{5.105}$$

where $d = 2c$, the minimum distance between symbols. The probability of a symbol error is simply

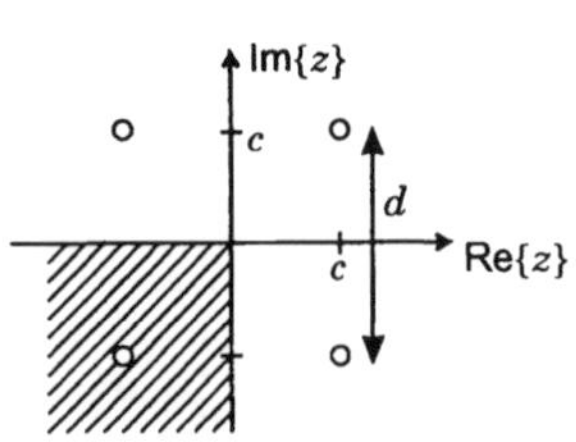

Fig. 5-34. The shaded region is the decision region corresponding to the symbol $-c - jc$.

$$\begin{aligned}\Pr[\text{symbol error}] &= 1 - \Pr[\text{correct}] \\ &= 2Q(d/2\sigma) - Q^2(d/2\sigma) \\ &= 2Q(\sqrt{E/N_0}) - Q^2(\sqrt{E/N_0}),\end{aligned} \tag{5.106}$$

where we substituted $d/2 = c = \sqrt{E_a/2}$. Typically the SNR is high enough that $Q^2(\cdot)$ is negligible relative to $Q(\cdot)$, so that:

$$\Pr[\text{symbol error}] \approx 2Q(\sqrt{E/N_0}). \tag{5.107}$$

This approximation is reasonable, because communication is likely to be useful only if the probability of error is small (see Problem 5-16). Here the coefficient "2" reflects the fact that every symbol has two nearest neighbors.

Example 5-36.

The 4-PSK constellation is $\mathcal{A} = \{\pm b, \pm jb\}$, where $b = \sqrt{E_a}$. Suppose the receiver rotates z by 45 degrees, yielding $z' = ze^{j\pi/4} = a' + n'$, where $a' = ae^{j\pi/4}$ is the rotated symbol and $n' = ne^{j\pi/4}$ is the rotated noise. But $a' \in \{\pm c \pm jc\}$ where $c = \sqrt{E_a/2}$, which we recognize this as the 4-QAM alphabet. Thus, the rotation converts 4-PSK to 4-QAM. The statistics of the rotated noise are the same as the statistics of the non-rotated noise, since the noise is circularly symmetric. We conclude that 4-PAM has the same performance as 4-QAM, namely (5.106).

Example 5-37.

The 16-QAM alphabet is $\mathcal{A} = \{\pm c \pm jc, \pm c \pm j3c, \pm 3c \pm jc, \pm 3c \pm j3c\}$ where $c = \sqrt{E_a/10}$, as illustrated in Fig. 5-35. The probability of error for 16-QAM can be found by a similar method. Consider the four *inside* points. Their decision regions are squares with sides equal to $d = 2c$. In this case, the probability of a correct decision is

$$\Pr[\text{correct} \mid a \text{ on the inside}] = [1 - 2Q(d/2\sigma)]^2 \tag{5.108}$$

so the probability of error is

$$\Pr[\text{error} \mid a \text{ on the inside}] = 4Q(d/2\sigma) - 4Q^2(d/2\sigma) \approx 4Q(d/2\sigma), \tag{5.109}$$

consistent with the fact that every interior point has four nearest neighbors. The probability of error for the corner symbols is similar to the probability of error for the 4-PSK signal,

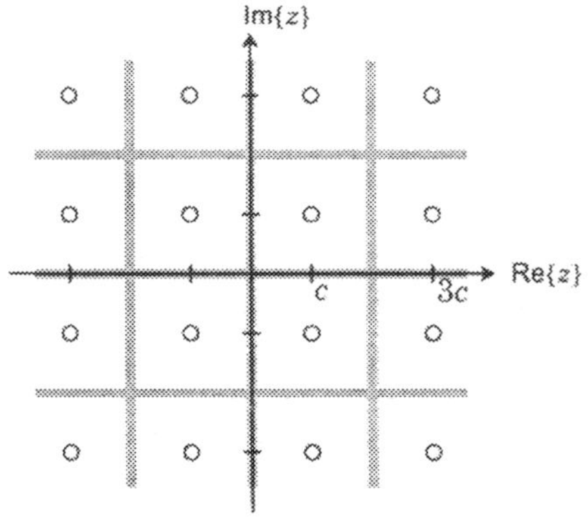

Fig. 5-35. The 16-QAM alphabet and the corresponding minimum-distance decision regions.

$$\Pr[\text{error} \mid a \text{ in the corner}] \approx 2Q(d/2\sigma)\,, \tag{5.110}$$

and

$$\Pr[\text{error} \mid a \text{ not inside or corner}] \approx 3Q(d/2\sigma)\ . \tag{5.111}$$

Assuming the symbols are equally likely, the total probability of error is

$$\begin{aligned}\Pr[\text{error}] &\approx \frac{4}{16} \times 4Q(d/2\sigma) + \frac{8}{16} \times 3Q(d/2\sigma) + \frac{4}{16}\,2Q(d/2\sigma) \\ &= 3Q(d/2\sigma) \\ &= 3Q\left(\sqrt{\frac{E}{5N_0}}\right).\end{aligned} \tag{5.112}$$

Generalizing the above analysis, it is easy to find an exact expression for the symbol-error probability for an M-QAM alphabet in the practical case when $M = 2^b$ and b is even, without making the assumption of high SNR (see Problem 5-19):

$$\Pr[\text{error}] = 4\left(1 - \frac{1}{\sqrt{M}}\right)Q\left(\sqrt{\frac{3E/N_0}{M-1}}\right) - 4\left(1 - \frac{1}{\sqrt{M}}\right)Q^2\left(\sqrt{\frac{3E/N_0}{M-1}}\right)\ . \tag{5.113}$$

As the size of the alphabet M gets large, a greater percentage of the points will be in the interior with four nearest neighbors, and so the coefficient of $Q(\cdot)$ tends towards 4.

Example 5-38.

The M-PSK alphabet is the set of M points equally spaced on a circle of radius $c = \sqrt{E_a}$: $\mathcal{A} = \{c, ce^{j2\pi/M}, ce^{j4\pi/M}, \ldots ce^{j2\pi(M-1)/M}\}$. The 16-PSK alphabet is shown in Fig. 5-36. The symmetry of the alphabet implies that the symbol error probability is the same regardless of which symbol is transmitted, so let us assume that the transmitted symbol is $a = \sqrt{E_a}$. The conditional pdf for z is then $\mathcal{CN}(\sqrt{E_a}, 2\sigma^2)$. The minimum-distance decision region for this symbol is the nonshaded region of the complex plane corresponding to angles less than π/M in magnitude. The error probability is the probability that the complex noise perturbs z to fall outside of this region. Although an exact expression cannot be found, it is easy to derive a tight bound. Let P_1 denote the probability that z crosses the upper threshold that separates c from its nearest upper neighbor. It can be expressed as $P_1 = \Pr[\text{Im}\{n'\} > d/2]$, where $n' = ne^{-j\pi/M}$ is a rotated version of the noise. Because the noise is circularly symmetric, the rotation does not change the noise statistics; therefore, P_1 reduces to $Q(d/2\sigma)$. In terms of the figure, we may interpret $Q(d/2\sigma)$ as the two-dimensional integral of the conditional pdf for z over the striped region above the upper threshold. Similarly, the probability that the noise crosses the lower threshold is also $Q(d/2\sigma)$, which represents the pdf integral over the shaded region below the lower threshold. Let P_2 denote the probability that z falls in the decision region for $-c$, which can be found by integrating the conditional pdf for z over the crosshatched region shown in the figure. Since this region is the intersection of the striped and shaded regions, we have:

$$\Pr[\text{error}] = 2Q(d/2\sigma) - P_2\,. \tag{5.114}$$

In practice P_2 will be negligible compared to $Q(d/2\sigma)$, and so by ignoring it we arrive at the following upper bound:

$$\Pr[\text{error}] < 2Q(d/2\sigma)$$
$$= 2Q\left(\sqrt{2E/N_0}\sin\frac{\pi}{M}\right), \tag{5.115}$$

where we used $d/2 = \sqrt{E_a}\sin(\pi/M)$. The bound is tight for $M \geq 4$.

5.5.2. Bandwidth and SNR Requirements of Passband PAM

In the previous section, we derived expressions for the probability of symbol error as a function of SNR = E/N_0 for a variety of modulation schemes. In this section we will compare the performance of these modulation schemes in terms of their bandwidth requirements (or equivalently spectral efficiency) and SNR requirements. To make the comparison fair, however, we must first account for the fact that different modulation schemes convey a different number of bits per symbol, and also have different bandwidth requirements. Towards this end, we first introduce E_b as the amount of signal energy received per source bit. In terms of the signal energy E received per transmitted symbol, it can be expressed as:

$$E_b = E/b\,, \tag{5.116}$$

where $b = \log_2 |\mathcal{A}|$ is the number of bits conveyed by a uniformly drawn symbol. The energy per bit leads to the following as a definition for the *per-bit signal-to-noise ratio*:

$$E_b/N_0\,. \tag{5.117}$$

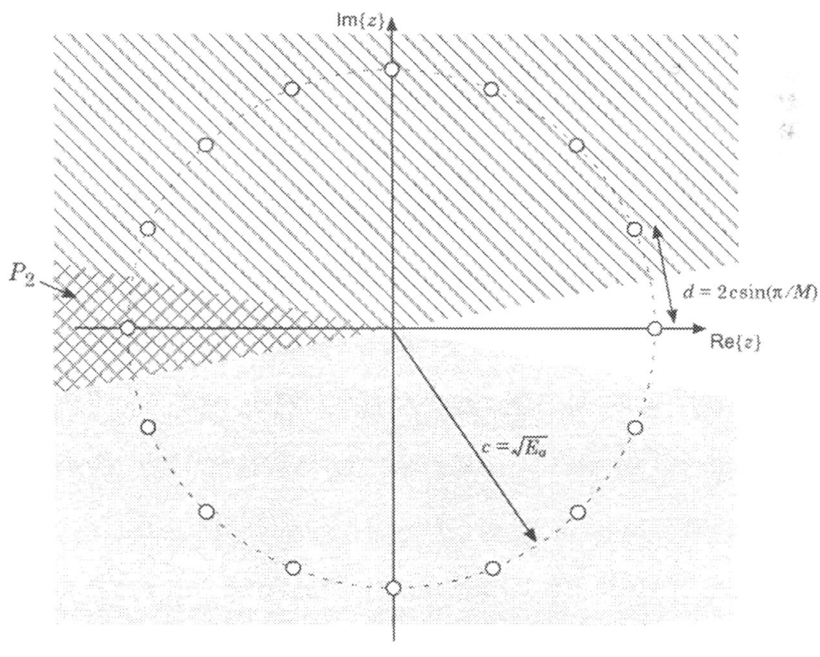

Fig. 5-36. The 16-PSK constellation.

In terms of the usual definition $SNR = E/N_0$, this SNR is $E_b/N_0 = SNR/b$. Alternatively, in terms of the received signal power P and bit rate R_b, the per-bit SNR is:

$$E_b/N_0 = P/(N_0 R_b) . \tag{5.118}$$

The second change we make is to convert symbol-error probability to bit-error probability. Bit errors are caused by symbol errors, but the exact relationship between their probabilities depends on the mapper that maps input bits into symbols. If the SNR is high enough, as it is for most useful communication systems, then a symbol is far more likely to be mistaken for one of its neighbors in the constellation than for more distant symbols (see for example Problem 5-17). Mappers often implement a *Gray mapping*, as illustrated in Fig. 5-37, in which nearest neighbors correspond to bit groups that differ by only one bit. Thus, the most probable symbol errors cause only a single bit error. Thus, with Gray mapping at high SNR, we have:

$$\Pr[\text{bit error}] \approx \frac{1}{b}\Pr[\text{symbol error}] , \tag{5.119}$$

where again $b = \log_2 |\mathcal{A}|$ is the bits per symbol. In the following we will use P_b to denote the bit-error probability.

In some cases, P_b can be calculated directly and exactly, without resorting to the approximation of (5.119).

Example 5-39.

Consider the Gray-mapped 4-PSK alphabet of Fig. 5-37(a). We can view 4-PSK as a rotated version of 4-QAM, which in turn can be viewed as a pair of BPSK alphabets in quadrature, each with energy $E_b = E/2$. The Gray mapping of Fig. 5-37(a) uses the first bit to determine the in-phase BPSK symbol, and the second bit to determine the quadrature BPSK symbol. Therefore, the *exact* bit-error probability for Gray-mapped 4-PSK is given by (5.101) with E replaced by E_b:

$$P_b = Q(\sqrt{2E_b/N_0}) . \tag{5.120}$$

In contrast, plugging (5.106) into the approximation of (5.119) yields:

$$P_b \approx Q(\sqrt{2E_b/N_0}) - \frac{1}{2}Q^2(\sqrt{2E_b/N_0}) . \tag{5.121}$$

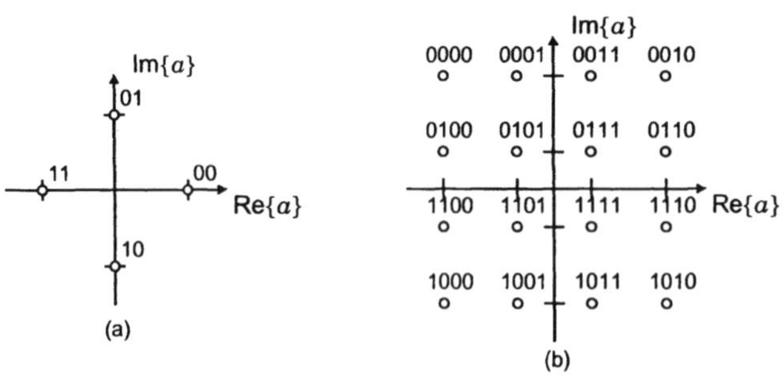

Fig. 5-37. Gray-mapped 4-PSK (a) and 16-QAM (b). Associated with each symbol is a set of source bits. Notice that only one bit differs between any two adjacent symbols. This minimizes the number of bit errors per nearest-neighbor symbol errors.

We see that the approximation of (5.119) erroneously includes a $Q^2(\cdot)$ term in this example.

After converting SNR to per-bit SNR, and after converting symbol-error probability to bit-error probability, the analysis results of the previous section for BPSK, M-PSK and M-QAM can be summarized as follows:

$$P_b^{\text{BPSK}} = Q(\sqrt{2E_b/N_0}), \tag{5.122}$$

$$P_b^{M\text{-QAM}} \approx \frac{4}{b}(1-2^{-b/2})Q\left(\sqrt{\frac{3bE_b/N_0}{2^b-1}}\right), \tag{5.123}$$

$$P_b^{M\text{-}PSK} \approx \frac{2}{b}Q\left(\sqrt{2bE_b/N_0}\sin\frac{\pi}{M}\right). \tag{5.124}$$

If we solve (5.123) for E_b/N_0, we arrive at an expression for the SNR per bit required by M-QAM to achieve a certain bit-error probability, namely:

$$E_b/N_0 = \Gamma\frac{2^b-1}{b} \quad \text{(QAM)}, \tag{5.125}$$

where
$$\Gamma = \frac{1}{3}\left(Q^{-1}\left(\frac{bP_b}{4(1-2^{-b/2})}\right)\right)^2. \tag{5.126}$$

In contrast, the Shannon limit of SNR per bit is also given by (5.125) but with $\Gamma = 1$. Thus, we may interpret Γ as the SNR gap from capacity. At $P_b = 10^{-6}$ and as b increases from 2 to 8, the gap Γ decreases roughly linearly from 8.8 dB to 8.4 dB. Similarly, solving (5.124) for E_b/N_0 leads to the SNR requirement for M-PSK:

$$E_b/N_0 = \frac{(Q^{-1}(bP_b/2))^2}{2b\sin^2\frac{\pi}{M}} \quad \text{(PSK)}, \tag{5.127}$$

where again $M = 2^b$.

These expressions quantify the power requirements of the different modulation schemes, but they say nothing about the *bandwidth* requirements. Let us define the *normalized* bandwidth requirement as the ratio of the required bandwidth to the achieved bit rate. With zero excess bandwidth, the bandwidth requirement for any passband PAM scheme is equal to the symbol rate. Therefore, the normalized bandwidth requirement is $1/\log_2|\mathcal{A}| = 1/b$. In Fig. 5-38(a) we plot the per-bit SNR requirement as a function of the normalized bandwidth requirement for several QAM and PSK modulation schemes. The key feature of this figure is the tradeoff between signal power and bandwidth: As the size of the alphabet grows large, the bandwidth requirement goes down while the power requirement goes up. This tradeoff is fundamental. The Shannon limit also exhibits the same tradeoff. For example, with unbounded bandwidth, the Shannon limit on the SNR per bit is –1.6 dB. As the bandwidth becomes more

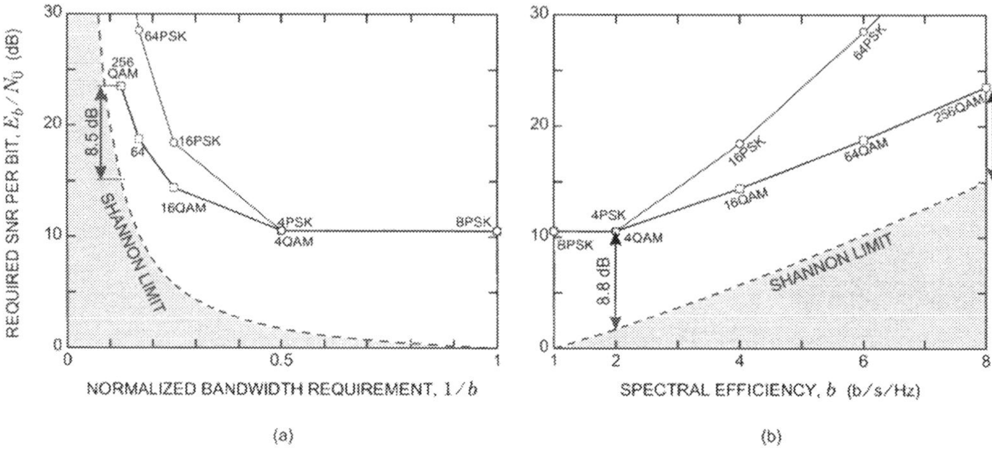

Fig. 5-38. Comparing M-QAM and M-PSK to the Shannon limit. In (a), we plot the required SNR per bit as a function of the normalized bandwidth requirement W/R_b, which reduces to $1/b$ with zero excess bandwidth. This plot clearly shows the tradeoff between SNR and bandwidth. In (b), we plot the required SNR per bit as a function of spectral efficiency b. Both figures assume $P_b = 10^{-6}$.

and more constrained, the SNR requirement grows larger. For example, when the bandwidth available is only 10% of the desired bit rate, the Shannon limit grows to 20.1 dB. Fig. 5-38(a) also clearly illustrates the superiority of QAM over PSK.

In Fig. 5-38(b) we plot the SNR requirement as a function of spectral efficiency b. Obviously, spectral efficiency is just the inverse of the normalized bandwidth requirement, so Fig. 5-38(b) and Fig. 5-38(a) convey the same information. In Fig. 5-38(b) we see that the gap to capacity for QAM (see (5.126)) is roughly constant as spectral efficiency grows; in contrast, the gap to capacity for PSK grows larger with increasing spectral efficiency.

5.6. Further Reading

The minimum-distance approach to receiver design is not standard in textbooks or the literature. The benefit of this approach is that a wealth of receiver structures have been quickly derived from a common design principle. It turns out that the principle of minimum-distance receiver design is optimal in a certain sense (defined in Chapter 7) for channels with additive white Gaussian noise. Thus, these receiver structures are usually derived from noise considerations, an approach that is more circuitous.

A good summary of the current state of the art of modulation for linear Gaussian-noise channels can be found in [16]. The design and optimization of passband PAM alphabets was considered in [4]–[9]. Higher-dimensional alphabets are proposed in [11]. The original Viterbi algorithm reference is [14], with a more tutorial paper following shortly thereafter [17]. A broader perspective is given in an excellent tutorial by D. Forney [18]. A probabilistic view of the Viterbi algorithm will be presented in Section 7.4. The Viterbi algorithm was originally derived to decode error-correcting codes (see Chapter 12). The proposal to use the Viterbi algorithm for ISI channels comes from Forney [15]. Omura [19] first pointed out that the Viterbi algorithm could be derived from the principles of dynamic programming, invented by Bellman [20]. A comprehensive coverage of its use in error-correcting codes can be found in Viterbi and Omura [21], which also has an extensive set of references.

Problems

Problem 5-1. Suppose that the rectangular pulse of Fig. 5-1 is used with the bandlimited channel of Example 5-17, where

$$W = 1/T. \tag{5.128}$$

Assume that the transmitted symbols have power $E[|a_k|^2] = E_a$, and assume that successive data symbols are uncorrelated. Sketch the power spectrum of the received signal $r(t)$. Describe qualitatively the distortion of the signal.

Problem 5-2. Consider a baseband PAM system using the raised-cosine pulses in (5.8). Assume the symbol sequence is white and normalized, so its power spectrum is

$$S_a(e^{j\theta}) = 1 \ . \tag{5.129}$$

Show that the transmit power is independent of T for any roll-off factor α.

Problem 5-3. Consider a channel that is ideally bandlimited to $|f| \leq 1500$ Hz. What is the maximum symbol rate using pulses with 50% excess bandwidth? Assume that the receive filter will be an ideal lowpass filter and that ISI is not tolerable.

Problem 5-4. In an example of a *partial response* system, a symbol sequence is generated as follows:

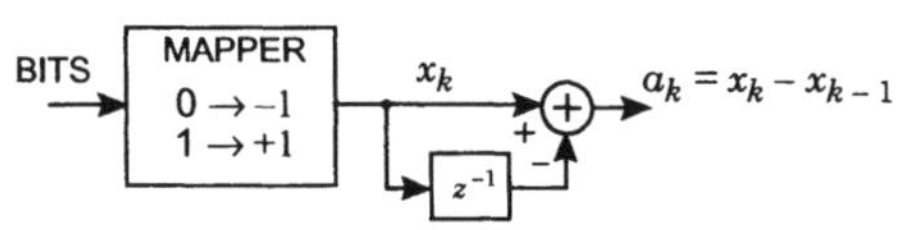

Assume the incoming bits are random, independent of one another, and zeros and ones are equally probable.

(a) Find $S_a(e^{j\theta})$, the power spectrum of the symbol sequence. Sketch it.

(b) Suppose the transmit pulse $g(t)$ is an ideal lowpass pulse with zero excess bandwidth. Find the power spectrum of the baseband PAM signal. A well-labeled, careful sketch is sufficient.

(c) Find the pulse $h(t)$ such that the transmitted signal

$$s(t) = \sum_{m=-\infty}^{\infty} a_m g(t - mT) \tag{5.130}$$

can be written as

$$s(t) = \sum_{m=-\infty}^{\infty} x_m h(t - mT) . \tag{5.131}$$

Assuming $g(t)$ from (b), does $h(t)$ satisfy the Nyquist criterion? Assuming a perfect channel, $b(t) = \delta(t)$, is there a stable receive filter $f(t)$ such that the overall pulse shape $p(t) = h(t) * b(t) * f(t)$ satisfies the Nyquist criterion?

Problem 5-5. Show that the horizontal eye opening for a pulse with zero excess bandwidth and binary antipodal signaling is zero. Assume there is no coding, so any sequence of symbols is permissible. A consequence of this is that the timing recovery for such a channel would have to be absolutely perfect.

Problem 5-6. In a baseband PAM system, assume $G(f) = \sqrt{T}$ and $B(f) = 1$, so the transmitter sends an impulse stream and the channel has infinite bandwidth. Further assume that the channel noise has power spectrum $S_N(f) = N_0/2$, and assume $S_a(e^{j\theta}) = 1$.

(a) Show that $E[|a_k|^2] = 1$.

(b) Show that the power spectrum of the transmitted signal is independent of T. *Hint:* Use the results of Appendix 3-A, where a random phase is introduced to make the transmit signal WSS.

(c) Find the receive filter $F(f)$ such that the output of the receive filter has a pulse shape that is an ideal zero-excess-bandwidth pulse shape.

(d) Find the SNR at the slicer with $p(t)$ given in part (c).

(e) Find the SNR at the slicer when the pulse $p(t)$ has the triangular spectrum of Fig. 5-3(b), given by

$$P(f) = \begin{cases} T - |f|T^2; & |f| < 1/T \\ 0; & \text{otherwise} \end{cases} . \tag{5.132}$$

Compare this SNR to that in part (d).

Problem 5-7. Suppose a baseband PAM signal with a complex-valued pulse $g(t)$ whose Fourier transform is shown in the following figure:

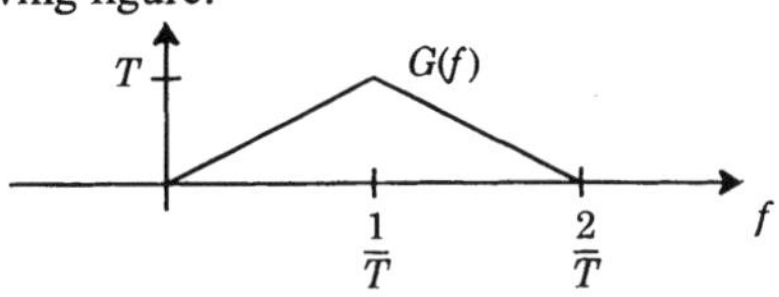

(a) Does this pulse satisfy the Nyquist criterion? Does $\text{Re}\{g(t)\}$ satisfy the Nyquist criterion?

(b) Suppose that the symbols a_k are uncorrelated, so that

$$S_a(e^{j2\pi fT}) = E_a . \tag{5.133}$$

Give the power spectrum of the PAM signal. Sketch it.

(c) Find $g(t)$.

Problem 5-8. Consider a channel where the equalized pulse $p(t) = g(t) * b(t) * f(t)$ is a raised-cosine pulse. Assume white noise with PSD $N_0/2$ and white symbols with PSD $S_a(e^{j\theta}) = E_a$. Find the SNR after the receive filter as a function of the excess bandwidth (for the range of zero to 100%) for the following three transmitted pulse shapes and an ideal bandlimited channel:

(a) The transmitted pulse is an impulse.

(b) The transmitted pulse is a raised-cosine pulse with the same excess bandwidth as desired at the receiver.

(c) The Fourier transform of the transmitted pulse is the square root of the Fourier transform of a raised-cosine pulse (tedious).

Problem 5-9. Suppose you are to design a digital communication system to transmit a speech signal sampled at 8 KHz with 8 bits per sample. Find the minimum bandwidth required for each of the following methods:

(a) Binary antipodal baseband PAM.

(b) Binary antipodal passband PAM.

(c) 4-PSK.

(d) 16-QAM.

Define bandwidth to cover positive frequencies only, as shown in the following figure:

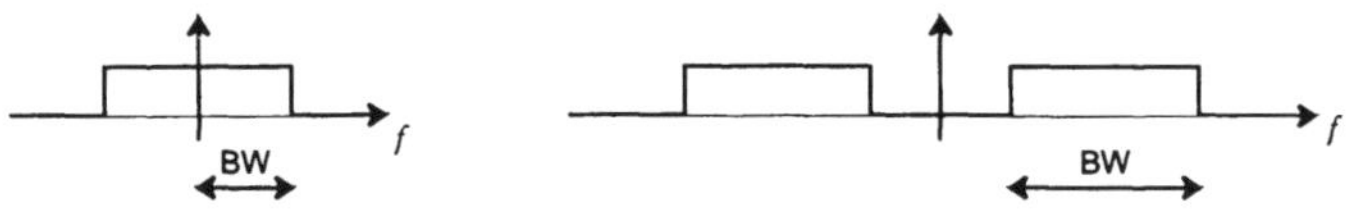

Problem 5-10. Consider a channel with bandwidth 10 KHz:

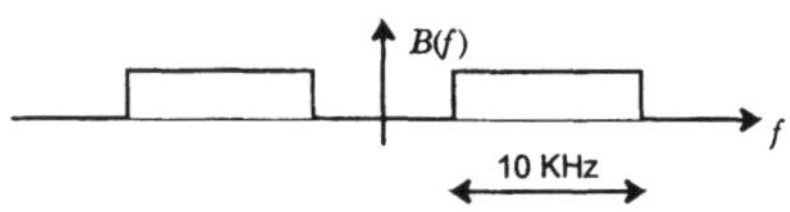

(a) What is the complex envelope $\tilde{b}(t)$ of this channel's impulse response?

(b) Let $p(t) = g(t) * \frac{1}{\sqrt{2}}\tilde{b}(t) * f(t)$, where $g(t)$ is the impulse response of the transmit filter and $f(t)$ is the impulse response of the receive filter. Find the maximum *bit rate* achievable with zero ISI using the following methods:

- 4-PSK with $p(t)$ being a minimum-bandwidth pulse,
- binary antipodal with $p(t)$ being a 50% excess-bandwidth raised-cosine pulse,
- 16-QAM with $p(t)$ being a 100% excess-bandwidth raised-cosine pulse,
- 16-QAM with $p(t)$ as shown below:

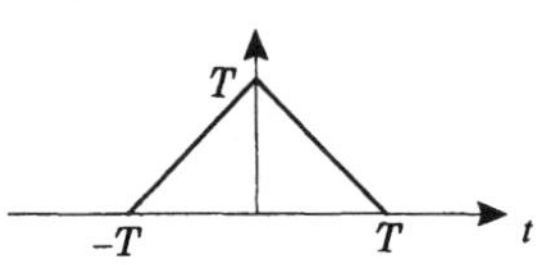

(c) Assuming the 4-PSK signal of part (b), give transfer functions for filters $g(t)$ and $f(t)$ (and justify). Assume an additive-white Gaussian noise channel.

Problem 5-11. Design a hardware configuration for mappers for the constellations in Fig. 5-14.

Problem 5-12.

(a) Give an example of a pulse $h(t)$ with time-duration that is exactly two symbol periods ($2T$) (and hence it is not bandlimited) and obeys the Nyquist criterion at the output of a matched filter, $\rho_h(k) = E_h\delta_k$.

(b) Repeat part (a) for three symbol periods ($3T$).

Problem 5-13.

(a) Show that the pulse autocorrelation obeys the symmetry relation $\rho_h(k) = \rho_h{}^*(-k)$.

(b) Show that the folded spectrum is non-negative real valued on the unit circle.

Problem 5-14. Define

$$S_{h,+}(z) = \sum_{k=0}^{\infty} \rho_h(k) z^{-k} , \tag{5.134}$$

and show that the folded spectrum is

$$S_h(z) = S_{h,+}(z) + S_{h,+}{}^*(1/z^*) - \rho_h(0) . \tag{5.135}$$

This gives a convenient way to calculate the folded spectrum.

Problem 5-15. Let $h_0(t)$ be a complex-valued pulse shape that has energy E_0 and that is orthogonal to all its translates by multiples of the symbol interval T. Let $F(z) = \sum_{k=0}^{\mu} f_k z^{-k}$ be a general FIR filter with memory μ, and define a pulse shape

$$h(t) = \sum_{k=0}^{\mu} f_k h_0(t - kT) . \tag{5.136}$$

(a) Show that

$$\rho_h(k) = E_0 f_k * f_{-k}{}^* , \tag{5.137}$$

and hence that

$$S_h(z) = E_0 F(z) F^*(1/z^*) . \tag{5.138}$$

(b) What is the pulse energy?

Problem 5-16. Compare $Q(d/2\sigma)$ and $Q^2(d/2\sigma)$ for values of $d = 2$ and $\sigma = 1/2$. Do it again for $\sigma = 1/4$. Is the approximation in (5.107) valid for these values of σ? You may use to approximate $Q(\cdot)$.

Problem 5-17. Consider the 4-PSK constellation $\{\pm\sqrt{E}, \pm j\sqrt{E}\}$ with unit energy, $E = 1$, as shown in Fig. 5-37. Assume that the real and imaginary parts of additive Gaussian noise are independent zero-mean Gaussian with variance $\sigma^2 = N_0/2 = 1/16$. Assuming that the transmitted symbol is -1, find a numerical value for the probability that the received sample is closer to j than to -1, and compare it to the probability that the received sample is closer to 1 than to -1. You may use Fig. 3-1 to estimate the probabilities.

Problem 5-18. Show that the probability of error for the 16-QAM constellation of Fig. 5-35 can be written

$$\Pr[\text{error}] = 3Q(d/2\sigma) - 2.25Q^2(d/2\sigma)\,. \tag{5.139}$$

Problem 5-19. Consider an M-QAM alphabet in the practical case when $M = 2^b$ and b is even. Show that the probability that the minimum-distance receiver makes a decision error is exactly given by (5.113), assuming additive white Gaussian noise.

Problem 5-20.

(a) Find the union bound on the probability of error for the 16-QAM constellation in Fig. 5-35. Assume $a_k = c + jc$ is actually transmitted.

(b) The CCITT V.29 standard for full-duplex transmission at 9600 b/s over voiceband channels uses the constellation shown in Fig. 5-39. Find the union bound on the probability of error. Assume $a_k = 1 + j$ is transmitted.

(a) Explain why the exact analysis technique that lead to (5.113) would be difficult to apply for the V.29 constellation.

(b) Find c in Fig. 5-35 so that the two constellations have the same power. Use the union bounds of parts (a) and (b) to compare their performance.

Problem 5-21. Show that the minimum-distance sequence detector of Fig. 5-22 reduces to that of Fig. 5-19 when there is no ISI.

Problem 5-22. Suppose we add a first-order rational causal allpass filter before the Euclidean norm in Fig. 5-23, or equivalently in both the forward paths. Combine the allpass filter with the precursor equalizer in the top path, and combine the allpass filter with the channel model filter in the bottom path. For simplicity, assume $\gamma^2 = 1$.

(a) Determine the resulting transfer functions of the precursor equalizer and discrete-time channel model filters.

(b) Show that it is not possible to use the allpass filter to turn the new precursor equalizer into a causal filter.

(c) Determine the impulse response of the new channel model filter.

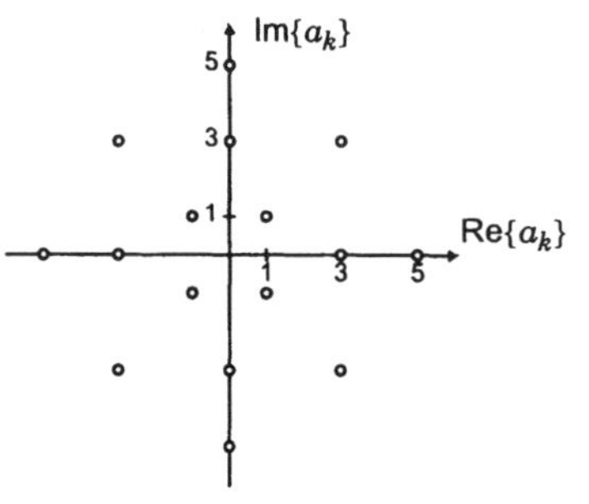

Fig. 5-39. The constellation for the CCITT V.29 standard for transmission at 9600 b/s over voiceband channels.

Problem 5-23. Consider a real-valued memoryless channel $Y = X + N$ where $N \sim \mathcal{N}(0,\ 0.1)$, and suppose the input is constrained to have unit energy, $E[X^2] \le 1$. If the input is selected randomly from an L-ary PAM alphabet $\mathcal{A} = \{\pm c, \pm 3c, \ldots, \pm(L-1)c\}$, how big can L be in order to maintain a bit-error probability of roughly 10^{-6} using minimum-distance detection?

Problem 5-24. Suppose we add a first-order rational causal allpass filter before the Euclidean norm in Fig. 5-24, or equivalently in both the forward paths. As in Problem 5-22, combine the allpass filter with the precursor equalizer and channel-model filter.

(a) Determine the transfer functions of the precursor equalizer and discrete-time channel model filters.

(b) Show that it is possible to turn the new precursor equalizer filter in the upper path into a causal filter by choosing the allpass filter appropriately. What is the allpass filter, and what are the resulting transfer functions of the precursor equalizer and channel-model filters?

(c) Determine the impulse response of the new channel model filter for the general case of a.

(d) Repeat c. for the particular allpass filter of b.

Problem 5-25. For a possibly complex-valued WSS stationary random process $S(t)$, define a cosine-modulated version

$$Z(t) = \sqrt{2}\cos(2\pi f_c t + \Theta)S(t) \quad , \tag{5.140}$$

where Θ is uniformly distributed over $[0, 2\pi)$ and is independent of $S(t)$. Show that with this uniformly distributed carrier phase, $Z(t)$ is WSS and find its power spectrum. Further, show that without the random phase $Z(t)$ is not WSS, and explain why not.

Problem 5-26. Show that $X(t) = \sqrt{2}\,\mathrm{Re}\{e^{j(2\pi f_c t + \Theta)}S(t)\}$ is WSS for any WSS $S(t)$ when Θ is uniformly distributed on $[0, 2\pi)$.

Problem 5-27. Let $Y(t)$ be a zero-mean WSS real-valued random process, and let $\hat{Y}(t)$ be its Hilbert transform. Show that the signal

$$X(t) = \sqrt{2}\,\mathrm{Re}\{e^{j2\pi f_c t}(Y(t) + j\hat{Y}(t))\} \tag{5.141}$$

is WSS and find its power spectrum.

References

1. T. Noguchi, Y. Daido, and J. Nossek, "Modulation Techniques for Microwave Digital Radio," *IEEE Communications Magazine*, Vol. 24, no. 10, p. 21, Oct. 1986.
2. D. Taylor and P. Hartmann, "Telecommunications by Microwave Digital Radio," *IEEE Communications Magazine*, Vol.24, no. 8, p. 11, Aug. 1986.
3. J. M. Wozencraft and I. M. Jacobs, *Principles of Communication Engineering*, Wiley, New York 1965.

4. C. R. Cahn, "Combined Digital Phase and Amplitude Modulation Communication Systems," *IRE Transactions on Communications Systems*, Vol. CS-8, pp. 150-154, 1960.

5. J. C. Hancock and R. W. Lucky, "Performance of Combined Amplitude and Phase-Modulated Communication Systems," *IRE Trans. Communications Systems*, Vol. CS-8, pp. 232-237, 1960.

6. C. N. Campopiano and B. G. Glazer, "A Coherent Digital Amplitude and Phase Modulation Scheme," *IRE Transactions on Communications Systems*, Vol. CS-10, pp. 90-95, 1962.

7. R. W. Lucky and J. C. Hancock, "On the Optimum Performance of *N*-ary Systems Having Two Degrees of Freedom," *IRE Trans. Communications Systems*, Vol. CS-10, pp. 185-192, 1962.

8. R. W. Lucky, *Digital Phase and Amplitude Modulated Communications Systems*, Purdue University, Lafayette, IN, 1961.

9. G. J. Foschini, R. D. Gitlin, and S. B. Weinstein, "Optimization of Two-Dimensional Signal Constellations in the Presence of Gaussian Noise," *IEEE Trans. Comm.*, Vol. 22, no. 1, Jan. 1974.

10. G. D. Forney, Jr., R. G. Gallager, G. Lang, F. M. Longstaff, S. U. Qureshi, "Efficient Modulation for Band-Limited Channels," *IEEE J. Selected Areas Comm.*, Vol. 2, No. 5 Sep. 1984.

11. A. Gersho and V. B. Lawrence, "Multidimensional Signal Constellations for Voiceband Data Transmission," *IEEE Journal on Selected Areas in Communications,* Vol. 2, no. 5, Sep. 1984.

12. I. S. Gradshteyn, I. M. Ryzhik, *Table of Integrals, Series, and Products*, Academic Press, 1980.

13. I. Korn, *Digital Communications*, Van Nostrand Reinhold, New York, 1985.

14. A. J. Viterbi, "Error Bounds for Convolutional Codes and an Asymptotically Optimum Decoding Algorithm," *IEEE Trans. on Information Theory*, Vol. IT-13, pp. 260-269, April 1967.

15. G. D. Forney, Jr., "Maximum-Likelihood Sequence Estimation of Digital Sequences in the Presence of Intersymbol Interference," *IEEE Trans Info. Theory*, Vol. 18, pp. 363-378 May 1972.

16. G. D. Forney, Jr., and G. Ungerboeck, "Modulation and Coding for Linear Gaussian Channels," *IEEE Transactions on Information Theory*, Vol. 44, No. 6, pp. 2384-2415, October 1998.

17. A. J. Viterbi, "Convolutional Codes and their Performance in Communication Systems," *IEEE Trans. on Communication Tech.*, Vol. COM-19, pp. 751-772, Oct. 1971.

18. G. D. Forney, Jr., "The Viterbi Algorithm," *Proceedings of the IEEE*, Vol. 61 (3), March 1973.

19. J. K. Omura, "On the Viterbi Decoding Algorithm," *IEEE Trans. on Information Theory*, Vol. IT-15, pp. 177-179, 1969.

20. R. E. Bellman, *Dynamic Programming*, Princeton University Press, Princeton, NJ, 1957.

21. A. J. Viterbi and J. K. Omura, *Principles of Digital Communication and Coding*, McGraw-Hill, 1979.

6

Advanced Modulation

In Chapter 5 we described transmitter and receiver design using PAM. In this chapter we extend to other modulation schemes. We begin by considering a general form of modulation called M-ary modulation, in which one of M signals is transmitted every signaling interval. In this general setting, we present the correlation and projection receivers as practical means for implementing the minimum-distance receiver, and we present a union-bound approximation for the resulting probability of error in the presence of AWGN.

We then specialize to orthogonal modulation, for which the M candidate signals are equal energy and orthogonal. Orthogonal modulation includes frequency-shift keying and pulse-position modulation as special cases. We examine the generalized Nyquist criterion, which lower-bounds the bandwidth needed to accommodate M orthogonal pulses without ISI. We show that orthogonal modulation has good power efficiency but poor spectral efficiency, the opposite of the large-alphabet PAM schemes considered in the previous chapter. Orthogonal modulation is then combined with PAM to give orthogonal pulse-amplitude modulation (OPAM). Two practical examples of this combination, multicarrier modulation and code-division multiplexing, are then described. The relationship between bandwidth and signal dimensionality will be explored with the aid of the Landau-Pollak theorem. Finally, we introduce the rate-normalized SNR as a means for comparing the performance of different modulation schemes with eachother, and also comparing them to the Shannon limit.

6.1. M-ary Modulation

In baseband PAM, symbols $\{a_k\}$ modulate the amplitude of a pulse train according to:

$$s(t) = \sum_{k=-\infty}^{\infty} a_k g(t-kT) . \tag{6.1}$$

A single pulse shape $g(t)$ is used in one signaling interval, and its amplitude is modulated by a data symbol. In this chapter we generalize the PAM model to *M-ary modulation* by allowing the pulse shape in any signaling interval to be chosen from a set of M possibilities, $\{g_i(t);\ 0 \le i \le M-1\}$, to represent $\log_2 M$ bits of information. The transmitted signal is then:

$$s(t) = \sum_{k=-\infty}^{\infty} g_{a_k}(t-kT) , \tag{6.2}$$

where a_k takes on values in the set $\{0, 1, \ldots, M-1\}$. The data symbol thus indexes *which pulse* is transmitted in the k-th symbol interval, rather than the *amplitude* of the pulse that is transmitted.

Example 6-1.
In Chapter 5 the 8-PAM signal set was defined as $\{ag(t) : a \in \mathcal{A}\}$ with alphabet $\mathcal{A} = \{\pm 1, \pm 3, \pm 5, \pm 7\}$. In terms of the new notation of (6.2), the 8-PAM signal set is $g_0(t) = -7g(t)$, $g_1(t) = -5g(t), \ldots, g_7(t) = 7g(t)$, as sketched below for a rectangular pulse:

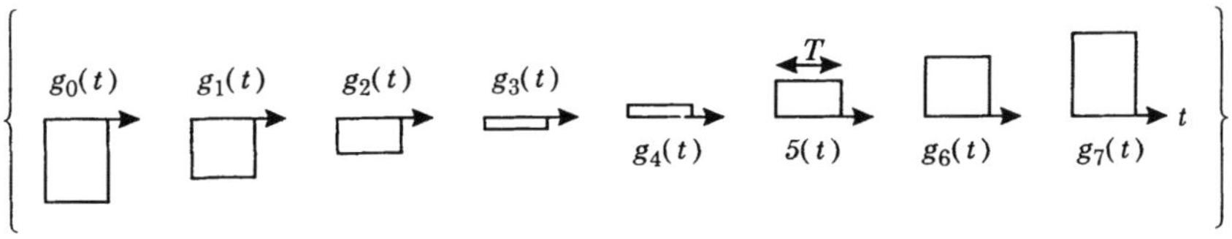

6.1.1. Baseband Equivalent Model

For passband systems, the pulses $\{g_i(t);\ 0 \le i \le M-1\}$ often represent the complex envelopes of the transmitted signal. In that case, the transmitted passband signal will be

$$x(t) = \sqrt{2}\,\mathrm{Re}\{e^{j2\pi f_c t} s(t)\} . \tag{6.3}$$

An alternative viewpoint is to define the passband equivalent pulses

$$\hat{g}_i(t) = \sqrt{2}\,\mathrm{Re}\{e^{j2\pi f_c t} g_i(t)\}. \tag{6.4}$$

These can then be used to form directly the passband signal

$$x(t) = \sum_{k=-\infty}^{\infty} \hat{g}_{a_k}(t-kT) . \tag{6.5}$$

Both interpretations will be useful.

Exercise 6-1.

Show that the correlation between two passband pulses is equal to the real part of the correlation between the corresponding complex envelopes:

$$\int_{-\infty}^{\infty} \hat{g}_i(t)\,\hat{g}_j(t)\,dt = \mathrm{Re}\left\{\int_{-\infty}^{\infty} g_i(t) g_j^*(t)\,dt\right\}, \tag{6.6}$$

where $\hat{g}_i(t)$ is defined by (6.4). This implies, among other things, that the energy of a pulse is equal to the energy of its complex envelope, and that two pulses are orthogonal if and only if their complex envelopes are orthogonal.

6.1.2. One-Shot Minimum-Distance Detection

A one-shot M-ary transmitter selects the transmitted signal from an M-ary signal set $\{g_0(t), \ldots g_{M-1}(t)\}$. If the n-th pulse $g_n(t)$ is transmitted, the receiver observation waveform is:

$$r(t) = h_n(t) + n(t)\,, \tag{6.7}$$

where $h_n(t)$ is the n-th *received pulse* and where $n(t)$ is noise. In terms of the Fourier transform of the channel impulse response $b(t)$, the received pulse is:

$$H_n(f) = G_n(f)B(f + f_c)\,. \tag{6.8}$$

The model (6.7) and (6.8) applies to both the baseband and passband cases. In the baseband case, $g_n(t)$ is real-valued and transmitted directly with no upconversion, so we set the carrier frequency to zero in (6.8), yielding $h_n(t) = g_n(t) * b(t)$. In the passband case where $g_n(t)$ is the complex envelope of the transmitted signal, the waveform $r(t)$ in (6.7) represents the received signal after downconversion, and (6.8) is equivalent to $h_n(t) = g_n(t) * \tilde{b}(t)/\sqrt{2}$. Since the model (6.7) is valid for both the baseband and passband case, we will focus on the more general passband case in the following.

Correlation Receiver

In Section 5.3 we introduced the minimum-distance detector in the context of PAM; it chooses the symbol that best describes the observation in a minimum-distance (or minimum error energy) sense. This strategy generalizes in a natural way to the case of general M-ary modulation. Given the observation (6.7), the minimum-distance receiver decides on the signal $h_i(t)$ that is closest to $r(t)$ in a minimum-distance sense, minimizing:

$$J = \int_{-\infty}^{\infty} |\,r(t) - h_i(t)\,|^2\,dt\,. \tag{6.9}$$

Expanding the integrand into three terms and integrating each separately yields:

$$J = E_r - 2\mathrm{Re}\left\{\int_{-\infty}^{\infty} r(t)\,h_i^*(t)\,dt\right\} + E_i\,, \tag{6.10}$$

where E_r is the energy of $r(t)$, and where E_i is the energy of the i-th received pulse:

$$E_r = \int_{-\infty}^{\infty} |r(t)|^2 dt, \qquad E_i = \int_{-\infty}^{\infty} |h_i(t)|^2 dt. \tag{6.11}$$

Let us introduce $y_i = \text{Re}\{\langle r(t), h_i(t)\rangle\}$ as the real correlation between the observation and the i-th received pulse:

$$y_i = \text{Re}\left\{\int_{-\infty}^{\infty} r(t)\, h_i^*(t) dt\right\}. \tag{6.12}$$

Because the first term in (6.10) is independent of the i-th pulse, the minimum-distance receiver can equivalently minimize the second two terms, or equivalently *maximize* the following:

$$J_i = y_i - \frac{1}{2} E_i. \tag{6.13}$$

This defines the correlation receiver for the an isolated pulse of a general M-ary signal set. A block diagram is shown in Fig. 6-1. Intuitively, the receiver correlates the observation with each of the possible received signals $\{h_0(t), \ldots h_{M-1}(t)\}$, then subtracts half the energy of the corresponding pulse, yielding J_i. The minimum-distance receiver compares the numbers $J_0, \ldots J_{M-1}$ that result and selects the largest.

There are two common variations of the correlation receiver shown in Fig. 6-1. First, the correlations of (6.12) can be calculated directly using a correlator (Section 5.3.1) instead of a matched filter. Second, exploiting (6.6), they can be calculated directly in terms of the real

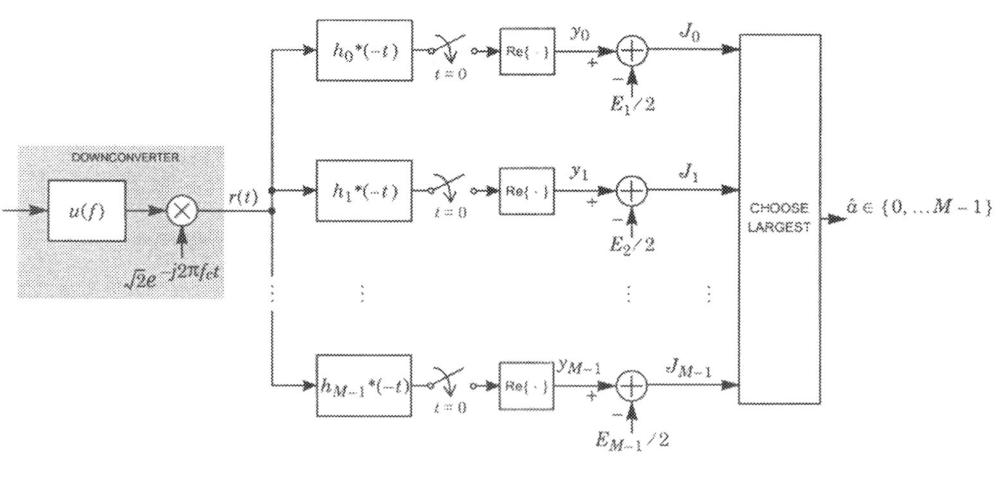

Fig. 6-1. The correlation receiver, which implements the minimum-distance receiver for an isolated pulse of general passband M-ary modulation. The downconverted signal is correlated against the M possible received pulses using a bank of sampled matched filters. Energies are subtracted from each correlation and the largest of the results determines the minimum-distance decision.

pulses themselves instead of in terms of the complex envelopes. This would be appropriate for baseband modulation where there is no upconverter at the transmitter. These two variations are illustrated in Fig. 6-2, which shows a correlation receiver for M-ary modulation implemented using correlators.

Projection Receiver

The complexity of the correlation receiver is dominated by the number M of correlations it must compute, either using a correlator or a matched filter. Since M can be very high in some applications ($M = 512$ is not unheard of), there is a need to reduce complexity. This section derives an alternative method for implementing the minimum-distance receiver called the projection receiver. We will rely heavily on the signal-space ideas introduced in Section 2.6.

Let $\mathcal{S} = \text{span}\{h_0(t), \ldots h_{M-1}(t)\}$ denote the *signal space*, and let $\hat{r}(t)$ denote the projection of $r(t)$ onto $\mathcal{S}$. Then because the projection error $r(t) - \hat{r}(t)$ is orthogonal to everything residing in $\mathcal{S}$, including $\hat{r}(t) - h_i(t)$, the minimum-distance cost function of (6.9) reduces to:

$$J = \int_{-\infty}^{\infty} | r(t) - \hat{r}(t) + \hat{r}(t) - h_i(t) |^2 dt$$

$$= \int_{-\infty}^{\infty} | r(t) - \hat{r}(t) |^2 dt + \int_{-\infty}^{\infty} | \hat{r}(t) - h_i(t) |^2 dt . \quad (6.14)$$

Because the first term is independent of the candidate decision $h_i(t)$, the minimum-distance receiver can equivalently minimize:

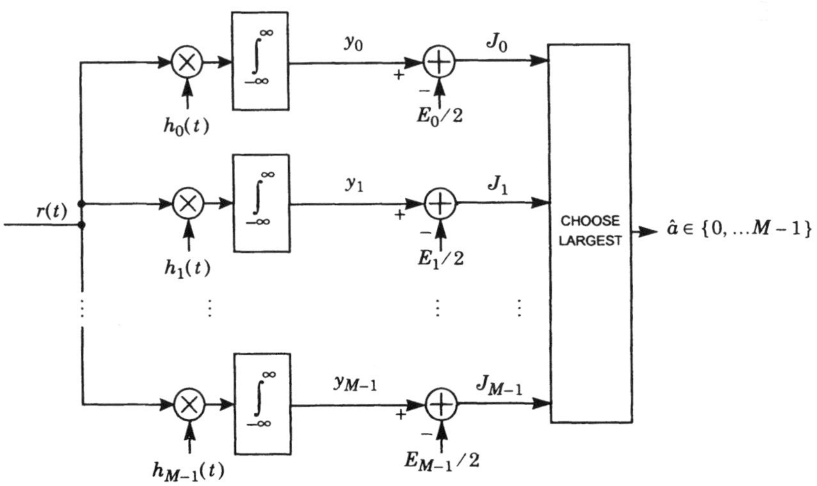

Fig. 6-2. The correlation receiver for the special case of baseband M-ary modulation, where the received signal and received pulses are real.

$$J' = \int_{-\infty}^{\infty} | \hat{r}(t) - h_i(t) |^2 dt . \tag{6.15}$$

Let $\{\phi_1(t), \dots \phi_N(t)\}$ be an orthonormal basis for the signal space $\mathcal{S}$, as described in Section 2.6. For example, the basis may be the result of applying the Gram-Schmidt procedure to the signal set $\{h_0(t), \dots h_{M-1}(t)\}$. Let $\boldsymbol{h}_i = [h_{i,1}, h_{i,2}, \dots h_{i,N}]^T$ denote the vector of projection coefficients for $h_i(t)$, where:

$$h_{i,j} = \int_{-\infty}^{\infty} h_i(t)\phi_j^*(t)dt . \tag{6.16}$$

From the Parseval's relationship of (2.99), the minimum-distance cost of (6.15) reduces to:

$$J' = \| \boldsymbol{r} - \boldsymbol{h}_i \|^2 , \tag{6.17}$$

where $\boldsymbol{r} = [r_1, \dots r_N]^T$ is the projection vector for $r(t)$, with:

$$r_j = \int_{-\infty}^{\infty} r(t)\phi_j^*(t)dt . \tag{6.18}$$

The new cost function (6.17) suggests a new way of implementing the minimum distance receiver, called the *projection receiver*. First, the received waveform is projected onto the signal space basis functions, yielding an N-dimensional observation vector $\boldsymbol{r}$. Then the cost (6.17) is minimized. Intuitively, the receiver "rounds" the projection vector $\boldsymbol{r}$ to the nearest signal vector in the N-dimensional complex space. A block diagram of the projection receiver is shown in Fig. 6-3.

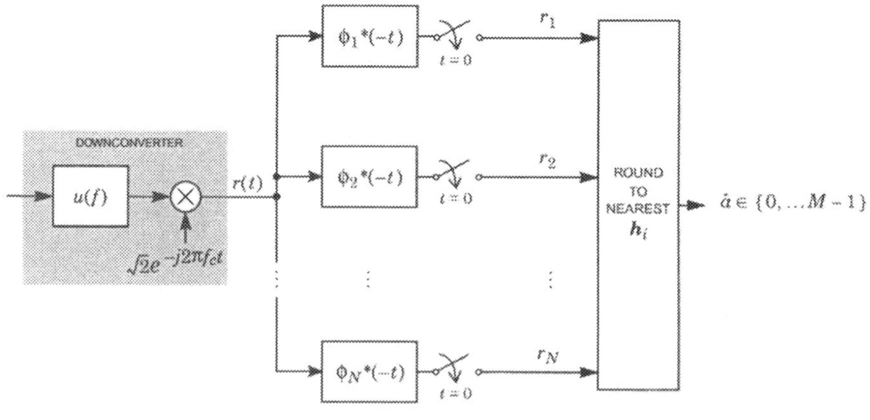

Fig. 6-3. The projection receiver for an isolated pulse of passband M-ary modulation.

The complexity of the projection receiver is roughly governed by N, the signal space dimension, while the complexity of the correlation receiver is governed by M, the size of the signal set. Since $N \le M$, the projection receiver will never be more complex than the correlation receiver. In some cases the complexity savings of the projection receiver is significant.

Example 6-2.

In concept one could implement the minimum-distance receiver for 16-QAM using the correlation receiver of Fig. 6-1, but the complexity would be high, since there would be sixteen sampled matched filters. The projection receiver is much simpler. In particular, since all elements of the 16-QAM signal set take the form $ah(t)$, an application of the Gram-Schmidt procedure leads to a one-dimensional signal space spanned by $\phi(t) = h(t)/\sqrt{E_h}$. So the projection receiver of Fig. 6-3 reduces to a downconverter, a single matched filter, a sampler, and a slicer. In fact, this is precisely the same receiver structure as the one-shot minimum-distance receiver derived in the previous chapter and shown in Fig. 5-19.

6.2. Probability of Error

We will now calculate bounds on the probability of error for a general M-ary modulation format, assuming white Gaussian noise and a minimum-distance receiver.

6.2.1. Performance in AWGN

Let us look closer at the j-th projection coefficient of (6.18), assuming that the i-th pulse was received:

$$\begin{aligned} r_j &= \int_{-\infty}^{\infty} h_i(t)\phi_j^*(t)dt + \int_{-\infty}^{\infty} n(t)\phi_j^*(t)dt \\ &= h_{i,j} + n_j , \end{aligned} \tag{6.19}$$

where

$$n_j = \int_{-\infty}^{\infty} n(t)\phi_j^*(t)dt . \tag{6.20}$$

Therefore, the projection vector is given by:

$$\boldsymbol{r} = \boldsymbol{h}_i + \boldsymbol{n} , \tag{6.21}$$

where $\boldsymbol{n} = [n_1, \ \dots \ n_N]^T$ is a vector of noise samples.

Properties of The Gaussian Noise Vector

Fact:

Let $n(t)$ be the complex envelope of real white-Gaussian noise with PSD $N_0/2$, and let n_i denote the correlation of (6.20) between $n(t)$ and the i-th element of an orthonormal basis $\{\phi_1(t), \dots \phi_N(t)\}$. When the bandwidths of the basis functions are all less than the carrier frequency, the components of the vector $\boldsymbol{n} = [n_1, \ \dots, \ n_N]^T$ are *i.i.d.* $\mathcal{CN}(0, N_0)$.

The bandwidth restriction is nearly always met in practice and is not a serious constraint. Intuitively it requires only that the passband signal (after upconversion) not overlap d.c., which is usually the case.

Proof:

We have already seen that the complex envelope of white Gaussian noise is circularly symmetric and Gaussian (Section 3.2.8). The random variables $\{n_1, \ldots n_N\}$ defined by (6.20) are all linear functions of the same circularly symmetric Gaussian random process, and hence they form a circularly symmetric jointly Gaussian random vector. To completely specify its statistics we need only identify its mean and autocorrelation matrix. Since $n(t)$ has zero mean it is clear from (6.20) that the components of $\boldsymbol{n}$ have zero mean. The i,j-th element of the autocorrelation matrix for $\boldsymbol{n}$ is:

$$\begin{aligned} \mathrm{E}[n_i n_j^*] &= \mathrm{E}\left[\int_{-\infty}^{\infty} n(t)\phi_i^*(t)dt \int_{-\infty}^{\infty} n^*(\tau)\phi_j(\tau)d\tau\right] \\ &= \int_{-\infty}^{\infty}\int_{-\infty}^{\infty} \mathrm{E}[n(t)n^*(\tau)]\phi_i^*(t)\phi_j(\tau)dt\,d\tau \\ &= N_0\int_{-\infty}^{\infty}\int_{-\infty}^{\infty} \delta(t-\tau)\phi_i^*(t)\phi_j(\tau)dt\,d\tau \\ &= N_0\int_{-\infty}^{\infty} \phi_i^*(t)\phi_j(t)dt \\ &= N_0\delta_{i,j}\,. \end{aligned} \tag{6.22}$$

As shown in Fig. 6-3, we can realize (6.20) by sampling the output of a filter matched to the j-th basis function when the noise complex envelope $n(t)$ is the input. This interpretation is useful because it clearly indicates the implications of the bandwidth assumption. Specifically, although strictly speaking the PSD for the complex envelope $n(t)$ is $N_0 u(f+f_c)$, the assumption that the carrier frequency exceeds the bandwidth of the basis functions allows us to replace the input PSD $N_0 u(f+f_c)$ by N_0 without affecting the filter outputs. This explains the third equality in (6.22). The last equality follows from the fact that the basis set is orthonormal.

The result (6.22) implies that the autocorrelation matrix for $\boldsymbol{n}$ is diagonal, so that the components of $\boldsymbol{n}$ are uncorrelated. Since uncorrelated circularly symmetric Gaussian random variables are also independent, this completes the proof.

This fact has several important implications that will be exploited in this chapter. Specifically the noise vector $\boldsymbol{n}$ has the following properties:

- The set of $2N$ real-valued random variables $\{\mathrm{Re}\{n_1\}, \ldots \mathrm{Re}\{n_N\}, \mathrm{Im}\{n_1\}, \ldots \mathrm{Im}\{n_N\}\}$ are mutually independent, zero mean and Gaussian with variance $N_0/2$.
- The vector $\boldsymbol{n}$ is circularly symmetric and Gaussian with zero mean and $\mathrm{E}[\boldsymbol{n}\boldsymbol{n}^*] = N_0\mathbf{I}$.
- The components of $\boldsymbol{n}$ are uncorrelated, so that $\mathrm{E}[n_i n_j^*] = 0$ for $i \neq j$.
- The components of $\boldsymbol{n}$ are circularly symmetric, or $\mathrm{E}[n_i n_j] = 0$ for $1 \leq i, j \leq N$.
- A complex random variable defined as follows:

$$X = \langle \boldsymbol{n}, \boldsymbol{e} \rangle = \boldsymbol{e}^*\boldsymbol{n} \tag{6.23}$$

will be $\mathcal{CN}(0, N_0)$ for *any* vector $\boldsymbol{e}$ with unit length (satisfying $\| \boldsymbol{e} \| = 1$).

The last item in the list deserves further clarification. Clearly X is Gaussian, since it is a linear combination of independent Gaussian random variables. Further, it is circularly symmetric ($\mathrm{E}[X^2] = 0$), since it is a linear function of a circularly symmetric Gaussian vector. This implies that $\mathrm{Re}\{X\}$ and $\mathrm{Im}\{X\}$ are identically distributed and independent. To determine the statistics of X, we need only determine is its variance. Calculating this directly,

$$\mathrm{E}[\,|X|^2] = \mathrm{E}\Big[\sum_{i=1}^{N}\sum_{k=1}^{N} n_i n_k{}^* e_i{}^* e_k\Big] = \sum_{i=1}^{N} \mathrm{E}[\,|n_i|^2]\,|e_i|^2 = N_0 \sum_{i=1}^{N} |e_i|^2 = N_0\,. \tag{6.24}$$

Thus X has the same variance as the components of $\boldsymbol{n}$, namely N_0, and as a result, the real and imaginary parts of X each have variance $N_0/2$. This result can also be explained intuitively, since $\langle \boldsymbol{n}, \boldsymbol{e} \rangle$ is the projection of $\boldsymbol{n}$ on the span of a unit-magnitude vector $\boldsymbol{e}$, or the component of $\boldsymbol{n}$ in the direction of unit-vector $\boldsymbol{e}$. Since $\boldsymbol{n}$ has the same variance in each of its components, it stands to reason that the variance of the component of $\boldsymbol{n}$ in *any* direction has the same variance, not just in the direction of the principal axes.

Vector-Valued Signal in Vector-Valued Noise

The projection receiver converts the continuous-time minimum-distance criterion (6.9) into a vector-valued minimum distance criterion (6.17). The projection process leads to the N-dimensional complex vector $\boldsymbol{r}$ of (6.21), consisting of a known signal vector and an additive complex Gaussian noise vector:

$$\boldsymbol{r} = \boldsymbol{h}_i + \boldsymbol{n}\,, \tag{6.25}$$

where $\boldsymbol{r} = [r_1, r_2, \ldots, r_N]^T$ is the projection vector for the received signal, and $\boldsymbol{h}_i = [h_{i,1}, h_{i,1}, \ldots, h_{i,N}]^T$ is the projection vector for the i-th received pulse. We will refer to the set of M known such vectors $\{\boldsymbol{h}_1, \ldots \boldsymbol{h}_M\}$ as *signal vectors.*

Bounds on the Probability of Error

We would like to determine the probability that the minimum-distance receiver makes an incorrect decision, given additive white Gaussian noise. However, a closed-form solution is not known in general, and so we often must resort to bounding the probability of error.

We will first determine the *pairwise error probability* $P_{i\to j}$, defined as the probability that the received signal is closer to $\boldsymbol{h}_j$ than it is to $\boldsymbol{h}_i$, given that $\boldsymbol{h}_i$ was transmitted, for some $j \neq i$. In other words

$$P_{i\to j} = \Pr\Big[\, \|\boldsymbol{r} - \boldsymbol{h}_j\|^2 < \|\boldsymbol{r} - \boldsymbol{h}_i\|^2 \,|\, \boldsymbol{h}_i \text{ transmitted} \Big]. \tag{6.26}$$

Substituting $\boldsymbol{h}_i + \boldsymbol{n}$ for $\boldsymbol{r}$,

$$\begin{aligned} P_{i\to j} &= \Pr\Big[\, \|\boldsymbol{n} - (\boldsymbol{h}_j - \boldsymbol{h}_i)\|^2 < \|\boldsymbol{n}\|^2 \Big] \\ &= \Pr\Big[\, \|\boldsymbol{n}\|^2 + \|\boldsymbol{h}_j - \boldsymbol{h}_i\|^2 - 2\mathrm{Re}\{\langle \boldsymbol{n}, \boldsymbol{h}_j - \boldsymbol{h}_i \rangle\} < \|\boldsymbol{n}\|^2 \Big]. \end{aligned} \tag{6.27}$$

Cancelling the $\| \boldsymbol{n} \|^2$ term and dividing both sides of the inequality by

$$d_{i,j} = \| \boldsymbol{h}_j - \boldsymbol{h}_i \| , \tag{6.28}$$

the distance between $\boldsymbol{h}_i$ and $\boldsymbol{h}_j$, we get equivalently

$$P_{i \to j} = \Pr\left[\text{Re}\{\langle \boldsymbol{n}, (\boldsymbol{h}_j - \boldsymbol{h}_i)/d_{i,j} \rangle\} > d_{i,j}/2 \right]. \tag{6.29}$$

Since the vector $(\boldsymbol{h}_j - \boldsymbol{h}_i)/d_{i,j}$ has unit length, (6.23) implies that the Re{ · } term above is a Gaussian random variable with variance $\sigma^2 = N_0/2$. Therefore, the pairwise error probability is

$$P_{i \to j} = Q\left(\frac{d_{i,j}}{2\sigma} \right), \tag{6.30}$$

where $Q(\cdot)$ is the integral of the tail of the unit-variance Gaussian distribution (Chapter 3).

Using this result, we can determine the error probability for the binary case ($M = 2$). Since in this case each symbol conveys one bit, the error probability reduces to the bit-error probability.

Example 6-3.
If $M = 2$, and if $\boldsymbol{r} = \boldsymbol{h}_0 + \boldsymbol{n}$, then an error occurs if $\boldsymbol{r}$ is closer to $\boldsymbol{h}_1$ than to $\boldsymbol{h}_0$. This occurs with probability

$$\Pr[\boldsymbol{h}_1 \text{ chosen} \mid \boldsymbol{h}_0 \text{ transmitted}] = Q\left(\frac{d_{01}}{2\sigma} \right). \tag{6.31}$$

On the other hand, if $\boldsymbol{h}_1$ is transmitted, $\boldsymbol{h}_0$ will be chosen with the same probability. Therefore, regardless of the *a priori* probabilities, the error probability for any binary signal set $\{h_0(t), h_1(t)\}$ is $Q(d/2\sigma)$, where $d = \| \boldsymbol{h}_0 - \boldsymbol{h}_1 \| = (\int_{-\infty}^{\infty} | h_0(t) - h_1(t) |^2 dt)^{1/2}$ and $\sigma^2 = N_0/2$.

Example 6-4.
The result of the previous example makes it easy to compare the performance of different binary modulation schemes. For example, consider the three cases of binary antipodal signaling, on-off keying, and binary orthogonal signaling shown below:

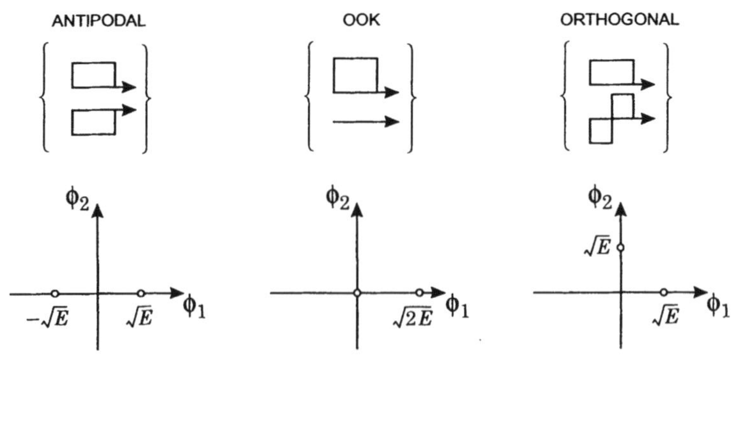

Assuming both signals are equally likely, the average energy is E in all cases. The distance for antipodal signaling is $2\sqrt{E}$, while in the other two cases the distance is $\sqrt{2E}$. Based on (6.31), the bit-error probability for antipodal signaling is thus $Q(\sqrt{2E/N_0})$, as compared to $Q(\sqrt{E/N_0})$ for OOK and orthogonal signaling. This implies that binary antipodal signaling enjoys a 3-dB advantage over OOK or binary orthogonal modulation.

The exact error probability for three or more signals ($M \geq 3$) can be difficult to calculate, since the minimum-distance decision boundary can be very complicated. We saw special cases in Chapter 5 where it is not too difficult. More generally, however, we can establish bounds on the error probability that are easy to apply. These bounds become tight as $\sigma \to 0$, and thus represent not only bounds, but also accurate approximations for small σ (small error probability). Since most digital communication systems operate at low error probability, these bounds are very useful.

The upper bound will be based on the *union bound* described in Chapter 3.1. For N events $\{\mathcal{E}_n, 1 \leq n \leq N\}$, the union bound is

$$\Pr[\bigcup_{n=1}^{N} \mathcal{E}_n] \leq \sum_{n=1}^{N} \Pr[\mathcal{E}_n] . \tag{6.32}$$

We begin by bounding the probability that the minimum-distance receiver makes an error when $\boldsymbol{h}_0$ is transmitted. If we define $\mathcal{E}_j$ as the event that $\boldsymbol{r} = \boldsymbol{h}_0 + \boldsymbol{n}$ is closer to $\boldsymbol{h}_j$ than $\boldsymbol{h}_0$, then the pairwise error probability $P_{0\to j}$ can be written as $P_{0\to j} = \Pr[\mathcal{E}_j]$. Then

$$\Pr[\text{error} \mid \boldsymbol{h}_0 \text{ transmitted}] = \Pr[\bigcup_{j=1}^{M-1} \mathcal{E}_j] \leq \sum_{j=1}^{M-1} \Pr[\mathcal{E}_j] . \tag{6.33}$$

But since

$$\Pr[\mathcal{E}_j] = P_{0\to j} = Q\left(\frac{d_{0j}}{2\sigma}\right), \tag{6.34}$$

we get

$$\Pr[\text{error} \mid \boldsymbol{h}_0 \text{ transmitted}] \leq \sum_{j=1}^{M-1} Q\left(\frac{d_{0j}}{2\sigma}\right) . \tag{6.35}$$

This is an upper bound on the probability of error, conditioned on $\boldsymbol{h}_0$ being transmitted.

It was shown in Chapter 3 that $Q(\cdot)$ is a very steep function of its argument for large arguments (corresponding to high SNR). This implies that the sum in (6.35) tends to be dominated by the term corresponding to the smallest argument. Define $d_{0,\min}$ as the smallest $d_{0,j}$, $1 \leq j \leq M-1$. Then the union bound of (6.35) can be approximated by

$$\Pr[\text{error} \mid \boldsymbol{h}_0 \text{ transmitted}] \approx K_0 Q\left(\frac{d_{0,\min}}{2\sigma}\right), \tag{6.36}$$

where K_0 is the number of signals that are distance $d_{0,\min}$ away from $\boldsymbol{h}_0$. We can no longer assert that (6.36) is an upper bound, since we have thrown away positive terms, making the right side smaller. However, for small σ, (6.36) remains an accurate approximation. It is also intuitive that the error probability would be dominated by the signals that are closest to $\boldsymbol{h}_0$, since the nearest signals are the ones most likely to be confused with $\boldsymbol{h}_0$.

A lower bound on error probability can also be established. Since $\bigcup_{j=1}^{M-1} \mathcal{E}_j$ contains $\mathcal{E}_m$ for any $1 \le m \le M-1$, we get that

$$\Pr[\bigcup_{j=1}^{M-1} \mathcal{E}_j] \ge \Pr[\mathcal{E}_m] = Q\left(\frac{d_{0,m}}{2\sigma}\right) . \tag{6.37}$$

Obviously, the bound is tightest when $d_{0,m} = d_{0,\min}$, since that will maximize the right side of (6.37). Thus, (6.37) acts as a lower bound on error probability. Combining it with the upper bound of (6.35), we arrive at:

$$Q\left(\frac{d_{0,\min}}{2\sigma}\right) \le \Pr[\text{error} \mid \boldsymbol{h}_0 \text{ transmitted}] \le \sum_{j=1}^{M-1} Q\left(\frac{d_{0,j}}{2\sigma}\right). \tag{6.38}$$

The approximation of (6.36) for the upper bound is seen to be only a factor of K_0 larger than the lower bound. This bound applies equally well for any other transmitted signal $\boldsymbol{h}_i$ with K_0 replaced by K_i and $d_{0,\min}$ replaced by $d_{i,\min}$, where $d_{i,\min}$ is the minimum distance from $\boldsymbol{h}_i$ to any other signal, and K_i is the number of signals at distance $d_{i,\min}$.

We are often interested in the overall probability of error P_e, defined as the probability that the wrong signal is chosen by the minimum-distance criterion. To calculate P_e, we must know $\{p_i, 0 \le i \le M-1\}$, the set of *a priori* probabilities of the M signals being transmitted. Then

$$P_e = \sum_{i=0}^{M-1} \Pr[\boldsymbol{h}_i \text{ not chosen} \mid \boldsymbol{h}_i \text{ transmitted}] \cdot p_i . \tag{6.39}$$

Substituting the union-bound approximation, P_e can be approximated as

$$P_e \approx \sum_{i=0}^{M-1} p_i K_i Q\left(\frac{d_{i,\min}}{2\sigma}\right) . \tag{6.40}$$

As before, (6.40) will be dominated by the terms with the smallest argument to $Q(\cdot)$. Thus,

$$P_e \approx K \cdot Q\left(\frac{d_{\min}}{2\sigma}\right) , \tag{6.41}$$

where K is a constant, called the *error coefficient*, and $d_{\min}$ is the minimum distance between any pair of signals. The error coefficient has the interpretation as the average number of signals at the minimum distance. Since K has a much milder impact on P_e and the argument of $Q(\cdot)$, the error probability at high SNR is dominated by the minimum distance $d_{\min}$.

6.3. Orthogonal Modulation

In this section we consider a special case of M-ary modulation called *orthogonal modulation*, in which the signal set is *orthogonal* and *equal energy*, meaning that

$$\int_{-\infty}^{\infty} g_i(t)g_j^*(t)dt = E_g\delta_{i-j}, \tag{6.42}$$

for some constant E_g. We will begin with two simplifications. First, we will avoid the issue of ISI by transmitting and receiving a single isolated pulse. (Later we will generalize the Nyquist criterion to design orthogonal signals that avoid ISI.) Second, we will assume the effect of the channel transfer function is benign, so that the corresponding received pulses $h_i(t)$ are also orthogonal and equal energy, namely:

$$\int_{-\infty}^{\infty} h_i(t)h_j^*(t)dt = E\delta_{i-j}, \tag{6.43}$$

where E is the energy of any of the received pulses. For practical applications, such as multicarrier and code-division multiple access, discussed below, this assumption is often valid. It is also obviously valid for channels with inherently flat frequency responses.

For reasons seen shortly, orthogonal signaling has poor spectral efficiency, and hence is rarely used when bandwidth is at a premium. Nonetheless, it is valuable as a starting point for more elaborate techniques that combine it with PAM (Section 6.4).

6.3.1. The Minimum-Distance Receiver for Orthogonal Modulation

The correlation and projection receivers of the previous section are easily adapted to the case of orthogonal modulation. When the pulse set is orthogonal, the Gram-Schmidt orthonormalization process has no impact other than to normalize each pulse to have unit energy. Therefore, the dimension of the signal space is equal to the size of the signal set, $N = M$. It follows that the projection receiver is no less complex than the correlation receiver. Because all signals have the same energy, subtracting the energies in Fig. 6-1 is not needed. Thus, with orthogonal modulation, the minimum-distance receiver simplifies to that shown in Fig. 6-4.

Intuitively, the minimum-distance receiver correlates the observation waveform with each of the possible received pulses. The fact that the received pulses $h_i(t)$ are orthogonal implies that when pulse i was transmitted, the i-th correlation y_i will be equal to $\sqrt{E}$ plus noise, while the other correlations $y_{k \neq i}$ will be noise only. So it makes intuitive sense that the maximum y_i should determine which pulse was transmitted. From a signal-space perspective, each y_i looks only in the direction of $h_i(t)$ in linear space by forming an inner product (cross-correlation) of $h_i(t)$ with the received signal.

Detection of an isolated pulse is only the beginning, of course. To detect a sequence of pulses, the matched-filter receiver in Fig. 6-4 can be modified so that samples are taken at multiples of T, rather than just once at $t = 0$. This will also be optimal in a minimum-distance sense if such sampling does not result in intersymbol interference.

6.3.2. Error Probability for Orthogonal Modulation

For orthogonal modulation, each signal projection vector is real and of the form

$$h_i = [0, 0, \dots 0, \sqrt{E}, 0, \dots, 0]^T \tag{6.44}$$

where the non-zero term is in the i-th position. Thus, every signal is the same distance from every other signal, namely $d = \sqrt{2E}$, so the minimum distance is $d_{\min} = \sqrt{2E}$. The geometric picture is sketched below for the case of $M = 3$:

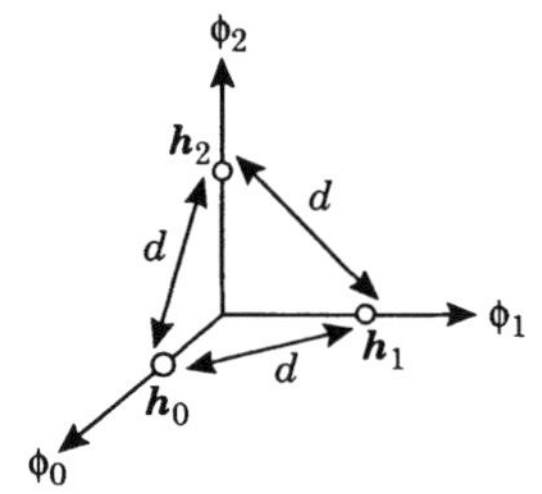

For M orthogonal signals, there are $M - 1$ other signals at the minimum distance, and the error probability is independent of which signal is transmitted. Using (6.41), the error probability is approximated by:

$$P_e \approx (M-1) \cdot Q\left(\frac{d_{\min}}{2\sigma}\right)$$

$$= (M-1) \cdot Q\left(\sqrt{\frac{E}{N_0}}\right) . \tag{6.45}$$

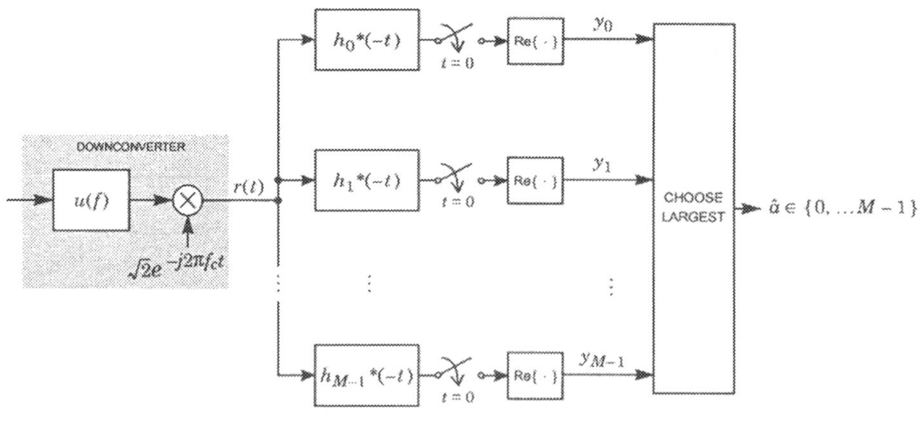

Fig. 6-4. The correlation receiver for an isolated pulse of passband orthogonal modulation. For the baseband case, there is no downconverter and there is no real-part operator.

The argument of $Q(\cdot)$ is not a function of M, but the error probability does increase slowly with M because of the factor $M-1$ multiplying $Q(\cdot)$. The basic trade-off is that, at the expense of more bandwidth and with only a minor penalty in error probability, the number of bits per symbol can be increased by increasing M. It is somewhat surprising that the required increase in bandwidth associated with increasing M does not result in a greater penalty in error probability, since increasing bandwidth is usually associated with allowing more noise into the receiver.

The exact error probability for orthogonal modulation can be calculated with relative ease because of the simplicity of the signal geometry. In particular, we showed in the previous section that the minimum-distance receiver simplifies, for this geometry, to the criterion

$$\max_{l}\{y_l\}\ , \tag{6.46}$$

or in words, the receiver chooses the largest correlation between the received signal and the orthogonal pulses. The error probability does not depend on which signal is transmitted, so assume that it is $\boldsymbol{h}_0$. In that case, all the y_l are independent real Gaussian random variables with variance $\sigma^2 = N_0/2$, and all are zero-mean except for y_0, which has mean $\sqrt{E}$. A correct decision is made if y_0 is larger than y_l, $1 \le l \le M-1$. Thus,

$$\Pr[\text{correct decision} \mid \boldsymbol{h}_0 \text{ transmitted}, y_0 = y] = \left(1 - Q\left(\frac{y}{\sigma}\right)\right)^{M-1} . \tag{6.47}$$

Since this error probability will be the same regardless of which signal is transmitted, we need only average (6.47) over the statistics of y_0, yielding:

$$\begin{aligned} \Pr[\text{error}] &= \Pr[\text{error} \mid \boldsymbol{h}_0 \text{ transmitted}] \\ &= 1 - \Pr[\text{correct decision} \mid \boldsymbol{h}_0 \text{ transmitted}] \\ &= 1 - \int_{-\infty}^{\infty} f_{y_0}(y)\left(1 - Q\left(\frac{y}{\sigma}\right)\right)^{M-1} dy, \end{aligned} \tag{6.48}$$

where $f_{y_0}(y)$ is the p.d.f. of a Gaussian random variable with mean $\sqrt{E}$ and variance σ^2,

$$f_{y_0}(y) = \frac{1}{\sqrt{2\pi\sigma^2}} e^{-(y-\sqrt{E})^2/(2\sigma^2)} . \tag{6.49}$$

The integral in (6.48) does not have a closed form solution, but is tabulated in [1] and plotted in Fig. 6-5 using solid lines. In contrast, the dashed lines in the figure were calculated using the approximation (6.45), which is seen to be very accurate at high SNR. Notice that for large SNR, the error probability is only weakly dependent on M, the number of orthogonal pulses, as predicted by the error probability approximation. Going from $M = 2$ to $M = 128$ results in less than a 2 dB penalty.

6.3.3. Examples of Orthogonal Modulation

Two canonical forms of orthogonal modulation are *frequency-shift keying (FSK)* and *pulse-position modulation (PPM)*. They are roughly duals of one another, since in FSK each signal gets a different slice of the frequency band, where as in PPM each signal gets a different slice of the signaling interval.

Frequency-Shift Keying

Example 6-5.

In *binary FSK*, two distinct carrier frequencies are used. An example is shown in Fig. 6-6, in which a lower-frequency pulse $h_0(t)$ is used to represent a binary "0" and a higher frequency pulse $h_1(t)$ is used to represent a binary "1". It is easy to show that these two pulse shapes are orthogonal if each contains an integral number of cycles in a symbol interval.

Because FSK is a special case of orthogonal modulation, we can use the correlation (or the equivalent matched filter) receiver.

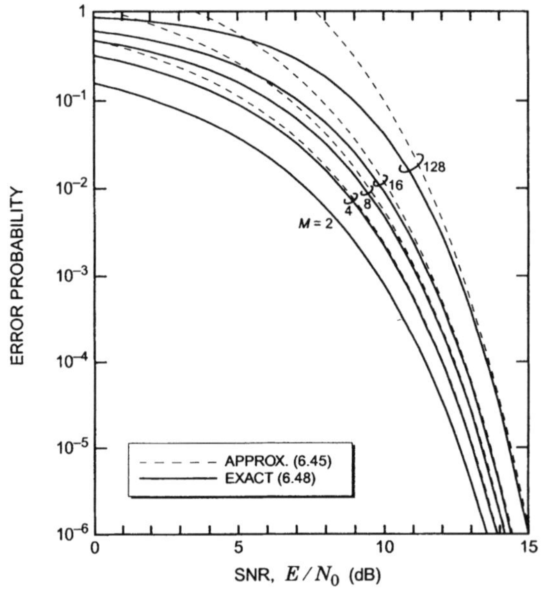

Fig. 6-5. Error probability for M orthogonal pulses with equal energy E and AWGN. The dashed lines were calculated using the approximation (6.45), which is seen to be accurate at high SNR.

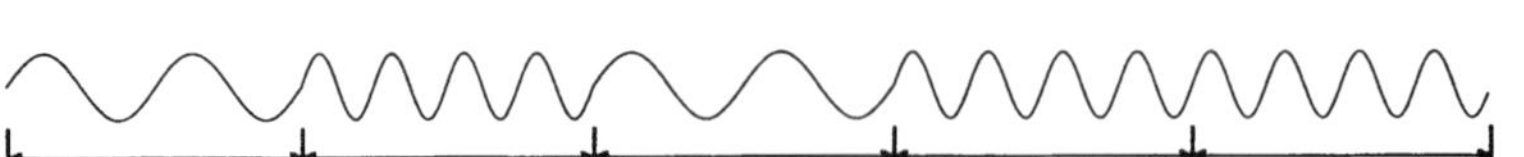

Fig. 6-6. An example of a binary FSK signal.

Example 6-6.
The binary FSK pulses of the previous example, properly normalized, can be written

$$h_0(t) = \sqrt{\frac{2E}{T}} \sin(2\pi f_0 t) w(t)$$
$$h_1(t) = \sqrt{\frac{2E}{T}} \cos(2\pi f_1 t) w(t), \tag{6.50}$$

where $w(t) = u(t) - u(t - T)$ is a rectangular *window* of duration T. The matched filters $f_i(t) = h_i(-t)$ have finite impulse response, but are non-causal. We can define causal matched filters,

$$f_0(t - T) = h_0(T - t),$$
$$f_1(t - T) = h_1(T - t). \tag{6.51}$$

For an isolated pulse, the sampling is delayed to $t = T$ instead of $t = 0$.

More generally, a set of M frequencies can be chosen, and the resulting pulse shapes will be orthogonal if each frequency is chosen with an integral number of cycles per symbol interval (this condition is sufficient, but not necessary).

Like all forms of orthogonal modulation, FSK is not spectrally efficient, but it does offer some advantages:

- *Incoherence.* For detection of passband PAM signals we have thus far assumed that the exact carrier frequency and phase is available at the receiver to perform the demodulation. However, as we will show in Chapter 15, carrier recovery is far from trivial, especially on certain channels where the received carrier phase varies rapidly. Examples of such channels are optical fiber and radio links to rapidly moving vehicles such as aircraft. Effective FSK receivers can be designed that make no attempt to recover the carrier phase. Such receivers are said to be *incoherent.*
- *Ease of implementation.* Very simple FSK modems are possible mainly because incoherent detection is feasible. For example, the receiver shown in Fig. 6-7 discriminates between transmitted frequencies simply by measuring the density of zero crossings. The performance of this receiver is difficult to analyze, partly because Gaussian noise on the channel does not imply Gaussian noise at the input to the slicer. (A similar receiver is successfully analyzed in [2].) Fortunately, as we saw, the correlation receiver is easier to analyze, and provides a lower bound on the probability of error for any suboptimal receiver.
- *Immunity from certain nonlinearities.* Most FSK modulation techniques result in a *constant envelope,* in which information is carried by the zero crossings of the signal

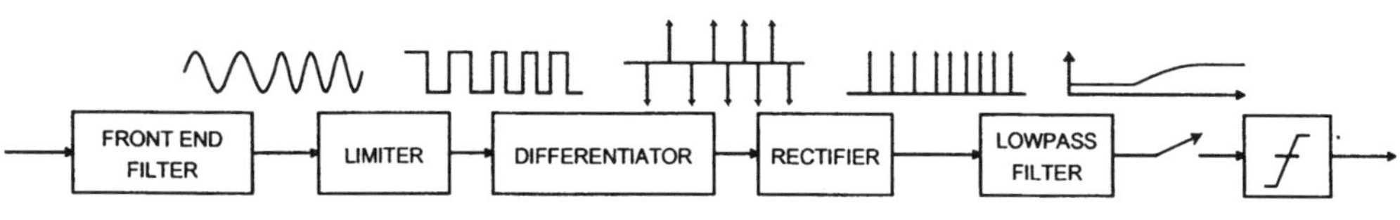

Fig. 6-7. A zero-crossing detector for binary FSK with the accompanying waveforms.

alone. Some channels, such as those using RF amplifiers operating at or near saturation, hard-limit the signal.

Example 6-7.

The CCITT V.21 300 b/s voiceband modem standard uses binary FSK with frequencies of 1080 ± 100 Hz for transmission in one direction and 1750 ± 100 Hz for transmission in the other direction. The motivation for using FSK in this case is the hardware simplicity of the receiver and the fact that no carrier or timing recovery is required. This standard was developed in an age of more expensive hardware built with discrete devices.

Example 6-8.

The transmit power of a satellite is limited, and more power can be generated if the transponder is operated in or near its nonlinear saturation region. The transponder is essentially a peak-power limiter, and a *constant envelope* modulation method such as FSK can achieve a higher average power for a given peak power.

FSK also has significant disadvantages, other than spectral inefficiency. Binary FSK suffers a 3 dB penalty in SNR required to achieve a given probability of error relative to binary antipodal PAM (recall Example 6-4). Because the basic FSK signal is not usually a linear function of the data, existing adaptive equalization techniques (Chapter 8) for linear modulation are not applicable. Compensation for channel distortion is therefore more difficult than for PAM signals. FSK is therefore primarily limited to channels such as fiber optics and satellite where frequency dispersion and selective fading are not a problem.

The rectangular window of Example 6-6 leads to an infinite bandwidth requirement. In contrast, very little bandwidth is required when a sinc-shaped window is used.

Exercise 6-2.

Let

$$h_n(t) = \sqrt{\frac{E}{T}}\left(\frac{\sin(\pi t/(2T))}{\pi t/(2T)}\right)\cos\left((n+\tfrac{1}{2})\frac{\pi t}{T}\right), \tag{6.52}$$

for $n \in \{0 \ldots, M-1\}$. Show that these pulses are ideally bandlimited to the frequency range $n/(2T) \le |f| < (n+1)/(2T)$, as shown in Fig. 6-8, so that the aggregate bandwidth occupied by the first M pulses is $M/(2T)$. Also show that they are orthogonal, so that $\langle h_i(t), h_j(t) \rangle = E\delta_{i,j}$.

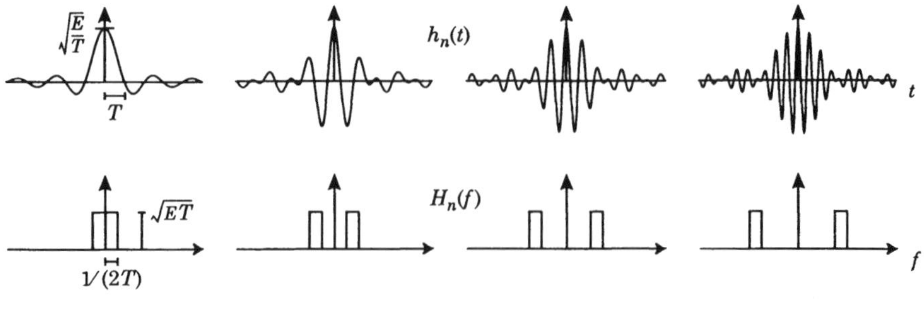

Fig. 6-8. Time domain (top) and frequency domain (bottom) plots of the pulses in (6.52) for $n = 1, 2, 3$, and 4.

Pulse-Position Modulation

Another example of orthogonal modulation is pulse-position modulation (PPM), widely used in optical communications. Roughly speaking, PPM divides up the signaling interval into M slots or *chips* of width T/M, and sends a pulse in exactly one of the slots to convey $\log_2 M$ bits of information. An example of a 4-ary PPM signal set with a rectangular pulse is sketched in Fig. 6-9(a). Nonrectangular shapes are also possible. A general M-ary PPM signal set is given by $\{g(t), g(t-T/M), \ldots g(t-(M-1)T/M)\}$. To avoid ISI and ensure orthogonality, the overall pulse shape after a chip-matched filter, namely $p(t) = g(t) * g^*(-t)$, must satisfy the Nyquist criterion when sampled at the chip rate M/T. The rectangular pulse works, but its bandwidth is too large for many applications. The pulse with minimal bandwidth that satisfies this criterion is $g(t) = \sqrt{EM/T}\sin(M\pi t/T)/(M\pi t/T)$. The 4-PPM signal set with this pulse is shown in Fig. 6-9(b). The bandwidth requirement in this case is $M/(2T)$, a factor of M larger than for the case of M-ary PAM.

Interestingly, comparing the minimum-bandwidth PPM signal set to the sinc-windowed FSK signal set of (6.52), we see that both require the same total bandwidth: $M/(2T)$. In the following we will see that this is the minimum bandwidth required for orthogonal signaling and no ISI.

6.3.4. The Generalized Nyquist Criterion

There is a fundamental lower bound on the bandwidth required by an orthogonal signal set, assuming that we wish to avoid ISI. The Nyquist criterion, discussed in Section 5.1.1, states that for baseband PAM with symbol rate T, the minimum signal bandwidth is $1/(2T)$ Hz. We can now generalize the Nyquist criterion and show that the minimum bandwidth of an orthogonal signal set of size M is $M/(2T)$ Hz. Thus, the requirement that there be M orthogonal pulses in the symbol interval increases the minimum bandwidth requirement by M.

Assume the receiver structure of Fig. 6-4, and for an isolated-pulse input, sample the matched-filter outputs at all integer multiples of T. To avoid ISI, if the signal input is pulse $h_i(t)$, then the samples at the output of the filter matched to $h_i(t)$ must satisfy the ordinary Nyquist criterion,

$$h_i(t) * h_i^*(-t)\,\big|_{t=kT} = \delta_k\,, \quad 0 \le i \le M-1\,. \tag{6.53}$$

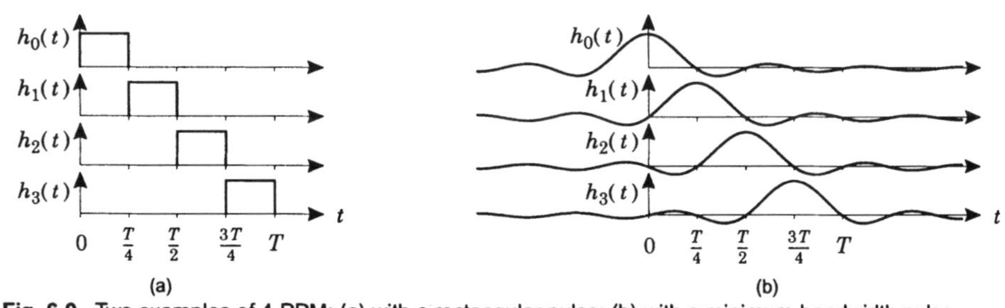

Fig. 6-9. Two examples of 4-PPM: (a) with a rectangular pulse; (b) with a minimum-bandwidth pulse.

In addition, to avoid crosstalk between pulses, if $h_i(t)$ is the input to a filter matched to pulse $h_j(t)$, for $j \neq i$, then the output sampled at $t = kT$ must be zero for all k,

$$h_i(t) * h_j^*(-t) \big|_{t = kT} = 0\,, \quad j \neq i, \quad -\infty < k < \infty\,. \tag{6.54}$$

These two conditions can be written together in a single compact form,

$$h_i(t) * h_j^*(-t) \big|_{t = kT} = \delta_k \delta_{j-i}\,. \tag{6.55}$$

We can express these conditions in terms of an equivalent frequency-domain criterion. Let $h_i(t)$ have Fourier transform $H_i(f)$. When we input $h_i(t)$ to a filter matched to $h_j(t)$, the output has Fourier transform $H_i(f)H_j^*(f)$. Sample this at $t = kT$, the discrete-time Fourier transform has to be unity for $j = i$ and zero for $j \neq i$, and hence

$$\frac{1}{T}\sum_{m=-\infty}^{\infty} H_i\left(f - \frac{m}{T}\right) H_j^*\left(f - \frac{m}{T}\right) = \delta_{j-i}\,. \tag{6.56}$$

Equation (6.56) is called the *generalized Nyquist criterion*. It is clear that a bandwidth of $M/(2T)$ is sufficient to satisfy this criterion, as demonstrated by both the sinc-windowed FSK signal set of Exercise 6-2 as well as the minimum-bandwidth PPM signal set of the previous section. In Appendix 6-A, we show that this bandwidth is also necessary.

A Set of Orthonormal Pulses with Excess Bandwidth

The ideally bandlimited pulse set in Fig. 6-8 is not realizable. Practical pulse sets designed by Chang [3] achieve close to the minimum bandwidth promised by the generalized Nyquist criterion. We will derive these pulses, relegating many details to Appendix 6-A.

Let $w(t)$ be a windowing pulse shape chosen such that $w(t) * w(-t)$ (the pulse shape at the output of a matched filter) satisfies the Nyquist criterion for symbol rate $1/(2T)$; that is,

$$\int_{-\infty}^{\infty} w(t) w(t - 2kT)\, dt = \delta_k\,. \tag{6.57}$$

The minimum bandwidth of $w(t)$ is $1/(4T)$, but to allow a gradual rolloff, assume that it has twice the minimum bandwidth, or $1/(2T)$. An example of a $|W(f)|^2$ satisfying these conditions is shown in Fig. 6-10(a). Note that only the magnitude of $W(f)$ is constrained by this condition, not the phase. We will choose the phase later to satisfy the generalized Nyquist criterion.

We can generalize (6.52) as follows. For $n \in \{0, \ldots, M-1\}$, define a set of pulse shapes

$$h_n(t) = w(t) \cos\left((n + \tfrac{3}{2}) \frac{\pi t}{T} \right)\,. \tag{6.58}$$

Exercise 6-3.

Show that as long as the window $w(t)$ satisfies (6.57), then for each n, the matched filter output $h_n(t) * h_n(-t)$ satisfies the ordinary Nyquist criterion (6.53).

To show that the $\{h_n(t)\}$ satisfy the generalized Nyquist criterion, we must verify that (6.56) holds. This requires some of the same machinery used to prove the minimum bandwidth of orthogonal modulation, so we defer the details to Appendix 6-A. In particular, in the appendix we show that (6.56) holds if $W(f)$, the Fourier transform of $w(t)$, satisfies

$$\mathrm{Re}\{W(f)W(1/(2T)-f)\}=0\ , \qquad 0 \le f \le 1/(4T)\ . \tag{6.59}$$

This can be satisfied by adjusting the phase of $W(f)$ to have a particular symmetry about $1/(4T)$, without regard to the magnitude. The magnitude can be chosen to satisfy (6.57) and simultaneously the phase can be chosen to satisfy (6.59). To see this, write $W(f)$ in terms of its magnitude and phase, $W(f)=A(f)e^{j\theta(f)}$. Then

$$\mathrm{Re}\{W(f)W(1/(2T)-f)\}=A(f)A(1/(2T)-f)\cos\Big(\theta(f)+\theta(1/(2T)-f)\Big)\ . \tag{6.60}$$

This function will be zero for all $0 \le f \le 1/(4T)$ if

$$\theta(f)+\theta(1/(2T)-f)=\pm\pi/2 \tag{6.61}$$

over the same range of f.

Example 6-9.

As an example of a phase function satisfying this constraint, let

$$\theta(f)=-\pi fT+\gamma(f) \tag{6.62}$$

where $\gamma(f)$ is any function that has odd symmetry about $1/(4T)$. Then,

$$\theta(f)+\theta(1/(2T)-f)=-\pi/2+\gamma(f)+\gamma\,(1/(2T)-f)\ =\ -\pi/2\ . \tag{6.63}$$

This phase function includes a linear-phase term, corresponding to a delay, plus another arbitrary phase term meeting the odd-symmetry constraint. There is thus considerable freedom in choosing this phase function.

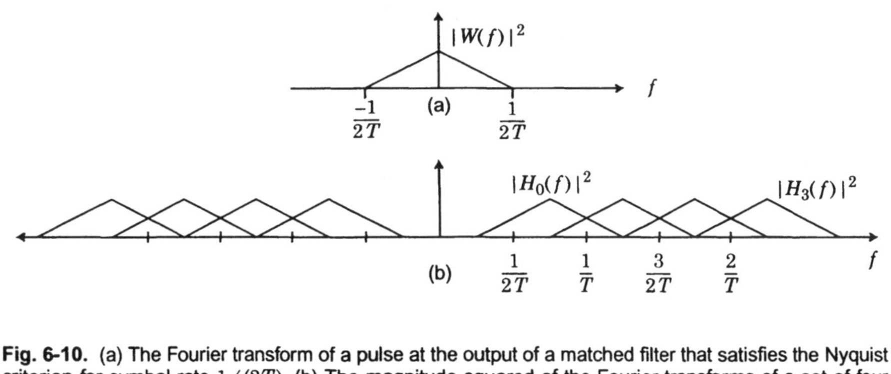

Fig. 6-10. (a) The Fourier transform of a pulse at the output of a matched filter that satisfies the Nyquist criterion for symbol rate $1/(2T)$. (b) The magnitude squared of the Fourier transforms of a set of four orthonormal pulses that satisfy the generalized Nyquist criterion at symbol rate $1/T$, but unlike the pulses in Fig. 6-8 overlap in the frequency domain.

To summarize, we have defined a class of pulse sets (6.58) that satisfy the generalized Nyquist criterion. The magnitude $|W(f)|$ has been chosen to force $w(t) * w(-t)$ to satisfy the ordinary Nyquist criterion at sampling rate $1/(2T)$, and the phase of $W(f)$ has been chosen to force the pulses to be orthogonal as required.

The pulses $\{h_0(t), \dots h_{M-1}(t)\}$ cover the frequency interval $[1/(4T), (2M+3)/(4T)]$, for a total aggregate bandwidth of $(M+1)/(2T)$. This is only slightly larger than the minimum bandwidth $M/(2T)$, for large M. This increase in bandwidth is the small price paid for achieving gradual rolloff.

6.3.5. Noise Immunity and Spectral Efficiency of Orthogonal Modulation

Gray mapping exploits the fact that some symbol errors are more likely than others, and is thus incompatible with orthogonal modulation, where *all* neighboring signals are equidistant. In fact, there is no room for optimization at all in the mapping of bits to orthogonal signals; all mappings will yield the same bit-error probability. Therefore, the relationship between bit-error and symbol-error probability of (5.120) cannot be applied to orthogonal modulation.

Suppose a particular signal is transmitted, and consider its $M-1$ equidistant neighbors, each labeled by a different block of $b = \log_2 M$ bits. Unlike a Gray mapping, where all nearest-neighbor labels would differ in only one bit, we have only $\binom{b}{1}$ neighbors differing in one bit, $\binom{b}{2}$ differing in two bits, and in general $\binom{b}{k}$ differing in k bits. Averaging $k\binom{b}{k}$ over $k \in \{1, \dots b\}$ yields $M/2$. Since the probability of choosing any one of the neighbors is $P_e/(M-1)$, where P_e is given in (6.48), we conclude that the bit-error probability for orthogonal signaling is

$$P_b = \frac{M/2}{M-1} P_e . \tag{6.64}$$

Since each pulse conveys $\log_2 M$ bits of information, the energy per bit for M-ary orthogonal modulation is $E_b = E/\log_2 M$. Therefore, from (6.45) we can approximate the bit-error probability by:

$$P_b \approx \frac{M}{2} Q\left(\sqrt{\log_2 M \frac{E_b}{N_0}}\right) . \tag{6.65}$$

Solving for E_b/N_0 leads to the following per-bit SNR requirement for M-ary orthogonal modulation:

$$E_b/N_0 = \frac{\left(Q^{-1}\left(\frac{P_b}{M/2}\right)\right)^2}{\log_2 M} . \tag{6.66}$$

The minimum bandwidth for ISI-free orthogonal modulation is $M/(2T)$ Hz. Consequently the best spectral efficiency is

$$\nu = \frac{\log_2 M}{T(M/2T)} = \frac{2\log_2 M}{M} . \tag{6.67}$$

Differentiating the spectral efficiency (6.67) with respect to M reveals that it achieves a maximum value of $\nu = 2/(e \cdot \log_e 2) = 1.07$ b/s/Hz when $M = e$. But since M must be an integer, the best we can do is $M = 3$, where $\nu = 1.057$ b/s/Hz. Both $M = 2$ and $M = 4$ give $\nu =$ 1 b/s/Hz, which is almost as good.

The performance of M-ary orthogonal modulation is illustrated in Fig. 6-11, where we plot the E_b/N_0 required for $P_b = 10^{-6}$ as a function of the bandwidth requirement normalized by the bit rate; *i.e.*, as a function of the inverse of the spectral efficiency. This same plot was made for QAM and PSK in the previous chapter (Fig. 5-38(a)). Also shown in the figure is the performance of QAM and PSK and the Shannon limit. Observe that, compared to 4-QAM, binary orthogonal modulation ($M = 2$) is 3 dB less power efficient and half as spectrally efficient. On the other hand, 4-ary orthogonal modulation is only about 0.2 dB less power-efficient than 4-QAM, but still half as bandwidth efficient. As M grows large, M-ary modulation becomes more power efficient but less bandwidth efficient. This is the opposite of QAM and PSK, which become less power-efficient but more bandwidth efficient as the alphabet gets large.

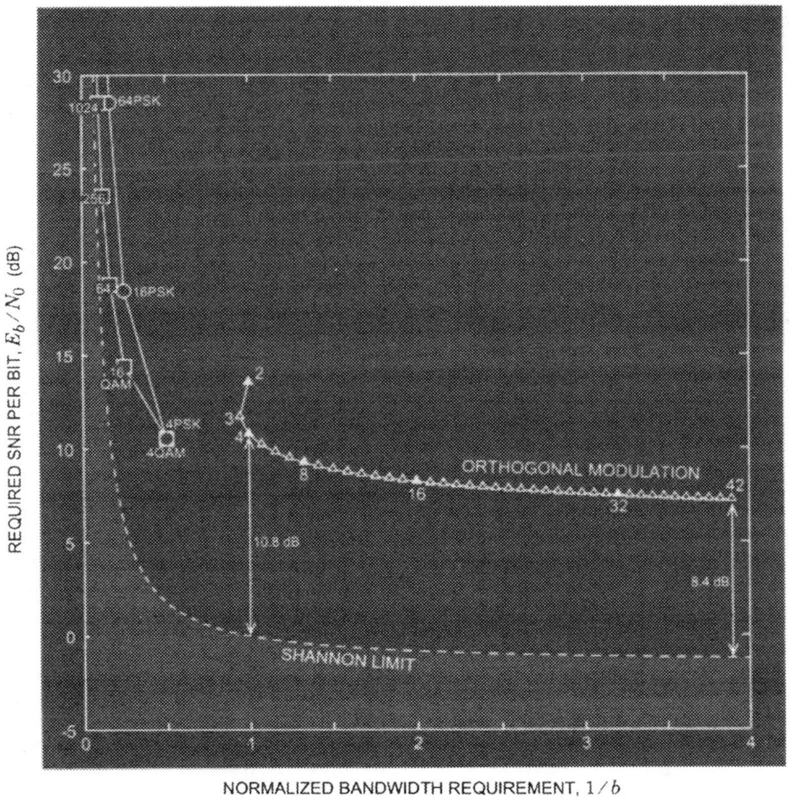

Fig. 6-11. Orthogonal modulation is more power-efficient than QAM but less bandwidth-efficient. (The x-axis is the bandwidth normalized by the bit rate, which is the inverse of the spectral efficiency.)

PAM can be much more spectrally efficient, assuming that the signal-to-noise ratio allows us to increase the number of bits/symbol, because that increase in bits/symbol comes without any impact on either the bandwidth or the symbol interval. On the other hand, orthogonal signaling is inherently less susceptible to noise, because we are in effect doing a binary rather than M-ary amplitude modulation of the transmitted pulse.

6.3.6. Practical Frequency-Shift Keying

In this section we look closer at FSK and examine two practical issues that arise: continuous-phase FSK, and incoherent detection.

Continuous-Phase Frequency-Shift Keying

As defined in Example 6-6, the FSK transmitted signal can have either discontinuous phase or continuous phase, depending on the signal design, as illustrated below:

Continuous phase, shown on the right, is desirable, since the high frequency components are reduced. This is important for a bandlimited channel, and particularly important when the channel is nonlinear.

Example 6-10.
In many applications, we would like to drive the transmitter RF amplifier hard enough into saturation that significant nonlinearity results. As shown in Fig. 6-12, a practical FSK transmitter will apply bandpass filtering prior to upconversion and amplification. Now there is unfortunately considerable interaction between phase transitions and the nonlinearity of the amplifier, since the bandpass filter will convert the phase transitions into an amplitude transient, which might be quite large. The power of the signal at RF will then have to be backed off in order to keep the amplitude of these transients within the peak power limitation, thereby reducing the average signal power and increasing the error rate at the receiver. In this case continuous phase is preferred.

For pulses of the form (6.50) to result in a continuous-phase FSK signal, each pulse must traverse an integer number of cycles, so

$$f_i T = M_i \tag{6.68}$$

where M_i is an integer. For maximum spectral efficiency, it also helps to minimize the *frequency separation* between signaling frequencies f_i. We want minimum frequency separation while maintaining phase continuity and orthogonality of pulses.

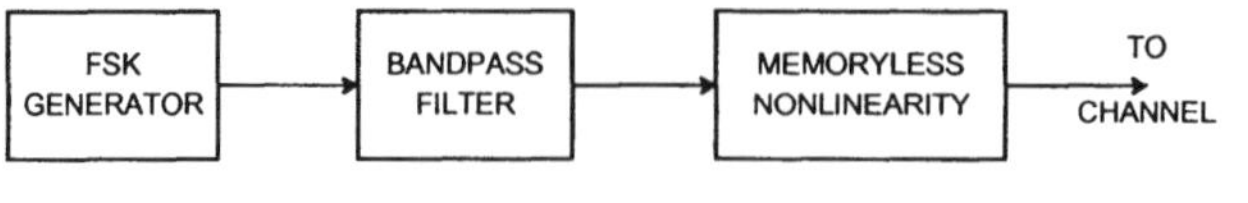

Fig. 6-12. Bandpass filtering before the nonlinearity of a power amplifier.

Example 6-11.

For the binary FSK pulses of (6.50), continuous phase requires

$$f_0 T = M_0 \quad \text{and} \quad f_1 T = M_1 \tag{6.69}$$

where M_0 and M_1 are integers. Minimum frequency separation $|f_0 - f_1|$ requires that $|M_0 - M_1| = 1$. One of the pulses must traverse exactly one more cycle than the other. Two such pulses are superimposed below:

The pulse frequencies satisfy

$$2f_d = |f_1 - f_0| = 1/T \tag{6.70}$$

where f_d is called the *peak deviation* from the nominal carrier frequency $f_c = (f_0 + f_1)/2$. Such pulses are easily shown to be orthogonal (see Problem 6-1). A smaller frequency separation than that given by (6.70), maintaining both phase continuity and orthogonality, but only when we generalize the M-ary model by adding *memory*; this will be explained in Section 6.5.1.

Exercise 6-4.

As in Example 6-6, assume the channel is benign, so that the received pulses are equal to the transmitted pulses given in (6.50). Assume as in Example 6-11 that

$$f_0 = M_0/T \quad \text{and} \quad f_1 = M_1/T, \tag{6.71}$$

where M_0 and M_1 are arbitrary integers. Show that with these frequencies, the pulses (6.50) satisfy (6.56), and hence satisfy the generalized Nyquist criterion.

For continuous-phase FSK signals with M pulses, Example 6-11 suggests that the frequencies of pulses must differ by integer multiples of $1/T$. If the pulses have frequencies $f_0 < f_1 < \ldots < f_{M-1}$, then

$$f_i - f_{i-1} = 1/T; \quad \text{for} \quad 1 \le i \le N-1 \tag{6.72}$$

is the minimum separation. This frequency spacing of $1/T$ between pulses is twice as large as that achieved by the ideally bandlimited orthogonal pulses of (6.52) and the Chang pulses of (6.58).

Example 6-12.

For binary FSK, a frequency spacing of $1/(2T)$, equivalent to the spacing of the Chang pulses in (6.58), is possible (see Problem 6-1). However, when more than two pulses of the form (6.50) are used, they will not in general be orthogonal with such spacing.

Incoherent Receivers

The matched-filter and correlation receivers require that the receiver accurately know the frequency and phase of the sinusoids used to form the FSK pulses. Such receivers are said to be *coherent*.

Exercise 6-5.

Consider the reception of a single isolated pulse,

$$g_0(t) = \sqrt{\tfrac{2}{T}}\sin(2\pi f_0 t)w(t)\ , \tag{6.73}$$

where $f_0 = M_0/T$ for some integer M_0, and $w(t)$ is a rectangular window over $[0, T)$, as in Example 6-6. Assume a benign channel so that $h_0(t) = g_0(t)$ is received. Suppose that the receiver cannot accurately determine the phase of the sinusoid in (6.73), so it uses the erroneous matched filter

$$f(t) = \sqrt{\tfrac{2}{T}}\sin(-2\pi f_0 t + \theta)w(-t)\ , \tag{6.74}$$

where θ is the phase error. If the receiver also cannot determine the frequency f_0 accurately, then θ may be time varying. Assume that if it is time varying, then it varies slowly enough to be considered constant over one sample interval. Show that the sampled output of the received filter is

$$g_0(t) * f(t)\Big|_{t=0} = \cos\theta\ . \tag{6.75}$$

Hence, if the phase is correct, $\theta = 0$, and the sampled output is unity. However, if the phase error is $\theta = \pi/2$, then the sampled output is zero! Hence, uncertainty in θ can significantly compromise the performance of the matched filter receiver.

One prime advantage of FSK is the ability to use an incoherent receiver, which does not require knowledge of the carrier phase and can tolerate small inaccuracies in the carrier frequency. The zero-crossing detector of Fig. 6-7 is an incoherent receiver, but the matched-filter and correlation receivers are coherent. The question arises whether the matched-filter receiver can be modified to operate incoherently. We will show here that it can, and in Chapter 7 we will show that a slightly more general structure is optimal if the carrier phase is a uniformly distributed random variable over $[0, 2\pi)$. Incoherent receivers for FSK perform nearly as well as coherent receivers, so the additional cost of a coherent receiver may not be justified.

We can arrive at an incoherent receiver by heuristic arguments. Matched filters for FSK signals are bandpass filters, although rather crude bandpass filters because of the relatively high sidelobes. Their magnitude frequency response is shown in Fig. 6-13. This suggests another intuitive argument for the matched-filter receiver; it simply measures the output of a bank of bandpass filters with center frequencies at each of the signaling frequencies, and selects the largest. Note however that the *magnitude* frequency response will not be affected

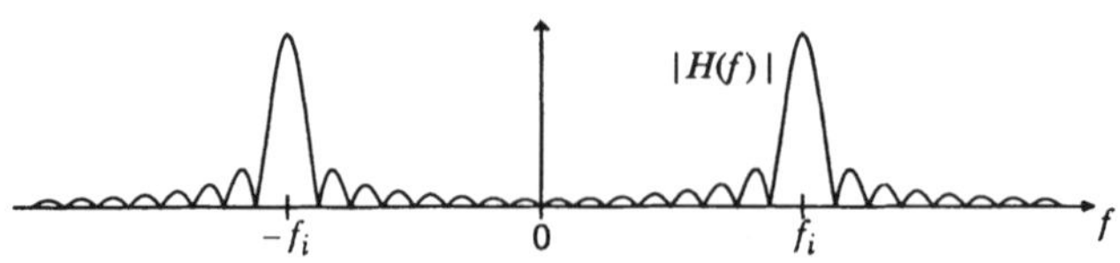

Fig. 6-13. The magnitude response of a typical matched filter for FSK receivers.

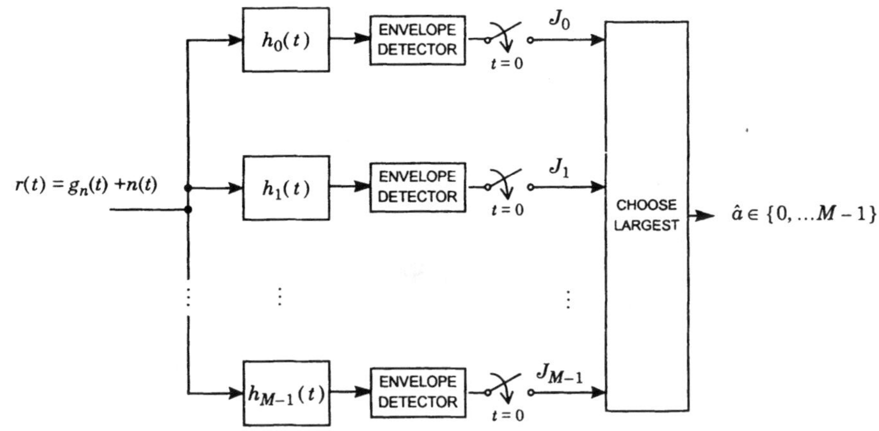

Fig. 6-14. An incoherent receiver for isolated pulse FSK signals. The filters $h_i(t)$ are bandpass filters centered at the signaling frequencies.

Fig. 6-15. a. An ideal envelope detector uses a gain of two and a phase splitter to get a complex signal. The magnitude of the phase splitter output is equal to the amplitude of the sinusoidal input (see Problem 6-5). b. An approximate envelope detector uses a peak detector and a lowpass filter.

by phase errors like those in Exercise 6-5. It will also be minimally affected by small frequency errors. This suggests that we could simply estimate the envelope of the filter output, as shown in Fig. 6-14. An envelope detector, shown in Fig. 6-15, simply finds the amplitude of a sinusoidal signal.

Ideally, all outputs of all the filters in Fig. 6-14 but the correct one will be zero. The output of the correct one will be a sinusoid, and the envelope detector will capture the level of the sinusoid. Moreover, unlike the matched filter receiver, the time at which we take the sample at the output of the envelope detector is not critical. The receiver can tolerate relatively large errors in the timing phase, although the average rate at which we take the samples needs to be precise in either case. Timing recovery is covered in Chapter 16.

6.4. Orthogonal Pulse-Amplitude Modulation (OPAM)

In Section 6.3 we saw that orthogonal modulation has a maximum spectral efficiency of about one bit/sec/Hz, which is poor in comparison to PAM. The reason that PAM has better spectral efficiency is that increasing the number of bits per symbol does not expand the bandwidth, as it does in orthogonal modulation. Fortunately, limiting ourselves to pure PAM or orthogonal signaling is not necessary. They can be combined to form *orthogonal pulse-amplitude modulation (OPAM)*. In this section we describe the general properties of OPAM transmitters and receivers. We then present two important special cases of OPAM: multicarrier modulation (both discrete-multitone and orthogonal frequency-division multiplexing) and code-division multiplex access.

6.4.1. OPAM: Combined PAM and Orthogonal Modulation

Orthogonal modulation and PAM can be combined by choosing a set of N orthonormal pulse shapes $\{g_n(t): n = 0, \ \ldots N-1\}$, amplitude-modulating each pulse shape with a different symbol $a_k^{(n)}$ from an alphabet of M symbols, and sending all of the pulses simultaneously, i.e., sending:

$$s(t) = \sum_{k=-\infty}^{\infty} \sum_{n=0}^{N-1} a_k^{(n)} g_n(t-kT) . \tag{6.76}$$

This combination of orthogonal modulation and PAM is called *orthogonal PAM (OPAM)*. A block diagram of an OPAM transmitter is shown in Fig. 6-16. In each symbol interval of length T, N symbols are simultaneously transmitted using N distinct pulses. Because the pulse shapes are orthogonal, the superposition of pulses can be sorted out at the receiver by a bank of matched filters. Note that $s(t)$ may represent the complex envelope, in which case the actual transmitted signal would be:

$$x(t) = \sqrt{2}\,\mathrm{Re}\{s(t)e^{j2\pi f_c t}\} . \tag{6.77}$$

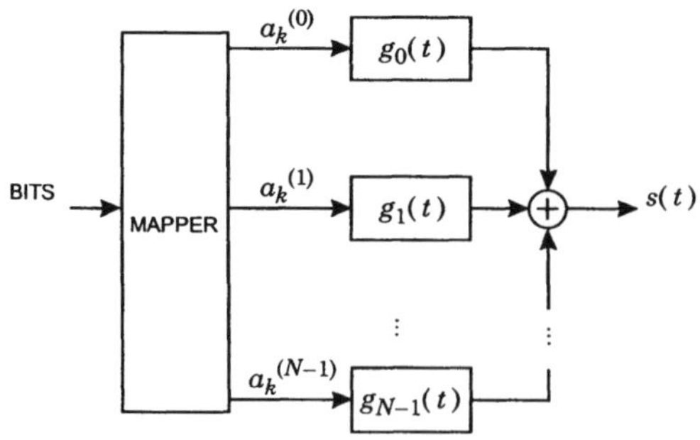

Fig. 6-16. An OPAM transmitter.

Example 6-13.

PAM is clearly a special case of (6.76) in which $N = 1$. For passband PAM, (6.76) represents the complex envelope of the transmitted signal. Interestingly, sometimes the passband PAM signal can be represented directly as an OPAM signal. Consider a passband PAM signal where the carrier frequency f_c has an integer number of cycles in one symbol interval, so that $f_c T$ is an integer. Then we can rewrite the quadrature representation (5.21) of a passband PAM signal as

$$x(t) = \sum_{k=-\infty}^{\infty} \text{Re}\{a_k\}g_0(t - kT) + \text{Im}\{a_k\}g_1(t - kT) , \tag{6.78}$$

where

$$g_0(t) = \sqrt{2}\cos(2\pi f_c t)g(t) \quad \text{and} \quad g_1(t) = -\sqrt{2}\sin(2\pi f_c t)g(t) . \tag{6.79}$$

These two pulses are orthogonal whenever the bandwidth of $g(t)$ is less than the carrier frequency, regardless of the particular shape of $g(t)$. Hence, for certain carrier frequencies, passband PAM can be viewed as an example of OPAM for $N = 2$.

Example 6-14.

Ordinary orthogonal signaling may also be viewed as a special case of OPAM, provided we add a constraint that the symbols $a_k^{(n)}$ take on values 0 or 1 in such a way that exactly one of the set $\{a_k^{(0)}, a_k^{(1)}, \ldots, a_k^{(N-1)}\}$ has value 1.

In the remainder of this section we focus on the unconstrained case where the modulating symbols $\{a_k^{(n)}\}$ are chosen independently.

In the following we will show that the spectral efficiency and power efficiency of OPAM is independent of N, and equivalent to that of PAM ($N = 1$).

Spectral Efficiency and Power Efficiency

Consider first *passband* OPAM, where (6.76) represents the complex envelope of the transmitted signal. The upconverter will double the bandwidth requirement from $W = N/(2T)$ to $W = N/T$. When the symbols are chosen independently from the same complex alphabet of size M, the symbol set $\{a_k^{(n)} : n = 0, \ldots, N-1\}$ can assume one of M^N values, communicating $\log_2 M^N$ bits of information per signaling interval T. Thus the best spectral efficiency is

$$\nu = \frac{\log_2 M^N}{WT} = \log_2 M . \tag{6.80}$$

Comparing this result to (5.36), we conclude that the spectral efficiency of OPAM is equivalent to passband PAM with the same alphabet size M, and independent of the dimensionality N.

For baseband signaling, where (6.76) represents the real-valued transmitted signal, the bandwidth requirement would be cut in half. But since the symbols would be real-valued, they would convey half as much information, so the overall spectral efficiency would be the same.

The probability of symbol error is easily derived for OPAM, because the orthogonality of the different pulses implies that they have no impact on eachother after a matched-filter receiver. Thus, the probability of symbol error can be calculated by considering just one of the pulses in isolation. This means that we can use directly the analysis of Chapter 5 for PAM. With M-ary QAM, for example, the probability of error for OPAM is given by (5.113):

$$P_e \approx 4\left(1 - \frac{1}{\sqrt{M}}\right)Q\left(\sqrt{\frac{3SNR}{M-1}}\right), \tag{6.81}$$

where we made the substitution $SNR = E/N_0$. This substitution is easily justified: Since E represents the average received energy for one of the pulses, it is a fraction $1/N$ of the total received energy PT per signaling interval, namely $E = PT/N$. With no excess bandwidth ($W = N/T$), it follows that $E/N_0 = P/(N_0W) = SNR$. The key point is that the symbol-error probability depends only on the SNR, and is independent of the dimensionality N. Assuming they both have the same SNR and the same alphabet, PAM and OPAM have the same error probability, independent of N.

Example 6-15.

Consider what happens when we base OPAM on the 4-ary PPM signal set Fig. 6-9(b). Each signaling interval of length T is divided into four chips of duration $T/4$. With OPAM, a different PAM symbol modulates each of the chip pulses. But this reduces to conventional PAM with symbol rate $4/T$! Using similar reasoning, it follows that *any* conventional PAM system may be viewed as OPAM based on N-ary PPM by simply grouping each block of N symbols together and treating it as a single block of OPAM symbols. Clearly, nothing can be gained or lost by this reinterpretation. It thus makes sense that the performance of OPAM will be independent of N.

Increasing the Symbol Interval

Suppose we have a fixed channel with bandwidth W. We are free to increase the dimensionality N of the signal set if we simultaneously increase the symbol interval T. Intuitively, we are compensating for the reduced symbol rate by increasing the number of symbols transmitted per symbol interval, thereby keeping the spectral efficiency fixed.

There are two fundamentally different ways to choose the set of orthogonal pulses as we increase T. One way is to hold the bandwidth of all the pulses fixed at W, and somehow make them orthogonal (a method for doing this will be discussed shortly). When we do this, the effects of non-ideal channel transfer functions can be mitigated, because we can make T so large that any time dispersion on the channel becomes insignificant relative to T. This effect is easy to visualize intuitively in the time domain, and will be further quantified in the frequency domain below.

The second way to choose orthogonal pulses as T increases is to make each pulse have bandwidth on the order of $1/(2T)$, satisfying the ordinary Nyquist criterion, and make them orthogonal by placing them at different center frequencies, spaced $1/(2T)$ apart. This was the approach used to design the idealized pulses (6.52) and the Chang pulses (6.58). Again, the effect of this is to mitigate the effects of non-ideal channels, because as T (and hence N) increases, the bandwidth of each pulse decreases (in inverse proportion). This implies that only a portion of the channel transfer function over a narrower and narrower bandwidth affects each pulse transmission. Eventually, for sufficiently large T, the channel transfer function will be essentially constant over the bandwidth of the pulse, and introduce insignificant ISI. Furthermore, the pulses will tend to retain their orthogonality because they are in approximately non-overlapping frequency bands, both at the channel input and output (the latter assuming the channel is linear and time-invariant). This is the approach taken in multicarrier modulation below, where we will further consider the effects of ISI.

An additional advantage of increasing T is that impulse noise phenomena that are highly localized in time will have less impact on pulses of greater time duration.

Increasing Bandwidth

The second way of increasing N is to hold the symbol interval T constant and increase the bandwidth W. This is not feasible on many media, if the additional bandwidth is not available. However, for radio, the medium itself has very broad bandwidth and is typically *frequency-division multiplexed (FDM)* by assigning different users to non-overlapping frequency bands. FDM is closely related to the multicarrier modulation mentioned above, except that in FDM the different users typically do not overlap one another in frequency.

An alternative approach to maintaining the separation of the users is to aggregate N users sharing a bandwidth W, where now W is N times as large as the nominal bandwidth required by each user in FDM. Then, rather than each user transmitting a pulse that does not overlap the bandwidth of the other users, all users are assigned pulses that occupy the full bandwidth W. The orthogonality of these pulses is assured by the techniques mentioned above. This approach to sharing a fixed bandwidth among a set of users, allowing their transmissions to completely overlap in frequency, is called *code-division multiple access (CDMA)*, and is elaborated further below. It is particularly advantageous in *cellular radio* systems, as discussed further in Chapter 17.

Receiver Design

A correlation receiver for OPAM is shown in Fig. 6-17. It shares some similarities with the correlation receiver for orthogonal modulation shown Fig. 6-4. As before, consider only a single signaling interval (which now carries N symbols $a^{(0)} \ldots a^{(N-1)}$), and let

$$s(t) = \sum_{n=0}^{N-1} a^{(n)} g_n(t) , \tag{6.82}$$

and assume the received signal is corrupted only by noise. In particular, the received pulse shapes $h_n(t)$ are identical to the transmitted pulse shapes. Instead of selecting one of N candidate pulses as done in Fig. 6-4, each pulse is assumed to carry independent data, and hence has its own slicer, as shown in Fig. 6-17. If the pulses satisfy the generalized Nyquist criterion, there is no crosstalk between pulses at the matched filter output, sampled at the appropriate time, so each slicer responds only to its corresponding pulse.

We will now give two important examples of OPAM modulation, *multicarrier modulation* and *code-division multiple access* (CDMA).

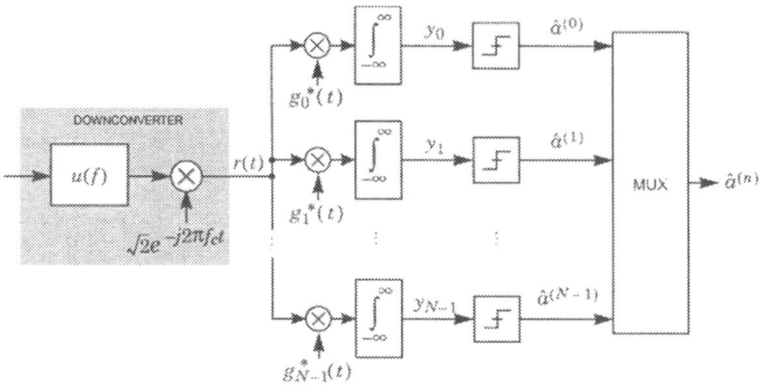

Fig. 6-17. A correlation receiver for OPAM.

6.4.2. Multicarrier Modulation

Multicarrier modulation is the generic term used for any OPAM scheme in which each of the orthogonal pulses is roughly localized in the frequency domain. It includes as special cases frequency-division multiplexing, orthogonal frequency-division multiplexing (OFDM), and discrete multitone transmission.

Our starting point for multicarrier modulation will be the following set of pulses:

$$g_n(t) = \frac{1}{\sqrt{T}} e^{j2\pi nt/T} w(t), \quad \text{for } n = 0, \dots N-1, \tag{6.83}$$

where $w(t) = u(t) - u(t-T)$ is a rectangular window over $[0, T)$. These pulses are similar to the FSK pulses of Example 6-6 with the same frequency separation $1/T$ used in Example 6-11. This frequency separation ensures continuous phase, but is twice as large as the frequency separation of the Chang pulses (6.58).

Exercise 6-6.

Show that the pulses in (6.83) are orthonormal.

Using these pulses, the OPAM signal $s(t)$ in (6.76) consists of superimposed finite-length signals modulated on different carriers. The technique is closely related to frequency-division multiplexing, although there is considerable overlap between neighboring frequency bands, as illustrated below:

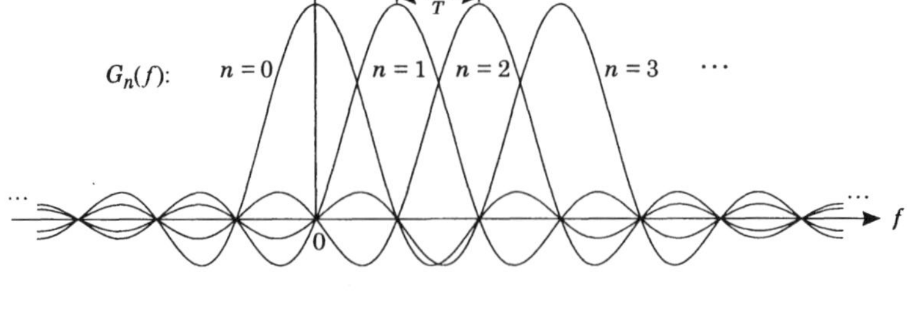

These pulses are of theoretical interest but are somewhat idealized and not very practical, primarily because the rectangular window in (6.83) prevents the signals from being bandlimited. The next section shows how these pulses can be modified to make them more practical.

A Better Set of Orthonormal Pulses

We will see that a key advantage of multicarrier modulation is that it can be implemented with low cost by applying a discrete-time signal to an ideal D/A converter. The pulses of (6.83) cannot be generated in this way because they are not bandlimited; the output of a D/A converter will always be ideally bandlimited to half the sample rate. In this section we introduce a new set of orthonormal pulses that are similar in spirit to those described above, but which lead to low-complexity implementation.

Let $\{\tilde{g}_n(t)\}$ be the result of passing the signals of (6.83) through the following back-to-back cascade of an ideal A/D and D/A converter:

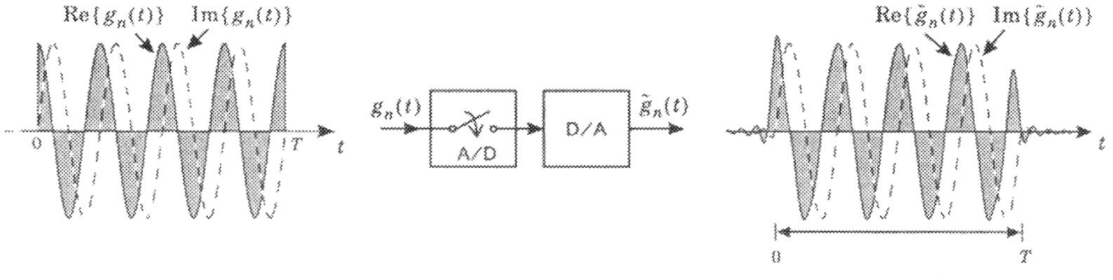

First, the pulse $g_n(t)$ is sampled at the rate N/T. Because of the rectangular window, only N of these samples will be nonzero. Then, these N samples are interpolated using an ideal reconstruction filter for this sample rate, which is bandlimited to half the sample rate.

The above diagram shows a specific example where $N = 32$ and $n = 4$, so that the sinusoid completes precisely four cycles during the signaling interval. We see that $\tilde{g}_4(t)$ looks very similar to $g_4(t)$ except at the transitions near $t = 0$ and $t = T$, where the bandlimited nature of $\tilde{g}_4(t)$ leads to small oscillations in the time domain. Mathematically, the new pulse set can be written as:

$$\tilde{g}_n(t) = \frac{1}{\sqrt{N}}\sum_{k=0}^{N-1} e^{j2\pi nk/N} p(t - kT/N), \quad \text{for } n = 0, \ldots N-1, \tag{6.84}$$

where $p(t)$ is an ideal unit-energy reconstruction filter for a sample rate of N/T:

$$p(t) = \sqrt{\frac{T}{N}}\frac{\sin(\pi Nt/T)}{\pi t}. \tag{6.85}$$

Multicarrier modulation (or more generally OPAM) based on the pulse set $\{\tilde{g}_n(t)\}$ of (6.84) is called *discrete-multitone modulation (DMT)* in the context of DSL applications, and *orthogonal-frequency-division multiplexing (OFDM)* in the context of wireless applications.

Remarkably, the transformation from $\{g_n(t)\}$ to $\{\tilde{g}_n(t)\}$ does not alter orthogonality: just like the originals, the new pulse set $\{\tilde{g}_n(t)\}$ satisfies the generalized Nyquist criterion.

Exercise 6-7.
Show that the set of pulses $\{\tilde{g}_n(t)\}$ defined by (6.84) are orthonormal.

An important feature of $\{\tilde{g}_n(t)\}$ is that they are bandlimited to half the sample rate. This property is illustrated in the following sketch, which compares the Fourier transforms of $g_n(t)$ and $\tilde{g}_n(t)$ when $n = 15$ and $N = 32$:

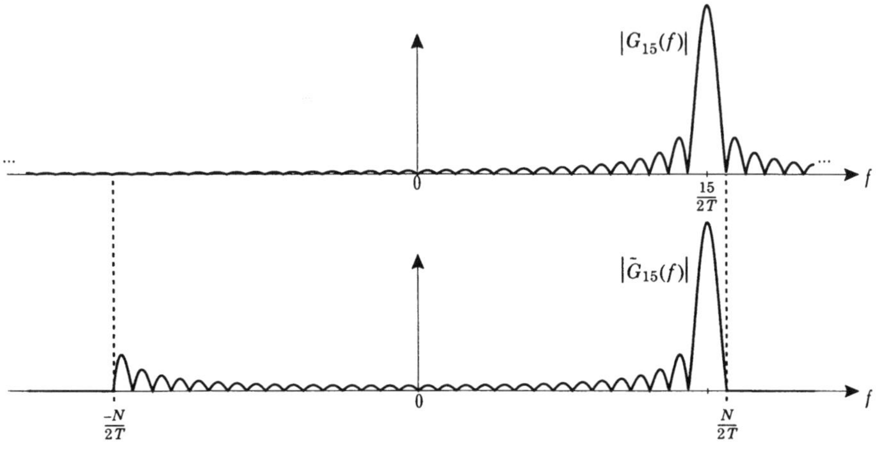

Both pulses have most of their energy concentrated at $15/T$, but unlike the infinite extent for $g_{15}(t)$, we see that $\tilde{g}_{15}(t)$ is strictly bandlimited to the range $|f| < N/(2T)$.

Using the new pulses $\{\tilde{g}_n(t)\}$, a one-shot OPAM transmitter sends:

$$s(t) = \sum_{n=0}^{N-1} a^{(n)} \tilde{g}_n(t) . \tag{6.86}$$

Substituting $\tilde{g}_n(t)$ from (6.84) and changing the order of summation leads to:

$$\begin{aligned} s(t) &= \sum_{n=0}^{N-1} a^{(n)} \frac{1}{\sqrt{N}} \sum_{k=0}^{N-1} e^{j2\pi nk/N} p(t - kT/N) \\ &= \sum_{k=0}^{N-1} \left\{ \frac{1}{\sqrt{N}} \sum_{n=0}^{N-1} a^{(n)} e^{j2\pi nk/N} \right\} p(t - kT/N) . \end{aligned} \tag{6.87}$$

We immediately recognize the term within the brackets as the k-th inverse DFT coefficient of $\{a^{(0)}, a^{(1)}, \ldots a^{(N-1)}\}$, call it s_k, so that the OPAM transmitter sends:

$$s(t) = \sum_{k=0}^{N-1} s_k p(t - kT/N) . \tag{6.88}$$

We further recognize (6.88) as an interpolation of $\{s_0, \ldots s_{N-1}\}$ with the ideal interpolation filter $p(t)$. Together, these two observations have an important implication: the multicarrier signal $s(t)$ of (6.86) can be generated by a cascade of an IFFT and a D/A converter with sample rate N/T, as sketched below:

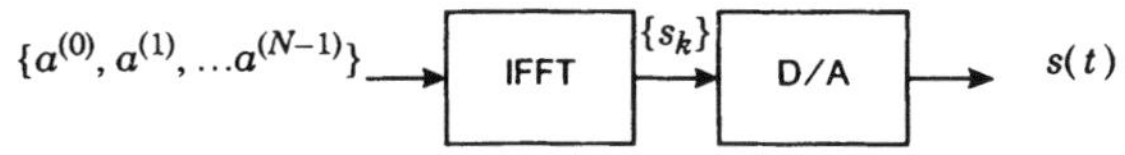

The complexity is extremely low. The above transmitter might look different from the general OPAM transmitter of Fig. 6-16 but the two are precisely equivalent when the pulse set is defined by (6.84).

The Minimum-Distance Receiver

The minimum-distance receiver also simplifies dramatically when the pulses of (6.84) are used. We consider the usual one-shot case in which the channel is benign; it adds noise but otherwise does not distort the signal, so that the received pulses are the same as the transmitted pulses. As illustrated in Fig. 6-17, the one-shot minimum-distance receiver for any OPAM scheme requires that the received signal $r(t)$ be correlated with each of the pulses $\{\tilde{g}_n(t)\}$. Substituting $\tilde{g}_n(t)$ from (6.84) and exchanging the order of summation and integration, the n-th correlation simplifies to:

$$\begin{aligned} y_n &= \int_{-\infty}^{\infty} r(t)\tilde{g}_n^*(t)dt \\ &= \int_{-\infty}^{\infty} r(t)\frac{1}{\sqrt{N}}\sum_{k=0}^{N-1} e^{-j2\pi nk/N} p(t-kT/N)dt \\ &= \frac{1}{\sqrt{N}}\sum_{k=0}^{N-1} e^{-j2\pi nk/N}\int_{-\infty}^{\infty} r(t)p(t-kT/N)dt\,. \end{aligned} \tag{6.89}$$

Let r_k denote the last integral above; it can be calculated by passing $r(t)$ through a filter $p(t)$ and sampling the output at time kT/N. Since $p(t)$ represents an ideal *antialiasing* filter for a sample rate of N/T, we can equivalently interpret r_k as the k-th output of an ideal A/D converter that includes an antialiasing filter. With this definition, the n-th correlation simplifies to:

$$y_n = \frac{1}{\sqrt{N}}\sum_{k=0}^{N-1} r_k e^{-j2\pi nk/N}\,. \tag{6.90}$$

We immediately recognize this as the n-th DFT coefficient of the set of samples $\{r_0, \ldots r_{N-1}\}$. Thus, we can implement the entire bank of correlators simultaneously using a simple cascade of an A/D converter and an FFT:

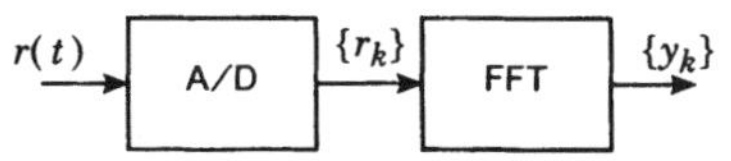

The A/D converter is understood to include the unit-energy antialiasing filter $p(t)$. This front-end might look different from that of the general OPAM correlator shown in Fig. 6-17, but the two are precisely equivalent when the pulse set is defined by (6.84).

Combating ISI with a Cyclic Prefix

So far we have assumed a benign channel that only adds noise. We now relax this assumption and examine the effects of ISI on the multicarrier signal. We will see that the ISI destroys the orthonormality between the pulses, but that it can be restored by a simple modification at the transmitter involving a cyclic prefix.

Let h_k denote the equivalent discrete-time impulse response between the transmitted samples s_k and received samples r_k. Let μ denote the memory of this impulse response, so that h_k is nonzero for $k \in \{0, \dots \mu\}$ only. Assume that $\mu \le N$. In other words, assume that the number N of carriers is large enough that one symbol interval (N sample periods) is longer than the impulse response of the channel. Thus, we can write the noise-free channel output as a finite convolution,

$$r_k = \sum_{i=0}^{N-1} h_i s_{k-i} . \tag{6.91}$$

If we attempted to use the DFT-based receiver just derived — despite the ISI — we would end up taking the DFT of the above linear convolution. It would be nice if the DFT of the convolution would reduce to the product of DFT's, but this property applies only to *circular* convolution, not linear convolution.

Let us briefly review the relationship between circular convolution and the DFT. The circular convolution of h_k and s_k is defined by:

$$x_k = \sum_{i=0}^{N-1} h_i s_{(k-i)_N} , \tag{6.92}$$

where $(\cdot)_N$ is a modulo operator that reduces its argument to the range $\{0, \dots N-1\}$. Let $H_n/\sqrt{N}$ denote the n-th DFT coefficient of $\{h_k\}$; with this normalization, and given our assumption that the channel is FIR with memory $\mu < N$, we may interpret H_n as the DTFT of the channel impulse response sampled at frequency $2\pi n/N$, namely:

$$H_n = H(e^{j2\pi n/N}) . \tag{6.93}$$

Also, as usual, let $a^{(n)}$ denote the n-th DFT coefficient of s_k. With this notation, taking the DFT of both sides of (6.92) leads to the product of DFT coefficients:

$$X_n = H_n a^{(n)} . \tag{6.94}$$

With a modification to the modulation format we can make the ordinary convolution equal to the circular convolution. All we need to do is precede the N samples $\{s_0, \dots s_{N-1}\}$ by a *cyclic prefix* $\{s_{-\mu}, \dots s_{-1}\}$ of length μ:

$$s_{-i} = s_{N-i}, \text{ for } 1 \le i \le \mu , \tag{6.95}$$

as illustrated in Fig. 6-18. Although these μ redundant symbols convey no new information, they have an important benefit: they make the ordinary convolution (6.91) equal to the circular convolution of (6.92). The receiver will discard the observations $\{r_{-\mu}, \dots r_{-1}\}$ corresponding to the cyclic prefix, which also eliminates the ISI from the previous signaling interval, and will base its decisions on the DFT of the remaining samples $\{r_0, \dots r_{N-1}\}$, which will now be:

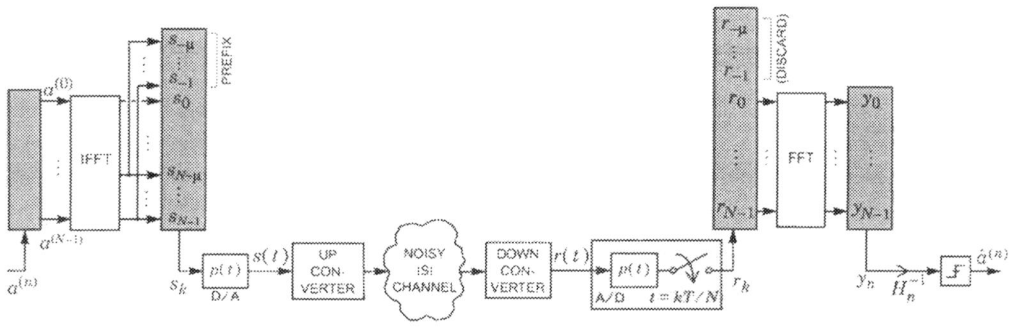

Fig. 6-19. A multicarrier modulation system using FFTs. The shaded boxes are serial-to-parallel or parallel-to-serial converters. In practice, the complex D/A and A/D converters are implemented using a pair of real D/A and A/D converters having the same sample rate.

$$y_n = H_n a^{(n)} . \tag{6.96}$$

In other words, neglecting noise, the n-th correlation will be the (desired) n-th symbol $a^{(n)}$ scaled by the complex constant H_n. There is no interference from other pulses; the effect of the dispersive channel is only to scale each of the N symbols by the channel frequency response evaluated at the corresponding frequency. The receiver can easily compensate by scaling the n-th correlation by $1/H_n$ before applying the slicer.

A complete multicarrier system based on FFT blocks is shown in Fig. 6-19. Although the picture only shows the one-shot case, it also applies to a continuous stream of pulses. The IFFT and FFT would be computed once per signaling interval, which has duration $T = (N + \mu)T'$ with the cyclic prefix (where T' is the sample period). Each such interval carries a superposition of N orthogonal pulses and N symbols.

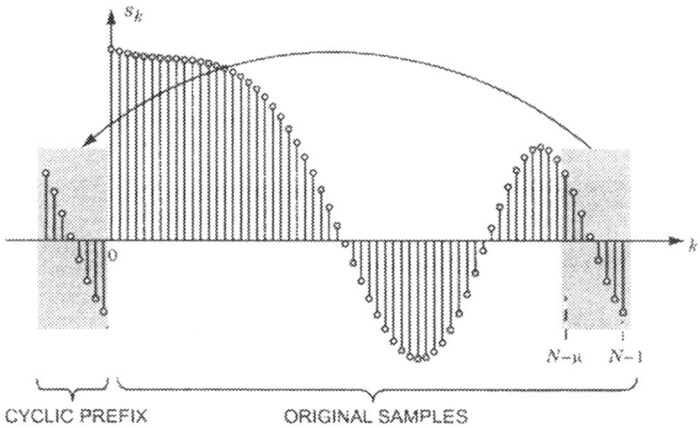

Fig. 6-18. A cyclic extension of the samples s_k restores orthogonality despite a dispersive channel.

Normalizing the slicer inputs by $1/H_n$ is not necessary for binary or 4-QAM alphabets, but it is critical for larger alphabets. In most practical situations, we do not know precisely the frequency response of the channel, but H_n can be estimated from observations of y_n. The mechanism used to estimate H_n is related to the *decision-directed* techniques used for adaptive equalization (Chapter 9) and carrier recovery (Chapter 15). Referring to Fig. 6-19, if the decisions $\{\hat{a}^{(n)}\}$ are all correct, then we can determine what the variables y_n, for $n = 0, \ldots, N-1$ should have been by computing an inverse DFT. Comparing what these variables should have been to the actual observation, using (6.96), we obtain an estimate of H_n for $n = 0, \ldots, N - 1$. These estimates can then be used for scaling the slicer inputs in subsequent signaling intervals.

It is sometimes useful to view the multicarrier system of Fig. 6-19 in terms of an equivalent passband PAM system having a higher symbol rate. Looking closer at Fig. 6-19, we recognize the inner channel from s_k to r_k as a special case of a conventional *single-carrier* PAM system for which:

- the "symbol" rate is N/T
- the k-th transmitted "symbol" is s_k
- the transmit pulse shape is $p(t)$, which has zero excess bandwidth
- the receive filter is not a matched filter but still provides sufficient statistics, because the lack of excess bandwidth implies that the baud-spaced samples avoids aliasing.

With this view, we may interpret multicarrier modulation as conventional single-carrier PAM with only one difference: there is extra *preprocessing* at the transmitter. Namely, each block of N information-bearing symbols $\{a^{(n)}\}$ is transformed into a block of $N + \mu$ transmitted symbols $\{s_k\}$ using an IFFT and a cyclic prefix.

Summary

Multicarrier modulation has many advantages. Foremost is that it can avoid the need for computationally intensive time-domain equalization (Chapter 8); the cyclic prefix prevents interference between pulses despite the presence of a dispersive channel. And an implementation based on D/A, A/D, and FFT blocks is extremely efficient. Also, as with many OPAM systems, the long signaling interval relative to a comparable PAM signal makes the multicarrier signal less sensitive to impulse noise that is highly localized in time. In wireless applications, multicarrier modulation together with error-control coding (Chapter 12) is an effective means for mitigating the effects of frequency-selective fading channels.

A significant advantage of multicarrier modulation is the ability to choose separately the alphabet for each carrier depending on its corresponding SNR. Of course this is possible only when the transmitter has knowledge of the channel frequency response. In particular, from (6.96) it is clear that for a fixed n, the sequence of symbols $a_k^{(n)}$ modulates the n-th carrier. The transmitter can adjust both the size and the energy of the n-th alphabet to match the n-th gain H_n, thus controlling the transmitted power spectrum and the noise immunity. In Chapter 8, we will find a specific shape for the power spectrum (based on waterpouring) that is ideal, in that it permits the performance to approach the theoretical capacity of the channel. Multicarrier signals can easily synthesize that shape. It is reasonable to dynamically adjust the choice of alphabet at each carrier frequency as the channel changes over time. Consequently, if a channel has a deep notch in its frequency response, or develops a deep notch due to frequency-selective multipath fading, for example, we would use a very simple binary antipodal alphabet at that frequency. We might even avoid using that frequency altogether. Correspondingly, at frequencies where the channel response is strong, we would use a large symbol alphabet. Such adaptive loading strategies are an integral part of DMT modems for DSL, but are less common in OFDM for wireless applications, because the time-varying nature of a wireless channel makes it difficult to acquire timely and accurate knowledge of the channel frequency response at the transmitter.

If N is large, and the symbols $a_k^{(n)}$ are appropriately random, then the probability distribution of a sample of the OPAM signal (6.76) will approach a Gaussian distribution, from the central limit theorem. Hence, a key disadvantage of multicarrier modulation is that the peak-to-average power ratio can be larger than with more conventional signals. This suggests that OPAM modulation is not advisable for either peak-power-limited channels or nonlinear channels. The linearity requirements of the power amplifier of a multicarrier transmitter are also more stringent than for a single-carrier system, increasing cost. However, the practical disadvantages of a Gaussian distribution becomes a theoretical advantage when we recall that, as shown in Chapter 4, a signal that achieves channel capacity has a Gaussian distribution.

The benefits of the cyclic prefix come at the price of a reduced throughput; of the $N + \mu$ transmitted symbols, only N convey information, leading to a rate-loss penalty of:

$$\frac{\mu}{N + \mu} . \tag{6.97}$$

This same fraction of signal energy is also wasted.

Example 6-16.

The IEEE 802.11a and ETSI Hiperlan 2 proposed wireless LAN standards for the 5 GHz band are based on a sample rate of 20 MHz, suitable for operation over a passband channel with a 20 MHz bandwidth. The number of tones is $N = 64$. The length of the cyclic prefix is $\mu = 16$ samples, which is large enough to accommodate a delay spread of at most 0.8 μs. Therefore, from (6.97), $16/(64 + 16)$ or 20% of the transmitted symbols are redundant and convey no information.

Example 6-17.

The ADSL standard [4] uses a prefix of length 32 with $N = 512$, leading to a penalty of 5.9%.

The overhead penalty of the cyclic prefix can be made small by choosing N large. Even without a cyclic prefix, if the impulse response of the channel is short compared to the symbol interval T, the difference between the circular convolution (6.92) and the ordinary convolution (6.91) might be small enough that (6.96) is close enough to be useful [5].

6.4.3. Spread Spectrum

Spread spectrum systems are PAM systems that deliberately use pulses with much more than the minimum bandwidth $1/(2T)$ required by the Nyquist criterion. Spread spectrum has a long history, mostly in secure military communications, as discussed by Scholtz [6]. More recently, a number of commercial applications have arisen, for example in digital cellular systems. A useful definition of spread-spectrum is [7]:

> Spread-spectrum is a means of transmission in which the signal occupies a bandwidth in excess of the minimum necessary to send the information; the band spread is accomplished by means of a code that is independent of the data, and a synchronized reception with the code at the receiver is used for de-spreading and subsequent data recovery.

The *spreading code* used in this definition will be defined shortly.

In Chapter 5 we saw that the SNR achieved with a matched-filter or correlation receiver depends on the energy in the received pulse $h(t)$, but not on its bandwidth. So from the perspective of SNR, there is no harm in using a pulse with a broad bandwidth, as long as a matched filter receiver is used. There are several reasons for using large bandwidth:

- Pulses with a broader spectrum are less sensitive to channel impairments that are highly localized in frequency. Such impairments arise, for example, with frequency-selective multipath fading.
- Spread spectrum signals are less vulnerable to *jamming*, in which a hostile party is trying to deliberately disrupt the communication.
- Spread spectrum signals can be concealed. By using very wide bandwidth pulses, these signals can be placed in regions of the spectrum already occupied by other signals, and in effect be masked by the other signals.
- Many spread spectrum users can share a common bandwidth without interfering much with one another.

Bandwidth and Probability of Error

Assume the channel noise is white and Gaussian. We have already seen that the bit-error probability is independent of the channel bandwidth; it depends only on the energy of the received pulse (Example 6-4). An intuitive explanation of this bandwidth independence is as follows. The bandwidth required for N orthogonal pulses satisfying the generalized Nyquist criterion is $W = N/(2T)$, which implies that $N = 2WT$. By definition, spread spectrum corresponds to $2WT \gg 1$, where this approximation becomes accurate. Since $h(t)$ only occupies one dimension, the matched filter captures only a fraction $1/2WT$ of the total noise in bandwidth W. While the variance of this total noise is proportional to W, on net, the noise variance at the output of matched filter is not dependent on W.

It is common in practice to characterize the signal-to-noise ratio (SNR) at the receiver input, in preference to the energy per bit and noise power spectral density. The received signal power is equal to the energy per symbol E times the symbol rate $1/T$, namely $P = E/T$. Furthermore, the total noise power within bandwidth W is $N_0 W$. The received SNR is defined as the ratio of signal power to noise power,

$$SNR = \frac{P}{N_0 W}. \tag{6.98}$$

Substituting into $Q(\sqrt{2E/N_0})$, we can express the bit-error probability for binary antipodal signaling as a function of SNR:

$$P_b = Q(\sqrt{2WT \cdot SNR}) . \tag{6.99}$$

If we keep SNR constant, then P_b decreases as the dimensionality $2WT$ increases. However, in order to keep SNR constant for a fixed N_0, P has to be increased in proportion to W. If P is kept fixed, then P_b is independent of W as stated earlier.

Generating Broadband Pulses

Spread spectrum requires ways to generate broadband pulses with controlled spectral properties. A whole family of pulse shapes $h(t)$, each with the same amplitude spectrum and different phase spectra, is conveniently generated using a *chip waveform* and *spreading sequence*. In this approach, the symbol interval T is divided into N subintervals or chips, each of duration $T_c = T/N$. The pulse $h(t)$ is formed from a PAM modulation of the chip pulses by some deterministic sequence $\{x_0, x_1, \ldots x_{N-1}\}$, called the *spreading sequence*, according to:

$$h(t) = \sum_{m=0}^{N-1} x_m h_c(t - mT_c) , \qquad H(f) = H_c(f) \sum_{m=0}^{N-1} x_m e^{-j2\pi f m T_c} . \tag{6.100}$$

The bandwidth of the resulting pulse will equal the bandwidth of $h_c(t)$. Typically, we choose $h_c(t)$ to satisfy the Nyquist criterion at the chip rate $1/T_c$, which requires a minimum bandwidth of $W = 1/(2T_c) = N/(2T)$; this causes a bandwidth expansion by a factor of $N = T/T_c$. The spectrum can be controlled to some degree by the spreading sequence, with precisely N degrees of freedom.

The chip waveform and spreading sequence can also be used to generate orthogonal pulses (see Problem 6-16). For example, such orthogonal pulses are required for CDMA systems.

ISI and Spread Spectrum

The preceding error probability calculation presumed no ISI. In fact, spread spectrum affords a degree of immunity to ISI. To understand this, we need to consider the pulse shape at the output of a receiver matched filter, and then the effect of channel dispersion.

Keeping the symbol interval T fixed, and increasing the bandwidth W, the ISI in the transmit pulse can be reduced. For a large $2WT$, a pulse bandlimited to W can be largely confined to an interval T, in the sense that a diminishing fraction of the pulse energy falls outside that interval as $2WT$ increases. (In fact, approximately $2WT$ orthogonal pulses can satisfy this condition simultaneously.) When $2WT$ is near unity, even a single pulse cannot come close to being time-limited to T. Thus, the *transmit* pulse can come much closer to being time-limited in a spread-spectrum system.

However, the *receive* pulse is affected by the channel; furthermore, we are interested in the ISI at the matched filter output rather than the channel output. (Recall that the matched filter is crucial to the operation of a spread spectrum system because of its power to suppress in-band noise.) Assume for the moment that the channel is ideal. The isolated-pulse output of the matched filter is then the pulse autocorrelation function, $\rho_h(t)$. The Fourier transform of that isolated pulse is $|H(f)|^2$, which by definition has a wide bandwidth W.

Example 6-18.

Suppose that $|H(f)|^2$ is constant over the bandwidth W and zero elsewhere. If we normalize the energy of $h(t)$ to unity, then $|H(f)|^2 = 1/2W$. The isolated pulse at the matched filter output is

$$\rho_h(t) = \text{sinc}(2\pi Wt) . \tag{6.101}$$

As W increases, the energy of this pulse concentrates in a shorter time duration. Furthermore, if $2WT$ is an integer, $\rho_h(t)$ always obeys the Nyquist criterion,

$$\rho_h(kT) = \text{sinc}(k\pi \cdot 2WT) = \delta_k . \tag{6.102}$$

This simple example illustrates two important points:

- For an ideal channel, the isolated pulse at the output of the matched filter is dependent only on the magnitude spectrum of $h(t)$, and is not dependent on the phase spectrum. Even though we have specified the magnitude spectrum in Example 6-18, there remains flexibility in choosing the phase.
- The time duration of isolated pulse at the output of the matched filter can be much shorter than the symbol interval, even though $h(t)$ completely fills the symbol interval. The greater W, the shorter this duration can be.

Pulses of the form of (6.100) can be designed to have a narrow autocorrelation function. The pulse autocorrelation (matched filter output isolated pulse) will conform to Example 6-18 if two sufficient (but not necessary) conditions are satisfied:

- The chip pulse $h_c(t) = \text{sinc}(\pi t/T_c)$ is an ideal LPF with bandwidth $W = 1/(2T_c)$, and
- The sequence $\{x_m\}$ is chosen to satisfy

$$\left|\sum_{m=0}^{N-1} x_m e^{-j2\pi f m T_c}\right|^2 = 1 \quad \text{for all } f . \tag{6.103}$$

Example 6-19.

A trivial case is $x_k = \delta_{k-L}$ for some $0 \le L \le N-1$. Regardless of L, (6.103) is satisfied. The choice of L affects the phase spectrum of $h(t)$, but not the magnitude spectrum. The problem with this choice is that the peak signal is very large in relation to E_h, creating practical difficulties on most channels, and especially on radio channels.

Example 6-20.

We can increase E_h for a given peak signal by choosing a sequence so that $|x_0|^2 = |x_1|^2 \ldots = |x_{N-1}|^2 = 1/N$. For such choices, (6.103) cannot be exactly satisfied. However, a good approximate approach is to force (6.103) to be satisfied at uniformly spaced frequencies, with spacing $1/(NT_c)$,

$$\left|\sum_{m=0}^{N-1} x_m e^{-j2\pi mn/N}\right|^2 = 1/N , \quad 0 \le n \le N-1 . \tag{6.104}$$

This is the condition that the DFT of $\{x_m\}$ have unit magnitude.

It is possible to come very close to satisfying the two conditions that $|x_m|^2 = 1$ and the DFT have a unit magnitude, by making $\{x_m\}$ a maximal-length shift register sequence (this will be shown in Chapter 12). The result is called *direct-sequence spread spectrum*. Spreading codes can also be used to design orthogonal modulation signal sets (see Problem 6-16). Unlike the FSK pulse sets described earlier, these can overlap one another completely in frequency.

Having established a method of designing a broadband $h(t)$ that has very narrow autocorrelation $\rho_h(t)$, the next question is the effect of channel dispersion. Assuming the channel has impulse response $b(t)$, the output of the matched filter becomes

$$h(t) * b(t) * h^*(-t) = \rho_h(t) * b(t) \ . \tag{6.105}$$

As before, the phase spectrum of $h(t)$ does not matter. The non-ideal channel increases the time duration of the matched filter output. However, since $\rho_h(t)$ can be kept very narrow when W is large, $\rho_h(t) * b(t)$ will have time duration approximately equal to the duration of $b(t)$. As long as the duration of $b(t)$ is smaller than the symbol interval T, the channel dispersion will not have a significant effect.

Example 6-21.

Spread spectrum is often used on radio channels, which suffer from multipath distortion. Suppose we take a two-path model,

$$b(t) = \delta(t) + \alpha\delta(t-\tau) \ , \tag{6.106}$$

where τ is the relative delay of the second path. For the $\rho_h(t)$ of Example 6-18, the isolated pulse output of the matched filter with this dispersive channel will be

$$f(t) = b(t) * \rho_h(t) = \text{sinc}(2\pi Wt) + \alpha\text{sinc}(2\pi W(t - \tau)) \ . \tag{6.107}$$

The symbol-rate samples of this isolated pulse are then

$$f(kT) = \delta_k + \alpha\text{sinc}(2\pi W(kT - \tau)) \ . \tag{6.108}$$

As W gets large, assuming that $|\tau| < T$, the ISI gets small. This is illustrated in Fig. 6-20. For $2WT = 5$ or 400% excess bandwidth, the multipath distortion sampled at the symbol interval can still be fairly large at the output of the matched filter. For $2WT = 128$, even a half-symbol

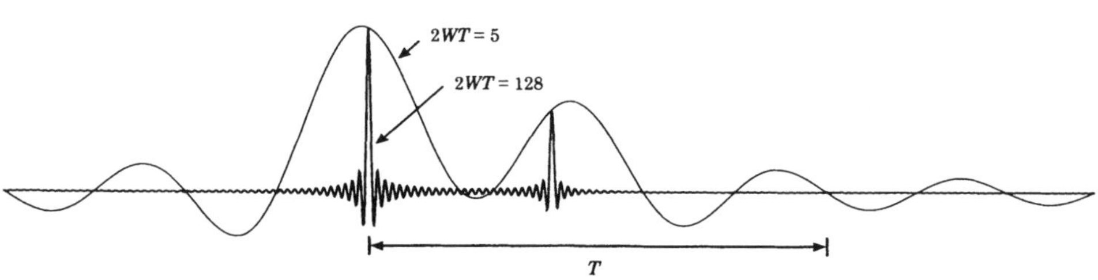

Fig. 6-20. The isolated pulses at the output of a matched filter for a two-path multipath channel with $\tau = 0.4T$ and $\alpha = 0.5$. Two bandwidth expansions are shown: $2WT = 5$ and $2WT = 128$. Note that the ISI will be small for $2WT = 128$ as long as delay spread τ is a little less than the symbol interval T, but for $2WT = 5$ ISI will be significant even for very small τ.

multipath spread results in a very small ISI, because the basic pulse shape at the matched filter output decays so rapidly.

This example illustrates the desirable effect of bandwidth expansion in terms of minimizing the effect of channel dispersion on the isolated pulse at the output of the matched filter. However, two caveats are in order:

- Spread spectrum with large $2WT$ is not *immune* to ISI; for example, consider what happens when $\tau = T$ in Example 6-21. In fact, if the multipath delay spread is $\tau = T$, ISI will be a problem no matter how large $2WT$ may be! Spread spectrum successfully mitigates channel dispersion only where the time-delay spread is smaller than a symbol interval.
- Fig. 6-20 illustrates that the timing recovery must be increasingly accurate as $2WT$ increases. Fortunately, the broader bandwidth of the signal is helpful in increasing the timing recovery accuracy as well.

Spread Spectrum and Jamming or Interference

We have shown that any bandwidth increase has no impact on the probability of error on white Gaussian noise channels, but it does help mitigate the effects of ISI. Another benefit of increasing bandwidth is that it improves immunity to jamming and broadband interference.

Assume that the noise on the channel is generated by a jammer. *Jamming* is a deliberate attempt to disrupt communication by generating a broadband interference signal. In practice, the jammer is limited in the power it can generate. The jammer generates bandlimited white noise with power P_J over the signal bandwidth W, with spectral density $N_0/2 = P_J/(2W)$. Since the jamming signal is white over the signal bandwidth, from the perspective of the receiver the error probability is the same as for white Gaussian noise (the presence or absence of out-of-band noise will be inconsequential). The receive SNR is now

$$SNR = \frac{P}{N_0 W} = \frac{P}{P_J} , \tag{6.109}$$

independent of bandwidth. The fact that the SNR is bandwidth-independent has profound implications, since from (6.99), P_b is now strongly dependent on the dimensionality $2WT$. In fact, as $2WT$ increases by expanding W, P_b decreases. For this reason, $2WT$ is called the *processing gain.*

Example 6-22.
If the processing gain is $2WT = 10^3$ (30 dB), the jammer power is effectively suppressed by 30 dB. That is, in going from $2WT \approx 1$ to $2WT = 10^3$ by expanding the bandwidth by 1000, 30 dB greater jammer power P_J can be tolerated with the same error probability.

The processing gain can also be interpreted in signal space. The pulse $h(t)$ defines a one-dimensional subspace, and our assumption is that the jammer does not know the direction of this subspace. Thus, the jammer must spread its power P_J evenly over all $2WT$ dimensions of the subspace of signals bandlimited to W Hz and time limited to a symbol interval T (this is equivalent to the jammer transmitting bandlimited white noise). The matched filter responds to the jammer noise in the direction of the signal only, and hence at the output of the matched filter the jammer power is reduced by $2WT$.

The foregoing presumes that the jammer cooperates, and transmits bandlimited white noise, or equivalently spreads its power evenly over all dimensions of the signal subspace. But clearly the jammer would do better to concentrate its power in the direction of the signal, because then there would be no processing gain! Increasing the signal bandwidth is beneficial only if the jammer does not know the direction of the signal, and therefore must spread its jamming power equally in all directions. If the jammer transmits a one-dimensional signal, the jammer power in the direction of signal vector $h(t)$ can fall anywhere between 100% (no processing gain) and 0% (infinite processing gain).

The use of the term "jammer" implies a military connotation, but in commercial microwave radio systems an important consideration is co-channel interference. This interference, for example between two satellites within the aperture of a single antenna, between a satellite and terrestrial radio system, or among different users of a terrestrial cellular radio system, has similar characteristics to jamming. If the signal and interferer both spread their bandwidth, keeping their total powers the same, then a processing gain results. In fact, if we take steps to actively reduce interference, the interferer can avoid transmitting in the one-dimension of the signal, resulting in an infinite processing gain! This is the principle behind the use of spread spectrum as a multiple access technique, as described below.

6.4.4. Code-Division Multiplexing

Another application of OPAM is *multiple access*, where distinct transmitter-receiver pairs share a single channel. Each transmitter-receiver pair would typically use only one of the orthogonal pulses. As long as the other transmitter-receiver pairs use different orthogonal pulses, the OPAM receiver structures given above are effective in separating the signals.

Reversing the order of summation and rewriting (6.76) in the form

$$s(t) = \sum_{n=0}^{N-1} u_n(t), \qquad u_n(t) = \sum_{k=-\infty}^{\infty} a_k^{(n)} g_n(t-kT), \tag{6.110}$$

we can now think of $s(t)$ as being the superposition of N PAM subchannel signals $\{u_n(t), 0 \le n \le N-1\}$, as shown in Fig. 6-21. Each of these PAM signals transmits an independent stream of data symbols $\{a_k^{(n)}, -\infty < k < \infty\}$ using its own distinctive pulse shape $g_n(t)$. All these N PAM signals can share the same channel as long as the pulses they use are orthogonal to one another. The matched filter in the receiver for one subchannel will not respond to the pulse shapes used by the other subchannels, as long as the set of pulses satisfies the general Nyquist criterion.

Example 6-23.

In multicarrier modulation, the pulses are chosen to be sinusoids of different frequencies. A multiple access scheme based on this set of pulses would be termed a *frequency-division multiple access (FDMA)* system.

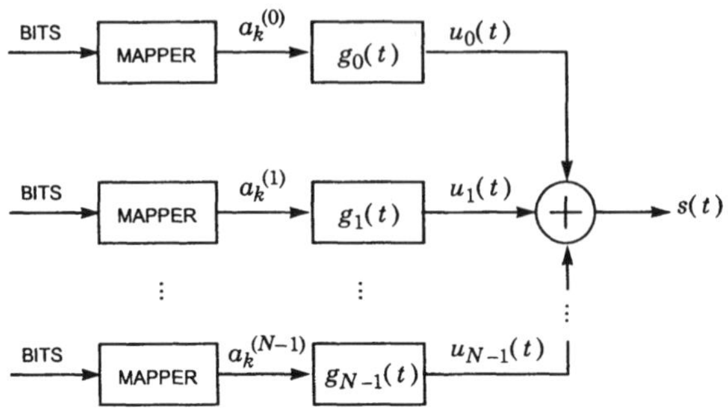

Fig. 6-21. In code-division multiple access (CDMA), N bit streams share the same channel by using N distinct orthogonal pulses $g_i(t)$.

Example 6-24.
An alternative is to choose a set of broadband pulses $g_n(t)$, each one of which fills the entire bandwidth of $|f| \le N/(2T)$. For this particular choice of pulses $g_n(t)$, (6.110) is known as *code-division multiple access (CDMA)*. The pulses used in CDMA are typically generated using a pseudorandom sequence, generated using a technique described in Chapter 12. For now, observe that these pulses can have broad bandwidth (the pseudorandom sequence ensures this) and will be orthogonal to one another.

Previously we described spread spectrum as a technique that can counter certain types of noise and interference signals by greatly expanding the bandwidth of the transmitted pulse. It has very poor spectral efficiency, but can be used in situations where spectral efficiency is not important. In CDMA, each PAM subchannel is itself a spread-spectrum signal.The motivation here is not to counter jamming signals so much as to allow other PAM signals (using different orthogonal pulse shapes) to share the same channel. Of course, it is also possible in CDMA to expand the bandwidth beyond $N/(2T)$, thereby gaining both multiple access and immunity to jamming signals of the type discussed in Section 6.4.3.

6.5. Modulation with Memory

The modulation techniques considered so far can be considered as memoryless, in the sense that the signal transmitted during the k-th signaling interval depends only on the k-th symbol (or, for OPAM, the k-th set of symbols), and in particular does not depend on previous symbols. In practice it can be advantageous to generalize the modulator so that previous symbols as well as the k-th symbol determine the transmitted signal. In this section we examine three common types of modulation with memory: continuous-phase modulation, minimum-shift keying, and differential modulation.

6.5.1. Continuous-Phase Modulation

A continuous-phase FSK transmitter can be implemented using a *voltage-controlled oscillator* (VCO, Chapter 14) driven by a baseband PAM signal, as shown in Fig. 6-22. The VCO frequency varies about a nominal carrier f_c and automatically maintains phase continuity. The output signal can be written

$$x(t) = K\cos\left[2\pi\left(f_c t + f_d\int_{-\infty}^{t} s(\tau)d\tau\right)\right]. \tag{6.111}$$

A signal of this form is called *continuous-phase modulation* (CPM). The baseband data signal is written

$$s(t) = \sum_{k=-\infty}^{\infty} a_k g(t - kT). \tag{6.112}$$

When the data signal $s(t)$ is normalized to peak at unity, $|s(t)| \le 1$, the factor f_d is the peak frequency deviation. When $g(t)$ is rectangular, CPM is called *continuous-phase FSK* (CPFSK). Other pulse shapes are also used and can lead to significantly improved performance.

The nonlinear relationship between the data signal $s(t)$ and the FSK signal is clearly seen in (6.111). Among other effects, this nonlinear relationship makes it very difficult to find an expression for the power spectrum of a general CPFSK signal. Suitable derivations and plots are given elsewhere (see for example [2]). The bandwidth of an FSK signal is usually wider than a passband PAM signal with the same symbol rate, except when *minimum shift keying (MSK)* is used. MSK, discussed in the following subsection, is characterized by a frequency deviation f_d that is half that predicted by (6.72), equivalent to the spacing of the Chang pulses (6.58).

Minimum Shift Keying

In Section 6.3 we determined that orthogonal FSK pulses have continuous phase if the frequency separation is equal to the symbol rate $1/T$. However, it is possible to reduce the frequency separation still further while maintaining orthogonality and phase continuity using the model in (6.111). When the frequency separation is half of that predicted by (6.72), the modulation technique is called minimum-shift keying (MSK).

MSK was invented by Doelz and Heald [8], and has been used in some microwave radio systems. It is sometimes called *fast FSK* (FFSK) because the spectral efficiency is higher than more traditional FSK signals. The separation in frequency between pulses is equivalent to that

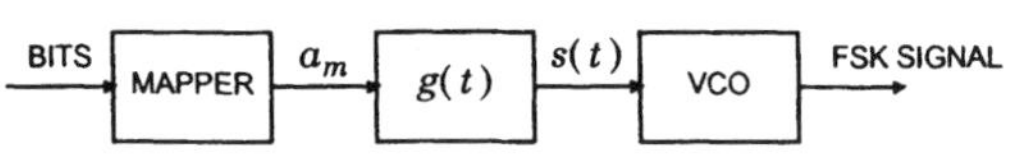

Fig. 6-22. An implementation of an FSK transmitter using a voltage controlled oscillator (VCO).

of the minimum-bandwidth pulses (6.52) and the Chang pulses (6.58). By introducing memory from one symbol to the next, it cleverly maintains phase continuity while decreasing the frequency separation. If the MSK pulses have frequencies $f_0 < f_1 < \ldots < f_{N-1}$, then

$$f_i - f_{i-1} = 1/(2T); \quad \text{for } 1 \le i \le N-1 . \tag{6.113}$$

We will illustrate MSK for the binary case in the following example.

Example 6-25.

Suppose that $|f_1 - f_2| = 1/(2T)$ in the binary FSK example. The peak frequency deviation is $f_d = 1/(4T)$. The higher frequency sinusoid traverses half a cycle more than the lower frequency sinusoid within one symbol interval. The *M*-ary model of (6.2) for FSK would use two pulses like those in Fig. 6-23(a), resulting in the signal shown in Fig. 6-23(b), which has discontinuous phase. Using the continuous-phase model of (6.111), however, the corresponding signals are shown in Fig. 6-24. The frequency separation is half the minimum frequency separation of Example 6-11, yet the pulses are orthogonal (see Problem 6-2).

The correlation or matched-filter receiver needs to be modified slightly to accommodate the MSK model. Notice from Fig. 6-24 that

$$g_i(t) = \pm \sin(2\pi f_i t) w(t) , \tag{6.114}$$

where $w(t)$ is a rectangular window over [0, T), as in Example 6-6. There are two signals with opposite polarity for each symbol. Since the signals with opposite polarities bear the same information, the matched-filter receiver should compare the *absolute value* of the sampled outputs, as shown in Fig. 6-25.

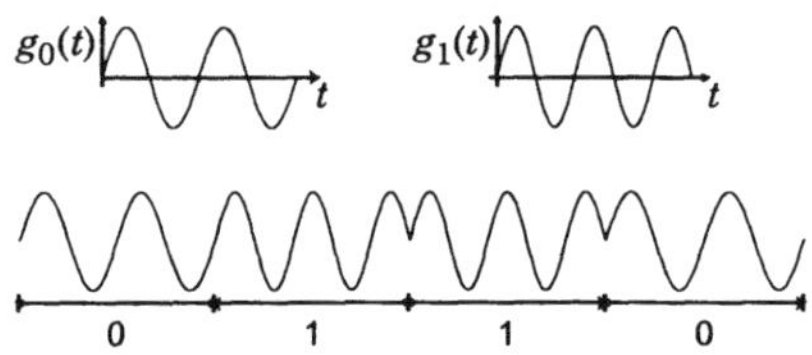

Fig. 6-23. (a) Two orthogonal FSK pulses with frequency separation $|f_1 - f_0| = 1/(2T)$. (b) An FSK signal with discontinuous phase.

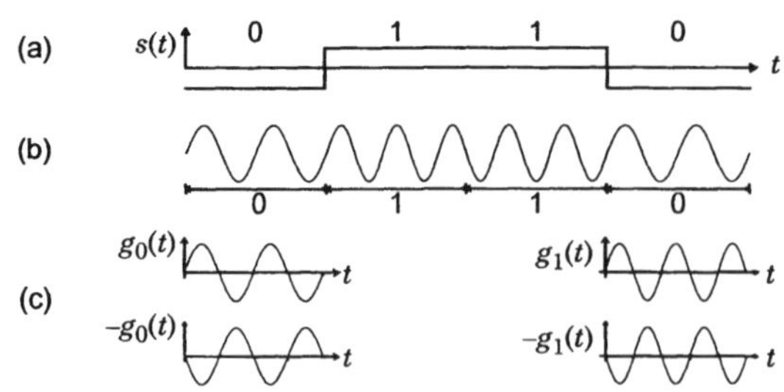

Fig. 6-24. This figure shows the relevant signals for continuous-phase binary MSK. The peak frequency deviation is $f_d = 1/(4T)$. (a) A data signal. (b) The corresponding MSK signal.(c) The pulses. Notice that each bit value has two possible pulses which are negatives of one another.

It should be emphasized at this point that MSK is not orthogonal modulation. From (6.114), the two possible transmitted pulses are $\sin(2\pi f_i t)$ and $-\sin(2\pi f_i t)$, which are not orthogonal. In fact, MSK can be thought of as a constrained form of OPAM.

The receiver in Fig. 6-25 throws away potentially useful information, since based on the history of the received signal, we could infer which of the two polarities would be expected. Making use of this information can reduce the probability of error. In the next section we will show how to take advantage of this additional information.

The pulses in (6.114) can be re-written

$$g_i(t) = \sin[2\pi f_c t + \pi b t/(2T) + \phi] w(t) , \tag{6.115}$$

where $b \in \{\pm 1\}$ determines the transmit frequency and $\phi \in \{0, \pi\}$ depending on which phase is being transmitted. The nominal carrier frequency is

$$f_c = (f_0 + f_1)/2 . \tag{6.116}$$

One representation for the transmitted MSK signal is therefore

$$x(t) = \sum_{k=-\infty}^{\infty} \sin[2\pi f_c t + \pi b_k t/(2T) + \phi_k] w(t - kT) , \tag{6.117}$$

where b_k is determined by the data and ϕ_k ensures phase continuity.

Exercise 6-8.

Show that to maintain phase continuity we need

$$\phi_k = \phi_{k-1} + (b_{k-1} - b_k)\pi k/2 \mod 2\pi . \tag{6.118}$$

Expression (6.118) explicitly shows the dependence of the phase in each symbol interval on the data.

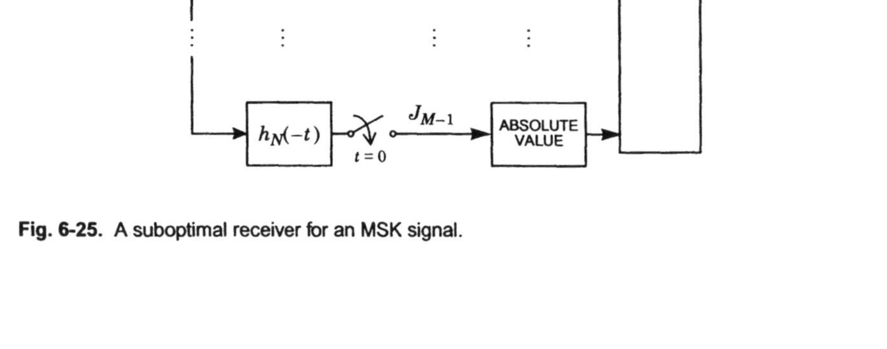

Fig. 6-25. A suboptimal receiver for an MSK signal.

The matched-filter receiver of Fig. 6-25 performs roughly as well with binary MSK as with FSK. In the next section we will show that the performance can be improved to make MSK better than FSK, in that the probability of error will be lower and the bandwidth will be smaller. This is accomplished by using the information thrown away by the absolute values in Fig. 6-25, the error probability of MSK becomes essentially equivalent to PSK.

An interesting property of MSK signals is that they can be interpreted as passband PAM signals where the quadrature component is delayed half a symbol interval with respect to the in-phase component. Such signals are called *offset keyed QAM* (OQAM or OK-QAM) or *offset keyed phase-shift keying* (OPSK or OK-PSK). MSK is a special case (see Problem 6-4).

6.5.2. Detection of CPM

We now consider the detection problem for a CPM signal defined by (6.111) and (6.112). It is common to normalize the pulse shape $g(t)$ so that it integrates to $1/2$. With this normalization, the CPM signal has a phase change of $a_k f_d \pi$ radians in one symbol interval, with respect to the carrier f_c.

MSK provides a constant envelope signal with considerably narrower bandwidth than a constant envelope PSK (using rectangular pulses). However, since for CPFSK $g(t)$ in (6.112) is rectangular, there is a discontinuity in the first derivative of the CP signal. This implies that with smoother choices for $g(t)$ we can significantly reduce the bandwidth by ensuring continuous first, or even second or third, derivatives. The simplest case, called *full-response CPM*, uses a $g(t)$ that is zero outside the interval $0 \le t \le T$. The second case, called *partial-response CPM*, uses a $g(t)$ that extends over several symbol intervals. The term partial response refers to the deliberate introduction of ISI for spectrum control. In fact, the spectral properties of partial-response CPM signals are considerably better than full-response CPM and CPFSK [9].

The evolution of a CPM signal over time can be compactly described using a *phase diagram*. The phase diagram plots the phase term from (6.111), namely

$$2\pi f_d \int_{-\infty}^{t} s(\tau) d\tau \,, \tag{6.119}$$

for all possible input symbols $\{a_k\}$.

Example 6-26.

The phase diagram for MSK is shown in Fig. 6-26.

Since phase is modulo 2π, the phase diagram is best viewed wrapped around a cylinder with circumference 2π. If the modulation index f_d is rational, then the phase diagram will have a finite number of points on this cylinder. In this case, it can be viewed as the trellis diagram for a Markov chain. Since the phase evolves according to a Markov chain, the Viterbi algorithm is commonly used in the detector to perform optimal sequence detection.

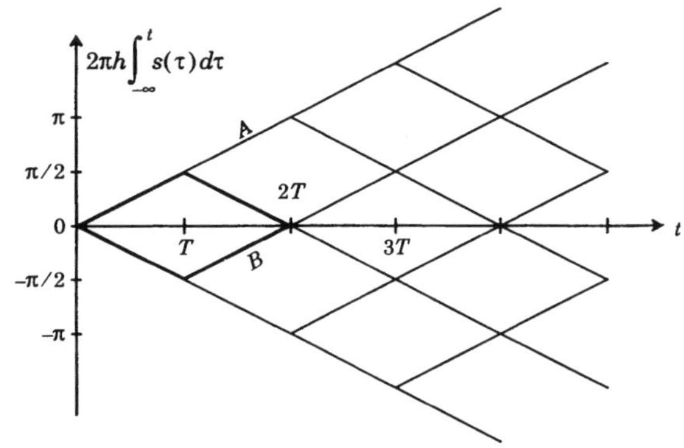

Fig. 6-26. A phase diagram for an MSK signal.

Example 6-27.
The receiver structure Fig. 6-25 for MSK signals performs roughly as well as the optimal matched-filter receiver for orthogonal FSK signals. However, by taking the absolute value of the sampled output of the matched filters, that structure discards useful information. The useful information that is discarded is easily seen in Fig. 6-26. The receiver in Fig. 6-25 makes no distinction between the path labeled "A" and the path labeled "B", which correspond to signals that are π radians apart, or antipodal. Obviously, making the distinction would be useful. Considering that the phase is modulo 2π, the phase diagram in Fig. 6-26 has only four states, best viewed as lying on a cylinder. The minimum-distance error event has length two and is the bold diamond shape in Fig. 6-26 (see Problem 6-15). Furthermore, the squared distance of this error event is twice the squared distance between the two orthogonal signaling pulses, so application of the Viterbi algorithm results in a 3 dB improvement over the receiver in Fig. 6-25. That receiver was shown to perform roughly as well as an optimal orthogonal FSK receiver, which is 3 dB worse than an optimal binary antipodal (2-PSK) signal of the same average power. Consequently, by using the Viterbi algorithm with MSK we are able to recover the 3 dB loss associated with FSK and match the performance of PSK.

Construction of the trellis and application of the Viterbi algorithm becomes more complicated for CPM signals other than MSK. The number of states depends on the modulation index f_d and the extent of the pulse $g(t)$.

6.5.3. Differential Encoding and DPSK

Most of the signal constellations we have described have the practical problem that they are rotationally invariant for some particular angles of rotation, typically multiples of $\pi/2$. By this we mean that if the constellation is rotated, there is no way that the receiver can distinguish it from a valid constellation, unless it knows the actual transmitted data symbols, which it does not. If this problem is not dealt with, the receiver may decide on a receive phase corresponding to a rotated constellation, with the result that the information bits will be incorrectly decoded. This problem can be eliminated by using *differential encoding*, in which the information is encoded by the *change* in constellation position between symbols rather than absolute position. Differential encoding can be mixed with absolute encoding. For the 16-

QAM constellation of Fig. 5-12, for example, it is common to differentially encode the quadrant (specified by two bits), while the point within the quadrant (specified by another two bits) is encoded absolutely.

Differential encoding is especially valuable on channels with rapid fading, such as the mobile radio channel. It is common on such channels to use PSK modulation, which makes the detection of data symbols insensitive to fluctuations in the amplitude of the received signal, since the slicer considers only the angle and not the amplitude of its input. However, the phase of such a channel can also vary rapidly, especially during deep amplitude fades. At the expense of some noise immunity, this phase fluctuation can be mitigated by using PSK with differential encoding and differential detection. This combination of PSK, differential encoding, and differential decoding is called *differential PSK (DPSK)*.

We will consider differential encoding only for the case of PSK, although it should be recognized that it is often used with other constellations as well. For PSK, the transmitted symbols are of the form $a_k = e^{j\phi_k}$ where the phases ϕ_k are chosen from some alphabet. The phase ϕ_k is determined by

$$\phi_k = \phi_{k-1} + \Delta_k , \tag{6.120}$$

where the difference in phase from one symbol to the next, Δ_k, carries the information, not the absolute phase ϕ_k.

Example 6-28.

In *differential binary PSK (DBPSK)*, one of two phases is transmitted. For this case, these two phases are π apart, and the coder can map a zero bit into $\Delta_k = 0$ (two successive transmitted phases are identical) and a one bit into $\Delta_k = \pi$ (two successive transmitted phases are π apart).

Example 6-29.

The IS-54 standard for digital cellular radio in North America transmits two bits per symbol, using is a form of *quadrature PSK (QPSK)*. However, rather than associating these two bits with four phases, in actuality eight equally-spaced phases are used, as shown in Fig. 6-27. At any given symbol (k) the data symbol assumes only one of four phases chosen from the sets $\{0, \pi/2, \pi, 3\pi/2\}$ (for odd-numbered symbols) and $\{\pi/4, 3\pi/4, 5\pi/4, 7\pi/4\}$ (for even-numbered symbols), where these two sets are offset by $\pi/4$ relative to one another. Two information bits are coded as a *change* in phase by one of the values $\{\pi/4, 3\pi/4, 5\pi/4, 7\pi/4\}$. The possible phase transitions from one symbol to another are shown in Fig. 6-27. The differential phase Δ_k is determined from the two input information bits in accordance with the following table:

Bits	Δ_k
1 1	$5\pi/4$
0 1	$3\pi/4$
0 0	$\pi/4$
1 0	$7\pi/4$

There are two choices in the receiver design when using differential encoding: *coherent* or *synchrodyne detection*, which attempts to learn and track the absolute phase of the received data symbols, and *differential detection*, which looks at only the change in phase from one symbol to the other, as illustrated in Fig. 6-27. Synchrodyne detection (Fig. 6-27(a)) is appropriate when differential encoding is used only to mitigate the rotational invariance of the signal constellation. Suppose the input samples to the detector in both cases are

$$y_k = e^{j(\phi_k + \theta_k)} + n_k \ , \tag{6.121}$$

where θ_k is some unknown phase rotation and n_k is the complex-valued additive noise. In the coherent case, we assume that $\theta_k = \theta$, where θ assumes certain discrete phases (e.g. any multiple of $\pi/4$ in Example 6-29) that allow the slicer to work properly. The y_k are applied to a conventional slicer designed for the ϕ_k, and it is assumed that the receiver estimates $\phi_k + \theta$ rather than ϕ_k. After the slicer, a difference operation forms an estimate of Δ_k, independent of θ, that directly represents the information bits.

The second alternative, differential detection, is shown in Fig. 6-27(b). This approach avoids tracking a rapidly varying channel phase. For this case, the statistic $y_k y_{k-1}^*$ is formed before the slicer. The slicer is designed to have the proper thresholds for $e^{j\Delta_k}$ rather than $e^{j\phi_k}$. There are two consequences of this:

- In the absence of noise, the input to the slicer is the proper phase Δ_k regardless of θ. This is valuable on channels with rapid phase variations, since it means that the carrier phase does not have to be tracked.
- There is an increase in the noise at the slicer input; this is the price paid for the insensitivity to phase rotation.

We will now verify these two properties. The slicer input is

$$y_k y_{k-1}^* = e^{j\Delta_k} e^{j(\theta_k - \theta_{k-1})} + e^{j(\phi_k + \theta_k)} n_{k-1}^* + e^{-j(\phi_{k-1} + \theta_{k-1})} n_k + n_k n_{k-1}^* \ . \tag{6.122}$$

Assume that the phase rotation θ_k does not change too much from one symbol to the next ($\theta_k \approx \theta_{k-1}$). This is a valid assumption as long as the symbol rate is high relative to the rate of phase change. With this assumption, the signal term at the slicer input is $e^{j\Delta_k}$, independent of θ_k. Looking at the noise terms, $n_k n_{k-1}^*$ is the product of two noise terms, and hence will typically be insignificant. The phase factors multiplying n_k and n_{k-1}^* do not affect their variance. Approximating these terms as independent, the total noise variance is now

$$E[\,|n_k + n_{k-1}^*|^2] = 4\sigma^2 \ , \tag{6.123}$$

or twice as large as in the coherent case. There is thus roughly a 3 dB penalty for differential detection (twice as much noise power). A more refined analysis that takes account of the

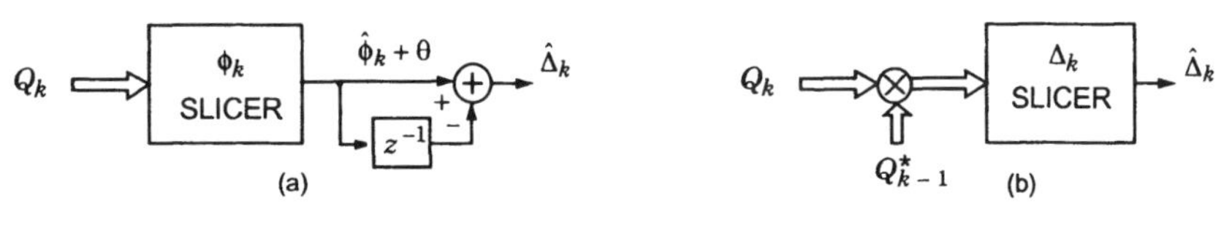

Fig. 6-27. Two detection techniques for DPSK. (a) Coherent, which requires an accurate phase reference, and (b) differential, which allows an arbitrary slowly-varying phase rotation of the data symbols.

correlation of the two noise terms reveals that the penalty is actually about 2.3 dB at high SNR.

6.6. Bandwidth and Signal Dimensionality

This chapter has introduced the important concept of the dimensionality of the subspace spanned by the set of known signals. In this section, we develop additional insight into the relationship between this dimensionality and the bandwidth required to accommodate this set of signals. Specifically, we will shed new light on the generalized Nyquist criterion by examining it from a different perspective. In Section 2.6 we defined the subspace of signals spanned by a set of M known signals, and observed that this subspace is finite dimensional (with dimension N). Our goal is to understand better the relationship between the bandwidth of the signals and the dimension of the subspace.

6.6.1. Landau-Pollak Theorem

No signal can be both time limited and bandlimited. A bandlimited signal is not time limited, in the sense that its energy cannot be totally confined to any finite interval of time, and a time-limited function is not bandlimited, in the sense that its energy cannot be totally confined to a finite band of frequencies. However, it is possible for functions to be bandlimited and *approximately* time limited, or time limited and *approximately* bandlimited. For example, consider a function $f(t)$ that is causal and bandlimited to W Hz, and also has finite energy E_f. Then $f(t)$ never goes precisely to zero beyond any fixed time t_0, but because it has finite energy it will decay gradually to zero. One way to measure the rate of that decay is to calculate the fraction $\varepsilon(t_0)$ of its energy outside the interval $[0, t_0]$, where $\varepsilon(t_0) < 1$. Specifically, let

$$\int_0^{t_0} |f(t)|^2 dt = E_f(1 - \varepsilon(t_0)) \ . \tag{6.124}$$

Since a fraction $\varepsilon(t_0)$ of the energy is outside the interval, a fraction $1 - \varepsilon(t_0)$ is within the interval. For a causal finite-energy function $f(t)$, as $t_0 \to \infty$, $\varepsilon(t_0) \to 0$. If we define the signal to be approximately time limited to an interval when less than a specific fraction ε of its energy is outside that interval, then we can always choose a large enough interval that the signal is approximately time limited.

Although the signal space of all finite-energy signals is infinite-dimensional, it is also true that the subset of such signals that are bandlimited to W Hz and approximately time limited to $[0, t_0]$ is approximately finite dimensional, with dimension $2Wt_0 + 1$. This statement is made rigorous by the *Landau-Pollak theorem* [10].

There exists a set of $2Wt_0 + 1$ orthonormal waveforms $\phi_i(t)$, such that for any (possibly complex-valued) finite-energy waveform $f(t)$ with energy E_f that is bandlimited to $|f| \le W$, for any constant $0 < \varepsilon < 1$, and for any t_0 sufficiently large that

$$\int_0^{t_0} |f(t)|^2 dt \ > \ E_f(1-\varepsilon) \ , \tag{6.125}$$

there exists a set of $2Wt_0 + 1$ coefficients f_i such that

$$\int_{-\infty}^{\infty} |f(t) - \sum_{i=0}^{2Wt_0} f_i \phi_i(t)|^2 \, dt \ < \ 12E_f\varepsilon \ . \tag{6.126}$$

We can state this theorem in words as follows. If less than a fraction ε of a bandlimited signal's energy is outside an interval $[0, t_0]$, then that signal can be approximated by a linear combination of a set of $2Wt_0 + 1$ orthonormal waveforms with an error which has energy less than a fraction 12ε of the signal's energy. Thus, the dimensionality of the subspace of all signals approximately time limited to t_0 and bandlimited to W is approximately $2Wt_0 + 1$, in the sense that a small fraction of the signal's energy is outside a signal subspace of dimension $2Wt_0 + 1$. As t_0 increases, the fraction of energy outside this subspace (which is also growing in dimensionality) gets smaller.

6.6.2. Relation to the Generalized Nyquist Criterion

In the generalized Nyquist criterion, we made no attempt to time-limit the pulse waveforms $h_n(t)$ to the symbol interval T. Thus, the Landau-Pollak theorem does not apply directly. However, the generalized Nyquist criterion and the Landau-Pollak theorem are connected, and consistent with one another, as we now show.

The key to forming the connection is to consider a sequence of L signaling intervals. Suppose $h_n(t)$, $0 \le n \le N-1$ is a set of pulses bandlimited to W Hz that satisfy the generalized Nyquist criterion. Generate an OPAM signal consisting of L signaling intervals:

$$s(t) = \sum_{k=0}^{L-1} \sum_{n=0}^{N-1} a_k^{(n)} h_n(t-kT) \ . \tag{6.127}$$

Since $s(t)$ is a linear combination of NL orthogonal waveforms $h_n(t-kT)$, $0 \le n \le N-1$, $0 \le k \le L-1$, it lies in an NL-dimensional subspace of signal space. It is also easy to show (see Problem 6-11) that under very mild conditions, $s(t)$ is approximately time limited to $[0, LT]$ in the sense that the fraction of the energy of $s(t)$ outside this interval goes to zero as $L \to \infty$. Thus, the Landau-Pollak theorem tells us that $s(t)$ can be approximated by $2WLT + 1$ orthonormal functions, with increasing accuracy as $L \to \infty$. This means that this dimensionality must be at least the actual dimensionality NL,

$$2WLT + 1 \ge NL \ , \qquad W \ge \frac{NL-1}{2LT} \ . \tag{6.128}$$

As $L \to \infty$, the Landau-Pollak theorem implies that the bandwidth required is $W \ge N/(2T)$ Hz, which is consistent with the generalized Nyquist criterion.

6.6.3. Impact of Signal Bandwidth on the Isolated Pulse

One impact of the Landau-Pollak theorem is that the parameter $2Wt_0$, the so-called *time-bandwidth product,* plays an important role for signals that are approximately time limited and bandlimited. For a bandlimited signal with bandwidth W, as $2Wt_0$ increases, a couple of things happen:

- The fraction of the signal energy confined to an appropriate time interval of duration t_0 will increase.
- The fraction of the signal energy falling outside a $2Wt_0+1$ dimensional subspace of signal space will decrease.

When $2Wt_0$ is small, the notion of a pulse being confined to an interval of duration t_0 is crude at best. However, as $2Wt_0$ gets large, we can design bandlimited pulses that are, for all practical purposes, confined to the interval of duration t_0.

The Landau-Pollak theorem considers a waveform with bandwidth W and requires us to find a sufficiently large time limit t_0 such that most of the energy of the waveform lies within $[0, t_0]$. An alternative approach is to hold t_0 fixed and increase the bandwidth W, allowing the waveform to be increasingly confined to $[0, t_0]$. The dual notions of increasing the bandwidth or the time interval both arise in digital communication.

Example 6-30.
In spread spectrum, a single pulse $h(t)$ is amplitude modulated for each data symbol. The Nyquist criterion says that a bandwidth of $1/(2T)$ is required if ISI is to be avoided. In fact, in spread spectrum the bandwidth W is much larger (often hundreds of times), so that $2WT$ is very large. In this case, it is possible to make the pulse $h(t)$ nearly time limited to the symbol interval T. This implies in turn that ISI is rarely a practical issue in spread spectrum systems. In fact, countering or avoiding ISI is often a motivation for using spread spectrum; the essential property is that the time-bandwidth product is very large. This issue was addressed in Section 6.4.3.

Example 6-31.
In orthogonal modulation (for example FSK), one of M orthogonal pulses is transmitted for each symbol. The minimum bandwidth W required to satisfy the generalized Nyquist criterion is $W = M/(2T)$, or $2WT = M$. If M is large (not two or three), a side effect of this larger bandwidth is that the pulses can be designed to be largely confined to the symbol interval T. One advantage of orthogonal modulation can thus be less susceptibility to ISI.

Example 6-32.
In multicarrier modulation, the usual perspective is that as the dimensionality of the signal set is increased, the bandwidth W is kept constant but the symbol interval T is increased. While the symbol rate is thereby reduced, a number of orthogonal pulses, each independently amplitude modulated with their own data symbols, can be transmitted simultaneously and separated out at the receiver by a bank of matched filters, keeping the spectral efficiency approximately constant. In fact, a maximum of $N = 2WT$ orthogonal pulses can be defined consistent with generalized Nyquist criterion, and with this maximum N, the spectral efficiency is not affected by increasing N and T. One side effect is that the pulses can be designed to be more confined to a symbol interval, in this case because of the increase in T for constant W. Again, the system can be less susceptible to ISI as a result. Often a short guard time between pulses will completely eliminate ISI.

6.6.4. Communication Link in a Network Context

While communication theorists tend to focus on two performance parameters, bit rate and error rate, a third parameter — the delay a communication link imparts — is of great importance when a link is incorporated in a network. This parameter is closely related to the Nyquist criterion and the Landau-Pollak theorem, and we now explore this relationship briefly.

Suppose a communication link imparts a total signal delay τ from input to output. For example, in free space τ equals the physical length of the link divided by the speed of light. (The link also imparts dispersion, which we ignore for simplicity.) Taking a snapshot of time and looking at the signal waveform as a function of distance along the link, we see τ seconds of the past transmitted waveform. The Landau-Pollak theorem tells us that the number of orthogonal pulses that can be in transit at any given time is thus approximately $2W\tau + 1 \approx 2W\tau$. This observation is important to the design of a network (and applications utilizing the network), because the bits represented by these pulses in transit are *latent*: they have been generated and transmitted, but have not yet arrived. A network engineer thinks of this in terms of latent bits rather than pulses, and says equivalently that the number of bits in transit equals the bit-rate-delay product $R_b\tau$, where R_b is the bit rate.

An often invoked metaphor for a network is a "superhighway." This is misleading because the way we increase the throughput of a highway (with a fixed number of lanes) is to increase the speed limit. On a communication link, the "speed limit" is fixed by nature (at some large fraction of the speed of light), and the way to increase throughput is to make the bits physically smaller and closer together: increasing the transmitted bandwidth and symbol rate is equivalent to deceasing the space occupied on the link by individual bits and pulses (or decreasing the size and spacing of the vehicles on the highway).[1]

From the point of view of applications, the latency of bits is less important than the latency of *messages*; that is, groups of bits that have application meaning only when taken as a group. For example, in a tightly coupled parallel computation, the latency of messages between processors (numerical values input to another part of the computation) directly affect the time it takes to complete an overall computation. The latency of a message on a communication link is the time that elapses between when that message is available to be transmitted until it arrives at the destination in its entirety. (Of course, networks have other sources of latency including queuing delay). This message latency has two components: the message transmit time (L/R_b seconds for an L-bit message) and the transit time τ. As we increase the "performance" of the link by increasing R_b (using the techniques in this book), the transmit time decreases but the transit time does not. Thus, there comes a point of diminishing returns (with respect to message latency) where the unchanging transit time dominates message latency; this is where the entire message occupies a small portion of the bits in transit on the link. The transit time thus comes to dominate the message latency performance of a link, a fact

1. Network engineers also commonly refer to the network "bandwidth," when they really mean "bit-rate throughput." This is confusing, because the two parameters are proportional, but the constant of proportionality (the spectral efficiency) can differ widely from unity.

that has profound implications to the future of networks and their applications. The only way to reduce this transit time is to make the communication link shorter, or equivalently the entire system smaller.

6.7. Capacity and Modulation

There are a number of ways of comparing different modulation techniques. The appropriate method depends on the context. Some of the measures of interest when making comparisons include:

- The *average transmitted power* is limited on most media, due to physical constraints or regulatory fiat. On some media, *peak transmitted power* is also of concern. Either of these limitations, together with the attenuation of the medium, will limit the received signal power. The *noise immunity* of the modulation system is measured by the minimum received power relative to the noise power required to achieve a given error probability.
- The *probability of error* is normally the basic measure of the fidelity of the information transmission. Of primary concern in some contexts is the probability of bit error, and in other contexts the probability of block error, for some appropriate block size.
- The *spectral efficiency* is of great concern on bandwidth-constrained media, such as radio. The spectral efficiency is the ratio of two parameters: the information bit rate available to the user and the bandwidth required on the channel.
- A measure that has not been discussed yet is the potential advantage of using coding techniques (Chapters 12 and 13). This advantage is different for modulation formats that are otherwise comparable. *Coding gain* is usually defined as the decrease in received signal power that could be accommodated at the same error probability if coding were used in conjunction with the modulation system.
- An important criterion in practice, although one we do not emphasize in this book, is the *implementation cost* or *design effort* required to realize a given modulation system.

The variety of measures for comparison of modulation systems makes it difficult to define one "standard measure of comparison." For example, for two modulation techniques, we can set the transmitted powers equal and compare the uncoded probability of error. However, the usefulness of this comparison will be compromised if the two modulation systems have different bandwidth requirements or provide different information bit rates for the same bandwidth.

In this section, we first illustrate how to make simple comparisons of uncoded modulation systems, addressing specifically baseband and passband PAM. Following this, a more sophisticated approach (based on a "rate-normalized SNR") is developed. This approach allows modulation systems to be compared against the fundamental limits of information theory (Chapter 4) and against one another in a way that is independent of bit rate or bandwidth requirements. The rate-normalized SNR takes into account most of the parameters mentioned — transmit power, noise power, and spectral efficiency — and summarizes them in a single universal error probability plot that further displays the available coding gain.

Comparisons will be made under the following assumptions:

- The channel is an ideal bandlimited channel with bandwidth W and additive white Gaussian noise with power spectral density $N_0/2$.
- The average transmit power is constrained to be P. There is no peak power limitation.
- Symbol error probability adequately reflects the performance of the system. Moreover, the union bound is sufficiently accurate as an approximation to error probability.

6.7.1. Error Probability of PAM

The simplest approach to comparing modulation systems is to calculate their error probability as a function of all the relevant parameters. A convenient approximate formula for the probability of symbol error (based on the union bound) for PAM is given by (6.41), which we repeat here,

$$P_e \approx K \cdot Q\left(\frac{d_{\min}}{2\sigma}\right), \tag{6.129}$$

where $d_{\min}$ is the minimum distance between any pair of alphabet symbols, K is the average number of nearest neighbors at that distance, and $\sigma^2 = N_0/2$. Let W denote the signal bandwidth, E denote the average signal energy, and T denote the symbol interval, so that the signal power is $P = E/T$. We will find it convenient to rewrite the error probability in terms of $\text{SNR} = P/(N_0 W)$ as follows:

$$P_e \approx K \cdot Q(\sqrt{2WT \cdot \eta_A \cdot SNR}), \tag{6.130}$$

where we have introduced:

$$\eta_A = \frac{d_{\min}^2}{4E_a}. \tag{6.131}$$

This formula applies to any PAM system, as long as the effect of ISI is ignored, and expresses P_e in terms of the received SNR, dimensionality $2WT$, and a parameter of the signal constellation, η_A. Often it is desired to express the probability in terms of the spectral efficiency, which is

$$\nu = \frac{\log_2 M}{WT}, \tag{6.132}$$

where M is the number of points in the signal constellation.

It is instructive to determine η_A for two standard constellation designs.

Example 6-33.

For a baseband PAM constellation with M equally spaced points, let M be even and let the constellation consist of odd integers. Thus, $\mathcal{A} = \{(2m - 1),\ 1 - M/2 \le m \le M/2\}$, and the minimum distance is $d_{\min} = 2$. Assuming all the points in the constellation are equally probable, the variance is (calculating the average-squared valued of only the positive points, taking advantage of symmetry)

$$E = \frac{2}{M}\sum_{m=1}^{M/2}(2m-1)^2 = \frac{M^2-1}{3}. \tag{6.133}$$

Substituting into (6.131), and using (6.132),

$$\eta_A = \frac{3}{M^2-1} = \frac{3}{2^{2WT\nu}-1}. \tag{6.134}$$

Example 6-34.

For a passband PAM $L \times L$ square constellation with $M = L^2$ points, and again using odd integers on each axis, the minimum distance is again $a_{\min} = 2$, and the variance for equally probable points can be shown to be $E = 2(M-1)/3$. Substituting into (6.131) and (6.132),

$$\eta_A = \frac{3/2}{M-1} = \frac{3/2}{2^{WT\nu}-1}. \tag{6.135}$$

In both the baseband and passband cases, as the number of points in the constellation increases, or equivalently the spectral efficiency increases, η_A gets smaller and as expected, from (6.130) we must increase SNR to hold P_e fixed.

It is useful to determine P_e when the highest possible symbol rate consistent with the Nyquist criterion is used, since this will maximize the SNR and minimize P_e. For the baseband case, the maximum symbol rate is $1/T = W$, and hence $2WT = 1$. For the passband case, a passband channel with bandwidth W corresponds to a complex baseband channel with bandwidth $W/2$, and thus the maximum symbol rate is $1/T = W$ or $2WT = 2$. Expressed in terms of SNR and ν, the resulting P_e is the same for both cases,

$$P_e \approx K \cdot Q\left(\sqrt{3 \cdot \frac{SNR}{2^\nu - 1}}\right). \tag{6.136}$$

This is a "universal" formula that applies to both baseband and passband PAM systems with square QAM constellations and the maximum possible symbol rate.

It is difficult to compare modulation systems that have different bit rates or spectral efficiencies. The universal formula for P_e in (6.136) gives a hint as to a possible approach, since it shows that, when expressed in terms of spectral efficiency, all baseband and passband square QAM constellations are equivalent. Another simplification of this formula is that P_e does not depend on SNR or ν individually, but only through the ratio $SNR/(2^\nu - 1)$. It is helpful, therefore, to define a new parameter SNR_{norm}, which is called the *rate-normalized SNR*, as

$$SNR_{\text{norm}} = \frac{SNR}{2^\nu - 1}. \tag{6.137}$$

Now, the P_e is a function of only two parameters, K and SNR_{norm},

$$P_e \approx K \cdot Q(\sqrt{3 \cdot SNR_{\text{norm}}}) . \tag{6.138}$$

Two square baseband or passband QAM constellations with the maximum symbol rate and the same SNR_{norm} will have approximately the same P_e.

The utility of (6.138) is that it expresses very succinctly P_e for a variety of PAM systems, including baseband, passband, and different bit rates and symbol rates. The simplicity of this result leads us to speculate that there may be something fundamental about this tradeoff between SNR and ν expressed in SNR_{norm}. Indeed there is, although we will have to take a diversion, calculating the capacity of the channel, to uncover it.

6.7.2. Capacity of the Ideal Gaussian Channel

Two approaches to comparing modulation systems operating over the same channel are first, to compare them directly, or second, to compare them against the fundamental limits of channel capacity (Chapter 4). The channel capacity tells us the maximum bit rate (or equivalently spectral efficiency) that can be obtained on the underlying channel. Comparing against capacity has a couple of benefits. First, it gives an indirect comparison of the systems against one another. Second, the comparison against fundamental limits gives us a valuable benchmark, because it indicates the maximum possible benefits of doing channel error-correction coding (Chapters 12 and 13).

The capacity of an ideal bandlimited Gaussian channel with additive white Gaussian noise will now be derived, and subsequently compared against the spectral efficiency of several modulation techniques operating over this same channel. A general way to compare a given modulation system against this capacity, based on the rate-normalized SNR already encountered in QAM modulation, will be uncovered.

The frequency response of an ideal channel with bandwidth W Hz is shown in Fig. 6-28 for two cases, baseband and passband. The convention is that the bandwidth of the channel is W in both cases. This implies that the baseband channel is equivalent to the passband channel for the specific carrier frequency $f_c = W / 2$. Intuitively, we would not expect the carrier frequency to affect the capacity of the channel, since the noise is white, and thus we expect the capacity of the two channels in Fig. 6-28 to be identical. In fact, that is the case.

We are interested in calculating the capacity C of these two channels, where capacity has the units bits/sec. Thus, C can be directly compared to the bit rate achieved by a given modulation system. The capacity is calculated under a transmitted power constraint, so that the

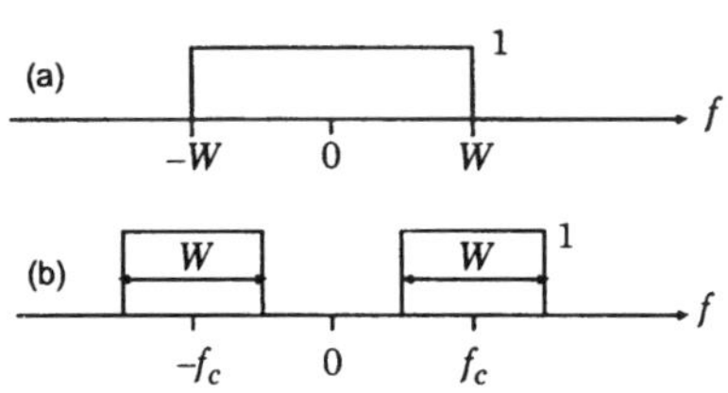

Fig. 6-28. An ideal bandlimited channel with bandwidth W for the (a) baseband, and (b) passband,

transmitted signal is constrained to have power P. Also of interest is the spectral efficiency ν, which has the units of bits/sec-Hz. We define ν_c as the spectral efficiency of a system operating at the limits of capacity, and thus

$$\nu_c = C/W. \tag{6.139}$$

The channel coding theorem (Chapter 4) says that a certain spectral efficiency ν_c can be achieved with transmit power P in the sense that an arbitrarily small probability of error can be achieved by *some* modulation and coding scheme. Further, it says that if you try to achieve a higher ν at this P, the probability of error is necessarily bounded away from zero for *all* modulation and coding schemes. The tradeoff between ν_c and P as quantified by the channel capacity theorem is thus a fundamental limit against which all modulation systems can be compared.

The capacity of the ideal channels in Fig. 6-28 with additive white Gaussian noise is simple to calculate using the capacity of a vector Gaussian channel (Chapter 4). We will do the calculation for the baseband case since it is slightly easier, although the passband case is also straightforward (Problem 6-20). Utilizing the Landau-Pollak theorem (Section 6.6.1) and an orthogonal expansion of the signal subspace, any transmitted signal with bandwidth W Hz can be approximately represented in a time interval of length T by $2WT$ orthonormal waveforms, with increasing accuracy as $T \to \infty$. From Section 6.1.2, the minimum-distance receiver will generate a set of $2WT$ decision variables, by (6.21),

$$\boldsymbol{r} = \boldsymbol{s} + \boldsymbol{n}, \tag{6.140}$$

where $\boldsymbol{s}$ is the $2WT$-dimensional vector of signal components, and $\boldsymbol{n}$ is a vector of Gaussian independent noise components, each component having variance $\sigma^2 = N_0/2$. In this case, since the signal, noise, and channel are real-valued, all vectors in (6.140) are real-valued.

The capacity of channel (6.140), consisting of an N-dimensional real-valued vector signal in real-valued vector Gaussian noise, with total signal variance E and noise variance per dimension $\sigma^2 = N_0/2$, is given by (4.36),

$$C_{VG} = \frac{N}{2}\log_2(1 + SNR), \qquad SNR = \frac{E/N}{\sigma^2}. \tag{6.141}$$

This is the capacity for a *single use* of the vector channel, or equivalently the capacity for the continuous-time channel over a time interval of length T. The signal-to-noise ratio SNR is defined as the ratio of the total signal variance to the total noise variance.

The constraint that the transmitted power is P implies that the average transmitted energy in time interval T must be PT, and thus

$$\mathrm{E}\left[\int_0^T s^2(t)\,dt\right] = \sum_{n=1}^{N} \mathrm{E}[s_n^2] = E = PT. \tag{6.142}$$

Defining C_T as the capacity for time interval T, with this power constraint,

$$C_T = WT \cdot \log_2(1 + SNR)\ , \qquad SNR = \frac{PT}{WTN_0} = \frac{P}{N_0 W}\ . \tag{6.143}$$

In this case, SNR can again be interpreted as signal-to-noise ratio, since the numerator P is the total signal power at the channel output, and the denominator is the total noise power within the signal bandwidth (the noise spectral density $N_0/2$ times the total bandwidth, which is $2W$ for positive and negative frequencies).

The capacity per unit time is

$$C = C_T/T = W \cdot \log_2(1 + SNR) \quad \text{bits/sec}\ . \tag{6.144}$$

This expression for the capacity of a bandlimited channel is known as the *Shannon limit.* Alternative proofs and interpretation of this result are given in [11][12].

Fundamental Limit in Spectral Efficiency

The spectral efficiency is the bit rate per unit time (capacity) divided by the bandwidth, and thus the maximum spectral efficiency predicted by the channel capacity is

$$\nu_c = C/W = \log_2(1 + SNR) \qquad \text{bits/sec -Hz}\ . \tag{6.145}$$

If ν is the spectral efficiency of any practical modulation scheme operating at signal-to-noise ratio SNR, then we must have $\nu \le \nu_c$.

Rate-Normalized Signal-to-Noise Ratio

Rewriting (6.145) in a different way, if a modulation system is operating at the limits of capacity with signal-to-noise ratio SNR and spectral efficiency ν_c, then

$$\frac{SNR}{2^{\nu_c} - 1} = 1\ . \tag{6.146}$$

This relation has a striking similarity to SNR_{norm} defined in (6.137), and SNR_{norm} was shown in (6.138) to largely determine P_e for a rectangular baseband or passband QAM constellation. The only difference is that ν, the spectral efficiency of the PAM modulator, is substituted for ν_c, the spectral efficiency at capacity limits. The combination of (6.138) and (6.146) suggests that SNR_{norm} is a fundamental and useful parameter of a modulation system [13]. In fact, since $\nu \le \nu_c$ for a system operating short of the capacity limit,

$$SNR_{\text{norm}} = \frac{2^{\nu_c} - 1}{2^{\nu} - 1} \ge 1\ . \tag{6.147}$$

This is another way of expressing the Shannon limit on the operation of a given modulation system; if the modulation system operates at signal-to-noise ratio SNR with spectral efficiency ν, and the corresponding $SNR_{\text{norm}} > 1$, then there is nothing fundamental preventing that system from having an arbitrarily small P_e. (If it has a large P_e, that is only because it is falling short of fundamental limits). Conversely, if $SNR_{\text{norm}} < 1$, the P_e of the system is necessarily bounded away from zero, because the parameters of the system (SNR and ν) are violating Shannon limits. In this case, the capacity theorem does not prevent P_e from being small, but it does guarantee that there is nothing we could do (like adding error-control coding) to make P_e arbitrarily small, short of changing the parameters SNR and/or ν. Thus, $SNR_{\text{norm}} > 1$ is the region where we want to operate on an ideal bandlimited white Gaussian noise channel.

It is useful to plot the relationship between SNR and SNR_{norm}, where both are expressed in dB, as in Fig. 6-29. Taking the logarithm of (6.137),

$$SNR_{\text{norm,dB}} = SNR_{\text{dB}} - \Delta SNR_{\text{dB}}\,, \qquad \Delta SNR_{\text{dB}} = 10 \cdot \log_{10}(2^{\nu} - 1)\,. \tag{6.148}$$

At large spectral efficiencies, the unity term can be ignored, and ΔSNR_{dB} approaches an asymptote of $\Delta SNR_{\text{dB}} \approx 3\nu$. Thus, for a hypothetical high spectral efficiency system operating at the limits of capacity, 3 dB of additional SNR is required to increase spectral efficiency by one bit/sec-Hz. At low spectral efficiencies, a larger increase in SNR is required. Remarkably, PAM systems with square QAM constellations operating at a constant $P_e > 0$ obey exactly the same tradeoff between ν and SNR, as indicated by (6.138). (Although the tradeoff is the same, they will require a higher absolute SNR to achieve a reasonable P_e, as will be seen shortly.)

For any modulation system, the gap (usually expressed in dB) between SNR_{norm} and unity (the minimum value of SNR_{norm}) is a measure of how far short of fundamental limits the modulation scheme falls. Specifically, it is a measure of how much the transmitted power (or equivalently the SNR) must be increased to achieve a given spectral efficiency, relative to the lower bound on transmitted power (or SNR) predicted by capacity. The usefulness of SNR_{norm} is that it summarizes SNR and ν in a single parameter, and the Shannon limit is very simply expressed in terms of SNR_{norm}.

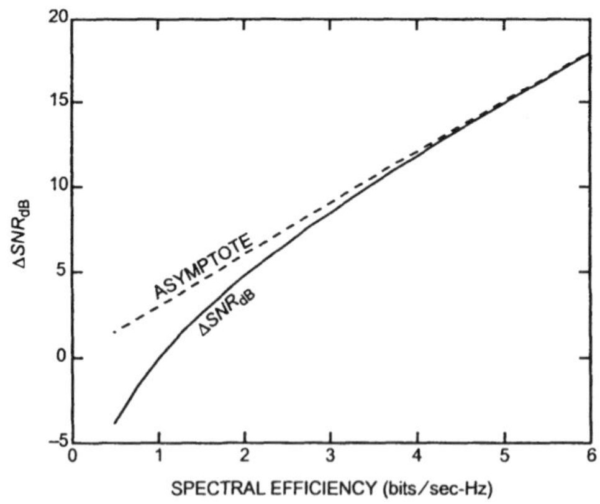

Fig. 6-29. The difference between SNR and SNR_{norm} in dB plotted against spectral efficiency. The "asymptote" is the same relationship ignoring the "1 +" term.

6.7.3. Using Normalized SNR in Comparisons

While $SNR_{\text{norm}} = 1$ corresponds to a hypothetical system operating at capacity, all practical modulation schemes will have a non-zero error probability for all values of SNR_{norm}. A useful way to characterize P_e is to parameterize it on SNR_{norm}, because SNR_{norm} expresses both SNR and ν in a single parameter, and because the Shannon limit is so simply characterized in terms of SNR_{norm}. In this section we illustrate how the symbol error probability P_e can be calculated as a function of SNR_{norm} for different modulation techniques. This relationship gives us several types of information:

- Comparisons can be made between different modulation techniques. For two modulation systems operating at the same P_e (as approximated by the union bound, and ignoring the effect of the error coefficient K), and at the same spectral efficiency, if the superior modulation system allows SNR_{norm} to be 3 dB lower, then it allows 3 dB lower transmit power. Alternatively, if the two systems are operating at the same SNR, then the superior system will operate at a spectral efficiency that is one bit/sec-Hz higher (asymptotically at high ν).
- Comparisons can be made between a modulation system and fundamental limits. At a given P_e and ν, the difference between SNR_{norm} and unity (usually expressed in dB) tells us how far the modulation system is operating from fundamental limits, in the sense that it is requiring a higher SNR or lower ν to achieve the same spectral efficiency. This quantifies, for example, the ultimate potential benefit of adding error-correction coding to the system (Chapters 12 and 13).
- Reasonable comparisons between modulation systems operating at different information bit rates and spectral efficiencies can be made. For example, we might want to compare two schemes utilizing the same bandwidth but having a different number of points in the constellation (and hence different spectral efficiency). Comparing them each against the Shannon limit is an indirect way of comparing them against each other.

We are interested in a wide range of error probabilities (some applications are more demanding than others), and thus it is useful to plot the functional relationship between P_e and SNR_{norm}, and compare to capacity ($SNR_{\text{norm}} = 1$). This will now be illustrated for several modulation systems.

Baseband and Passband PAM

Earlier in this chapter we found that the symbol-error probability for baseband and passband QAM (as well as for OPAM) was (6.130), where the parameter η_A is given by (6.134) (baseband case) and (6.135) (passband case). Expressing P_e in terms of SNR_{norm}, rather than SNR, (6.130) can be rewritten as

$$P_e \approx K \cdot Q(\sqrt{\gamma_A \cdot SNR_{\text{norm}}}), \tag{6.149}$$

where

$$\gamma_A = \eta_A(2^{\nu} - 1) . \tag{6.150}$$

This assumes that the bandwidth on the channel is the minimum consistent with the Nyquist criterion. For the baseband case, from (6.134),

$$\gamma_A = \frac{d_{\min}^2 (2^\nu - 1)}{4E} , \tag{6.151}$$

and for the passband case, from (6.135),

$$\gamma_A = \frac{d_{\min}^2 (2^\nu - 1)}{2E} . \tag{6.152}$$

Example 6-35.

For square QAM constellations, as shown in (6.138), the remarkably simple result is that $\gamma_A = 3$. This holds for all cases where the number of points in the constellation is even (baseband case) or the square of an even number (passband case).

For other PAM constellations, γ_A is a parameter of the constellation that is independent of scaling, but is a function of the geometry of the constellation. Remarkably, the P_e of any PAM constellation across a wide range of *SNR*'s, is accurately summarized in this single parameter γ_A. It can be determined directly from (6.151) or (6.152). The error coefficient K is also relevant, although much less important.

We can plot P_e vs. SNR_{norm} under different conditions, and get a set of universal rate-normalized curves. First, in Fig. 6-30, γ_A is held fixed (at $\gamma_A = 2$), and K is varied. This set of curves has several interesting interpretations. First, it shows how large an SNR_{norm} is required to achieve a given error probability for these assumed parameters. As expected, the required SNR_{norm} increases as P_e gets smaller. The Shannon limit dictates that $SNR_{\text{norm}} > 1$, or 0 dB. Since the channel capacity theorem guarantees the feasibility of achieving any (arbitrarily small) error probability, it is theoretically possible to achieve any point on the 0 dB SNR_{norm} axis; conversely, since $SNR_{\text{norm}} \geq 1$, the probability of error will be theoretically bounded

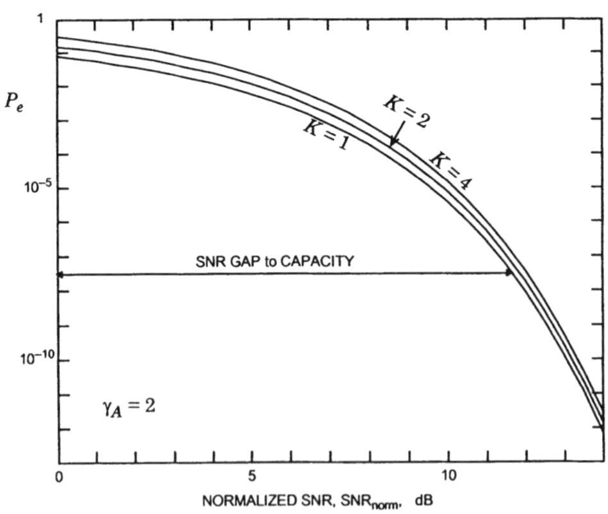

Fig. 6-30. A plot of P_e vs. SNR_{norm} for passband PAM assuming $\gamma_A = 2$ and three typical values of K. This illustrates that K has a relatively minor effect on the error probability.

away from zero at any point to the left of the 0 dB SNR_{norm} axis. In this sense, the 0 dB axis represents the limit on deliverable performance as dictated by Shannon limit. At a given P_e the horizontal distance between the 0 dB SNR_{norm} axis and the curve, labeled "SNR gap to capacity," represents the increase in SNR_{norm} required relative to capacity. Also, the horizontal distance between two curves represents the difference in SNR_{norm} required for two different signal constellations to achieve the same P_e. This gap can be made up in one of two ways: operate the system at a higher SNR, or at a lower ν.

By definition, the SNR gap to capacity goes to zero as $SNR_{\text{norm}} \to 1$. What may be surprising is that P_e can be small (like 10^{-1}) at this crossover point, or even for $SNR_{\text{norm}} < 1$. Doesn't the channel capacity theorem rule out any useful operation for $SNR_{\text{norm}} < 1$? Two points should be made about this behavior. First, since the error probability is based on the union bound, it is generally not wise to trust these quantitative results at low SNR (high P_e), except for modulation schemes for which the union bound is exact (such as binary antipodal signaling). Second, although it would be tempting to assert that the channel capacity tells us something specific about the error probability of any modulation scheme operating at $SNR_{\text{norm}} < 1$, in fact it only asserts that in this region the error probability is bounded away from zero. It does not tell us what that bound is. Thus, the channel capacity theorem does not rule out any non-zero error probability at the point where $SNR_{\text{norm}} = 1$.

In Fig. 6-30 the effect of K on SNR_{norm} is small, emphasizing that K has a relatively minor influence on P_e. The effect of γ_A is much more significant, as illustrated in Fig. 6-31. The major factor distinguishing different signal constellations is γ_A. We will calculate γ_A for a couple of cases to illustrate this.

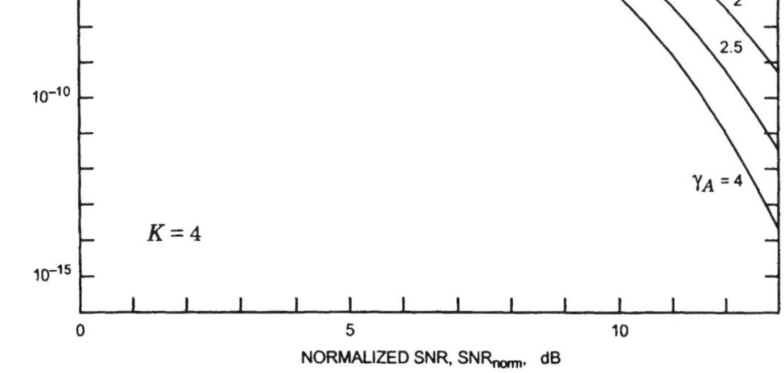

Fig. 6-31. A plot of P_e vs. SNR_{norm} for passband PAM, assuming $K = 4$ and different values of γ_A.

Example 6-36.

All rectangular QAM constellations are equivalent, in the sense that they require the same SNR_{norm} to achieve a given error probability. That tradeoff between SNR_{norm} and P_e is the $\gamma_A = 3$ curve in Fig. 6-31. For example, at an error rate of $P_e = 10^{-6}$, the SNR gap to capacity is about 9 dB, independent of the size of the constellation. However, at a fixed P_e, square QAM constellations do require different unnormalized SNR's, since for the passband case

$$SNR = SNR_{norm} \cdot (2^{\nu} - 1) = SNR_{norm} \cdot (M - 1) . \tag{6.153}$$

As M increases, the SNR must increase in proportion to $M-1$ because of the need to increase the signal power to maintain the same minimum distance. Looking at it another way, as the spectral efficiency ν increases, the SNR must be increased in proportion to $(2^{\nu} - 1)$.

Example 6-37.

For a PSK signal constellation, all the points fall on the unit circle, and thus $E = 1$ independent of the distribution of signal constellation points. It is straightforward to show that $d_{min} = 2\sin(\pi/M)$, and thus,

$$\gamma_A = 2(M-1)\sin^2(\pi/M) . \tag{6.154}$$

In this case, γ_A is strongly dependent on M, in contrast to rectangular QAM. This dependence is plotted in Fig. 6-32, where the largest γ_A is 3, the same as rectangular QAM, for $M = 3$ and $M = 4$. Thus, the SNR gap to capacity for PSK is the same for 3-PSK and 4-PSK as it is for rectangular QAM. The equivalence at $M = 4$ is obvious, since 4-PSK is in fact a square QAM constellation. Both 2 *-PSK* (binary antipodal) and M-PSK for $M > 4$ are inferior to rectangular QAM in the sense that they require a larger SNR_{norm} to achieve the same P_e (higher SNR at the same ν or lower ν at the same SNR). In the case of 2-PSK, which is equivalent to binary antipodal signaling, its gap is larger than QAM because it is a passband PAM system that fails to use the quadrature axis (we have shown previously that a *baseband* PAM binary antipodal constellation has $\gamma_A = 3$).

The SNR gap to capacity for PSK increases rapidly at a given P_e as M increases. Intuitively, this is because PSK does not efficiently pack the circularly-shaped constellation with a regular grid of points, and thus suffers in spectral efficiency as compared to square QAM constellations. We can also plot the P_e directly, as shown in Fig. 6-33, for different M, taking into account the corresponding K. $M = 4$ has the smallest gap (equivalent to rectangular QAM), and $M = 2$ and 8 have roughly the same gap (because the γ_A is about the same, as seen in Fig. 6-32). Choosing large values of M for PSK results in significantly poorer performance.

If it is desired to compare two constellations against one another analytically, the simplest approach is to set the arguments of $Q(\cdot)$ to be equal, which ignores the effect of K. In other words, compare their γ_A.

Example 6-38.

To compare rectangular QAM against PSK, set

$$3 \cdot SNR_{norm,QAM} = 2(M-1) \cdot \sin^2(\pi/M) \cdot SNR_{norm,PSK} \tag{6.155}$$

to yield

$$\frac{SNR_{\text{norm,PSK}}}{SNR_{\text{norm,QAM}}} = \frac{3}{2(M-1)\sin^2(\pi/M)} . \tag{6.156}$$

This relationship is plotted vs. M in Fig. 6-34 in dB. The penalty in SNR_{norm} for using PSK is shown as a function of M. For any M greater than four, PSK requires a higher SNR to achieve a similar error probability. All values of M are squares of even integers, which are the only square QAM constellation sizes.

Spread Spectrum

Our examples of the SNR gap to capacity for PAM thus far have presumed that the maximum feasible symbol rate in relation to channel bandwidth is used. In spread spectrum, a much lower symbol rate is used, and the SNR gap to capacity will expand accordingly. We can quantify this effect as follows. Considering the bandpass case, from substituting (6.135) into (6.130),

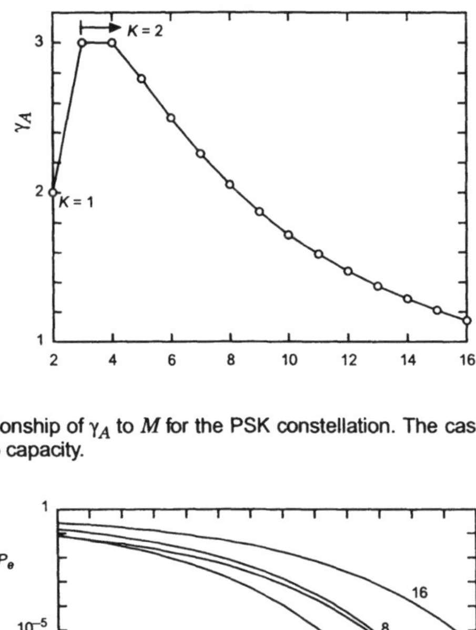

Fig. 6-32. The relationship of γ_A to M for the PSK constellation. The cases $M = 3$ and $M = 4$ have the smallest SNR gap to capacity.

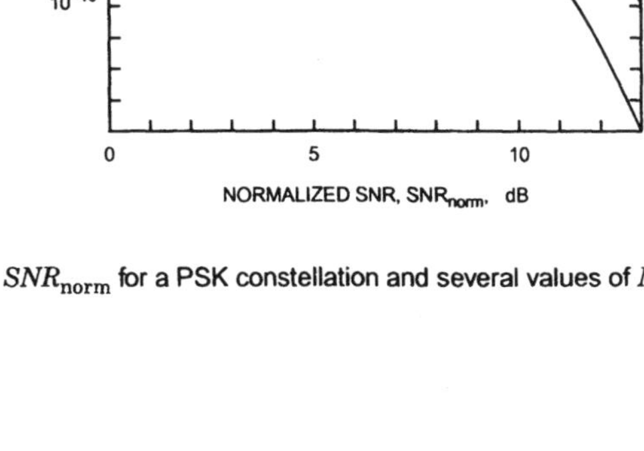

Fig. 6-33. P_e vs. SNR_{norm} for a PSK constellation and several values of M.

$$P_e \approx K \cdot Q(\sqrt{3\gamma_{SS} SNR_{\text{norm}}}) \tag{6.157}$$

where the additional factor is

$$\gamma_{SS} = \frac{WT \cdot (2^{\nu} - 1)}{2^{WT\nu} - 1} \ . \tag{6.158}$$

Since ν is a function of WT, it is useful to express γ_{SS} in terms of M, the number of points in the constellation,

$$\gamma_{SS} = \frac{WT \cdot (M^{1/(WT)} - 1)}{M - 1} \ . \tag{6.159}$$

For the maximum symbol rate, $2WT = 2$ and $\gamma_{SS} = 1$. More generally, however, $\gamma_{SS} < 1$, forcing SNR_{norm} to be larger for the same P_e and increasing the SNR gap to capacity. This implies that coding is more beneficial in spread spectrum systems.

Exercise 6-9.

Show that as $WT \to \infty$, $\gamma_{SS} \to (\log_e M)/(M-1)$.

The effect of γ_{SS} is to increase the SNR gap to capacity. Asymptotically, the gap is increased by $(M-1)/\log_e M$ for a signal constellation of size M. For $M = 4$, the smallest M for which the formula for γ_A is valid, the SNR gap to capacity is increased by $3/\log_e 4 = 2.16$ (3.3 dB).

Penalizing spread spectrum in its SNR gap to capacity, although understandable in terms of its reduced spectral efficiency, is unfair when we realize that multiple spread spectrum signals can coexist within the same bandwidth, as in code-division multiple access. The increase in SNR gap to capacity calculated here is for a *single* spread spectrum signal occupying the bandwidth. Also, these results are for white noise with fixed power spectral density on the channel, which is not always the context in which spread spectrum is used.

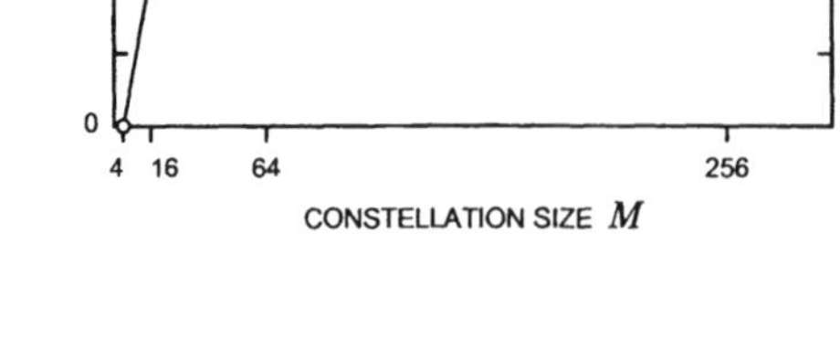

Fig. 6-34. The penalty in dB for PSK in comparison to QAM, vs. constellation size M, when M is an even power of 2.

Orthogonal Modulation

The symbol error probability for orthogonal modulation was found in (6.45) to be:

$$P_e \approx (M-1) \cdot Q\left(\sqrt{\frac{E}{N_0}}\right), \tag{6.160}$$

where $E = PT$ is the energy per signaling interval T, and P is the signal power. But the minimum bandwidth for M-ary orthogonal modulation is $W = M/(2T)$, so that:

$$\text{SNR} = \frac{P}{N_0 W} = \frac{2}{M} \cdot \frac{E}{N_0}. \tag{6.161}$$

Furthermore, as shown in Section 6.3, the spectral efficiency of orthogonal modulation is

$$\nu = \frac{\log_2 M^2}{M}, \qquad 2^{\nu} - 1 = M^{2/M} - 1 \ . \tag{6.162}$$

Thus, rewriting (6.160) in terms of $SNR_{\text{norm}} = \text{SNR}/(2^{\nu} - 1)$ yields:

$$P_e \approx (M-1) \cdot Q(\sqrt{\gamma_M SNR_{\text{norm}}}) \ , \tag{6.163}$$

where

$$\gamma_M = \tfrac{1}{2} M(M^{2/M} - 1) \ . \tag{6.164}$$

As M gets large, the factor γ_M increases monotonically, which reduces the error probability, but the multiplier $M - 1$ increases it. The resulting error probability is plotted for several values of M in Fig. 6-35; the net effect is that the gap to capacity decreases as M increases.

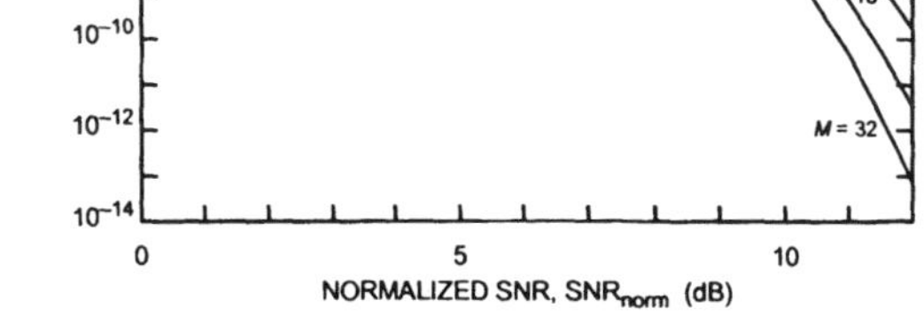

Fig. 6-35. P_e vs. SNR_{norm} for orthogonal modulation for several values of M, where M is the dimensionality.

From Fig. 6-35 it appears that orthogonal modulation gives similar performance to QAM, in the sense that the SNR gap is similar. However, as we will see in Chapter 12, there are known ways to use channel coding to "shrink the gap" in the case of PAM, but there is no way to do this directly with orthogonal modulation.

6.8. Further Reading

The geometric approach to estimating the error probability of modulation systems was originally inspired by the text by Wozencraft and Jacobs [14], which remains recommended reading. The approach to comparing modulation systems used here, based on the normalized signal-to-noise ratio, was heavily influenced by [13]. Spread-spectrum is covered in some detail in the digital communications textbooks by Proakis [15], Cooper and McGillem [16], and Ziemer and Peterson [17]. Our coverage has relied heavily on two excellent tutorial articles [7][18].

More detailed information about FSK can be obtained from Proakis [15] and Lucky, Salz, and Weldon [2], both of which derive the power spectrum of continuous-phase FSK signals. For tutorials on MSK, we recommend Pasupathy [19] and Haykin [20]. The relationship between MSK and QPSK modulation is described in detail by Gronemeyer and McBride [21] and Morais and Feher [22]. For a more general treatment of continuous-phase modulation, see [23]. An excellent treatment of noncoherent and partially coherent modulation is given by Simon, Hinedi, and Lindsey [24].

Multicarrier modulation dates back to the 1960s, with early work carried out by Chang [3], Saltzberg [25], Powers and Zimmerman [26], and Darlington [27]. The use of the DFT to construct the transmitted signal is described by Weinstein and Ebert [28] and Hirosaki [5]. Kalet [29] and Ruiz, Cioffi, and Kasturia [30] considered the design of the symbol sets and allocation of bit rates when the channel is distorted, with the latter applying coset coding (Chapter 13). The technique has been applied to the voiceband data channel [31], the high-speed digital subscriber loop (HDSL) channel [32][33], wireless LANs, and the magnetic recording channel [34].

Appendix 6-A.
The Generalized Nyquist Criterion

In this appendix, we first show that a bandwidth of $N/(2T)$ is required to satisfy the Nyquist criterion,

$$\frac{1}{T}\sum_{m=-\infty}^{\infty} H_n\left(f-\frac{m}{T}\right)H_l^*\left(f-\frac{m}{T}\right)=\delta_{l-n}. \tag{6.165}$$

Then we demonstrate a class of practical pulse shapes that satisfy the criterion with a bandwidth close to the minimum.

Proof of Necessity

In (6.52) and Fig. 6-8 we showed a pulse set with bandwidth $N/(2T)$ that meets the criterion. This shows that the a bandwidth of $N/(2T)$ is sufficient. We will now show that it is also necessary.

The left side of (6.165) is a periodic function of f with period $1/T$. Consequently, we need only verify (6.165) for $f \in [-1/(2T),\ 1/(2T)]$. (If $h_n(t)$ is real for all n, then we need only verify (6.165) for $\in [0,\ 1/(2T)]$. By conjugate symmetry of $H_n(f)$ it is automatically satisfied for the rest of the range.) Assume that all pulses $h_n(t)$ for each $0 \le n \le N-1$ lie within the frequency range $|f| \le K/(2T)$ for some arbitrary integer K (we already know that the generalized Nyquist criterion can be met with bandlimited pulses). Then the summation in (6.165) becomes finite, so for $f \in [-1/(2T),\ 1/(2T)]$,

$$\frac{1}{T}\sum_{m=-M_1}^{M_2} H_n\left(f-\frac{m}{T}\right)H_l^*\left(f-\frac{m}{T}\right) = \delta_{l-n}, \tag{6.166}$$

where M_1 and M_2 are integers that depend on K. Specifically, we want to make them as small as possible while maintaining the equivalence of (6.166) and (6.165). We require that the range $[f - M_1/T, f + M_2/T]$ be at least as large as the total bandwidth of the pulses $-K/(2T)$, $K/(2T)]$ for each $f \in [-1/(2T),\ 1/(2T)]$. If K is odd, then the smallest values are $M_1 = M_2 = (K-1)/2$, so there are K terms in the summation. If K is even, we can use different values of M_1 and M_2 in the ranges $f \in [-1/(2T),\ 0]$ and $f \in [0,\ 1/(2T)]$. For $f \in [-1/(2T),\ 0]$, we can use $M_1 = (K/2) - 1$ and $M_2 = K/2$. In the latter range we can use $M_1 = K/2$ and $M_2 = (K/2) - 1$. In all cases, the number of terms in the summation is $M_1 + M_2 + 1 = K$.

For each fixed f, define a vector $\mathbf{H}_n(f)$ consisting of the K terms in the summation, $H_n(f - m/T))$ for $m = -M_1, \ldots, M_2$. The dimensionality of $\mathbf{H}_n(f)$ is K. (If real-valued pulses are desired, then f can be restricted to the interval $f \in [0,\ 1/(2T)]$ with the constraint $H_n(-f) = H_n(f)^*$.) Now we can write (6.166) as

$$\frac{1}{T}\mathbf{H}_n(f)^T\mathbf{H}_l(f)^* = \delta_{l-n}, \quad -1/(2T) \le f \le /(2T), \quad n, l \in \{0, 1, \ldots, N-1\}, \tag{6.167}$$

where $\mathbf{H}_n(f)^T$ is the transpose of $\mathbf{H}_n(f)$. Thus, the generalized Nyquist criterion can be satisfied if, for each $f \in [-1/(2T),\ 1/(2T)]$, a set of N orthogonal equal-length K-dimensional Euclidean vectors $\mathbf{H}_n(f)$, $n \in \{0, 1, \ldots, N-1\}$, can be found. Clearly, N orthonormal vectors can be found if their dimensionality is at least N, or $K \ge N$, and cannot be found for smaller K. Thus, a bandwidth of $N/(2T)$ will suffice, as confirmed by the earlier example.

We have argued that for N orthonormal pulses bandlimited to an integer multiple of $1/(2T)$, the multiple must be $K \ge N$ to satisfy (6.166). Choosing the minimum value, $K = N$, we can now show that the entire bandwidth $[-N/(2T),\ N/(2T)]$ must be used. Thus we prove that the bandwidth cannot be reduced further from $N/(2T)$. Specifically, note that if there is any value of f in this interval where all N vectors are zero-valued, then (6.167) cannot be true. To see this, define the $N \times K$ matrix $\mathbf{H}(f)$ where row n is $\mathbf{H}_n(f)^T$, and note that (6.167) is equivalent to

$$\frac{1}{T}\mathbf{H}(f)\mathbf{H}(f)^* = \mathbf{I} \tag{6.168}$$

for all $f \in [-1/(2T),\ 1/(2T)]$, where $\mathbf{H}(f)^*$ is the conjugate transpose and $\mathbf{I}$ is the identity matrix. Hence $\mathbf{H}(f)$ must have full rank. Furthermore, when $N = K$, the matrix is square, so this implies that each component of the vectors $\mathbf{H}_n(f)$ must be non-zero for some $n \in \{0, 1, \ldots, N-1\}$, or else $\mathbf{H}(f)$ would have an all-zero column. This in turn implies that $H_n(f) \neq 0$ for some $n \in \{0, 1, \ldots, N-1\}$ for every $f \in [-N/(2T),\ N/(2T)]$. Thus, we arrive at the following theorem:

> **Theorem 6-1.** The minimum aggregate bandwidth required to satisfy the generalized Nyquist criterion is $N/(2T)$. More precisely, if a set of pulses is constrained to lie within a bandwidth of $N/(2T)$, then the pulses collectively fill this bandwidth, leaving no gaps. If the bandwidth is lowpass (it need not be), then for every $f \in [-N/(2T),\ N/(2T)]$, $H_n(f) \neq 0$ for some $n \in \{0, 1, \ldots, N-1\}$. Conversely, there exist sets of pulses satisfying the generalized Nyquist criterion with aggregate bandwidth $N/(2T)$.

For $N > 1$, the minimum-bandwidth set of pulses is not unique. To see this, note that if $\boldsymbol{H}(f)$ satisfies (6.168), then $\mathbf{UH}(f)$ will also satisfy (6.168) for any *unitary* matrix $\mathbf{U}$ (see Problem 6-7). (A matrix $\mathbf{U}$ is unitary if $\mathbf{U}^{-1} = \mathbf{U}^*$, where $\mathbf{U}^*$ is the conjugate transpose.)

For the four pulses shown in Fig. 6-8, if we simply number them left to right, then in the range $f \in [0,\ 1/(2T)]$,

$$\mathbf{H}(f) = \sqrt{T}\begin{bmatrix} 0 & 0 & 1 & 0 \\ 0 & 1 & 0 & 0 \\ 0 & 0 & 0 & 1 \\ 1 & 0 & 0 & 0 \end{bmatrix}, \tag{6.169}$$

which satisfies (6.168). In the range $f \in [-1/(2T),\ 0]$ the matrix is similar, but not identical. These pulses, which are non-overlapping in the frequency domain, illustrate the possibilities, but are not practical due to the discontinuity in the frequency response at the band edge. What is needed is a set of pulses that have gradual rolloff, and thus overlap one another in the frequency domain.

A Practical Pulse Set With Minimum Bandwidth

The pulse set in (6.52) is not practical to implement because the pulses are ideally bandlimited. In Section 6.3.4 we generalized (6.52) to

$$h_n(t) = w(t)\cos((n + 3/2)\pi t/T)\ , \tag{6.170}$$

where $w(t)$ is a pulse with $|W(f)|$ chosen so that $w(t) * w(-t)$ satisfies the ordinary Nyquist criterion for symbol rate $1/(2T)$, half the desired symbol rate [3]. We allow the bandwidth of $w(t)$ to be as high as $1/(2T)$, twice its theoretical minimum, allowing gradual rolloff in the frequency domain. While the magnitude of $W(f)$ is specified so that each pulse satisfies the ordinary Nyquist criterion, we will now show how to select the phase of $W(f)$ so that pulses $h_n(t)$ are mutually orthogonal and satisfy the generalized Nyquist criterion.

We need to show that (6.165) is satisfied for $l \neq n$. Since $W(f)$ is bandlimited to $1/(2T)$, $h_n(t)$ and $h_{n+1}(t)$ have a 50% overlap in the frequency domain, but $h_n(t)$ and $h_l(t)$ do not overlap for $|n-l| \geq 2$. Thus (6.165) holds trivially for $|n-l| \geq 2$.

The only case left is $l = n + 1$. We will now show that, for any $w(t) * w(-t)$ satisfying the Nyquist criterion (with respect to symbol rate $1/(2T)$), the phase of $W(f)$ can be chosen such that $h_n(t)$ and $h_{n+1}(t)$ satisfy (6.165). $|H_n(f)|^2$ and $|H_{n+1}(f)|^2$ are plotted in the Fig. 6-36(b), where without loss of generality it is assumed that n is even. The interval $f \in [0, 1/(2T)]$ is highlighted, as are all translates of this interval by $1/T$ that overlap the spectra in question. We see immediately that there are two cases to consider, $f \in [0, 1/(4T)]$ and $f \in [1/(4T), 1/(2T)]$. In the first case, at all translates by $1/T$, one or the other of $\mathbf{H}_n(f)$ and $\mathbf{H}_{n+1}(f)$ is zero, and thus their inner product must be zero. For the second case, both $\mathbf{H}_n(f)$ and $\mathbf{H}_{n+1}(f)$ have non-zero coordinates for exactly two of the translates by kT, and for those two translates the vectors are, using the fact that $W^*(-f) = W(f)$ (since $w(t)$ is assumed real-valued), for the interval $f \in [1/(4T), 1/(2T)]$,

$$\mathbf{H}_n(f) = [W^*(3/(4T) - f),\ W(f - 1/(4T))] \tag{6.171}$$

$$\mathbf{H}_{n+1}(f) = [W(f - 1/(4T)),\ W^*(3/(4T) - f)]\ . \tag{6.172}$$

All the zero components of both vectors are omitted, since they will not affect the inner product between the two vectors. The inner product between $\mathbf{H}_n(f)$ and $\mathbf{H}_{n+1}(f)$ will be zero if

$$\mathrm{Re}\{W(f - 1/(4T))W(3/(4T) - f)\} = 0\ , \qquad f \in [1/(4T), 1/(2T)]\ . \tag{6.173}$$

Changing variables, this can be written

$$\mathrm{Re}\{W(f)W(1/(2T) - f)\} = 0\ , \qquad f \in [0, 1/(4T)]\ . \tag{6.174}$$

This condition can be satisfied by adjusting the phase of $W(f)$ to have a particular symmetry about $1/(4T)$. An example showing how to do this is included in Section 6.3.4.

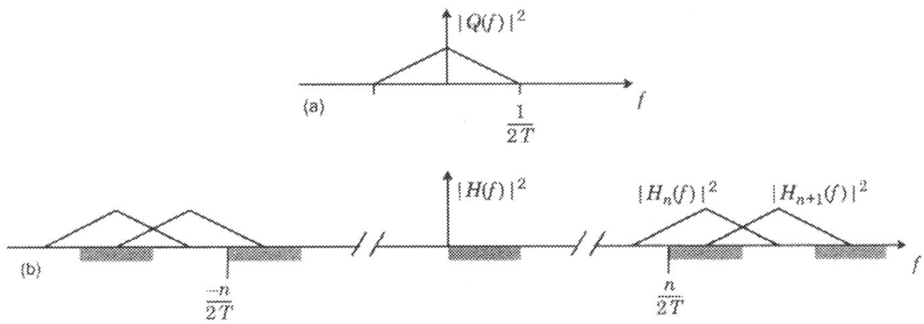

Fig. 6-36. An orthonormal pulse set overlapping in the frequency domain. a. The magnitude-squared of a pulse satisfying the Nyquist criterion at the output of a matched filter, with respect to rate $1/(2T)$. b. The magnitude-squared of two pulses overlapping one another in the frequency domain.

Problems

Problem 6-1.

(a) Show that the binary FSK pulses given in (6.50) are orthogonal when

$$f_2 - f_1 = 1/T \tag{6.175}$$

and $f_1 + f_2 = K/T$ for some integer K.

(b) Show that they are also orthogonal when

$$f_2 - f_1 = 1/(2T) \tag{6.176}$$

and $f_1 + f_2 = K/(2T)$ for some integer K.

Problem 6-2. Consider the binary CPFSK with pulses given by (6.114) and shown in Fig. 6-24(c). Assume the nominal carrier frequency satisfies

$$w_c = \frac{f_0 + f_1}{2} = \frac{K}{4T}, \tag{6.177}$$

for some integer K. Show that the frequency spacing of MSK signals (6.113) is the minimum frequency spacing that results in orthogonal pulses.

Problem 6-3. Consider designing an MSK transmission system with $N = 8$ pulses of different frequencies. Suppose that the lowest frequency is $f_0 = 10$ MHz. Also suppose that the symbol interval is $T = 1$ μs. Find f_1 through f_7.

Problem 6-4. In this problem we show that an MSK signal can be described as an offset-keyed PSK signal with half sinusoidal pulse shapes.

(a) Show that (6.117) can be written

$$x(t) = \cos(2\pi f_c t)\sum_{k=-\infty}^{\infty} I_k \sin\left(\frac{\pi t}{2T}\right) w(t - kT)$$
$$+ \sin(2\pi f_c t)\sum_{m=-\infty}^{\infty} Q_k \cos\left(\frac{\pi t}{2T}\right) w(t - kT), \tag{6.178}$$

where $Q_k = \cos(\phi_k)$ and $I_k = b_k\cos(\phi_k)$. This is close to a passband PAM form, but more work is required.

(b) Show that if k is even then $Q_k = Q_{k-1}$, and if k is odd then $I_k = I_{k-1}$. *Hint*: Use (6.118).

(c) Use parts (a) and (b) to show that

$$x(t) = \cos(2\pi f_c t)\sum_{k \text{ even}} p(t - kT)(-1)^{k/2} I_k$$
$$+ \sin(2\pi f_c t)\sum_{k \text{ odd}} p(t - kT)(-1)^{(k+1)/2} Q_k , \tag{6.179}$$

where $$p(t) = \sin\left(\frac{\pi t}{2T}\right)\Big(w(t) + w(t - T)\Big) \tag{6.180}$$

is one half of one cycle of a sinusoid. This is a passband PAM signal with pulse shape $p(t)$ (which extends over $2T$). Notice however that since one of the summations is over even k and the other is over odd k, the in-phase and quadrature parts of the signal are offset from one another by T. The symbol rate is $1/(2T)$, the in-phase symbols are $(-1)^{k/2}I_k$ for even k, and the quadrature symbols are $(-1)^{(k+1)/2}Q_k$ for odd k.

Problem 6-5. Show that if $A\cos(2\pi f_c t)$ is fed into an ideal phase splitter to produce a complex output signal, that the magnitude of the complex output signal is the constant $A/2$. Hence, the amplitude of a sinusoid (its *envelope*) can be found using the structure in Fig. 6-15(a).

Problem 6-6. Consider the OPAM modulation of (6.76). Suppose you are to achieve a total bit rate of 19.2 kb/s using $N = 128$ distinct orthonormal pulses. Assume each pulse is modulated using a 4-PSK constellation. Find the symbol interval T.

Problem 6-7. We are to design a real-valued set of orthonormal pulses for orthogonal modulation. In the range $0 \le f < 1/(2T)$, define the matrix of (6.168) to be

$$\boldsymbol{H}(f) = \frac{\sqrt{T}}{2}\begin{bmatrix} 1 & 1 & 0 & 0 \\ -1 & 1 & 0 & 0 \\ 0 & 0 & 1 & 1 \\ 0 & 0 & -1 & 1 \end{bmatrix}. \tag{6.181}$$

This clearly satisfies (6.168). Sketch $H_n(f)$ for $n = 0, 1, 2, 3$, assuming $M_1 = 2$ and $M_2 = 1$ in (6.166). Find expressions for $h_n(t)$ for $n = 0, 1, 2, 3$. How does the bandwidth efficiency compare to that of the pulses in Fig. 6-8?

Problem 6-8.

(a) Verify that the 3×3 DFT matrix

$$\boldsymbol{D} = \frac{1}{\sqrt{3}}\begin{bmatrix} 1 & 1 & 1 \\ 1 & e^{-j2\pi/3} & e^{-j4\pi/3} \\ 1 & e^{-j4\pi/3} & e^{-j8\pi/3} \end{bmatrix} \tag{6.182}$$

satisfies

$$\mathbf{DD^*} = \mathbf{I} \tag{6.183}$$

where $\mathbf{I}$ is the identity matrix.

(b) Use the property in part (a) to find a 3×3 matrix $\mathbf{H}(f)$ that satisfies (6.168) for $f \in [-1/(2T), 1/(2T))$.

(c) Use $\mathbf{H}(f)$ from part (b) to design an orthogonal modulation pulse set that satisfies the generalized Nyquist criterion. Assume that in defining $\mathbf{H}(f)$, we use $M_1 = 0$ and $M_2 = 2$ in (6.166). Give both time and frequency domain expressions or detailed sketches (whichever is more convenient) for the three pulses.

(d) Is your pulse set real-valued in time? What is the spectral efficiency if these pulses are used for orthogonal modulation over a real channel? What is the spectral efficiency if the pulses are used for OPAM (combined orthogonal modulation and PAM), assuming an alphabet size of M? How does the spectral efficiency compare to what you would get using the ideal bandlimited pulses of Fig. 6-8 for OPAM? How does it compare to ideal baseband and passband PAM? Are the pulses practical?

Problem 6-9. Define $d_{i,j} = \| \mathbf{S}_i - \mathbf{S}_j \|$. Show that $\mathbf{Y} = \mathbf{S}_i + \mathbf{E}$ is closer to $\mathbf{S}_j$ than to $\mathbf{S}_i$ if and only if

$$\mathrm{Re}\{\langle \mathbf{E}, \mathbf{U} \rangle\} > \frac{d_{i,j}}{2} \tag{6.184}$$

where $\mathbf{U}$ is a unit vector in the direction of $(\mathbf{S}_j - \mathbf{S}_i)$.

Problem 6-10. Calculate the minimum distance $d_{\min}$ for the following cases:

(a) Slicer design, with the data-symbol alphabet 4-PSK and magnitude unity.

(b) Isolated pulse PAM, where the pulse energy is σ_h and the data symbol is chosen from the same alphabet as in a.

(c) Orthogonal modulation with pulse energy σ_h^2.

(d) PAM with ISI for two data symbols ($K = 2$) and a pulse autocorrelation function $\rho_h(k) = \alpha^{|k|}$ for a real-valued $0 \le \alpha < 1$.

Problem 6-11. Assume that in (6.127), the causal orthogonal bandlimited pulses $h_n(t)$ are chosen such that their tail energy is bounded by

$$\int_{t_0}^{\infty} h_n^2(t)\,dt \le \frac{\alpha}{t_0^2}, \tag{6.185}$$

for some constant α. Thus, the energy falls off at least as the square of time. Show that the fraction of the energy in $s(t)$ falling outside the interval $[0, KT]$ goes to zero as $K \to \infty$. Thus, the signal becomes approximately time limited to $[0, KT]$ as $K \to \infty$.

Problem 6-12. Assume a benign passband channel, $B(f) = 1$, with real additive Gaussian noise with power spectrum $S_N(f) = N_0/2$. The transmit filter produces a 100% excess-bandwidth raised-cosine pulse. The transmit power cannot be greater than unity. The complex envelope $f(t)$ of the receive filter is an ideal low-pass filter that permits the 100% excess-bandwidth pulse to get through undistorted. The constellation is 16-QAM. Find the probability of error as a function of N_0 and T.

Problem 6-13. Consider the constellation in the following figure:

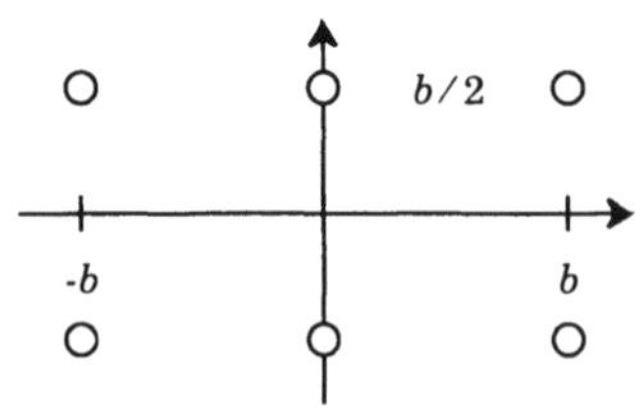

Assume that

- The inner two symbols each have probability $1/4$.
- The outer four symbols each have probability $1/8$.
- The noise in each dimension is independent and Gaussian with variance σ^2.

(a) Design a coder for this constellation that achieves these probabilities if the input bits are equally likely to be zero or one.

(b) Find the exact probability of error as a function of b.

(c) Find the signal power as a function of b.

(d) Give the probability of error as function of the SNR. Use an approximation from Fig. 3-1 to find the probability of error when $SNR = 10$ dB.

(e) Give approximations for the probability of error. Compute the approximate probabilities of error when $SNR = 10$ dB.

Problem 6-14. Suppose that more than two dimensions are available for our alphabet. Consider an alphabet where the symbols are vertices of an M dimensional hypercube, shown for $M = 3$ in the following figure:

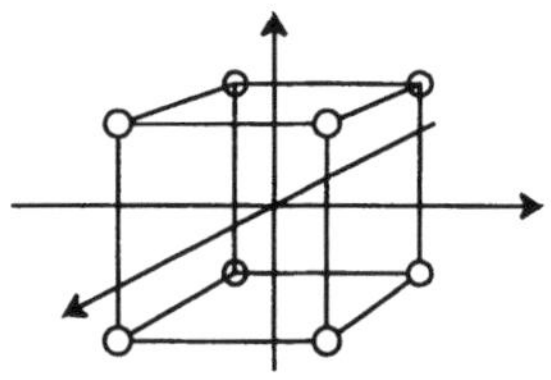

Assume that all the symbols are equally likely and the noise in each dimension is independent with variance σ^2.

(a) What are the decision regions?

(b) What is the probability of error as a function of M and the minimum distance d between points in the signal constellation.

Problem 6-15. Consider the following MSK signal:

$$x(t) = \cos\left[2\pi f_c t + \pi\int_{-\infty}^{t} s(\tau)d\tau + \frac{\pi}{2}\right], \tag{6.186}$$

where $s(t)$ is given by (6.112), $a_k \in \{\pm 1\}$, and $g(t)$ is a rectangular pulse over $[0, T)$ that integrates to $1/2$.

(a) Using Fig. 6-26 as a starting point, draw a trellis with a finite number of states that describes the phase evolution of the MSK signal.

(b) Show that the minimum-distance error event is the error event of length one. Find its distance.

(c) Compare the optimal sequence detector performance to that of the receiver in Fig. 6-25.

Problem 6-16. Show that if the translates of the chip waveform $h_c(t)$ are mutually orthogonal, then $N = 2WT$ pulses of the form of (6.100) can be made mutually orthogonal by choice of the spreading sequences. Specify the required properties of the spreading sequence.

Problem 6-17. Consider the spreading sequence $\{x_0, x_1, \ldots x_{N-1}\}$ in (6.100) to be the impulse response of a causal, discrete-time FIR filter. Condition (6.103) suggests that we would like this filter to be an allpass filter. Use the results of Section 2.5.4 to show that the only FIR filters that are allpass have impulse response δ_{k-L}, for some integer L. Thus, (6.103) can be exactly satisfied only for the trivial choice of spreading sequence in Example 6-19.

Problem 6-18. Consider a spread-spectrum system operating in $N = 2WT$ dimensional signal space, where the isolated pulse signal is chosen randomly. Let a set of orthonormal basis functions for this space be $\phi_i(t)$, $1 \le i \le N$. The one-dimensional binary antipodal transmitted signal is chosen to be

$$\pm s(t) = \pm \sum_{i=1}^{N} s_i \phi_i(t) \tag{6.187}$$

and a jammer generates a similar signal

$$x(t) = \sum_{i=1}^{N} x_i \phi_i(t)\,, \tag{6.188}$$

where the s_i and x_i are mutually independent zero-mean random variables. The signal components are chosen to have variance $\mathrm{E}[s_i^2] = E/N$ where E is the average energy per bit, and the jammer chooses the x_i variances to satisfy the constraint

$$\sum_{i=1}^{N} \mathrm{E}[x_i^2] = E_J\,, \tag{6.189}$$

where E_J is the constrained jamming signal energy per symbol interval. The receiver is told the s_i, and applies a matched filter to the received signal.

(a) Determine the signal and noise random variables at the output of the matched filter.

(b) Define the SNR at the matched filter output as the ratio of the mean-signal squared to the noise variance. Show that this SNR has a processing gain of N independently of how the jammer distributes its energy among the signal components. (This result is due to the fact that the signal is truly random and unknown to the jammer.)

Problem 6-19. Consider a fiber optic communication link on which the group velocity is half the speed of light.

(a) For each one meter of distance and at a bit rate of 10 Gb/s, how many bits are in transit?

(b) At what distance (in meters) does the transmit time of a 1 megabyte message equal the transit time?

Problem 6-20. Verify that the capacity of the passband channel in Fig. 6-28(b) is given by (6.144). Use a signal space argument with complex rather than real signals, noise, and vectors.

Problem 6-21. One way to view digital communication is as a way in which to exchange channel bandwidth for improved noise immunity. For example, in analog modulation systems, FM modulation is thought of primarily as a way to achieve higher post-demodulation SNR in exchange for increased bandwidth. In this problem we quantify this tradeoff for digital communication systems, using SNR for a given channel capacity as a measure of noise immunity. Given two white Gaussian noise channels as in Fig. 6-28, with bandwidths W_1 and W_2, where $W_2 > W_1$, suppose they have the same channel capacity. Find a relation for the SNRs (in dB) required for the two channels as a function of the bandwidth expansion factor W_2/W_1. Interpret this relation. You may assume large SNR.

Problem 6-22. It is common to use binary PSK in spread spectrum modulation. Find the SNR gap to capacity for 2-PSK spread spectrum as a function of WT. What is the increase in the SNR gap to capacity asymptotically as $WT \rightarrow \infty$, expressed in dB?

References

1. S.W.Golomb, *Digital Communications with Space Applications*, Prentice Hall, N.J. (1964).
2. R. W. Lucky, J. Salz, and E. J. Weldon, Jr., *Principles of Data Communication*, McGraw-Hill Book Co., New York (1968).
3. R. W. Chang, "Synthesis of Band-Limited Orthogonal Signals for Multichannel Data Transmissions," *Bell System Technical Journal* (Aug. 1966).
4. *American National Standards Institute*, "Network and customer installation interfaces — Asymmetrical digital subscriber line (ADSL) metallic interface," ANSI Standard T1E1, pp. 413 1995, August 18, 1995.
5. B. Hirosaki, "An Orthogonal Multiplexed QAM System Using the Discrete Fourier Transform," *IEEE Transactions on Communications*, pp. 982-989 (July 1981).
6. R. A. Scholtz, "The Origins of Spread-Spectrum Communications," *IEEE Trans. Communications*, Vol. COM-30 (5), p. 822 (May 1982).
7. R. L. Pickholtz, D. L. Schilling, and L. B. Milstein, "Theory of Spread-Spectrum Communications – A Tutorial," *IEEE Trans. Commun.*, Vol. COM-30 (5), p. 855 (May 1982).
8. M. L. Doelz and E. H. Heald, "Minimum-Shift Data Communications Systems," U. S. Patent No. 2,977,417, (March 28, 1961).
9. T. Aulin, N. Rydbeck, and C.-E. W. Sundberg, "Continuous Phase Modulation - Part II: Partial Response Signaling," *IEEE Trans. on Communications*, Vol. COM-29 (3), (March 1981).
10. H. J. Landau and H. O. Pollak, "Prolate Spheroidal Wave Functions, Fourier Analysis, and Uncertainty, III: The Dimension of the Space of Essentially Time- and Band-Limited Signals," *Bell System Technical Journal*, Vol. 41, p. 1295 (1962).
11. C. E. Shannon and W. Weaver, *The Mathematical Theory of Communication*, University of Illinois Press, Urbana, Illinois (1963).
12. C. E. Shannon, "Communication in the Presence of Noise," Proc. IRE, Vol. 37, pp. 10-21 (January 1949).
13. G. D. Forney, Jr and M. V. Eyuboglu, "Combined Equalization and Coding Using Precoding," *IEEE Communications Magazine*, (Dec. 1991).
14. J. M. Wozencraft and I. M. Jacobs, *Principles of Communication Engineering*, Wiley, New York (1965).
15. J. G. Proakis, *Digital Communications*, Fourth Edition, McGraw-Hill Book Co., New York (2001).

16. G. R. Cooper and C. D. McGillem, *Modern Communications and Spread Spectrum*, McGraw-Hill Book Co., New York (1986).

17. R. E. Ziemer and R. L. Peterson, *Digital Communications and Spread Spectrum Systems*, Macmillan, New York (1985).

18. C. E. Cook and H. S. Marsh, "An Introduction to Spread Spectrum," *IEEE Communications Magazine*, p.8, (March 1983).

19. S. Pasupathy, "Minimum Shift Keying: A Spectrally Efficient Modulation," *IEEE Communications Magazine,* Vol. 17, No. 4 (July 1979).

20. S. Haykin, *Communications Systems, Second Edition,* John Wiley & Sons, Inc. (1983).

21. S. Gronemeyer and A. McBride, "MSK and Offset QPSK Modulation," *IEEE Transactions on Communications*, Vol. 24, No. 8 (Aug. 1976).

22. D. H. Morais and K. Feher, "Bandwidth Efficiency and Probability of Error Performance of MSK and Offset QPSK Systems," *IEEE Transactions on Comm.,* Vol. 27, No. 12 (Dec. 1979).

23. J. B. Anderson, T. Aulin, and C.-E. Sundberg, *Digital Phase Modulation*, Plenum, New York, NY, 1986.

24. M. K. Simon, S. M. Hinedi, and W. C. Lindsey, Digital Communication Techniques, Prentice Hall, Englewood Cliffs, New Jersey, 1995.

25. B. R. Saltzberg, "Performance of an Efficient Parallel Data Transmission System," *IEEE Transactions on Communications*, Vol. 15, pp. 805-811 (Dec. 1967).

26. E. Powers and M. Zimmerman, "TADIM — A Digital Implementation of a Multichannel Data Modem," *Proceedings of ICC* (1968).

27. S. Darlington, "On Digital Single-Sideband Modulators," *IEEE Transactions on Circuit Theory,* Vol 17, pp. 409-414 (Aug. 1970).

28. S. B. Weinstein and P. M. Ebert, "Data Transmission by Frequency Division Multiplexing Using the Discrete Fourier Transform," *IEEE Trans. Communications*, Vol. 19, No. 5 (Oct. 1971).

29. I. Kalet, "The Multitone Channel," *IEEE Trans. Communications*, Vol 37, No. 2, pp. 119-124 (Feb. 1989).

30. A. Ruiz, J. Cioffi, and S. Kasturia, "Discrete Multiple Tone Modulation with Coset Coding for the Spectrally Shaped Channel," *IEEE Trans. Comm.,* Vol. 40, No. 6, pp. 1012-1029, June 1992.

31. J. A. C. Bingham, "Multicarrier Modulation for Data Transmission: An Idea Whose Time Has Come," *IEEE Communications Magazine,* pp. 5-14 (May 1990).

32. J. W. Lechleider, "The Optimum Combination of Block Codes and Receivers for Arbitrary Channels," *IEEE Transactions on Communications*, Vol. 38, No. 5 (May 1990).

33. S. Kasturia, J. T. Aslanis, and J. M. Cioffi, "Vector Coding for Partial Response Channels," *IEEE Transactions on Information Theory,* Vol. 36, No. 4, pp. 741-762 (July 1990).

34. E. Feig, "Practical Aspects of DFT-Based Frequency Division Multiplexing for Data Transmission," *IEEE Transactions on Communications*, Vol. 38, No. 7, pp. 929-932 (July 1990).

7

Probabilistic Detection

A fundamental problem in digital communications is the corruption of the transmitted signal by noise. The minimum-distance philosophy for receiver design is reasonably robust in the presence of noise, but two questions arise: when is it optimal? And what should be done when it is not? In this chapter we start with the statistics of the noise and develop a theory of optimal detection for both discrete-time and continuous-time channels. With this theory, we identify the circumstances under which the minimum-distance receiver is optimal. Moreover, we take a systematic approach to receiver design based on a probabilistic characterization of the noise that is applicable to a wide range of applications beyond the classical model of additive white Gaussian noise, including those for which minimizing distance is not optimal. The probabilistic tools developed in this chapter play an important role in iterative decoding of error-control codes (Chapter 12).

The general approach to deriving optimal receivers is to model the relationship between the transmitted and received signals by a joint probability distribution. Based on the noisy observation (the received signal corrupted by noise), we wish to *estimate* or *detect* the input signal. We use the term *estimation* when the transmitted signal (or more generally the quantity to be estimated) is a continuous-valued random variable, as is often the case in an analog communication system, and the term *detection* when the transmitted signal is discrete-valued (even if the received signal is continuous-valued). The primary distinction is that in detection we can often recover the signal exactly with high probability, a restatement of the regeneration principle of Chapter 1. In estimation, by contrast, we must be satisfied with a recovered signal that may be more accurate than the observation but will not be exact. In this chapter we study only detection, although very similar techniques can be applied to the estimation problems of analog communications.

This chapter will adopt the notation that random quantities will be upper case (such as X) whereas deterministic quantities will be lower case (such as x). This will help us distinguish between a random variable and a particular outcome of that random variable, an important distinction in the analysis of this chapter.

In order to address the detection problem, we need a statistical model for the received signal. Before the data symbols arrive at the detector, they are processed by a transmitter, pass through a channel, and are further processed by the front end of the receiver. Some of this processing is deterministic, such as any filtering functions, and some is random, such as additive noise on the channel. In this chapter we call the deterministic portion *signal generation* and a random component *noise generation*. The model is shown in Fig. 7-1. The input X_k is a discrete-time and discrete-valued random process. It is not only discrete-valued but has a *finite* number of possible values, each of which is a function of the source bits.

Example 7-1.

Suppose X_k is a data symbol sequence. Then the signal generator could be a discrete-time equivalent transmit filter and channel transfer function, representing ISI, and the noise generator could be *additive Gaussian noise*, independent of X_k.

Example 7-2.

Suppose X_k is a bit sequence. The signal generator could be a coder that produces another bit sequence, and the noise generator could be modeled as a *binary symmetric channel* (BSC), which randomly inverts some of the bits.

The receiver uses the *observation* Y_k to make a *decision* about X_k. The observation can be either discrete or continuous-valued. Since each X_k has a finite number of possible values, the detector must make a decision from among a finite number of alternatives.

Two related detection methods are covered: *maximum likelihood (ML)* and *maximum a-posteriori probability (MAP)*. MAP detection, also called *Bayesian* detection, is optimal in the sense that it minimizes the probability of error. ML detection is a special case of MAP detection for the special case when all the possible inputs are equally likely.

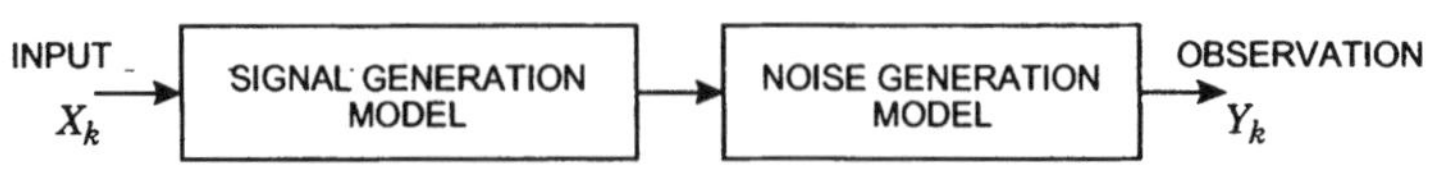

Fig. 7-1. A signal X_k to be transmitted is processed deterministically (signal generation) and stochastically (noise generation).

We begin with simple signal generation models and progress to more realistic (and more complicated) models. We address the two basic noise generators of Example 7-1 and Example 7-2 — additive Gaussian noise and the BSC. We begin with the detection of a single real-valued data symbol, where the input is a real-valued data symbol and the signal generator is trivial. Then we progress to the detection of vector-valued inputs, which applies to the case of complex-valued signal constellations among others. The next step is to derive the optimal detector in additive Gaussian noise, for both the discrete-time and continuous-time cases. Up to this point the optimal detectors have been defined for the detection of a single data symbol, and the next extension is to ISI, where it is shown that the minimum-distance receiver design of Section 5.4 is optimal in additive Gaussian noise. The Viterbi algorithm will be reviewed from a probabilistic viewpoint. We then introduce the BCJR algorithm for calculating a posteriori probabilities after an ISI channel with white Gaussian noise. The BCJR has many other applications in digital communication, including the detection of convolutional and trellis codes (Chapters 12 and 13). We then relax the known-signal assumption by allowing the carrier phase to be unknown and random. Finally, the detection of a shot-noise signal with known intensity, characteristic of fiber optic systems, is considered.

7.1. Detection of a Single Real-Valued Symbol

In this section, we consider the simplest case, where the input is a single random variable X (a single data symbol A) rather than a random process, and the signal generator passes this symbol directly through to the noise generator without modification. The data symbol has as a sample space the alphabet $\mathcal{A}$, as discussed in Chapter 5. Noise generators that arise in practice for this case result in either a discrete-valued observation Y or a continuous-valued observation. We give examples of both cases in the following subsections, at the same time illustrating the ML and MAP detectors.

7.1.1. Discrete-Valued Observations

Some noise generators result in discrete-valued observations Y. In order to design a detector, we must know the discrete distribution of Y conditioned on knowledge of the data symbol, $p_{Y|A}(y \mid \hat{a})$, as this completely specifies the noise generator. The *maximum-likelihood* (ML) detector chooses $\hat{a} \in \mathcal{A}$ to maximize the *likelihood* $p_{Y|A}(y \mid \hat{a})$, where y is the observed outcome of Y.

Example 7-3.
Suppose that we have additive discrete noise N, so that $Y = A + N$. Assume A and N are independent and take on values zero and one according to the flip of two fair coins. There are three possible observations, $y = 0,\ 1,$ or 2. The likelihoods for the observation $y = 0$ are

$$p_{Y|A}(0 \mid \hat{a}) = \begin{cases} 0.5; & \text{for } \hat{a} = 0 \\ 0\ ; & \text{for } \hat{a} = 1 \end{cases} \tag{7.1}$$

so if the observation is $y = 0$, the ML detector selects $\hat{a} = 0$. If the observation is $y = 1$, then the likelihoods are equal

$$p_{Y|A}(1 \mid \hat{a}) = \begin{cases} 0.5; & \text{for } \hat{a} = 0 \\ 0.5; & \text{for } \hat{a} = 1 \end{cases} \tag{7.2}$$

so the ML detector selects either zero or one (it could choose randomly, for example). If the observation is $y = 2$, the ML detector selects $\hat{a} = 1$.

The advantage of ML detection is that the likelihood $p_{Y|A}(y \mid \hat{a})$ is easily computed for each possible $\hat{a}$ knowing only the statistics of the noise generator and not the statistics of the data symbol.

The *maximum a-posteriori probability* (MAP) detector maximizes the *posterior probability* $p_{A|Y}(\hat{a} \mid y)$. This should be more intuitively appealing than maximizing the likelihood. In fact, we will see that the MAP receiver minimizes the probability of error, and hence is optimal for any application that prefers correct decisions over incorrect. We can write the posterior probability in terms of the likelihood using Bayes' rule (3.32)

$$p_{A|Y}(\hat{a} \mid y) = \frac{p_{Y|A}(y|\hat{a})p_A(\hat{a})}{p_Y(y)} . \tag{7.3}$$

A MAP detector therefore needs to know the probabilities $p_A(\cdot)$ of the symbols. These probabilities are called the *prior probabilities*, the *a priori probabilities*, or more simply, just the *priors*. Since Bayes' rule is used to compute the posterior probability, MAP detection is also called *Bayesian detection*. Since the denominator does not depend on the detector decision $\hat{a}$, maximizing the posterior probability is the same as maximizing the numerator, namely $p_{Y|A}(y \mid \hat{a})p_A(\hat{a})$. If $p_A(\hat{a})$ is constant for all $\hat{a} \in \mathcal{A}$, then maximizing the posterior probability is the same as maximizing the likelihood $p_{Y|A}(y \mid \hat{a})$, and MAP reduces to ML.

Example 7-4.

In Example 7-3, $p_A(\hat{a}) = 0.5$, for each possible $\hat{a}$, so MAP and ML are equivalent.

Example 7-5.

If we knew *a priori* in Example 7-3 that one of the coins was biased (unfair), say

$$p_A(0) = 0.75 , \qquad p_A(1) = 0.25 , \tag{7.4}$$

then the MAP detector would not generally give the same result as the ML detector. It is easily shown that if the observation is $y = 0$ or $y = 2$, the ML or MAP detections are the same and are correct with probability one. If the observation is $y = 1$, however, the detectors are not the same. In this case, the likelihoods are the same, as specified in (7.2), so the ML detector can arbitrarily select a decision. However, when $y = 1$, the posterior probabilities are

$$p_{A|Y}(\hat{a} \mid 1) = \begin{cases} 0.75; & \text{for } \hat{a} = 0 \\ 0.25; & \text{for } \hat{a} = 1 \end{cases} \tag{7.5}$$

so the MAP detector always gives $\hat{a} = 0$ when the observation is $y = 1$.

Whenever we are detecting a signal based on a noisy observation there is of course the possibility of making an error or mistake. The *probability of error*

$$P_e = \Pr[\hat{a} \neq a] \tag{7.6}$$

is a good measure of the quality of our detector design.

Example 7-6.

In Example 7-3, when the coin flip is fair, the MAP and ML detectors are identical, so their probability of error is identical. We can compute that probability of error. The observation $Y = A + N$ takes on values in $\{0, 1, 2\}$ with probabilities $\{0.25, 0.5, 0.25\}$, respectively. If the observation is $y = 0$, then both detectors give the result $\hat{a} = 0$, and no error occurs. Similarly, if $y = 2$, both detectors give $\hat{a} = 1$, and no error occurs. If the observation is $y = 1$, however, both detectors are unable to choose a unique maximum. If the detectors select $\hat{a}$ to randomly equal zero or one, then they will be wrong half the time. Since the $p_Y(1) = 0.5$, the total probability of error is $P_e = 0.25$.

In the case of equal prior probabilities for A, ML and MAP detectors give identical results, and their probability of error is identical. This is no longer true if the priors are different. In this case, a MAP detector will always have a lower probability of error than a ML detector.

Example 7-7.

Assume that the data symbol A in Example 7-3 is generated by an unfair coin, just as in Example 7-5. We again assume a noisy observation $Y = A + N$, where N is a fair coin. If the observation is either $y = 0$ or $y = 2$, then both detectors are correct, and no error is made. If $y = 1$ and the ML detector arbitrarily selects a decision, it will make errors with probability 0.5 each time it observes $y = 1$. This occurs with probability $p_Y(1) = 0.5$, so the total probability of error is 0.25. The MAP detector, however, will do much better, because when $y = 1$ it always selects $\hat{a} = 0$. This is incorrect for only ¼ of such observations, and $p_Y(1) = 0.5$, so the probability of error has been reduced to 0.125.

In fact, the MAP detector minimizes the probability of error. To show this, note the probability of a correct decision is

$$\Pr[\text{correct decision}] = \Sigma_{y \in \mathcal{Y}} \Pr[\text{correct decision} \mid Y = y] p_Y(y) \,. \tag{7.7}$$

Since $p_Y(y) \geq 0$, the probability of a correct decision is maximized if the decision for each observation y maximizes Pr[correct decision | $Y = y$]. This is what the MAP receiver does since the probability of a correct decision is

$$\Pr[\text{correct decision} \mid Y = y] = p_{A \mid Y}(\hat{a} \mid y) \,, \tag{7.8}$$

and this is precisely the posterior probability maximized by the MAP detector.

In summary, ML and MAP are different detection techniques, but yield the same result when the prior probabilities are equal. If the prior probabilities are different, the MAP detector will yield a lower (and indeed the minimum) probability of error. If we have no information about the prior probabilities, we usually assume they are equal, and MAP reduces to ML.

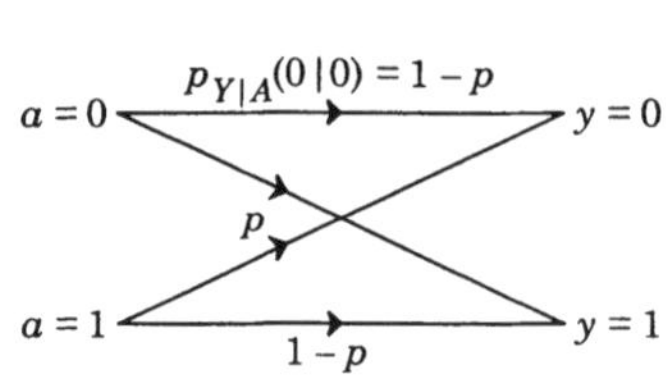

Fig. 7-2. A schematic diagram of the binary symmetric channel (BSC), which flips bits with probability p.

The most common noise generation model in digital communications that results in a discrete-valued observation is the *binary symmetric channel (BSC)*. This channel applies when the input data symbol is binary, and the conditional probabilities are shown in Fig. 7-2. The noisy observation differs from the binary input symbol with probability p, which is called the error probability or *crossover probability* of the BSC. The ML and MAP detectors for a BSC are derived in Problem 7-1.

7.1.2. Continuous-Valued Observations

A common noise generation model corrupts the input data symbol A by *additive continuous-valued* noise N. The observation $Y = A + N$ is then a continuous-valued random variable. The MAP detector selects $\hat{a}$ to maximize $p_{A|Y}(\hat{a} \mid y)$ while the ML detector selects $\hat{a}$ to maximize $p_{Y|A}(y \mid \hat{a})$. Given an observation $Y = y$, to find the MAP detector we use the mixed form of Bayes' rule (3.31) to write

$$p_{A|Y}(\hat{a} \mid y) = \frac{f_{Y|A}(y|\hat{a})p_A(\hat{a})}{f_Y(y)} . \tag{7.9}$$

The denominator is independent of the decision, so we need only maximize the numerator. This is illustrated by example.

Example 7-8.
Suppose that A takes on value ± 1 according to the flip of an unfair coin, so that $p_A(+1) = 0.75$ and $p_A(-1) = 0.25$. Suppose further that we observe $Y = A + N$ where N is a continuous-valued random variable with the p.d.f. given in Fig. 7-3(a). We can derive the MAP detector and its probability of error. We need to select $\hat{a}$ to maximize $f_{Y|A}(y \mid \hat{a})p_A(\hat{a})$. In Fig. 7-3(b), this quantity is shown for the two possible values of $\hat{a}$ as a function of the observation. From this figure it is immediately evident that the MAP detector will set $\hat{a} = 1$ if $y > -0.5$, otherwise $\hat{a} = -1$. An error never occurs if $A = +1$. An error occurs one third of the time that -1 is transmitted, because one third of the time the observation will be greater than -0.5. Hence, the probability of error is $p_A(-1)/3 = 1/12$. This is the area of the shaded region Fig. 7-3(b).

Example 7-9.
We now find the ML detector and its probability of error for the same scenario as in the previous example. In Fig. 7-3(c) we plot the likelihoods for the two possible decisions as a function of the observations. The likelihoods are equal in the region $-0.5 < y < 0.5$, so the ML detector can make its decision arbitrarily. A legal ML detector sets its threshold at -0.5, and has performance identical to that of the MAP detector. But another legal ML detector sets its threshold at zero (halfway

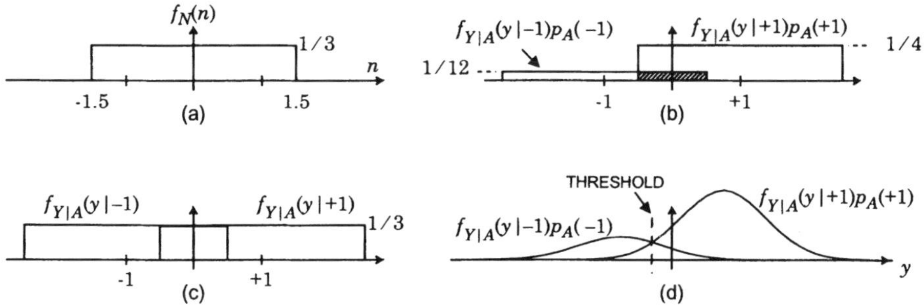

Fig. 7-3. (a) The p.d.f. of uniformly distributed noise. (b) For $\hat{a} = 1$, the MAP criterion $f_{Y|A}(y \mid \hat{a})p_A(\hat{a})$ is plotted. Note that the MAP detector will prefer $\hat{a} = 1$ for any observation $y > -0.5$. (c) The likelihoods as a function of the observation y are plotted for $\hat{a} = \pm 1$. Note that the ML detector has no preference when the observation is $-0.5 < y < 0.5$. (d) The MAP criterion is plotted assuming additive Gaussian noise.

between the two possibilities); this detector will make an error $1/6$ of the time for each possible transmission, so the probability of error is $1/6$.

The most common distribution for additive noise in digital communications is Gaussian, rather than uniform as in previous examples. The principle of the detectors is the same.

Example 7-10.

In Fig. 7-3(d) we show the functions $f_{Y|A}(y \mid \pm 1)p_A(\pm 1)$ as functions of the observations assuming additive Gaussian noise. For the MAP detector, the threshold is selected where these curves cross. For the ML detector, the threshold is zero.

In the next section we will consider the additive Gaussian noise case for the more general situation where the signal and noise are vector-valued.

7.2. Detection of a Signal Vector

Many of the communication channels we describe in this book can be modeled as noise corrupting a vector-valued signal. Although typical channels accept only scalar-valued signals, a convenient vector communication model can often be obtained using the technique shown in Fig. 7-4. A vector of transmitted symbols is converted to a sequence of scalars for transmission over an additive noise channel, and then reconverted to a vector at the channel output. In effect we have taken a finite sequence of samples and modeled them as a vector.

Example 7-11.

In Section 6.1 we saw that the minimum-distance receiver for a continuous-time channel can be implemented using a projection receiver, which first transforms the continuous-time received signal into a projection vector of dimension N. The projection receiver then proceeds to minimize distance for the vector-valued channel.

A general model for the situation is as follows. The signal generator accepts an input X and maps it into a vector signal **S** with dimension N.

Example 7-12.

The signal generator might take a set of N consecutive data symbols as the signal vector, $\mathbf{S} = [A_1, \ldots, A_N]^T$.

The observation is a vector $\mathbf{Y}$ with the same dimension as the signal. The noise generator is specified by the conditional distribution of the observation given the signal, $f_{\mathbf{Y}|\mathbf{S}}(\mathbf{y} \mid \mathbf{s})$. The detector decides which signal vector $\hat{\mathbf{s}}$ from among all the possible signal vectors was actually transmitted based on the observation. A common characteristic of the noise generator is *independent noise components*, by which we mean precisely that

$$f_{\mathbf{Y}|\mathbf{S}}(\mathbf{y} \mid \mathbf{s}) = \prod_{k=1}^{N} f_{Y_k|S_k}(y_k | s_k), \tag{7.10}$$

or in words, given knowledge of the signal vector, each component of the noise generation is independent of the others.

In the following subsections we will consider first the ML detector (for which the signal generator does not need to be statistically characterized) and then the MAP detector.

7.2.1. ML Detection

The ML detector chooses the signal vector $\hat{\mathbf{s}}$ from among all the possibilities in order to maximize the conditional probability $f_{\mathbf{Y}|\mathbf{S}}(\mathbf{y} \mid \hat{\mathbf{s}})$, a probability given directly by the noise generation model. It is simple in that the statistics of the signal $\mathbf{S}$ need not be taken into account. Two important examples are the additive Gaussian noise generator and the BSC noise model.

Example 7-13.

Consider the AWGN channel:

$$\mathbf{Y} = \mathbf{S} + \mathbf{N}, \tag{7.11}$$

where the signal vector $\mathbf{S}$ is randomly chosen from a set of M possibilities, and where the complex-valued noise vector $\mathbf{N}$ is circularly symmetric Gaussian with uncorrelated (and hence independent) components with variance $2\sigma^2$. As discussed in Section 6.1, this is the model for the projection vector with M-ary signaling in additive white Gaussian noise, (6.25). Given a

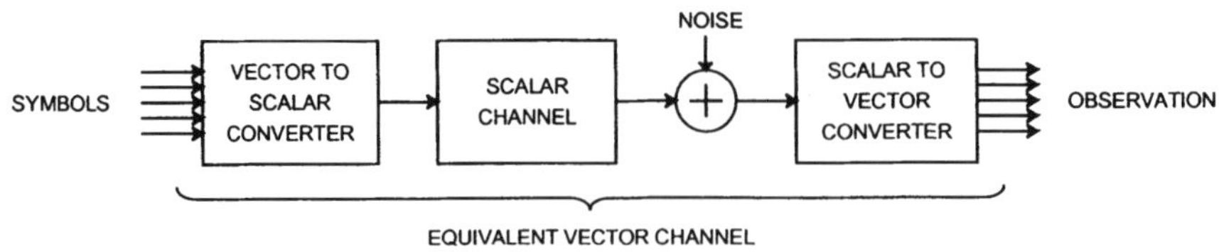

Fig. 7-4. A scalar channel plus some signal processing can sometimes be modeled as a vector channel.

particular outcome $\boldsymbol{s}$ for S, the received signal **Y** is a complex-valued Gaussian vector with mean equal to $\boldsymbol{s}$, and hence has the probability density function

$$f_{\mathbf{Y}|\mathbf{S}}(\boldsymbol{y}|\boldsymbol{s}) = f_{\mathbf{N}}(\boldsymbol{y} - \boldsymbol{s}) \ . \tag{7.12}$$

Hence the ML detector selects $\hat{\boldsymbol{s}}$ to maximize $f_{\mathbf{N}}(\boldsymbol{y} - \hat{\boldsymbol{s}})$. Since the components of **N** are assumed Gaussian and independent, from (3.49) we obtain

$$f_{\mathbf{N}}(\boldsymbol{n}) = \prod_{k=1}^{N} f_{N_k}(n_k) \ . \tag{7.13}$$

Since, for a complex Gaussian random variable with independent real and imaginary components,

$$f_{N_k}(n) = \frac{1}{2\pi\sigma^2} e^{-|n|^2/2\sigma^2}, \tag{7.14}$$

it follows that

$$f_{\mathbf{N}}(\boldsymbol{n}) = \frac{1}{(2\pi\sigma^2)^N} e^{-\|\boldsymbol{n}\|^2/2\sigma^2} \ . \tag{7.15}$$

Since the exponential is a monotonic function of its exponent, maximizing $f_{\mathbf{N}}(\boldsymbol{y} - \hat{\boldsymbol{s}})$ is equivalent to minimizing $\|\boldsymbol{y} - \hat{\boldsymbol{s}}\|^2$. In other words, the ML detector reduces to the minimum-distance detector for the special case of a white-Gaussian noise vector channel.

Example 7-14.

For the binary symmetric channel (BSC) of Fig. 7-2 with independent noise components, the components of the signal vector are binary, as are the components of the observation. The conditional probability for one channel use is

$$p_{Y_k|S_k}(y|\hat{s}) = \begin{cases} p; & y \neq \hat{s} \\ 1-p; & y = \hat{s} \end{cases} . \tag{7.16}$$

Define the metric $d_H(\hat{\boldsymbol{s}}, \boldsymbol{y})$ as the number of components in which the two binary vectors $\hat{\boldsymbol{s}}$ and $\boldsymbol{y}$ disagree. This metric $d_H(\cdot, \cdot)$ arises frequently in coding theory where is it called the *Hamming distance* or *Hamming metric* after R. Hamming of Bell Laboratories, who did pioneering work on algebraic coding in the 1950's. If the vectors have dimension N, then the joint conditional distribution is

$$p_{\mathbf{Y}|\mathbf{S}}(\boldsymbol{y} \mid \hat{\boldsymbol{s}}) = p^{d_H(\hat{\boldsymbol{s}}, \boldsymbol{y})}(1-p)^{N - d_H(\hat{\boldsymbol{s}}, \boldsymbol{y})} = (1-p)^N \left(\frac{p}{1-p}\right)^{d_H(\hat{\boldsymbol{s}}, \boldsymbol{y})} \ . \tag{7.17}$$

When $p < \frac{1}{2}$ as is usual, the ML detector chooses $\hat{\boldsymbol{s}}$ to minimize $d_H(\hat{\boldsymbol{s}}, \boldsymbol{y})$. In other words, it chooses the signal vector closest to the observation vector in Hamming distance.

The two important noise generators with independent additive noise components, the additive Gaussian noise and the BSC, result in a similar ML detector: choose that signal vector which is the closest to the observation vector. The only difference between the two cases is the manner in which we measure "distance;" in the first case we use Euclidean distance and in the other we use Hamming distance. In both cases, the sample space $\mathcal{Y}$ can be divided into *decision regions*. The i-th decision region is the set of all $y \in \mathcal{Y}$ closer to s_i (in Euclidean or Hamming distance) than to any other s_j, $j \neq i$. In other words, it is the set of all observations that will lead to the decision $\hat{s} = s_i$.

Example 7-15.
Let the signal s be a two-dimensional vector from the set

$$\left\{ \begin{bmatrix} -1 \\ 0 \end{bmatrix}, \begin{bmatrix} 1 \\ 0 \end{bmatrix}, \begin{bmatrix} 0 \\ 1 \end{bmatrix}, \begin{bmatrix} 0 \\ -1 \end{bmatrix} \right\}. \quad (7.18)$$

The signal space and decision regions are shown below:

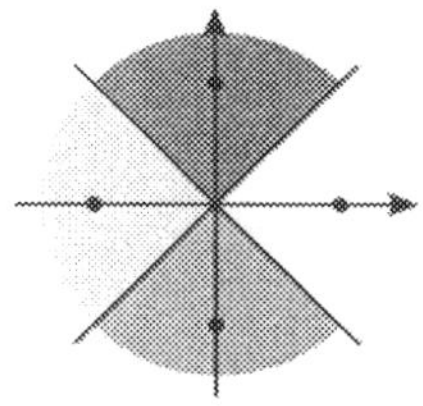

The ML detector in both cases is simple and intuitive, and has the feature that the noise variance σ^2 or crossover probability p need not be known by the detector. This feature is desirable, since in many digital communications systems the noise variance or error probability may be dependent on factors such as the distance of transmission, and therefore difficult to know in advance.

7.2.2. MAP Detector

The MAP detector is somewhat more complicated and requires extra knowledge about the noise statistics; for the Gaussian case it requires knowledge of the noise variance, and for the BSC case it requires knowledge of the crossover probability. Assume that the prior probabilities $p_S(\hat{s})$ are known for each possible signal $\hat{s}$. Given the observation y, the MAP detector selects $\hat{s}$ to maximize the posterior probability

$$p_{S|Y}(\hat{s} \mid y) = \frac{f_{Y|S}(y|\hat{s})p_S(\hat{s})}{f_Y(y)} . \quad (7.19)$$

The denominator is independent of $\hat{s}$, so the MAP detector equivalently selects $\hat{s}$ to maximize the numerator, $f_{Y|S}(y|\hat{s})p_S(\hat{s})$. If the prior probabilities $p_S(\hat{s})$ are equal (the signals are equally likely), then the MAP detector reduces to the ML detector, regardless of the channel parameters.

Example 7-16.

For the additive Gaussian noise channel of Example 7-13, the MAP detector maximizes

$$f_N(y - \hat{s})p_S(\hat{s}) = \left(\frac{1}{(2\pi\sigma^2)^N} e^{-\|y - \hat{s}\|^2/2\sigma^2}\right) p_S(\hat{s}). \tag{7.20}$$

Taking the natural logarithm, a monotonic function, we see that this is equivalent to *minimizing*

$$\| y - \hat{s} \|^2 - 2\sigma^2 \log p_S(\hat{s}). \tag{7.21}$$

The sample space for **Y** can again be divided into decision regions, but the boundaries of the regions are not determined strictly on the basis of Euclidean distance if the prior probabilities are not equal, and unfortunately depend on the noise variance σ^2 and the prior probabilities (see Problem 7-3 and Problem 7-4).

A similar result occurs in the case of the BSC of Example 7-14; namely, the detector does not merely minimize the Hamming distance but rather takes into account the error probability p and the prior probabilities if they are not equal (see Problem 7-2).

In practice, it is often the case that the prior probabilities or the noise variance are not known with sufficient accuracy to implement a MAP detector. When the MAP detector is used, the decision boundaries can be set for the worst case noise variance σ^2, thereby giving the optimal performance in the worst case. In more critical applications the noise variance can be estimated.

7.2.3. Probability of Error for BSC ML Detector

The probability of error was considered in Section 6.2 for the Gaussian noise case and minimum-distance receiver design. But we just showed that the ML detector in AWGN reduces to the minimum-distance detector. Thus, the error probability determined in Section 6.2, namely $P_e \approx KQ(d_{min}/2\sigma)$, applies directly to the ML detector in AWGN. In this section we extend this result to the BSC channel. Basically all that differs is the definition of distance and the definition of the $Q(\cdot)$ function.

Two-Signal Case

Consider two binary vectors s_i and s_j that differ in d components (the Hamming distance is d). Suppose s_i is transmitted through a BSC noise generator. The conditional probability of the received bits, given the transmitted bits, has the form given by (7.17). From Example 7-14, the ML detector will choose s_j instead of s_i if the received bits y are closer in Hamming distance to s_j than to s_i. Any error that occurs in a component for which the bits in s_i and s_j are the same will not affect ML detection, since it will impact the Hamming distances to both s_j and s_i equally. Pessimistically assume that a detection error is always made if the received bits y are equidistant from s_i and s_j. A detection error then occurs if more than t errors occur in the d bits that differ, where

$$t = \left\lfloor \frac{d-1}{2} \right\rfloor = \begin{cases} (d-1)/2; & \text{if } d \text{ is odd} \\ (d/2)-1; & \text{if } d \text{ is even} \end{cases}. \tag{7.22}$$

The number of errors in d components is binomial, so the probability that the ML detector prefers $\mathbf{s}_j$ over $\mathbf{s}_i$ (i.e., the pairwise error probability $P_{i \to j}$) is:

$$Q(d, p) = \sum_{i=t+1}^{d} \binom{d}{i} p^i (1-p)^{d-i} . \tag{7.23}$$

We choose the notation $Q(d, p)$ for this sum in order to emphasize the similarity to the $Q(\cdot)$ function used in the Gaussian case. By convention, whenever a function Q has two arguments, we mean (7.23), and whenever it has one argument, we mean (3.38).

Example 7-17.

Suppose that $\mathbf{s}_i = [0\ 0\ 0\ 0\ 0\ 0]$ and $\mathbf{s}_j = [1\ 1\ 0\ 1\ 1\ 1]$. The Hamming distance is $d = 5$. If $\mathbf{s}_i$ is transmitted over a BSC with error probability p, the ML detector will prefer $\mathbf{s}_j$ if three or more of the received bits in the five differing positions are changed by the channel. The probability of this occurring is

$$Q(5, p) = \binom{5}{3} p^3 (1-p)^2 + \binom{5}{4} p^4 (1-p) + \binom{5}{5} p^5$$
$$= 10p^3(1-p)^2 + 5p^4(1-p) + p^5 . \tag{7.24}$$

Three or More Signals

In Section 6.2, we derived an upper bound (the union bound) and a lower bound on the error probability for M-ary modulation in AWGN. That same bound applies to the ML detector for the BSC with a redefinition of distance and the definition of the $Q(\cdot)$ function. Just as these bounds are tight for small σ in the Gaussian case, they become tight for small p in the BSC case.

$Q(d, p)$ is monotonic in d for the BSC, just as $Q(d/2\sigma)$ is monotonic in d for the Gaussian channel. Consequently, exactly the same bounds apply, where $d_{\min}$ is Euclidean distance for the Gaussian channel, and $d_{\min}$ is Hamming distance for the BSC. The conclusion is that for small p, the error probability is

$$P_e \approx K \cdot Q(d_{\min}, p) \tag{7.25}$$

where $d_{\min}$ is the minimum Hamming distance among all pairs of transmitted signal vectors, and where K is the average number of neighbors at the minimum distance.

7.3. Known Signals in Gaussian Noise

In this section we consider the problem of designing the ML detector for one of M signals in additive Gaussian noise, first for the discrete-time case and then for the continuous-time case. The discrete-time case for white noise follows in straightforward fashion from the results of Section 7.2. Our main concern will be with extending this result to nonwhite Gaussian noise, and subsequently to continuous time. For generality, we will treat the case of complex-valued noise and received signals. Subsequently we will apply the results directly to the baseband and passband signals of specific interest.

7.3.1. Discrete-Time Received Signal

We first consider a discrete-time received signal of the form

$$Y_k = s_k^{(m)} + N_k \ , \quad 0 \le k < \infty \ , \tag{7.26}$$

where $\{s_k^{(1)}, \dots s_k^{(M)}\}$ is a set of M known signals, $s_k^{(m)}$ is the m-th of those signals, and N_k is additive zero-mean Gaussian noise. All quantities in (7.26) are assumed to be complex-valued.

White Noise Case

Assume the noise is white and Gaussian with variance $2\sigma^2$ and circularly symmetric and hence is also stationary. This is similar to the vector signal case considered in Section 7.2, except that the number of components in the vector is countably infinite. We can handle this by using the previous structure for the N-dimensional vector and allowing $N \to \infty$. The ML detector then calculates the Euclidean distance between the received signal and each known signal. Hence, it calculates

$$J_m = \sum_{k=0}^{\infty} | Y_k - s_k^{(m)} |^2$$
$$= \sum_{k=0}^{\infty} | Y_k |^2 + \sum_{k=0}^{\infty} | s_k^{(m)} |^2 - 2\text{Re}\{ \sum_{k=0}^{\infty} Y_k s_k^{(m)*} \} \ , \tag{7.27}$$

for $1 \le m \le M$, and then chooses the m for which J_m is *minimum*. The first term is not a function of m, so it can be ignored. Thus, the ML criterion is equivalent to *maximizing*

$$R_m = \text{Re}\{ \sum_{k=0}^{\infty} Y_k s_k^{(m)*} \} - \tfrac{1}{2} E_m \ , \qquad E_m = \sum_{k=0}^{\infty} | s_k^{(m)} |^2 \ , \tag{7.28}$$

where E_m is the energy of the m-th signal. This detector correlates the received signal with each of the known signals $\{s_k^{(1)}, \dots s_k^{(M)}\}$, and then takes the real part of the result. This is repeated for each m, and after subtracting half the energy, the decision is the m for which difference is maximum. This interpretation of the receiver is shown in Fig. 7-5(a). This is a discrete-time version of the correlation receiver structure that was seen earlier in Chapter 6 (Fig. 6-1), except that here we are operating on discrete-time signals.

An equivalent way to generate R_m is to apply the received signal to a filter with impulse response $s_{-k}^{(m)*}$, which has transfer function $S_m^*(1/z^*)$, and sample the output at $k = 0$. This interpretation of the receiver is shown in Fig. 7-5(b). The filter $S_m^*(1/z^*)$ is a discrete-time version of the matched filter first encountered in Chapter 5. As in the continuous-time case, the discrete-time matched filter is anticausal. If it happens to be FIR, then it can be realized as a causal filter plus a delay. If it is IIR, then it can only be approximated in practice.

Nonwhite Noise Case

We now consider wide-sense stationary circularly symmetric Gaussian noise with power spectrum $S_N(z)$ that is not necessarily white. The receiver of Fig. 7-5 is the ML detector for white noise only, so it cannot be directly applied to the nonwhite noise. A useful trick is to first apply the received signal to a *noise-whitening filter.*

The power spectrum $S_N(z)$ must be non-negative real on the unit circle. In Section 2.5.5 we derived the spectral factorization of a rational $S_N(z)$ that is real-valued and non-negative on the unit circle,

$$S_N(z) = \gamma^2 M(z) M^*(1/z^*) \tag{7.29}$$

where $M(z)$ is a monic and loosely minimum-phase transfer function. This factorization generalizes to non-rational $S_N(z)$ as well. Assume that $S_N(z)$ has no zeros on the unit circle. In that case, $1/M(z)$ is causal, stable, and strictly minimum-phase. As illustrated in Fig. 7-6, the received signal can be passed through the strictly minimum-phase whitening filter $1/\gamma M(z)$, which produces noise at the output that is unit-variance white and Gaussian. The whitening filter also affects the signal, and the Z-transform of the new signal is $S_m(z)/\gamma M(z)$, where $S_m(z)$ is the Z-transform of $\{s_k^{(m)},\ 0 \le k < \infty\}$. The matched-filter receiver of Fig. 7-5

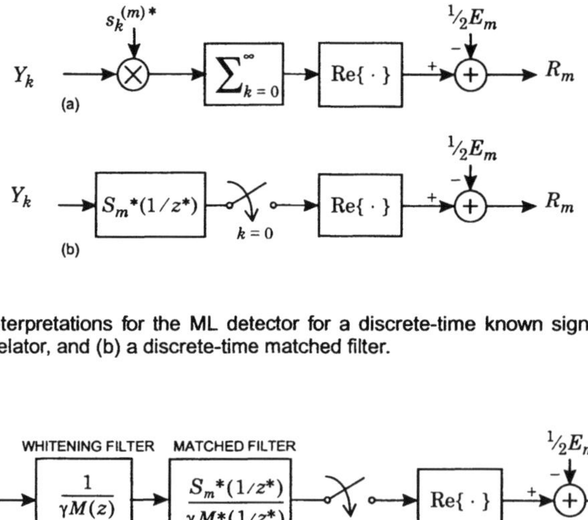

Fig. 7-5. Two interpretations for the ML detector for a discrete-time known signal in white Gaussian noise. (a) A correlator, and (b) a discrete-time matched filter.

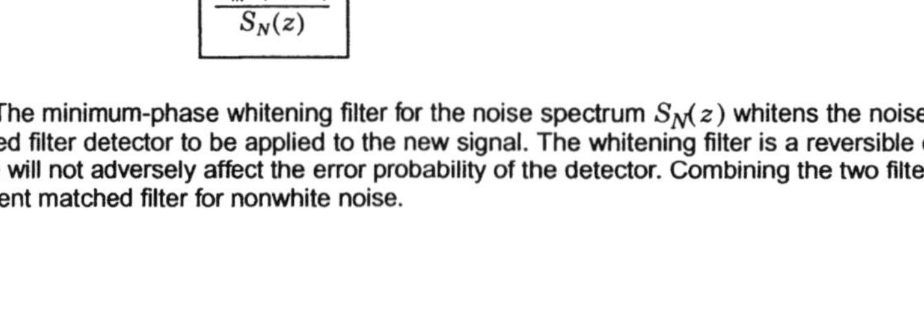

Fig. 7-6. The minimum-phase whitening filter for the noise spectrum $S_N(z)$ whitens the noise, allowing the matched filter detector to be applied to the new signal. The whitening filter is a reversible operation, and hence will not adversely affect the error probability of the detector. Combining the two filters, we get an equivalent matched filter for nonwhite noise.

can be applied to this whitened received signal, taking into account the new signal spectrum. Since linear filtering preserves the circular symmetry of Gaussian noise, the white noise is circularly symmetric, and thus the samples of the noise are mutually independent. As also shown in Fig. 7-6, the whitening and matched filters can be combined into a single filter, equivalent to that in Fig. 7-5 except that the transfer function is normalized by the noise spectrum $S_N(z)$. This normalization is appropriate for a matched filter in nonwhite noise, and the result is an ML detector.

7.3.2. Continuous-time Reception

Suppose we have a continuous-time received signal,

$$Y(t) = s_m(t) + N(t), \quad 0 \le t < T, \tag{7.30}$$

where $\{s_1(t), \ldots, s_M(t)\}$ is a set of known signals over the interval $[0, T)$. The noise $N(t)$ is assumed to be a zero-mean circularly symmetric stationary Gaussian process with power spectrum $S_N(f)$ and autocorrelation function $R_N(\tau)$. For reasons of mathematical tractability, we address a finite (but arbitrarily large) time interval $0 \le t < T$.

To develop the ML detector as we have in the discrete-time case, some new techniques have to be introduced. Our approach is to turn this into a problem equivalent to the discrete-time case by using a signal-space expansion of the random process $N(t)$, $0 \le t < T$, in terms of a countable set of functions,

$$N(t) = \sum_{i=1}^{\infty} N_i \phi_i(t), \qquad 0 \le t < T \tag{7.31}$$

where the functions are orthonormal in signal space,

$$\int_0^T \phi_i(t)\phi_j^*(t)\,dt = \delta_{i-j} \,. \tag{7.32}$$

Equality in (7.31) is in the sense of mean-square convergence of the series on the right to the process on the left. This expansion is most useful if the coefficients are themselves uncorrelated,

$$\mathrm{E}[N_i N_j^*] = \sigma_i^2 \delta_{i-j} \,. \tag{7.33}$$

Under quite general conditions, a set of orthonormal functions $\{\phi_i(t)\}$ can be found such that (7.33) is satisfied (even for the case where $N(t)$ is not wide-sense stationary as we assume here). The resulting expansion is known as the *Karhunen-Loeve expansion*.

First, taking the inner product of both sides of (7.31) with $\phi_j(t)$,

$$N_j = \int_0^T N(t)\phi_j^*(t)\,dt \,. \tag{7.34}$$

Since N_j is a linear function of a Gaussian process, it is a Gaussian random variable, and further it is circularly symmetric since $N(t)$ is assumed circularly symmetric. This circular symmetry together with (7.33) implies that the N_j are statistically independent. In Appendix 7-A, it is shown that a necessary and sufficient condition on $\{\phi_j(t)\}$ for (7.33) to be satisfied is

$$\int_0^T R_N(t-\tau)\phi_j(\tau)\,d\tau = \sigma_j^2\phi_j(t)\,, \qquad 1 \le j < \infty, \quad 0 \le t < T. \tag{7.35}$$

Although the left side of (7.35) looks like a convolution, the equality is valid only for the finite time interval $0 \le t < T$, so it is in fact not a convolution. This is an *integral equation*, and, in analogy to similar matrix equations, $\phi_j(t)$ is called an *eigenfunction* of $R_N(t)$ with corresponding *eigenvalue* σ_j^2.

The question arises as to whether there exists a set of complete orthonormal eigenfunctions and corresponding non-zero eigenvalues $\{\phi_j(t), \sigma_j, 1 \le j < \infty\}$ satisfying (7.35). This is considered in some detail by Van Trees [1], where it is confirmed that they do exist under rather general conditions. For our purposes, it suffices that the power spectrum $S_N(f)$ be non-zero for all f. Fortunately, in the following we don't actually have to find the eigenfunctions satisfying (7.35); it suffices to know that they exist.

Now returning to the original detection problem of (7.30), the approach is to expand the received signal in the same set of orthonormal functions that arise out of the Karhunen-Loeve expansion of $N(t)$,

$$Y(t) = \sum_{i=1}^{\infty} Y_i\phi_i(t)\,, \qquad Y_i = s_i^{(m)} + N_i\,, \qquad 0 \le t < T\,. \tag{7.36}$$

The coefficients N_i are uncorrelated, circularly symmetric (and hence independent) Gaussian random variables. Their variances are not necessarily equal, $E[\,|N_i|^2] = \sigma_i^2$, and the $s_i^{(m)}$ are the coefficients of $s_m(t)$ with respect to the orthonormal basis functions $\phi_i(t)$, namely

$$s_i^{(m)} = \int_0^T s_m(t)\phi_i^*(t)\,dt\,. \tag{7.37}$$

The Karhunen-Loeve expansion turns the continuous-time received signal into an equivalent discrete-time received signal, at least in a mathematical if not literal sense (since the i in Y_i is not time, but an index over the signal-space basis). The continuous-time received signal $Y(t)$ is represented on the finite time interval $0 \le t < T$ by the countable set of random variables $\{Y_i, 1 \le i < \infty\}$. We can apply the earlier discrete-time results to this equivalent representation, with the slight complication that the noise samples do not all have equal variance. Assuming that the eigenvalues are all non-zero, this problem is easily circumvented by normalizing the samples by dividing both sides by the known standard deviation,

$$\frac{Y_i}{\sigma_i} = \frac{s_i^{(m)}}{\sigma_i} + \frac{N_i}{\sigma_i}, \qquad 1 \le i < \infty\,. \tag{7.38}$$

The N_i/σ_i are all unit-variance Gaussian random variables. The normalization of (7.38) can be considered a form of whitening, similar to the whitening filter applied in Fig. 7-6.

The normalized representation of (7.38) satisfies all the conditions assumed for the discrete-time case, namely a set of known discrete-time signals with additive white noise variables. We can therefore apply the earlier detector to the normalized received signal Y_i/σ_i, where the known signal component is $s_i^{(m)}/\sigma_i$. Thus, the ML detector minimizes

$$J_m = \sum_{i=1}^{\infty} \left| \frac{Y_i}{\sigma_i} - \frac{s_i^{(m)}}{\sigma_i} \right|^2 = \sum_{i=1}^{\infty} \frac{|Y_i - s_i^{(m)}|^2}{\sigma_i^2} \tag{7.39}$$

over all possible signals $1 \le m \le M$. As in the discrete-time case, the first term $\sum_{i=1}^{\infty} |Y_i|^2$ will be independent of m, so the ML detector equivalently maximizes the decision variable

$$R_m = \mathrm{Re}\left\{ \sum_{i=1}^{\infty} \frac{Y_i s_i^{(m)*}}{\sigma_i^2} \right\} - \tfrac{1}{2} E_m , \qquad E_m = \sum_{i=1}^{\infty} \frac{|s_i^{(m)}|^2}{\sigma_i^2} . \tag{7.40}$$

In Appendix 7-A, this result is related to the original continuous-time signals. In particular, defining a function $g_m(t)$ that satisfies the integral equation

$$\int_0^T R_N(t-\tau) g_m(\tau) d\tau = s_m(t) , \qquad 1 \le m \le M , \quad 0 \le t < T , \tag{7.41}$$

then (7.40) can be written as

$$R_m = \mathrm{Re}\left\{ \int_0^T Y(t) g_m^*(t) dt \right\} - \tfrac{1}{2} E_m , \qquad E_m = \int_0^T s_m(t) g_m^*(t) dt . \tag{7.42}$$

Example 7-18.
If the additive noise is white, so that its autocorrelation is $R_N(\tau) = N_0 \delta(\tau)$, then $N_0 g_m(t) = s_m(t)$. In that case, the receiver simply correlates with each of the known signals $s_m(t)$, $1 \le m \le M$.

The significance of (7.42) cannot be overstated. It shows that the infinite-dimensional continuous-time received signal can be reduced to a finite set of M decision variables R_m, $1 \le m \le M$, where M is the number of known signals. R_m consists of a correlation against $g_m(t)$, as shown in Fig. 7-7(a). As in the discrete-time case, the correlation detector is equivalent to the continuous-time matched filter detector of Fig. 7-7(b). For the special case of white noise, Example 7-18 establishes that the matched filters in Fig. 7-7 are matched to the set of signal waveforms $\{s_1(t), \ldots s_M(t)\}$.

If we let $T \rightarrow \infty$, we get a different interpretation and a better understanding of $g_m(t)$. In that case, assuming that $g_m(t)$ is causal, (7.41) approaches a convolution equation $R_N(t) * g_m(t) = s_m(t)$, or $G_m(f) = S_m(f)/S_N(f)$. This limiting case has the interpretation shown in Fig. 7-7(c). In the white noise case, the matched filter has impulse response $s_m^*(t)$ and transfer function $S_m^*(f)$.

In the nonwhite noise case, a different interpretation takes advantage of the fact that $S_N(f)$ is positive real-valued. It can be factored as the product of two identical filters,

$$S_N(f) = S_N^{1/2}(f) \cdot S_N^{1/2}(f) . \tag{7.43}$$

Thus, in Fig. 7-7(c) the matched filter has been divided into two parts: a whitening filter $1/S_N^{1/2}(f)$ that has white noise at the output, and a filter matched to the signal at the whitening filter output. The signal component at the output of the whitening filter is $S_m(f)/S_N^{1/2}(f)$, and the second filter is matched to this new signal. The factorization into whitening and matched filtering is similar to the discrete-time case (Fig. 7-6). E_m is the energy of the signal $s_m(t)$ *after* it passes through the whitening filter, making it consistent with the white-noise case.

7.3.3. Sufficient Statistics

In (7.42), define the M complex-valued decision variables

$$V_m = \int_0^T Y(t) g_m^*(t)\, dt , \quad 1 \le m \le M , \tag{7.44}$$

as labeled in Fig. 7-7. The ML detector first calculates these M decision variables, corresponding to the M signals, and then chooses m to maximize $R_m = \text{Re}\{V_m\} - E_m/2$. The ML detector is thus summarizing the continuous-time received signal $\{Y(t), 0 \le t < T\}$ by the M random variables $\{V_1, \ldots, V_M\}$. In the process, it is clearly throwing away a lot of information about $\{Y(t), 0 \le t < T\}$. The information that is being thrown away is considered *irrelevant* by the ML detector.

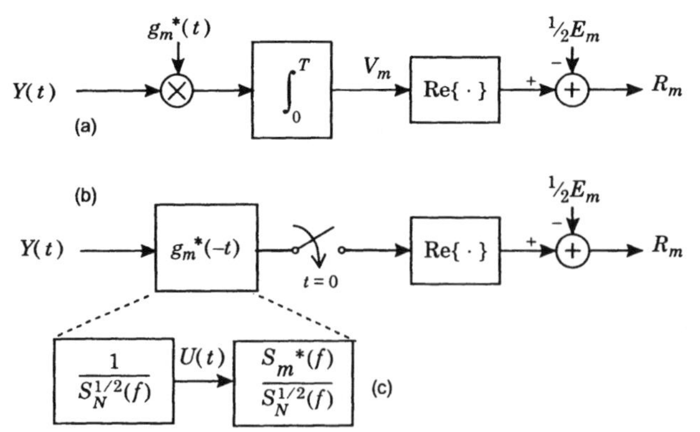

Fig. 7-7. The ML detector for a continuous-time known signal in additive Gaussian noise. (a) The correlation receiver. (b) The matched filter receiver. (c) The matched filter in the limit as $T \to \infty$.

The overall goal of this subsection is summarized in Fig. 7-8. Starting with the received signal $\{Y(t), 0 \le t < T\}$, the Karhunen-Loeve expansion coefficients $\{Y_i\}$ give an equivalent representation. This countable set of random variables is much easier to deal with analytically, but remains impractical for implementation because of the infinite number of variables. However, the ML detector further reduces the received signal to M decision variables $\{V_1, \ldots, V_M\}$, where M is the number of known signals. For purposes of implementation, this finite set of decision variables is a dramatic improvement. Conceptually, however, if the dimensionality of the subspace spanned by the signals is less than M, then based on the experience of Chapter 6 we would expect that a number of decision variables equal to the dimension of the subspace would suffice. In fact, it will now be shown that the received signal can be represented by a set of N *sufficient statistics* $\{U_1, \ldots, U_N\}$, where $N \le M$, and N will be defined shortly. The sufficient statistics summarize $\{Y(t), 0 \le t < T\}$ for purposes of detection of $\{s_1(t), \ldots s_M(t)\}$.

The N sufficient statistics can be used for purposes of ML detection. This reduces the number of decision variables that must be dealt with in the implementation of the ML detector. Remarkably, the sufficient statistics can be relied upon for the detection of the known signals $\{s_1(t), \ldots, s_M(t)\}$ for *any* criterion of optimality, not just the ML criterion. For example, they would serve equally well as the starting point for MAP detection.

The intuitive basis for sufficient statistics is that they retain all the information in the received signal that is relevant to the detection of the known signals $\{s_1(t), \ldots, s_M(t)\}$ and discard only information that is irrelevant. As we will also show below, the sufficient statistics $\{U_1, \ldots, U_N\}$ can be obtained from the ML decision variables $\{V_1, \ldots, V_M\}$ as pictured in Fig. 7-8, and therefore all the information in the sufficient statistics must also be included in the ML decision variables. Thus, the ML decision variables are themselves sufficient statistics, and could be used by any detection criterion (not just the ML criterion) without compromising performance.

Our starting point is to define a new set of signals $\{f_1(t), \ldots, f_M(t)\}$ that are related to the original set $\{s_1(t), \ldots, s_M(t)\}$ by the Fourier transform

$$F_m(f) = \frac{S_m(f)}{S_N^{1/2}(f)} , \tag{7.45}$$

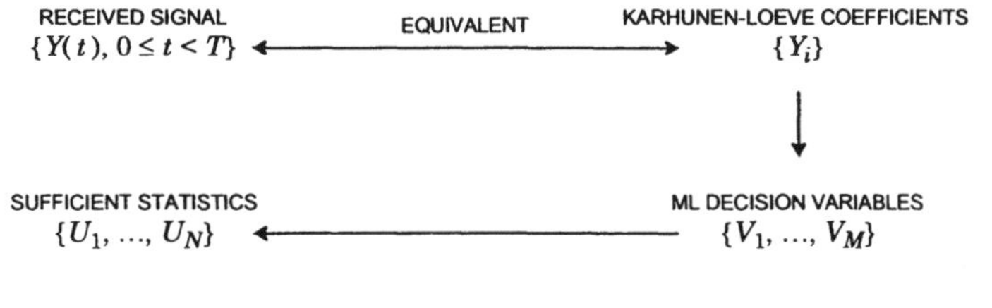

Fig. 7-8. The progression from the received signal $Y(t)$ through a progression of decision variables, all of which retain all relevant information for detection of the known signals.

where $S_m(f)$ is the Fourier transform of the pulse $s_m(t)$. The new signals are easy to understand in terms of Fig. 7-7(c): $f_m(t)$ is the response of the whitening filter to $s_m(t)$. In other words, whereas $\{s_1(t), \ldots, s_M(t)\}$ is signal set for the original nonwhite-noise channel, $\{f_1(t), \ldots, f_M(t)\}$ is the signal set for the equivalent white-noise channel model after the whitening filter. The second half of the matched filter in Fig. 7-7(c) is matched to $f_m(t)$.

Assume the $\{f_1(t), \ldots, f_M(t) : 0 \le t < \infty\}$ span a subspace S_f of signal space of dimension $N \le M$, and let $\{\psi_k(t), 1 \le k < \infty\}$ be a complete set of orthonormal functions chosen so that the first N, $\{\psi_1(t), \ldots \psi_N(t)\}$, serve as a basis for S_f. Then we can write

$$f_m(t) = \sum_{k=1}^{N} F_k^{(m)} \psi_k(t) \,, \qquad F_k^{(m)} = \int_0^\infty f_m(t) \psi_k^*(t) dt \,. \tag{7.46}$$

Let $U(t)$ denote the output of the whitening filter in Fig. 7-7(c). In other words, whereas $Y(t)$ is the original observation for the nonwhite channel model, $U(t)$ is the observation for the equivalent white-noise model. In the following, we will represent $U(t)$ in terms of the basis $\{\psi_k(t), 1 \le k < \infty\}$, and show that only the first N coordinates are relevant to detecting $\{s_1(t), \ldots, s_M(t)\}$.

The components of $U(t)$ with respect to the new basis are

$$U_k = \int_0^\infty U(t) \psi_k^*(t) dt \,, \qquad 1 \le k \le \infty \,. \tag{7.47}$$

Substituting for $U(t)$ from

$$U(t) = f_m(t) + W(t) \tag{7.48}$$

where $W(t)$ is white noise with unit variance,

$$U_k = \int_0^\infty f_l(t) \psi_k^*(t) dt + \int_0^\infty W(t) \psi_k^*(t) dt$$

$$= \begin{cases} F_k^{(m)} + W_k, & 1 \le k \le N \\ W_k, & k > N \end{cases} . \tag{7.49}$$

The noise components W_k are mutually independent because of the white noise component in $U(t)$ and the orthonormality of the $\psi_k(t)$. Only the first N components of U_k depend on the signal.

The $\{U_1, \ldots, U_N\}$ of (7.47) are sufficient statistics for the received signal $Y(t)$. This means they contain all the relevant information in $\{Y(t), 0 \le t < T\}$ for purposes of detection of known signals from the set $\{s_1(t), \ldots, s_M(t)\}$ with respect to *any* criterion of optimality, not just the ML criterion. In fact, the MAP detector, or any other detector can start by calculating these sufficient statistics. The justification for this sufficient statistic property is as follows:

- Only the first N components of U_k have a signal component.
- The remaining components are statistically independent of the first N components. Thus, they do not contain any information about the first N components.

- Thus $\{U_k, N+1 \le k < \infty\}$ is irrelevant to the detection of the signal, regardless of what criterion of optimality is used.

The procedure used to arrive at this smaller set of decision variables is identical to the expansion in Section 6.1.2, except there we expanded $\{s_1(t), \ldots, s_M(t)\}$ in terms of an N-dimensional basis rather than the whitening-filter-adjusted signal set $\{f_1(t), \ldots f_M(t)\}$. (Note that due to the whitening filter, the signal sets $\{s_1(t), \ldots s_M(t)\}$ and $\{f_1(t), \ldots f_M(t)\}$ may actually have a different dimensionality.) Thus, (7.46) and (6.16) are essentially the same, except for the intervening whitening filter.

Example 7-19.

As an example of the utility of the sufficient statistic argument, consider the reception of a single PAM pulse, where $s_m(t) = A_m \cdot h(t)$ and the data symbol A_m assumes one of M values $1 \le m \le M$. For this case the received signal is one-dimensional, since all signals are linear multiples of a single waveform $h(t)$. Thus, if the additive noise is white and Gaussian, any detector can first form a sufficient statistic for the received signal

$$V_1 = \int_0^\infty Y(t)h^*(t)\,dt\ , \tag{7.50}$$

or if the noise is nonwhite, it may apply the received signal to a matched filter with transfer function $H^*(f)/S_N(f)$ and sample at time $t = 0$. The resulting *single* random variable summarizes the received signal for purposes of detection with respect to *any* criterion.

It can be shown (Problem 7-23) that $\{U_1, \ldots, U_N\}$ can be obtained from the larger set of decision variables $\{V_1, \ldots, V_M\}$ by a simple linear transformation. This dependence is shown in Fig. 7-8. It follows that $\{V_1, \ldots, V_M\}$ must also be a set of sufficient statistics, since $\{U_1, \ldots, U_N\}$ could not contain any relevant information about $\{Y(t), 0 \le t < T\}$ not present in $\{V_1, \ldots, V_M\}$. It is usually advantageous to use $\{U_1, \ldots, U_N\}$ rather than $\{V_1, \ldots, V_M\}$ since it has fewer decision variables, and furthermore the Gaussian noise components in $\{U_1, \ldots, U_N\}$ are independent as established in (7.47). This latter property makes it easier to develop the remainder of the receiver structure based on the actual optimality criterion.

Based on the sufficient statistic results, and given a criterion of optimality, an optimal receiver structure can be developed as follows:

- The receiver front end calculates the N (or M) sufficient statistics. One of these finite sets of decision variables replaces the received signal for purposes of the design of the remainder of the receiver.
- The statistics of the sufficient statistics are established for the particular noise and the set of known signals.
- The remainder of the receiver structure is determined by the criterion of optimality as applied to the sufficient statistics and their statistical properties.

In summary we conclude that, generally speaking, the minimum-distance receiver structure of the previous chapter is optimal with respect to the maximum-likelihood criterion, provided that the noise is stationary and Gaussian, and provided that a noise-whitening filter is used.

7.3.4. Optimal Detectors for PAM with ISI

In Chapter 5 a receiver for PAM with ISI was derived using a minimum-distance criterion. The receiver front end uses a whitened-matched filter (WMF). In geometric terms, this filter projects the received signal onto an orthonormal basis for the signal space. The receiver then minimizes distance in discrete time. In this section, we extend these results in several ways:

- We show that the WMF output is a set of sufficient statistics for the received signal when the noise is Gaussian. Thus, a receiver designed with respect to *any* criterion of optimality can share this same WMF front end.
- We show that the minimum-distance receiver design of Chapter 5 is, as expected, the ML detector for a set of M^L known signals consisting of a sequence of L data symbols. As a result, we call this receiver the *ML sequence detector (MLSD)*.
- We extend these results to nonwhite noise, and in particular show that the WMF can be reformulated for this case.

WMF Outputs as Sufficient Statistics

As in Chapter 5, assume that the received signal consists of a finite sequence of L data symbols, each with the same alphabet with size M, modulating a basic complex baseband pulse $h(t)$,

$$\hat{Y}(t) = \sqrt{2}\,\mathrm{Re}\left\{\sum_{k=0}^{L-1} a_k h(t-kT) e^{j2\pi f_c t}\right\} + N(t). \tag{7.51}$$

For the moment assume that the noise is white and Gaussian with PSD $S_N(f) = N_0/2$. Then we can consider (7.51) as consisting of a signal portion drawn from a set of $M = |\mathcal{A}|^L$ known signals, together with additive white Gaussian noise. In Section 7.3, we established that a set of sufficient statistics can be generated by downconverting to yield the complex envelope $Y(t)$ of the received signal, and then correlating with each possible complex baseband signal,

$$\int_{-\infty}^{\infty} Y(t) \sum_{k=0}^{L-1} a_k^* h^*(t-kT)dt = \sum_{k=0}^{L-1} a_k^* U_k \tag{7.52}$$

where

$$U_k = \int_{-\infty}^{\infty} Y(t)h^*(t-kT)\,dt, \qquad 0 \le k \le L-1. \tag{7.53}$$

This correlation has to be repeated for all $M = |\mathcal{A}|^L$ sequences of data symbols $\{a_0, \ldots, a_{L-1}\}$.

In practice, calculating $M = |\mathcal{A}|^L$ correlations is not feasible as L gets large, but fortunately, (7.52) can be generated from the L decision variables $\{U_0, \ldots, U_{L-1}\}$ in (7.53). These L variables summarize the received signal from the perspective of calculating (7.52). The $\{U_0, \ldots, U_{L-1}\}$ are themselves sufficient statistics. This reduces the number of sufficient statistics from $|\mathcal{A}|^L$ down to just L. These L sufficient statistics are the outputs of L correlators against $h^*(t-kT)$, $0 \le k \le L-1$. As shown in Fig. 5-19, these L correlators can be replaced by a *single* matched filter, matched to $h(t)$, followed by a sampler at $t = kT$ for

$0 \le k \le L-1$. Thus, we conclude that a receiver structure consisting of a downconverter followed by a filter matched to the complex baseband pulse $h(t)$ followed by a symbol-rate sampler generates a set of sufficient statistics for the received signal detection.

This result is easily generalized to nonwhite noise using the results of Section 7.3.2, specifically the asymptotic results as $T \to \infty$. In this case, the output of the downconverter is first whitened by filter $1/S_{\tilde{N}}^{1/2}(f)$, and the set of known signals is replaced by the set of known signals as modified by this whitening filter. Equivalently, the matched filter can be replaced by a filter matched to $h(t)$ normalized by $S_{\tilde{N}}(f)$. The resulting front end that generates the sufficient statistics $\{U_0, \ldots, U_{L-1}\}$ is shown in Fig. 7-9(a). The noise samples at the sampler output are not white, but rather have power spectrum

$$S_n(e^{j2\pi fT}) = \frac{1}{T}\sum_{k=-\infty}^{\infty} \frac{|H(f-m/T)|^2}{S_{\tilde{N}}(f-m/T)} . \tag{7.54}$$

This is similar to the folded $S_h(z)$ of Chapter 5, (5.58), except that $|H(f)|^2$ is normalized by $S_{\tilde{N}}(f)$. As in Chapter 5, we can invoke a minimum-phase spectral factorization to write

$$S_n(z) = \gamma_n^2 M_n(z) M_n^*(1/z^*) \tag{7.55}$$

where γ_n^2 is a positive constant and $M_n(z)$ is a monic minimum-phase transfer function.

In Fig. 7-9(a) a maximum-phase whitening filter is added to the output of the sampled matched filter, yielding an output noise process N_k that is white, Gaussian, and circularly symmetric, with variance $1/\gamma_n^2$. The resulting discrete-time channel model from input symbols to WMF outputs is shown in Fig. 7-9(b).

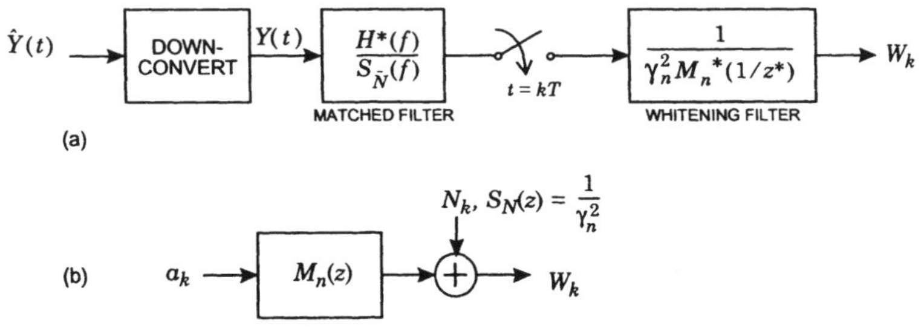

Fig. 7-9. A whitened-matched filter for PAM and nonwhite noise. (a) Complete front end of the receiver, and (b) equivalent discrete-time model.

Example 7-20.

The front end of Fig. 7-9 reduces to the WMF derived in Section 5.4.3 when the channel noise is white, $S_N(f) = N_0/2$. For this case, $S_n(e^{j2\pi fT}) = S_h(e^{j2\pi fT})/N_0$, where $S_h(z)$ is the folded spectrum, and thus:

$$S_n(z) = \frac{S_h(z)}{N_0} = \frac{\gamma^2}{N_0} M(z)M^*(1/z^*) \ . \tag{7.56}$$

The N_k thus have variance $1/\gamma_n^2 = N_0/\gamma^2$.

A receiver for detection of the data symbols can be safely based on the WMF front end of Fig. 7-9(a) regardless of what criterion of optimality is applied. This result is quite remarkable when we consider that symbol-rate sampling at the matched filter output is generally at less than the Nyquist rate, so aliasing of both noise and signal is inherent in this sampling. This aliasing will not compromise the performance of the receiver as long as the filter before the sampling is a matched filter.

Maximum-Likelihood Sequence Detector

The sufficient statistic argument allows us to use the front end of Fig. 7-9(a) for any detection criterion, so we will now apply it to the ML detector. With the WMF front end, the equivalent discrete-time model of Fig. 7-9(b) can be used as a starting point for application of the ML criterion. In particular, the ML detector for this equivalent discrete-time model was developed in Section 7.3. It chooses the sequence of data symbols that minimizes the Euclidean distance

$$\min_{\{a_0, \ldots, a_{L-1}\}} \sum_{k=0}^{\infty} | W_k - \textstyle\sum_{l=0}^{L-1} a_l m_{k-l} |^2 \ . \tag{7.57}$$

This is precisely the minimum-distance receiver design of Chapter 5, and thus that receiver design is equivalent to the detector using the criterion of maximizing the likelihood of the received signal conditional on a sequence of data symbols $\{a_0, \ldots, a_{L-1}\}$. Since an entire sequence of data symbols is detected at once, this detector is called the *maximum-likelihood sequence detector (MLSD)*. If all sequences are equally likely, the MLSD minimizes the probability of making one or more errors in a sequence of data symbols. That is, the criterion penalizes the detector equally for making any number of detection errors.

We have now derived the MLSD criterion of (7.57) in two ways. First, in Section 5.4 it was shown that the discrete-time criterion of (7.57) is equivalent to a continuous-time minimum-distance receiver design, in the sense that both criteria will choose the same sequence of data symbols. Second, in this section, using the argument that the WMF forms a sufficient statistic for the received signal detection, and also using the white noise property of the WMF output, we have shown that the criterion of (7.57) is optimal in the ML sense. Combining these two facts, we arrive at the conclusion that minimum-distance receiver design is optimal in the ML sense for PAM with ISI on the white Gaussian noise channel.

7.4. ML Sequence Detection with the Viterbi Algorithm

The Viterbi algorithm was presented in Chapter 5 as an efficient technique for solving the *minimum-distance* sequence detection problem. We now show how it can be generalized to solve the *maximum-likelihood* sequence detection problem for *any* Markov signal generator and *any* noise generator with independent noise components.

Let $\Psi = [\Psi_0, \dots \Psi_{L+\mu}] \in \{0, \dots Q-1\}^{L+\mu+1}$ denote the sequence of states of a finite-state machine from time $k = 0$ to $L + \mu$, and let the vector ψ denote an outcome of this random vector. Similarly let the vector $\mathbf{Y} = [Y_0, \dots Y_{L+\mu-1}]$ denote the corresponding vector of noisy observations. (There is one fewer observation than states because observations correspond to transitions between states). Then given a particular observation y, the *MAP sequence detector* selects the vector ψ that maximizes the posterior probability $p_{\Psi|\mathbf{Y}}(\psi \mid y)$. Note that the criterion is to maximize the *a posteriori* probability of the *whole sequence* of states, rather than a single state, and hence the term *sequence* detector.

In this section we omit the pdf subscripts when there is no ambiguity, writing $p(\psi \mid y)$ instead of $p_{\Psi|\mathbf{Y}}(\psi \mid y)$, for example. The shorthand will greatly simplify notation. Furthermore, we will use the notation $f(y)$ to denote the probability density function for $\mathbf{Y}$, which implies that $\mathbf{Y}$ is continuous-valued, as in the case of additive Gaussian noise. If it is discrete-valued, as in the BSC, then simply replace $f(y)$ with $p(y)$.

The MAP sequence detector can equivalently maximize the product $p(\psi \mid y)f(y)$ because $f(y)$ is not dependent on our choice ψ. From the mixed form of Bayes' rule (3.31), we can equivalently maximize $f(y \mid \psi)p(\psi)$. We look at the second factor first.

Exercise 7-1.

Use Bayes' rule and the Markov property to show that

$$p(\psi) = p(\psi_0) \prod_{k=0}^{L+\mu-1} p(\psi_{k+1} \mid \psi_k) \,. \tag{7.58}$$

This is intuitive because the a priori probability of a given state trajectory ψ is equal to the product of the probabilities of the corresponding state transitions and the probability of the initial state. Since we assume the initial state is known, $p(\psi_0) = 1$.

We now look at the first factor. Because of the independent noise components assumption,

$$f(y \mid \psi) = \prod_{k=0}^{L+\mu-1} f(y_k \mid \psi) \,. \tag{7.59}$$

Furthermore, since Y_k depends on only two of the states in ψ, we can write

$$f(y \mid \psi) = \prod_{k=0}^{L+\mu-1} f(y_k \mid \psi_k, \psi_{k+1}) \,. \tag{7.60}$$

Putting these results together, the goal of the MAP sequence detector is to find the state sequence ψ, or equivalently the path through the trellis, that maximizes

$$f(y\,|\,\psi)p(\psi) = \prod_{k=0}^{L+\mu-1}\Big[f(y_k\,|\,\psi_k,\,\psi_{k+1})p(\psi_{k+1}\,|\,\psi_k)\Big]\,. \tag{7.61}$$

We can interpret this quantity as a *path metric* equal to the product of branch metrics, where the term in the square brackets represents the branch metric for the transition (ψ_k, ψ_{k+1}). The MAP detector then calculates the path metric for each path through the trellis and finds the path with the largest path metric.

To make things more precise we will adopt the notation of Section 5.4.4. Assume there are Q states labeled 0, … $Q-1$. Assume that the state is known to start at zero at time zero, and end at zero at time $L+\mu$. Let $a^{(p,\,q)}$ and $s^{(p,\,q)}$ denote, respectively, the unique input symbol and unique signal generator output associated with a valid transition from state $p \in \{0, 1, \ldots Q-1\}$ to state $q \in \{0, 1, \ldots Q-1\}$. In other words, following the usual convention of labeling state transitions by the input/output pair, the label for a transition from state p to state q is $(a^{(p,\,q)}, s^{(p,\,q)})$. This notation will be illustrated by example.

Example 7-21.

Consider a sequence of OOK symbols $a_k \in \{0, 1\}$ over an ISI channel with transfer function $H(z) = 1 + 0.5z^{-1}$, for which one stage of the trellis is shown below:

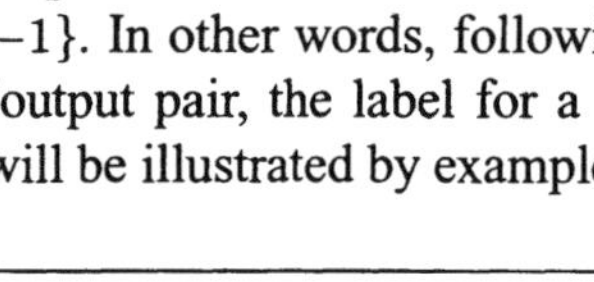

(This same example was considered in Example 5-27.) The input and output labels of the trellis stage shown above indicate that:

$$\begin{array}{llll} a^{(0,0)} = 0\,, & a^{(0,1)} = 1\,, & a^{(1,0)} = 0\,, & a^{(1,1)} = 1\,, \\ s^{(0,0)} = 0\,, & s^{(0,1)} = 1\,, & s^{(1,0)} = 0.5\,, & s^{(1,1)} = 1.5\,. \end{array} \tag{7.62}$$

In terms of this notation, $p(\psi_{k+1}=q\,|\,\psi_k=p)$ reduces to $p_{A_k}(a^{(p,\,q)})$, the a priori probability that the k-th symbol is $a^{(p,\,q)}$, and also $f(y_k\,|\,\psi_k,\,\psi_{k+1})$ reduces to $f_{Y_k|S_k}(y_k\,|\,s^{(p,\,q)})$, where S_k denotes the k-th signal generator output. Thus, (7.61) suggests that the branch metric for a transition from state p to q at time k should be:

$$\gamma_k(p,\,q) = f_{Y_k|S_k}(y_k\,|\,s^{(p,\,q)})p_{A_k}(a^{(p,\,q)})\,. \tag{7.63}$$

If it is impossible to transition from state p to state q, the corresponding branch metric reduces to zero, because the a priori probability of a symbol causing such a transition is evidently zero. Often the branch metric can be significantly simplified. For example, if all symbols are equally likely then the a priori factor $p(a^{(p,\,q)})$ is a constant and can be omitted. Alternatively, if we do not know these a priori probabilities, we can assume a uniform distribution and again omit this term. In either case the MAP sequence detector reduces to the *ML sequence detector.*

The Viterbi algorithm of Section 5.4.4 can be used to efficiently find the path with the largest metric, with only a minor modification to handle the multiplicative branch metric. The modified algorithm will be called the *multiplicative* Viterbi algorithm. (We will see that it shares many similarities with the BCJR algorithm of the next section.) Define the *survivor* for state p at time k as the partial path beginning at time zero and state zero and ending at state p at time k with maximal metric, and call this maximal metric $\alpha_k(p)$. The Viterbi algorithm exploits the fact that survivors at time $k+1$ will build on survivors at time k. In particular, the partial path metrics at time $k+1$ are related to the partial path metrics at time k by:

$$\alpha_{k+1}(q) = \max_p\{\alpha_k(p)\gamma_k(p, q)\} . \tag{7.64}$$

Since the state is known to be zero at time zero, the α metrics are initialized according to $[\alpha_0(0), \ldots \alpha_0(Q-1)] = [1, 0, \ldots 0]$, or more succinctly, $\alpha_0(p) = \delta_p$. The multiplicative Viterbi algorithm uses this recursion to recursively determine the survivors for $k \in \{1, 2, \ldots L+\mu-1\}$. The last survivor (for state zero at time $L+\mu$) is then the decision path.

In practice the complexity of this recursion is reduced by tracking the negative of the logarithm of the survivor metric, a monotonic function. Taking the negative logarithm of (7.64) allows us to replace the branch metrics by their negative logarithm, replace the multiplication by addition, and replace maximization by minimization. With these changes, (7.64) reduces to the familiar add-compare-select recursion of Section 5.4.4.

Example 7-22.
Consider the case of complex additive Gaussian noise with $Y_k = S_k + N_k$, where S_k is the k-th output of the signal generator. In this case the branch metric of (7.63) reduces to

$$\gamma_k(p, q) = \frac{1}{2\pi\sigma^2}e^{-|y_k - s^{(p,q)}|^2/(2\sigma^2)}p_{A_k}(a^{(p, q)}) . \tag{7.65}$$

The negative of the logarithm of the branch metrics is proportional to

$$|y_k - s^{(p, q)}|^2 - 2\sigma^2 \log p_{A_k}(a^{(p, q)}) + 2\sigma^2 \log(2\pi\sigma^2) . \tag{7.66}$$

The last term is common to all transitions and can be omitted. In the special case when all symbols are equally likely, the second term is also a constant and can be omitted, yielding an effective branch metric equal to the squared Euclidean distance. Hence we have rederived the result from Section 7.2 in another way! We have also shown how it can be generalized to accommodate a priori information about the transmitted symbols, namely, by using (7.66).

Example 7-23.
Consider a finite-state signal generator that produces binary vectors $\{\mathbf{S}_k\}$ of length N, and assume a BSC noise generator with crossover probability p, so that (7.63) reduces to:

$$\begin{aligned}\gamma_k(p, q) &= f(y_k \,|\, \mathbf{s}^{(p, q)})p_{A_k}(a^{(p, q)}) \\ &= \left(\frac{p}{1-p}\right)^{d_H(y_k,\, \mathbf{s}^{(p,q)})} p_{A_k}(a^{(p, q)})(1-p)^N\end{aligned} \tag{7.67}$$

where $a^{(p,q)}$ and $\mathbf{s}^{(p,q)}$ are the unique inputs and outputs associated with a transition from p to q. The negative logarithm of this branch metric is proportional to:

$$d_H(y_k, \boldsymbol{s}^{(p,\,q)}) - \log p_{A_k}(a^{(p,\,q)}) / \log(\tfrac{p}{1-p}) - N\log(1-p) / \log(\tfrac{p}{1-p}) . \qquad (7.68)$$

The last term is common to all transitions and can be ignored. In the special case when all input symbols are equally likely, the second term is also a constant and can be omitted, yielding an effective branch metric equal to $d_H(y_k, \boldsymbol{s}^{(p,\,q)})$, the Hamming distance. Thus, whereas Euclidean distance is appropriate for the AWGN channel, Hamming distance is appropriate for the BSC channel.

7.5. A Posteriori Probability Detection with BCJR

In this section we summarize the Bahl, Cocke, Jelinek, Raviv (BCJR) algorithm [2] for *a posteriori probability (APP)* detection of symbols corrupted by an ISI channel and AWGN.

Although the BCJR and Viterbi algorithms share many similarities, they differ in a fundamental way because they compute very different quantities. The Viterbi algorithm computes *hard* decisions by performing sequence detection. The BCJR algorithm computes *soft* information about the data symbols in the form of *a posteriori* probabilities for each of the transmitted symbols. When applied to PAM over an ISI channel, for example, the Viterbi algorithm is an example of a hard-output equalizer, whereas the BCJR algorithm is an example of a soft-output equalizer. Of course, this soft information could be converted to hard decisions by choosing each decision so as to maximize its a posteriori probability; the resulting decisions would be *individually* optimal in the sense that they have minimal probability of being in error. Thus, whereas the Viterbi algorithm produces the *most likely symbol sequence*, which minimizes the probability of a sequence error, maximizing the APP's as calculated by the BCJR algorithm produces the *sequence of most likely symbols*, which minimizes the average symbol-error rate. The distinction is subtle and is not by itself enough to warrant much interest in BCJR. Indeed, the hard-decision performance of BCJR is only marginally better than Viterbi.

The real value of BCJR is it that the APP's are valuable in their own right; they may be used to estimate the reliability of a decision, and are especially useful when interacting with other modules in a receiver. For example, so-called *soft decoding* of error-correction decodes base on APP's can significantly outperform hard decoding based on hard decisions. Furthermore, APP's are needed for implementing iterative receivers based on turbo processing.

The BCJR algorithm shares many similarities with the multiplicative Viterbi algorithm. Both are based on the same trellis, both assign the same multiplicative branch metric to transitions, and both progress through the trellis recursively. However, rather than making one pass through the trellis from start to finish, as is done in the Viterbi algorithm, the BCJR algorithm makes two passes: one forward pass from start to finish, and then a second backward pass from finish to start. Therefore, roughly speaking, the complexity of the BCJR algorithm is twice that of the Viterbi algorithm.

To be concrete, we consider a signal generator with Q states labeled $\{0, \ldots\ Q-1\}$ and L inputs symbols $\{a_0, \ldots\ a_{L-1}\}$ drawn independently from an alphabet $\mathcal{A}$. For example, the symbols might be drawn from a complex PAM alphabet and the signal generator might be an FIR filter (representing ISI) with memory $\log_{|\mathcal{A}|} Q$. The input symbols are not necessarily drawn uniformly from $\mathcal{A}$. Our presentation of the BCJR algorithm will rely on the same trellis structure used by the Viterbi algorithm (Section 5.4.4). As in Section 5.4.4, we make the simplifying assumption that an idle symbol is input for all time $k < 0$ and $k \geq L$, so that the state of the signal generator is known to be zero at both time $k = 0$ and time $L + \mu$. The trellis consists of $L + \mu$ stages, the first L being due to the data symbols, while the remaining μ are due to the idle symbols. Let $y = [y_0, \ldots\ y_{L+\mu-1}]$ denote the set of observations, one for each trellis stage. There are $|\mathcal{A}|$ branches emanating from all nodes in the first L stages, one for each possible input symbol.

Simply stated, the aim of the APP detector is to compute $p_{A_k}(a \mid y)$ for each $a \in \mathcal{A}$, and for all $k \in \{0, \ldots\ L-1\}$. In the context of the trellis diagram, these probabilities are easily computed once the *a posteriori transition probabilities* $\sigma_k(p, q) = \Pr[\psi_k = p, \psi_{k+1} = q \mid y]$ are known for each state transition (or branch) in the trellis. The BCJR algorithm provides a computationally efficient method for finding these state transition probabilities.

The key to the BCJR algorithm is a decomposition of the a posteriori probability for a transition at time k into three separable factors: the first depending only on the "past" observations $y_{l<k} = \{y_l : l < k\}$, the second depending on only the "present" observation y_k, and the third depending only on the "future" observations $y_{l>k} = \{y_l : l > k\}$. This decomposition is illustrated in Fig. 7-10. We can derive this decomposition through the following series of straightforward equalities:

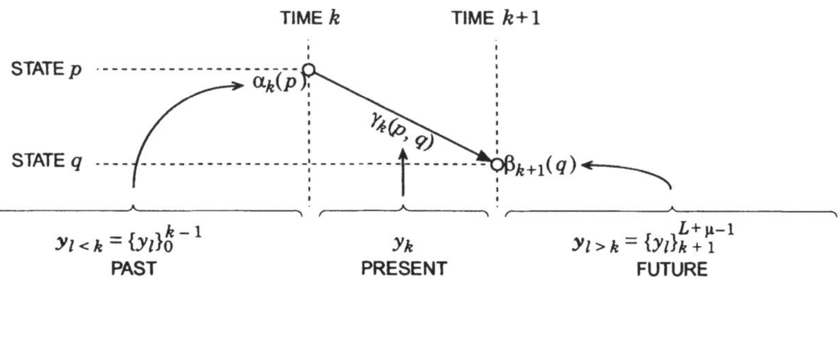

Fig. 7-10. A transition from state p to state q at the k-th stage of the trellis. Associated with the starting state is the quantity $\alpha_k(p)$, which depends only on the past; associated with the ending state is the quantity $\beta_{k+1}(q)$, which depends only on the future, and associated with the branch is the metric $\gamma_k(p, q)$, which depends only on the present. The a posteriori probability of this particular transition is proportional to the product $\alpha k(p)\gamma_k(p, q)\beta_{k+1}(q)$.

$$\begin{aligned}\sigma_k(p, q) &= f(\psi_k = p, \psi_{k+1} = q, y)/f(y)\\ &= f(\psi_k = p, \psi_{k+1} = q, y_{l<k}, y_k, y_{l>k})/f(y)\\ &= f(y_{l>k} \mid \psi_k = p, \psi_{k+1} = q, y_{l<k}, y_k) f(\psi_k = p, \psi_{k+1} = q, y_{l<k}, y_k)/f(y). \quad (7.69)\end{aligned}$$

Because of the Markov property of the finite-state machine model, knowledge of the state at time $k + 1$ supersedes knowledge of the state at time k, and it also supersedes knowledge of y_k and $y_{l<k}$, so that (7.69) reduces to:

$$\begin{aligned}\sigma_k(p, q) &= f(y_{l>k} \mid \psi_{k+1} = q) f(\psi_k = p, \psi_{k+1} = q, y_{l<k}, y_k)/f(y)\\ &= f(y_{l>k} \mid \psi_{k+1} = q) f(\psi_{k+1} = q, y_k \mid \psi_k = p, y_{l<k}) f(\psi_k = p, y_{l<k})/f(y). \quad (7.70)\end{aligned}$$

Again, exploiting the Markov property, this simplifies (with a reordering of terms) to:

$$\begin{aligned}\sigma_k(p, q) &= \underbrace{f(\psi_k = p, y_{l<k})}\underbrace{f(\psi_{k+1} = q, y_k \mid \psi_k = p)}\underbrace{f(y_{l>k} \mid \psi_{k+1} = q)}\underbrace{/f(y)} \quad (7.71)\\ &= \alpha_k(p) \times \gamma_k(p, q) \times \beta_{k+1}(q) \;/\; f(y).\end{aligned}$$

This equation defines $\alpha_k(p)$, $\beta_{k+1}(q)$, and $\gamma_k(p, q)$. We immediately recognize $\gamma_k(p, q)$ as being identical to the multiplicative Viterbi branch metric (7.63) from Section 7.4. Although the quantity $\alpha_k(p)$ defined above is not identical to the partial path metric $\alpha_k(p)$ of Section 7.4, we purposefully use the same notation so as to emphasize their similarities.

Observe that $\alpha_k(p)$ is a probability measure for state p at time k that depends only on the past observations $y_{l<k}$. On the other hand, $\beta_{k+1}(q)$ is a probability measure for state q at time $k + 1$ that depends only on the future observations $y_{l>k}$. And finally, $\gamma_k(p, q)$ is a branch metric for a transition from state p at time k to state q at time $k + 1$, and it depends only on the present (k-th) observation y_k. The implication of (7.71) is that, as illustrated in Fig. 7-10, the a posteriori probability for a particular transition can be calculated (to within a constant factor) by simply multiplying the α quantity for the starting state by the branch metric by the β quantity for the ending state.

The ultimate goal is to calculate the a posteriori probabilities for the symbols a_k, and not for the transitions. Fortunately, it is easy to translate from one to the other. Specifically, let $S_a \subseteq \{0, \ldots Q - 1\}^2$ denote the set of integer pairs (p, q) for which a state transition from p to q corresponds to an input symbol of a. We illustrate this notation by a simple example.

Example 7-24.

Let us return to the OOK example with ISI channel $H(z) = 1 + 0.5z^{-1}$, for which one stage of the trellis is:

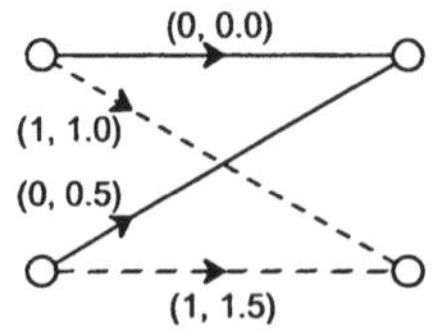

The transitions corresponding to an input of a one are shown as dashed lines. Therefore, in terms of the above notation, $\mathcal{S}_0 = \{(0, 0), (1, 0)\}$ identifies the solid transitions, and $\mathcal{S}_1 = \{(0,1), (1,1)\}$ identifies the dashed transitions.

The a posteriori probability that the k-th symbol is a particular $a \in \mathcal{A}$ can be easily computed by simply adding all of the transition probabilities in stage k corresponding to that a:

$$p_{A_k \mid \mathbf{Y}}(a \mid \boldsymbol{y}) = \sum_{(p,q) \in \mathcal{S}_a} \Pr[\psi_k = p, \psi_{k+1} = q \mid \boldsymbol{y}] \tag{7.72}$$

$$= \frac{1}{f(\boldsymbol{y})} \sum_{(p,q) \in \mathcal{S}_a} \alpha_k(p)\gamma_k(p, q)\beta_{k+1}(q) \,. \tag{7.73}$$

In practice, the quantity $f(\boldsymbol{y})$ in the denominator is ignored because it is common to all posteriori probabilities. If desired, its impact can be *implicitly* accounted for by first calculating (7.73) without $f(\boldsymbol{y})$, and then normalizing the results so that the a posteriori pdf for each of the transmitted symbols sums to unity, $\sum_{a \in \mathcal{A}} p_{A_k \mid \mathbf{Y}}(a \mid \boldsymbol{y}) = 1$.

The branch metric is determined by the statistics of both the noise and the source. For example, the branch metric for the AWGN channel and BSC are given by (7.65) and (7.67), respectively. There are many applications in which all symbols are equally likely, in which case the a priori probability factor $p_{A_k}(a)$ is a constant, independent of a. In such applications, the BCJR branch metric is equivalent to the usual Euclidean-distance or Hamming-distance metric. However, the coding of symbols before transmission (for example, using source coding or channel coding) can make some symbols more likely than others, and exploiting this knowledge can be beneficial.

That the transition APP can be decomposed into the form $\alpha_k(p)\gamma_k(p, q)\beta_{k+1}(q)$ is interesting in theory, but to make this decomposition valuable in practice we need a computationally efficient method for computing the $\{\alpha_k(p)\}$ and $\{\beta_k(p)\}$ values. Fortunately, they can be computed recursively with very low complexity. Specifically, it is easy to show (see Appendix 7-C) that the metrics $\{\alpha_k(p)\}$ can be calculated recursively according to:

$$\alpha_{k+1}(q) = \sum_{p=0}^{Q-1} \alpha_k(p)\gamma_k(p, q) \,. \tag{7.74}$$

Remarkably, this BCJR recursion is very similar to the Viterbi recursion of (7.64); the only difference is that the maximum operator $\max_p\{\,\cdot\,\}$ of the Viterbi recursion has been replaced by a summation operator $\Sigma_p\{\,\cdot\,\}$. Thus, whereas the Viterbi recursion compares contributions from several branches and selects one, the BCJR recursion accepts contributions from all branches. This difference is not always significant, however, because the exponential form of the branch metric (7.65) implies that the maximum is often a good approximation to the sum.

As shown in Appendix 7-C, the metrics $\{\beta_k(q)\}$ can be also be calculated recursively, using the exact same recursion as above but working *backwards*, starting at the end of the trellis and moving to the beginning, according to:

$$\beta_k(p) = \sum_{q=0}^{Q-1} \gamma_k(p, q)\beta_{k+1}(q) \,. \tag{7.75}$$

The BCJR algorithm can now be summarized as follows.

1. Just as for the Viterbi algorithm, calculate the branch metrics (γ) for all branches in the trellis. (For example, use (7.65) for the AWGN channel, or use (7.67) for the BSC.)
2. Calculate forward metrics (α) using (7.74) for $q \in \{0, \dots Q-1\}$, as $k = \{0, 1, \dots L+\mu-1\}$, with initialization $\alpha_0(p) = \delta_p$.
3. Calculate backward metrics (β) using (7.75) for $p \in \{0, \dots Q-1\}$, as $k = \{L+\mu-1, \dots 1, 0\}$, with initialization $\beta_{L+\mu}(p) = \delta_p$.
4. Calculate the a posteriori probabilities using (7.73) for each $a \in \mathcal{A}$, $k \in \{0, \dots L-1\}$.

Interestingly, steps 1 and 2 are just the multiplicative version of the Viterbi algorithm with $\max_p\{\cdot\}$ replaced by $\Sigma_p\{\cdot\}$. We need only add steps 3 and 4 to realize the BCJR algorithm.

7.5.1. Special Case: Binary Alphabet

If the input alphabet is binary, $\mathcal{A} = \{\pm 1\}$, then the a posteriori probabilities $p_{A_k|\mathbf{Y}}(a\,|\,\mathbf{y})$ are completely characterized by the scalar $p_{A_k|\mathbf{Y}}(1\,|\,\mathbf{y}) = 1 - p_{A_k|\mathbf{Y}}(-1\,|\,\mathbf{y})$, or equivalently by the ratio $p_{A_k|\mathbf{Y}}(1\,|\,\mathbf{y})/p_{A_k|\mathbf{Y}}(-1\,|\,\mathbf{y})$, or equivalently by its logarithm:

$$\lambda_k = \log\left(\frac{p_{A_k|\mathbf{Y}}(+1\,|\,\mathbf{y})}{p_{A_k|\mathbf{Y}}(-1\,|\,\mathbf{y})}\right), \tag{7.76}$$

which from (7.73) can be expressed as:

$$\lambda_k = \log\left(\frac{\sum_{(p,q)\in\mathcal{S}_1} \alpha_k(p)\gamma_k(p,q)\beta_{k+1}(q)}{\sum_{(p,q)\in\mathcal{S}_{-1}} \alpha_k(p)\gamma_k(p,q)\beta_{k+1}(q)}\right). \tag{7.77}$$

The BCJR algorithm for binary alphabets thus uses (7.77) to produce the a posteriori log-likelihood ratios $\lambda_0, \lambda_1, \dots, \lambda_{L-1}$, one for each transmitted symbol. If desired, the decisions that minimize the probability of being wrong can then be formed using a simple quantizer:

$$\hat{a}_k = \text{sign}(\lambda_k). \tag{7.78}$$

This is *the* minimum-probability-of-error detector. As mentioned earlier, however, the hard decisions produced by quantizing the a posteriori probabilities are not significantly more reliable than those produced by the Viterbi algorithm. Thus, in practice, the additional complexity of the BCJR algorithm is only justified when the a posteriori probabilities themselves are desired.

7.5.2. Normalization

Implementation of the forward and backward recursions on a finite-precision computer can lead to numerical underflow as the values for α and β become small. To prevent such numerical problems, the forward and backward recursions are commonly replaced by the following *normalized* versions:

$$\alpha_{k+1}'(q) = A_k \sum_{p=0}^{Q-1} \alpha_k'(p)\gamma_k(p, q)\,, \tag{7.79}$$

$$\beta_k'(p) = B_k \sum_{q=0}^{Q-1} \gamma_k(p, q)\beta_{k+1}'(q)\,, \tag{7.80}$$

where the constants A_k and B_k are chosen to prevent α and β from becoming too small. A common choice is to choose A_k and B_k so that $\sum_p \alpha_k'(p) = \sum_p \beta_k'(p) = 1$ for each k, or else so that $\alpha_k'(0) = \beta_k'(0) = 1$ for each k (the latter strategy having the advantage of reducing slightly the storage requirements). It is easy to see that the net effect of these normalizations is:

$$\alpha_k'(p) = (\Pi_{i=0}^{k} A_i)\alpha_k(p), \qquad \beta_k'(p) = (\Pi_{i=k}^{L+\mu-1} B_i)\beta_k(p)\,. \tag{7.81}$$

Thus, using α_k' and β_k' in place of α_k and β_k in (7.72) would only scale the a posteriori probabilities for the k-th symbol by the constant factor $(\Pi_{i=0}^{k} A_i)(\Pi_{i=k+1}^{L+\mu-1} B_i)$. This factor is of no consequence, because it applies equally to all symbols in the alphabet, and it can easily be accounted for by rescaling the a posteriori pdf to ensure that it sums to unity. For the same reason, all of the branch metrics within a given stage k of the trellis can be scaled by the same constant without affecting the APP calculations. This constant can change from one stage to the next. For example, the $1/(2\pi\sigma^2)$ factor in (7.65) and the $(1-p)^N$ factor in (7.67) can be ignored in practice.

Example 7-25.
Suppose a sequence of $L = 3$ OOK symbols $\{a_0, a_1, a_2\} \in \{0, 1\}^3$ are transmitted over an ISI channel with memory $\mu = 1$ and AWGN. The trellis has $L+\mu-1 = 4$ stages, as shown below:

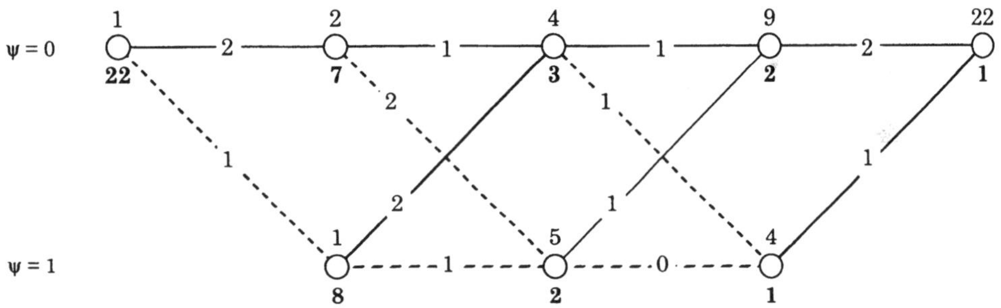

The transitions caused by inputs of *one* are shown as dashed lines. Suppose the *a priori* probabilities $\{p_{A_i}(a_i)\}$ and the relevant observations $\{y_0, y_1, y_2, y_3\}$ are such that the branch metrics defined by (7.65) are as shown above. (Almost surely these branch metrics would *not* be integers, but we assume integers to simplify our calculations.) In other words, suppose step 1 of the BCJR algorithm has already been performed. The aim of this example is to illustrate the remaining steps of the BCJR algorithm.

Associated with each node are two metrics. We write the forward metric α above the node, and the backward metric β below the node, in bold. According to step 2, the forward metric is unity at state zero at time zero, so $\alpha_0(0) = 1$ is written above the root node. The forward recursion dictates that

$\alpha_1(0) = \alpha_0(0)\gamma_0(0, 0) = 1 \times 2 = 2$, which is written above the state 0, time 1 node. Similarly, it dictates that $\alpha_1(1) = \alpha_0(0)\gamma_0(0, 1) = 1 \times 1 = 1$, which is written above the state 1, time 1 node. Then we move on to the next stage. For example, according to the forward recursion,

$$\begin{aligned}\alpha_2(1) &= \alpha_1(0)\gamma_0(0, 1) + \alpha_1(1)\gamma_0(1, 1) \\ &= 2 \times 2 + 1 \times 1 = 5 ,\end{aligned} \tag{7.82}$$

which explains why a "5" is written above the node at time 2, state 1.

After calculating all of the forward metrics, step 3 tells us to repeat the same procedure *backwards*. The state is known to be zero after the last stage, so $\beta_{L+1}(0) = \mathbf{1}$ is written below the last node. For example, the "**7**" written below the node at time 1, state 0 is there because the backward recursion implies that:

$$\begin{aligned}\beta_1(0) &= \gamma_1(0, 0)\beta_2(0) + \gamma_1(0, 1)\beta_2(1) \\ &= 1 \times \mathbf{3} + 2 \times \mathbf{2} = \mathbf{7} .\end{aligned} \tag{7.83}$$

Observe that the last backward metric agrees with the last forward metric, $\beta_0(0) = \alpha_{L+1}(0) = 22$. They will always agree.

Once the forward and backward metrics are known, the a posteriori log-likelihood ratios defined in (7.76) can be calculated according to (7.77). The results are:

$$\lambda_0 = \log\left(\frac{1 \times 1 \times 8}{1 \times 2 \times 7}\right) = \log\left(\frac{4}{7}\right) > 0 , \tag{7.84}$$

$$\lambda_1 = \log\left(\frac{2 \times 2 \times 2 + 1 \times 1 \times 2}{2 \times 1 \times 3 + 1 \times 2 \times 3}\right) = \log\left(\frac{5}{6}\right) > 0 , \tag{7.85}$$

$$\lambda_2 = \log\left(\frac{4 \times 1 \times 1 + 5 \times 0 \times 1}{4 \times 1 \times 2 + 5 \times 1 \times 2}\right) = \log\left(\frac{2}{9}\right) < 0 . \tag{7.86}$$

The numerators are the sum of the $\alpha \times \gamma \times \beta$ products for the dashed transitions, which correspond to an input of one, while the denominators are the sum of the $\alpha \times \gamma \times \beta$ products for the solid transitions, which correspond to an input of zero. The signs of the a posteriori LLR's indicate that the symbol-wise MAP decisions are $\hat{a}_0 = 1$, $\hat{a}_1 = 1$, and $\hat{a}_2 = 0$.

7.6. Symbol-Error Probability for MLSD

In this section we relate the probability of *sequence* error to the probability of *symbol* error for the ML sequence detector, focusing primarily on the case of continuous-time passband PAM with real additive white Gaussian noise with PSD $N_0/2$.

By interpreting a sequence of PAM pulses as a single signal, we can apply directly the general M-ary analysis results of Section 6.1. The fact that we have found a computationally efficient algorithm for making the ML decision will not change that error probability.

However, in this context, the error probability is really the probability of a *sequence error.* By sequence error, we mean *one or more* errors in the detection of the entire sequence of data symbols. Specifically, from (6.41), the sequence error probability is approximately

$$P_e \approx K \cdot Q\left(\frac{d_{\min}}{2\sigma}\right) . \tag{7.87}$$

In this expression, $\sigma^2 = N_0/2$, and $d_{\min}$ is the minimum Euclidean distance between two distinct sequences of data symbols, that is, the minimum distance between any signal $s(t) = \sum_{k=0}^{L-1} a_k h(t-kT)$ and any other signal $s'(t) = \sum_{k=0}^{L-1} a_k' h(t-kT)$:

$$d_{\min}^2 = \min_{\{a_k\} \neq \{a_k'\} \in \mathcal{A}^L} \int_{-\infty}^{\infty} |s(t) - s'(t)|^2 dt . \tag{7.88}$$

However, caution is in order. As the length L of the sequence gets large, the number of distinct paths with minimum distance from the correct path also gets large, invalidating the approximations in Section 6.1 that led to (7.87). In fact, as L gets large, the probability of sequence error usually approaches unity! Fortunately, the probability of *symbol* error of the minimum-distance sequence detector stays small even if L goes to infinity. The symbol-error probability of the minimum-distance sequence detector will be explored below.

7.6.1. Error Events

For purposes of determining the symbol error probability, it is useful to introduce the concept of an *error event.* Let $\{\psi_k\}$ be the correct state sequence and $\{\hat{\psi}_k\}$ be the sequence selected by the Viterbi algorithm. Over a long time, $\{\psi_k\}$ and $\{\hat{\psi}_k\}$ will typically diverge and remerge several times. Each distinct separation is called an error event, which is therefore defined as a correct path through the trellis paired with an error path that begins and ends with the correct state. By definition, the error path does not share any intermediate states with the correct state sequence. The *length* of an error event is the number of intermediate (incorrect) nodes in the path.

Example 7-26.
Examples of error events of length one and two are shown in Fig. 7-11 for a two-state trellis. The assumed correct state trajectory is shown by dashed lines, and the error event by solid lines. There are error events of unbounded length, although as we will see, the probability of the longer events will usually (but not always) be negligibly small.

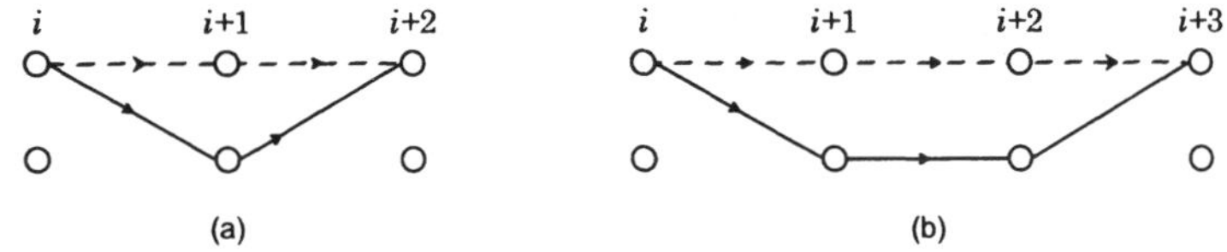

Fig. 7-11. When the correct state sequence ψ and the detected state sequence $\hat{\psi}$ diverge and remerge, we have an error event. Two error events are shown here. (a) The shortest error event for the two-state trellis in Example 7-24. (b) The next longest error event. In both cases we have assumed the correct state trajectory is all zeros, shown with the dashed lines.

An error event has one or more *symbol errors*, which are incorrect symbols or bits that result from taking an incorrect path through the trellis. In Appendix 7-B we show that the probability of symbol error is dominated by the probability of the minimum distance error event at high SNR. For the Gaussian noise case,

$$\Pr[\text{symbol error}] \approx C \cdot Q(d_{\min}/2\sigma) \tag{7.89}$$

and for the BSC case

$$\Pr[\text{symbol error}] \approx C \cdot Q(d_{\min},\ p)\ , \tag{7.90}$$

where C is some constant between P and R given by (7.157) and (7.149) respectively, and $Q(\cdot\,,\cdot)$ is defined by (7.23). As long as P and R are reasonably close to unity, we need not be too concerned with this multiplicative constant.

The following procedure will find the distance of any particular error event for either the Gaussian or BSC cases. Assume a correct state sequence, and label each branch in the trellis with its squared distance from the corresponding branch of the correct state sequence. This would be the branch metric if the channel were noiseless. The correct state sequence will have branch metrics that are zero, and normally all the branches not on the correct path will have a non-zero branch metric. For each possible error event, we can find the distance of that error event very simply by computing its path metric.

Example 7-27.
Continuing the ISI Example 7-24, Fig. 7-12 shows the trellis labeled with the branch metrics assuming a noiseless channel and an all-zeros transmitted sequence. The path metric for each path through the trellis is now the square of the Euclidean distance of that path from the correct all-zeros path. The error event of length one is easily seen to have Euclidean distance $\sqrt{1.25}$, and the error event of length two has distance $\sqrt{3.5}$. Longer error events have still greater distances. Obviously, the error event of length one is much more probable than longer error events. It is easy to show by exhaustive search that all possible correct paths through the trellis are at least distance $\sqrt{1.25}$, and none has smaller distance (see Exercise 7-2 below), so $\sqrt{1.25}$ is the minimum distance for all possible correct paths. It is shown in Appendix 7-B that $C = 1$, so

$$\Pr[\text{symbol error}] \approx Q(\sqrt{1.25}/2\sigma)\ . \tag{7.91}$$

Exercise 7-2.
Completing Example 7-27, show that for each possible correct path through the trellis, the minimum-distance error event has length one and distance $\sqrt{1.25}$.

In Example 7-27 we found the minimum distance by inspection. This will not be so easy in general. Fortunately, it turns out that the Viterbi algorithm can itself be used to find the

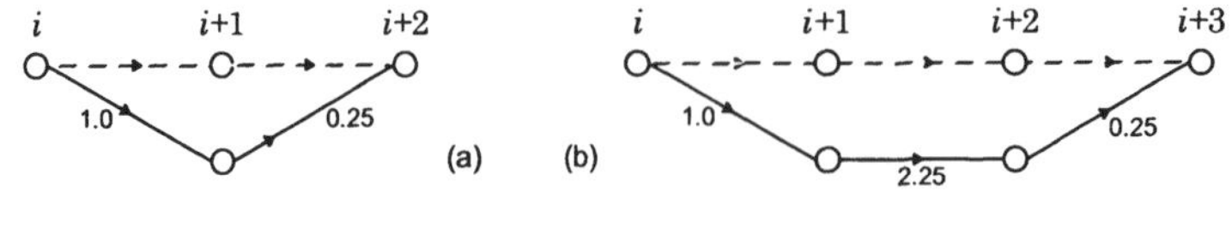

Fig. 7-12. The trellis of Fig. 7-11 labeled with the distances from the correct branches, assuming the correct branches are the dashed ones.

minimum-distance error event for a given correct state sequence! Using the trellis diagram labeled with the same branch metrics as above, we can begin on the correct path and use the Viterbi algorithm to find the survivor at each stage, excluding the correct path (the only one with zero path metric). Each survivor at a node on the correct path corresponds to an error event, and the path metric is the distance of this error event from the correct path. At each stage of the algorithm, keep track of the minimum-distance error events recorded thus far; when all survivors have a partial path metric greater than this minimum, the minimum-distance error event has been found for one assumed correct path through the trellis.

As shown in Appendix 7-B, it is the global minimum distance $d_{\min}$ that dominates the probability of error, not the minimum distance from one particular path through the trellis. Fortunately, usually we do not need to examine all possible correct paths to find the minimum distance. Better techniques are available for both the ISI examples (discussed shortly), and the BSC examples (discussed in Chapter 12). In both cases we exploit symmetry based on linearity, although the nature of the linearity is quite different in each case.

7.6.2. Calculating the Minimum-Distance for ISI

We now examine the problem of calculating the $d_{\min}$ of (7.88) for PAM over an ISI channel. In Section 5.4.3 we showed that a PAM signal of the form $s(t) = \sum_{k=0}^{L-1} a_k h(t - kT)$ can be expanded in terms of the WMF basis $\{\phi(t - kT)\}$, where the expansion coefficients are $\gamma a_k * m_k$, and where m_k and γ are defined by the factorization $S_h(z) = \gamma^2 M(z) M^*(1/z^*)$. Therefore, from Parseval's relationship, (7.88) simplifies to:

$$d_{\min}^2 = \min_{\{a_k\} \neq \{a_k'\} \in \mathcal{A}^L} \sum_{k=0}^{\infty} \left| \sum_{n=0}^{L-1} \gamma a_n m_{k-n} - \sum_{n=0}^{L-1} \gamma a_n' m_{k-n} \right|^2 dt \tag{7.92}$$

$$= \min_{\{\varepsilon_0 \neq 0, \varepsilon_1, \ldots \varepsilon_{L-1}\}} \gamma^2 \sum_{k=0}^{\infty} \left| \sum_{n=0}^{L-1} \varepsilon_n m_{k-n} \right|^2 , \tag{7.93}$$

where $\varepsilon_k = a_k - a_k'$ is called an *error symbol*. The minimization is over all sequences of $\{a_k\}$ and $\{a_k'\}$ that are not equal; that is, that differ for at least one k. Therefore, the minimization is over the *error alphabet* $\mathcal{A} - \mathcal{A}$, with the constraint that at least one of the ε_k's must be nonzero. By time invariance, we can limit attention to error events that begin with an error at time $k = 0$; that is, the minimization of (7.93) is over all $\{\varepsilon_k\}$ such that $\varepsilon_0 \neq 0$.

The minimum-distance problem of (7.93) can be formulated in a form that can be solved by the Viterbi algorithm, but only for the case where $M(z)$ is an FIR filter, where an FSM model holds. Assuming $m_k = 0$ for $k > \mu$, the convolution sum can be reversed,

$$d_{\min}^2 = \min_{\{\varepsilon_0 \neq 0, \varepsilon_1, \ldots \varepsilon_{L-1}\}} \sum_{k=0}^{L+\mu-1} \left| \sum_{i=0}^{\mu} m_i \varepsilon_{k-i} \right|^2 . \tag{7.94}$$

Two desirable things happen in the FIR case as formulated in (7.94). First, the summation over k becomes finite, although in practice this is not too helpful because we are interested in very large values of L. Second, the minimization can be formulated as the minimization of the path metric in a trellis. Toward this end, define a state

$$\Psi_k = [\varepsilon_{k-1}, \ldots, \varepsilon_{k-2}, \varepsilon_{k-\mu}] \tag{7.95}$$

and a trellis diagram for the progression of this state. Labeling each branch of this trellis with the corresponding branch metric $|\sum_{i=0}^{\mu} m_i \varepsilon_{k-i}|^2$, the minimization problem becomes one of finding the non-zero path through this trellis with minimum path metric. The minimization is over all error symbol sequences starting with $\varepsilon_0 \neq 0$, and hence over all paths through the trellis where the starting state is [0, 0, …, 0] and the first branch is non-zero.

This formulation of the minimum-distance problem has its price, as well as a major advantage. The price is that the alphabet of error symbols is larger than the alphabet of the data symbols, increasing the number of states in the trellis. In general if the data symbol alphabet has size M, the error symbol alphabet can be as large as M^2, although normally it is smaller.

Example 7-28.
If the data symbols are real-valued, M is odd, and the data-symbol alphabet is all integers in the range $[-(M-1)/2, (M-1)/2]$, then the error symbols (difference between two data symbols) can assume all values in the range $[-(M-1), (M-1)]$, or $(2M-1)$ distinct values. This is generally much smaller than M^2.

The advantage of this formulation is that the branch metric is a function of each branch in isolation, not pairs of branches (correct path and error path), greatly reducing the number of alternatives that have to be considered.

The Viterbi algorithm can now be applied to finding d_{min} in a simple example (where it is hardly needed).

Example 7-29.
For the ISI channel of Example 7-24, where $H(z) = 1 + 0.5z^{-1}$, the branch metric is given by $(\varepsilon_k + 0.5\varepsilon_{k-1})^2$. If the data symbols are binary, with alphabet $\{0, 1\}$, the error ε_k is ternary, with alphabet $\{0, \pm 1\}$. The corresponding three-state trellis diagram is shown in Fig. 7-13(a), and the

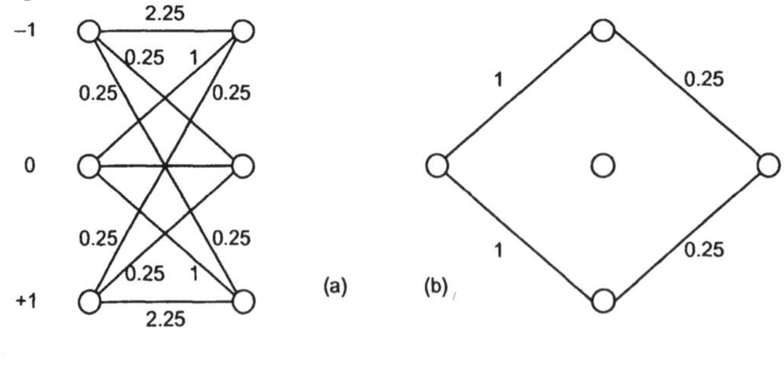

Fig. 7-13. a. The trellis diagram for finding the minimum distance for the ISI example of Example 7-29. b. The two minimum-distance error events.

two minimum-distance error events are shown in Fig. 7-13(b). Each error event has a single symbol error.

To complete the probability-of-error analysis it is helpful to estimate the error coefficient C in (7.89) or (7.90) by finding P and R, where $P \le C \le R$. Recall that P is the probability that there is an error event starting at a fixed time i with distance $d_{\min}$, and R is given by (7.149), or

$$R = \sum_{e \in B} w(e)\Pr[\psi] \tag{7.96}$$

where B is the set of error events with minimum distance, $w(e)$ is the number of symbol errors in the error event e, and ψ is the correct state trajectory.

Example 7-30.

Continuing Example 7-29, at any given time k, exactly one minimum-distance error event can start at that time, regardless of the correct state trajectory. If $a_k = 0$, which occurs with probability ½, then the top error event is possible, corresponding to $\varepsilon_k = -1$. If $a_k = 1$, which also occurs with probability ½, then the bottom error event in Fig. 7-13(b) is possible, corresponding to $\varepsilon_k = +1$. Consequently, $P = 1$. Since the error event includes one symbol error, $w(e) = 1$ for each $e \in B$, and since there is only one minimum distance error event possible for each correct path through the trellis, $R = 1$. Consequently, $C = 1$ and Pr[symbol error] $\approx Q(\sqrt{1.25}/2\sigma)$.

Example 7-31.

Consider a channel with response $H(z) = 1 - z^{-1}$ and a binary alphabet $\{0, 1\}$. The trellis is shown in Fig. 7-14(a). What is interesting about this channel is that two of the branches not on the correct zero path have zero branch metric. As a result, there are an infinite number of error events at the minimum distance of $\sqrt{2}$, corresponding to any sequence of consecutive errors of the same polarity. First note that since every correct path has at least one minimum-distance error event, $P =$ 1. To find R, consider the contribution of an error event with m consecutive errors. This error event can only occur if m consecutive data symbols have the same polarity as the error, and the $(m + 1)$ symbol has the opposite polarity. This event has probability $2^{-(m+1)}$. This error event contributes m symbol errors, and there are two such error events, so

$$R = 2\sum_{m=1}^{\infty} m 2^{-(m+1)} = 2\,. \tag{7.97}$$

So C may be as large as 2. The behavior of this particular channel is analogous to that of a *catastrophic convolutional encoder*, discussed further in Chapter 12.

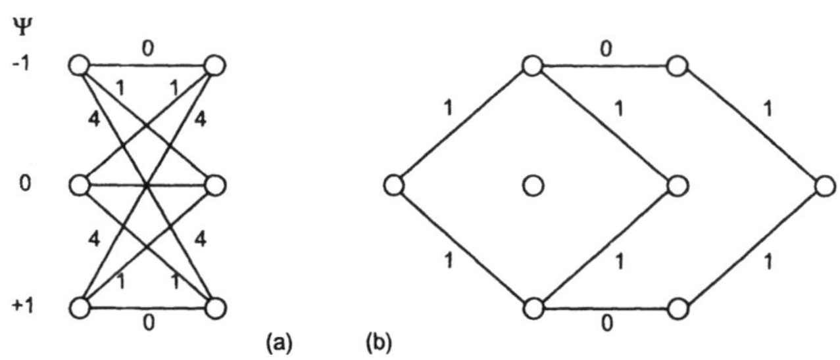

Fig. 7-14. a. The trellis for the channel of Example 7-31. (b) Four of the minimum-distance error events. There are actually an infinite number, corresponding to any sequence of consecutive errors with the same polarity.

In retrospect, the infinite number of minimum-distance error events of Example 7-31 is not unexpected, since a sequence of consecutive ones and a sequence of consecutive zeros both result in the same channel output sequence (all zeros). Thus, it is difficult for the ML sequence detector to distinguish the two cases; if it happens to make an error at the beginning, which is the only place where they differ, it will get the entire sequence wrong.

It may seem that an infinite number of error sequences must be considered in finding the minimum distance. In fact Fig. 7-14 illustrates that minimum-distance error events can be infinite in length, although this is unusual. However, as long as P and R are moderate, it is adequate to find just the minimum distance, and necessarily not the number of error events at the minimum distance. If our goal is limited to finding the minimum distance, *there are only a finite number of error events that need be considered.* This statement follows from the following observation: if an error event passes through the same state twice (other than the all-zero path), then it need not be considered in searching for the minimum-distance error event. This is because the shorter error event obtained by removing that portion of the path between the two passes through the same state must have a path metric at least as small. This rule allows us to remove from consideration many error events. In particular, if the trellis has N states, then only error events that pass through the $N-1$ non-zero states at most once need be considered, and such error events are of length at most $N-1$.

Example 7-32.

For a symmetric three-state trellis, corresponding to binary data symbols and memory $\mu = 1$, it suffices to consider the error events shown below:

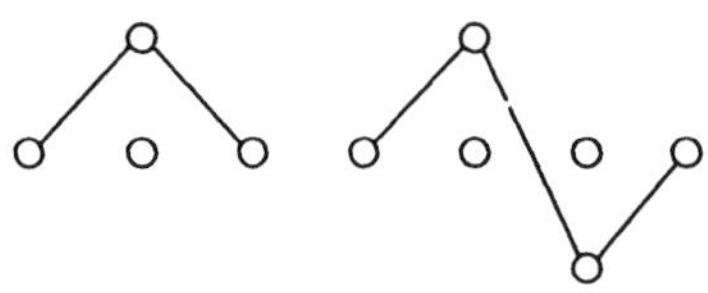

By symmetry all branches that are vertical mirror images of one another have the same branch metric. There are thus two error events not shown (the mirror images) that need not be considered since they have the same path metrics. Of the two error events shown, the second will always have a path metric at least as large as the first, because of the mirror-image symmetry of the branch metrics and because of the extra branch in the middle. Thus, for a symmetric three-state trellis, the error event of length one *always* has the minimum distance. A word of caution is in order, however. From Example 7-31 we know that there can be many error events with this minimum distance.

For a trellis with a large number of states, the number of error events that must be considered, although finite, is still large. In this case, the minimum-distance error events can be found using the Viterbi algorithm, and it is wise to use a computer.

7.7. Incoherent Detection

In Chapter 6, FSK was presented as a modulation technique suitable for transmission over channels that cause rapidly varying carrier phase. One of the major advantages of FSK is the ability to *incoherently* detect a signal, without deriving the carrier phase. Intuitively, this can be accomplished by realizing a set of bandpass filters, one centered at each of the known signal frequencies, and measuring the power at the output of each filter. The question arises, however, as to the optimal detection technique where the carrier phase is unknown. We will now derive the *optimal incoherent detector*, applying directly the results of Section 7.3.

Assume that the carrier phase is random, with the goal of rederiving the ML detector. The complex envelope of the received signal is now of the form

$$Y(t) = e^{j\Theta}\tilde{s}_m(t) + N(t), \quad 1 \le m \le M, \tag{7.98}$$

where Θ is assumed independent of the signal, and the noise $N(t)$ is white, circularly symmetric and Gaussian. In the absence of any other relevant information, we can assume that Θ is uniformly distributed over the interval $[0, 2\pi)$; this is also the most tractable choice analytically, leading to a simple result. The general approach to determining the ML detector is to first condition on knowledge of $\Theta = \theta$, and then average over θ.

Assume that the $\{\tilde{s}_1(t), \dots \tilde{s}_M(t)\}$ span a subspace of dimension N, and that this subspace has an orthonormal basis $\{\psi_1(t), \dots \psi_N(t)\}$. If the carrier phase is known to be $\Theta = \theta$, then a sufficient statistic is

$$V_n = \int_0^\infty Y(t)\psi_n^*(t)\,dt, \quad 1 \le n \le N, \tag{7.99}$$

as in (7.44). Incorporating phase θ in the calculation of the sufficient statistic would simply multiply V_n by $e^{j\theta}$, but not add any additional information. Substituting (7.98) into (7.99), and observing that the $2f_c$ term will integrate to zero, we can express the sufficient statistics as an N-dimensional vector,

$$\mathbf{V} = e^{j\theta}\,\tilde{\mathbf{S}}_m + \mathbf{N}, \tag{7.100}$$

where $\tilde{\mathbf{S}}_m$ is a vector of the coefficients of $\tilde{s}_m(t)$ with respect to the orthonormal basis and $\mathbf{N}$ is a vector of independent circularly symmetric Gaussian random variables. The effect of the unknown carrier phase θ is to shift the signal component of $\mathbf{V}$ by phase θ.

To determine the ML detector, we must determine the probability density function of $\mathbf{V}$ conditioned on the m-th signal being transmitted. As the first step, we find the p.d.f. of $\mathbf{V}$ conditioned on both m and the phase Θ, $f_{\mathbf{V}|M,\Theta}(v \mid m,\theta)$. This is a multi-dimensional Gaussian density function, given by

$$f_{\mathbf{V}|M,\Theta}(v \mid m,\theta) = \frac{1}{(2\pi\sigma^2)^N} e^{-\|v - e^{j\theta}\tilde{\mathbf{S}}_m\|^2/2\sigma^2}, \tag{7.101}$$

where $\sigma^2 = N_0/2$ is the variance of the real or imaginary part of the Gaussian noise. This formidable expression will be made yet more formidable by finding $f_{\mathbf{V}|M}(v \mid m)$ by integrating out the dependence on θ. But do not despair — the end result is simple! It is useful to derive first the following simple result.

Exercise 7-3.

Define the modified Bessel function of zero order,

$$I_0(x) = \frac{1}{2\pi}\int_{-\pi}^{\pi} e^{x\cos(\theta)}d\theta \tag{7.102}$$

for a real-valued x. Show that for a complex-valued z,

$$I_0(|z|) = \frac{1}{2\pi}\int_{-\pi}^{\pi} e^{\mathrm{Re}\{e^{j\theta}z^*\}}d\theta\,. \tag{7.103}$$

Hint: Write z in polar coordinates.

Using this result, we can find the marginal density of the received signal by integrating against the density function of Θ,

$$f_{V|M}(v\,|\,m) = \int_{-\pi}^{\pi} f_{V|M,\Theta}(v\,|\,m,\theta)f_\Theta(\theta)\,d\theta$$
$$= \frac{1}{(2\pi\sigma^2)^N}\exp\left(\frac{-1}{2\sigma^2}(\|v\|^2 + \|\tilde{S}_m\|^2)\right)I_0\left(\frac{|\langle v, \tilde{S}_m\rangle|}{\sigma^2}\right). \tag{7.104}$$

When each signal has the same energy so that $\|\tilde{S}_m\|$ is a constant, the exponential term in (7.104) is independent of m. Therefore, from the monotonicity of $I_0(x)$, the ML receiver selects m to maximize

$$J_m = |\langle v, \tilde{S}_m\rangle| = \left|\sum_{n=1}^{N} v_n \tilde{s}_n^{(m)*}\right| = \left|\int_{-\infty}^{\infty} Y(t)\tilde{s}_m^*(t)dt\right|. \tag{7.105}$$

This is the simple form that was promised. A receiver structure to compute J_m is shown in Fig. 7-15. Instead of correcting the matched filter output for the phase, as would be done if the phase were known, the receiver simply determines the magnitude of the matched filter output, throwing away any phase information.

Example 7-33.

For binary FSK, $\tilde{s}_1(t) = e^{-j2\pi f_d t}w(t)$ and $\tilde{s}_2(t) = e^{j2\pi f_d t}w(t)$, where $2f_d$ is the deviation between the two signals, and where $w(t) = u(t) - u(t-T)$ is a rectangular window. In terms of the passband signal $\hat{Y}(t)$, the optimal receiver calculates the quantity:

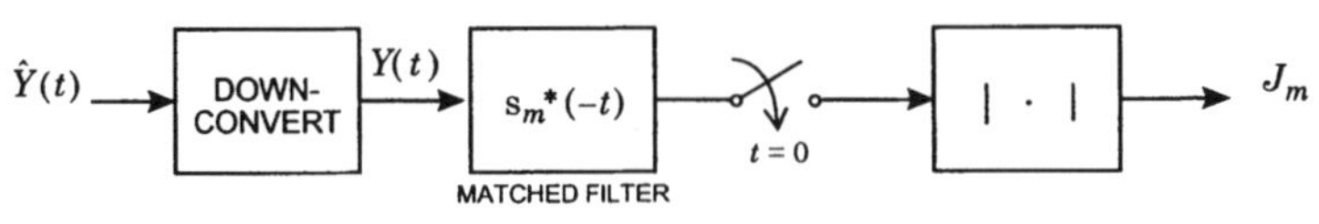

Fig. 7-15. The optimal incoherent receiver uses a matched filter (a correlator could be used also) and throws away phase information.

$$\left| \int_0^T \hat{Y}(t) e^{-j2\pi(f_c \pm f_d)t} dt \right|^2 =$$

$$\left(\int_0^T \hat{Y}(t)\cos(2\pi(f_c \pm f_d)t)\, dt \right)^2 + \left(\int_0^T \hat{Y}(t)\sin(2\pi(f_c \pm f_d)t)\, dt \right)^2 \tag{7.106}$$

as shown in Fig. 7-16. Since the signal energies are the same, and since the Bessel function is monotonic, J_1 and J_2 given by (7.106) are compared and the signal corresponding to the maximum chosen. Intuitively, since the phase of the signal is unknown, we must correlate against two quadrature sinusoids, since we are then assured of a strong correlation for any signal phase for one or the other sinusoid phases. This receiver is also equivalent to passing the received signal through two filters

$$\text{bg}(t) = e^{j2\pi(f_c \pm f_d)t} w(-t). \tag{7.107}$$

These are roughly bandpass filters centered at $f_c + f_d$ and $f_c + f_d$, the two transmitted frequencies. The filter outputs are each followed by an envelope detector. This structure was previously shown in Fig. 6-14, where it was justified on intuitive grounds.

The calculation of the probability of error for an incoherent detector is rather involved, and each case is best treated individually. The starting point is substituting (7.100) into (7.105), so that the decision variable becomes, conditioned on transmitted signal m,

$$J_i = | e^{j\theta} \langle \tilde{\mathbf{S}}_m, \tilde{\mathbf{S}}_i \rangle + \langle \mathbf{N}, \tilde{\mathbf{S}}_i \rangle | . \tag{7.108}$$

We can illustrate the probability of error calculation using FSK.

Example 7-34.

For FSK the two signals are orthogonal (with the proper choice of frequency deviation), and thus $\langle \tilde{\mathbf{S}}_1, \tilde{\mathbf{S}}_2 \rangle = 0$. Hence, the decision variables of (7.108) become, assuming $\tilde{\mathbf{S}}_1$ is transmitted,

$$J_1 = | e^{j\theta} \| \tilde{\mathbf{S}}_1 \|^2 + \langle \mathbf{N}, \tilde{\mathbf{S}}_1 \rangle |, \qquad J_2 = | \langle \mathbf{N}, \tilde{\mathbf{S}}_2 \rangle | . \tag{7.109}$$

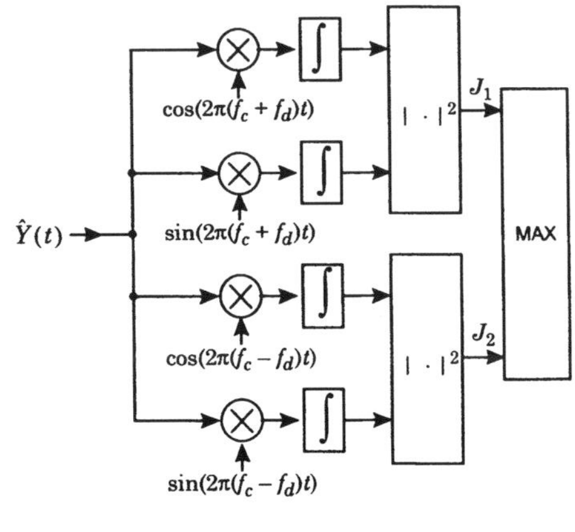

Fig. 7-16. The optimal incoherent receiver for a binary FSK signal using correlators.

The probability of error becomes, conditioned on $\tilde{\mathbf{S}}_1$ transmitted,

$$\Pr[\text{error} \mid \tilde{\mathbf{S}}_1 \text{ transmitted}] = \Pr\Big[\, |\langle \mathbf{N}, \tilde{\mathbf{S}}_2 \rangle| \;>\; | e^{j\theta} \| \tilde{\mathbf{S}}_1 \|^2 + \langle \mathbf{N}, \tilde{\mathbf{S}}_1 \rangle | \,\Big]. \tag{7.110}$$

This probability is in fact independent of θ, and by symmetry is the same as the error probability conditioned on $\tilde{\mathbf{S}}_2$ being transmitted. The two random variables on the left and right sides are not independent. The evaluation of the result is rather involved, and leads to an expression in terms of a tabulated function known as Marcum's Q function.

For more examples of the calculation of the probability of error, the interested reader is referred to [3][4].

7.8. Shot Noise Signal with Known Intensity

For most digital communication media the additive Gaussian noise model considered earlier in this chapter is quite appropriate. The major exception is the optical-fiber channel under some circumstances. The *signal* at the output of an optical detector is actually *shot noise* or a *filtered Poisson process*. The randomness of the signal itself is therefore a contributor to errors, as is thermal noise introduced in receiver preamplifiers. The appropriate detection technique thus depends on the situation:

- When thermal noise is the dominant impairment, the white Gaussian noise detector is appropriate, since taking account of the shot noise nature of the signal will have little impact.
- When thermal noise is insignificant, as in a homodyne or heterodyne optical receiver, the shot noise nature of the signal will be dominant, and the optimal detector should take it into account.

In this section we consider optimal ML detection of a shot-noise signal with time-varying known intensity, approximating the case where thermal noise is insignificant. When the shot-noise has high intensity (roughly speaking, when there are a large number of received photons per bit), we showed in Section 3.4.5 that the signal could be accurately approximated as a deterministic signal (the mean value of the shot noise) plus additive Gaussian noise. Unfortunately, the variance of the latter is time-varying (see (3.218)), and hence this noise is non-stationary and the previous results do not apply even approximately to this case. By resolving the detection problem, we will show that the ML detector correlates against not the signal intensity as one might expect from the Gaussian case, but rather against the *logarithm* of the signal intensity. We will give a simplified derivation of this result under specific assumptions, to avoid getting overly involved in the details.

The current output of the photodetector is a shot noise process with intensity (average number of photoelectrons generated per unit time) proportional to the incident optical power. Assume that one of M signals is received on the interval $[0, T]$, where the current intensity for the m-th signal is $\lambda_m(t) + \lambda_{\text{dark}}$ where λ_{dark} is the dark current that is always present, regardless of the signal. Our first simplifying assumption is that $h(t)$, the response of the receiver circuitry (photodetector biasing circuitry and preamplifier) to a single photoelectron assumes the particular shape of Fig. 7-17(a). This shape simplifies the calculation, because the photodetector output becomes a *modified counting process*, as shown in Fig. 7-17(b). We call it a counting process since the current $I(t)$ at time t is equal to the count of the number of photoelectrons in the interval of time $[t - T_h, t]$.

For the assumed pulse shape, the current waveform is not strictly bandlimited and we cannot sample without aliasing. However, it is a reasonable approximation to sample the current every T_h seconds, yielding

$$K_n = I(nT_h) \, , \tag{7.111}$$

which equals precisely the number of photoelectrons since the last sample; that is, the number of photoelectrons generated over the interval $[(n-1)T_h, nT_h]$. We choose this sampling rate for two reasons:

- The successive samples are statistically independent, since the arrival of photoelectrons follows a Poisson process and hence the number of arrivals in non-overlapping intervals are independent. If we sampled any faster, we would lose this independence, complicating the results to follow.
- This is the minimum sampling rate that avoids throwing away useful information about the signal, since with less frequent samples there would be photoelectrons that would be missed because they did not affect any sample.

In summary, the samples K_n are independent and Poisson distributed with parameter

$$\Gamma_{n\,|\,m} = \int_{(n-1)T_h}^{nT_h} \lambda_m(t)dt \; + T_h\lambda_{\text{dark}} \, . \tag{7.112}$$

If we assume that the response time T_h of the photodetector will be much faster than the rate of change in the intensity, as will almost always be the case, then we get the approximation

$$\Gamma_{n\,|\,m} \approx T_h\lambda_m(nT_h) + T_h\lambda_{\text{dark}} . \tag{7.113}$$

To determine the ML detector, we must calculate the probability of the received samples K_n conditioned on the m-th signal having been transmitted. The probability of one sample $I(nT_h) = K_n$ is the Poisson distribution of (3.155),

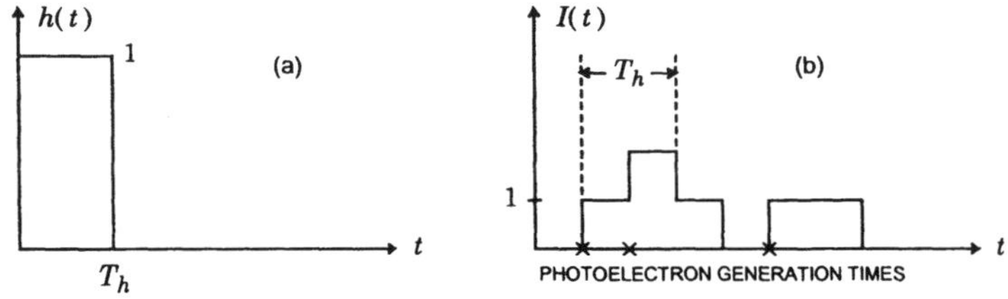

Fig. 7-17. a. The assumed shape of $h(t)$, the response of the detector to a single photoelectron, for calculation of the optimal receiver processing. (b) An example of the sample function of the shot noise photodetector current resulting from this pulse shape.

$$p_{K_n|M}(k_n | m) = \frac{\Gamma_{n|m}^{k_n} e^{-\Gamma_{n|m}}}{k_n!} , \tag{7.114}$$

and since the samples are independent, the joint probability of L samples is the product of the probabilities,

$$p_{K_1, K_2, \ldots, K_L|M}(k_1, \ldots, k_L | m) = \prod_{n=1}^{L} \frac{\Gamma_{n|m}^{k_n} e^{-\Gamma_{n|m}}}{k_n!} , \tag{7.115}$$

where L is the number of samples required to cover the signaling interval $[0, T]$, $T = LT_h$. For convenience, we calculate the logarithm of this probability,

$$\log p_{K_1, K_2, \ldots, K_L|M}(k_1, \ldots, k_L | m) = \sum_{n=1}^{L} (k_n \log\Gamma_{n|m} - \Gamma_{n|m} - \log(k_n!)) . \tag{7.116}$$

Since the logarithm is monotonic, the ML detector chooses the signal m for which this quantity is maximum. In (7.116), the only quantity that is a function of the signal is $\Gamma_{n|m}$, and hence the last term can be discarded. We can make a couple of further simplifications. First, from (7.112),

$$\sum_{n=1}^{L} \Gamma_{n|m} = E_m + LT_h\lambda_{\text{dark}} , \tag{7.117}$$

where

$$E_m = \int_0^T \lambda_m(t)\, dt \tag{7.118}$$

is the integral of a quantity proportional to the incident power on the detector and hence is proportional to the total energy incident on the detector for the m-th signal. The second term in (7.117) can be ignored since it is signal independent. Also, the first term in (7.116) can be approximated by an integral, since from (7.113)

$$\begin{aligned} \sum_{n=1}^{L} k_n \log_e\Gamma_{n|m} &\approx \sum_{n=1}^{L} i(nT_h)\log(T_h(\lambda_m(nT_h) + \lambda_{\text{dark}})) \\ &\approx \frac{1}{T_h}\int_0^T i(t)\log\left[T_h\lambda_{\text{dark}}\left(1 + \frac{\lambda_m(t)}{\lambda_{\text{dark}}}\right)\right] dt . \end{aligned} \tag{7.119}$$

Finally, we can ignore the signal-independent T_h and $T_h\lambda_{\text{dark}}$ terms, so that the ML detector equivalently finds the signal that satisfies

$$\max_{m \in \{1, \ldots, M\}} \left\{ \Xi_m = \int_0^T I(t)\log\left(1 + \frac{\lambda_m(t)}{\lambda_{\text{dark}}}\right) dt - E_m \right\}. \tag{7.120}$$

For high intensity, where the signal current is large relative to the dark current,

$$\Xi_m \approx \int_0^T I(t)\log\lambda_m(t)\, dt - E_m , \tag{7.121}$$

case, where the dark current is larger than the signal, we can approximate $\log(1 + \varepsilon)$ by ε, and thus the ML detector reduces to a correlation against the signal as in the Gaussian case,

$$\Xi_m \approx \int_0^T I(t)\lambda_m(t)\, dt - E_m \,. \tag{7.122}$$

Example 7-35.

If the signal intensity $\lambda(t)$ is constant over an interval $[0, T]$, the ML detector simply integrates the current over that interval. Hence the ML detector reduces to an "integrate and dump" detector.

7.9. Further Reading

The Karhunen-Loeve expansion is described in [1]. A good overview of the Viterbi algorithm is given in [5]. The Viterbi algorithm was originally developed in the context of error-control coding as a means for determining the most likely transmitted codeword. It was later applied to the problem ML sequence detection for mitigating ISI [6]. Similarly, the BCJR algorithm for APP decoding was originally developed as a means for the optimal bit-by-bit decoding of error-correcting codes [2], although a similar algorithm appeared earlier for mitigating ISI [7]. The BCJR algorithm arises in a wide range of statistical inference applications, such as speech recognition, where it is known as the forward-backward algorithm [8].

Appendix 7-A. Karhunen-Loeve Expansion

In this Appendix, we derive some of the detailed results for the Karhunen-Loeve expansion used in Section 7.3.2.

Integral Equation

It is necessary to show that (7.35) imposes necessary and sufficient conditions for (7.33) to be satisfied. Assuming that (7.33) holds, after substituting for N_i and N_j from (7.34) and exchanging the order of expectation and integration,

$$\sigma_i^2\delta_{i,j} = \int_0^T \phi_i^*(t) \int_0^T R_N(t - \tau)\phi_j(\tau)\, d\tau\, dt \,. \tag{7.123}$$

A sufficient condition for this to be satisfied is (7.35), as can be established by substituting (7.35) into (7.123). To show that (7.35) is a necessary condition, we assume that (7.31) and (7.33) are valid, and show that this implies (7.35). Multiplying (7.31) by N_n^* and taking the expected value, we get

$$E[N(t)N_n^*] = E[\sum_{i=1}^{\infty} N_i N_n^* \phi_i(t)] = \sigma_n^2 \phi_n(t), \quad 0 \le t \le T. \tag{7.124}$$

Similarly, multiplying the conjugate of (7.34) by $N(t)$ and taking the expectation,

$$E[N(t)N_n^*] = E[N(t)\int_0^T N^*(\tau)\phi_n(\tau)\, d\tau] = \int_0^T R_N(t-\tau)\phi_n(\tau)\, d\tau, \quad 0 \le t \le T. \tag{7.125}$$

Equating these two results establishes (7.35).

Derivation of the Continuous-time Whitening Filter

We now derive (7.42). Define

$$g_m(t) = \sum_{i=1}^{\infty} \frac{s_i^{(m)}}{\sigma_i^2} \cdot \phi_i(t) \tag{7.126}$$

and then (7.41) follows directly by multiplying both sides of (7.126) by $R_N(\tau - t)$ and integrating. Similarly,

$$\int_0^T s_m(t) g_m^*(t)\, dt = \sum_{i=1}^{\infty} \frac{s_i^{(m)*}}{\sigma_i^2} \int_0^T s_m(t)\phi_i^*(t)\, dt = \sum_{i=1}^{\infty} \frac{|s_i^{(m)}|^2}{\sigma_i^2} = E_m. \tag{7.127}$$

Sufficient Statistic Argument

In Section 7.3.3 we derived a set of sufficient statistics $\{U_1, \ldots U_N\}$ for the received signal $Y(t)$, $0 \le t \le T$, by letting $T \to \infty$ and using intuitive arguments. Here we derive these sufficient statistics carefully for finite T using the Karhunen-Loeve expansion. The results remain valid as $T \to \infty$. Define

$$f_m(t) = \sum_{i=1}^{\infty} \frac{s_i^{(m)}}{\sigma_i^2} \phi_i(t), \quad 1 \le m \le L, \quad 0 \le t \le T, \tag{7.128}$$

where $\{\phi_i(t),\ \sigma_i^2,\ 1 \le i < \infty\}$ are the eigenfunctions and eigenvalues of (7.35) and the $\{s_i^{(m)}\}$ are given by (7.37). As $T \to \infty$, (7.128) approaches the same definition as (7.46). The complete orthonormal basis $\{\psi_k(t),\ 1 \le k < \infty\}$ is chosen so that the first N members are a basis for the subspace $\mathcal{S}_f$ spanned by the $\{f_1(t), \ldots f_M(t)\}$ in (7.128). Using (7.46),

$$F_k^{(m)} = \int_0^T f_m(t)\psi_k^*(t)\, dt = \sum_{i=1}^{\infty} \frac{s_i^{(m)}}{\sigma_i^2} \psi_{k,i}^*, \tag{7.129}$$

where

$$\psi_{k,i} = \int_0^T \psi_k(t)\phi_i^*(t)\, dt. \tag{7.130}$$

Since the $\phi_i(t)$ are a complete orthonormal set, and (7.130) are the components of $\psi_k(t)$ in terms of the $\phi_i(t)$, it follows that

$$\psi_k(t) = \sum_{i=1}^{\infty} \psi_{k,i}\phi_i(t) , \tag{7.131}$$

and thus

$$\delta_{k,j} = \int_0^T \psi_k(t)\psi_j^*(t)\, dt = \sum_{i=1}^{\infty}\sum_{l=1}^{\infty} \psi_{k,i}\psi_{j,l}^* \int_0^T \phi_i(t)\phi_l^*(t)\, dt$$
$$= \sum_{i=1}^{\infty} \psi_{k,i}\psi_{j,i}^* . \tag{7.132}$$

Thus, the discrete-time sequences $\{\psi_{k,i},\ 1 \le k \le \infty\}$ for $1 \le i < \infty$ are orthonormal.

Returning to the received signal of (7.38), and forming the inner product of both sides with $\psi_{k,i},\ 1 \le i < \infty$, thus expressing it in terms of a new basis $\{\psi_{k,i},\ 1 \le k < \infty\}$,

$$U_k = \sum_{i=1}^{\infty} \frac{Y_i}{\sigma_i}\psi_{k,i}^* = \sum_{i=1}^{\infty} \frac{s_i^{(m)}}{\sigma_i}\psi_{k,i}^* + \sum_{i=1}^{\infty} \frac{N_i}{\sigma_i}\psi_{k,i}^* . \tag{7.133}$$

The first term on the right side of (7.133) is $F_k^{(m)}$, and the second is a noise term W_k. All that remains to establish (7.49) is to show that W_k is white, which follows from

$$\mathrm{E}[W_k W_j^*] = \mathrm{E}\Big[\sum_{i=1}^{\infty}\sum_{l=1}^{\infty} \frac{N_i N_l^*}{\sigma_i \sigma_l}\psi_{k,i}^*\psi_{j,l}\Big] = \sum_{i=1}^{\infty} \psi_{j,i}\psi_{k,i}^* = \delta_{i,j} . \tag{7.134}$$

Since W_k is a linear function of a circularly symmetric process N_k, it is circularly symmetric, and (7.134) implies that the W_k are mutually independent.

We can express the sufficient statistics in terms of continuous-time signals as follows. Substituting from (7.130) in (7.133),

$$U_k = \sum_{i=1}^{\infty} \frac{Y_i}{\sigma_i} \int_0^T \psi_k^*(t)\phi_i(t)\, dt = \int_0^T U(t)\psi_k^*(t)\, dt \tag{7.135}$$

where

$$U(t) = \sum_{i=1}^{\infty} \frac{Y_i}{\sigma_i}\phi_i(t) , \qquad 0 \le t \le T , \tag{7.136}$$

is the output of the whitening filter. This confirms (7.47).

Appendix 7-B. Bit-Error Probability for Sequence Detectors

We showed in Section 7.6 that the probability of sequence error is easy to obtain using the vector channel results from Section 6.2, but is often useless because the probability approaches unity as the sequence gets large. Instead of the probability of sequence error, we can compute the probability that an *error event* begins at a particular time. This effectively normalizes the probability of sequence error per unit time. Error events are defined in Section 7.6.1.

We are most often interested however in the probability of a *bit* or *symbol* error rather than an error event. In this appendix we derive the error event probability and a general expression for the probability of bit or symbol error that does not depend on linearity in the system. We then show that for the additive Gaussian white noise case the probability of error is approximately $C \cdot Q(d_{\min}/2\sigma)$, where C is a constant that we can easily bound, and $d_{\min}$ is the distance of the minimum distance error event. For the BSC channel case, the probability of error is approximately $C \cdot Q(d_{\min}, p)$, where $Q(\cdot,\cdot)$ is defined by (7.23).

After the sequence detector selects a path through the trellis, the receiver must translate this path into its corresponding bit sequence (recall the one-to-one mapping between incoming bit sequences and state trajectories). Several bit or symbol errors may occur as a consequence of each error event. Let E denote the set of all error events starting at time i. Each element e of E is characterized by both a correct path ψ and an incorrect path $\hat{\psi}$ that diverge and remerge some time later. We make a stationarity assumption that $\Pr[e]$ is independent of i, the starting time of the error event. This will of course not be true if the trellis is finite, but if it is long relative to the length of the significant error events then the approximation is accurate. Each error event causes one or more *detection errors*, where a detection error at time k means that in_k at stage k of the trellis is incorrect. For the ISI example, each X_k is a symbol A_k, so a detection error is the same as a symbol error. For the binary coding examples, each X_k is a set of one or more bits. Define

$$c_m(e) = \begin{cases} 1; & \text{if } e \text{ has a detection error in position } e \text{ (from the start } i) \\ 0; & \text{otherwise} \end{cases} \tag{7.137}$$

This function characterizes the sample times corresponding to detection errors in error event e.

Example 7-36.

Consider Example 7-24, the ISI example. Let e_1 denote the error event of Fig. 7-11(a), which assumes that the correct state trajectory ψ consists of zero states. From we the figure of Example 7-24 we see that this error event causes decisions $\hat{x}_i = 1$ and $\hat{x}_{i+1} = 0$. Since $x_i = 0$ and $x_{i+1} = 0$ are the correct decisions,

$$c_m(e_1) = \delta_m . \tag{7.138}$$

The probability of a particular error event e starting at time i and causing a detection error at time k is

$$c_{k-i}(e)\Pr[e]\ . \tag{7.139}$$

Since the error events in E are disjoint (if one occurs no other can occur),

$$\Pr[\text{detection error at time } k] = \sum_{i=-\infty}^{k} \Sigma_{e \in E} \Pr[e] c_{k-i}(e)\ . \tag{7.140}$$

Exchanging the order of summation, assuming this is legitimate,

$$\Pr[\text{detection error at time } k] = \Sigma_{e \in E} \Pr[e] \sum_{i=-\infty}^{k} c_{k-i}(e)\ . \tag{7.141}$$

By a change of variables,

$$w(e) = \sum_{i=-\infty}^{k} c_{k-i}(e) = \sum_{m=0}^{\infty} c_m(e) \tag{7.142}$$

which is the total number of detection errors in e. Thus

$$\Pr[\text{detection error}] = \Sigma_{e \in E} \Pr[e] w(e)\ , \tag{7.143}$$

where we note that the dependence on k has disappeared. Hence, the probability of a detection error at any particular time is equal to the expected number of detection errors caused by error events starting at any fixed time i. In retrospect this result is not unexpected, since from the perspective of time k, the probability of a detection error at that time must take into account all error events starting at times prior to k.

The probability of the error event e depends on the probabilities of both the correct and incorrect paths ψ and $\hat{\psi}$ that make up e,

$$\Pr[e] = \Pr[\psi]\Pr[\hat{\psi} \mid \psi]\ . \tag{7.144}$$

It is usually difficult to find exact expressions for $\Pr[\hat{\psi} \mid \psi]$, but bounds are often easy. The reason is easy to see in the simple example of Fig. 7-18, where we assume there are only three possible trajectories. In Fig. 7-18(a) the ML decision regions for the three signals are shown. These decision regions lie in a M-dimensional space that we schematically represent on the two-dimensional page. Now suppose that ψ is the actual trajectory, corresponding to signal $\boldsymbol{s}$ in Fig. 7-18(a). The region corresponding to the detection of $\hat{\psi}$ is shown in Fig. 7-18(b). The

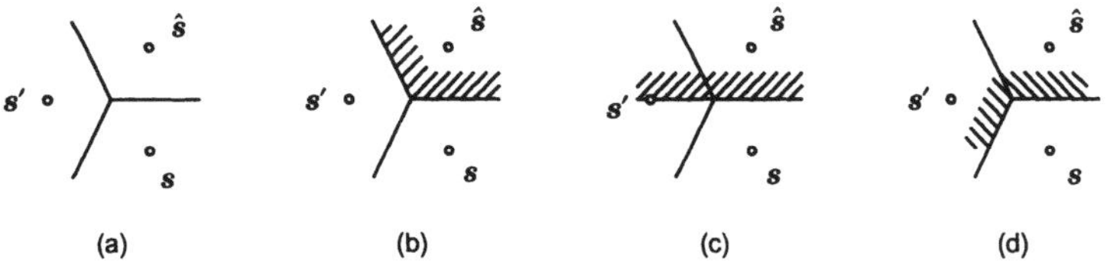

Fig. 7-18. a. Three signals corresponding to state trajectories ψ, $\hat{\psi}$ and ψ' where ψ is the actual state trajectory. (b) If received signal (with noise) is in the shaded region, the ML detector will choose trajectory $\hat{\psi}$. (c) The decision region for $\hat{\psi}$ if there were only two signals, ψ and $\hat{\psi}$. d. If the received signal (with noise) is in the shaded region, the ML detector will generate an error event.

probability of the noise carrying us into this region is very difficult to calculate, especially as the number of possible trajectories gets large. However, this probability is easy to upper bound by using the larger decision region of Fig. 7-18(c), which ignores the possibility of any trajectory other than ψ and $\hat{\psi}$.

For the additive white Gaussian noise model, the probability of the region in Fig. 7-18(c) is

$$\Pr[\hat{\psi} \mid \psi] \le Q(d(\hat{\psi}, \psi)/2\sigma) \tag{7.145}$$

where $d(\hat{\psi}, \psi)$ is the Euclidean distance between transmitted signals $\hat{\boldsymbol{s}}$ and $\boldsymbol{s}$ corresponding to state trajectories $\hat{\psi}$ and ψ. For the BSC,

$$\Pr[\hat{\psi} \mid \psi] \le Q(d(\hat{\psi}, \psi), p) \tag{7.146}$$

where $d(\hat{\psi}, \psi)$ is now a Hamming distance and $Q(\cdot\,, \cdot)$ is defined by (7.23). The bound is precisely the probability that the received signal is closer to the signal $\hat{\boldsymbol{s}}$ corresponding to $\hat{\psi}$ than it is to the signal $\boldsymbol{s}$ corresponding to the correct path ψ. Combining (7.143), (7.144), and (7.145), for the Gaussian case

$$\Pr[\text{detection error}] \le \textstyle\sum_{e \in E} w(e)\Pr[\psi]Q(d(\hat{\psi}, \psi)/2\sigma)\ . \tag{7.147}$$

This can be written

$$\Pr[\text{detection error}] \le \textstyle\sum_{e \in B} w(e)\Pr[\psi]Q(d_{\min}/2\sigma) + \text{other terms} \tag{7.148}$$

where B is the subset of error events in E that have distance $d_{\min}$, and the "other terms" all have arguments to the $Q(\cdot)$ function larger than $d_{\min}/2\sigma$. At high SNR these other terms become insignificant and the upper bound on Pr[detection error] approaches $RQ(d_{\min}/2\sigma)$, where

$$R = \textstyle\sum_{e \in B} w(e)\Pr[\psi]\ . \tag{7.149}$$

For the BSC case, just replace $Q(d_{\min}/2\sigma)$ with $Q(d_{\min}, p)$.

Example 7-37.

In Example 7-36 we considered the error event e_1 shown in Fig. 7-11(a), and found that $w(e_1) = 1$ and the distance is $\sqrt{1.25}$. This is the minimum distance, and it occurs for the eight error events shown in Fig. 7-19, each of which also have $w(e) = 1$. Consequently, (7.149) becomes

$$R = \textstyle\sum_{e \in B} \Pr[\psi] \tag{7.150}$$

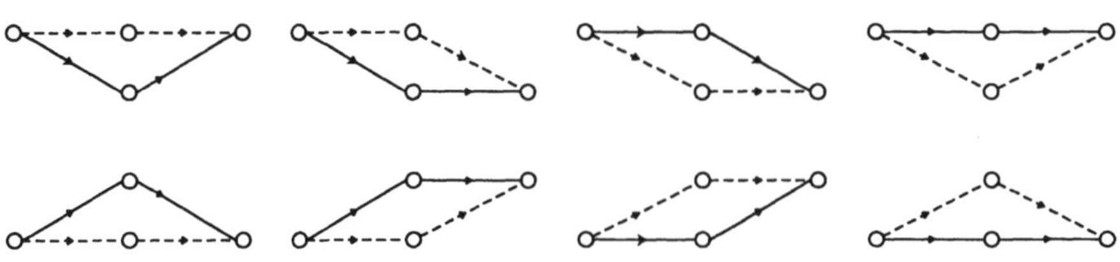

Fig. 7-19. Eight error events that have the same probabilities.

where B is the set of error events shown in Fig. 7-19. Assuming all possible actual paths are equally likely, and noting that only three successive states of each ψ in Fig. 7-19 are specified, $\Pr[\psi] = 2^{-3} = 1/8$ for each one. Hence $R = 1$ and (7.148) becomes

$$\Pr[\text{detection error}] \le Q(\sqrt{1.25}/2\sigma) + \text{other terms} . \tag{7.151}$$

We can get an idea of the magnitude of the "other terms" by considering the set of second most probable error events, one of which is shown in Fig. 7-11(b). It has length two, distance $d = \sqrt{3.5}$ from the correct path, weight $w(e_2) = 2$, and $\Pr[\psi] = 1/16$, so its contribution to the sum is $0.125Q(\sqrt{3.5}/2\sigma)$. Because of the exponential decrease in $Q(\cdot)$, this term is orders of magnitude smaller than the first for small σ. Consequently, we conclude that the "other terms" in (7.151) can safely be ignored.

A *lower* bound on the probability of detection error for the additive white Gaussian noise model can also be found. Combining this with the upper bound in (7.147) leads to an accurate estimate of Pr[detection error]. Returning to Fig. 7-18, we want to use the decision region of Fig. 7-18(c) to somehow obtain a lower bound. Shown in Fig. 7-18(d) is the decision region corresponding to *any error event* conditioned on actual state sequence ψ. The probability of any error event is evidently lower bounded by calculating the probability of the smaller decision region in Fig. 7-18(c). Thus, we see that in order to determine a lower bound, we must start with the probability of *any* error event, rather than the probability of a particular error event.

Since $w(e) \ge 1$ for all error events e, then from (7.143)

$$\Pr[\text{detection error}] \ge \textstyle\sum_{e \in E} \Pr[e] = \Pr[\text{an error event}] . \tag{7.152}$$

Now consider a particular actual path ψ through the trellis. For this path, let $d_{\min}(\psi)$ denote the distance of the minimum distance error event (either Euclidean or Hamming). Of course, $d_{\min}(\psi) \ge d_{\min}$, where $d_{\min}$ is the minimum distance error event over all possible actual state sequences ψ. As in Fig. 7-18(c), if ψ is the actual state sequence, the probability of an error event is lower bounded by Fig. 7-18(c). Obviously to make this bound strongest, we want to choose a particular error event that is closest to ψ, one of those at distance $d_{\min}(\psi)$. Hence, for the Gaussian case

$$\Pr[\text{an error event} \mid \psi] \ge Q(d_{\min}(\psi)/2\sigma) . \tag{7.153}$$

Combining this with (7.152) we get

$$\Pr[\text{detection error} \mid \psi] \ge Q(d_{\min}(\psi)/2\sigma) . \tag{7.154}$$

Consequently,

$$\Pr[\text{detection error}] \ge \textstyle\sum_{\psi} \Pr[\psi] Q(d_{\min}(\psi)/2\sigma) . \tag{7.155}$$

If we omit some terms in this summation, the bound will still be valid since the terms are all non-negative. Thus, let us retain only those state sequences ψ for which $d_{\min}(\psi) = d_{\min}$,

$$\Pr[\text{detection error}] \ge \textstyle\sum_{\psi \in A} \Pr[\psi] Q(d_{\min}/2\sigma) , \tag{7.156}$$

where A is the set of actual paths ψ that have a minimum distance error event, and d_{min} is that minimum distance. Define

$$P = \Sigma_{\psi \in A} \Pr[\psi] , \tag{7.157}$$

the probability that a randomly chosen ψ has an error event starting at any fixed time with distance d_{min} (or is consistent with a minimum distance error event). Then

$$\Pr[\text{detection error}] \ge PQ(d_{min}/2\sigma) . \tag{7.158}$$

In retrospect this lower bound is intuitive, since we would expect that every state sequence consistent with a minimum-distance error event will result in a probability of error event at least as large as $Q(d_{min}/2\sigma)$, and each error event will result in at least one detection error. For the common case where all possible paths ψ through the trellis are consistent with a minimum-distance error event, $P = 1$. This is true in Example 7-37.

Combining our upper and lower bounds,

$$PQ(d_{min}/2\sigma) \le \Pr[\text{detection error}] \le RQ(d_{min}/2\sigma) , \tag{7.159}$$

where the upper bound is approximate since some terms were thrown away. We conclude that at high SNR

$$\Pr[\text{detection error}] \approx C \cdot Q(d_{min}/2\sigma) \tag{7.160}$$

for some constant C between P and R. The BSC case is identical,

$$\Pr[\text{detection error}] \approx C \cdot Q(d_{min}, p) \tag{7.161}$$

where $Q(\cdot , \cdot)$ is defined in (7.23).

Example 7-38.

Continuing Example 7-37, note that $P = R = 1$. Hence

$$\Pr[\text{detection error}] \approx Q(\sqrt{1.25}/2\sigma) . \tag{7.162}$$

For this example, each detection error causes exactly one bit error, so Pr[detection error] = Pr[bit error]. Hence, with the sequence detector we get approximately the same probability of error as for an isolated pulse and a matched filter receiver (see Problem 7-9).

In general, a single detection error may cause more than one bit error. Suppose each input to the Markov chain X_k is determined by n source bits (and hence comes from an alphabet of size 2^n). Then each detection error causes at least one and at most n bit errors. Hence we can write

$$\frac{1}{n} \Pr[\text{detection error}] \le \Pr[\text{bit error}] \le \Pr[\text{detection error}] . \tag{7.163}$$

Typically we make the pessimistic assumption that

$$\Pr[\text{bit error}] \approx \Pr[\text{detection error}] . \tag{7.164}$$

Appendix 7-C. BCJR Forward/Backward Recursions

We first derive the forward recursion of (7.74), by using a series of straightforward and self-explanatory equalities. Starting with the definition of $\alpha_k(p)$ in (7.71), we have:

$$\begin{aligned}\alpha_{k+1}(q) &= f(\psi_{k+1} = q, y_{l<k+1}) \\ &= f(\psi_{k+1} = q, y_k, y_{l<k}) \\ &= \sum_{p=0}^{Q-1} f(\psi_{k+1} = q, y_k, \psi_k = p, y_{l<k}) \\ &= \sum_{p=0}^{Q-1} f(\psi_{k+1} = q, y_k \mid \psi_k = p, y_{l<k}) f(\psi_k = p, y_{l<k}) \\ &= \sum_{p=0}^{Q-1} f(\psi_{k+1} = q, y_k \mid \psi_k = p) f(\psi_k = p, y_{l<k}) \\ &= \sum_{p=0}^{Q-1} \alpha_k(p)\gamma_k(p, q) \,. \end{aligned} \tag{7.165}$$

The derivation of the backward recursion (7.75) is similar. Starting with the definition of $\beta_{k+1}(q)$ in (7.71), we have

$$\begin{aligned}\beta_k(p) &= f(y_{l>k-1} \mid \psi_k = p) \\ &= f(y_{l>k}, y_k \mid \psi_k = p) \\ &= \sum_{q=0}^{Q-1} f(y_{l>k}, y_k, \psi_{k+1} = q \mid \psi_k = p) \\ &= \sum_{q=0}^{Q-1} f(y_{l>k} \mid y_k, \psi_{k+1} = q, \psi_k = p) f(y_k, \psi_{k+1} = q \mid \psi_k = p) \\ &= \sum_{q=0}^{Q-1} f(y_{l>k} \mid \psi_{k+1} = q) f(y_k, \psi_{k+1} = q \mid \psi_k = p) \\ &= \sum_{q=0}^{Q-1} \gamma_k(p, q)\beta_{k+1}(q) \,. \end{aligned} \tag{7.166}$$

Problems

Problem 7-1. Suppose a binary symbol $A \in \{0, 1\}$ with a priori probabilities $p_A(0) = q$ and $p_A(1) = 1 - q$ is transmitted through the BSC of Fig. 7-2. The observation $Y \in \{0, 1\}$ is also binary; it equals A with probability $1 - p$.

(a) Find the ML detection rule. Assume $p < \frac{1}{2}$.

(b) Find the probability of error of the ML detector as a function of p and q.

(c) Assume $p = 0.2$ and $q = 0.9$. Find the MAP detector and its probability of error. Compare this probability of error to that in part (b).

(d) Find the general MAP detector for arbitrary p and q.

(e) Find the conditions on p and q such that the MAP detector always selects $\hat{a} = 0$.

Problem 7-2. Consider the vector detection problem for the BSC of Example 7-14. Specify the MAP detector for some given prior probabilities for signal vectors.

Problem 7-3. Consider a random variable $X \in \{-3, -1, +1, +3\}$ with a priori probabilities $p_X(\pm 3) = 0.1$ and $p_X(\pm 1) = 0.4$. Given an observation y of the random variable $Y = X + N$, where N is a zero mean Gaussian random variable with variance σ^2, independent of X, find the decision regions for a MAP detector. Now suppose $\sigma^2 = 0.25$ and $y = 2.1$. What is the decision?

Problem 7-4. Assume the random variable X is from the alphabet $\mathcal{X} = \{x_1, x_2\}$ with nonuniform a priori probabilities $p_X(x_1) \neq p_X(x_2)$. Define the random variable $Y = X + N$, where N is a zero mean Gaussian random variable with variance σ^2, independent of X. Give an expression for the MAP decision boundary between x_1 and x_2.

Problem 7-5. Consider M vectors each a distance d_{min} from the other vectors. Assume an ML detector will be used to distinguish between these vectors.

(a) Give an example for $M = 3$ of such a set of vectors where d_{min} is a Euclidean distance of $\sqrt{2}$.

(b) Give an example for $M = 3$ of such a set of vectors where d_{min} is a Hamming distance of 2.

(c) Use the union bound to find an upper bound on the probability of error for your two examples, assuming additive white Gaussian noise for (a) and a BSC for (b). First give the bound assuming s_1 is transmitted, then give the bound without this assumption.

Problem 7-6. Suppose you are given N observations $x_1, \ldots, x_N$ of the zero mean independent Gaussian random variables $X_1, \ldots, X_N$. Assume that the random variables have the same (unknown) variance σ^2. What is the ML estimator for the variance?

Problem 7-7. Given a Gaussian channel with independent noise components, one of the following four equally likely signals is transmitted: $(1, 1)$, $(1, -1)$, $(-1, 1)$, $(-1, -1)$. Determine the *exact* probability of error of an ML detector for Gaussian noise with variance σ^2.

Problem 7-8.

(a) Repeat Problem 7-7 for a BSC with error probability p and four equally likely signals: (000000), (000111), (111000), (111111).

(b) What is this error probability when $p = 0.1$? Compare to the minimum distance approximation.

Problem 7-9. Suppose that a symbol A from the alphabet $\mathcal{A} = \{0, 1\}$ is transmitted through the LTI system with impulse response

$$h_k = \delta_k + \tfrac{1}{2}\delta_{k-1} \, . \qquad (7.167)$$

and corrupted by additive white Gaussian noise with variance σ^2.

(a) Determine the structure of the ML detector.

(b) Calculate the probability of error for the ML detector.

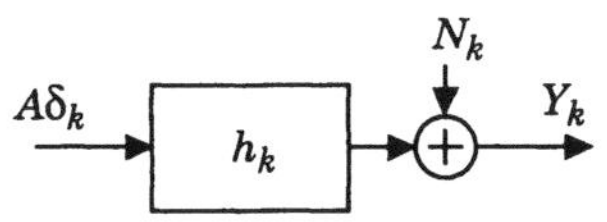

Fig. 7-20. A simple case of a discrete-time channel with intersymbol interference.

Problem 7-10. Consider the system in Fig. 7-20. Assume N_k is a sequence of independent zero mean Gaussian random variables with variance σ^2. Assume the symbol alphabet is $\mathcal{A} = \{0, 1\}$ and that the channel impulse response is

$$h_k = \delta_k - 0.5\delta_{k-1} + 0.1\delta_{k-2} . \tag{7.168}$$

(a) Derive the matched filter receiver.

(b) Give an expression for the probability of error.

(c) Now suppose $\mathcal{A} = \{-1, 0, +1\}$. Repeat part (a).

Problem 7-11.

(a) Determine the optimal incoherent receiver structure for passband PAM modulation, where the data symbols assume only two values $A_0 = 0$ and $A_0 = 1$. This is known as amplitude shift keying (ASK).

(b) Discuss the conditions under which passband PAM can be successfully demodulated using an incoherent receiver.

(c) Find an expression for the probability of error of the type derived in Example 7-34. You do not need to evaluate this expression.

Problem 7-12.

(a) Derive a discrete-time channel model analogous to Fig. 7-9 where instead of a matched filter, a general filter $F^*(f)$ is used, the Gaussian noise added on the channel has power spectrum $S_N(f)$, and symbol-rate sampling is used at the output of the filter.

(b) Determine a model for the special case where the matched filter optimized for the particular noise spectrum is used.

(c) As a check, verify that your model reduces to Fig. 7-9 when the filter is a matched filter and the noise is white.

Problem 7-13. Suppose that for a channel with additive white noise, the response at the output of the whitened matched filter is $M(z)$. Find an example of a received pulse $h(t)$ that results in this response.

Problem 7-14. Find the appropriate branch metric for the ML sequence detector for the following signal and noise generators. Let the signal generator output be non-negative, $S_k \geq 0$ for all k. Assume the noise generator outputs conditioned on a particular signal are independent and identically distributed Poisson random variables, and the k-th output has mean value αS_k. (This model is roughly equivalent to what might be encountered on a fiber optics channel.)

Problem 7-15. Repeat Problem 7-14 for the following. The signal generator output is binary, assuming the values $\{0, 1\}$, and the channel is the binary-erasure channel of Problem 4-8, with independent channel uses.

Problem 7-16. Consider transmitting bits X_k (zeros and ones) over a channel with additive white Gaussian noise. Assume that $X_k = 0$ for $k < 0$ and $k \geq K$. Suppose $K = 3$ and the observation sequence is $y_0 = 0.6$, $y_1 = 0.9$, $y_2 = 1.3$, and $y_3 = 0.3$.

(a) Find the ML decision sequence $\hat{x}_k$ assuming that the additive noise is the only degradation (no ISI) and that X_k are i.i.d.

(b) Suppose you are told that the ISI channel of $H(z) = 1 + 0.5z^{-1}$ is being used. Draw the trellis for the Markov model and label the transition weights. What is the ML detection of the incoming bit sequence?

Problem 7-17. Consider the following ISI model with AWGN:

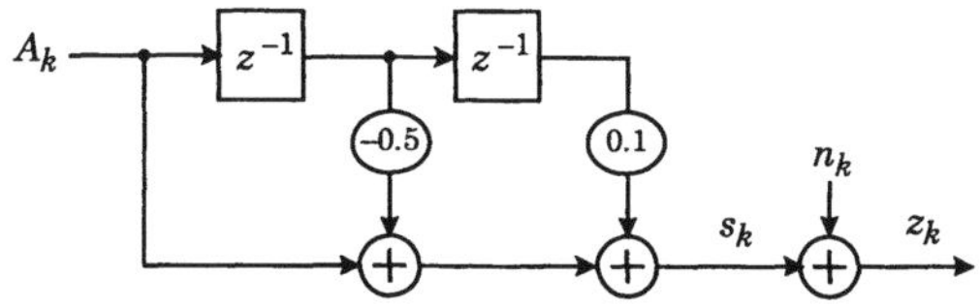

Assume A_k is equally likely to be 0 or 1, and the A_k are independent for all k. Assume additive Gaussian white noise with variance σ^2.

(a) Model the system as a shift register process and draw the state transition diagram. Label the arcs with the input/output pair (A_k, S_k).

(b) Draw one stage of a trellis and label with the input/output pairs (A_k, S_k).

(c) Assume $\psi_k = 0$ for $k \leq 0$ and $k \geq 5$. Suppose the observation sequence is $y_0 = 0.5$, $y_1 = -0.2$, $y_2 = 0.9$, $y_3 = 1.2$, and $y_4 = 0.1$. Draw a complete trellis with branch weights labeled.

(d) Use the Viterbi algorithm to find the ML decision sequence.

(e) Assuming that the correct state trajectory is $\psi_k = 0$ for all k, find the minimum-distance error event and its distance.

(f) Argue convincingly that the minimum distance found in part (e) is the minimum distance for any correct state trajectory.

Problem 7-18. Consider a response $M(z) = 1 + m_1 z^{-1} + m_2 z^{-2}$ at the output of a whitened matched filter. Assume the data symbols have alphabet $\{0, 1\}$.

(a) Draw the trellis diagram that can be used to find the minimum distance, and label each transition with the appropriate branch metric.

(b) Specify a finite set of error events that it suffices to consider in searching for the minimum distance.

(c) Give an example of a channel such that the minimum-distance error event is of length less than $(1 + m_1^2 + m_2^2)$.

Problem 7-19. For a response $M(z) = 1 + dz^{-1} + dz^{-2}$ at the whitened matched filter output, find the minimum distance as a function of $0 \leq d \leq 1$ assuming binary signaling from alphabet $\{0, 1\}$.

Problem 7-20. For the response $M(z) = 1 + z^{-1}$, find the set of minimum-distance error events and the resulting bound on the probability of symbol error assuming binary signaling with alphabet $\{0, 1\}$. Can you give an intuitive explanation for the error events you find?

Problem 7-21. A sequence of three symbols a_0, a_1, a_2 from the binary alphabet $\{\pm 1\}$, with a priori probabilities satisfying $p_{A_0}(1) = 2p_{A_1}(1) = p_{A_2}(1) = 1/2$, is transmitted across an ISI channel with transfer function $H(z) = 1 - z^{-1}$. Use the BCJR algorithm to find the a posteriori log-likelihood ratios λ_0, λ_1, and λ_2 of (7.76) when the observation after an AWGN channel with real variance $\sigma^2 = 0.5$ is $y = [y_0, \ldots y_3] = [1, 0, -1, 0]$. Assume that the idle symbol is -1, and that the state is constrained to be idle at times $k = 0$ and $k = 3$.

Problem 7-22.

(a) Apply the Chernov bound to the problem of detecting a binary shot noise signal with two known intensity functions as well as dark current. In particular, assume that the shot noise signal has filtering function $f(t)$, which is a composite of the photodetector response and any post-filtering, and that this signal is directly sampled at $t = 0$ and applied to a slicer.

(b) Minimize the upper bound with respect to $f(t)$, and show that the resulting detector correlates the logarithm of the known intensity against the delta-function shot noise signal.

(c) Evaluate the upper bound for the solution of (b)

Problem 7-23. Using the results of Appendix 7-A, develop the following relationship between $\{V_1, \ldots, V_M\}$ and $\{U_1, \ldots, U_N\}$, which serves to establish that $\{V_1, \ldots, V_M\}$ is also a set of sufficient statistics,

$$V_l = \sum_{k=1}^{N} F_k^{(m)*} U_k, \quad 1 \le m \le M. \tag{7.169}$$

References

1. H. L. Van Trees, *Detection, Estimation, and Modulation Theory*, Part I, Wiley, New York (1968).
2. L. Bahl, J. Cocke, F. Jelinek, and J. Raviv, Optimal Decoding of Linear Codes for Minimizing Symbol Error Rate, *IEEE Trans. Info. Theory*, Vol. 20, pp. 284-287, March 1974.
3. J. G. Proakis, *Digital Communications*, Fourth Edition, McGraw-Hill Book Co., New York (2001).
4. 3 S. Benedetto, E. Biglieri, and V. Castellani, *Digital Transmission Theory*, Prentice-Hall, Inc., Englewood Cliffs, NJ (1987).
5. G. D. Forney, Jr., "The Viterbi Algorithm," *Proceedings of the IEEE*, Vol. 61 (3), (March 1973).
6. G. D. Forney, Jr., "Maximum-Likelihood Sequence Estimation of Digital Sequences in the Presence of Intersymbol Interference," *IEEE Trans. on Information Theory*, Vol. IT-18, pp. 363-378 (May 1972).

7. R. W. Chang and J. C. Hancock, "On Receiver Structures for Channels Having Memory," *IEEE Trans. Inform. Theory*, vol. IT-12, pp. 463-468, Oct. 1966.

8. L. A. Rabiner, "A Tutorial on Hidden Markov Models and Selected Applications in Speech Recognition, *Proceedings of the IEEE*, Vol. 77, No. 2, pp. 257 -286, Feb. 1989.

8

Equalization

In PAM, intersymbol interference (ISI) results from linear amplitude and phase distortion in the channel that broadens the pulses and causes them to interfere with one another. The Nyquist criterion specifies a condition on the received pulses under which there is no ISI. Generally this or a similar condition is not satisfied unless we *equalize* the channel, meaning roughly that we filter to compensate for the channel distortion. Unfortunately, any equalization of amplitude distortion also enhances or amplifies any noise introduced by the channel, called *noise enhancement*. There is therefore a tradeoff between accurately minimizing intersymbol interference and minimizing the noise at the slicer input. Ultimately, of course, our goal is to minimize the probability of error.

In Chapter 7 we derived the maximum-likelihood sequence detector (MLSD), which is the optimal receiver for detecting a sequence of data symbols under the maximum-likelihood criterion. Rather than filtering the ISI to eliminate it entirely, the MLSD filters to ensure that the ISI is causal and monic; the resulting front-end structure is identical to the whitened matched filter (WMF) described in Section 5.4. The detection is performed by comparing the WMF output sequence against what that output sequence would be for each feasible sequence of input data symbols in the absence of noise; it chooses the data symbols that best match the WMF output according to a Euclidean distance measure.

Both MLSD (implemented using the Viterbi algorithm) and the symbol-wise MAP detector (implemented using the BCJR algorithm) are based on the same trellis structure. The complexity of both algorithms is governed by the number of states, which grows exponentially with the memory of the channel. Therefore, neither algorithm is feasible when the channel impulse response has many coefficients. In this chapter we describe several simpler strategies for mitigating ISI that are applicable to channels with large memory: linear equalization, decision-feedback equalization, and transmitter precoding.

In Section 8.1, the optimal equalization structures are derived under a *zero-forcing (ZF) criterion*, which means that we completely eliminate the ISI at the slicer input. The starting point is the WMF, which is used as the front-end of the receiver of Chapters 5 and 7, and relies on the sufficient statistic argument. In Section 8.2 these results are generalized by relaxing the assumption that the WMF is used, and also by considering an alternative *mean-square error (MSE) criterion*. The MSE criterion reduces noise enhancement by allowing residual ISI at the slicer input, and attempts to minimize the sum of the ISI and noise. In Section 8.3 we demonstrate that the equalization, including the matched filter in the WMF, can be implemented in discrete time if the sampling rate is increased to some multiple of the symbol rate, as is often done in practice. Section 8.4 describes briefly a practical equalizer filter structure called the *transversal filter*, which is just another name for a direct-form FIR filter. The transversal filter plays a central role in the realization of *adaptive equalization* as described in Chapter 9. Finally, in Section 8.5 we consider the channel capacity of a channel with ISI, and derive *Price's result*, which suggests that the DFE receiver structure does not compromise channel capacity at high SNR. In other words, with appropriate channel coding, the trellis-based equalizers of the previous chapter are not only high in complexity, but also unnecessary.

In this chapter we assume that the linear distortion of the channel is exactly known. Furthermore, we do not constrain the complexity of the receiver or its filter structures. Both of these assumptions are relaxed in Chapter 9, where we describe constrained-complexity filters, known as *adaptive equalizers*, that adapt to unknown or changing channel conditions.

Notation

Some of the expressions in this chapter become fairly complicated, so will often use a simplified notation in which the frequency variable is suppressed. For a transfer function $F(z)$, which becomes $F(e^{j\theta})$ when evaluated on the unit circle, we will write simply F. Similarly, the reflected transfer function $F^*(1/z^*)$, which becomes $F^*(e^{j\theta})$ when evaluated on the unit circle, will be written in shorthand notation as F^*.

Example 8-1.

Given a transfer function $S(z)$ that is non-negative real on the unit circle, the minimum-phase spectral factorization can be written in four ways,

$$S(z) = \gamma^2 M(z) M^*(1/z^*) , \qquad S(e^{j\theta}) = \gamma^2 |M(e^{j\theta})|^2 , \tag{8.1}$$

$$S = \gamma^2 M M^* , \qquad S = \gamma^2 |M|^2 . \tag{8.2}$$

We will often use in this chapter the shorthand notations of (8.2).

Also, in this chapter the arithmetic and geometric means of a function will arise frequently, and we will use a shorthand notation for these means. Given a *non-negative real-valued* function $f(x)$ and a subset X of the x axis, define $|X|$ as the total size (measure) of the set X. For example, if $X = \{ x: |x| \le x_0 \}$, then $|X| = 2x_0$. The *arithmetic mean* of $f(x)$ over X is defined as

$$\langle f \rangle_{A,X} = \frac{1}{|X|} \int_X f(x) dx \,, \tag{8.3}$$

which has the interpretation as the average value of $f(x)$ over the interval X. Similarly, define a *geometric mean* of $f(x)$ over X

$$\langle f \rangle_{G,X} = \exp\left\{ \frac{1}{|X|} \int_X \log f(x) \, dx \right\} . \tag{8.4}$$

The geometric mean is independent of the base of both the exponential and the logarithm, as long as they agree. So, for example, we could write $\langle f \rangle_{G,\,X} = 2^{\langle \log_2 f \rangle_{A,X}}$.

Jensen's inequality implies that $E[\log f(Y)] \le \log E[f(Y)]$ for any random variable Y and any nonnegative function $f(\cdot)$, with equality if and only if $f(Y)$ is deterministic. By interpreting Y as a random variable uniformly distributed over the set X, Jensen's inequality implies that the geometric mean is always less than or equal to the arithmetic mean:

$$\langle f \rangle_{G,X} \le \langle f \rangle_{A,X} \tag{8.5}$$

with equality if and only if $f(x)$ is a constant over the set X.

Example 8-2.

The constant γ^2 in the geometric mean formula of (8.2) is, from (2.69),

$$\gamma^2 = \langle S \rangle_{G,[-\pi,\,\pi)}, \tag{8.6}$$

where the independent variable in this case is θ rather than x.

The arithmetic and geometric means have several useful properties. For both types of mean, the mean of a real nonnegative constant is itself,

$$\langle a \rangle_{A,X} = \langle a \rangle_{G,X} = a \,. \tag{8.7}$$

Similarly,

$$\langle a \cdot f \rangle_{A,\,X} = a \cdot \langle f \rangle_{A,\,X} \,, \qquad \langle a \cdot f \rangle_{G,\,X} = a \cdot \langle f \rangle_{G,\,X} \,. \tag{8.8}$$

The arithmetic mean also obeys the distributive law,

$$\langle a \cdot f + a \cdot g \rangle_{A,\,X} = a \cdot \langle f \rangle_{A,X} + b \cdot \langle g \rangle_{A,X} \,, \tag{8.9}$$

but the geometric mean does not. Conversely, the geometric mean obeys the multiplicative laws

$$\langle f \cdot g \rangle_{G,X} = \langle f \rangle_{G,X} \cdot \langle g \rangle_{G,X}, \qquad \langle \frac{f}{g} \rangle_{G,X} = \frac{\langle f \rangle_{G,X}}{\langle g \rangle_{G,X}}, \tag{8.10}$$

which the arithmetic mean does not.

8.1. Optimal Zero-Forcing Equalization

In Chapter 7 we derived receivers for PAM with ISI based on optimal detection criteria. In particular, the MLSD offers a computational load that is constant with time. However, in practice the MLSD is rarely used to equalize ISI for a couple of reasons. First, the complexity of trellis-based equalizer grows exponentially with the memory in the channel, making them impractical for all but channels with very short impulse responses. Second, in high performance data communication systems, error-control coding is almost always used (Chapters 12 and 13), and surprisingly it has been shown that coding can be used to approach channel capacity in the presence of ISI without trellis-based equalization.

In this section, we will first review the results of Chapter 7 as they apply to PAM with ISI, and in the process develop some useful bounds on the performance of receivers in the presence of ISI. Following this background, we will describe three practical and widely used equalization techniques — linear equalization, decision-feedback equalization, and transmitter precoding — and compare their performance to one another and to the MLSD.

8.1.1. Background Results

The results of Chapter 7 will prove very useful as a starting point for the derivation of optimal equalizer structures for ISI. Not only does Chapter 7 establish a canonical front end for all receivers, including those based on equalization, but it also establishes some useful bounds on the performance of such receivers.

In Section 7.3 we established that the whitened matched filter (WMF), first encountered in Section 5.4 as an embodiment of the minimum-distance receiver design criterion, develops a set of sufficient statistics for the received signal. When the noise on the channel is white and Gaussian, the WMF can be used as the front end of a receiver designed according to any criterion of optimality. This result is particularly valuable because of the desirable properties of the WMF output:

- It is a sampled data signal, with sampling rate equal to the symbol rate.
- The equivalent discrete-time channel model is causal and minimum-phase, the latter implying that the energy of the impulse response is maximally concentrated in the early samples (Problem 2-29).
- The equivalent additive noise is Gaussian, circularly symmetric and white, implying that the samples of this complex-valued noise are independent.

These properties greatly simplify the derivation of the remainder of the receiver for any particular criterion. It was shown in Section 7.3 that the WMF can be generalized to nonwhite

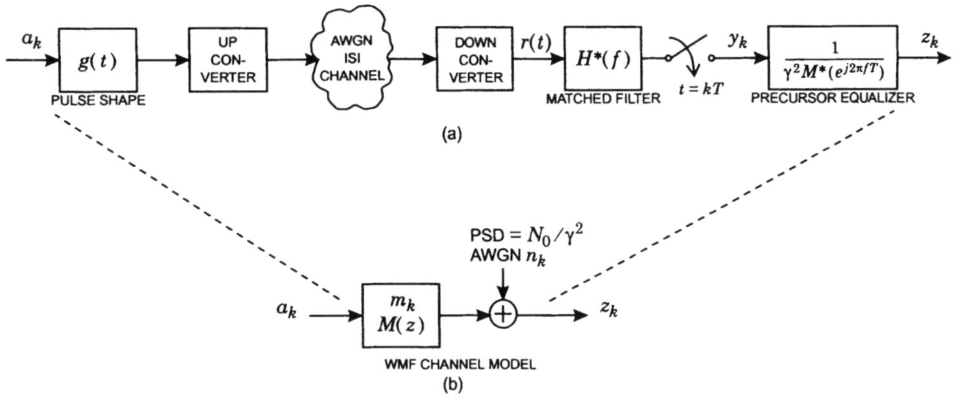

Fig. 8-1. A passband PAM transmitter, channel, and receiver front end (a). The transmitter consists of a transmit filter and upconverter; the channel has ISI and adds white Gaussian noise with PSD $N_0/2$; the receiver consists of a downconverter, a MF, a sampler, and a precursor equalizer. The equivalent WMF channel model (b) is a causal monic minimum-phase filter, with additive white and circularly symmetric Gaussian noise.

channel noise, and that the equivalent channel model is essentially the same. In this section we will assume the channel noise is white.

A passband PAM system together with a WMF receiver front end is shown in Fig. 8-1(a). In the following we review the properties of the equivalent channel model for both the sampled MF front end and the WMF front end.

Sampled MF Channel Model

Let us first derive the equivalent impulse response between the transmitted data symbols a_k and the sampled MF output y_k. This can be done by setting $a_k = \delta_k$, so that the noiseless downconverter output is an isolated pulse $h(t)$. The equivalent impulse response at the output of the sampler is then $\rho_h(k)$, the sampled autocorrelation function of $h(t)$. Taking the Z transform, the equivalent transfer function is $S_h(z)$, the folded spectrum.

Next, let us derive the noise power spectrum after the sampled MF. When the original passband noise is white and Gaussian with power spectrum $N_0/2$, the downconverted noise can be modeled as circularly symmetric and Gaussian with power spectrum N_0. (See Section 3.2.8). This implies that the power spectrum after the MF is also circularly symmetric and Gaussian with power spectrum $N_0 |H(f)|^2$. Since sampling a random process aliases its power spectrum according to Section 3.2.5, it follows that the power spectrum of the discrete-time noise after the sampled MF is $N_0 S_h(z)$, where $S_h(e^{j2\pi fT})$ is the folded spectrum of (5.66).

Together, the above results imply that the following is an equivalent model for the sampled MF output when the original channel adds white Gaussian noise with power spectrum $N_0/2$:

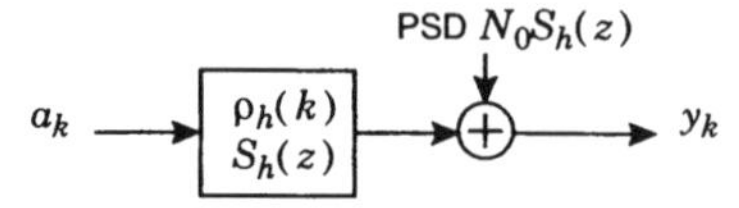

SAMPLED MF CHANNEL MODEL

Interestingly, the equivalent transfer function and the equivalent noise PSD are identical to within a constant. This channel model has two drawbacks: the impulse response is two-sided (and hence noncausal), and the noise is not white. Both drawbacks are rectified by the WMF.

WMF Channel Model

Let us now derive the equivalent channel model after the noise-whitening filter of Fig. 8-1(a), or in other words, after the WMF. We have already seen that the data symbols see a transfer function of $S_h(z)$ at the output of the sampler, which is nonnegative real on the unit circle and thus admits a minimum-phase spectral factorization (Section 2.5):

$$S_h(z) = \gamma^2 M(z) M^*(1/z^*) \,. \tag{8.11}$$

Therefore, the equivalent transfer function after the noise whitener $1/\gamma^2 M^*(1/z^*)$ is simply $M(z)$. This equivalent response is causal. For this reason, the filter $1/\gamma^2 M^*(1/z^*)$ is known as a *precursor equalizer*, because, among other things, it eliminates the noncausal portion of the ISI, called the precursor. In other words, it turns a two-sided impulse response $\rho_h(k)$ at the sampled MF output into a causal response m_k at the WMF output.

We have already seen that the noise after the sampled MF is circularly symmetric Gaussian with power spectrum $N_0 S_h(z)$. Therefore, the precursor equalizer changes the power spectrum of the noise to:

$$S_n(z) = N_0 S_h(z) \cdot \frac{1}{\gamma^2 M^*(1/z^*)} \cdot \frac{1}{\gamma^2 M(z)} = \frac{N_0}{\gamma^2} \,. \tag{8.12}$$

The noise after the WMF is white. This explains why the precursor equalizer is also known as a *noise whitener*, and which also explains the *whitened* MF terminology. Furthermore, the samples of this noise are statistically independent, since it is white and circularly symmetric. In addition, the real and imaginary parts are statistically independent, and have the same variance, namely

$$\sigma^2 = \frac{N_0/2}{\gamma^2} = \frac{N_0}{2} \langle S_h^{-1} \rangle_G$$

$$= \frac{N_0}{2} \cdot \exp\left\{ \frac{-1}{2\pi} \int_{-\pi}^{\pi} \log S_h(e^{j\theta}) \, d\theta \right\}, \tag{8.13}$$

where we made use of (2.69).

Therefore, the equivalent channel model for the WMF is that shown in Fig. 8-1(b). The equivalent transfer function is $M(z)$, which is causal and monic, and the noise is white and circularly symmetric Gaussian with power spectrum N_0/γ^2.

It is not necessary to evaluate the integral of (8.13) for rational folded spectra, since the constant γ^2 can be read directly from the rational spectrum. This is illustrated by two examples.

Example 8-3.
Let the received pulse be $h(t) = \sqrt{2E_h/\tau}\, e^{-t/\tau}u(t)$, where $u(t)$ is the unit step function, $\tau > 0$, and E_h is the pulse energy. Calculating the pulse autocorrelation directly yields:

$$\rho_h(k) = E_h \alpha^{|k|}, \qquad \alpha = e^{-T/\tau}, \tag{8.14}$$

so that the folded spectrum is first-order all-pole and rational:

$$S_h(z) = E_h(1-\alpha^2) \cdot \frac{1}{1-\alpha z^{-1}} \cdot \frac{1}{1-\alpha z} . \tag{8.15}$$

This is written in the form of a minimum phase spectral factorization, since the first term $M(z) = 1/(1-\alpha z^{-1})$ is minimum-phase and monic ($M(\infty) = 1$). Because each term is monic, we deduce that

$$\gamma^2 = E_h(1-\alpha^2), \tag{8.16}$$

without the need to evaluate the geometric-mean integral.

Example 8-4.
Let the received pulse be $h(t) = h_0(t) + \alpha h_0(t-T)$, where $h_0(t)$ is a pulse shape with energy E_0 that is orthogonal to its translates by kT. Then the autocorrelation of this pulse is $\{\ldots\ 0, \alpha E_0, (1+\alpha^2)E_0, \alpha E_0, 0, \ldots\}$, so that the folded spectrum is:

$$S_h(z) = E_0(1+\alpha z^{-1})(1+\alpha z) . \tag{8.17}$$

Again, this is written in the form of a monic minimum-phase spectral factorization, and we identify $\gamma^2 = E_0$.

The existence of the spectral factorization of (8.11) was established in Section 2.5 only for rational folded spectra. Fortunately, it holds more generally, in fact for any folded spectrum $S_h(z)$ such that both $S_h(e^{j\theta})$ and $\log S_h(e^{j\theta})$ are integrable on the interval $[-\pi, \pi)$. This is known as the *Paley-Wiener condition*. From (8.13), the integrability of $\log S_h(e^{j\theta})$ is required for γ^2 to be well defined.

Example 8-5.
If $S_h(e^{j\theta}) = 0$ over any interval of frequencies, then $\log S_h(e^{j\theta}) = -\infty$ over that same interval, and thus $\log S_h(e^{j\theta})$ is not integrable. Intuitively, this is obvious since for this condition the noise at the output of the sampled matched filter will have a vanishing power spectrum over an interval of frequencies, and clearly it cannot then be whitened.

We cannot guarantee the existence of a spectral factorization of (8.11) unless the folded spectrum vanishes on the unit circle only at discrete points, and even then it must vanish in such a way that its logarithm is integrable. Fortunately, rational folded spectra can vanish at only a finite set of points on the unit circle (at most the number of zeros), and their logarithm is always integrable.

The existence of the WMF in Fig. 8-1 depends on the spectral factorization of $S_h(z)$, which depends in turn on the integrability of $S_h(e^{j\theta})$. The WMF exists for rational folded spectra without poles on the unit circle, but not for all non-rational folded spectra. The receiver design techniques described in this section do not apply to folded spectra for which the WMF does not exist.

Very useful for the sequel is the observation that the energy in the received pulse $h(t)$ can be related to γ^2 and $M(z)$ through the time-domain version of (8.11):

$$\rho_h(k) = \gamma^2 \cdot m_k * m_{-k}^* \tag{8.18}$$

and hence

$$E_h = \rho_h(0) = \gamma^2 \sum_{k=0}^{\infty} |m_k|^2 \ . \tag{8.19}$$

In the following sections, we will use the WMF of Fig. 8-1(a) as the front end of the receiver. For purposes of designing the equalizer structures, it is appropriate to use the channel model of Fig. 8-1(b) as the starting point.

Probability of Error

It was shown in Section 6.2 that when a minimum-distance receiver design criterion is used for the equivalent channel model of Fig. 8-1(b), the error probability is approximately

$$P_e \approx K \cdot Q(\sqrt{\Gamma}/2) \ , \qquad \Gamma = d_{\text{min}}^2 / \sigma^2 \ , \tag{8.20}$$

where $K \geq 1$ is the error coefficient, equal to the average number of signals at the minimum distance, d_{min} is the minimum Euclidean distance between pairs of known signals, and $\sigma^2 = N_0/(2\gamma^2)$ is the variance of the real or imaginary parts of the additive noise. The interpretation of the terminology *known signals* depends on the context. We will see two examples below: the isolated-pulse case, and the sequence case.

It was observed that the argument of $Q(\cdot)$ has much greater impact on the error probability than the multiplicative constant K. Therefore, in this chapter we will compare modulation techniques primarily through their value of Γ, rather than comparing the error probability directly. Since Γ is the primary parameter of a particular receiver design impacting its probability of error, we will call it the *figure of merit* for the receiver equalization or detection technique. Generally, receivers with a higher figure of merit will have a lower error probability (although K must be taken in account to be definitive).

Matched-Filter Bound

The WMF leads directly to two upper bounds on the figure of merit Γ in the presence of ISI. The first bound is the matched filter bound, which presumes that a single data symbol is transmitted. That is, the input to the equivalent channel of Fig. 8-1(b) is $a_0\delta_k$ for some isolated data symbol a_0. The output of the channel is then

$$z_k = a_0 m_k + n_k \,. \tag{8.21}$$

This is the case of a received discrete-time signal in additive white Gaussian noise. The ML detector is shown in Section 7.3 to be a matched filter with transfer function $M^*(1/z^*)$ followed by a sampler at $k = 0$ and a slicer for a scaled version of the data symbol a_0. Actually, this discrete-time matched filter is (within a constant) the inverse of the equalizer in the WMF, and thus the ML detector reverts to the output of the *continuous-time* matched filter $H^*(f)$ sampled at time $t = 0$. Since the continuous-time received signal in this case is $a_0 h(t)$, this is obvious in retrospect. The original matched filter was adequate without the need for the precursor equalizer.

The set of known signals in (8.21) is $a_0 m_k$ for all a_0 in the symbol alphabet. The minimum distance for this set of known signals is

$$d_{\min}^2 = \min \sum_{k=0}^{\infty} |a_0^{(1)} m_k - a_0^{(2)} m_k|^2 = a_{\min}^2 \sum_{k=0}^{\infty} |m_k|^2 \,, \tag{8.22}$$

where $a_0^{(1)}$ and $a_0^{(2)}$ are two different data symbols, the minimization is over all such pairs such that $a_0^{(1)} \neq a_0^{(2)}$, and $a_{\min}$ is the minimum distance within the data-symbol alphabet. From (8.19), this can be written in the form

$$d_{\min}{}^2 = \frac{a_{\min}^2 E_h}{\gamma^2} \,. \tag{8.23}$$

Thus, from (8.20) we get a figure of merit of

$$\Gamma_{\text{MF}} = \frac{d_{\min}^2}{\sigma^2} = a_{\min}^2 \cdot \frac{2E_h}{N_0} \,. \tag{8.24}$$

The constant $2E_h/N_0$ is a kind of signal-to-noise ratio, equal to the received pulse energy divided by the power spectrum of the white noise.

In the presence of ISI, the receiver will generally not achieve a figure of merit of Γ_{MF} because the matched-filter receiver does not take ISI into account. However, Γ_{MF} represents a very useful benchmark, since the difference between the actual receiver figure of merit (always smaller than Γ_{MF}) and Γ_{MF} is a measure of the severity of the ISI, as well as the effectiveness of our methods for countering its effects.

Figure of Merit for the MLSD

Another upper bound on the figure of merit is Γ_{MLSD}, the figure of merit for the MLSD. No receiver based on equalization or other techniques for countering ISI can perform better than the MLSD. Thus, Γ_{MLSD} represents an upper bound on the figure of merit. Moreover, any gap between Γ_{MLSD} and Γ_{MF} is due to ISI and not to shortcomings in the receiver design. This difference is a fundamental measure of the severity of the ISI. Surprisingly, in many cases this difference is zero, meaning that within a multiplicative constant the error probability of the MLSD is essentially the same as the MF bound even in the presence of ISI.

The figure of merit for the MLSD is given by

$$\Gamma_{MLSD} = \frac{2d_{min}^2}{N_0/\gamma^2} = \frac{2d_{min}^2\gamma^2}{N_0}\ , \tag{8.25}$$

where

$$d_{min}^2 = \min_{\{\varepsilon_0 \neq 0,\, \varepsilon_1, \ldots \varepsilon_{L-1}\}} \sum_{k=0}^{L+\mu-1} \left| \sum_{i=0}^{\mu} m_i \varepsilon_{k-i} \right|^2 \ . \tag{8.26}$$

The calculation of this minimum distance is described in Section 7.6.

A pair of very useful bounds on Γ_{MLSD} can be developed simply. The first bound shows that the MLSD has a figure of merit less than or equal to the matched-filter bound. This follows from the simple observation that if we perform the minimization over a set of restricted error events, then this cannot match the minimum distance. In particular, in (8.26) constrain ε_k to be of the form $\varepsilon_k = \varepsilon_0\delta_k$, so that

$$\begin{aligned} d_{min}^2 &\le \min_{\varepsilon_0 \neq 0} \sum_{k=0}^{L+\mu-1} |\varepsilon_0 m_k|^2 = \min_{\varepsilon_0 \neq 0} |\varepsilon_0|^2 \sum_{k=0}^{\infty} |m_k|^2 \\ &= a_{min}^2 \sum_{k=0}^{\infty} |m_k|^2 = a_{min}^2 E_h/\gamma^2, \end{aligned} \tag{8.27}$$

where we have used (8.19). Equation (8.27) leads to the desired upper bound on Γ_{MLSD},

$$\Gamma_{MLSD} \le \frac{a_{min}^2 E_h/\gamma^2}{N_0/(2\gamma^2)} = a_{min}^2 \cdot \frac{2E_h}{N_0} = \Gamma_{MF}\ . \tag{8.28}$$

It is possible for $\Gamma_{MLSD} = \Gamma_{MF}$, even in the presence of ISI. This will occur whenever one of the minimum-distance error events is the single error ε_0, since in that case (8.27) becomes an equality.

Example 8-6.

Suppose the minimum-phase channel response is $M(z) = 1 + \alpha z^{-1}$. Assume the original data symbols are taken from the alphabet $\{0, 1\}$, so that the error symbols have alphabet $\{0, \pm 1\}$. For the given $M(z)$, the error-event trellis has three states, as shown in Fig. 8-2. Shown are the only two possible error events, and the shorter error event is *always* the minimum-distance event because its path metric is smaller by $(1-\alpha)^2$. The minimum distance error event has only a single non-zero error ε_0. Hence, $\Gamma_{MLSD} = \Gamma_{MF}$ for this channel, and the MLSD achieves the same figure of merit as the matched filter bound. We can verify this in another way, because $d_{min}^2 = 1 + \alpha^2$, and hence

$$\Gamma_{MLSD} = \frac{d_{min}^2}{N_0/(2\gamma^2)} = \frac{1+\alpha^2}{N_0} \cdot \frac{2E_h}{1+\alpha^2} = \frac{2E_h}{N_0} = \Gamma_{MF}\ . \tag{8.29}$$

In (8.29) we have used the fact that $\gamma^2 = E_h/(1+\alpha^2)$ and also that $a_{\min} = 1$ for this alphabet.

A lower bound on Γ_{MLSD} follows from the inequality

$$\sum_{k=0}^{L+\mu-1}\left|\sum_{i=0}^{\mu} m_i\varepsilon_{k-i}\right|^2 \ge \left|\sum_{i=0}^{\mu} m_i\varepsilon_{-i}\right|^2 = |\varepsilon_0 m_0|^2 = |\varepsilon_0|^2 . \qquad (8.30)$$

For any candidate error sequence, the left-hand side of (8.29) is greater than or equal to the right side. Thus, for the error sequence that minimizes the left side (and achieves $d^2_{\min}$), $|\varepsilon_0|^2$ is not greater. If we minimize $|\varepsilon_0|^2$ over ε_0 (yielding $a^2_{\min}$), we get an even smaller (or equal) quantity. Thus, it follows that,

$$d^2_{\min} \ge a^2_{\min} \, , \qquad \Gamma_{\text{MLSD}} \ge a^2_{\min} \cdot \frac{2\gamma^2}{N_0} . \qquad (8.31)$$

This bound will prove useful in comparing the MLSD to other receivers later. (The lower bound is actually the figure of merit of the decision-feedback equalizer considered later.) If there is ISI ($M(z) \ne 1$), and $M(z)$ is an FIR filter (as it must be for application of the VA), then this inequality is strict (Problem 8-2).

8.1.2. Zero-Forcing Linear Equalizer

An obvious sub-optimal receiver eliminates ISI using a linear filter called a *linear equalizer*. For example, we can choose the receive filter so as to satisfy the Nyquist criterion at the slicer input. However, such a filter is not unique. The Nyquist criterion specifies neither the equalized pulse at the slicer input nor the receive filter. Even if the excess bandwidth of the equalized pulse is chosen, the pulse and the corresponding receive filter are not unique. The question is, among those satisfying the Nyquist criterion, which receive filter minimizes the noise at the slicer or equivalently minimizes the probability of symbol error?

The optimal receive filter under the constraint that the Nyquist criterion must be satisfied at the slicer input can now be determined easily because the WMF can be used as a front end for *any* criterion of optimality, including this one. Placing a discrete-time symbol-rate filter at the WMF output, the filter that will satisfy the Nyquist criterion is unique because of the

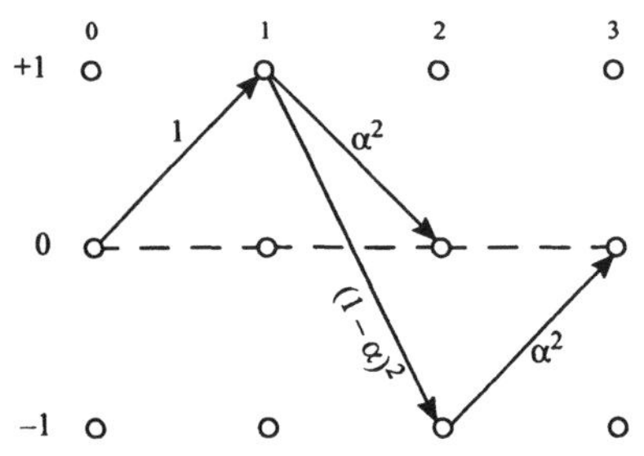

Fig. 8-2. The trellis for determining the minimum distance in Example 8-6.

symbol-rate sampling. That filter, called the *zero-forcing linear equalizer (LE-ZF)*, is the inverse filter $M^{-1}(z)$, assuming the discrete-time equivalent channel model for the WMF of Fig. 8-1(b), since the Nyquist criterion is equivalent to an overall unity transfer function for the WMF plus equalizer. The term "zero-forcing" refers to the fact that we are constraining the ISI to be entirely absent, a constraint that will be relaxed in Section 8.2. Since $M(z)$ is a monic causal minimum-phase filter, the equalizer must have all its poles inside or on the unit circle; hence it is a causal filter, although it is not necessarily minimum phase or stable, because of the possibility of poles on the unit circle. We will see that when $M(z)$ has zeros on the unit circle, the LE-ZF is not useful since the noise at the slicer input would have to have infinite variance (the figure of merit would be zero).

Combining the equalizer $M^{-1}(z)$ with the WMF discrete-time equalizer in Fig. 8-1(a), we show the resulting LE receiver structure in Fig. 8-3. Among all receiver structures consisting of a downconverter, receiver filter, and equalizer constrained to eliminate ISI at the slicer input, this one is optimal (minimizes error probability). The combined discrete-time equalizer has transfer function $S_h^{-1}(z)$, the inverse of the folded spectrum.

Figure of Merit for the LE-ZF

The figure of merit for the LE is easily determined from the fact that the noise at the WMF output is white and circularly symmetric with variance N_0/γ^2. The power spectrum of the noise at the output of the LE is therefore (suppressing the z and $1/z^*$ variables):

$$S_v = \frac{N_0}{\gamma^2} \cdot \frac{1}{M} \cdot \frac{1}{M^*} = \frac{N_0}{S_h} \,. \tag{8.32}$$

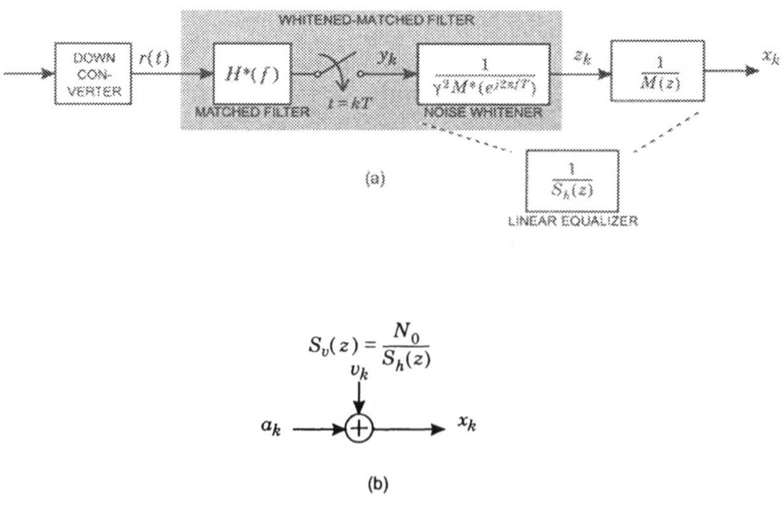

Fig. 8-3. The zero-forcing linear equalizer. (a) It consists of a matched filter, symbol-rate sampler, and a discrete-time equalizer with transfer function that is the inverse of the folded spectrum. (b) The equivalent channel model consists of an ideal channel plus nonwhite Gaussian noise.

The input to the slicer is the current data symbol plus an additive Gaussian noise sample. To find the variance of this noise, we need only integrate the power spectrum above, yielding:

$$2\sigma_v^2 = N_0 \cdot \langle S_h^{-1} \rangle_A \ . \tag{8.33}$$

The minimum distance is that of the data symbol, $a_{\min}$, and thus the figure of merit is

$$\Gamma_{\text{ZF-LE}} = \frac{2a_{\min}^2}{N_0} \cdot \langle S_h^{-1} \rangle_A^{-1} \ . \tag{8.34}$$

For rational spectra, it is easier to calculate $\Gamma_{\text{ZF-LE}}$ by finding the coefficient of z^0 in $S_h^{-1}(z)$ than it is to evaluate this integral.

Example 8-7.
For the first-order all-pole received pulse of Example 8-3, the inverse of the folded spectrum is:

$$S_h^{-1}(z) = \frac{(1-\alpha z^{-1})(1-\alpha z)}{(1-\alpha^2)E_h} \tag{8.35}$$

so that the coefficient of z^0 is $(1+\alpha^2)(1-\alpha^2)^{-1}/E_h$. Thus, the figure of merit is

$$\Gamma_{\text{ZF-LE}} = a_{\min}^2 \frac{2E_h}{N_0} \cdot \frac{1-\alpha^2}{1+\alpha^2} = \Gamma_{\text{MF}} \cdot \frac{1-\alpha^2}{1+\alpha^2} \ . \tag{8.36}$$

Note that $\Gamma_{\text{ZF-LE}} \to 0$ as $|\alpha| \to 1$, which is the case where the channel pole approaches the unit circle.

Example 8-8.
For the first-order all-zero received pulse $h(t) = h_0(t) + \alpha h_0(t-T)$ of Example 8-4, the inverse of the folded spectrum is

$$S_h^{-1}(z) = \frac{1+\alpha^2}{E_h} \cdot \frac{1}{(1-\alpha z^{-1})(1-\alpha z)} \tag{8.37}$$

and when expanded in z^{-k} has a coefficient of z^0 of $(1+\alpha^2)(1-\alpha^2)^{-1}/E_h$. As a result, the figure of merit is

$$\Gamma_{\text{ZF-LE}} = a_{\min} \frac{2E_h}{N_0} \cdot \frac{1-\alpha^2}{1+\alpha^2} = \Gamma_{\text{MF}} \cdot \frac{1-\alpha^2}{1+\alpha^2} \ . \tag{8.38}$$

Note that $\Gamma_{\text{ZF-LE}} \to 0$ as $|\alpha| \to 1$, because the channel zero approaches the unit circle and the LE-ZF cannot equalize it.

Bound on Figure of Merit

An upper bound on $\Gamma_{\text{ZF-LE}}$ that is useful for later comparisons can be developed. Let f_k denote the impulse response of the equalizer $M^{-1}(z)$. This response is causal, and since $M^{-1}(\infty) = 1$, it is also monic ($f_0 = 1$). Since the input noise to this LE-ZF is white, we can express the output variance in terms of this impulse response as:

$$2\sigma_v^2 = \frac{N_0}{\gamma^2} \cdot \sum_{k=0}^{\infty} |f_k|^2 \geq \frac{N_0}{E_h}, \tag{8.39}$$

with equality if and only if $M(z) = 1$ (there is no ISI). From this it follows that

$$\Gamma_{\text{ZF-LE}} = \frac{a_{\min}^2}{\sigma_v^2} \leq a_{\min}^2 \cdot \frac{2\gamma^2}{N_0}. \tag{8.40}$$

The right side of (8.40) is the figure of merit for the zero-forcing decision-feedback equalizer, considered later in Section 8.1.3.

Existence of the LE-ZF

The existence of the LE-ZF is not guaranteed, but depends on the folded spectrum $S_h(z)$. The first requirement is the existence of the WMF, or in other words that $\log S_h(e^{j\theta})$ be integrable. As we saw earlier, the WMF is precluded if, for example, $S_h(e^{j\theta})$ vanishes on an interval. The LE-ZF has the more stringent requirement that the filter $M^{-1}(z)$ be stable, and that the noise variance at its output be finite. The problematic case is where $M(z)$ has one or more zeros on the unit circle, in which case $M^{-1}(z)$ is a well-defined filter but is not stable, and the variance of the noise at the slicer is infinite. This can be seen from (8.39) above, because when $M^{-1}(z)$ has a pole on the unit circle, f_k does not decay to zero as $k \to 0$, and thus $\sum_{k=0}^{\infty} |f_k|^2 = \infty$.

In summary, when $S_h(z)$ is rational, the LE-ZF will not be useful (in the sense that the slicer noise variance would be infinite) whenever $S_h(z)$ has zeros on the unit circle. Intuitively, this is because the LE-ZF, in the process or adding gain to compensate for channel attenuation, cannot compensate for even an algebraic zero in the frequency response of the channel. Note that under these conditions the WMF *does* exist, so that for example the MLSD can be implemented.

When $S_h(z)$ is not rational, the precise condition for the LE-ZF to be useful is that the inverse folded spectrum $S_h^{-1}(e^{j\theta})$ be integrable, which guarantees a finite noise variance at the slicer input.

8.1.3. Zero-Forcing Decision-Feedback Equalizer

The *zero-forcing decision-feedback equalizer (DFE-ZF)* is a nonlinear receiver structure that offers a performance intermediate between the LE and the MLSD. In the absence of channel coding (Section 12), the DFE is frequently used since it offers a good compromise between performance and implementation complexity.

The DFE-ZF follows from the observation that the WMF equivalent channel model of Fig. 8-1(b) is monic and causal. Thus, the residual ISI at this point is called *postcursor ISI*, meaning that the interference at the WMF output is only from *past* data symbols. This is distinguished from *precursor ISI*, where the interference is from future symbols. The distinguishing feature of postcursor ISI is that if we know the past data symbols, we can *cancel* the ISI by subtracting a replica of the ISI from the WMF output. If we are making symbol-by-symbol decisions using a slicer, then in fact we do have *estimates* of the past data symbols, namely the slicer output. In the DFE, these estimates are used to cancel the postcursor ISI at the slicer input.

The DFE is related to the LE in Fig. 8-4. In Fig. 8-4(a), the LE is shown in a different form, where the equalizer filter $M^{-1}(z)$ is realized as a filter $M(z) - 1$ placed in a feedback loop. It is easily shown that the transfer function of this feedback system is $M^{-1}(z)$. The feedback loop is also realizable; because $M(z)$ is causal and monic, $M(z) - 1$ is a strictly causal response with no zero-delay tap (such a tap could not be realized in the feedback loop!). This filter is stable as long as $M(z)$ has no zeros on the unit circle (it is strictly minimum-phase).

The DFE, first suggested by Austin [1][2], is shown in Fig. 8-4(b). There are two equivalent ways of looking at this receiver structure. One follows from the observation that, in the absence of noise, the output of the LE (input to the slicer) in Fig. 8-4(a) is precisely the data symbol a_k. This property is not affected by moving the slicer *inside* the feedback loop,

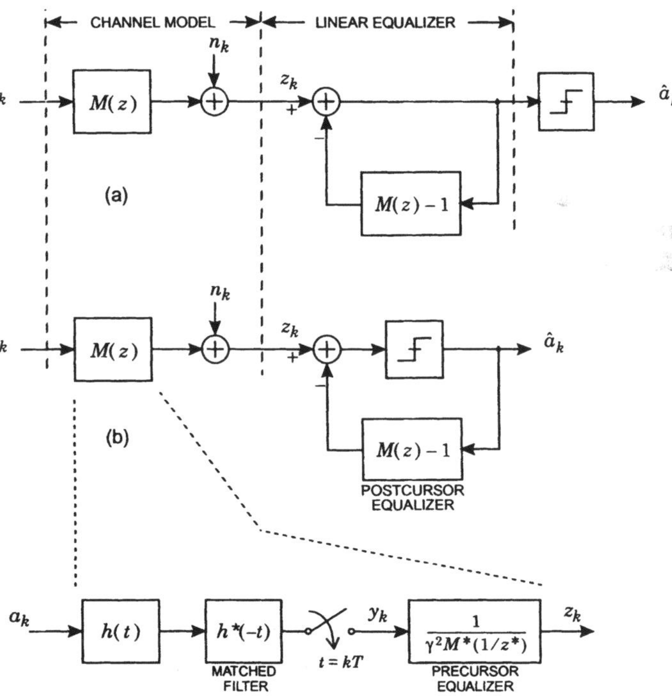

Fig. 8-4. The zero-forcing decision-feedback equalizer (DFE-ZF). (a) A different realization of the LE-ZF, in which the filter $M^{-1}(z)$ is implemented with a feedback filter. (b) The DFE-ZF, in which the slicer is moved inside the feedback loop, reducing the noise enhancement. (c) Expansion of the channel model. The front end of the DFE-ZF, identical to the WMF, consists of a matched filter and precursor equalizer, where the latter also doubles as a whitening filter.

since in the absence of noise the slicer has no effect. The effect on the *noise* of moving the slicer inside the feedback loop is beneficial, however, since the effect of the slicer is to remove any noise that would otherwise recirculate back through the feedback loop to the input to the slicer (this will be verified mathematically momentarily). Moving the slicer inside the feedback loop also has a beneficial stabilizing effect; even if $G(z)$ has zeros on the unit circle and the LE-ZF filter is not stable, the nonlinear filter of Fig. 8-4(b) is stable, since the output of the slicer is always bounded! Thus, the DFE-ZF exists and is a stable system whenever the WMF exists.

The second viewpoint, ISI cancellation, is illustrated in Fig. 8-5. The channel model of Fig. 8-4(c) is shown in a different way in Fig. 8-5(b). As illustrated in Fig. 8-4(c), the front end of the DFE-ZF is identical to the WMF, including a matched filter, symbol-rate sampler, and maximum-phase whitening filter (labeled *precursor equalizer*). In the context of the DFE-ZF, that whitening filter plays another, more important role, that of equalizing the response to be causal. This is why it is also called a precursor equalizer. If there is any ISI at all, then there is precursor ISI because of the symmetry about $t = 0$ of the matched filter response. The precursor ISI is eliminated by the precursor equalizer. The resulting model $M(z)$ is monic and causal, and hence can be viewed as shown in Fig. 8-5(b). Since the output of the slicer is an estimate $\hat{a}_k$ of the current data symbol a_k, if $\hat{a}_k = a_k$, then the postcursor equalizer exactly cancels the ISI introduced by the model in Fig. 8-5(b).

An example is shown in Fig. 8-6. The output of the sampled matched filter has both precursor and postcursor ISI, but the precursor equalizer eliminates the precursor ISI. The postcursor ISI is cancelled by the *postcursor equalizer* $M(z) - 1$ using the symbol estimates generated by the slicer. Since the ISI at the output of the precursor equalizer has transfer function $M(z) - 1$, a replica of this ISI can be generated using the feedback filter. This argument depends on the assumption that all decisions are correct. In fact, when the slicer makes incorrect decisions, the ISI correction becomes flawed for future decisions. This phenomenon is known as *error propagation*, and is discussed later.

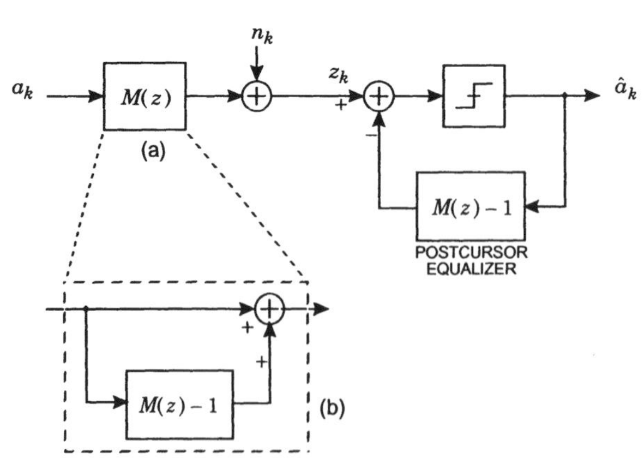

Fig. 8-5. (a) Another view of the zero-forcing decision-feedback equalizer (DFE-ZF). (b) An expansion of the channel model that shows clearly that this model introduces postcursor ISI that is cancelled by the DFE. $M(z)$ is causal and monic, so $M(z) - 1$ is strictly causal.

Figure of Merit of the DFE-ZF

The DFE performance is easily calculated assuming that all past decisions are correct. In that case, the signal component at the slicer input is the current data symbol a_k, free of ISI, and hence the minimum distance is the data-symbol alphabet minimum distance $a_{\min}$. The noise at the slicer input is precisely the noise at the WMF output, which has variance N_0/γ^2. The figure of merit is therefore

$$\Gamma_{\text{ZF-DFE}} = \frac{a_{\min}^2}{N_0/(2\gamma^2)} = \frac{2a_{\min}^2\gamma^2}{N_0} \ . \tag{8.41}$$

Using (8.13) this can be expressed directly in terms of the folded spectrum as

$$\Gamma_{\text{ZF-DFE}} = \frac{2a_{\min}^2}{N_0} \cdot \langle S_h \rangle_G = \frac{2a_{\min}^2}{N_0} \cdot \langle S_h^{-1} \rangle_G^{-1} \ . \tag{8.42}$$

The second form of (8.42) is written in a form to emphasize the comparison to (8.34); the two equations are similar except that the arithmetic mean in (8.34) is replaced by the geometric mean in (8.42). For rational spectra, $\Gamma_{\text{ZF-DFE}}$ is easily calculated directly from the spectrum

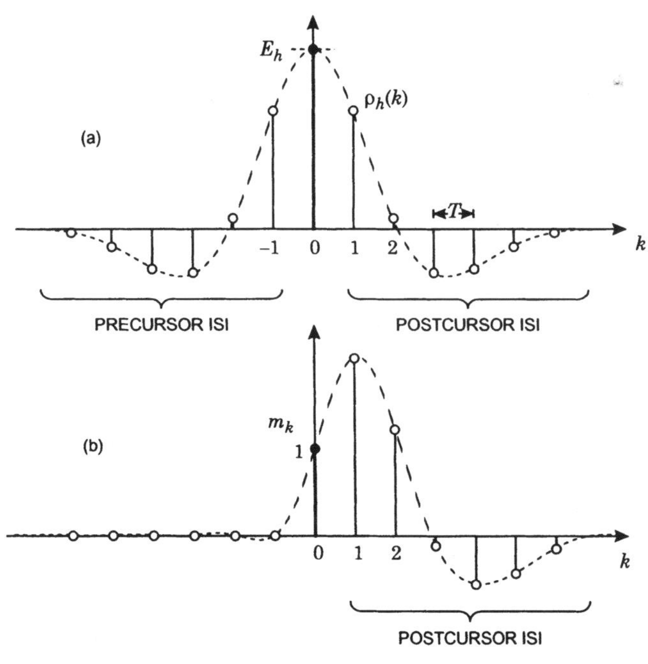

Fig. 8-6. Example pulse shapes in a DFE. (a) The pulse shape at the output of the sampled matched filter is always symmetric about $k = 0$. The *cursor* at time zero is marked by a thick line for emphasis. (b) After the precursor equalizer (whitening filter) the pulse is causal and minimum phase; that is, it has postcursor ISI only. The postcursor equalizer cancels the tail.

without using (8.42). However, (8.42) is very useful for non-rational spectra, or to avoid performing a spectral factorization.

Example 8-9.

Let us consider again the first-order all-pole pulse in Example 8-7, but this time for the ZF-DFE. Inspection of the folded spectrum of (8.35) reveals that $\gamma^2 = E_h(1 - \alpha^2)$. This illustrates the ease with which γ^2 can be calculated without evaluating a geometric mean integral. Then the figure of merit is

$$\Gamma_{\text{ZF-DFE}} = a_{\text{min}}^2 \cdot \frac{2E_h}{N_0} \cdot (1 - \alpha^2) . \tag{8.43}$$

As in the ZF-LE, $\Gamma_{\text{ZF-DFE}} \to 0$ as $|\alpha| \to 1$, because the channel pole is approaching the unit circle. Also, from Example 8-7, $\Gamma_{\text{ZF-DFE}} / \Gamma_{\text{ZF-LE}} = 1 + \alpha^2$, so that as expected the DFE-ZF always has a larger figure of merit than the LE-ZF, by as much as 3 dB.

Example 8-10.

Repeating the first-order all-zero pulse of Example 8-8 for the DFE-ZF, in this case by inspection $\gamma^2 = E_h/(1 + \alpha^2)$, and hence the figure of merit is

$$\Gamma_{\text{ZF-DFE}} = a_{\text{min}}^2 \cdot \frac{2E_h}{N_0} \cdot \frac{1}{1 + \alpha^2} . \tag{8.44}$$

In contrast to the ZF-LE, the DFE-ZF is well behaved as $|\alpha| \to 1$. The DFE-ZF suffers at most a 3 dB penalty relative to the MF bound, whereas the LE-ZF may suffer an arbitrarily large penalty.

Two bounds, (8.31) and (8.40), establish that the figure of merit of the DFE-ZF falls between the LE-ZF and the MLSD,

$$\Gamma_{\text{ZF-LE}} \le \Gamma_{\text{ZF-DFE}} \le \Gamma_{\text{MLSD}} . \tag{8.45}$$

These are strict inequalities, unless there is no ISI, in which case they become equalities. (An additional inequality is that $\Gamma_{\text{MLSD}} \le \Gamma_{\text{MF}}$, which can be an equality even in the presence of ISI.) The intuitive reason the DFE-ZF performs better than the LE-ZF is that postcursor ISI is cancelled without noise enhancement, since the slicer removes noise fed back through the postcursor equalizer filter.

Existence of the DFE-ZF

Since the DFE-ZF uses the WMF as a front end, it exists whenever the WMF exists; that is, both $S_h(e^{j\theta})$ and $\log S_h(e^{j\theta})$ must be integrable. Roughly speaking, the folded spectrum $S_h(e^{j\theta})$ cannot vanish on an interval, although algebraic zeros (characteristic of rational spectra) are permissible. Thus, another advantage of the DFE-ZF over the LE-ZF is that the DFE-ZF has a finite noise variance at the slicer input even when the folded spectrum has zeros on the unit circle, whereas the slicer input noise will have infinite variance for the LE-ZF under the same conditions. This is a desirable side effect of performing precursor equalization, but not postcursor equalization, using a linear filter.

Optimality of the WMF as the DFE-ZF Front End

We based the DFE on the WMF, without considering alternative filters. The basic concept of canceling postcursor ISI with a postcursor equalizer does not depend on the details of the impulse response of the precursor equalizer, except that the equivalent discrete-time isolated-pulse response up to and including the precursor equalizer must be monic and causal, so that the ISI is in fact postcursor.

As shown in Fig. 8-7, any filter $E(z)$ can be placed at the output of the WMF, changing the equivalent transfer function from $M(z)$ to $M(z)E(z)$, as long as the feedback filter is changed from $M(z) - 1$ to $M(z)E(z) - 1$. $E(z)$ must be chosen to meet the two constraints above:

- Since $M(z)$ is causal, $M(z)E(z)$ will be causal if and only if $E(z)$ is causal.
- Since $M(z)$ is monic ($M(\infty) = 1$), $M(z)E(z)$ will be monic ($M(\infty)E(\infty) = 1$) if and only if $E(z)$ is monic ($E(\infty) = 1$).

For a filter $E(z)$ with impulse response e_k meeting these constraints, since the noise at the input to $E(z)$ is white with power spectrum N_0/γ^2, the variance of the output (input noise to the slicer) will have variance

$$\frac{N_0}{\gamma^2} \sum_{k=0}^{\infty} |e_k|^2 \geq \frac{N_0}{\gamma^2}, \tag{8.46}$$

with equality if and only if $E(z) = 1$. This establishes that the original configuration of Fig. 8-4 is optimal in the sense of minimizing the noise variance at the slicer input.

In Section 5.4.2, it was pointed out that the minimum-distance receiver design (which is equivalent to the MLSD) does not require a minimum-phase spectral factorization. Equivalently, adding an allpass filter to the WMF and to the equivalent channel model will not change the minimum-distance criterion. This would be equivalent to making $E(z)$ a causal monic filter with a constant magnitude response in Fig. 8-7. We have just shown, however, that the minimum-phase spectral factorization *is* critical to the DFE, because it minimizes the noise at the slicer input. Adding a monic and causal filter $E(z)$ with constant magnitude $|E(e^{j\theta})|$ to the WMF will increase the noise variance without affecting the signal level at the slicer.

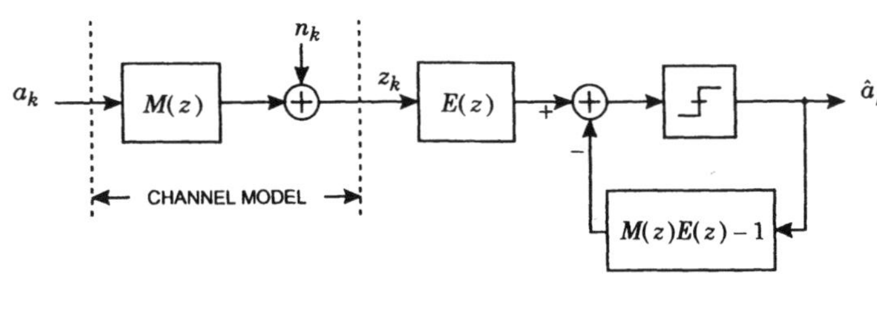

Fig. 8-7. An arbitrary DFE-ZF precursor equalizer can be realized by concatenating a monic and causal filter $E(z)$ with the WMF front end.

An intuitive explanation for this difference between the MLSD and the DFE is as follows: the DFE relies on the first sample of the equivalent impulse response to make the decision on the data symbol, and throws away the remainder of the received signal energy in the postcursor ISI. The minimum-phase channel model maximizes the energy in the first sample (see Problem 2-29), and is thus important to the performance of the DFE. The MLSD, on the other hand, uses *all* the energy in the equivalent channel impulse response, and hence will not be affected by the relative distribution of that energy between the first sample and the postcursor ISI.

In actual practice, the minimum-phase spectral factorization is helpful in the implementation of the MLSD as well. As shown in Section 5.4.4, the use of the Viterbi algorithm to realize the DFE requires that the channel response be FIR, and the computational complexity of the algorithm increases exponentially with the number of taps in the FIR channel model. It is often necessary to approximate an IIR response $M(z)$ with an FIR filter, and the number of taps required for a good approximation will generally be minimized when the spectral factorization is minimum-phase, since again that concentrates the energy in the low-delay coefficients.

Linear Predictor Interpretation of DFE-ZF

The optimal DFE-ZF structure can be derived in a slightly different way, one which lends additional insight. This alternative derivation illustrates a connection between optimal linear prediction (Section 3.2.3) and the WMF and DFE, and also explains in another way why the LE results in more noise enhancement than the DFE. The connection is illustrated in Fig. 8-8, where a DFE is placed at the output of a LE-ZF rather than WMF, and consists of a linear predictor $E(z)$, which introduces postcursor ISI, and a DFE postcursor equalizer. The key observation is that the noise at the output of the LE-ZF is not white, as demonstrated by (8.32), unless of course there is no ISI ($M(z) = 1$). Since the noise samples are correlated, we can take advantage of this correlation, and apply an optimal linear prediction error filter $E(z)$ as shown in Fig. 8-8. Like all linear prediction error filters, $E(z)$ is monic and causal. Thus, while $E(z)$ introduces ISI (which was absent at the output of the LE-ZF), this is postcursor ISI and can be canceled by a DFE-ZF feedback filter without enhancing the noise.

As shown in Section 3.2.3, the optimal $E(z)$ is the monic minimum-phase whitening filter. The noise power spectrum at the input to the predictor is proportional to $1/M(z)M^*(1/z^*)$, and the monic minimum-phase portion is thus $M^{-1}(z)$. The optimal whitening filter is the inverse of this, $M(z)$, which also happens to be the inverse of the LE-ZF filter $M^{-1}(z)$. Thus, the optimal predictor and the LE-ZF filter cancel one another, the optimal front end for the DFE is the WMF, and the optimal postcursor equalizer is $E(z) - 1 = M(z) - 1$.

This establishes in a different way that the WMF is a front end that generates white noise at its output, and among all such front ends is the one that minimizes the noise variance. It also explains in a compelling way the enhanced performance of the DFE-ZF relative to the LE-ZF. The DFE-ZF takes advantage of the correlation of noise samples at the output of the LE-ZF to reduce the noise variance at the slicer input, and cancels the resulting postcursor ISI using a DFE postcursor equalizer.

DFE Error Propagation

One obvious potential problem with the DFE is that any decision errors at the output of the slicer will cause a corrupted estimate of the postcursor ISI to be generated by the postcursor equalizer. The result is that a single error causes a reduction in the margin against noise for a number of future decisions. This phenomenon is called *error propagation*, and results in an error rate greater than would be predicted on the basis of SNR calculations alone. Error propagation is explored further in Appendix 8-A, where it is shown that the benefit of reduced noise enhancement usually far outweighs the effect of error propagation.

8.1.4. Transmitter Precoding

DFE error propagation can be avoided by using *transmitter precoding*. Transmitter precoding is sometimes called *Tomlinson-Harashima coding*, in honor of its co-inventors [3][4][5]. This technique has also been called *generalized partial response* [6], since the ordinary partial response invented earlier [7] is a special case.

The idea of precoding is to move the cancellation of the postcursor ISI to the transmitter, where the past transmitted symbols are known without the possibility of errors. However, this means that the postcursor ISI impulse response $M(z)$ must be known precisely at the transmitter. In practice, in most situations this impulse response must be estimated in the receiver using adaptive filter techniques (Chapter 9), and passed back to the transmitter in order to use transmitter precoding. This is feasible on channels that are time-invariant or slowly time-varying, but is not feasible on channels (such as mobile radio) that are rapidly time-varying.

The basis of transmitter precoding is the observation that the channel model through the WMF, $M(z)$, and LE equalizer $M^{-1}(z)$, both linear and time-invariant, can be reversed without compromising the requirement that the Nyquist criterion be satisfied at the slicer input, as shown in Fig. 8-9(a). The LE equalizer $M(z)$ can be put in the transmitter. There are two benefits to this:

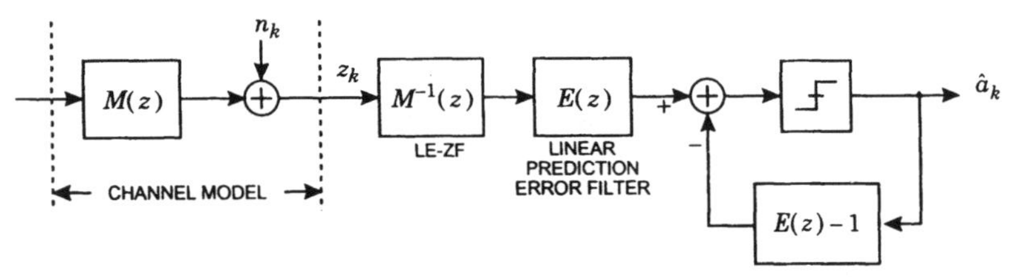

Fig. 8-8. Showing the connection between the optimal DFE precursor equalizer and optimal linear prediction. The DFE precursor equalizer can be obtained by applying a linear predictor to the nonwhite noise at the LE-ZF output.

- Since the receiver is now the WMF followed directly by a slicer, the noise at the slicer input is that of the WMF; that is, it is the same as for the DFE. Thus, even though the receiver uses linear equalization, it does not suffer the noise enhancement of linear equalization because the equalization is done *prior* to the channel, where the noise is introduced.
- The error probability may actually be slightly lower than with the DFE, because the postcursor ISI cancellation is done in the transmitter and there is no possibility of error propagation as there is in the DFE.

In the transmitter-based linear equalizer of Fig. 8-9(a), the transmitted data symbols x_k are the original data symbols a_k filtered by $M^{-1}(z)$. This filter is simply placed between the original data symbols and the pulse-amplitude modulator, so that the transmitted signal takes the form $\sum_k x_k g(t - kT)$. When the transmit filter, channel, and WMF front end in the receiver are taken into account, the channel model of Fig. 8-9(a) results.

Simply doing the equalization in the transmitter as shown in Fig. 8-9(a), is not advisable, however, because it increases the average and peak power in the transmitted signal. If the impulse response of $M^{-1}(z)$ is f_k, which is causal and monic, then the peak transmitted sample is increased by a factor of $\sum_{k=0}^{\infty} |f_k|^2 > 1$. Likewise, if the data symbols are independent and identically distributed, then the average power of the transmitted symbols is multiplied or enhanced by a factor $\sum_{k=0}^{\infty} |f_k|^2 > 1$. Not surprisingly, when the equalizer was at the receiver, it resulted in noise enhancement. By moving it to the transmitter, we induce signal enhancement. If we penalize the system for these increases in transmitted peak and average power, then we end up back where we started: a transmitter-based linear equalizer performs no better or worse than a receiver-based linear equalizer.

Modification to Reduce the Transmitted Power

Fortunately, there is a simple solution that substantially reduces these peak and average power penalties, and in fact makes them go away entirely in the limit of large data-symbol constellations. This approach is easiest to understand in the one-dimensional case, so assume the data symbols a_k are drawn from the L-ary alphabet (where L is even) $\{-(L-1), -(L-3), \ldots, -3, -1, 1, 3, \ldots, (L-3), (L-1)\}$. That is, the data symbols are chosen among all odd integers in the range $(-L, L)$. Consider the modification of the transmitter-equalizer shown in Fig. 8-9(b), in which an additional term $2L \cdot i_k$ is added to the feedback, where the sequence of integers $\{i_k, -\infty < k < \infty\}$ is yet to be determined. Define an *expanded symbol* c_k,

$$c_k = a_k + 2L \cdot i_k \ . \tag{8.47}$$

Since a_k is an odd integer and $2L \cdot i_k$ is an even integer, their sum is odd, and the alphabet of the expanded symbol c_k is the set of *all* odd integers, not those limited to the range $(-L, L)$. Both the original data symbol alphabet and the expanded data symbol alphabet are illustrated in Fig. 8-10. The original data symbol a_k can be recovered from the expanded symbol c_k by reducing it modulo $2L$,

$$a_k = c_k \text{ modulo } 2L \ . \tag{8.48}$$

By (x modulo $2L$), we mean specifically the unique value of $x + 2L \cdot m$ when the integer m is chosen such that $-L < x + 2L \cdot m \le L$. The *precoded symbol* x_k is transmitted, where

$$x_k = a_k + 2L \cdot i_k - v_k \ . \tag{8.49}$$

Fig. 8-9(b) is equivalent to Fig. 8-9(a) with the data symbol a_k replaced by the expanded symbol c_k of (8.47). Thus, in the receiver, the data symbol a_k can be recovered by first applying an *expanded slicer*, which detects the expanded symbol c_k, and then reducing the result modulo $2L$. The expanded slicer simply applies thresholds to detect a data symbol

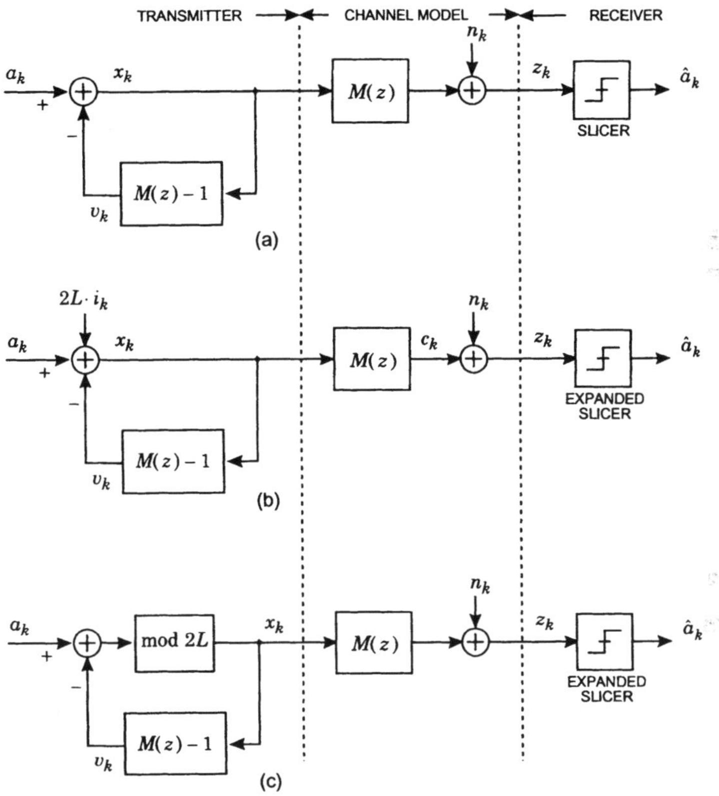

Fig. 8-9. Derivation of the transmitter precoder. (a) The LE-ZF, realized as a feedback filter as in Fig. 8-4(a), can be placed in the transmitter rather than the receiver. (b) For the data-symbol alphabet specified in the text, an arbitrary sequence of samples $2L \cdot i_k$ can be added to the data symbols, as long as the receiver slicer is appropriately expanded. (c) The transmitted power can be minimized by choosing $2L \cdot i_k$ to yield an equivalent modulo $2L$ operation.

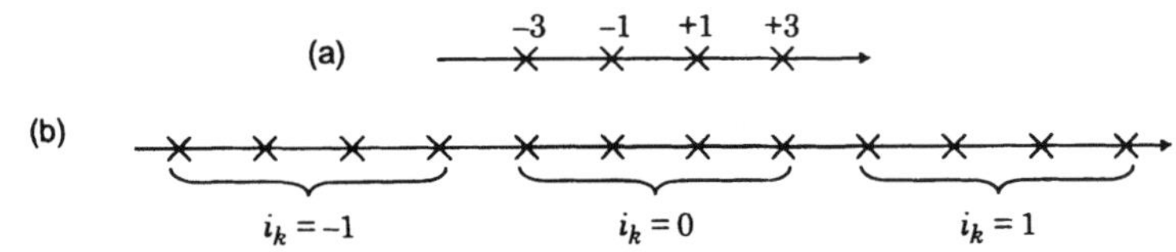

Fig. 8-10. Data symbol alphabets for transmitter precoding: (a) The original data symbols a_k for $L = 4$, and (b) the expanded data symbols c_k, shown for three values of i_k. (In general, i_k can assume larger values as well.)

expected to be any odd integer, as opposed to the original slicer that expects an odd integer in the range $(-L, L]$. With the expanded slicer, the system of Fig. 8-9(b) works just like Fig. 8-9(a) for any sequence of integers i_k. Furthermore, the error probability is essentially the same, since the minimum distance is unchanged (the distance between odd integers is $d_{\min} = 2$ in either case). There will be a slight increase in error probability due to end effects, as the original constellation is bounded, and the two outer symbols are therefore detected with a slightly lower error probability, a property lost with the expanded constellation.

The final step is to choose the sequence of integers i_k. For this purpose, observe from Fig. 8-9(b) and (8.49) that a term $2L \cdot i_k$ appears directly in the precoded symbols x_k. Since our original goal was to reduce the peak or average power penalty, i_k should be chosen to minimize the magnitude of each x_k. In fact i_k can always be chosen to limit x_k to the range $(-L, L]$, which is equivalent to reducing it modulo $2L$; choosing any other i_k results in a precoded symbol with a larger magnitude. This transmitter precoding approach, shown in Fig. 8-9(c), results in a small increase in the range of the precoded symbol alphabet, as the original data symbols are limited in magnitude to $L - 1$, and the precoded symbols are limited to L. For large L, this is only a slight increase in transmitted power.

The operation of the precoder can be better understood with the following simple (and accurate, for large L) model that predicts the statistics of the precoded symbols for an idealized statistical model of the original data symbols. Assume that the data symbols A_k are independent uniformly distributed random variables on $(-L, L]$. Of course, they actually have a discrete distribution, but this approximation becomes more accurate as L increases. This approximation is an important analytical tool in coding theory, and has been called the *continuous approximation* [8]. We will now show that under these conditions, the precoded symbols x_k are also independent and uniform-distributed on $(-L, L]$. First, in Fig. 8-9(c),

$$x_k = (a_k - v_k) \text{ modulo } 2L \ . \tag{8.50}$$

The first observation is that X_k is uniformly distributed and independent of v_k, regardless of the channel response $M(z)$.

Exercise 8-1.

Show that $P_{x_k|v_k}(x \mid v)$ is a uniform distribution on $(-L, L]$. Since the particular outcome v has no bearing on this conditional distribution, x_k is statistically independent of v_k.

It follows readily that X_k is independent of $\{v_k, v_{k-1}, \ldots\}$. Obviously x is dependent on a_k, but it is independent of $\{a_{k-1}, a_{k-2}, \ldots\}$, since the a_k's are independent. Since x_{k-l} is a function of a_{k-l} and v_{k-l}, it follows that x_k is independent of x_{k-l} for $l \geq 1$.

The approximation that the x_k are independent uniformly distributed random variables implies that the statistics of the precoded symbols are very similar to the statistics of the original data symbols, if the latter are independent identically distributed and are approximately equally likely to assume the L values in their alphabet. In particular, the variance $\mathrm{E}[x_k^2]$ of the precoded symbols is approximately $L^2/3$, the variance of a uniform random variable. The original data symbols have variance $(L^2 - 1)/3$, if they are equally likely. This approximation suggests that for large L, the average power of the precoded symbols x_k is approximately the same as the average power of the original symbols. Thus, the modulo operation accomplishes the objective of reducing the peak and average transmitted power to approximately that of the original signal.

In the absence of channel coding, transmitter precoding does not offer a substantial advantage over the DFE, since all it accomplishes is to eliminate error propagation, which turns out to be a minor problem in practice. Transmitter precoding has the major disadvantage that it requires precise knowledge of the channel in the transmitter. For these reasons, it was not used until recently. However, it has recently become important as one of three available equalization methods for obtaining the best performance in combination with channel coding on channels with intersymbol interference (Chapter 13). The DFE, while widely used in uncoded systems, is fundamentally incompatible with channel coding because of the requirement for immediate symbol decisions to cancel postcursor ISI. Channel decoding inevitably introduces a multi-symbol delay in the detection and decision process.

8.2. Generalized Equalization Methods

In Section 8.1, two important equalization techniques were introduced, linear and decision-feedback equalization. The structures developed were optimal under the criterion of minimizing the noise at the slicer input subject to the constraint that there be no ISI. There are several directions that these results can be generalized, and all are important in practice:

- Remove the assumption that the front end of the receiver consist of a matched filter followed by a symbol-rate sampler. While this structure was shown to be optimal for Gaussian channel noise, it is problematic on many real channels because it requires knowledge of the channel transfer function and presumes that this transfer function is not changing with time. There are adaptive techniques for dealing with these problems (Chapter 9), but they typically work in discrete time. Thus, practical receivers for unknown or time-varying channels typically do not use front-end matched filtering, although as shown in Section 8.3 the equivalent operation can be performed in discrete time if the sampling rate is increased.
- Generalize the criterion of optimality to allow for residual ISI at the slicer. By allowing residual ISI, we can reduce the variance of the noise, and usually there is a net advantage in SNR at the slicer.
- Constrain the complexity of the equalization filters, so that we can make them practical to implement and control the implementation cost.

In this section, we will deal with two of these issues. First we will not assume a matched-filter receiver front end. Second, we will define a *mean-square error (MSE)* criterion that takes into account ISI as well as noise. We will design optimal equalizers under both the zero-forcing (ZF) and MSE criteria. In this section we will stick with symbol-rate sampling, relaxing that assumption in Section 8.3, and we will not constrain the equalizer complexity, leaving that consideration to Chapter 9 where adaptive equalization is covered.

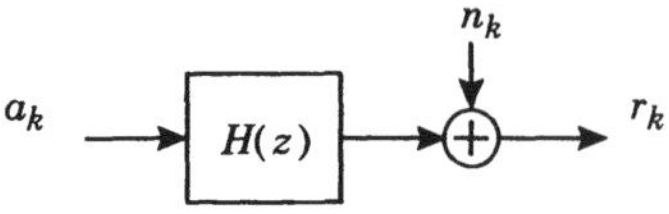

Fig. 8-11. A discrete-time channel model sampled at the symbol rate. The channel is not necessarily causal or monic, and the noise is not necessarily white.

8.2.1. Preliminaries

Before deriving optimal equalizer structures, we will choose a basic discrete-time channel model that does not presume front-end matched filtering, as in Section 8.1, and also define the MSE criterion.

Data-Symbol and Channel Model

The noise, signal, and channel model we consider here is shown in Fig. 8-11. Unlike for the ZF criterion, to design receivers according to the MSE criterion we must assume a statistical model for the data symbols, so that $\{a_k\}$ is viewed as a random process. The complex-valued data symbols a_k and additive complex noise $\{n_k\}$ are both assumed to be zero-mean wide-sense stationary discrete-time random processes with power spectra

$$S_a = \gamma_a^2 \cdot M_a M_a^* \tag{8.51}$$

$$S_n = \gamma_n^2 \cdot M_n M_n^* \tag{8.52}$$

where the minimum-phase spectral factorizations of Section 2.5.2 have been introduced, and M_a and M_n are loosely minimum-phase, causal, monic transfer functions. In case these are white, the power spectra reduce to γ_a^2 or γ_n^2 respectively. We also make the reasonable assumption that the noise and data symbols are uncorrelated and independent. The channel H, presumed to be rational, introduces *dispersion* or *intersymbol interference (ISI)*. It is no longer assumed that H is non-negative real on the unit circle, as it would be with a MF front end, nor is it minimum phase and monic, as it would be with a WMF front end. For purposes of analysis, it is convenient to decompose the rational H canonically in the manner of (2.52),

$$H = H_0 \cdot z^r \cdot H_{\text{min}} \cdot H_{\text{max}} \cdot H_{\text{zero}} \tag{8.53}$$

where H_0 is a complex constant, H_{min} is monic, causal, and minimum-phase, H_{max} is monic, anticausal, and maximum-phase, and H_{zero} contains all the zeros on the unit circle, and is causal and monic. Generally a flat delay in the channel is of no consequence, so we will assume that $r = 0$.

Example 8-11.

When the receiver front-end is a matched filter followed by symbol-rate sampling, $H = S_h$. For this case, $H_{\text{min}} = H_{\text{max}}^*$; $H_0 = \gamma^2$, and H_{zero} assumes a particular form where zeros come in pairs.

Example 8-12.
Given the channel

$$H(z) = \frac{(1-0.1z^{-1})(1-2z^{-1})(1-z^{-1})}{1-0.5z^{-1}}, \tag{8.54}$$

the minimum-phase and unit-circle zero terms are all in monic form, but a bit of manipulation is required on the maximum-phase term,

$$(1-2\,z^{-1}) = -2z^{-1}(1-0.5z)\,, \tag{8.55}$$

and hence we can identify

$$H_0 = -2, \quad r = -1, \quad H_{\min}(z) = \frac{1-0.1z^{-1}}{1-0.5z^{-1}},$$
$$H_{\max}(z) = 1-0.5z, \quad H_{\text{zero}}(z) = 1-z^{-1}\,. \tag{8.56}$$

Then we can set $r = 0$, in effect dealing with the channel $zH(z)$ rather than $H(z)$.

Physical Constraints on the Channel Model

Physical channels will never have a left-sided component to the impulse response. Thus we would expect $H_{\max}$ to be an FIR filter, implying that it has poles only at $z = \infty$, because there are no poles in H outside the unit circle (except possibly at $z = \infty$). Channel poles outside the unit circle, while mathematically feasible, can be ruled out by physical considerations.

Example 8-13.
The equivalent channel response for the WMF is S_h, which is non-negative real on the unit circle. Excluding poles at $z = 0$ and $z = \infty$, if S_h has any additional poles at all, then it will have poles outside the unit circle. The practical problem here is that if the received pulse $h(t)$ is right sided but does not have finite support (it decays to zero but never reaches zero), the matched filter will be left-sided and will not have finite support, and hence is not physically realizable. Under this condition, the matched filter can only be approximated by a filter with a finite-support impulse response, and the resulting response will again be right-sided. For this practical approximation, unlike the ideal matched filter case, H can have poles inside the unit circle, but not outside.

While we expect $H_{\max}$ to be FIR, it is not necessarily unity. That is, channel zeros outside the unit circle are feasible, and common if we use the WMF front end.

Example 8-14.
A broadband channel model for mobile radio, where the delay spread is large, consists of independent Rayleigh fading channels with different delays. Assume that there are two resolvable paths, so the impulse response of the channel is $c_1h(t) + c_2h(t-\tau)$ where c_1 and c_2 are independent Gaussian random variables. (Actually, c_1 and c_2 are random processes, but we are interested in them at some fixed time.) The τ is the differential delay of the two paths, and $h(t)$ is the impulse response of the ideal channel, typically dominated by the transmit and receive filters, and chosen to satisfy the Nyquist criterion. If the output of the receive filter is sampled at the symbol rate, then the discrete-time channel has impulse response $h_k = c_1\delta_k + c_2h(kT-\tau)$. Just to

illustrate what can happen, for simplicity assume that τ is a multiple of the symbol interval T, $\tau = mT$, so that

$$h_k = c_1\delta_k + c_2\delta_{k-m}\ , \qquad H(z) = c_1 + c_2 z^{-m}\ . \tag{8.57}$$

This channel model has poles only at $z = 0$, and the m zeros satisfy

$$z^m = \frac{-c_2}{c_1}\ , \qquad |z| = \left|\frac{c_2}{c_1}\right|^{1/m}\ . \tag{8.58}$$

Hence all m zeros have the same radius $|c_2|/|c_1|$. The channel is minimum-phase if and only if $|c_2| \le |c_1|$. Thus, the channel will have zeros outside the unit circle, and not be minimum-phase, if the greater delay path has a higher strength than the lesser delay path. If the lesser delay path is the direct path to the transmitter, this is unlikely to happen. However, if both paths are reflected, as during a deep shadowing, then the channel could easily not be minimum-phase. Furthermore, due to the fading, the channel is likely to alternate between minimum and non-minimum phase.

As we will see shortly, these zeros outside the unit circle, if they exist, are quite problematic for both the LE and DFE, because they require equalization filters with poles outside the unit circle. Practically speaking these zeros outside the unit circle can only be approximately equalized, and accurate equalization requires high-complexity filters. The situation described in Example 8-14 is particularly difficult for equalization, because the channel can be minimum- and non-minimum-phase at different times. As we will see, the desired structure for the equalizer when the channel model has zeros outside the unit circle is much different than when it is minimum-phase.

Mean-Square Error

In this section we want to allow residual ISI at the slicer input, in which case there is not only noise but ISI at the slicer input. The simple $Q(\cdot)$ formula for the probability of error no longer applies. Rather than calculate the exact probability of error, which is difficult, we will compare and optimize equalizers using an MSE criterion. The MSE is simply the variance of the error between the slicer input and the actual data symbol. We denote it by ε^2, and define it as

$$\varepsilon^2 = \mathrm{E}[\,|e_k|^2]\ , \qquad e_k = y_k - a_k\ , \tag{8.59}$$

where a_k is the data symbol, assumed to be a random variable, and y_k is the input sample to the slicer. The expectation is with respect to both the data-symbol statistics and the noise statistics. To calculate the MSE, it is necessary to know the power spectrum of the additive channel noise, S_n, the power spectrum of the data symbols, S_a (assuming the data symbols are modeled as a wide-sense stationary random process), and the equivalent channel response H.

In Section 8.1 we used a "figure of merit" Γ to compare different equalization techniques, where the probability of error was directly related to Γ through the $Q(\cdot)$ function. It is natural to define a quantity similar to the figure of merit, where we replace the Gaussian noise variance in the denominator by half the MSE (half because the MSE is the variance of the complex-valued error, and σ^2 is the variance of only the real or imaginary component),

$$\Gamma = \frac{a_{\min}^2}{\varepsilon^2/2} \ . \tag{8.60}$$

The philosophy here is that the residual *ISI* is a source of "noise" similar to the additive Gaussian noise, so we simply add its variance. In fact, on some channels the ISI might be approximately Gaussian, by the central limit theorem, and then a good approximation to the error probability would be $K \cdot Q(\frac{1}{2}\sqrt{\gamma})$. Even without having to make this approximation, minimizing ε^2 is a reasonable criterion for the design of the equalizer.

In this section we will simply use ε^2 as a measure of the effect of ISI and noise, and design and compare different receiver structures based on this measure. Since e_k is not Gaussian, this approach is not equivalent to minimizing the error probability, but because of its simplicity, it is widely applied in practice. As we will see in Chapter 9, this criterion is also widely used for the adaptation of equalizers as well.

8.2.2. Linear Equalizer

The linear equalizer (LE) applies the channel output to an equalizer filter *C*, as shown in Fig. 8-12, the purpose of which is to reduce or eliminate the ISI. The error e_k is the difference between the slicer input and the current data symbol, which we want to be as small as possible. As is conventional, the error is shown as the difference between the slicer input and output, called the slicer error, which will be the same as e_k as defined in (8.59) only if the slicer makes a correct decision. In this section we will assume that the slicer always makes a correct decision for purposes of analysis of the MSE and design of the equalizer filters (this assumption becomes relevant only in the case of the DFE). With this assumption, the slicer error can be generated in the convenient form of Fig. 8-12(a).

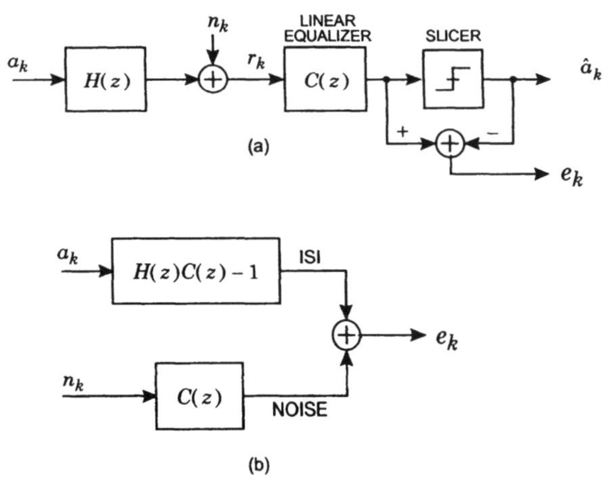

Fig. 8-12. A linear equalizer receiver. (a) The receiver uses an equalizer filter *C* and slicer, generating error e_k. (b) Equivalent way of generating the error assuming that decisions are correct.

Assuming that the slicer decisions are correct, $\hat{a}_k = a_k$, then the filters in Fig. 8-12(a) can be combined as shown in Fig. 8-12(b). This illustrates clearly that e_k consists of two components: the output of the top filter is the residual ISI after equalization, and the output of the bottom filter is the noise component at the slicer input. There is generally a tradeoff between these two components of error. Minimizing the ISI enhances the noise, so if we are willing to accept more ISI after equalization we can reduce the noise enhancement. Designing C in accordance with the MSE criterion represents one way to specify the desired tradeoff between noise enhancement and ISI.

The power spectrum of the slicer error evaluated on the unit circle, from Fig. 8-12(b) and the assumption of independence between the data symbols and noise, is

$$S_e = S_a \,|\, HC - 1 \,|^2 + S_n \,|\, C \,|^2 \,. \tag{8.61}$$

The *mean-square error (MSE)*, which is the variance of the slicer error, $E[\,|\, e_k \,|^2]$, is the integral of the power spectrum given by (8.61),

$$\varepsilon^2 = \langle S_e \rangle_A \,. \tag{8.62}$$

We will now determine the equalizer filter C for two cases: First, we will constrain the ISI to be zero (the ZF criterion) and then we will not constrain the ISI (the MSE criterion).

Zero-Forcing Criterion

In Section 8.1 we constrained the ISI to be zero at the slicer. We now repeat that design for a general channel model H. The equalizer is simply chosen to force the ISI component of the slicer error to zero (hence the name zero-forcing), or

$$C = \frac{1}{H} = \frac{1}{H_0 \cdot H_{\text{min}} \cdot H_{\text{max}} \cdot H_{\text{zero}}} \,. \tag{8.63}$$

We can make several observations:

- C does not depend on either the data-symbol or noise statistics. The MSE will depend on the noise spectrum, although not on the data-symbol spectrum because the ISI is forced to zero.
- This equalizer cannot be stable unless $H_{\text{zero}} = 1$; that is, the channel has no zeros on the unit circle.
- H_{min}^{-1} is a readily implemented causal minimum-phase filter.
- H_{max}^{-1} is an anticausal maximum-phase filter. In the practical case where H_{max} is an FIR filter (since poles would result in an impractical left-sided channel impulse response), H_{max}^{-1} is an all-pole IIR anticausal filter, which is impractical to implement and can at best be approximated.

To summarize, a minimum-phase channel is relatively straightforward to equalize. When the channel has a maximum-phase component, even if that component is FIR, the LE-ZF suddenly becomes impractical to implement and can at best be approximated. This approximation can be an FIR filter, but will often require a high order to be accurate.

The only component of slicer error is the noise, which has, from (8.61) and (8.63), variance

$$\varepsilon^2_{\text{ZF-LE}} = \langle S_n / |H|^2 \rangle_A \ . \tag{8.64}$$

For white noise and rational H, a convenient way to evaluate $\varepsilon^2_{\text{ZF-LE}}$ without the need to integrate is to note that it is γ_n^2 times the coefficient of z^0 in an expansion of $(HH^*)^{-1}$.

Example 8-15.

For white noise ($S_n(z) = \gamma_n^2$) and a first-order FIR channel, $H(z) = 1 - cz^{-1}$ for some complex-valued c, the LE-ZF is $C(z) = 1/(1 - cz^{-1})$. When the channel is minimum-phase ($|c| < 1$) the equalizer impulse response is $c_k = c^k u_k$, where u_k is the unit step function, a stable causal response. When the channel is not minimum-phase ($|c| > 1$), the impulse response is anticausal and IIR, $c_k = -c^k u_{-k-1}$. The noise is processed only by $C(z)$, and hence has power spectrum $\gamma_n^2 C(z)C^*(1/z^*)$. The MSE can be evaluated from the partial fraction expansion

$$C(z)C^*(1/z^*) = \frac{1}{H(z)H^*(1/z^*)} = \frac{1}{1-|c|^2}\left(\frac{cz^{-1}}{1-cz^{-1}} + \frac{1}{1-c^*z}\right) . \tag{8.65}$$

For the minimum-phase case, the first term in (8.65) starts at $k = 1$, and does not contribute to the z^0 term. The second term does, and $\varepsilon^2_{\text{ZF-LE}} = \gamma_n^2/(1 - |c|^2)$. For the non-minimum-phase case, rewrite (8.65) as

$$C(z)C^*(1/z^*) = \frac{1}{H(z)H^*(1/z^*)} = \frac{1}{|c|^2 - 1}\left(\frac{1}{1-c^{-1}z} + \frac{c^*z}{1-c^{*-1}z^{-1}}\right) . \tag{8.66}$$

In this case, only the first term contributes to z^0, and $\varepsilon^2_{\text{ZF-LE}} = \gamma_n^2/(|c|^2 - 1)$. In both cases $\varepsilon^2_{\text{ZF-LE}} \to \infty$ as $|c| \to 1$. As noted in Section 8.1.2, the LE-ZF is not useful when the channel model has zeros on the unit circle.

Example 8-16.

We can now rederive the results of Section 8.1 by setting $H = M$ and setting $S_n = N_0/\gamma^2$, where the spectral factorization of S_h is $S_h = \gamma^2 MM^*$. This corresponds to the symbol-rate discrete-time channel model for the WMF front end. In this case, the LE-ZF is $C = M^{-1}$ and the MSE is

$$\varepsilon^2_{\text{ZF-LE}} = \frac{1}{2\pi}\int_{-\pi}^{\pi} \frac{N_0/\gamma^2}{|M|^2}\, d\theta = N_0 \cdot \langle S_h^{-1} \rangle_A \, . \tag{8.67}$$

Equation (8.67) is consistent with (8.34). For this case, it *appears* that the equalizer is relatively simple to implement because the channel model is strictly minimum-phase. However, that is *not* the case, because the trouble is hidden in the WMF, which includes a strictly maximum-phase filter $1/M^*$. That WMF filter can only be approximated in practice if M has any zeros, since the WMF will then have poles outside the unit circle. The continuous-time matched filter may also be problematic if the received pulse $h(t)$ is causal and has unbounded support.

Mean-Square Error Criterion

The second method for designing the equalizer is to minimize the mean-square error (MSE) $E[|e_k|^2]$, taking into account both the ISI and noise components. The resulting equalizer is known as the mean-square error linear equalizer (LE-MSE). Since C is not constrained in complexity (this assumption will be removed in Chapter 9), C can be chosen independently at each frequency. The MSE can therefore be minimized by minimizing S_e given by (8.61) at each frequency by judicious choice of C.

First note that the power spectrum of the signal plus noise at the channel output in Fig. 8-11 is

$$S_r = S_a |H|^2 + S_n \ . \tag{8.68}$$

It will turn out that this power spectrum plays a crucial role in the equalizer design.

Exercise 8-2.
Show that by completing the square, (8.61) can be written as

$$S_e = S_r |C - S_a S_r^{-1} H^*|^2 + S_a S_n S_r^{-1} \ . \tag{8.69}$$

Since all terms in (8.69) are positive, it can be minimized at each frequency by forcing the first term to zero,

$$C = S_a S_r^{-1} H^* \tag{8.70}$$

which is shown in Fig. 8-13. We recognize that the term H^* is a discrete-time MF, and separate it out. Again, we can make some important observations:

- The reason that the LE-ZF does not contain a MF is that the equalizer would simply find the inverse of this filter, making it pointless. However, the LE-MSE includes a discrete-time matched filter.
- The MF is typically difficult to realize when the channel response has poles, as noted previously. Those poles will, for a physically meaningful channel and continuous-time receive filter, be inside the unit circle, which will place poles outside the unit circle in the MF. The resulting anticausal IIR filter can only be approximated by a realizable filter.
- The equalizer $S_a S_r^{-1}$, without the MF included, is non-negative real on the unit circle. Excluding poles at $z = 0$ and $z = \infty$, if it has any additional poles, then some of these must be outside the unit circle, making it unrealizable.
- If, as is typically the case, $S_n \neq 0$ on the unit circle (the channel noise is non-zero at all frequencies), then the equalizer $S_a S_r^{-1}$ can have no poles on the unit circle, and hence

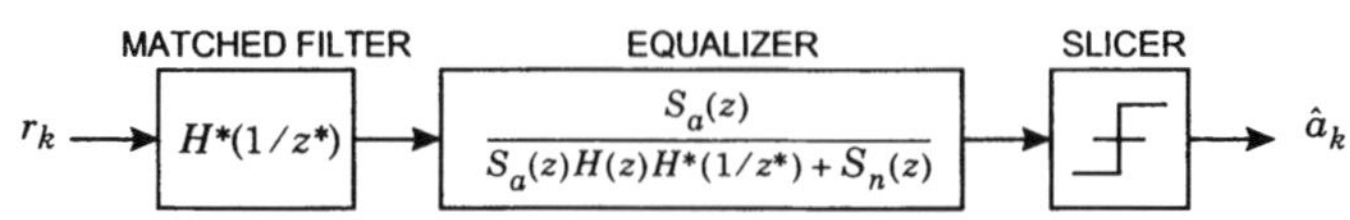

Fig. 8-13. The LE-MSE receiver, which minimizes the mean-square slicer error, consists of a MF and equalizer.

is stable. Similarly, since the channel H is assumed to be stable and have no poles on the unit circle, then the matched filter H^* will also be stable. Hence, for this case the LE-MSE is stable, in contrast to the LE-ZF, which will have poles on the unit circle if H has zeros on the unit circle. However, it is possible for the LE-MSE to have poles outside the unit circle as noted above, in which case it can only be approximated in practice even though it is mathematically well defined as a stable filter.

- As $S_n \to 0$, the LE-MSE approaches the LE-ZF, which is expected since in that case the LE-MSE will ignore the noise and focus on minimizing the ISI.

To summarize, the practicality of the LE-MSE is quite distinct from the LE-ZF. The LE-ZF has difficulty with non-minimum-phase channels, in the sense that the equalizer can only be approximated, and that approximation may have relatively high complexity. The LE-MSE has a similar difficulty with any channel with poles, except at $z = 0$ and $z = \infty$, due to the MF portion. The LE-ZF is not stable for channels with zeros on the unit circle. On the other hand, the LE-MSE generally has no difficulty with non-minimum-phase channels, where there are zeros outside the unit circle, or with channels with zeros on the unit circle.

Assuming (8.70), the power spectrum of the slicer error is the last term in (8.69), and hence the MSE is

$$\varepsilon^2_{\text{MMSE-LE}} = \langle\, S_n / (\,|H|^2 + S_n S_a^{-1})\,\rangle_A \,. \tag{8.71}$$

Comparing with (8.64), the extra $S_n S_a^{-1}$ term in the denominator ensures that $\varepsilon^2_{\text{MMSE-LE}} \le \varepsilon^2_{\text{ZF-LE}}$, since the integrand must be smaller at some frequencies in the MSE case. Furthermore, if $S_r S_a^{-1} \ne 0$ on the unit circle, which will typically be the case, then $\varepsilon^2_{\text{MMSE-LE}}$ is guaranteed to be finite regardless of the channel H. Thus the LE-MSE is guaranteed to be stable, although it may not have a right-sided impulse response. This nice property follows intuitively from the fact that this equalizer can avoid infinite noise enhancement (which plagues the LE-ZF) by allowing some residual ISI at the slicer.

Example 8-17. ---

Continuing Example 8-16, where the receiver front end is a WMF, assume that the data symbols are white ($S_a = \gamma_a^2$). Then the equalizer is

$$C = \frac{S_a M^*}{S_a |M|^2 + S_n} = \frac{M^*}{|M|^2 + N_0/(\gamma^2\gamma_a^2)} \,, \tag{8.72}$$

and we see that $C \to M^{-1}$ (the LE-ZF solution) as $N_0/(\gamma^2\gamma_a^2) \to 0$ (the high-SNR case). The MSE is given in integral form by

$$\varepsilon^2_{\text{MMSE-LE}} = \langle N_0/(S_h + N_0/\gamma_a^2)\rangle_A \,, \tag{8.73}$$

which approaches the ZF-LE as $N_0 \to 0$.

8.2.3. Decision-Feedback Equalizer

To determine the optimal DFE under an MSE criterion, we will draw heavily on the connection between the DFE and linear prediction established in Section 8.1.3. The opportunity for improving on the MSE of the LE comes from the fact that the slicer error samples are correlated, and hence can be reduced by a linear prediction error filter. The LE slicer error e_k from Fig. 8-12(b) is reproduced in Fig. 8-14(a), with the addition of a linear prediction filter E, which is constrained to be causal and monic. We know that the new slicer error e_k' will have a smaller variance than the LE slicer error e_k if E is chosen properly, but the question is whether the slicer error configuration of Fig. 8-14(a) corresponds to a practical DFE configuration. Fortunately, the DFE configuration of Fig. 8-14(b) is equivalent, assuming that decisions are correct ($\hat{a}_k = a_k$). Furthermore, the feedback filter is realizable, because the postcursor equalizer in the feedback loop, $(E - 1)$, is a strictly causal filter; that is, the zero-delay coefficient is zero since E is monic. Thus, the output of this filter, subtracted from the slicer input, is a function of *past* decisions only, as it must be to avoid zero delay in the feedback loop.

We know from the properties of optimal prediction in Section 3.2.3 that if the prediction error filter is properly chosen, then the slicer error e_k' in Fig. 8-14 will have a variance no larger than e_k, the LE slicer error, for the same choice of C. Furthermore, we know that the DFE slicer error must be white for a properly chosen prediction error filter E. Thus, the addition of the feedback filter in Fig. 8-14 must be beneficial, in the sense of reducing the mean-square slicer error, when compared to a LE using the same C.

Having defined the DFE structure, it remains to determine the optimal equalizers (precursor equalizer CE and postcursor equalizer $E - 1$). It is much simpler to determine the optimal C and E, and then infer the precursor and postcursor equalizers. In fact, for any C, E is chosen as the optimal linear prediction error filter. Once E is so determined, it is simple to determine the optimal C.

As in the case of the LE, there are two alternative approaches: the zero-forcing DFE (DFE-ZF) forces the ISI to zero at the slicer input, while the mean-square DFE (DFE-MSE) minimizes the variance of the slicer error. It turns out to be easy to show that C is the same for the LE-ZF and DFE-ZF, and likewise for the LE-MSE and DFE-MSE. Thus, for either criterion the only difference between the LE and DFE is the addition of the linear prediction error filter in the DFE. In both cases, the linear prediction filter results in a white slicer error process, although in the ZF case that process is Gaussian and in the MSE case it is not (because of the residual ISI).

Let us first assume some filter C, and choose the optimal linear predictor E. For a given C, the power spectrum of the LE slicer error is given by (8.61), and a monic minimum-phase spectral factorization of this spectrum can be performed,

$$S_e = \varepsilon_{DFE}^2 \cdot M_e M_e^* \tag{8.74}$$

where ε^2_{DFE} is the variance of the innovations e_k', and hence is the mean-square DFE slicer error. The prediction error filter is the monic minimum-phase whitening filter $E = M_e^{-1}$. A convenient formula for ε^2_{DFE} is given by (2.69),

$$\varepsilon^2_{DFE} = \langle S_e \rangle_G . \tag{8.75}$$

Naturally, S_e, M_e, and E all depend on C, which depends in turn on the criterion used.

Zero-Forcing Criterion

Considering first the DFE-ZF, assume that there are no zeros on the unit circle ($H_{\text{zero}} = 1$) so that the LE-ZF is stable (we will relax this assumption momentarily). The LE-ZF is then $C = H^{-1}$, and the only component of the LE slicer error is the noise passing through C. The slicer error spectrum is

$$S_e = \frac{S_n}{HH^*} = \frac{\gamma_n^2 M_n M_n^*}{|H_0|^2 \cdot H_{\text{min}} H_{\text{max}} H_{\text{min}}^* H_{\text{max}}^*} \tag{8.76}$$

and we can immediately see that $\varepsilon^2_{\text{ZF-DFE}} = \gamma_n^2 / |H_0|^2$. A more universal formula follows from (8.75),

$$\varepsilon^2_{\text{ZF-DFE}} = \langle S_n / |H|^2 \rangle_G . \tag{8.77}$$

Furthermore, the prediction error filter (monic minimum-phase whitening filter) is:

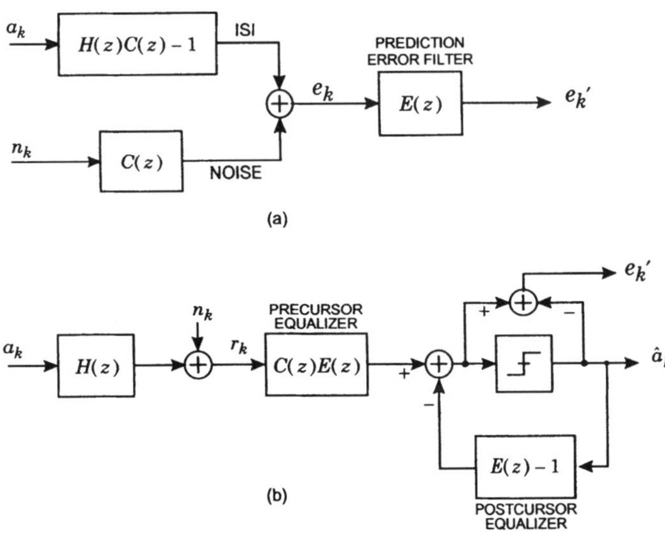

Fig. 8-14. The decision-feedback equalizer (DFE). (a) Adding a linear prediction filter to Fig. 8-12(b) to whiten the slicer error. (b) An equivalent DFE structure assuming that slicer decisions are correct.

$$E = \frac{H_{\min} H_{\max}{}^*}{M_n}, \tag{8.78}$$

thereby determining the precursor equalizer,

$$CE = H^{-1} \cdot \frac{H_{\min} H_{\max}{}^*}{M_n} = \frac{1}{H_0} \cdot \frac{H_{\max}{}^*}{H_{\max}} \cdot M_n^{-1}. \tag{8.79}$$

A few observations:

- The term $M_n{}^{-1}$ is a minimum-phase noise-whitening filter. This ensures that the noise at the slicer input is white. This is expected, since the ISI is completely eliminated by the precursor and postcursor equalizers, and thus the noise is the only component of slicer error. The slicer error will always be white for an optimal DFE.
- The $H_{\max}{}^*/H_{\max}$ term is an allpass filter. With a phase-only filtering, the precursor ISI can be eliminated, and there is no noise enhancement due to frequency-dependent precursor equalizer gain. In contrast, the LE eliminates *both* precursor and postcursor ISI at the expense of noise enhancement.
- The allpass filter does not modify the noise spectrum at the slicer input, allowing the noise-whitening filter to do its job without interference.
- The response of the channel plus precursor equalizer is E, which is minimum-phase. The role of the allpass filter is thus to convert the maximum-phase channel component to minimum-phase by reflecting poles and zeros inside the unit circle. Another interpretation of this result is that among all causal responses with the same magnitude Fourier transform, the minimum-phase response has maximum energy near $k = 0$ (Problem 2-28). Thus, in this sense the minimum-phase response minimizes the energy of the ISI which must be canceled by the postcursor equalizer, minimizing the signal energy that is thrown away by canceling it in the postcursor equalizer.
- If the channel is minimum-phase and the noise is white, no precursor equalizer at all is required, aside from a flat gain! Thus, the DFE-ZF is like the LE-ZF, in that it finds minimum-phase channels much easier to deal with. In fact, except for the noise-whitening filter required when the noise is not white, the precursor equalizer is unnecessary!
- If there is a maximum-phase component $H_{\max}$, then in a practical sense it must be an FIR filter (all its poles are at $z = \infty$). In this case the allpass filter component of the precursor equalizer has poles outside the unit circle, and hence can only be approximated by a (relatively high-complexity) FIR filter. Thus, the DFE-ZF has a similar difficulty as the LE-ZF with non-minimum-phase channels.

Example 8-18.

Consider the same channel as in Example 8-15, $H(z) = 1 - cz^{-1}$. For the minimum-phase case, $|c| < 1$. We identify $H_0 = 1$ and $H_{\max}(z) = 1$, and hence no precursor equalizer is needed. Because there is no precursor equalizer, the slicer error is $\varepsilon^2_{\text{ZF-DFE}} = \gamma_n^2$, and there is no noise enhancement. The postcursor equalizer, $E(z) - 1 = -cz^{-1}$, simply cancels the single ISI sample. Note that nothing special happens as $|c| \to 1$, in sharp contrast to the LE-ZF, which is not stable in that case.

Example 8-19.

When we repeat Example 8-18 for the non-minimum-phase case, $|c| > 1$, we will obtain a much different answer! Rewriting $H(z) = -cz^{-1}(1 - c^{-1}z)$ we can ignore the z^{-1} term, and identify $H_0 = -c$ and $H_{max}(z) = 1 - c^{-1}z$. The MSE is thus $\varepsilon^2_{ZF\text{-}DFE} = A_z^2 \,/\, |c|^2$, and thus the slicer error is smaller than in the minimum-phase case (since the receiver is basing its decision on the larger delayed sample c). Since $E(z) = 1 - c^{*-1}z^{-1}$, the postcursor equalizer becomes $E(z) - 1 = -c^{*-1}z^{-1}$, a single tap as in the minimum-phase case. The big difference is in the precursor equalizer, which is the allpass filter

$$C(z)E(z) = \frac{1}{c} \cdot \frac{1 - c^{*-1}z^{-1}}{1 - c^{-1}z} . \tag{8.80}$$

This precursor equalizer is anticausal, since it has a pole outside the unit circle, and can at best only be approximated by a realizable filter. There is thus a wide gap between the complexity of the precursor equalizer for the minimum-phase and non-minimum-phase channels. When $|c|$ crosses unity, in principle nothing bad happens (in contrast to the LE-ZF), but in practice the structure of the precursor equalizer changes dramatically. This can present implementation difficulties, and is a real problem for example on broadband Rayleigh fading channels (Example 8-14).

It should be noted from these two examples that the minimum-phase solution will work in the maximum-phase case, in the sense of eliminating the ISI component of the slicer error, but the penalty paid will be a larger MSE slicer error (γ_n^2 rather than $\gamma_n^2/|c|^2$). This statement is more generally true: as long as the maximum-phase channel component has poles only at $z = 0$ or $z = \infty$, which is normally expected on physical grounds, zero ISI at the slicer can be ensured by a postcursor equalizer only (if an appropriate delay is added to the channel) because the channel has a right-sided impulse response. However, a penalty in MSE is paid for not equalizing to a minimum-phase response.

Example 8-20.

Continuing Example 8-16, assume that the front end is a WMF. Then $C = 1/M$, and the power spectrum of the LE slicer error is

$$S_e = \frac{N_0/\gamma^2}{|M|^2} = \frac{N_0/\gamma^2}{MM^*} . \tag{8.81}$$

The optimal predictor is thus the inverse of the minimum-phase portion, $E = M$, and the forward equalizer is $CE = 1$; that is, as expected, the precursor equalizer is actually the WMF. The MSE is given by

$$\varepsilon^2_{ZF\text{-}DFE} = \frac{N_0}{\gamma^2} = N_0 \cdot \langle S_h^{-1} \rangle_G , \tag{8.82}$$

consistent with (8.42).

One of the practically important properties of the DFE-ZF is that, in contrast to the LE-ZF, it works perfectly well in the presence of zeros on the unit circle. To see this, assume that $H_{zero} \neq 1$, but design the precursor equalizer as before, turning the maximum-phase component into a minimum-phase component. The response of channel plus precursor equalizer is then EH_{zero} rather than E. Since this response is still causal and monic, the ISI can still be canceled by a strictly causal postcursor equalizer $EH_{\text{zero}} - 1$. In effect we have included H_{zero} in the minimum-phase component of the channel, which is customary and introduces no mathematical difficulties.

Example 8-21.
Replace $H(z)$ in Example 8-19 by $H(z) = (1 - cz^{-1})(1 - z^{-1})$ for $|c| > 1$. Retaining the same allpass precursor equalizer, the isolated pulse response at the precursor equalizer output is

$$(1 - c^{*-1})(1 - z^{-1}) = 1 - (c^* + 1)c^{*-1}z^{-1} + c^{*-1}z^{-2} \tag{8.83}$$

and the ISI can be canceled with postcursor equalizer $(-(c^* + 1)c^{*-1}z^{-1} + c^{*-1}z^{-2})$.

Mean-Square Error Criterion

The DFE-MSE optimal filters are almost as easy to determine. The first observation is that (8.75) is monotonically increasing in S_e at each frequency, and hence will be minimized by choosing C at each frequency to minimize S_e. Thus C for the DFE-MSE is precisely the same as C for the LE-MSE, since C was designed to meet the same criterion. C is given by (8.70) and S_e is given by the last term in (8.69). To find the whitening filter E, first do a minimum-phase spectral factorization of S_r,

$$S_r = S_a HH^* + S_n = \gamma_r^2 \cdot M_r M_r^* , \tag{8.84}$$

$$\gamma_r^2 = \langle S_a |H|^2 + S_n \rangle_G , \tag{8.85}$$

and thus from (8.69)

$$S_e = \frac{S_a S_n}{S_r} = \frac{\gamma_a^2 \gamma_n^2}{\gamma_r^2} \cdot \frac{M_a M_a^* M_n M_n^*}{M_r M_r^*} = \varepsilon^2_{\text{MMSE-DFE}} \cdot M_e M_e^* . \tag{8.86}$$

It follows that

$$\varepsilon^2_{\text{MMSE-DFE}} = \frac{\gamma_a^2 \gamma_n^2}{\gamma_r^2} = \langle S_n / (|H|^2 + S_n S_a^{-1}) \rangle_G , \tag{8.87}$$

and

$$E = \frac{1}{M_e} = \frac{M_r}{M_a M_n} . \tag{8.88}$$

The optimal precursor equalizer is

$$CE = \frac{\gamma_a^2}{\gamma_r^2} \cdot H^* \cdot \frac{M_a^*}{M_r^*} \cdot M_n^{-1} . \tag{8.89}$$

As in the LE-MSE, this solution includes a matched filter H^*, and as in the DFE-ZF it includes a noise-whitening filter M_n^{-1}. The remaining term is, in contrast to the DFE-ZF, not an allpass filter. It is straightforward to verify that this MSE solution approaches the ZF solution as

$S_n \rightarrow 0$ (Problem 8-8), and it follows from (8.77) and (8.87) that $\varepsilon^2_{\text{MMSE-DFE}} \le \varepsilon^2_{\text{ZF-DFE}}$ because the integrand is smaller at some frequencies for the MSE criterion.

Example 8-22.

Continuing Example 8-16, assume that the front end is the WMF. The power spectrum of the equivalent channel output is

$$S_r = S_a \, |M|^2 + \frac{2N_0}{\gamma^2} = \gamma_r^2 M_r M_r^* \, . \tag{8.90}$$

The only simplification is that the noise-whitening filter is not required, since the noise at the WMF output is already white ($M_n = 1$), and hence the postcursor equalizer is $E = M_r / M_a$ and the precursor equalizer is

$$CE = \frac{\gamma_a^2}{\gamma_r^2} \cdot \frac{M M_a^*}{M_r^*} \, . \tag{8.91}$$

The MSE is

$$\varepsilon^2_{\text{MMSE-DFE}} = \langle N_0 / (S_h + N_0 S_a^{-1}) \rangle_G \, . \tag{8.92}$$

Unbiased Mean-Square Error

If we calculate the transfer function from the data symbol A_k to the slicer error for the DFE-MSE design, it is, from Fig. 8-14,

$$(HC - 1)E = \frac{-\gamma_n^2}{\gamma_r^2} \cdot \frac{M_n^*}{M_a M_r^*} \, . \tag{8.93}$$

The striking property of this solution is that, since $M_n^* / M_a M_r^*$ is monic, the slicer error has a component proportional to the current data symbol a_k, namely $(-\gamma_n^2 / \gamma_r^2) \cdot a_k$. Thus, the total component of the current symbol at the slicer input is this error plus a_k, or

$$\frac{\gamma_r^2 - \gamma_n^2}{\gamma_r^2} \cdot a_k \, . \tag{8.94}$$

While this minimizes the MSE, the error probability can be made smaller by removing this bias in the amplitude of the data symbol reaching the slicer. (Recall that the slicer error is not Gaussian, so minimizing the MSE is not precisely the same as minimizing the error probability.) Thus, the performance can be improved by scaling the slicer input by a factor which removes this bias, this factor being $\gamma_r^2 / (\gamma_r^2 - \gamma_n^2)$. No other modifications are necessary, as the precursor and postcursor equalizers remain the same (since the gain is added after ISI cancellation). The resulting design is called the *unbiased DFE-MSE (MMSE-DFE-U)* [9].

The MSE of the MMSE-DFE-U, $\varepsilon^2_{\text{MMSE-DFE,U}}$, must be larger than $\varepsilon^2_{\text{MMSE-DFE}}$. We can find out what it is from Fig. 8-15, which shows the relationship between the slicer error of the DFE-MSE, e_k', and the MMSE-DFE-U, e_k''. First, the DFE-MSE slicer input sample y_k is obtained by adding the data symbol to e_k', and then the slicer input is multiplied by the adjustment factor, and then the new slicer error is obtained by subtracting the data symbol. From this figure,

$$(\gamma_r^2 - \gamma_n^2)\cdot e_k'' = \gamma_r^2\cdot e_k' + \gamma_n^2\cdot a_k\,. \tag{8.95}$$

Finding the MSE from this relationship is easy for the particular case where the data symbols are zero-mean and independent ($M_a = 1$). For this case, a_k is independent of e_k'', because the latter has no component of a_k, and a_k is independent of all the other components that make up e_k'' (the noise and the other data symbols). Thus, we can easily calculate the variance,

$$\gamma_r^4\cdot\varepsilon^2_{\text{MMSE-DFE}} = (\gamma_r^2 - \gamma_n^2)^2\cdot\varepsilon^2_{\text{MMSE-DFE,U}} + \gamma_n^4\cdot\gamma_a^2\,, \tag{8.96}$$

which, with the aid of (8.87) is easily solved for $\varepsilon^2_{\text{MMSE-DFE,U}}$,

$$\varepsilon^2_{\text{MMSE-DFE,U}} = \frac{\gamma_a^2\gamma_n^2}{\gamma_r^2 - \gamma_n^2}\,. \tag{8.97}$$

The increase in MSE due to removing the bias is a factor of $\gamma_r^2/(\gamma_r^2 - \gamma_n^2)$. Furthermore, it is readily shown that

$$\frac{\gamma_a^2}{\varepsilon^2_{\text{MMSE-DFE}}} = \frac{\gamma_a^2}{\varepsilon^2_{\text{MMSE-DFE,U}}} + 1\,, \tag{8.98}$$

and since the two ratios are in the form of signal-to-noise ratios (SNR's), we see that the SNR of the MMSE-DFE-U is indeed smaller than the SNR of the DFE-MSE, in fact by exactly unity.

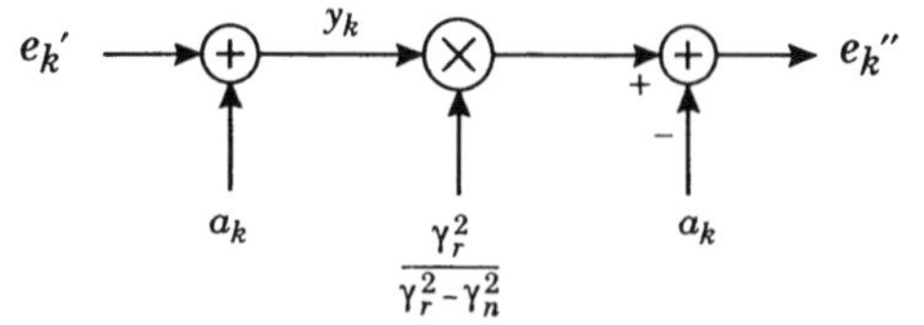

Fig. 8-15. The slicer error e_k'', for the unbiased DFE-MSE (MMSE-DFE-U) obtained from e_k', the slicer error for the DFE-MSE.

The MMSE-DFE-U has two analytical disadvantages: the MSE is larger than for the DFE-MSE, and unlike the DFE-MSE the slicer error is not white. However, we do expect it to have a lower error probability. From a theoretical point of view, the MMSE-DFE-U is related in a remarkable canonical way to the capacity of the discrete-time channel, as will be shown in Section 8.5.

It has also been shown [9] that among all DFE equalizers that are unbiased, the MMSE-DFE-U achieves the minimum MSE. Since the DFE-ZF is also an unbiased DFE equalizer, it follows that

$$\varepsilon^2_{\text{MMSE-DFE}} \le \varepsilon^2_{\text{MMSE-DFE,U}} \le \varepsilon^2_{\text{ZF-DFE}} \,. \tag{8.99}$$

8.2.4. Maximum-Likelihood Sequence Detector

We demonstrated in Section 8.1 that the MLSD and the DFE-ZF share the same WMF front-end filtering. In Section 8.2.3 we generalized the DFE-ZF to channels that do not necessarily have a matched-filter front end. The Viterbi algorithm can be applied at the output of the DFE-ZF precursor equalizer, in place of the WMF. This is illustrated in Fig. 8-16. Fig. 8-16(a) shows a configuration in which the Viterbi algorithm is applied at the output of the precursor equalizer, and Fig. 8-16(b) shows the equivalent channel model to the input of the Viterbi algorithm. This equivalent channel model displays all the characteristics needed to apply the Viterbi algorithm; the equivalent channel is causal and monic and the noise samples are Gaussian and independent.

The only restriction in using the Viterbi algorithm in detecting the data symbol sequence a_k under an ML criterion is that E be an FIR filter, so that the channel model is a finite-state machine. This requires the following conditions:

- $H_{max}{}^*$ must be an FIR filter, implying that, excluding poles at $z = 0$ and $z = \infty$, H_{max} have no poles. As we have discussed, this requirement is also necessary for physical realizability of the channel model.
- H_{min} must be an FIR filter, implying that, excluding poles at $z = 0$ and $z = \infty$, the channel model H must have no poles at all.
- Excluding zeros at $z = 0$ and $z = \infty$, M_n must have no zeros, implying that the noise spectrum S_n must be an all-pole spectrum (such a noise process is known as *autoregressive*).

It should be emphasized that the configuration of Fig. 8-16(a) is not the ML sequence detector applied to the continuous-time received signal (in the sense derived in Chapter 7) unless of course the channel model H was obtained from the WMF receiver front end. However, if we do not use a WMF front end, the Viterbi algorithm applied as in Fig. 8-16(a) is still a MLSD with respect to that discrete-time channel model, even though it is suboptimal with respect to the continuous-time model.

What do we do when E is not FIR? The VA remains useful in this case, even though it is not precisely an ML detector. There are two possible approximations we can use:

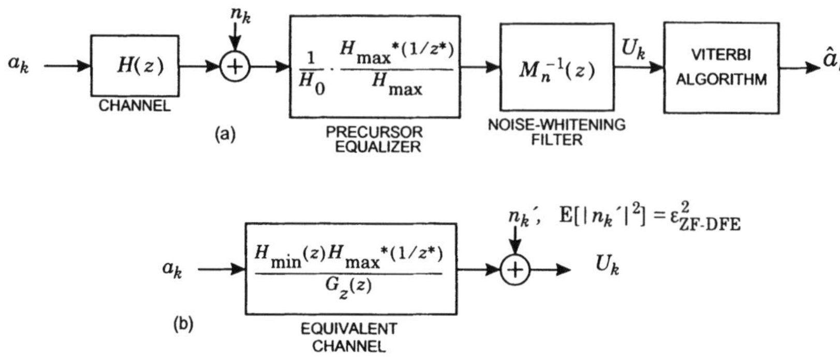

Fig. 8-16. (a) The Viterbi algorithm as applied to the detection of data symbols in the presence of ISI. It must be assumed that the isolated pulse response at the precursor equalizer output is FIR. (b) Equivalent model, in which the channel is causal and the additive noise n_k' is white with variance $\varepsilon^2_{ZF\text{-}DFE}$.

- The precursor equalizer can be modified to result in an FIR response, in which case the noise samples at the VA input will not be precisely white.
- The channel model can be truncated to an FIR response assumed in the realization of the VA, in which case there will be some residual ISI not considered in the VA.

Either of these approximations become more accurate as the number of taps in the artificially assumed FIR response becomes larger.

8.3. Fractionally Spaced Equalizer

One of the surprising properties of the sampled MF and the WMF is the symbol-rate sampling. Assuming that the PAM signal design uses some excess bandwidth, as it must for practical reasons, this sampling rate is less than would be required to avoid aliasing distortion. It seems strange that aliasing distortion would be deliberately introduced, and indeed there are some practical problems that result. Thus, it is common to use a higher sampling rate in a receiver front end and equalization, using an architecture known as the *fractionally spaced equalizer (FSE)* [10]. This fractionally spaced approach applies to all the receiver design strategies we have considered: the LE and DFE, WMF, and MLSD. A properly designed FSE is equivalent to the matched filter plus symbol-rate sampling, but offers considerable advantages [11][12] [13]. First we will outline two of the problems addressed by the FSE, and then derive the FSE structure in two ways — in the time-domain and in the frequency-domain.

Aliasing and Sampling Phase

From a practical perspective, the assumptions under which the sampled matched filter was derived are problematic. There are several assumptions underlying this structure that are often violated in practice:

- The receiver sampling phase is precisely known relative to the symbol interval.

- The channel response $h(t)$ is known precisely. This is partially overcome with the adaptive equalization techniques of Chapter 9, but the practical difficulty is the continuous-time matched filter, which is not easily adapted.
- The discrete-time filters, such as the linear equalizer or DFE precursor equalizer, have unconstrained complexity.

To see the effect of a sampling phase error, assume that the input isolated pulse is delayed in time by an unknown time t_0 [14]. The complex-valued baseband pulse is then $h(t - t_0)$ rather than $h(t)$, but the sampling times $t = kT$ at the matched filter output are unchanged.

Exercise 8-3.
Verify that, with this time delay, the white-noise folded spectrum of an isolated pulse after sampling is

$$S_{h,t_0}(e^{j2\pi fT}) = \frac{1}{T}e^{j2\pi ft_0}\sum_{m=-\infty}^{\infty}\left|H\left(f-\frac{m}{T}\right)\right|^2 e^{j2\pi mt_0/T} . \tag{8.100}$$

While $S_h(e^{j2\pi fT})$ is a non-negative real-valued function, $S_{h,t_0}(e^{j2\pi fT})$ is no longer necessarily either real-valued or non-negative. In fact, $S_{h,t_0}(e^{j2\pi fT})$ can have spectral nulls or near-nulls that are not present in $S_h(e^{j2\pi fT})$. An equalizer that compensates for these nulls can produce considerably more noise enhancement than when $t_0 = 0$.

Example 8-23.
When there is less than 100% excess bandwidth, we get the simpler relation

$$S_{h,t_0}(e^{j2\pi fT}) = e^{j2\pi ft_0}\,|H(f)|^2\left(1 + \alpha e^{-j2\pi t_0/T}\right) \quad , \quad 0 \le f \le 1/(2T) \tag{8.101}$$

where

$$\alpha = \left|\frac{H(f-1/T)}{H(f)}\right|^2 . \tag{8.102}$$

If the channel falls off monotonically, then α will generally be less than unity. The term in parentheses in (8.101) has squared magnitude

$$\left|1 + \alpha e^{-j2\pi t_0/T}\right|^2 = 1 + \alpha^2 + 2\alpha\cos(2\pi t_0/T) , \tag{8.103}$$

which has minimum value $(1-\alpha)^2$ when $t_0 = T/2$. The folded spectrum has a near-null at the frequency for which α is maximum, and the depth of the null depends on t_0, with the worst case when $t_0 = T/2$. The folded spectrum depends on t_0, and for some values of t_0 the noise enhancement can be much worse than for others.

When equalizer filters are made adaptive (Chapter 9), their complexity must constrained. The symbol-rate folded spectrum in a passband QAM channel can have rapid transitions near the band edges that are difficult to equalize using a constrained complexity equalizer [15].

A receiver front end that moves the matched filter to discrete time, which requires a higher than symbol-rate sampling, is equivalent to the sampled matched filter under idealized assumptions, but is superior under more realistic assumptions. This alternative receiver structure will be derived first in the time domain, and then in the frequency domain.

Time Domain Derivation

Assuming that the noise spectrum is white, the matched filter following demodulation equivalently correlates against the received pulse $h^*(t)$. For simplicity, assume that the channel has an excess bandwidth of less than 100%, so that $h(t)$ is bandlimited to $1/T$ Hz. The sampling theorem (Section 2.3) tells us that the received pulse $h(t)$ can be expanded in terms of its samples at twice the symbol rate, $h_m = h(mT/2)$,

$$h(t) = \sum_{m=-\infty}^{\infty} h_m \text{sinc}\left(\frac{2\pi}{T}(t - mT/2)\right), \tag{8.104}$$

where $\text{sinc}(x) = \sin(x)/x$. We can replace $h(t)$ by (8.104) in calculating the folded spectrum, resulting in a sampled matched filter output of

$$\sum_{m=-\infty}^{\infty} h_m^* s_{2k+m}, \tag{8.105}$$

where

$$s_n = \int_{-\infty}^{\infty} r(t) e^{-j2\pi f_c t} \text{sinc}\left(\frac{2\pi}{T}(t - nT/2)\right) dt. \tag{8.106}$$

$\{s_n, -\infty < i < \infty\}$ is evidently another sufficient statistic for the received signal since the matched filter output can be deduced from it using (8.105).

The sufficient statistics $\{s_n\}$ in (8.106) are simply the sampled output of an ideal lowpass filter, where the filter bandwidth is $1/T$ as shown in Fig. 8-17(a). This lowpass filter rejects all out-of-band noise components, and especially the out-of-band noise that would otherwise alias in-band after sampling. Since the output has bandwidth less than $1/T$, by the sampling theorem it can be sampled at rate $2/T$ (twice the symbol rate) without loss of information. The calculation of the second sufficient statistic in (8.105) can be thought of as applying a matched

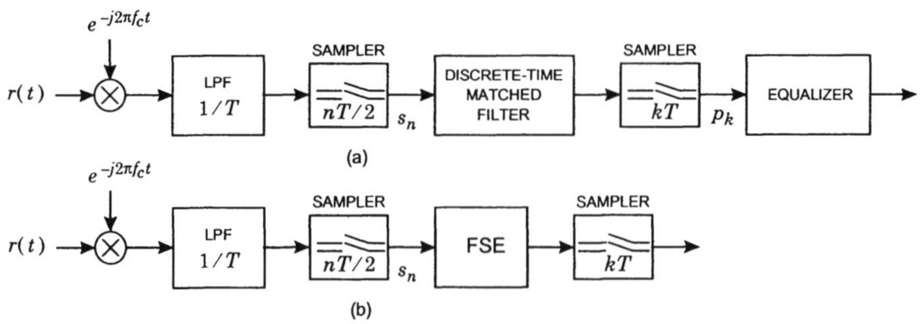

Fig. 8-17. Fractionally spaced equalizer for 100% excess bandwidth. a. Realization with separate discrete-time matched filter and equalizer. b. Realization where the discrete-time matched filter and equalizer are combined.

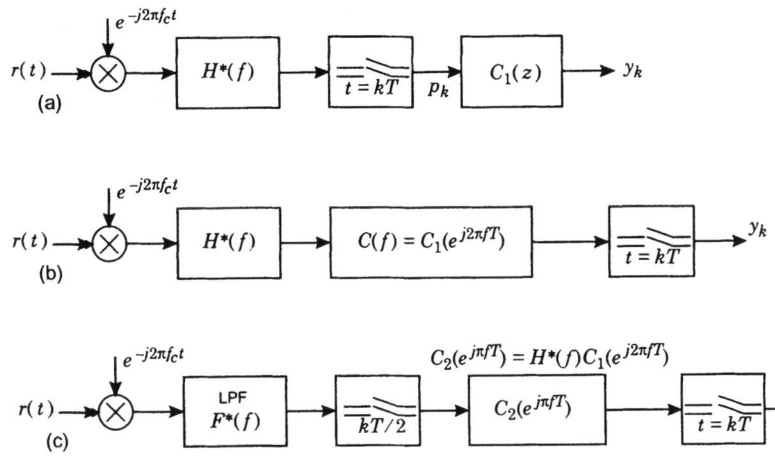

Fig. 8-18. FSE interpretation in the frequency domain. a. The starting point is the conventional matched filter, symbol-rate sampler, and equalizer $C_1(z)$. b. The equalizer implemented in continuous time. c. The matched filter and equalizer implemented in discrete time with sampling rate equal to twice the symbol rate.

filter to the lowpass filter output in discrete time. The output of this matched filter can be sampled at the symbol rate as before, so that this discrete-time matched filter also performs a decimation (by a factor of two) implicitly. This is followed by a symbol-rate equalizer, for example a linear equalizer or a precursor equalizer.

The two discrete-time filters in Fig. 8-17(a) can be combined as in Fig. 8-17(b), where the resulting combined matched filter and equalizer is called a *fractionally spaced equalizer (FSE)*. It is a filter that has an input sampling rate equal to twice the symbol rate, and output sampling rate equal to the symbol rate. We can think of it as a filter with input and output sampling rates equal to twice the symbol rate, where every second output sample is not used and hence need not be calculated. The FSE structure shown in Fig. 8-17(b) is applicable to any of the design strategies considered previously, including the sampled matched filter, WMF, LE, DFE, and MLSD. An FSE is assumed in Fig. 5-21.

Exercise 8-4.

Determine a formula for the output of the FSE before decimation in terms of the received pulse $h(t)$ and equalizer $C(z)$, and thereby determine the transfer function of the FSE.

Frequency Domain Derivation

It is instructive to rederive the FSE entirely in the frequency domain. The conventional matched filter $H^*(f)$, symbol-rate sampler, and a discrete-time equalizer $C_1(z)$ are shown in Fig. 8-18(a). The first step in the derivation of the FSE is to implement the discrete-time filter equivalently in continuous time as a filter with transfer function $C_1(e^{j2\pi fT})$. This moves the symbol-rate sampler after the equalizer, but does not change the output.

Exercise 8-5.

Show mathematically that Fig. 8-18(a) and Fig. 8-18(b) are equivalent.

Assume again that the excess bandwidth is less than 100% (this can be relaxed, see Problem 8-9). Then the combined filter $H^*(f)C_1(e^{j2\pi fT})$ has bandwidth less than $1/T$, and hence can be implemented as a discrete-time filter with sampling rate equal to $2/T$, or twice the symbol rate. This is shown in Fig. 8-18(c); the channel output is first filtered by an antialiasing filter that eliminates all noise and signal components above the symbol rate $1/T$, a twice symbol-rate sampler, and a discrete-time filter that realizes both the matched filter and the precursor equalizer. The output of this discrete-time filter is resampled at the symbol rate, implying that this is a decimating filter (Problem 8-10).

Interpretation

The symbol-rate sampling after a matched filter usually introduces aliasing. In both the time and frequency domains, we have seen that the FSE uses a discrete-time filter at a sampling rate such that there is no aliasing before filtering. The result is that the discrete-time filter can be designed to adjust for the sampling phase in the matched filtering portion of its response, thereby eliminating the effect of sampling phase on noise enhancement. In practice, we will find in Chapter 9 that the discrete-time FSE can adapt automatically to compensate for the sampling phase, thereby reducing the effect of sampling phase on noise enhancement. In addition, the limited degrees of freedom of a finite discrete-time filter are more effective when deployed before decimation to the symbol rate, because the aliased sidebands can be filtered independently.

It should be remembered that a twice-symbol-rate FIR FSE with the same number of coefficients as a symbol-rate discrete-time filter has an impulse response that will span half the time interval. In spite of this, experience has shown that the FSE will perform as well for the same number of coefficients for all channel conditions, and noticeably better for channels with severe band-edge delay distortion [12].

We do not need to redo the equalization designs of Sections 8.1 and 8.2, because of the equivalence to the FSE shown in Fig. 8-18. Thus, a simple approach is to design the equalizers according to the theory of Sections 8.1 and 8.2, and then transfer the resulting design to the fractionally spaced implementation using the equivalence property. Note that fractionally spaced equalizers can be used for the LE, the WMF, and the precursor equalizer of the DFE. However, the postcursor equalizer of the DFE will always use symbol-rate sampling, because it operates at the same sampling rate as the slicer.

8.4. Transversal Filter Equalizers

In previous sections of this chapter, we have considered equalizer designs without regard to implementation complexity. In practice any equalizer that is built must be realizable, which means that it must have a rational transfer function. Rational transfer functions come in two basic types, finite impulse response (FIR) or infinite impulse response (IIR). FIR filters can be implemented even if they are not causal, if an additional delay is permissible. Such a delay will be permissible in the feedforward, but not feedback, path. IIR filters can be implemented as stable and causal filters as long as their poles are inside the unit circle.

If the folded spectrum $S_h(z)$ happens to be rational, then all the equalizer designs are rational. However, it is very unusual for a channel to have *precisely* a rational transfer function, so we must resort to an approximation. Moreover, we saw in Section 8.2 that equalizer designs resulting from optimization often have poles outside the unit circle, when the discrete-time channel model has a maximum-phase component. These equalizers can only be approximated, usually by an FIR filter. The usual approximation would be something like

$$C(z) = \sum_{k=-N}^{N} c_k z^{-k} \quad , \tag{8.107}$$

which is noncausal. In fact an equalizer of the form of (8.107) is realizable if the output is delayed by N symbol intervals, since

$$z^{-N}C(z) = \sum_{k=0}^{2N} c_{k-N} z^{-k} \tag{8.108}$$

is causal. Of course any physically realizable system must be strictly causal, but we have assumed a non-causal matched filter, which again can be approximated by a causal filter plus delay.

In digital communication, an FIR digital filter is usually given the special name *transversal filter*, a terminology that originated in early continuous-time realizations using analog delay lines. The coefficients c_k of the filter are usually called the *tap weights* or *tap coefficients* of the filter. In the case of (8.107) there are precisely $2N + 1$ taps for the filter, which gives us $2N + 1$ degrees of freedom in the design of the filter.

Optimization of equalizer designs based on a constrained-complexity filter such as (8.107) is feasible, although quite different techniques than those used earlier in this chapter are necessary. The equalizer transfer function can no longer be chosen independently at each frequency as was done earlier. This implies that the zero-forcing criterion is no longer feasible, since a constrained-complexity filter may not be able to force the ISI to zero. Thus, a minimum MSE criterion is often used, and the MSE is minimized over the choice of filter tap coefficients (considered as a $(2N + 1)$-dimensional vector). An example of this type of optimization will arise in Chapter 9, as a first step in the design of adaptive equalizers.

8.5. ISI and Channel Capacity

In this chapter we have established the effect of ISI on the error probability of several receiver designs, including both equalization and the MLSD. Another related question is the effect of ISI on channel capacity. While the error probability results allow us to predict the impact of ISI when channel coding is not used, the channel capacity results allow us to assess the effect of ISI in the presence of channel coding. In Section 6.7 we did a similar development for ideal channels without ISI, and found the performance of different modulation techniques in relation to the fundamental limits of capacity in the absence of ISI. It was found that a given modulation technique at a given probability of error displayed an "SNR gap to capacity," which is the difference between the normalized SNR required to achieve that error probability and the normalized SNR that the channel capacity theorem says is required for a vanishingly small error probability. Our purpose here is to extend those results to channels with ISI.

Since implementation of processing techniques is easier in discrete time, we must choose a sampling rate and transmit and receive filters that effectively transmit a discrete-time signal over a continuous-time channel. If the continuous-time channel is not bandlimited, then this conversion to discrete time sometimes involves a reduction in capacity, although, as we will see, that is more likely to occur at high SNR. For the most part, we will apply our discrete-time techniques to continuous-time channels that are strictly bandlimited, because this case is analytically much simpler. (At the end of the section, some differences in the conclusions for channels that are not strictly bandlimited will be mentioned.) For strictly bandlimited channels we will find some surprising conclusions, including:

- ISI always reduces the capacity of a channel. This is surprising because we have seen that ISI does not always appreciably increase the error probability for PAM, for example on some channels with mild ISI when the MLSD achieves a figure of merit equal to the matched filter bound.
- On channels with ISI, we will find a generalized definition of normalized *SNR*, SNR_{norm}, that has the same interpretation as SNR_{norm} in Section 6.7; namely, $SNR_{\text{norm}} \geq 1$, with equality when the modulation technique is operating at capacity limits. The difference between SNR_{norm} and unity represents an "SNR gap to capacity", or increase in transmitted power or SNR relative to the minimum transmitted power or SNR at which it is feasible to operate, as quantified by the channel capacity.
- When we use PAM in conjunction with the equalization strategies considered in Section 8.1, we will quantify the SNR gap. For the DFE-ZF, we will find that this SNR gap approaches a constant at high SNR that is *independent of the channel response.* Since an ISI-free channel is a special case of a channel with ISI, this asymptotic SNR

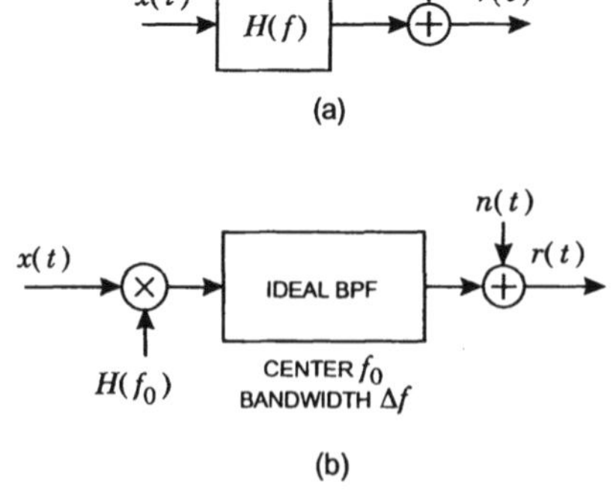

Fig. 8-19. Continuous-time channel models for the calculation of capacity. (a) A continuous-time channel with additive Gaussian noise. (b) A small subband of the channel.

gap is, at high SNR, the same on channels with ISI as on ideal channels. This implies that channel coding (Chapter 12) has approximately the same potential for improvement in the normalized SNR at a given error probability on channels with ISI as on ideal channels. This does *not* imply that ISI is benign with respect to channel capacity, since it does reduce channel capacity as noted above, but rather it *does* imply that the gap between coded modulation systems and capacity can be closed essentially to the same extent on channels with ISI as on ideal channels, at least at high SNR. Furthermore, it suggests that the DFE-ZF, and not the MLSD, is a good receiver structure to use as a starting point.

- The SNR gap for the MMSE-DFE-U is fixed, independent of the ISI, at all SNR's (not only at high SNR as with the DFE-ZF), as long as the transmit filter is optimized. This suggests that the unbiased DFE-MSE may be a good canonical receiver structure for coded modulation systems at low SNR.

8.5.1. Continuous-Time Water Pouring

As in Section 6.7, the first step will be to determine the capacity of the continuous-time additive Gaussian noise channel. We do this first for a general channel, and then apply the results to the special case of a passband channel.

Suppose, as pictured in Fig. 8-19, that we have a real-valued channel with transfer function $H(f)$ and additive real-valued Gaussian noise $n(t)$. While we are primarily interested in the white noise case, the following development is no harder if we assume that the noise has a general power spectrum $S_n(f)$. In Section 6.7 we developed the channel capacity for a similar situation, except the channel was an ideal channel with bandwidth B Hz. We were then able to derive the fundamental limits to the spectral efficiency of this ideal channel. Our purpose here is to extend these results to channels with ISI, and hence general $H(f)$ and $S_n(f)$.

The earlier ideal channel results can be applied to the present situation by dividing bandwidth into small bins of width Δf, where Δf is small enough that the channel transfer function is approximately constant over a range of Δf. (Of course, later we will take $\Delta f \to 0$, at which time the approximation will become precise.) One of these subbands, centered at frequency f_0, is shown in Fig. 8-19(b). This subband is modeled by a flat gain of $H(f_0)$ together with an ideal (unit gain) real-valued bandpass filter with bandwidth Δf centered at frequency f_0. Even though the additive noise is not white, it can be considered as white with power spectral density $S_n(f_0)$ within the bandwidth of interest in the subchannel. Since each of these subbands can be used in principle by a separate and independent digital communication system, the total capacity is the aggregate capacity of each of these subbands.

Suppose there is a constraint on the average transmit power $E[x^2(t)] = P$. Furthermore, let $S_x(f)$ be an appropriate input power spectrum that meets this constraint. Choosing $S_x(f)$ is equivalent to deciding how to distribute the available input power across frequencies, or across subbands. Our approach is to determine the capacity of each subband, assuming its input power is constrained by $S_x(f)$ (at the frequency of that subband), giving us an expression for the total capacity as a function of $S_x(f)$. Subsequently, the $S_x(f)$ that maximizes this total capacity will be found, subject to the constraint that $E[x^2(t)] = P$.

The capacity of one subband can be determined as follows. The constraint on the power into the subband is $2S_x(f_0) \cdot \Delta f$, taking into account both positive and negative frequencies. This power and the gain factor $H(f_0)$ is equivalent to an ideal (unity transfer function) subchannel with input power constrained to $2S_x(f_0)\,|H(f_0)|^2 \cdot \Delta f$. The capacity of this subchannel is given by (6.144),

$$C(f_0) = \Delta f \cdot \log_2\left(1 + \frac{2S_x(f_0)|H(f_0)|^2 \Delta f}{2S_n(f_0)\Delta f}\right)$$

$$= \Delta f \cdot \log_2\left(1 + \frac{S_x(f_0)|H(f_0)|^2}{S_n(f_0)}\right). \tag{8.109}$$

The total capacity is then the sum of the subchannel capacities, which becomes an integral in the limit of $\Delta f \rightarrow 0$,

$$C = \int_0^\infty \log_2\left(1 + \frac{S_x(f)|H(f)|^2}{S_n(f)}\right) df$$

$$= \frac{1}{2}\int_{-\infty}^{\infty} \log_2\left(1 + \frac{S_x|H|^2}{S_n}\right) df \quad \text{bits/sec.} \tag{8.110}$$

In this and many subsequent expressions the frequency variable is suppressed for compactness.

Since capacity in each subband is achieved by a wide-sense stationary Gaussian input process, total capacity is achieved if $x(t)$ is wide-sense stationary Gaussian with power spectrum $S_x(f)$. What remains is to translate an overall input power constraint into an optimal power spectrum $S_x(f)$. Our desire is to maximize C under the constraints

$$P = \int_{-\infty}^{\infty} S_x \, df \,, \qquad S_x \geq 0 \,, \tag{8.111}$$

where P is the transmitted power. Using a Lagrange multiplier approach, we can equivalently choose S_x as a function of λ to maximize

$$\frac{1}{2}\int_{-\infty}^{\infty} \log_2\left(1 + \frac{S_x|H|^2}{S_n}\right) df + \lambda \int_{-\infty}^{\infty} S_x df \,, \tag{8.112}$$

and then choose λ to meet the power constraint (8.111). Writing this as a single integral and differentiating the integrand with respect to S_x at each frequency and setting that derivative to zero, we get that S_x is of the form $(1/2\lambda - S_n/|H|^2)$. Taking into account that S_x must be positive, the result is

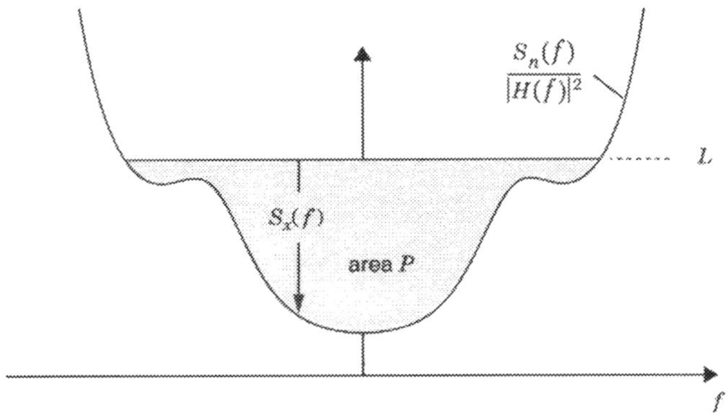

Fig. 8-20. An illustration of water pouring for the optimization of the transmit power spectrum in achieving capacity. A "bowl" shaped like $S_n(f)/|H(f)|^2$ is formed, and water with volume P is poured into it up to level L. The capacity-achieving transmit spectrum at each frequency is the depth of the water.

$$S_x = \begin{cases} L - \frac{S_n}{|H|^2}, & f \in F \\ 0, & \text{otherwise} \end{cases} \tag{8.113}$$

where the constant $L = 1/2\lambda$ is chosen to meet the power constraint and F is the set of frequencies for which $L > S_n/|H|^2$. F is called the *water-pouring band*. This method of determining the transmit spectrum is called *water pouring*, and we will call the resulting transmit spectrum S_x the *water-pouring spectrum* [16].

The approach of splitting up the overall channel into subchannels also points out that the input random process that achieves capacity is a wide-sense stationary (and hence strictly stationary) real-valued Gaussian process with power spectral density $S_x(f)$. Thus, communication systems that wish to approach capacity have to somehow ensure that the transmit spectrum approximates the water-pouring spectrum, and also that the distribution is approximately Gaussian.

This solution is easiest to understand pictorially, as in Fig. 8-20. A "bowl" $S_n/|H|^2$ is formed, and water is poured into the bowl up to level L until the total volume of water equals the transmit power constraint. The transmit spectrum vs. frequency is then the depth of the water. Water pouring concentrates the transmit power at those frequencies where the ratio $|H|^2/S_n$ is relatively large; that is, the channel has relatively little attenuation or the noise power spectrum is relatively small.

Passband Channel

Observe that the water-pouring approach applies equally well to passband as well as baseband channels.

Example 8-24.

For an ideal passband white Gaussian noise channel with bandwidth B, $S_n/|H|^2 = N_0/2$ is a constant within the channel bandwidth. The bowl (actually two bowls) have infinite-slope sides, as illustrated in Fig. 8-21, because $|H|^2 = 0$ outside the channel bandwidth. Thus, the water-pouring band is precisely the full passband bandwidth. The volume in one of the bowls must be $P/2$, which implies that the depth of the water must be $P/2B$; that is, $S_x = P/2B$ over the bandwidth of the channel. The capacity is then

$$C = B \cdot \log_2(1 + SNR)\ , \qquad SNR = P/(N_0B)\ , \tag{8.114}$$

consistent with (6.144).

A formula for capacity in terms of a complex baseband equivalent channel can be derived for a passband channel. Assume the channel H is strictly bandlimited to $f_0 - B/2 \le |f| \le f_0 + B/2$. Then the capacity integral of (8.110) can be safely limited to this frequency range, since the water-pouring band must fall in this range. Thus, (8.110) can be written as the sum of two integrals, one for positive and one for negative frequencies, or equivalently twice the positive-frequency integral due to the symmetry of $|H|$ about $f = 0$ (since the channel H is real-valued). Thus,

$$C = \int_{f_0 - B/2}^{f_0 + B/2} \log_2\left(1 + \frac{S_x(f)|H(f)|^2}{S_n(f)}\right) df$$

$$= \int_{-B/2}^{B/2} \log_2\left(1 + \frac{S_x(f + f_0)|H(f + f_0)|^2}{S_n(f + f_0)}\right) df\ . \tag{8.115}$$

When water pouring is performed at passband, it is clear from symmetry that half the power will be at positive frequencies, and thus we can formulate water pouring at baseband using $P/2$ in place of P,

$$S_x(f + f_0) = \begin{cases} L - \dfrac{S_N(f + f_0)}{|H(f + f_0)|^2}, & f \in F \\ 0, & \text{otherwise} \end{cases} \tag{8.116}$$

where L is chosen such that

$$\int_{-B/2}^{B/2} S_x(f + f_0)\ df = P/2\ . \tag{8.117}$$

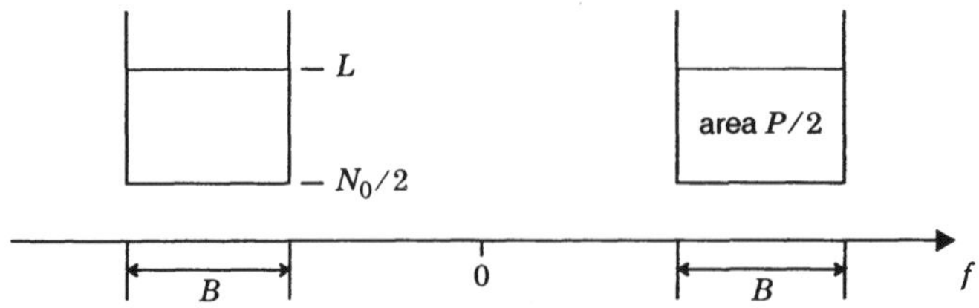

Fig. 8-21. Illustration of the water pouring procedure for the ideal bandpass white Gaussian noise channel.

Also, F must be a subset of $[-B/2, B/2]$. The complex baseband channel does not necessarily have spectra symmetric about $f = 0$ (the baseband-equivalent channel and noise are complex-valued), and thus the water-pouring band F is not necessarily symmetric about $f = 0$.

8.5.2. Discrete-Time Water Pouring

When a continuous-time channel is strictly bandlimited, and we derive a discrete-time channel from it using a sufficiently high sampling rate, then the discrete-time channel will have the same capacity as the underlying continuous-time channel. This will also be true of channels which are not bandlimited, but for which the transfer function approaches zero as frequency gets large. We will now form the connection between the channel capacities of discrete- and continuous-time channels, leading to a formula for the capacity of a discrete-time channel. We will gain the ability to compare the capacity of this channel with the performance of PAM in conjunction with receiver design techniques covered earlier in this chapter.

Assume that $H(f)$ is a real-valued continuous-time channel, which may or may not be bandlimited and may be baseband or passband. A discrete-time channel is derived in Fig. 8-22(a) by sampling at rate $2B = 1/T$ at both input and output with ideal LPF transmit and receive filters each with gain $\sqrt{T}$, to ensure that their impulse response has unit energy. The resulting discrete-time system is pictured in Fig. 8-22(b). We will now relate the discrete-time transfer function and power spectra to the similar quantities for the continuous-time channel.

Signal Transfer Function

Since the receive filter is bandlimited to half the sampling rate, there is no aliasing distortion in the final sampling operation. The overall gain of T in the transmit and receive filters makes up for the factor of $1/T$ in the conversion from continuous to discrete time, and hence the discrete-time and continuous-time transfer functions are directly related,

$$H(e^{j2\pi fT}) = H(f)\,, \qquad |f| \le B\,. \tag{8.118}$$

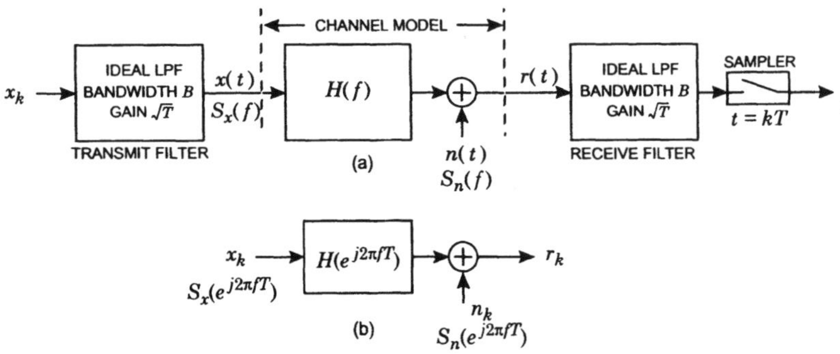

Fig. 8-22. The way in which a discrete-time system can be obtained from a continuous-time system employing the sampling theorem. (a) Deriving the discrete-time channel by sampling at the input and output. (b) The resulting discrete-time channel model.

Transmit Power Constraint

The discrete- and continuous-time power spectra are related by

$$S_x(f) = S_x(e^{j2\pi fT}) , \qquad |f| \le B , \tag{8.119}$$

since there is a factor of $1/T$ in the conversion from discrete- to continuous-time (see Appendix 3-A) compensated by the $\sqrt{T}$ gain of the filter. The relation

$$\mathrm{E}[x_k^2] = T\int_{-1/(2T)}^{1/(2T)} S_x(e^{j2\pi fT})df = T\int_{-1/(2T)}^{1/(2T)} S_x(f)df = T\cdot \mathrm{E}[x^2(t)] \tag{8.120}$$

implies that a continuous-time power constraint $P = \mathrm{E}[x^2(t)]$ is equivalent to a discrete-time constraint on the energy of a single sample, $PT = \mathrm{E}[x_k^2]$. Furthermore, any continuous-time power spectrum $S_x(f)$ can be generated, *providing that it is bandlimited* to B Hz. Thus, if capacity for the continuous-time channel demands a transmit power spectrum wider than this, in accordance with water pouring, then it cannot be generated by the configuration of Fig. 8-22(a) for sampling rate $2B$. Conversely, if the bandwidth of the continuous-time water-pouring band is less than B, then the discrete-time channel capacity will equal the continuous-time channel capacity, with the transmit power spectrum chosen appropriately.

Output Noise Spectrum

When the power spectrum of the noise on the continuous-time channel is $S_n(f)$, then the noise at the output of the receive filter is bandlimited to B Hz and multiplied by T. The sampling operation does not introduce aliasing distortion, and multiplies the power spectrum by $1/T$, and thus after sampling

$$S_n(e^{j2\pi fT}) = S_n(f) . \qquad |f| \le B . \tag{8.121}$$

Capacity of the Discrete-Time System

The capacity of the discrete-time system with input power constraint $\mathrm{E}[x_k^2] = E$ will equal the capacity of the continuous-time system with input power constraint $\mathrm{E}[x^2(t)] = E/T$, providing that the resulting water-pouring band of the continuous-time system is contained in $|f| \le B$. This will always be true if the continuous-time system is strictly bandlimited to B. Therefore, by assuming that $H(f)$ is bandlimited to B and satisfies (8.118) for $|f| \le B$, we can calculate the capacity of the discrete-time system indirectly by calculating the capacity of the continuous-time system with input power constraint P. Thus, the capacity of a real-valued discrete-time baseband channel $H(e^{j2\pi fT})$ with additive real-valued noise $S_n(e^{j2\pi fT})$ is

$$C = \frac{1}{2}\int_{-1/(2T)}^{1/(2T)} \log_2\left(1 + \frac{S_x(e^{j2\pi fT})|H(e^{j2\pi fT})|^2}{S_n(e^{j2\pi fT})}\right) df , \qquad \text{bits/sec} , \tag{8.122}$$

where $S_x(e^{j2\pi fT})$ is given by the water-pouring formula

$$S_x(e^{j2\pi fT}) = \begin{cases} L - \dfrac{S_n(e^{j2\pi fT})}{|H(e^{j2\pi fT})|^2}, & f \in F \\ 0, & \text{otherwise} \end{cases} \tag{8.123}$$

where L is chosen such that the transmit power $P = \mathrm{E}[x_k^2]/T$ is

$$P = \int_{-1/(2T)}^{1/(2T)} S_x(e^{j2\pi fT})\, df \ . \tag{8.124}$$

Passband Case

We can use the capacity of a continuous-time passband channel, given by (8.115), to derive the capacity of a discrete-time complex-valued baseband channel derived from that passband channel. The connection is the upconversion and downconversion steps illustrated in Fig. 8-23. The resulting complex baseband channel is shown in Fig. 8-23(b). As shown in Chapter 2, the equivalent baseband channel is $H(f + f_c)$, and the equivalent baseband power spectrum is $S_n(f) = 2S_{\tilde{n}}(f + f_c)$, where $S_{\tilde{n}}(f)$ is the power spectrum of the real passband noise. The factor of two in S_n can be interpreted intuitively in two ways:

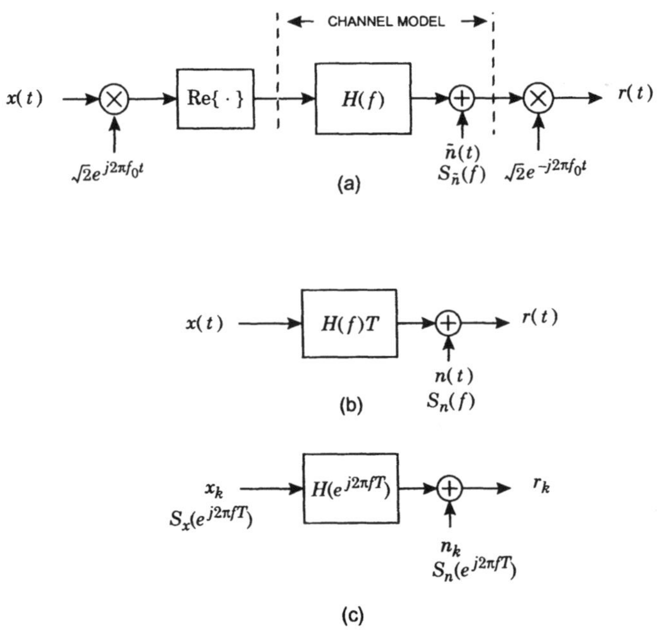

Fig. 8-23. The derivation of a complex baseband model from a passband channel $B(f)$. (a) A continuous-time bandpass channel. (b) An equivalent complex baseband channel derived using upconverter and downconverter. (c) A discrete-time complex baseband channel derived by applying the sampling and transmit/receive filter of Fig. 8-22 to the baseband channel of (b).

- The noise power spectrum doubles due to spreading the same noise over half the bandwidth at baseband compared to passband.
- The noise is divided between the real and imaginary parts in equal variance (due to circular symmetry of the complex Gaussian noise), each part with the same power spectrum (referred to baseband) as in the passband case, but half the bandwidth.

Finally, due to the $\sqrt{2}$ factor in the transmitter, the relation between discrete-time and continuous-time signal power spectra is $S_x(e^{j2\pi fT}) = 2S_x(f)T$, which normalizes the total power to be the same at passband and baseband, and also results in the power constraint, from (8.117),

$$P = \int_{-1/(2T)}^{1/(2T)} S_x(e^{j2\pi fT})\, df\,, \tag{8.125}$$

corresponding to power constraint $P = \mathrm{E}[\,|x_k|^2]/T$.

Substituting these values into (8.115), the capacity of the complex baseband discrete-time channel is

$$C = \int_{-1/(2T)}^{1/(2T)} \log_2\left(1 + \frac{S_x(e^{j2\pi fT})|H(e^{j2\pi fT})|^2}{S_n(e^{j2\pi fT})}\right) df \ \text{ bits/sec}\,, \tag{8.126}$$

where the functional form of the transmit spectrum is, from (8.116),

$$S_x(e^{j2\pi fT}) = \begin{cases} 2L - \dfrac{S_n(e^{j2\pi fT})}{|H(e^{j2\pi fT})|^2}, & f \in F \\ 0, & \text{otherwise} \end{cases} \tag{8.127}$$

and $2L$ is chosen to meet constraint (8.125).

Example 8-25.

When the noise on the continuous-time channel is white with power spectrum $N_0/2$, $S_n = N_0$. It is convenient to define a normalized transmit spectrum,

$$S_x' = \frac{S_x}{N_0} = \begin{cases} L' - |H|^{-2}, & f \in F \\ 0, & \text{otherwise} \end{cases} \tag{8.128}$$

in which case L' is chosen to meet the power constraint of (8.125),

$$\frac{P}{N_0 B} = T\int_{-1/(2T)}^{1/(2T)} S_x'\, df \tag{8.129}$$

and the capacity of (8.126) becomes

$$C \cdot T = T\int_{-1/(2T)}^{1/(2T)} \log_2(1 + S_x' \cdot |H|^2)\, df\ . \tag{8.130}$$

Of course, P/N_0B is familiar as the SNR of the ideal white Gaussian noise channel. These relations are convenient for calculating the capacity for this white noise case. If the variable of integration is changed from f to $\nu = fT$, the factors of T multiplying the integral conveniently go away. For example,

$$C \cdot T = \int_{-1/2}^{1/2} \log_2(1 + S_x'(e^{j2\pi\nu}) \cdot |H(e^{j2\pi\nu})|^2) \, d\nu \ . \tag{8.131}$$

Since $1/T$ is the sampling rate in Hz, the quantity $C \cdot T = C/B$ is the capacity per sample. Alternatively, it is the capacity as measures in bits/sec/Hz.

Example 8-26.

Consider a single-zero channel $H(z) = (1 - cz^{-1})/\sqrt{1 + |c|^2}$, normalized to have a unit-energy impulse response. The capacity is plotted in Fig. 8-24 for three cases: the actual channel with $|c| = 0.99$, the ideal channel ($c = 0$), and a formula derived later that applies asymptotically at high SNR. For our present purposes, the interesting comparison is between the ideal channel and the channel with ISI. At low SNR, the channel with ISI actually has a higher capacity than the ideal channel, because the water-pouring spectrum can be concentrated at frequencies where the channel transfer function has gain. At high SNR, there is a modest penalty in capacity due to ISI.

Example 8-27.

The capacity of the single-pole channel $H(z) = \sqrt{1 + |c|^2}/(1 - cz^{-1})$ for $|c| = 0.99$ is plotted in Fig. 8-25. This channel has much more severe ISI than the single-zero channel, and as a result there is a much greater penalty in capacity due to ISI.

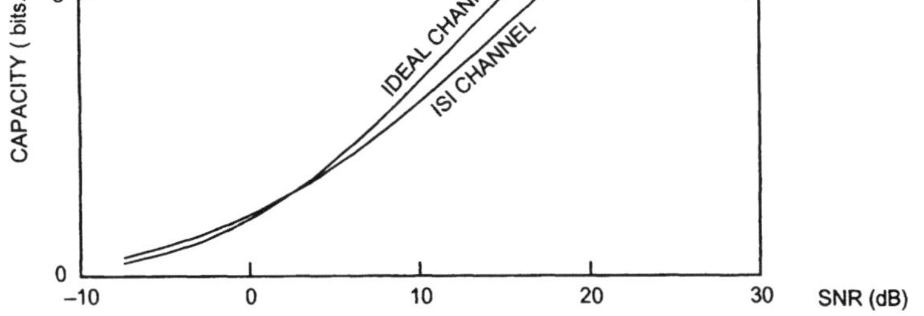

Fig. 8-24. The capacity per sample is plotted against SNR P/N_0B for a single-zero channel with $|c| = 0.99$. The capacity is independent of the angle of c.

8.5.3. Relation to Arithmetic and Geometric Means

Once the water-pouring band F is determined, the capacity formulas can all be simplified in a useful way in terms of the arithmetic and geometric means. This gives a compact formula for the capacity, with the caveat that the formula is not self-contained, in that water-pouring must be performed first to determine F. Although this applies to both the continuous-time and discrete-time cases, we are primarily interested in discrete-time here. We will do the baseband case first, followed by the complex baseband equivalent to a passband channel.

Baseband

Assume we know F. From (8.123), we can calculate the total power for the water-pouring spectrum directly by integrating S_x, and restricting the integral to F,

$$\frac{P}{|F|} = \langle S_x \rangle_{A,F} = L - \langle S_n / |H|^2 \rangle_{A,F}, \tag{8.132}$$

which gives us an equation for L in terms of the power constraint P. Also, substituting (8.123) into (8.122), where the integral is restricted to $f \in F$,

$$C = \frac{1}{2} \int_F \log_2\left(\frac{L|H|^2}{S_n}\right) df. \tag{8.133}$$

This is directly related to the geometric mean,

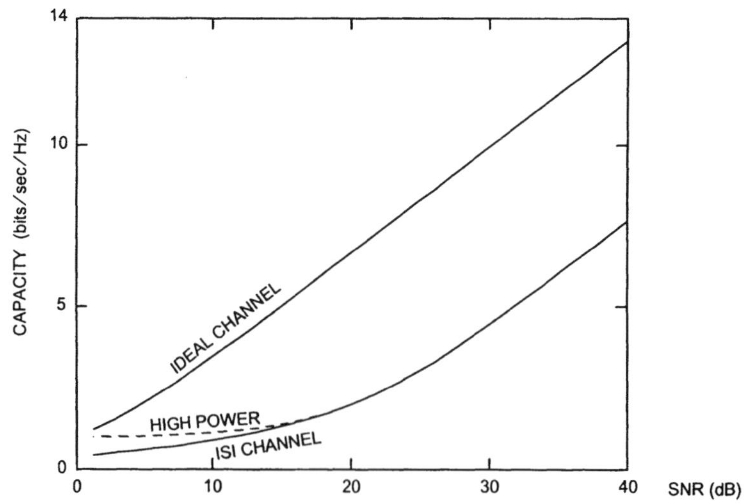

Fig. 8-25. The capacity per sample plotted against SNR P/N_0B for a single-pole channel with $|c| = 0.99$. The capacity is independent of the angle of c.

$$C = \frac{|F|}{2}\log_2\langle L|H|^2/S_n\rangle_{A,F} = \frac{|F|}{2}\log_2\left(\frac{L}{\langle S_n/|H|^2\rangle_{G,F}}\right). \tag{8.134}$$

Finally, substituting from (8.132),

$$C = \frac{|F|}{2}\log_2\left(\frac{P/|F| + \langle S_n/|H|^2\rangle_{A,F}}{\langle S_n/|H|^2\rangle_{G,F}}\right) \quad \text{bits/sec.} \tag{8.135}$$

Example 8-28.

When $S_n = N_0/2$ is white noise, the capacity becomes

$$C = \frac{|F|}{2}\log_2\left(\frac{SNR + \langle|H|^{-2}\rangle_{A,F}}{\langle|H|^{-2}\rangle_{G,F}}\right) \quad \text{bits/sec,} \tag{8.136}$$

where $SNR = P/(|F|N_0/2)$ is the channel *input* signal-to-noise ratio within the water-pouring band F ($|F|$ is the total bandwidth of the water-pouring band in Hz, and when multiplied by $N_0/2$ we get the total noise within this bandwidth). Note that the signal-to-noise ratio at the channel *output* is dependent on the channel transfer function. Equation (8.136) bears a striking relationship to (8.114), and of course when there is no ISI ($H = 1$ for $f \in F$) it reduces to (8.114).

Once the water-pouring band F is determined, the only parameters of the channel that need be known are the arithmetic and geometric means of $S_n/|H|^2$ over F. All channels with the same water-pouring band, and the same arithmetic and geometric means over that bandwidth, have the same capacity. This result is striking in view of the fact that the arithmetic and geometric means (over the entire bandwidth) determine the performance of the equalizers considered in Section 8.2 as well.

Complex Baseband Case

The only differences in the complex baseband case are the replacement of L by $2L$ in the water-pouring formula (which doesn't change the result since it is just another constant), the fact that the noise is complex, and the removal of the factor of $1/2$. Thus, the capacity for the complex baseband case is similar,

$$C = |F|\cdot\log_2\left(\frac{P/|F| + \langle S_n/|H|^2\rangle_{A,F}}{\langle S_n/|H|^2\rangle_{G,F}}\right) \quad \text{bits/sec.} \tag{8.137}$$

The biggest difference is a capacity that is a factor of two larger, due to the complex signals; that is, the capacity *per dimension* stays the same. Of course, other differences in practice include a 50% smaller bandwidth (compensating for the higher dimensionality), and spectra that are generally not symmetric about $f = 0$.

8.5.4. Spectral Efficiency, SNR, and Normalized SNR

Thus far in this section, the capacity of both continuous- and discrete-time channels with additive Gaussian noise have been determined. These results will now be applied to characterizing the performance of digital communication systems operating over such channels. The results of Section 6.7, in which the operation of specific modulation techniques were related to the fundamental limits of capacity, will be extended to channels with ISI. This will be done by defining a normalized signal-to-noise ratio, SNR_{norm}, which is a generalization of the SNR_{norm} defined in Section 6.7, with the same interpretation that $SNR_{\text{norm}} \geq 1$ with equality if the system operates at the fundamental capacity limits.

For the remainder of this section, we will limit attention to the complex baseband channel shown in Fig. 8-23(c). In typical applications, this channel will have been derived from an underlying continuous-time channel by transmit and receive filtering and sampling, but we will not concern ourselves with how that happened. The ISI on the channel is represented by the transfer function $H(e^{j2\pi fT})$, and the additive Gaussian noise has power spectrum $S_n(e^{j2\pi fT})$. When the channel is used for digital communication with PAM modulation, x_k will be replaced by the data symbols a_k, and the channel output r_k will be applied to one of the receiver structures considered in Section 8.2, such as the LE, DFE, or MLSD. Our concern is with the performance of the latter receivers, as measured by P_e, and how that performance relates to the fundamental capacity limits for the discrete-time channel of Fig. 8-23(c).

Since capacity is normally expressed in bits/sec, or in terms of spectral efficiency, we need to know the sampling rate in Fig. 8-23(c), which is the symbol rate as well. Define the symbol rate as $B_0 = 1/T$, where T is the symbol interval. In typical applications, the complex baseband channel of Fig. 8-23(c) would be associated with an underlying continuous-time passband channel. If we start with a passband channel with nominal bandwidth B, the highest possible symbol rate is $B_0 = B$. The complex baseband channel has bandwidth $B/2$, and the highest symbol rate is double this, or B.

Spectral Efficiency

The spectral efficiency depends on the bit rate and the bandwidth of the continuous-time channel that supports this bit rate. For the latter, we will assume that the symbol rate B_0 has been chosen to be as high as possible, and thus the underlying continuous-time channel has bandwidth B_0. If the continuous-time channel actually has a higher bandwidth than this, and we are using it inefficiently by choosing a lower symbol rate than necessary, then the spectral efficiency will actually be lower than that calculated here. With this assumption, the spectral efficiency is

$$\nu_c = \frac{C}{B_0} = \frac{|F|}{B_0} \cdot \log_2\left(\frac{P/|F| + \langle S_n/|H|^2\rangle_{A,F}}{\langle S_n/|H|^2\rangle_{G,F}}\right) \quad \text{bits/sec-Hz.} \tag{8.138}$$

When the water-pouring band is the full bandwidth of the channel, then the factor $|F|/B_0 = 1$.

Signal-to-Noise Ratio

The *signal-to-noise ratio (SNR)* is defined as the ratio of the average signal power to the total noise power. This SNR applies only to the discrete-time channel; if the underlying continuous-time channel were considered instead, the result might be different because of the bandlimiting effects of the receive filter, etc. We will now calculate the SNR when the input spectrum is chosen according to water pouring; such a spectrum can achieve capacity. There are two natural places to define the SNR; at the input to the channel, or at the output. We will focus on the input to the channel; this is simpler because the signal power at the channel output depends on the details of S_x (not just its integral) as well as the channel H. The channel-output SNR is relegated to Problem 8-14.

At the input to the channel, the signal power is, by definition, P. Although the channel model defines the noise spectrum as S_n at the output of the channel H, this is equivalent to a noise with power spectrum $S_n/|H|^2$ at the channel input.

$$\int_F S_n/|H|^2 \, df = |F| \cdot \langle S_n/|H|^2 \rangle_{A,F} , \tag{8.139}$$

and the input SNR is then

$$SNR_{\text{in}} = \frac{P/|F|}{\langle S_n/|H|^2 \rangle_{A,\,F}} . \tag{8.140}$$

This SNR does not depend on the water-pouring spectrum, but rather depends only on the water-pouring band F and the total signal power P (this is an advantage of using channel-input SNR). For the case where $|F| = B_0$, SNR_{in} can also be interpreted as the signal energy per sample divided by the noise variance; in other words, in this case SNR_{in} coincides with the usual definition of SNR.

Normalized SNR

Equation (8.138), for the maximum spectral efficiency $B_0 = B$, can be rewritten as

$$\frac{P_S/|F|}{2^{\nu_c B_0/|F|} \langle S_n/|H|^2 \rangle_{G,\,F} - \langle S_n/|H|^2 \rangle_{A,\,F}} = 1 . \tag{8.141}$$

This is an alternative way to define the maximum achievable spectral efficiency ν_c for a given set of channel parameters (F, $\langle S_n/|H|^2 \rangle_{G,F}$, and $\langle S_n/|H|^2 \rangle_{A,F}$); it is simply a generalization of (6.146) for the ideal channel. This motivates the definition of a *rate-normalized SNR*, SNR_{norm}, that is a direct generalization of (6.137). For a system operating over channel H, with output noise spectrum S_n, with input power P, and achieving a spectral efficiency ν, define

$$SNR_{\text{norm}} = \frac{P/|F|}{2^{\nu B_0/|F|} \langle S_n/|H|^2 \rangle_{G,\,F} - \langle S_n/|H|^2 \rangle_{A,\,F}} . \tag{8.142}$$

This expression is simply (8.141) with the limiting spectral efficiency ν_c replaced by the actual achieved spectral efficiency ν. It is more complicated than it looks at first glance, since in general F (and hence $|F|$, $\langle S_n/|H|^2\rangle_{A,F}$, and $\langle S_n/|H|^2\rangle_{G,F}$) is a complicated function of P. However, substituting for P from (8.141),

$$SNR_{\text{norm}} = \frac{2^{\nu_c B_0/|F|}\langle S_n/|H|^2\rangle_{G,F} - \langle S_n/|H|^2\rangle_{A,F}}{2^{\nu B_0/|F|}\langle S_n/|H|^2\rangle_{G,F} - \langle S_n/|H|^2\rangle_{A,F}} \geq 1 \ , \tag{8.143}$$

with equality if the system is operating at the fundamental limits of capacity. That is, the relation $\nu \leq \nu_c$ is equivalent to the bound $SNR_{\text{norm}} \geq 1$. Note that (8.143) does not require that the system occupy the water-pouring band F, but rather applies to any system operating with transmit power P and spectral efficiency ν.

We can express SNR_{norm} in terms of the SNR_{in} by substituting for P from (8.140),

$$SNR_{\text{norm}} = \frac{SNR_{\text{in}}}{(2^{\nu B_0/|F|}\langle S_n/|H|^2\rangle_{G,F}/\langle S_n/|H|^2\rangle_{A,F}) - 1} \ , \tag{8.144}$$

a relation that is similar to, and a generalization of, the ideal channel case in (6.137).

Example 8-29.
For a complex baseband discrete-time channel derived from a continuous-time ideal passband channel with bandwidth B and white noise $N_0/2$, $S_n = N_0$ and $H = 1$, and the water-pouring band will be the full bandwidth, $F = [-B/2, B/2]$. Hence the arithmetic and geometric means will both be N_0, and $SNR_{\text{norm}} = SNR_{\text{in}}/(2^{\nu} - 1)$. Furthermore, $SNR_{\text{in}} = P/N_0B$, and this expression for SNR_{norm} is the same definition as (6.137). Unlike that simpler expression, (8.144) is a function of the ISI on the channel through the parameters F, $\langle S_n/|H|^2\rangle_{G,F}$, and $\langle S_n/|H|^2\rangle_{A,F}$.

It is important to note that in SNR_{norm}, F depends on P (through water pouring), and hence the other factors ($|F|$, $\langle S_n/|H|^2\rangle_{A,F}$, and $\langle S_n/|H|^2\rangle_{G,F}$) also depend on P. Thus, unlike in the ideal channel case, SNR_{norm} is in general not a linear function of P. As in Section 6.7, SNR_{norm} expressed in dB is related to the "SNR gap to capacity." In Section 6.7, that SNR gap to capacity was defined as the increase in transmitted power (or equivalently SNR) required to achieve a give P_e, relative to the Shannon limit on power at the same spectral efficiency. Unfortunately that same simple interpretation does not apply to the ISI case considered here, because SNR_{norm} is not linearly related to the transmit power P. However, based on SNR_{norm} we could still infer the SNR gap to capacity by taking into account the precise nonlinear relationship between P and SNR_{norm}. Fortunately, at high SNR this is not necessary, and the simpler interpretation of Section 6.7 applies, as we now show.

Capacity at High SNR

On some (but not all) channels H the water-pouring bandwidth eventually becomes the full bandwidth $|F| = B_0$ at high signal powers. Two cases are illustrated in Fig. 8-26 (for the white noise case). If $|H|^2$ is non-zero for all $|f| \leq B_0/2$ as in Fig. 8-26(a), the sides of the bowl become infinitely steep at the band edge, and for sufficiently high signal powers $|F| = B_0$. In contrast, on a channel with a zero at the band edge, as shown in Fig. 8-26(b), or for that matter a zero anywhere else, $|F| < B_0$ at all finite (even if large) signal powers.

If P ever becomes large enough that $|F| = B_0$, a major simplification occurs, in that the parameters $\langle S_n/|H|^2 \rangle_{A,F}$ and $\langle S_n/|H|^2 \rangle_{G,F}$ are constants that are not a function of P and SNR_{in} is proportional to P. In that case, SNR_{norm} has the same simple interpretation as that in Section 6.7; namely, it is the increase in signal power (or equivalently SNR_{in}) required to achieve a given P_e, relative to the Shannon limit.

Example 8-30.
If the noise on a passband channel is white, $S_n = N_0$, and $|F| = B_0$, then from (8.142)

$$SNR_{\text{norm}} = \frac{P/(N_0 B_0)}{2^{\nu}\langle |H|^{-2}\rangle_{G,F} - \langle |H|^{-2}\rangle_{A,F}} \,. \tag{8.145}$$

The only term on the right that is a function of P is P itself. Thus, SNR_{norm} and P are directly proportional.

Example 8-31.
In both Fig. 8-24 and Fig. 8-25, the high SNR approximation, obtained by setting $|F| = B_0$ and replacing the arithmetic and geometric means by their values over the full Nyquist bandwidth, are plotted. These asymptotes are useful at high SNR, especially in the single-pole channel case. However, the single-zero case illustrates that very high signal powers may be need to approach the asymptote, because H will be very small in the vicinity of a zero near the unit circle. For the case of $|c| = 1$, where there is a null in H at the angle of c, the arithmetic mean will not even exist and the

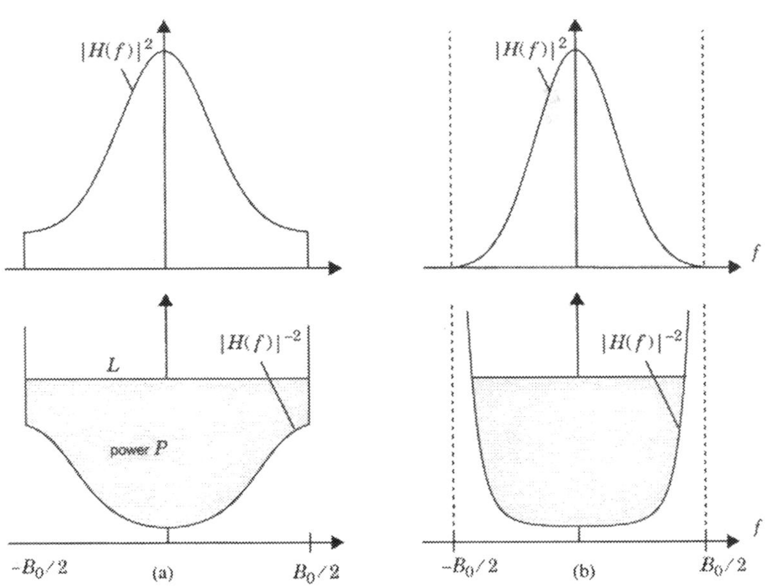

Fig. 8-26. Illustration of water pouring on channels with an without zeros. (a) When H is non-zero for all $|f| \le B_0/2$, the bowl has infinitely steep sides at the band edge. (b) When H has a zero at the band edge, the sides of the bowl have finite slope, and water pouring never fills the bandwidth. Similar behavior will occur if H has a zero anywhere in-band.

water-pouring band will never fill the Nyquist interval. In this case, there is no meaningful high-power asymptote.

Whenever a channel has nulls within its bandwidth, the water-pouring band F will always exclude a small frequency interval around these nulls. This is fortunate, because otherwise $\langle S_n/|H|^2 \rangle_{A,F}$ would be infinite! If $|H|^{-2}$ is zero over an interval of frequencies, then F will exclude that interval, which is fortunate since otherwise $\langle S_n/|H|^2 \rangle_{G,F}$ and $\langle S_n/|H|^2 \rangle_{A,F}$ would both be infinite. In fact, the water-pouring procedure ensures that $|H|^2 > 0$ over F, and therefore both $\langle S_n/|H|^2 \rangle_{G,F}$ and $\langle S_n/|H|^2 \rangle_{A,F}$ are bounded for all finite powers P.

However, nulls in $|H|^2$ do imply that $|F| < B_0$ for all P. Thus, the simple interpretation of SNR_{norm} never occurs, because $\langle S_n/|H|^2 \rangle_{G,F}$ and $\langle S_n/|H|^2 \rangle_{A,F}$ are functions of P at all power levels and SNR never becomes precisely proportional to P. On the other hand, for simple isolated nulls, the interval excluded from the water-pouring band will shrink to zero, and we would not expect the nulls to have a large effect at high P. This issue has to be addressed carefully, and would take us too far afield here [17][18].

Remarkably, at high SNR, provided that $|F| = B_0$, the capacity of the discrete-time channel is directly related to the MSE's of the LE-ZF and DFE-ZF given by (8.64) and (8.87),

$$C = B_0 \cdot \log_2\left(\frac{\varepsilon^2_{\text{ZF-LE}} + P/B_0}{\varepsilon^2_{\text{ZF-DFE}}} \right) \quad \text{bits/sec} . \tag{8.146}$$

At high SNR, the capacity of a discrete-time channel can be predicted by knowing $\varepsilon^2_{\text{ZF-LE}}$ and $\varepsilon^2_{\text{ZF-DFE}}$ alone. The normalized SNR becomes

$$SNR_{\text{norm}} = \frac{P/B_0}{2^{\nu}\varepsilon^2_{\text{ZF-DFE}} - \varepsilon^2_{\text{ZF-LE}}} . \tag{8.147}$$

8.5.5. Relationship of Capacity to Unbiased DFE-MSE

The formula for capacity of (8.146), which applies to many channels at high signal powers, can be replaced by an even simpler formula,

$$C = B_0 \cdot \log_2(1 + SNR_{\text{MMSE-DFE-U}}) , \qquad SNR_{\text{MMSE-DFE-U}} = \frac{E_a}{\varepsilon^2_{\text{MMSE-DFE,U}}} , \tag{8.148}$$

where $\varepsilon^2_{\text{MMSE-DFE,U}}$ is the MSE of the unbiased DFE-MSE designed in Section 8.2 and E_a is the alphabet energy. Remarkably, this formula holds at *all* signal powers, whereas (8.146) holds only at high power. To understand this result, the design of the unbiased DFE-MSE that leads to (8.148) needs to be clarified:

- Unlike the ZF case, the statistics of the data symbols matter to the MSE criterion, so they must be specified. The input symbols a_k are chosen to be zero-mean, stationary, and mutually independent (and hence white).
- A discrete-time transmit filter $G(z)$ is added, filtering the data symbols before they reach the channel, in order to control the shape of the spectrum of the signal at the

discrete-time channel input subject to an average power constraint. (Note that the underlying continuous-time channel will normally also include a continuous-time transmit filter, which is not affected.)

- The unbiased MSE between the equalizer output and the data symbols is minimized by choice of the receiver precursor and postcursor filters, taking into account $G(z)$.
- The MSE is minimized by the choice of $G(z)$, which remarkably turns out to result in a channel-input spectrum exactly to the water-pouring spectrum that achieves capacity.

This result demonstrates a close canonical relationship between the MMSE-DFE-U and channel capacity. In particular, (8.148) is in precisely the functional form of the capacity for an ideal white noise channel, except that the SNR of the MMSE-DFE-U is substituted for the SNR of the ideal channel. The optimal $G(z)$ for the MMSE-DFE-U results in a transmit spectrum reaching the channel that is precisely the water-pouring spectrum for that channel. Furthermore, (8.148) holds at all SNR, and is not just an asymptotic result. Thus, the SNR of the MMSE-DFE-U, for an optimized transmit filter that results in the water-pouring spectrum at the channel input, bears the same relationship to capacity on channels with ISI as the simple SNR plays in the absence of ISI. This result was derived [9] in a more general continuous-time channel context. The derivation we give here is simpler but also less general.

The derivation of (8.148) is based on the configuration of Fig. 8-27. The input data symbols a_k are zero-mean and white with energy E_a. Furthermore, we assume that they are mutually independent. The data symbols are put through $G(z)$, which allows us to generate any input power spectrum to the channel $E_a|G|^2$. Since the effects of channel noise could be eliminated by choosing a transmit filter with a very large gain, we constrain the power at the transmit filter output to be P, or in other words set $PT = E_a\langle |G|^2\rangle_A$. A MMSE-DFE-U equalizer is put at the output of the channel, generating output samples y_k, which would then be applied to a slicer.

The earlier analysis of Section 8.2 considered a similar situation, and determined the MSE assuming the data symbols are white and independent. That analysis did not include any discrete-time transmit filter $G(z)$, but that shortcoming is trivially overcome by recognizing that the transmit filter can simply be combined with the channel H, and those earlier results can be applied with H replaced by GH. We first minimize the MSE for any given transmit filter G, and subsequently choose G to minimize the MSE under the transmit power constraint. From (8.98) and (8.87),

$$\frac{E_a}{\varepsilon^2_{\text{MMSE-DFE-U}}} + 1 = \frac{E_a}{\varepsilon^2_{\text{MMSE-DFE}}} = \frac{\gamma_r^2}{\gamma_n^2} = \frac{\langle S_r\rangle_G}{\langle S_n\rangle_G} = \langle S_r/S_n\rangle_G \,, \tag{8.149}$$

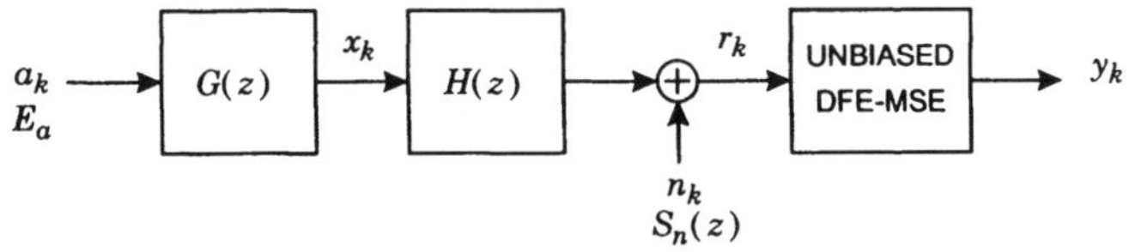

Fig. 8-27. The configuration used for deriving the canonical relationship between the MMSE-DFE-U and channel capacity.

where

$$S_r = E_a\,|\,GH\,|^2 + S_n\ . \tag{8.150}$$

If we define $S_x = E_a\,|\,G\,|^2$ as the input power spectrum to the channel, we get that

$$S_r = S_x\,|\,H\,|^2 + S_n\ , \tag{8.151}$$

and

$$\frac{E_a}{\varepsilon^2_{\text{MMSE-DFE-U}}} + 1 = \langle\, 1 + S_x\,|\,H\,|^2/S_n\,\rangle_G\ . \tag{8.152}$$

This gives us the minimum MSE for any given transmit filter G. The final step is to choose G to minimize the MSE, subject to the constraint that the power in S_x is P. We can relate this optimization back to capacity, since from (8.126),

$$2^{C/B_0} = \langle\, 1 + S_x\,|\,H\,|^2/S_n\,\rangle_G\,. \tag{8.153}$$

The right side of (8.153) is maximized when S_x is the water-pouring spectrum. Thus, maximizing the SNR of (8.152) is mathematically equivalent to finding the capacity, and the transmit spectrum S_x that maximizes (8.152) is the water-pouring spectrum. When S_x is determined by water pouring, we can set (8.152) and (8.153) equal,

$$\frac{E_a}{\varepsilon^2_{\text{MMSE-DFE-U}}} + 1 = 2^{C/B_0}\,, \tag{8.154}$$

thereby establishing (8.148).

8.5.6. Impact of ISI on Capacity

The effect of channel H and noise S_n on the capacity can be quantified by setting the capacity of the channel with ISI equal to the capacity of an ideal channel and solving for the relationship between the transmitted power required in both cases. This comparison is most meaningful if we assume that the noise is white in both cases, $S_n = N_0$, where we get

$$\frac{P_{\text{ISI}}}{N_0|F|} = \langle\,|\,H\,|^{-2}\,\rangle_{G,F}\left(\frac{P_{\text{IDEAL}}}{N_0B_0} + 1\right)^{B_0/|F|} - \langle\,|\,H\,|^{-2}\,\rangle_{A,F}\,. \tag{8.155}$$

P_{ISI} can be larger or smaller than P_{IDEAL}. For example, if the channel has a large gain, the channel with ISI may require less transmitted power in spite of the ISI. To separate out the effects of ISI, we can normalize the channel response, for example by setting the received isolated pulse energy E_h equal to that of the ideal channel.

As the transmitted power increases, $|\,F\,| \to B_0$, and (8.155) simplifies to

$$SNR_{\text{in,ISI}} = \langle\,|\,H\,|^{-2}\rangle_G(SNR_{\text{in,IDEAL}} + 1) - \langle\,|\,H\,|^{-2}\,\rangle_A\,. \tag{8.156}$$

Asymptotically, at high SNR, the effect of the ISI is to increase the required SNR_{in} by a factor of $\langle |H|^{-2} \rangle_G$.

Example 8-32.
Given a minimum-phase causal IIR channel with unit-energy impulse response and a single pole,

$$H(z) = \frac{\sqrt{1-|c|^2}}{1-cz^{-1}} \tag{8.157}$$

for $|c| < 1$, the means are

$$\langle |H|^{-2} \rangle_A = \frac{1+|c|^2}{1-|c|^2}, \quad \langle |H|^{-2} \rangle_G = \frac{1}{1-|c|^2}. \tag{8.158}$$

Thus, the capacity at high SNR is

$$CT = \log_2(SNR \cdot (1-|c|^2) + (1+|c|^2)), \tag{8.159}$$

where $SNR = P/N_0B_0$. This formula can be used for conveniently illustrating the effect of pole radius on capacity, as shown in Fig. 8-28. As $|c| \to 1$, the ISI gets worse and there is a rapid drop-off in capacity. Asymptotically at high SNR, the channel with ISI requires a larger SNR by a factor of $1/(1-|c|^2)$ to achieve the same capacity as the ideal channel ($c = 0$). For example, in the case plotted in Fig. 8-25, $|c| = 0.99$, and the SNR must be larger by 20 dB, which is the asymptotic difference between the "ideal channel" and "actual channel" curves in Fig. 8-25.

8.5.7. Performance of the LE and DFE

Using the results from the previous section, the performance of the linear and decision-feedback equalizers is easily related to capacity limits by calculating SNR_{norm} at a given error probability, and comparing to unity. This comparison is simple at high signal power, if the

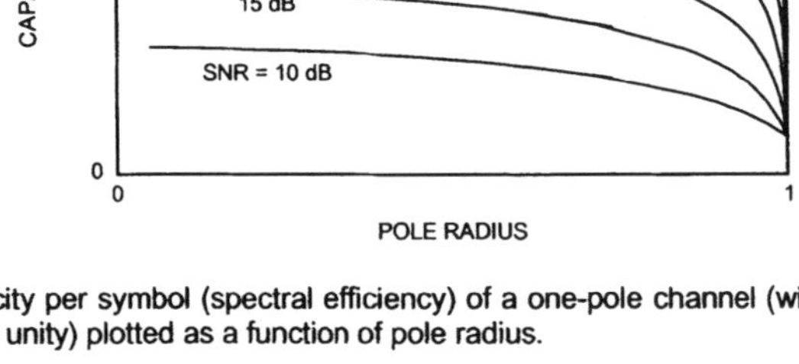

Fig. 8-28. The capacity per symbol (spectral efficiency) of a one-pole channel (with impulse response energy normalized to unity) plotted as a function of pole radius.

water pouring bandwidth becomes the full bandwidth of the channel, so we will restrict ourselves to this case. Generally this corresponds to the case where both the LE-ZF and DFE-ZF exist, further simplifying matters.

For a zero-forcing equalizer, if the transmit symbol is $\alpha \cdot a_k$ (the factor α allows us to vary the transmit power), then the input to the slicer is $\alpha \cdot a_k$ plus additive Gaussian noise with variance ε^2 for MSE ε^2 (where $\varepsilon^2 = \varepsilon^2_{\text{ZF-LE}}$ or $\varepsilon^2 = \varepsilon^2_{\text{ZF-DFE}}$). Since the Gaussian noise is circularly symmetric, the real and imaginary parts have equal variance $\sigma^2 = \varepsilon^2/2$. In the DFE-ZF case, this presumes optimistically that the slicer has made no past decision errors within the memory of the postcursor equalizer. The minimum distance is $\alpha \cdot a_{\text{min}}$, where a_{min} is the minimum distance over the symbol alphabet. When the data symbols are independent and each have variance $\sigma_A{}^2$, the transmitted energy per symbol is $PT = \alpha^2 E_a$. If K is the average number of symbols at the minimum distance, the error probability is approximately

$$P_e \approx K \cdot Q\left(\frac{d_{\text{min}}}{2\sigma}\right) = K \cdot Q\left(\sqrt{\frac{\alpha^2 a^2_{\text{min}}}{2\varepsilon^2}}\right) = K \cdot Q\left(\sqrt{\frac{a^2_{\text{min}} PT}{2E_a \varepsilon^2}}\right). \tag{8.160}$$

This gives us an accurate estimate of the error probability for zero-forcing equalizers, but we are interested in their performance in relation to capacity. If the PAM system operates at minimum bandwidth (zero excess bandwidth), then the spectral efficiency is $\nu = \log_2 M$. Furthermore, if the input noise is white, at sufficiently high power levels the water-pouring band becomes the entire Nyquist interval, and the normalized signal-to-noise ratio becomes, from (8.147), generalized to a complex baseband channel,

$$SNR_{\text{norm}} = \frac{PT}{2^{\nu}\varepsilon^2_{\text{ZF-DFE}} - \varepsilon^2_{\text{ZF-LE}}}, \tag{8.161}$$

and substituting for PT in (8.160),

$$P_e \approx K \cdot Q(\sqrt{\gamma_A \gamma_{\text{ISI}} SNR_{\text{norm}}}), \tag{8.162}$$

where

$$\gamma_A = \frac{(M-1)a^2_{\text{min}}}{2E_a} \tag{8.163}$$

is a property of the symbol constellation (already defined for the ideal channel case in Section 6.7), and

$$\gamma_{\text{ISI}} = \frac{M\varepsilon^2_{\text{ZF-DFE}} - \varepsilon^2_{\text{ZF-LE}}}{(M-1)\varepsilon^2} \tag{8.164}$$

is a function of the choice of equalizer structure (through ε^2) and also the size of the constellation M. We saw in Section 6.7 that for square QAM constellations, $\gamma_A = 3$ independent of M, the size of the constellation. Thus, the effect of ISI on the SNR gap to capacity is embodied in the term γ_{ISI}, which is also a function of M, the constellation size.

Equation (8.162) is a remarkably simple characterization of SNR_{norm} that applies at high SNR to discrete-time channels H with additive white noise $S_n = N_0$. It reflects the nature of the ISI on the channel through only two parameters, the MSE of the LE-ZF and DFE-ZF. To reiterate, it applies only when the water-pouring bandwidth fills the entire Nyquist bandwidth. The effect of the ISI is reflected in the factor γ_{ISI}, which will generally be less than unity, requiring SNR_{norm} to be larger to achieve the same error probability. Thus, the difference between γ_{ISI} and unity is a measure of the increase in the SNR gap to capacity due to ISI, for the particular equalizer structure used. Of course, when there is no ISI, $\varepsilon^2_{\text{ZF-DFE}} = \varepsilon^2_{\text{ZF-LE}} = \varepsilon^2$, and $\gamma_{\text{ISI}} = 1$ and the SNR gap to capacity reduces to that displayed for the ideal channel in Section 6.7.

For large signal constellations M (large ν), γ_{ISI} approaches an asymptote

$$\Gamma_{\text{ISI}} \to \frac{\varepsilon^2_{\text{ZF-DFE}}}{\varepsilon^2} . \tag{8.165}$$

There are two cases of interest:

- When we use the LE-ZF, then for large M

$$\gamma_{\text{ISI}} \to \frac{\varepsilon^2_{\text{ZF-DFE}}}{\varepsilon^2_{\text{ZF-LE}}} , \tag{8.166}$$

and the SNR gap to capacity is increased, relative to the ideal channel case, by precisely the MSE penalty of the LE-ZF relative to the DFE-ZF.

- If we use the DFE-ZF, then for large M

$$\gamma_{\text{ISI}} \to 1 , \tag{8.167}$$

and there is no increase in the SNR gap to capacity due to ISI.

This last fact is a remarkable conclusion first observed by Price [19]. It says that the SNR gap to capacity at high signal powers is independent of the ISI, and in particular is the same for channels with ISI and for ideal channels, as long as the DFE-ZF is used as the receiver structure. Thus, channel coding has potentially the same benefit (the same gap to close) on channels with ISI as on ideal channels. This statement has to be qualified, however, with many caveats:

- The water-pouring band must fill the entire baseband bandwidth at high signal powers; that is, the channel transfer function must be non-zero. (However, it has been shown that the result is unchanged if the channel has isolated zeros, for example at the band edge.)
- While this statement is true for general symbol-rate sampled discrete-time channels, it applies to continuous-time channels only if they can be represented by an equivalent discrete-time channel through transmit/receive filtering and sampling without information loss. That is, they must be *strictly bandlimited.* (It has been shown that this asymptotic result does not apply to many channels that are not strictly bandlimited, as will be discussed later.)
- It applies asymptotically only at high signal powers and large signal constellations.

- Our approximation to error probability assumes that all past decisions are correct. In fact, the error probability of the DFE-ZF will be higher, due to error propagation. As shown in Appendix 8-A, the effect on error probability is to increase the factor K, so it is not a major effect.

Price's result suggests that, at least in principle, channel coding techniques designed for the ideal Gaussian channel should have approximately the same potential benefit (reduction in signal power for the same P_e) on channels with ISI, independent of the nature of the ISI, at high signal powers. This follows from the fact that the SNR gap to capacity for the DFE-ZF is independent of the nature of the ISI, including the case of no ISI, at high signal powers. This result also suggests that the DFE-ZF is a good receiver structure on which to base a channel coding system for channels with ISI, since channel coding techniques should be able to close the SNR gap to capacity to the same extent from a starting point of the DFE-ZF as on the ideal channel. In particular, it shows that the MLSD receiver, which is optimal with respect to probability of error in the absence of channel coding, is not required at high signal powers in the presence of channel coding to obtain good performance. This result is surprising in light of the fact that the DFE-ZF arbitrarily cancels a part of the received signal energy, the part residing in the postcursor ISI, while the MLSD uses this energy to advantage.

There are some obstacles to applying Price's result in practice, since it depends on the ability to cancel the postcursor ISI using past decisions. This depends on the constellation being discrete, and further depends on immediate decisions, both of which are incompatible with available channel coding techniques. Fortunately, ways have been found to circumvent these obstacles, as discussed in Section 13.4.

It is tempting to apply a similar analysis to the MMSE-DFE-U. Since it is canonically related to capacity, we might expect its SNR gap to capacity to be independent of ISI at all signal powers (as long as the transmit filter is designed in accordance with water pouring). However, we cannot estimate the error probability of the MMSE-DFE-U simply, since there is residual ISI at the slicer (and hence the slicer error is not Gaussian); further, the Gaussian noise component of the slicer error is not white. Thus, the extent to which the MMSE-DFE-U may or may not close the SNR gap to capacity at low signal powers, relative to the DFE-ZF, remains an open question.

8.6. Further Reading

For further details on the design of optimal detectors in various circumstances arising in digital communication, see Wozencraft and Jacobs [20] or Proakis [21]. An excellent treatise on equalization can be found in [22] or [15]. The literature on the subject is very extensive, and those references have extensive bibliographies. On the Viterbi algorithm for sequence detection, the original paper by Forney [23] and a later tutorial article [24] are highly recommended. An excellent tutorial description of transmitter precoding, Price's result, and the capacity of channels with ISI is [25]. A nonlinear equalizer structure using decision-aided cancellation not covered here is the *intersymbol interference canceler* suggested by Proakis [26] and elaborated by Gersho and Lim [27]. Simplifications of the full-blown MLSD for ISI channels have been explored in several articles [28][29][30].

Appendix 8-A. DFE Error Propagation

In this appendix we consider the nature of error propagation in the DFE, finding in particular upper bounds on the error probability that demonstrate that the effects are usually insignificant compared to the benefits of reduced noise enhancement.

We can model the error propagation phenomenon by removing the assumption that $a_k = \hat{a}_k$, and rewriting the slicer input as

$$y_k = a_k + \sum_{j=1}^{\infty} g_j a_{k-j} - \sum_{j=1}^{\infty} g_j \hat{a}_{k-j} + n_k \tag{8.168}$$

where n_k is the complex-valued noise at the slicer input. This can be rewritten as

$$y_k = a_k + v_k + z_k \tag{8.169}$$

where the middle term is the residual ISI due to incorrect cancellation of postcursor ISI samples given by

$$v_k = \sum_{j=1}^{N} g_j w_{k-j} \,, \qquad w_k = a_k - \hat{a}_k \,, \tag{8.170}$$

and a finite number N taps has been assumed.

For purposes of understanding this phenomenon further, specialize to the baseband case with binary antipodal signaling, so that $a_k = \pm 1$ and w_k assumes the values $\{\pm 2, 0\}$. For this case the slicer will apply a threshold at zero, and we can easily calculate the probability for both types of error at time k. The first type of error occurs when $a_k = 1$ and the slicer input is negative, and results in $w_k = 2$,

$$\Pr[w_k = 2] = p \cdot \Pr[1 + v_k + n_k < 0] \tag{8.171}$$

where $p = \Pr[a_k = 1]$. Similarly the probability that $w_k = -2$ is

$$\Pr[w_k = -2] = (1 - p) \cdot \Pr[-1 + v_k + n_k > 0] \tag{8.172}$$

Exercise 8-6.

Show that if n_k is a real-valued zero-mean Gaussian random variable, and the data symbols are equally likely, $p = 1/2$, then for any ISI v_k

$$\Pr[w_k \neq 0] < 1/2 \,. \tag{8.173}$$

The intuition behind this result is that no matter how big the ISI, it actually *reduces* the error probability for one polarity of data symbol, and if the symbols are equally probable then the error probability can be no worse than ½ for any residual ISI.

Of course the conclusion that the error probability is no worse than ½ is of little value since we could flip a coin in the receiver and do just as well. In order to get stronger results, assume that the data symbols a_k and the noise samples n_k are independent. From Section 8.6 we know that the noise samples will be independent as $N \rightarrow \infty$ for the optimal forward filter design, and should be approximately true for N sufficiently large. Then we can see that $\Pr[w_k]$ depends only on w_{k-j}, $1 \le j \le N$, and therefore w_k is a Markov chain (Section 3.3) with 3^N states. While the steady-state probability of the states of this chain can be calculated in principle [2], we can easily develop a simple model for this chain that gives an upper bound on the error probability as well as considerable insight into the error propagation phenomenon [31].

Assume that in the absence of ISI the error probability is $P_{e,0}$,

$$P_{e,0} = \Pr[w_k \ne 0 \mid w_{k-j} = 0, \ 1 \le j \le N] \, . \tag{8.174}$$

Further, make the worst case assumption that if there is residual ISI, the error probability is ½,

$$\Pr[w_k \ne 0 \mid w_{k-j} \ne 0 \text{ for some } j \in \{1, 2, \ldots, N\}] = \tfrac{1}{2} \, . \tag{8.175}$$

Define a Markov chain X_k, where X_k is a count of the number of successive correct decisions that have been made up to but not including time k. That is, $X_k = n$ if $w_{k-1} = w_{k-2} = \ldots = w_{k-n} = 0$ and $w_{k-n-1} \ne 0$. We now get the following model for an error propagation event. If $X_k \ge N$ there is no residual interference because there have been no errors made within the memory of the FIR DFE feedback filter. Assume that $X_k \ge N$ but that an error is made anyway at time k, $w_k \ne 0$, due to the additive noise, resulting in $X_{k+1} = 0$. Then according to our worst-case model, errors will be made thereafter with probability ½ until such time that N correct decisions in a row have been made, at which time we revert to the state of zero residual ISI and the error probability returns to $P_{e,0}$. Suppose that on average it takes K time increments until N correct decisions in a row have been made. We call K the average length of an error propagation event. Since the error probability is ½ during the event, error propagation results in an average of $K/2$ errors for every error due to the random noise. Thus, the error probability taking into account ISI is

$$P_e = (\tfrac{1}{2}K + 1) \, P_{e,0} \, . \tag{8.176}$$

In actuality we expect that the error probability will be less than this, since the error probability during an error event will in fact be less than ½. This bound can be strengthened and made more rigorous as in [31].

This logic, even though worst-case, gives considerable insight into the mechanism of error propagation. It demonstrates that error events will terminate whenever we make N correct decisions in a row, and that this happens in relatively short order because the error probability is no worse than ½ no matter how large the ISI gets. This argument depends strongly on the assumption of equally probable (random) data, and in fact there are worst-case data sequences for which the error event can persist much longer than we have predicted here. This suggests that it is important to insure that the data is random when the DFE is used, and also suggests that it is desirable to keep N as small as possible consistent with obtaining most of the benefit of the DFE.

We can determine K in (8.176) by observing that K is the average number of tosses of a fair coin before we obtain N heads in a row. This was determined in Problem 3-13 to be

$$K = \frac{1 - 1/2^N}{1/2^{N+1}} = 2(2^N - 1) \ . \tag{8.177}$$

Substituting into (8.176), we get

$$P_e = 2^N \cdot P_{e,0} \ , \tag{8.178}$$

and the error probability is multiplied by a factor of 2^N due to error propagation. If N is fairly modest, this error multiplication is much more than offset by the benefit of the DFE in reducing $P_{e,0}$. For example, $N = 3$ results in an order of magnitude increase in error probability due to error propagation, and the DFE must reduce $P_{e,0}$ by only about half a dB to result in a net reduction in error probability.

Problems

Problem 8-1. Extend Example 8-6 to a channel with two zeros, and the same alphabet.

(a) Sketch one stage of the trellis, showing only the structure of the trellis (allowable branches) and not bothering to label the branch metrics.

(b) Sketch the error event corresponding to a single error; that is, sketch the one which, if it is the minimum-distance error event, implies that $\Gamma_{MLSD} = \Gamma_{MF}$.

(c) Sketch two other error events which are relatively short, and are thus possible candidates as the minimum-distance error event.

Problem 8-2. Show that the inequality in (8.31) is strict when $M(z)$ is FIR and $M(z) \neq 1$. *Hint:* You will need to use the fact that only a finite number of error events need be considered in calculating the minimum distance.

Problem 8-3. Given a channel with transfer function $H(z) = 1/(1 - cz^{-1})$, with complex-valued pole location c such that $|c| \neq 1$, verify the following:

(a) $\langle |H|^{-2} \rangle_A = 1 + |c|^2$.

(b) $\langle |H|^{-2} \rangle_G = 1$ for $|c| < 1$ and $\langle |H|^{-2} \rangle_G = |c|^2$ for $|c| > 1$.

(c) Note that the geometric mean is everywhere smaller, as expected.

Problem 8-4. Consider the channel of Problem 8-3:

(a) Find the transfer function of the LE-ZF, assuming the noise is white.

(b) Find $\varepsilon^2_{\text{ZF-LE}}$ for both the minimum-phase and maximum-phase cases.

(c) Interpret the result as a function of $|c|$.

(d) Discuss the practicality of this channel model.

Problem 8-5. Repeat Problem 8-4, except normalize the channel impulse response such that the energy in the impulse response is unity for all c. Be sure to treat the minimum- and maximum-phase cases separately, and explain the results intuitively.

Problem 8-6. For the channel model of Problem 8-3 and white noise:

(a) When $|c| < 1$, find the precursor and postcursor equalizers and the resulting MSE.

(b) Repeat (a) for $|c| > 1$.

(c) Interpret the results of (a) and (b), and compare to the results of Problem 8-4.

Problem 8-7. As in Problem 8-5, modify the results of Problem 8-6 by normalizing the channel impulse response to unit energy.

Problem 8-8. Show that $\varepsilon^2_{\text{MMSE-DFE}}$ of (8.87) approaches $\varepsilon^2_{\text{ZF-DFE}}$ as $S_n \to 0$, and further that the precursor and postcursor equalizers for the MSE criterion approach those of the DFE-ZF.

Problem 8-9. Derive the structure of a FSE for the following circumstances, using a sampling rate that is no higher than necessary:

(a) The received pulse has excess bandwidth less than 200%.

(b) The received pulse has excess bandwidth less than 50%.

Problem 8-10. Assuming that the FSE of Fig. 8-18(c) is implemented as a non-FIR discrete-time filter, derive a time-domain expression for the output y_k in terms of the samples x_k at the output of the antialiasing lowpass filter $F(f)$.

Problem 8-11. Derive a formula for the capacity of the continuous-time passband channel in terms of arithmetic and geometric means for an equivalent complex baseband channel model.

Problem 8-12. Suppose the method for deriving a discrete-time baseband channel from a continuous-time baseband channel of Fig. 8-22 is replaced by a more practical system, where the transmit filter is replaced by a general filter $\sqrt{T}\,G(f)$, and the receive filter is replaced by a general filter $\sqrt{T}\,F(f)$. Also do not assume that the filters G, H, or F are necessarily bandlimited.

(a) Derive a formula for the capacity of the resulting discrete-time channel.

(b) Derive the capacity when the receive filter is chosen to be a filter matched to the received pulse.

(c) Under what conditions can you be certain that the capacity of the discrete-time channel of (a) has the same capacity as the underlying continuous-time channel?

(d) Repeat (c) for the receive filter of (b).

(e) Repeat (d) when a discrete-time precursor equalizer is added to turn the receiver front end into a WMF.

Problem 8-13. Show that the discrete-time capacity formulas for the baseband and passband cases both reduce to the capacity of (8.114) when the noise is white and the discrete-time channel is ideal.

Problem 8-14. In the chapter, the normalized SNR was expressed in terms of the channel-input SNR. In this problem, we will reformulate it in terms of the channel-output SNR.

(a) Derive an expression for the SNR at the output of the channel within the water-pouring band, SNR_{out}, where the signal power is derived for the water-pouring signal spectrum.

(b) Utilizing the result of (a), express SNR_{norm} in terms of SNR_{out}.

Problem 8-15. With the MMSE-DFE-U, we found benefit in including a transmit filter, and optimizing that filter resulted in the water-pouring filter at the channel input. The question arises as to the benefit of a transmit filter to the DFE-ZF. Include a transmit filter G in the DFE-ZF, assuming the data symbols are white ($M_a = 1$) and the transmit power is constrained to P, $PT = E_a\langle |G|^2\rangle_A$. Under these conditions, prove that the optimal transmit filter, with the goal of minimizing the MSE, is a flat filter. That is, the trivial transmit filter is optimal.

Problem 8-16. Generalize the results of Problem 8-15 by allowing the transmit data symbols to have a general power spectrum S_a.

(a) Show that the optimal transmit filter for the DFE-ZF is not flat, but is in fact a whitening filter for the transmit symbols.

(b) Find the resulting MSE.

(c) Show that the MSE is always smaller when the transmit symbols are not white and the optimal transmit filter is used, than when the symbols are white with the same variance.

Problem 8-17. Since $\varepsilon^2_{\text{MMSE-DFE}} \le \varepsilon^2_{\text{ZF-DFE}}$, it might appear from (8.165) that a DFE-MSE could potentially operate at a smaller gap to capacity on a channel with ISI than on the ideal channel (that is $\gamma_{\text{ISI}} > 1$). Explain why this conclusion would be wrong.

References

1. M. E. Austin, *Decision-Feedback Equalization for Digital Communication Over Dispersive Channels*, M.I.T. Lincoln Laboratory, Lexington, Mass (August 1967).

2. C. A. Belfiore and J. H. Park, "Decision Feedback Equalization," *Proceedings of the IEEE*, Vol. 67 (8), (Aug. 1979).

3. M. Tomlinson, "New Automatic Equalizer Employing Modulo Arithmetic," *Electronic Letters*, Vol. 7, (March 1971).

4. H. Harashima and H. Miyakawa, "A Method of Code Conversion for a Digital Communication CHannel with Intersymbol Interference," *Transactions Institute Electronic Communication Engineering*, (Japan), vol. 52-A, June 1969.

5. H. Harashima and H. Miyakawa, "Matched-Transmission Technique for Channels with Intersymbol Interference," *IEEE Trans. on Communications*, Vol. COM-20, p. 774 (Aug. 1972).

6. D. G. Messerschmitt, "Generalized Partial Response for Equalized Channels with Rational Spectra," *IEEE Trans. on Communications*, Vol. COM-23 (11), p. 1251 (Nov. 1975).

7. A. Lender, "Correlative Level Coding for Binary-Data Transmission," *IEEE Spectrum*, Vol. 3, p. 104 (Feb. 1966).

8. G. D. Forney, Jr, "Coset Codes – Part I: Introduction and Geometrical Classification," *IEEE Trans. Information Theory*, Vol. IT-34, p. 1123 (1988).

9. J. M. Cioffi, G. P. Dudevoir, M. Vedat Eyuboglu, and G. D. Forney, Jr, MMSE Decision-Feedback Equalizers and Coding. Part I: General Results. to appear.

10. R. W. Lucky, "Signal Filtering with the Transversal Equalizer," *Proc. Seventh Annual Allerton Conference on Circuits and System Theory*, p. 792 (Oct. 1969).

11. R. D. Gitlin and S. B. Weinstein, "Fractionally Spaced Equalization: An Improved Digital Transversal Equalizer," *Bell System Technical Journal*, Vol. 60 (2), (Feb. 1981).

12. S. U. H. Qureshi and G. D. Forney, Jr., "Performance Properties of a T/2 Equalizer," *NTC '77 Proceedings*, 1977.

13. G. Ungerboeck, "Fractional Tap-Spacing and Consequences for Clock Recovery in Data Modems," *IEEE Trans. on Communications*, (Aug. 1976).

14. J. E. Mazo, "Optimum Timing Phase for an Infinite Equalizer," *Bell System Technical Journal*, Vol. 54 (1), (Jan. 1975).

15. S.U.H.Qureshi, "Adaptive Equalization," pp. 640 in *Advanced Digital Communications Systems and Signal Processing Techniques*, ed. K. Feher, Prentice-Hall, Englewood Cliffs, N.J. (1987).

16. R. Gallager, *Information Theory and Reliable Communication*, John Wiley and Sons, Inc., New York (1968).

17. J. R. Barry, E. A. Lee, and D. G. Messerschmitt, "Capacity Penalty Due to Ideal Tail-Canceling Equalization," *IEEE Trans. on Information Theory*, vol. 42, no. 4, pp. 1062-1071, July 1996.

18. J. R. Barry, E. A. Lee, and D. G. Messerschmitt, "Capacity Penalty Due to Ideal Zero-Forcing Decision-Feedback Equalization," *Proceedings Int. Conf. Communications*, (June 1993).

19. R. Price, "Nonlinearly Feedback-Equalized PAM vs. Capacity for Noisy Filter Channels," Proc. 1972 *IEEE International Conf. Communications*, pp. 22-12 (June 1972).

20. J. M. Wozencraft and I. M. Jacobs, *Principles of Communication Engineering*, Wiley, New York (1965).

21. J. G. Proakis, *Digital Communications*, Fourth Edition, McGraw-Hill Book Co., New York (2001).

22. J. G. Proakis, "Advances in Equalization for Intersymbol Interference," *Advances in Communication Systems*, Vol. 4, (1975).

23. G. D. Forney, Jr., "Maximum-Likelihood Sequence Estimation of Digital Sequences in the Presence of Intersymbol Interference," *IEEE Trans. on Information Theory*, Vol. IT-18, pp. 363-378 (May 1972).

24. G. D. Forney, Jr., "The Viterbi Algorithm," *Proceedings of the IEEE*, Vol. 61 (3), (March 1973).

25. M. V. Eyuboglu and G. D. Forney, Jr, "Trellis Precoding: Combined Coding, Precoding, and Shaping for Intersymbol Interference Channels," *IEEE Trans. Information Theory*, (March 1992).

26. J. G. Proakis, "Adaptive Nonlinear Filtering Techniques for Data Transmission," *IEEE Symposium on Adaptive Processes, Decision, and Control*, p. XV.2.1 (1970).

27. A. Gersho and T. L. Lim, "Adaptive Cancellation of Intersymbol Interference for Data Transmission," *Bell System Technical Journal*, Vol. 70, pp. 1997-2021 (Nov. 1981).

28. D. D. Falconer and F. R. Magee, Jr, "Adaptive Channel Memory Truncation for maximum Likelihood Sequence Estimation," *Bell System Technical Journal*, Vol. 52, p. 1541 (Nov. 1973).

29. W. U. Lee and F. S. Hill, "A Maximum-Likelihood Sequence Estimator with Decision Feedback Equalization," *IEEE Trans. on Communications*, Vol. COM-25 (9), pp. 971-979 (Sep. 1977).

30. K. Wesolowski, "An Efficient DFE & ML Suboptimum Receiver for Data Transmission Over Dispersive Channels Using Two-Dimensional Signal Constellations," *IEEE Trans. on Communications*, Vol. COM-35, (3), pp. 336-339 (March 1987).

31. D. L. Duttweiler, J. E. Mazo, and D. G. Messerschmitt, "Error Propagation in Decision-Feedback Equalizer," *IEEE Trans. on Information Theory*, Vol. IT-20, pp. 490-497 (Jul. 1974).

9

Adaptive Equalization

In Chapter 8 we derived a set of receiver structures that counter intersymbol interference under the assumptions of a known channel and unconstrained implementation complexity. The resulting structures are impractical for most applications in the exact form we derived them for several reasons. First, assumption of a known received pulse shape is unrealistic, particularly for channels such as the digital subscriber loop (with bridged taps), radio channel (with selective fading), and voiceband data channel, where there are significant variations in the channel affecting the reception. Thus, the received pulse shape is not actually known in advance for these channels, and is sometimes varying during actual transmission. Second, the receiver structures we derived usually have an infinite number of coefficients, and cannot be realized. Third, our optimizations did not take into account significant impairments such as timing jitter and timing offset, which must be considered in the design of receive filtering.

Timing offset and jitter will be considered in Chapter 16. In this chapter we will address the problem of estimating the actual channel isolated pulse response, and automatically adjusting an equalizer to equalize this channel. This is known as *adaptive equalization*, and was first proposed and analyzed by R.W. Lucky in 1965 [1][2][3], building on earlier work in adaptive filtering by B. Widrow and M.E. Hoff, Jr. [4]. Our general approach will be to define practical filter structures similar to those found to be optimal in Chapter 8, and then arrange to adapt the parameters of those structures to the actual channel characteristics.

The simplest form of adaptive equalizer is shown in Fig. 9-1 in block diagram form. The received signal is applied to a receive filter. Since the channel is not assumed to be known, the receive filter is usually *not* a matched filter, although it may be a compromise approximation. Rather, it is more likely to be a lowpass filter which simply rejects all out-of-band noise. The output of the receive filter is sampled, usually at the symbol rate or twice the symbol rate, and applied to an *adaptive equalizer*. The adaptive equalizer may be realized as a *finite transversal filter*, which is a version of the transversal filter encountered in Section 8.4, with a finite number of taps or coefficients. The object is to *adapt* the coefficients to minimize the noise and intersymbol interference at the output, which is applied to a slicer to make the decisions on the data symbols. The adaptation of the equalizer is driven by an *error signal*, which indicates to the equalizer the direction that the coefficients must be moved to more accurately represent the data symbols at the slicer input.

In the steady state, the adaptation of the equalizer is *decision directed*, meaning that the receiver decisions are used to generate the error signal. In the absence of intersymbol interference and noise, the slicer input would precisely equal the transmitted data symbols, and the slicer output would equal the slicer input. Thus, there would be no error, and the error signal at the adaptive equalizer would be zero. This would tell the equalizer that no adjustment of coefficients is necessary. If there were noise alone at the slicer input, but no intersymbol interference, the error signal would be non-zero, but would average to zero resulting in no net change in the coefficients. But when there is intersymbol interference, the resulting error signal can be used to adjust the coefficients so as to reduce that intersymbol interference. A more detailed block diagram for the passband PAM case was shown in Fig. 5-21.

The adaptive equalizer thus uses the regenerative effect (Chapter 1) to advantage: since the slicer regenerates a noise- and intersymbol interference-free representation of the transmitted data symbols, a comparison of these symbols with the slicer input can be used to adjust the equalizer. Of course, the slicer makes occasional errors, but due to the long averaging time of the equalizer coefficient adjustment algorithm these errors have no significant effect.

Decision-directed equalizer adjustment is effective in tracking slow variations in the channel response. It is often, however, not effective during initial acquisition since the intersymbol interference can be so bad as to cause a very high error rate initially. For this reason the initial acquisition of the equalizer is often accomplished by using a *training signal*. In this mode of operation, the transmitter generates a data symbol sequence known to the

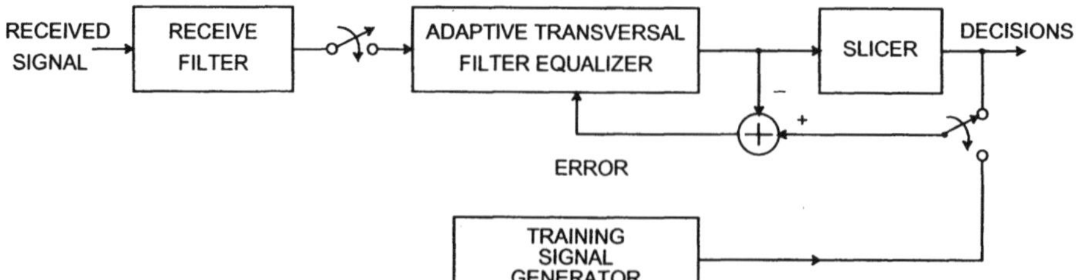

Fig. 9-1. Block diagram of an adaptive equalizer.

receiver. The receiver therefore substitutes this known training signal in place of the slicer output, as shown in Fig. 9-1. Once an agreed period of time has elapsed, the slicer output is substituted and actual data transmission begins.

9.1. Constrained-Complexity Equalizers

Before we can adapt an equalizer, we must specify the *structure* of a suitable filter that has *finite degrees of freedom* or *constrained complexity*, and only then can we design an algorithm for adapting the parameters of that structure. The constrained complexity transversal filter defined in Section 8.4 is suitable for that purpose. Using this structure, in this section we resolve the optimization of the filter coefficients under the mean-square error (MSE) criterion for the linear equalizer (LE) case. This requires the solution of a set of linear equations; we discuss a specific method to obtain this solution called the *mean-square error gradient (MSEG) algorithm*. This method is of interest because it leads directly to an *adaptation algorithm* in Section 9.2.

9.1.1. Equalizer Structure

A block diagram of a complete adaptive decision-feedback equalizer system is shown in Fig. 9-2. The coefficients of the equivalent discrete-time channel, $H(z)$, will in general be complex-valued, and the additive noise N_k will also be complex-valued. The channel response and receive filter are reflected in the equivalent discrete-time response $H(z)$. This channel model assumes that the demodulation has been performed prior to equalization, as was derived in Chapter 8, which is known as a *baseband adaptive equalizer*. In practice a *passband equalizer* is usually used, as discussed in Section 9.5, but we discuss the baseband case first since it is easier to understand and also models the important baseband channel case. We have also assumed symbol-rate sampling in the equalizer; the fractionally spaced case (Section 9.4) will be deferred to Section 9.4.

The DFE consists of a precursor equalizer and a postcursor equalizer; the postcursor equalizer is absent in the LE ($D(z) = 0$). The error signal between slicer input and output,

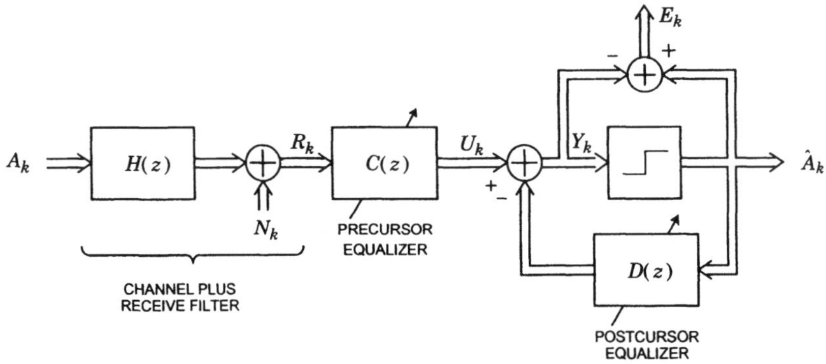

Fig. 9-2. An adaptive decision-feedback equalizer. A linear equalizer results when $D(z) = 0$.

$$E_k = \hat{A}_k - Y_k \ , \tag{9.1}$$

will be used to adapt the precursor and postcursor equalizers in the decision-directed mode; during the training mode the actual transmitted data symbol A_k is substituted for the decision $\hat{A}_k$. In the analysis of the equalizer, we assume that there are no decision errors, and thus use A_k in place of $\hat{A}_k$. This assumption is justified by experience, which confirms that as long as the error rate is below ten percent or so there is no appreciable effect on equalizer operation.

Linear Equalizer (LE) Structure

For the LE in Chapter 8 the precursor equalizer is non-causal, even with unconstrained complexity, and thus it is natural to assume that the finite transversal filter has taps symmetrically spaced around the zero-delay tap. If we assume the total number of coefficients N is odd, and let $L = (N-1)/2$, then this equalizer would be of the form

$$C(z) = \sum_{m=-L}^{L} c_m z^{-m} \tag{9.2}$$

where c_m, $-L \le m \le L$ are the N coefficients of the precursor equalizer. This filter is not causal, but can always be made causal at the expense of an additional delay through the equalizer.

Decision-Feedback Equalizer (DFE) Structure

For the DFE, in Chapter 8 the unconstrained complexity precursor equalizer was anticausal in order to cancel precursor intersymbol interference, so we can assume an anticausal N coefficient equalizer of the same form,

$$C(z) = \sum_{m=-(N-1)}^{0} c_m z^{-m} . \tag{9.3}$$

The postcursor equalizer must be a strictly causal filter with M taps, *viz.*

$$D(z) = \sum_{m=1}^{N} d_m z^{-m} . \tag{9.4}$$

The number of feedforward coefficients N is allowed to be different from the number of postcursor equalizer coefficients M.

9.1.2. Minimum MSE Solution

We now return to the same problem solved in Chapter 8, the known channel case, and re-derive the optimal equalizer for constrained complexity, using the *minimum mean-square error (MSE)* criterion. We also make the assumption that all signals are wide-sense stationary. The solution will lead directly to a method of adapting the equalizer in Section 9.2. We specialize to the LE case here, and defer the DFE until Section 9.4.

In Chapter 8, the ZF criterion forced the intersymbol interference to zero, while the MSE criterion allowed intersymbol interference in order to reduce the noise variance at the slicer input. For the constrained complexity transversal filter, intersymbol interference is inevitable since there are insufficient degrees of freedom to get completely rid of it. Thus the ZF criterion is a little different — it now forces the intersymbol interference to zero for only a finite set of precursors and postcursors. The MSE criterion is much the same as before — it minimizes the MSE at the slicer input. Since intersymbol interference is inevitable, the constrained complexity MSE criterion makes more sense and is usually the one used. Therefore, we will consider only that criterion and will leave the ZF criterion to the problems.

For the transversal filter it is appropriate to use vector and matrix notation. Throughout this chapter we use $\boldsymbol{x}^T$ to denote a matrix or vector transpose, $\bar{\boldsymbol{x}}$ to denote a complex conjugate, and $\boldsymbol{x}^*$ to denote a conjugate (or Hermitian) transpose, so that $\boldsymbol{x}^* = \bar{\boldsymbol{x}}^T$. Define a column vector of transversal filter coefficients

$$\boldsymbol{c} = [c_{-L}, \ldots, c_L]^T \tag{9.5}$$

where $L = (N-1)/2$, and N is the number of equalizer taps, assumed odd. Also define a vector of past and future input samples to the equalizer,

$$\boldsymbol{r}_k = [R_{k+L}, \ldots, R_k, \ldots, R_{k-L}]^T. \tag{9.6}$$

If the slicer output is assumed to equal the actual transmitted data symbols (no decision errors), then the error signal is

$$E_k = A_k - Y_k \,, \qquad\qquad Y_k = \boldsymbol{c}^T \boldsymbol{r}_k \,. \tag{9.7}$$

In general all these quantities are complex-valued, although for the baseband channel they are all real-valued. Assume that all the random processes are wide-sense stationary with known statistics, and design the equalizer coefficient vector $\boldsymbol{c}$ to minimize the MSE, defined as $\mathrm{E}[\,|E_k|^2]$.

Exercise 9-1.

Explicitly evaluate the mean-square error and show that it is equal to

$$\begin{aligned}\mathrm{E}[\,|E_k|^2] &= \mathrm{E}[\,|A_k|^2] - 2\mathrm{Re}\{\boldsymbol{c}^*\mathrm{E}[A_k\bar{\boldsymbol{r}}_k]\} + \boldsymbol{c}^*\mathrm{E}[\bar{\boldsymbol{r}}_k\boldsymbol{r}_k^T]\boldsymbol{c} \\ &= \mathrm{E}[\,|A_k|^2] - 2\mathrm{Re}\{\boldsymbol{c}^*\boldsymbol{\alpha}\} + \boldsymbol{c}^*\boldsymbol{\Phi}\boldsymbol{c},\end{aligned} \tag{9.8}$$

where α and Φ are defined as

$$\boldsymbol{\alpha} = \mathrm{E}[A_k\bar{\boldsymbol{r}}_k]\,, \tag{9.9}$$

$$\boldsymbol{\Phi} = \mathrm{E}[\bar{\boldsymbol{r}}_k\boldsymbol{r}_k^T] = \begin{bmatrix} \phi_0 & \phi_{-1} & \cdots & \phi_{-(N-1)} \\ \phi_1 & & & \\ \phi_2 & & \ddots & \\ \vdots & & & \\ \phi_{N-1} & & & \phi_0 \end{bmatrix}, \qquad \phi_m = \mathrm{E}[R_{k+m}R_k^*]\,. \tag{9.10}$$

Exercise 9-2.

Suppose the channel model of Fig. 9-2 is based on a receiver front end consisting of a downconverter, a receive filter, and a baud-rate sampler. Assume that the transmitted symbols are zero-mean and uncorrelated with energy E_a, and assume the passband noise is white and Gaussian with PSD $N_0/2$. Show that the autocorrelation function of the equalizer input is

$$\phi_m = E_a \Sigma_k h_{k+m} h_k^* + N_0 \rho_f(m) , \tag{9.11}$$

$\rho_f(k)$ is the autocorrelation of the receive filter impulse response $f(t)$.

The MSE is not a function of time due to the wide-sense stationarity assumption. The autocorrelation function ϕ_m is just a simplified notation for $R_R(m)$, and Φ is an *autocorrelation matrix* for the sampled data signal.

Exercise 9-3.

Show that Φ has the following important properties:

(a) Φ is a *Hermitian matrix*, i.e.

$$\Phi^* = \Phi . \tag{9.12}$$

(b) Φ is a *Toeplitz matrix*, i.e. the (i, j) element is a function of $i - j$ [5][6].

(c) Φ is a positive semidefinite matrix, i.e. the *Hermitian form* $\boldsymbol{x}^*\Phi\boldsymbol{x}$ is real-valued for any vector $\boldsymbol{x}$ and is also non-negative,

$$\boldsymbol{x}^*\Phi\boldsymbol{x} \geq 0. \tag{9.13}$$

In most but not all applications it can be assumed that this autocorrelation matrix is positive definite, and hence nonsingular. Instances where it is singular will be discussed in Section 9.4, but for the time being assume it is nonsingular.

Our goal is to find the vector $\mathbf{c}$ that minimizes (9.8). There are several ways to accomplish this; in the spirit of illustrating useful techniques we will demonstrate two approaches. The first approach is to express the MSE in a form for which the minimum MSE solution is obvious.

Example 9-1.

As an aid to intuition, it is often helpful to specialize to the degenerate case of a single real-valued coefficient c. This simplifies the equations dramatically, and yet reveals many of the interesting properties. If the input signals A_k and R_k are real-valued, then (9.8) becomes

$$\mathrm{E}[E_k^2] = \mathrm{E}[A_k^2] - 2\alpha c + \phi_0 c^2 = \mathrm{E}[A_k^2] - \alpha^2/\phi_0 + \phi_0(c - \alpha/\phi_0)^2 \tag{9.14}$$

where $\alpha = \mathrm{E}[A_k R_k]$ and $\phi_0 = \mathrm{E}[R_k^2]$. Since the square term is non-negative, the MSE is minimized if it is zero, and the optimal coefficient is $c_{\text{opt}} = \alpha/\phi_0$.

Exercise 9-4.

Verify by multiplying it out that

$$\mathrm{E}[\,|E_k|^2] = \mathrm{E}[\,|A_k|^2] - \boldsymbol{\alpha}^*\Phi^{-1}\boldsymbol{\alpha} + (\Phi^{-1}\boldsymbol{\alpha} - \mathbf{c})^*\Phi(\Phi^{-1}\boldsymbol{\alpha} - \mathbf{c}) . \tag{9.15}$$

is equivalent to (9.8).

Since Φ is positive semidefinite, the last term in (9.15) is non-negative (from (9.13)) and is minimized by the choice

$$\mathbf{c}_{\text{opt}} = \Phi^{-1}\boldsymbol{\alpha} . \tag{9.16}$$

Since only the last term in (9.15) depends on $\boldsymbol{c}$, (9.16) also minimizes the mean-square error, which has a resultant minimum value

$$\xi_{\min} = \mathrm{E}[\,|A_k|^2] - \alpha^*\Phi^{-1}\alpha \;=\; \mathrm{E}[\,|A_k|^2] - \alpha^*\boldsymbol{c}_{\text{opt}} \,. \tag{9.17}$$

Finding $\boldsymbol{c}_{\text{opt}}$ from (9.16) requires the solution of a system of linear equations. Under our nonsingular Φ assumption, the solution of these equations is unique.

We now have a useful formula for the MSE, (9.15), which we can write in the form

$$\mathrm{E}[\,|E_k|^2] \;=\; \xi_{\min} + (\boldsymbol{c} - \boldsymbol{c}_{\text{opt}})^*\Phi(\boldsymbol{c} - \boldsymbol{c}_{\text{opt}}) \;. \tag{9.18}$$

This consists of a term independent of the coefficient vector $\boldsymbol{c}$, and a second term that is a Hermitian form in a vector $(\boldsymbol{c} - \boldsymbol{c}_{\text{opt}})$ with matrix Φ.

Another way to find the optimal solution is to take the gradient of the MSE with respect to the coefficient vector, and set that gradient to zero to find $\boldsymbol{c}_{\text{opt}}$. For the baseband channel case, where everything is real-valued, this is straightforward. In the complex-valued case we have to be more careful because derivatives of some innocuous-looking complex-valued functions do not exist (for example, the derivative of z^*, the conjugate of a complex variable z, does not exist anywhere!). Fortunately, with the MSE we are dealing with a real-valued function of a complex vector $\boldsymbol{c}$, which makes life simpler because we can consider the MSE to be a real-valued function of two real-valued vectors (the real and imaginary parts of $\boldsymbol{c}$).

If we define the real and imaginary parts of the complex quantities in (9.8),

$$\boldsymbol{c} = \boldsymbol{c}_{\mathrm{R}} + j\boldsymbol{c}_{\mathrm{I}} \,, \qquad \alpha = \alpha_{\mathrm{R}} + j\alpha_{\mathrm{I}} \,, \qquad \Phi = \Phi_{\mathrm{R}} + j\Phi_{\mathrm{I}} \,, \tag{9.19}$$

and consider the real-valued function $\mathrm{E}[\,|E_k|^2]$ to be a function of two real-valued vectors $\boldsymbol{c}_{\mathrm{R}}$ and $\boldsymbol{c}_{\mathrm{I}}$.

Exercise 9-5.

(a) Show that as a consequence of the Hermitian property,

$$\Phi_{\mathrm{R}} = \Phi_{\mathrm{R}}{}^T \,, \qquad \Phi_{\mathrm{I}} = -\Phi_{\mathrm{I}}{}^T \,. \tag{9.20}$$

(b) Show that

$$\nabla_{\boldsymbol{c}_{\mathrm{R}}}\boldsymbol{c}^*\Phi\boldsymbol{c} = 2\Phi_{\mathrm{R}}\boldsymbol{c}_{\mathrm{R}} - 2\Phi_{\mathrm{I}}\boldsymbol{c}_{\mathrm{I}} \,, \qquad \nabla_{\boldsymbol{c}_{\mathrm{I}}}\boldsymbol{c}^*\Phi\boldsymbol{c} = 2\Phi_{\mathrm{R}}\boldsymbol{c}_{\mathrm{I}} + 2\Phi_{\mathrm{I}}\boldsymbol{c}_{\mathrm{R}} \,, \tag{9.21}$$

$$\nabla_{\boldsymbol{c}_{\mathrm{R}}}\mathrm{Re}\{\boldsymbol{c}^*\alpha\} = \alpha_{\mathrm{R}} \,, \qquad \nabla_{\boldsymbol{c}_{\mathrm{I}}}\mathrm{Re}\{\boldsymbol{c}^*\alpha\} = \alpha_{\mathrm{I}} \,, \tag{9.22}$$

where $\nabla_{\boldsymbol{x}}$ is the gradient with respect to vector $\boldsymbol{x}$.

(c) Show that if we define a gradient of a real-valued function with respect to a complex vector $\boldsymbol{c}$ as

$$\nabla_{\boldsymbol{c}} = \nabla\boldsymbol{c}_{\mathrm{R}} + j\nabla\boldsymbol{c}_{\mathrm{I}} \tag{9.23}$$

then

$$\nabla_{\boldsymbol{c}}\boldsymbol{c}^*\Phi\boldsymbol{c} = 2\Phi\boldsymbol{c} \;, \qquad \nabla_{\boldsymbol{c}}\mathrm{Re}\{\boldsymbol{c}^*\alpha\} = \alpha \;. \tag{9.24}$$

Given the results of the exercise, we can find $\boldsymbol{c}_{\text{opt}}$ by taking the gradients of $\mathrm{E}[\,|E_k|^2]$ from (9.8) with respect to $\boldsymbol{c}_\mathrm{R}$ and $\boldsymbol{c}_\mathrm{I}$ and setting them to zero to find the real and imaginary parts of $\boldsymbol{c}_{\text{opt}}$. In view of our definition of a gradient with respect to $\boldsymbol{c}$, this is equivalent to

$$\nabla_{\boldsymbol{c}}\mathrm{E}[\,|E_k|^2] = 2\Phi\boldsymbol{c} - 2\alpha = 0\,, \tag{9.25}$$

which yields the same solution as (9.16).

Orthogonality Principle

If we calculate the crosscorrelation between the slicer error signal and the input signal to the equalizer, assuming the equalizer coefficient vector is optimal,

$$\mathrm{E}[E_k\bar{\boldsymbol{r}}_k] = \mathrm{E}[(A_k - \boldsymbol{c}_{\text{opt}}{}^T\boldsymbol{r}_k)\bar{\boldsymbol{r}}_k] = \mathrm{E}[A_k\bar{\boldsymbol{r}}_k - \bar{\boldsymbol{r}}_k\boldsymbol{r}_k{}^T\boldsymbol{c}_{\text{opt}}] = \alpha - \Phi\boldsymbol{c}_{\text{opt}} = 0\,. \tag{9.26}$$

This result is known as the *orthogonality principle.* This principle is the first inkling of a possible approach to adapting the equalizer. In particular, it tells us a way in which the filter can tell if the coefficients are optimal; namely, the different delays of the sampled data signal at the input should be orthogonal to the slicer error. If this condition is not met, then the orthogonality principle doesn't necessarily tell us which direction to move the coefficients to bring them closer to the optimum, but we will find such a method in the next section.

9.1.3. The MSE Gradient Algorithm

The previous results indicate that a system of linear equations must be solved in order to find the optimal MSE coefficient vector. Any number of numerical techniques could be used to solve these equations, but we now focus on the MSE gradient (MSEG) algorithm. This method is of interest because it leads directly to an adaptive algorithm for equalizer adjustment, the *stochastic gradient (SG) algorithm*, in Section 9.2. Further, understanding of the convergence properties of the MSEG algorithm is a prerequisite to the understanding of the SG algorithm.

The MSEG algorithm defines a sequence of coefficient vectors that is guaranteed to converge to $\boldsymbol{c}_{opt}$, assuming that a unique optimum exists. As a starting point to the derivation, R_k is again assumed to be wide-sense stationary with nonsingular autocorrelation matrix (9.10). The output MSE given by (9.8) is a quadratic form in the coefficient vector and therefore has a unique global minimum. The MSE can be viewed as a surface in (N+1)-dimensional space (since it is a function of N coefficients). The quadratic nature of this surface makes it simple to adjust the weights iteratively to minimize the MSE by descending along the MSE surface. This is the MSE gradient algorithm.

Since the algorithm is iterative in nature, a notation for the coefficient vector that reflects this is needed. Thus, call the j-th iteration of the coefficient vector $\boldsymbol{c}_j$. Given the present coefficient vector $\boldsymbol{c}_j$, by subtracting off a term proportional to the error gradient, $\nabla_{\boldsymbol{e}}\mathrm{E}[\,|E_k|^2]$, the resultant tap vector should be closer to $\boldsymbol{c}_{\text{opt}}$. This is because the gradient of the error is a vector in the direction of maximum increase of the error. Moving a short distance in the opposite (negative) direction of the gradient should therefore reduce the error. On the other hand, moving too far in that direction might actually overshoot the minimum, and result in instability.

The MSEG algorithm is illustrated in Fig. 9-3 for the order two case ($N = 2$). Because of the quadratic nature of the MSE, the contours of constant mean-square error are elliptical. The negative of the gradient points in the direction of maximum decrease of the mean-square error. When the step size is small, the mean-square error is reduced at each step of the algorithm, and approaches the minimum of (9.17) asymptotically. When the step size is too large, as shown in Fig. 9-3, the mean-square error can actually increase, and the algorithm becomes unstable.

The MSEG algorithm is explicitly

$$\boldsymbol{c}_{j+1} = \boldsymbol{c}_j - \frac{\beta}{2} \nabla_{\boldsymbol{c}_j} \mathrm{E}[\,|E_k|^2] \,, \tag{9.27}$$

where β is a small adaptation constant or step size that controls the size of the change in $\boldsymbol{c}_j$ at each update. The division by two is included to avoid a factor of two in the subsequent adaptation algorithm. From (9.25), this algorithm becomes

$$\boldsymbol{c}_{j+1} = \boldsymbol{c}_j + \beta(\alpha - \Phi \boldsymbol{c}_j) = (\mathbf{I} - \beta\Phi)\boldsymbol{c}_j + \beta\alpha \,, \tag{9.28}$$

where $\mathbf{I}$ is the identity matrix. Hopefully, if this algorithm is simply iterated from some arbitrary initial guess $\boldsymbol{c}_0$ it will converge to $\boldsymbol{c}_{\text{opt}}$ of (9.16).

Example 9-2.

Continuing Example 9-1, in this case (9.28) becomes

$$c_{j+1} = (1 - \beta\phi_0)c_j + \beta\alpha \,, \tag{9.29}$$

where c_j is the j-th iteration of a single real-valued coefficient c. It is simple to find a formula for c_j from (9.29), but even easier to subtract the optimal coefficient from Example 9-1 from both sides to obtain

$$(c_{j+1} - c_{\text{opt}}) = (1 - \beta\phi_0)(c_j - c_{\text{opt}}) = (1 - \beta\phi_0)^j(c_0 - c_{\text{opt}}) \,, \tag{9.30}$$

which demonstrates that $c_j \to c_{\text{opt}}$ as long as $|1 - \beta\phi_0| < 1$. Thus, the step size of the algorithm must be in the range $0 < \beta < 2/\phi_0$ for there to be convergence. The convergence is exponential in the iteration index. The fastest convergence is for $\beta = 1/\phi_0$, in which case the MSEG algorithm converges in a single iteration!

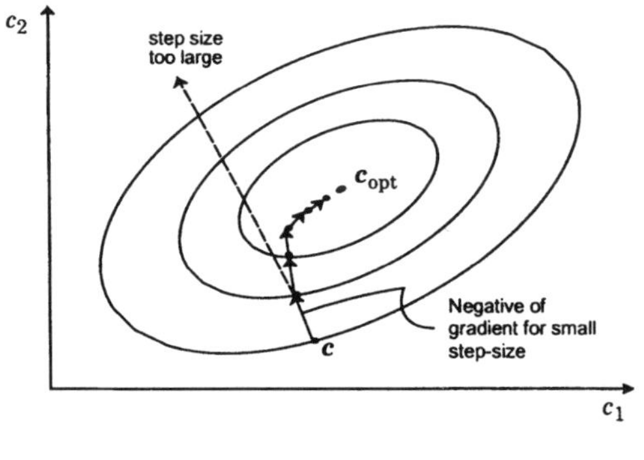

Fig. 9-3. The elliptical contours of equal MSE, and an illustration of the MSEG algorithm.

This simple example is readily extended to the general case. If c_{opt} from (9.16) is subtracted from both sides of (9.28) and

$$q_j = c_j - \Phi^{-1}\alpha \tag{9.31}$$

is defined as the error between the actual and optimal coefficient vector, then

$$q_{j+1} = (\mathbf{I} - \beta\Phi)q_j = (\mathbf{I} - \beta\Phi)^{j+1}q_0 . \tag{9.32}$$

The question becomes whether this error converges to zero.

The behavior of (9.32) depends critically on the eigenvalues of the matrix Φ, which we denote by $\lambda_1,\ldots,\lambda_n$, and which are explored in the following exercise.

Exercise 9-6.

Let **A** be a Hermitian matrix. Then establish the following facts:

(a) The eigenvalues of **A** are real-valued.

(b) Let λ_i and λ_j be two distinct (real-valued) eigenvalues of **A**. Show that the associated eigenvectors v_i and v_j are orthogonal; that is, $v_i^* v_j = 0$.

(c) Assume for simplicity that the eigenvalues of **A** are all distinct, and that the eigenvectors are normalized to unit length ($v_i^* v_i = 1$). Define the matrix

$$\mathbf{V} = [v_1, v_2, \cdots, v_N] \tag{9.33}$$

called the *modal matrix*. Show that this modal matrix is *unitary*, i.e.

$$\mathbf{V}^{-1} = \mathbf{V}^*. \tag{9.34}$$

(d) Show that

$$\mathbf{A} = \mathbf{V}\Lambda\mathbf{V}^* \tag{9.35}$$

where Λ is a diagonal matrix of eigenvalues,

$$\Lambda = \mathrm{diag}(\lambda_1, \lambda_2, \ldots, \lambda_N) . \tag{9.36}$$

(e) Show that if **A** is positive definite (semi-definite) then its eigenvalues are all positive (non-negative).

The results of the exercise generalize to the case where the eigenvalues are not distinct [7]. In particular, if an eigenvalue is repeated with multiplicity m, then there are m corresponding linearly independent eigenvectors. These can always be constructed (by the Gram-Schmidt procedure, for example) to be mutually orthogonal and orthogonal to all the other eigenvectors. Thus, the decompositions of (9.35) apply to a general Hermitian matrix. More precisely, for any Hermitian matrix Φ there exists a unitary matrix V such that the diagonalizing transformation of (9.35) holds.

Exercise 9-7.

These results can be applied to our autocorrelation matrix Φ, which is Hermitian from Exercise 9-3. Let $\{\lambda_1, \lambda_2, \ldots, \lambda_N\}$ be the real-valued non-negative eigenvalues of Φ and let $\{v_1, v_2, \ldots, v_N\}$

be a set of associated eigenvectors chosen to be mutually orthogonal. Establish that Φ can be written in the form

$$\Phi = \sum_{i=1}^{n} \lambda_i v_i v_i^* \ . \tag{9.37}$$

This is known as a *spectral decomposition* of the matrix.

Exercise 9-8.

Show that the eigenvectors of the matrix $(\mathbf{I} - \beta\Phi)$ are the same as the eigenvectors of Φ, and that the eigenvalues are $(1 - \beta\lambda_i)$, $1 \le i \le n$, and thus establish the following decomposition, known as a *modal decomposition*,

$$(\mathbf{I} - \beta\Phi)^j = \sum_{i=1}^{n} (1 - \beta\lambda_i)^j v_i v_i^* \ . \tag{9.38}$$

The i-th term in (9.38) is known as the *i-th mode of the convergence*.

From (9.38) and (9.32), the error vector $\boldsymbol{q}_j$ obeys a trajectory that is the sum of N modes, the i-th of which is proportional to $(1 - \beta\lambda_i)^j$. The speed of convergence of each of these modes is governed by β. If β is made too large, then one or more of the $(1 - \beta\lambda_i)$ terms will be larger than unity in magnitude, and the error vector in (9.38) will actually increase in size with time. This is quite consistent with the intuitive behavior exhibited in Fig. 9-3 since the large β causes an overshoot of the minimum and actually increases the error.

This acceptable range of β can be investigated further if we order the eigenvalues from smallest to largest, denoting the smallest as λ_{min} and the largest as λ_{max}. Then the $(1 - \beta\lambda_i)$ term that governs how large β can get is the one corresponding to the largest eigenvalue, and hence the condition for $\boldsymbol{q}_j$ decaying exponentially to zero is

$$0 < \beta < \frac{2}{\lambda_{\text{max}}} \ . \tag{9.39}$$

This determines the largest value of β, but of more interest is the β corresponding to the fastest convergence of the MSEG algorithm. For a fixed β, the speed of convergence of the algorithm can be considered to be dominated by the slowest converging mode in (9.38). This slowest mode corresponds to the largest value of $|1 - \beta\lambda_i|$. The two extreme cases are plotted in Fig. 9-4, where this term is calculated for λ_{min} and λ_{max}. The corresponding curves for the other eigenvalues lie in between these two curves. The value of β that results in the fastest convergence is the point labeled β_{opt} in the figure. Choice of any other value of β results in a slower convergence of the mode corresponding to either the maximum or minimum eigenvalue. This optimal value of β is easily shown to be

$$\beta_{\text{opt}} = \frac{2}{\lambda_{\text{min}} + \lambda_{\text{max}}} \ , \tag{9.40}$$

and for this choice of β the modes corresponding to both minimum and maximum eigenvalues converge at the same rate, namely proportional to

$$\left(\frac{\lambda_{max}/\lambda_{min} - 1}{\lambda_{max}/\lambda_{min} + 1}\right)^j . \tag{9.41}$$

The quantity in the parenthesis is plotted in Fig. 9-5 as a function of the parameter $\lambda_{max}/\lambda_{min}$. This parameter, the ratio of largest to smallest eigenvalue, is called the *eigenvalue spread*. The eigenvalue spread has a minimum value of unity, and can be arbitrarily large. The larger the eigenvalue spread of the autocorrelation matrix, the slower the convergence of the MSEG algorithm. As seen in Fig. 9-5, the convergence becomes arbitrarily slow as the eigenvalue spread approaches infinity since the quantity in parentheses in (9.41) approaches unity.

The important role of the eigenvalues can be further quantified based on the following exercise.

Exercise 9-9.
Show that

$$E[\,|E_k|^2] - \xi_{min} = (\boldsymbol{c}_k - \boldsymbol{c}_{opt})^*\boldsymbol{\Phi}(\boldsymbol{c}_k - \boldsymbol{c}_{opt}) = \sum_{i=1}^{n} \lambda_i \,|(\boldsymbol{c}_k - \boldsymbol{c}_{opt})^*\boldsymbol{v}_i|^2 , \tag{9.42}$$

using expansion (9.38).

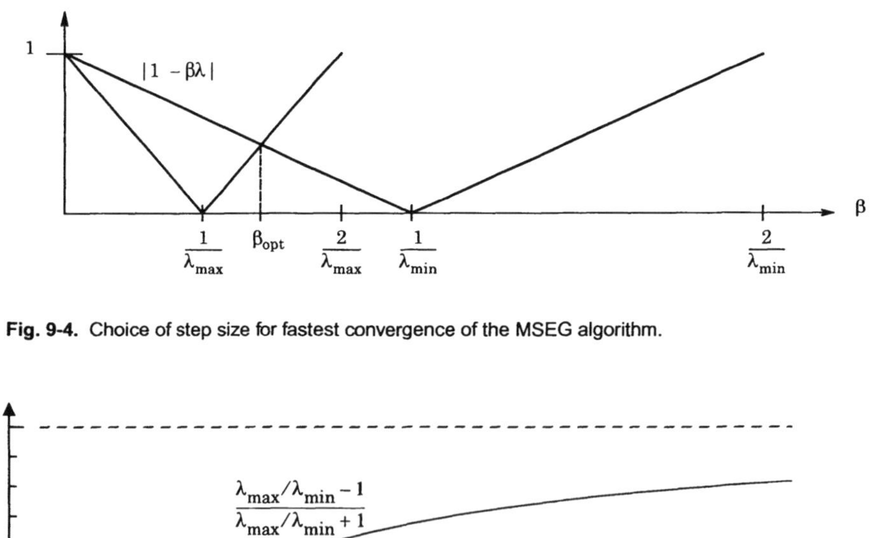

Fig. 9-4. Choice of step size for fastest convergence of the MSEG algorithm.

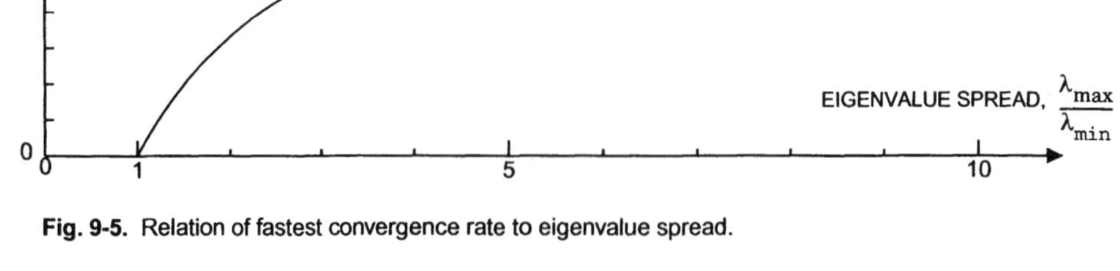

Fig. 9-5. Relation of fastest convergence rate to eigenvalue spread.

Equation (9.42) decomposes the MSE into its components in the direction of each of the eigenvectors, and shows that the component in that direction is proportional to the corresponding eigenvalue. The MSE therefore increases most rapidly in the direction of the eigenvector corresponding to λ_{max} and most slowly in the direction corresponding to λ_{min}. The largest acceptable step size β is determined by the maximum eigenvalue, since the gradient will be the largest in the direction of the eigenvector corresponding to the largest eigenvalue, and the correction of the MSEG algorithm will thus be largest in that direction. As β gets larger, that is the direction in which the increment in the algorithm will first be so large as to actually increase the MSE.

The β_{opt} of (9.40) is optimized to speed the convergence of the coefficient vector. An alternative approach is to choose β to maximize the rate of convergence of the MSE. From (9.42), the MSE at time $k = 1$ can be expressed in terms of the coefficient error $\boldsymbol{q}_1$ as

$$E[\,|E_1|^2] - \xi_{min} = \sum_{i=1}^{n} \lambda_i |\boldsymbol{q}_1^* \boldsymbol{v}_i|^2 , \tag{9.43}$$

which can then be expressed in terms of $\boldsymbol{q}_0$ using (9.38) as

$$E[\,|E_1|^2] - \xi_{min} = \sum_{i=1}^{n} \lambda_i (1 - \beta\lambda_i)^2 |\boldsymbol{q}_0^* \boldsymbol{v}_i|^2 . \tag{9.44}$$

To get the maximum reduction in MSE from step $k = 0$ to step $k = 1$, we would minimize (9.44) with respect to β. Unfortunately, this requires some assumption about the initial coefficient error $\boldsymbol{q}_0$. However, the largest contributor to the MSE is likely to be the term in the sum corresponding to the largest eigenvalue, λ_{max}. A reasonable strategy is therefore to choose β to immediately force that term to zero. This is achieved by choosing $\beta = 1/\lambda_{max}$. The conclusion is that (9.40) is not necessarily the best choice for step size, if the criterion is to most quickly reduce the MSE as opposed to reduce the norm of the coefficient error vector. The step size that speeds the convergence of the MSE depends on the initial coefficient error vector, but $\beta = 1/\lambda_{max}$ is a reasonable choice.

The convergence of the MSEG algorithm can be interpreted graphically by plotting the contours of equal mean-square error as in Fig. 9-6. Equation (9.42) illustrates that the contours of equal mean-square error are elliptical in shape, with the principal axes in the direction of the eigenvectors. The eccentricities of the ellipses are directly related to the relative sizes of the eigenvalues. This is illustrated in Fig. 9-6 for the $N = 2$ case. It is assumed that $\lambda_2 > \lambda_1$, in which case the mean-square error increases more rapidly in the direction of $\boldsymbol{v}_2$. The direction of the two orthogonal eigenvectors is shown. The major axis of the ellipse is in the direction of $\boldsymbol{v}_1$, and the minor axis in the direction of $\boldsymbol{v}_2$.

The case where the eigenvalue spread is small (eigenvalues approximately equal) is shown in Fig. 9-6(a); the ellipse is close to being a circle. A larger eigenvalue spread is shown in Fig. 9-6(b); there the ellipse is more eccentric.

For a small eigenvalue spread, the gradient correction is always nearly in the direction of the minimum mean-square error, and the length of the gradient vector is always approximately the same. For a larger eigenvalue spread, the direction of the negative gradient can be quite different from the direction of the minimum, although for small steps the mean-square error still gets smaller. Since each step does not go directly toward the minimum, the number of required steps will be increased for some starting conditions. More importantly, the length of the gradient vector will be much smaller in the direction of the major axis of the ellipse, since the MSE is not varying as rapidly in that direction. The step size is therefore governed by the largest eigenvalue, so that the steps do not overshoot in the direction of the corresponding eigenvector, which is the minor axis of the ellipse. A step size that maintains stability along the minor axis results in very small increments in the direction of the major axis.

An intuitive interpretation of Fig. 9-4 also follows from Fig. 9-6. Consider the case where the starting coefficient vector is on the minor axis of the ellipse, so that convergence is in the direction of the eigenvector corresponding to the largest eigenvalue. If β is chosen to be smaller than $1/\lambda_{max}$, then each step of the algorithm in this direction does not overshoot the minimum, and the MSE gets smaller. When $\beta = 1/\lambda_{max}$, the algorithm converges to the minimum in one iteration. When β is greater than $1/\lambda_{max}$, the algorithm overshoots on each iteration, but as long as β is smaller than $2/\lambda_{max}$ the MSE still decreases and the algorithm converges. It is advantageous from the point of view of maximizing the worst-case convergence rate to choose $\beta = \beta_{opt}$, a choice that results in the algorithm overshooting the minimum in the direction of the eigenvector corresponding to λ_{max} in order that the algorithm converge faster in the direction of the eigenvector corresponding to λ_{min}.

Since the eigenvalue spread plays such an important role in the adaptation speed, it is instructive to relate it to the power spectral density of the wide-sense stationary random process R_k, the samples of the data waveform. It is a classical result of Toeplitz form theory [6] that the eigenvalues of (9.10) are bounded by

$$\min_{\theta} S(e^{j\theta}) < \lambda_i < \max_{\theta} S(e^{j\theta}) \,, \tag{9.45}$$

where $S(e^{j\theta})$ is the power spectral density of the reference random process defined as the Fourier transform of the autocorrelation function (the elements of the matrix),

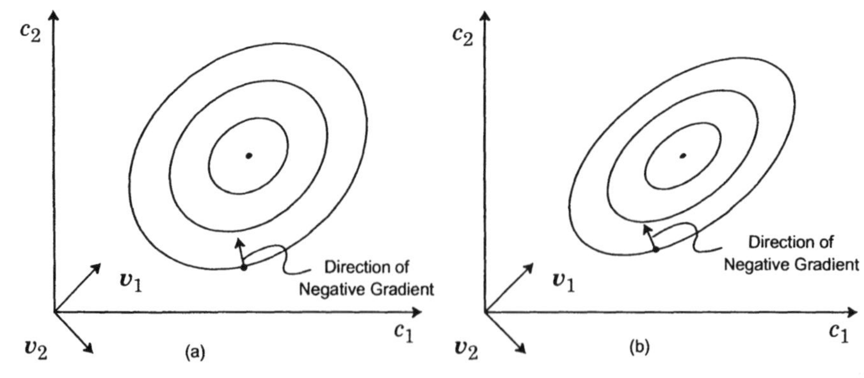

Fig. 9-6. Effect of eigenvalue spread on convergence. a. Small eigenvalue spread. b. Larger eigenvalue spread.

$$S(e^{j\theta}) = \sum_{k=-\infty}^{\infty} \phi_k e^{-jk\theta} . \tag{9.46}$$

While the eigenvalues depend on the order of the matrix, n, as $n \to \infty$

$$\lambda_{min} \to \min_{\theta} S(e^{j\theta}) , \qquad \lambda_{max} \to \max_{\theta} S(e^{j\theta}) . \tag{9.47}$$

See [6] for more precise statements of these results. This interesting relationship between the eigenvalues and the power spectrum is explored in Problem 9-4.

It follows that the spectra that result in slow convergence of the MSEG algorithm are those for which the ratio of the maximum to the minimum of the spectrum is large, and spectra that are almost flat (have an eigenvalue spread near unity) result in fast convergence. The intuition behind this result is that a large eigenvalue spread is related to a large correlation among the input samples, which in turn slows convergence because of interactions between the convergence of the different coefficients of the transversal filter.

Since the modes of convergence of the MSEG algorithm are all of the form of γ^j, where γ is a positive real number less than unity and j is the iteration number, the error in decibels can be determined by taking the logarithm of the square (Problem 9-8),

$$10 \cdot \log_{10}(\gamma^{2j}) = [10 \cdot \log_{10}(\gamma^2)] \cdot j , \tag{9.48}$$

and thus the error expressed in decibels decreases linearly with iteration number (the constant factors multiplying these exponentially decaying terms give a constant factor in decibels). The convergence of a MSEG algorithm is thus often expressed in units of *dB per iteration*, which is the number of decibels of decrease in the error power per iteration.

9.2. Adaptive Linear Equalizer

On many practical channels, one cannot pretend to know the autocorrelation matrix Φ, and hence the known-channel solution of Section 9.1 is not applicable.

Example 9-3.
On the voiceband telephone channel, there is significant variation in the amplitude and phase of the channel transfer function from one call to another. On a terrestrial microwave channel, under normal conditions the channel is nearly ideal, but there can be conditions under which there is considerable variation in the transfer function due to selective fading.

The technique is to modify the MSEG algorithm, by a simple trick, to allow adaptation. The resulting algorithm is known as the is the *stochastic gradient (SG)* algorithm. The approach taken in the SG algorithm is to substitute a time average for the ensemble average in the MSE solution. This adaptation algorithm is also sometimes called the LMS adaptive transversal filter. The term LMS stands for *least-mean square*, although the algorithm does not provide an exact solution to the problem of minimizing the mean-square error but rather only approximates the solution. This approximation is the price paid for not requiring that the channel be known or stationary.

If we don't know the channel, then we cannot calculate the expectation in (9.8). However, we can, if we choose, calculate the modulus-squared error without the expectation,

$$|E_k|^2 = |A_k|^2 - 2\text{Re}\{A_k \boldsymbol{c}^* \bar{\boldsymbol{r}}_k\} + \boldsymbol{c}^* \bar{\boldsymbol{r}}_k \boldsymbol{r}_k^T \boldsymbol{c} . \tag{9.49}$$

Using Exercise 9-5, and using the fact that $\bar{\boldsymbol{r}}_k \boldsymbol{r}^T$ is a Hermitian matrix, we can take the gradient of this expression,

$$\nabla_{\boldsymbol{c}} |E_k|^2 = -2\bar{\boldsymbol{r}}_k(A_k - \boldsymbol{r}_k^T \boldsymbol{c}) = -2E_k \bar{\boldsymbol{r}}_k . \tag{9.50}$$

Because we are dealing with well-behaved quadratic functions, and the gradient and expectation are linear operators, they can be interchanged. The expectation of the gradient in (9.50) is the same as the gradient of $\text{E}[|E_k|^2]$,

$$\text{E}[\nabla_{\boldsymbol{c}} |E_k|^2] = \nabla_{\boldsymbol{c}} \text{E}[|E_k|^2]. \tag{9.51}$$

Therefore, (9.50) is an *unbiased estimator* of the gradient in the MSEG algorithm derived in Section 9.1. The SG algorithm substitutes this "noisy" or "stochastic" gradient for the actual gradient in the algorithm of (9.28). What results is

$$\boldsymbol{c}_{k+1} = \boldsymbol{c}_k - \tfrac{\beta}{2} \nabla_{\boldsymbol{c}}[|E_k|^2] \big|_{\boldsymbol{c} = \boldsymbol{c}_k} , \tag{9.52}$$

or

$$\begin{aligned} \boldsymbol{c}_{k+1} &= \boldsymbol{c}_k + \beta E_k \bar{\boldsymbol{r}}_k \\ &= [\mathbf{I} - \beta \bar{\boldsymbol{r}}_k \boldsymbol{r}_k^T] \boldsymbol{c}_k + \beta A_k \bar{\boldsymbol{r}}_k . \end{aligned} \tag{9.53}$$

There is another rather subtle but important difference between this SG algorithm and the algorithm of (9.28). In the former algorithm, the index j corresponded to the iteration number for the iterative algorithm for solving a system of linear equations. In (9.53) on the other hand, the iteration number k corresponds to the sample number (or time index) of the data waveform at the input to the equalizer. Thus each iteration corresponds to a new sample. The algorithm is in effect performing a time average in order to estimate the gradient.

It is not surprising to note the similarity between the SG algorithm of (9.53) and the MSEG algorithm of (9.28). The former substitutes the stochastic matrix $\bar{\boldsymbol{r}}_k \boldsymbol{r}_k^T$ for Φ and the stochastic vector $A_k \bar{\boldsymbol{r}}_k$ for the vector α. In each case the deterministic matrix or vector corresponds to the ensemble average of the stochastic matrix or vector for the stationary case.

A realization of adaptation algorithm (9.53) is illustrated in Fig. 9-7. What is shown is a single filter coefficient and how it contributes both to the transversal filter structure as well as how it is adapted. Denote the j-th coefficient c_j at time k by $[\boldsymbol{c}_k]_j$. Then the adaptation algorithm of (9.53) can be rewritten for the j-th component as

$$[\boldsymbol{c}_{k+1}]_j = [\boldsymbol{c}_k]_j + \beta E_k R_{k-j}^* , \quad -L \le j \le L . \tag{9.54}$$

Specifically, the input sample to the equalizer R_{k-j} is taken from the output of the same unit delay as is used for multiplication by c_j, conjugated, and multiplied by the j-th coefficient at time k. The resultant value contributes to the summation which is subtracted from the data symbol A_k to obtain the error sample E_k. In accordance with this equation, $[\boldsymbol{c}_k]_j$ is obtained by cross-correlating (time averaging) the estimation error E_k with the delayed input R_{k-j}. This cross-correlation consists of taking the product of E_k with R^*_{k-j} and step size β and accumulating the result.

This algorithm is not surprising in light of the orthogonality principle of (9.26). In particular, when all the filter coefficients are optimal, the orthogonality principle says that the input to the accumulator in Fig. 9-7 will average to zero. Under these conditions, the output of the accumulator will maintain the same average value (namely the optimal filter coefficient). What the orthogonality principle does not tell us directly is that when the coefficients are not optimal, the non-zero average of the accumulator input is of the correct sign so as to move each coefficient toward the optimum.

Example 9-4.

Continuing Example 9-1 for a single real-valued coefficient c, denote this coefficient at time k as c_k, and then (9.54) becomes

$$c_{k+1} = c_k + \beta E_k R_k \,, \qquad E_k = A_k - c_k R_k \,. \tag{9.55}$$

Now suppose that $c_k = c_{\text{opt}} + \Delta$ for some $\Delta > 0$. Then the correction term in the algorithm is

$$E_k R_k = (A_k - c_{\text{opt}} R_k) R_k - \Delta R_k^2 \,. \tag{9.56}$$

The first term in (9.56) has an average value of zero by the orthogonality principle, and the second term is *always* negative, decreasing the coefficient on average as desired. (In spite of the fact that the second term gives the correction a bias in the right direction, the correction is stochastic because of the first term, and hence will sometimes go in the wrong direction.)

This example is easily generalized. The term in the product $E_k R^*_{k-j}$ which depends on $[\mathbf{c}_k]_j$ is $-[\mathbf{c}_k]_j \, |R_{k-j}|^2$. Consider for example the case where the real part of $[\mathbf{c}_{\text{opt}}]_j$ is positive. Since $|R_{k-j}|^2$ is positive real-valued, if the real part of $[\mathbf{c}_k]_j$ is too large then the real part of this gradient term is more negative than it would be if the coefficient were optimal. This makes the average correction to the real part of $[\mathbf{c}_k]_j$ negative and on average the real part of $[\mathbf{c}_{k+1}]_j$ is smaller than $[\mathbf{c}_k]_j$. The same logic applies to the imaginary part of $[\mathbf{c}_k]_j$.

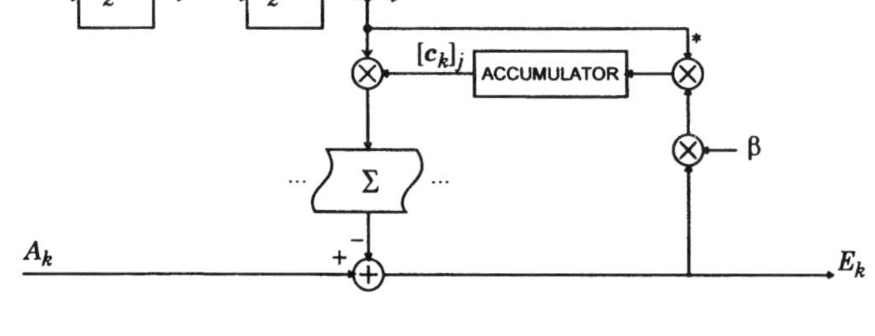

Fig. 9-7. SG algorithm for one coefficient. Although signals are shown with single lines, in general all signals and coefficients are complex-valued.

This explanation does not answer the important question as to the effect of the interaction between the adaptation of the different coefficients. This interaction occurs because all the coefficients affect the error E_k, which in turn affects the coefficient adaptation. We now address this question.

9.2.1. Convergence of the SG Algorithm

One difference between the SG algorithm of (9.53) and the MSEG algorithm of (9.28) is that in (9.28) the coefficient vector follows a deterministic and predictable trajectory, while the trajectory in (9.53) is random or stochastic. The cause of this random fluctuation is the use of the time average in place of ensemble average, or alternatively the use of the input samples in place of the ensemble averages.

A useful method for analyzing the convergence behavior of an adaptive filtering algorithm is to assume (often unrealistically) that the input samples can be modeled as a wide-sense stationary random process with known statistics; that is, return to the assumptions of Section 9.1. The coefficient vector can then be expected to converge in some sense to be determined to the MSE solution. The speed of this convergence, while not directly applicable to a case where the input statistics are actually changing with time, is a good indication of the convergence performance of the algorithm.

If the assumption is made that the input can be modeled as a wide-sense stationary random process, we would first like to find the average trajectory of the coefficient vector in (9.53). If the step size β is small, the coefficient vector will vary slowly, since the update at each sample time is proportional to β. The input random process will therefore vary rapidly relative to the coefficient vector. If we take the expectation of (9.53) with respect to the statistics of $\boldsymbol{r}_k$, it is therefore a good approximation to assume that the coefficient vector $\boldsymbol{c}_k$ is a constant with respect to this expectation, and hence

$$\begin{aligned} \boldsymbol{c}_{k+1} &= \mathrm{E}[(\mathbf{I} - \beta \bar{\boldsymbol{r}}_k \boldsymbol{r}_k^T)\boldsymbol{c}_k] + \beta \mathrm{E}[A_k \bar{\boldsymbol{r}}_k] \\ &\approx (\mathbf{I} - \beta \Phi)\boldsymbol{c}_k + \beta \alpha \,, \end{aligned} \tag{9.57}$$

where the expectation is only with respect to the input process. Although the coefficient vector is varying slowly, it is still random, and hence we must still take the expectation of both sides with respect to the ensemble of coefficient vectors,

$$\mathrm{E}[\boldsymbol{c}_{k+1}] \approx (\mathbf{I} - \beta \Phi)\mathrm{E}[\boldsymbol{c}_k] + \beta \alpha \,. \tag{9.58}$$

This approximate average trajectory precisely obeys the earlier deterministic MSEG algorithm. It can be asserted without further analysis that within the accuracy of this approximation the average trajectory of the SG converges to the optimal coefficient vector under the same condition that guarantees convergence of the MSEG algorithm, and the nature of the convergence is identical to that discussed in Section 9.1.2.

This does not mean, however, that any particular coefficient vector trajectory itself converges to the optimum, but only that the average of all trajectories converges to the optimum. In fact, the coefficient vector does not converge to the optimum. Even after convergence of the coefficient vector in the mean-value sense, the difference equation (9.53) still has a stochastic driving term, and therefore the coefficient vector continues to fluctuate about the optimal coefficient vector randomly. The larger the value of the step size β, the larger this fluctuation. The size of this fluctuation will be considered in a moment. Keeping it reasonably small generally requires a much smaller step size than the value given by (9.40).

These considerations make it important to calculate some measure of the variation of the coefficient vector about this optimum. In analogy to (9.31), define an error vector between the actual coefficient vector at time k and the optimal vector as

$$\boldsymbol{q}_k = \boldsymbol{c}_k - \boldsymbol{c}_{\text{opt}}. \tag{9.59}$$

Exercise 9-10.

Substitute into the SG algorithm of (9.53) to show that the update equation for the error vector is

$$\boldsymbol{q}_{k+1} = \Gamma_k \boldsymbol{q}_k + \beta D_k \bar{\boldsymbol{r}}_k , \tag{9.60}$$

where Γ_k is a stochastic matrix

$$\Gamma_k = \mathbf{I} - \beta \bar{\boldsymbol{r}}_k \boldsymbol{r}_k^T , \tag{9.61}$$

and D_k is the error signal for a transversal filter with optimal coefficients,

$$D_k = A_k - \boldsymbol{r}_k^T \boldsymbol{c}_{\text{opt}} . \tag{9.62}$$

This interesting relationship demonstrates again that the coefficient vector can never reach its optimum because this stochastic equation has a driving term which never goes to zero (except in the degenerate case where the optimal filter yields a zero error signal).

One measure of how well the SG algorithm is working would be the Euclidean norm of the coefficient error vector,

$$\|\boldsymbol{q}_k\|^2 = \boldsymbol{q}_k^* \boldsymbol{q}_k. \tag{9.63}$$

This is the appropriate measure to use if the accuracy with which the algorithm approximates the optimal coefficients is the primary concern. If, on the other hand, we are interested in how well the adaptive filter does its job, as manifested by the size of the error signal (which after all is what causes incorrect decisions in the slicer), then we are interested in the size of E_k rather than the error vector.

Actually, as one might expect, these two measures are closely related. If the coefficient vector is fixed, then from (9.15),

$$\text{E}[\,|E_k|^2] = \xi_{\min} + \boldsymbol{q}^* \Phi \boldsymbol{q} , \tag{9.64}$$

where $\xi_{\min}$ is the minimum MSE that would result from the use of the optimal coefficient vector $\boldsymbol{c}_{opt}$ ($\boldsymbol{q} = \mathbf{0}$). The second term in (9.64) we call the *excess MSE*, or that MSE over and above the minimum possible due to the non-optimal coefficient vector. If the step size β is very small, then within the time frame of significant variation in the input random process R_k we would expect very little change in $\boldsymbol{q}$. Thus, we can substitute the time-varying coefficient vector for the fixed vector in (9.64) to accurately find how the MSE varies with time,

$$\mathrm{E}[\,|E_k|^2] \approx \xi_{\min} + \boldsymbol{q}_k{}^*\Phi\boldsymbol{q}_k\,. \tag{9.65}$$

In order to find the average MSE, we must average (9.65) over the ensemble of coefficient error vectors, or

$$\mathrm{E}[\,|E_k|^2] \approx \xi_{\min} + \mathrm{E}[\boldsymbol{q}_k{}^*\Phi\boldsymbol{q}_k]\,. \tag{9.66}$$

This equation establishes the link that we were looking for between the coefficient error vector and the filter output MSE.

Rather than analyze the dynamics of (9.66) in general, we will specialize to an input process R_k with uncorrelated zero-mean samples. For this case, the autocorrelation matrix is diagonal,

$$\Phi = \phi_0\mathbf{I}\,, \qquad \phi_0 = \mathrm{E}[\,|R_k|^2]\,. \tag{9.67}$$

The average MSE as a function of time from (9.66) becomes a simpler expression

$$\mathrm{E}[\,|E_k|^2] = \xi_{\min} + \phi_0\mathrm{E}[\|\boldsymbol{q}_k\|^2] \tag{9.68}$$

that relates directly to the Euclidean norm of the coefficient error vector. Estimation of this norm is somewhat tedious, so we defer it to Appendix 9-A. Derived there is a difference equation in the error vector norm

$$\mathrm{E}[\|\boldsymbol{q}_{k+1}\|^2] = \gamma\cdot\mathrm{E}[\|\boldsymbol{q}_k\|^2] + \beta^2N\phi_0\xi_{\min}\,, \qquad \gamma = 1 - 2\beta\phi_0 + \beta^2N\phi_0{}^2\,. \tag{9.69}$$

There is only a single mode of adaptation due to the assumption of a white input spectrum (and therefore there is only one distinct eigenvalue). This relation demonstrates both the speed with which the error vector norm decreases with time as the filter is adapting and the asymptotic error vector norm, which is non-zero due to the continued stochastic driving term even after nominal convergence. This latter contribution to excess MSE is related to the minimum MSE for a fixed-coefficient filter because this error appears in the error signal driving adaptation even after nominal convergence of the coefficient vector.

The condition for stability of the filter, in the sense that the error vector norm decreases with time, is that

$$|\gamma| < 1. \tag{9.70}$$

The quantity γ is plotted in Fig. 9-8, where we see several interesting properties. Starting with zero, as we increase the step size β the speed of convergence increases, until a maximum speed is reached at

$$\beta_{\mathrm{opt}} = \frac{1}{N\phi_0}\,. \tag{9.71}$$

Continuing to increase the step size slows convergence, until eventually we reach instability at twice the optimal step size. The condition for stability is

$$0 < \beta < \frac{2}{N\phi_0} = 2\beta_{\mathrm{opt}}\,. \tag{9.72}$$

This condition is considerably more stringent than the condition for convergence of the average coefficient vector (9.39). This implies that we could get a situation where the average coefficient vector is converging but the norm of the error vector is diverging. Since this is unacceptable, (9.72) is the most stringent condition.

Since convergence is exponential, it is instructive to define a time constant τ as the number of samples required for the MSE to decrease by a factor of e^{-1}, $\gamma^{\tau} = 1/e$; solving for τ we get, in the range of small β,

$$\tau \approx \frac{1}{2\beta\phi_0}, \tag{9.73}$$

with a shortest time constant corresponding to β_{opt} of $\tau \approx N/2$. The important conclusion here is that the best rate of convergence is dependent on the number of filter coefficients — the more coefficients, the longer it takes for the coefficients to converge. This is not surprising in view of the inevitable interaction of the coefficients. The more coefficients there are, the more "noise" is introduced into the adaptation of each coefficient by the simultaneous adaptation of the other coefficients.

Aside from the rate of convergence, the other parameter of interest is the asymptotic error in the filter coefficients after convergence. The stationary point in (9.69) is

$$\mathrm{E}[\|\boldsymbol{q}_k\|^2] \to \frac{N\beta}{2 - N\beta\phi_0}\xi_{min} \qquad \text{as} \quad k \to \infty . \tag{9.74}$$

This is plotted in Fig. 9-9, where we see that the asymptotic error increases as the step size increases until it blows up at twice the optimal step size. At the optimal step size (optimal in terms of rate of convergence of MSE, not the asymptotic MSE), the error is

$$\mathrm{E}[\|\boldsymbol{q}_k\|^2] \to \frac{1}{\phi_0}\xi_{min}. \tag{9.75}$$

In view of (9.66), the asymptotic MSE is

$$\mathrm{E}[|E_k|^2] \to \xi_{min} + \xi_{min} = 2\xi_{min}. \tag{9.76}$$

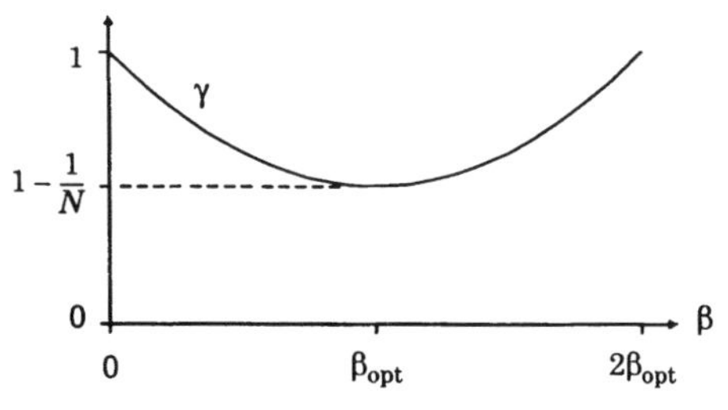

Fig. 9-8. A plot of γ vs. β characterizing convergence of the MSE for a white input spectrum. A small N ($N = 2$) is plotted to exaggerate the curve.

Thus, for the fastest convergence, the total MSE is twice the minimum MSE for a fixed-coefficient filter, with half that MSE attributable to the asymptotic wandering of the filter coefficients about their optimal value.

There is an important tradeoff between speed of convergence and asymptotic MSE. If the goal is to minimize the asymptotic MSE, then choose as small a step size as possible. Generally what limits how small we can make the step size is the number of bits of precision in the arithmetic we use to implement the SG algorithm. In this case we can get the asymptotic MSE as close to the minimum MSE for a fixed-coefficient filter as we like at the expense of higher precision arithmetic. On the other hand, if our goal is to maximize the rate of convergence, then the step size should be chosen to be β_{opt}, with the penalty that the asymptotic MSE is twice as large as the minimum possible value.

A similar analysis to what we have done here can be applied to the general input spectrum case [8][9][10][11]. Surprisingly, the results are essentially the same as for the white input case we have considered here. Thus, the large effect of eigenvalue spread on the convergence of the average coefficient vector does not extend to the convergence of the MSE. The reason for this can be seen in the expression for excess MSE given in (9.42). In the direction of eigenvectors corresponding to small eigenvalues, the MSE does not change very rapidly, as manifested in the λ_i term. This implies that the coefficient vector tends to have larger excursions in this direction. Small eigenvalues thus cause problems in the convergence of the coefficient vector in the direction of the corresponding eigenvectors. However, these same excursions do not cause as large an impact on the resulting MSE precisely because of the small eigenvalue. Thus, if the goal is to accurately estimate the optimal coefficient vector, then small eigenvalues are a problem, but if the goal is to minimize the MSE of the adaptive filter they do not present nearly as great a problem. Even in the latter case, however, they can lead to numerical problems, as discussed in Section 9.4.

9.2.2. Common Modifications

Several modifications are commonly made to the SG algorithm as we have derived it.

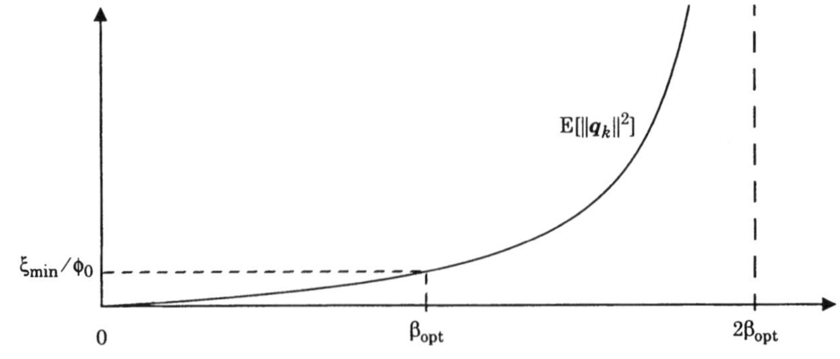

Fig. 9-9. Asymptotic average Euclidean norm squared of the filter coefficients after convergence as a function of the step size.

Normalization of Step Size

The SG algorithm displays an undesirable dependence of speed of convergence on input signal power. This can be seen from (9.53); if the input signal is increased in size by a factor γ, then this is equivalent to increasing the step size β by the factor γ^2 to $\beta\gamma^2$. Another way to see this is that the optimal step size of (9.71) requires that the step size β should be inversely proportional to the input signal power ϕ_0. The speed of convergence and asymptotic MSE of the SG algorithm is strongly affected by the size of the input signal. A serious consequence is that if the input signal grows too large the adaptation algorithm becomes unstable.

The standard solution to this difficulty is to normalize the step size of the algorithm. The size of the updates can be kept approximately the same size on average if the update is normalized by an estimate of the input signal power, which is equivalent to choosing a step size in (9.53) equal to

$$\beta_k = \frac{a}{\sigma_k^2 + b}, \tag{9.77}$$

where β_k is the step size at time k, a and b are some appropriately chosen constants, and σ_k^2 is an estimate of the input signal power at time k. The purpose of the b in the denominator is to prevent β_k from becoming too large (causing instability) when the input signal power becomes very small.

As an example of how the input signal power can be estimated, we can use an exponentially weighted time average of the input signal power,

$$\sigma_k^2 = (1-\alpha)\sum_{j=0}^{\infty} \alpha^j |R_{k-j}|^2 \tag{9.78}$$

where α is an appropriately chosen constant and the $(1 - \alpha)$ factor normalizes the estimate to be an unbiased estimate of the input signal power. The reason for choosing this estimate is that it can be written recursively as

$$\sigma_k^2 = \alpha\sigma_{k-1}^2 + (1-\alpha)|R_k|^2. \tag{9.79}$$

Gear-Shift Algorithms

There is a tradeoff between asymptotic MSE and speed of convergence of the SG algorithm. Speed of convergence is important in two contexts. First, if we start the equalizer up on an unknown channel, then we would like to have rapid convergence of the equalizer. Second, if there is any variation of the channel, we would like the equalizer to track this variation. In many applications of adaptive equalization, variation of the channel is quite slow.

Example 9-5.

In voiceband data modems, there is no mechanism to cause any significant variation of the impulse response of the channel once a telephone connection is established. Thus, any changes are minor and occur over a long time period.

On these channels, a very small step size would be desirable after convergence of the equalizer to insure a small asymptotic excess MSE, but a larger step size is needed during initial convergence. The solution is a *gear-shift* algorithm, in which the step size is initially larger, and shifted to a smaller value after a sufficient period of time for convergence to have occurred.

There are examples of channels in which the tracking capability of the equalizer is important, in which case a larger excess MSE must be accepted in order to gain this tracking capability.

Example 9-6.

In a microwave radio system, selective fading can vary fairly rapidly. Therefore, the time constant of the MSE adaptation should be small relative to the time of significant variation of the fading. In a mobile radio system, the variations are even faster, making speed of adaptation the most important factor.

9.3. Adaptive DFE

Just as the coefficients of a transversal filter equalizer can be adapted, so too can the coefficients of a decision feedback equalizer. We will follow the model of the linear equalizer, and consider first the MSE solution for a finite precursor and postcursor equalizer filter in the DFE, followed by derivation of the stochastic gradient algorithm (we can dispense with the MSE gradient algorithm now that we know the principle of the stochastic gradient).

We will assume the form of the DFE shown in Fig. 9-2 in which the precursor equalizer is anticausal with N coefficients and the postcursor equalizer is causal with M coefficients. The slicer input is given by the relation

$$Y_k = \sum_{i=-(N-1)}^{0} c_i R_{k-i} - \sum_{i=1}^{M} d_i \hat{A}_{k-i}, \tag{9.80}$$

which is illustrated in Fig. 9-10. The figure has been drawn to illustrate an interesting interpretation of the finite DFE as a two-sided transversal filter, with non-causal coefficients to

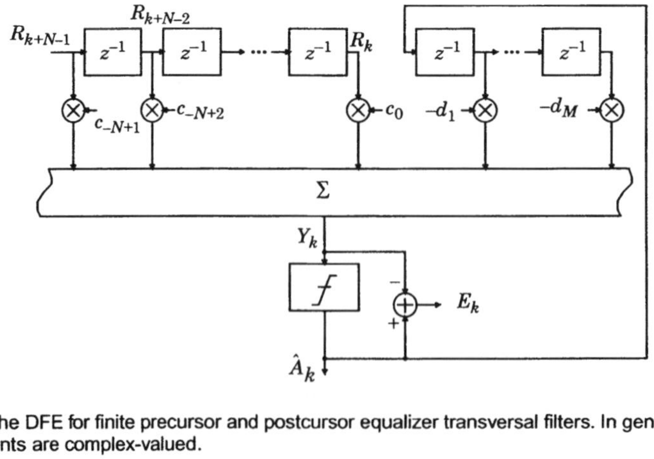

Fig. 9-10. The DFE for finite precursor and postcursor equalizer transversal filters. In general all signals and coefficients are complex-valued.

cancel precursor intersymbol interference and causal coefficients to cancel postcursor intersymbol interference. This is similar to the LE, with the important difference that the input to the causal portion of the filter is the decisions rather than the output of the precursor equalizer filter. This difference will obviously change the desired tap coefficients as well as reduce the noise enhancement due to equalization.

Exercise 9-11.
The coefficients of *both* the causal and non-causal portions of the DFE equalizer will be different from the corresponding coefficients of the LE. Why?

9.3.1. MSE Solution

It is instructive to find the optimal finite equalizers using the MSE criterion. We have to re-solve this problem, first considered in Chapter 8, since the filters now have constrained complexity. It is permissible to assume that there are no decision errors in calculating the output of the postcursor equalizer. The resulting filter, combining the channel model of Fig. 9-2 and the equalizer of Fig. 9-10, is shown in Fig. 9-11(a). After combining the filter blocks, we get the configuration of Fig. 9-11(b).

The objective is to minimize $E[|E_k|^2]$ over the choice of the postcursor equalizer filter coefficients and the precursor equalizer filter coefficients. The former are easier to find. Assuming the noise and data symbols to be uncorrelated, $E[|E_k|^2]$ in Fig. 9-11(b) is the sum of two terms, one for the noise and the other for the intersymbol interference, where the noise term is independent of $D(z)$. Hence, we can minimize just the intersymbol interference term over the postcursor equalizer coefficients. This term is of the form

$$\sum_m v_m A_{k-m} , \tag{9.81}$$

where

$$v_k = \begin{cases} \sum_{i=-(N-1)}^{0} c_i p_{k-i} - d_k , & k \in \{1, 2, \ldots, M\} \\ \sum_{i=-(N-1)}^{0} c_i p_{k-i} , & \text{otherwise} \end{cases} \tag{9.82}$$

It is evident that as long as the data symbols are uncorrelated with one another, $E[|E_k|^2]$ will be minimized by choosing $D(z)$ to eliminate the first M intersymbol interference samples; that is, force $v_m = 0,\ 1 \le m \le M$. Hence, the optimal postcursor equalizer coefficients are

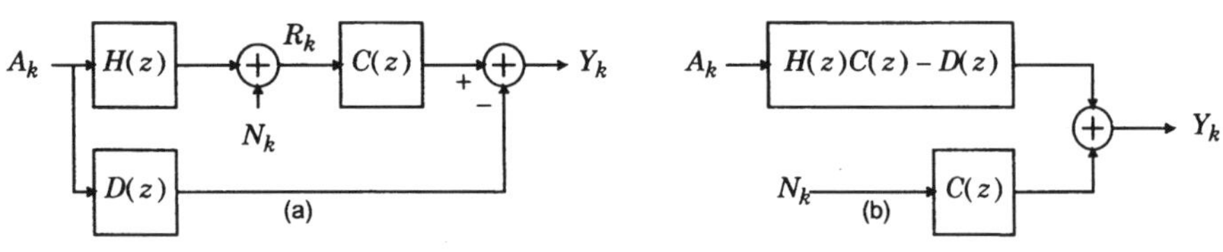

Fig. 9-11. (a) Equivalent of channel, precursor equalizer, and postcursor equalizer assuming no decision errors. (b) Simplification after combining filters.

$$d_m = \sum_{i=-(N-1)}^{0} c_i p_{m-i}, \quad m \in \{1, 2, \dots, M\}. \tag{9.83}$$

This says, not surprisingly, that there is no benefit to leaving any postcursor intersymbol interference after the postcursor equalization within the memory of the filter, since this cannot reduce the noise.

Having found the postcursor equalizer coefficients in terms of the precursor equalizer coefficients, we substitute this solution and then minimize over the precursor equalizer coefficients c_k.

Exercise 9-12.

Assume the data symbols are mutually uncorrelated with mean zero and variance E_a. Define a vector of precursor equalizer coefficients

$$\boldsymbol{c} = [c_{-(N-1)}, \dots, c_0]^T \tag{9.84}$$

and show that the optimal $\boldsymbol{c}$ satisfies (9.16) where the matrix Φ has elements given by (9.11) except that the summation is missing the terms $m = 1$ to $m = M$ and the vector α is given by

$$\alpha^T = E_a[p_{-(N-1)}, \dots, p_0]. \tag{9.85}$$

9.3.2. Stochastic Gradient Algorithm

The derivation of a stochastic gradient algorithm for the DFE is a simple extension of the LE case. First we define an augmented vector of N+M filter coefficients,

$$\boldsymbol{v} = [c_{-(N-1)}, \dots, c_0, -d_1, \dots, -d_M]^T \tag{9.86}$$

and an augmented input signal vector

$$\boldsymbol{w}_k = [R_{k+(N-1)}, \dots, R_k, A_{k-1}, \dots, A_{k-M}]^T. \tag{9.87}$$

Then the DFE slicer error can be expressed as

$$E_k = \hat{a}_k - \boldsymbol{v}_k{}^T \boldsymbol{w}_k. \tag{9.88}$$

This is identical to the LE case with $\boldsymbol{c}$ replaced by $\boldsymbol{v}$ and $\boldsymbol{p}$ replaced by $\boldsymbol{w}$, and hence we can immediately infer that the SG algorithm is, from (9.53),

$$\boldsymbol{v}_{k+1} = \boldsymbol{v}_k + \beta E_k \overline{\boldsymbol{w}}_k. \tag{9.89}$$

9.4. Fractionally Spaced Equalizer

Thus far in this chapter we have considered the adaptation of a linear equalizer with sample rate equal to the symbol rate. In practice it would be more common to adapt a *fractionally spaced equalizer (FSE)* discussed in Section 8.3 for several reasons. First, incomplete knowledge of the channel makes it impossible to realize a matched filter directly. The FSE structure allows us, in effect, to adapt the matched filter as well as the equalizer.

Second, the FSE is less influenced by sampling phase, as discussed in Section 8.3. Third, for implementation, the separate matched filter has the effect of doubling the loss of the channel, greatly increasing the gain necessary in the transversal filter equalizer and increasing its dynamic range requirements.

If we start with an FSE and use a MSE criterion for adaptation, within the constraints of the finite degrees of freedom the resulting filter will perform both the matched filtering and equalizer functions. The adaptation is also a straightforward extension of the earlier case. For example, if the FSE sampling rate is twice the symbol rate, we can think of the FSE as a transversal filter that operates at twice the symbol rate but simply fails to calculate every second output sample. The adaptation is based on the output samples that are calculated. However, the FSE does suffer from one subtle difficulty, which is a form of numerical ill-conditioning. We will address this issue in the following subsection.

9.4.1. Conditions for Unique MSE Solution

The existence of a unique solution to the MSE problem, as well as the arguments for the convergence of the gradient and SG algorithms, depended on the nonsingularity of the input autocorrelation matrix Φ . The singular case corresponds to one or more zero eigenvalues. From the argument in Section 9.1, the case of concern is where the spectrum of the reference input vanishes at some frequency, since (9.47) would then predict that one (or more) eigenvalues would approach zero as $N \rightarrow \infty$. That the vanishing of the input spectrum would cause problems is not surprising, since the equalizer transfer function in the regions of zero spectrum does not affect the MSE, so the filter coefficients would obviously not be unique.

The FSE displays this ill-conditioning problem, because the bandwidth of the input data signal is deliberately made less than half the sampling rate. Although there will likely be noise components at all frequencies, they may be small and still not prevent some eigenvalues from being very small. Thus, we pay a price for the reduced sensitivity to sampling phase, and other benefits of the FSE, in adversely affecting the convergence properties of the equalizer. The small eigenvalues lead in particular to a problem with coefficient saturation, as will now be detailed.

9.4.2. Coefficient Drift

In any implementation, analog or digital, there will be a maximum value that a filter coefficient can assume. As one or more eigenvalues get very small, it becomes more likely that this maximum value will be inadequate, and the proper operation of the filter comes into question. This point is illustrated in Fig. 9-12. The region of allowed filter coefficients for a two-coefficient filter is usually a square centered at the origin as constrained by implementation considerations. As an eigenvalue approaches zero, the sensitivity of the mean-square error in the direction of the corresponding eigenvector to the filter coefficients becomes very small. Even after convergence of the filter coefficients to the optimum, there will continue to be a fluctuation of the filter coefficients about that optimum, with the adaptation algorithm continually bringing the coefficients back toward the optimum. The fluctuation of the coefficients in the direction of least sensitivity (the direction of the eigenvector corresponding to the minimum eigenvalue) will tend to be larger. Since the coefficients can drift relatively freely in this direction, the phenomenon is called *coefficient drift*. As the eigenvalue gets smaller, the probability of coefficient drift taking the coefficients out of the allowed region gets large.

There are several possible solutions to this problem. If the coefficients drift to the edge of the allowed region, then further adaptation could cause an overflow. A simple solution is to saturate the coefficients when they reach the edge, in effect constraining the adaptation to the allowed region. A less attractive solution is to inject a small component of white noise at the input to the adaptation algorithm, thereby increasing the smallest eigenvalue (Problem 9-14). Obviously this degrades performance, since this noise component will also appear at the slicer input.

A third solution, which also degrades performance, is to introduce a *coefficient leakage* that tends to force the coefficients toward the origin [12]. This leakage can be obtained by changing the criterion that the adaptation algorithm is minimizing to

$$E[\,|E_k|^2] + \mu\|\mathbf{c}\|^2 , \tag{9.90}$$

where μ is another small constant. Instead of simply minimizing the MSE, the criterion tries in addition to minimize the length of the coefficient vector. Like the added white noise, this criterion results in some compromise in the asymptotic mean-square error and coefficient vector, which are no longer optimal in the sense of (9.16) (Problem 9-15).

9.5. Passband Equalization

The equalization techniques that we have discussed thus far are based on a baseband channel model, in which it has been assumed that demodulation has been performed prior to equalization. It is possible to perform equalization prior to demodulation; this is called *passband equalization*. This is an important extension of the results thus far. Passband equalization is much more common than baseband equalization, for reasons that will be

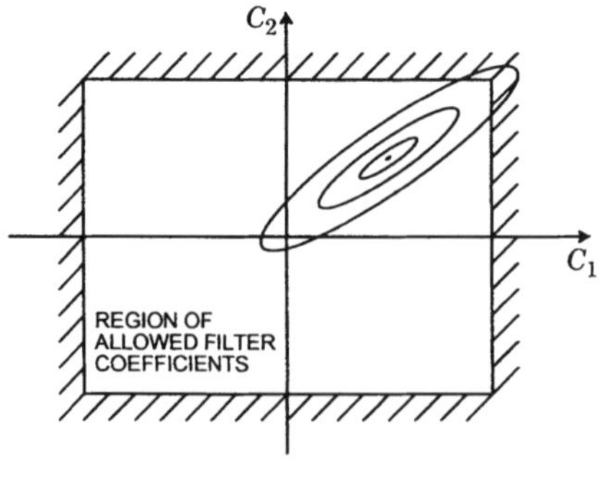

Fig. 9-12. Contours of equal MSE for a large eigenvalue spread.

elaborated in Section 15. Basically these reasons relate to the difficulty in placing the adaptive equalizer in the carrier recovery loop. Passband equalization mitigates these difficulties by allowing us to place the entire demodulation structure after the equalization. Passband equalization was proposed in a seminal paper by Falconer in 1976 [13]. Passband fractionally spaced equalization was assumed in Fig. 5-21.

The first step in deriving the passband equalizer structure is to derive a new channel model assuming that demodulation is not performed in the receiver front end. Such a channel model is shown in Fig. 9-13(a). We show the usual QAM transmitter, a channel impulse response $b(t)$, a receive filter $f(t)$, a phase splitter to generate the analytic signal, and a demodulator with frequency f_1. The baseband channel model that we have considered thus far corresponds to $f_1 = f_c$. The case where there is no demodulation corresponds to $f_1 = 0$. Other values of f_1 are possible; an important example would be where f_1 was chosen to *nominally* equal f_c, but with a small and unknown frequency offset due to the fact that f_c and f_1 are generated by independent oscillators.

If we define

$$\Delta f = f_c - f_1 , \tag{9.91}$$

then the channel model of Fig. 9-13(b) can be derived.

Exercise 9-13.
Show that the model of Fig. 9-13(b) follows from Fig. 9-13(a), where

$$h(t) = g(t) * ((b(t) * f(t))e^{-j2\pi f_c t} . \tag{9.92}$$

By a simple manipulation the model of Fig. 9-13(b) becomes

$$\Sigma_k A_k h(t - kT) e^{-j2\pi\Delta f t} = \Sigma_k \tilde{A}_k \tilde{h}(t - kT) , \tag{9.93}$$

where

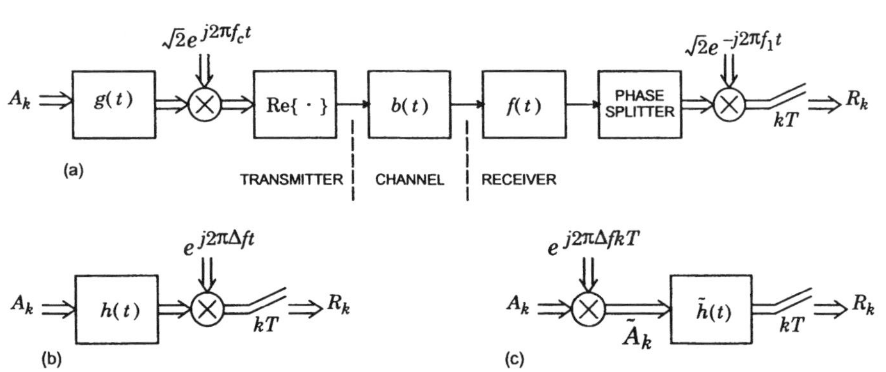

Fig. 9-13. A passband channel model. (a) The transmitter, channel, and receiver assuming that demodulation uses frequency f_1 rather than f_c. (b) Equivalent channel model consisting of baseband filter and modulator. (c) Equivalent channel model consisting of modulator and passband filter.

$$\tilde{A}_k = A_k e^{j2\pi\Delta fkT} \tag{9.94}$$

is called the *rotated data symbol* and

$$\tilde{h}(t) = h(t)e^{j2\pi\Delta ft} \tag{9.95}$$

is called the *passband channel response*. This relation corresponds to the model shown in Fig. 9-13(c). Since $h(t)$ is a baseband filter, centered at d.c., $\tilde{h}(t)$ is a passband filter, centered at frequency Δf. One way to view this model is as a modulation operation followed by passband filter, as opposed to a baseband filter followed by modulation in Fig. 9-13(b).

For purposes of equalization, we can think of the channel as consisting of a passband filter $\tilde{h}(t)$ driven by the rotated data symbols $\tilde{A}_k$. This logic leads to the passband equalizer structure shown in Fig. 9-14(a). The sampled channel output is fed through a passband equalizer $C(z)$, the purpose of which is to invert the response $\tilde{h}(t)$ to yield a good estimate of the rotated symbols $\tilde{A}_k$. If we wanted to replicate the baseband equalizer structure, we would build a slicer appropriate for the rotated symbols. The simplest way to do this is to take a slicer appropriate for the non-rotated symbols, and precede that slicer by the reverse rotation and follow it by the rotation, as shown in the figure. Since the purpose of the passband equalizer is to generate a good estimate of the rotated symbols, we must use an error signal for adaptation which is the difference between the equalizer output and the rotated data symbol, as shown. We call this error signal $\tilde{E}_k$, since it is a rotated version of the error E_k between input and output of the non-rotated slicer,

$$\tilde{E}_k = e^{j2\pi\Delta fkT}E_k \,. \tag{9.96}$$

This is made clearer by the equivalent equalizer structure shown in Fig. 9-14(b). We can think of this structure as realizing a slicer and error generator appropriate for the non-rotated data symbols, whereas the equalizer works in a rotated data symbol world. The rotators simply convert between the two worlds.

The convergence of the adaptive passband equalizer follows directly from the baseband case, since the two are equivalent except for two facts:

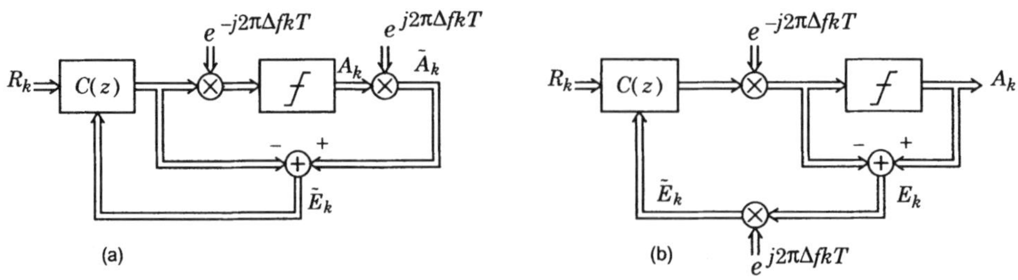

Fig. 9-14. Passband equalizer structure. a. Direct generation of rotated error signal. b. Rotation of slicer error signal.

- The passband equalizer is driven by rotated data symbols rather than the non-rotated symbols.
- The passband equalizer is inverting the passband channel response $\tilde{h}(t)$ rather than the baseband response $h(t)$.

The relationship between the statistics of the non-rotated and rotated data symbols is easily developed.

Exercise 9-14.

Show that the power spectrum of the rotated data symbols is given by

$$S_{\tilde{A}}(e^{j2\pi fT}) = S_A(e^{j2\pi(f-\Delta f)T}) . \tag{9.97}$$

Hence, if the non-rotated symbols are white (uncorrelated), then so are the rotated symbols.

The convergence properties of the passband equalizer are therefore the same as those of the baseband equalizer when the transmitted data symbols are uncorrelated.

The passband equalizer can be used in several ways:

- We can phase-lock the demodulation carrier to the incoming carrier, making $f_1 = f_c$ ($\Delta f = 0$). This raises problems to be discussed in Chapter 15, because it puts the equalizer delay into the carrier recovery loop, slowing the tracking capability of that loop.
- We can choose $f_1 = 0$ ($\Delta f = f_c$). This is the passband equalizer case.
- We can choose f_1 to nominally equal f_c, but without phase locking with a carrier recovery loop. For this case Δf will be small but unknown. We can think of this case as a baseband equalizer with a small frequency offset and phase compensation at the output. In this case $\tilde{h}(t)$ is not a passband function, although it has been shifted in frequency by a small amount.

In either of the second two cases, a carrier recovery loop driving the rotators at the equalizer output is required, as discussed in Chapter 15. Case 3 could therefore be described as a baseband equalizer with a phase-tracking carrier loop. The relative merits of the second two realizations is considered in Problem 9-17.

9.6. Further Reading

The tutorial article by Qureshi on adaptive equalization is highly recommended reading [10][11], as is the recent book by Proakis [14]. The early treatise by Lucky, Salz, and Weldon is somewhat dated but still recommended reading [3]. There are several books on the general topic of adaptive filtering [9][15][16].

We have not discussed the adaptation of a ML sequence detector (Viterbi algorithm). This is not straightforward; the issues are addressed at some length in [11].

Several methods to speed the convergence of the adaptive equalizers have been omitted. These include an alternative structure called the *lattice filter* [9][17][18], and a class of adaptation algorithms called the *least-squares (LS)* algorithms [9]. There are many versions of the LS algorithms, including those based on both transversal filter [19] and lattice filter realizations [20].

Appendix 9-A. SG Algorithm Error Vector Norm

In this appendix we approximate the expected value of the norm of the error vector $\boldsymbol{q}_k$ for an input random process consisting of zero-mean independent samples.

An update for $\boldsymbol{q}_k$ is given in (9.60). By direct calculation,

$$\|\boldsymbol{q}_{k+1}\|^2 = \boldsymbol{q}_{k+1}{}^*\boldsymbol{q}_{k+1} = \boldsymbol{q}_k{}^*\Gamma_k{}^*\Gamma_k\boldsymbol{q}_k + 2\beta\mathrm{Re}\{D_k{}^*\boldsymbol{r}_k^T\Gamma_k\boldsymbol{q}_k\} + \beta^2\,|D_k|^2\|\boldsymbol{r}_k\|^2 . \tag{9.98}$$

Since by assumption D_k is the error of an optimal fixed-coefficient equalizer, by the orthogonality principle it is uncorrelated with the vector of input samples $\boldsymbol{r}_k$. If we also assume it is independent,

$$\mathrm{E}[D_k{}^*\boldsymbol{r}_k^T\Gamma_k\boldsymbol{q}_k] = \mathrm{E}[D_k{}^*]\mathrm{E}[\boldsymbol{r}_k^T\Gamma_k\boldsymbol{q}_k] \tag{9.99}$$

and noting that $\mathrm{E}[D_k] = 0$, fortuitously the expectation of the middle term is zero.

Assuming that the error vector $\boldsymbol{q}_k$ is changing very slowly, we can assume it is a constant with respect to the expectation over the input random vector $\boldsymbol{r}_k$, and thus the mean value of the error vector norm versus time is

$$\|\boldsymbol{q}_{k+1}\|^2 = \boldsymbol{q}_k{}^*\mathrm{E}[\Gamma_k{}^*\Gamma_k]\boldsymbol{q}_k + \beta^2\mathrm{E}[\,|D_k|^2]\mathrm{E}[\|\boldsymbol{r}_k\|^2]. \tag{9.100}$$

In this expectation, we have

$$\mathrm{E}[\,|D_k|^2] = \xi_{\min} \;, \qquad \mathrm{E}[\|\boldsymbol{r}_k\|^2] = N\phi_0 \;. \tag{9.101}$$

To evaluate the expectation of the first term, we write it out explicitly,

$$\begin{aligned}\mathrm{E}[\Gamma_k{}^*\Gamma_k] &= \mathbf{I} - 2\beta\Phi + \beta^2\mathrm{E}[(\boldsymbol{r}_k^T\bar{\boldsymbol{r}}_k)(\bar{\boldsymbol{r}}_k\boldsymbol{r}_k^T)] \\ &= \mathbf{I} - 2\beta\Phi + \beta^2\mathrm{E}[\|\boldsymbol{r}_k\|^2(\bar{\boldsymbol{r}}_k\boldsymbol{r}_k^T)] ,\end{aligned} \tag{9.102}$$

where $\Phi = \phi_0\mathbf{I}$ by assumption. The last term is difficult since it involves fourth-order statistics, which we will have to approximate (unless the R_k are Gaussian, in which case exact evaluation is possible). In terms of the original input process, the (m, n) element of this matrix (indexed from the center) is

$$E[\sum_{j=-L}^{L} |R_{k+j}|^2 R_{k+m}^* R_{k+n}] \ . \tag{9.103}$$

Because of the assumed independence and zero mean of the R_k, this expectation is zero for $m \neq n$; therefore, the matrix is diagonal. When $m = n$, it reduces to

$$E[\left[\sum_{j=-L}^{L} |R_{k+j}|^2 |R_{k+m}|^2\right] = E[|R_{k+m}|^4] + \sum_{j \neq m} E[|R_{k+j}|^2]E[|R_{k+m}|^2]$$

$$= \eta_a + (N-1)\phi_0^2 , \tag{9.104}$$

where

$$\eta_a = E[|R_{k+m}|^4] \ . \tag{9.105}$$

We will approximate η_a as the square of the second moment,

$$\eta_a \approx \phi_0^2 \ . \tag{9.106}$$

The second term is precisely $(N-1)\phi_o^2$; hence, the entire sum is approximately $N\phi_0^2$. We can get some idea of the accuracy of this approximation from considering the Gaussian example.

Exercise 9-15.
Assume that R_k is a complex-valued Gaussian random variable with independent identically-distributed real and imaginary parts. Show that

$$\eta_a = 2\phi_0^2 \ . \tag{9.107}$$

Hence the approximation above has the correct dependence on ϕ_0 but is off by a factor of two.

Substituting the approximation of (9.106) into (9.104), we get

$$E[\Gamma_k^* \Gamma_k] = (1 - 2\beta\phi_0 + \beta^2 N\phi_0^2)\mathbf{I} , \tag{9.108}$$

and the result of (9.69) is established. The error in this approximation will generally be small when N is reasonably large. The approximation is necessary since the statistics are governed by the very complicated intersymbol interference, and the fourth moment is therefore very difficult to evaluate explicitly.

Problems

Problem 9-1. For a linear predictor of input process R_k, define a vector of prediction coefficients

$$\mathbf{f} = [f_1, \ldots, f_N]^T \tag{9.109}$$

and a vector of past data samples,

$$\mathbf{r}_k = [R_{k-1}, \ldots, R_{k-N}]^T . \tag{9.110}$$

Then the prediction error of an N-th order predictor is

$$E_k = R_k - \mathbf{f}^T \boldsymbol{r}_k. \tag{9.111}$$

Find the optimal set of coefficients and the resultant minimum MSE.

Problem 9-2. Rederive (9.16) directly from orthogonality principle (9.26).

Problem 9-3. Derive the orthogonality principle for a linear predictor.

Problem 9-4. This problem will attempt to make plausible the relationship between the eigenvalues of an autocorrelation matrix and the power spectrum displayed in (9.47) Let $\boldsymbol{\Phi}$ be a $(2L + 1) \times (2L + 1)$ autocorrelation matrix, and let the components of an eigenvector of this matrix be

$$\boldsymbol{v} = [v_{-L}, v_{-L+1}, \ldots, v_L]^T \ . \tag{9.112}$$

(a) Show that the eigenvector and associated eigenvalue λ satisfies the relationship

$$\sum_{i=-L}^{L} \phi_{j-i} v_i = \lambda v_j \,, \quad -L \le j \le L \ . \tag{9.113}$$

(b) Let $L \to \infty$ and take the Fourier Transform to show that

$$S(e^{j\theta}) V(e^{j\theta}) = \lambda V(e^{j\theta}) \,, \tag{9.114}$$

where $S(e^{j\theta})$ is the power spectrum defined by (9.46) and

$$V(e^{j\theta}) = \sum_{k=-\infty}^{\infty} v_k e^{-jk\theta} \ . \tag{9.115}$$

(c) Where $S(e^{j\theta})$ is a single valued function (that is, it doesn't assume the same value at two different frequencies), argue that the infinite eigenvectors have components that are samples of a complex exponential ($e^{j\theta_0 k}$), with corresponding eigenvalues equal to the power spectrum at the same frequency.

(d) Use these results to argue the validity of (9.47).

Problem 9-5. Consider an input wide-sense stationary random process with autocorrelation function $\phi_m = \alpha^{|m|}$.

(a) Find the power spectrum of this random process.

(b) Find the asymptotic minimum and maximum eigenvalues of the autocorrelation matrix.

(c) Find, as a function of α, the eigenvalues and eigenvectors of the 2×2 autocorrelation matrix.

(d) Find the eigenvalue spread of the autocorrelation matrix as predicted by approximate relation (9.47) and compare to the results of c.

(e) Find, as a function of α and as $N \to \infty$, the step size β and resulting dominant mode of convergence of the MSEG algorithm. Interpret this result intuitively.

Problem 9-6. Show that for the MSEG algorithm, the error vector is given by

$$\mathbf{c}_j - \mathbf{c}_{\text{opt}} = \sum_{i=1}^{N} (1 - \beta\lambda_i)^j (\boldsymbol{v}_i^T (\mathbf{c}_0 - \mathbf{c}_{\text{opt}})) \boldsymbol{v}_i \,, \tag{9.116}$$

and interpret this equation.

Problem 9-7. Using the results of Problem 9-6 and (9.38) show that the excess MSE is given, for the MSEG algorithm, by the relation

$$\mathrm{E}[E_k^2] - \xi_{\min} = \sum_{i=1}^{N} \lambda_i (1-\beta\lambda_i)^{2j} (\boldsymbol{v}_i^* (\boldsymbol{c}_0 - \boldsymbol{c}_{\text{opt}}))^2 , \tag{9.117}$$

and interpret this equation.

Problem 9-8. Consider the dominant mode of the MSEG algorithm.

(a) The excess MSE as a function of time expressed in decibels is approximately given by $\gamma_1 - \gamma_2 \cdot j$. Find the constants γ_1 and γ_2.

(b) Evaluate these constants for the particular case where β is very small, and discuss the tradeoff between speed of convergence and step size for this case. Thus, the MSE expressed in decibels decreases linearly with time.

Problem 9-9. For an input process Y_k with mean value $\mu \neq 0$, show that the minimum-MSE first-order predictor is

$$\hat{Y}_k = \rho Y_{k-1} + (1-\rho)\mu , \tag{9.118}$$

where ρ is a normalized covariance, defined as

$$\rho = \frac{\phi_1 - \mu^2}{\phi_0 - \mu^2} . \tag{9.119}$$

Problem 9-10. A real-valued input WSS process has power spectral density

$$\Phi(z) = \frac{A}{(1-\alpha z)(1-\alpha z^{-1})} , \quad 0 < \alpha < 1 \tag{9.120}$$

and the region of convergence includes the unit circle.

(a) Find the autocorrelation function ϕ_m.

(b) Find the predictor coefficients for a minimum MSE N-th order predictor.

Problem 9-11. In the MSEG algorithm, in place of a fixed step size β, use a variable step size β_j in the determination of $\boldsymbol{c}_j$.

(a) Show that the error vector at iteration j is given by

$$\boldsymbol{q}_j = \sum_{l=1}^{n} \prod_{i=1}^{j} (1-\beta_i\lambda_l)(\boldsymbol{v}_l^* \boldsymbol{q}_0) \cdot \boldsymbol{v}_l . \tag{9.121}$$

(b) Show that you can force the error vector to zero in precisely N iterations by proper choice of the step sizes assuming you know the eigenvalues of the matrix (but that is all you need to know).

Problem 9-12. For the signal power estimation algorithm of (9.78), assume that R_k is a real-valued zero-mean white Gaussian process.

(a) Show that ${\sigma_k}^2$ is an unbiased estimator of the signal power.

(b) Find the variance of this estimate. Interpret how this variance depends on step size α.

Problem 9-13. Assume an FSE with sampling rate equal to twice the symbol rate.

(a) Write the equations describing the input-output relationship of the FSE.

(b) Find the SG algorithm that adapts this equalizer structure.

Problem 9-14. Assume that a white noise component with variance σ^2 is added to the input of an adaptive filter. Quantify the effect of this noise on the coefficient drift as follows:

(a) What is the new set of eigenvalues?

(b) What is the new eigenvalue spread? What is the effect of σ?

(c) For the MSE solution, what is the effect of σ on the MSE?

Problem 9-15. Suppose the MSE criterion is modified to minimize (9.90). Find the coefficient vector $\mathbf{c}_\mu$ which minimizes this quantity, and show that the error between this solution and the coefficient vector $\mathbf{c}_{\text{opt}}$ of (9.16) is given by

$$\mathbf{c}_\mu - \mathbf{c}_{\text{opt}} = -\mu \sum_{i=1}^{n} \frac{\mathbf{v}_i^{*\prime}\boldsymbol{\alpha}}{\lambda_i(\lambda_i + \mu)} \mathbf{v}_i \,, \tag{9.122}$$

and that the resulting excess MSE of (9.8) is given by

$$E[{E_k}^2] - \xi_{\min} = \mu^2 \sum_{i=1}^{n} \frac{|\mathbf{v}_i^{*\prime}\boldsymbol{\alpha}|^2}{\lambda_i(\lambda_i + \mu)^2} \,. \tag{9.123}$$

How does this MSE increase as we vary μ?

Problem 9-16. Continuing Problem 9-15:

(a) Find the MSE gradient algorithm which iteratively minimizes this error.

(b) Find the criterion on the step size of this algorithm which guarantees stability.

(c) Find the step size which maximizes the rate of convergence.

(d) Investigate how the maximum rate of convergence can be altered by the choice of μ, particularly where the eigenvalue spread is large. Discuss the tradeoff between "excess MSE" and rate of convergence.

(e) How do these results apply to the stochastic gradient algorithm of (9.53)?

Problem 9-17. Consider a voiceband data modem with the following characteristics: carrier frequency 1800 Hz, symbol rate 2400 Hz, excess bandwidth 10%. Assume that all sampling rates used in the receiver are an integer multiple of the symbol rate and are chosen to be as small as possible. Draw a block diagram of a receiver using a fractionally-spaced linear equalizer, labeling the sampling rates at each point.

(a) Assume a baseband equalizer with phase-locked demodulation at the receiver front end.

(b) Assume a passband equalizer.

(c) Assume a baseband equalizer with a demodulator at the front end, but not phase-locked to the receive carrier.

(d) Which of these realizations appears to be more attractive? Why?

Problem 9-18. For the same conditions as in Problem 9-17b, draw a block diagram of a DFE with a passband precursor equalizer labeling the sampling rates at each point.

References

1. R. W. Lucky, "Automatic Equalization for Digital Communications," *Bell System Technical Journal*, Vol. 44, pp. 547-588 (Apr. 1965).
2. R. W. Lucky and H. R. Rudin, "An Automatic Equalizer for General-Purpose Communication Channels," *Bell System Technical Journal*, Vol. 46, p. 2179 (Nov. 1967).
3. R. W. Lucky, J. Salz, and E. J. Weldon, Jr., *Principles of Data Communication*, McGraw-Hill Book Co., New York (1968).
4. B. Widrow and M. Hoff, Jr., "Adaptive Switching Circuits," *IRE WESCON.*, pp. 96-104 (1960).
5. R. M. Gray, "On the Asymptotic Eigenvalue Distribution of Toeplitz Matrices," *IEEE Trans. on Information Theory*, Vol. IT-18, pp. 725-730 (Nov. 1972).
6. U. Grenander and G. Szego, *Toeplitz Forms and Their Applications*, UC Press (1958).
7. R. Bellman, *Introduction to Matrix Analysis*, McGraw-Hill, New York (1960).
8. G. Ungerboeck, "Theory on the Speed of Convergence in Adaptive Equalizers for Digital Communication," *IBM J. Res. and Develop.*, pp. 546-555 (Nov. 1972).
9. M. L. Honig and D. G. Messerschmitt, *Adaptive Filters: Structures, Algorithms, and Applications*, Kluwer Academic Publishers, Boston (1984).
10. S. U. H. Qureshi and G. D. Forney, Jr., "Performance Properties of a T/2 Equalizer," *NTC '77 Proceedings*, 1977.
11. S. U. H.Qureshi, "Adaptive Equalization," pp. 640 in *Advanced Digital Communications Systems and Signal Processing Techniques*, ed. K. Feher, Prentice-Hall, Englewood Cliffs, N.J. (1987).
12. R. D. Gitlin, H. C. Meadors, Jr., and S. B. Weinstein, "The Tap-Leakage Algorithm: An Algorithm for the Stable Operation of a Digitally Implemented, Fractionally Spaced Adaptive Equalizer," *Bell System Technical Journal*, Vol. 61 (8), (Oct. 1982).
13. D. D. Falconer, "Jointly Adaptive Equalization and Carrier Recovery in Two-Dimensional Digital Communication Systems," *Bell System Technical Journal*, Vol. 53 (3), (March 1976).
14. J. G. Proakis, *Digital Communications*, Fourth Edition, McGraw-Hill, New York (2001).

15. C. F. N. Cowan and P. M. Grant, *Adaptive Filters*, Prentice-Hall, (1985).

16. B. Widrow and S. D. Stearns, *Adaptive Signal Processing*, Prentice Hall, (1985).

17. J. Makhoul, "Stable and Efficient Lattice Methods for Linear Prediction," *IEEE Trans. on ASSP*, Vol. ASP-25, pp. 423-428 (Oct. 1977).

18. E. H. Satorius and S. T. Alexander, "Channel Equalization Using Adaptive Lattice Algorithms," *IEEE Trans. on Communications*, Vol. COM-27, p. 899 (Jun. 1979).

19. D. D. Falconer and L. Ljung, "Application of Fast Kalman Estimation to Adaptive Equalization," *IEEE Trans. on Communications*, Vol. COM-26, pp. 1439-1446 (Oct. 1978).

20. B. Friedlander, "Lattice Filters for Adaptive Processing," *Proc. IEEE*, Vol. 70, No. 8, pp. 829-867, 1982.

10

MIMO Communications

Prior chapters were concerned with communication across a *single-input single-output (SISO)* channel for which the channel input and channel output were scalar-valued signals. In this chapter we study the problem of communicating across a *multiple-input multiple-output (MIMO)* channel, for which the channel input and output are vector-valued signals. In doing so, this chapter adds another impediment to the mix. Not only must a receiver contend with ISI and noise, but also with *interference* between the inputs. Depending on the application, this interference may be referred to as co-channel interference, adjacent-channel interference, crosstalk, or multiuser interference. Whatever its name, the presence of this interference is what distinguishes this chapter from previous chapters.

At one level, a MIMO channel is a straightforward extension of an ISI channel to higher dimensions. Indeed, as we will see, many useful communication strategies for MIMO channels are obviously extensions of ISI equalization strategies to the case of a matrix-valued channel. However, this view can be misleading because it fails to recognize that MIMO channels possess unique characteristics that have no counterpart in ISI channels. For example, because the different inputs to a MIMO channel can represent n different users, it follows that:

- the interfering symbols might have much higher energy than the desired symbols
- there are $n!$ ways to arrange the order in which the n input symbols are detected

In contrast, neither is possible in scalar ISI channels, where instead, the interfering symbols have the same energy as the desired symbols, and where an equalizer detects the symbols in the same order in which they were transmitted. As we will see, MIMO channels have many such unique properties that call for new transmission and detection strategies.

As special cases, this chapter also considers *multiple-input single-output (MISO)* and *single-input multiple-output (SIMO)* channels, as can arise when either the transmitter or receiver uses an antenna array, for example. Antenna arrays are commonly used in wireless channels as a means for providing diversity against multipath fading, and thus MIMO channels frequently arise in the context of wireless channels. For this reason, there will be more of a focus on wireless channels in this chapter than in previous chapters.

There are two classes of applications for which the MIMO model arises: *single-user* channels and *multiuser* channels. The class of multiuser channels can be further decomposed into two types: *multiple-access* channels and *broadcast* channels.

- In a *single-user* MIMO channel, the emitters are collocated at a single transmitter, and the sensors are collocated at a single receiver, facilitating coordination at both ends. A single-user system is also known as a *point-to-point* system. The n inputs originate from the same source and are often highly correlated. The receiver aims to detect all inputs simultaneously.
- In a *multiple-access* channel, or multipoint-to-point channel, there are multiple transmitters (or *users*) that are geographically isolated from one another and for which there is little or no coordination. Specifically, each user has its own distinct message to convey, so that the symbols emitted from different transmitters are statistically independent. A receiver whose goal is to detect all inputs is a *centralized multiuser detector*, whereas a receiver whose goal is to detect only one of the inputs is *decentralized.* In a cellular setting, the uplink channel (from portables to the basestation) is an example of a multiple-access channel. The basestation receiver is often centralized.
- In a *broadcast* channel, or point-to-multipoint channel, there are multiple receivers (or *users*) that are geographically distinct and for which there is no coordination. The transmitter, which consists of one or more coordinated emitters, sends distinct messages to the multiple users simultaneously. The downlink (from the basestation to the portables) is an example of a broadcast channel. The receivers are generally decentralized.

Example 10-1.

A home gains access to the telephone network through a twisted pair of wires. Before the central office, the twisted pairs from neighboring homes are typically bundled into a single cable with a protective sheath, as illustrated in Fig. 10-1(a). The mutual interference among the twisted pairs within the cable leads to a single-user n-input n-output channel model.

Example 10-2.

As shown in Fig. 10-1(b), a wireless transmitter with an array of n antenna elements and a receiver array with m elements constitute a single-user MIMO channel with n inputs and m outputs.

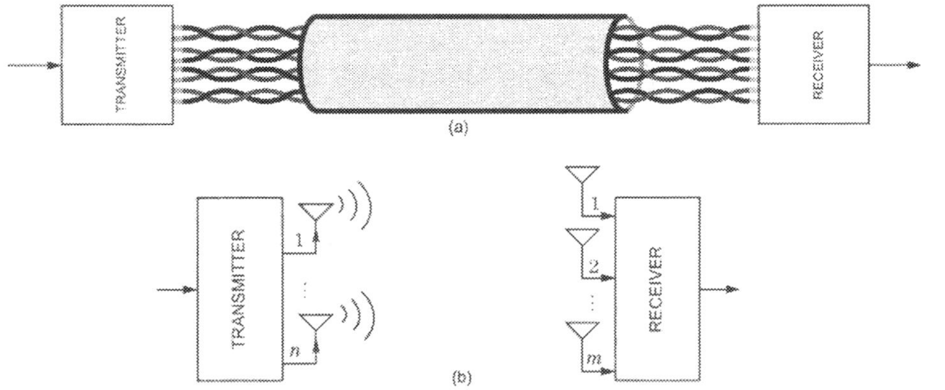

Fig. 10-1. Examples of single-user MIMO channels: (a) Mutual interference or crosstalk among the twisted pairs bundled within a single cable leads to a MIMO channel model; (b) a wireless transmitter with an array of n antennas communicating to a receiver with an array of m antennas leads to an n-input m-output channel model.

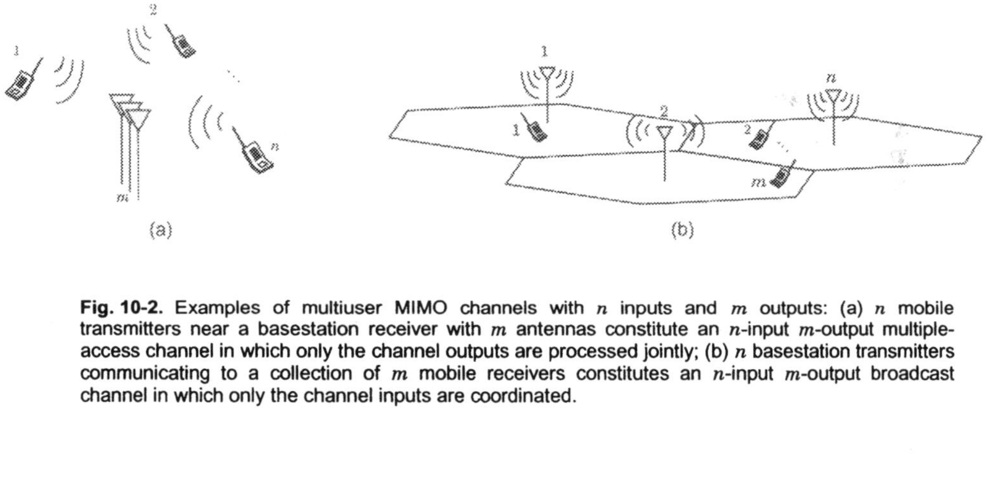

Fig. 10-2. Examples of multiuser MIMO channels with n inputs and m outputs: (a) n mobile transmitters near a basestation receiver with m antennas constitute an n-input m-output multiple-access channel in which only the channel outputs are processed jointly; (b) n basestation transmitters communicating to a collection of m mobile receivers constitutes an n-input m-output broadcast channel in which only the channel inputs are coordinated.

Example 10-3. ————————————————————

The wireless channels illustrated in Fig. 10-2(a) and Fig. 10-2(b) are examples of multiple-access and broadcast multiuser channels, respectively, that take a MIMO model.

For our purposes, the three categories of MIMO channels are distinguished by the amount of coordination at each end: a multiple-access channel coordinates only at the receiver, a broadcast channel coordinates only at the transmitter, and a single-user channel coordinates at both ends. Throughout this chapter, whenever possible, we will focus our discussion on a general MIMO channel model that encompasses all three categories.

10.1. Basics of MIMO Systems

This chapter considers continuous-time and discrete-time signals that are vector-valued, as well as filters that are matrix-valued. In this section we briefly summarize the notation and properties of such signals and systems.

10.1.1. Deterministic Signals

In this chapter, column vectors will be bold and lower case, such as $\boldsymbol{a}$, and matrices will be bold and uppercase, such as $\mathbf{A}$. We use the notation $\mathbf{A}^T$ to denote a transpose, $\bar{\mathbf{A}}$ to denote a complex conjugate, and $\mathbf{A}^*$ to denote a conjugate (or Hermitian) transpose, so that $\mathbf{A}^* = \bar{\mathbf{A}}^T$. When $\mathbf{A}$ is invertible, we use $\mathbf{A}^{-*}$ as a shorthand for $(\mathbf{A}^{-1})^*$. The rank of a matrix is the number of linearly independent columns. A *real* matrix satisfies $\bar{\mathbf{A}} = \mathbf{A}$, a *symmetric* matrix satisfies $\mathbf{A}^T = \mathbf{A}$, and a *Hermitian* matrix satisfies $\mathbf{A}^* = \mathbf{A}$.

A continuous-time vector-valued signal $\boldsymbol{s}(t) = [s_1(t), \ldots, s_n(t)]^T$ of dimension n is a column vector whose components are continuous-time scalar signals. A linear-time-invariant n-input m-output filter is characterized by an impulse response matrix $\mathbf{H}(t)$ of dimension $m \times n$, whose component $h_{ij}(t)$ in row i and column j is the response at the i-th output to an impulse at the j-th input, assuming all other inputs are silent. The output of a MIMO filter with impulse response matrix $\mathbf{H}(t)$ and input $\boldsymbol{s}(t)$ is given by the convolution integral:

$$\boldsymbol{y}(t) = \int_{-\infty}^{\infty} \mathbf{H}(\tau)\boldsymbol{s}(t-\tau)\,d\tau\,. \tag{10.1}$$

We define the Fourier transform of a vector-valued signal $\boldsymbol{s}(t)$ as $\mathbf{S}(f) = [S_1(f), \ldots, S_n(f)]^T$, where $S_i(f)$ is the Fourier transform of $s_i(t)$. Similarly, the Fourier transform of $\mathbf{H}(t)$ is $\mathbf{H}(f)$, where the component $H_{ij}(f)$ of $\mathbf{H}(f)$ in row i and column j is the Fourier transform of $h_{ij}(t)$. Although we will use the same notation $\mathbf{H}(\cdot)$ for both the time and frequency domain, the ambiguity is easily resolved by considering the context and the argument of $\mathbf{H}(\cdot)$. Taking the Fourier transform of both sides of (10.1) yields $\mathbf{Y}(f) = \mathbf{H}(f)\mathbf{S}(f)$.

Discrete-time signals and systems will use an analogous notation. A discrete-time vector-valued signal $\boldsymbol{x}_k = [x_k^{(1)}, \ldots, x_k^{(n)}]^T$ of dimension n is a vector whose i-th component is a scalar sequence $x_k^{(i)}$. An LTI n-input m-output filter is characterized by a matrix-valued impulse response $\mathbf{H}_k$ of dimension $m \times n$, whose component $h_k^{(i,j)}$ in row i and column j is the response at the i-th output to an impulse δ_k at the j-th input. The matrix sequence $\mathbf{H}_k$ will be referred to as an impulse response, even though it would be more accurate to call it a matrix of impulse responses. The output of a MIMO filter with impulse response $\mathbf{H}_k$ and input $\boldsymbol{x}_k$ is:

$$\boldsymbol{y}_k = \sum_{l=-\infty}^{\infty} \mathbf{H}_l \boldsymbol{x}_{k-l}\,. \tag{10.2}$$

The Z-transform of $\boldsymbol{x}_k$ is $\mathbf{X}(z) = [X_1(z), \ldots X_n(z)]^T$, where $X_i(z) = \sum_{k=-\infty}^{\infty} x_k^{(i)} z^{-k}$ is the Z-transform of $x_k^{(i)}$. Rather than viewing $\mathbf{X}(z)$ as a vector of Z transforms, we can equivalently view it as the Z-transform of the vector-valued sequence $\boldsymbol{x}_k$, using:

$$\mathbf{X}(z) = \sum_{k=-\infty}^{\infty} \boldsymbol{x}_k z^{-k}. \tag{10.3}$$

When $z = e^{j\theta}$ or $z = e^{j2\pi fT}$, $\mathbf{X}(z)$ reduces to the Fourier transform $\mathbf{X}(e^{j\theta})$ or $\mathbf{X}(e^{j2\pi fT})$. Similarly, the Z-transform of $\mathbf{H}_k$ is $\mathbf{H}(z)$, where the component $H_{ij}(z)$ of $\mathbf{H}(z)$ in row i and column j is the Z-transform of $h_k^{(i,j)}$, or equivalently:

$$\mathbf{H}(z) = \sum_{k=-\infty}^{\infty} \mathbf{H}_k z^{-k}. \tag{10.4}$$

Taking the Z transform or Fourier transform of (10.2) yields $\mathbf{Y}(z)=\mathbf{H}(z)\mathbf{X}(z)$ or $\mathbf{Y}(e^{j\theta})=\mathbf{H}(e^{j\theta})\mathbf{X}(e^{j\theta})$. For this reason, $\mathbf{H}(z)$ is referred to as the *transfer function* of the MIMO filter.

A discrete-time MIMO filter with impulse response $\mathbf{H}^*_{-k}$ is said to be *matched* to the filter with impulse response $\mathbf{H}_k$. If the original filter has dimension $m \times n$, its matched filter has dimension $n \times m$. The Z-transform of the matched filter $\mathbf{H}^*_{-k}$ is $\mathbf{H}^*(1/z^*)$, as is easily verified from (10.4). The transfer function $\mathbf{H}^*(1/z^*)$ is said to be the *parahermitian conjugate* of $\mathbf{H}(z)$. A transfer function $\mathbf{H}(z)$ that satisfies $\mathbf{H}^*(1/z^*) = \mathbf{H}(z)$ is said to be *parahermitian* [1]. Clearly, a matrix can only be parahermitian if it is square. If $\mathbf{H}(z)$ is parahermitian then $\mathbf{H}(e^{j\theta})$ is Hermitian for all θ.

A MIMO filter is *stable* if its transfer function is stable. A transfer function $\mathbf{H}(z)$ is stable if bounded inputs produce bounded outputs, which is true if and only if the entries $H_{ij}(z)$ are stable; *i.e.*, their regions of convergence include the unit circle.

The rank of a transfer function $\mathbf{H}(z)$ will generally depend on the value of z. For a rational MIMO transfer function (whose elements are rational scalar transfer functions), the rank of $\mathbf{H}(z)$ will be the same – the so-called *normal rank* – for all but a finite set of points in the z plane. The largest possible value for the normal rank is the minimum of m and n.

The notion of poles and zeros that is so useful for scalar filters does not readily extend to MIMO filters. A *transmission zero* of a MIMO transfer function $\mathbf{H}(z)$ can be defined as a value of z for which the rank of $\mathbf{H}(z)$ is smaller that its normal rank. For the special case when $\mathbf{H}(z)$ is square with full normal rank, the transmission zeros reduce to the conventional zeros of the scalar $\det\mathbf{H}(z)$. Observe that a transmission zero at $z = z_0$ does not imply that $\mathbf{H}(z_0) = \mathbf{0}$; it only implies that $\mathbf{H}(z_0)$ has a smaller rank than at other values of z, so that there exists a constant vector $\boldsymbol{u}$ such that an input of the form $z_0^k\boldsymbol{u}$ produces a zero output $\mathbf{0}$. In this sense, a transmission zero has associated with it a *direction* (or more precisely, a subspace) as well as a frequency.

There is no direct relationship between a transmission zero of $\mathbf{H}(z)$ and the zeros of the components of $\mathbf{H}(z)$.

Example 10-4.

The rank of

$$\mathbf{H}(z) = \begin{bmatrix} 1 & z \\ z & 1 \end{bmatrix}$$

is two for all values of z in the complex plane except at $z = \pm 1$, where it has rank one. Hence, its normal rank is two, and its transmission zeros are at $z = \pm 1$. In contrast, the components of $\mathbf{H}(z)$ have a zero only at $z = 0$.

The poles of $\mathbf{H}(z)$ are defined as the union of all of the poles of $H_{ij}(z)$. Unlike the scalar case, a particular value of z can be both a pole and a transmission zero, without cancellation!

Example 10-5.

The stable transfer function $\mathbf{H}(z) = \begin{bmatrix} 2+z^{-1} & 3+z^{-1} \\ 1 & \frac{1}{2+z^{-1}} \end{bmatrix}$ has both a pole and a transmission zero at $z = -1/2$.

Space-Time Causality

A *causal* system is a system that does not require knowledge of future inputs to produce a given output. There are three ways to define causality for a MIMO filter, depending on how "future" is defined. All three are valid extensions of causality from scalar sequences to vector sequences, in the sense that all reduce to the conventional definition when the filter has a single input and a single output.

Given a MIMO filter with input $\boldsymbol{x}_k$ and output $\boldsymbol{y}_k$, one way to extend causality to MIMO systems is to define a causal system as one in which the output $\boldsymbol{y}_k$ depends only on $\{\boldsymbol{x}_k, \boldsymbol{x}_{k-1}, \boldsymbol{x}_{k-2}, \ldots\}$. For an LTI system, this is equivalent to the condition that $\mathbf{H}_k = \mathbf{0}$ for $k < 0$. This definition does not order the components of a particular $\boldsymbol{x}_k$, but instead implicitly assumes that they occur simultaneously. The past and future with respect to a particular vector $\boldsymbol{x}_k$ are distinguished in Fig. 10-3(a).

A less obvious but equally valid definition for causality would order the components of a vector sequence so that space takes precedence over time, according to $\{\{\ldots x_{-1}^{(1)}, x_0^{(1)}, x_1^{(1)} \ldots\}, \{\ldots x_{-1}^{(2)}, x_0^{(2)}, x_1^{(2)} \ldots\}, \ldots \{\ldots x_{-1}^{(n)}, x_0^{(n)}, x_1^{(n)} \ldots\}\}$. In other words, we could declare that $x_k^{(i)}$ comes before $x_l^{(j)}$ when $i < j$, regardless of k and l, and that $x_k^{(i)}$ comes before $x_l^{(i)}$ when $k < l$. From the perspective of a particular component $x_k^{(i)}$, the past and future are illustrated in Fig. 10-3(b). With this ordering, all of the matrix coefficients $\{\mathbf{H}_k\}$ of a causal LTI filter would be lower triangular, and the entire transfer function $\mathbf{H}(z)$ of a causal

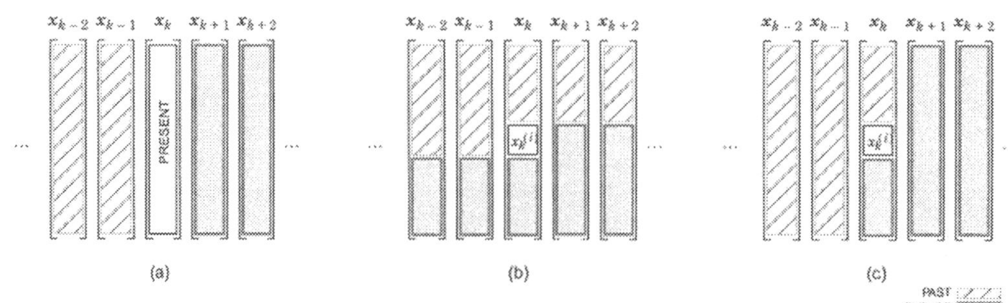

Fig. 10-3. Three ways to order the elements of a vector-valued sequence.

filter would be lower triangular. Unlike the diagonal elements, the elements below the diagonal would not necessarily be causal in the scalar sense. This version of causality can be useful in some theoretical treatments of multiuser detection [2].

The most convenient definition of causality for our purposes is to order the set of all components $\{x_k^{(i)}$: for all i, $k\}$ so that time takes precedence over space. In particular, from the perspective of $x_k^{(i)}$, the "past" includes not only $\{\boldsymbol{x}_{k-1}, \boldsymbol{x}_{k-2} \dots \}$ but also $\{x_k^{(1)}, \dots x_k^{(i-1)}\}$, and the "future" includes not only $\{\boldsymbol{x}_{k+1}, \boldsymbol{x}_{k+2} \dots \}$ but also $\{x_k^{(i+1)}, \dots x_k^{(n)}\}$. These relationships are sketched in Fig. 10-3(c). With this ordering, the transfer function $\mathbf{H}(z)$ of a causal LTI filter is generally not lower triangular, although the zero-th coefficient $\mathbf{H}_0$ is. Furthermore, all components of $H_{ij}(z)$ will be causal in the scalar sense. This leads to the following definition for space-time causality, which we adopt in the remainder of the chapter.

Definition. $\mathbf{H}(z)$ is *space-time causal* if and only if (1) $\mathbf{H}_k = \mathbf{0}$ for $k < 0$, and (2) $\mathbf{H}_0$ is lower triangular.

The value of this definition will become clear later in the chapter, when we describe decision-feedback detectors. Suffice it to say that this definition leads to a unified view of decision-feedback detection that encompassed DFE for scalar ISI channels and successive interference cancellation for multiuser channels.

Another way to understand the triangular constraint is to consider a *scalar* (time-varying) filter sandwiched between a parallel-to-serial converter and a serial-to-parallel converter, as shown below, where the input and output switches are synchronized to move in unison:

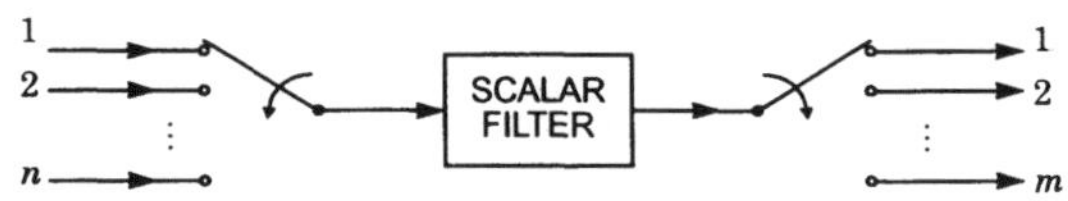

This arrangement could be used to emulate an arbitrary LTI MIMO system. However, if we constrain the scalar filter to be *causal*, then this arrangement can only emulate a *space-time causal* system for which $\mathbf{H}_0$ is lower triangular.

Example 10-6.

The transfer function $\mathbf{H}_1(z) = \begin{bmatrix} 1 & z^{-1} \\ 1 & 1 \end{bmatrix}$ is space-time causal, but $\mathbf{H}_2(z) = \begin{bmatrix} 1 & 1 \\ z^{-1} & 1 \end{bmatrix}$ is not.

A transfer function $\mathbf{H}(z)$ is *strictly space-time causal* if $\mathbf{H}(z)$ is space-time causal and $\mathbf{H}_0$ is *strictly* lower triangular, with zeros on the diagonal as well as above. The output of a strictly space-time-causal filter depends only on the past. A transfer function $\mathbf{H}(z)$ is space-time *anticausal* if its parahermitian conjugate (or equivalently, its matched filter) $\mathbf{H}^*(1/z^*)$ is space-time causal.

If $\mathbf{H}(z)$ is space-time causal then $H_{ij}(z)$ will be causal for $i \geq j$ and $H_{ij}(z)$ will be strictly causal for $i < j$. It follows that if $\mathbf{H}(z)$ is space-time causal and stable, the poles of all of its components $\{H_{ij}(z)\}$ will be inside the unit circle.

A space-time causal or anticausal filter $\mathbf{H}(z)$ is *monic* if $\mathbf{H}_0$ has ones on the diagonal. The transfer function $\mathbf{H}_1(z)$ of Example 10-6 is monic. A constant triangular matrix $\mathbf{H}_0$ is also said to be monic when it has ones on the diagonal. So a space-time causal or anticausal $\mathbf{H}(z)$ is monic if $\mathbf{H}_0$ is monic.

Example 10-7.

Consider the linear feedback system shown in Fig. 10-4(a). Just as in the scalar case, such a system can be implemented only when the feedback filter $\mathbf{B}(z)$ is strictly causal. If the ordering of Fig. 10-3(a) were used, so that all components of a particular output $\boldsymbol{y}_k$ were calculated simultaneously, then we would require that $\mathbf{B}_k = \mathbf{0}$ for $k \le 0$. If the ordering of Fig. 10-3(b) were used, so that the first output sequence $\{y_k^{(1)}\}$ were computed in its entirety (for all $k \in \{-\infty, \ldots \infty\}$) before computing any element of the second output sequence $\{y_k^{(2)}\}$, then we would require that $\mathbf{B}_k$ be strictly lower triangular for all $k \in \{-\infty, \ldots \infty\}$. Equivalently, we would require that $\mathbf{B}(z)$ be strictly lower triangular, with diagonal elements that are strictly causal in the scalar sense, but whose elements below the diagonal would not necessarily be causal in the scalar sense. But we will be using the ordering of Fig. 10-3(c), so that $\{y_k^{(1)}, \ldots y_k^{(i-1)}\}$ as well as $\{\boldsymbol{y}_{k-1}, \boldsymbol{y}_{k-2} \ldots\}$ are available when computing $y_k^{(i)}$; in this case, we require that $\mathbf{B}(z)$ be strictly space-time causal in the sense that $\mathbf{B}_k = \mathbf{0}$ for $k < 0$ and $\mathbf{B}_0$ be strictly lower triangular. With this constraint, the linear feedback system of Fig. 10-4(a) is implementable, and the transfer function mapping $\boldsymbol{x}_k$ to $\boldsymbol{y}_k$ is given by:

$$\mathbf{C}(z) = (\mathbf{I} + \mathbf{B}(z))^{-1}\mathbf{F}(z) \, . \tag{10.5}$$

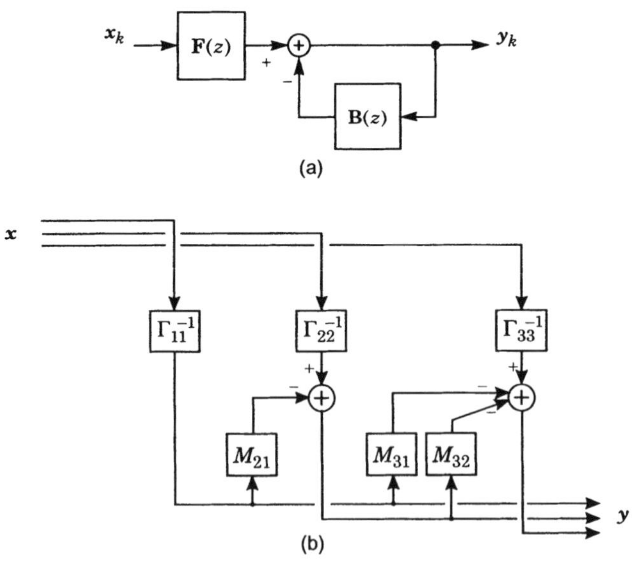

Fig. 10-4. A general linear feedback system is shown in (a). A special case is shown in (b), where $\mathbf{F}(z) = \Gamma^{-1}$ and $\mathbf{B}(z) = \mathbf{M} - \mathbf{I}$; this system implements the inverse of a lower-triangular matrix $\mathbf{G}$, where Γ is the diagonal part of $\mathbf{G}$ and $\mathbf{M} = \Gamma^{-1}\mathbf{G}$ is monic and lower triangular. This is the *backsubstitution* method for inverting a triangular matrix.

In particular, by choosing $\mathbf{B}(z) = \mathbf{M}(z) - \mathbf{I}$ and $\mathbf{F}(z) = \Gamma^{-1}$, we can use the feedback system to implement $\mathbf{G}(z)^{-1}$, the inverse of a space-time causal filter $\mathbf{G}(z) = \Gamma\mathbf{M}(z)$, where Γ is the diagonal part of $\mathbf{G}_0$, and where $\mathbf{M}(z)$ is monic and space-time causal. A special case is shown in Fig. 10-4(b), where the inverse of a constant 3×3 triangular matrix is implemented using linear feedback. This method for implementing the inverse of a triangular matrix is known as back substitution.

Definition. A square filter $\mathbf{H}(z)$ is *minimum phase* if and only if both $\mathbf{H}(z)$ and $\mathbf{H}^{-1}(z)$ are space-time causal and stable.

10.1.2. Random Signals and MIMO Systems

A random vector sequence $\boldsymbol{x}_k = [x_k^{(1)}, \ldots, x_k^{(n)}]^T$ of dimension n is a vector of scalar random sequences, or equivalently, a sequence of random vectors. The *autocorrelation function* $\mathbf{R}(d, k)$ is an $n \times n$ function defined by:

$$\mathbf{R}(d, k) = \mathrm{E}[\boldsymbol{x}_{k+d}\boldsymbol{x}_k^*] . \tag{10.6}$$

It can be viewed as the correlation matrix for the pair random vectors $\boldsymbol{x}_{k+d}$ and $\boldsymbol{x}_k$. Without the expectation operator, the product of the column vector $\boldsymbol{x}_{k+d}$ and row vector $\boldsymbol{x}_k^*$ would have rank one. With the expectation operator, however, the rank can be as high as n. A random sequence is *wide-sense stationary (WSS)* if its autocorrelation function is independent of time k, in which case (10.6) is written as $\mathbf{R}(d)$. If $\boldsymbol{x}_k$ is WSS, the covariance of $\boldsymbol{x}_{k+d}$ and $\boldsymbol{x}_k$ depends only on the time lag d. The *power spectral density (PSD) matrix*, or power spectrum, of a WSS random sequence is defined as the Z-transform $\mathrm{S}(z)$ or Fourier transform $\mathrm{S}(e^{j\theta})$ of its autocorrelation function, when it exists:

$$\mathrm{S}(z) = \sum_{d=-\infty}^{\infty} \mathbf{R}(d)z^{-d} . \tag{10.7}$$

Its inverse is given by:

$$\mathbf{R}(k) = \frac{1}{2\pi}\int_{-\pi}^{\pi} \mathrm{S}(e^{j\theta})e^{jk\theta}\, d\theta . \tag{10.8}$$

A vector random process is said to be *spatially white* if its PSD is proportional to the identity matrix, $\mathrm{S}(z) = \phi(z)\mathbf{I}$ for some valid scalar PSD $\phi(z)$. It is said to be *temporally white* if its PSD is a constant, $\mathrm{S}(z) = \mathrm{S}$, independent of z. A vector random process that is both spatially white and temporally white is said to be *white*; its PSD is of the form $\mathrm{S}(z) = \phi\mathbf{I}$ for some real positive constant ϕ.

Example 10-8.
If $\{x_k^{(1)}, \ldots x_k^{(n)}$, for all $k\}$ are independent and uniformly distributed over $\{\pm 1\}$, then the autocorrelation function of the vector random sequence $\boldsymbol{x}_k = [x_k^{(1)}, \ldots, x_k^{(n)}]^T$ is:

$$\mathbf{R}(d, k) = \mathrm{E}[\boldsymbol{x}_{k+d}\boldsymbol{x}_k^*] = \delta_d\mathbf{I} . \tag{10.9}$$

Since this is independent of k, $\boldsymbol{x}_k$ is WSS with $\mathbf{R}(d) = \delta_d\mathbf{I}$. Taking the Z-transform, the PSD of $\boldsymbol{x}_k$ is $\mathbf{S}(z) = \mathbf{I}$. Hence, $\boldsymbol{x}_k$ is white (both temporally and spatially).

Exercise 10-1.

The PSD $\mathbf{S}(z)$ and autocorrelation function $\mathbf{R}(d)$ of a vector random process $\boldsymbol{x}_k$ have many useful properties that will be exploited in this chapter. Most of these properties are direct extensions of their scalar counterparts. Show that:

(a) $E[\,|x_k^{(i)}|^2] = R_{ii}(0) = \frac{1}{2\pi}\int_{-\pi}^{\pi} S_{ii}(e^{j\theta})\,d\theta$.

(b) $E[\|\boldsymbol{x}_k\|^2] = \text{tr}\{E[\boldsymbol{x}_k\boldsymbol{x}_k^*]\} = \text{tr}\{\mathbf{R}(0)\} = \frac{1}{2\pi}\int_{-\pi}^{\pi} \text{tr}\{\mathbf{S}(e^{j\theta})\}d\theta = \frac{1}{2\pi}\int_{-\pi}^{\pi} \sum_{i=1}^{n} S_{ii}(e^{j\theta})\,d\theta$.

(c) $\mathbf{R}^*(-d) = \mathbf{R}(d)$. Thus, if we interpret $\mathbf{R}(d)$ as a filter impulse response, it is matched to itself.

(d) $\mathbf{S}^*(1/z^*) = \mathbf{S}(z)$. Thus, a PSD is parahermitian.

(e) For any angle θ, $\boldsymbol{u}^*\mathbf{S}(e^{j\theta})\boldsymbol{u} \ge 0$ for all vectors $\boldsymbol{u}$.

(f) For any angle θ, the eigenvalues $\{\lambda_i(\theta)\}$ of $\mathbf{S}(e^{j\theta})$ are nonnegative, $\lambda_i(\theta) \ge 0$ for all i.

A matrix $\mathbf{S}(e^{j\theta})$ satisfying property (e) is said to be *positive semidefinite (PSD)*. Hence, one can say that a PSD is PSD! A matrix $\mathbf{S}(e^{j\theta})$ satisfying property (e) is said to be *positive definite* if $\boldsymbol{u}^*\mathbf{S}(e^{j\theta})\boldsymbol{u} = 0$ implies that $\boldsymbol{u} = \mathbf{0}$; in this case, all of its eigenvalues in (f) become strictly positive.

We say that a given transfer function $\mathbf{S}(z)$ is a *valid PSD* if one could construct a WSS random sequence with $\mathbf{S}(z)$ as its PSD. In terms of the properties of Exercise 10-1, $\mathbf{S}(z)$ is a valid PSD if and only if it satisfies properties (d) and (e). In the scalar case, (d) reduces to the requirement that the power spectrum be real, and (e) reduces to the requirement that it cannot be negative.

Example 10-9.

Consider the following two matrices:

$$\mathbf{S}_1(z) = \begin{bmatrix} 1 & z^{-1} \\ z & 2 \end{bmatrix}, \qquad \mathbf{S}_2(z) = \begin{bmatrix} 1 & z^{-1} \\ z^{-1} & 2 \end{bmatrix}. \tag{10.10}$$

The matrix $\mathbf{S}_1(z)$ is a valid PSD. The matrix $\mathbf{S}_2(z)$ is not a valid PSD because it does not satisfy (d). The matrix $\mathbf{S}_3(z) = -\mathbf{S}_1(z)$ is not a valid PSD, because it satisfies (d) but not (e).

Filtering a Random Sequence

If a WSS random sequence $\boldsymbol{x}_k$ with PSD $\mathbf{S}_{\boldsymbol{x}}(z)$ is passed through a MIMO filter with transfer function $\mathbf{H}(z)$, then the output $\boldsymbol{y}_k$ will also be WSS, and its PSD will be given by:

$$\mathbf{S}_{\boldsymbol{y}}(z) = \mathbf{H}(z)\mathbf{S}_{\boldsymbol{x}}(z)\mathbf{H}^*(1/z^*)\,. \tag{10.11}$$

On the unit circle, this reduces to:

$$\mathbf{S}_{\boldsymbol{y}}(e^{j\theta}) = \mathbf{H}(e^{j\theta})\mathbf{S}_{\boldsymbol{x}}(e^{j\theta})\mathbf{H}^*(e^{j\theta})\,. \tag{10.12}$$

It is important to emphasize that, as in (10.11), the three matrices in this expression *do not commute*, and must be written in this order. This is in contrast to the scalar case, where (10.12) reduces to $S_x(e^{j\theta})|H(e^{j\theta})|^2$.

To demonstrate (10.11), we begin with the definition (10.6) for the autocorrelation function of the filter output, and substitute the convolution sum of (10.2):

$$\begin{aligned}
\mathbf{R}_y(d) &= \mathrm{E}[\boldsymbol{y}_{k+d}\boldsymbol{y}_k^*] \\
&= \mathrm{E}\,[\textstyle\sum_{l=-\infty}^{\infty}\mathbf{H}_l\boldsymbol{x}_{k+d-l}\sum_{i=-\infty}^{\infty}\boldsymbol{x}_{k-i}^*\mathbf{H}_i^*\,] \\
&= \textstyle\sum_{l=-\infty}^{\infty}\mathbf{H}_l\sum_{j=-\infty}^{\infty}\mathbf{R}_x(d-j-l)\mathbf{H}_{-j}^* \quad (\text{where } j=-i) \\
&= \mathbf{H}_d \otimes \mathbf{R}_x(d) \otimes \mathbf{H}_{-d}^* \quad (\text{where } \otimes \text{ denotes convolution})\,. \qquad (10.13)
\end{aligned}$$

Taking the Z-transform of (10.13) yields (10.11). The above convolution operator is not commutative.

10.1.3. Matrix Spectral Factorization

Suppose you are given a valid PSD $\mathbf{S}(z)$ of dimension $n \times n$. How would you go about creating a random sequence with $\mathbf{S}(z)$ as its PSD? Alternatively, if you were given a random sequence $\boldsymbol{x}_k$ with PSD $\mathbf{S}(z)$, how would you go about whitening it, or predicting future samples based on a linear combination of past samples? Answers to questions such as these are based on a *minimum-phase spectral factorization* of the PSD matrix $\mathbf{S}(z)$ [3][4].

Theorem 10-1. *(Matrix Spectral Factorization.)* If $\mathbf{S}(z)$ is a valid $n \times n$ spectral matrix that is rational and nonsingular on the unit circle, there exists a *unique right* factorization:

$$\mathbf{S}(z) = \mathbf{M}^*(1/z^*)\Gamma^2\mathbf{M}(z) \quad (\text{right})\,, \qquad (10.14)$$

where $\mathbf{M}(z)$ is *monic* and *minimum phase*, and where Γ^2 is a diagonal matrix with positive diagonal components. There also exists a *unique left* factorization:

$$\mathbf{S}(z) = \mathbf{M}(z)\Gamma^2\mathbf{M}^*(1/z^*) \quad (\text{left})\,, \qquad (10.15)$$

where again $\mathbf{M}(z)$ is monic and minimum phase, and Γ^2 is positive definite and diagonal. In either case, Γ^2 satisfies:

$$\det\Gamma^2 = \exp\left\{\frac{1}{2\pi}\int_{-\pi}^{\pi}\log\det\mathbf{S}(e^{j\theta})\,d\theta\right\}. \qquad (10.16)$$

At times it will be convenient to combine the factors Γ and $\mathbf{M}(z)$ into one, by defining the minimum-phase filter $\mathbf{G}(z) = \Gamma\mathbf{M}(z)$ for the right factorization, and $\mathbf{G}(z) = \mathbf{M}(z)\Gamma$ for the left factorization, so that the right and left factorizations reduce to $\mathbf{S}(z) = \mathbf{G}^*(1/z^*)\mathbf{G}(z)$ and $\mathbf{S}(z) = \mathbf{G}(z)\mathbf{G}^*(1/z^*)$, respectively. We should emphasize that the $\mathbf{M}(z)$ of (10.14) will not be the same as the $\mathbf{M}(z)$ of (10.15). They are given the same name only because they are both monic and minimum phase. This causes no confusion in practice because a given application will require that only one of the factorizations be performed. A numerical technique for factoring spectral matrices is explored in Problem 10-1.

The above matrix spectral factorization reduces to well-known factorizations in two limiting cases. In the scalar case, when $n = 1$, it reduces to the minimum-phase spectral factorization of Chapter 2. Alternatively, in the memoryless case when $\mathbf{S}(z)$ is a constant, independent of z, the matrix spectral factorization reduces to the *Cholesky decomposition*. (We will examine the Cholesky decomposition in more detail in Section 10.3.4.)

Example 10-10.

The spectral matrix:

$$\mathbf{S}(z) = \begin{bmatrix} 10 & 30 \\ 41 & 83 \end{bmatrix} z + \begin{bmatrix} 105 & 21 \\ 21 & 426 \end{bmatrix} + \begin{bmatrix} 10 & 41 \\ 30 & 83 \end{bmatrix} z^{-1} \tag{10.17}$$

can be factored as both $\mathbf{S}(z) = \mathbf{G}_L(z)\mathbf{G}_L^*(1/z^*)$ and $\mathbf{S}(z) = \mathbf{G}_R^*(1/z^*)\mathbf{G}_R(z)$, where

$$\mathbf{G}_L(z) = \begin{bmatrix} 10 & 0 \\ 1 & 20 \end{bmatrix} + \begin{bmatrix} 1 & 2 \\ 3 & 4 \end{bmatrix} z^{-1}, \text{ and } \mathbf{G}_R(z) = \begin{bmatrix} 10.078 & 0 \\ 0.564 & 19.845 \end{bmatrix} + \begin{bmatrix} 0.908 & 3.834 \\ 1.512 & 4.183 \end{bmatrix} z^{-1}. \tag{10.18}$$

We are now ready to address the realization problem that was used to motivate the spectral factorization at the beginning of this section. Specifically, to generate a random sequence $\{\boldsymbol{x}_k\}$ with a given PSD $\mathbf{S}(z)$, one need only generate a white vector sequence $\{\boldsymbol{u}_k\}$ with i.i.d. unit-variance components, so that $\mathbf{S}_{\boldsymbol{uu}}(z) = \mathbf{I}$, and then pass $\boldsymbol{u}_k$ through a filter whose transfer function $\mathbf{G}(z)$ is defined by the left factorization (10.15) of the given PSD, namely $\mathbf{S}(z) = \mathbf{G}(z)\mathbf{G}^*(1/z^*)$. Then (10.11) implies that the filter output will have the desired PSD.

10.1.4. Linear Prediction

In this section we extend the linear prediction results of Section 3.2.4 to vector-valued random sequences. We first consider the space-only problem of linear prediction for a temporally white random process. Afterwards we consider the general space-time prediction problem.

Linear Prediction in Space

Consider the problem of *linear prediction* for a single random vector $\boldsymbol{x}$ of dimension n, whose mean is zero and whose autocorrelation matrix $\mathbf{R}_{\boldsymbol{xx}} = E[\boldsymbol{xx}^*]$ is full rank. As just explained, such a matrix admits a unique (Cholesky) factorization $\mathbf{R}_{\boldsymbol{xx}} = \mathbf{M}\Gamma^2\mathbf{M}^*$, where $\mathbf{M}$ is monic and lower triangular, and Γ^2 is diagonal with real and positive diagonal elements. Let us form an estimate $\hat{x}_i$ of its i-th component as a linear combination of its past components $\{x_1, \dots x_{i-1}\}$, namely $\hat{x}_i = \sum_{j=1}^{i-1} P_{i,j}x_j$. In particular, because the first component of $\boldsymbol{x}$ has no past, we have $\hat{x}_1 = 0$.The vector of estimates $\hat{\boldsymbol{x}} = [\hat{x}_1, \dots \hat{x}_n]^T$ can be expressed as:

$$\hat{\boldsymbol{x}} = \mathbf{P}\boldsymbol{x}, \tag{10.19}$$

where $\mathbf{P}$ is a *strictly* lower-triangular matrix of prediction coefficients $\{P_{i,j}\}$. The problem is to choose the prediction coefficients so as to minimize the sum mean-squared prediction error $E[\|\boldsymbol{x} - \hat{\boldsymbol{x}}\|^2] = \text{tr}\mathbf{R}_{\boldsymbol{ee}}$, where $\mathbf{R}_{\boldsymbol{ee}} = E[\boldsymbol{ee}^*]$ is the autocorrelation matrix of the prediction error $\boldsymbol{e} = \boldsymbol{x} - \hat{\boldsymbol{x}} = (\mathbf{I} - \mathbf{P})\boldsymbol{x}$. Specifically, we have:

$$\begin{aligned} E[\|\boldsymbol{x} - \hat{\boldsymbol{x}}\|^2] &= \mathrm{tr}\mathbf{R}_{ee} \\ &= \mathrm{tr}\{(\mathbf{I} - \mathbf{P})\mathbf{R}_{xx}(\mathbf{I} - \mathbf{P})^*\} \\ &= \mathrm{tr}\{(\mathbf{I} - \mathbf{P})\mathbf{M}\Gamma^2\mathbf{M}^*(\mathbf{I} - \mathbf{P})^*\} \\ &= \mathrm{tr}\{(\mathbf{I} + \mathbf{D})\Gamma^2(\mathbf{I} + \mathbf{D})^*\}, \end{aligned} \tag{10.20}$$

where we have introduced $(\mathbf{I} + \mathbf{D}) = (\mathbf{I} - \mathbf{P})\mathbf{M}$. Since $\mathbf{M}$ is monic and lower triangular, and since $\mathbf{P}$ is strictly lower triangular, the product $(\mathbf{I} - \mathbf{P})\mathbf{M}$ must be monic and lower triangular. Therefore, the matrix $\mathbf{D}$ must be strictly lower triangular. Multiplying (10.20) out yields:

$$E[\|\boldsymbol{x} - \hat{\boldsymbol{x}}\|^2] = \mathrm{tr}\{\Gamma^2 + \mathbf{D}\Gamma^2 + \Gamma^2\mathbf{D}^* + \mathbf{D}\Gamma^2\mathbf{D}^*\} \tag{10.21}$$

$$= \mathrm{tr}\{\Gamma^2\} + \mathrm{tr}\{\mathbf{D}\Gamma^2\mathbf{D}^*\}, \tag{10.22}$$

where we exploited the fact that the trace of a strictly triangular matrix is zero. The second trace in (10.22) is nonnegative, because it is the trace of a Hermitian matrix, which has nonnegative eigenvalues. Hence, we can do no better than to force the second term to zero by choosing $\mathbf{D} = \mathbf{0}$, yielding a minimum MSE of $\mathrm{tr}\{\Gamma^2\}$. Thus, solving $(\mathbf{I} + \mathbf{D}) = (\mathbf{I} - \mathbf{P})\mathbf{M}$ for $\mathbf{P}$ with $\mathbf{D} = \mathbf{0}$ yields the solution to the linear prediction problem:

$$\mathbf{P} = \mathbf{I} - \mathbf{M}^{-1}. \tag{10.23}$$

Since $\mathbf{M}$ is lower triangular and monic, it is invertible, and $\mathbf{P}$ is strictly lower triangular.

Example 10-11.

Consider a random vector $\boldsymbol{x} = [x_1, x_2, x_3]^T$ whose autocorrelation matrix is:

$$\mathbf{R}_{xx} = \begin{bmatrix} 16 & 8 & 4 \\ 8 & 20 & 10 \\ 20 & 10 & 21 \end{bmatrix} = \underbrace{\begin{bmatrix} 1 & 0 & 0 \\ 1/2 & 1 & 0 \\ 1/4 & 1/2 & 1 \end{bmatrix}}_{\mathbf{M}} \underbrace{\begin{bmatrix} 16 & 0 & 0 \\ 0 & 16 & 0 \\ 0 & 0 & 16 \end{bmatrix}}_{\Gamma^2} \underbrace{\begin{bmatrix} 1 & 1/2 & 1/4 \\ 0 & 1 & 1/2 \\ 0 & 0 & 1 \end{bmatrix}}_{\mathbf{M}^*}, \tag{10.24}$$

where we have shown the unique left Cholesky factorization $\mathbf{R}_{xx} = \mathbf{M}\Gamma^2\mathbf{M}^*$. Suppose we want to predict x_2 given x_1, and also x_3 given x_1 and x_2. From (10.23), the optimal linear predictor is $\hat{\boldsymbol{x}} = \mathbf{P}\boldsymbol{x}$, where:

$$\mathbf{P} = \mathbf{I} - \mathbf{M}^{-1} = \mathbf{I} - \begin{bmatrix} 1 & 0 & 0 \\ 1/2 & 1 & 0 \\ 1/4 & 1/2 & 1 \end{bmatrix}^{-1} = \begin{bmatrix} 0 & 0 & 0 \\ 1/2 & 0 & 0 \\ 0 & 1/2 & 0 \end{bmatrix}. \tag{10.25}$$

It follows that $\hat{x}_2 = 0.5x_1$, and $\hat{x}_3 = 0.5x_2$. The resulting MSE is $\mathrm{tr}\{\Gamma^2\} = 48$.

Linear Prediction in Space and Time

In this section we generalize the above result to solve the *space-time* linear prediction problem. Consider a WSS random sequence $\boldsymbol{x}_k$ of dimension n whose PSD $\mathbf{S}_{xx}(z)$ is left-factored according to $\mathbf{S}_{xx}(z) = \mathbf{M}(z)\Gamma^2\mathbf{M}^*(1/z^*)$, where $\mathbf{M}(z)$ is monic and minimum phase, and where Γ^2 is a positive-definite diagonal matrix. Let us form an estimate $\hat{x}_k^{(i)}$ of its i-th component at time k as a linear combination of its past components $\{x_k^{(1)}, \dots x_k^{(i-1)}\}$ and $\{\boldsymbol{x}_{k-1}, \boldsymbol{x}_{k-2}, \dots\}$. In vector form, $\hat{\boldsymbol{x}}_k = [\hat{x}_k^{(1)}, \dots \hat{x}_k^{(n)}]^T$ can be expressed as:

$$\hat{\boldsymbol{x}}_k = \sum_{l=0}^{\infty} \mathbf{P}_l \boldsymbol{x}_{k-l}, \tag{10.26}$$

where $\{\mathbf{P}_k\}$ is a strictly space-time causal MIMO filter of prediction coefficients with transfer function $\mathbf{P}(z)$, so that $\mathbf{P}_k = \mathbf{0}$ for $k < 0$, and $\mathbf{P}_0$ is strictly lower triangular. The MMSE linear predictor chooses $\mathbf{P}(z)$ to minimize $\mathrm{E}[\|\boldsymbol{x}_k - \hat{\boldsymbol{x}}_k\|^2] = \mathrm{tr}\frac{1}{2\pi}\int_{-\infty}^{\infty}\mathbf{S}_{ee}(e^{j\theta})d\theta$, where $\mathbf{S}_{ee}(z)$ is the PSD of the prediction error $\boldsymbol{e}_k = \boldsymbol{x}_k - \sum_{l=0}^{\infty}\mathbf{P}_l\boldsymbol{x}_{k-l}$. But:

$$\begin{aligned}
\mathbf{S}_{ee}(z) &= (\mathbf{I} - \mathbf{P}(z))\mathbf{S}_{xx}(z)(\mathbf{I} - \mathbf{P}^*(1/z^*)) \\
&= (\mathbf{I} - \mathbf{P}(z))\mathbf{M}(z)\Gamma^2\mathbf{M}^*(1/z^*)(\mathbf{I} - \mathbf{P}^*(1/z^*)) \\
&= \mathbf{E}(z)\Gamma^2\mathbf{E}^*(1/z^*),
\end{aligned} \tag{10.27}$$

where we have introduced $\mathbf{E}(z) = (\mathbf{I} - \mathbf{P}(z))\mathbf{M}(z)$. Being the product of two monic and space-time causal transfer functions, $\mathbf{E}(z) = \sum_{k=0}^{\infty}\mathbf{E}_k z^{-k}$ is also monic and space-time causal, with $\mathbf{E}_0$ monic and lower triangular. The MSE can now be expressed as:

$$\begin{aligned}
\mathrm{E}[\|\boldsymbol{x} - \hat{\boldsymbol{x}}\|^2] &= \mathrm{tr}\frac{1}{2\pi}\int_{-\pi}^{\pi}\mathbf{S}_{ee}(e^{j\theta})d\theta \\
&= \mathrm{tr}\frac{1}{2\pi}\int_{-\pi}^{\pi}\mathbf{E}(e^{j\theta})\Gamma^2\mathbf{E}^*(e^{j\theta})d\theta \\
&= \mathrm{tr}\sum_{k=0}^{\infty}\mathbf{E}_k\Gamma^2\mathbf{E}_k^* \\
&= \mathrm{tr}\{\mathbf{E}_0\Gamma^2\mathbf{E}_0^*\} + \mathrm{tr}\{\textstyle\sum_{k=1}^{\infty}\mathbf{E}_k\Gamma^2\mathbf{E}_k^*\}.
\end{aligned} \tag{10.28}$$

The second trace is nonnegative, because it is the trace of a semi-positive definite matrix, and hence we can do no better than to force it to zero by choosing $\mathbf{E}_k = \mathbf{0}$ for $k \geq 1$. Given a monic constraint, the first term is minimized by choosing $\mathbf{E}_0 = \mathbf{I}$, as proved in the last section; hence, the $\mathbf{E}(z)$ that minimizes MSE is the identity matrix. Using $\mathbf{E}(z) = (\mathbf{I} - \mathbf{P}(z))\mathbf{M}(z)$, we find that the best prediction filter is:

$$\mathbf{P}(z) = \mathbf{I} - \mathbf{M}(z)^{-1}. \tag{10.29}$$

This solution reduces to the space-only solution (10.23) when the random process is temporally white, and it reduces to the time-only solution of (3.76) when $\boldsymbol{x}_k$ is a scalar. The MSE after the best linear predictor is $\mathrm{E}[\|\boldsymbol{x}_k - \hat{\boldsymbol{x}}_k\|^2] = \mathrm{tr}\{\Gamma^2\}$. Using the arithmetic-geometric mean inequality and (10.16), this MSE is bounded by:

$$\mathrm{MSE}_{\min} = \mathrm{tr}\{\Gamma^2\} \geq n\cdot\exp\left\{\frac{1}{2\pi n}\int_{-\pi}^{\pi}\log\det\mathbf{S}_{xx}(e^{j\theta})\,d\theta\right\}. \tag{10.30}$$

This expression might be useful for a PSD that cannot be factored analytically, or even for an experimentally measured PSD.

10.2. The Gaussian MIMO Channel

The bulk of this chapter concerns the general MIMO channel shown in Fig. 10-5. There are n *emitters* or *users* that produce the channel inputs, and there are m *sensors* that produce the channel outputs. The complex envelope of the j-th emitted signal is $\sum_k a_k^{(j)} g_j(t - kT)$, where $\{a_k^{(j)}\}$ is the complex symbol sequence for the j-th emitter, and $g_j(t)$ is the transmit pulse shape for the j-th emitter. Let $r_i(t)$ denote the complex envelope of the waveform observed at the i-th receiver sensor. It will be convenient to represent the n channel input sequences by the vector-valued sequence $\boldsymbol{a}_k = [a_k^{(1)}, \ldots a_k^{(n)}]^T$, and similarly the m output waveforms by the vector-valued observation $\boldsymbol{r}(t) = [r_1(t), \ldots r_m(t)]^T$. Then $\boldsymbol{r}(t)$ is related to $\{\boldsymbol{a}_k\}$ according to:

$$\boldsymbol{r}(t) = \sum_{k=-\infty}^{\infty} \mathbf{H}(t - kT)\boldsymbol{a}_k + \boldsymbol{n}(t) , \tag{10.31}$$

where $\mathbf{H}(t)$ is an $m \times n$ matrix of impulse responses whose $(i, j)^{th}$ component $h_{ij}(t)$ is the received response at the i-th output due to an impulse at the j-th input. As in previous chapters,

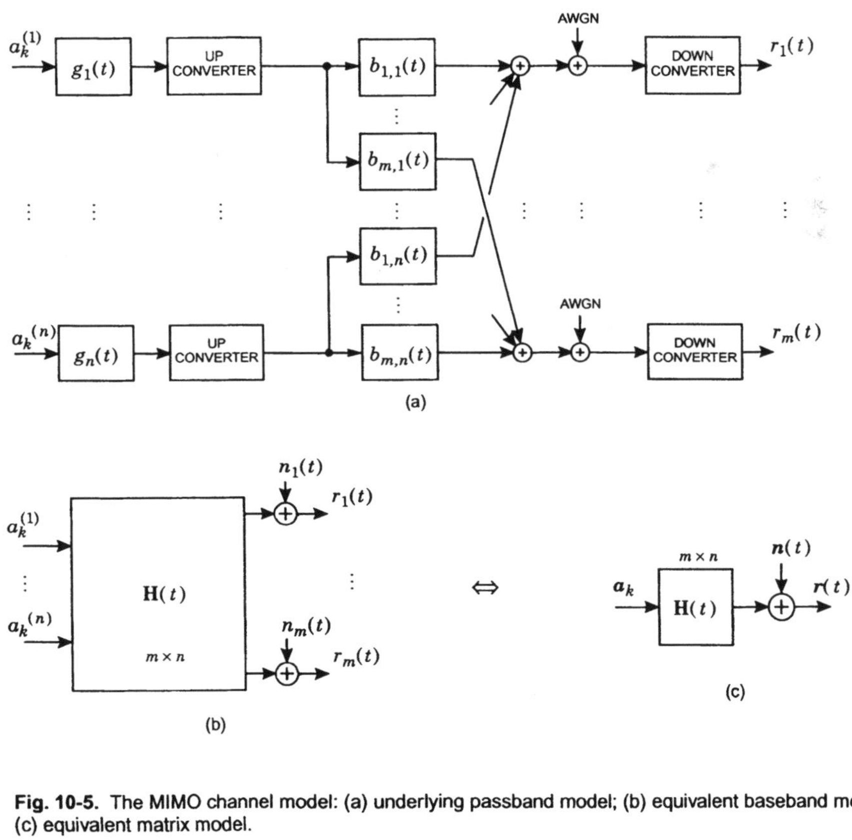

Fig. 10-5. The MIMO channel model: (a) underlying passband model; (b) equivalent baseband model; (c) equivalent matrix model.

$h_{ij}(t)$ is generally complex-valued, taking into account the effects of the transmitter pulse shape $g_j(t)$ as well as the passband channel response $b_{ij}(t)$, according to:

$$H_{ij}(f) = G_j(f)B_{ij}(f + f_c). \tag{10.32}$$

We assume that the noise vector $\boldsymbol{n}(t) = [n_1(t), \ldots n_m(t)]^T$ has independent complex-symmetric components, each white and Gaussian, satisfying $E[n_i(t+\tau)n_j^*(t)] = N_0\delta_{ij}\delta(\tau)$. Implicit in this model is the assumption that all transmitters are complex PAM with the same symbol rate $1/T$. However, the inputs are not assumed to be synchronous. The relative timing offsets of the different inputs are not shown explicitly in (10.31) because they have been absorbed into the definitions of the different components of $\mathbf{H}(t)$.

The equation (10.31) is reminiscent of the standard complex baseband model for PAM-based systems that has dominated the book, but with the important distinction that the input and output are vectors, and the received pulse shape has been replaced by a matrix of received pulses.

10.2.1. The Matrix Matched Filter

Of course, a special case of the MIMO channel is the SISO channel, which was studied extensively in the previous chapters. In the SISO case we saw that there were two ways to formulate the receiver optimization problem. One was to find the most likely sequence of transmitted symbols, which was solved by the Viterbi algorithm. The other was to find the sequence of most likely symbols, which was solved by quantizing the soft outputs of the BCJR algorithm. In the MIMO setting, there are *three* ways to formulate the receiver optimization problem:

- The *joint maximum-likelihood sequence detector (JMLSD)* chooses the entire sequence $\{\boldsymbol{a}_k\}$ that maximizes the likelihood function $f(\boldsymbol{r}(t)\,|\,\{\boldsymbol{a}_k\})$.
- The *vector-by-vector ML* detector chooses each $\boldsymbol{a}_k$ so as to maximize $f(\boldsymbol{r}(t)\,|\,\boldsymbol{a}_k)$.
- The *symbol-by-symbol ML* detector chooses each element $a_k^{(i)}$ of each $\boldsymbol{a}_k$ to maximize $f(\boldsymbol{r}(t)\,|\,a_k^{(i)})$.

In the following we focus on the JML sequence detector — loosely called the ML sequence detector — because it is the easiest to implement.

Given that the noise is white and Gaussian, the JML sequence detector will choose the symbol sequence $\{\boldsymbol{a}_k\}$ that minimizes:

$$J = \int_{-\infty}^{\infty} \| \boldsymbol{r}(t) - \sum_{k=-\infty}^{\infty} \mathbf{H}(t - kT)\boldsymbol{a}_k \|^2 \, dt \, , \tag{10.33}$$

which reduces to:

$$J = \int_{-\infty}^{\infty} \| \boldsymbol{r}(t) \|^2 \, dt - 2\sum_k \mathrm{Re}\left\{ \boldsymbol{a}_k^* \int_{-\infty}^{\infty} \mathbf{H}^*(t - kT)\boldsymbol{r}(t) dt \right\} + \int_{-\infty}^{\infty} \| \sum_k \mathbf{H}(t - kT)\boldsymbol{a}_k \|^2 \, dt \, . \tag{10.34}$$

Only the first two terms depend on $\boldsymbol{r}(t)$, but the receiver may ignore the first term because it is independent of $\{\boldsymbol{a}_k\}$. Hence, everything about $\boldsymbol{r}(t)$ that the receiver needs to know in order to recover the input symbols is captured by the integral in the second term, namely:

$$\boldsymbol{y}_k = \int_{-\infty}^{\infty} \mathbf{H}^*(t - kT)\boldsymbol{r}(t)dt \,. \tag{10.35}$$

In other words, the set $\{\boldsymbol{y}_k\}$ is a *sufficient statistic* for determining the JML sequence.

The receiver may calculate the vector $\boldsymbol{y}_k$ by passing $\boldsymbol{r}(t)$ through a *matrix matched-filter (MMF)* with impulse response $\mathbf{H}^*(-t)$, and sampling the output at time kT, as shown in Fig. 10-6(a). The sampled-MF transforms the underlying continuous-time channel to an equivalent discrete-time channel with input $\boldsymbol{a}_k$ and output $\boldsymbol{y}_k$, as shown in Fig. 10-6(b).

The MMF has two key properties. First, it transforms the continuous-time observation into a discrete-time sequence that is sufficient for recovering the channel input. Second, it transforms the underlying $m \times n$ continuous-time channel into an equivalent discrete-time channel model that is *square* with dimension $n \times n$. Interestingly, this dimension does not depend on the number of receiver sensors m.

The impulse response of the sampled MF channel of Fig. 10-6(b) is easily derived. Substituting (10.31) into (10.35), we find that:

$$\boldsymbol{y}_k = \sum_{l=-\infty}^{\infty} \int_{-\infty}^{\infty} \mathbf{H}^*(t - kT)\mathbf{H}(t - lT)dt\,\boldsymbol{a}_l + \boldsymbol{n}_k \,, \tag{10.36}$$

or equivalently:

$$\boldsymbol{y}_k = \sum_{l=-\infty}^{\infty} \mathbf{S}_{k-l}\boldsymbol{a}_l + \boldsymbol{n}_k \,, \tag{10.37}$$

where

$$\mathbf{S}_k = \int_{-\infty}^{\infty} \mathbf{H}^*(t - kT)\mathbf{H}(t)\,dt \tag{10.38}$$

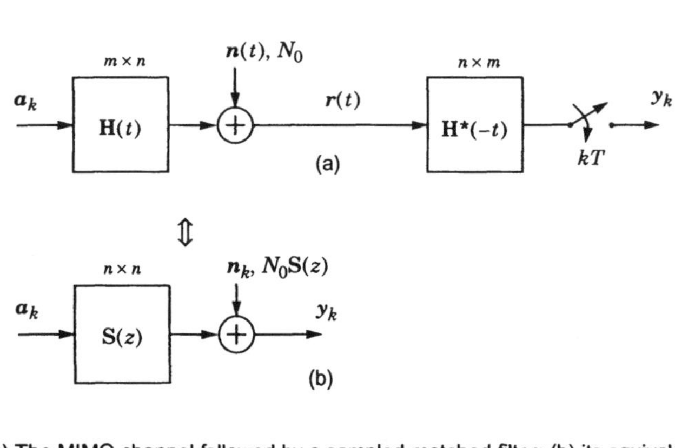

Fig. 10-6. (a) The MIMO channel followed by a sampled matched filter; (b) its equivalent model. In (b), the channel response $\mathrm{S}(z)$ and the noise PSD $N_0\mathrm{S}(z)$ are proportional to one another.

and
$$\boldsymbol{n}_k = \int_{-\infty}^{\infty} \mathbf{H}^*(t - kT)\boldsymbol{n}(t)\, dt\,. \tag{10.39}$$

We may interpret (10.38) as a generalization of the sampled-autocorrelation function $\rho_h(k)$ of (5.65) to MIMO channels. Since $\mathbf{H}^*(t)$ has dimension $n \times m$ and $\mathbf{H}(t)$ has dimension $m \times n$, each $\mathbf{S}_k$ has dimension $n \times n$. The Z-transform of $\mathbf{S}_k$ is $\mathbf{S}(z) = \Sigma_k \mathbf{S}_k z^{-k}$, and the Fourier transform is:

$$\mathbf{S}(e^{j2\pi fT}) = \tfrac{1}{T}\Sigma_k \mathbf{H}^*(f - \tfrac{k}{T})\mathbf{H}(f - \tfrac{k}{T})\,. \tag{10.40}$$

Since $\mathbf{S}(z)$ is parahermitian and positive semidefinite, it is a valid PSD. We refer to $\mathbf{S}(z)$ and $\mathbf{S}(e^{j2\pi fT})$ as the *folded spectrum*, since it is a MIMO generalization of the scalar folded spectrum (5.66).

The filtered and sampled noise sequence $\boldsymbol{n}_k$ of (10.39) has autocorrelation matrix:

$$\begin{aligned}
\mathbf{R}_{\boldsymbol{n}}(m) &= \mathrm{E}[\boldsymbol{n}_{k+m}\boldsymbol{n}_k^*] \\
&= \mathrm{E}\left[\int_{-\infty}^{\infty} \mathbf{H}^*(t - kT - mT)\boldsymbol{n}(t)\,dt \int_{-\infty}^{\infty} \boldsymbol{n}(\tau)^*\mathbf{H}(\tau - kT)\,d\tau\right] \\
&= N_0\int_{-\infty}^{\infty} \mathbf{H}^*(t - kT - mT)\mathbf{H}(t - kT)\,dt \;=\; N_0\mathbf{S}_m\,.
\end{aligned} \tag{10.41}$$

Taking the Z-transform, we find that the power-spectrum matrix for the noise $\boldsymbol{n}_k$ is $\mathbf{S}_{\boldsymbol{n}}(z) = N_0\mathbf{S}(z)$. Interestingly, it is proportional to the equivalent channel transfer function.

10.2.2. Separating Space From Time

The spatial and temporal properties of the MIMO impulse response $\mathbf{H}(t)$ can be separated in the *narrowband* case, for which the symbol rate is sufficiently small that the frequency response of the channel can be modeled as a constant over the band of frequencies occupied by the signal. In this case, assuming that all n transmitters use the same pulse shape $g(t)$, the MIMO impulse response of (10.31) reduces to:

$$\mathbf{H}(t) = g(t)\mathbf{H}\,, \tag{10.42}$$

where $\mathbf{H}$ is a constant matrix. Therefore, the matched filter $\mathbf{H}^*(-t)$ can be decomposed into the cascade of a space-only memoryless matrix $\mathbf{H}^*$ followed by a bank of time-only matched-filters $g(-t)$. This is illustrated in Fig. 10-7. The matrix $\mathbf{H}^*$ is called a *linear* or *spatial combiner*, and can be interpreted as a spatial matched filter.

Example 10-12.
Consider a single-user single-antenna narrowband transmitter with complex envelope $s(t) = \Sigma_k a_k g(t - kT)$, and a receiver with two antenna elements, producing the pair of observations $r_1(t) = h_1 s(t) + n_1(t)$ and $r_2(t) = h_2 s(t) + n_2(t)$. The MIMO impulse response is given by (10.42) with $\mathbf{H} = [h_1, h_2]^T$. As shown in Fig. 10-8, the MF receiver first applies the spatial combiner $\mathbf{H}^*$, which produces $y(t) = h_1^* r_1(t) + h_2^* r_2(t)$, and then applies $y(t)$ to a scalar temporal MF $g(-t)$. In this setting, the spatial combiner $\mathbf{H}^*$ is commonly referred to as a *maximal-ratio combiner*, because it is the combiner that maximizes SNR.

The decision to perform spatial processing before temporal processing above is somewhat arbitrary; when the channel is narrowband as given by (10.42), the MMF can just as well be decomposed into a temporal sampled MF bank followed by a spatial combiner **H***. But a receiver that performs spatial combining first will require only a single RF downconverter in practice, and thus has a complexity advantage, because downconverters can be expensive to implement. On the other hand, a receiver that performs temporal filtering first could implement combining after the samplers, making it easier to track a time-varying or unknown channel using digital processing. The price paid for this added flexibility is the need for two downconverters and two A/D converters instead of one of each. Interestingly, because the output of the spatial combiner provides sufficient statistics, so must its input. Hence, the MF bank alone transforms the continuous-time model of (10.31) into an equivalent discrete-time model. This model will also be *memoryless* if $|G(f)|^2$ is a Nyquist pulse.

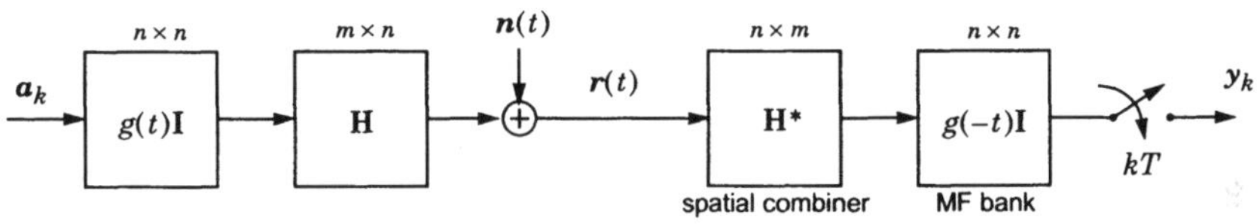

Fig. 10-7. In the narrowband case, the spatial and temporal processing of the MF can be separated.

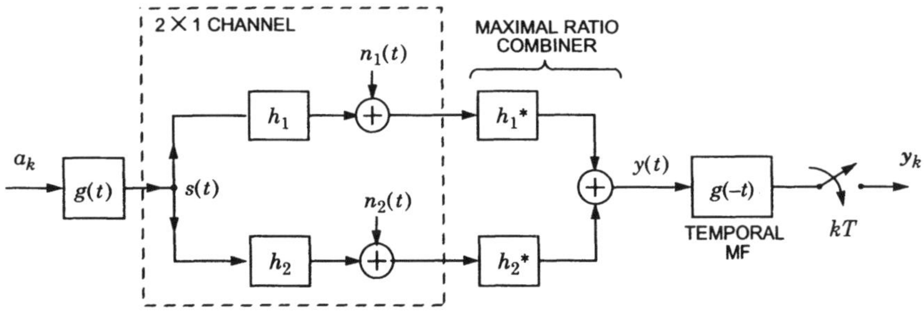

Fig. 10-8. The MMF for a narrowband SIMO system reduces to a maximal-ratio combiner followed by a single temporal MF. The 2 × 1 channel model within the dotted box implicitly includes a pair of down converters, once for each receiver antenna. Therefore, the figure shows combining after downconversion. In practice the combining can be performed at RF, before downconversion, allowing implementation with only a single downconverter.

10.2.3. Linear Separability

The Nyquist criterion for scalar PAM signals states that, to avoid ISI, the Fourier transform of the pulse shape must alias to a constant. This result may be generalized to MIMO systems as follows [5]. If sampling $\boldsymbol{r}(t) = \Sigma_l \mathbf{H}(t - lT)\boldsymbol{a}_l$ at the symbol rate is to reproduce $\boldsymbol{a}_k$, so that $\boldsymbol{r}(kT) = \boldsymbol{a}_k$, without any ISI and without any co-channel interference, then $\mathbf{H}(t)$ must satisfy:

$$\mathbf{H}(kT) = \delta_k \mathbf{I} , \tag{10.43}$$

or equivalently:

$$\frac{1}{T}\sum_k \mathbf{H}(f - \tfrac{k}{T}) = \mathbf{I} , \tag{10.44}$$

where $\mathbf{H}(f)$ is the Fourier transform of $\mathbf{H}(t)$. These two equations represent the time-domain and frequency-domain manifestations of the *generalized Nyquist criterion*, which we first encountered in Section 6.3.4.

The Nyquist criterion has important implications. Consider first a SISO receiver that observes $r(t) = \Sigma_l a_l h(t - lT)$ in the absence of noise. An implication of the Nyquist criterion is that, in order for the receiver to be able to recover the transmitted symbols through linear filtering and symbol-rate sampling, the pulse shape $h(t)$ must be transformable into a Nyquist pulse through linear filtering, or equivalently, the folded spectrum $S(e^{j\theta})$ must be nonzero for all θ. From this we concluded that the bandwidth must be at least $1/(2T)$.

The question becomes more interesting in the MIMO case. Consider a MIMO receiver that observes $\boldsymbol{r}(t) = \Sigma_l \mathbf{H}(t - lT)\boldsymbol{a}_l$ in the absence of noise. Under what conditions can the receiver recover the transmitted symbols through linear MIMO filtering and symbol-rate sampling alone? The answer is the same: Only when the folded spectrum $\mathbf{S}(e^{j\theta})$ is invertible for all θ. But the implications are more profound, because it leads to an important relationship between the number of users (or more generally, number of channel inputs) n, the signal bandwidth W, and the number of receiver sensors m.

Let W be the bandwidth of all of the channel inputs, so that all components of $\mathbf{H}(f)$ are nonzero for $|f| \le W$ and zero for $|f| > W$. Let $1/T$ be the common symbol rate of all of the inputs. Then the number of inputs n that can be linearly separated must satisfy:

$$n \le m \lfloor W2T \rfloor , \tag{10.45}$$

where m is the number of receiver sensors. This result is a direct consequence of the generalized Nyquist criterion, and is proved in Appendix 10-A. It is useful to interpret the term $\lfloor W2T \rfloor$ as a *bandwidth expansion factor*, since it is the integer part of the ratio of the actual bandwidth W to the minimum bandwidth $1/(2T)$. For the SISO case with $m = n = 1$, (10.45) reduces to the familiar requirement that $W \ge 1/(2T)$.

The implications of (10.45) are that a receiver can linearly separate n channel inputs only when the product of the bandwidth expansion factor and the number of receiver sensors is at least n. With less than 100% excess bandwidth, the burden falls entirely on array size, and we must have $n \le m$. (This agrees with the well-known array-processing adage that an array with

m elements can cancel $m - 1$ narrowband interferers.) At the other extreme, with only a single receiver sensor, the burden falls entirely on the bandwidth expansion factor; in this case, (10.45) becomes $n \le \lfloor W2T \rfloor$, which implies that we must have $W \ge n/(2T)$. Hence, to accommodate n users, their bandwidths must be expanded by a factor of n. This is the principle behind spread-spectrum multiple-access communications. The implication of (10.45) is that by doubling the number of receiver sensors, we halve the required bandwidth expansion factor.

10.2.4. Joint ML Sequence Detection

Let us return to the problem of joint ML sequence detection posed in Section 10.2.1, this time with the aim of developing an efficient implementation. To get started, we briefly review our progress so far. From (10.34), the joint ML sequence detector for the channel of (10.31) chooses the decision sequence $\{\boldsymbol{a}_k\}$ so as to minimize:

$$J' = -2\sum_k \mathrm{Re}\{\boldsymbol{a}_k{}^*\boldsymbol{y}_k\} + \int_{-\infty}^{\infty} \| \sum_k \mathbf{H}(t - kT)\boldsymbol{a}_k \|^2 \, dt \,, \tag{10.46}$$

where $\boldsymbol{y}_k$ is defined in (10.35) as the output of the MMF $\mathbf{H}^*(-t)$ at time kT. The equivalent discrete-time MIMO channel mapping $\boldsymbol{a}_k$ to $\boldsymbol{y}_k$ is called the sampled-MF channel, and is shown in Fig. 10-6(b):

$$\boldsymbol{y}_k = \sum_l \mathbf{S}_l \boldsymbol{a}_{k-l} + \boldsymbol{n}_k \,, \tag{10.47}$$

where $\mathbf{S}_k$ is the sampled-autocorrelation function defined in (10.38), with Z-transform $\mathbf{S}(z)$, and where the PSD of the noise $\boldsymbol{n}_k$ is $N_0\mathbf{S}(z)$. If the cascade of the channel $\mathbf{H}(t)$ and the MMF $\mathbf{H}^*(-t)$ satisfies the generalized Nyquist criterion (10.44), then $\mathbf{S}(z)$ reduces to the identity matrix, and (10.47) reduces to a memoryless channel without crosstalk.

The Space-Time Whitened-Matched Filter

The key step to developing an efficient implementation of the ML sequence detector is the *space-time whitened-matched filter (WMF)*, which can be realized by appending a discrete-time filter after the sampled-matched filter. The WMF is often motivated as a means for achieving white Gaussian noise, which permits application of the Viterbi algorithm. However, this view of the WMF neglects to address its optimality with respect to the original ML criterion, as applied to the underlying continuous-time channel. In this section we follow the *geometric* approach of Section 5.4.3: we identify the WMF with an orthonormal basis for the signal space. When the MIMO channel reduces to a SISO channel, the space-time WMF presented here reduces to the scalar WMF of Section 5.4.

Consider the folded spectrum $\mathbf{S}(z)$, where $\mathbf{S}_k$ is defined in (10.38). Since $\mathbf{S}(z)$ is a valid PSD, it admits a unique right spectral factorization, according to Theorem 10-1:

$$\mathbf{S}(z) = \mathbf{G}^*(1/z^*)\mathbf{G}(z) \,, \tag{10.48}$$

where $\mathbf{G}(z) = \sum_{k=0}^{\infty} \mathbf{G}_k z^{-k}$ is space-time causal, with $\mathbf{G}_0$ lower triangular. When $\mathbf{S}(z)$ is constant, (10.48) reduces to a Cholesky factorization. Let us define an $m \times n$ filter $\boldsymbol{\Phi}(t)$ by its Fourier transform:

$$\mathbf{\Phi}(f) = \mathbf{H}(f)\mathbf{G}^{-1}(e^{j2\pi fT}) . \tag{10.49}$$

Clearly, this is a generalization of the scalar filter (5.80) to the MIMO case. In the following we will define the space-time WMF as a filter matched to $\mathbf{\Phi}(t)$. First, we show how $\mathbf{\Phi}(t)$ relates to the signal space.

Theorem 10-2. The set of translates $\{\mathbf{\Phi}(t - kT)\}$ forms an orthonormal basis for the signal space $\mathcal{S}$ spanned by $\{\mathbf{H}(t - kT)\}$, that is, the set of signals of the form $\{\sum_k \mathbf{H}(t - kT)\boldsymbol{x}_k\}$.

Proof. To demonstrate orthonormality, we must show that the following inner product:

$$\mathbf{D}_k = \int_{-\infty}^{\infty} \mathbf{\Phi}^*(t - kT)\mathbf{\Phi}(t)\, dt \tag{10.50}$$

reduces to $\mathbf{D}_k = \delta_k \mathbf{I}$. We may view $\mathbf{D}_k$ as a sampled version of $\mathbf{P}(t) = \mathbf{\Phi}^*(-t) \otimes \mathbf{\Phi}(t)$, which has Fourier transform $\mathbf{P}(f) = \mathbf{\Phi}^*(f)\mathbf{\Phi}(f)$. Hence, the Fourier transform of $\mathbf{D}_k = \mathbf{P}(kT)$ is:

$$\begin{aligned} \mathbf{D}(e^{j2\pi fT}) &= \frac{1}{T}\sum_k \mathbf{P}(f - \tfrac{k}{T}) \\ &= \frac{1}{T}\sum_k \mathbf{G}^{-*}(e^{j2\pi fT})\mathbf{H}^*(f - \tfrac{k}{T})\mathbf{H}(f - \tfrac{k}{T})\mathbf{G}^{-1}(e^{j2\pi fT}) \\ &= \mathbf{G}^{-*}(e^{j2\pi fT})\left\{\frac{1}{T}\sum_k \mathbf{H}^*(f - \tfrac{k}{T})\mathbf{H}(f - \tfrac{k}{T})\right\}\mathbf{G}^{-1}(e^{j2\pi fT}) \\ &= \mathbf{G}^{-*}(e^{j2\pi fT})\mathbf{S}(e^{j2\pi fT})\mathbf{G}^{-1}(e^{j2\pi fT}) = \mathbf{I} , \end{aligned} \tag{10.51}$$

where we used the fact that $e^{j2\pi(f - k/T)T} = e^{j2\pi fT}$ for all k. Taking the inverse transform yields $\mathbf{D}_k = \delta_k \mathbf{I}$, which proves that the elements of $\{\mathbf{\Phi}(t - kT)\}$ are orthonormal.

To demonstrate that $\{\mathbf{\Phi}(t - kT)\}$ forms a basis for the space $\mathcal{S}$ spanned by $\{\mathbf{H}(t - kT)\}$, we need only show that any element of $\mathcal{S}$ may be expressed as a right-linear combination of $\{\mathbf{\Phi}(t - kT)\}$. But if $\boldsymbol{s}(t) \in \mathcal{S}$ then there exists a sequence $\{\boldsymbol{a}_k\}$ such that $\boldsymbol{s}(t) = \sum_k \mathbf{H}(t - kT)\boldsymbol{a}_k$. In other words, we may generate $\boldsymbol{s}(t)$ by passing $\{\boldsymbol{a}_k\}$ through a filter with transfer function $\mathbf{H}(f)$. But from (10.49), we may decompose $\mathbf{H}(f)$ into the cascade of a discrete-time filter $\mathbf{G}(z)$ followed by a continuous-time filter $\mathbf{\Phi}(f)$. If we define $\boldsymbol{b}_k = \sum_l \mathbf{G}_l \boldsymbol{a}_{k-l}$ as the output of the discrete-time filter, we may express $\boldsymbol{s}(t)$ as a linear combination of $\{\mathbf{\Phi}(t - kT)\}$, namely $\boldsymbol{s}(t) = \sum_k \mathbf{\Phi}(t - kT)\boldsymbol{b}_k$, which proves that $\{\mathbf{\Phi}(t - kT)\}$ spans $\mathcal{S}$. This completes the proof.

Having established that $\{\mathbf{\Phi}(t - kT)\}$ is an orthonormal basis for the signal space, the JML sequence detector can, without loss of information, project the continuous-time observation onto this basis, yielding:

$$\boldsymbol{r}_k = \int_{-\infty}^{\infty} \mathbf{\Phi}^*(t - kT)\boldsymbol{r}(t)\, dt , \quad \text{for all } k . \tag{10.52}$$

This projection may be implemented using the *space-time whitened matched filter*, which is just a filter matched to $\mathbf{\Phi}(t)$. Such an implementation is shown in Fig. 10-9(a), where $\boldsymbol{r}(t)$ is passed through a filter matched to $\mathbf{\Phi}(t)$, and the output is sampled at time kT. From (10.49), the space-time WMF has Fourier transform:

$$\Phi^*(f) = \mathbf{G}^{-*}(e^{j2\pi fT})\mathbf{H}^*(f). \tag{10.53}$$

Hence, as shown in Fig. 10-9(b), we may implement $\mathbf{G}^{-*}(e^{j2\pi fT})$ in discrete time, after the sampler. Thus, the projection of (10.52) can be implemented by filtering the output of a conventional sampled matched filter.

As we have seen earlier (see (10.41)), the PSD of the noise $\boldsymbol{n}_k$ at the output of the sampled matched filter is $\mathbf{S}_{\boldsymbol{n}}(z) = N_0\mathbf{S}(z)$, where $\mathbf{S}(z)$ is the folded spectrum. Therefore, the noise $\boldsymbol{w}_k$ after the discrete-time filter $\mathbf{G}^{-*}(1/z^*)$ is white:

$$\begin{aligned}\mathbf{S}_{\boldsymbol{w}}(z) &= \mathbf{G}^{-*}(1/z^*)\mathbf{S}_{\boldsymbol{n}}(z)\mathbf{G}^{-1}(z) \\ &= \mathbf{G}^{-*}(1/z^*)N_0\mathbf{S}(z)\mathbf{G}^{-1}(z) \\ &= N_0\mathbf{I}\,,\end{aligned} \tag{10.54}$$

where we exploited the spectral factorization of (10.48). For this reason, we may interpret the discrete-time filter $\mathbf{G}^{-*}(1/z^*)$ after the sampler as a *noise-whitening filter*.

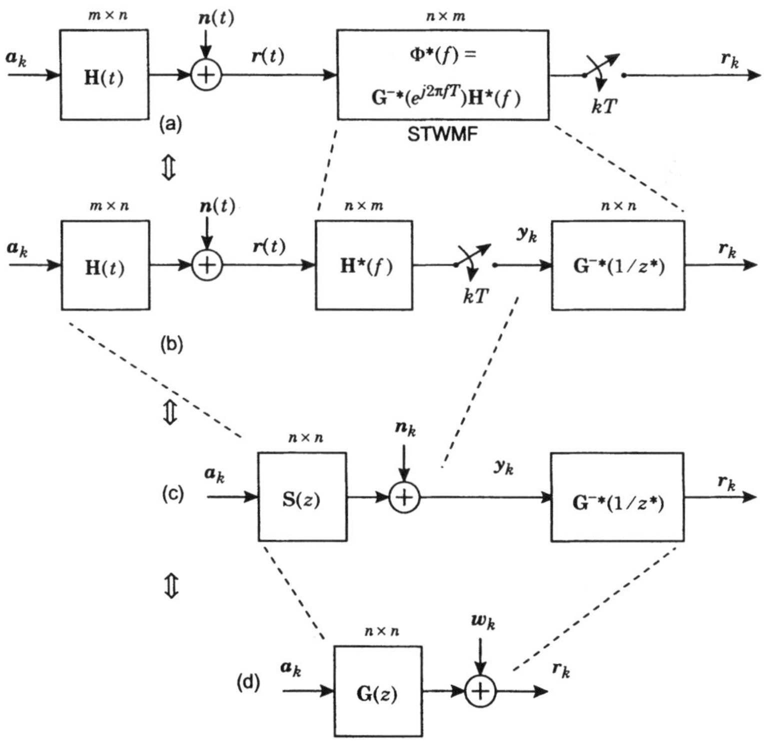

Fig. 10-9. The space-time WMF for MIMO channels can be implemented using a sampled matched filter followed by a discrete-time noise-whitening filter. It transforms a continuous-time white-noise channel (a), (b) into an equivalent discrete-time white-noise channel (c), (d) whose transfer function $\mathbf{G}(z)$ is causal.

The filter $\mathbf{G}^{-*}(1/z^*)$ not only whitens the noise, it also has a desirable impact on the desired signal. Specifically, since the transfer function between the channel input $\boldsymbol{a}_k$ and the sampled-MF output y_k is $\mathbf{S}(z)$, it follows that the transfer function between the channel input and the space-time WMF output is $\mathbf{G}^{-*}(1/z^*)\mathbf{S}(z)$, which reduces to $\mathbf{G}(z)$ from (10.48). Thus, the space-time WMF maps the underlying continuous-time white-noise channel, which is not necessarily space-time causal, into an equivalent discrete-time white-noise channel that is space-time causal.

Minimizing Discrete-Time Distance

An important fact that makes joint MLSD feasible in some circumstances is that the original minimum-distance criterion as applied to the continuous-time observation is equivalent to minimizing distance in discrete time, after the space-time WMF. We now demonstrate this result, where we use the notation $\|\cdot\|^2$ in two different ways, depending on whether its argument is a continuous-time vector signal or a discrete-time vector sequence:

$$\| \boldsymbol{r}(t)\|^2 = \sum_{i=1}^{n}\int_{-\infty}^{\infty}|r_i(t)|^2 dt\,, \qquad \| \boldsymbol{r}_k\|^2 = \sum_{i=1}^{n}\sum_{k=-\infty}^{\infty}|r_k^{(i)}|^2\,. \tag{10.55}$$

With this notation, the original joint MLSD cost, as applied to $\boldsymbol{r}(t)$, can be written as:

$$\begin{aligned} J &= \| \boldsymbol{r}(t) - \textstyle\sum_{l=-\infty}^{\infty}\mathbf{H}(t-lT)\boldsymbol{a}_k\|^2 \\ &= \left\| \boldsymbol{r}(t) - \hat{\boldsymbol{r}}(t) + \left\{ \hat{\boldsymbol{r}}(t) - \textstyle\sum_{l=-\infty}^{\infty}\mathbf{H}(t-lT)\boldsymbol{a}_l \right\}\right\|^2 \end{aligned} \tag{10.56}$$

where we have introduced $\hat{\boldsymbol{r}}(t)$, the projection of $\boldsymbol{r}(t)$ onto the signal space $\mathcal{S}$ spanned by $\{\mathbf{H}(t-kT)\}$. The term within the braces $\{\,\cdot\,\}$ lies within $\mathcal{S}$, and hence it can be expressed as a linear combination of the basis $\{\mathbf{\Phi}(t-kT)\}$, namely:

$$\left\{ \hat{\boldsymbol{r}}(t) - \sum_{l=-\infty}^{\infty}\mathbf{H}(t-lT)\boldsymbol{a}_l\right\} = \sum_{k=-\infty}^{\infty}\mathbf{\Phi}(t-kT)\left\{\boldsymbol{r}_k - \sum_{m=-\infty}^{\infty}\mathbf{G}_{k-m}\boldsymbol{a}_m\right\}. \tag{10.57}$$

By the projection theorem, this signal is orthogonal to the projection error $\boldsymbol{r}(t) - \hat{\boldsymbol{r}}(t)$. Furthermore, from Parseval's identity we find that its energy is given by:

$$\left\| \hat{\boldsymbol{r}}(t) - \sum_{l=-\infty}^{\infty}\mathbf{H}(t-lT)\boldsymbol{a}_l\right\|^2 = \left\|\boldsymbol{r}_k - \sum_{m=-\infty}^{\infty}\mathbf{G}_{k-m}\boldsymbol{a}_m\right\|^2. \tag{10.58}$$

Therefore, (10.56) reduces to:

$$J = \left\|\boldsymbol{r}(t) - \hat{\boldsymbol{r}}(t)\right\|^2 + \left\|\boldsymbol{r}_k - \sum_{m=-\infty}^{\infty}\mathbf{G}_{k-m}\boldsymbol{a}_m\right\|^2. \tag{10.59}$$

Because the first term is independent of $\{\boldsymbol{a}_k\}$, only the second term need be minimized. We recognize $\boldsymbol{r}_k$ as the result of passing $\boldsymbol{r}(t)$ through the space-time WMF, and we recognize the summation as a convolution of the candidate symbol sequence $\boldsymbol{a}_k$ with the WMF channel response $\mathbf{G}_k$. Thus, we have shown that the minimum-distance receiver operating on the continuous-time waveform reduces to a minimum-distance receiver operating on the space-time WMF output.

Suppose that $\mathbf{G}(z)$ is FIR with finite memory μ, so that $\mathbf{G}(z) = \sum_{k=0}^{\mu} \mathbf{G}_k z^{-k}$. Under this condition, the WMF output $\boldsymbol{r}_k$ is the output of a finite-state machine perturbed by white-Gaussian noise, and hence the ML sequence $\{\boldsymbol{a}_k\}$ may be determined using the Viterbi algorithm of Section 5.4.4. The state at time k may be defined as:

$$\psi_k = [\boldsymbol{a}_{k-1}, \boldsymbol{a}_{k-2}, \ldots, \boldsymbol{a}_{k-\mu}], \tag{10.60}$$

which can take on any of $|\mathcal{A}|^{n\mu}$ values, where $\mathcal{A}$ is the alphabet used by each of the n channel inputs. Since the complexity of the Viterbi algorithm is governed by the number of states, the JMLSD is practical only for small values of $|\mathcal{A}|$, n, and μ.

Example 10-13.

A single-input channel with memory 3 and alphabet size 4 has 64 states, which is manageable. But a four-input channel under the same conditions would have over 16 million states!

When the JML detector is impractical, suboptimal detectors with reduced complexity must be used. As described in the following section, practical strategies for MIMO detection include linear detection, decision-feedback detection, and multistage detection. So as to simplify their presentation, the next section will temporarily restrict its attention to a simple *memoryless* discrete-time model. Once we understand the basic principles for this special case we will be better prepared to tackle the more general continuous-time problem, which is addressed in Section 10.4.

10.3. Memoryless MIMO Channels

In this section we focus on the special case of a *memoryless* discrete-time channel, which embodies much of the essence of the general MIMO communications problem while avoiding complicated notation. Specifically, we consider the following n-input m-output channel:

$$\begin{aligned} \boldsymbol{r} &= \textstyle\sum_{i=1}^{n} a_i \boldsymbol{h}_i + \boldsymbol{n} \\ &= \mathbf{H}\boldsymbol{a} + \boldsymbol{n}\,, \end{aligned} \tag{10.61}$$

where $\boldsymbol{r} = [r_1, \ldots r_m]^T$ is the channel output, $\mathbf{H} = [\boldsymbol{h}_1, \ldots \boldsymbol{h}_n]$ is the $m \times n$ channel matrix whose i-th column is $\boldsymbol{h}_i$, $\boldsymbol{a} = [a_1, \ldots a_n]^T$ is the channel input, and $\boldsymbol{n} = [n_1, \ldots n_m]^T$ is the noise.

We will make the assumption that the n channel inputs are independent and uniformly chosen from the same discrete and complex alphabet. The inputs are thus uncoded. Furthermore, we will assume that the n columns $\boldsymbol{h}_1, \ldots \boldsymbol{h}_n$ of $\mathbf{H}$ are linearly independent, so that the rank of $\mathbf{H}$ is n. This assumption implies that there are at least as many channel outputs as channel inputs, $m \geq n$, so that $\mathbf{H}$ is a *square* or *tall* matrix. As usual, we will also assume white Gaussian noise, so that the real and imaginary components of $\boldsymbol{n}$ are i.i.d. zero-mean

Gaussian with variance $N_0/2$. The independence of the inputs and noise allows us to use the model (10.61) independently for each signaling interval, and hence without loss of generality we may focus on a single signaling interval.

Example 10-14.
The model of (10.61) can arise from the *narrowband* model of (10.42), where all transmitters use the same pulse shape $g(t)$, and the channel has a flat frequency response, so that $\mathbf{H}(t) = g(t)\mathbf{H}$. Specifically, suppose $g(t)$ is a unit-energy square-root Nyquist pulse, $\int g(t)g(t-kT)\,dt = \delta_k$. If the receiver applies a bank of filters matched to $g(t)$, one for each of the m receiver sensors, and samples their outputs at multiples of T, the m outputs will be related to the n channel inputs by (10.61).

Example 10-15.
The memoryless model also arises in *synchronous CDMA*, where n independent users transmit simultaneously to a receiver with a single sensor, leading to a $1 \times n$ continuous-time channel $\mathbf{H}(t) = [h_1(t), \ldots\ h_n(t)]$. Assume that each $h_i(t)$ is nonzero over the interval $[0, T)$ only. Let $\{\phi_1(t), \ldots\ \phi_m(t)\}$ denote an orthonormal basis for the span of $\{h_1(t), \ldots\ h_n(t)\}$. Then a bank of m filters matched to $\phi_1(t), \ldots\ \phi_m(t)$, which provides sufficient statistics, leads to (10.61). Interestingly, the dimensions change; a $1 \times n$ continuous-time channel becomes an $m \times n$ discrete-time channel.

The last example illustrates that the number of rows m in $\mathbf{H}$ may not be equal to the number of receiver sensors. This is a common occurrence. It is best to think of the number of rows m as an *effective* number of sensors that may be greater than the actual number of physical sensors.

The columns of $\mathbf{H}$ may be interpreted as the *signatures* for the channel inputs. In Example 10-14, the i-th column $\boldsymbol{h}_i$ represents the *spatial* signature for the i-th transmitter, whereas in Example 10-15, it represents the *spreading* signature for the i-th user.

10.3.1. Joint ML Detection

On a memoryless MIMO channel there are two ways to formulate the maximum-likelihood receiver optimization problem:

- The *joint-ML (JML)* detector chooses $\boldsymbol{a}$ so as to maximize $f(\boldsymbol{r} \mid \boldsymbol{a})$.
- The *individual-ML (IML)* detector chooses each element a_i of $\boldsymbol{a}$ to maximize $f(\boldsymbol{r} \mid a_i)$.

If $\hat{\boldsymbol{a}}$ denotes a vector of decisions, the JML detector minimizes the probability of vector error $\Pr[\hat{\boldsymbol{a}} \neq \boldsymbol{a}]$, whereas the IML detector minimizes the probability of symbol error $\Pr[\hat{a}_i \neq a_i]$. In the following we will focus on the JML detector because it is simpler, and also because its performance is not significantly different from the IML detector.

Because of the white Gaussian noise, the joint likelihood function is given by:

$$f(\boldsymbol{r} \mid \boldsymbol{a}) = \frac{1}{(\pi N_0)^m} \exp\left\{\frac{-1}{N_0}\|\boldsymbol{r} - \mathbf{H}\boldsymbol{a}\|^2\right\}. \tag{10.62}$$

Hence, the ML decision $\hat{\boldsymbol{a}}$ is also the minimum-distance decision, minimizing:

$$\|\boldsymbol{r} - \mathbf{H}\boldsymbol{a}\|^2 = \|\boldsymbol{r}\|^2 - 2\mathrm{Re}\{\boldsymbol{a}^*\mathbf{H}^*\boldsymbol{r}\} + \boldsymbol{a}^*\mathbf{H}^*\mathbf{H}\boldsymbol{a}\,. \tag{10.63}$$

Since the first term is independent of $\boldsymbol{a}$, we can equivalently choose $\boldsymbol{a}$ so as to minimize the following metric:

$$J(\boldsymbol{a}) = -2\text{Re}\{\boldsymbol{a}^*\boldsymbol{y}\} + \boldsymbol{a}^*\mathbf{H}^*\mathbf{H}\boldsymbol{a} , \tag{10.64}$$

we have introduced $\boldsymbol{y} = \mathbf{H}^*\boldsymbol{r}$. We may interpret $\boldsymbol{y}$ as the result of applying the spatial MMF $\mathbf{H}^*$ to $\boldsymbol{r}$. The MF is especially advantageous when $m \gg n$, because it converts the original m-dimensional observation into an n-dimensional observation that is a sufficient statistic. Any subsequent processing can then be performed with lower complexity.

A brute-force solution to the JML detection problem would perform an *exhaustive search*, whereby (10.64) would be computed for all possible $\boldsymbol{a}$, and the lowest-metric $\boldsymbol{a}$ would win. The number of candidate input vectors is $|\mathcal{A}|^n$, which can be prohibitively large.

Example 10-16.

If each of ten inputs uses a 64-QAM alphabet, there are 64^{10} possibilities. A receiver capable of one billion calculations of (10.64) per second would require 36 years to complete an exhaustive search. All of this to recover only 60 bits! This would explode to 150000 years if the number of inputs were to grow from ten to twelve.

In Section 10.3.8 we describe the *sphere detector*, which organizes its search in a clever way so as to avoid the full complexity of an exhaustive search. We delay further discussion on this topic until after our discussion on decision-feedback detection, which will greatly simplify its presentation.

When JML detection is impractical, we must resort to suboptimal alternatives with reduced complexity.

10.3.2. Genie-Aided Receivers and the Matched-Filter Bound

Before considering reduced-complexity receivers, we first consider a hypothetical receiver that bounds the performance of all receivers. Consider the output $\boldsymbol{y} = \mathbf{H}^*\boldsymbol{r}$ of the MMF:

$$\begin{aligned} \boldsymbol{y} &= \mathbf{H}^*\boldsymbol{r} , \\ &= \mathbf{H}^*(\mathbf{H}\boldsymbol{a} + \boldsymbol{n}) \\ &= \mathbf{R}\boldsymbol{a} + \boldsymbol{w} , \end{aligned} \tag{10.65}$$

where we have introduced the matrix:

$$\mathbf{R} = \mathbf{H}^*\mathbf{H} . \tag{10.66}$$

This matrix will play a central role in our analysis. It can be viewed as a matrix of crosscorrelations, since its (i, j)-th component is equal to the correlation between the i-th and j-th column of $\mathbf{H}$, namely $R_{ij} = \boldsymbol{h}_i^*\boldsymbol{h}_j$. The autocorrelation matrix of the Gaussian noise $\boldsymbol{w} = \mathbf{H}^*\boldsymbol{n}$ after the MMF is $\text{E}[\boldsymbol{w}\boldsymbol{w}^*] = N_0\mathbf{R}$. Let $\mathbf{R}^d = \text{diag}\{\, \|\boldsymbol{h}_1\|^2, \ldots \|\boldsymbol{h}_n\|^2\}$ denote the diagonal part of $\mathbf{R}$. In terms of this diagonal matrix, we may rewrite the MMF output as:

$$\boldsymbol{y} = \underbrace{\mathbf{R}^d\boldsymbol{a}}_{\text{desired}} + \underbrace{(\mathbf{R} - \mathbf{R}^d)\boldsymbol{a}}_{\text{interference}} + \boldsymbol{w} \; . \tag{10.67}$$

The matrix $\mathbf{R} - \mathbf{R}^d$ has zeros on the diagonal. The i-th component of $\boldsymbol{y}$ can be written as $y_i = R_{ii}a_i + x_i + w_i$, where $x_i = \sum_{j \neq i} R_{ij}\, a_j$ represents the contribution to the i-th MMF output from the interfering symbols $\{a_{j \neq i}\}$. Hence, the first and second vectors in (10.67) contain the desired and interference terms, respectively.

A *genie-aided* receiver for the i-th symbol a_i is a receiver that somehow knows the interfering symbols $\{a_{j \neq i}\}$. Consider a bank of such genie-aided receivers, one for each symbol a_1 through a_n. Since the interference term is deterministic and known to each receiver, the best it can do is to subtract it off. Hence, the bank of genie-aided receivers would subtract the interference term from (10.67), yielding:

$$\boldsymbol{z} = \boldsymbol{y} - (\mathbf{R} - \mathbf{R}^d)\boldsymbol{a}\ , \tag{10.68}$$

which reduces to $\boldsymbol{z} = \mathbf{R}^d\boldsymbol{a} + \boldsymbol{w}$. Since the interference has been eliminated, the i-th receiver need only quantize z_i to arrive at its genie-aided decision.

The genie-aided receiver is the gold standard against which practical receivers are compared. Its performance is a bound for any practical receiver, and is commonly called the *matched-filter bound (MFB)*, the *single-user* bound, or the *perfect interference-cancellation* bound. Since the i-th component of (10.68) is $z_i = \|\boldsymbol{h}_i\|^2 a_i + w_i$, its SNR and MSE are:

$$SNR_{MFB,i} = \frac{\|\boldsymbol{h}_i\|^2}{N_0}\ , \qquad MSE_{MFB,i} = \frac{N_0}{\|\boldsymbol{h}_i\|^2}\ , \tag{10.69}$$

assuming $E[\,|a_i|^2] = 1$. Simply stated, the MFB is the performance of the MF receiver for user i in the absence of other interferers. Like the MFB for ISI channels, the MFB is often not attainable by a practical receiver, even by the joint ML receiver.

10.3.3. Linear MIMO Detection

The simplest detector for memoryless MIMO channels is the *linear detector*, illustrated in Fig. 10-10. While in theory it does not always perform as well as more sophisticated detectors, in some applications it can be preferred due to its ease of implementation and robustness to nonidealities. A *decentralized* linear detector makes a decision about the i-th symbol by quantizing the decision statistic $y_i = \boldsymbol{w}_i{}^*\boldsymbol{r}$, where $\boldsymbol{w}_i$ is an $m \times 1$ vector of combining coefficients. A *centralized* linear detector is a bank of independent decentralized detectors operating in parallel; it produces a vector of decisions by independently quantizing the

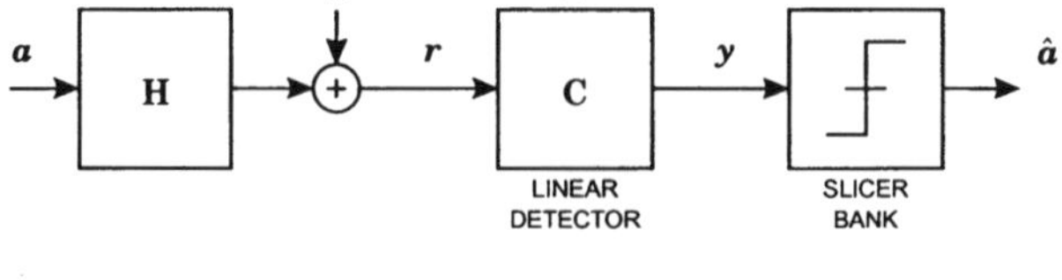

Fig. 10-10. A centralized linear detector applied to the memoryless MIMO channel of (10.61).

components of the vector $y = Cr$ for some $n \times m$ matrix $\mathbf{C}$, whose i-th row is w_i^*. In the sequel we will often focus on the more general case of centralized detection, knowing that we need only discard all but one of the rows of $\mathbf{C}$ to produce a decentralized detector.

Like the linear equalizer of Chapter 8, the linear detector for MIMO channels comes in two varieties: zero-forcing (ZF) and minimum mean-squared error (MMSE). A third possibility is a matched-filter (MF) detector, which essentially ignores interference. We briefly consider the MF detector first.

Matched-Filter Detection, the Near-Far Problem, and Power Control

The decentralized *MF detector* for user i chooses $w_i = h_i$. This would be optimal in the absence of the other interfering columns $\{h_{j \neq i}\}$, but it can be grossly suboptimal when they are present. Nevertheless, the MF detector has the advantage that a decentralized receiver need only know a single column of $\mathbf{H}$. In the context of multiuser detection for CDMA, the MF receiver is known as the *conventional receiver*; in this case, the columns of $\mathbf{H}$ represent the signatures of the different users, so that a decentralized MF detector need only know the signature of the desired user. Its performance is close to optimal in the special case when the columns of $\mathbf{H}$ happen to be nearly orthogonal; that is, when the off-diagonal elements of $\mathbf{R} = \mathbf{H}^*\mathbf{H}$ are negligible compared to the diagonal elements. This condition is rarely met in practice, with the possible exception of synchronous CDMA with power control, as described below.

A challenging problem in multiuser systems is the so-called *near-far problem*. Essentially, the near-far problem occurs whenever the energies of the interfering signals are much greater than the energy of the desired signal. It can arise whenever a nearby (and hence powerful) interferer inhibits a receiver from detecting the information transmitted by a far-away (and hence weak) desired user, as sketched below:

This phenomenon is peculiar to multiuser systems. It is not present in single-user ISI channels, because there the interfering symbols have the same energy as the desired symbol. Hence, this is one example of a MIMO phenomenon that has no counterpart in scalar ISI channels.

The MF or conventional receiver is particularly susceptible to the near-far problem. Consider the synchronous CDMA system of Example 10-15. Let $A_j = \| h_j \|$ denote the received amplitude for the j-th user. In practice the different signatures are designed to have a low normalized correlation, so that $| h_j^* h_i | / (A_i A_j) \ll 1$. Nevertheless, practical constraints prevent these correlations from being zero. Regardless of how small these nonzero correlations are, there will always exist a set of interfering amplitudes that will make the interference dominate over the desired signal, rendering the conventional MF receiver useless.

A classical approach to the near-far problem is to couple the MF receiver with a *power control* strategy, in which the power emitted by each transmitter is adjusted so that the received amplitudes for all users are equal: $A_j = A$ for all j. Power control is a form of zero-forcing linear MIMO equalization at the transmitters, and can be effective in some multiuser applications. However, in *single-user* MIMO applications for which the different channel inputs originate from the same transmitter, power control can be grossly suboptimal. Indeed, a capacity-achieving MIMO transmitter will follow a water-pouring strategy — the opposite of power control [6].

Zero-Forcing Linear Detection

A *zero-forcing (ZF) linear* detector chooses the matrix **C** of Fig. 10-10 so as to eliminate interference completely, regardless of noise enhancement. Specifically, a ZF linear detector chooses **C** so that $\mathbf{CH} = \mathbf{I}$. Such a matrix will always exist, given our assumption that there are at least as many channel outputs as inputs, and that the columns of **H** are linearly independent. When the channel has just as many outputs as inputs, so that **H** is a square matrix, then the ZF linear detector is unique: $\mathbf{C} = \mathbf{H}^{-1}$. When the channel has more outputs than inputs, $m > n$, then there are an infinite number of matrices **C** satisfying $\mathbf{CH} = \mathbf{I}$. In such cases we define *the* ZF linear detector as the unique such **C** that also minimizes the resulting $\text{MSE} = E[\|\mathbf{C}\boldsymbol{r} - \boldsymbol{a}\|^2]$.

The ZF linear detector is easily derived once we decompose **C** into the product $\mathbf{C} = \mathbf{WH}^*$, where **W** is an $n \times n$ matrix to be determined, and $\mathbf{H}^*$ is an MMF. We may do this because the MMF is known to give sufficient statistics, whether the criterion be ML or MMSE. Now, our constraint $\mathbf{CH} = \mathbf{I}$ becomes $\mathbf{WH}^*\mathbf{H} = \mathbf{I}$. Because **H** has full column rank, $\mathbf{H}^*\mathbf{H}$ is invertible; therefore, we must have $\mathbf{W} = (\mathbf{H}^*\mathbf{H})^{-1}$. Hence, the ZF linear detector is given by the $n \times m$ matrix:

$$\mathbf{C} = (\mathbf{H}^*\mathbf{H})^{-1}\mathbf{H}^*. \tag{10.70}$$

This matrix is also known as the *Moore-Penrose pseudoinverse* of **H**. When **H** is itself invertible, **C** reduces to $\mathbf{H}^{-1}$.

The decentralized ZF linear detector has a useful geometric interpretation. With respect to a given symbol index i, the desired symbol modulates the i-th column $\boldsymbol{h}_i$ of **H**, while the interfering symbols modulate the remaining columns $\{\boldsymbol{h}_{j \neq i}\}$. In this context, the *interference subspace* is defined as the span of the interfering columns $\{\boldsymbol{h}_{j \neq i}\}$. We can decompose $\boldsymbol{h}_i$ as the sum of a vector within the interference subspace and another orthogonal to it, according to:

$$\boldsymbol{h}_i = \hat{\boldsymbol{h}}_i + (\boldsymbol{h}_i - \hat{\boldsymbol{h}}_i)\,, \tag{10.71}$$

where the first term $\hat{\boldsymbol{h}}_i$ is the projection of $\boldsymbol{h}_i$ onto the interference subspace, and the second term $(\boldsymbol{h}_i - \hat{\boldsymbol{h}}_i)$ is the projection error. This decomposition is illustrated below:

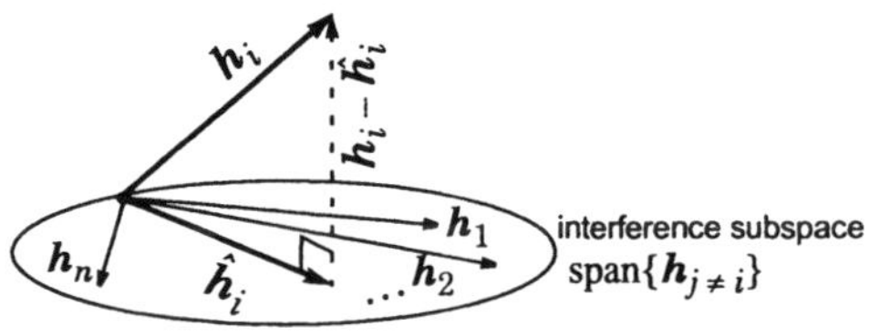

The ZF detector must discard the first term $\hat{\boldsymbol{h}}_i$ of (10.71) in order to satisfy the zero-interference constraint. In doing so, the ZF detector throws away that fraction of the signal that lies in the interference subspace. The corresponding loss of signal energy can lead to a significant penalty in performance. Scaling the ZF detector output to compensate for this loss in signal energy results in an amplification of the noise, or *noise enhancement.*

After discarding $\hat{\boldsymbol{h}}_i$, the ZF linear detector keeps only the remaining term in (10.71), so that its coefficient vector is proportional to the projection error: $\boldsymbol{w}_i = (\boldsymbol{h}_i - \hat{\boldsymbol{h}}_i)/G^2$. The projection theorem of Chapter 2 ensures that the projection error will be orthogonal to the interfering columns $\{\boldsymbol{h}_{j \neq i}\}$. The constant $G^2 = \|\boldsymbol{h}_i - \hat{\boldsymbol{h}}_i\|^2$ is chosen to force $\boldsymbol{w}_i{}^*\boldsymbol{h}_i = 1$, which compensates for the signal loss.

A systematic way of deriving the decentralized ZF linear detector is to apply the Gram-Schmidt orthonormalizaton procedure to the columns of $\mathbf{H}$ arranged so that the desired column $\boldsymbol{h}_i$ is last. Dividing the last basis vector by G will give the desired ZF linear detector $\boldsymbol{w}_i$.

In the context of multiuser detection, the ZF linear detector is called the *decorrelating* detector. Its importance stems from its optimality with respect to near-far resistance and ML detection in the presence of unknown signal amplitudes. Both attributes are best understood after we modify our channel model to explicitly account for signal amplitudes. If we define $A_i = \|\boldsymbol{h}_i\|$ as the received *amplitude* for the i-th user, then the channel model (10.61) reduces to:

$$\begin{aligned} \boldsymbol{r} &= \textstyle\sum_{i=1}^{n} a_i A_i \boldsymbol{s}_i + \boldsymbol{n} \\ &= \mathbf{S}\mathbf{A}\boldsymbol{a} + \boldsymbol{n}\,, \end{aligned} \tag{10.72}$$

where $\boldsymbol{s}_i = \boldsymbol{h}_i / \|\boldsymbol{h}_i\|$ is the unit-length signature for the i-th input. In terms of matrices, we have effectively decomposed the channel matrix $\mathbf{H}$ into the product $\mathbf{H} = \mathbf{SA}$, where $\mathbf{A} = \text{diag}(A_1, \dots A_n)$ is a diagonal matrix of signal amplitudes, with $A_i = \|\boldsymbol{h}_i\|$, and where $\mathbf{S} = [\boldsymbol{s}_1, \dots \boldsymbol{s}_n]$ is a matrix of unit-length signatures.

When the receiver has perfect knowledge of the received amplitudes, the joint ML detector of Section 10.3.1 is optimal. At the other extreme, however, when the receiver knows the signatures but has no a priori knowledge of the received amplitudes, and the alphabet is binary, the decorrelator is optimal. This is explored in the following example.

Example 10-17.

Consider the model of (10.72), where the receiver knows $\mathbf{S}$ exactly but has no *a priori* information about $\mathbf{A}$. This models the scenario in which the receiver knows the signatures of all transmitters, but it does not know the distance to each user. In this case, the receiver could solve the *joint* estimation problem of finding the symbols $\boldsymbol{a}$ and amplitudes $\mathbf{A}$ that jointly maximize the likelihood function $f(\boldsymbol{r} \mid \boldsymbol{a}, \mathbf{A})$, or equivalently that minimize:

$$\|\boldsymbol{r} - \mathbf{SA}\boldsymbol{a}\|^2 = \|\boldsymbol{r}\|^2 + (\mathbf{A}\boldsymbol{a} - \mathbf{R_S}^{-1}\boldsymbol{y})^*\mathbf{R_S}(\mathbf{A}\boldsymbol{a} - \mathbf{R_S}^{-1}\boldsymbol{y}) - \boldsymbol{y}^*\mathbf{R_S}^{-1}\boldsymbol{y}, \tag{10.73}$$

where we have introduced $\boldsymbol{y} = \mathbf{S}^*\boldsymbol{r}$ as the output of a bank of unit-energy matched filters, and $\mathbf{R_S} = \mathbf{S}^*\mathbf{S}$ as a matrix of signature correlations. The equality in (10.73) follows by completing the square and is easily verified. The first and last term do not depend on $\boldsymbol{a}$ and $\mathbf{A}$; hence, the joint estimator need only minimize the middle term. For the special case of the BPSK alphabet $\{\pm 1\}$, we can force the middle term to zero by choosing $\mathbf{A}\boldsymbol{a} = \mathbf{R_S}^{-1}\boldsymbol{y}$, which leads to the solutions $\hat{A}_i = |z_i|$ and $\hat{a}_i = \text{sign}(z_i)$ where $\boldsymbol{z} = \mathbf{R_S}^{-1}\boldsymbol{y}$. Hence, the decorrelator solves the joint maximum-likelihood estimation problem. The solution is more complicated when the alphabets are complex.

The performance of the ZF linear detector is easily evaluated. Since $\boldsymbol{r} = \mathbf{H}\boldsymbol{a} + \boldsymbol{n}$, the output $\boldsymbol{y} = \mathbf{C}\boldsymbol{r}$ of the ZF detector of (10.70) reduces to:

$$\begin{aligned} \boldsymbol{y} &= (\mathbf{H}^*\mathbf{H})^{-1}\mathbf{H}^*(\mathbf{H}\boldsymbol{a} + \boldsymbol{n}) \\ &= \boldsymbol{a} + \tilde{\boldsymbol{n}}, \end{aligned} \tag{10.74}$$

where the noise $\tilde{\boldsymbol{n}}$ is no longer white, but has autocorrelation matrix $N_0\mathbf{CC}^* = N_0\mathbf{R}^{-1}$, where $\mathbf{R} = \mathbf{H}^*\mathbf{H}$. Hence, the MSE for the i-th symbol is:

$$\text{MSE}_i = N_0(\mathbf{R}^{-1})_{ii}\,. \tag{10.75}$$

It is not hard to show that this expression can also be written as (see Problem 10-10):

$$\text{MSE}_i = N_0\|\boldsymbol{h}_i - \hat{\boldsymbol{h}}_i\|^{-2}, \tag{10.76}$$

where $\hat{\boldsymbol{h}}_i$ is the projection of $\boldsymbol{h}_i$ onto the subspace spanned by $\{\boldsymbol{h}_{j \neq i}\}$.

To quantify the gap in performance between any particular receiver and the MFB of the genie-aided receiver, three performance metrics are commonly used. The *multiuser efficiency* may be defined for our purposes as the ratio $MSE_{MFB,i}/MSE_i$. (See [7] for a more precise definition.) It is a number between zero and one that measures how close the detector comes to the MFB. The *asymptotic multiuser efficiency (AME)* is the limit of the multiuser efficiency as $N_0 \to 0$. The performance of a multiuser receiver is usually a strong function of the interfering amplitudes $\{A_{j \neq i}\}$. The *near-far resistance (NFR)* is defined by:

$$NFR_i = \min_{\{A_{j \neq i}\}} AME_i\,. \tag{10.77}$$

In words, the near-far resistance is the AME in the worst-case scenario when the interfering amplitudes happen to be just the right values to minimize the AME.

From (10.76), the multiuser efficiency for the ZF linear detector is:

$$MSE_{MFB,i}/MSE_i = \frac{\|\boldsymbol{h}_i - \hat{\boldsymbol{h}}_i\|^2}{\|\boldsymbol{h}_i\|^2} = 1 - \frac{\|\hat{\boldsymbol{h}}_i\|^2}{\|\boldsymbol{h}_i\|^2}. \tag{10.78}$$

The ZF linear detector achieves the MFB only when $\boldsymbol{h}_i$ is orthogonal to the interference subspace; i.e., when $\hat{\boldsymbol{h}}_i$ is zero. Since (10.78) is independent of N_0, it also describes the AME of the ZF linear detector. Furthermore, since (10.78) is independent of the interfering amplitudes, the NFR of the ZF linear detector is also given by (10.78).

It is a surprising result that the NFR of the JML detector is also given by (10.78). In other words, the ZF linear detector is optimally near-far resistant. Of course, for most interfering amplitudes the JML detector is superior to the ZF linear detector, but if the interfering amplitudes are worst-case, the JML detector can do no better than the ZF linear detector. A rigorous demonstration of this result would take us too far afield, but the gist is as follows, assuming binary alphabets, $a_i \in \{\pm 1\}$. First, the probability that the JML detector makes an error for the i-th symbol is roughly $Q(d_{\min}/\sqrt{2N_0})$ at high SNR, where:

$$d_{\min} = \min_{\{\boldsymbol{a},\, \tilde{\boldsymbol{a}}:a_i \neq \tilde{a}_i\}} \|\mathbf{SA}\boldsymbol{a} - \mathbf{SA}\tilde{\boldsymbol{a}}\| . \tag{10.79}$$

In contrast, the error probability for the ZF linear detector is $Q(\sqrt{2/MSE_i})$. We can define an *effective asymptotic* MSE for the JML detector by equating the two Q functions, yielding $MSE_i^{JML} = 4N_0/d_{\min}^2$. But $d_{\min}$ of (10.79) reduces to:

$$\begin{aligned} d_{\min} &= \min_{\{\boldsymbol{a},\, \tilde{\boldsymbol{a}}:a_i \neq \tilde{a}_i\}} \|A_i(a_i - \tilde{a}_i)\boldsymbol{s}_i + \textstyle\sum_{j \neq i} A_j(a_j - \tilde{a}_j)\boldsymbol{s}_j\| \\ &= \min_{a_i \in \{\pm 1\},\, e_j \in \{0, \pm 1\}} \|2A_i a_i \boldsymbol{s}_i - 2\textstyle\sum_{j \neq i} A_j e_j \boldsymbol{s}_j\| \\ &= \min_{e_j \in \{0, \pm 1\}} 2\|\boldsymbol{h}_i - \textstyle\sum_{j \neq i} A_j e_j \boldsymbol{s}_j\| , \end{aligned} \tag{10.80}$$

where we have introduced $e_j = (\tilde{a}_j - a_j)/2 \in \{0, \pm 1\}$. If (10.80) is then minimized over the set of interfering amplitudes, the sets $\{e_j\}$ and $\{A_j\}$ will be such that the term $\sum_{j \neq i} A_j e_j \boldsymbol{s}_j$ in (10.80) reduces to the projection $\hat{\boldsymbol{h}}_i$ of $\boldsymbol{h}_i$ onto the span of $\{\boldsymbol{h}_{j \neq i}\}$, in which case:

$$d_{\min} = 2\|\boldsymbol{h}_i - \hat{\boldsymbol{h}}_i\| . \tag{10.81}$$

Hence, the worst-case effective MSE for the JML detector is $4N_0/d_{\min}^2 = N_0\|\boldsymbol{h}_i - \hat{\boldsymbol{h}}_i\|^{-2}$, which is exactly the same as the MSE of (10.76) for the ZF linear detector.

Despite its optimality in the face of worst-case interfering amplitudes, the ZF detector can perform very poorly in other circumstances, as shown in the following example.

Example 10-18.

Consider the memoryless channel

$$\mathbf{H} = \begin{bmatrix} 1 & \varepsilon \\ \varepsilon & 0 \end{bmatrix}.$$

When ε is small, the second column is nearly collinear with the first. In order to eliminate the interference from the second user, the ZF detector for the first user would discard the first component of the channel output, even though it contains most of the desired signal energy, and even though the interference is relatively benign. When ε is small, it would be much better to ignore altogether the interference, and use an MF detector instead. To quantify the performance difference, we may calculate the respective outputs of the ZF linear detector (which uses the first row of $\mathbf{H}^{-1}$) and the MF detector (which uses the first row of $\mathbf{H}^*$) as:

$$y_{ZF} = a_1 + n_2/\varepsilon\,, \qquad y_{MF} = (1+\varepsilon^2)a_1 + \varepsilon a_2 + n_1 + \varepsilon n_2\,. \tag{10.82}$$

Comparing the two, the ZF detector has zero interference but amplified noise, while the MF detector has a small amount of interference and negligible noise amplification. Assuming 4-QAM alphabets of $\{\pm 1 \pm j\}/\sqrt{2}$, taking the sign of the real and imaginary components of y_{ZF} and y_{MF} will produce respective error probabilities of:

$$BER_{\text{ZF}} = Q\left(\frac{\varepsilon}{\sqrt{N_0}}\right), \qquad BER_{\text{MF}} = \tfrac{1}{2}Q\left(\frac{1-\varepsilon+\varepsilon^2}{\sqrt{N_0(1+\varepsilon^2)}}\right) + \tfrac{1}{2}Q\left(\frac{1+\varepsilon+\varepsilon^2}{\sqrt{N_0(1+\varepsilon^2)}}\right). \tag{10.83}$$

From these expressions we can conclude that the MF detector outperforms the ZF linear detector by roughly $1/\varepsilon^2$ in SNR when ε and N_0 are small; e.g., by 40 dB when $\varepsilon = 10^{-2}$ and BER $= 10^{-6}$.

Minimum-Mean-Squared Error Linear Detection

A drawback of the ZF linear detector is its insistence on forcing the interference to zero, regardless of the interference strength. As demonstrated in Example 10-18, the ZF detector will discard any desired signal energy that lies in the interference subspace, even if the interfering columns of $\mathbf{H}$ happen to have much less energy than the desired column $\boldsymbol{h}_i$. The signal energy lost in this way is effectively a form of noise enhancement. A better strategy is to choose $\mathbf{C}$ so as to balance the lost signal energy with the increased interference; it is much better to accept some residual interference if it allows you to capture more of the desired signal energy.

The *MMSE linear detector* chooses $\mathbf{C}$ so as to minimize the sum MSE $= \text{E}[\|\mathbf{C}\boldsymbol{r} - \boldsymbol{a}\|^2]$ directly, without the additional zero-forcing constraint that $\mathbf{CH} = \mathbf{I}$. Unlike the ZF detector, which minimizes interference but neglects noise, and unlike the MF detector, which minimizes noise but neglects interference, the MMSE detector achieves an optimal balance of noise enhancement and interference suppression. The MSE can also be expressed as MSE $= \text{tr}\{\mathbf{R}_e\}$, where:

$$\begin{aligned}\mathbf{R}_e &= \text{E}[(\mathbf{C}\boldsymbol{r} - \boldsymbol{a})(\mathbf{C}\boldsymbol{r} - \boldsymbol{a})^*] \\ &= \mathbf{C}\mathbf{R}_r\mathbf{C}^* + \mathbf{I} - \mathbf{H}^*\mathbf{C}^* - \mathbf{C}\mathbf{H} \\ &= (\mathbf{C} - \mathbf{H}^*\mathbf{R}_r^{-1})\mathbf{R}_r(\mathbf{C} - \mathbf{H}^*\mathbf{R}_r^{-1})^* + \mathbf{I} - \mathbf{H}^*\mathbf{R}_r^{-1}\mathbf{H} \qquad (10.84) \\ &= (\mathbf{C} - \mathbf{H}^*\mathbf{R}_r^{-1})\mathbf{R}_r(\mathbf{C} - \mathbf{H}^*\mathbf{R}_r^{-1})^* + N_0(\mathbf{H}^*\mathbf{H} + N_0\mathbf{I})^{-1}. \qquad (10.85)\end{aligned}$$

We have introduced $\mathbf{R}_r = \text{E}[\boldsymbol{r}\boldsymbol{r}^*]$, the received autocorrelation matrix. In deriving the above, we have assumed that the channel inputs are uncorrelated with unit energy, so that $\text{E}[\boldsymbol{a}\boldsymbol{a}^*] = \mathbf{I}$. (The columns of $\mathbf{H}$ can be scaled to account for an alphabet that does not have unit energy.) Only the first term of (10.85) depends on $\mathbf{C}$, and we can do no better than to make this term zero. Hence, the MMSE solution is $\mathbf{C} = \mathbf{H}^*\mathbf{R}_r^{-1}$, which can be rewritten in two equivalent ways:

$$\mathbf{C} = \mathbf{H}^*(\mathbf{H}\mathbf{H}^* + N_0\mathbf{I})^{-1} \tag{10.86}$$

$$= (\mathbf{H}^*\mathbf{H} + N_0\mathbf{I})^{-1}\mathbf{H}^* . \tag{10.87}$$

The equivalence is easily verified. Although the difference between (10.86) and (10.87) is subtle at first glance, closer inspection yields three important differences: the MF comes last in (10.86) but first in (10.87); the identity matrix has dimension $m \times m$ in (10.86) but $n \times n$ in (10.87); and the product $\mathbf{HH}^*$ in (10.86) becomes $\mathbf{H}^*\mathbf{H}$ in (10.87). The first form (10.86) of the MMSE detector is often preferred in analysis. However, the last form (10.87) can be implemented with lower complexity because the matrix inverse has a smaller dimension. It also obviously reduces to the ZF linear detector of (10.70) when the noise is zero.

After the MMSE linear detector, the autocorrelation of the error signal is given by the last term in (10.85). Hence, the MSE $E[\,|y_i - a_i|^2]$ for the i-th symbol is given by:

$$MSE_i = N_0(\tilde{\mathbf{R}}^{-1})_{ii} , \tag{10.88}$$

where we have introduced $\tilde{\mathbf{R}} = \mathbf{H}^*\mathbf{H} + N_0\mathbf{I}$. In the absence of noise, $\tilde{\mathbf{R}}$ reduces to $\mathbf{R}$, and the performance of the MMSE linear detector reduces to that of the ZF linear detector (10.75).

The MMSE detector of (10.86) and (10.87) is centralized. A decentralized detector for the i-th input takes the form $y_i = \boldsymbol{w}_i^*\boldsymbol{r}$, where $\boldsymbol{w}_i$ is the i-th column of $\mathbf{C}^*$, namely:

$$\begin{aligned} \boldsymbol{w}_i &= (\mathbf{HH}^* + N_0\mathbf{I})^{-1}\boldsymbol{h}_i \\ &= (\boldsymbol{h}_i\boldsymbol{h}_i^* + \Sigma_{j \neq i}\boldsymbol{h}_j\boldsymbol{h}_j^* + N_0\mathbf{I})^{-1}\boldsymbol{h}_i . \end{aligned} \tag{10.89}$$

From this expression we observe that, in two extreme cases, the MMSE detector approaches the MF detector. First, in the limit as the noise variance goes to infinity, then $\boldsymbol{w}_i \to \boldsymbol{h}_i/R$ with $R = N_0$. Second, if the noise variance is fixed but the interfering amplitudes go to zero, $\|\boldsymbol{h}_{j \neq i}\| \to 0$, then $\boldsymbol{w}_i \to \boldsymbol{h}_i/R$ with $R = \|\boldsymbol{h}_i\|^2 + N_0$.

Example 10-19.

The decentralized MMSE linear detector for the first input to the channel $\mathbf{H} = \begin{bmatrix} 2 & A_2 \\ 1 & 0 \end{bmatrix}$ ranges from the MF detector to the ZF linear detector as A_2 ranges from zero to infinity.

Example 10-20.

Consider again the channel from Example 10-18. The decentralized linear MMSE detector is the first row of $(\mathbf{H}^*\mathbf{H} + N_0\mathbf{I})^{-1}\mathbf{H}^*$, which can be expressed as $\boldsymbol{w}^* = [1, d]/A$, where:

$$d = \varepsilon\left(1 + \frac{\varepsilon^2}{N_0}\right), \quad \text{and} \quad A = 1 + \varepsilon^4/N_0 + 2\varepsilon^2 + N_0 . \tag{10.90}$$

If we fix ε and let $N_0 \to 0$, the MMSE detector approaches the ZF detector: $\boldsymbol{w}^* \to [0, \varepsilon^{-1}]$, which is the first row of $\mathbf{H}^{-1}$. On the other hand, if we fix N_0 and let $\varepsilon \to 0$, the MMSE detector approaches the MF detector: $\boldsymbol{w}^* \to [1, \varepsilon]/(1 + N_0)$, which is proportional to the first row of $\mathbf{H}^*$. Thus, for certain values of ε and N_0, the MMSE detector may resemble the ZF detector or the MF detector. Evaluating $y = \boldsymbol{w}^*\boldsymbol{r}$ in the general case leads to:

$$y_{\text{MMSE}} = \Big((1 + d\varepsilon)a_1 + \varepsilon a_2 + n_1 + dn_2\Big) / A\,. \tag{10.91}$$

Assuming 4-QAM alphabets of $\{\pm 1 \pm j\}/\sqrt{2}$, taking the sign of the real and imaginary parts of y_{MMSE} yields an overall bit-error probability of:

$$BER_{\text{MMSE}} = \tfrac{1}{2}Q\Big(\frac{1 - \varepsilon + d\varepsilon}{\sqrt{N_0(1 + d^2)}}\Big) + \tfrac{1}{2}Q\Big(\frac{1 + \varepsilon + d\varepsilon}{\sqrt{N_0(1 + d^2)}}\Big)\,. \tag{10.92}$$

This expression is plotted below for $\varepsilon = 10^{-2}$, along with BER_{ZF} from (10.83):

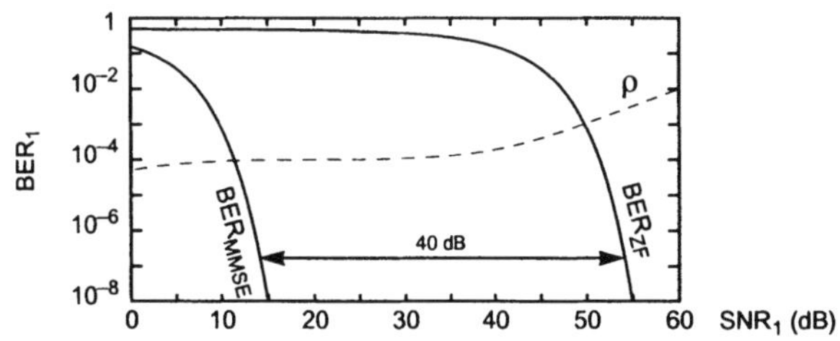

Again, like the MF, the MMSE detector is seen to outperform the ZF detector by $1/\varepsilon^2$ when ε and N_0 are small, or by 40 dB when $\varepsilon = 10^{-2}$. The dashed curve shows $\rho = \text{MSE}_{\text{MMSE}}/\text{MSE}_{\text{ZF}}$.

The previous example conveys an important message — the MMSE detector can significantly outperform the ZF detector, *even at high SNR*. At first glance this is a surprising result, because it seems to contradict two earlier results, namely:

- The MMSE detector coefficients of (10.87) approach the ZF detector coefficients of (10.70) as $N_0 \to 0$.
- The MSE performance of (10.88) for the MMSE detector approaches that of the ZF detector (10.75) as $N_0 \to 0$.

There is no contradiction, except that the above two limiting results are somewhat misleading. The value of SNR that is required to make the limits valid may be beyond the SNR range of interest. For example, the second result above implies that the ratio $\rho = \text{MSE}_{\text{MMSE}}/\text{MSE}_{\text{ZF}}$ satisfies $\rho \to 1$ as $N_0 \to 0$. While true in theory, the dotted curve in the previous example shows how the ratio actually behaves as a function of SNR for this example. We see that ρ does grow with SNR, but only slowly; ρ is only 10^{-2} when SNR = 60 dB. Eventually we will have $\rho \to 1$, but not until SNR is very high. Specifically, to achieve $\rho \ge 0.99$ requires SNR $\ge$ 100 dB. By then, (3.43) indicates that the BER of even the ZF detector will be infinitessimal: $BER_{ZF} < 10^{-217147}$.

Example 10-21.

Consider a receiver with an array of $m = 4$ omnidirectional antennas. If the antennas are arranged in a line with uniform half-wavelength spacing, the SIMO channel response to a narrowband transmitter with angle of incidence θ is given by the so-called *steering vector* [8]:

$$\mathbf{v}(\theta) = [1,\ e^{j\pi\sin(\theta)},\ e^{j2\pi\sin(\theta)},\ e^{j3\pi\sin(\theta)}]^T . \tag{10.93}$$

As illustrated in Fig. 10-11, suppose there is one desired user and three interferers with angles of incidence 15°, 30°, 70°, and 80°, respectively. Furthermore, suppose a reflection from the desired user arrives with angle of incidence 50° and attenuation 0.9, and that the second and third interferers are attenuated by 0.5 and 0.6, respectively. Then the MIMO model of (10.61) applies, with $\mathbf{H} = [\boldsymbol{h}_1, \boldsymbol{h}_2, \boldsymbol{h}_3, \boldsymbol{h}_4]$ and:

$$\boldsymbol{h}_1 = \mathbf{v}(15°) + 0.9\mathbf{v}(50°)\,, \quad \boldsymbol{h}_2 = \mathbf{v}(30°)\,, \quad \boldsymbol{h}_3 = 0.5\mathbf{v}(70°)\,, \quad \boldsymbol{h}_4 = 0.6\mathbf{v}(80°)\,.$$

A decentralized linear detector takes the form $y = \boldsymbol{c}^*\boldsymbol{r}$. In this context, given $\boldsymbol{c}$, the *antenna array pattern* is the response $|\boldsymbol{c}^*\mathbf{v}(\theta)|$ as a function of θ, and it measures the antenna gain or attenuation as a function of angle of incidence. The array patterns for the following four detectors are shown in Fig. 10-11 using a polar plot, assuming $N_0 = 0.08$ and a 4-QAM alphabet $\{\pm 1 \pm j\}/\sqrt{2}$:

- The *beamsteering* detector is $\boldsymbol{c} = \mathbf{v}(15°)$; it is matched to the LOS path only, and ignores both the reflected path of the desired signal and the interference.

- The maximal-ratio combiner (or MMF) is $\boldsymbol{c} = \boldsymbol{h}_1$; it is matched to all contributions from the desired signal, including the multipath, but ignores the interference.

- The zero-forcing detector is $\boldsymbol{c} = (\mathbf{HH}^*)^{-1}\boldsymbol{h}_1$, or equivalently the first column of $\mathbf{H}^{-*}$; it captures as much signal energy as possible, while steering nulls in the directions of all interferers. From the figure we see that the main lobe of the ZF detector is not very close to the desired signal, but it is as close as possible while satisfying the ZF constraint.

- The MMSE detector is $\boldsymbol{c} = (\mathbf{HH}^* + N_0\mathbf{I})^{-1}\boldsymbol{h}_1$. By relaxing the ZF constraint, the MMSE detector is able to accept more energy from the LOS path, and also steer a secondary beam to the multipath component, while very nearly nulling the interferers.

The output constellations clearly illustrate the superiority of the MMSE detector over the others.

A *centralized* MMSE detector is able to separate the signals from multiple narrowband users, even when they transmit at the same time, they are in the same vicinity, and they occupy the same bandwidth; this concept is known as *space-division multiple access (SDMA)*. The only requirement is that the spatial signatures $\{\boldsymbol{h}_i\}$ induced by the different users be linearly independent, a condition that is easily met in practice when the users are spatially separated and are subject to multipath propagation.

Example 10-22.

Consider three independent narrowband users transmitting simultaneously and in the same band to a receiver using a uniform linear array with 8 elements, and suppose the first two users are subject to multipath propagation, so that the channel matrix of (10.61) is $\mathbf{H} = [\boldsymbol{h}_1, \boldsymbol{h}_2, \boldsymbol{h}_3]$, where:

$$\boldsymbol{h}_1 = \mathbf{v}(85°) + 0.9\mathbf{v}(50°)\,, \quad \boldsymbol{h}_2 = \mathbf{v}(70°) + 0.8\mathbf{v}(120°), \quad \boldsymbol{h}_3 = 0.5\mathbf{v}(160°)\,.$$

The geometry of this channel is illustrated in Fig. 10-12. Let $\mathbf{c}_1^*$ and $\mathbf{c}_2^*$ denote the first and second row of the centralized linear MMSE detector of (10.87), assuming $N_0 = 0.08$. The array

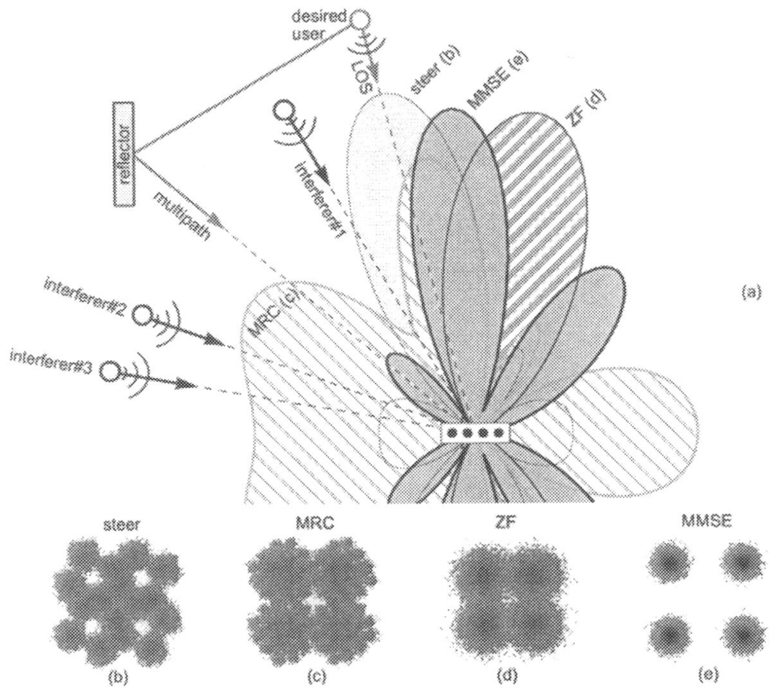

Fig. 10-11. (a) Channel geometry and array patterns for Example 10-21. Also shown are the corresponding output constellations for (b) beamsteering, (c) MRC, (d) ZF, and (e) MMSE linear detectors.

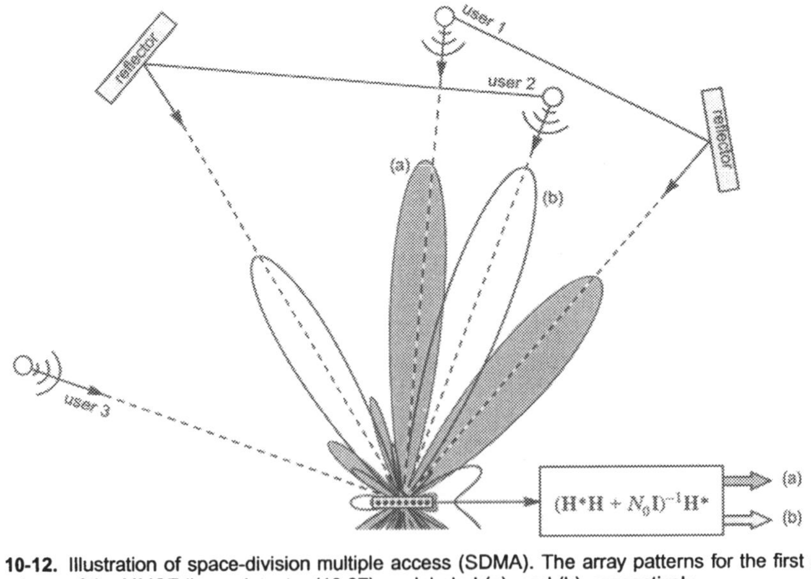

Fig. 10-12. Illustration of space-division multiple access (SDMA). The array patterns for the first and second row of the MMSE linear detector (10.87) are labeled (a), and (b), respectively.

pattern for $\mathbf{c}_1$ and $\mathbf{c}_2$ are shown in Fig. 10-12. Observe that the array pattern for $\mathbf{c}_1$ steers beams towards the LOS and reflected paths of user 1, while very nearly steering nulls towards the contributions from user 2 and user 3. Meanwhile, $\mathbf{c}_2$ steers beams towards the LOS and reflected paths of user 2, while nulling the contributions from users 1 and 3.

10.3.4. Decision-Feedback Detection for MIMO Channels

In this section we describe the *zero-forcing decision-feedback (ZF-DF) detector*, a nonlinear method for MIMO detection invented by Duel-Hallen [9]. The ZF-DF detector can be viewed as a modification of the conventional ZF-DFE for ISI channels that operates in the *spatial* domain instead of the time domain. It is a powerful technique that can significantly outperform a linear detector. The ZF-DF detector is an example of a more general technique called *successive interference cancellation (SIC)* [10]. In array-to-array MIMO wireless links, the ZF-DF detector is known as the *BLAST nulling and cancelling detector* [11][12], while in packet transmission applications it is known as a *generalized* DFE (GDFE) [13].

In the next five sections we will present five equivalent views of the DF detector: (1) an SIC view; (2) a matrix-valued DFE view; (3) a Gram-Schmidt view; (4) a Cholesky whitened-matched filter view; and (5) a linear prediction view. Each interpretation is valid in its own right, and each conveys additional insight into the workings of the DF detector. Although all five views apply to both the MMSE and ZF realizations of the DF detector, we will focus on the ZF version so as to simplify our presentation; the MMSE version will be considered separately in Section 10.3.5.

The SIC View

A DF detector for the memoryless channel $\boldsymbol{r} = \mathbf{H}\boldsymbol{a} + \boldsymbol{n}$ can be described as follows. First, a decision $\hat{a}_1$ about the first symbol is made by slicing the decision statistic $z_1 = \boldsymbol{w}_1{}^*\boldsymbol{r}$, where the coefficients of the vector $\boldsymbol{w}_1$ are chosen to suppress the interference from the interfering symbols $a_2, \dots a_n$. In other words, $\boldsymbol{w}_1{}^*$ is the decentralized ZF linear detector given by the first row of (10.70). This is the *nulling* step. The decision may be written as $\hat{a}_1 = Q\{\boldsymbol{w}_1{}^*\boldsymbol{r}\}$, where $Q\{z\}$ is the element of the alphabet closest to z. Second, the contribution to $\boldsymbol{r}$ from a_1 is reconstructed and subtracted off, yielding $\boldsymbol{r}_2 = \boldsymbol{r} - \hat{a}_1\boldsymbol{h}_1$. This is the *cancellation* step. When the decision is correct, the result is $\boldsymbol{r}_2 = \mathbf{H}_2\boldsymbol{a}_2 + \boldsymbol{n}$, where $\mathbf{H}_2 = [\boldsymbol{h}_2, \dots \boldsymbol{h}_n]$ is what is left after the first column of $\mathbf{H}$ is removed, and $\boldsymbol{a}_2 = [a_2, \dots a_n]^T$ is what is left after the first symbol of $\boldsymbol{a}$ is removed.

The cancellation produces a residual channel model $\boldsymbol{r}_2 = \mathbf{H}_2\boldsymbol{a}_2 + \boldsymbol{n}$ that is equivalent to the original, but with one less input, so we can follow a similar procedure to detect a_2. The process can repeat in a recursive fashion for symbols a_3, a_4, etc., until all n symbols have been detected. The detector can be summarized succinctly by the following recursion:

$$\boldsymbol{r}_{i+1} = \boldsymbol{r}_i - Q\{\boldsymbol{w}_i{}^*\boldsymbol{r}_i\}\,\boldsymbol{h}_i\,, \tag{10.94}$$

which is iterated for $i = 1$ to $n - 1$, with $\boldsymbol{r}_1 = \boldsymbol{r}$ initially. A block diagram of the DF detector is shown in Fig. 10-13.

The nulling vectors $\{w_1, \dots w_n\}$ for the ZF-DF detector are easily specified. Under the assumption of correct decisions, the i-th stage of the detector will face an effective channel of the form $r_i = \mathbf{H}_i a_i + n$, where $\mathbf{H}_i = [h_i, \dots h_n]$ is what is left of the original matrix after the first $i-1$ columns have been removed, and $a_i = [a_i, \dots a_n]^T$ is what is left after the first $i-1$ symbols of a are removed. The role of the i-th nulling vector is to extract the desired symbol while rejecting the contributions from the undetected symbols $a_{i+1}, \dots a_n$. In other words, the i-th nulling vector must satisfy the *zero-forcing condition*:

$$w_i^* \mathbf{H}_i = [1, \underbrace{0, 0, \dots 0}_{n-i}] \; ; \tag{10.95}$$

The minimum-norm (and hence minimum-MSE) solution is to apply the ZF linear detector (10.70) to the effective channel $\mathbf{H}_i$. Specifically, the i-th nulling vector for the ZF-DF detector is the first row of $\mathbf{C}_i = (\mathbf{H}_i^* \mathbf{H}_i)^{-1} \mathbf{H}_i^*$, the Moore-Penrose pseudoinverse of $\mathbf{H}_i$.

The Matrix View

In this section we develop a model of the DF detector based on matrices. We begin with Fig. 10-14, which shows an alternative implementation of the DF detector of Fig. 10-13. The two versions are precisely equivalent. The only difference is that the new detector cancels the interference from already detected symbols *after* each nulling vector w_i, whereas the original implementation of Fig. 10-13 performed the cancellation on the channel output r directly, *before* the nulling vectors. Subtracting the vector h_1 before the nulling filter w_2^* is the same as subtracting the scalar $w_2^* h_1$ after. Thus, choosing the coefficient b_{21} in Fig. 10-14 according to $b_{21} = w_2^* h_1$ ensures that the input to the second slicer of the new detector is identical to that in the original detector. Similarly, by choosing all of the cancellation coefficients $\{b_{ij}\}$ according to $b_{ij} = w_i^* h_j$, we ensure that Fig. 10-14 is equivalent to Fig. 10-13.

The new version becomes conceptually simple when we adopt a matrix notation. Let $y_i = w_i^* r$ denote the output of the i-th nulling vector in Fig. 10-14, before the cancellation. If we collect the n outputs into a single vector $y = [y_1, \dots y_n]^T$, we can express the n nulling operations by the single equation:

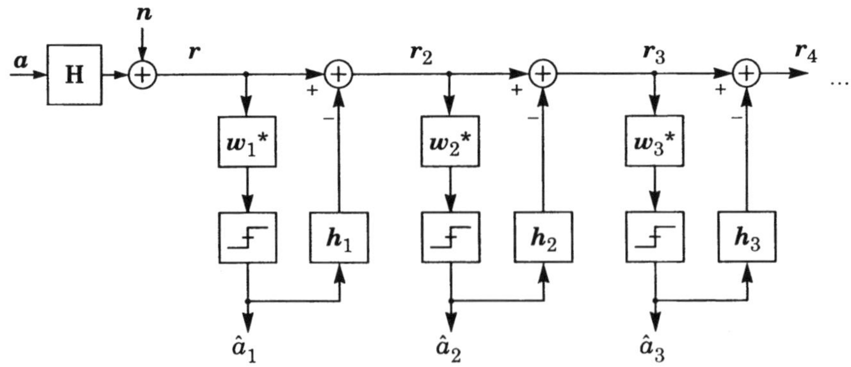

Fig. 10-13. The MIMO DF detector consists of n stages, one for each channel input. Each stage consists of four steps: nulling, detection, reconstruction, and cancellation.

$$y = \mathbf{F}r, \qquad \text{where } \mathbf{F} = \begin{bmatrix} w_1{}^* \\ w_2{}^* \\ \vdots \\ w_n{}^* \end{bmatrix}. \tag{10.96}$$

The matrix **F** will be called the *forward* filter; its i-th row is $w_i{}^*$.

The cancellation step can also be expressed in matrix form. As shown in Fig. 10-14, $b_{21}\hat{a}_1$ is subtracted from the second component of y, producing z_2. Similarly, $b_{31}\hat{a}_1 + b_{32}\hat{a}_2$ is subtracted from the third component of y, producing z_3. In general, $z_i = y_i - \sum_{j=1}^{i-1} b_{ij}\hat{a}_j$. In terms of the decision vector $\hat{a} = [\hat{a}_1, \ldots \hat{a}_n]^T$, the cancellation steps may be summarized by the single equation:

$$z = y - \mathbf{B}\hat{a}, \tag{10.97}$$

where **B** is a *strictly* lower triangular matrix (having zeros on and above the diagonal), with b_{ij} in row i and column $j < i$. With our definition of causality, **B** is a *strictly causal* spatial filter.

Combining (10.96) and (10.97), the vector z of slicer inputs can be written in the compact form:

$$z = \mathbf{F}r - \mathbf{B}\hat{a}. \tag{10.98}$$

This is the matrix view of the DF detector, as illustrated in Fig. 10-15. Observe the similarities between Fig. 10-15 and the DFE of Chapter 8 (such as the ZF-DFE of Fig. 8-14). In both cases there is a forward filter whose purpose is to suppress interference due to future (as-yet undetected) symbols, and in both cases there are decisions feeding a strictly causal feedback filter, whose purpose is to suppress interference due to past (already detected) symbols. The key difference here is that the symbols are ordered spatially rather than temporally, leading to a forward filter **F** and feedback filter **B** that are spatial rather than temporal.

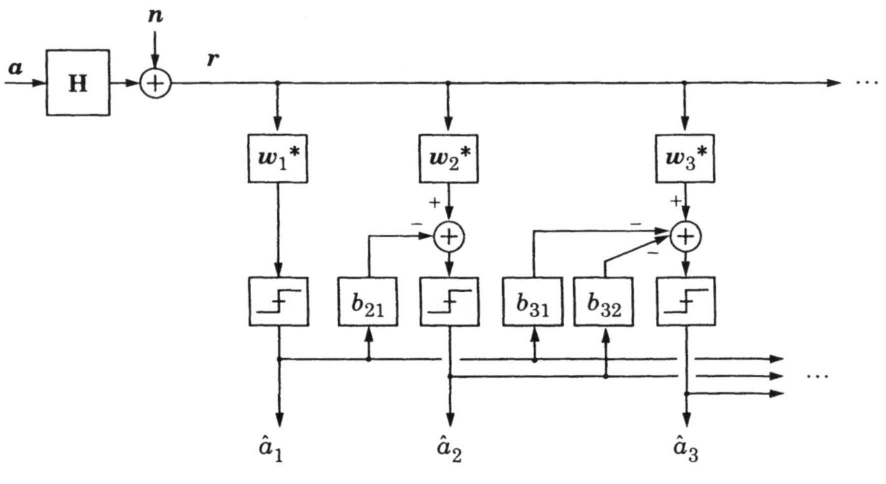

Fig. 10-14. An alternative but equivalent implementation of the DF detector of Fig. 10-13.

At first glance it might appear that the feedback system in Fig. 10-15 cannot be implemented because of a causality problem; it appears that to calculate $\hat{\boldsymbol{a}}$, the receiver must know $\boldsymbol{z}$, and to calculate $\boldsymbol{z}$, it must know $\hat{\boldsymbol{a}}$. However, there is no problem when the feedback is interpreted properly. Rather than trying to calculate the entire decision vector all at once, the components are calculated one by one, in keeping with the space-ordered notion of causality, beginning with the first. Because the feedback filter **B** is strictly causal (strictly lower triangular), the i-th component of the slicer input z_i depends only on the previous decisions $\{\hat{a}_1, \ldots \hat{a}_{i-1}\}$. The feedback system of Fig. 10-15 is thus self-consistent and can be implemented whenever the feedback filter is strictly triangular.

Originally we defined the feedback filter **B** as the strictly lower triangular matrix whose entry in row i and column $j < i$ is $b_{ij} = \boldsymbol{w}_i^* \boldsymbol{h}_j$. In terms of the forward filter **F** and channel **H**, the feedback filter can equivalently be defined as the strictly lower triangular part of the cascade **FH**:

$$\mathbf{B} = \{\mathbf{FH}\}_{\text{lower}} . \tag{10.99}$$

This is not surprising. Regardless of how the forward filter is chosen, the cascade of it with the channel yields an effective channel of **FH** as seen by the decision device and feedback filter. Because of the causality constraint, the best that **B** can do is to completely cancel any interference due to past decisions. The result of (10.99) is directly analogous to the result of (9.83) in Chapter 9, where it was shown that – in the context of temporal ISI channels – the optimal feedback filter is always the strictly causal part of the channel-forward-filter cascade, regardless of how the forward filter was chosen.

We have seen three equivalent block diagrams of the same DF detector. The first two (Fig. 10-13 and Fig. 10-14) may be preferable from an implementation standpoint, because they explicitly indicate the order in which the decisions are made. However, the last (Fig. 10-15) is often preferred for its compact form, conceptual simplicity, and its amenability to analysis.

We previously identified the i-th nulling vector for the ZF-DF detector as the first row of $(\mathbf{H}_i^* \mathbf{H}_i)^{-1} \mathbf{H}_i^*$, where $\mathbf{H}_i = [\boldsymbol{h}_i, \ldots \boldsymbol{h}_n]$. The following section uses geometric intuition to develop a more concise and conceptually simple specification of the nulling vectors.

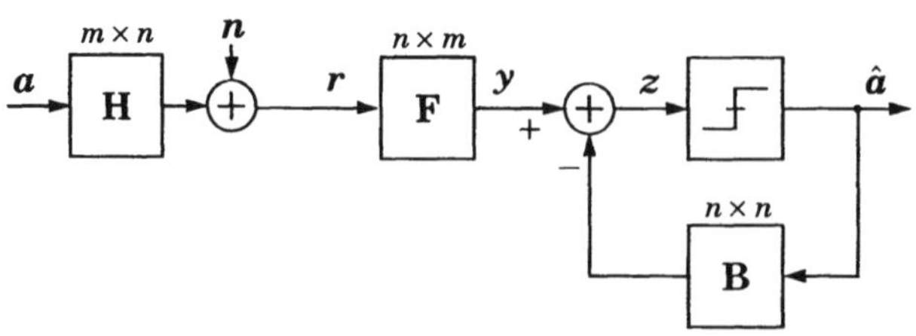

Fig. 10-15. A matrix-valued view of the DF detector that is equivalent to Fig. 10-13 and Fig. 10-14.

The Gram-Schmidt View

The zero-forcing nulling vectors $\{w_i\}$ are closely tied to a Gram-Schmidt (GS) basis for the columns of **H**. Hence, let us review the GS procedure for finding an orthonormal basis for the columns of **H**. Rather than beginning with the first column, we will perform the GS procedure in reverse order, beginning with the last column h_n.

Let $q_n = h_n / G_{n,n}$ be the n-th basis vector, where $G_{n,n} = \|h_n\|$ ensures a unit length. Following the GS procedure, we next compute $G_{n,n-1}q_n$, the projection of h_{n-1} onto the space spanned by q_n, where the expansion coefficient is $G_{n,n-1} = q_n{}^*h_{n-1}$. The projection error $e_{n-1} = h_{n-1} - G_{n,n-1}q_n$ will be orthogonal to q_n, so we need only normalize it to arrive at the next basis function, $q_{n-1} = e_{n-1}/G_{n-1,n-1}$, where $G_{n-1,n-1} = \|e_{n-1}\|$. (Our assumption that **H** is full rank ensures that the projection error will never be zero.) Repeating this procedure iteratively, the j-th basis vector is given by $q_j = e_j / G_{jj}$, where $e_j = h_j - \sum_{i=j+1}^{n} G_{i,j}q_i$ and $G_{i,j} = q_i{}^*h_j$, with j decreasing from n to 1. Note that $G_{jj} = \|e_j\|$ can also be expressed as $G_{jj} = q_j{}^*h_j$.

The above procedure produces n vectors $q_1 \ldots q_n$ that are unit length, orthogonal, and span the same space spanned by the columns of **H**. In particular, we can express each column of **H** in terms of the basis vectors. The last column is $h_n = G_{n,n}q_n$; it depends only on the last basis vector. The second-to-last column is $h_{n-1} = G_{n-1,n-1}q_{n-1} + G_{n,n-1}q_n$; it depends only on the last two basis vectors. If we express all of the columns of **H** as linear combinations of the basis vectors, and write the result in matrix form, we get the following important decomposition.

Gram-Schmidt (GS) Decomposition. An $m \times n$ matrix **H** that has rank n can be factored *uniquely* according to

$$\left[\begin{array}{c} \\ h_1, h_2, \ldots h_n \\ \\ \end{array}\right] = \left[\begin{array}{c} \\ q_1, q_2, \ldots q_n \\ \\ \end{array}\right] \begin{bmatrix} G_{1,1} & 0 & \cdots & 0 \\ G_{2,1} & G_{2,2} & \ddots & \vdots \\ \vdots & & \ddots & 0 \\ G_{n,1} & & \cdots & G_{n,n} \end{bmatrix} \iff \mathbf{H} = \mathbf{QG}\,, \tag{10.100}$$

where **Q** is an $m \times n$ matrix whose columns are unit length and orthogonal and span the range of **H**, and where **G** is an $n \times n$ lower triangular matrix with real and positive diagonal elements.

As we have seen, the GS decomposition **H** = **QG** follows directly from application of the GS procedure to the columns of **H**, from last to first. The diagonal elements of **G** are real and positive because G_{ii} is defined as the norm of a projection error vector. The GS decomposition is essentially equivalent to the so-called *truncated QR (orthogonal-triangular)* decomposition, except that the triangular matrix in (10.100) is lower triangular instead of upper triangular, and the diagonal components of **G** are necessarily real and positive.

The $m \times n$ matrix **Q** has a special property — it satisfies **Q*Q** = **I**. We may think of **Q** as the first n columns of an $m \times m$ unitary matrix. It will not satisfy **QQ*** = **I**, and hence it is not a unitary matrix, unless the channel is square ($m = n$).

Before relating the zero-forcing nulling vectors $\{\boldsymbol{w}_i\}$ to the GS decomposition of (10.100), we first point out an important fact: the i-th ZF nulling vector is expressible as a linear combination of the column set $\{\boldsymbol{h}_i, \ldots \boldsymbol{h}_n\}$, or in other words:

$$\boldsymbol{w}_i \in \text{Span}\{\boldsymbol{h}_i, \ldots \boldsymbol{h}_n\} . \quad (10.101)$$

To establish this result, observe that $\boldsymbol{w}_i$ does not need to contend with interference from symbols a_1 through a_{i-1}, because their impact has presumably already been cancelled. Hence, $\boldsymbol{w}_i$ sees an effective channel $\mathbf{H}_i = [\boldsymbol{h}_i, \ldots \boldsymbol{h}_n]$, where the columns of $\mathbf{H}$ corresponding to a_1 through a_{i-1} have been removed. A MF is known to provide sufficient statistics, and thus $\boldsymbol{w}_i{}^*$ can without loss of performance be implemented as the cascade of an MF followed by another filter $\boldsymbol{c}^* = [c_i{}^*, \ldots c_n{}^*]$, so that $\boldsymbol{w}_i{}^* = \boldsymbol{c}^* \mathbf{H}_i^*$. This implies that $\boldsymbol{w}_i = \sum_{j=i}^{n} c_j \boldsymbol{h}_j$, which is clearly a linear combination of $\{\boldsymbol{h}_i, \ldots \boldsymbol{h}_n\}$.

Now let us relate the zero-forcing nulling vectors to the GS basis $\{\boldsymbol{q}_i\}$, beginning with the last one. From (10.95), $\boldsymbol{w}_n$ is the minimum-norm vector satisfying $\boldsymbol{w}_n{}^* \boldsymbol{h}_n = 1$. There are no future symbols, so this is the only constraint. Since $\boldsymbol{w}_n \in \text{Span}\{\boldsymbol{h}_n\}$ from (10.101), the solution must be $\boldsymbol{w}_n = \boldsymbol{h}_n / \|\boldsymbol{h}_n\|^2$, or equivalently $\boldsymbol{w}_n = \boldsymbol{q}_n / \|\boldsymbol{h}_n\|$. Since $G_{n,n} = \|\boldsymbol{h}_n\|$, we can also write this as $\boldsymbol{w}_n = \boldsymbol{q}_n / G_{n,n}$.

Having found $\boldsymbol{w}_n$, we move on to $\boldsymbol{w}_{n-1}$, which is known to lie in the span of $\{\boldsymbol{h}_{n-1}, \boldsymbol{h}_n\}$. To satisfy $\boldsymbol{w}_{n-1}{}^* \boldsymbol{h}_n = 0$, we must have $\boldsymbol{w}_{n-1} \propto \boldsymbol{q}_{n-1}$, and to satisfy $\boldsymbol{w}_{n-1}{}^* \boldsymbol{h}_{n-1} = 1$, we must have $\boldsymbol{w}_{n-1} = \boldsymbol{q}_{n-1} / G_{n-1,n-1}$. Next we seek $\boldsymbol{w}_{n-2} \in \text{Span}\{\boldsymbol{h}_{n-2}, \boldsymbol{h}_{n-1}, \boldsymbol{h}_n\}$. Again, to satisfy $\boldsymbol{w}_{n-2}{}^* \boldsymbol{h}_{n-1} = \boldsymbol{w}_{n-2}{}^* \boldsymbol{h}_n = 0$ we must have $\boldsymbol{w}_{n-2} \propto \boldsymbol{q}_{n-2}$, and to satisfy $\boldsymbol{w}_{n-2}{}^* \boldsymbol{h}_{n-2} = 1$, we must have $\boldsymbol{w}_{n-2} = \boldsymbol{q}_{n-2} / G_{n-2,n-2}$.

By now it is clear that we will have $\boldsymbol{w}_i = \boldsymbol{q}_i / G_{ii}$ for all i. Thus, except for a scaling constant, *the zero-forcing nulling vectors are just the basis vectors found by applying the Gram-Schmidt procedure to the columns of* $\mathbf{H}$ *in reverse order.*

To see why the GS procedure produces the right zero-forcing nulling vectors, consider Fig. 10-16. With respect to a given i, the *interference subspace* for a *DF detector* is the span of $\{\boldsymbol{h}_{i+1}, \dots \boldsymbol{h}_n\}$. Let us decompose the column $\boldsymbol{h}_i$ corresponding to the desired symbol a_i according to:

$$\boldsymbol{h}_i = \hat{\boldsymbol{h}}_i + (\boldsymbol{h}_i - \hat{\boldsymbol{h}}_i) , \tag{10.102}$$

where $\hat{\boldsymbol{h}}_i$ is the projection of $\boldsymbol{h}_i$ onto the interference subspace. The component of $\boldsymbol{r}$ in the direction of $\hat{\boldsymbol{h}}_i$ is discarded by a ZF nulling vector, because it will contain contributions from at least one interfering symbol. The second term is the projection error, and it is orthogonal to the interference subspace. The GS procedure forms the i-th basis vector $\boldsymbol{q}_i$ by normalizing the projection error $\boldsymbol{h}_i - \hat{\boldsymbol{h}}_i$. Thus, by choosing $\boldsymbol{w}_i$ proportional to $\boldsymbol{q}_i$, we ensure that $\boldsymbol{w}_i$ is orthogonal to the interference subspace, and hence orthogonal to $\{\boldsymbol{h}_{i+1}, \dots \boldsymbol{h}_n\}$. (This is most of the zero-forcing constraint of (10.95).) Since $G_{ii} = \|\boldsymbol{h}_i - \hat{\boldsymbol{h}}_i\| = \boldsymbol{q}_i^*\boldsymbol{h}_i$, the scaling factor in $\boldsymbol{w}_i = \boldsymbol{q}_i / G_{ii}$ ensures that the desired symbol is passed by $\boldsymbol{w}_i^*$ with unity gain, so that $\boldsymbol{w}_i^*\boldsymbol{h}_i = 1$. (This is the rest of the zero-forcing constraint of (10.95).)

Since the forward filter is defined as the $n \times m$ matrix whose i-th row is $\boldsymbol{w}_i^*$, we find that the forward filter can now be expressed in compact form:

$$\mathbf{F} = \Gamma^{-1}\mathbf{Q}^*, \tag{10.103}$$

where we have introduced the diagonal matrix $\Gamma = \text{diag}\{G_{11}, G_{22} \dots G_{nn}\}$, which has real and positive diagonal elements. The implication of (10.103) is that we can view the forward filter $\mathbf{F}$ as the cascade of $\mathbf{Q}^*$, which projects the original m-dimensional observation onto an n-dimensional subspace, and also implements a change of basis or a rotation of coordinates, followed by a diagonal matrix Γ^{-1}. The change of basis will not change the statistics of the noise, so that the white Gaussian noise before $\mathbf{Q}^*$ will still be white and Gaussian after. After the diagonal matrix Γ^{-1}, the different noise components will remain independent but with different noise variances. In particular, the real and imaginary parts of the i-th noise component will have variance $(N_0/2)/G_{ii}^2$.

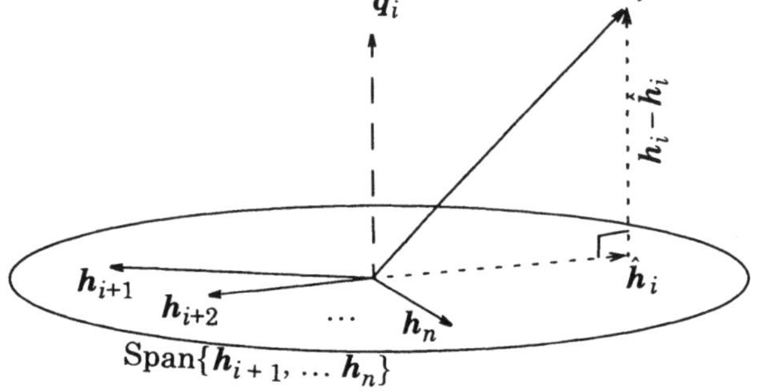

Fig. 10-16. Decomposing $\boldsymbol{h}_i$ into the sum of $\hat{\boldsymbol{h}}_i$, which is the projection of $\boldsymbol{h}_i$ onto the interference subspace, and $(\boldsymbol{h}_i - \hat{\boldsymbol{h}}_i)$, which is the projection error.

The Cholesky View and the Whitened-Matched Filter

We gain further insight into the DF detector by relating the forward filter **F** to a *whitened-matched filter (WMF)*, as derived from a Cholesky decomposition. Consider the Hermitian matrix $\mathbf{R} = \mathbf{H}^*\mathbf{H}$, which may be interpreted as the cascade of the channel and its matched filter. In this sense, **R** is a spatial version of the folded spectrum of (5.66). Just as the minimum-phase spectral factorization expresses the folded spectrum as a product of a minimum-phase filter and its match, the Cholesky decomposition expresses a Hermitian matrix as a product of a minimum-phase spatial filter **G** and its match.

Cholesky Decomposition. A nonsingular $n \times n$ matrix **R** that is positive definite and Hermitian (satisfying $\mathbf{R}^* = \mathbf{R}$) can be factored *uniquely* according to

$$\mathbf{R} = \mathbf{G}^*\mathbf{G} = \mathbf{M}^*\Gamma^2\mathbf{M}\,, \tag{10.104}$$

where **G** is lower triangular with real and positive diagonal elements $\{G_{ii} > 0\}$. We will find it convenient to further decompose **G** according to $\mathbf{G} = \Gamma\mathbf{M}$, where $\Gamma = \text{diag}\{G_{11}, G_{22} \ldots G_{nn}\}$ is a diagonal matrix consisting of the diagonal part of **G**, and $\mathbf{M} = \Gamma^{-1}\mathbf{G}$ is lower triangular and monic (having ones on its diagonal).

We do not have to prove this result from scratch, because it is essentially a corollary of the GS decomposition (10.100). With our standing assumption that **H** has full column rank, $\mathbf{R} = \mathbf{H}^*\mathbf{H}$ is nonsingular and Hermitian. Substituting $\mathbf{H} = \mathbf{QG}$ from (10.100) into $\mathbf{R} = \mathbf{H}^*\mathbf{H}$ immediately yields (10.104). In this case, the **G** matrix in the Cholesky decomposition is precisely the same as the **G** matrix of the GS decomposition. For the more general case where **R** is positive definite and Hermitian but not given in the form $\mathbf{H}^*\mathbf{H}$, we need only apply the GS procedure to the square-root matrix $\mathbf{H} = \Lambda^{1/2}\mathbf{V}^*$, where $\mathbf{R} = \mathbf{V}\Lambda\mathbf{V}^*$ is an eigendecomposition (see (9.35)).

Before relating the DF detector to the Cholesky decomposition, let us first decompose the $n \times m$ forward filter into the product $\mathbf{F} = \mathbf{CH}^*$, so that we may implement it using the cascade of a $n \times m$ matched filter $\mathbf{H}^*$ followed by an $n \times n$ filter **C**. Such a decomposition does not hinder performance, since the matched filter is known to give sufficient statistics. In situations where there are many more channel outputs than inputs ($m \gg n$), this decomposition is also beneficial because it simplifies the filter design problem — rather than optimizing an $n \times m$ filter **F**, we need only optimize an $n \times n$ filter **C**.

What choice of **C** maximizes SNR, subject to a zero-forcing constraint? Combining our earlier result that $\mathbf{F} = \Gamma^{-1}\mathbf{Q}^*$ with the decompositions $\mathbf{F} = \mathbf{CH}^*$ and $\mathbf{H} = \mathbf{QG}$ yields:

$$\mathbf{C} = \Gamma^{-1}\mathbf{G}^{-*} = \Gamma^{-2}\mathbf{M}^{-*}\,. \tag{10.105}$$

The second form of **C** is particularly illuminating because of its close match with the ZF-DFE of Section 8.1.3, where the forward filter after a sampled matched filter was shown to be the anticausal filter $C(z) = \gamma^{-2}(M^*(1/z^*))^{-1}$. Here, $\mathbf{C} = \Gamma^{-2}\mathbf{M}^{-*}$ is also anticausal (upper triangular). We interpreted $C(z)$ as a noise-whitening filter, and the cascade of the sampled matched filter and $C(z)$ was termed the whitened-matched filter (WMF). Here we have the analogous result for spatial channels: we may interpret the cascade of the MF and $\mathbf{G}^{-*}$ as a *spatial* whitened-matched filter; it produces sufficient statistics and white noise. From the GS decomposition, the spatial WMF $\mathbf{G}^{-*}\mathbf{H}^*$ reduces to the projection matrix $\mathbf{Q}^*$. The relationship between the spatial WMF and the ZF-DF detector is illustrated in Fig. 10-17.

The ZF-DF forward filter $\mathbf{F} = \Gamma^{-1}\mathbf{Q}^*$ can also be understood by considering its impact on the channel; using the GS decomposition, the transfer function of the channel-forward-filter cascade can be written as:

$$\begin{aligned} \mathbf{FH} &= (\Gamma^{-1}\mathbf{Q}^*)(\mathbf{QG}) \\ &= \Gamma^{-1}\mathbf{G} \\ &= \mathbf{M} . \end{aligned} \quad (10.106)$$

Thus, the ZF-DF forward filter transforms the channel into a monic, causal, and minimum phase transfer function. From (10.99), the feedback filter can now be expressed as

$$\mathbf{B} = \{\mathbf{M}\}_{\text{lower}} = \mathbf{M} - \mathbf{I} . \quad (10.107)$$

The optimal feedback filter is thus the strictly causal part of the monic and minimum-phase factor $\mathbf{M}$. (Compare this result with the optimal feedback filter of the temporal ZF-DFE of Section 8.1.3, which was shown to be $M(z) - 1$.)

The MSE performance of the ZF-DF detector is easy to evaluate. The forward filter $\mathbf{F} = \Gamma^{-1}\mathbf{Q}^*$ transforms the autocorrelation of the channel noise from $N_0\mathbf{I}$ to $N_0\Gamma^{-2}$. Since the feedback filter does not affect the noise, the MSE for symbol i (under the assumption of correct decision feedback) is:

$$MSE_i^{\text{DF}} = \frac{N_0}{\Gamma_{ii}^2} . \quad (10.108)$$

If we define $\hat{\boldsymbol{h}}_i^{\text{DF}}$ and $\hat{\boldsymbol{h}}_i^{\text{L}}$ as the projection of $\boldsymbol{h}_i$ onto the subspaces spanned by $\{\boldsymbol{h}_{i+1}, \ldots \boldsymbol{h}_n\}$ and $\{\boldsymbol{h}_{j \neq i}\}$, respectively, then the MSE for the ZF-DF and ZF-linear detectors become:

$$MSE_i^{\text{DF}} = N_0\|\boldsymbol{h}_i - \hat{\boldsymbol{h}}_i^{\text{DF}}\|^{-2} , \qquad MSE_i^{\text{L}} = N_0\|\boldsymbol{h}_i - \hat{\boldsymbol{h}}_i^{\text{L}}\|^{-2} . \quad (10.109)$$

Since $\hat{\boldsymbol{h}}_i^{\text{L}}$ will always be as close or closer to $\boldsymbol{h}_i$ than $\hat{\boldsymbol{h}}_i^{\text{DF}}$, the DF detector is always as good or better than the linear detector.

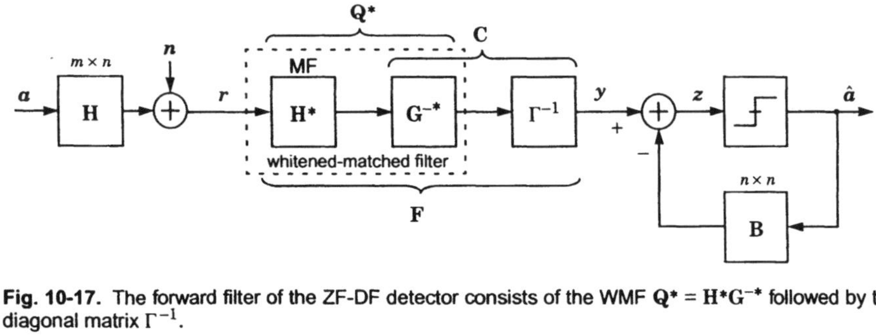

Fig. 10-17. The forward filter of the ZF-DF detector consists of the WMF $\mathbf{Q}^* = \mathbf{H}^*\mathbf{G}^{-*}$ followed by the diagonal matrix Γ^{-1}.

The Linear-Prediction View

A particularly useful way to interpret the DF detector is to begin with a *linear* detector and then apply linear prediction to reduce the noise variance. In particular, consider the centralized ZF linear detector of (10.70), which computes $\boldsymbol{y} = \mathbf{C}\boldsymbol{r}$ with $\mathbf{C} = \mathbf{R}^{-1}\mathbf{H}^*$ and $\mathbf{R} = \mathbf{H}^*\mathbf{H}$. Since $\boldsymbol{r} = \mathbf{H}\boldsymbol{a} + \boldsymbol{n}$, the output of the linear detector will be:

$$\boldsymbol{y} = \boldsymbol{a} + \tilde{\boldsymbol{n}}\,, \tag{10.110}$$

where the noise $\tilde{\boldsymbol{n}}$ is no longer white; instead, its autocorrelation matrix is $N_0\mathbf{CC}^* = N_0\mathbf{R}^{-1}$.

Using linear prediction, we can exploit the correlation of the noise to reduce its variance. For example, if we knew the first $i-1$ samples $\{\tilde{n}_1, \ldots \tilde{n}_{i-1}\}$ of the noise, we could form an estimate $\hat{n}_i$ of $\tilde{n}_i$ and subtract this estimate from y_i before any subsequent processing. This process is complicated by the fact that we do not have access to $\tilde{n}_i$ directly but rather the sum $a_i + \tilde{n}_i$. Thus, if we apply a linear prediction error filter $\mathbf{I} - \mathbf{P}$ to $\boldsymbol{y}$ we would reduce the noise variance at the expense of reintroducing interference. Fortunately, however, the causality of the prediction process implies that the resulting interference will be strictly causal, and can thus be eliminated using decision feedback. In particular, the output of the prediction error filter is:

$$(\mathbf{I} - \mathbf{P})\boldsymbol{y} = \boldsymbol{a} - \mathbf{P}\boldsymbol{a} + \boldsymbol{e}\,, \tag{10.111}$$

where $\boldsymbol{e} = \tilde{\boldsymbol{n}} - \hat{\boldsymbol{n}}$ is the effective noise with reduced variance after prediction. Since $\mathbf{P}$ is strictly lower triangular, we can add $\mathbf{P}\hat{\boldsymbol{a}}$ to (10.111) using decision feedback. In this way we get the benefit of zero interference as well as reduced noise variance. This view of the DF detector is shown in Fig. 10-18(a). An equivalent implementation is shown in Fig. 10-18(b), where the prediction filter and feedback filter are implemented using a single filter.

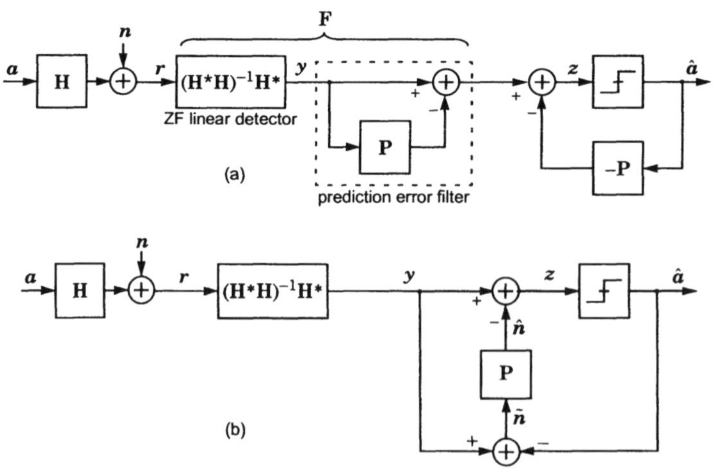

Fig. 10-18. The ZF-DF detector implemented using linear prediction: (a) with a separate prediction filter and feedback filter; (b) with a combined prediction filter and feedback filter.

The optimal prediction filter can be found by first performing a Cholesky decomposition of the autocorrelation matrix of $\tilde{\boldsymbol{n}}$, according to $N_0\mathbf{R}^{-1} = \mathbf{M}^{-1}(N_0\Gamma^{-2})\mathbf{M}^{-*}$, and then applying (10.23). This leads to an optimal prediction-error filter of $\mathbf{I} - \mathbf{P} = \mathbf{M}$. The cascade of the ZF linear detector and the prediction error filter $\mathbf{I} - \mathbf{P}$ reduces to an effective forward filter of:

$$\mathbf{F} = (\mathbf{I} - \mathbf{P})\mathbf{C} = \mathbf{M}\mathbf{R}^{-1}\mathbf{H}^* = \mathbf{M}(\mathbf{M}^{-1}\Gamma^{-2}\mathbf{M}^{-*})\mathbf{H}^* = \Gamma^{-2}\mathbf{M}^{-*}\mathbf{H}^* = \Gamma^{-1}\mathbf{Q}^*, \tag{10.112}$$

which agrees with (10.103). The corresponding feedback filter is $\mathbf{B} = -\mathbf{P} = \mathbf{M} - \mathbf{I}$, which agrees with (10.107). Thus, the linear-prediction approach yields precisely the same forward and feedback filters derived earlier.

The LP view of the DF detector is useful because it illustrates vividly the relationship between the ZF linear and ZF-DF detectors. Since the feedback filter has no impact on the first component, it is clear that the first slicer operates directly on the output of the ZF linear detector. Under the assumption of correct decisions, it is also clear that the remaining decisions will be more reliable than after a linear detector, because they are subject to a reduced noise variance.

10.3.5. The MMSE-DF Detector

In this section we derive the MMSE-DF detector. It has the same form as the ZF-DF detector shown in Fig. 10-15, but we relax the requirement that the forward filter completely force to zero the interference from future symbols; instead, the MMSE-DF detector chooses its forward and feedback filters so as to minimize the sum MSE $E[\|\boldsymbol{z} - \boldsymbol{a}\|^2]$, under the assumption that the decisions are correct. To aid analysis, we adopt the matrix view of the DF detector. Keep in mind that the rows of the forward filter are the nulling vectors of Fig. 10-13.

Our derivation will be aided by the block diagram of Fig. 10-19. Fig. 10-19(a) shows how the DF detector of Fig. 10-15 simplifies when the decisions are assumed to be correct. The diagram of Fig. 10-19(b) is equivalent, but the contributions to the error signal $\boldsymbol{e} = \boldsymbol{z} - \boldsymbol{a}$ of the channel input $\boldsymbol{a}$ and noise $\boldsymbol{n}$ have been separated. Specifically, the symbols $\boldsymbol{a}$ see a transfer function $\mathbf{FH} - (\mathbf{I} + \mathbf{B})$, and the noise sees a transfer function $\mathbf{F}$. If we decompose the forward filter $\mathbf{F}$ into the product $\mathbf{F} = (\mathbf{I} + \mathbf{B})\mathbf{C}$, then the equivalent diagram of Fig. 10-19(c) results. We may decompose $\mathbf{F}$ in this way because $\mathbf{I} + \mathbf{B}$ is monic and triangular and thus always invertible. By expressing $\mathbf{F}$ in terms of $\mathbf{C}$ and $\mathbf{B}$, we have changed the design problem from one of finding the best $\mathbf{F}$ and $\mathbf{B}$ to one of finding the best $\mathbf{C}$ and $\mathbf{B}$. The benefit of this decomposition is that $\mathbf{C}$ and $\mathbf{B}$ may now be optimized *separately*. Indeed, from Fig. 10-19(c) we see that only the last block depends on $\mathbf{B}$. And because $\mathbf{I} + \mathbf{B}$ is a monic and causal filter, it can be interpreted as a linear-prediction error filter. Hence, optimizing $\mathbf{B}$ reduces to a linear prediction problem; regardless of how $\mathbf{C}$ is chosen, the best $\mathbf{B}$ will be determined by a Cholesky factorization of the autocorrelation matrix of the intermediate error signal ε. Specifically, as shown in Section 10.1.4, the optimal $\mathbf{B}$ satisfies $(\mathbf{I} + \mathbf{B}) = \mathbf{M}_\varepsilon^{-1}$, where $\mathbf{R}_\varepsilon = \mathbf{M}_\varepsilon \Gamma_\varepsilon^2 \mathbf{M}_\varepsilon^*$ is a Cholesky decomposition of the autocorrelation matrix of the intermediate error signal ε.

The MSE after this predictor will be $E[\|\boldsymbol{e}\|^2] = \mathrm{tr}\{\Gamma_\varepsilon^2\}$. It remains now only to choose $\mathbf{C}$ so as to minimize $\mathrm{tr}\{\Gamma_\varepsilon^2\}$. From Fig. 10-19(c), the autocorrelation matrix for ε is:

$$\mathbf{R}_\varepsilon = (\mathbf{CH} - \mathbf{I})(\mathbf{CH} - \mathbf{I})^* + N_0 \mathbf{CC}^*. \tag{10.113}$$

This is exactly the same as (10.84), the autocorrelation matrix for the error signal after a *linear* detector. Therefore, we can use our prior results for MMSE linear detection here: the best $\mathbf{C}$ is the MMSE linear detector given by (10.86) or (10.87).

With this choice for $\mathbf{C}$, the autocorrelation matrix of the residual error signal is the last term in (10.85), or $\mathbf{R}_\varepsilon = N_0 \tilde{\mathbf{R}}^{-1}$, where we have introduced the new matrix:

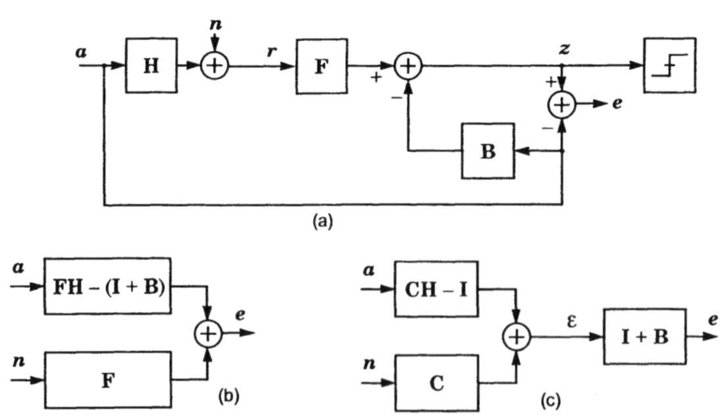

Fig. 10-19. Three equivalent block diagrams relating the channel input and noise to the error of a DF detector: (a) after assuming correct decisions; (b) after separating signal from noise; (c) after factoring out the term **I** + **B**.

$$\tilde{\mathbf{R}} = \mathbf{H}^*\mathbf{H} + N_0\mathbf{I} \,. \tag{10.114}$$

It will be Hermitian and positive definite even when $\mathbf{H}$ is rank-deficient, as long as the noise is nonzero. Hence, we can factor it using a Cholesky decomposition:

$$\tilde{\mathbf{R}} = \tilde{\mathbf{M}}^*\tilde{\Gamma}^2\tilde{\mathbf{M}} \,, \tag{10.115}$$

where $\tilde{\mathbf{M}}$ is monic and lower triangular, and $\tilde{\Gamma}$ is diagonal with real and positive diagonal elements. In terms of this decomposition, the residual error autocorrelation reduces to:

$$\mathbf{R}_\varepsilon = N_0\tilde{\mathbf{M}}^{-1}\tilde{\Gamma}^{-2}\tilde{\mathbf{M}}^{-*} \,. \tag{10.116}$$

Since $\tilde{\mathbf{M}}^{-1}$ is itself a monic and lower triangular matrix, and since $N_0\tilde{\Gamma}^{-2}$ is a positive diagonal matrix, (10.116) itself defines a Cholesky decomposition of $\mathbf{R}_\varepsilon$. The best feedback filter $\mathbf{B}$ will be such that $(\mathbf{I} + \mathbf{B})$ cancels $\tilde{\mathbf{M}}^{-1}$; or in other words:

$$\mathbf{B} = \tilde{\mathbf{M}} - \mathbf{I} \,. \tag{10.117}$$

Combining (10.87) and (10.117), we find that the forward filter of the MMSE-DF detector is:

$$\begin{aligned} \mathbf{F} &= (\mathbf{I} + \mathbf{B})\mathbf{C} \\ &= \tilde{\mathbf{M}}(\mathbf{H}^*\mathbf{H} + N_0\mathbf{I})^{-1}\mathbf{H}^* \\ &= \tilde{\mathbf{M}}\tilde{\mathbf{R}}^{-1}\mathbf{H}^* \\ &= \tilde{\mathbf{M}}\tilde{\mathbf{M}}^{-1}\tilde{\Gamma}^{-2}\tilde{\mathbf{M}}^{-*}\mathbf{H}^* \\ &= \tilde{\Gamma}^{-2}\tilde{\mathbf{M}}^{-*}\mathbf{H}^* = \tilde{\Gamma}^{-1}\tilde{\mathbf{G}}^{-*}\mathbf{H}^* \,. \end{aligned} \tag{10.118}$$

Equations (10.117) and (10.118) define the filters of the MMSE-DF detector. Compared to the corresponding filters of the ZF-DF detector, we see that the MMSE filters differ only in that $\tilde{\mathbf{M}}$ has replaced $\mathbf{M}$, and $\tilde{\Gamma}$ has replaced Γ. Equivalently, whereas the ZF filters were based on a Cholesky factorization of $\mathbf{H}^*\mathbf{H}$, the MMSE filters are based on a factorization of $\mathbf{H}^*\mathbf{H} + N_0\mathbf{I}$. In the limit of low noise, the MMSE-DF detector approaches the ZF DF detector. From (10.116), the MSE for the i-th output of the MMSE-DF detector is:

$$MSE_i = \frac{N_0}{\tilde{\Gamma}_{ii}^2} \,. \tag{10.119}$$

An important difference between the MMSE and ZF detectors is the impact of the forward filter. In (10.106) we saw that the ZF forward filter transforms the channel into a monic and causal transfer function, $\mathbf{F}_{ZF}\mathbf{H} = \mathbf{M}$. In contrast, the MMSE forward filter transforms the channel into something that is neither monic nor causal:

$$\begin{aligned} \mathbf{FH} &= \tilde{\Gamma}^{-2}\tilde{\mathbf{M}}^{-*}\mathbf{H}^*\mathbf{H} \\ &= \tilde{\Gamma}^{-2}\tilde{\mathbf{M}}^{-*}(\mathbf{H}^*\mathbf{H} + N_0\mathbf{I} - N_0\mathbf{I}) \\ &= \tilde{\Gamma}^{-2}\tilde{\mathbf{M}}^{-*}(\tilde{\mathbf{M}}^*\tilde{\Gamma}^2\tilde{\mathbf{M}} - N_0\mathbf{I}) \\ &= \tilde{\mathbf{M}} - N_0\tilde{\Gamma}^{-2}\tilde{\mathbf{M}}^{-*}. \end{aligned} \tag{10.120}$$

The first term is monic and causal, but the second term is not. Thus, the second term in (10.120) will result in residual interference that will not be cancelled by the feedback filter. This interference is scaled by the noise variance and is often small relative to the desired signal.

Like the temporal DFE of Chapter 8, the output of the MMSE-DF detector is slightly biased. If we assume correct decisions are fed back, the input to the slicer after an MMSE-DF detector is given by:

$$\begin{aligned} \boldsymbol{z} &= \mathbf{F}\boldsymbol{r} - \mathbf{B}\hat{\boldsymbol{a}} \\ &= \mathbf{F}(\mathbf{H}\boldsymbol{a} + \boldsymbol{n}) - \mathbf{B}\boldsymbol{a} \\ &= (\mathbf{F}\mathbf{H} - \mathbf{B})\boldsymbol{a} + \mathbf{F}\boldsymbol{n} \\ &= \left(\tilde{\mathbf{M}} - N_0\tilde{\Gamma}^{-2}\tilde{\mathbf{M}}^{-*} - (\tilde{\mathbf{M}} - \mathbf{I})\right)\boldsymbol{a} + \mathbf{F}\boldsymbol{n} \\ &= (\mathbf{I} - N_0\tilde{\Gamma}^{-2}\tilde{\mathbf{M}}^{-*})\boldsymbol{a} + \mathbf{F}\boldsymbol{n}\,. \end{aligned} \tag{10.121}$$

In particular, the coefficient in z_i of the desired symbol a_i is $(1 - N_0/\tilde{\Gamma}_{ii}^2)$, which from (10.119) can also be written as $(1 - MSE_i)$. This factor represents a bias that will be small when the noise is small, but for best performance (in terms of error probability, not MSE) the receiver should nonetheless divide z_i by this factor before the slicer.

We should point out that the matrix view of the DF detector used in this section was for convenience only. We can at any time revert back to the nulling-slicing-cancellation view used in Fig. 10-13 or Fig. 10-14 by interpreting the i-th row of the forward filter $\mathbf{F}$ in (10.118) as the i-th nulling vector $\boldsymbol{w}_i^*$. Of course, unlike the ZF nulling vectors, the "nulling" vectors of the MMSE-DF detector do not completely reject the interference, they only attenuate it.

10.3.6. Order of Detection

Up to this point we have assumed that the DF detector detects the symbols in the *natural* order of first a_1, then a_2, and so on. However, this is not the only ordering possible. The receiver may detect the symbols in any order it desires. An important question is: of the $n!$ possible orderings, which is best? The order in which the symbols are detected is an important degree of freedom at the receiver's disposition that can have a huge impact on performance. (This is another example of a MIMO issue that has no counterpart in scalar ISI channels.)

The ZF-DF detector does not treat each user the same. The first symbol derives no benefit from DF, and is in fact detected using a ZF linear detector. In contrast, under the assumption of correct decisions, the last symbol achieves the performance of the perfect interference canceller or MFB. On fading channels (see Chapter 11), the diversity benefit varies from one symbol to the next. Specifically, as explained in Section 11.4, the first symbol sacrifices diversity in order to null the interferers, and achieves a diversity order of only $m - n + 1$. On a square channel, it has no diversity at all. In contrast, the last symbol achieves full diversity order m.

Let i_j denote the index of the j-th symbol to be detected, so that the desired ordering is specified by an ordered set $\{i_1, i_2, \ldots i_n\}$ that is a permutation of the integers $\{1, 2, \ldots n\}$. Analytically, the ordering of users is captured by a permutation matrix **P**, where the j-th column of **P** is the i_j-th column of an $n \times n$ identity matrix, so that $\boldsymbol{a}' = \mathbf{P}^T\boldsymbol{a}$ contains the elements of $\boldsymbol{a}$ in the desired order. The permutation matrix is orthonormal, satisfying $\mathbf{P}\mathbf{P}^T = \mathbf{I}$. Substituting $\boldsymbol{a} = \mathbf{P}\boldsymbol{a}'$ into the channel model yields:

$$\begin{aligned} \boldsymbol{r} &= \mathbf{H}\boldsymbol{a} + \boldsymbol{n} \\ &= \mathbf{H}\mathbf{P}\boldsymbol{a}' + \boldsymbol{n} \\ &= \mathbf{H}'\boldsymbol{a}' + \boldsymbol{n}\,, \end{aligned} \tag{10.122}$$

where we have introduced $\mathbf{H}' = \mathbf{HP}$. We recognize this as equivalent to our original channel model, but with $\boldsymbol{a}$ and $\mathbf{H}$ replaced by $\boldsymbol{a}'$ and $\mathbf{H}'$, respectively. Since $\boldsymbol{a}'$ has the same statistics as $\boldsymbol{a}$, we conclude that the main impact of changing the order of detection is to change the channel from $\mathbf{H}$ to $\mathbf{H}' = \mathbf{HP}$, which can also be expressed as:

$$\mathbf{H}' = \begin{bmatrix} \boldsymbol{h}_{i_1}, \boldsymbol{h}_{i_2}, \ldots \boldsymbol{h}_{i_n} \end{bmatrix}. \tag{10.123}$$

Hence, changing the order of detection is equivalent to rearranging the columns of **H**.

The probability that any of the symbol decisions is incorrect will be dominated by the largest MSE_i, or equivalently by the smallest G_{ii}. Hence, we would like the ordering to ensure that $\min\{G_{ii}\}$ is as large as possible. The ordering also impacts the probability of error propagation. If an early symbol (say the first) is detected incorrectly, then the cancellation step of its stage will not reduce interference but will instead increase it, thus making it very likely that the decisions to follow will also be incorrect. These observations lead to the following two-step heuristic for choosing the ordering:

- Identify the ordering or orderings that maximize the worst-case SNR.
- Of these, choose the ordering that maximizes $J_i = \sum_{k=1}^{i} G_{kk}^2$ for all $i \in \{1, \ldots n\}$.

The *BLAST ordering procedure* is a recursive algorithm in which the symbol index i_k to be detected at stage k is chosen from the set of undetected symbols so as to minimize its MSE, as summarized by the following pseudocode [12]:

```
for k = 1 : n ,
    Let i_k = row of pseudoinverse of H with smallest norm, excluding rows {i_1, ... i_(k-1)}      (10.124)
    Replace the i_k-th column of H by all zeros
end
```

It is a remarkable fact that this procedure attains both heuristic goals outlined above: it maximizes the worst-case SNR, and it maximally concentrates $\{G_{kk}\}$ at small k. This is a local and greedy algorithm, in the sense that each symbol index is chosen without consideration of its impact on future symbols. Nevertheless, it is globally optimal. The complexity of the above procedure is high because it requires repeated computations of a matrix pseudoinverse; a reduced-complexity algorithm for finding the optimal ordering is explored in Problem 10-12.

Example 10-23.

Consider a two-input channel $\mathbf{H} = [\boldsymbol{h}_1, \boldsymbol{h}_2]$, and let $A_1 = \|\boldsymbol{h}_1\|$, $A_2 = \|\boldsymbol{h}_2\|$, and $\rho = \boldsymbol{h}_2{}^*\boldsymbol{h}_1/(A_1A_2)$, where the Cauchy-Schwartz inequality implies that $|\rho| \le 1$. The natural ordering would lead us to perform a Cholesky factorization of $\mathbf{H}^*\mathbf{H}$, and the permuted ordering would lead us to perform a Cholesky factorization of $[\boldsymbol{h}_2, \boldsymbol{h}_1]^*[\boldsymbol{h}_2, \boldsymbol{h}_1]$, yielding respective Cholesky factors of:

$$\mathbf{G} = \begin{bmatrix} A_1\sqrt{1-|\rho|^2} & 0 \\ A_1\rho & A_2 \end{bmatrix} \text{ and } \mathbf{G}' = \begin{bmatrix} A_2\sqrt{1-|\rho|^2} & 0 \\ A_2\rho & A_1 \end{bmatrix}. \tag{10.125}$$

Whether the receiver should swap the order of detection or not depends on whether $\mathbf{G}$ or $\mathbf{G}'$ contains the minimum diagonal element. Comparing the two, we conclude that the detector should swap only if the second input has more energy than the first, or $A_2 > A_1$. This simple result holds in general only for the case when there are two inputs; with more than two inputs, the optimal ordering cannot be found simply by sorting the signal energies. (See Example 10-24.) Observe from (10.125) that neither $\mathbf{G}$ nor $\mathbf{G}'$ will have its diagonal elements in decreasing order when $(1-\rho^2)^{1/2} < A_2/A_1 < (1-\rho^2)^{-1/2}$, for example, when $\rho = 0.1$ and $A_1 = A_2$.

Example 10-24.

Consider a three-input channel $\mathbf{H} = [\boldsymbol{h}_1, \boldsymbol{h}_2, \boldsymbol{h}_3]$, where $\boldsymbol{h}_1 = [1, 0, 1, 0]^T$, $\boldsymbol{h}_2 = [1, 1, 0, 0]^T$, and $\boldsymbol{h}_3 = [1, 0, 1, 1]^T$. There are $3! = 6$ possible orderings, which lead to the following diagonal elements for $\mathbf{G}$:

	123	132	213	231	312	321
G_{11}:	0.77	0.77	1.22	1.22	1.	1.
G_{22}:	1.29	1.58	0.82	1.	1.22	1.22
G_{33}:	1.73	1.41	1.73	1.41	1.41	1.41

The last three permutations are preferable over the others, because they maximize the worst-case G_{ii}. Of these, the 231 permutation is preferred over the other two, because its G_{11} is as big as possible, which minimizes the probability of error propagation. None of the permutations results in the condition $G_{11} > G_{22} > G_{33}$. Observe that, although the third input has more energy than the first two, the optimal ordering does not detect it first. Thus, we have a counterexample to the proposition that the optimal detection order can be found by sorting signal energies.

The following example demonstrates that an optimally ordered DF detector can outperform a naturally ordered DF detector by 5 dB.

Example 10-25.

Consider a two-input two-output memoryless MIMO channel with AWGN and independent *Rayleigh fading*, meaning that the channel coefficients $\{h_{ij}\}$ are i.i.d. zero-mean circularly symmetric Gaussian random variables. Both inputs are chosen independently and uniformly from a 4-QAM alphabet. In Fig. 10-20 we show the performance of the ZF-DF and MMSE-DF detectors, both without sorting and with optimal sorting, as found using Monte-Carlo simulations after averaging over one million independently generated channel, symbol, and noise realizations. The figure illustrates that sorting improves performance by 2 dB with ZF-DF detection, and it improves

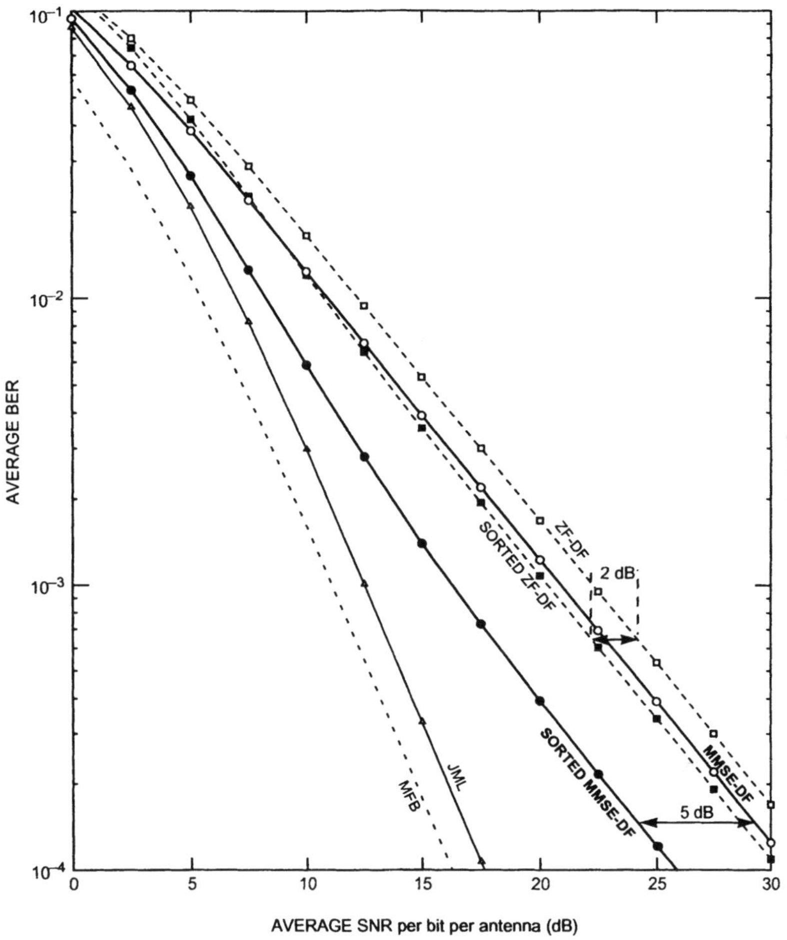

Fig. 10-20. A comparison of DF detectors, with and without sorting, for Example 10-25. The advantage of sorting is 2 dB for the ZF-DF detector, and it is 5 dB for the MMSE-DF detector. For comparison purposes, the performance of the joint ML detector and MFB are also shown.

performance by 5 dB with MMSE-DF detection. Clearly, the order in which the symbols are detected has a huge impact on performance.

Although the main purpose of this example was to demonstrate the benefits of sorting, it also illustrates the clear superiority of MMSE over ZF. Specifically, the gain of MMSE over ZF is 4.4 dB with sorting.

10.3.7. Multistage Parallel-Interference Cancellation

The DF detector of the previous section cancels interference from the symbols one by one, in sequential order. For this reason, it is characterized as a *serial* or *successive* interference cancellation (SIC) method. The performance is a strong function of the order in which the symbols are detected. In this section we describe the *parallel-interference canceller (PIC)* or *multistage detector* [14], a related MIMO detector in which the interference from all symbols is cancelled simultaneously, or in parallel.

The PIC is easily motivated as a straightforward attempt at achieving the MFB performance of the genie-aided receiver. In fact, the PIC uses the same cancellation technique used by the genie-aided receiver of Section 10.3.2 (see (10.68)), except that a vector of *tentative* decisions $\tilde{\boldsymbol{a}}$ is used in place $\boldsymbol{a}$:

$$\boldsymbol{z} = \mathbf{H}^*\boldsymbol{r} - \mathbf{B}\tilde{\boldsymbol{a}} \ , \tag{10.126}$$

where we have introduced $\mathbf{B} = \mathbf{R} - \mathbf{R}^d$. This defines the PIC receiver. Comparing it to the DF detector of (10.98), we see the same basic structure, but with three important differences:

- The forward filter $\mathbf{F} = \mathbf{H}^*$ is the MMF.
- The "feedback" filter $\mathbf{B} = \mathbf{R} - \mathbf{R}^d$ is neither lower triangular nor upper triangular; in general, it has zeros only on the diagonal.
- The input to the feedback filter is a vector of tentative decisions, not final decisions.

The tentative decisions $\tilde{\boldsymbol{a}}$ may be found using any of the detectors described in the previous sections, including the linear and DF detectors, either ZF or MMSE. Thus, we may think of the PIC detector as additional processing that may be appended to another detector in an attempt to improve its performance. The ultimate performance of the PIC detector will depend on the reliability of the tentative decisions. Correct tentative decisions would imply that the PIC receiver would attain the MFB. (Of course, if correct decisions could be guaranteed then

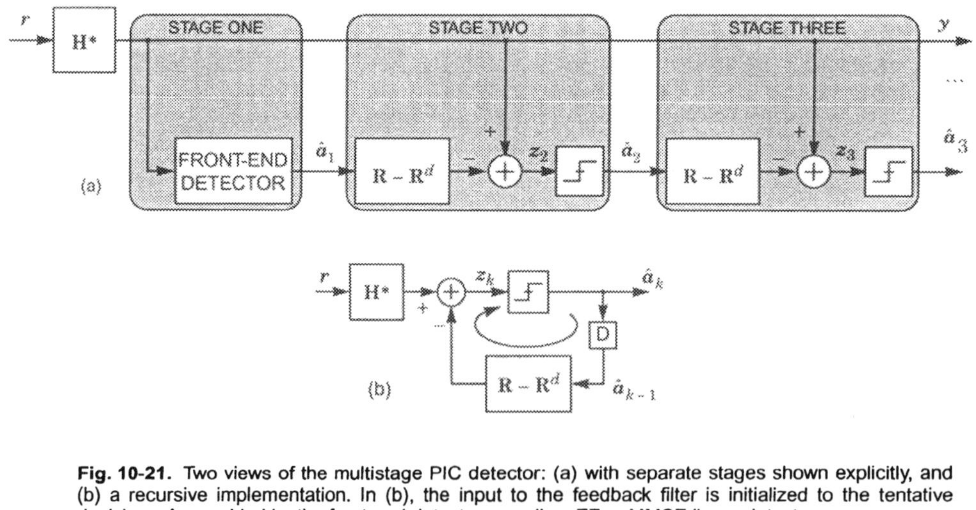

Fig. 10-21. Two views of the multistage PIC detector: (a) with separate stages shown explicitly, and (b) a recursive implementation. In (b), the input to the feedback filter is initialized to the tentative decisions $\hat{\boldsymbol{a}}_1$ provided by the front-end detector, usually a ZF or MMSE linear detector.

there would be no room for improvement, and hence no need for the PIC structure in the first place.)

The PIC naturally lends itself to a recursive or *multistage* implementation, where final decisions from stage k are used as tentative decisions for stage $k + 1$. The hope is that each iteration will improve the reliability of the decisions. Specifically, repeated application of (10.126) yields the recursion:

$$\boldsymbol{z}_{k+1} = \mathbf{H}^*\boldsymbol{r} - \mathbf{B}\hat{\boldsymbol{a}}_k , \tag{10.127}$$

where $\hat{\boldsymbol{a}}_k$ is found by quantizing $\boldsymbol{z}_k$, and where $\hat{\boldsymbol{a}}_1$ is initialized by a conventional linear or DF detector.

Two equivalent views of the multistage PIC detector are shown in Fig. 10-21. In Fig. 10-21(a), the different stages are shown explicitly. In Fig. 10-21(b), the structure of the DF detector is used, and the recursion of (10.127) is implied.

The multistage PIC detector of (10.127) is aggressive because it attempts to completely cancel all interference, despite the fact that the decisions at the initial stage may not be reliable. The reliability of the decisions does not necessarily improve from one stage to the next; in some circumstances, the detector can enter a limit cycle in which a sequence of decision vectors repeats periodically [7]. Divsalar *et al.* proposed a less aggressive multistage detector that aims to only partially cancel interference during the initial stages, leading to a *partial PIC (PPIC)* detector that adapts the slicer input according to [15]:

$$\boldsymbol{z}_{k+1} = p_k(\mathbf{H}^*\boldsymbol{r} - \mathbf{B}Q\{\boldsymbol{z}_k\}) + (1 - p_k)\boldsymbol{z}_k , \tag{10.128}$$

where the step size p_k is a number between zero and one that increases with each iteration. Typically, $\boldsymbol{z}_1$ is initialized using the linear MMSE detector of (10.87). If p_k is fixed at zero, then the PPIC detector does not deviate from the linear MMSE detector. On the other hand, if $p_k = 1$, the PPIC detector reverts to the conventional PIC of (10.127). A common strategy is to choose p_k as an increasing function of k, so that as the decisions get more and more reliable, the detector approaches the PIC.

10.3.8. Tree-Based Sphere Detection

In this section we return to the problem of joint ML detection, as described in Section 10.3.1. The complexity of an exhaustive search is exponential in the number of channel inputs n, since it enumerates all $|\mathcal{A}|^n$ possibilities. We might be discouraged by the fact that the joint ML detection problem is NP-hard, meaning that all known solutions have a worst-case complexity that is exponential in n. But that is a pessimistic view of the problem, because it focuses on the worst-case complexity. In practice, the complexity is a random variable that depends on the statistics of the channel input and noise, and the *average* complexity is often more relevant than its worst-case value. In this section we describe a tree-based algorithm for realizing the JML detector called a *sphere detector* [16] that can be significantly more efficient than an exhaustive search, attaining polynomial-time complexity in some circumstances [17].

Recall that the JML detector chooses its decision $\hat{\boldsymbol{a}}$ so as to minimize the total cost:

$$\begin{aligned} J(\boldsymbol{a}) &= \|\boldsymbol{r} - \mathbf{H}\boldsymbol{a}\|^2 \\ &= \|\boldsymbol{y} - \mathbf{G}\boldsymbol{a}\|^2 \\ &= \sum_{i=1}^{n} | y_i - \sum_{j=1}^{i} G_{ij} a_j |^2, \end{aligned} \tag{10.129}$$

where we have introduced $\boldsymbol{y} = \mathbf{Q}^*\boldsymbol{r}$, the WMF output, where $\mathbf{H} = \mathbf{QG}$ is the Gram-Schmidt decomposition of (10.100), so that $\mathbf{G}$ is lower triangular with positive real diagonal elements.

The sphere detector derives its name from the geometric picture shown in Fig. 10-22, which depicts each of the $|\mathcal{A}|^n$ possible *candidate* vectors $\{\mathbf{G}\boldsymbol{a}\}$ as they might appear in n-dimensional complex space, and the WMF output $\boldsymbol{y}$ (marked by $\times$), which is one of the candidates perturbed by white Gaussian noise. The JML problem is to find the candidate closest to $\boldsymbol{y}$. The sphere detector avoids the complexity of an exhaustive search by only considering candidates that fall within a given hypersphere centered at $\boldsymbol{y}$. The radius of the sphere must be chosen carefully. If it is too small, there will be no candidates inside, and if it is too big, there will be little benefit over an exhaustive search. A common choice is to use the decision $\tilde{\boldsymbol{a}}$ from a reduced-complexity detector — such as a DF detector — to define the radius $\sqrt{J}$ of the sphere, according to $J = \|\boldsymbol{y} - \mathbf{G}\tilde{\boldsymbol{a}}\|^2$. This guarantees that there is at least one candidate within the sphere, namely $\mathbf{G}\tilde{\boldsymbol{a}}$. Furthermore, at high SNR, the effectiveness of the DF detector ensures that there will not be many other candidates within the sphere. In the following we describe a tree-based algorithm that restricts the search to candidates within the sphere.

We can evaluate the cost of (10.129) by assigning metrics to the branches of a tree, and then summing branch metrics along a path from the root to a leaf of the tree. In Fig. 10-23 we show such a tree for the special case of $n = 3$ inputs and a binary alphabet $\mathcal{A} = \{\pm 1\}$. The starting point is the left-most node, or *root*. The tree consists of three stages, or in general n stages, one for each input (and hence one for each WMF output). The root is connected to $|\mathcal{A}|$ child nodes, one for each possible value for a_1, and each of these is connected to $|\mathcal{A}|$ child

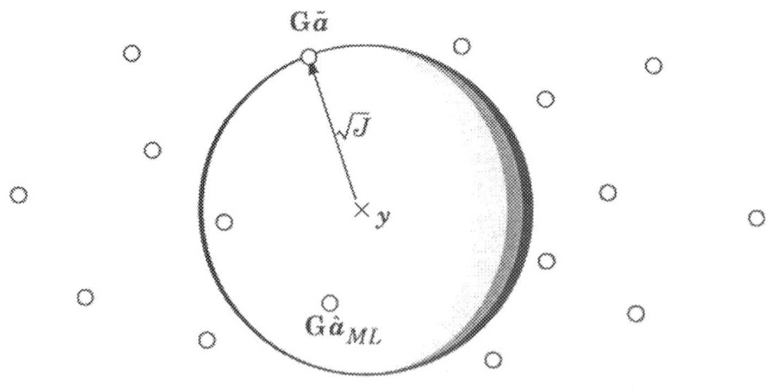

Fig. 10-22. The n-dimensional candidates $\{\mathbf{G}\boldsymbol{a}\}$, marked by ○, and the WMF output $\boldsymbol{y}$, marked by $\times$.

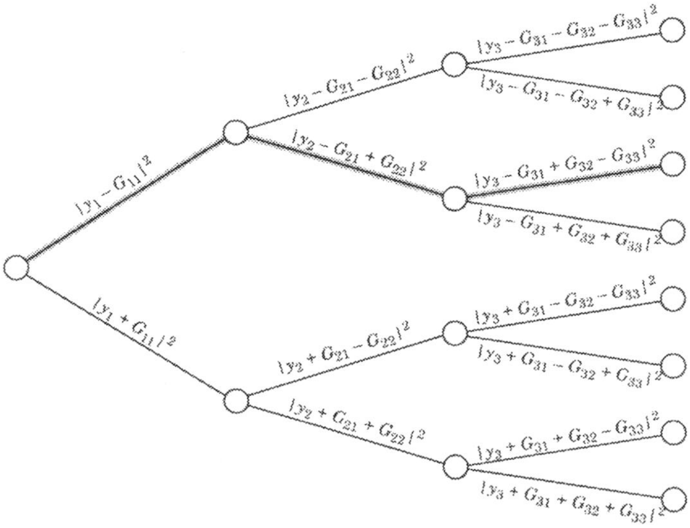

Fig. 10-23. A tree for the case of 3 binary inputs. An upper branch corresponds to an input of $a_i = 1$, and a lower branch to $a_i = -1$. Each branch has been labeled by the branch metric of (10.130).

nodes, depending on a_2, and so on. Thus, for each possible $\boldsymbol{a}$ there is a unique path through the tree that begins at the root and ends at one of the right-most nodes, or *leaf* nodes. For example, the path corresponding to $[a_1\ a_2\ a_3] = [1, -1, 1]$ has been highlighted.

In terms of this tree, we may interpret (10.129) as the sum of n *branch metrics*, one for each branch in a path from the root to a leaf node, where the metric for a branch in the i-th stage with path history $\{a_1, a_2, \dots a_i\}$ is defined as:

$$|y_i - \textstyle\sum_{j=1}^{i} G_{ij} a_j|^2 . \tag{10.130}$$

We see that the metric for a branch at the i-th stage depends only on the i-th observation y_i. With these labels, we can define the *cost* of any node as the sum of the branch metrics between it and the root. In terms of the geometry of Fig. 10-22, we may interpret the cost of a leaf node as the squared distance between the corresponding candidate and the WMF output.

Once we construct a tree and label its branches, the JML detection problem can be reformulated as a problem of finding the path through the tree with the lowest cost. A brute-force search would consider all $|\mathcal{A}|^n$ possible paths, one for each leaf node. At the other extreme is the ZF-DF detector, which can be interpreted as a *greedy* and short-sighted algorithm that travels through the tree, beginning at the root and ending at a leaf node, and at each node always takes the branch of least cost. The DF detector is simple and fast, but it does not always yield the ML solution. This is because an early branch with a low cost may lead to a subtree with a high cost, as illustrated below:

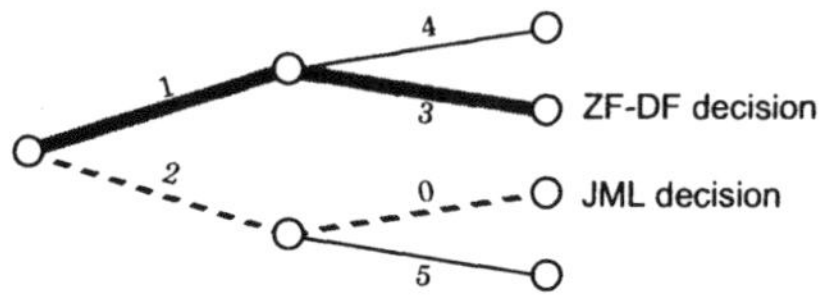

In the above example, the ZF-DF detector would choose the grey path, yielding a cost of 4, whereas the ML detector would choose the dashed path, yielding a cost of only 2.

The sphere detector efficiently organizes its search by exploiting the structure of the tree. It begins by using any low-complexity detector, such as a DF detector, to arrive at a tentative decision $\tilde{\boldsymbol{a}}$ with cost $J = \|\boldsymbol{y} - \mathbf{G}\tilde{\boldsymbol{a}}\|^2$. The key to the sphere detector is the observation that, once the cost of a particular node exceeds J, there is no need to extend this node any further, since additional branches will only add to the cost. We can thus eliminate this node *and all of its children* from further consideration.

There are two ways to implement a sphere detector. The first is a breadth-first search that mimics the Viterbi algorithm, moving through the tree one stage at a time, and discarding those nodes whose cost exceeds J (because such nodes necessarily lead to a leaf node that is outside the sphere). The nodes that are not discarded are said to be *survivors*.

A Breadth-First (Viterbi-Like) Sphere Detector

1. Define J as the cost (10.129) for any decision $\tilde{\boldsymbol{a}}$, perhaps that of a DF detector.
   Initialize the set of survivors $\mathcal{S}_0$ to the root node, and initialize its cost to zero.
2. For each tree stage $i \in \{1, \ldots n\}$:

   Define the i-th set of survivors $\mathcal{S}_i$ as the children of the previous survivors in $\mathcal{S}_{i-1}$, but excluding those children whose partial path metric exceeds J.
3. The leaf node in $\mathcal{S}_n$ having the smallest cost determines the ML decision $\hat{\boldsymbol{a}}_{ML}$.

Rather than pursuing all possible paths in parallel, a faster implementation arranges its search in a best-first manner by pursuing first the most promising paths. A leaf node will thus be reached faster, allowing us to decrease the threshold J, which ultimately allows us to discard more paths early on. When J decreases, the sphere radius of Fig. 10-22 shrinks. To facilitate the description of the algorithm, let $\eta(\mathcal{S})$ denote the node in $\mathcal{S}$ whose cost is smallest.

A Best-First Sphere Detector

1. Initialize J as the cost (10.129) for any decision $\tilde{\boldsymbol{a}}$, perhaps that of a DF detector.
   Initialize the set of survivors $\mathcal{S}$ to the root node, and initialize its cost to zero.
2. While $\eta(\mathcal{S})$ is not a leaf node:

   Eliminate $\eta(\mathcal{S})$ from $\mathcal{S}$, replacing it by those of its children whose cost is less than J.
   If any of these children are leaf nodes, reset J to the cost of the leaf node that is smallest.
3. The leaf node $\eta(\mathcal{S})$ having the smallest cost determines the ML decision $\hat{\boldsymbol{a}}_{ML}$.

This best-first algorithm is essentially equivalent to a form of sequential decoding for convolutional codes known as a *stack decoder*. In discrete optimization problems it is characterized as a *best-first branch-and-bound* technique [18]. A disadvantage of this approach is the potentially large memory requirements needed to store the surviving nodes along with their costs, as well as the computational complexity of finding the node with the smallest cost. The memory requirement is actually a random variable that depends on the particular realization of the channel input and noise random variables. A practical implementation would allocate enough memory to make the probability of overflow sufficiently small, and then resort to a suboptimal decision whenever overflow does occur.

As an alternative, one can formulate a local *depth-first* search technique with almost no storage requirements and no sorting, analogous to the *Fano algorithm* for sequential decoding. A Fano-like sphere detector steps forward through the tree from one node to the next, choosing the untraversed branch having the smallest cost, and adding branch metrics along the way, until it gets to a leaf node. The algorithm then backs up to the previous node — subtracting the branch metric — and repeats, updating the threshold J each time it gets to a low-cost leaf node, and backing up to a previous node whenever either all forward moves would increase the cost above the threshold J, or all forward moves have already been pursued. The search ends when no more moves are possible.

The complexity of the sphere detector is governed by the number of nodes it visits, a random variable whose probability mass function is a strong function of SNR. As SNR increases, the mass becomes more and more concentrated near the minimal value n. In contrast, the DF detector always visits exactly n nodes, regardless of SNR, and an exhaustive search always visits every one of the $(|\mathcal{A}|^{n+1} - |\mathcal{A}|)/(|\mathcal{A}| - 1)$ nodes in the tree. While the wort-case complexity of the sphere detector can be very high, the average complexity can be extremely small, sometimes comparable to that of the DF detector.

Example 10-26.

Consider a 16-input, 16-output memoryless channel in AWGN whose coefficients $\{h_{ij}\}$ are i.i.d. zero-mean complex Gaussian random variables, and assume the inputs are independent uncoded 16-QAM symbols. An exhaustive search would have to consider $16^{16} = 2^{64}$ leaf nodes, a near impossibility. Let N denote the number of nodes visited by a depth-first sphere detector, assuming the inputs are ordered optimally according to the BLAST ordering procedure. The probability mass function for N is easy to estimate using simulation. Two examples are shown in Fig. 10-24. When the SNR is 18 dB, as shown in Fig. 10-24(a), the sphere detector visits $\overline{N} = 3995$ nodes on average, which is about 250 times bigger than the 16 nodes visited by the DF detector, but over one million times smaller than an exhaustive search. When the SNR is 24 dB, as shown in Fig. 10-24(b), the sphere detector visits only $\overline{N} = 48.6$ nodes on average, barely three times as big as the DF detector. In fact, from the figure we see that the sphere detector is most likely to visit 18 nodes, just two more than the DF detector. Thus, the JML performance of an exhaustive search is often attainable with complexity comparable to the DF detector.

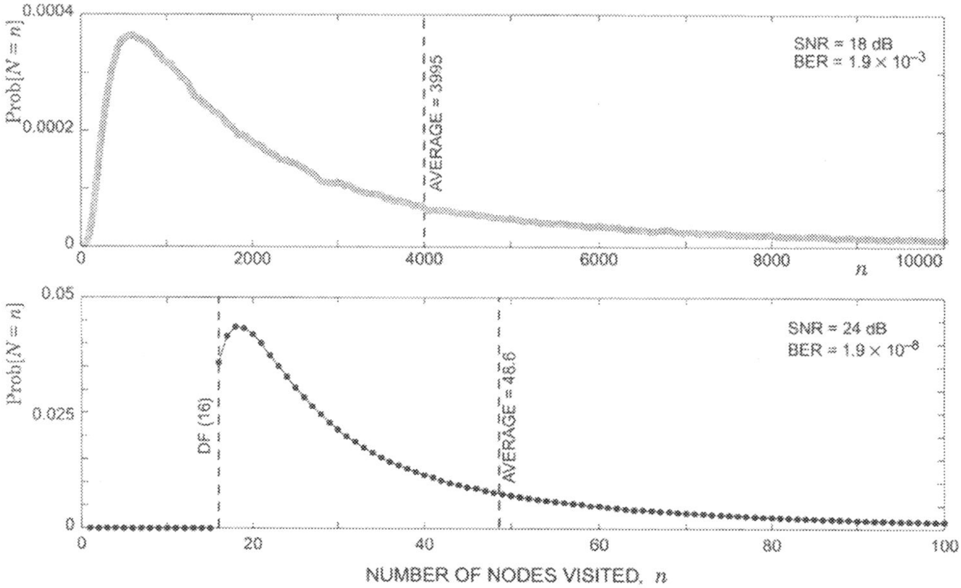

Fig. 10-24. Estimated probability mass functions for N, the number of nodes visited by the depth-first sphere detector, at (a) SNR = 18 dB and (b) SNR = 24 dB, assuming a 16 × 16 Rayleigh-fading channel with 16-QAM inputs. These results were found by simulating the sphere detector T times, with independent noise, channel, and symbol realizations for each trial, then estimating the pmf for N according to $\text{Prob}[N = n] = I_n / T$, where I_n is the number of trials for which n nodes were visited. The number T of trials was 2.5×10^5 for (a) and 5×10^6 for (b). The area under the tail beyond 10000 in (a) and beyond 100 in (b) is about 8% and 9%, respectively.

10.3.9. Performance Comparison

In this section we briefly compare the performance of many of the detection methods described in previous sections. Rather than specializing to a particular channel matrix **H**, which might not be representative for an application of interest, we will instead average performance over an ensemble of randomly generated channel matrices. Specifically, we consider a two-input two-output channel whose coefficients $\{h_{ij}\}$ are i.i.d. zero-mean complex Gaussian random variables. Both inputs are chosen independently and uniformly from a unit-energy 4-QAM alphabet $\{\pm 1 \pm j\}/\sqrt{2}$. The results are found using Monte-Carlo simulations after averaging over one million independent channel realizations, and are shown in Fig. 10-25 in the form of average BER (for both channel inputs) versus average SNR.

In this example, we see that the MMSE-linear detector outperforms the ZF-linear detector by 1.8 dB. This gap jumps significantly for DF detectors: the MMSE-DF detector is 4.4 dB better than the ZF-DF detector. Comparing DF detection to linear detection, we see that the MMSE-DF detector is 6.4 dB better than the MMSE-linear detector. (The difference is less dramatic for ZF detectors: the ZF-DF detector outperforms the ZF-linear detector by only

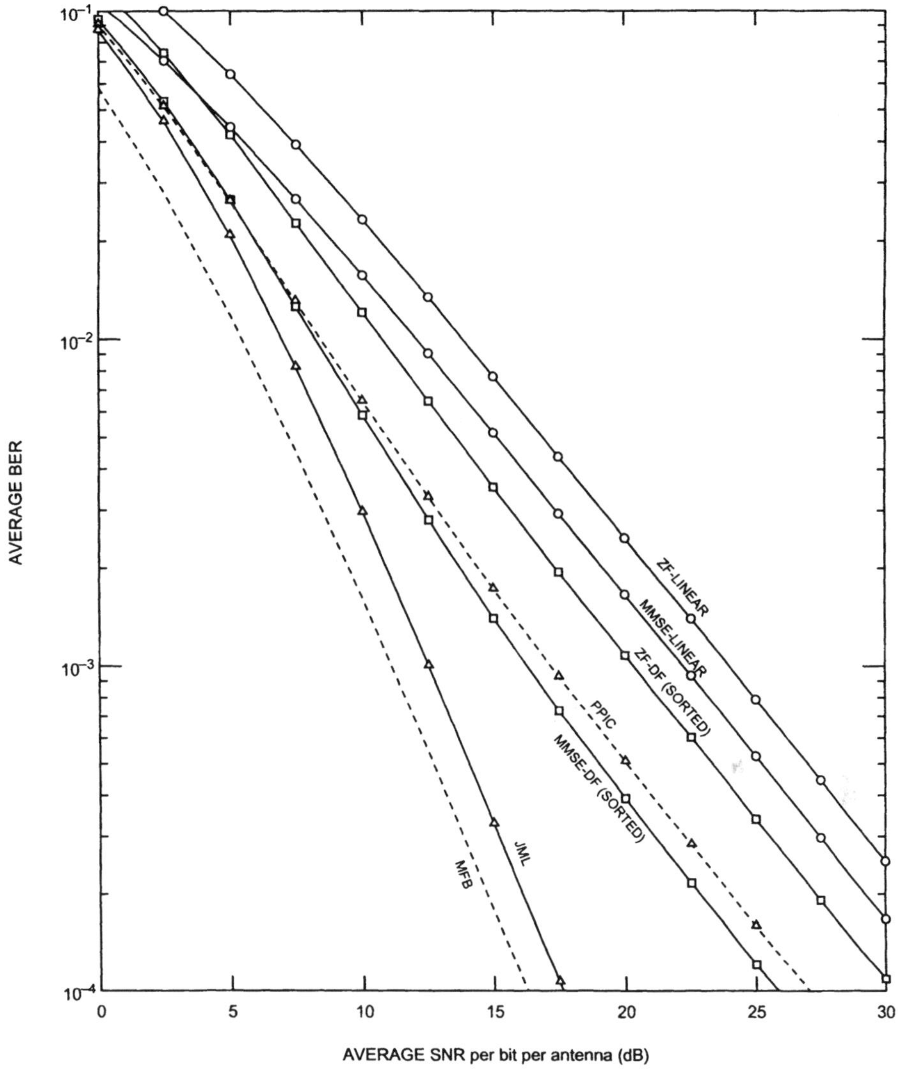

Fig. 10-25. BER averaged over one million random two-input two-output Rayleigh fading channels.

3.6 dB.) The PPIC detector described in Section 10.3.7 (with $\{p_1, \ldots p_7\} = \{0.1, 0.2, \ldots 0.6, 1\}$) is 3.2 dB better than the ZF-DF detector, but 1.2 dB worse than the MMSE-DF detector. Of the suboptimal schemes considered, the optimally sorted MMSE-DF detector offers the best performance, but is still 8.2 dB worse than the JML detector at BER = 10^{-4}. The JML detector of Section 10.3.8 falls 1.3 dB short of the MFB.

10.4. MIMO Detection with Channel Memory

In this section we show how the linear and decision-feedback detectors of the previous section — presented in the context of the one-shot memoryless channel — generalize to the continuous-time MIMO channel with memory of (10.31).

10.4.1. Linear Detection

In its most general form, a linear MIMO detector for the $m \times n$ channel of (10.31) consists of a continuous-time receive filter $\mathbf{F}(t)$ of dimension $n \times m$, a symbol-rate sampler, and a bank of slicers, as shown in Fig. 10-26(a). Because the sampled-MF projects $\boldsymbol{r}(t)$ onto the signal space spanned by $\{\mathbf{H}(t - kT)\}$, we can — without loss of generality — decompose $\mathbf{F}(t)$ into the cascade of the MF $\mathbf{H}^*(-t)$, a sampler, and discrete-time filter $\mathbf{C}(z)$, as shown in Fig. 10-26(b). By projecting onto the signal space, the receiver discards only noise components that are statistically independent of the desired symbols, and hence irrelevant to their detection. Therefore, the optimal receiver filter $\mathbf{F}(t)$ with respect to any reasonable optimization criterion — whether it be with respect to minimizing MSE, minimizing BER or some other criterion — will have a frequency response given by $\mathbf{F}(f) = \mathbf{C}(e^{j2\pi fT})\mathbf{H}^*(f)$, and can be implemented as shown in Fig. 10-26(b). By decomposing the receiver filter in this way, the receiver filter is completely specified by the discrete-time filter $\mathbf{C}(z)$. In Fig. 10-26(c) we have replaced the cascade of the channel, the matched filter, and the sampler by the equivalent model: a discrete-time filter with transfer function $\mathbf{S}(z)$ equal to the folded spectrum, and noise with PSD equal to $N_0\mathbf{S}(z)$.

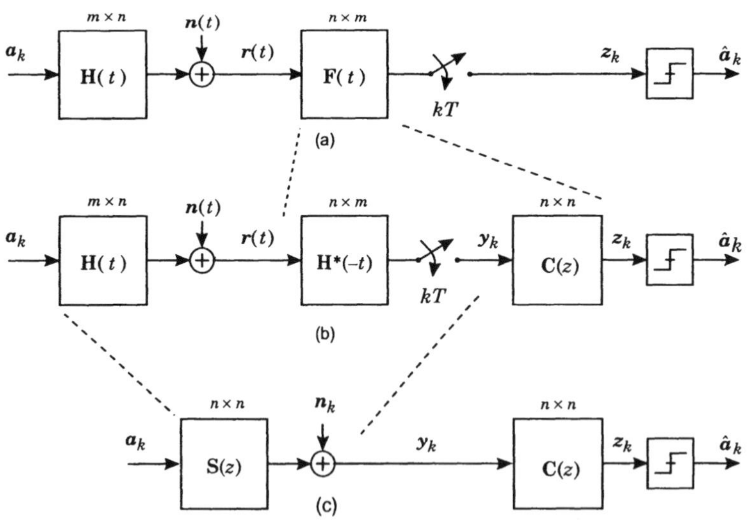

Fig. 10-26. (a) A linear detector; (b) a linear detector that is based on a sampled-MF front end. In (c), the cascade of the channel and sampled-MF is replaced by its equivalent model.

ZF Linear Detector

The zero-forcing linear detector chooses $\mathbf{C}(z)$ so as to eliminate interference completely. Assuming that its inverse is stable, we have no choice but to choose $\mathbf{C}(z) = \mathbf{S}^{-1}(z)$. The output is given by:

$$z_k = a_k + \tilde{n}_k \,, \tag{10.131}$$

which is free of interference, and where the PSD of the filtered noise $\tilde{n}_k$ is:

$$\begin{aligned} \mathbf{S}_{\tilde{n}}(z) &= \mathbf{C}(z)\mathbf{S}_n(z)\mathbf{C}^*(1/z^*) \\ &= \mathbf{S}^{-1}(z)\{N_0\mathbf{S}(z)\}\mathbf{S}^{-*}(1/z^*) \\ &= N_0\mathbf{S}^{-1}(z) \,, \end{aligned} \tag{10.132}$$

where we exploited the fact that $\mathbf{S}(z) = \mathbf{S}^*(1/z^*)$. Therefore, the MSE $E[\,|\,z_k^{(i)} - a_k^{(i)}\,|^2]$ for the i-th symbol can be expressed as:

$$\text{MSE}_i = N_0 \frac{1}{2\pi}\int_{-\pi}^{\pi} (\mathbf{S}^{-1}(e^{j\theta}))_{ii}\, d\theta \,. \tag{10.133}$$

MMSE Linear Detector

The MMSE linear detector chooses $\mathbf{C}(z)$ of Fig. 10-26 so as to minimize the MSE sum $E[\|z_k - a_k\|^2]$ directly, without forcing the interference to zero. As discussed in Section 10.3 for memoryless channels, the MMSE linear detector achieves an optimal balance between noise enhancement and interference suppression. The MSE sum can be expressed as:

$$\text{MSE} = \text{tr}\left\{\frac{1}{2\pi}\int_{-\pi}^{\pi} \mathbf{S}_e(e^{j\theta})\, d\theta\right\}, \tag{10.134}$$

where $\mathbf{S}_e(z)$ is the PSD of the error signal $e_k = z_k - a_k$. From Fig. 10-28(b) we see that the symbols contribute to e_k through a filter with transfer function $\mathbf{C}(z)\mathbf{S}(z) - \mathbf{I}$, whereas the noise n_k contributes to e_k through a filter with transfer function $\mathbf{C}(z)$. Therefore, the PSD of the error signal is (using $\mathbf{C}$ and $\mathbf{C}^*$ as shorthand for $\mathbf{C}(z)$ and $\mathbf{C}^*(1/z^*)$, respectively):

$$\mathbf{S}_e = (\mathbf{CS} - \mathbf{I})(\mathbf{S}^*\mathbf{C}^* - \mathbf{I}) + \mathbf{C}\{N_0\mathbf{S}\}\mathbf{C}^* \tag{10.135}$$

$$= (\mathbf{C} - \tilde{\mathbf{S}}^{-1})\,\tilde{\mathbf{S}}\mathbf{S}(\mathbf{C} - \tilde{\mathbf{S}}^{-1})^* + N_0\,\tilde{\mathbf{S}}^{-1}, \tag{10.136}$$

where we have introduced:

$$\tilde{\mathbf{S}}(z) = \mathbf{S}(z) + N_0\mathbf{I} \,. \tag{10.137}$$

The last equality in (10.136) follows from completing the square and is easily verified. Only the first term in (10.136) depends on $\mathbf{C}$, and we can do no better than to make this term zero. Hence, the MMSE solution is $\mathbf{C}(z) = \tilde{\mathbf{S}}^{-1}(z)$. With this choice, the MSE performance is:

$$\text{MSE}_i = N_0 \frac{1}{2\pi}\int_{-\pi}^{\pi} (\tilde{\mathbf{S}}^{-1}(e^{j\theta}))_{ii}\, d\theta \,. \tag{10.138}$$

As $N_0 \to 0$, $\tilde{\mathbf{S}} \to \mathbf{S}$, and the MMSE detector approaches the ZF detector.

10.4.2. MMSE-DF Detector

In this section we describe the MIMO DF detector for the general MIMO channel of (10.31), as shown in Fig. 10-27. This detector is a straightforward extension of the memoryless DF detector considered in Section 10.3.4. The detector consists of a sampled MMF, which provides sufficient statistics, followed by a forward filter $\mathbf{F}(z)$, whose function is to mitigate interference from future (undetected) symbols. Interference due to past (already detected) symbols is reconstructed using a strictly space-time causal feedback filter $\mathbf{B}(z)$ and then subtracted, yielding the slicer input $\boldsymbol{z}_k$.

We now derive the MMSE-DF detector, which chooses its forward and feedback filters so as to minimize the sum MSE $E[\|\boldsymbol{z}_k - \boldsymbol{a}_k\|^2]$, under the assumption that the decisions are correct. The derivation deviates from that of Section 10.3.5 because the filters under consideration have memory, and also because they operate on the sampled MF output, and are thus subject to noise that is not white. In Fig. 10-28(a) we show how the block diagram of Fig. 10-27 simplifies when the decisions are correct, and when the contributions to the error signal $\boldsymbol{e}_k = \boldsymbol{z}_k - \boldsymbol{a}_k$ from the channel input $\boldsymbol{a}_k$ and noise $\boldsymbol{n}_k$ have been separated. Specifically, the symbols $\boldsymbol{a}_k$ see a transfer function $\mathbf{F}(z)\mathbf{S}(z) - (\mathbf{I} + \mathbf{B}(z))$, and the noise sees a transfer function $\mathbf{F}(z)$. If we decompose the forward filter according to $\mathbf{F}(z) = (\mathbf{I} + \mathbf{B}(z))\mathbf{C}(z)$, then the equivalent diagram of Fig. 10-28(b) results. This diagram shows that only the last block depends on $\mathbf{B}(z)$. Furthermore, because $\mathbf{I} + \mathbf{B}(z)$ is a monic and space-time causal filter, it can be interpreted as a linear-prediction error filter. Hence, optimizing $\mathbf{B}(z)$ reduces to a space-time linear prediction problem: regardless of how $\mathbf{C}(z)$ is chosen, the best $\mathbf{B}(z)$ will be

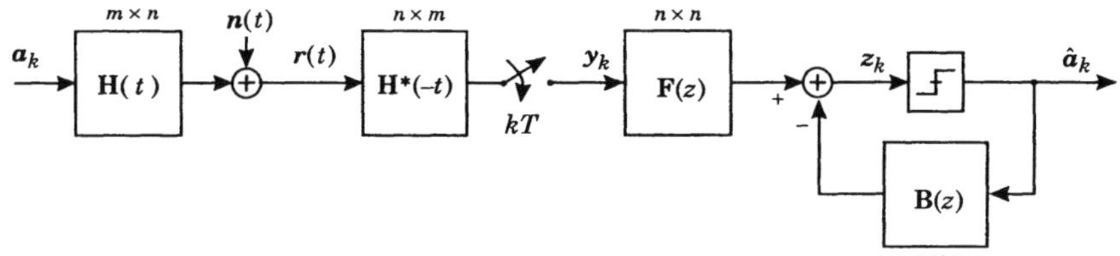

Fig. 10-27. A MIMO DF detector applied to the continuous-time MIMO channel of (10.31).

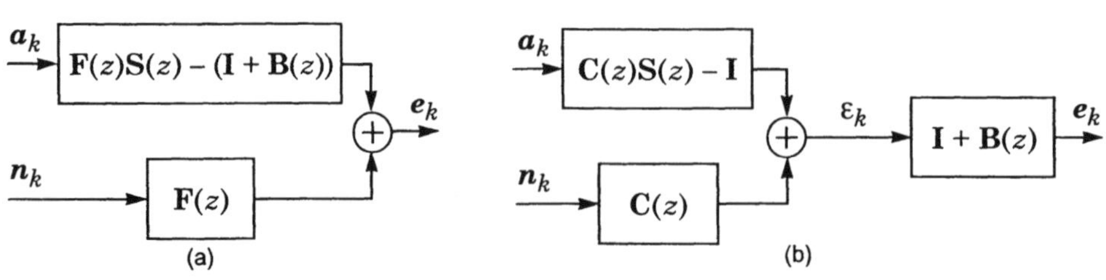

Fig. 10-28. (a) A block diagram showing how the channel input and noise contribute to the error of a MIMO DF detector, assuming correct decisions; (b) after factoring out the term $\mathbf{I} + \mathbf{B}(z)$.

determined by a spectral factorization of the PSD of the intermediate error signal ε_k. Specifically, as shown in Section 10.1.4, the optimal $\mathbf{B}(z)$ satisfies $(\mathbf{I} + \mathbf{B}(z)) = \mathbf{M}_\varepsilon(z)^{-1}$, where $\mathbf{M}_\varepsilon(z)$ is defined by the spectral factorization $\mathbf{S}_\varepsilon(z) = \mathbf{M}_\varepsilon(z)\Gamma_\varepsilon{}^2\mathbf{M}_\varepsilon^*(1/z^*)$ of the PSD of ε_k.

After the linear prediction error filter is optimized, the MSE sum is $\text{E}[\|\boldsymbol{e}_k\|^2] = \text{tr}\{\Gamma_\varepsilon{}^2\}$. We now must find the $\mathbf{C}(z)$ that minimizes $\text{tr}\{\Gamma_\varepsilon{}^2\}$. From Fig. 10-28(b), the PSD for ε_k is:

$$\mathbf{S}_\varepsilon(z) = (\mathbf{C}(z)\mathbf{S}(z) - \mathbf{I})(\mathbf{S}^*(1/z^*)\mathbf{C}^*(1/z^*) - \mathbf{I}) + N_0\mathbf{C}(z)\mathbf{S}(z)\mathbf{C}^*(1/z^*)\,. \tag{10.139}$$

Recognizing this as the error PSD (10.135) after a linear detector, we immediately conclude that the solution is the MMSE linear detector, $\mathbf{C}(z) = \tilde{\mathbf{S}}^{-1}(z)$, where $\tilde{\mathbf{S}}(z) = \mathbf{S}(z) + N_0\mathbf{I}$. With this choice for $\mathbf{C}(z)$, the PSD of the residual error signal is the last term in (10.136), namely:

$$\mathbf{S}_\varepsilon(z) = N_0\,\tilde{\mathbf{S}}(z)^{-1}. \tag{10.140}$$

Observe that $\tilde{\mathbf{S}}(z) = \mathbf{S}(z) + N_0\mathbf{I}$ is itself a valid PSD and it will be full rank even when $\mathbf{S}(z)$ is not, provided that the noise is nonzero. Hence, it admits a spectral factorization:

$$\tilde{\mathbf{S}}(z) = \tilde{\mathbf{M}}^*(1/z^*)\tilde{\Gamma}^2\tilde{\mathbf{M}}(z)\,, \tag{10.141}$$

where $\tilde{\mathbf{M}}(z)$ is monic and space-time causal, and $\tilde{\Gamma}$ is diagonal with real and positive diagonal elements. In terms of this factorization, the residual error PSD reduces to:

$$\mathbf{S}_\varepsilon(z) = N_0\tilde{\mathbf{M}}(z)^{-1}\tilde{\Gamma}^{-2}\tilde{\mathbf{M}}^{-*}(1/z^*)\,. \tag{10.142}$$

Since $\tilde{\mathbf{M}}(z)^{-1}$ is itself monic and space-time causal, and since $N_0\tilde{\Gamma}^{-2}$ is a positive diagonal matrix, (10.142) itself defines a spectral factorization of $\mathbf{S}_\varepsilon(z)$, namely $\mathbf{S}_\varepsilon(z) = \mathbf{M}_\varepsilon(z)\Gamma_\varepsilon{}^2\mathbf{M}_\varepsilon^*(1/z^*)$ with $\mathbf{M}_\varepsilon(z) = \tilde{\mathbf{M}}(z)^{-1}$ and $\Gamma_\varepsilon{}^2 = N_0\tilde{\Gamma}^{-2}$. As shown in Section 10.1.4, the best feedback filter $\mathbf{B}(z)$ will cause the space-time prediction error filter $(\mathbf{I} + \mathbf{B}(z))$ to cancel $\tilde{\mathbf{M}}(z)^{-1}$:

$$\mathbf{B}(z) = \tilde{\mathbf{M}}(z) - \mathbf{I}\,. \tag{10.143}$$

This defines the feedback filter. Combining the result $\mathbf{C}(z) = \tilde{\mathbf{S}}^{-1}(z)$ with the definition $\mathbf{F}(z) = (\mathbf{I} + \mathbf{B}(z))\mathbf{C}(z)$, we find that the forward filter of the MMSE-DF detector is:

$$\mathbf{F}(z) = (\mathbf{I} + \mathbf{B}(z))\tilde{\mathbf{S}}^{-1}(z) \tag{10.144}$$

$$= \tilde{\mathbf{M}}(z)\tilde{\mathbf{M}}(z)^{-1}\tilde{\Gamma}^{-2}\tilde{\mathbf{M}}^{-*}(1/z^*) \tag{10.145}$$

$$= \tilde{\Gamma}^{-2}\tilde{\mathbf{M}}^{-*}(1/z^*)\,. \tag{10.146}$$

Equations (10.143) and (10.146) define the filters of the MMSE-DF detector.

As expected, the solutions (10.143) and (10.146) reduce to the scalar solutions of Chapter 8 when the channel has only one input and one output, namely $B(z) = \tilde{M}(z) - 1$ and $F(z) = 1/(\tilde{\gamma}^2\tilde{M}^*(1/z^*))$. Furthermore, the solutions reduce to the memoryless solutions of Section 10.3.5 when the folded spectrum $\mathbf{S}(z) = \mathbf{R}$ is a constant, independent of z.

From (10.142), the MSE for the i-th output of the MMSE-DF detector is:

$$MSE_i = \frac{N_0}{\tilde{\Gamma}_{ii}^2} . \tag{10.147}$$

In the absence of noise, the MMSE-DF detector reduces to the MIMO ZF-DF detector. In this case, $\tilde{\mathbf{S}}(z)$ reduces to $\mathbf{S}(z)$, so that the forward and feedback filters are also given by (10.143) and (10.146), but with $\mathbf{M}(z)$ and Γ in place of $\tilde{\mathbf{M}}(z)$ and $\tilde{\Gamma}$, respectively.

Example 10-27.

Consider the continuous-time channel of (10.31) with two inputs and one output, so that $\mathbf{H}(t) = [h_1(t), h_2(t)]^T$, where $h_1(t)$ and $h_2(t)$ are as sketched below [7]:

This model might arise in an *asynchronous CDMA* application, where $h_1(t)$ and $h_2(t)$ are the signatures assigned to user 1 and user 2, respectively. The two pulses would be orthogonal if they were transmitted synchronously, but the second pulse is delayed by an amount αT, which leads to multiuser interference. Assume that the normalized delay parameter α is between 0 and 1, and assume that both signatures are normalized to have unit energy (i.e., perfect power control). The components of the 2×2 sampled autocorrelation matrix $\mathbf{S}_k$, as defined by (10.38), are:

$$S_k^{(1,1)} = \int_{-\infty}^{\infty} h_1(t - kT)h_1(t)\, dt = \delta_k$$

$$S_k^{(1,2)} = \int_{-\infty}^{\infty} h_1(t - kT)h_2(t)\, dt = \rho\delta_k - \rho\delta_{k-1}$$

$$S_k^{(2,1)} = \int_{-\infty}^{\infty} h_2(t - kT)h_1(t)\, dt = \rho\delta_k - \rho\delta_{k+1}$$

$$S_k^{(2,2)} = \int_{-\infty}^{\infty} h_2(t - kT)h_2(t)\, dt = \delta_k , \tag{10.148}$$

where the correlation parameter $\rho \in [0, \sqrt{3}/4]$ depends on $\alpha \in [0, 1]$ according to:

$$\rho = \int_{-\infty}^{\infty} h_1(t)h_2(t)\,dt = \sqrt{3}\,\alpha(1 - \alpha) . \tag{10.149}$$

Therefore, the folded spectrum is:

$$\mathbf{S}(z) = \begin{bmatrix} 0 & 0 \\ -\rho & 0 \end{bmatrix} z + \begin{bmatrix} 1 & \rho \\ \rho & 1 \end{bmatrix} + \begin{bmatrix} 0 & -\rho \\ 0 & 0 \end{bmatrix} z^{-1} . \tag{10.150}$$

Observe that the folded spectrum reduces to a constant (the identity matrix) when $\alpha = 0$, i.e., when the two waveforms are synchronized. It is easy to verify that the spectral factorization of the folded spectrum is given by $\mathbf{S}(z) = \mathbf{M}^*(1/z^*)\Gamma^2\mathbf{M}(z)$, where:

$$\mathbf{M}(z) = \begin{bmatrix} 1 & 0 \\ \rho/\gamma^2 & 1 \end{bmatrix} + \begin{bmatrix} 0 & -\rho/\gamma^2 \\ 0 & 0 \end{bmatrix} z^{-1}, \quad \Gamma = \begin{bmatrix} \gamma & 0 \\ 0 & \gamma \end{bmatrix}, \text{ and } \gamma = \sqrt{\frac{1}{2} + \sqrt{\frac{1}{4} - \rho^2}}. \tag{10.151}$$

Let us consider the ZF-DF detector, which has the same structure as the MMSE-DF detector, but chooses its filters under the assumption that the noise is zero, even when it is not. Therefore, from (10.143) and (10.146), the forward and feedback filters are given by:

$$\mathbf{F}(z) = \Gamma^{-2}\mathbf{M}^{-*}(1/z^*) = \frac{1}{\gamma^2 - \rho z}\begin{bmatrix} 1 & -\beta \\ -\beta z & 1 \end{bmatrix}$$

$$\mathbf{B}(z) = \mathbf{M}(z) - \mathbf{I} = \begin{bmatrix} 0 & -\beta z^{-1} \\ \beta & 0 \end{bmatrix}, \tag{10.152}$$

where we have introduced $\beta = \rho/\gamma^2 \in [0, 0.5]$. An implementation of the ZF-DF detector is shown in Fig. 10-29. The receiver front end consists of an MMF $\mathbf{H}^*(-t)$, which reduces to a pair of scalar matched filters, followed by a pair of samplers. The forward and feedback filters of (10.152) are shown explicitly. In practice, the noncausal filters $1/(\gamma^2 - \rho z)$ and $z/(\gamma^2 - \rho z)$ would have to be delayed sufficiently to make them approximately causal; this delay is not shown. Recall that for the memoryless DF detector, the first-user decision reduces to that of a linear detector. From Fig. 10-29 we see that this result is no longer true for asynchronous channels (channels with memory). Indeed, the user 1 decision at time k is based in part on the user 2 decision at time $k - 1$. The MSE for both users is N_0/γ^2, while the MFB attained by the genie-aided detector is $MSE_{\text{MFB}} = N_0$. Thus, the MSE of the ZF-DF detector exceeds the MFB by a factor of γ^{-2}. This penalty ranges from 0 dB to 1.25 dB as α ranges from 0 to $1/2$.

In contrast, the ZF-linear detector would take the form of Fig. 10-26 with $\mathbf{C}(z) = \mathbf{S}(z)^{-1}$:

$$\mathbf{C}(z) = \mathbf{S}(z)^{-1} = \begin{bmatrix} 1 & \rho(1 - z^{-1}) \\ \rho(1 - z) & 1 \end{bmatrix}^{-1} = \frac{1}{1 - \rho^2(1 - z^{-1})(1 - z)}\begin{bmatrix} 1 & -\rho(1 - z^{-1}) \\ -\rho(1 - z) & 1 \end{bmatrix}. \tag{10.153}$$

Therefore, from (10.138), the MSE for both users is given by:

$$MSE_{\text{ZFL}} = N_0 \frac{1}{2\pi}\int_{-\pi}^{\pi} \frac{1}{1 - \rho^2|1 - e^{-j\theta}|^2}\, d\theta = \frac{2N_0}{\xi - \rho^4/\xi}, \tag{10.154}$$

where:

$$\xi = \frac{1 - \rho^2}{2} + \sqrt{\left(\frac{1 - \rho^2}{2}\right)^2 - \rho^2}.$$

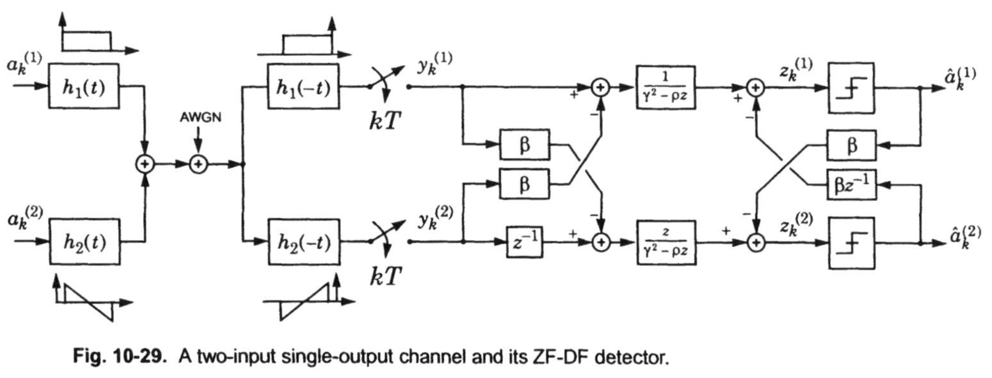

Fig. 10-29. A two-input single-output channel and its ZF-DF detector.

Relative to the MFB, the penalty is $(\xi - \rho^4/\xi)^{-1}$, which ranges from 0 dB to 1.42 dB as α ranges from 0 to $1/2$. It is at most 0.17 dB worse than the ZF-DF detector.

10.5. Further Reading

The theory of spectral factorization for stationary vector processes was developed by Wiener *et al.* in 1957 [3][4]. Practical factorization techniques can be found in [19][20]. The capacity of the Gaussian MIMO channel was derived in [21]. Verdu provides comprehensive coverage of multiuser detection [7]. The ML sequence detector for MIMO channels was proposed in [22][23] and improved upon in [24]. The linear MMSE detector for MIMO channels was proposed in [25]. A DF detector for 2×2 channels with analytic symmetry was developed in [26]. Linear detection for the 2×2 dually polarized radio channel was proposed in [27]. Salz generalized the linear detector to arbitrary dimension, and also optimized the transmit filters [28]. A DF detector for the 2×2 dually polarized channel was proposed by Kavehrad and Salz [29]. This detector was extended to arbitrary dimensions by Duel-Hallen [30]. A joint optimization of the transmitter and receiver filters for a MIMO channel with DF detection can be found in [31]. Related to the successive and parallel interference cancellation methods is a hybrid method that adopts the PIC recursion but updates only a subset of symbols at each stage, perhaps even just one, as opposed to all [32][33]. A feature of the sphere detector not explored here is its ability to provide soft outputs [34]. Adaptive implementations of multiuser detectors are described in [7][35][36].

Appendix 10-A. Proof of Separability Result (10.45)

In this appendix we prove the result (10.45), which states that, given an n-input m-output system with bandwidth W and symbol rate $1/T$, the receiver can linearly separate the n inputs only if $n \le m\lfloor W2T \rfloor$. We first observe that linear separability is possible only when the folded spectrum $\mathbf{S}(e^{j\theta})$ is full rank for all θ; this is because the MMF outputs are sufficient statistics, so separation is possible only if we can invert the folded spectrum.

We now must show that if $\mathbf{S}(e^{j\theta})$ is full rank for all θ, then we must have $n \le m\lfloor W2T \rfloor$. Recall from (10.40) that the folded spectrum is:

$$\mathbf{S}(e^{j2\pi fT}) = \frac{1}{T}\sum_k \mathbf{H}^*(f - \tfrac{k}{T})\mathbf{H}(f - \tfrac{k}{T}) . \tag{10.155}$$

We first claim that there exists a frequency f such that the number of nonzero terms in the summation of (10.155) is $\lfloor W2T \rfloor$. The frequency at which the number of nonzero terms is smallest will be $f = 0$ when $\lfloor W2T \rfloor$ is odd, and it will be $f = 1/(2T)$ when $\lfloor W2T \rfloor$ is even. Regardless, the number of nonzero terms is $\lfloor W2T \rfloor$.

We treat the case $m > n$ and $m \le n$ separately. We assume that $\mathbf{H}(f)$ is full rank for $|f| \le W$. If $m > n$, then $\mathbf{H}^*(f)\mathbf{H}(f)$ has rank n, and we need only one term ($\lfloor W2T \rfloor \ge 1$) in the summation in (10.155) to equal the rank of the left-hand side. But $\lfloor W2T \rfloor \ge 1$ implies $n \le m\lfloor W2T \rfloor$ when $m > n$. On the other hand, if $m \le n$, then each of the $\lfloor W2T \rfloor$ nonzero terms in (10.155) has rank m. The right-hand side (which is a sum of $\lfloor W2T \rfloor$ rank-m matrices) can equal the left-hand side (which is a rank n matrix) only if $m\lfloor W2T \rfloor \ge n$. Regardless of whether $m \le n$ or $m > n$, we always require that $m\lfloor W2T \rfloor \ge n$.

Problems

Problem 10-1. A conceptually simple numerical technique for performing the spectral factorization in (10.15) is the *Bauer method*, which creates an $m(k + 1) \times m(k + 1)$ matrix $\mathbf{R}_k$ of the form [20]:

$$\mathbf{R}_k = \begin{bmatrix} \mathbf{S}_0 & \mathbf{S}_{-1} & \mathbf{S}_{-2} & \cdots \mathbf{S}_{-k} \\ \mathbf{S}_1 & \mathbf{S}_0 & \mathbf{S}_{-1} & \vdots \\ \mathbf{S}_2 & \mathbf{S}_1 & \ddots & \\ \vdots & \ddots & & \mathbf{S}_{-1} \\ \mathbf{S}_k & \cdots & \mathbf{S}_1 & \mathbf{S}_0 \end{bmatrix},$$

where $\mathbf{S}(z) = \sum_k \mathbf{S}_k z^{-k}$ is the $m \times m$ PSD to be factored. To find the minimum-phase factor $\mathbf{G}(z) = \sum_k \mathbf{G}_k z^{-k}$ such that $\mathbf{S}(z) = \mathbf{G}(z)\mathbf{G}^*(1/z^*)$, let $\mathbf{R}_k = \mathbf{L}_k\mathbf{L}_k^*$ denote a Cholesky factorization of the above matrix, where $\mathbf{L}_k$ is lower triangular with real nonnegative diagonal elements. Under very mild conditions on $\mathbf{S}(z)$ [20], the last block row of $\mathbf{L}_k$ approaches $[\mathbf{G}_k, \ldots \mathbf{G}_1, \mathbf{G}_0]$ as $k \to \infty$. Convergence can be very fast, so that k need not be large to accurately estimate $\mathbf{G}(z)$.

(a) Use the Bauer method with $k = 1$ to approximate the factor $\mathbf{G}(z)$ in $\mathbf{S}(z) = \mathbf{G}(z)\mathbf{G}^*(1/z^*)$, assuming:

$$\mathbf{S}(z) = \begin{bmatrix} 64.01 & 1.6z + 8 + 0.2z^{-1} \\ 0.2z + 8 + 1.6z^{-1} & 0.2z + 5.04 + 0.2z^{-1} \end{bmatrix}. \tag{10.156}$$

(b) How close is the part (a) approximation to the true solution?

Problem 10-2. Find the transfer function $\mathbf{H}(z)$ of the following two-input three-output filter:

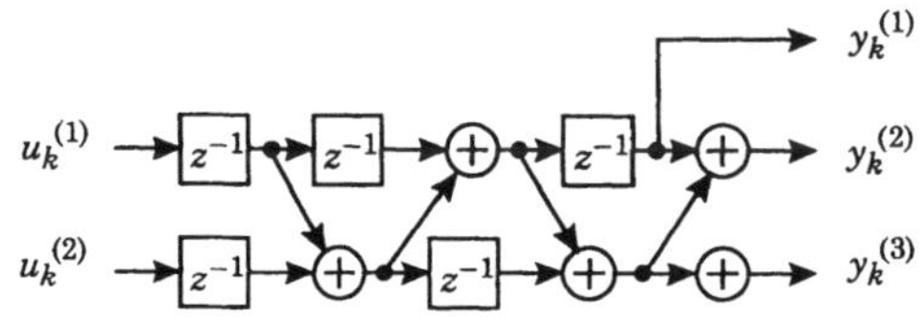

Problem 10-3. Assume that $u_k^{(1)}$ and $u_k^{(2)}$ are *i.i.d.* $\mathcal{N}(0, 1)$ in Problem 10-2, and let us define a two-dimensional random vector sequence $\boldsymbol{x}_k = [y_k^{(1)}, y_k^{(3)}]$ by selecting the first and last outputs of the above system.

(a) Find the autocorrelation function $\mathbf{R}_{\boldsymbol{x}}(m)$ for $\boldsymbol{x}_k$.

(b) Find the power spectrum $\mathbf{S}_{\boldsymbol{x}}(z)$ for $\boldsymbol{x}_k$.

Problem 10-4. Let $\boldsymbol{X}_k$ be a temporally white random vector sequence that is not spatially white, so that its power spectrum is a constant but not diagonal. In particular, suppose the power spectrum of $\boldsymbol{X}_k$ is $\mathbf{S}_X(z) = \mathbf{R}_{xx}$, where $\mathbf{R}_{xx}$ is specified in (10.24). Consider the following system:

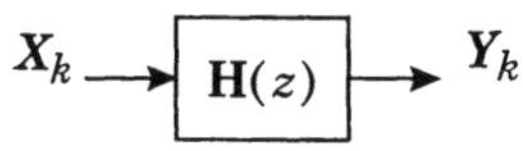

(a) Find a transfer function $\mathbf{H}(z)$ so that the output power spectrum $\mathbf{S}_Y(z)$ is given by (10.17).

(b) Is the answer to part (a) unique?

Problem 10-5. Consider a single-input double-output channel with received impulse responses $h_1(t) = u(t-T)$ and $h_2(t) = u(t-2T)$, where $u(t)$ is the unit step. Find the folded spectrum $\mathbf{S}(z)$.

Problem 10-6. Consider a single-input double-output channel. Find and sketch specific examples of received impulse responses $h_1(t)$ and $h_2(t)$ so that the folded spectrum is given by:

$$\mathbf{S}(z) = \begin{bmatrix} 9 & 3z \\ 3z^{-1} & 3 \end{bmatrix}. \tag{10.157}$$

Problem 10-7. Let $\mathbf{A}(z) = \sum_k \mathbf{A}_k z^{-k}$ and $\mathbf{B}(z) = \sum_k \mathbf{B}_k z^{-k}$ be stable transfer functions of dimension $m \times n$ and $p \times n$, respectively. Prove or disprove the following proposition:

$$\frac{1}{2\pi}\int_{-\pi}^{\pi} \mathbf{A}(e^{j\theta})\mathbf{B}^*(e^{j\theta})d\theta = \sum_{k=-\infty}^{\infty} \mathbf{A}_k \mathbf{B}_k^* \ . \tag{10.158}$$

Problem 10-8. Let $\boldsymbol{x}_k = [x_k^{(1)}, x_k^{(2)}]^T$ be a random process with PSD given by (10.17). Let $\hat{x}_k^{(1)} = p_1 x_{k-1}^{(1)} + p_2 x_{k-1}^{(2)}$ be a linear-prediction estimate of $x_k^{(1)}$. Find the prediction coefficients p_1 and p_2 that minimize the prediction error variance.

Problem 10-9. Consider a *real*-valued channel of the form $\boldsymbol{r} = a_1\boldsymbol{h}_1 + a_2\boldsymbol{h}_2 + \boldsymbol{n}$, which is a special case of (10.61) with only two inputs. Assume a_1 and a_2 are independently and uniformly chosen from the binary alphabet $\{\pm 1\}$, and assume that the noise components are real, independent and Gaussian with variance σ^2. This problem compares the jointly optimal (JML) detector, which jointly chooses a_1 and a_2 to maximize the conditional pdf $f(\boldsymbol{r} \mid a_1, a_2)$, to the individually optimal (IML) detector for the first input, which chooses a_1 to maximize $f(\boldsymbol{r} \mid a_1)$. Let $y_1 = \boldsymbol{h}_1^T\boldsymbol{r}$, let $y_2 = \boldsymbol{h}_2^T\boldsymbol{r}$, and let $\rho = \boldsymbol{h}_2^T\boldsymbol{h}_1$.

(a) Show that the JML decision for a_1 can be expressed as [7]:

$$\hat{a}_1^{\text{JML}} = \text{sign}\left\{ y_1 + \tfrac{1}{2}g_0(y_2 - \rho) - \tfrac{1}{2}g_0(y_2 + \rho) \right\}, \tag{10.159}$$

where we have introduced the nonlinearity $g_0(x) = |x|$.

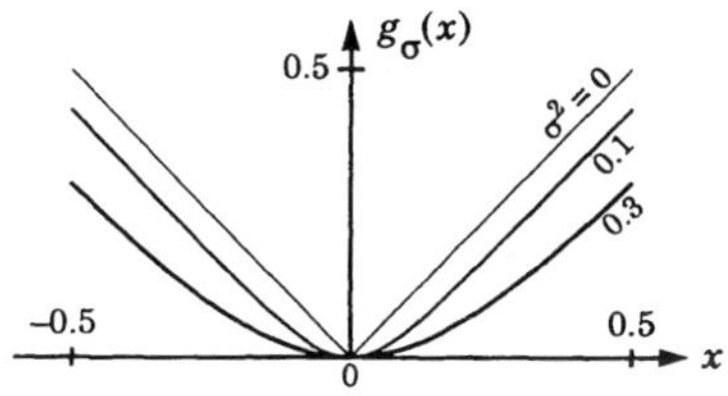

Fig. 10-30. The function $g_\sigma(x)$ for Problem 10-9.

(b) Show that the IML decision for a_1 is also given by (10.159), but with $g_0(x) = |x|$ replaced by $g_\sigma(x) = \sigma^2 \log\cosh(x/\sigma^2)$ [7]. This nonlinearity is sketched in Fig. 10-30.

(a) Show that $g_\sigma(x) \to |x|$ as $\sigma \to 0$. It follows that $\hat{a}_1^{\text{IML}}$ and $\hat{a}_1^{\text{JML}}$ usually agree at high SNR.

(b) Find a value for y_1 such that $\hat{a}_1^{\text{IML}} \neq \hat{a}_1^{\text{JML}}$, assuming $y_2 = 1$, $\rho = 1$ and $\sigma = 1$.

Problem 10-10. Show that (10.75) equals (10.76).

Problem 10-11. Show that (10.86) equals (10.87).

Problem 10-12. Consider the linear-prediction view of the DF detector shown in Fig. 10-31, where $\boldsymbol{c}_i$ is the i-th row of the pseudoinverse of $\mathbf{H}$. With this view we can derive an alternative algorithm for finding the optimal ordering that is significantly less complex than the BLAST procedure of (10.124).

(a) Show that the optimal choice for the first symbol index is $i_1 = \underset{i \in \{1, \dots n\}}{\arg\min} \| \boldsymbol{c}_i \|^2$.

(b) Once i_1 is chosen according to part (a), show that the optimal choice for i_2 satisfies:

$$i_2 = \underset{i \neq i_1,\ p}{\arg\min} \| \boldsymbol{c}_i - p\boldsymbol{c}_{i_1} \|^2 . \tag{10.160}$$

(c) Once $\{i_1, \dots i_{k-1}\}$ are optimally chosen, show that the optimal i_k is given by:

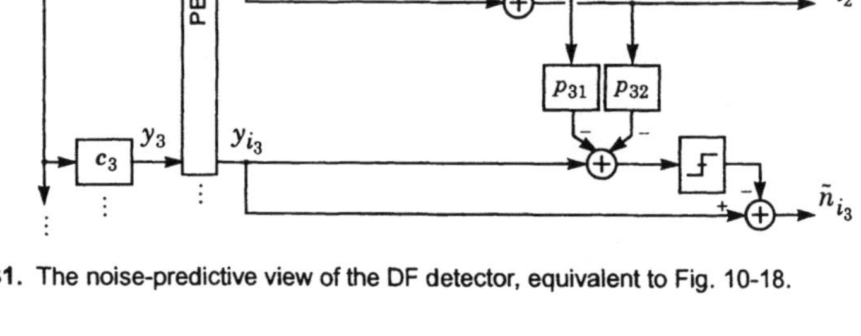

Fig. 10-31. The noise-predictive view of the DF detector, equivalent to Fig. 10-18.

$$i_k = \underset{i \notin \{i_1, \ldots i_{k-1}\}}{\arg\min} \| c_i - \hat{c}_i \|^2 , \tag{10.161}$$

where $\hat{c}_i$ is the projection of c_i onto the subspace spanned by $\{c_{i_1}, \ldots c_{i_{k-1}}\}$. The recursion (10.161) can be implemented with $O(n^3)$ complexity using a modified version of the Gram-Schmidt procedure [37], as opposed to the $O(n^4)$ complexity of (10.124).

Problem 10-13. Consider the three-input memoryless channel $\mathbf{H} = [\boldsymbol{h}_1, \boldsymbol{h}_2, \boldsymbol{h}_3]$ of Example 10-24, where $\boldsymbol{h}_1 = [1, 0, 1, 0]^T$, $\boldsymbol{h}_2 = [1, 1, 0, 0]^T$, and $\boldsymbol{h}_3 = [1, 0, 1, 1]^T$. Suppose the inputs are *i.i.d.* uniformly chosen from $\{\pm 1\}$, and the channel adds real white-Gaussian noise having the identity as an autocorrelation matrix.

(a) Assuming the best-first sphere detector is used with the natural ordering, find a *numerical value* for the probability that the first node extended (besides the root node) is *not* a part of the actual transmitted path.

(b) Repeat part (a) based on the 231 BLAST ordering, as described in Example 10-24.

(c) Based on a comparison of (a) and (b), argue qualitatively why the BLAST ordering reduces the complexity of the sphere detector tree search.

References

1. P Vaidyanathan, *Multirate Systems and Filter Banks*, Prentice Hall, Englewood Cliffs, 1993.
2. T. Guess and M. Varanasi, "Deriving Optimal Successive Decoders for the Asynchronous CDMA Channel Using Information Theory," *Conference of Information Sciences and Systems*, Princeton University, March 16-17, 2000.
3. N. Wiener and P. Masani, "The Prediction Theory of Multivariate Stochastic Processes I: The Regularity Condition," *Acta Math.*, vol. 98, no. 1, pp. 111-150, 1957.
4. N. Wiener and P. Masani, "The Prediction Theory of Multivariate Stochastic Processes II: The Linear Predictor," *Acta Math.*, vol. 99, pp. 93-137, 1958.
5. D. A. Schnidman, "A Generalized Nyquist Criterion and an Optimum Linear Receiver for a Pulse Modulation System," *Bell System Technical Journal*, pp. 2163-2177, November 1967.
6. R. Gallager, *Information Theory and Reliable Communication*, John Wiley and Sons, Inc., New York, 1968.
7. S. Verdú, *Multiuser Detection*, Cambridge University Press, 1998.
8. D. H. Johnson and D. E. Dudgeon, *Array Signal Processing*, Prentice-Hall, 1993.
9. A. Duel-Hallen, "Decorrelating Decision-Feedback Multiuser Detector for Synchronous Code-Division Multiple Access Channel," *IEEE Trans. Comm.*, Vol. 41, No. 2, pp. 285-290, Feb. 1993.
10. A. J. Viterbi, "Very Low Rate Convolution Codes for Maximum Theoretical Performance of Spread-Spectrum Multiple-Access Channels, *IEEE JSAC*, Vol. 8, No. 4, pp. 641-649, May 1990.

11. G. D. Golden, G. J. Foschini, R. A. Valenzuela, and P. W. Wolniansky, "Detection Algorithm and Initial Laboratory Results using the V-BLAST Space-Time Communication Architecture," *Electronics Letters*, Vol. 35, No. 1, pp. 14-15, 1999.

12. G. J. Foschini, G. D. Golden, R. Valenzuela, and P. Wolniansky, "Simplified Processing for Wireless Communication at High Spectral Efficiency," *IEEE JSAC*, Vol. 17, No. 11, pp. 1841-1852, 1999.

13. J. M. Cioffi and G. D. Forney, "Generalized Decision-Feedback Equalization for Packet Transmission with ISI and Gaussian Noise," in *Communication, Computation, Control and Signal Processing*, A. Paulraj, et al., Eds., Kluwer, 1997, pp. 79–127.

14. M. Varanasi and B. Aazhang, "Multistage Detection in Asynchronous Code-Division Multiple-Access Communications," *IEEE Trans. Comm.*, vol. 38, pp. 509-519, April 1990.

15. D. Divsalar, M. Simon and D. Raphaeli, "Improved Parallel Interference Cancellation for CDMA," *IEEE Trans. Comm.*, vol. 46, no. 2, pp. 258-268, Feb. 1998.

16. M. O. Damen, A. Chkief, and J. C. Belfiore, "Lattice Code Decoder for Space-Time Codes," *IEEE Communication Letters*, pp. 161-163, May 2000.

17. B. Hassibi and H. Vikalo, "On the Expected Complexity of Sphere Decoding," *35th Asilomar Conf. Sig., Syst. and Comp.*, November 4-7, 2001.

18. A. Grama and V. Kumar, "A Survey of Parallel Search Algorithms for Discrete Optimization Problems," *ORSA Journal of Computing*, vol. 7, no. 4, pp. 365-385, 1995.

19. M. C. Davis, "Factoring the Spectral Matrix," *IEEE Trans. Auto. Control*, pp. 296-305, October 1963.

20. D. C. Youla and N. N. Kazanjian, "Bauer-Type Factorization of Positive Matrices and the Theory of Matrix Polynomials Orthogonal on the Unit Circle," *IEEE Trans.n Circuits and Systems*, vol. CAS-25, no. 2, pp. 57-69, February 1978.

21. L. H. Brandenburg and A. D. Wyner, "Capacity of the Gaussian Channel with Memory: The Multivariate Case," *The Bell System Tech. Journal*, vol. 53, no. 5, pp. 745-778, May-June 1974.

22. W. Van Etten, "Maximum Likelihood Receiver for Multiple Channel Transmission Systems," *IEEE Trans. Comm.*, pp. 276-283, February 1976.

23. W. Van Etten, "An Optimum Linear Receiver for Multiple Channel Digital Transmission Systems," *IEEE Trans. Comm.*, pp. 828-834, August 1975.

24. S. Verdú, "Minimum Probability of Error for Asynchronous Gaussian Multiple-Access Channels," *IEEE Trans. Info. Theory*, Vol. 32, pp. 85-96, January 1986.

25. A. R. Kaye and D. George, "Transmission of Multiplexed PAM Signals over Multiple Channel and Diversity Systems," *IEEE Trans. Comm.Techn.*, vol. 18, no. 5, pp. 520-526, October 1970.

26. D. D. Falconer and G. J. Foschini, "Theory of Minimum Mean-Square-Error of QAM Systems Employing Decision Feedback Equalization," *Bell System Technical Journal*, vol. 52, no. 10, pp. 1821-1849, December 1973.

27. N. Amitay and J. Salz, "Linear Equalization Theory in Digital Data Transmission Over Dually Polarized Fading Radio Channels," *AT&T Bell Laboratories Technical Journal*, vol. 63, no. 10, pp. 2215-2259, December 1984.

28. J. Salz, "Digital Transmission Over Cross-Coupled Linear Channels," *AT&T Technical Journal*, vol. 64, no. 6, pp. 1147, July-August 1985.

29. M. Kavehrad and J. Salz, "Cross-Polarization Cancellation and Equalization in Digital Transmission Over Dually Polarized Multipath Fading Channels," *AT&T Technical Journal*, vol. 64, no. 10, pp. 2211-2245, December 1985.

30. A. Duel-Hallen, "Equalizers for Multiple Input/Multiple Output Channels and PAM Systems with Cyclostationary Input Sequences," *IEEE JSAC*, vol. 10, no. 3, pp. 630-639, April 1992.

31. J. Yang and S. Roy, "On Joint Transmitter and Receiver Optimization for Multiple-Input Multiple-Output (MIMO) Transmission Systems," *IEEE Trans. Comm.*, vol. 42, no. 12, December 1994, pp. 3221-3231.

32. Y. Sun, "Local Maximum Likelihood Multiuser Detection," *Proc. Conference on Information Science and Systems*, Baltimore, pp. 7-12, March 2001.

33. G. Barriac and U. Madhow, "PASIC: A New Paradigm for Low Complexity Multiuser Detection," *Proc. Conference on Information Science and Systems*, Baltimore, March 2001.

34. B. M. Hochwald and S. ten Brink, "Achieving Near-Capacity on a Multiple-Antenna Channel," submitted to *IEEE Trans. Comm.*, July 2001.

35. M. L. Honig and H. V. Poor, "Adaptive Interference Suppression," in *Wireless Communications: Signal Processing Perspectives* (H. V. Poor and G. W. Wornell, Eds.), Prentice-Hall, 1998.

36. M. Honig and M. K. Tsatsanis, "Adaptive Techniques for Multiuser CDMA Receivers," *IEEE Signal Processing Magazine*, vol. 17, no. 3, pp. 49–61, May 2000.

37. D. W. Waters and J. Barry, "Noise-Predictive Decision-Feedback Detection for Multiple-Input Multiple-Output Channels," *IEEE Int. Symp. Advances Wireless Comm.*, Victoria, Sept. 2002.

11

Fading and Diversity

The previous chapter examined MIMO communications from an abstract point of view, with an eye towards all types of MIMO applications. In contrast, this chapter specializes to wireless MIMO applications, which suffer from not only additive noise and multiuser interference but also from *multipath fading*.

A common theme running throughout this chapter is *diversity*, a strategy for mitigating the effects of multipath fading. Diversity is available whenever multiple, independently fading channels link the transmitter and receiver. Such multiple channels naturally occur in MIMO applications for which the transmitter or receiver use an antenna array. In fact, the hope for diversity is often what motivates the use of an antenna array in the first place. In this chapter we will define diversity more precisely, and we will describe various means for extracting diversity from a receiver antenna array. We will quantify the benefits of diversity for the case of Rayleigh fading. The last half of the chapter will be devoted to emerging techniques for achieving diversity with an antenna array at the *transmitter* instead of the receiver, such as delay-diversity transmission and space-time codes.

11.1. Types of Diversity

The effect of multiple paths of propagation at the receiver is a random multiplicative channel gain that can vary rapidly with time over a wide range of amplitudes and phases. In the Rayleigh-fading model, the complex gain is a random variable with a circularly symmetric zero-mean Gaussian distribution, so that its amplitude has a Rayleigh distribution, and its phase has a uniform distribution. Two examples of Rayleigh-fading channels are illustrated in Fig. 11-1.

Diversity is a powerful technique for mitigating the effects of fading. The basic idea behind diversity is to simultaneously transmit across multiple channels that are fading *independently*; that way, it is very unlikely that all channels fade simultaneously. If the probability that any one channel fades is p, the probability that m independent channels fade simultaneously is p^m. For example, the two channels shown in Fig. 11-1 are fading independently, and it can be seen that the two channels rarely fade at the same time.

Three important forms of diversity are *time*, *frequency*, and *space diversity*.

- *Time diversity* exploits the time-varying nature of the fading channel, as illustrated in Fig. 11-1. For example, a transmitter might send the same symbol at three different times, with the delay between symbols chosen large enough to ensure that the three transmissions experience independent fading. In other words, the transmitter might employ a $(3, 1)$ *repetition code* followed by a block interleaver. More generally, exploiting time diversity requires some form of error-control coding (Chapter 12) and interleaving; the redundancy of the error-control code spreads the message symbols over multiple coded symbols, and the interleaver breaks up bursts of errors in an attempt to make the coded symbols experience independent fading. A slowly varying channel would require a very long interleaver, and the latency that results would be unacceptable

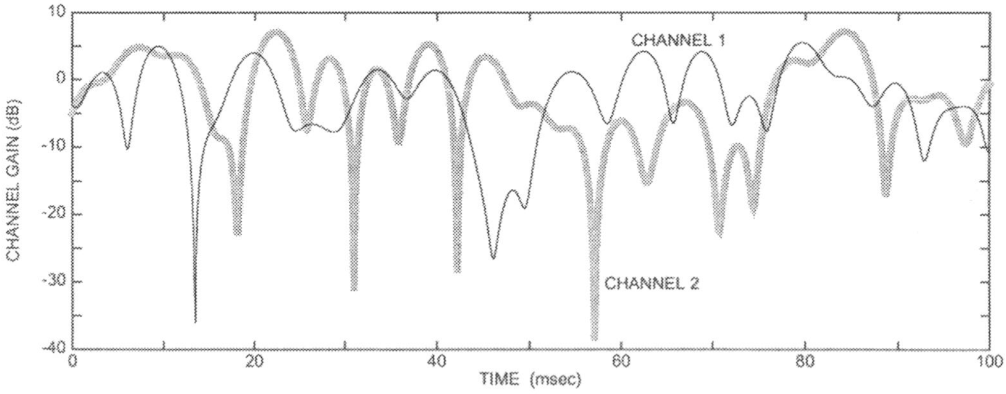

Fig. 11-1. Two independent Rayleigh-fading amplitudes, as generated from Jake's model [2], assuming 40 uniformly spaced angles of incidence with independent random phases and a maximum Doppler frequency of 100 Hz. The two channels are never in a deep fade at the same time. Instead, the minimum of the two gains is never smaller than –7 dB over the range shown.

in some applications. Hence, time diversity is not always available, especially in extreme cases such as a static environment (a user stopped at a red light, for example).

- *Frequency diversity* exploits the frequency-selective nature of multipath fading channels, and in particular, the fact that the fading correlation at two different frequencies tends towards zero as the frequency separation increases. Being the dual of time diversity, frequency diversity is not always available; just as time diversity diminishes as the channel variation slows, and disappears altogether for a static environment, so does frequency diversity diminish as the delay spread decreases, and disappears altogether for a flat-fading channel. Frequency diversity can be exploited using a combination of OFDM and coding, spread-spectrum modulation and RAKE reception, or even conventional equalization (such as DFE) for the case of a single-carrier QAM system whose bandwidth is already large relative to the inverse of the delay spread.
- *Spatial diversity* exploits the fact that the fading experienced by multiple antennas at a receiver will be independent when the distance between antennas is sufficiently large. The required separation distance depends on the details of the propagation environment, and can range from as little as one half wavelength (for rich scattering environments with paths arriving from all angles of incidence) to many wavelengths (for the case of a strong LOS component and a narrow range of angles of incidence). Compared to time and frequency diversity, spatial diversity has the advantage that it can be achieved without the latency of interleaving and without the bandwidth expansion of coding. Further, it does not require additional signal processing such as OFDM, DFE, or RAKE reception, although it does increase implementation complexity somewhat.

All forms of diversity can be analyzed in a similar manner, but for simplicity, the remainder of this chapter will focus almost exclusively on spatial diversity.

11.2. Receiver Diversity

Consider the wireless link with receiver diversity shown in Fig. 11-2(a), consisting of a transmitter with a single antenna and a receiver with m antennas. In the narrowband case, this system is a special case of (10.61) with $n = 1$ input, leading to the following memoryless single-input multiple-output (SIMO) model:

$$\boldsymbol{r} = a\boldsymbol{h} + \boldsymbol{n} , \tag{11.1}$$

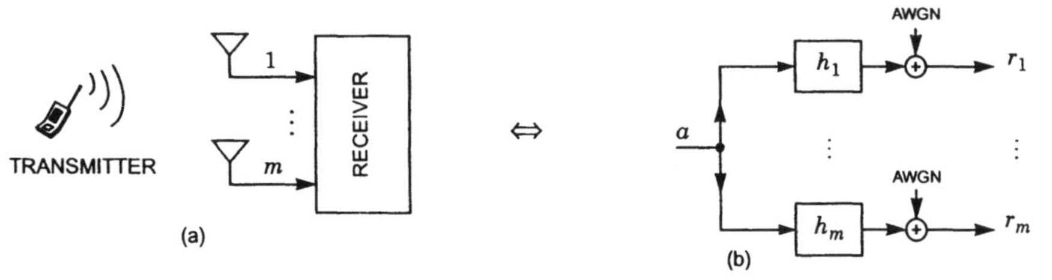

Fig. 11-2. (a) A wireless link with receiver diversity; (b) its equivalent model for the narrowband case.

where a is the transmitted symbol chosen from a finite alphabet, where $\boldsymbol{h} = [h_1, \ldots h_n]^T$ is a vector of channel gains, and where $\boldsymbol{n}$ is the noise. Here, h_i denotes the complex channel gain from the transmitter to the i-th receiver antenna. A block diagram representation of this model is shown in Fig. 11-2(b).

The fact that the channel is fading does not change the receiver design problem; we can build on the extensive results of Chapter 11. Specifically, given knowledge of $\boldsymbol{h}$, the receiver that minimizes the probability of error is the ML detector, which — for a SIMO channel — can be implemented by slicing the output of an MMF. However, the complexity of estimating and tracking $\boldsymbol{h}$ and implementing the MMF can be high when the channel varies rapidly. For this reason, suboptimal strategies are sometimes used. The most common strategies for combining the outputs of the antennas — a process referred to as *diversity combining* — are listed below:

- *Switched combining* — The receiver switches to one of its antennas, say antenna i, and stays with it until its channel gain $|h_i|$ falls below a predetermined or adaptive threshold, at which time it switches blindly to the next antenna.
- *Selection combining* — The receiver selects the antenna whose channel gain $|h_i|$ is largest at any instant. Mathematically, the decision statistic is:

$$y = r_{i_{\max}}, \qquad \text{where } i_{\max} = \arg\max_i\{\,|h_i|\,\}\,. \tag{11.2}$$

  This performs better than switched combining, but it requires that all m channel gains be estimated simultaneously, as opposed to only one.
- *Equal-gain combining* — Let α_i and θ_i denote the amplitude and phase of h_i, respectively, so that $h_i = \alpha_i e^{j\theta_i}$. In equal-gain combining, all antennas contribute to the decision statistic according to:

$$y = \Sigma_i e^{-j\theta_i} r_i\,. \tag{11.3}$$

  In other words, the phases of the different gains are adjusted so that the contributions add constructively. This performs almost as well as maximal-ratio combining, without requiring that the amplitudes $\{\alpha_i\}$ be estimated.
- *Maximal-ratio combining* — This is just another name for the MMF in the context of a memoryless SIMO channel. As in equal-gain combining, the phases are adjusted to ensure constructive addition, but then the different gains are scaled so as to maximize the SNR after combining, leading to:

$$y = \Sigma_i \alpha_i e^{-j\theta_i} r_i = \Sigma_i h_i^* r_i = \boldsymbol{h}^* \boldsymbol{r}\,. \tag{11.4}$$

  This performs the best, but requires the estimation of all of the complex channel gains.

The above strategies work best when there are no interferers. In the presence of interferers, we should add the ZF and MMSE linear detectors to the list; these detectors may be viewed as a means for diversity combining that take into account not only the fading but also the interference.

11.3. Performance Analysis for Rayleigh Fading

The aim of this section is to quantify the benefits of receiver diversity. Our starting point is the model $\boldsymbol{r} = a\boldsymbol{h} + \boldsymbol{n}$ of (11.1), where for simplicity we specialize to the 4-QAM alphabet, so that a is chosen uniformly from $\{\pm 1 \pm j\}\sqrt{E_b}$. The alphabet is scaled so that the energy per symbol is $2E_b$. Since 4-QAM conveys two bits of information, the energy *per bit* is E_b. Furthermore, we assume i.i.d. Rayleigh fading on the m subchannels, so that the channel gains $\{h_i\}$ are i.i.d. zero-mean unit-variance circularly symmetric complex Gaussian random variables, satisfying $\mathrm{E}[\,|h_i|^2] = 1$. Thus, the random vector $\boldsymbol{h}$ is completely characterized by its diagonal autocorrelation matrix $\mathrm{E}[\boldsymbol{h}\boldsymbol{h}^*] = \mathbf{I}$. Finally, we assume AWGN, so that the real and imaginary parts of the components of $\boldsymbol{n}$ are independently Gaussian with zero mean and variance $N_0/2$.

The next section analyzes the performance of a receiver that uses maximal-ratio combining. In Section 11.3.2, we analyze the performance with selection combining.

11.3.1. Performance with Maximal-Ratio Combining

Given knowledge of $\boldsymbol{h}$, the detector that minimizes the probability of error is the ML detector, which slices the output y of an MMF, and thus performs maximal-ratio combining:

$$\begin{aligned} y &= \boldsymbol{h}^*\boldsymbol{r} \\ &= \|\boldsymbol{h}\|^2 a + \tilde{n} \ , \end{aligned} \tag{11.5}$$

where $\tilde{n} = \boldsymbol{h}^*\boldsymbol{n}$ is a complex Gaussian random variable satisfying $\mathrm{E}[|\tilde{n}|^2] = N_0\|\boldsymbol{h}\|^2$. We can view (11.5) as the equation for an *effective* SISO channel whose instantaneous SNR per bit is:

$$\begin{aligned} \gamma_{\text{eff}} &= \|\boldsymbol{h}\|^2 \frac{E_b}{N_0} \\ &= \sum_{i=1}^{m} \gamma_i \ , \end{aligned} \tag{11.6}$$

where we have introduced the *instantaneous SNR per bit for the i-th antenna*:

$$\gamma_i = |h_i|^2 \frac{E_b}{N_0} \ . \tag{11.7}$$

The i.i.d. assumption implies that the *average* SNR per bit is the same for each antenna, namely:

$$\mathrm{E}[\gamma_i] = E_b/N_0 \ . \tag{11.8}$$

The effective SNR of (11.6) is sometimes called the *postdetection* SNR, so as to distinguish it from the predetection SNR of E_b/N_0. Given $\boldsymbol{h}$, the instantaneous BER for the effective SISO channel of (11.5) with Gray mapping is:

$$\Pr[\text{error} \mid \boldsymbol{h}] = Q(\sqrt{2\gamma_{\text{eff}}}) \ . \tag{11.9}$$

Because γ_{eff} is a random variable that depends on the channel, the above conditional probability is also random. To find the average BER, this expression must be averaged over the pdf for γ_{eff}:

$$\text{BER} = \int_0^{\infty} f_{\gamma_{\text{eff}}}(t) Q(\sqrt{2t})\, dt\,. \tag{11.10}$$

Since $\{h_i\}$ are i.i.d. complex Gaussian, each γ_i has a central chi-square distribution with two degrees of freedom, and $\gamma_{\text{eff}} = \sum_i \gamma_i$ has a chi-square distribution with $2m$ degrees of freedom:

$$f_{\gamma_{\text{eff}}}(t) = \frac{1}{(m-1)!(E_b/N_0)^m} t^{m-1} e^{-tN_0/E_b}\,, \quad \text{for } t \ge 0\,. \tag{11.11}$$

Repeated integration by parts of (11.10) with (11.11) leads to the following closed-form expression for the average BER on Rayleigh-fading channels with diversity order m:

$$\text{BER} = p^m \sum_{k=0}^{m-1} \binom{m-1+k}{k} (1-p)^k\,, \tag{11.12}$$

where we have introduced:

$$p = \frac{1}{2} - \frac{1}{2}\left(1 + \frac{1}{E_b/N_0}\right)^{-1/2}. \tag{11.13}$$

In Fig. 11-3 we plot the average BER of (11.12) versus the average SNR per bit per antenna for $m \in \{1, 2, 3, 4, 6, 10\}$.

The parameter p has a useful interpretation — it is the average BER *without* diversity. This is because a SISO fading channel without diversity is just a special case of (11.1) with $m = 1$, in which case (11.12) reduces to:

$$\text{BER} = p. \tag{11.14}$$

Of course, the performance of a SISO fading channel is of great practical interest, since diversity is not always available. Taking a closer look at (11.13) we see that, at high SNR, the BER without diversity is well-approximated by:

$$\text{BER} = p \approx \frac{1}{4E_b/N_0}\,. \tag{11.15}$$

Hence, without diversity, the asymptotic BER on a Rayleigh-fading channel is inversely proportional to SNR, which implies a very slow decay as SNR increases; see the curve labeled *no diversity* in Fig. 11-3. A BER near 10^{-9} requires an astronomical SNR of 84 dB. This is in stark contrast to a static channel, for which BER decreases *exponentially* with SNR, and for which a BER of 10^{-9} requires an SNR of only 12.5 dB. Roughly speaking, the poor performance without diversity is due to the fact that the deep fades of Fig. 11-1 dominate the overall average BER, so that a low average BER is achieved only when the SNR is high enough that the instantaneous BER is always low, even during a deep fade.

Returning to the general diversity case of $m > 1$, we can derive a simplified expression when the SNR is high. Specifically, if E_b/N_0 is so large that the term $(1-p)$ can be approximated by 1, then (11.12) simplifies to:

$$\text{BER} \approx K_m p^m, \qquad \text{where } K_m = \sum_{k=0}^{m-1}\binom{m-1+k}{k} = \binom{2m-1}{m}. \tag{11.16}$$

We see that the impact of order-m diversity is to raise the BER without diversity — namely p — to the power of m. Furthermore, since $p \approx (4E_b/N_0)^{-1}$ when E_b/N_0 is large, (11.16) is asymptotically:

$$\text{BER} \approx K_m\left(\frac{1}{4E_b/N_0}\right)^m. \tag{11.17}$$

Hence, when BER is plotted versus E_b/N_0 on a log-log scale, the curve approaches a straight line at high SNR, with an asymptotic slope proportional to m. A higher diversity leads to a steeper slope. The straight-line asymptotes predicted by (11.17) are shown as dashed lines in Fig. 11-3, and they are seen to be accurate at high SNR, especially when m is small.

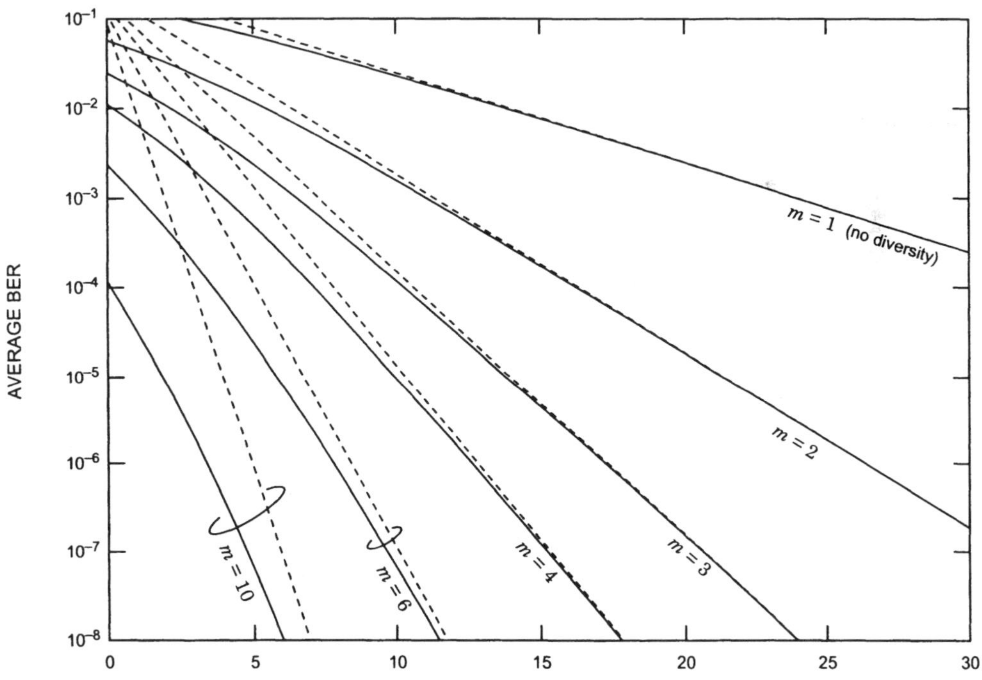

Fig. 11-3. Performance of a ML detector with order-m diversity for 4-QAM over a SIMO Rayleigh-fading channel with AWGN. The solid lines are exact, from (11.12). The dashed lines are the straight-line asymptotes of (11.17).

Comparing the curves labeled $m = 1$ and $m = 2$ in Fig. 11-3 at a BER of 10^{-3}, we see that a receiver with order-two diversity requires 13 dB less SNR than a receiver with no diversity at this BER. The improvement is less dramatic but still significant as m increases from 2 to 3.

The expression (11.17) shows that — with respect to average BER — the impact of order-m diversity on a fading channel is to raise the SNR to the power of m. In contrast, the impact of order-m diversity on a *static* (non-fading) channel is to only multiply the SNR by m, a much less dramatic improvement.

11.3.2. Performance with Selection Combining

In this section we perform analysis similar to that of the previous section, but assuming that the receiver performs selection combining instead of maximal-ratio combining. We will find that, as expected, selection combining performs somewhat worse than maximal-ratio combining, but it nevertheless achieves full diversity order. The channel model will be the same, namely 4-QAM with i.i.d. Rayleigh fading and AWGN, as described at the beginning of Section 11.3.

The selection-combining strategy dictates that the receiver select the antenna whose gain is largest in magnitude, and discard the observations from the other antennas. The index of the selected antenna is $i_{\max} = \arg\max_i\{ |h_i| \}$, so that the selected observation is:

$$y = h_{i_{\max}} a + n_{i_{\max}} . \qquad (11.18)$$

This is an effective SISO channel whose instantaneous (postdetection) SNR per bit is XE_b/N_0, where we have introduced the random variable

$$X = |h_{i_{\max}}|^2. \qquad (11.19)$$

Thus, given a particular channel $\boldsymbol{h}$, the bit-error rate with Gray-mapped 4-QAM is

$$\Pr[\text{error} \mid \boldsymbol{h}] = Q(\sqrt{2XE_b/N_0}) . \qquad (11.20)$$

Under the standard Rayleigh assumption that $\{h_i\}$ are i.i.d. zero-mean unit-variance complex Gaussian random variables, the cdf of X is straightforward to derive:

$$\begin{aligned} F_X(x) &= \Pr[|h_{i_{\max}}|^2 < x] \\ &= \Pr[(|h_1|^2 < x) \text{ and } (|h_2|^2 < x) \text{ and } \cdots \text{ and } (|h_m|^2 < x)] \\ &= \Pr[|h_1|^2 < x]\Pr[|h_2|^2 < x] \cdots \Pr[|h_m|^2 < x] \\ &= F_{|h_i|^2}(x)^m \\ &= (1 - e^{-x})^m u(x). \end{aligned} \qquad (11.21)$$

The pdf of X is thus:

$$\begin{aligned} f(x) &= \frac{d}{dx} F_X(x) \\ &= m(1 - e^{-x})^{m-1} e^{-x} u(x) \end{aligned}$$

$$= \sum_{k=1}^{m} (-1)^{(k-1)} \binom{m}{k} k e^{-kx} u(x) \ . \tag{11.22}$$

The average BER can now be found by averaging $Q(\sqrt{2XE_b/N_0})$ over this pdf, yielding:

$$BER = \sum_{k=1}^{m} (-1)^{(k-1)} \binom{m}{k} \int_0^\infty k e^{-kx} Q(\sqrt{2XE_b/N_0})\, dx \ . \tag{11.23}$$

The k-th integral can be interpreted as the average BER when the SNR has an exponential distribution with mean $E_b/(kN_0)$, or in other words, the average BER with no diversity and a path loss of $1/k$. We have already evaluated an integral of this form and need not do it again; the k-th integral reduces to the quantity p defined in (11.13), but with E_b/N_0 decreased by a factor of $1/k$. Based on this observation, the BER of (11.23) reduces to:

$$BER = \sum_{k=1}^{m} (-1)^{(k-1)} \binom{m}{k} \left\{ \frac{1}{2} - \frac{1}{2}\left(1 + \frac{k}{E_b/N_0}\right)^{-1/2} \right\}.$$

$$= \frac{1}{2} \sum_{k=0}^{m} (-1)^{k} \binom{m}{k} \left(1 + \frac{k}{E_b/N_0}\right)^{-1/2} \tag{11.24}$$

$$\approx \frac{A_m}{(E_b/N_0)^m}, \tag{11.25}$$

where the coefficient A_m satisfies the recursion $A_m = (m - \frac{1}{2})A_{m-1}$ with initialization $A_1 = \frac{1}{4}$. Specifically, it takes on the values of $A_m = \frac{1}{4}, \frac{3}{8}, \frac{15}{16}, \ldots$ for $m = 1, 2, 3$, etc. The approximation (11.25) is asymptotically tight, and it can be derived using a Taylor series expansion (Problem 11-3). The asymptotic expression clearly shows that selection combining achieves a diversity order of m, just like maximal-ratio combining. The performance is not quite as good as with maximal-ratio combining, however; comparing (11.24) with (11.17), we see that, when compared to maximal ratio combining, the BER with selection combining is asymptotically larger by a factor of 2, 6, and 24 for $m = 2, 3$, and 4, respectively.

11.4. The Diversity-Interference Trade-Off

There are two key benefits to an antenna array at the receiver:

- the ability to mitigate interference
- the ability to mitigate small-scale fading (diversity)

However, it is important to realize that these two benefits cannot be fully attained simultaneously using linear processing. An antenna array with linear processing that mitigates interference has a diminished capacity to mitigate fading, and likewise an antenna array that mitigates fading has a diminished capacity to mitigate interference. There is a fundamental trade-off between mitigating interference and mitigating fading, at least for the case of linear detection, as summarized by the following conservation result.

Conservation Theorem: The diversity order of a decentralized ZF linear receiver for an n-input m-output ($m \geq n$) Rayleigh flat-fading MIMO channel is $m - n + 1$. Equivalently, since the aim of such a receiver is to null the contributions from the $n - 1$ interferers, this can be summarized as:

$$\text{diversity order} + \#\text{interferers} = \#\text{receive antennas} . \tag{11.26}$$

In particular, if the channel is square ($m = n$), so that the number of antennas is equal to the number of users, then the linear receiver derives no diversity gain. Nulling the interferers exhausts all degrees of freedom of the array. At the other extreme, if there are no interferers, then all degrees of freedom are available to mitigate fading, and the diversity order is equal to the number of antennas. A proof of the theorem is given in Appendix 11-A.

The conservation theorem applies to the ZF linear detector only. In contrast, the JML detector is capable of simultaneously mitigating interference *and* achieving diversity. This is illustrated for the two-input, two-output Rayleigh channel in Fig. 10-25, where we see that the slope of the JML curve is twice as steep as that of the ZF linear detector. In this case, the JML detector achieves diversity order two, whereas the linear detector achieves diversity order one.

Although the conservation theorem applies to the MMSE detector in the limit as the noise variance goes to zero, the *effective* diversity order achieved by an MMSE detector can be higher than that of the ZF detector, provided that the SNR is sufficiently small. Specifically, the MMSE diversity order will range from $m - n + 1$ to m, depending on the strength of the interfering signals relative to the desired signal. This is because an MMSE detector does not try to null those interferers that are already below the noise floor; instead, it simply ignores them. Roughly speaking, the diversity order achieved by an MMSE detector is $m - n_{\text{eff}} + 1$, where n_{eff} is the number of significant interferers. This notion is explored in the following example.

Example 11-1.

Consider a two-user wireless Rayleigh-fading channel and a receiver with two antennas, where the user 2 signal is attenuated by a factor of α relative to that of user 1, so that the channel model is:

$$\mathbf{H} = \begin{bmatrix} h_{11} & \alpha h_{12} \\ h_{21} & \alpha h_{22} \end{bmatrix}, \tag{11.27}$$

where $\{h_{ij}\}$ are i.i.d. zero-mean complex Gaussian random variables. Both users use the same 4-QAM alphabet. The effective diversity order achieved by the MMSE linear detector for user 1 ranges from two to one as the attenuation constant α ranges from zero to infinity. This is illustrated in Fig. 11-4, where the average user-1 BER after a MMSE detector is shown as a function of average user-1 SNR for $\alpha \in \{0.01, 0.1, 0.2, 0.3, 0.4, 1, 10\}$, assuming 4-QAM alphabets for both users. The curves in Fig. 11-4 are labeled by the *signal-to-interference ratio* $SIR = \alpha^{-2}$, which is the average power of the desired user relative to that of the interferer. When the desired user is 20 dB weaker than the interferer, the linear MMSE detector is essentially identical to the linear ZF detector, yielding a diversity order of unity (i.e., no diversity). When the two users have equal power, the MMSE detector outperforms the ZF detector by about 1.3 dB. This gain of MMSE over ZF increases as the interferer weakens, until an SIR of 40 dB leads to essentially full diversity order of two, which is the MFB. At low SNR and high SIR, the MMSE curve appears to have a steeper slope and thus a higher diversity order. Eventually, however, as SNR grow sufficiently high, the MMSE slope approaches the ZF slope.

The main point of the previous example was to show that the MMSE detector can achieve a higher effective diversity order than that of the ZF detector. But it also reinforces our earlier result (see Example 10-20) that the performance of the MMSE detector does not always approach that of the ZF linear detector, even at high SNR, despite the fact that the coefficients of the MMSE detector approach those of the ZF detector at high SNR. This can be explained in part by the fading nature of the channel in this example. Regardless of how small the noise variance is, there is a nonzero probability that the channel coefficients will be even smaller, i.e., the channel will experience a deep fade. The performance during a deep fade tends to dominate the overall average BER performance, and during a deep fade, the MMSE and ZF filters differ substantially.

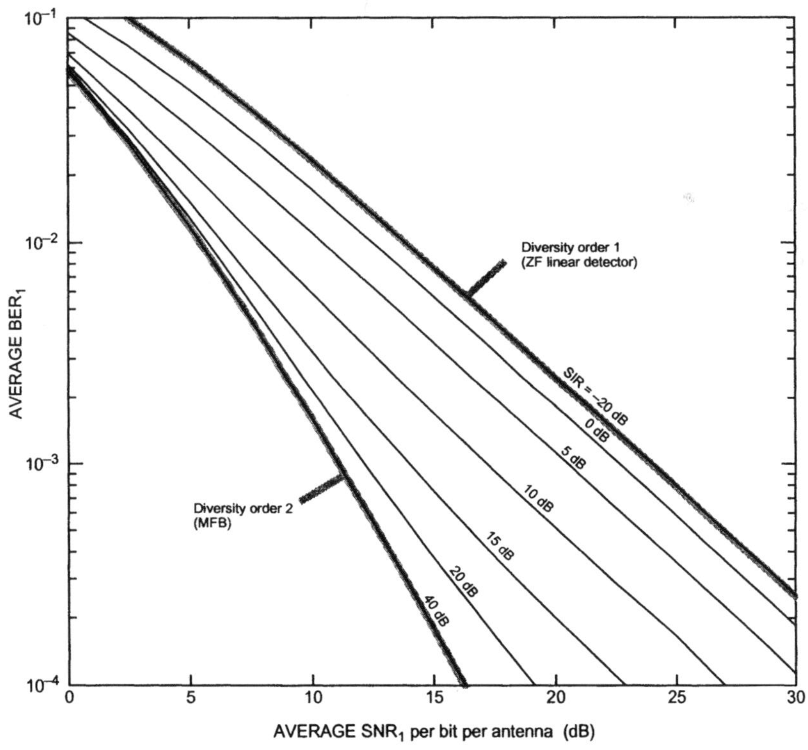

Fig. 11-4. The average BER for user 1 after a linear MMSE detector for the 2×2 channel of Example 11-1, assuming i.i.d. Rayleigh fading, AWGN, and 4-QAM. The MMSE curves were found using Monte Carlo simulations over 10^7 trials, whereas the lower and upper gray curves were from (11.12) and (11.13), respectively.

The conservation theorem has an important consequence in the context of ZF-DF detection. Suppose a ZF-DF detector is applied to a memoryless n-input m-output Rayleigh-fading channel. Refer back to the nulling and cancelling view of the ZF-DF detector, as illustrated in Fig. 10-13, with nulling vectors $\{w_1, \ldots w_n\}$. Because the first nulling vector must null the $n - 1$ remaining symbols, it provides a diversity order of $m - n + 1$. The second nulling vector, however, need only null the $n - 2$ undetected symbols, since the effects of the first symbol were subtracted off. Hence, it provides a diversity order of $m - n + 2$. In general, from (11.28), the diversity order for detecting the i-th symbol in a ZF-DF detector is:

$$\text{diversity order}_i = m - n + i \,. \tag{11.28}$$

In particular, because the last symbol detected sees no interference at all, it enjoys a full diversity order of m. This result relies on the assumption of correct decisions; in practice, when the effects of error propagation are taken into account, the diversity order can be smaller than (11.28) [1].

Example 11-2.
Consider DF detection applied to a 2-input 2-output Rayleigh fading channel. In this case, (11.28) tells us that the first symbol derives no diversity benefit, whereas the second symbol enjoys a diversity order of two. The BER for the first input will thus be much higher than that for the second. When the overall BER is computed as the average over both inputs, the lower BER dominates, and hence the overall diversity order is only one. This claim is confirmed by two previous results: Fig. 10-28, which applies to the 2×2 Rayleigh channel of Example 10-25, and Fig. 10-25, which applies to the similar example of Section 10.3.9. In both figures, we see that the slope of the BER-vs.-SNR curve is twice as steep for the JML detector as for any of the DF detectors. Furthermore, from Fig. 10-25 we observe that, although the DF detectors outperform the linear detectors, the DF and linear detectors have the *same* diversity order. Of all of the detectors considered, only the JML detector achieves a diversity order of two. It should be mentioned that it is possible to recoup some of the diversity order lost by the DF detector through the use of error-correction coding, since the different diversity orders seen by the two inputs is essentially a form of time diversity.

11.5. Transmit Diversity

Unlike the receiver diversity schemes of the previous section, which exploited an antenna array at the receiver, a *transmit diversity* scheme uses linear or nonlinear processing to spread information across multiple antennas at the transmitter. Transmit diversity schemes can be classified as either indirect or direct. Indirect schemes include delay diversity and frequency-offset diversity. Direct schemes include closed-loop precompensation, space-time block codes, and space-time trellis codes.

11.5.1. Delay Diversity and Frequency-Offset Diversity

An *indirect* transmit diversity scheme transforms the spatial diversity of the transmitter array into either temporal diversity or frequency diversity, so that it can be harvested at the receiver using conventional single-antenna techniques. Examples of indirect transmit diversity include delay diversity and frequency-offset diversity.

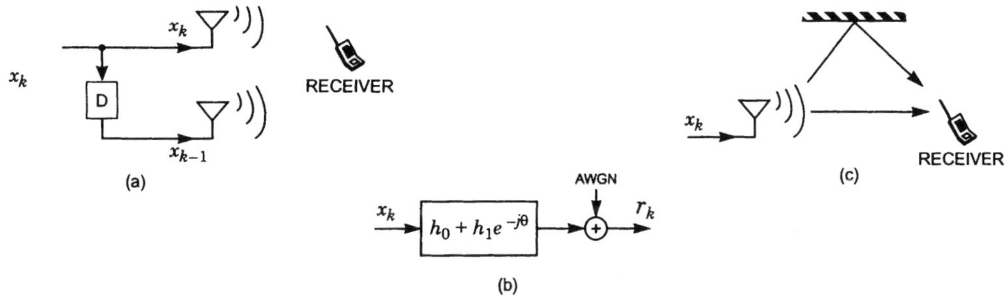

Fig. 11-5. Illustration of delay diversity: (a) flat-fading MISO model; (b) equivalent block diagram representation; (c) equivalent SISO channel with ISI.

The basic idea behind delay diversity is to transform a flat-fading channel into a frequency-selective (ISI) channel, so that conventional equalization methods such as DFE or MLSD can be used to extract diversity. The simplest form of delay diversity is shown in Fig. 11-5, where a narrowband transmitter with two antennas communicates to a receiver with one antenna across a flat-fading channel. The symbols transmitted from the first antenna are again transmitted from the second antenna, after being delayed by one signaling interval, so that if $\{x_k\}$ are the information symbols, the vector of symbols transmitted during the k-th signaling interval is:

$$\boldsymbol{a}_k = \begin{bmatrix} x_k \\ x_{k-1} \end{bmatrix}. \tag{11.29}$$

Substituting this into (10.61) with $n = 2$ inputs and $m = 1$ output leads to

$$r_k = h_0 x_k + h_1 x_{k-1} + n_k , \tag{11.30}$$

as shown schematically in Fig. 11-5(b). This is the equation for a *SISO* channel that has frequency-selective fading, as might arise from a LOS path and an independently fading reflected path having the same energy, as illustrated in Fig. 11-5(c). When the transmitter has n antennas, the delay-diversity strategy can be generalized by letting $\boldsymbol{a}_k = [x_k, x_{k-1}, \dots x_{k-n+1}]^T$ be the vector of symbols transmitted at time k, which leads to an equivalent SISO channel with n independently fading propagation paths. Conventional equalization techniques such as DFE or MLSD will not only recover the transmitted symbols, but they will also extract the full diversity benefit of order n [3].

Example 11-3.

Let us explicitly derive the BER of a ZF-DFE applied to the delay-diversity scheme of Fig. 11-5. As illustrated in Fig. 11-6, the optimal DFE filters depends on whether the effective ISI channel $H(z) = h_0 + h_1 z^{-1}$ is minimum phase. If $|h_0| > |h_1|$, the channel is minimum phase, in which case the optimal ZF-DFE has an identity forward filter and $h_1 z^{-1}$ as a feedback filter, as sketched in Fig. 11-6(a). On the other hand, If $|h_0| < |h_1|$, the ZF-DFE would consist of an all-pass forward filter $\tilde{H}(z)/H(z)$ that transforms the channel into its minimum-phase cousin $\tilde{H}(z) = h_1 + h_0 z^{-1}$, followed by a feedback filter of $h_0 z^{-1}$, as sketched in Fig. 11-6(b). The all-pass filter effectively swaps h_0 with h_1. In either case, the gain with the smaller magnitude will always be in the feedback filter, whereas the gain with the larger magnitude will always scale the slicer input, so that the slicer input (assuming correct decisions) is given by:

$$z_k = h_{\max} x_k + n_k , \tag{11.31}$$

where $h_{\max} = h_0$ when $|h_0| > |h_1|$, and $h_{\max} = h_1$ when $|h_1| > |h_0|$. The expression (11.31) is essentially identical to that produced by a *receiver* array using the *selection combining* strategy of (11.2). In fact, there is a 3-dB difference between the two schemes. With a single transmit antenna and 4-QAM, the alphabet energy is $2E_b$ (see Section 11.3), but since the delay-diversity scheme has two transmit antennas, the alphabet energy is only E_b for each. In other words, the 4-QAM alphabet for this example is $x_k \in \{\pm 1 \pm j\}\sqrt{E_b/2}$. Thus, the average BER can be found by cutting the SNR in half for the selection diversity result of (11.24), with $m = 2$:

$$BER = 2\left\{\frac{1}{2} - \frac{1}{2}\left(1 + \frac{2}{E_b/N_0}\right)^{-1/2}\right\} - \left\{\frac{1}{2} - \frac{1}{2}\left(1 + \frac{4}{E_b/N_0}\right)^{-1/2}\right\} \tag{11.32}$$

$$\approx \frac{3/2}{(E_b/N_0)^2} . \tag{11.33}$$

The last approximation comes from (11.25) and is asymptotically tight. Since the SNR is squared in (11.33), we conclude that the combination of a two-antenna delay-diversity transmitter and an ideal ZF-DFE achieves diversity order two.

A second example of indirect transmit diversity is frequency-offset diversity, which transforms the spatial diversity into *time* diversity. This can be accomplished, for example, by forcing each transmit antenna to use a slightly different carrier frequency. More generally, the upconverter for the i-th antenna could include an extra factor of $e^{j\theta_i(t)}$, so that its output is related to its input $\tilde{s}_i(t)$ by:

$$s_i(t) = \sqrt{2}\,\mathrm{Re}\{\tilde{s}_i(t) e^{j(2\pi f_0 t + \theta_i(t))}\} . \tag{11.34}$$

When all transmit antennas send the same symbol, so that $\tilde{s}_i(t) = ag(t)$, then the above strategy leads to an effective SISO channel whose time-varying gain is given by:

$$\Sigma_i h_i e^{j\theta_i(t)} . \tag{11.35}$$

This gain can vary rapidly, even when the channel coefficients $\{h_i\}$ are static. Thus, by artificially inducing time variations in the channel, the above strategy is able to convert a static

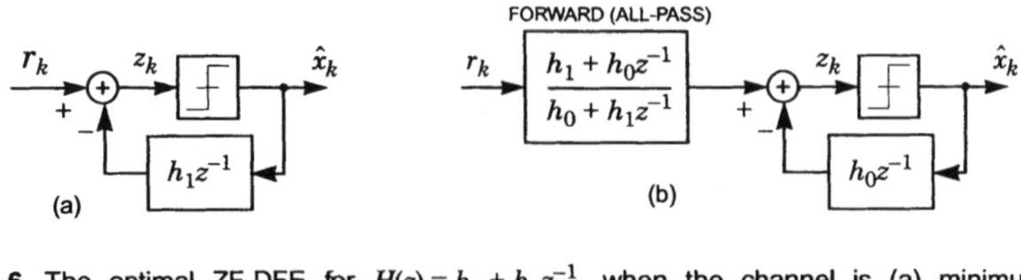

Fig. 11-6. The optimal ZF-DFE for $H(z) = h_0 + h_1 z^{-1}$, when the channel is (a) minimum-phase ($|h_0| > |h_1|$) and (b) non-minimum phase ($|h_0| < |h_1|$). There is no forward filter in (a), and the forward filter in (b) is the all-pass filter that essentially swaps h_1 for h_0 and h_0 for h_1.

fading channel into a time-selective fading channel. The resulting diversity can be harvested using error-correction coding and interleaving, as discussed at the beginning of this chapter.

11.5.2. Closed-Loop Transmit Diversity

In contrast to the indirect transmit diversity schemes of the previous section, a *direct* transmit diversity scheme leaves the spatial diversity in the spatial domain, and does not require any additional processing like coding or interleaving. In this section we describe the closed-loop form of direct transmit diversity for the case when the transmitter has knowledge of the channel [4]. Consider the MISO fading channel of Fig. 11-7(a), consisting of a narrowband transmitter with n antennas and a receiver with a single antenna. When all transmit antennas use the same pulse shape and carrier frequency, the equivalent baseband model is:

$$r = \mathbf{H}\boldsymbol{a} + \tilde{n} \ , \tag{11.36}$$

where $\mathbf{H} = [h_1, \ \dots \ h_n]$ is a row vector and h_i is the complex Rayleigh-fading channel gain from the i-th transmit antenna to the receiver, where $\boldsymbol{a} = [a_1, \ \dots \ a_n]^T$ and a_i is the complex symbol transmitted from the i-th antenna, and where $\tilde{n}$ is complex Gaussian noise satisfying $\text{E}[|\tilde{n}|^2] = N_0$. A block diagram representation of this model is shown in Fig. 11-7(b).

Despite the lack of an antenna array at the receiver, the transmitter can ensure diversity gain by sending the same complex symbol — say x — from each antenna, but adjusting the phases so that they add constructively at the receiver, and adjusting the amplitudes so as to maximize the SNR at the receiver. With this strategy, the channel input may be expressed as:

$$\boldsymbol{a} = x\boldsymbol{w} \ , \tag{11.37}$$

where $\boldsymbol{w} = [w_1, \ \dots \ w_n]^T$ and w_i is the complex weight for the i-th transmit antenna. In the absence of multipath, the transmitter might choose the weights so as to steer a beam towards the receiver, a strategy referred to as *transmit beamforming*. More generally, subject to a unit-norm constraint, the optimal choice for the i-th weight is $w_i = h_i^* / \|\mathbf{H}\|$, which leads to a vector of combining weights $\boldsymbol{w} = \mathbf{H}^* / \|\mathbf{H}\|$ that resembles a matched filter at the transmitter. (See Problem 11-1.) With this choice for $\boldsymbol{w}$, the receiver output reduces to:

$$r = \|\mathbf{H}\| x + \tilde{n} \ . \tag{11.38}$$

Fig. 11-7. (a) A wireless link with transmit diversity; (b) its equivalent narrowband model.

This is just a scalar multiple of the output (11.5) obtained using conventional receive diversity. Hence, the analysis for receiver diversity applies directly to the case of transmit diversity, and we immediately conclude that a transmitter with n antennas may achieve a diversity order of n by using matched-filter precompensation at the transmitter. Clearly, this is possible only when the transmitter knows the channel values $\{h_i\}$, which is difficult in rapidly varying mobile applications [5].

11.5.3. The Alamouti Space-Time Block Code

This section describes the *Alamouti space-time block code*, a simple and effective method for achieving spatial diversity when the transmitter has an array with two antennas. This is an example of a *direct* and *open-loop* method for transmit diversity because, unlike the indirect methods described earlier, it does not transform spatial diversity into temporal or frequency diversity, and unlike closed-loop methods, it does not require channel knowledge at the transmitter.

To send a pair of complex symbols $\{x_1, x_2\}$ chosen from some QAM or PSK constellation, a transmitter with two antennas that uses the Alamouti space-time block code requires two signaling intervals. During the first, x_1 and x_2 are transmitted from the first and second antennas, respectively, and during the second, $-x_2{}^*$ and $x_1{}^*$ are transmitted from the first and second antennas, respectively. In other words, the pair of information symbols $\boldsymbol{x} = \{x_1, x_2\}$ are transformed into the *space-time codeword* or *codematrix*:

$$\mathbf{A}(\boldsymbol{x}) = \begin{bmatrix} x_1 & -x_2{}^* \\ x_2 & x_1{}^* \end{bmatrix} \qquad (11.39)$$

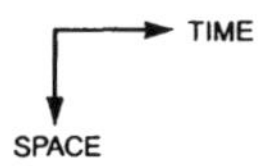

where the first column identifies the symbols transmitted during the first signaling interval, and the second column identifies the symbols transmitted during the second interval. The rows of the space-time codeword correspond to the spatial dimension, while the columns correspond to the time dimension. In the following we show that the JML detector for this code is simple to implement, and that it achieves a (maximal) diversity order of twice the number of receiver antennas. Furthermore, we will see that the encoder does not incur a penalty in capacity when the receiver has a single antenna. One can thus cleanly separate the functions of diversity and channel coding. Because of these advantages, the Alamouti code has been incorporated in the IEEE 802.11a and IEEE 802.16a wireless standards and also a third-generation standard for wideband CDMA [6].

The observations during the first and second signaling interval at a receiver with a single antenna are given by, respectively:

$$r_1 = [h_1 \quad h_2]\begin{bmatrix} x_1 \\ x_2 \end{bmatrix} + n_1\,, \qquad \text{and} \qquad r_2 = [h_1 \quad h_2]\begin{bmatrix} -x_2{}^* \\ x_1{}^* \end{bmatrix} + n_2\,, \qquad (11.40)$$

where n_1 and n_2 represent the additive noise during the two signaling intervals. We have assumed that the channel does not change over the span of two signaling intervals. Equivalently, combining r_1 and $r_2{}^*$ (note the conjugation) into a single vector yields:

$$\underbrace{\begin{bmatrix} r_1 \\ r_2^* \end{bmatrix}}_{\boldsymbol{r}} \underset{=}{=} \underbrace{\begin{bmatrix} h_1 & h_2 \\ h_2^* & -h_1^* \end{bmatrix}}_{\mathbf{H}_{\text{eff}}} \underbrace{\begin{bmatrix} x_1 \\ x_2 \end{bmatrix}}_{\boldsymbol{x}} \underset{+}{+} \underbrace{\begin{bmatrix} n_1 \\ n_2^* \end{bmatrix}}_{\boldsymbol{n}} . \tag{11.41}$$

Observe that the Alamouti code transforms a *MISO* (two-input single-output) channel into an effective *MIMO* (two-input two-output) channel with transfer function $\mathbf{H}_{\text{eff}}$. Since n_2 and n_2^* are identically distributed, the complex noise vector $\boldsymbol{n}$ is white and Gaussian with the usual distribution: zero-mean with covariance matrix $\text{E}[\boldsymbol{n}\boldsymbol{n}^*] = N_0\mathbf{I}$. The conjugation of r_2 may be unexpected, but it serves a useful function: it makes the columns of $\mathbf{H}_{\text{eff}}$ *orthogonal.* Specifically, $\mathbf{H}_{\text{eff}}^*\mathbf{H}_{\text{eff}}$ is proportional to the identity matrix, namely $\mathbf{H}_{\text{eff}}^*\mathbf{H}_{\text{eff}} = \|\mathbf{H}\|^2\mathbf{I}$, where $\|\mathbf{H}\|^2 = |h_1|^2 + |h_2|^2$ is the squared norm of the underlying MISO channel $\mathbf{H} = [h_1, h_2]$.

The receiver that minimizes the probability of detecting either x_1 or x_2 erroneously is the joint ML detector of Section 10.3.1, which chooses $\boldsymbol{x} = [x_1, x_2]^T$ so as to minimize:

$$\|\boldsymbol{r} - \mathbf{H}_{\text{eff}}\boldsymbol{x}\|^2 . \tag{11.42}$$

Passing $\boldsymbol{r}$ through the MMF yields the following sufficient statistics:

$$\begin{aligned} \boldsymbol{y} &= \mathbf{H}_{\text{eff}}^*\boldsymbol{r} \\ &= \mathbf{H}_{\text{eff}}^*(\mathbf{H}_{\text{eff}}\boldsymbol{x} + \boldsymbol{n}) \\ &= \|\mathbf{H}\|^2\boldsymbol{x} + \tilde{\boldsymbol{n}} , \end{aligned} \tag{11.43}$$

where we exploited the orthogonality condition, $\mathbf{H}_{\text{eff}}^*\mathbf{H}_{\text{eff}} = \|\mathbf{H}\|^2\mathbf{I}$, and where we introduced $\tilde{\boldsymbol{n}} = \mathbf{H}_{\text{eff}}^*\boldsymbol{n}$. The fact that the columns of $\mathbf{H}_{\text{eff}}$ are orthogonal has two key implications. First, there is no crosstalk between x_1 and x_2 after the MMF, as shown by (11.43). In other words, the MMF *diagonalizes* the effective channel. Second, the noise after the MMF is white:

$$\begin{aligned} \text{E}[\tilde{\boldsymbol{n}}\tilde{\boldsymbol{n}}^*] &= \mathbf{H}_{\text{eff}}^*\text{E}[\boldsymbol{n}\boldsymbol{n}^*]\mathbf{H}_{\text{eff}} \\ &= \mathbf{H}_{\text{eff}}^*(N_0\mathbf{I})\mathbf{H}_{\text{eff}} = N_0\|\mathbf{H}\|^2\mathbf{I} . \end{aligned} \tag{11.44}$$

Together, these two results dramatically simplify the implementation of the JML detector. Specifically, since the MMF produces no crosstalk and independent noise, the JML decisions may be calculated by applying the pair of outputs of the MMF to a pair of *independent*, scalar slicers. This amounts to the computation and comparison of at most $2|\mathcal{A}|$ scalar distances, as opposed to the $|\mathcal{A}|^2$ vector distances implied by (11.42).

Comparing (11.43) to (11.5), we immediately see that the MMF output for a two-input single-output channel with Alamouti transmit-diversity coding is statistically *identical* to the MMF (or maximal-ratio combiner) output (11.5) for a single-input two-output channel with *receive* diversity and no space-time coding. Hence, the Alamouti scheme leads to order-two diversity, without an array at the receiver, without channel knowledge at the transmitter, and with only simple linear combining and slicing at the receiver. Because of the equivalence, we need not analyze (11.43) anew, but can apply the analysis of Section 11.3 directly, with one minor modification: the Alamouti transmit-diversity scheme suffers a 3 dB penalty relative to a receive-diversity scheme. This penalty arises because, to achieve the same error probability, the energy of x_i in (11.43) must be identical to the energy of a in (11.5), which implies that the total signal energy emitted from an Alamouti-encoded transmitter — having two antennas — is twice that emitted by an uncoded single-antenna transmitter. Thus, from (11.17), the average BER with Alamouti encoding, 4-QAM and i.i.d. Rayleigh fading at high SNR is:

$$\text{BER} \approx \frac{3/4}{(E_b/N_0)^2}. \tag{11.45}$$

As a reminder, $E_b = \text{E}[\,|x_i|^2]\text{E}[\,|h_i|^2]$ is the average received energy per bit.

When the receiver uses m antennas instead of one, the JML detector is still easily implementable using linear processing and scalar slicing, and it still achieves maximal diversity order, namely $2m$. To see why, consider a two-input m-output channel with $m \times 2$ channel matrix $\mathbf{H} = [\boldsymbol{h}_1, \boldsymbol{h}_2]$, and suppose the transmitter uses the Alamouti code of (11.39). If $\boldsymbol{r}_1$ and $\boldsymbol{r}_2$ denote the m-dimensional receiver observations during the two signaling intervals, respectively, the $2m$-dimensional vector $\boldsymbol{r} = [\boldsymbol{r}_1{}^T, \boldsymbol{r}_2{}^*]^T$ is related to the information symbol vector $\boldsymbol{x} = [x_1, x_2]^T$ by:

$$\boldsymbol{r} = \mathbf{H}_{\text{eff}}\boldsymbol{x} + \boldsymbol{n}\ , \quad \text{where } \mathbf{H}_{\text{eff}} = \left[\begin{array}{c|c} \boldsymbol{h}_1 & \boldsymbol{h}_2 \\ \hline \bar{\boldsymbol{h}}_2 & -\bar{\boldsymbol{h}}_1 \end{array}\right], \tag{11.46}$$

and where the AWGN satisfies $\text{E}[\boldsymbol{n}\boldsymbol{n}^*] = N_0\mathbf{I}$. Once again we see that the columns of the $2m \times 2$ matrix $\mathbf{H}_{\text{eff}}$ are orthogonal, so that the MMF — which is known to provide sufficient statistics — diagonalizes the channel:

$$\begin{aligned} \boldsymbol{y} &= \mathbf{H}_{\text{eff}}{}^*\boldsymbol{r} \\ &= \|\mathbf{H}\|^2\boldsymbol{x} + \tilde{\boldsymbol{n}}\ , \end{aligned} \tag{11.47}$$

where $\text{E}[\tilde{\boldsymbol{n}}\tilde{\boldsymbol{n}}^*] = N_0\|\mathbf{H}\|^2\mathbf{I}$. This is precisely the same as the case of $m = 1$, see (11.43), except that the *underlying* channel $\mathbf{H}$ has dimension $m \times 2$ instead of 1×2, and thus the squared Frobenius norm $\|\mathbf{H}\|^2 = \|\boldsymbol{h}_1\|^2 + \|\boldsymbol{h}_2\|^2$ has a chi-squared distribution with $4m$ degrees of freedom instead of 4. It follows that independent slicing of the outputs of the MMF will yield a diversity order of $2m$, the maximal possible for a two-input m-output channel.

Interestingly, although the Alamouti code applied to a two-input m-output channel achieves full diversity for any value of m, it is only capacity-preserving when $m = 1$. In other words, the capacity of the channel (11.47) after Alamouti coding is strictly less than the capacity of the underlying channel whenever $m > 1$, even though the two capacities are equal when $m = 1$.

Example 11-4.

The cascade of an Alamouti encoder, a static $m \times 2$ AWGN channel $\mathbf{H}$, and an MMF $\mathbf{H}_{\text{eff}}{}^*$ results in an effective scalar (SISO) channel whose capacity is easily calculated, assuming the information symbols are constrained to satisfy $\text{E}[\,|x|^2] = E$. Let $\{\lambda_1, \lambda_2\}$ denote the eigenvalues of $\mathbf{H}^*\mathbf{H}$. They are real, nonnegative, and satisfy $\lambda_1 + \lambda_2 = \|\mathbf{H}\|^2$. From (11.47), the capacity with Alamouti coding is then $C_A = \log_2(1 + \|\mathbf{H}\|^2E/N_0)$ bits per signaling interval. In contrast, the capacity of the underlying channel without knowledge of the channel at the transmitter is [7]

$C = \log_2(1 + \lambda_1 E/N_0) + \log_2(1 + \lambda_2 E/N_0)$. When $m = 1$, we have $\lambda_1 = \|\mathbf{H}\|^2$ and $\lambda_2 = 0$, so that $C_A = C$. However, when $m > 1$, and assuming the two columns of $\mathbf{H}$ are linearly independent so that both eigenvalues are nonzero (which happens with probability one with i.i.d. Rayleigh fading), we have $C_A < C$, since then:

$$\begin{aligned} 2^C &= (1 + \lambda_1 E/N_0)(1 + \lambda_2 E/N_0) \\ &= 1 + \lambda_1 E/N_0 + \lambda_2 E/N_0 + \lambda_1\lambda_2(E/N_0)^2 \\ &> 1 + \lambda_1 E/N_0 + \lambda_2 E/N_0 \\ &= 2^{C_A} \, . \end{aligned} \tag{11.48}$$

The previous example illustrates a key benefit of the Alamouti code — it allows the transmitter to *separate* the spatial processing from the temporal processing, without penalty. Specifically, to approach the capacity of the underlying two-input single-output channel, the transmitter need only choose a channel code that approaches capacity on the equivalent SISO channel, perhaps one based on turbo or LDPC codes, without regard for the spatial dimension of the channel. Likewise, the Alamouti encoder and MMF combiner can ignore the fact that the information symbols are coded. At least on the two-input single-output channel, the channel code need not be designed jointly with the space-time code.

11.5.4. Space-Time Block Codes

The Alamouti space-time code, which maps information symbols $\{x_1, x_2\}$ to the space-time codeword $\mathbf{A}$ of (11.39), has three key features: it does not require channel knowledge at the transmitter, it makes JML detection easy to implement, and it provides maximal diversity order. More generally, a *space-time block code* is a mapping of K complex information symbols $\{x_1, \ldots x_K\}$ to a codeword $\mathbf{A}$ with dimension $n \times L$, such that the L columns of $\mathbf{A}$ are transmitted during L consecutive signaling intervals. The aim is to achieve maximal diversity with minimal complexity at the receiver. Preferably, JML detection may be implemented using linear processing only, and a full diversity order of mn is achieved. The *rate* of a space-time code is the average number of information symbols $\{x_i\}$ conveyed per signaling interval, and is equal to the ratio K/L. For example, since the Alamouti code requires two signaling intervals to transmit two information symbols, its rate is unity.

The design of space-time codes is often based on the union bound for word-error probability. Consider a space-time code consisting of $|\mathcal{A}|^K$ codewords, one for each possible sequence of input symbols $\{x_1, \ldots x_K\}$. Assume that the codewords are normalized so that a randomly chosen column has unit energy, and that $\mathrm{E}[|h_{ij}|^2] = E$, which implies that E/N_0 is the average SNR per receive antenna. If all codewords are equally likely, the union bound for the average probability that the ML detector makes a decision error is:

$$P_e \le \frac{1}{|\mathcal{A}|^K} \sum_{i=1}^{|\mathcal{A}|^K} \sum_{j \ne i}^{|\mathcal{A}|^K} P_{i \to j} \tag{11.49}$$

$$\le (|\mathcal{A}|^K - 1) \max_{i \ne j} P_{i \to j} \, , \tag{11.50}$$

where $P_{i \to j} = \Pr[\|\mathbf{R} - \mathbf{H}\mathbf{A}_j\|^2 < \|\mathbf{R} - \mathbf{H}\mathbf{A}_i\|^2 \,|\, \mathbf{A}_i \text{ transmitted}]$ is the *pairwise-error probability*, namely, the probability that an ML receiver would prefer $\mathbf{A}_j$ over $\mathbf{A}_i$, given that $\mathbf{A}_i$ was transmitted. The looser bound of (11.50) results when each pairwise probability is replaced by its worst-case value. It is shown in Appendix 11-B that, under the assumption of i.i.d. static Rayleigh fading and AWGN, the pairwise error probability is related to the average SNR per antenna by:

$$P_{i \to j} \le \left(\frac{\eta E}{4N_0}\right)^{-mr}, \tag{11.51}$$

where r is the rank of the following $n \times n$ matrix:

$$\mathbf{B} = (\mathbf{A}_i - \mathbf{A}_j)(\mathbf{A}_i - \mathbf{A}_j)^*, \tag{11.52}$$

and where $\eta = (\lambda_1 \lambda_2 \ldots \lambda_r)^{1/r}$ is the geometric mean of the nonzero eigenvalues of $\mathbf{B}$.

Because of the exponent mr in (11.51), each pairwise error probability has an effective diversity order of mr. The diversity order of the total error probability will be dominated by the diversity order of the worst-case pairwise error probability. Therefore, the *diversity order* of the space-time code is $mr_{\min}$, where $r_{\min}$ is the minimum value for the rank of $\mathbf{B}$ over all distinct codewords $\mathbf{A}_i$ and $\mathbf{A}_j$. This relationship leads to the *rank criterion*, a simple litmus test that helps guide the design of full-diversity space-time codes:

> **Rank Criterion** — A space-time code achieves maximal diversity order (namely mn) if and only if the $n \times L$ difference matrix $\mathbf{A}(\boldsymbol{x}) - \mathbf{A}(\boldsymbol{x}')$ has rank n whenever $\boldsymbol{x} \ne \boldsymbol{x}'$.

The factor $\eta = (\lambda_1 \lambda_2 \ldots \lambda_r)^{1/r}$ in (11.51) can be viewed as a *pairwise coding gain*, since its impact is to amplify the SNR; when error probability is plotted versus SNR, the coding gain shifts the curve to the left, but does not change its slope. The *coding gain* $\eta_{\min}$ for the code as a whole is the minimum pairwise coding gain over all distinct codeword pairs for which $r = r_{\min}$. When comparing two candidate space-time codes, both having the same diversity order, the one with the higher coding gain is preferred.

The Alamouti code is an example of a *linear* and *orthogonal* space-time block code; it is linear because the real and imaginary components of the codeword matrix $\mathbf{A}(\boldsymbol{x})$ are linear combinations of the real and imaginary components of the information symbols $\{x_i\}$, and it is orthogonal because the rows of $\mathbf{A}$ are orthogonal with equal norm, namely $\|\boldsymbol{x}\|$. Linearity implies that the difference $\mathbf{A}(\boldsymbol{x}) - \mathbf{A}(\boldsymbol{x}')$ reduces to $\mathbf{A}(\boldsymbol{e})$, where $\boldsymbol{e} = \boldsymbol{x} - \boldsymbol{x}'$. Orthogonality has two important implications. First, it implies that simple linear processing is sufficient for JML detection. Second, it implies that $\mathbf{B}$ of (11.52) is $\mathbf{B} = \mathbf{A}(\boldsymbol{e})\mathbf{A}(\boldsymbol{e})^* = \|\boldsymbol{e}\|^2\mathbf{I}$, which clearly satisfies the rank criterion. Hence, space-time block codes that are linear and orthogonal automatically achieve full diversity. Furthermore, because the matrix of (11.52) reduces to $\mathbf{B} = \|\boldsymbol{x} - \boldsymbol{x}'\|^2\mathbf{I}$ for linear orthogonal codes, its eigenvalues are all equal, and the coding gain $\eta_{\min} = (\lambda_1 \lambda_2 \ldots \lambda_r)^{1/r}$ reduces to $\min_{\boldsymbol{x} \ne \boldsymbol{x}'} \|\boldsymbol{x} - \boldsymbol{x}'\|^2$, the squared minimum Euclidean distance between the information symbol vectors.

Example 11-5.
The Alamouti code of (11.39) is clearly linear and clearly orthogonal. Hence, from the above discussion, it follows that the Alamouti code achieves a diversity order of $2m$ with JML detection for any number of receive antennas m. (This same result was derived earlier, but in a different way.) Furthermore, with a 4-QAM alphabet normalized so that $\|\boldsymbol{x}\|^2 = 1$, the geometric mean $\eta = \|\boldsymbol{x} - \boldsymbol{x}'\|^2$ satisfies $\eta \in \{1, 2, 3, 4\}$, depending on $\boldsymbol{x}$ and $\boldsymbol{x}'$. The coding gain is thus $\eta_{\min} = 1$.

It has been shown that, within the class of linear and orthogonal space-time block codes, a rate-one code exists only when there are $n = 2$ transmit antennas [8]. In this sense, the Alamouti code is unique. For the special cases of $n = 3$ and $n = 4$ antennas, the highest-rate linear and orthogonal space-time block codes have rate $3/4$. Linear orthogonal codes are known for all $n \geq 5$ as well, but none with rate greater than $1/2$ [8].

Example 11-6.
Tarokh *et al.* [8] proposed the following rate-$3/4$ space-time block code that maps three information symbols $\{x_1, x_2, x_3\}$ to four antennas over four signaling intervals:

$$\mathbf{A}_1(\boldsymbol{x}) = \frac{1}{2}\begin{bmatrix} 2x_1 & -2x_2^* & \sqrt{2}x_3^* & \sqrt{2}x_3^* \\ 2x_2 & 2x_1^* & \sqrt{2}x_3^* & -\sqrt{2}x_3^* \\ \sqrt{2}x_3 & \sqrt{2}x_3 & -x_1 - x_1^* + x_2 - x_2^* & x_2 + x_2^* + x_1 - x_1^* \\ \sqrt{2}x_3 & -\sqrt{2}x_3 & -x_2 - x_2^* + x_1 - x_1^* & -x_1 - x_1^* - x_2 + x_2^* \end{bmatrix}. \quad (11.53)$$

Similarly, Tirkkonen and Hottinen proposed the following somewhat simpler rate-$3/4$ code [9]:

$$\mathbf{A}_2(\boldsymbol{x}) = \begin{bmatrix} x_1 & -x_2^* & -x_3^* & 0 \\ x_2 & x_1^* & 0 & x_3^* \\ x_3 & 0 & x_1^* & -x_2^* \\ 0 & -x_3 & x_2 & x_1 \end{bmatrix}. \quad (11.54)$$

Both examples are clearly linear, and it is easily verified that rows of $\mathbf{A}$ are equal-norm and orthogonal. Thus, the rank criterion is satisfied, so that full diversity is achieved. Furthermore, the JML detector reduces to a combination of linear processing and scalar slicing at the receiver, greatly simplifying detection. If there are three antennas instead of four, a suitable rate-$3/4$ space-time block code can be constructed from either of the above codes by deleting one of the rows of $\mathbf{A}$.

11.5.5. Space-Time Trellis Codes

In later chapters we will see that error-correction codes — which mitigate noise — can be classified as either block codes or trellis codes. Similarly, space-time codes — which primarily mitigate fading — can be classified in the same way. The space-time block codes of the previous section are analogous to block codes, mapping a fixed block of information symbols to a fixed block of antenna outputs. We now consider *space-time trellis codes*, which map an arbitrary number of information symbols to antenna outputs according to a finite-state machine. They are extensions of convolutional codes and trellis-coded modulation for the case of multiple transmit antennas. Beyond the above analogy, however, there is a philosophical difference that distinguishes space-time block codes from space-time trellis codes. Whereas space-time block codes aim to mitigate fading only, with little or no coding gain, space-time trellis codes aim to jointly mitigate fading and noise, by providing both diversity gain and coding gain. In other words, whereas space-time block codes *separate* the tasks of mitigating fading and mitigating noise, space-time trellis codes perform the tasks jointly.

The simplest example of a space-time trellis code is the delay-diversity technique described earlier, as illustrated in Fig. 11-5(a), and repeated here for convenience:

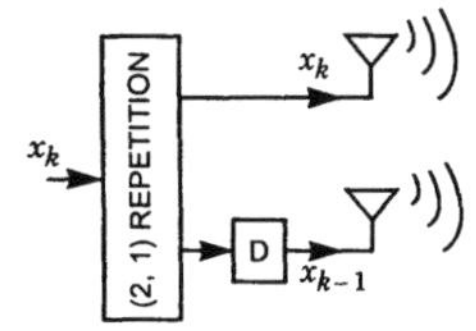

This transmitter can be modeled as a finite-state machine whose state at time k is x_{k-1}, the previous input symbol, and a sequence of input symbols uniquely defines a path through a trellis, where the number of trellis states is equal to the size of the input alphabet. The receiver can implement ML sequence detection using the Viterbi algorithm.

If we fix the number of input symbols to K, the delay-diversity transmitter defines an equivalent rate-$K/(K+1)$ space-time *block* code, as follows:

$$\mathbf{A}(\boldsymbol{x}) = \begin{bmatrix} x_1 & x_2 & x_3 & \cdots & x_{K-1} & x_K & 0 \\ 0 & x_1 & x_2 & x_3 & \cdots & x_{K-1} & x_K \end{bmatrix}. \tag{11.55}$$

The two zeros in the first and last columns may be replaced by any *idle* symbol chosen from the same alphabet used by $\{x_i\}$; they serve to initialize and terminate the trellis to a known state. Having transformed the delay-diversity scheme into a space-time block code, we can now use the rank criterion to determine its diversity order. This is clearly a linear code, and hence the difference $\mathbf{A}(\boldsymbol{x}) - \mathbf{A}(\boldsymbol{x}')$ between two distinct codewords reduces to $\mathbf{A}(\boldsymbol{e})$, where at least one element of $\boldsymbol{e} = \boldsymbol{x} - \boldsymbol{x}'$ is nonzero. It is easy to verify that the columns of $\mathbf{A}(\boldsymbol{e})$ containing the first and last nonzero element of $\boldsymbol{e}$ are linearly independent; therefore, the rank criterion is satisfied, and delay diversity achieves full diversity.

Example 11-7.

A slightly more efficient rate-1 encoder results by moving x_K from the last column to the first:

$$\mathbf{A}(\boldsymbol{x}) = \begin{bmatrix} x_1 & x_2 & x_3 & \cdots & x_{K-1} & x_K \\ x_K & x_1 & x_2 & x_3 & \cdots & x_{K-1} \end{bmatrix}, \tag{11.56}$$

thereby eliminating the idle symbols. This is an example of a *tail-biting* code because it begins and ends in the same state. However, since the rank of the difference matrix will be one when all elements of $\boldsymbol{x} - \boldsymbol{x}'$ are identical, the rank criterion is not satisfied, and hence this code does not achieve full diversity.

The motivation for space-time trellis codes is simple — to improve the performance of delay diversity, without increasing complexity. A good example of such an improvement can be described for the special case of the 8-PSK alphabet, $\mathcal{A} = \{e^{jl\pi/4}\}$ with $l \in \{0, 1, \ldots 7\}$. If we identify symbols by their integer labels l, the (2, 1) repetition code can be expressed as

$C = \{00,\ 11,\ 22,\ 33,\ 44,\ 55,\ 66,\ 77\}$; its minimum Euclidean distance is $d_{\min} = (4 - 2\sqrt{2})^{1/2} \approx 1.082$. In contrast, the *optimal* (2, 1) block code of 8-PSK symbols (maximizing $d_{\min}$) is $C' = \{00,\ 15,\ 22,\ 37,\ 44,\ 51,\ 66,\ 73\}$; it achieves a maximal minimum Euclidean distance of $d_{\min} = 2$. The optimal code obeys a simple rule: if l_k is the index of the k-th information symbol, the parity symbol index is $5l_k$ (modulo 8). If we use this code in place of the repetition code, but keep the structure of the delay-diversity transmitter, we arrive at the two-antenna 8-state 8-PSK space-time trellis code of [10]:

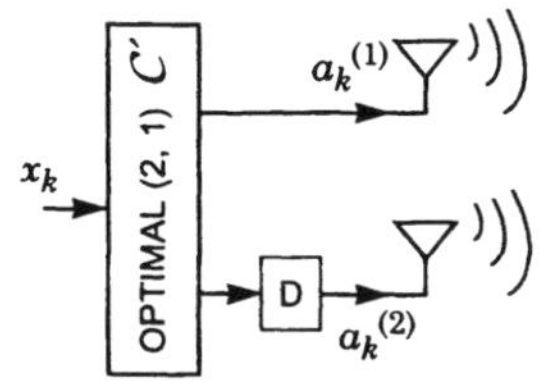

It has better performance than the delay-diversity scheme, but with exactly the same decoding complexity. The space-time codeword that results has the form:

$$\begin{bmatrix} l_1 & l_2 & l_3 & \cdots & \\ 0 & 5l_1 & 5l_2 & 5l_3 & \cdots \end{bmatrix}, \tag{11.57}$$

so that the length-two codewords from C' appear diagonally within the space-time codeword.

Both the delay-diversity scheme and the space-time trellis code described above satisfy the rank criterion, and hence achieve full diversity, but the latter performs better because its coding gain $\eta_{\min}$ is larger. Even better performance can be achieved when the encoder operates on information *bits* instead of complex symbols, and when the encoder is designed to simultaneously maximize diversity order and coding gain.

Example 11-8.

The best-known 4-state 4-PSK space-time trellis code for two transmit antennas and multiple receiver antennas consists of a convolutional code with generator polynomial:

$$\mathbf{G}(D) = \begin{bmatrix} 0 & 2 \\ 2 & 3 \end{bmatrix} + \begin{bmatrix} 1 & 2 \\ 2 & 0 \end{bmatrix} D \tag{11.58}$$

followed by a pair of 4-PSK mappers [11], but with arithmetic performed modulo-4 instead of modulo-2, as sketched in Fig. 11-8(a). The delay-diversity scheme with 4-PSK can also be viewed in this way, except its generator is:

$$\mathbf{G}_{\mathrm{dd}}(D) = \begin{bmatrix} 2 & 1 \\ 0 & 0 \end{bmatrix} + \begin{bmatrix} 0 & 0 \\ 2 & 1 \end{bmatrix} D. \tag{11.59}$$

In either case, the encoder accepts a pair of bits $\boldsymbol{b}_k = [b_k^{(1)},\ b_k^{(2)}]^T$ during the k-th signaling interval, which together with the encoder state $\psi_k = [b_{k-1}^{(1)},\ b_{k-1}^{(2)}]^T$, defines the two symbols $a_k^{(1)}$ and $a_k^{(2)}$ transmitted simultaneously from the two antennas. One stage of the trellis is shown in Fig. 11-8(b). It has the same spectral efficiency and complexity as 4-QAM with two-antenna delay diversity, but its performance is significantly better (see Fig. 11-10).

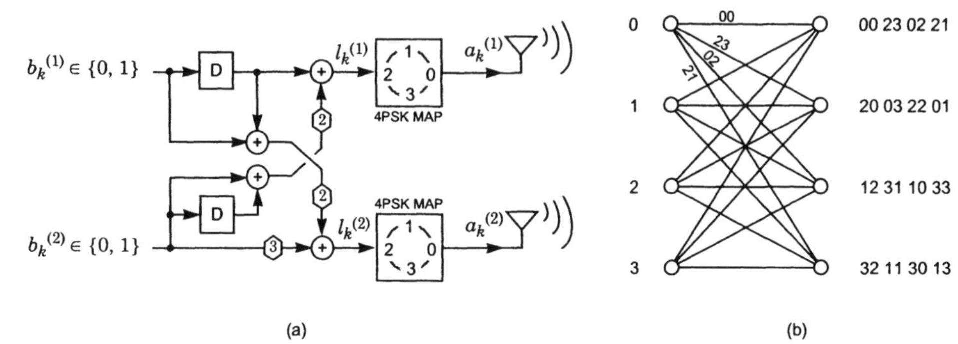

Fig. 11-8. A space-time trellis code for 4-PSK and two transmit antennas [11]. The encoder structure is shown in (a). The mappers convert an integer $l \in \{0,1,2,3\}$ into a 4-PSK symbol a according to $a = \exp(jl\pi/2)$. One trellis stage is shown in (b), where the state labels are $2b_{k-1}^{(1)} + b_{k-1}^{(2)}$, and the branch labels are $\{l_k^{(1)}, l_k^{(2)}\}$.

The 4-state space-time trellis code of the previous example was constrained to use a 4-PSK alphabet. In doing so it conveys two bits per signaling interval, so that its spectral efficiency is 2 bits/sec/Hz. We can achieve the same spectral efficiency with less SNR by relaxing the constraint that the alphabet be 4-PSK. A particularly effective strategy for two transmit antennas is to *concatenate* an outer code with an inner Alamouti space-time block code. Since the Alamouti code transforms the MIMO channel into an effective SISO channel, a good conventional code — such as a trellis code — may be used as an outer code to achieve good overall performance. In order to achieve the same spectral efficiency, a trellis code uses a larger alphabet.

Example 11-9.

Consider the *concatenated* strategy illustrated in Fig. 11-9. The inner Alamouti space-time block code transforms the $m \times 2$ underlying channel into an effective SISO channel, as seen by the outer coder and decoder. The outer code is chosen as a four-state 8-PSK trellis code; this makes for a fair comparison, because it has the same complexity (4 states) and the same spectral efficiency (2 b/s/Hz) as the space-time trellis code of Fig. 11-8. The performance of this concatenated strategy is shown in Fig. 11-10, where average word-error rate is plotted versus average SNR, assuming two transmit antennas and $m = 1, 2,$ and 4 receive antennas over a Rayleigh block-fading channel. For comparison purposes, the performance of two-antenna delay diversity with 4-PSK and the space-time trellis code of Fig. 11-8 are also shown. The concatenated scheme consistently outperforms the delay-diversity scheme by 2.4 dB, regardless of the number of receive antennas. The performance of the space-time trellis code, however, depends strongly on the number of receive antennas. With only one receive antenna, the space-time trellis code performs about the same as delay diversity; with $m = 2$ receive antennas, it is 1 dB better, and with $m = 4$ receive antennas, it is about 2 dB better, and roughly as good as the concatenated scheme.

We will see much more about trellis codes in Chapter 13.

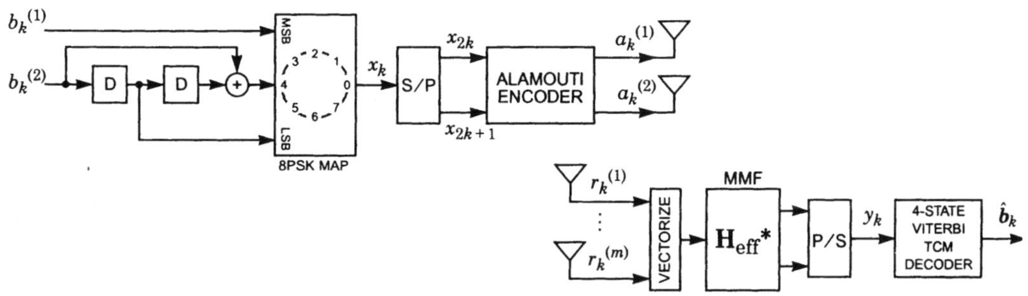

Fig. 11-9. A concatenated space-time coded system. The transmitter consists of an outer trellis code and an inner Alamouti space-time block code. At the receiver, the MMF and parallel-to-serial converter transform the Alamouti-encoded MIMO channel into an effective SISO channel, so that a conventional 4-state Viterbi-based TCM decoder may be used.

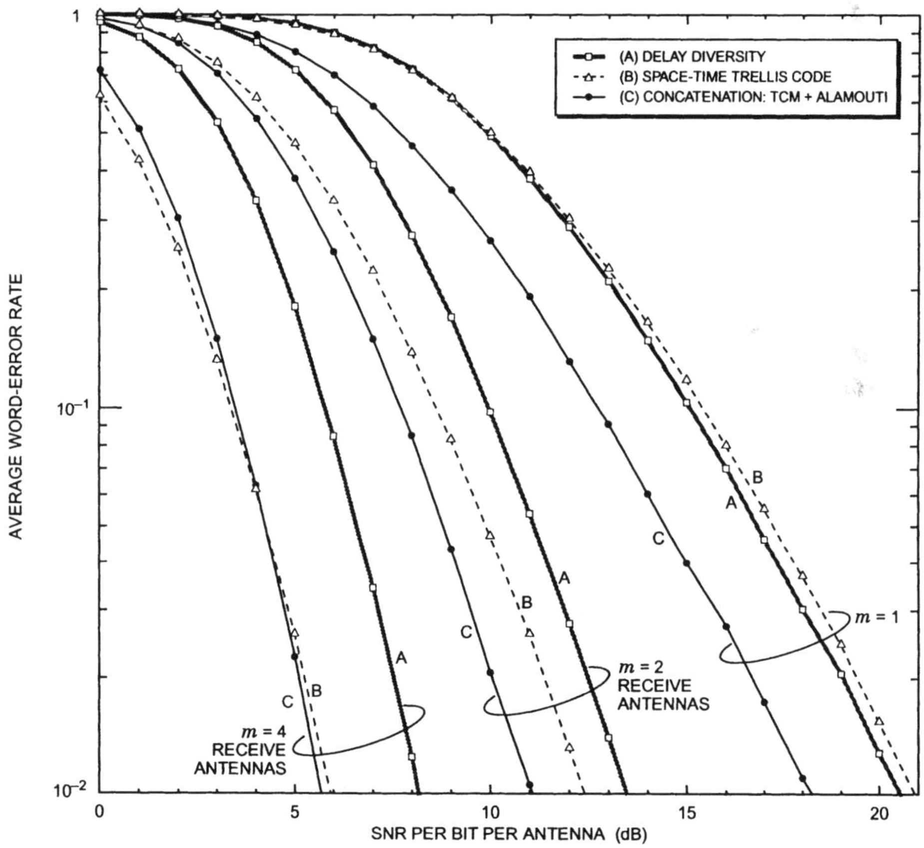

Fig. 11-10. A performance comparison of three alternative space-time coding techniques: (A) 4-PSK with delay diversity (11.59), (B) 4-PSK with $G(D)$ from (11.58), and (C) trellis-coded 8-PSK with Alamouti space-time block coding, as shown in Fig. 11-9. All three strategies achieve 2 b/s/Hz with two transmit antennas, and the complexity of the 4-state Viterbi algorithm at the receiver is the same for all three cases. The word length is 131 symbols, and the results are averaged over independent Rayleigh fading channels that are constant over the duration of one codeword, and independent from one codeword to the next.

11.6. Layered Space-Time Modems

Multiple antennas at a transmitter can provide diversity against fading using the transmit-diversity strategies of Section 11.5. These strategies are most appropriate when the receiver has only one or two antennas, in which case receive diversity alone may be insufficient to provide adequate protection against multipath fading. Alternatively, when the receiver already has many antennas, so that there is minimal benefit to supplementing the receive diversity with additional transmit diversity, the multiple transmit antennas may be used to increase the data rate instead, a concept known as *spatial multiplexing*. The basic idea is to use the multiple transmit antennas to simultaneously transmit multiple independent streams of symbols, without spatial redundancy.

The simplest form of spatial multiplexing is known as *V-BLAST* (*vertical Bell Labs layered space-time*) transmission, where *independent* substreams are transmitted from each antenna. From the receiver's point of view, each transmitting antenna represents an *independent user*, and hence the receiver can use centralized multiuser detection techniques to simultaneously recover the symbols transmitted from all antennas. A V-BLAST transmitter is sketched below:

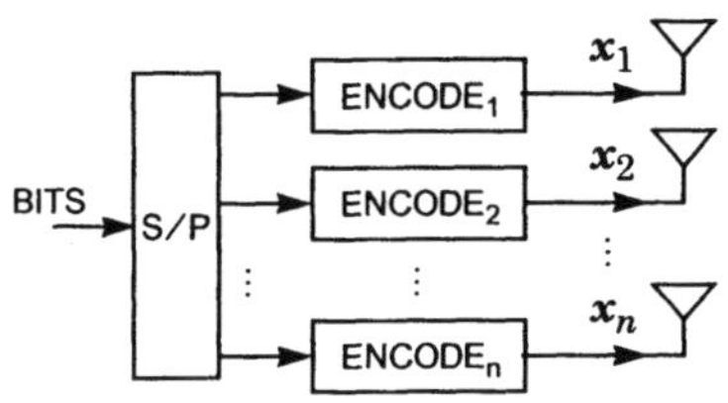

The source bits are demultiplexed into n independent substreams, one for each transmitting antenna. Each substream is coded independently using a conventional one-dimensional code of length L, such as a trellis code or LDPC code, and then transmitted from its own dedicated antenna. The complexity of the transmitter is reduced because the bit rate seen by each encoder is a fraction $1/n$ of the overall bit rate. If we define $\boldsymbol{x}_i = [x_i^{(1)}, \ldots x_i^{(L)}]$ as the i-th coded substream, corresponding to the i-th transmit antenna, then this strategy results in the following space-time codeword:

A = 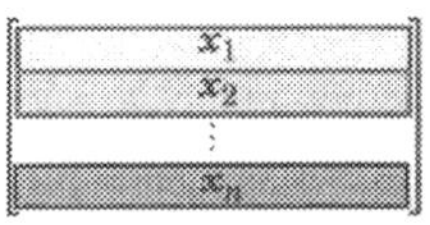 . (11.60)

In contrast to prior sections, the codeword is organized by rows instead of columns. A receiver with m antennas will observe the following $m \times L$ matrix over L signaling intervals:

$$\begin{bmatrix} \boldsymbol{r}_1 \\ \boldsymbol{r}_2 \\ \vdots \\ \boldsymbol{r}_n \end{bmatrix} = \mathbf{H} \begin{bmatrix} \boldsymbol{x}_1 \\ \boldsymbol{x}_2 \\ \vdots \\ \boldsymbol{x}_n \end{bmatrix} + \begin{bmatrix} \overline{\boldsymbol{n}}_1 \\ \boldsymbol{n}_2 \\ \vdots \\ \boldsymbol{n}_n \end{bmatrix}, \tag{11.61}$$

or

$$\mathbf{R} = \mathbf{HA} + \overline{\mathbf{N}}, \tag{11.62}$$

where $\boldsymbol{r}_i = [r_i^{(1)}, \ldots r_i^{(L)}]$ represents the L observations at receiver antenna i, and where $\overline{\boldsymbol{n}}_i = [\overline{n}_i^{(1)}, \ldots \overline{n}_i^{(L)}]$ is the corresponding AWGN satisfying $E[|\overline{n}_i^{(l)}|^2] = N_0$. A BLAST receiver can use ZF-DF or MMSE-DF detection; for simplicity we describe the ZF-DF detector. If $\mathbf{H} = \mathbf{QG}$ is a Gram-Schmidt decomposition of $\mathbf{H}$, then application of the WMF $\mathbf{Q}^*$ yields:

$$\mathbf{Y} = \mathbf{Q}^*\mathbf{R} = \mathbf{GA} + \mathbf{N}\,, \tag{11.63}$$

where $\mathbf{G}$ is lower triangular, and where $\mathbf{N}$ has the same statistics as $\overline{\mathbf{N}}$. The first row of $\mathbf{Y}$ is $\boldsymbol{y}_1 = G_{11}\boldsymbol{x}_1 + \boldsymbol{n}_1$, with no interference from the other substreams $\{\boldsymbol{x}_2, \ldots \boldsymbol{x}_n\}$; thus, we may pass $\boldsymbol{y}_1$ to the first substream decoder to arrive at the decision codeword $\hat{\boldsymbol{x}}_1$. If the decision is correct, subtracting $G_{21}\hat{\boldsymbol{x}}_1$ from the second row of $\mathbf{Y}$ yields $\boldsymbol{z}_2 = G_{22}\boldsymbol{x}_2 + \boldsymbol{n}_2$, which can be decoded to yield $\hat{\boldsymbol{x}}_2$. This process can be repeated for each of the n substreams, as illustrated in Fig. 11-11. This detector is the same in spirit as the ZF-DF detector of Section 10.3.4, but with one important distinction: the coding is taken into account. In particular, the detector of Fig. 11-11 decodes the entire first row of $\mathbf{A}$ before making any decisions about the second row of $\mathbf{A}$. This is in contrast to a ZF-DF detector that ignores coding, which would make *tentative* decisions about $\mathbf{A}$ column by column, and would perform decoding only after the decision-feedback process had been completed.

A drawback of V-BLAST is that, because the different antennas send independent information, there is no possibility for transmit diversity. If there is to be diversity, it must come from the receiver array. Furthermore, with ZF-DF detection, the SNR seen by the i-th substream is proportional to $G_{ii}^2 = \|\boldsymbol{h}_i - \hat{\boldsymbol{h}}_i\|^2$, where $\hat{\boldsymbol{h}}_i$ is the projection of $\boldsymbol{h}_i$ onto the subspace spanned by $\{\boldsymbol{h}_{i+1}, \ldots \boldsymbol{h}_n\}$. The problem is that some of these G_{ii}'s can be very small. This might happen because all of the gains associated with the i-th transmitter are small (in which case $\boldsymbol{h}_i$ has a small norm), or perhaps because $\boldsymbol{h}_i$ is nearly expressible as a linear combination of $\{\boldsymbol{h}_{i+1}, \ldots \boldsymbol{h}_n\}$. For the special case of Rayleigh fading, G_{ii}^2 is a chi-square random variable with $2(m - n + i)$ degrees of freedom (see (11.28) and (11.67)), so G_{11} has the

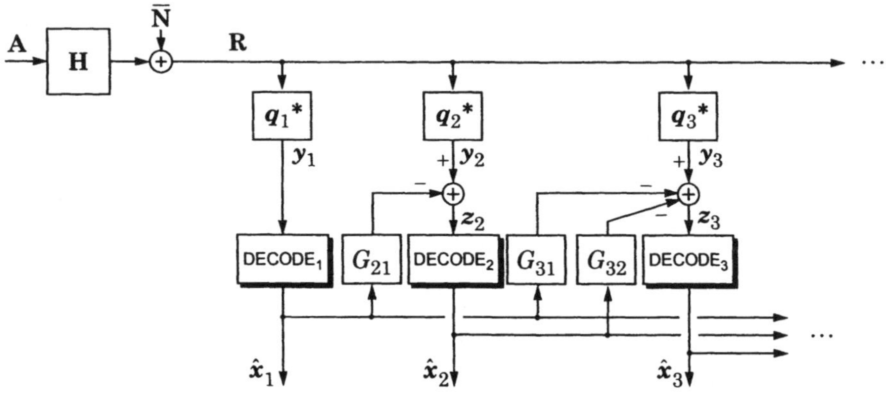

Fig. 11-11. The ZF-DF detector for a coded V-BLAST transmitter. Unlike the ZF-DF detector of Fig. 10-14, the decision devices are now decoders, which produce codewords of length L as outputs.

smallest diversity benefit and is the most likely to be small. In any case, if any one of the substreams sees a bad subchannel, its poor performance will dominate the overall error rate performance.

The diagonal-BLAST (D-BLAST) strategy was proposed by Foschini as a means for overcoming the drawbacks of V-BLAST [12]. The basic idea is simple: rather than assigning each substream to a dedicated antenna, the assignment is periodically cycled, with a dwell time of W symbol periods, as sketched below:

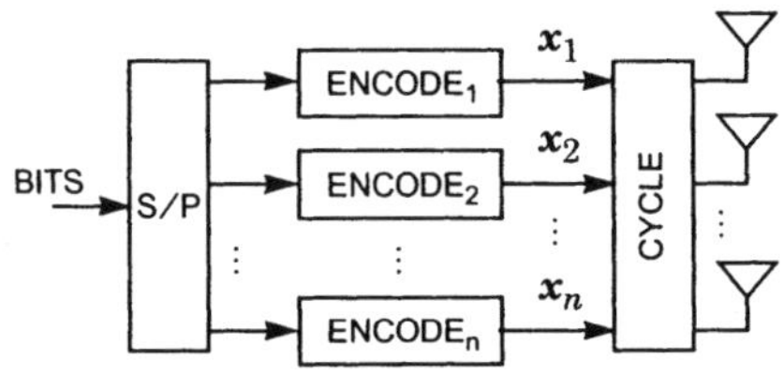

This effectively decomposes the i-th coded substream of length L is into $J = L/W$ subblocks, $\boldsymbol{x}_i = [\boldsymbol{x}_i^{(1)}, \ldots \boldsymbol{x}_i^{(J)}]$, where each $\boldsymbol{x}_i^{(j)}$ represents a block of W coded symbols. The coded substreams thus fill the space-time codeword in a block-diagonal fashion, as illustrated below for the case of $n = 3$ antennas:

$$\mathbf{A} = \begin{bmatrix} x_1^{(1)} & x_2^{(1)} & x_3^{(1)} & x_1^{(4)} & x_2^{(4)} & x_3^{(4)} & \cdots & x_2^{(J-2)} & x_3^{(J-2)} & 0 & 0 \\ 0 & x_1^{(2)} & x_2^{(2)} & x_3^{(2)} & x_1^{(5)} & x_2^{(5)} & x_3^{(5)} & \cdots & x_2^{(J-1)} & x_3^{(J-1)} & 0 \\ 0 & 0 & x_1^{(3)} & x_2^{(3)} & x_3^{(3)} & x_1^{(6)} & x_2^{(6)} & x_3^{(6)} & \cdots & x_2^{(J-1)} & x_3^{(J)} \end{bmatrix}. \quad (11.64)$$

Comparing this space-time codeword with the delay-diversity scheme of (11.55), we see that delay diversity is a special case of D-BLAST for which the encoders are repetition codes and the dwell time is one symbol period. Comparing to (11.57), we see that the two-antenna 8-state 8-PSK space-time trellis code of [10] can also be viewed as a special case of D-BLAST.

Without cycling, the i-th substream is obliterated when the i-th gain G_{ii} is small. The advantage of cycling is that it spreads the bad gains evenly across all substreams. By giving all substreams access to all transmit antennas, cycling ensures that all substreams encounter a channel with the same statistics, namely, one with a time-varying gain that cycles through all $\{G_{11}, \ldots G_{nn}\}$. Cycling converts the spatial diversity potential of the transmitter array into time diversity potential. Sometimes the channel is good, sometimes it is bad. The error-correction capabilities of the code are able to compensate for the bad times.

A receiver for D-BLAST based on DF detection can be implemented in two ways. It can use tentative decisions and DF to detect $\mathbf{A}$ column by column, and decode a substream only after all of its subblocks have been detected. At the expense of increased complexity, improved performance can be achieved by detecting the elements of $\mathbf{A}$ diagonal by diagonal (substream by substream), using decision feedback after decoding to cancel previously detected diagonals, and using linear processing to mitigate interference from undetected diagonals.

Appendix 11-A. Proof of Conservation Theorem

The conservation theorem of (11.26) is easily proved. It applies to the flat-fading Rayleigh MIMO channel, where the $m \times n$ matrix $\mathbf{H}$ has i.i.d. zero-mean unit-variance complex Gaussian entries, satisfying $E[\,|h_{ij}|^2] = 1$. Without loss of generality, assume user 1 is the user of interest. As discussed in Section 10.3.3, the ZF linear detector is proportional to the projection error $\boldsymbol{h}_1 - \hat{\boldsymbol{h}}_1$, where $\hat{\boldsymbol{h}}_1$ is the projection of $\boldsymbol{h}_1$ onto the interference space spanned by the interfering columns $\{\boldsymbol{h}_2, \dots \boldsymbol{h}_n\}$.

The key to the proof is the following thought experiment. Suppose we were to make the following two changes to the channel model:

- Zero the interfering columns $\{\boldsymbol{h}_2, \dots \boldsymbol{h}_n\}$
- Replace the first column $\boldsymbol{h}_1$ by $\boldsymbol{h}_1 - \hat{\boldsymbol{h}}_1$

Since the ZF linear detector is orthogonal to the interference subspace, its output would not be affected by the above changes. But these changes would transform the original MIMO channel into the SIMO fading channel of (11.1) with an effective $m \times 1$ transfer function of $\boldsymbol{h}_1 - \hat{\boldsymbol{h}}_1$, and in terms of this new channel model, the ZF linear detector would actually be the MF detector, or maximal-ratio combiner. Hence, just as in (11.5), its output can expressed as

$$\begin{aligned} y &= (\boldsymbol{h}_1 - \hat{\boldsymbol{h}}_1)^* \boldsymbol{r} \\ &= \|\boldsymbol{h}_1 - \hat{\boldsymbol{h}}_1\|^2 a_n + N, \end{aligned} \tag{11.65}$$

where N is a complex zero-mean Gaussian RV satisfying $E[\,|N|^2] = N_0 \|\boldsymbol{h}_1 - \hat{\boldsymbol{h}}_1\|^2$.

To establish the conservation theorem, we need only show that $\|\boldsymbol{h}_1 - \hat{\boldsymbol{h}}_1\|^2$ has a chi-square distribution with $2(m - n + 1)$ degrees of freedom. Let $\mathbf{Q}_1 = [\boldsymbol{q}_1, \dots \boldsymbol{q}_{n-1}]$ contain an orthonormal basis for the interference subspace, as found by applying the Gram-Schmidt procedure to $\{\boldsymbol{h}_2, \dots \boldsymbol{h}_n\}$, and let $\mathbf{Q}_2 = [\boldsymbol{q}_n, \dots \boldsymbol{q}_m]$ contain a completion of that basis that does not exploit knowledge of $\boldsymbol{h}_1$, so that $\mathbf{Q} = [\mathbf{Q}_1, \mathbf{Q}_2]$ is a unitary matrix that is statistically independent of $\boldsymbol{h}_1$. The projection error $\boldsymbol{h}_1 - \hat{\boldsymbol{h}}_1$ will be orthogonal to $\{\boldsymbol{q}_1, \dots \boldsymbol{q}_{n-1}\}$, and hence it may be expressed as a linear combination of the columns of $\{\boldsymbol{q}_n, \dots \boldsymbol{q}_m\}$, namely:

$$\boldsymbol{h}_1 - \hat{\boldsymbol{h}}_1 = \sum_{i=n}^{m} g_i \boldsymbol{q}_i, \tag{11.66}$$

where we have introduced $g_i = \boldsymbol{q}_i^* \boldsymbol{h}_1$ for $i \in \{n, \dots m\}$. Therefore,

$$\|\boldsymbol{h}_1 - \hat{\boldsymbol{h}}_1\|^2 = \sum_{i=n}^{m} |g_i|^2. \tag{11.67}$$

Since $\boldsymbol{g} = [g_n, \dots g_m]^T = \mathbf{Q}_2^* \boldsymbol{h}_1$ is a linear transformation of a jointly Gaussian vector, it is itself jointly Gaussian, and is thus characterized by its autocorrelation matrix:

$$E[\boldsymbol{g}\boldsymbol{g}^*] = \mathbf{Q}_2^* E[\boldsymbol{h}_1 \boldsymbol{h}_1^*] \mathbf{Q}_2 = \mathbf{Q}_2^* \mathbf{I} \mathbf{Q}_2 = \mathbf{I}. \tag{11.68}$$

We see that the components $\{g_i\}$ are i.i.d. with precisely the same distribution as the components $\{h_i\}$ of the original Rayleigh-fading channel. Therefore, the diversity order is equal to the number of terms in the sum (11.67), namely $m - n + 1$. This completes the proof.

Appendix 11-B. Bound on Pairwise Error Probability

In this appendix we derive the bound of (11.51). We assume that the codeword columns are normalized so that $E[\|\boldsymbol{a}_l\|^2] = 1$, and that $\{h_{ij}\}$ are i.i.d. zero-mean unit-variance Gaussian random variables satisfying $E[|h_{ij}|^2] = E$, so that E/N_0 is the average SNR per receive antenna. The pairwise-error probability $P_{i\to j}$ is the probability that $\mathbf{HA}_i + \mathbf{N}$ is closer to $\mathbf{HA}_j$ than to $\mathbf{HA}_i$, where the elements of $\mathbf{N}$ are i.i.d. zero-mean complex Gaussian random variables satisfying $E[|n_{ij}|^2] = N_0$. In this appendix, as usual, the norm of a matrix is assumed to be the Frobenius norm, defined by $\|\mathbf{H}\|^2 = \|\boldsymbol{h}_1\|^2 + \ldots + \|\boldsymbol{h}_n\|^2$ when $\mathbf{H} = [\boldsymbol{h}_1, \ldots \boldsymbol{h}_n]$. The pairwise-error probability may be expressed as:

$$\begin{aligned} P_{i\to j} &= \Pr[\|\mathbf{H}(\mathbf{A}_i - \mathbf{A}_j) + \mathbf{N}\| < \|\mathbf{N}\|] \\ &= E\big[\Pr[\|\mathbf{H}(\mathbf{A}_i - \mathbf{A}_j) + \mathbf{N}\| < \|\mathbf{N}\| \mid \mathbf{H}]\big] \\ &= E\left[Q\left(\frac{\|\mathbf{H}(\mathbf{A}_i - \mathbf{A}_j)\|}{\sqrt{2N_0}}\right)\right] \\ &= E\left[Q\left(\sqrt{\frac{X}{2N_0}}\right)\right] \\ &\le E\left[\exp\left(\frac{-X}{4N_0}\right)\right] = \psi_X\left(\frac{-1}{4N_0}\right), \end{aligned} \qquad (11.69)$$

where we have introduced the random variable $X = \|\mathbf{H}(\mathbf{A}_i - \mathbf{A}_j)\|^2$ and its characteristic function $\psi_X(s) = E[e^{sX}]$. The inequality follows from (3.43).The key step is to find $\psi_X(s)$.

Let r denote the rank of the $n \times n$ Hermitian matrix $\mathbf{B} = (\mathbf{A}_i - \mathbf{A}_j)(\mathbf{A}_i - \mathbf{A}_j)^*$, whose eigendecomposition is $\mathbf{B} = \mathbf{U}\Lambda\mathbf{U}^*$, where $\mathbf{U}$ is unitary, and where Λ is a diagonal matrix with real, nonnegative, nonincreasing diagonal elements $\{\lambda_1, \ldots \lambda_r, 0, \ldots 0\}$. In terms of this decomposition, and by exploiting the identity $\|\mathbf{M}\|^2 = \mathrm{tr}\{\mathbf{MM}^*\}$, we may write X as:

$$X = \|\mathbf{H}(\mathbf{A}_i - \mathbf{A}_j)\|^2 = \mathrm{tr}\{\mathbf{HBH}^*\} = \mathrm{tr}\{\mathbf{H}(\mathbf{U}\Lambda\mathbf{U}^*)\mathbf{H}^*\} = \|\mathbf{HU}\Lambda^{1/2}\|^2. \qquad (11.70)$$

Because $\mathbf{H}$ is Gaussian and $\mathbf{U}$ is unitary, $\mathbf{HU}$ and $\mathbf{H}$ are identically distributed. Hence, if we were to eliminate $\mathbf{U}$ from (11.70), we would not change its probability distribution. In other words, X has the same probability distribution as:

$$X' = \|\mathbf{H}\Lambda^{1/2}\|^2 = \lambda_1\|\boldsymbol{h}_1\|^2 + \ldots + \lambda_r\|\boldsymbol{h}_r\|^2 . \qquad (11.71)$$

Therefore,

$$\begin{aligned}\psi_X(s) &= \mathrm{E}[\exp(s\Sigma_i\lambda_i\|\boldsymbol{h}_i\|^2)]\\ &= \prod_{i=1}^{r}\mathrm{E}[\exp(s\lambda_i\|\boldsymbol{h}_i\|^2)]\\ &= \prod_{i=1}^{r}\psi_{Y_i}(s\lambda_i)\\ &= \prod_{i=1}^{r}(1-s\lambda_i E)^{-m}\,,\end{aligned} \tag{11.72}$$

where we introduced $Y_i = \|\boldsymbol{h}_i\|^2$, a chi-square random variable with $2m$ degrees of freedom and mean mE, and exploited the fact that its characteristic function is $\psi_{Y_i}(s) = (1-sE)^{-m}$. Substituting (11.72) into (11.69) yields:

$$\begin{aligned}P_{i\to j} &\le \prod_{i=1}^{r}\left(1+\frac{\lambda_i E}{4N_0}\right)^{-m}\\ &< \prod_{i=1}^{r}\left(\frac{\lambda_i E}{4N_0}\right)^{-m}\\ &= \left(\frac{\eta E}{4N_0}\right)^{-mr},\end{aligned} \tag{11.73}$$

where we have introduced $\eta = (\lambda_1\lambda_2 \ldots \lambda_r)^{1/r}$, the geometric mean of the nonzero eigenvalues of $\mathbf{B} = (\mathbf{A}_i - \mathbf{A}_j)(\mathbf{A}_i - \mathbf{A}_j)^*$.

Problems

Problem 11-1. For the MISO model of (11.36), show that the combining weight vector $\boldsymbol{w}$ in (11.37) that maximizes the SNR at the receiver, subject to the constraint that $\|\boldsymbol{w}\| = 1$, is $\boldsymbol{w} = \mathbf{H}^*/\|\mathbf{H}\|$.

Problem 11-2. Let $\boldsymbol{n} = [n_1, \ldots, n_4]^T$ with $\{n_i\}$ i.i.d. $\mathcal{CN}(0, N_0)$. Let $\hat{\boldsymbol{n}}$ denote the projection of $\boldsymbol{n}$ onto the subspace spanned by $\boldsymbol{x} = \frac{1}{2}[1,-1,1,-1]^T$ and $\boldsymbol{y} = \frac{1}{2}[1,1,-1,-1]^T$. Characterize the pdf of $\|\hat{\boldsymbol{n}}\|^2$.

Problem 11-3. Derive (11.25) from (11.24). The following two identities may be helpful [13]:

$$(1+x)^{-1/2} = \sum_{k=0}^{\infty}\frac{(2k-1)!!}{(2k)!!}(-x)^k\,,$$

$$\sum_{k=0}^{m}(-1)^k k^n\binom{m}{k} = \begin{cases}0 & \text{if } n\in\{0,1,2,\ldots,m-1\}\\ (-1)^m m! & \text{if } n = m\end{cases} \tag{11.74}$$

The double factorial notation is short for $(2k-1)!! = 1\cdot3\cdot5\cdots(2k-1)$ and $(2k)!! = 2\cdot4\cdot6\cdots(2k)$.

Problem 11-4. Consider the SIMO channel of (11.1), assuming i.i.d. Rayleigh fading, AWGN and 4-QAM. As described at the beginning of Section 11.3, the received average SNR per bit per antenna is E_b/N_0. However, the effective (or postdetection) average SNR per bit is generally larger.

(a) Show that maximal-ratio combining increases the average postdetection SNR by a factor of m.

(b) Show that selection combining increases the average postdetection SNR by a factor of $\sum_{k=1}^{m} \frac{1}{k}$.

Problem 11-5. Suppose a transmitter with two antennas somehow knows the response of the narrowband channel between it and a receiver with a single antenna. Instead of using matched-filter precompensation, suppose the transmitter uses selection instead, *i.e.*, only transmits from the antenna with the largest gain. Find an approximate expression for the bit-error probability as a function of E_b/N_0 for Gray-mapped 4-QAM with AWGN and i.i.d. Rayleigh fading.

Problem 11-6. An OFDM transmitter that was originally designed for use with only a single transmit antenna, suddenly acquires a second transmit antenna. To avoid a total redesign of the transmitter, it would be nice if delay diversity could be used, so that the signal that was originally transmitted from the first antenna will be transmitted again from the second antenna, only delayed relative to the first. Describe the changes that will be necessary at the receiver in order for this delay-diversity scheme to achieve the benefit of transmit diversity.

Problem 11-7. Consider the following set of space-time codes:

$$\mathbf{A}_1 = \begin{bmatrix} x_1 & x_2{}^* \\ x_2 & x_1{}^* \end{bmatrix}, \mathbf{A}_2 = \begin{bmatrix} x_1 & -x_2 \\ x_2 & x_1 \end{bmatrix}, \mathbf{A}_3 = \begin{bmatrix} x_1{}^* & -x_2 \\ x_2{}^* & x_1 \end{bmatrix}, \mathbf{A}_4 = \begin{bmatrix} x_1 & x_2 \\ x_2 & x_1 \end{bmatrix}. \tag{11.75}$$

(a) Find the diversity order achieved by each, assuming that the information symbols $\{x_i\}$ are independently and uniformly selected from a 4-QAM alphabet, and assuming the receiver has a single antenna.

(b) For those that achieve full diversity in (a), find the coding gain.

References

1. N. Prasad and M. K. Varanasi, "Analysis of Decision-Feedback Detection for MIMO Rayleigh Fading Channels and Optimum Allocation of Transmitter Powers and QAM Constellations," in *Proc. Allerton Conf. on Comm. Control, and Comput.*, Monticello, IL, October 2001.

2. G. L. Stuber, *Principles of Mobile Communication, Second Edition*, Kluwer Academic Publishers, 2001.

3. J. H. Winters, "The Diversity Gain of Transmit Diversity in Wireless Systems with Rayleigh Fading," *IEEE Trans. Vehic.Techn.*, Vol. 47, No. 1, pp. 119-123, Feb. 1998.

4. D. Gerlach and A. Paulraj, "Adaptive Transmitting Antenna Arrays with Feedback," *IEEE Signal Processing Letters*, vol. 1, no. 10, pp. 150-152, October 1994.

5. B. Raghothaman, G. Mandyam, and R. T. Derryberry, "Performance of Closed Loop Transmit Diversity with Feedback Delay," *34th Asilomar Conf. Sig., Syst. and Comp.*, pp. 102 -105, 2000.

6. TSG RAN WG1 (Radio Layer 1) Specifications, "Transmitter Diversity Solutions for Multiple Antennas," http://www.3gpp.org/ftp/Specs/archive/25_series/25.869/.

7. G. J. Foschini and M. Gans, "On Limits of Wireless Communications in a Fading Environment when using Multiple Antennas," *Wireless Personal Comm.*, vol. 6, no. 3, pp. 311-355, Mar. 1998.

8. V. Tarokh, H. Jafarkhani, and A. R. Calderbank, "Space-Time Block Codes from Orthogonal Designs," *IEEE Trans. Info. Theory*, Vol. 45, No. 5, pp. 1456-1467, July 1999.

9. O. Tirkkonen and A. Hottinen, "Square-Matrix Embeddable Space-Time Block Codes for Complex Signal Constellations," *IEEE Trans. Info. Theory*, Vol. 48, pp. 384-395, Feb. 2002.

10. V. Tarokh, N. Seshadri, and A. R. Calderbank, "Space-Time Codes for High Data Rate Wireless Communication: Performance Criterion and Code Construction," *IEEE Trans. Info. Theory*, Vol. 44, No. 2, pp. 744–765, March 1998.

11. J. Yuan, Z. Chen, B. Vucetic, and W. Firmanto, "Performance Analysis and Design of Space-Time Coding on Fading Channels," preprint, submitted to *IEEE Trans. Comm.*, 2000.

12. G. J. Foschini, "Layered Space-Time Architecture for Wireless Communications in a Fading Environment when using Multi-Element Antenna," *Bell Labs Tech. J.*, vol. 1, pp. 41-59, 1996.

13. I. S. Gradshteyn and I. M. Ryzhik, *Tables of Integrals, Series, and Products*, Fourth Edition, Academic Press, 1980.

12

Error Control

Error-control coding is the name given to the process of converting source bits into transmitted symbols so as to make possible reliable communications despite the presence of noise. As shown in Fig. 12-1, a *channel coder* precedes the mapping of bits to symbols at the transmitter. The channel coder constrains the symbol sequence $\{a_k\}$ so that only a strict subset of all possible symbol sequences can be transmitted. There is thus redundancy in the coded sequence, which can be exploited at the receiver to improve the robustness to noise.

Even the most rudimentary communication system will use some form of error-control coding. For example, error-control coding can be used to facilitate the *detection* of errors at the receiver; the receiver may then repeatedly request retransmissions until no errors are detected. A more ambitious goal is error *correction*, where the receiver not only detects an error but also corrects it. Instead of correcting errors after they occur, a still more ambitious goal is to *prevent* errors before they occur, by a combination of detection and decoding known as soft

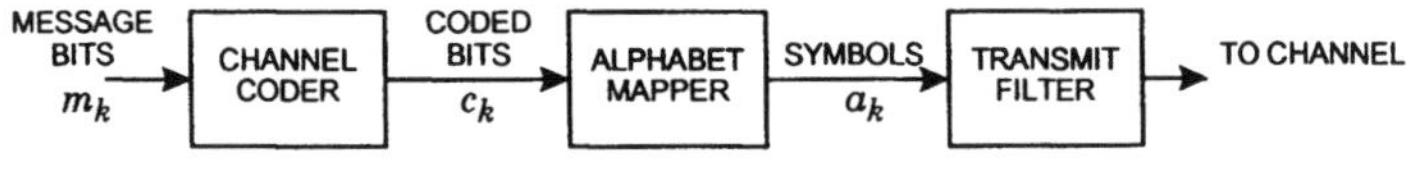

Fig. 12-1. A channel coder translates source bits into coded bits to protect them from noise.

decoding. The ultimate aim of error-control coding is to close the gap between the performance of uncoded modulation and the Shannon limit, allowing a practical system to communicate reliably at a rate close to the Shannon capacity. To be concrete we will examine the subject for the special case of the AWGN channel.

In previous chapters we saw a fundamental tradeoff between bandwidth and signal power. Orthogonal modulation schemes were seen to be appropriate when bandwidth is more plentiful than power, whereas large-alphabet PAM was seen to be the better choice when power is more plentiful than bandwidth. There is a similar dichotomy in choosing an appropriate error-control strategy. Specifically, let us distinguish between the *low-SNR regime*, where bandwidth is relatively plentiful so that the target spectral efficiency is 1 b/s/Hz or lower, and the *high-SNR regime*, where bandwidth is relatively scarce and the target spectral efficiency is higher than 1 b/s/Hz.[1] In the low-SNR regime, *binary error-control coding* in combination with binary modulation is nearly optimal. Binary coding is the subject of this chapter. On the other hand, the high-SNR regime calls for *signal-spacing coding* using larger PAM alphabets, as described in the next chapter.

There is an important fundamental difference between binary coding and signal-space coding. The binary coding strategies described in this chapter fix the alphabet size and increases the symbol rate by inserting extra transmitted symbols which depend deterministically on the information bits. The higher symbol rate implies a lower spectral efficiency. Because bandwidth is proportional to the symbol rate, binary coding increases the bandwidth requirement. In contrast, signal-space coding of Chapter 13 fixes the symbol rate and increases the size of the alphabet. The bandwidth requirement does not increase. Signal-space coding is most appropriate for bandwidth-limited media where the target spectral efficiency is high.

Two fundamentally different types of *decoding* are used, *hard* and *soft*, as illustrated in Fig. 12-2. With hard decoding, the receiver first makes hard decisions about the transmitted symbols using a memoryless slicer. The hard decoder operates on these hard decisions. Since not all bit patterns are permitted by the code, the decoder can *detect* or *correct* bit errors. From the perspective of the coder, the channel is binary and makes transmission errors. Often a binary symmetric channel (BSC) noise generation model is used (Fig. 7-2).

A soft decoder, by contrast, makes direct decisions about the information bits without making intermediate decisions about the transmitted symbols. As shown in Fig. 12-2, a soft decoder operates directly on the continuous-valued samples of the received signal. Instead of correcting errors after they occur, as done by a hard decoder, a soft decoder *prevents* errors by combining slicing with channel decoding. We can think of soft decoding as a combination of slicing and removing redundancy.

---

1. The value of $SNR = P/(N_0W)$ is natural choice for distinguishing power-limited channels from bandwidth-limited channels, since SNR is proportional to power and inversely proportional to bandwidth. Furthermore, SNR also determines the Shannon limit on spectral efficiency, namely $\log_2(1 + SNR)$.

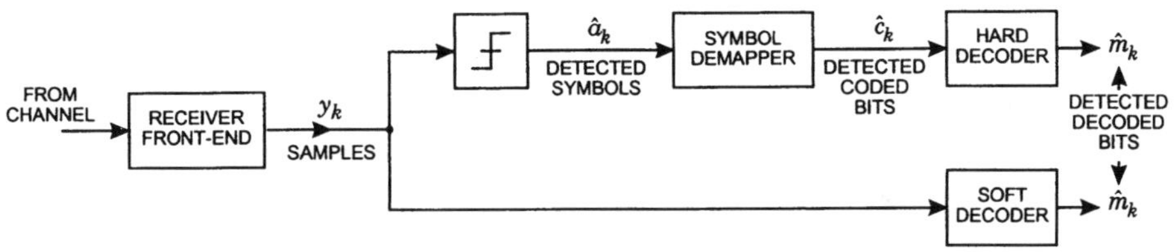

Fig. 12-2. In hard decoding (upper branch), decisions are made about the incoming symbols by a slicer, the symbols are decoded into bits by a demapper, and the channel decoder maps these coded bits into uncoded bits. A soft decoder (lower branch) operates directly on continuous-valued samples of the incoming signal rather than on the detected bits.

Although more complex to implement, soft decoding performs better than hard decoding because it makes use of information that the slicer would otherwise throw away. For example, if the slicer input is halfway between slicer levels, the hard decoder would be forced to make a decision, whereas the soft decoder would make note of the uncertainty and incorporate that uncertainty in the final decision. When we have no control over the modulation and demodulation process, however, hard decoding is the only option. For example, error-control coding is sometimes applied to an *existing* digital communication system that has already been designed and implemented, but for which the error rate is too large for our intended purpose. This scenario is illustrated in Fig. 12-3. The error rate can be decreased using binary coding at the transmitter and hard decoding at the receiver, at the expense of a lower information rate.

Within the class of binary codes we will consider *block codes* and *convolutional codes.* Both hard and soft decoding are used for block and convolutional codes, while only soft decoding is used for signal-space codes (Chapter 13). A block code maps blocks of k source bits into blocks of n coded bits where $n > k$. Such a block code is said to have *code rate* k/n, where the terminology refers to the fraction of the total bit rate devoted to information bits. A convolutional coder also produces coded bits at a higher rate than the source bits, but it does so without dividing the source bits into blocks. Instead, a convolutional encoder acts on a *stream* of message bits that in theory might last forever. In both cases there are more coded bits than source bits, and the coded bits have redundant information about the source bits.

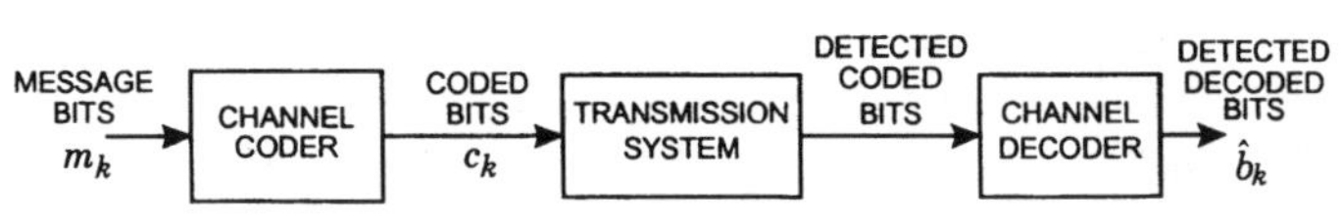

Fig. 12-3. A channel coder and hard decoder added to an existing digital communication system.

A new class of codes called *turbo-like codes* has emerged in the last decade that promises both performance near the Shannon limit as well as low-complexity implementation. Examples of turbo-like codes include serially and parallel-concatenated turbo codes, repeat-accumulate codes, and low-density parity-check codes. They behave very differently from the algebraic and convolutional codes that came before. For many decades the usual approach to code design has been to maximize minimize distance, which will in turn minimize the value of the largest $Q(\cdot)$ function in the union bound. This strategy has two flaws: it ignores other distances that may contribute nonnegligible $Q(\cdot)$ terms, and more importantly, it ignores the *number* of neighbors at that minimum distance; in other words, it ignores the error coefficient that multiplies the $Q(\cdot)$ function. The problem is that the union bound is not very tight when the operating SNR is near the Shannon limit. The turbo-like codes described later in this chapter can approach capacity not because they have a large minimum distance (in fact their minimum distance can be relatively small), but because they have a good *distance spectrum*.

The remainder of this chapter is organized as follows. In Section 12.1 we examine the optimality of binary coding when the spectral efficiency is low. In Section 12.2 and Section 12.3 we describe binary block codes and convolutional codes, respectively. In Section 12.4 we describe low-density parity-check codes, and in Section 12.5 we describe turbo codes, which include repeat-accumulate codes and turbo equalization as special cases.

Error correction coding is a large subject, and in this book we hope to convey the most important concepts and leave the details to the extensive and excellent literature on the subject.

12.1. The Capacity Penalty of Binary Coding

The capacity of an AWGN channel is achieved only when the input symbols are chosen according to a Gaussian distribution. However, we will now argue that there is very little penalty in capacity when the input alphabet is constrained to be binary antipodal, provided that the system is operating in the low-SNR regime where the target spectral efficiency is small.

To understand the near-optimality of binary coding at low spectral efficiencies, consider the Shannon capacity of the real-valued memoryless AWGN channel

$$r = a + n\,, \tag{12.1}$$

where the noise is real, zero-mean and Gaussian with variance $N_0/2$. In Chapter 4 we learned that the capacity of this channel, subject to an energy constraint of $E = \mathrm{E}[a^2]$, is:

$$C = \tfrac{1}{2}\log_2(1 + \mathrm{SNR}) \text{ bits per real symbol}\,, \tag{12.2}$$

where $SNR = 2E/N_0$.

The above model directly applies to baseband PAM, but it also applies to *one dimension* (real or imaginary) for passband PAM. Either way, a power constraint of P on the underlying continuous-time channel implies a per-dimension energy constraint of $E = P/(2W)$, assuming zero excess bandwidth. (For baseband, $E = PT$ where $W = 1/(2T)$, while for passband, the energy is split in half for each dimension, $E = PT/2$, where $W = 1/T$.) This implies that the SNR of the discrete-time channel (12.1) is identical to the SNR of the continuous-time channel, $SNR = P/N_0W = 2E/N_0$.

Since spectral efficiency is the number of bits conveyed by *two* real symbols, or equivalently by one complex symbol, we can invert the above capacity equation and solve it for SNR as a function of the spectral efficiency $b = 2C$:

$$SNR = 2^b - 1 \ . \tag{12.3}$$

This is the Shannon limit on SNR. Any practical and reliable system achieving a spectral efficiency of b (b/s/Hz) must have an SNR at least as large as $2^b - 1$. For example, the SNR must be at least 0 dB for a spectral efficiency of 1 b/s/Hz.

Recall that the *energy per bit* is defined as the ratio of signal power to bit rate, which is $E_b = P/(bW)$ when b is the spectral efficiency in b/s/Hz. It follows that $E_b/N_0 = SNR/b$, so that the Shannon limit on E_b/N_0 is:

$$E_b/N_0 = (2^b - 1)/b \ . \tag{12.4}$$

In Fig. 12-4 we plot this E_b/N_0 requirement as a function of the spectral efficiency. It achieves a minimum value of log(2) = –1.59 dB in the limit as the spectral efficiency goes to zero, which might happen when the channel bandwidth is unlimited. On the other hand, when the target bit rate and channel bandwidth coincide, so that the spectral efficiency is one, the per-bit SNR requirement is exactly 0 dB. The shaded region below this limit is unattainable.

Also shown in Fig. 12-4 is a plot of the E_b/N_0 requirement as a function of the mutual information between the channel output and input when the input is constrained to be chosen uniformly from the *binary* antipodal alphabet $\{\pm\sqrt{E}\}$. This mutual information has no closed-form solution but it can be calculated numerically using (4.25), or equivalently, it can be expressed in terms of a zero-mean unit-variance Gaussian random variable U as [1]:

$$C_b = 1 - \mathrm{E}[\log_2(1 + e^{-2U\sqrt{SNR} - 2SNR})] \text{ bits per real symbol} \ . \tag{12.5}$$

The key observation from the figure is that the binary-input constraint has a negligible impact on the E_b/N_0 requirement when the spectral efficiency is lower than about 0.5 b/s/Hz. Even at 1 b/s/Hz, the penalty is only 0.19 dB. The penalty grows quickly thereafter. At the modest spectral efficiency of 1.9 b/s/Hz, the penalty due to a binary alphabet is over 2.5 dB, and the penalty grows without bound as the spectral efficiency approaches 2 b/s/Hz.

Example 12-1.

The communication link from a deep-space probe to an earth-based receiver is a classic example of a low-SNR, power-limited channel. Signal power is scarce because of the great distance between transmitter and receiver. In contrast, the bandwidth is plentiful, especially relative to the modest data rates required. For example, the spectral efficiency for the Mars Pathfinder and Jupiter Cassini deep-space probes is less than 0.2 b/s/Hz [2]. Not surprisingly, in light of Fig. 12-4, deep-space communications relies exclusively on binary coding.

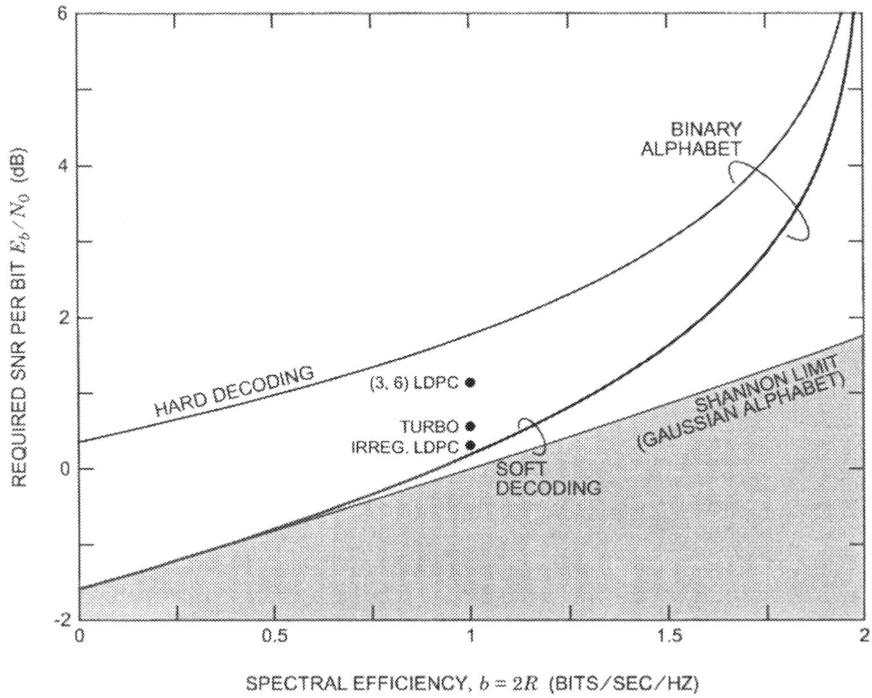

Fig. 12-4. Capacity of the AWGN with various constraints on the input and output. The lower curve is the Shannon limit $(2^b - 1)/b$, which is achieved when the channel inputs have a Gaussian distribution. The next curve above is the Shannon limit when the input alphabet is constrained to be binary. The penalty due to binary inputs disappears at low spectral efficiency. The uppermost curve constrains the input alphabet to be binary and also constrains the receiver to perform hard decoding. The extra penalty due to hard decoding is seen to range from 2 dB down to 1 dB as the binary code rate ranges from 0 to 0.98. The three circles mark the performance of specific codes to be described later in the chapter.

At this point we should point out that when we say *binary modulation* in this chapter we really mean binary modulation *per dimension*, in the sense that the inputs to the real-valued channel model of (12.1) are binary. In this sense, 4-QAM with alphabet $\{\pm 1 \pm j\}$ is actually a form of binary modulation, since both the real and the imaginary components of the transmitted symbols are binary, and the two components independently experience an identical channel. We can thus constrain our discussion to real binary modulation without loss of generality.

When the transmitted symbols are binary, one might be tempted to insert a one-bit quantizer at the front-end of the receiver, which would memorylessly convert each noisy observation into a *hard* decision about which symbol was transmitted. The remainder of the receiver would then be easier to implement because it would operate only on bits. This quantization would transform the binary-input real-output channel into a binary-input binary-output channel which, because of the symmetry of the Gaussian noise, is actually a BSC with crossover probability $p = Q(\sqrt{SNR})$. A decoder that operates on these binary decisions is said to be a *hard decoder*, while in contrast, a decoder that operates on the original real-valued samples is a *soft decoder.*

In theory we can quantify the penalty due to hard decoding by evaluating the BSC capacity $1 + p\log_2 p + (1 - p)\log_2(1 - p)$ from (4.18) using $p = Q(\sqrt{SNR})$. The results are represented by the uppermost curve in Fig. 12-4. The capacity penalty due to hard decoding — on top of the binary-input penalty — is seen to range from 2 dB down to 1 dB as the spectral efficiency increases from zero to 2 b/s/Hz.

The performance of three representative turbo-like codes is illustrated in Fig. 12-4. All three codes have rate 1/2 (unity spectral efficiency) and block lengths of one million bits, and the points in the figure identify the value of E_b/N_0 required to achieve a bit-error rate of 10^{-6}, as described in [3]. The uppermost point (labeled *(3, 6) LDPC*) is a regular low-density parity-check code [4], described in Section 12.4, which falls within 1 dB of the Shannon limit. The point below that (labeled *turbo*) is a turbo code [5], described in Section 12.5.1. The lowermost point (labeled *irregular LDPC*) is an irregular LDPC code, described in Section 12.4, which is only 0.13 dB shy of the Shannon limit for binary inputs.

12.2. Binary Linear Block Codes

An (n, k) binary block coder maps blocks of k source bits, or *message* bits, into blocks of n coded bits, where $n > k$. The n coded bits depend only on the k source bits, so the coder is said to be *memoryless*. Assume that the k source bits are collected in a shift register, as shown in Fig. 12-5, and the n coded bits are serialized to be sent to the symbol mapper. The block coder circuitry itself often consists only of modulo-two adders (exclusive-or gates).

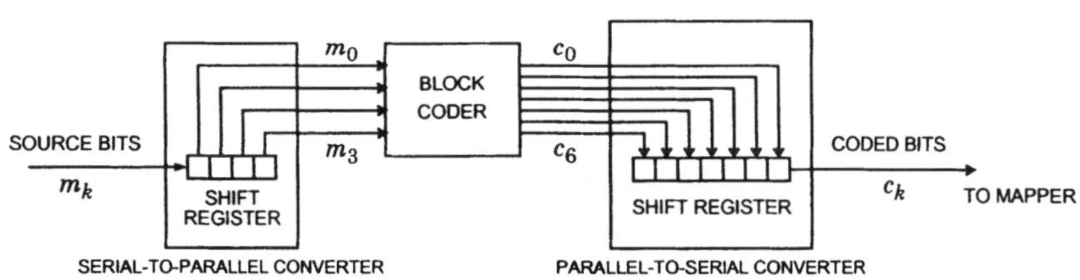

Fig. 12-5. A block coder collects k incoming bits in a shift register, codes them, and shifts them out serially to the symbol mapper. The example shown is an $(n, k) = (7, 4)$ block code.

Example 12-2.

A *single-parity-check coder* takes a block of k message bits $[m_0, \ldots m_{k-1}]$ and appends an extra bit, called a parity bit c_k, yielding a codeword of the form $\boldsymbol{c} = [m_0, \ldots m_{k-1}, c_k]$. Furthermore, the extra bit c_k is chosen so that the total number of ones in the codeword is even, which occurs if and only if the extra bit is the modulo-two summation of the message bits:

$$c_k = m_0 \oplus \cdots \oplus m_{k-1} \; , \tag{12.6}$$

where $\oplus$ denotes modulo-two addition.

A general *linear code* or *parity-check code* is a code for which all coded bits are modulo-two summations of subsets of the message bits. This implies the existence of a *generator matrix* $\mathbf{G}$ with k rows and n columns, consisting of zeros and ones such that the coded bits $\boldsymbol{c}$ are related to the message bits by:

$$\boldsymbol{c} = \boldsymbol{m}\mathbf{G} \; , \tag{12.7}$$

where the addition that occurs in the matrix multiplication is modulo-two.

Example 12-3.

The generator matrix for the single-parity-check code of Example 12-2 is

$$\mathbf{G} = \begin{bmatrix} 1 & 0 & 0 & \ldots & 0 & 1 \\ 0 & 1 & 0 & & 0 & 1 \\ 0 & 0 & 1 & & 0 & 1 \\ \vdots & & & \ddots & & \vdots \\ 0 & 0 & 0 & \cdots & 1 & 1 \end{bmatrix} . \tag{12.8}$$

Codes like that of Example 12-3 which have a generator matrix of the form

$$\mathbf{G} = [\mathbf{I}_k \,|\, \mathbf{P}] \; , \tag{12.9}$$

where $\mathbf{I}_k$ is the k-dimensional identity matrix, are called *systematic*. The first k coded bits are exactly the message bits, so that a systematic codeword takes the form $\boldsymbol{c} = [\boldsymbol{m}, \boldsymbol{x}]$, where $\boldsymbol{x}$ is a set of $n - k$ parity bits. The single-parity-check code of Example 12-3 is systematic.

Example 12-4.

A more elaborate systematic parity-check code is the (7, 4) *Hamming code*, which has generator matrix

$$\mathbf{G} = \begin{bmatrix} 1 & 0 & 0 & 0 & 1 & 0 & 1 \\ 0 & 1 & 0 & 0 & 1 & 1 & 1 \\ 0 & 0 & 1 & 0 & 1 & 1 & 0 \\ 0 & 0 & 0 & 1 & 0 & 1 & 1 \end{bmatrix} . \tag{12.10}$$

For a particular binary row vector $\boldsymbol{m}$, $\boldsymbol{c} = \boldsymbol{m}\mathbf{G}$ is called a *codeword*. A *code* is the set of all possible codewords. Contrast this with an *encoder*, which is a rule for mapping message bits to coded bits. It is thus possible that two distinct encoders generate the same code. Every codeword is a modulo-two summation of rows of the generator matrix, with the zero vector being a degenerate example. Clearly, therefore, the modulo-two sum of any two codewords is a codeword, and parity-check codes are said to be *closed* under modulo-two summation. This is a desirable property for codes, and is elaborated in Appendix 12-A. In fact, it is shown in the appendix that all parity-check codes are *linear subspaces* over the binary field, or just *linear* for short, and that all linear block codes are parity-check codes with some generator matrix. Furthermore, all linear block codes are equivalent to a systematic code, obtained by reordering the coded bits. In the following we specialize to linear codes and exploit their special properties. This constraint does not cost us much, since linear codes are sufficient to approach the capacity of symmetric binary-input channels like the BSC and AWGN channels considered here [6].

An important goal is to compare the performance of soft decoding to hard decoding. The basic difference between the two cases is that in the soft decoding case the detector chooses the codeword closest to the reception in Euclidean distance, and in the hard decoding case the detector uses Hamming distance. Furthermore, we will find, not unexpectedly in view of the analysis in Chapter 7, that the probability of error is dominated by the two codewords that are closest in Euclidean or Hamming distance at high SNR. To compare the two approaches, we need to find a relationship between Euclidean and Hamming distance between a given pair of codewords; this depends on the symbol mapper, that is, the mapping from coded bits c_k into symbols a_k in Fig. 12-1. The Euclidean distance is measured in terms of the symbols, whereas the Hamming distance is measured in terms of coded bits. Comparisons below assume that binary antipodal signaling is used, so that the symbol mapper maps $\{0, 1\}$ into $\{\pm\sqrt{E}\}$.

Exercise 12-1.

Show that for a binary antipodal alphabet, the Hamming distance and Euclidean distance between a pair of codewords are related by

$$d_E = 2\sqrt{E d_H}\,. \tag{12.11}$$

When the code is linear, it is shown in Appendix 12-A that the minimum Hamming distance between codewords is equal to the minimum Hamming weight (number of ones) among all the nonzero codewords,

$$d_{H,\min} = \min_{\mathbf{c} \in C,\, \mathbf{c} \neq \mathbf{0}} w_H(\mathbf{c})\,. \tag{12.12}$$

Example 12-5.

In the single-parity-check code of Example 12-2, all codewords have an even-valued Hamming weight. Thus, the smallest Hamming weight among all nonzero codewords is two. This is also the minimum Hamming distance between codewords. Thus, the minimum Euclidean distance with binary antipodal signaling is $d_{E,\min} = 2\sqrt{2E}$.

12.2.1. Performance of Soft Decoders

The value of error-control coding is often quantified by the *coding gain*, defined as the amount by which the signal power may be decreased with coding to achieve the same bit-error probability as an uncoded system. Implicit in this comparison is the assumption that both systems are operating at the same information bit rate over the same channel with the same noise power spectrum $N_0/2$. Furthermore, for the binary codes of this chapter, both the coded and uncoded systems are assumed to use binary antipodal signaling.

Because signal power is proportional to E_b/N_0 when the bit rate and noise power spectrum are fixed, we can equivalently define coding gain as the decrease in the E_b/N_0 required to achieve the desired bit-error probability because of the code:

$$\text{coding gain} = \frac{(E_b/N_0)^{\text{required}}_{\text{uncoded}}}{(E_b/N_0)^{\text{required}}_{\text{coded}}} \,. \tag{12.13}$$

A coding gain of 3 dB, for example, indicates that the coded system can achieve the same performance as the uncoded system with half as much signal power. The coding gain is generally a function of the desired bit-error probability.

For the soft decoding case, the input to the decoder is the sample stream before it is applied to a slicer. Assume the equivalent channel is a discrete-time additive Gaussian noise channel with independent noise components and no ISI. A codeword $\boldsymbol{c}$ is transmitted as a vector $\boldsymbol{a}$ with components chosen from $\{\pm\sqrt{E}\}$, and the noise samples have variance $\sigma^2 = N_0/2$. Critical to our analysis is the fact that the energy E per symbol decreases with the code rate $R = k/n$ according to:

$$E = RE_b \,, \tag{12.14}$$

where E_b is the energy per bit. This relationship can be partially justified by examining the units: [energy per symbol] = [bits per symbol]·[energy per bit]. But it is more instructive to attach a physical explanation to this relationship. A code of rate R will result in more coded bits than message bits, requiring that the symbol rate be *increased* by a factor of $1/R$ in order to maintain the same bit rate. Therefore, to avoid an increase in signal power, the energy per symbol must *decrease* by a factor of R. In other words, with a fixed power constraint, the energy available per pulse decreases as we increase the rate at which we send pulses.

The received samples y_k can be collected into the vector

$$\boldsymbol{y} = \boldsymbol{a} + \boldsymbol{n} \,, \tag{12.15}$$

where $\boldsymbol{n}$ is a vector of i.i.d. Gaussian random variables. This detection problem is familiar from Chapter 7, where we showed that the ML detector selects the valid signal vector $\hat{\boldsymbol{a}}$ closest in Euclidean distance to the observed vector $\boldsymbol{y}$. Hence, the union-bound approximation gives

$$\Pr[\text{block error}] \approx KQ\left(\frac{d_{E,\min}}{2\sigma_c}\right)$$

$$= KQ\left(\sqrt{2Rd_{\min}\frac{E_b}{N_0}}\right) , \tag{12.16}$$

where $d_{E,\min} = 2\sqrt{Ed_{\min}} = 2\sqrt{RE_b d_{\min}}$, and where $d_{\min}$ is the minimum *Hamming* distance of the code. The coefficient K is the average number of codewords with Hamming distance $d_{\min}$ from a given codeword.

In contrast to the above expression for a coded system, the bit-error probability for an uncoded binary antipodal system has a simple closed-form solution, namely:

$$\Pr[\text{bit error, uncoded}] = Q\left(\sqrt{\frac{2E_b}{N_0}}\right). \tag{12.17}$$

In fact, the union-bound estimate becomes exact in this case, since (12.16) reduces to (12.17) for the uncoded case where $R = d_{\min} = K = 1$. At high SNR, the probability of error for a block of k uncoded bits is then:

$$\Pr[\text{block error, uncoded}] \approx kQ\left(\sqrt{\frac{2E_b}{N_0}}\right). \tag{12.18}$$

We can now directly compare the block-error performance of the coded and uncoded systems. If we ignore the constant multipliers of $Q(\cdot)$, which is reasonable at high E_b/N_0, we can equate the arguments of the $Q(\cdot)$ in the coded and uncoded cases, yielding:

$$Rd_{\min}(E_b/N_0)^{\text{required}}_{\text{coded}} = (E_b/N_0)^{\text{required}}_{\text{uncoded}} . \tag{12.19}$$

This implies that the *asymptotic coding gain* is simply the product of the code rate and minimum Hamming distance:

$$\text{asymptotic coding gain} = Rd_{\min} . \tag{12.20}$$

This gain is *asymptotic* because it only applies in the limit of large E_b/N_0, or equivalently small probability of error. The true coding gain is usually somewhat less.

Example 12-6.

For the $(n, n-1)$ single-parity-check code (Example 12-2), the code rate is $R = (n-1)/n$, and the minimum Hamming distance is $d_{\min} = 2$, so that the asymptotic coding gain is:

$$Rd_{\min} = 2(n-1)/n . \tag{12.21}$$

With $n = 3$, for example, the asymptotic coding gain is $4/3$ or 1.25 dB. In Fig. 12-6(a), we compare the bit-error probability P_b with ML decoding of the (3, 2) single-parity-check code to the bit-error probability (12.17) of an uncoded system. The true coding gain increases from 0.72 dB at $P_b = 10^{-3}$ to 0.89 dB at $P_b = 10^{-5}$. The 1.25 dB predicted by the asymptotic coding gain is seen to be somewhat optimistic at these values of P_b.

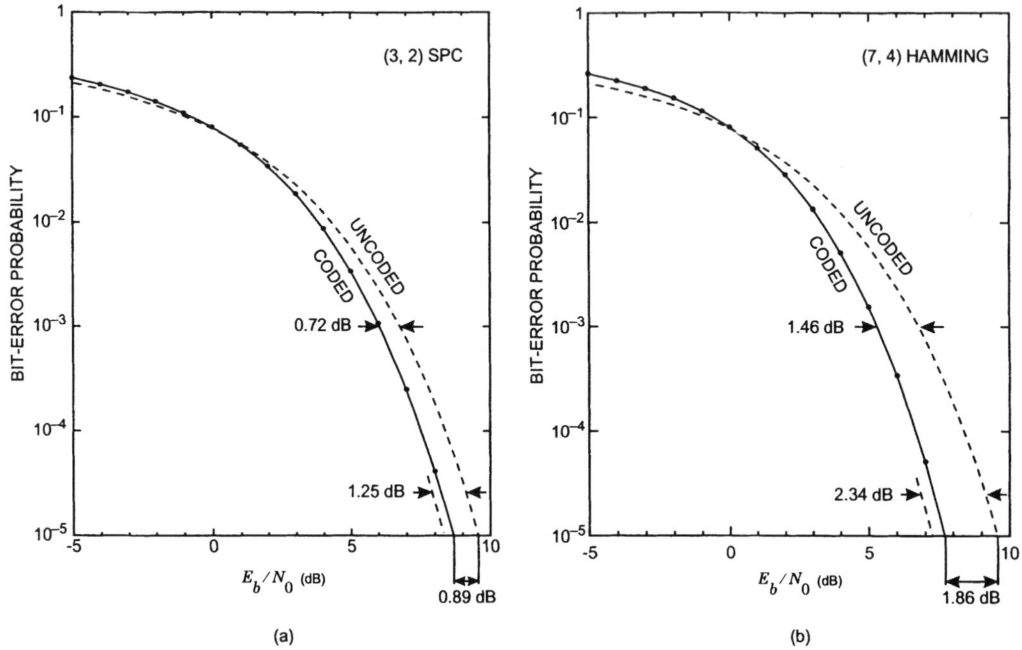

Fig. 12-6. Illustration of coding gain with soft decoding for (a) the (3, 2) single-parity-check code, and (b) the (7, 4) Hamming code, assuming AWGN with binary signaling. Also shown is the bit-error probability for an uncoded system. In (a), we see that the coding gain for the (3, 2) single-parity-check code is 0.89 dB at $P_b = 10^{-5}$, which is close to the 1.25 dB predicted by the asymptotic coding gain. The coding gain for the (7, 4) Hamming code is 1.86 dB at $P_b = 10^{-5}$, about 0.5 dB short of the 2.34 dB asymptotic coding gain.

Example 12-7.

For the (7, 4) Hamming code (Example 12-4), the code rate is $R = 4/7$, and the minimum Hamming weight of all nonzero codewords is $d_{\min} = 3$. This is easily seen by considering all possible linear combinations of the rows of the generator matrix. Hence, the asymptotic coding gain with soft decoding is:

$$Rd_{\min} = \frac{12}{7} = 2.34 \text{ dB}. \tag{12.22}$$

The bit-error probability of the (7, 4) Hamming code with soft decoding is shown in Fig. 12-6(b). The coding gain grows from 1.46 dB at $P_b = 10^{-3}$ to is 1.86 dB at $P_b = 10^{-5}$. The 2.34 dB predicted by the asymptotic coding gain becomes more and more accurate as the bit-error probability gets smaller. More precise comparisons can be made. However, it is clear that the asymptotic coding gain is accurate to within a fraction of a dB for reasonable probabilities of error.

The soft decoder can be expensive to implement because computing the distance between the observed vector and each possible codeword is usually impractical for large n. Fortunately, practical algorithms have been developed [7][8][9][10][11].

12.2.2. Performance of Hard Decoders

A hard decoder operates on the output of the slicer. If the discrete-time channel up to the slicer has a noise generation model with independent noise components, then after the slicer the equivalent binary channel will usually be a BSC (Fig. 7-2). With a code of rate R and binary antipodal signaling in AWGN, the crossover probability is:

$$p = Q\left(\sqrt{2R\frac{E_b}{N_0}}\right). \tag{12.23}$$

Clearly, this crossover probability *increases* as the code rate decreases. In other words, the introduction of a code makes the "raw" bit-error rate worse. A useful code will more than compensate for this increase, resulting in an overall bit-error rate after decoding that is smaller.

Let $\boldsymbol{c}$ denote the transmitted codeword and let $\boldsymbol{r}$ denote the corresponding bits emerging from the BSC. In this case we showed in Chapter 7 that the ML detector selects the codeword $\hat{\boldsymbol{c}}$ closest in Hamming distance to $\boldsymbol{r}$.

Example 12-8.

For the (7, 4) Hamming code, if $\boldsymbol{c} = 0000000$ is transmitted and $\boldsymbol{r} = 0001010$ is received, the ML detected codeword is $\hat{\boldsymbol{c}} = 0001011$, which is closer in Hamming distance than the all-zero codeword.

The question that arises now is how many bit errors can be corrected by an ML detector for a given code. It is clear that if $\boldsymbol{r}$ is closer to $\boldsymbol{c}$ than to any other codeword then any errors in $\boldsymbol{r}$ will be corrected by the ML detector. Certainly if $\boldsymbol{r}$ has fewer than

$$t = \left\lfloor \frac{d_{H,\min} - 1}{2} \right\rfloor \tag{12.24}$$

errors then those errors can be corrected, where $\lfloor \cdot \rfloor$ denotes the "floor" function, or the greatest integer less than or equal to the argument. This value t appeared before (see (7.22)).

Example 12-9.

In the single-parity-check code of Example 12-2 and Example 12-3, each codeword has an even number of ones, so $d_{H,\min} = 2$. The code can correct

$$t = \left\lfloor \frac{d_{H,\min} - 1}{2} \right\rfloor = 0 \tag{12.25}$$

bit errors. Hence this code is not useful at all for hard error correction. It can detect any odd number of bit errors. Note that with soft decoding, this code is useful for reducing the error rate at a given signal level, but not with hard decoding.

Example 12-10.

The (7, 4) Hamming code has $d_{H,\min} = 3$ and $t = 1$, and hence can correct all single bit errors.

It is difficult to quantify analytically the coding gain when hard decoding is used, but we can get a rough estimate by making several approximations. In Section 7.2 we derived upper and lower bounds on the probability of error for vectors of bits transmitted over a BSC. We found the following union-bound approximation (7.25) for the probability of a block error:

$$\Pr[\text{block error}] \approx K\sum_{i=t+1}^{d_{\min}} \binom{d_{\min}}{i} p^i(1-p)^{d_{\min}-i}, \tag{12.26}$$

where $d_{\min}$ is the minimum Hamming distance of the code. For small p, the first term dominates, so that this can be approximated by:

$$\Pr[\text{block error}] \approx K\binom{d}{t+1} p^{t+1}. \tag{12.27}$$

Using $p = Q(\sqrt{2RE_b/N_0})$ and $Q(x) \approx e^{-x^2/2}$ yields:

$$\Pr[\text{block error}] \approx K\binom{d}{t+1} e^{-R(t+1)E_b/N_0}. \tag{12.28}$$

Using the same approximation $Q(x) \approx e^{-x^2/2}$ for the *uncoded* case yields:

$$\Pr[\text{block error, uncoded}] \approx kQ(\sqrt{2E_b/N_0}) \approx ke^{-E_b/N_0}. \tag{12.29}$$

Comparing the uncoded and coded cases and ignoring the multiplying coefficients leads to an asymptotic coding gain for hard decoding of:

$$\text{asymptotic coding gain}_{(\text{hard})} = R(t+1). \tag{12.30}$$

This is almost half the asymptotic coding gain of $Rd_{\min}$ that arises from soft decoding. This approximate analysis suggests that hard ML decoding should perform almost 3 dB worse than soft ML decoding. In reality the difference is usually closer to 1.5 dB or 2 dB.

Example 12-11.

Like all Hamming codes, the (15, 11) Hamming code has $d_{\min} = 3$ and $t = 1$, and hence can correct all single bit errors. The asymptotic coding gains for hard and soft decoding are thus:

$$R(t+1) = \frac{22}{15} = 1.66 \text{ dB (hard)}, \qquad Rd_{\min} = \frac{33}{15} = 3.42 \text{ dB (soft)}. \tag{12.31}$$

The anticipated difference is thus 1.76 dB. The bit-error probability for this code with hard decoding over the BSC can be found analytically through a straightforward but tedious counting exercise, yielding:

$$P_{b,\text{hard}} = 21p^2 + 119p^3 + 392p^4 + 1036p^5 + 2093p^6 + 3067p^7 + 3368p^8 + \\ + 2912p^9 + 1967p^{10} + 973p^{11} + 336p^{12} + 84p^{13} + 15p^{14} + p^{15}, \tag{12.32}$$

where $p = Q(\sqrt{2\frac{11}{15}E_b/N_0})$. The bit-error probability with soft decoding can be found via simulations. The results are shown in Fig. 12-7, which reveals that the actual coding gain for hard and decoding at $P_b = 10^{-5}$ are:

$$\text{coding gain} = 1.2 \text{ dB (hard)}, \qquad \text{coding gain} = 2.6 \text{ dB (soft)}. \tag{12.33}$$

In particular, we see that the soft decoder outperforms the hard decoder by 1.4 dB.

The union-bound estimate is not necessary for some codes. For any code we can be *certain* of correcting up to t bit errors, but *some* error patterns with more bit errors may be correctable also, unless the code is a so-called *perfect* code. A perfect binary code has the property that all bit patterns of length n are within Hamming distance t of one and only one codeword. All Hamming codes are perfect. The performance of the ML detector is easy to determine for perfect codes. With independent noise components, the probability of m bit errors in a block of n bits has a binomial distribution. For perfect codes, a block decoding error is sure to occur if more than t bit errors occur, so

$$\Pr[\text{block error}] = \sum_{m=t+1}^{n} \binom{n}{m} p^m (1-p)^{n-m}$$
$$= 1 - \sum_{m=0}^{t} \binom{n}{m} p^m (1-p)^{n-m} . \tag{12.34}$$

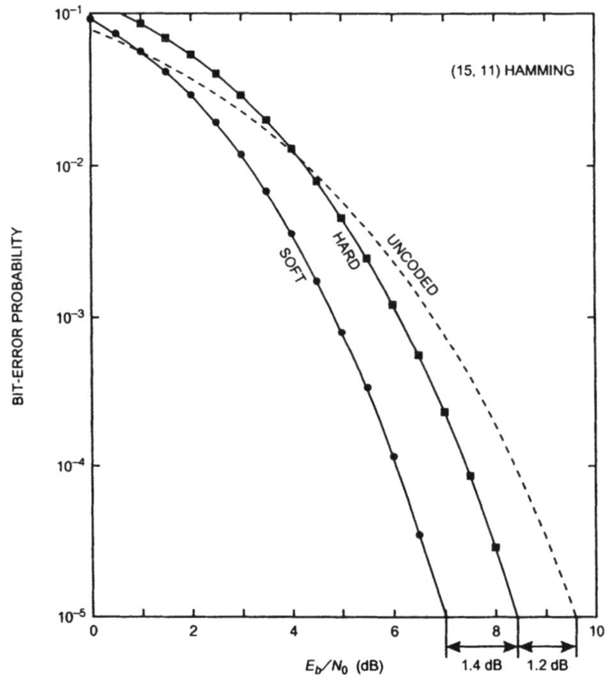

Fig. 12-7. A comparison of bit-error probability for the (15, 11) Hamming code with soft and hard decoding on the AWGN channel with binary signaling. At $P_b = 10^{-5}$, the coding gain is 1.2 dB with hard decoding, and 2.6 dB with soft decoding.

Example 12-12.

The (7, 4) Hamming code is a perfect code, so all seven-bit patterns are either a codeword or one bit distant from exactly one codeword. It follows that

$$\Pr[\text{block error}] = 1 - (1-p)^7 - 7p(1-p)^6 . \tag{12.35}$$

The usefulness of (12.34) is limited to perfect codes. For those codes that are not perfect, (12.34) is an upper bound,

$$\Pr[\text{block error}] \le \sum_{m=t+1}^{n} \binom{n}{m} p^m (1-p)^{n-m} , \tag{12.36}$$

because *some* error patterns with more than t bits errors will be corrected by the ML decoder. This bound often gives a good estimate. Many practical codes are *quasiperfect*, meaning that although some error patterns with $t + 1$ bit errors are corrected, none with $t + 2$ or more are corrected. For these we can get a lower bound:

$$\Pr[\text{block error}] \ge \sum_{m=t+2}^{n} \binom{n}{m} p^m (1-p)^{n-m} . \tag{12.37}$$

Together these upper and lower bounds lead to good estimates of the performance of codes that are sometimes easier to use than the bounds of Section 7.2.

Many other bounds, both tighter and looser, are known. We refer the interested reader to the extensive coding literature.

12.2.3. Parity-Check Matrix

ML soft and hard decoders find the codeword closest (in Euclidean or Hamming distance) to the received block. Direct implementation becomes difficult for large k and n, since there are 2^k distances that need to be computed and compared. Fortunately, for hard decoding, efficient techniques have evolved. The basic approach is to design the code to have a rich *algebraic* structure, and then exploit that structure in the decoding process. Recently, algebraic techniques have also been applied to soft decoding. Although we cannot give a comprehensive treatment of decoding techniques, we can at least illustrate some of the most important concepts.

Consider a systematic linear (n, k) binary block code, which has a generator matrix of the form

$$\mathbf{G} = [\, \mathbf{I}_k \mid \mathbf{P} \,] . \tag{12.38}$$

Given a row vector $\boldsymbol{m}$ of k bits, the corresponding codeword is $\boldsymbol{c} = \boldsymbol{m}\mathbf{G}$, a row vector which can be written

$$\boldsymbol{c} = [\, \boldsymbol{m} \,,\, \boldsymbol{x} \,] , \tag{12.39}$$

where $\boldsymbol{x} = \boldsymbol{m}\mathbf{P}$ is a row vector with $n - k$ parity-check bits. Note that since $\boldsymbol{x} = \boldsymbol{m}\mathbf{P}$,

$$\boldsymbol{m}\mathbf{P} \oplus \boldsymbol{x} = \mathbf{0} , \tag{12.40}$$

where the modulo-two addition is performed element-wise. This can be written

$$[\,\boldsymbol{m}\,,\ \boldsymbol{x}\,]\begin{bmatrix}\mathbf{P}\\ \mathbf{I}_{n-k}\end{bmatrix}=\mathbf{0}\,, \tag{12.41}$$

or

$$\boldsymbol{c}\mathbf{H}^T=\mathbf{0} \tag{12.42}$$

where

$$\mathbf{H}=[\,\mathbf{P}^T\mid\mathbf{I}_{n-k}\,]\ . \tag{12.43}$$

$\mathbf{H}$ is called a *parity-check matrix*, because it can be used to test if a vector $\boldsymbol{c}$ is a codeword by checking (12.42).

Example 12-13.

The parity-check matrix for the $(k+1, k)$ single-parity-check code of Example 12-2 and Example 12-3 is

$$\mathbf{H}=[1\ \ 1\ \ \ldots\ \ 1\ \ 1]\ . \tag{12.44}$$

As we already knew, we can check a bit vector to see if it is a codeword by summing (modulo-two) all of the bits and checking to see if the sum is zero.

Example 12-14.

A parity-check matrix for the (7, 4) Hamming code of Example 12-4 is

$$\mathbf{H}=\begin{bmatrix}1&1&1&0&1&0&0\\0&1&1&1&0&1&0\\1&1&0&1&0&0&1\end{bmatrix}. \tag{12.45}$$

An example of a codeword is $\boldsymbol{c}_1=[0\ 1\ 1\ 0\ 0\ 0\ 1]$, which satisfies $\boldsymbol{c}_1\mathbf{H}^T=\mathbf{0}$. By contrast, $\boldsymbol{c}_2=[0\ 1\ 1\ 0\ 1\ 0\ 1]$ is not a codeword, since $\boldsymbol{c}_2\mathbf{H}^T=[1\ 0\ 0]$.

Although we have shown only how to get a parity-check matrix given the generator matrix of a systematic code, it is possible to find one for any linear block code. In fact, the parity-check matrix can be a compact and useful representation of the code.

For a received vector $\boldsymbol{r}$, calculation of the distance to every codeword can be avoided by using a parity-check matrix. Write

$$\boldsymbol{r}=\boldsymbol{c}\oplus\boldsymbol{e} \tag{12.46}$$

where $\boldsymbol{c}$ is the transmitted codeword and $\boldsymbol{e}$ is the error pattern. Define the *syndrome* to be

$$\boldsymbol{s}=\boldsymbol{r}\mathbf{H}^T=\boldsymbol{c}\mathbf{H}^T+\boldsymbol{e}\mathbf{H}^T=\boldsymbol{e}\mathbf{H}^T. \tag{12.47}$$

Thus the syndrome depends only on the error pattern $\boldsymbol{e}$ and not on the transmitted codeword. The syndrome is zero if and only if $\boldsymbol{e}$ is a codeword (recall that $\mathbf{0}$ is always a codeword of a linear code). Efficient decoders use the syndrome to represent the error pattern, which can then be corrected.

12.2.4. Hamming Codes

Quite a variety of block codes have been developed, each with its advantages and disadvantages. We will describe a very small subset, beginning with Hamming codes.

Hamming codes of length n are designed to correct any single-bit error within a block of n bits. This is possible only if the syndrome for each such error pattern is distinct and nonzero. (The all-zero syndrome is already reserved for the all-zero codeword.) Equivalently, in light of (12.47), this can happen only if the columns of the parity-check matrix **H** are nonzero and distinct. If m is the number of rows in **H**, then there can be at most $2^m - 1$ distinct and nonzero columns in **H**. When all such columns are used, the resulting code is known as a *Hamming code*. The (7, 4) Hamming code is one example for which $m = 3$. For any positive integer m, there exists a Hamming code with parameters

$$(n, k) = (2^m - 1,\ 2^m - 1 - m)\ . \tag{12.48}$$

The rate of the code approaches unity as m grows large.

Example 12-15.
We have already seen the parity-check matrix for the (7, 4) Hamming code:

$$\mathbf{H} = \begin{bmatrix} 1\,1\,1\,0\,1\,0\,0 \\ 0\,1\,1\,1\,0\,1\,0 \\ 1\,1\,0\,1\,0\,0\,1 \end{bmatrix}. \tag{12.49}$$

The columns consist of all possible nonzero three-bit vectors, here arranged in systematic form.

Once **H** is found, the generator matrix **G** for a systematic code can be found by comparing (12.43) and (12.38).

Exercise 12-2.
Show that every Hamming code has $d_{H,\min} = 3$. Consequently, all Hamming codes can correct single errors ($t = 1$).

Hamming codes are perfect codes with minimum distance 3, meaning that every length n bit pattern has distance 0 or 1 from exactly one codeword.

12.2.5. Cyclic Codes

Many practical block codes are *cyclic codes*, which have rich algebraic properties that lead to efficient encoding and decoding techniques. Some of the flavor of this algebra can be obtained from the discussion of Galois fields in Appendix 12-A. See Section 8.6 for some further reading on this topic.

An (n, k) linear block code is said to be *cyclic* if any cyclic shift of a codeword produces another codeword. All Hamming codes can be put into cyclic form.

Exercise 12-3.
Verify that all cyclic shifts of 1000101, namely

1100010 0110001 1011000 0101100 0010110 0001011 1000101

are codewords of the (7, 4) Hamming code.

The algebraic properties of cyclic codes permit collapsing the information contained by the generator matrix into a single polynomial, not surprisingly called the *generator polynomial.* Manipulations of this polynomial representation are powerful, permitting the synthesis of good codes and efficient coding and decoding techniques.

12.2.6. BCH and Reed-Solomon Codes

BCH codes, named after the inventors, Bose, Ray-Chaudhuri, and Hocquenghem, are a large class of multiple-error-correcting codes invented around 1960. For any positive integers m and t, there is a t-error-correcting binary BCH code with

$$n = 2^m - 1\,, \qquad k \ge n - mt\,. \tag{12.50}$$

In order to correct t errors, it is clear that the minimum Hamming distance is bounded by

$$d_{H,\min} \ge 2t + 1\,. \tag{12.51}$$

BCH codes are important primarily because practical and efficient decoding techniques have been found [13], and because of the flexibility in the choice of parameters (n and k).

An important class of nonbinary BCH codes are *Reed-Solomon* codes, in which the symbols are blocks of bits. Their importance is again the existence of practical decoding techniques, as well as their ability to correct bursts of errors.

12.2.7. Maximal-Length Shift Register Codes

In order to give a taste of cyclic codes without getting involved with the algebraic techniques that are required for a general treatment, consider a class of codes called *maximal-length shift register codes*. They are practically much less important the BCH and Reed-Solomon codes, but can be described without introducing any new techniques. Maximal-length shift registers are described in Appendix 12-B.

Example 12-16.
A maximal-length feedback shift register with $m = 4$ stages is shown in Fig. 12-8.

A block coder using a circuit such as that in Fig. 12-8 operates as follows: source bits are divided into blocks of length m. The shift register is loaded with these bits and clocked $2^m - 1$ times. The output from the circuit is then regarded as a codeword of length $2^m - 1$. The result is an $(n, k) = (2^m - 1, m)$ block code. By picking the appropriate output from the circuit, the code is easily made systematic.

Example 12-17.

A systematic linear (15, 4) block code is generated by the system illustrated in Fig. 12-8 if the output codeword is $c_k = X_{k-4}$. The first four bits out are the source bits.

The rate of maximal-length shift register codes is

$$k/n = m/(2^m - 1)\,, \tag{12.52}$$

which becomes very small as m becomes large. This limits the usefulness of these codes.

Exercise 12-4.

Show that a systematic maximal-length shift register code is a parity check code, and hence is linear.

Exercise 12-5.

Show that the Hamming weight of all nonzero codewords of a maximal-length shift register code is 2^{m-1}, where m is the length of the shift register.

Example 12-18.

In the (15, 4) maximal-length shift register code, all nonzero codewords have Hamming weight 8. Since the code is linear, the minimum Hamming distance is 8, and the code can correct up to 3 bit errors.

Maximal-length shift register codes are cyclic codes. This is easily seen by examining a state transition diagram of a maximal-length shift register.

Example 12-19.

A state transition diagram for the $m = 4$ maximal-length shift register is shown in Fig. 12-9. From its circular structure we see that the initial condition determines where in the circle to start. Consequently, every nonzero codeword is a cyclic shift of every other nonzero codeword.

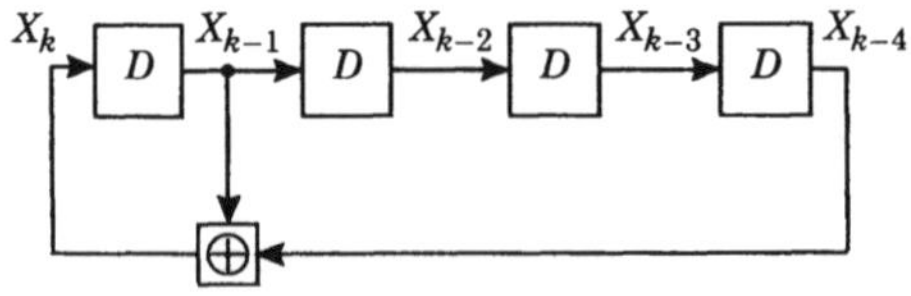

Fig. 12-8. A feedback shift register with $m = 4$ stages. If the shift register is loaded with an initial 4-bit pattern, then as the shift register is clocked, the output will be a periodic bit sequence with period $2^m - 1 = 15$. This example is a maximal-length feedback shift register.

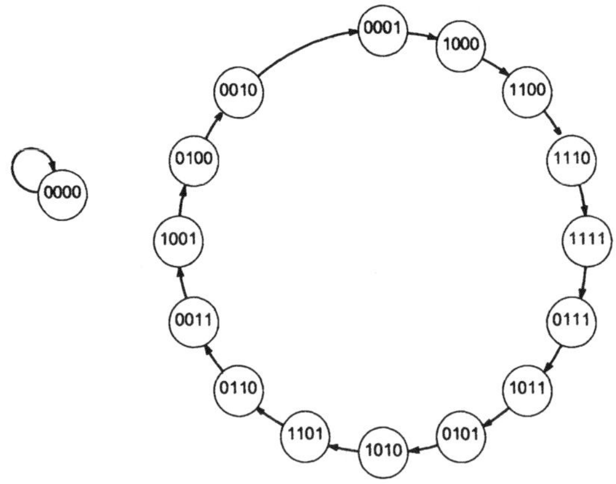

Fig. 12-9. A state transition diagram for the circuit in Fig. 12-8. From its circular structure we see that the block code that it generates is cyclic.

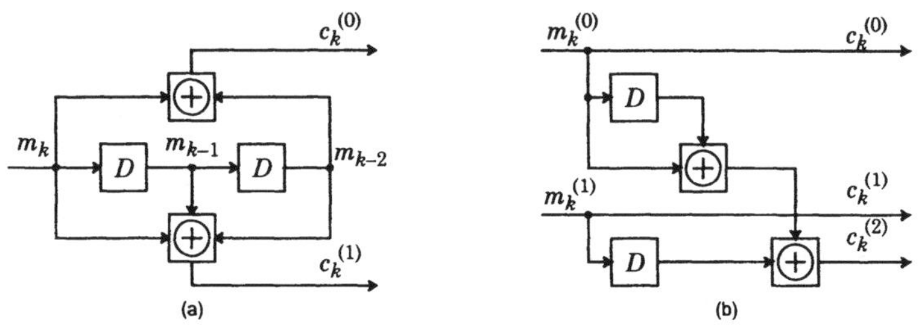

Fig. 12-10. Two convolutional coders. (a) A rate-$1/2$ convolutional coder denoted conv($1/2$). (b) A rate-$2/3$ convolutional coder denoted conv($2/3$). The coder in (b) is systematic because the source bits appear directly in the coded bits.

12.3. Convolutional Codes

A *convolutional coder* is a finite-memory system (rather than a memoryless system, as in the case of the block coder). The name refers to the fact that the added redundant bits are generated by modulo-two convolutions.

Example 12-20.

Two convolutional coder examples are shown in Fig. 12-10. We will use these example coders often in this section.

Convolutional codes are often preferred to block codes, primarily because they are conceptually and practically simpler, and their performance generally exceeds that of good block codes when complexity is constrained [31]. Their good performance is attributable in part to the availability of practical soft decoding techniques. Convolutional codes become particularly simple to describe once block codes are understood.

In describing convolutional codes it is convenient to adopt the *D-transform* notation. Given any binary sequence b_k (deterministic or random), the modulo-two D-transform is

$$b(D) = \ldots \oplus b_{-1}D^{-1} \oplus b_0 \oplus b_1 D \oplus b_2 D^2 \oplus \ldots \tag{12.53}$$

where $\oplus$ denotes modulo-two addition. In other words, it is just like a Z-transform, except that the additions are modulo-two and the symbol D is used instead of z^{-1}. Now any convolution of two sequences

$$c_k = g_k * b_k \tag{12.54}$$

can be written in the "D-domain" as

$$cD) = g(D)b(D) . \tag{12.55}$$

Just as for block codes, linear convolutional coders are constructed using modulo-two adders with the addition of delay elements. A convolutional coder can be described using a *generator matrix* where instead of the entries being zero or one, the entries are polynomials in D with coefficients that are either zero or one.

Example 12-21.

The generator matrix for the conv(1 / 2) coder of Fig. 12-10(a) is

$$\mathbf{G}(D) = [1 \oplus D^2,\ 1 \oplus D \oplus D^2] , \tag{12.56}$$

and the generator matrix for conv(2 / 3) in Fig. 12-10(b) is

$$\mathbf{G}(D) = \begin{bmatrix} 1 & 0 & 1 \oplus D \\ 0 & 1 & D \end{bmatrix} . \tag{12.57}$$

The entries in the generator matrix are transfer functions. Define the row vectors

$$\boldsymbol{m}(D) = [m^{(0)}(D), \ldots, m^{(k-1)}(D)] \tag{12.58}$$

$$\mathbf{c}(D) = [c^{(0)}(D), \ldots, c^{(n-1)}(D)] \tag{12.59}$$

where $m^{(i)}(D)$ and $c^{(i)}(D)$ are modulo-two D-transforms of $m_k^{(i)}$ and $c_k^{(i)}$ (see Fig. 12-10). The convolutional coder is defined by the matrix D-transform relation

$$\mathbf{c}(D) = \boldsymbol{m}(D)\mathbf{G}(D) . \tag{12.60}$$

The notation for a general convolutional coder is summarized in Fig. 12-11. Just as with block codes, a convolutional code has a *parity-check matrix*.

Example 12-22.

Comparing (12.57) with (12.38) and (12.43), a parity-check matrix for conv(2 / 3) is

$$\mathbf{H}(D) = [\ 1 \oplus D, D, 1\ ] . \tag{12.61}$$

Exercise 12-6.

Verify that for conv(2/3) and $\mathbf{H}(D)$ given by (12.61),

$$\mathbf{c}(D)\mathbf{H}^T(D) = \mathbf{0} \tag{12.62}$$

for all code sequences $\mathbf{c}(D)$.

Just as with block codes, the parity-check matrix is a compact specification of the code. If the code is systematic, the generator matrix is easy to derive from the parity-check matrix, or vice versa. If the code is not systematic, then some modulo-two algebra may be required.

Example 12-23.

Consider again conv(1/2). From (12.56) and (12.60) we know that

$$\begin{aligned} c^{(0)}(D) &= (1 \oplus D^2)m(D) \\ c^{(1)}(D) &= (1 \oplus D \oplus D^2)m(D). \end{aligned} \tag{12.63}$$

Multiply both sides of the first equation by $(1 \oplus D \oplus D^2)$ and of the second equation by $(1 \oplus D^2)$ and notice that the two right hand sides are equal. Hence the left hand sides are equal,

$$c^{(0)}(D)(1 \oplus D \oplus D^2) = c^{(1)}(D)(1 \oplus D^2)\ , \tag{12.64}$$

or

$$c^{(0)}(D)(1 \oplus D \oplus D^2) \oplus c^{(1)}(D)(1 \oplus D^2) = 0\ . \tag{12.65}$$

Hence a parity-check matrix is

$$\mathbf{H}(D) = [1 \oplus D \oplus D^2,\ 1 \oplus D^2]\ . \tag{12.66}$$

The *memory* μ of a nonrecursive convolutional code may be defined as the maximum degree of the polynomials in the generator matrix (the maximum length of any impulse response):

$$\mu = \max_{i,j}\ [\deg(g_{ij}(D))]\ . \tag{12.67}$$

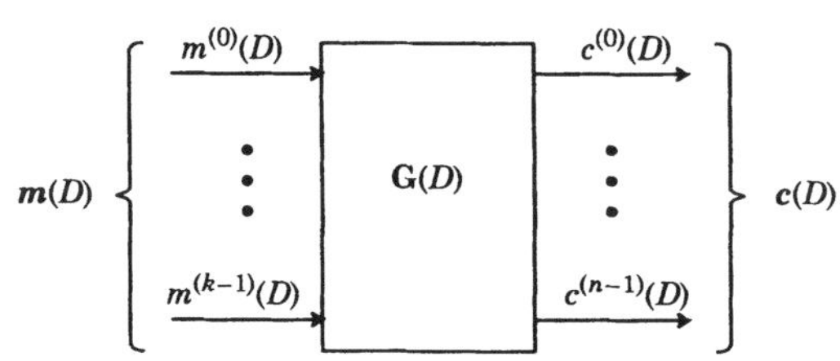

Fig. 12-11. A general convolutional coder. The modulo-two D-transform input and output sequences are shown, as is the generator matrix.

The *constraint length* of a nonrecursive convolutional code is $\mu + 1$. The conv(1/2) coder in Fig. 12-10(a) has memory $\mu = 2$ and constraint length 3, while conv(2/3) in Fig. 12-10(b) has memory $\mu = 1$ and constraint length 2. Note that both encoders have only two memory elements. (These definitions of memory and constraint length are common but not universal in the literature.)

Given a parity-check matrix, it is often easy to design a systematic encoder.

Example 12-24.

Given the parity-check matrix for conv(2/3) discussed previously, namely

$$\mathbf{H}(D) = [1 \oplus D,\ D,\ 1]\ , \tag{12.68}$$

we will now derive the encoder shown in Fig. 12-10(b). From (12.62) we know that

$$(1 \oplus D)c^{(0)}(D) \oplus Dc^{(1)}(D) \oplus c^{(2)}(D) = 0\ . \tag{12.69}$$

To get a systematic code, assign

$$c^{(0)}(D) = m^{(0)}(D) \qquad \text{and} \qquad c^{(1)}(D) = m^{(1)}(D)\ . \tag{12.70}$$

Then note from (12.69) that

$$c^{(2)}(D) = (1 \oplus D)c^{(0)}(D) \oplus Dc^{(1)}(D)\ , \tag{12.71}$$

or

$$c^{(2)}(D) = (1 \oplus D)m^{(0)}(D) \oplus Dm^{(1)}(D)\ . \tag{12.72}$$

In the time domain this is

$$c_k^{(2)} = m_k^{(0)} \oplus m_{k-1}^{(0)} \oplus m_{k-1}^{(1)}\ , \tag{12.73}$$

which is implemented in Fig. 12-10(b).

Given a parity-check matrix, it is not always quite so simple to find a systematic implementation.

Example 12-25.

Consider the parity-check matrix for conv(1/2) given in (12.66). The encoder given in Fig. 12-10(a) is not systematic, but there is an encoder that is systematic and has the same parity-check matrix. In all important respects the resulting code will be equivalent. From (12.62),

$$(1 \oplus D \oplus D^2)c^{(0)}(D) \oplus (1 \oplus D^2)c^{(1)}(D) = 0\ . \tag{12.74}$$

Again, to get a systematic code, set $c^{(0)}(D) = m(D)$. From (12.74),

$$(1 \oplus D \oplus D^2)m(D) = (1 \oplus D^2)c^{(1)}(D)\ . \tag{12.75}$$

Given this equation we can construct the system shown in Fig. 12-12(a) that generates $c_k^{(1)}$ from m_k. It is a cascade of two linear subsystems, the order of which can be reversed to get the implementation in Fig. 12-12(b). Those readers familiar with the design of recursive digital filters will recognize this procedure.

The above generalizes so that any rate-$1/2$ *nonrecursive, nonsystematic* convolutional encoder of the form $\mathbf{G}(D) = [g_1(D)\ g_2(D)]$ can be transformed into a *recursive, systematic* convolutional encoder $\mathbf{G}'(D) = [1, g_2(D)/g_1(D)]$ that generates the same set of codewords. The encoder of Fig. 12-12 is an example of a recursive systematic encoder. The encoder is recursive because of the feedback path. In signal-processing parlance, a recursive encoder is an infinite-impulse response (IIR) filter. As we will see in Section 12.5, such encoders play a crucial role in turbo codes.

For every k input bits, n bits are produced by the coder, so the *rate* of the convolutional coder is $R = k/n$. As with block codes, the rate is defined as the ratio of the input bit rate to output bit rate. In the performance calculations to follow, assume that the output bit stream is transmitted by a binary symbol alphabet, requiring an increase in symbol rate due to the coding.

The representation of convolutional codes using generator matrices highlights their similarity to block codes. However, to do ML detection of the coded sequence, it is more convenient to represent convolutional codes as Markov chains. It is clear from Fig. 12-10 that a convolutional code is a *shift register process*. If the input sequence $\boldsymbol{m}_k$ is i.i.d., then the output $\boldsymbol{c}_k$ of the coder is a Markov chain, and we can define the state of the Markov chain to be the bits stored in the delay elements in the coder.

Example 12-26.

For conv$(1/2)$, define the state of the Markov chain at time k to be the previous two message bits:

$$\Psi_k = [m_{k-1}, m_{k-2}] \,. \tag{12.76}$$

The state transition diagram is shown in Fig. 12-14, and the trellis diagram is shown in Fig. 12-14. Recall the trellis illustrates the progression through states over time. The transitions in the trellis in Fig. 12-14(b) are labeled with the (input, output) pairs (m_k, $[c_k^{(0)}, c_k^{(1)}]$). In Fig. 12-14(c) they are

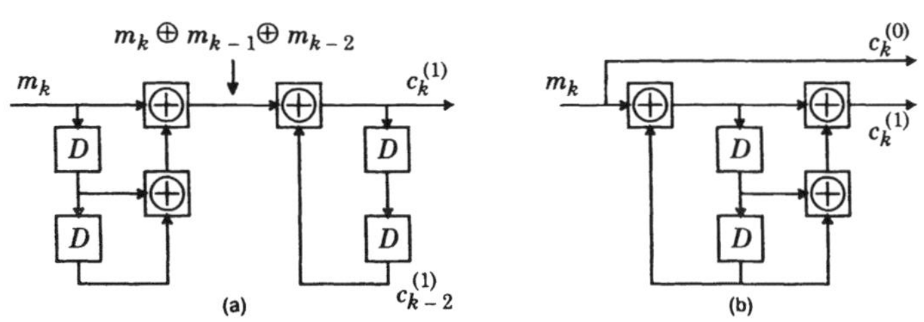

Fig. 12-12. Two versions of a systematic coder that has the same parity-check matrix as conv$(1/2)$. The one on the right is obtained by inverting the order of the two stages in the one on the left and combining the two delay lines into one. These are examples of *recursive-systematic* encoders.

labeled with the transmitted symbols instead of the coded bits, assuming binary antipodal signaling.

Example 12-27.

For conv(2/3) the state is

$$\Psi_k = [m_{k-1}^{(0)}, m_{k-1}^{(1)}] \ . \tag{12.77}$$

The state transition diagram for this code is fully connected, as shown in Fig. 12-15.

If the noise generation model has independent noise components, as we usually assume, then we can use the Viterbi algorithm (Section 7.4) for ML detection of the coded signal. In the soft decoding case with Gaussian noise a Euclidean distance branch metric is appropriate, while in

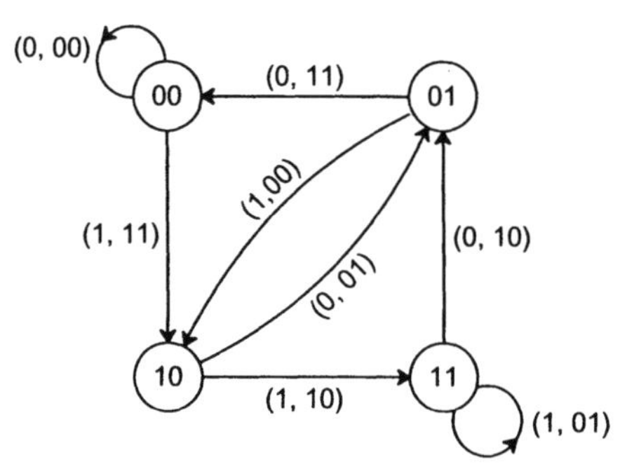

Fig. 12-13. The state transition diagram of conv(1/2) of Fig. 12-10(a). The arcs are labeled (b_k, [$c_k^{(0)}$, $c_k^{(1)}$]), where b_k is the input bit that triggers the transition and $c_k^{(0)}$ and $c_k^{(1)}$ are the outputs produced.

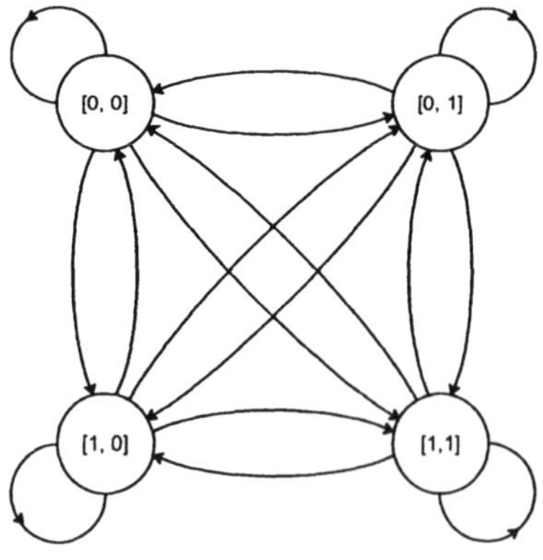

Fig. 12-15. A state transition diagram for a Markov chain modeling the convolutional coder of Fig. 12-10(b).

the hard decoding case with a BSC it is Hamming distance. Alternatively, we can perform APP decoding using the BCJR algorithm (Section 7.5). We will see in Section 12.5 that a key building block of turbo codes are rate-1/2 recursive systematic convolutional encoders similar to the one shown in Fig. 12-12(b).

12.3.1. Performance of Soft Decoders

A coded transmission system with a soft decoder is shown in Fig. 12-16. Assume additive white Gaussian noise with variance $\sigma^2 = N_0/2$, and binary antipodal signaling with alphabet $\mathcal{A} = \{\pm\sqrt{RE_b}\}$.

Example 12-28.
For conv(1/2) of Fig. 12-10(a), the coded system transmits at twice the symbol rate, so there is half as much energy available for each symbol, namely:

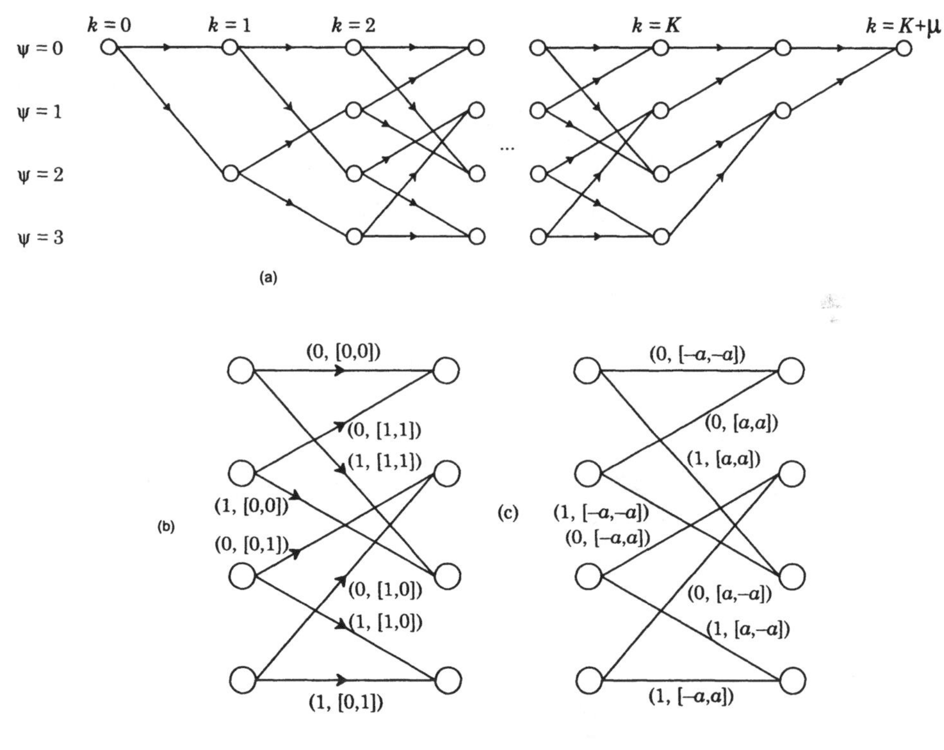

Fig. 12-14. (a) A four-state trellis illustrating all possible state transitions of the Markov chain in Fig. 12-13, which represents conv(1/2), assuming the initial and termination states are zero. (b) One stage of the trellis is shown with the transitions labeled with $(m_k, [c_k^{(0)}, c_k^{(1)}])$, where m_k is the input bit that triggers the transition and $c_k^{(0)}$ and $c_k^{(1)}$ are the outputs produced. (c) The branches of the trellis labeled with the transmitted symbols rather than the bits out of the coder for binary antipodal signaling with levels $\pm a$, where $a = \sqrt{E}$.

$$E = \tfrac{1}{2}E_b \,. \tag{12.78}$$

We will see that the power of the code more than makes up for this attenuation. Otherwise the coding scheme would not be useful!

For the coded system with soft decoder, instead of labeling trellis branches with output bits, they should be labeled with output *symbols*.

Example 12-29.

For conv(1/2) of Fig. 12-10(a), the trellis is shown in Fig. 12-14(c). Every transition is triggered by an input bit and produces a pair of output symbols. Given the corresponding pair of noisy observations y_k and y_{k+1}, and for a branch corresponding to symbols $[\hat{a}_k,\ \hat{a}_{k+1}]$, the ML detector computes the branch metric

$$d^2 = (y_k - \hat{a}_k)^2 + (y_{k+1} - \hat{a}_{k+1})^2 \,. \tag{12.79}$$

The path metric, the sum of branch metrics, is the square of the Euclidean distance between the observation and the transmitted symbols for that path.

The ML detector for a soft decoder selects the path through the trellis with the minimum path metric, as shown in Section 7.4. From Section 7.6, the probability of symbol error at high SNR is

$$\Pr[\text{symbol error}] \approx CQ(d_{\min}/2\sigma) \,, \tag{12.80}$$

where $d_{\min}$ is the minimum Euclidean distance of an error event and C is a constant between P and R given in Appendix 7-B.

Example 12-30.

For the trellis of Fig. 12-14(c), the error event with the minimum distance is shown in Fig. 12-17 when the correct path is the all-zero sequence. The square of the Euclidean distance between the correct path and the error event is

$$d_{\min}^2 = 20E \,. \tag{12.81}$$

Every possible correct path through the trellis has exactly one minimum distance error event, and each such error event e has one symbol error, $w(e) = 1$, so $P = C = R = 1$, and

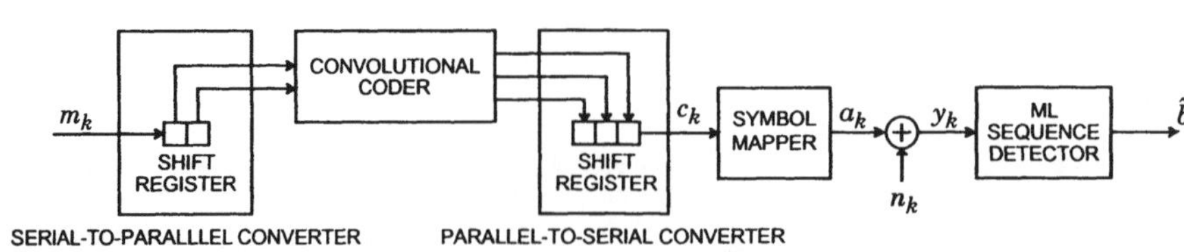

Fig. 12-16. A coded transmission system with a soft decoder (the ML sequence detector).

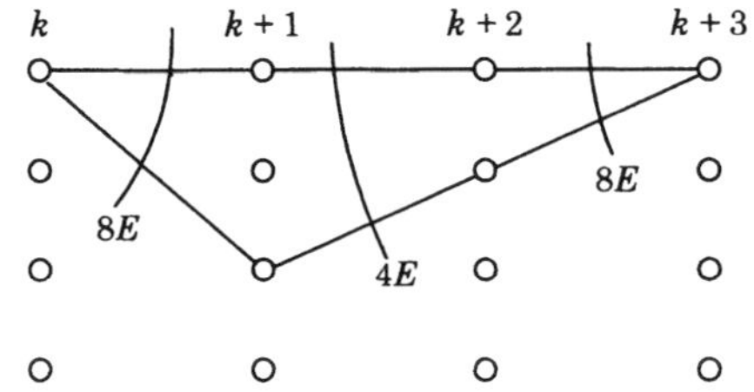

Fig. 12-17. Minimum-distance error event for conv(1/2).

$$\Pr[\text{symbol error}] \approx Q\left(\sqrt{\frac{10E}{N_0}}\right) = Q\left(\sqrt{\frac{5E_b}{N_0}}\right). \tag{12.82}$$

Each symbol represents one bit, so this is also the bit-error probability. In contrast, the bit-error probability for the uncoded case is given by (12.17). Setting the arguments of $Q(\cdot)$ equal for the coded and uncoded case, we see that the coding gain is:

$$\frac{5}{2} = 4 \text{ dB}. \tag{12.83}$$

This is a significant improvement over the uncoded system, and also is considerably better than the specific block codes considered in Section 12.2. The approximations are valid when the probability of error is low.

The above coding gain is nothing more than the product of the code rate ($R = 1/2$) and the minimum-Hamming distance ($d_{\min} = 5$) of the convolutional code. If we consider only high SNR so that we may ignore the error coefficients, the asymptotic coding gain for a general convolutional codes is $Rd_{\min}$, just as it was for binary block codes.

In Example 12-30, finding C was easy because P and R were both equal to one. In general, however, finding P and R is more difficult. Furthermore, finding $d_{\min}$ can be tedious. The general technique described in Section 7.6.2, which uses the Viterbi algorithm to find $d_{\min}$, works for all cases. If the code is linear (Appendix 12-A), then the task is greatly simplified because only one actual path through the trellis must be considered. In this case, either the Viterbi algorithm technique of Section 7.6.2 or the signal flow graph technique of Appendix 12-C can be used. Fortunately, exhaustive searches for encoders with maximal free distance for a given memory have already been performed, and the results are tabulated in the literature [44][45].

12.3.2. Performance of Hard Decoders

Next we compare the hard decoder to both the soft decoder and the uncoded system. For the hard decoder, the channel and receiver front end can again be modeled as a BSC. In this case the appropriate branch metric is the Hamming distance between the received bits and the transmitted bits corresponding to that branch. Each branch has a set of L output bits associated with it ($L = 2$ in conv(1/2) and $L = 3$ in conv(2/3)). The Hamming distance branch metric is an integer between 0 and L. The path metric is the sum of these branch metrics. We can use the Viterbi algorithm to implement the ML detector, which chooses the path through the trellis with minimum path metric. The analysis is similar to the Gaussian noise case, but $Q(\cdot)$ is replaced by $Q(\cdot, \cdot)$ given by (7.23). Suppose the codeword $\boldsymbol{c}$ is transmitted and the channel error probability is p. Then the probability that the received bits are closer in Hamming distance to another codeword $\boldsymbol{r}$ is $Q(d, p)$, where d is the Hamming distance between $\boldsymbol{c}$ and $\boldsymbol{r}$.

Example 12-31.

For the coder conv(1/2) in Fig. 12-10(a), the minimum-distance error event has length $K = 2$ and is the same as for the soft decoder shown in Fig. 12-17. We will bound the probability that this particular error event begins at some time $k = i$. Write the noise-free input codeword as

$$\boldsymbol{c} = [\boldsymbol{c}_i,\ \boldsymbol{c}_{i+1},\ \boldsymbol{c}_{i+2}] \tag{12.84}$$

where each $\boldsymbol{c}_k$ is a pair of bits determined by a state transition. (We can consider codewords of finite length only because we are considering an error event of finite length.) Assume a zero state trajectory, $\psi_k = 0$ for all k, so $\boldsymbol{c} = [(0,0), (0,0), (0,0)]$. The minimum distance error event in Fig. 12-17 has a corresponding codeword $\boldsymbol{r} = [(1,1), (0,1), (1,1)]$ and is Hamming distance five from $\boldsymbol{c}$. This is exactly the situation described in Example 7-17! The observation $\boldsymbol{r}$ will be closer to $\boldsymbol{r}$ than to $\boldsymbol{c}$ if three or more bits are changed in the five positions in which the two codewords differ. The probability of this occurring is given by (7.24), so

$$\Pr[\text{this error event}] \le Q(5, p) = 10p^3(1-p)^2 + 5p^4(1-p) + p^5 . \tag{12.85}$$

This is the same bound given in (7.146).

Appendix 7-B shows that when p is small, the probability of a detection error is approximately $CQ(d_{\min}, p)$ (see (7.161)) for some constant C between P and R given by (7.157) and (7.150). As in the Gaussian case, the situation is simple if every possible actual path through the trellis has exactly one minimum distance error event, and that minimum distance error event has exactly one detection error. In this case, $P = R = C = 1$.

Example 12-32.

The minimum distance error event in Example 12-31 has exactly one detection error (the first detected bit of the three erroneous stages will be incorrect, see Fig. 12-14(b)). Furthermore, since the code is linear, every actual path through the trellis also has exactly one minimum distance error event with exactly one detection error. From (7.161) we can assert

$$\Pr[\text{bit error}] = \Pr[\text{detection error}] \approx 10p^3(1-p)^2 + 5p^4(1-p) + p^5 . \tag{12.86}$$

It is assumed that other error events are far less likely than the minimum distance error event. Compare (12.86) to the probability of bit error p of an uncoded system. If $p = 0.1$, then the uncoded system has a probability of bit error of 0.1, while the coded system has probability of bit error approximately 0.0086.

In general, evaluating $Q(d_{\min}, p)$ exactly can be tedious. Fortunately, for p close to zero, the first term in the summation in (7.23) will dominate, so

$$Q(d_{\text{min}}, p) \approx \binom{d_{\text{min}}}{t+1} p^{t+1}(1-p)^{d-t+1} \approx \binom{d_{\text{min}}}{t+1} p^{t+1}\,, \tag{12.87}$$

where $t = \left\lfloor \frac{d_{\text{min}} - 1}{2} \right\rfloor$.

Example 12-33.

Continuing Example 12-32, using (12.87) we get

$$\Pr[\text{bit error}] \approx 10p^3\,, \tag{12.88}$$

as long as p is close to zero. In order to compare hard and soft decoders, we must relate the error probability of the BSC to the SNR on the channel prior to the slicer. To make a rough approximation, we can determine the E_b/N_0 required to achieve a *particular* probability of error, say 10^{-5}. Solving

$$10p^3 = 10^{-5} \tag{12.89}$$

for p we find $p \approx 0.01$. Hence p is close to zero and our approximation is valid for this probability of error. To achieve a crossover probability of $p \approx 0.01$ on an additive Gaussian white noise channel with binary antipodal signaling and a code rate of $1/2$ it is necessary that

$$Q(\sqrt{2(\tfrac{1}{2})E_b/N_0}) \approx 0.01\,, \quad (E_b/N_0)_{\text{coded}} = 7.3 \text{ dB}\,. \tag{12.90}$$

To achieve the same error probability without coding would require

$$Q(\sqrt{2E_b/N_0}) \approx 10^{-5}, \quad (E_b/N_0)_{\text{uncoded}} = 9.6 \text{ dB}\,. \tag{12.91}$$

The coding gain with hard decoding is thus about 2.3 dB, far short of the 4 dB coding gain predicted for soft decoding.

Although our analysis has been limited to simple examples, we conclude that just as with block codes, hard decoders for convolutional codes yield less coding gain than do soft decoders. As with block codes, this does not mean that hard decoders are not used. Their implementation may be simpler, they can improve existing transmission systems with minimal modification, and the gain on channels with other than Gaussian white noise may be better.

12.4. Low-Density Parity-Check Codes

As the name suggests, *low-density parity-check (LDPC)* codes are block codes defined by a parity-check matrix that is sparse. They were first proposed in 1962 by Gallager [4][14], along with an elegant iterative decoding scheme whose complexity grows only linearly with block length. Despite their promise, LDPC codes were largely forgotten for several decades until they were rediscovered by MacKay and Neal [15]. Today the value of LDPC codes is widely recognized. Their remarkable performance ensures that they will not be forgotten again. In contrast to many codes that were invented well after 1962, LDPC codes offer both better performance *and* lower decoding complexity. In fact, it is an irregular LDPC code (with block length 10^7) that currently holds the distinction of being the world's best-performing rate-$1/2$ code, falling only 0.04 dB short of the Shannon limit [16].

12.4.1. Parity-Check Codes

It is useful to think of a parity-check code of length N as a code whose codewords all satisfy a set of M linear parity-check constraints. Such a code is uniquely defined by its $M \times N$ parity-check matrix $\mathbf{H}$, whose M rows specify each of the M constraints. For example, if the first constraint specifies that bits 3 and 7 must be equal, then the first row of $\mathbf{H}$ contains a 1 in position 3 and 7 and zeros elsewhere. The parity-check code is the set of binary vectors satisfying all constraints, i.e., the set of binary vectors $\boldsymbol{c}$ satisfying $\boldsymbol{c}\mathbf{H}^T = \mathbf{0}$. Each linearly independent constraint cuts the number of valid codewords in half. Thus, if $r = \mathrm{rank}(\mathbf{H}) \le M$ is the number of linearly independent rows in $\mathbf{H}$, then the number of codewords is 2^{N-r}, and the code dimension is $K = N - r$. Because each codeword of length N conveys K information bits, the code rate is K/N.

A low-density parity-check (LDPC) code is defined by a parity-check matrix that is sparse [4].

Definition: A *regular* (j, k) *LDPC matrix* is an $M \times N$ binary matrix having exactly j ones in each column and exactly k ones in each row, where $j < k$ and both are small compared to N.

The "low density" terminology stems from the fact that the fraction of ones in a regular (j, k) LDPC matrix is k/N, which approaches zero as $N \to \infty$. An *irregular* [3] LDPC matrix is still sparse, but not all rows and columns contain the same number of ones. To simplify our discussion we will focus on regular LDPC matrices in this section.

By definition, every parity-check equation of a regular LDPC code involves exactly k bits, and every bit is involved in exactly j parity-check equations. The restriction $j < k$ is needed to ensure that more than just the all-zero codeword satisfies all of the constraints, or equivalently, to ensure a nonzero code rate. Indeed, the total number of ones in $\mathbf{H}$ is $Mk = Nj$, since there are M rows, each containing k ones, and there are N columns, each containing j ones. The code rate $R = 1 - M/N$ is then $R = 1 - j/k$, assuming the M rows are linearly independent. The need for $j < k$ is thus clear. For best performance, $j \ge 3$ is also required [4].

An $M \times N$ regular (j, k) LDPC matrix can often (but not always) be conveniently expressed in terms of the following shorter parity-check matrix $\mathbf{H}_0$:

$$\mathbf{H}_0 = \begin{bmatrix} \underbrace{1\;1\;1\ldots1}_{k} & & & \\ & \underbrace{1\;1\;1\ldots1}_{k} & & \\ & & \ddots & \\ & & & \overbrace{1\;1\;1\ldots1}^{k} \end{bmatrix}. \tag{12.92}$$

It has $N/k = M/j$ rows and N columns. The m-th row contains ones in columns $(m-1)k + 1$ through mk and zeros elsewhere. By itself, $\mathbf{H}_0$ would define a code consisting of N/k *independent* single-parity check constraints, with the first row constraining the parity of the first block of k coded bits, the next row constraining the parity of the second block of k bits, etc. In this code, every parity check involves k bits, and each bit is involved in one and only one parity check. Hence, $\mathbf{H}_0$ alone defines a $(1, k)$ regular LDPC code. However, the performance of this code would be poor. In fact, because the first two columns of $\mathbf{H}_0$ are linearly dependent (or equivalently because 0000... 0 and 1100... 0 are both valid codewords), the minimum distance for the code would be two.

We can construct a regular (j, k) LDPC matrix by stacking j column permutations of $\mathbf{H}_0$ one atop another:

$$\mathbf{H} = \begin{bmatrix} \pi_1(\mathbf{H}_0) \\ \pi_2(\mathbf{H}_0) \\ \\ \pi_j(\mathbf{H}_0) \end{bmatrix}. \tag{12.93}$$

Here $\pi_i(\mathbf{H}_0)$ denotes a matrix whose columns are a permuted version of the columns of $\mathbf{H}_0$. Since each row of $\mathbf{H}_0$ has k ones, each row of $\mathbf{H}$ also has k ones. Similarly, since each column of $\mathbf{H}_0$ contains a single one, each column of $\mathbf{H}$ contains j ones. We remark that not all valid (j, k) regular LDPC matrices $\mathbf{H}$ can be expressed in the form of (12.93). We also remark that the building block $\mathbf{H}_0$ in (12.92) was chosen somewhat arbitrarily; any column permutation of $\mathbf{H}_0$ could have been used in its place. For example, we can equivalently use $\mathbf{H}_0 = [\mathbf{I}\ \mathbf{I}\ \mathbf{I} \ldots \mathbf{I}]$ in place of (12.92), where $\mathbf{H}_0$ is a concatenation of k identity matrices $\mathbf{I}_{N/k}$.

A proper choice of the permutations will allow the minimum distance of the code defined by $\mathbf{H}$ to increase beyond two. The prospect of designing j different permutations of length N may at first seem daunting, especially for large N. However, Gallager proved that a totally random choice will on average produce an excellent code [14]. In particular, if each permutation is chosen independently and uniformly from the set of all $N!$ possible permutations, then the expected minimum distance that results will increase linearly with N. A code with such a property is said to be "good." The idea of designing codes randomly did not originate with Gallager, but dates back to Shannon's original work [17]. The beauty of Gallager's design is that, unlike Shannon's random codes, we will see that it is possible to decode Gallager's codes with complexity that grows only linearly with N.

Example 12-34.

The following is an example of a LDPC matrix with word length $N = 20, j = 3$, and $k = 4$ [4]:

$$\mathbf{H} = \left[\begin{array}{cccccccccccccccccccc}
1&1&1&1&0&0&0&0&0&0&0&0&0&0&0&0&0&0&0&0\\
0&0&0&0&1&1&1&1&0&0&0&0&0&0&0&0&0&0&0&0\\
0&0&0&0&0&0&0&0&1&1&1&1&0&0&0&0&0&0&0&0\\
0&0&0&0&0&0&0&0&0&0&0&0&1&1&1&1&0&0&0&0\\
0&0&0&0&0&0&0&0&0&0&0&0&0&0&0&0&1&1&1&1\\ \hline
1&0&0&0&1&0&0&0&1&0&0&0&1&0&0&0&0&0&0&0\\
0&1&0&0&0&1&0&0&0&1&0&0&0&0&0&0&1&0&0&0\\
0&0&1&0&0&0&1&0&0&0&0&0&0&1&0&0&0&1&0&0\\
0&0&0&1&0&0&0&0&0&0&1&0&0&0&1&0&0&0&1&0\\
0&0&0&0&0&0&0&1&0&0&0&1&0&0&0&1&0&0&0&1\\ \hline
1&0&0&0&0&1&0&0&0&0&0&1&0&0&0&0&0&1&0&0\\
0&1&0&0&0&0&1&0&0&0&1&0&0&0&0&1&0&0&0&0\\
0&0&1&0&0&0&0&1&0&0&0&0&1&0&0&0&0&0&1&0\\
0&0&0&1&0&0&0&0&1&0&0&0&0&1&0&0&1&0&0&0\\
0&0&0&0&1&0&0&0&0&1&0&0&0&0&1&0&0&0&0&1
\end{array}\right]. \tag{12.94}$$

The horizontal lines separate $\mathbf{H}_0$ and its two permutations. We see that each column contains $j = 3$ ones, and each row contains $k = 4$ ones. The corresponding permutations are $\pi_1 = [1\ 2\ \ldots\ 20]$, $\pi_2 = [1\ 5\ 9\ 13\ 2\ 6\ 10\ 17\ 3\ 7\ 14\ 18\ 4\ 11\ 15\ 19\ 8\ 12\ 16\ 20]$, and $\pi_3 = [1\ 5\ 9\ 13\ 17\ 2\ 6\ 10\ 14\ 18\ 7\ 3\ 11\ 15\ 19\ 8\ 16\ 4\ 12\ 20]$. It can be shown that the tenth row of $\mathbf{H}$ is the sum of the first nine rows, and that the fifteenth row is the sum of rows one through five and rows eleven through fourteen. Thus rows ten and fifteen are linearly dependent on the remaining rows. It can be shown that the remaining 13 rows are linearly independent. Hence, the rank of $\mathbf{H}$ is 13, the dimension of the LDPC code is $K = 7$, and the code rate is $K/N = 0.35$.

The previous parity-check matrix has 15 rows, but only 13 of them are independent. We could define a new full-rank parity-check matrix $\tilde{\mathbf{H}}$ by eliminating the redundant rows from $\mathbf{H}$. Then $\tilde{\mathbf{H}}$ would describe the same code as $\mathbf{H}$, since $\mathbf{c}\mathbf{H}^T = \mathbf{0}$ if and only if $\mathbf{c}\tilde{\mathbf{H}}^T = \mathbf{0}$. However, the number of ones in k columns of $\tilde{\mathbf{H}}$ would decrease each time a redundant row is removed, so that $\tilde{\mathbf{H}}$ would no longer obey the regularity property of a regular LDPC matrix. For this reason it is sometimes convenient to describe an LDPC code by a rank-deficient but regular LDPC matrix, as opposed to its more efficient but irregular full-rank equivalent.

Any parity-check code (including an LDPC code) may be specified by a *Tanner graph* [18][19], which is essentially a visual representation of the parity check matrix $\mathbf{H}$. Recall that an $M \times N$ parity-check matrix $\mathbf{H}$ defines a code in which the N bits of each codeword satisfy a set of M parity-check constraints. The Tanner graph contains N "bit" nodes, one for each bit, and M "check" nodes, one for each of the parity checks. The bit nodes are depicted using circles, while the check nodes are depicted using squares. The check nodes are connected to the bit nodes they check. Specifically, a branch connects check node m to bit node n if and only if the m-th parity check involves the n-th bit, or more succinctly, if and only if $H_{m,n} = 1$. This means that $\mathbf{H}$ is the adjacency matrix for the graph. The graph is said to be *bipartite* because there are two distinct types of nodes, bit nodes and check nodes, and there can be no direct connection between any two nodes of the same type.

Example 12-35.

The Tanner graph associated with the 15×20 LDPC matrix of (12.94) is shown in Fig. 12-18. The bit nodes are represented by the $N = 20$ circles at the top, while the check nodes are represented by the $M = 15$ squares at the bottom. The first (left-most) five check nodes correspond to $\mathbf{H}_0$, the second five to $\pi_2(\mathbf{H}_0)$, and the last five to $\pi_3(\mathbf{H}_0)$.

For the special case of a (j, k) regular LDPC code, each bit is involved in j parity checks. Hence, the number of branches emanating from a bit node is always j. Similarly, because each parity check involved k bits, the number of branches emanating from each check node is always k. Observe that the graph of Fig. 12-18 satisfies these properties.

The value of the Tanner graph will become clear in the next section, where we describe a decoding algorithm for LDPC codes. There, the graph will be used to describe a parallel implementation of a decoder, with the different nodes representing separate processors, and edges representing communication between processors.

An *irregular LDPC* code is a generalization of LDPC codes for which the parity-check matrix is still sparse, but for which not all bit nodes have the same degree, and/or not all check nodes have the same degree. The design of LDPC matrices, regular or irregular, is beyond our scope. Good LDPC codes have been designed using random constructions. For irregular LDPC codes, a fruitful approach is to optimize the distribution of the node degrees according to the theory of *density evolution* (see Section 12.4.3) [49][50][16]. Unfortunately, randomly generated LDPC codes suffer from high *encoding* complexity. The fundamental problem is that a randomly generated sparse parity-check matrix will almost surely lead to a dense generator matrix. Therefore, for a fixed code rate, the number of computations required to multiply a message of length K by a generator matrix of dimension $K \times N$ grows as N^2 for large N. Under certain constraints on the distribution of node degrees, some optimized LDPC codes can be encoded with complexity that grows only linearly with N [51]. An alternative approach is to constrain the parity-check matrix in such a way that the encoder has low complexity [52]. We will have to wait until our discussion on repeat-accumulate codes (Section 12.5.2) to see an LDPC code that can be both encoded and decoded with complexity that grows only linearly in N [34].

12.4.2. Decoding Parity-Check Codes

Before describing a decoding strategy for LDPC codes we need to establish some notation and background results. The probability distribution for a binary random variable $c \in \{0, 1\}$ is uniquely specified by the single parameter $p_c(1) = \Pr[c = 1]$, since it must be that $p_c(0) = 1 - p_c(1)$. Alternatively, the probability distribution is uniquely specified by the logarithm of the ratio:

$$\lambda = \log \frac{p_c(1)}{p_c(0)}. \tag{12.95}$$

To recover $p_c(1)$ from λ, we can solve this equation for $p_c(1)$, yielding $p_c(1) = 1/(1 + e^{-\lambda})$. The sign of λ indicates the most likely value for c; λ is positive when 1 is more likely than 0, and λ is negative when 0 is more likely than 1. Moreover, the magnitude $|\lambda|$ is a measure of

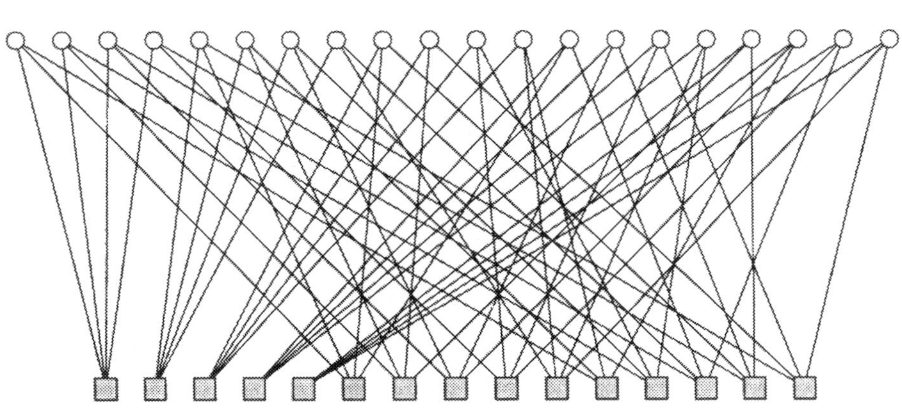

Fig. 12-18. The Tanner graph for the regular LPDC matrix of (12.94).

certainty. At one extreme, if $\lambda = 0$ then 0 and 1 are equally likely. At the other extreme, if $\lambda = \infty$ then $c = 1$ with probability 1, and $\lambda = -\infty$ implies $c = 0$.

Given a random bit $c \in \{0, 1\}$, let r denote an observation whose pdf depends on c according to $f(r \mid c)$. When c is fixed and $f(r \mid c)$ is viewed as a function of r, it is called a *conditional pdf.* On the other hand, when r is fixed, then $f(r \mid c)$ as a function of c is called the *likelihood function*.

Before making an observation, the *a priori* probabilities for c are $p_c(0)$ and $p_c(1)$. After making an observation, these probabilities change to the *a posteriori* probabilities (APP) $p_{c|r}(0 \mid r) = \Pr[c = 0 \mid r]$ and $p_{c|r}(1 \mid r) = \Pr[c = 1 \mid r]$. Because of Bayes rule, the a posteriori probability is proportional to the likelihood function:

$$p_c(1 \mid r) = f(r \mid c = 1)p_c(1)/f(r)\ . \tag{12.96}$$

Hence, the logarithm of the ratio of a posteriori probabilities can be expressed as:

$$\log\frac{p_{c|r}(1 \mid r)}{p_{c|r}(0 \mid r)} = \log\frac{f(r \mid c = 1)}{f(r \mid c = 0)} + \log\frac{p_c(1)}{p_c(0)}. \tag{12.97}$$

The first term on the right-hand side is called the *log-likelihood ratio (LLR)*. Strictly speaking, the second term on the right-hand side is a log-probability ratio, and the left-hand side is a log-APP ratio. However, with an abuse of notation, the second term on the right-hand side is more commonly called the *a priori LLR*, and the left-hand side is called the *a posteriori LLR*. If c is equally likely to be zero or one, then the *a priori* LLR is zero, and the *a posteriori* LLR is equal to the LLR.

The Tanh Rule

Let $\phi = \sum_{i=1}^{n} c_i$ denote the *parity* (modulo-two summation) of a set of n bits $\{c_1, \dots c_n\}$, so that $\phi = 0$ if there are an even number of ones, and $\phi = 1$ if there are an odd number. If the bits are independent, the LLR for the parity obeys the *tanh* rule [20][16].

Exercise 12-7.

(The Tanh Rule.) Let $\{c_1, \dots c_n\}$ be a set of independent bits with a priori probabilities defined by the LLR's $\lambda_i = \log(p_{c_i}(1)/p_{c_i}(0))$. Show that the LLR $\lambda_\phi = \log(p_\phi(1)/p_\phi(0))$ for the parity $\phi = \sum_{i=1}^{n} c_i$ is related to $\{\lambda_i\}$ by the so-called *tanh rule*:

$$\tanh\left(\frac{-\lambda_\phi}{2}\right) = \prod_{i=1}^{n}\tanh\left(\frac{-\lambda_i}{2}\right). \tag{12.98}$$

Solution. See Appendix 12-D.

Solving (12.98) for λ_ϕ yields the following equivalent relationship:

$$\lambda_\phi = -2\tanh^{-1}\left(\prod_{i=1}^{n}\tanh\left(\frac{-\lambda_i}{2}\right)\right). \tag{12.99}$$

Alternatively, if we treat the signs and magnitudes of the LLR's separately, then Appendix 12-D shows that (12.99) can equivalently be expressed as [4]:

$$\lambda_\phi = \sigma_\phi \cdot f\left(\sum_{i=1}^{n} f(|\lambda_i|)\right), \tag{12.100}$$

where $\sigma_\phi = -\prod_{i=1}^{n} \text{sign}(-\lambda_i)$, and where we have introduced the special function:

$$f(x) = \log\frac{e^x + 1}{e^x - 1} = -\log(\tanh(x/2)) \ . \tag{12.101}$$

In hardware implementations, (12.100) has advantages over the tanh rule because it involves the sum of n terms instead of the product n terms. Nevertheless, the tanh rule is preferred in analysis because of its conceptual simplicity.

The function $f(x)$ has some interesting properties. As shown in Fig. 12-19, it is positive and monotonically decreasing for $x > 0$, with $f(0) = \infty$ and $f(\infty) = 0$. Furthermore, $f(x)$ is its own inverse! That is, $f(f(x)) = x$ for all $x > 0$. This property is easily verified by direct substitution.

Let $\hat{c}_i \in \{0, 1\}$ denote the most likely value for the i-th bit, namely $\hat{c}_i = 1$ if $\lambda_i > 0$, else $\hat{c}_i = 0$. The sign of λ_ϕ, which indicates the most likely value for ϕ, is completely determined by the parity $\hat{\phi} = \sum_{i=1}^{n} \hat{c}_i$ of $\{\hat{c}_i\}$, since:

$$\sigma_\phi = -\prod_{i=1}^{n} \text{sign}(-\lambda_i) = -(-1)^{\hat{\phi}} . \tag{12.102}$$

This is to be expected, since the parity is most likely even when an even number of λ_i's are positive, and odd when an odd number are positive.

On the other hand, the magnitude of λ_ϕ, which measures the certainty that ϕ is its most likely value, is given by $|\lambda_\phi| = f(\Sigma_i f(|\lambda_i|))$. Suppose the k-th bit c_k is equally likely to be 0 or 1, so that $\lambda_k = 0$. The k-th term in the sum $\Sigma_i f(|\lambda_i|)$ is then infinity, so that the entire sum is infinite. Because $f(\infty) = 0$, it follows that (12.100) reduces to $\lambda_\phi = 0$ whenever any one of the bits has zero log-likelihood ratio. This makes intuitive sense, since if one bit is equally likely to be zero or one, then the parity of the entire vector is equally likely to be zero or one,

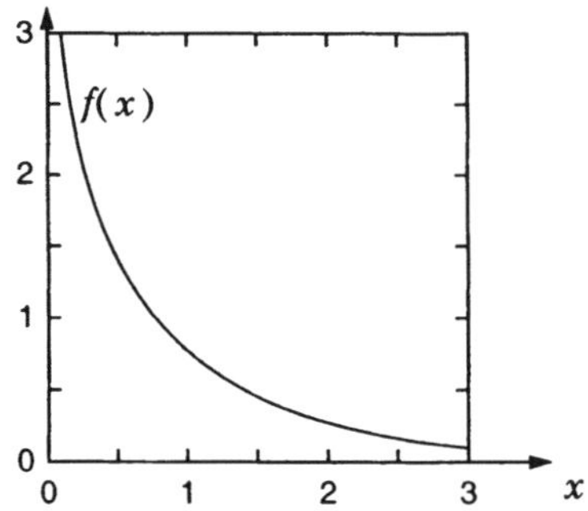

Fig. 12-19. The function $f(x)$ of (12.101).

regardless of the probabilities of the remaining bits. More generally, whenever one bit is significantly less certain than the others, so that the summation is dominated by $f(|\lambda_{\min}|)$, where $|\lambda_{\min}| = \min_i\{|\lambda_i|\}$, then the magnitude of λ_ϕ simplifies to:

$$\begin{aligned} |\lambda_\phi| &= f(\Sigma_i f(|\lambda_i|)) \\ &\approx f(f(|\lambda_{\min}|)) \\ &= |\lambda_{\min}|. \end{aligned} \tag{12.103}$$

The certainty in the parity of a vector can thus be approximated by the certainty of the least certain bit. Substituting (12.103) into (12.100) yields the following approximation:

$$\lambda_\phi \approx -(-1)^{\hat{\phi}} |\lambda_{\min}|. \tag{12.104}$$

In the next section we will show how (12.99) can be used in the decoding problem, but a lower-complexity approximation would use (12.104) in place of (12.99).

The Decoding Problem

Let us consider the problem of decoding a code with parity-check matrix **H**, given that the channel adds white Gaussian noise, so that the receiver observation $\boldsymbol{r} = [r_1 \ldots r_N]$ is related to the transmitted codeword $\boldsymbol{c} = [c_1 \ldots c_N]$ by:

$$\boldsymbol{r} = 2\boldsymbol{c} - 1 + \boldsymbol{n}, \tag{12.105}$$

where the components of the noise vector $\boldsymbol{n}$ are independent zero-mean Gaussian random variables with variance σ^2. (This implies an antipodal bit-to-symbol mapping $0 \to -1$, $1 \to 1$.)

The detector that minimizes the probability of error for the n-th bit would calculate the *a posteriori* LLR:

$$\begin{aligned} \lambda_n &= \log\frac{\Pr[c_n = 1|\boldsymbol{r}]}{\Pr[c_n = 0|\boldsymbol{r}]} \\ &= \log\frac{\Pr[c_n = 1|r_n, \{r_{i\neq n}\}]}{\Pr[c_n = 0|r_n, \{r_{i\neq n}\}]}, \end{aligned} \tag{12.106}$$

and then decide $\hat{c}_n = 1$ if $\lambda_n > 0$, and $\hat{c}_n = 0$ otherwise. Applying Bayes rule, the numerator in (12.106) can be written as:

$$\begin{aligned} \Pr[c_n = 1|r_n, \{r_{i\neq n}\}] &= \frac{f(r_n, c_n = 1, \{r_{i\neq n}\})}{f(r_n, \{r_{i\neq n}\})} \\ &= \frac{f(r_n|c_n = 1, \{r_{i\neq n}\}) f(c_n = 1, \{r_{i\neq n}\})}{f(r_n|\{r_{i\neq n}\}) f(\{r_{i\neq n}\})} \\ &= \frac{f(r_n|c_n = 1)\Pr[c_n = 1|\{r_{i\neq n}\}]}{f(r_n|\{r_{i\neq n}\})}. \end{aligned} \tag{12.107}$$

The last equality exploits the fact that, given c_n, r_n is independent of $\{r_{i\neq n}\}$. The denominator of (12.106) can be similarly expressed. Hence, (12.106) simplifies to:

$$\lambda_n = \log\frac{f(r_n|c_n = 1)}{f(r_n|c_n = 0)} + \log\frac{\Pr[c_n = 1|\{r_{i\neq n}\}]}{\Pr[c_n = 0|\{r_{i\neq n}\}]}$$

$$= \underbrace{\frac{2}{\sigma^2} r_n}_{\text{intrinsic}} + \underbrace{\log \frac{\Pr[c_n = 1 \,|\, \{r_{i \neq n}\}]}{\Pr[c_n = 0 \,|\, \{r_{i \neq n}\}]}}_{\text{extrinsic}}, \tag{12.108}$$

where we used the fact that $f(r_n \,|\, c_n) = (2\pi\sigma^2)^{-1/2} \exp(-(r_n - 2c_n + 1)^2 / (2\sigma^2))$ for the AWGN channel.

The first term in (12.108) represents the contribution from the n-th channel observation, and is called the *intrinsic* information, while the second term represents the contribution from the *other* observations, and is called the *extrinsic* information [21]. Interestingly, the contributions are combined by simply adding. Further, the intrinsic information is proportional to the n-th channel observation. The constant of proportionality $2/\sigma^2$ is called the *channel reliability* [20].

Exercise 12-8.
Consider a binary symmetric channel (BSC) with input and output alphabet $\{\pm 1\}$, as opposed to the usual alphabet $\{0, 1\}$. Show that the *a posteriori* LLR is again given by (12.108), except that the channel reliability is $\log\left(\frac{1-p}{p}\right)$ instead of $2/\sigma^2$, where p is the crossover probability.

Before we can simplify (12.108) further, we must examine more closely the structure of the Tanner graph. Consider Fig. 12-20, where we draw the graph of an LDPC code from the perspective of the n-th bit node, and where we have rearranged the bit nodes and check nodes

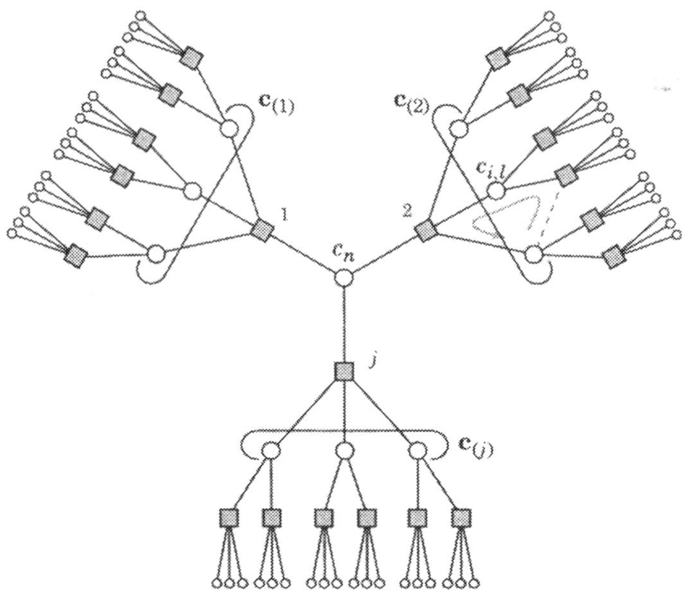

Fig. 12-20. The graph for an irregular LDPC code, from the perspective of the n-th bit node c_n. Removing any one of the edges incident on c_n would create two separate graphs. In fact, if the dotted edge did not exist, then all edges would have this property, and the graph would form a *tree*. This particular code is not regular, because some bit nodes (for example, the leaf nodes) are involved in only one parity-check equation, while others (for example, the n-th bit node) are involved in more than one.

so as to avoid crossing edges. The n-th bit is involved in exactly j parity checks, numbered 1 through j, and each of the checks involve $k-1$ other bits. As shown in the figure, let $\mathbf{c}_{(i)} = [c_{i,2} \ldots c_{i,k}]$ denote the set of bits involved in the i-th of the j parity check equations, excluding c_n.

A *cycle* is a path through the graph that begins and ends at the same bit node. The length of the cycle is the number of edges traversed. Because the graph is bipartite, the minimum length of a cycle is four. For example, for the graph shown in Fig. 12-20, the existence of a cycle depends on the dotted edge. With the dotted edge in place, the graph contains a single cycle of length four. However, if we were to remove the dotted edge, the graph would have no cycles. A graph without cycles is called a *tree*. A cycle-free graph has the following properties:

- removing any edge creates two separate subgraphs.
- there is a unique path connecting any pair of bit nodes.
- As a special case of the above, from the point of view of the bit node c_n: Every bit node is reachable from c_n through one and only one of the edges incident on c_n.
- If bit nodes c_j and c_k are reachable from c_n through different edges, then c_j and c_k are conditionally independent, when the n-th observation is excluded: $\Pr[c_j, c_k \mid \{r_{i \neq n}\}] = \Pr[c_j \mid \{r_{i \neq n}\}]\Pr[c_k \mid \{r_{i \neq n}\}]$.

Let $\phi(\mathbf{c})$ denote the parity of a set of bits $\mathbf{c}$. Because the j parity constraints of the code ensure that $c_n = \phi(\mathbf{c}_{(i)})$ for all $i = 1 \ldots j$, we can rewrite (12.108) as:

$$\lambda_n = \frac{2}{\sigma^2} r_n + \log \frac{\Pr[\phi(\mathbf{c}_{(i)}) = 1 \text{ for } i = 1 \ldots j \mid \{r_{i \neq n}\}]}{\Pr[\phi(\mathbf{c}_{(i)}) = 0 \text{ for } i = 1 \ldots j \mid \{r_{i \neq n}\}]}. \tag{12.109}$$

If the graph is cycle-free, the vectors $\mathbf{c}_{(1)}, \mathbf{c}_{(2)}, \ldots, \mathbf{c}_{(j)}$ are conditionally independent given $\{r_{i \neq n}\}$, and furthermore, the components of $\mathbf{c}_{(i)}$ are themselves conditionally independent given $\{r_{i \neq n}\}$. Hence, (12.109) reduces to:

$$\begin{aligned} \lambda_n &= \frac{2}{\sigma^2} r_n + \log \frac{\prod_{i=1}^{j} \Pr[\phi(\mathbf{c}_{(i)}) = 1 \mid \{r_{i \neq n}\}]}{\prod_{i=1}^{j} \Pr[\phi(\mathbf{c}_{(i)}) = 0 \mid \{r_{i \neq n}\}]} \\ &= \frac{2}{\sigma^2} r_n + \sum_{i=1}^{j} \log \frac{\Pr[\phi(\mathbf{c}_{(i)}) = 1 \mid \{r_{i \neq n}\}]}{\Pr[\phi(\mathbf{c}_{(i)}) = 0 \mid \{r_{i \neq n}\}]} \qquad (12.110) \\ &= \frac{2}{\sigma^2} r_n + \sum_{i=1}^{j} \lambda_{\phi(\mathbf{c}_{(i)})}, \qquad (12.111) \end{aligned}$$

where because the bits are conditionally independent, each $\lambda_{\phi(\mathbf{c}_{(i)})}$ satisfies the tanh rule of (12.99). In particular, if we introduce

$$\lambda_{i,l} = \log \frac{\Pr[c_{i,l} = 1 \mid \{r_{i \neq n}\}]}{\Pr[c_{i,l} = 0 \mid \{r_{i \neq n}\}]}, \tag{12.112}$$

then substituting (12.99) into (12.111) yields:

$$\lambda_n = \frac{2}{\sigma^2} r_n - 2\sum_{i=1}^{j} \tanh^{-1}\left(\prod_{l=2}^{k} \tanh\left(\frac{-\lambda_{i,l}}{2}\right)\right). \tag{12.113}$$

With the aid of the Tanner graph of Fig. 12-20, we may interpret (12.111) in terms of messages passed from node to node. Suppose the bit node associated with $c_{i,l}$ passes the "message" $\lambda_{i,l}$ to the i-th check node. In turn, the i-th check node collects the $k-1$ incoming messages from the other bits $\boldsymbol{c}_{(i)}$ involved (beside c_n), computes the *a posteriori* LLR $\lambda_{\phi(\mathbf{c}(i))}$ for their parity, and passes this "message" to the n-th bit node. Finally, the n-th bit node computes λ_n according to (12.111) by summing all of the incoming messages and adding $(2/\sigma^2)r_n$.

The Message-Passing Algorithm

But how to calculate $\lambda_{i,l}$? The key result is that $\lambda_{i,l}$ can be calculated *recursively*, again using an equation of the form (12.111), as long as the graph is cycle-free.

Consider the bit node labeled $c_{i,l}$ in Fig. 12-20. Obviously it is dependent on c_n, since both take part in the same parity-check equation. However, when conditioned on $\{r_{i \neq n}\}$, $c_{i,l}$ is *independent* of c_n. Even stronger, when we exclude the n-th observation r_n, we also exclude all of the observations that *pass through* r_n. This means that, for Fig. 12-20, two-thirds of the other observations are excluded when r_n is excluded. Only the remaining third that are in the same subgraph as $c_{i,l}$ are relevant in computing $\lambda_{i,l}$. In effect, by excluding r_n we are cutting the graph on the edges leaving bit-node n, producing j disjoint graphs. Since the resulting graphs are disjoint, they can be treated independently, and since they are all cycle-free, we can directly apply (12.111) to calculate the a posteriori LLR's. This recursion may then progress through the tree, until the leaves of the tree are reached.

The *message-passing algorithm* is a decoding technique in which messages are passed from node to node through the Tanner graph. The nodes act as independent processors, collecting incoming messages and producing outgoing messages. There is no global control over the timing or the content of the messages; instead, the bit and check nodes follow a common *local* rule: Send a message as soon as all necessary incoming messages have been received. When the graph is cycle-free, the message-passing algorithm is a recursive algorithm that always converges to the true a posteriori log-likelihood ratios defined by (12.106) after a finite number of messages have been passed.

However, most (if not all) "good" codes have cycles in their Tanner graphs. With cycles it no longer makes sense to have each node wait for all incoming messages before sending a message, because those nodes involved in a cycle would end up waiting forever. Instead it is common to modify the message-passing algorithm as described below. When applied to codes with cycles, this modified version of the message-passing algorithm is no longer an exact, recursive solution to the APP decoding problem but is instead an iterative and approximate solution. Fortunately, even when the graph has cycles, the message-passing algorithm performs remarkably well, and its complexity is extremely low.

The iterative message-passing decoder for a binary code with parity-check matrix **H** can be summarized concisely in terms of the index sets $\mathcal{M}_n = \{m: H_{m,n} = 1\}$ and $\mathcal{N}_m = \{n: H_{m,n} = 1\}$, as follows. Let $u_{m,n}^{(l)}$ denote an "upward" message from check-node m to bit-node n during the l-th iteration, and let $\lambda_n^{(l)}$ denote an estimate of the n-th a posteriori LLR (12.106) after l iterations. The message-passing decoder is:

Initialize —

- $u_{m,n}^{(0)} = 0,$ for all $m \in \{1, \dots M\}$ and $n \in \mathcal{N}_m$;
- $\lambda_n^{(0)} = \frac{2}{\sigma^2} r_n,$ for all $n \in \{1, \dots N\}$.

Iterate — for iteration counter $l = 1, 2, \dots l_{\max}$:

- *(Check-node update)*
  for $m \in \{1, \dots M\}$ and $n \in \mathcal{N}_m$:

$$u_{m,n}^{(l)} = -2\tanh^{-1}\left(\prod_{i \in \mathcal{N}_m - n} \tanh\left(\frac{-\lambda_i^{(l-1)} + u_{m,i}^{(l-1)}}{2} \right) \right) ; \quad (12.114)$$

- *(Bit-node update)*
  for $n \in \{1, \dots N\}$:

$$\lambda_n^{(l)} = \frac{2}{\sigma^2} r_n + \sum_{m \in \mathcal{M}_n} u_{m,n}^{(l)} . \quad (12.115)$$

In practice the algorithm can stop iterating as soon as all of the parity checks are satisfied. This algorithm can be interpreted in terms of passing messages through the Tanner graph. At the zero-th iteration, the message passed from the n-th bit node to each of its participating check nodes is $(2/\sigma^2)r_n$. The m-th check node collects its incoming messages, and passes as an outgoing message the LLR for the parity of the bits involved in the m-th check, excluding the recipient. Each bit node receives j messages, and the n-th bit node produces a new estimate for λ_n by adding $(2/\sigma^2)r_n$ to the summation of all of the incoming messages, as dictated by (12.111). The process then repeats: the check nodes pass their new estimates for $\{\lambda_n\}$ to the check nodes, excluding the contribution from the recipient, and so on. As illustrated in Fig. 12-21, the messages sent across an edge are a simple function of all messages incoming on the *other* edges.

The message-passing view of the iterative decoder makes it possible to realize a parallel implementation based on the Tanner graph, in which each of the M check nodes is a separate processor, and each of the N bit nodes is simply a summing node. Such an implementation makes feasible the use of LDPC code and iterative decoding at extremely high bit rates [22]. The memory requirements would be minimal, although layout of such a decoder would be difficult when N is large. At the other extreme is a fully serialized implementation, consisting

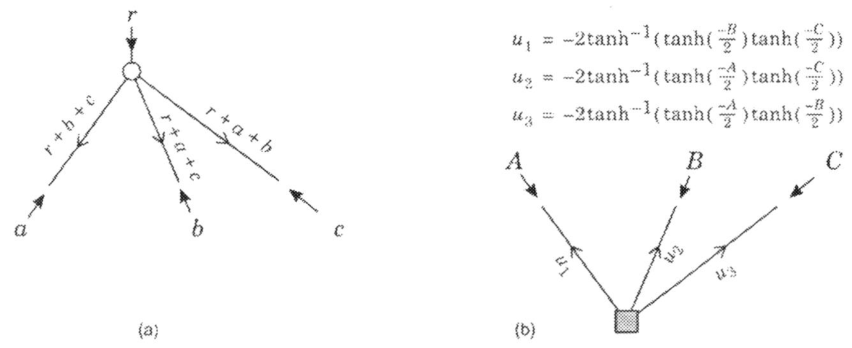

Fig. 12-21. (a) The message sent by a bit node down a particular edge is simply the sum of all messages incoming on the *other* edges, including the intrinsic information from the channel observation. (b) The message u_i sent by a check node up a particular edge is the tanh rule applied to the incoming messages on the *other* edges.

of a single check-node processor that computes each of the Mk check-to-bit messages one by one. The memory requirements can be significant. A software implementation works essentially this way.

Example 12-36.

Simulation results for the message-passing decoder of (12.114) and (12.115) are shown in Fig. 12-22, assuming the (3, 4) regular LDPC code of (12.94). The code rate is $R = 0.35$. After 1000 iterations, the code requires 6 dB to achieve a BER of 10^{-4}. In contrast, an uncoded BPSK system requires 8.4 dB to achieve the same BER; hence, this simple length-20 code offers a coding gain of 2.4 dB. The performance is far (about 6.4 dB) from the Shannon limit of –0.4 dB for a binary code with rate $R = 0.35$. The figure shows that 30 iterations is enough to come within 0.2 dB of a receiver that performs 3000 iterations.

Example 12-37.

Even though the Tanner graph for the (7, 4) Hamming code has cycles, the message-passing decoder is still effective. In Fig. 12-23, we show the BER performance after 7 iterations, and see an improvement of more than 1.5 dB when compared to a hard ML decoder.

12.4.3. Density Evolution

Analysis of the convergence properties of the message-passing decoder is difficult when the block length N is finite, but it simplifies considerably if we allow N to tend towards infinity [49][50][16]. Roughly speaking, for a fixed iteration number l, the probability that a randomly chosen bit node from a randomly constructed LDPC graph is part of a cycle of length less than l tends towards zero as $N \to \infty$. This fact allows us to ignore cycles as $N \to \infty$, in which case the messages incident on any node will be independent.

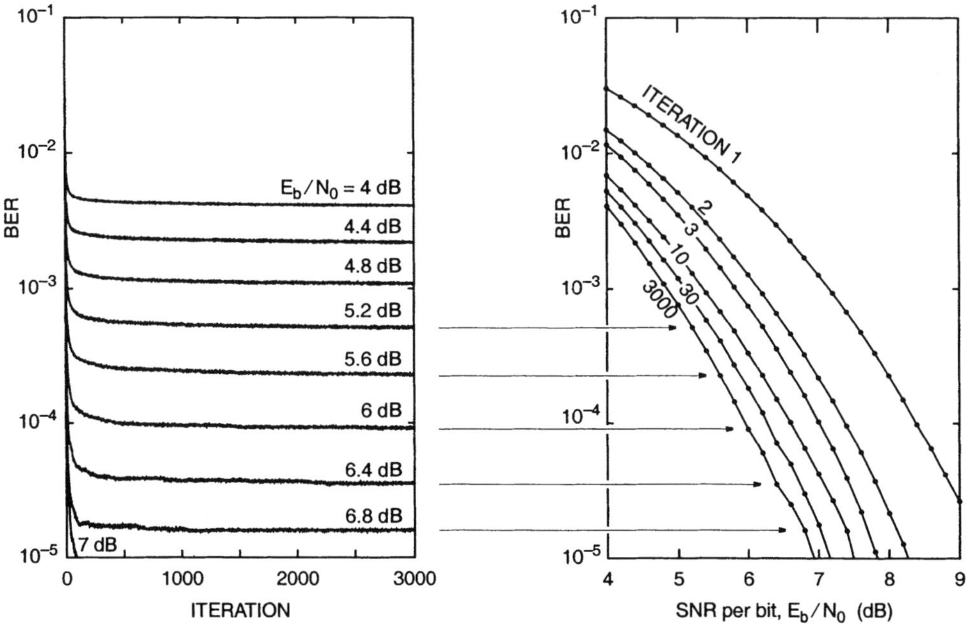

Fig. 12-22. Simulation results for the (3,4) regular LDPC code of (12.94), with block length 20 and rate R = 0.35, decoded using the message-passing algorithm of (12.114) and (12.115).

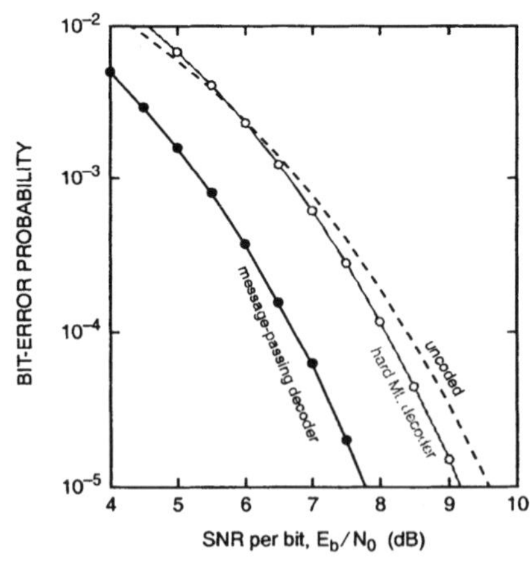

Fig. 12-23. Performance results for the (7, 4) Hamming code over an AWGN channel. The message-passing decoder outperforms the hard decoder by more than 1.5 dB after 7 iterations.

Density evolution is a technique for tracking the pdf's of the messages in the Tanner graph of an LDPC code, under the assumption that $N \to \infty$. By symmetry of the code, the channel, and the decoding algorithm, we may assume without loss of generality that the all-zero codeword was transmitted. (To avoid minus signs and thus simplify notation we also assume a $0 \to 1$, $1 \to -1$ bit-to-symbol mapping in this section, in contrast to our prior discussion. Therefore, the initial message sent from the n-th bit node for the AWGN channel is:

$$\lambda_n^{(0)} = (2/\sigma^2)(1 + \mathcal{N}(0, \sigma^2)) . \tag{12.116}$$

Observe that the variance of $\lambda_n^{(0)}$ is twice its mean:

$$\lambda_n^{(0)} \sim \mathcal{N}(m, 2m) , \tag{12.117}$$

where $m = (2/\sigma^2)$. A Gaussian random variable with this property is said to be *consistent.*

Although the initial messages are Gaussian, subsequent messages are not. Nevertheless, to simplify analysis even further, it is convenient to assume that all messages have a *consistent* Gaussian pdf. In this way, the problem of tracking an infinite-dimensional pdf collapses to one of tracking only a single parameter: the mean. In the following we briefly summarize the Gaussian approximation to density evolution [23][24][25][26]. We will assume that the LDPC code is regular with bit-node degree j and check-node degree k, although the analysis can be extended to irregular codes.

Let $u^{(l)}$ denote a randomly chosen upward (check-to-bit) message after l iterations, as calculated by a message-passing decoder using (12.114), and let μ_l denote its mean, $\mu_l = \mathrm{E}[u^{(l)}]$. Then, according to the tanh rule, this message satisfies:

$$\tanh\left(\frac{u^{(l)}}{2}\right) = \prod_{i=1}^{k-1} \tanh\left(\frac{v_i^{(l)}}{2}\right), \tag{12.118}$$

where $v_1^{(l)}, \ldots v_{k-1}^{(l)}$ are the relevant downward (bit-to-check) message after l iterations, each with mean $m_l = \mathrm{E}[v_i^{(l)}]$. We will assume that both $u^{(l)}$ and $v^{(l)}$ are consistent Gaussian random variables, so that $u^{(l)} \sim \mathcal{N}(\mu_l, 2\mu_l)$ and $v^{(l)} \sim \mathcal{N}(m_l, 2m_l)$. Taking the expectation of both sides of (12.118), and exploiting the independence of $\{v_i^{(l)}\}$, we have:

$$\psi(\mu_l) = \psi(m_l)^{k-1} , \tag{12.119}$$

where we have introduced the function:

$$\psi(m) = \mathrm{E}\left[\tanh\left(\tfrac{1}{2}\mathcal{N}(m, 2m)\right)\right] . \tag{12.120}$$

This function is shown in Fig. 12-24, where it is seen to have behavior similar to $\tanh(m/2)$. In particular, the function is invertible.

We now relate m_l of (12.119) to μ_{l-1}. The i-th downward message $v_i^{(l)}$ in (12.118) can be expressed as:

$$v_i^{(l)} = v_i^{(0)} + \sum_{n=1}^{j-1} u_n^{(l-1)} , \tag{12.121}$$

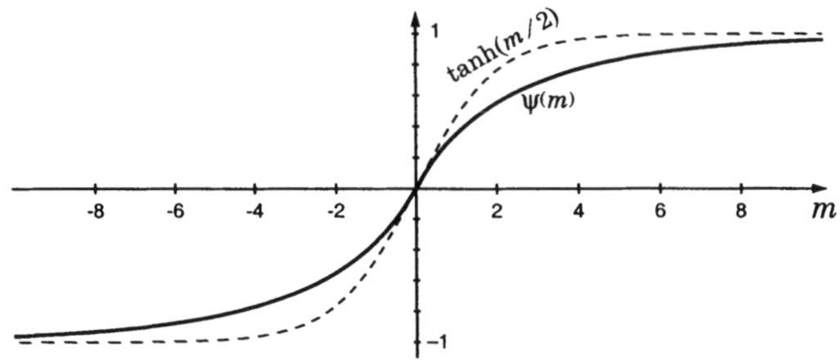

Fig. 12-24. The function $\psi(m)$ is roughly a stretched version of the function $\tanh(m/2)$.

where $v_i^{(0)}$ is the *initial* downward message (which is a consistent Gaussian with mean $2/\sigma^2$), and where $u_1^{(l-1)}, \ldots u_{j-1}^{(l-1)}$ are the relevant upward messages impinging on the bit node, which are mutually independent, each with mean μ_{l-1}. Thus, averaging (12.121) yields:

$$m_l = 2/\sigma^2 + (j-1)\mu_{l-1} \,. \tag{12.122}$$

Substituting (12.122) into (12.119) yields

$$\psi(\mu_l) = \psi\left(2/\sigma^2 + (j-1)\mu_{l-1}\right)^{k-1}. \tag{12.123}$$

Inverting $\psi(\cdot)$ leads to the following recursive relationship for the mean μ_l of a randomly chosen upward message after l iterations:

$$\mu_l = \psi^{-1}\left(\psi\left(2/\sigma^2 + (j-1)\mu_{l-1}\right)^{k-1}\right). \tag{12.124}$$

As dictated by the message-passing decoder, the initial upward message is zero, so that $\mu_0 = 0$. With this initialization, (12.124) can be iterated to determine the evolution of μ_l. The dynamics of this recursion are completely determined by three parameters: the bit-node degree j, the check-node degree k, and the SNR through $2/\sigma^2$.

Example 12-38.
Let $j = 4$, $k = 6$, and suppose we fix the SNR per bit at $E_b/N_0 = 1.72$ dB. (Recall that $E_b/N_0 = 1/(2R\sigma^2)$, where the rate is $R = 1 - j/k = 1/3$ in this example.) Then by iterating (12.124) we find that the mean μ_l approaches a constant of $\mu_l \to 0.375$. In Fig. 12-25, we plot (using a dashed line) the pdf of a consistent Gaussian pdf with mean 0.375. We see that there is a significant probability that the randomly chosen message will be negative, even after a large number of iterations. This implies that the probability of decoding error is nonzero.

Example 12-39.
Continuing the above example, suppose we again fix $j = 4$ and $k = 6$, but increase the SNR per bit from 1.72 dB to 1.73 dB. By iterating (12.124), we find that this small increase in SNR has a huge impact on μ_l, causing $\mu_l \to \infty$ as $l \to \infty$. The solid curves in Fig. 12-25 show the pdf of a consistent Gaussian random variable with the corresponding mean μ_l for iteration $l \in \{630, 631,$

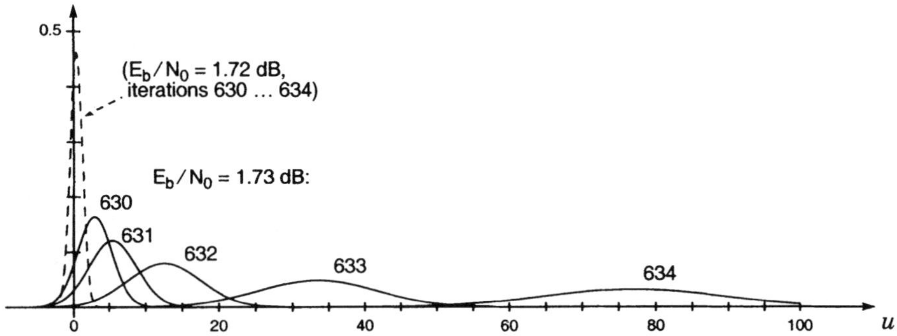

Fig. 12-25. The pdf of an upward message u as a function of E_b/N_0 and iteration.

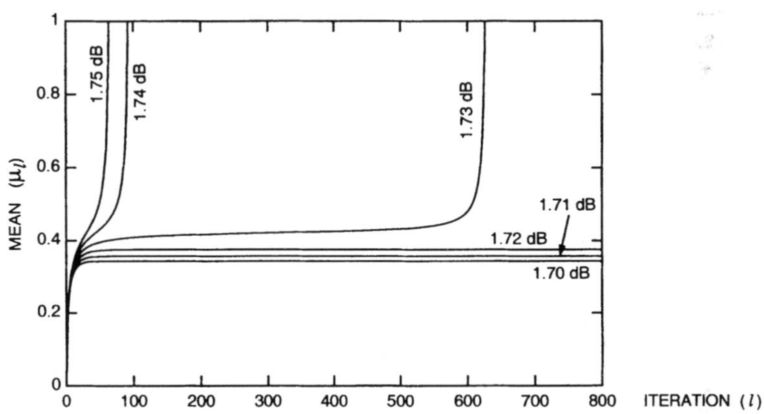

Fig. 12-26. The results of density evolution based on the Gaussian approximation for a (4, 6) regular LDPC code. Here, the mean of a randomly selected message is plotted as a function of iteration, for several different values of E_b/N_0. For E_b/N_0 = 1.73 dB and larger, the mean approaches infinity, which implies that the error probability approaches zero.

632, 633, 634}. As the mean gets larger, the probability of a negative message diminishes, until an infinite mean implies a vanishingly small probability of decoding error.

For any set of values of j, k, and σ^2, the recursion (12.124) always converges. Furthermore, there exists a SNR *threshold* γ such that $\mu_l \to \infty$ as $l \to \infty$ for all $E_b/N_0 > \gamma$, indicating that the decoding error probability approaches zero, while μ_l converges to a finite number for all $E_b/N_0 < \gamma$ as $l \to \infty$, indicating that the decoding error probability is nonzero.

Example 12-40.

Continuing the previous example, in Fig. 12-26 we plot the mean μ_l as a function of iteration l for different values of E_b/N_0, as found by iterating (12.124) with $j = 4$ and $k = 6$. The figure clearly illustrates the threshold phenomenon, with a threshold value near $\gamma = 1.73$ dB. The Shannon limit is –0.50 dB for rate-1/3 binary codes. In contrast, the results of Fig. 12-26 suggest that a sufficiently long (4, 6) regular LDPC code can achieve vanishingly small error probability at $E_b/N_0 =$ 1.73 dB. Hence, the gap to capacity is 2.2 dB. This gap can be reduced to negligible levels using irregular LDPC codes [16][3].

12.5. Turbo Codes

Although they were invented only a decade ago, turbo codes are already a part of many commercial standards, including the digital-video broadcasting return link (DVB-RCS), third-generation wireless standards 3GPP (wideband-CDMA) and 3GPP2 (CMDA 2000), as well as deep-space telemetry [22]. The idea behind turbo codes is to construct a powerful error-control code out of simple building blocks. Two or more easily decodable codes are concatenated either in parallel or in serial, separated by an interleaver that rearranges the bits in a pseudorandom fashion. The receiver consists of two or more low-complexity decoders, one for each of the component codes, that share information so that together they approximate a joint decoder. In Section 12.5.1 we describe parallel-concatenated turbo codes. In Section 12.5.2 we describe serially concatenated turbo codes, which include as special cases repeat-accumulate codes and turbo equalization.

12.5.1. Parallel Concatenated Codes

The original turbo code of Berrou *et al.* [5] uses a pair of simple-to-decode encoders and an interleaver, connecting them in *parallel* as shown in Fig. 12-27. First, a long block of message bits $\boldsymbol{m}$ is encoded using a rate-$1/2$ recursive systematic convolutional encoder with generator $\mathbf{G}(D) = [1, g_2(D)/g_1(D)]$, producing a systematic codeword of the form $[\boldsymbol{m}, \boldsymbol{x}_1]$. Next, the message bits are rearranged (or shuffled, or permuted) according to a pseudorandom

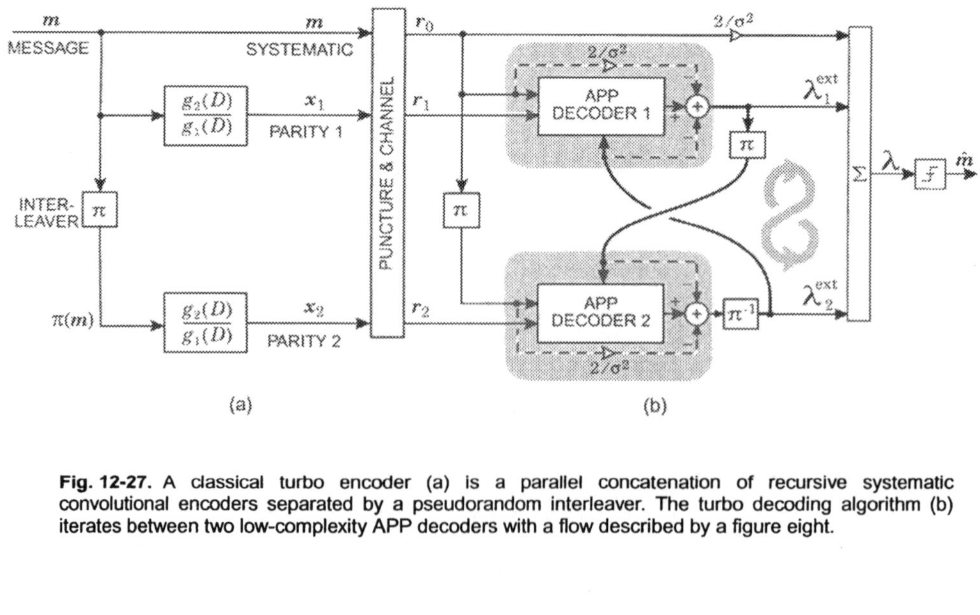

Fig. 12-27. A classical turbo encoder (a) is a parallel concatenation of recursive systematic convolutional encoders separated by a pseudorandom interleaver. The turbo decoding algorithm (b) iterates between two low-complexity APP decoders with a flow described by a figure eight.

interleaver π and then encoded a second time using the same generator, yielding a second set of parity bits $\boldsymbol{x}_2$. The output codeword is the original message along with the two sets of parity bits, $\boldsymbol{c} = [\boldsymbol{m}, \boldsymbol{x}_1, \boldsymbol{x}_2]$, so that the overall rate is nominally $1/3$. Let $\boldsymbol{r} = [\boldsymbol{r}_0, \boldsymbol{r}_1, \boldsymbol{r}_2]$ denote the corresponding vector of observations after an AWGN channel. To achieve a code rate higher than $1/3$ it is common to puncture (discard) a fraction of the parity bits. The component encoders are generally identical. The codes need not be systematic but they must be recursive for best performance [27].

True APP or ML decoding of the turbo code is not feasible because the memory of the encoder is governed by the length of the interleaver, which might be thousands or tens of thousands of bits. Instead, a turbo decoder approximates the APP decoder by decoding the two constituent codes separately but cooperatively, as shown in Fig. 12-27(b).

First, ignoring $\boldsymbol{r}_2$, the receiver performs APP decoding for the first encoder based on the relevant observations $\boldsymbol{r}_0$ and $\boldsymbol{r}_1$. The result is a set of a posteriori LLR's $\{\lambda_1, \ldots \lambda_K\}$, one for each message bit, given observations $\boldsymbol{r}_0$ and $\boldsymbol{r}_1$ but not $\boldsymbol{r}_2$:

$$\lambda_n = \log\frac{\Pr[m_i = 1 \mid \boldsymbol{r}_0, \boldsymbol{r}_1]}{\Pr[m_i = 0 \mid \boldsymbol{r}_0, \boldsymbol{r}_1]} \ . \tag{12.125}$$

Because the code is systematic, the n-th bit m_n is transmitted directly across the channel; let $r_0^{(n)}$ denote the corresponding observation. Then, according to (12.108), this a posteriori LLR decomposes into:

$$\lambda_n \ = \ \frac{2}{\sigma^2}\, r_0^{(n)} \ + \ \log\frac{\Pr[m_n = 1 \mid \{r_0^{(i \neq n)}\}, \boldsymbol{r}_1]}{\Pr[m_n = 0 \mid \{r_0^{(i \neq n)}\}, \boldsymbol{r}_1]} . \tag{12.126}$$

Let us use $\lambda_{1,\,\mathrm{ext}}^{(n)}$ to denote the *extrinsic* LLR, defined as the second term above; it represents the contribution to the n-th LLR that was not evident from the systematic observation $r_0^{(n)}$ alone. In other words, taken as a whole, the set of extrinsic LLR's $\{\lambda_{1,\,\mathrm{ext}}^{(n)}\}$ represents the *incremental* information added by the first set of parity observations, beyond what can be learned from observing $\boldsymbol{r}_0$ only.

Next, the extrinsic information about the message bits is isolated by subtracting the systematic LLR from the APP decoder output. This extrinsic information is then passed to the second APP decoder, which uses it as *a priori* information for estimating the LLR's based on the observations $\boldsymbol{r}_0$ and $\boldsymbol{r}_2$, ignoring $\boldsymbol{r}_1$. The resulting LLR's for the unpermuted message can be written as the sum:

$$\frac{2}{\sigma^2}\, r_0^{(n)} + \lambda_{1,\,\mathrm{ext}}^{(n)} + \lambda_{2,\,\mathrm{ext}}^{(n)} \ , \tag{12.127}$$

which implicitly defines the second-decoder *extrinsic* LLR $\lambda_{2,\,\mathrm{ext}}^{(n)}$ as the contribution to the second-decoder LLR's from the second set of parity observations, beyond what can be learned from the systematic and a priori information. As before, the extrinsic information quantifies the difference between the LLR's based on systematic and a priori information only and the LLR's based on systematic, a priori, and parity observations.

The turbo-decoding algorithm proceeds iteratively: decoder 2 passes its extrinsic information to decoder 1, which uses it as a priori information and then passes new extrinsic information to decoder 2, and so on. The flow of information follows a figure-eight pattern, as illustrated in Fig. 12-27. The algorithm can stop after a predetermined number of iterations, typically 10–20, or earlier if the magnitudes of the LLR's indicate reliable decisions. The final estimates of the LLR's for the message bits can be found by adding the systematic LLR's to the two extrinsic LLR's, as shown in the figure. If hard decisions are desired, these LLR's can be quantized.

If we are willing to define a new APP decoder block that produces only extrinsic information as its output, then the block diagram of Fig. 12-27 simplifies. Specifically, we need not explicitly perform the subtractions indicated by the dashed lines in the figure; they can be implicitly moved inside the APP decoder block to define a new extrinsic APP decoder. The two shaded regions surrounding each APP decoder define the extrinsic APP decoders. The advantage of showing the subtractions explicitly is that the APP block of Fig. 12-27 is precisely identical to the BCJR algorithm of Chapter 7. Thus, once the BCJR algorithm of Chapter 7 is understood, a turbo decoder is easily implemented by simply connecting two BCJR blocks as shown in the figure.

In pseudocode, the turbo-decoding algorithm can be described succinctly as follows:

$$\lambda_0 = \frac{2}{\sigma^2} \boldsymbol{r}_0$$
$$\lambda_{2,\text{ext}} = [0, \ldots 0]$$
for i = 1 to 20,
$$\quad \lambda_{1,\text{ext}} = \text{bcjr}(\boldsymbol{r}_0, \boldsymbol{r}_1, \lambda_{2,\text{ext}}) - \lambda_0 - \lambda_{2,\text{ext}}$$
$$\quad \lambda_{2,\text{ext}} = \pi^{-1}(\text{bcjr}(\pi(\boldsymbol{r}_0), \boldsymbol{r}_2, \pi(\lambda_{1,\text{ext}}))) - \lambda_0 - \lambda_{1,\text{ext}}$$
end
$$\hat{\boldsymbol{m}} = \text{sign}(\lambda_0 + \lambda_{1,\text{ext}} + \lambda_{2,\text{ext}})$$

The first two inputs to the function bcjr(· , · , ·) specify the channel observations and the third specifies the a priori information; this function can be implemented as described in Section 7.5 with outputs calculated according to (7.77).

The minimum distance $d_{\min}$ of turbo codes is generally small. This means that there exists some low-weight message vectors for which both parity vectors have low weight. However, the pseudorandom interleaver has the important benefit that such message vectors are very rare. In other words, the number of nearest neighbors with minimum distance $d_{\min}$ is small. Therefore, although a small minimum distance implies that some $Q(\cdot)$ terms in the union bound are large, the interleaver ensures that the corresponding error coefficients are small. In fact, the number of nearest neighbors is inversely proportional to the interleaver length (message length), a phenomenon known as *interleaver gain* [28]. Because of this, turbo codes are most effective when the block length is long.

Example 12-41.

The original turbo code had a block length of $2^{17} = 131072$ bits and came closer to the Shannon limit than any practical code had before [5]. The encoder was as shown in Fig. 12-27 with

$g_1(D) = 1 + D^4$ and $g_2(D) = 1 + D + D^2 + D^3 + D^4$, which implies that each APP decoder has 16 states. Alternating parity bits were punctured to increase the code rate to $1/2$. After 18 iterations of the iterative turbo-decoding procedure, a bit-error probability of 10^{-5} was achieved at $E_b/N_0 = 0.7$ dB, only 0.5 dB from the Shannon limit for binary inputs.

Consider a single rate-$1/n$ convolutional encoder with memory μ and K input bits. To *terminate* the trellis (*i.e.*, force the state to zero at time $K + \mu$) requires that a sequence of μ *pad* bits be fed into the encoder immediately after the message bits. For nonrecursive encoders the pad bits are all zero. However, for recursive encoders the pad bits needed to terminate the trellis are nonzero and are a deterministic function of the message bits. In the context of a parallel-concatenated turbo code there is a complication: it is very unlikely that the same set of pad bits will terminate both trellises. Instead, each encoder will require its own set of pad bits. This poses no problems to the receiver, which decodes each encoder separately anyway, and so expects to see two sets of pad bits, one for each encoder. To ensure that both sets of pad bits are transmitted, the first set of pad bits can be appended to the parity bits generated by the first encoder, and the second set of pad bits can be appended to the second set of parity bits. Thus, the encoder outputs $\boldsymbol{x}_1$ and $\boldsymbol{x}_2$ in the figure are each of length $K + 2\mu$, with $K + \mu$ representing parity bits and the remaining μ representing pad bits. The overall rate of the unpunctured terminated trellis code is then $K/(3K + 4\mu)$, which is essentially $1/3$ as long as $\mu \ll K$.

12.5.2. Serially Concatenated Codes

Unlike the parallel concatenation of codes described above, serially concatenated codes have a long history, dating back to the 1960's [29], and are still used today in many applications [30]. The archetypical concatenated code consists of a convolutional inner code with low-enough complexity that ML soft decoding using the Viterbi algorithm is feasible and practical. The error rate need only be in the range 10^{-2} to 10^{-3} [31]. Then, a high-rate Reed-Solomon code is used as an outer code to push down the error rate to an acceptable level with very little redundancy overhead. Since the errors after a Viterbi algorithm tend to cluster in bursts defined by error events that span multiple trellis stages, Reed-Solomon codes are an ideal choice as an outer code because of their excellent burst-error correcting capabilities. A regular symbol-wise interleaver can be inserted between the inner and outer code to spread longs error bursts into separate smaller bursts.

Example 12-42. ———

Concatenated codes are a mainstay on NASA spacecraft. For several decades the standard coding strategy consisted of a (255, 233) 8-bit Reed-Solomon outer code, a row-column interleaver, and a rate-$1/2$, 64-state convolutional code. Minor variations of this code were used on many spacecraft, including the Voyager (1977), Galileo (1989), and Cassini (1997) missions [30].

Serial-concatenated turbo codes are similar to classical concatenated codes, in that they both consist use an outer coder and an inner coder separated by an interleaver. However, there are three distinguishing features of a serial-concatenated turbo code. First, the interleaver is not regular but pseudorandom; second, the outer code must be easily APP decodable and thus is not a Reed-Solomon code; and third, the intent is to decode the concatenation iteratively using a pair of APP decoders. The inner code must be recursive for best performance, while the outer code need not be. On the other hand, the outer code should have a large minimum distance, preferably an odd minimum distance [32].

In Fig. 12-28(a) we show a generic serial-concatenated turbo encoder. In Fig. 12-28(b) we illustrate how the turbo-decoding algorithm can be applied to a serial concatenation of codes. The turbo decoder iterates between a pair of APP decoders, one for the inner code and another for the outer code. Unlike the parallel-concatenated turbo code, however, the information exchanged between the two decoders is not LLR's for the message bits, but rather, LLR's for the bits after the outer code but before the interleaver, i.e., the bits labeled $\boldsymbol{c}$ in the figure.

Although the BCJR algorithm of Chapter 7 was originally described so as to compute the message-bit LLR's, it is easily modified to compute those for the coded bits instead. Specifically, consider a rate-$1/2$ convolutional code, so that each stage of the trellis represents two coded bits. The a posteriori LLR for each one can be calculated following the basic approach of adding up a posteriori state transition probabilities, as specified by (7.77). To calculate the LLR's for the first coded bit, however, the numerator would sum over the transitions for which the first *coded bit* (not the message bit) is one, and the denominator would sum over the remaining transitions. This process would be repeated for the second coded bit, so that (7.77) would be calculated twice for each stage of the trellis.

There are two further subtleties in turbo decoding for serial-concatenation codes. First, the outer decoder has no direct access to the channel observations. (In contrast, both APP decoders for parallel-concatenated turbo codes had access to the channel observations.) Instead, the outer decoder is fed only a priori information. Second, unlike the a priori information fed to the APP decoders of a turbo decoder for parallel concatenation, the a priori information is regarding the *coded bits* and not the message bits. Fortunately, in the binary case these issues are easily accounted for by creating *effective* observations $\{\bar{r}_i\}$ from the a priori LLR's $\{\lambda_i^a\}$ according to $\bar{r}_i = \frac{\sigma^2}{2}\lambda_i^a$. These effective observations may then be used directly in the branch metric calculations of (7.65), along with uniform a priori probabilities for the message bits.

Repeat-Accumulate Codes

The repeat-accumulate (RA) code is the simplest nontrivial example of a serial-concatenated turbo code [54]. The outer code is a $(n, 1)$ repetition code, the simplest code capable of correcting a bit error. It has a low rate but a large minimum distance. The inner code is a rate-1 recursive convolutional code with generator $1/(1 + D)$; it is known as an accumulator because its output is the modulo-two summation or accumulation of the inputs. This is the simplest recursive coder imaginable. The inner and outer code are separated by a pseudorandom interleaver. For example, the following is a rate-$1/3$ RA encoder:

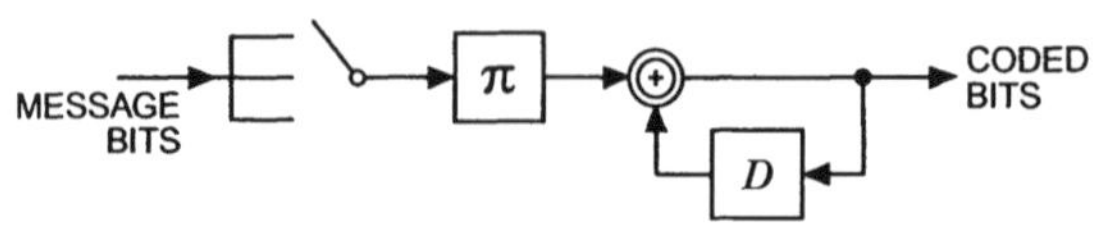

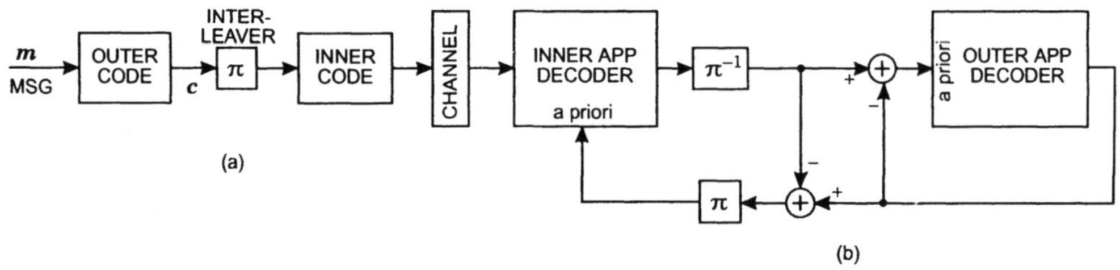

Fig. 12-28. A serial-concatenated turbo coder (a) consists of an outer code, a pseudorandom interleaver, and a recursive inner code. A turbo decoder (b) in this case iterates between a pair of APP decoders.

Individually, the three pieces of a RA code are themselves poor codes: the repetition code, the interleaver, and the accumulator by themselves each have zero coding gain. However, the RA code as a whole has remarkably good performance.

In principle one could decode an RA code using the generic structure of Fig. 12-28(b), for which the outer APP decoder would be trivial while the inner APP decoder would be based on a two-state BCJR algorithm. However, complexity can be significantly reduced by using a message-passing decoder instead. Consider for a moment a *systematic* RA code in which the original message bits are transmitted in addition to the accumulator output. Such a code is clearly linear, since the concatenation of linear codes is always linear, and the interleaver does not change this. Therefore, it can be represented by a parity-check matrix **H**, or equivalently by a Tanner graph.

Example 12-43. The Tanner graph for a systematic version of an RA code with a (3, 1) repetition outer code is sketched in Fig. 12-29 for the case of only two message bits. Unlike the Tanner graphs of Section 12.4, we have moved the bit nodes for the message bits down below the check nodes, to help distinguish them from the bit nodes for the parity bits at top. This graph is easily justified. The accumulator forces each output to be the modulo-two sum of the previous output and the input. Each of the check nodes enforce this constraint. The first (leftmost) check node constrains the first parity bit to be equal to the first interleaved bit, which happens to also be the first message bit in this example. Each succeeding check node constrains the parity bit to be equal to the modulo-two sum

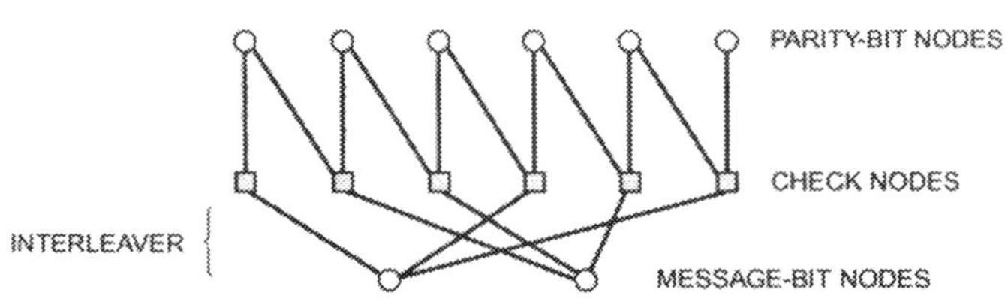

Fig. 12-29. The Tanner graph for a systematic RA code that results from the cascade of a (3, 1) repetition code, an interleaver $\pi = [1\ 4\ 6\ 2\ 3\ 5]$, and an accumulator, for the special case of two message bits.

of the previous parity bit and the corresponding accumulator input. The interleaver defines the connections between the message bit nodes and the check nodes. Observe that this graph is not regular: neglecting end effects, the message bit nodes have degree three, the parity bit nodes have degree two, and the check nodes have degree three.

Once the RA code is represented by a graph, the message-passing decoder of Section 12.4 is immediately applicable. To accommodate a nonsystematic RA code for which the systematic bits are not transmitted, we need only set the corresponding channel observations to zero. In other words, we can view the RA code as a *punctured* version of a systematic RA code. The Tanner graph and decoding algorithm would not change, except that the channel observations for the punctured bits would be zero.

An important advantage of the RA code over randomly generated LDPC codes is the low complexity of the encoder. Despite their conceptual simplicity and the ease with which they can be both encoded and decoded, RA codes perform surprisingly well.

Example 12-44. When decoded in the traditional manner, the Galileo concatenated code consisting of a rate-$1/4$ convolutional inner code with 2^{14} states and a (255, 223) 8-bit outer Reed-Solomon code can achieve a bit-error probability of 10^{-5} at $E_b/N_0 = 0.8$ dB. According to McEliece, a rate-$1/4$ RA code with message length 4096 can achieve the same bit-error probability at the same E_b/N_0, but using message passing the decoding complexity is 100 times smaller [33].

An *irregular* repeat-accumulate (IRA) coder has the same basic structure as the RA coder, but instead of repeating all K message bits an equal number of times, each of the first block of f_2K bits is repeated two times, each of the next block of f_3K bits is repeated three times, etc., where f_q denotes the fraction of bits repeated q times. The check nodes for the graph shown in Example 12-59 have total degree three but lower degree one. A further generalization is to increase the check-node lower degree to $a \geq 1$. The Tanner graph for a generic IRA graph is thus illustrated in Fig. 12-30.

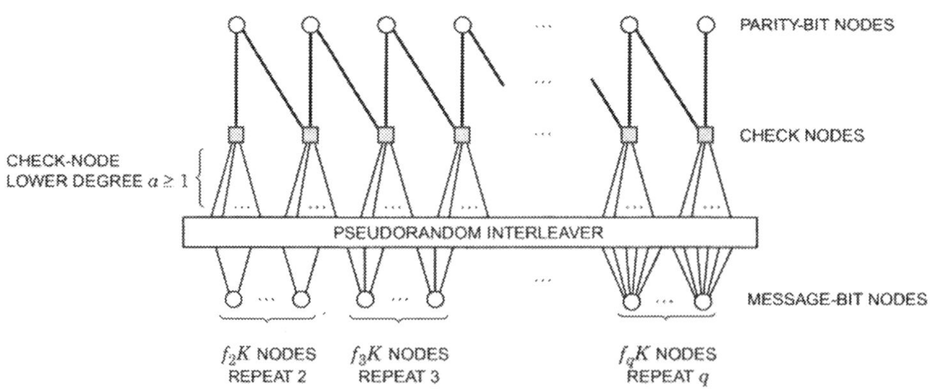

Fig. 12-30. The Tanner graph for an irregular repeat-accumulate code [34].

The code parameters $\{a, f_i\}$ can be optimized using density evolution so as to minimize the SNR threshold [34]. The resulting codes have similar performance to comparable irregular LDPC codes, outperforming turbo codes. In fact, from the Tanner graph we see that a systematic IRA code is a special case of an irregular LDPC code. However, IRA codes have the important advantage that their encoder can be implemented with low complexity. Specifically, for large block lengths N, the encoder complexity grows only linearly with N.

Example 12-45. A systematic IRA encoder is easy to implement in software as well as hardware. Using MATLAB, for example, one can encode a binary message vector `msg` with only a single line:

```
c = [msg, rem(cumsum(sum(msg(Pi))), 2)];
```
(12.128)

where `Pi` is a matrix of integers representing both the repetitions and the permutation, with the integers 1 through f_2K appearing twice, integers $f_2K + 1$ through $(f_2 + f_3)K$ appearing three times, etc. The placement of these integers within the matrix would be random.

Turbo Equalization

The concept of turbo decoding for serial-concatenated codes extends beyond the realm of just error-control coding. For example, consider the scenario in which a sequence of symbol are encoded by a classical (non-concatenated) error-control encoder, perhaps a simple convolutional encoder, interleaved to disperse the impact of nonstationary noise, and then transmitted across a channel that introduces ISI. A traditional receiver would first mitigate the ISI using an equalizer, ignoring the presence of the error-control code, and then perform decoding based on the equalizer outputs, ignoring the presence of the ISI.

Without interleaving, one might consider the possibility of performing the tasks of equalization and error-control decoding *jointly*, since the cascade of the error-control coder and the ISI channel may be viewed as a single finite-state machine, with a number of states equal to the product of the number of states of the encoder and the number of states of the ISI channel (assuming binary modulation). However, the interleaver would cause the number of states to grow exponentially with the length of the interleaver, making a joint equalizer and decoder impractical.

Fortunately, a simple change of view suggests a low-complexity alternative: by viewing the ISI channel as an inner "coder", the principles of turbo decoding for serial-concatenated codes can be applied directly to this scenario. The result is *turbo equalization*: an iterative method for approximating the joint equalizer and decoder that follows the turbo principle [53]. The block diagram of a turbo equalizer is identical to the serial turbo decoder of Fig. 12-28, except that the inner APP "decoder" is now the APP *equalizer* based on the BCJR algorithm (Section 7.5).

12.6. Historical Notes and Further Reading

In 1948, Shannon started the field of coding theory by demonstrating that through coding it is theoretically possible to achieve error-free transmission on noisy communication channels (Chapter 4). His results did not bound decoding complexity and delay, but nevertheless suggested that improvements in error probability could be achieved using practical codes. The first block codes, introduced in 1950 by Hamming, were capable of correcting single errors in hard-decoding system, but fell disappointingly short of the capacity predicted by Shannon. The next major advance emerged only after another 10 years, with the BCH codes in 1959 and the closely-related Reed-Solomon codes in 1960. The strong algebraic properties of these codes led to efforts to find efficient coding and decoding algorithms. Convolutional codes appeared in 1957. Initially, they lacked the algebraic properties relating to distance in block codes, but this shortcoming has been partially remedied. During the 1970s, these avenues of research continued, but the next major breakthrough did not occur until 1982 with the description of trellis codes (Chapter 13). The introduction of turbo codes in 1993 dramatically changed the way coding is looked at today.

A classic coding theory book, concentrating on block codes with algebraic (hard) decoding, is that of Berlekamp [35], who will convince any reader of the beauty of the subject. More recent comprehensive texts are Blahut [36] and Wicker [37]. A standard textbook that is still used, 35 years after its publication, is Gallager [6], which includes an extensive discussion of the performance of various block codes. A comprehensive treatment of block codes is MacWilliams and Sloane [38]. A particularly useful paper on convolutional codes is Massey [39]. McEliece [40] gives a very readable treatment of both information theory and coding. The chapter on convolutional codes is an excellent introduction to the subject. More detailed information about convolutional codes can be obtained from Forney [41], where equivalent realizations of codes are discussed. The second edition of a classic book by Peterson, written with Weldon, discusses efficient encoding and decoding of cyclic codes [13]. A more advanced text with a comprehensive treatment of convolutional codes is Viterbi and Omura [42]. Wilson gives a readable explanation of the interaction between coding and modulation [43]. Tables of good convolutional codes can be found in [44][45].

Appendix 12-A. Linear Codes

In order to treat linear codes in a general way, it is helpful to review some definitions from algebra. Linear codes turn out to be important enough to justify this digression.

A *ring* R is a set with two operations defined; we will call these operations *addition* and *multiplication* although they may bear little resemblance to ordinary addition and multiplication. To emphasize that they may be different, we denote addition by $\oplus$ and multiplication by " $\cdot$ ".

The most important feature of the addition operation is that adding any two operands in R produces a result in R.

Example 12-46.
The set of real numbers $\Re$ is a ring, and addition is ordinary. The sum of any two real numbers is a real number. The set of binary digits $Z_2 = \{0,1\}$ is a ring, where addition is modulo-two.

As with ordinary addition, $\oplus$ must be associative

$$(r_1 \oplus r_2) \oplus r_3 = r_1 \oplus (r_2 \oplus r_3) \tag{12.129}$$

and commutative

$$r_1 \oplus r_2 = r_2 \oplus r_1 \,. \tag{12.130}$$

Furthermore, the ring must have an *additive identity*, denoted "0", such that $r \oplus 0 = r$ for any r in R. Finally, every element r must have an *additive inverse* which when added to r produces the zero element.

Example 12-47.
The additive identity in $\Re$ is zero, and the additive inverse of any $r \in \Re$ is $-r$. The additive identity in Z_2 is zero, and the additive inverse of $z \in Z_2$ is z itself.

The multiplication operation " $\cdot$ " is similarly defined to operate on two elements in the ring to produce an element in the ring. Like ordinary multiplication it must be associative, but need not be commutative. It must however be distributive over addition,

$$r_1 \cdot (r_2 \oplus r_3) = r_1 \cdot r_2 \oplus r_1 \cdot r_3 \qquad (r_1 \oplus r_2) \cdot r_3 = r_1 \cdot r_3 \oplus r_2 \cdot r_3 \,. \tag{12.131}$$

To summarize, a ring has (1) closure under $\oplus$ and " $\cdot$ ", (2) associativity for $\oplus$ and " $\cdot$ ", (3) distributivity of " $\cdot$ " over $\oplus$ (12.131), (4) commutativity of $\oplus$, (5) an additive identity 0, and (6) an additive inverse $-r$.

Exercise 12-9.
Show from the above properties that $r \cdot 0 = 0$ for all $r \in R$.

A *field* is a ring where

- There is a *multiplicative identity* in the ring, denoted 1, such that $r \cdot 1 = r$,

- Multiplication is commutative ($r_1 \cdot r_2 = r_2 \cdot r_1$), and
- There is a *multiplicative inverse* $1/r \in R$ for every element of the field except the additive identity 0. The multiplicative inverse satisfies $r \cdot (1/r) = 1$.

Example 12-48.
Both $\Re$ and Z_2 from Example 12-46 are fields, but the set of positive integers Z_∞ is not because there is not a multiplicative inverse for all elements of the field. The set of all integers (positive and negative) is also not a field, but unlike Z_∞ it is a ring.

A field with a finite number of elements is called a *finite field*, or a *Galois field*. It turns out that all finite fields have p^m elements, where p is a prime number and m is any integer.

Example 12-49.
There is no field with six elements, but there is a field with three elements. Let $Z_3 = \{0, 1, 2\}$, $\oplus$ be modulo three addition, and " $\cdot$ " be modulo three multiplication. It is easy to verify that all of the above properties of fields apply, with the possible exception of the multiplicative inverse. We can verify that the multiplicative inverse exists for all elements from the following table:

element	inverse
0	none
1	1
2	2

Exercise 12-10.
Verify that $Z_5 = \{0, 1, 2, 3, 4\}$ is a field with addition and multiplication modulo 5. List the multiplicative inverses.

For both Z_3 and Z_5, the number of elements is prime. If the number of elements is not prime, but is $q = p^m$ for $m > 1$ and p prime, then multiplication in the field is more complicated than simple modulo-q multiplication.

Exercise 12-11.
Verify that multiplication in the field $Z_4 = \{0, 1, 2, 3\}$ is not simple modulo-four multiplication. Construct a multiplication table that satisfies the requirements for multiplication in a field.

Fortunately, in this book we are primarily interested in Z_2, so we need not get distracted by these complications. Suffice it to say that it is possible to define addition and multiplication so that $GF(q)$ is a field for any $q = p^m$, p prime.

All finite fields with q elements are equivalent and are denoted $GF(q)$. Hence any two-element field is equivalent to Z_2, or $GF(2)$.

A *vector space* $V_n(GF(q))$ over the field $GF(q)$ is a set of n-tuples of elements from the field. These are much like the vector spaces of Section 2.6, the only difference being that they are defined over Galois fields rather than the fields of real or complex numbers. Addition ($\oplus$) of two vectors is defined element-wise, and must produce a vector in the vector space (in other words the vector space is *closed under vector addition*). Vector addition is commutative and associative, and there is an additive identity (the zero vector) and an additive inverse (the negative of a vector). A vector can be multiplied element-wise by a scalar in the field, and the vector space is closed under scalar multiplication. Scalar multiplication is associative and distributive, from the properties of multiplication in the field.

Using these definitions, we first study the linear block codes, and then turn to convolutional codes.

Block Codes

An (n, k) block code is a set of 2^k vectors of length n in $V_n(GF(2))$. The *Hamming distance* $d_H(\mathbf{c}_1\ \mathbf{c}_2)$ between $\mathbf{c}_1$ and $\mathbf{c}_2$ in $V_n(GF(2))$ is the number of differing bits between $\mathbf{c}_1$ and $\mathbf{c}_2$. The *Hamming weight* $w_H(\mathbf{c})$ of $\mathbf{c}$ in $V_n(GF(2))$ is the number of ones in $\mathbf{c}$. Clearly

$$d_H(\mathbf{c}_1, \mathbf{c}_2) = w_H(\mathbf{c}_1 \oplus \mathbf{c}_2) \tag{12.132}$$

because $\mathbf{c}_1 \oplus \mathbf{c}_2$ has a component equal to one only in positions where $\mathbf{c}_1$ and $\mathbf{c}_2$ differ. To determine the performance of a code we are interested in finding

$$d_{H,\min} = \min_{\substack{\mathbf{c}_1, \mathbf{c}_2 \in C \\ \mathbf{c}_1 \neq \mathbf{c}_2}} d_H(\mathbf{c}_1, \mathbf{c}_2)\ , \tag{12.133}$$

which from (12.132) is

$$d_{H,\min} = \min_{\substack{\mathbf{c}_1, \mathbf{c}_2 \in C \\ \mathbf{c}_1 \neq \mathbf{c}_2}} w_H(\mathbf{c}_1 \oplus \mathbf{c}_2)\ . \tag{12.134}$$

To find this minimum distance it appears that we have to search over all pairs of codewords $\mathbf{c}_1$ and $\mathbf{c}_2$. Fortunately for linear codes this is not necessary.

An (n, k) *linear* block code C is a k-dimensional *subspace* of $V_n(GF(2))$. By *subspace* we mean that C itself is a vector space, and hence is closed under vector addition. I.e., if $\mathbf{c}_1$ and $\mathbf{c}_2$ are in C, then $\mathbf{c}_1 \oplus \mathbf{c}_2 \in C$. Hence (12.134) becomes

$$d_{H,\min} = \min_{\substack{\mathbf{c} \in C \\ \mathbf{c} \neq \mathbf{0}}} w_H(\mathbf{c})\ . \tag{12.135}$$

To find the minimum Hamming distance in a linear code we need only find the minimum Hamming weight in the linear code. Equivalently, set $\mathbf{c}_1 = \mathbf{0}$ and let $\mathbf{c}_2$ vary over all codewords in (12.134). In other words, when trying to find the probability of the most likely decoding error, assume that the zero vector is transmitted and then consider the probability of decoding to a non-zero codeword.

A *basis* of a vector space is a set of vectors such that every element in the vector space can be expressed as a linear combination of the vectors in the basis. Since the vector space is closed under scalar multiplication and vector addition, every linear combination of basis vectors must be in the vector space. We can use this fact to show that every linear block code can be generated using a generator matrix as shown in (12.7). Simply let the rows of the generator matrix $\boldsymbol{G}$ be a set of basis vectors for the code. Then any codeword may be written $\boldsymbol{m}\mathbf{G}$, where $\boldsymbol{m}$ is a binary vector of length k. It can also be shown that any linear block code has a systematic generator matrix (simply permute the columns of $\mathbf{G}$).

Associated with any (n, k) linear code C is a *dual code* $C^{\perp}$. The dual code is an $(n, n-k)$ linear code consisting of the set of vectors orthogonal to all vectors in C. (The vector $\boldsymbol{x}$ is orthogonal to $\boldsymbol{c}$ if $\boldsymbol{x}\boldsymbol{c}^T = 0$, where $\boldsymbol{c}^T$ is the transpose of $\boldsymbol{c}$.) If $\mathbf{G}$ is the generator matrix and $\mathbf{H}$ is a parity-check matrix for C, then $\mathbf{H}$ is a generator matrix for $C^{\perp}$, and $\mathbf{G}$ is a parity-check matrix.

Exercise 12-12.

Show that the rows of $\mathbf{H}$ are orthogonal to all codewords in C. Then show that $\mathbf{G}$ is a parity-check matrix for the code generated by $\mathbf{H}$.

Convolutional Codes

To study the linearity of convolutional codes we use the definition from (12.60), reproduced here,

$$\boldsymbol{c}(D) = \boldsymbol{m}(D)\mathbf{G}(D) , \tag{12.136}$$

where $\boldsymbol{m}(D)$ is a k-tuple of polynomials in D, $\boldsymbol{c}(D)$ is an n-tuple of polynomials in D, and $\mathbf{G}(D)$ is a $k \times n$ matrix of polynomials in D. Each polynomial has coefficients in $GF(2)$.

Exercise 12-13.

Let F_D be the set of polynomials in D with coefficients in $GF(2)$. Verify that F_D is a ring.

If we augment F_D to include rational polynomials in D, then it is a field as well. Define a vector space $V_n(F_D)$ of n-tuples of polynomials in F_D. The codewords $\boldsymbol{c}(D)$ are in $V_n(F_D)$. Furthermore, from (12.136), $\boldsymbol{c}(D)$ is formed by linear combinations of the rows of $\mathbf{G}(D)$. The rows of $\mathbf{G}(D)$, which are also in $V_n(F_D)$, form a basis for the code. The code is therefore a subspace and hence linear.

To find $d_{H,\min}$ for the code, define the Hamming distance $d_H(\boldsymbol{c}_1(D), \boldsymbol{c}_2(D))$ between two codewords to be the total number of differing coefficients in the polynomials $\boldsymbol{c}_1(D)$ and $\boldsymbol{c}_2(D)$. The Hamming weight $w_H(\boldsymbol{c}(D))$ of $\boldsymbol{c}(D)$ is the total number of non-zero coefficients in $\boldsymbol{c}(D)$. Then just as with block codes,

$$d_H(\boldsymbol{c}_1(D), \boldsymbol{c}_2(D)) = w_H(\boldsymbol{c}_1(D) \oplus \boldsymbol{c}_2(D)) . \tag{12.137}$$

Since the code is linear, $\boldsymbol{c}_1(D) \oplus \boldsymbol{c}_2(D)$ is a codeword and

$$d_{H,\min} = \min_{\substack{\boldsymbol{c}(D) \in C \\ \boldsymbol{c}(D) \neq \mathbf{0}}} w_H(\boldsymbol{c}(D)) . \tag{12.138}$$

We conclude that linear convolutional codes behave like linear block codes in that the minimum Hamming distance is the minimum Hamming weight. Put another way, we can safely assume the transmitted sequence is all zero and find the probability of the code sequence that is closest in Hamming distance to the zero sequence.

Linearity in Signal Space

From the above results we can find the minimum Hamming distance for linear block and convolutional codes with relative ease. The performance of hard decoders is determined primarily by this distance. However, the performance of soft decoders and signal space codes (e.g. trellis codes, covered in Chapter 13) is determined by the minimum *Euclidean* distance between coded *symbol* sequences, rather than *Hamming* distance between *bit* sequences. Often, when considering symbol sequences, linearity does not apply, so we *cannot* safely assume that the zero codeword is transmitted and find the minimum distance from it. An important exception is when the symbol alphabet is binary, meaning that the alphabet has only two symbols. Assume the codeword $\boldsymbol{c}_i$ (a binary vector) with corresponding symbols $\boldsymbol{a}_i$ is transmitted. Assume binary signaling, with symbols selected from the set $\mathcal{A} = \{a, b\}$. The Euclidean distance between $\boldsymbol{a}_i$ and $\boldsymbol{a}_j$ is

$$d_E(\boldsymbol{a}_i, \boldsymbol{a}_j) = |a - b| \sqrt{d_H(\boldsymbol{c}_i, \boldsymbol{c}_j)} \tag{12.139}$$

where $d_H(\boldsymbol{c}_i, \boldsymbol{c}_j)$ is the Hamming distance.

Example 12-50.

Two codewords of a (3, 2) simple parity-check code are $\boldsymbol{c}_1 = [1, 0, 1]$ and $\boldsymbol{c}_2 = [1, 1, 0]$. Then $d_H(\boldsymbol{c}_1, \boldsymbol{c}_2) = 2$. If binary antipodal signaling with $\mathcal{A} = \{\pm 1\}$ is used, then $\boldsymbol{a}_1 = [1, -1, 1]$ and $\boldsymbol{a}_2 = [1, 1, -1]$. The Euclidean distance is $d_E(\boldsymbol{a}_1, \boldsymbol{a}_2) = 2\sqrt{2}$, in agreement with (12.139).

Hence the minimum Euclidean distance between a transmitted sequence $\boldsymbol{a}_i$ and any other sequence $\boldsymbol{a}_j$ is

$$d_{E,\min} = |a - b| \sqrt{d_{H,\min}} \tag{12.140}$$

From the above arguments, if the code is linear, $d_{H,\min}$ can be found by assuming the zero codeword is being transmitted and finding the Hamming distance of all other codewords from the zero codeword.

When the signaling is not binary, things are not so simple. It is still not necessary to consider all possible pairs of codewords, however, in order to find the minimum Euclidean distance between pairs of codewords. Some straightforward simplifications are possible. Again, let $\boldsymbol{a}_i$ and $\boldsymbol{a}_j$ be two blocks of symbols corresponding to two codewords. Then

$$d_E(\boldsymbol{a}_i, \boldsymbol{a}_j) = w_E(\boldsymbol{e}) \, , \tag{12.141}$$

where $w_E(\boldsymbol{e})$ is the *Euclidean weight* of the vector $\boldsymbol{e} = \boldsymbol{a}_i - \boldsymbol{a}_j$ (the square root of the sum of the squares of the elements). Now to find $d_{E,\min}$, observe

$$d_{E,\min} = \min_{\boldsymbol{e} \in \Omega_{\boldsymbol{e}}} w_E(\boldsymbol{e}) \, . \tag{12.142}$$

It is important to note, however, that $\Omega_{\boldsymbol{e}}$ is not the set of permitted symbol sequences.

Example 12-51.

Continuing the previous example, note that

$$w_E(\boldsymbol{a}_1 - \boldsymbol{a}_2) = w_E([0, -2, +2]) = 2\sqrt{2} \, , \tag{12.143}$$

in agreement with previous results. The vector $\boldsymbol{e} = [0, -2, 2]$ does not consist of symbols.

The set Ω_e is generally significantly smaller than the set of all possible pairs of codewords, so the search space for $d_{E,\min}$ is reduced. This is essentially the same technique used in Section 7.6.2 in the determination of the minimum distance of ISI sequences. The complexity can be reduced further by observing that $w_E(e) = w_E(|e|)$ and searching over $\Omega_{|e|}$.

Example 12-52.
Continuing Example 12-51, $\Omega_{|e|}$ is

$$\Omega_{|e|} = \{[0, 2, 2], \quad [2, 0, 2], \quad [2, 2, 0]\} \ , \tag{12.144}$$

so again $d_{E,\min} = 2\sqrt{2}$. A similar reduction of the search space is used in Section 7.6.2 to find the minimum distance between transmitted signals in an ISI channel.

Appendix 12-B. Maximal-Length Feedback Shift Registers

Pseudorandom sequences are generated by a *feedback shift register* as pictured in Fig. 12-32. This device is governed by the relation

$$x_k = h_1 \cdot x_{k-1} \oplus \ldots \oplus h_n \cdot x_{k-n} , \tag{12.145}$$

where the summation is modulo-two, the output x_k is binary assuming the values "0" and "1", and similarly the coefficients of the shift register are binary. The zero coefficients correspond to no feedback tap, whereas the one coefficients correspond to the direct connection of the shift register output to the modulo-two summation.

In this appendix we will consider the properties of a periodic sequence generated by the shift register circuit of Fig. 12-32 with generator polynomial $h(D)$. While a full treatment of this problem requires some sophisticated mathematics, we can understand most of the properties of this generator using only elementary concepts. We will limit ourselves here to generator polynomials having binary coefficients, and all arithmetic will be modulo-two.

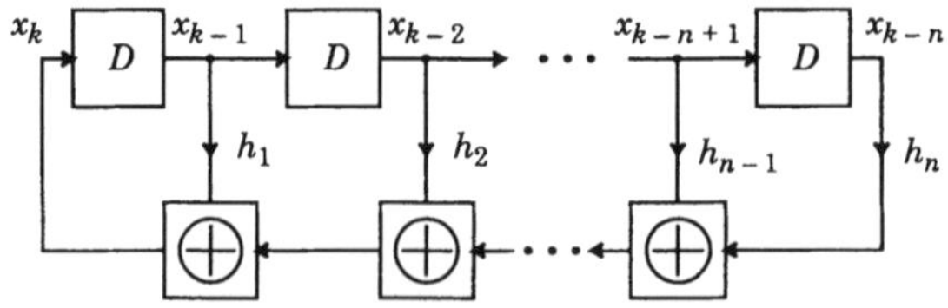

Fig. 12-31. A linear feedback shift register with binary input. The coefficients are binary, and the summation is modulo-two.

Example 12-53.

As an illustration of polynomial arithmetic over GF(2), multiplying the polynomials $(1 \oplus D)$ and $(1 \oplus D \oplus D^2)$,

$$(1 \oplus D)(1 \oplus D \oplus D^2) = 1 \oplus D \oplus D \oplus D^2 \oplus D^2 \oplus D^3 = 1 \oplus D^3 . \tag{12.146}$$

We have used the fact that, for example,

$$D \oplus D = (1 \oplus 1)D = 0 \cdot D = 0 . \tag{12.147}$$

We know that n-th order polynomials with real-valued coefficients always have n roots, but only if we allow those roots to be complex-valued. In general a polynomial with real coefficients cannot always be factored into a product of lower-order polynomials with real coefficients (for example $X^2 + 1$). Similarly, a GF(2) polynomial cannot always be factored into two or more polynomials with GF(2) coefficients.

Example 12-54.

Continuing Example 12-53, the polynomial $(1 \oplus D \oplus D^2)$ cannot be factored into the product of two first order polynomials over GF(2). In fact, the only first-order polynomials over GF(2) are D and $(1 \oplus D)$, and the reader can readily verify that they cannot be factors of $(1 \oplus D \oplus D^2)$.

A polynomial that has no factors other than itself and 1 is called an *irreducible polynomial over GF(2)*. In the sequel we will assume that the generator polynomial $h(D)$ is irreducible.

Returning to the feedback shift-register, the state $(x_{k-1}, \ldots, x_{k-n})$ can assume at most 2^n distinct values. From this fact, and other properties of the register, we can discern the following properties:

- If the state of the shift-register is all-zero (00 … 0) at any time, then it must always be all-zero. Thus, we must ensure that this state is never visited unless we are satisfied with a complicated circuit that just generates all-zeros at the output.
- If the state ever stays the same from one time increment to the next, then it will forever be the same. Thus, if the output is to be interesting (anything but all-zeros or all-ones), then we must ensure that the state always changes upon every time increment.
- The sequence of states must be periodic. Since there are only 2^n distinct states, the sequence of states must always return to an initial state, after which the sequence of states repeats. Since the output x_k is a function of the state, it must also be periodic.

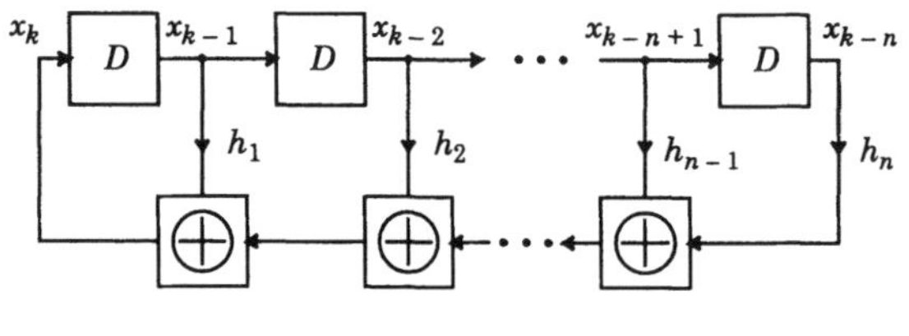

Fig. 12-32. A linear feedback shift register with binary input. The coefficients are binary, and the summation is modulo-two.

- Combining the first three, the maximum period of the states and outputs must be $(2^n - 1)$ time increments. This maximum period would correspond to a periodic sequence of states which change at every time increment and which cycle through every state except the all-zero state.

A feedback shift-register is called *maximal-length* if the period of the output is $r = 2^n - 1$.

Example 12-55.

Consider a shift-register base on the polynomial $h(D) = 1 \oplus D \oplus D^2$. From Example 12-54 this generator polynomial is irreducible. We can verify that the shift register is maximal-length, that is has period $2^2 - 1 = 3$. Starting with state (0,1), the following table specifies the state and output vs. time for four cycles:

x_k	x_{k-1}	1
1	0	1
1	1	0
0	1	1
1	0	1

Note that the state has returned to its initial value at the fourth time increment, and therefore the shift-register will continue with the same sequence of states. Also note that if we initialized the state with any of the other two values, the same sequence of states would result, but we would just start at a different point in the sequence.

We could easily envision that the period of a shift-register sequence could be less than $2^n - 1$ in length.

Exercise 12-14.

Show that the shift-register sequence corresponding to polynomial $h(D) = 1 \oplus D^2$ has period one or two depending on the initial state.

We would like to have some criterion to establish when a generator polynomial corresponds to a maximal-length shift-register sequence. When an irreducible polynomial $h(D)$ of degree n does not divide any polynomial $(1 \oplus D^m)$ for $m < 2^n - 1$, it is said to be *primitive*. A shift-register sequence is maximal-length if and only if the generator polynomial is primitive [55].

Example 12-56.

We can verify that the generator polynomial of Example 12-55, $h(D) = 1 \oplus D \oplus D^2$, is primitive. This is because it obviously does not divide $(1 \oplus D^2)$, while it does divide $(1 \oplus D^3)$ since

$$(1 \oplus D \oplus D^2)(1 \oplus D) = 1 \oplus D^3 \qquad (12.148)$$

from Example 12-53.

Fortunately, there exist primitive polynomials of all orders. The polynomials with *minimum weight*, that is with the minimum number of shift-register taps, of all orders up to $n = 34$ are listed in Table 12-1.

Example 12-57.

A maximal-length shift-register of order 12 can be found from Table 12-1. The octal entry is "10123," which corresponds to binary "1000001010011" and hence polynomial

$$h(D) = 1 \oplus D \oplus D^4 \oplus D^6 \oplus D^{12} . \tag{12.149}$$

Hence the shift-register is characterized by difference equation

$$x_k = x_{k-1} \oplus x_{k-4} \oplus x_{k-6} \oplus x_{k-12} . \tag{12.150}$$

An interesting property of maximal-length sequences is that if we look at n-bit segments of the sequence, we will see all possible n-bit words, with the exception of the all-zero word. This follows from the fact that the state of the shift-register passes through all possibilities except all-zeros, and the state is equal to the past n bits of the output. The maximal-length sequence therefore satisfies a minimal condition for "randomness," since we would expect to see all combinations of bits (except the all-zero) in such a sequence.

The output of a maximal-length shift register is often called a *pseudorandom sequence.* This is because, even though the sequence is deterministic and periodic, it displays many of the properties of a random sequence (analogous for example to a numerical algorithm for random number generation). We can see these properties reflected in the *relative frequency* and in the *autocorrelation function.*

The relative frequency of observing particular sequences of i bits in a maximal-length sequence is close to the probability of observing the i bits in an i.i.d. random sequence as long as $i \le n$, since all possible sequences of n bits occur once in one period of $2^n - 1$ bits, with the exception of the all-zero sequence.

Table 12-1. Minimal weight primitive polynomials of orders two through 34 [55]. Each entry in the table is an octal number, which when converted to binary specifies the coefficients of the polynomial $h(D)$. The most significant (left-most) bit is $h_n = 1$ and the least significant (right-most) bit is $h_0 = 1$.

Order	Polynomial	Order	Polynomial
2	7	19	2000047
3	13	20	4000011
4	23	21	10000005
5	45	22	20000003
6	103	23	40000041
7	211	24	100000207
8	435	25	200000011
9	1021	26	400000107
10	2011	27	1000000047
11	4005	28	2000000011
12	10123	29	4000000005
13	20033	30	10040000007
14	42103	31	20000000011
15	100003	32	40020000007
16	210013	33	100000020001
17	400011	34	201000000007
18	1000201		

Exercise 12-15.
Show that the relative frequency of any particular sequence of $i \leq n$ bits in the maximal-length sequence is

$$\frac{2^{n-i}}{2^n - 1} \approx 2^{-i} \tag{12.151}$$

for the case where the i bits do not constitute the all-zero sequence, and for the all-zero sequence of i bits

$$\frac{2^{n-i} - 1}{2^n - 1} \approx 2^{-i}. \tag{12.152}$$

The approximations apply to large n, and hence for this case the sequence looks random on a relative frequency basis as long as we don't observe blocks of bits greater than n.

The autocorrelation function can be determined using the *cycle-and-add property* of the maximal-length sequence [56]. This property says that if we modulo-two add the maximal-length sequence to itself, where one of the sequences has been shifted in time, we get another version of the same sequence shifted in time,

$$x_k \oplus x_{k+l} = x_{k+j} \tag{12.153}$$

for $l \in \{0, 1, \ldots, r-1\}$, where j depends on l. Of course, when $l = 0$ the sum is the all-zero sequence (this is a degenerate case of a maximal-length sequence). The cycle-and-add property follows from the fact that if $h(D)X(D) = 0$, then obviously $(1 \oplus D^l)h(D)X(D) = 0$, and therefore $(1 \oplus D^l)X(D)$ must also have generator polynomial $h(D)$. Many interesting properties can be derived from (12.153).

Example 12-58.
Since $x_k \oplus x_{k+l} = 0$ if and only if $x_k = x_{k+l}$, it follows that $x_k = x_{k+l}$ for precisely $(r-1)/2$ values of k within one period $k \in \{0, 1, \ldots, r-1\}$, and $x_k \neq x_{k+l}$ for precisely $(r+1)/2$ values of k. Again, $r = 2^n - 1$ is the length of the sequence.

In terms of the autocorrelation, we are usually interested in the autocorrelation of a binary antipodal sequence s_k obtained by mapping $x_k = 0$ into $s_k = -1$ and $x_k = 1$ into $s_k = +1$. We will call this new sequence the *binary antipodal maximal-length sequence*. The autocorrelation function of this sequence is defined as

$$R_s(l) = \frac{1}{r}\sum_{k=0}^{r-1} s_k s_{k+l}. \tag{12.154}$$

This is a *time-average* autocorrelation function averaged over one period of the sequence. Of course it is a periodic function of l, and hence we need only be concerned with the value for $l \in \{0, 1, \ldots, r-1\}$. Similarly, we can define a time-average mean value of the sequence as

$$\mu_s = \frac{1}{r}\sum_{k=0}^{r-1} s_k. \tag{12.155}$$

Exercise 12-16.

Using the relative frequency property, show that

$$\mu_s = \frac{1}{r} \tag{12.156}$$

which approaches zero as n (and hence r) gets large.

Exercise 12-17.

Use the cycle-and-add property to show that the autocorrelation function is given by

$$R_s(l) = \begin{cases} 1, & l \neq 0 \\ -1/r, & l \in \{1, 2, \ldots, r-1\} \end{cases} . \tag{12.157}$$

Hence, when r is large, the time-average autocorrelation function approaches zero except at multiples of the period. Except for the periodicity, this approaches the autocorrelation of a white sequence, and hence is another indication of the pseudo-random property.

Using this time-average autocorrelation, we can infer another important property of the binary antipodal maximal-length sequence; namely, its harmonic structure. Since this sequence is periodic, we can expand it using a DFT,

$$s_k = \frac{1}{r}\sum_{m=0}^{r-1} S_m e^{j2\pi mk/r}, \tag{12.158}$$

where

$$S_m = \sum_{k=0}^{r-1} s_k e^{-j2\pi mk/r}, \qquad m \in \{0, 1, \ldots, r-1\} . \tag{12.159}$$

We can easily relate the harmonics of the sequence to the autocorrelation function.

Exercise 12-18.

(a) Show that

$$\sum_{k=0}^{r-1} s_{k+l}\, e^{-j2\pi mk/r} = e^{j2\pi mk/r}\sum_{k=0}^{r-1} s_k\, e^{-j2\pi mk/r}. \tag{12.160}$$

(b) Show that

$$\frac{1}{r}\,|S_m|^2 = \sum_{l=0}^{r-1} R_s(l)\, e^{-j2\pi mk/r}. \tag{12.161}$$

(c) Evaluate this DFT to show that

$$|S_m|^2 = \begin{cases} 1, & m = 0 \\ r+1, & m \in \{1, 2, \ldots, r-1\} \end{cases} . \tag{12.162}$$

Hence, the harmonics of the sequence are all equal to one another in magnitude, except for the d.c. component, which is relatively small. This resembles the power spectrum of a white sequence, and this property makes these sequences desirable as spreading sequences in spread spectrum (Section 6.4.3).

Appendix 12-C. Path Enumerators

To estimate the performance of convolutional codes we need to find the most probable error event. For simple examples it can be found by inspection, but this method is seriously prone to error. In Section 7.6.2 we described a general technique that requires assuming an actual state trajectory and then using the Viterbi algorithm to find the minimum-distance error event for that state trajectory. In general, all possible actual state trajectories must be considered. In Section 7.6.2 we showed that this is not necessary for ISI examples, because we can exploit the linearity of the signal generation model. Fortunately, it is also not necessary for linear codes, which turn out to be even simpler than the ISI case. As shown in Appendix 12-A, for linear codes it is sufficient to consider only one actual path through the trellis. Hence, we can use the Viterbi algorithm to find the minimum-distance error event for any assumed actual path through the trellis, as detailed in Section 7.6.2, and the distance will be the global minimum distance. As indicated in the Appendix 12-A, this will work for linear codes using a hard decoder, or a soft decoder with binary antipodal signaling. In this appendix, we give an alternative technique using signal flow graphs (as in Section 3.3). Although this is not necessarily simpler than using the Viterbi algorithm, the technique can be used to simultaneously compute essentially all the information about the error events, such as their length, the number of bit errors in each one, and the number of error events at each distance. It gives much more information than just the minimum distance.

The general technique requires an assumption that the correct state trajectory remains in the zero state. This is why the technique is restricted to linear codes! An error event is therefore a path that leaves the zero state and later returns. We can enumerate all such paths, as shown in the following example.

Example 12-59.
Consider the convolutional coder in Fig. 12-38 (this coder is studied in Problem 12-10). The state transition diagram is shown in Fig. 12-33. The branches in the diagram are labeled with the variable z raised to a power equal to the Hamming distance of that branch from the zero branch. For

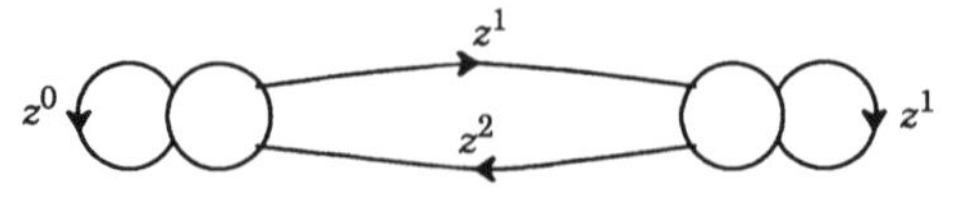

Fig. 12-33. The state transition diagram for the convolutional coder in Problem 12-10.

instance, the label $z^0 = 1$ follows from the fact that this branch *is* the zero branch, and hence its distance is zero. We are interested in the distance of all paths from the zero state back to the zero state. To find it, break the zero state into two states as shown in Fig. 12-34. If we view this graph as a signal flow graph, then the gain from the 0_1 state to the 0_2 state is a polynomial in z that enumerates the weights of all possible paths from the zero state back to itself. By inspection that polynomial is

$$T(z) = z^3 + z^4 + z^5 + \ldots \tag{12.163}$$

It is easy to see that error event with the minimum Hamming distance has Hamming distance 3, the error event with the second smallest Hamming distance has distance 4, etc. The polynomial $T(z)$ is called a path-enumerator polynomial. It can be expressed more compactly as

$$T(z) = \frac{z^3}{1-z} \; . \tag{12.164}$$

It can be readily verified that this is the same as (12.163) by carrying out the long division. The same technique can be used to enumerate the Euclidean distances of error events in order to determine the performance of a soft decoder as long as the alphabet is binary (see Problem 12-16).

The convolutional coder considered above does not have nearly as much coding gain as others of comparable complexity that we have considered. Unfortunately, for most useful linear codes, constructing the path enumerator polynomial by inspection can be difficult. Fortunately, a computer program or well-documented techniques from the literature can be used (for example, Mason's gain formula [46][47][48] is useful).

Example 12-60.

The convolutional coder state transition diagram of Fig. 12-13 is modified as shown in Fig. 12-35. The branch weights are again the variable z raised to the power of the Hamming distance of the branch from the zero branch. It is possible to show that

$$T(z) = \frac{z^5}{1-2z} \; . \tag{12.165}$$

By long division

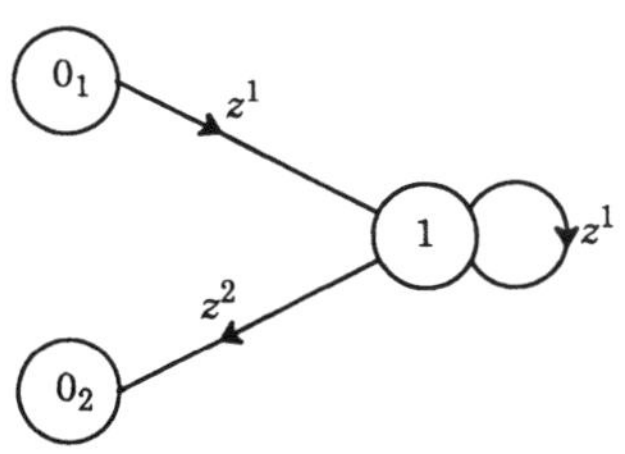

Fig. 12-34. To enumerate the paths from the zero state back to itself, break the zero state in two as shown. The weights on the arcs are a variable z raised to the power of the Hamming distance from the zero path.

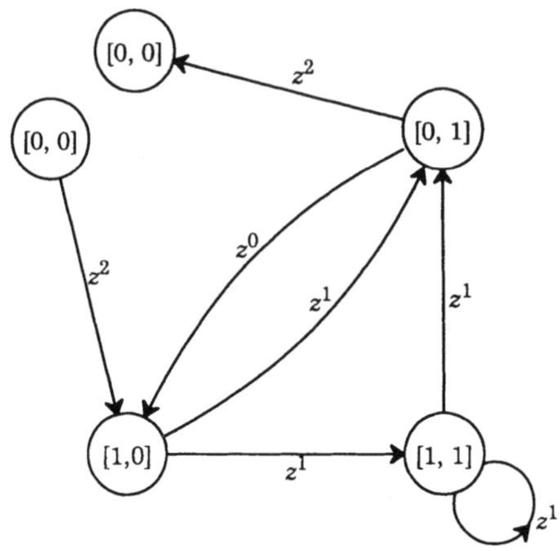

Fig. 12-35. The state transition diagram of Fig. 12-13 is modified for the purpose of enumerating the paths from the [0, 0] state back to itself. The branch weights are equal to z raised to the Hamming weight of the branch.

$$T(z) = z^5 + 2z^6 + 4z^7 + 8z^8 + \ldots \tag{12.166}$$

This says that there is one error event with Hamming distance 5 (which we knew already), two error events with Hamming distance 6, four with distance 7, etc.

The path enumerator technique can be used to obtain a variety of information about a code in addition to the distances of the error events. The extension is simple. In Problem 12-17 we show how to enumerate the number of bit errors in the error events and the length of the error events.

Appendix 12-D. Derivation of the Tanh Rule

Here we derive the multiplicative and additive forms ((12.100) and (12.100)) of the tanh rule.

Multiplicative Form

The parity $\phi = \Sigma_{i=1}^{n} c_i$ (modulo-two sum) of a set of n bits $\{c_1, \ldots c_n\}$ can be expressed as:

$$\phi = \tfrac{1}{2}\Big(1 - \prod_{i=1}^{n}(1 - 2c_i)\Big), \tag{12.167}$$

since an even number of ones in $\{c_i\}$ yields $\phi = 0$, while an odd number yields $\phi = 1$. The probability that the parity is odd is then the expected value of ϕ:

$$\Pr[\phi = 1] = E[\phi] \tag{12.168}$$

$$= \tfrac{1}{2}\Big(1 - E\Big[\prod_{i=1}^{n}(1-2c_i)\Big]\Big) \tag{12.169}$$

$$= \tfrac{1}{2}\Big(1 - \prod_{i=1}^{n}(1-2E[c_i])\Big) \quad \text{(because of independence)} \tag{12.170}$$

$$= \tfrac{1}{2}\Big(1 - \prod_{i=1}^{n}\Big(1 - \frac{2e^{\lambda_i}}{1+e^{\lambda_i}}\Big)\Big) \tag{12.171}$$

$$= \tfrac{1}{2}\Big(1 - \prod_{i=1}^{n}\frac{1-e^{\lambda_i}}{1+e^{\lambda_i}}\Big) \tag{12.172}$$

$$= \tfrac{1}{2}\Big(1 - \prod_{i=1}^{n}\tanh(-\lambda_i/2)\Big). \tag{12.173}$$

Since $\Pr[\phi = 0] = 1 - \Pr[\phi = 1]$, the LLR for ϕ becomes:

$$\lambda_\phi = \log\frac{\Pr[\phi = 1]}{\Pr[\phi = 0]} \tag{12.174}$$

$$= \log\left(\frac{1 - \Pi_i \tanh(-\lambda_i/2)}{1 + \Pi_i \tanh(-\lambda_i/2)}\right). \tag{12.175}$$

Using $\tanh(-\lambda/2) = (1 - e^{\lambda})/(1 + e^{\lambda})$, and letting $\Pi = \prod_{i=1}^{n}\tanh(-\lambda_i/2)$, we get:

$$\tanh(-\lambda_\phi/2) = \frac{1 - \left(\frac{1-\Pi}{1+\Pi}\right)}{1 + \left(\frac{1-\Pi}{1+\Pi}\right)} = \frac{(1+\Pi)-(1-\Pi)}{(1+\Pi)+(1-\Pi)} = \Pi = \prod_{i=1}^{n}\tanh(-\lambda_i/2), \tag{12.176}$$

which proves the tanh rule of (12.98).

Additive Form

We now derive the additive form of the tanh rule, namely (12.100). Substituting $-\lambda_i = \text{sign}(-\lambda_i)\cdot|\lambda_i|$ into (12.98) yields the pair of equations:

$$\text{sign}(-\lambda_\phi) = \prod_{i=1}^{n}\text{sign}(-\lambda_i), \tag{12.177}$$

$$\tanh\left(\frac{|\lambda_\phi|}{2}\right) = \prod_{i=1}^{n}\tanh\left(\frac{|\lambda_i|}{2}\right). \tag{12.178}$$

Taking –log(·) of both sides of (12.178) yields:

$$f(|\lambda_\phi|) = \sum_{i=1}^{n} f(|\lambda_i|), \tag{12.179}$$

where $f(x)$ is defined by (12.101). Because $f(f(x)) = x$ for all $x > 0$, we can apply $f(\cdot)$ to both sides of (12.179), yielding:

$$|\lambda_\phi| = f\left(\sum_{i=1}^{n} f(|\lambda_i|)\right). \tag{12.180}$$

Combining (12.177) and (12.180) yields (12.100).

Problems

Problem 12-1. An $(n, 1)$ *repetition code* is one where each bit is repeated n times.

(a) Compute the minimum Hamming distance $d_{H,\min}$ of such a code as a function of n.

(b) How does this coding technique compare with the (7, 4) Hamming code with a hard decoder?

Problem 12-2. Consider a linear block code C with parity-check matrix $\mathbf{H}$ and minimum Hamming distance $d_{H,\min}$ between codewords.

(a) Show that $d_{H,\min}$ is equal to the minimum number of columns of $\boldsymbol{H}$ that can be added to produce $\mathbf{0}$.

(b) Use part (a) to show that for all linear block codes

$$d_{H,\min} \leq n - k + 1 \ . \tag{12.181}$$

Problem 12-3. Given the generator matrix for a (7,3) linear block code

$$\mathbf{G} = \begin{bmatrix} 0 & 0 & 1 & 1 & 1 & 0 & 1 \\ 0 & 1 & 0 & 0 & 1 & 1 & 1 \\ 1 & 0 & 0 & 1 & 1 & 1 & 0 \end{bmatrix}, \tag{12.182}$$

(a) Construct the generator matrix of an equivalent systematic code.

(b) Find the parity-check matrix $\mathbf{H}$.

(c) Construct a table of all possible syndromes $\boldsymbol{s}$ and find the error pattern $\hat{\boldsymbol{e}}$ most likely to have resulted in that syndrome.

(d) What is the relationship of this code to the (7, 4) Hamming code?

(e) Find $d_{H,\min}$. How many bit errors in a block can be reliably corrected?

(f) Find the codeword $\boldsymbol{c} = \boldsymbol{m}\mathbf{G}$ for $\boldsymbol{m} = [1\ 0\ 1]$ and verify that $\boldsymbol{c}\mathbf{H}^T = \mathbf{0}$.

Problem 12-4.

(a) Give the generator and parity-check matrices in systematic form for the (15,11) Hamming code.

(b) Find a parity-check matrix for a non-systematic (15,11) Hamming code such that the syndrome of any bit pattern with one bit error can be interpreted as a binary number that identifies the position of the bit error.

Problem 12-5. Compare the performance of a (15, 11) Hamming code with a soft decoder to an uncoded system with the same source bit rate. Assume that both the coded and uncoded systems use a binary antipodal symbol alphabet $\{\pm a\}$. You may ignore the constant multiplying the $Q(\cdot)$ function.

Problem 12-6. Estimate the power advantage of the (15,4) maximal-length shift register code with a soft decoder. You may ignore the constants in front of the $Q(\cdot)$ term. To get the power advantage, compare to an uncoded system that also uses binary antipodal signaling, and a sufficiently lower symbol rate that the source bit rate is the same. Assume additive white Gaussian noise on the channel.

Problem 12-7. Consider the non-systematic convolutional coder in Fig. 12-36(a).

(a) Find the parity-check matrix.

(b) Show that Fig. 12-36(b) has the same parity-check matrix.

Problem 12-8. Consider transmitting bits m_k over a BSC channel using the convolutional coder conv(1/2) of Fig. 12-10(a). Assume that $m_k = 0$ for $k < 0$ and $k \geq K$. Suppose $K = 3$ and the observation sequence is $\{0,1, 0, 1, 1, 1, 1, 0, 0, 0, ...\}$. Draw the trellis for the Markov model and label the transition weights. What is the ML estimate of the incoming bit sequence?

Problem 12-9. Consider the rate-1/2 convolutional coder shown in Fig. 12-37.

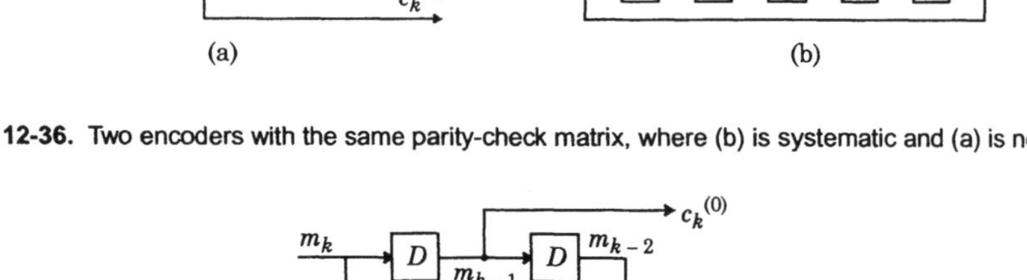

Fig. 12-36. Two encoders with the same parity-check matrix, where (b) is systematic and (a) is not.

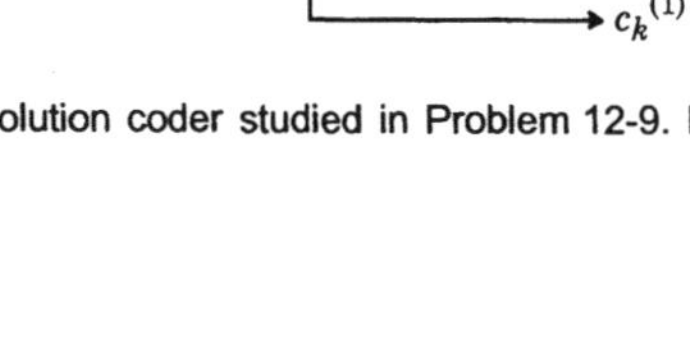

Fig. 12-37. A rate-1/2 convolution coder studied in Problem 12-9. It is not as good as the coder in Fig. 12-10(a).

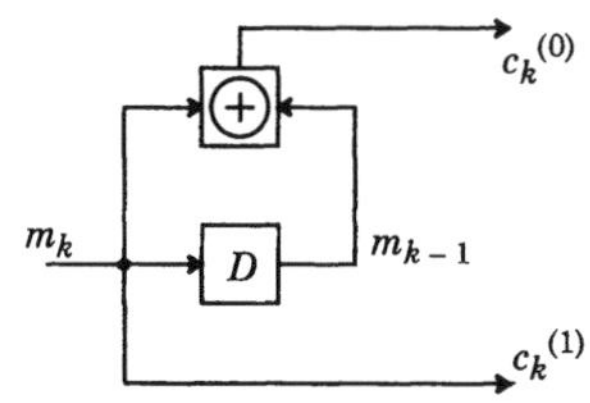

Fig. 12-38. A convolutional coder studied in Problem 12-10 and Problem 12-17.

(a) Draw the state transition diagram and trellis with each transition labeled with $(m_k,\ [c_k^{(0)},\ c_k^{(1)}])$.

(b) Assume that $c_k^{(0)}$ and $c_k^{(1)}$ are interleaved on a BSC with probability p of flipping a bit. Find the error event with the minimum Hamming distance. Find the probability of the error event and compare with the probability computed in (12.85).

(c) Assume that $c_k^{(0)}$ and $c_k^{(1)}$ are to be interleaved over an additive white Gaussian noise channel. Assume A_i is the symbol sequence chosen from the alphabet $\{-a,\ +a\}$. Assume an ML soft decoder. Estimate the probability of error event and compare it to the uncoded system and to conv(1/2) with soft decoding whose performance is approximated in (12.82). Give the comparison in dB.

Problem 12-10. Consider the convolutional coder shown in Fig. 12-38.

(a) Find the Hamming distance of the minimum distance error event and give an upper bound (like that in (12.85)) for the probability of this error event using a hard decoder.

(b) Assuming binary antipodal signaling with alphabet $\{\pm a\}$, find the error event with the minimum Euclidean distance. Estimate the coding gain using a soft decoder, assuming that this error event dominates. You may neglect the constant multiplying the $Q(\cdot)$ function.

Problem 12-11. Is the code consisting of the following codewords linear?

$$0010 \quad 0100 \quad 1110 \quad 1000 \quad 1010 \quad 1100 \quad 0110\ . \tag{12.183}$$

Problem 12-12. List the codewords of the dual of the (7, 4) Hamming code. What are n, k, and $d_{H,\min}$?

Problem 12-13. Consider the (3, 1) repetition code with binary antipodal signaling over an AWGN channel. Find numerical values for the three observations r_0, r_1, and r_2 such that the hard ML decision differs from the soft ML decision. If no such values exist, explain why.

Problem 12-14. Let $\mathbf{G}(D)$ be a matrix with one row and $n = 2^{\mu+1} - 1$ columns consisting of all possible nonzero polynomials of order μ or less. (For example, when $\mu = 1$, $\mathbf{G}(D) = [1,\ D,\ 1 + D]$.) This defines a family of rate-$1/n$ convolutional codes, parameterized by the encoder memory μ. Find a closed-form expression for the minimum Hamming distance $d_{\min}$ of the convolutional code for an arbitrary μ, expressed as a function of μ.

Problem 12-15. Find the minimum Hamming distance $d_{\min}$ between all distinct binary codewords generated by the rate-1/4 convolutional encoder with generator polynomial:

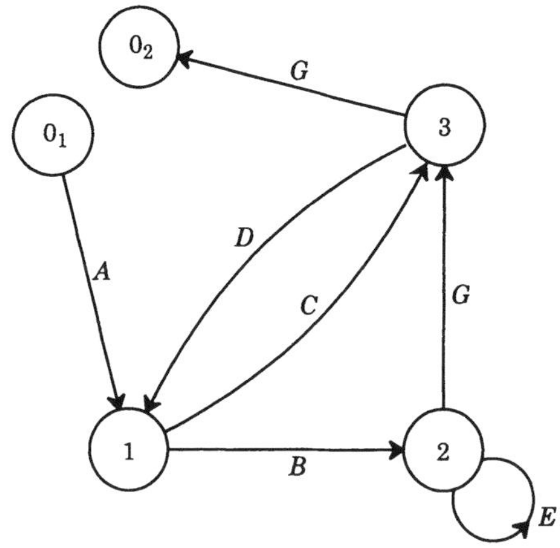

Fig. 12-40. A signal flow graph with branch weights shown.

$$G(D) = [1 + D + D^2, 1 + D + D^2, 1 + D + D^2, 1 + D^2] \quad (12.184)$$

Problem 12-16. Assuming binary antipodal signaling for the convolutional coder of Example 12-59, find an enumerator polynomial for the *squares* of the Euclidean distances.

Problem 12-17. In this problem we show how to use the path enumerator polynomials to find the distances, number of bit errors, and lengths of all error events simultaneously (for linear codes). Consider the convolutional coder in Fig. 12-38. To find just the Hamming distances of the error events we use the broken state transition diagram of Fig. 12-34. To find the number of bit errors and lengths of the error events we use the labels in Fig. 12-39.

(a) Find an expression for the gain from 0_1 to 0_2. It will be a polynomial $T(x, y, z)$ in x, y, and z.

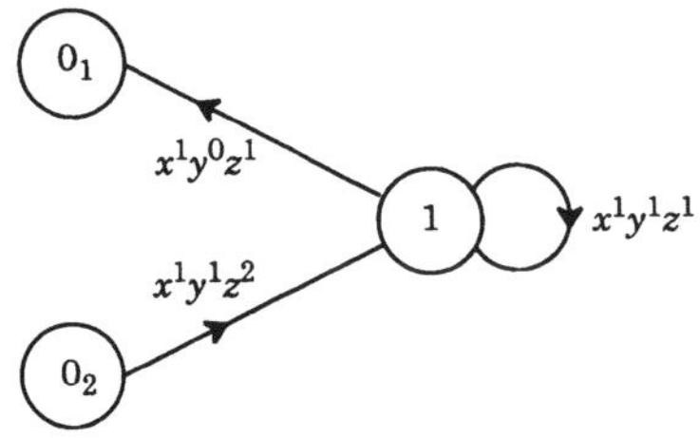

Fig. 12-39. The state transition diagram of Fig. 12-34 has been modified so that the exponent of x denotes the length of the branch (always one) and the exponent of y denotes the number of bit errors that occur if that branch is selected by the decoder (assuming the zero path is correct). The exponent of z still shows the Hamming weight of the branch.

(b) What is the distance and number of bit errors in the error event with length four (i.e. the error event that traverses five incorrect branches, or four incorrect states, before returning to the zero state)?

(c) Suppose you are told that the gain from 0_1 to 0_2 in Fig. 12-40 is

$$T = \frac{ACG(1-E)+ABFG}{1-DC-E-BFD+EDC} \,. \tag{12.185}$$

For the convolutional coder of Example 12-60, determine the number of distance 6 error events and their lengths, and the number of length 4 error events and their distances.

Problem 12-18. Consider the (3, 1) repetition code.

(a) Show that the Tanner graph has no cycles.

(b) Show that the message-passing algorithm converges to the true a posteriori LLR's after two iterations.

Problem 12-19. Does there exist a positive integer m for which the Tanner graph for the $(2^m - 1, 2^m - 1 - m)$ Hamming code has no cycles? Explain.

References

1. S. Shamai (Shitz), L. H. Ozarow, and A. D. Wyner, "Information Rates for a Discrete-Time Gaussian Channel with Intersymbol Interference and Stationary Inputs," *IEEEE Transactions on Information Theory,* Vol. 37, No. 6, pp. 1527-1539, November 1991.

2. D. J. Costello, Jr., J. Hagenauer, H. Imai, and S. B. Wicker, "Applications of Error-Control Coding," *IEEE Trans. on Information Theory,* Vol. 44, No. 6, pp. 2531-2560, October 1998.

3. T. J. Richardson, A. Shokrollahi, and R. Urbanke, "Design of Capacity-Approaching Low-Density Parity-Check Codes, *IEEE Trans. Info. Theory,* vol. 47, no. 2, pp. 619–637, Feb. 2001.

4. R. G. Gallager, "Low-Density Parity-Check Codes," *IRE Trans. Inform. Theory,* vol. IT-8, pp. 21–28, January 1962.

5. C. Berrou, A. Glavieux, and P. Thitimajshima, "Near Shannon Limit Error-Correcting Coding and Decoding: Turbo Codes," *IEEE Int. Conf. on Commun.*, Geneva, pp. 1064 1070, May 1993.

6. R. Gallager, *Information Theory and Reliable Communication*, Wiley, New York (1968).

7. D. Chase, "A Class of Algorithms for Decoding Block Codes with Channel Measurement Information," *IEEE Trans. on Information Theory,* Vol. IT-18, pp. 170-182 (Jan. 1972).

8. G. D. Forney, Jr., "Generalized Minimum Distance Decoding," *IEEE Trans. on Information Theory,* Vol. IT-12, pp. 125-131 (April 1966).

9. S. Wainberg and J. K. Wolf, "Algebraic Decoding of Block Codes Over a q-ary Input, Q-ary Output Channel, $Q > q$," *Information and Control,* Vol. 22, pp. 232-247, April 1973.

10. E. J. Weldon, Jr., "Decoding Binary Block Codes on Q-ary Output Channels," *IEEE Trans. on Information Theory*, Vol. IT-17, pp. 713-718 (Nov. 1971).

11. J. K. Wolf, "Efficient Maximum Likelihood Decoding of Linear Block Codes Using a Trellis," *IEEE Trans. on Information Theory*, Vol. IT-24, pp. 76-81 (Jan. 1978).

12. J. G. Proakis, *Digital Communications*, Fourth Edition, McGraw-Hill, New York (2001).

13. W. Peterson and E. Weldon, *Error-Correcting Codes*, 2nd Ed., M.I.T. Press, Cambridge, Mass. (1972).

14. R. G. Gallager, *Low-Density Parity-Check Codes*, MIT Press, Cambridge, MA, 1963.

15. D. J. C. MacKay and R. M. Neal, "Near Shannon-Limit Performance of Low-Density Parity-Check Codes," *Electronics Letters*, vol. 32, pp. 1645–1646, Aug. 1996.

16. S.-Y. Chung, G. D. Forney, Jr., T. J. Richardson, and R. Urbanke, "On the Design of Low-Density Parity-Check Codes within 0.0045 dB of the Shannon Limit," *IEEE Commun. Letters*, vol. 5, pp. 58-60, February 2001.

17. C. E. Shannon, "A Mathematical Theory of Communication," *Bell System Technical Journal*, vol. 27, pp. 379–423 and 623–656, July / Oct. 1948.

18. R. M. Tanner, "A Recursive Approach to Low Complexity Codes," *IEEE Trans. Inform. Theory*, vol. IT-27, pp. 533-547, September 1981.

19. F. R. Kschischang, B. J. Frey, and H.-A. Loeliger, "Factor Graphs and the Sum- Product Algorithm," *IEEE Trans. Information Theory*, vol. 47, no. 2, pp. 498-519, February 2001.

20. J. Hagenauer, E. Offer, and L. Papke, "Iterative Decoding of Binary Block and Convolutional Codes," *IEEE Trans. Information Theory*, vol. 42, pp. 429–445, March 1996.

21. C. Berrou, A. Glavieux, and P. Thitimajshima, "Near Shannon Limit Error- Correcting Coding and Decoding: Turbo Codes," *IEEE Intern. Conf. Commun.*, pp. 1064–70, Geneva, May 1993.

22. A. J. Blanksby and C.J. Howland, "A 690-mW 1-Gb/s 1024-bit, rate-1/2 Low-Density Parity-Check Code Decoder," *IEEE J. Solid-State Circuits*, vol.37, No.3, pp. 404-12, March 2002.

23. S.-Y. Chung, R. Urbanke, and T. J. Richardson, "Gaussian Approximation for Sum-Product Decoding of Low-Density Parity-Check Codes," in *Proc. Int. Symp. Inform. Theory*, page 318, Sorrento, Italy, June 2000.

24. S.-Y. Chung, "On the Construction of Some Capacity-Approaching Coding Schemes," *Ph.D. Dissertation*, Massachusetts Institute of Technology, Cambridge, MA, 2000.

25. S.-Y. Chung, T. J. Richardson, and R. Urbanke, "Analysis of Sum-Product Decoding of Low-Density Parity-Check Codes using a Gaussian Approximation," *IEEE Trans. Information Theory*, vol. 47, no. 2, pp. 657-670, February 2001.

26. H. El Gamal and A. R. Hammons, Jr., "Analyzing the Turbo Decoder using the Gaussian Approximation," in *Proc. Int. Symp. Inform. Theory*, page 319, Sorrento, Italy, June 2000.

27. D. J. Costello Jr., P. C. Massey, O. M. Collins, and O. Y. Takeshita, "Some Reflections on the Mythology of Turbo Codes," *Proc. 3rd ITG Conference on Source and Channel Coding*, Munich, Germany, pp. 157-160, January 2000.

28. S. Benedetto and G. Montorsi, "Unveiling turbo codes: Some results on parallel concatenated coding schemes," *IEEE Trans. Inform. Theory*, vol. 42, pp. 409 428, Mar. 1996.

29. G. D. Forney, Jr., Concatenated Codes. Cambridge, MA: MIT Press, 1966.

30. S. B. Wicker, V. K. Bhargava, *Reed-Solomon Codes and their Applications*, IEEE Press, 1994.

31. G. D. Forney, Jr. and G. Ungerboeck, "Modulation and Coding for Linear Gaussian Channels," *IEEE Trans. Information Theory*, Vol. 44, No. 6, pp. 238-2415, 1998.

32. S. Benedetto, D. Divsalar, G. Montorsi, and F. Pollara, "Serial concatenation of interleaved codes: Performance analysis, design, and iterative decoding," *IEEE Trans. Inform. Theory*, vol. 44, pp. 909-926, May 1998.

33. R. J. McEliece, "Achieving the Shannon Limit: a Progress Report," Plenary Lecture, *Allerton Conference on Communication, Control, and Computer*, 2000.

34. H. Jin, A. Khandekar, and R. McEliece, "Irregular Repeat-Accumulate Codes," *Proceedings 2nd International Symposium on Turbo codes and Related Topics*, Brest, France, pp. 1-8, Sept. 2000.

35. E. Berlekamp, *Algebraic Coding Theory*, McGraw-Hill Book Co., New York (1968).

36. R. E. Blahut, *Theory and Practice of Error Control Codes*, Addison-Wesley, (1983).

37. S. B. Wicker, *Error Control Systems for Digital Communication and Storage*, Prentice-Hall, (1995).

38. F. J. MacWilliams and J. J. Sloane, *The Theory of Error Correcting Codes*, North-Holland, New York (1977).

39. J. Massey, "Error Bounds for Tree Codes, Trellis Codes, and Convolutional Codes with Encoding and Decoding Procedures," in *Coding and Complexity*, ed. G. Longo, Springer-Verlang, New York (1977).

40. R. J. McEliece, *The Theory of Information and Coding*, Addison Wesley Pub. Co. (1977).

41. G. D. Forney, Jr., "Convolutional Codes I: Algebraic Structure," *IEEE Trans. on Information Theory*, Vol. IT-16, pp. 720-738 (Nov. 1970).

42. A. J. Viterbi and J. K. Omura, *Principles of Digital Communication and Coding*, McGraw-Hill (1979).

43. S. G. Wilson, *Digital Modulation and Coding*, Prentice Hall, Upper Saddle River, NJ, 1996.

44. K. J. Larsen, "Short Convolutional Codes with Maximal Free Distance for Rates 1/2, 1/3 and 1/4," *IEEE Trans. Information Theory*, Vol. IT-19, pp. 371-372 (1973).

45. D. G. Daut, J. W. Modestino, and L. D. Wismer, "New Short Constraint Length Convolutional Code Construction for Selected Rational Rates," *IEEE Trans. Information Theory*, Vol. IT-28, pp. 794-800 (1982).

46. C. L. Phillips and R. D. Harbor, *Feedback Control Systems*, Prentice-Hall (1988).

47. S. J. Mason, "Feedback Theory – Further Properties of Signal Flow Graphs," *Proc. IRE*, Vol. 44 (7), p. 920 (July 1956).

48. B. C. Kuo, *Automatic Control Systems*, Prentice-Hall, Englewood Cliffs, N.J. (1962).

49. M. Luby, M. Mitzenmacher, A. Shokrollahi, D. Spielman, "Analysis of Low-Density Codes and Improved Designs using Irregular Graphs," *IEEE Trans. Info. Theory*, Vol. 47, pp. 585-98, 2001.

50. T. Richardson and R. Urbanke, "The Capacity of Low-Density Parity-Check Codes Under Message-Passing Decoding," *IEEE Trans. Information Theory*, Vol. 47, pp. 599-618, 2001.

51. T. Richardson and R. Urbanke, "Efficient Encoding of Low-Density Parity-Check Codes," *IEEE Trans. Information Theory*, Vol. 47, pp. 638-656, 2001.

52. Y. Kou, S. Lin, and M. Fossorier, "Low-Density Parity-Check Codes Based on Finite Geometries: A Rediscovery and New Results," *IEEE Trans. Information Theory*, Vol. 47, No. 7, pp. 2711-2636, November 2001.

53. J. M. Tuchler, R. Hoetter, and A. C. Singer, "Turbo Equalization: Principles and New Results," *IEEE Trans. Communications*, Vol. 50, No. 5, pp. 754-767, May 2002.

54. D. Divsalar, H. Jin, and R. J. McEliece, "Coding Theorems for Turbo-Like Codes," *36th Allerton Conf. on Communication, Control, and Computing*, Allerton, Illinois, pp. 201-210, Sept. 1998.

55. W. Peterson and E. Weldon, *Error-Correcting Codes*, 2nd Ed., M.I.T. Press, Cambridge, Mass (1972).

56. S. W. Golomb, *Shift Register Sequences*, Holden-Day, San Francisco (1967).

13

Signal-Space Coding

The binary error-control schemes of Chapter 12 were designed specifically for binary modulation formats. They are ideally suited for the low-SNR regime where the target spectral efficiency is small. This chapter treats the problem of error control when the target spectral efficiency is large. In this scenario, one might be tempted to concatenate a binary encoder with an independently designed QAM symbol mapper, but the resulting performance will usually be poor.

The optimal receiver in white Gaussian noise uses soft decoding, a minimum-distance detection strategy based on a Euclidean signal-space distance rather than a Hamming distance. The resulting error probability is accurately predicted by the signal-space minimum distance. That minimum distance is affected in turn by how we do the mapping of bits to symbols. In this chapter, we introduce a superior approach to designing codes for soft decoding. We directly consider the geometry of signal sets within signal space, with the goal of maximizing the minimum distance, while meeting power and bandwidth constraints. This does not mean that the algebraic techniques emphasized in Chapter 12 (groups, fields, etc.) must be abandoned, but rather that these algebraic tools should be used in conjunction with geometric considerations.

Codes designed by choosing signal sets in signal space with desirable geometric (as well as algebraic) properties are called *signal-space codes*. They are advantageous on channels where spectral efficiency is at a premium (such as radio or voiceband telephone channels). To be practical, these signal-space codes must have a regular geometric and algebraic structure that can be used to simplify their implementation.

The key to obtaining large coding gains is to design codes in a subspace of signal space with high dimensionality, where a larger minimum distance in relation to signal power can be obtained. This is the essence of the channel coding theorem of Chapter 4. This high dimensionality does not necessarily imply a large bandwidth. For example, if we group together a large number of successive symbols in a PAM system, the resulting "vector" symbol is multidimensional. In other words, the dimensionality $2Bt_0$ can be increased for fixed bandwidth B by increasing the time interval t_0, making it multiple symbol intervals.

Our starting point will be to consider the problem of designing a signal constellation (in N-dimensional Euclidean space) that has a large minimum distance in relation to its average power. This is identical to the problem of designing baseband or passband signal constellations considered in Chapter 5 for $N = 1$ and $N = 2$, except that higher dimensionality is desired. We will show that it is advantageous to extend this design to $N > 2$, as illustrated in Fig. 13-1(a) in the context of passband PAM modulation. A sequence of $N/2$ two-dimensional (complex-valued) transmitted symbols can be considered as a single point in an N dimensional constellation. Each member of the constellation alphabet (called a codeword) is a vector in N-dimensional Euclidean space. Analogous to the mapper in Chapter 5, but generalized to N dimensions, a set of K input bits are used to choose one of 2^K codewords in the multidimensional constellation. That codeword is transmitted serially as $N/2$ data symbols over a passband PAM modulation system.

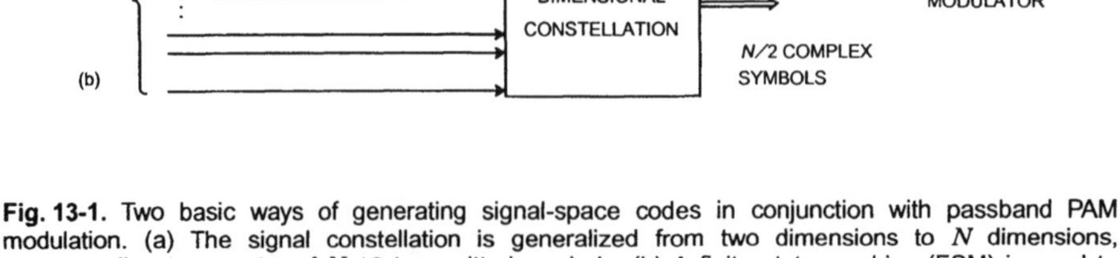

Fig. 13-1. Two basic ways of generating signal-space codes in conjunction with passband PAM modulation. (a) The signal constellation is generalized from two dimensions to N dimensions, corresponding to a vector of $N/2$ transmitted symbols. (b) A finite state machine (FSM) is used to choose a vector from an N-dimensional constellation.

An example of a multidimensional constellation is a *lattice code*, which is a generalization of some of the rectangular constellations of Chapter 5, and has geometric and algebraic structure that makes it practical to synthesize, analyze, and implement. Although lattice codes can approach capacity for large N, the exponential increase in the number of codewords with N rules out their use for large N.

A way to achieve significant coding gain without the implementation complexity of lattice codes is to extend the dimensionality of the transmitted signal by basing it on a finite state machine (FSM). As shown in Fig. 13-1(b), a group of K input bits is divided into two groups. The first group drives an FSM, which introduces redundancy by generating more bits at its output than there are at its input. The FSM output, together with the second group of input bits, specifies one from among a set of codewords in an N-dimensional signal constellation. The extra bits produced by the FSM implies an inherent increase in the number of points in the constellation. As in the multidimensional constellation, the codewords are transmitted serially as $N/2$ complex symbols. Not only are significant coding gains possible this way, but the implementation of the receiver maximum-likelihood detector (Chapter 7) can be based on the Viterbi algorithm, greatly reducing the complexity of soft decoding.

In Section 13.1 we will consider the design of multidimensional signal constellations, followed by trellis codes based on the FSM approach in Section 13.2. A generalization of the trellis code, the coset code, is introduced in Section 13.3. Finally, Section 13.4 discusses the combination of signal-space coding with ISI.

13.1. Multidimensional Signal Constellations

When passband PAM systems were designed in Chapter 5, a signal constellation was designed for a complex-valued data symbol, and this same constellation was used for each successive symbol. In Chapter 5, several intuitive design approaches for improving signal constellations were described, two of which are illustrated in Fig. 13-2. Three constellations are shown in Fig. 13-2 with the same minimum distance and 256 points. Thus, the three constellations will have the same spectral efficiency and, to accurate approximation, the same

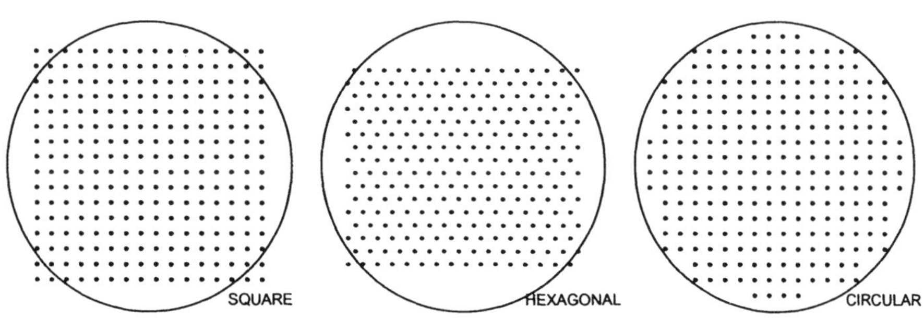

Fig. 13-2. A square QAM constellation, and two alternative constellations that illustrate shaping and coding gain. All three constellations have the same minimum distance, and each has 256 points.

error probability. Where they differ is in the variance of the data symbols, which (assuming equally probable points) are clearly different. Using the square QAM constellation as a reference, the other two constellations illustrate two basic approaches to improving constellation design.

The first idea is to change the shape or outline of the constellation without changing the relative positioning of points (on a regular square grid). The "circular" constellation approximates a circular shape. This circular constellation will have a lower variance than the square constellation, because every point that is moved from outside the circle to the inside will make a smaller contribution to the variance as a result. On these same grounds, a circular shape will have the lowest variance of any shaping region for a square grid of points. The resulting reduction in signal power is called *shaping gain.*

Significantly, shaping the constellation changes the marginal density of the real part or imaginary part of the data symbol. This is illustrated in Fig. 13-3, where the marginal density of the real-valued component of the complex symbol is compared for the square and circular constellations, assuming the points in the signal constellation are equally likely. For the circularly shaped constellation, the one-dimensional marginal density becomes nonuniform, even though the two-dimensional density *is* uniform.

A second approach to improving a constellation, also illustrated in Fig. 13-2, is to change the relative spacing of points in the constellation. The hexagonal constellation, in which points fall on a grid of equilateral triangles, also reduces the variance for the same minimum distance. (Alternatively, we could keep the variance constant, in which case the hexagonal constellation would have a larger minimum distance than the square constellation.) This decrease in power for the same minimum distance or increase in minimum distance for the same power through changing the relative spacing of the points is called *coding gain.*

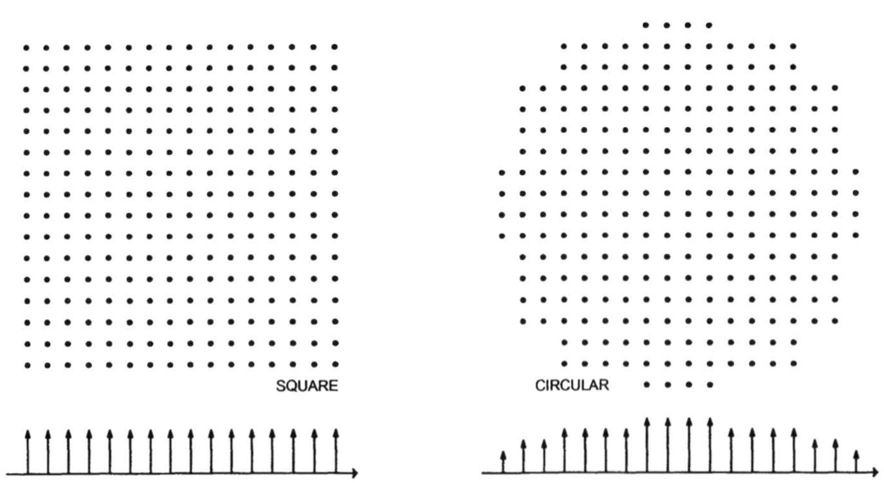

Fig. 13-3. The marginal probability density functions of one dimension for a an unshaped (square) and shaped (circular) two-dimensional constellation. The two-dimensional density is assumed to be uniform.

Coding and shaping gain can be combined, for example by changing the points in the circularly shaped constellation to a hexagonal grid while retaining the circular shaping.

If we define a constellation for a single data symbol a_k, as was done in Chapter 5, then it will be one-dimensional for the baseband case, and two-dimensional for the passband (complex) case. The square constellation of Fig. 13-2 is simply the Cartesian product of a pair of identical one-dimensional constellations. However, introducing either shaping gain or coding gain, or both, implies the two-dimensional constellation is no longer the Cartesian product of one-dimensional constellations. The advantages of shaping and coding gain imply that it is preferable to design a two-dimensional constellation directly as an alphabet of points in two-dimensional space, rather than taking the "lazy" approach of forming a Cartesian product of one-dimensional constellations.

Neither shaping nor coding gain is feasible in one dimension, but both are available in two dimensions. If going from one to two dimensions is beneficial, could it be that moving to even higher dimensions is a good idea? In this section, we will introduce the third fundamental idea in constellation design, the ***multidimensional signal constellation***. Taking the passband case, consider a sequence of complex-valued data symbols $\{a_k, -\infty < k < \infty\}$. A subset of $N/2$ successive symbols $\{a_k, a_{k+1}, \ldots, a_{k+N/2-1}\}$ (where of course N is even), can be considered as a vector in N-dimensional real-valued Euclidean space. Our convention is that a data symbol drawn from this N-dimensional constellation is transmitted once every $N/2$ symbol intervals. When we design a two-dimensional constellation, and choose the $N/2$ successive symbols to be an arbitrary sequence of two-dimensional symbols drawn from that constellation, the resulting N-dimensional constellation is a Cartesian product of $N/2$ two-dimensional constellations. An alternative is to design an N-dimensional constellation that is not constrained to have this Cartesian-product structure. This is then called an *N-dimensional signal constellation*. Whenever $N > 2$, it is called a multidimensional signal constellation.

Greater shaping and coding gains can be achieved with a multidimensional constellation than with a two-dimensional constellation. In Section 6.7 it was shown that two-dimensional constellations, on a Gaussian channel, suffer an "SNR gap to capacity." This gap can be closed completely with a multidimensional constellation as $N \to \infty$. This result is a straightforward application of the capacity theorem (Chapter 4) if the multidimensional constellation is not constrained, since it is simply a general channel code anticipated by the channel capacity theorem. Significantly, as will be cited below, there exist constellations that have an imposed structure (that are a multidimensional generalization of the square and hexagonal constellations) that can also completely close the gap as $N \to \infty$.

Practically speaking, significant shaping and coding gains can be achieved for modest N. However, multidimensional constellations suffer from a complexity that increases exponentially with dimensionality. To mitigate this, multidimensional constellations can be used in conjunction with trellis codes, as described in Section 13.3. Multidimensional constellations also serve to further our understanding of the structure of signal-space codes, and particularly the relationship between coding and shaping gain.

13.1.1. Lattice Codes

The two-dimensional constellations of Fig. 13-2 are special cases of *lattice codes*. Let the dimension of the constellation be N, and let $\{x_1, \ldots, x_I\}$ be a set of I linearly independent basis vectors in N-dimensional Euclidean space. Of course, we must have that $I \le N$. Consider the set of points in N-dimensional Euclidean space that can be expressed in the form

$$x = \sum_{i=1}^{I} k_i \cdot x_i , \quad (13.1)$$

where the $\{k_1, \ldots, k_I\}$ are integers. This countably infinite set of points is called a *lattice*, and is denoted by Λ. The regularly spaced set of points in a lattice is an appropriate choice for a multidimensional signal constellation for three reasons:

- Since the probability of error is determined by the minimum-distance, it is advantageous to choose regular arrays of points as in a lattice, where all points are equidistant from their neighbors.
- From an implementation perspective, the points in the constellation can be described and manipulated in terms of the vector of integers, using fixed-point integer arithmetic.
- The lattice has a convenient geometric and algebraic structure. Algebraically it is a *group*, meaning that it is closed with respect to vector summation and difference. One implication of this is that the zero vector must be a member of any lattice. This algebraic structure can be exploited in both implementation and in deriving various properties of the lattice.

A *lattice code* is made up of three constituents. First, there is the lattice Λ. Second, there is a translation vector a, and a translation of the lattice $\Lambda + a$ consisting of all points in the lattice translated by a. The motivation for this translation is that a lattice is constrained to have a point at zero (corresponding to all-zero integers), and the translation frees us from the constraint, for example to minimize the transmitted power. Third, there is a finite region S, with the convention that the lattice code is the intersection of Λ with S. This results in a finite number of points in the lattice code, and also gives another degree of control over the transmitted power. To display all these parameters at once, we write the lattice code in the form $(\Lambda + a) \cap S$.

Example 13-1.

The two-dimensional hexagonal constellation of Fig. 13-2 can be described as a lattice code, as illustrated in Fig. 13-4. The first step is to choose two basis vectors corresponding to the sides of an equilateral triangle,

$$x_1 = [d, 0] , \qquad x_2 = [d/2, \sqrt{3}d/2] . \quad (13.2)$$

In this case, d will be the minimum distance for the code. The second step is to choose some translation vector a ($a = 0$ shown). The third step is to choose a region S, shown as a circle (two-dimensional sphere). Finally, the lattice code consists of all points on the lattice that fall within S.

This example shows that many of the constellations illustrated in Chapter 5 (with the notable exception of the AM-PM constellations) can be formulated as lattice codes. The greatest significance of the lattice code formulation is that it readily extends to higher dimensions. Since it is difficult to draw or visualize dimensions higher than three, the mathematical structure of lattice codes becomes particularly important.

There are two properties of a lattice that influence its effectiveness as a code:

- The *minimum distance* $d_{\min}(\Lambda)$ of points in the lattice relates directly to the error probability of the lattice code (for an additive Gaussian noise channel at high SNR).
- The *fundamental volume* $V(\Lambda)$ is the volume of N-dimensional space associated with each lattice point. It is the inverse of the number of lattice points per unit volume. The fundamental volume is important because it relates directly to the number of lattice points within a given region S, and hence affects the spectral efficiency. (The spectral efficiency is affected by other factors as well, such as S, the PAM pulse bandwidth, etc.)

There is a direct relationship between the minimum distance and fundamental volume; increasing one tends to also increase the other. That is, for fixed transmit power, increasing the minimum distance tends to reduce the spectral efficiency.

13.1.2. Normalized SNR and Error Probability

For purposes of this chapter, we can assume that a symbol-rate discrete-time channel has been derived by demodulation and the sampled matched filter (Chapter 5). If the continuous-time channel has no ISI (the pulse autocorrelation $\rho_h(k) = \delta_k$), then the noise samples at the matched filter output will be white and Gaussian. Thus, assume the equivalent discrete-time channel is

$$Y_k = X_k + N_k , \tag{13.3}$$

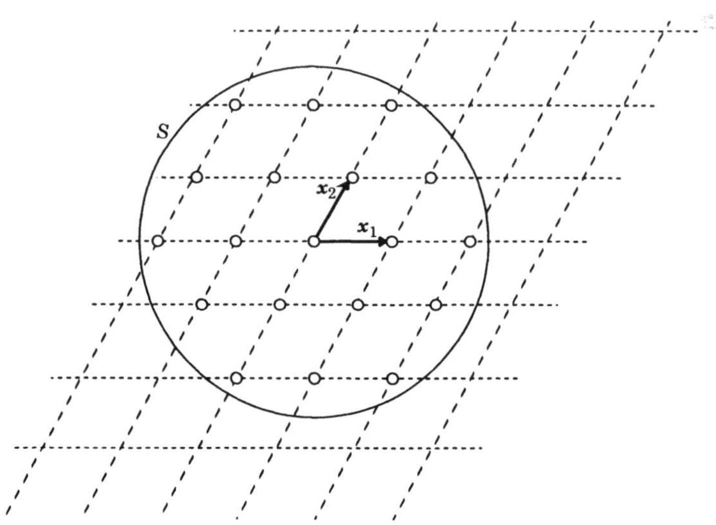

Fig. 13-4. Illustration of a two-dimensional lattice code, where S is a circle (two-dimensional sphere). The lattice points fall at the intersection of the dotted lines.

where X_k and Y_k are complex-valued inputs and outputs and N_k is circularly symmetric white Gaussian noise. Define $\sigma^2 = N_0/2$ as the noise variance per dimension, so that the noise energy (per two dimensions) is $E[\,|N_k|^2] = 2\sigma^2 = N_0$. Further, define E as the upper bound on the input *signal energy per two dimensions* for the input to this discrete-time channel,

$$E[\,|X_k|^2] \le E\,. \tag{13.4}$$

The *spectral efficiency* ν is defined as the number of bits of information communicated per complex sample, or the *bits per two dimensions*.

Example 13-2.
A passband PAM system operating at the maximum symbol rate relative to the bandwidth of the underlying channel will have a spectral efficiency ν bits/sec-Hz, since the symbol rate is equal to the bandwidth of the channel. Thus, for this idealized continuous-time channel, and only for this channel, the spectral efficiency as defined in Section 5.2.3 (in bits/sec-Hz), and the spectral efficiency of the discrete-time channel of (13.3) (in bits per two dimensions), will be numerically equal.

The Shannon limit on spectral efficiency for the channel of (13.3), ν_c, is given by (4.36) for $N = 2$,

$$\nu_c = \log_2(1 + SNR)\ , \qquad SNR = E/N_0\,. \tag{13.5}$$

Equivalently, (13.5) can be written in the form

$$\frac{SNR}{2^{\nu_c} - 1} = 1. \tag{13.6}$$

As in Chapter 6, a rate-normalized signal-to-noise ratio can be defined for a system operating over channel (13.3) with signal-to-noise ratio SNR and spectral efficiency ν,

$$SNR_{\text{norm}} = \frac{SNR}{2^{\nu} - 1}\,. \tag{13.7}$$

SNR_{norm} has the interpretation that the Shannon limit $\nu \le \nu_c$ is equivalent to $SNR_{\text{norm}} > 1$.

If complex symbols $X_k = A_k$ are transmitted over (13.3), then from Chapter 7 the error probability of a minimum-distance receiver is accurately approximated by the union bound as

$$P_e = C \cdot Q(d_{\text{min}}/2\sigma)\ , \tag{13.8}$$

where d_{min} is the two-dimensional Euclidean minimum distance between known signals. If (13.3) is used to transmit a vector signal, then P_e is likewise given by (13.8), except d_{min} is now the minimum Euclidean distance between vector signals, and the error coefficient C may be changed.

The error probability can be expressed in terms of the SNR_{norm}, ν, and P_e by expressing the squared argument of $Q(\,\cdot\,)$ as

$$\frac{d_{\text{min}}^2}{4\sigma^2} = \frac{d_{\text{min}}^2(2^{\nu} - 1)}{2E} \cdot \frac{E}{2\sigma^2(2^{\nu} - 1)} = \gamma \cdot SNR_{\text{norm}}\ , \tag{13.9}$$

where (13.5) and (13.7) have been used to define SNR_{norm}, and

$$\gamma = \frac{d_{\min}^2(2^{\nu} - 1)}{2E} . \tag{13.10}$$

Given this definition,

$$P_e \approx C \cdot Q(\sqrt{\gamma \cdot SNR_{\text{norm}}}) . \tag{13.11}$$

This expression for γ is similar to that derived in Chapter 6, except that it applies to vector signals, ν is measured by bits per two dimensions, and E is energy per two dimensions.

As shown in Fig. 6-31, the parameter γ relates directly to the SNR gap to capacity, and in particular the larger γ, the lower the error probability and the smaller the SNR gap to capacity. It was also shown in Chapter 6 that $\gamma = 3$ for a square two-dimensional QAM constellation, independent of the constellation size.

13.1.3. Coding and Shaping Gains for Lattice Codes

We will now show that for lattice codes with a large number of points, γ can be approximated in a particularly simple and insightful way. This will allow us to extend the concepts of shaping and coding gains as illustrated in two dimensions in Fig. 13-2 to multidimensional constellations.

The Continuous Approximation

The spectral efficiency and energy of a constellation can be expressed in terms of the parameters of the lattice and the shaping region. Considering first the spectral efficiency, if the volume of an N-dimensional shaping region S is defined as $V(S)$, then the number of points in $(\Lambda + \boldsymbol{a}) \cap S$ falling within S is accurately approximated by $V(S) / V(\Lambda)$, especially as this ratio gets large. Then for a given $(\Lambda + \boldsymbol{a}) \cap S$,

$$\nu \approx \frac{2}{N}\log_2\frac{V(S)}{V(\Lambda)}, \tag{13.12}$$

since the points are divided over $N/2$ complex symbols.

To calculate the energy a constellation, the starting point is the probability density of the data symbols, which consists of delta functions at the constellation points. When the number of points in $(\Lambda + \boldsymbol{a}) \cap S$ is large, then the points in the lattice within region S are closely spaced at regular intervals, and their probability density can reasonably be approximated as continuous rather than discrete. When the signal points are equally likely, then the appropriate continuous density is uniform over the region S. In calculating the energy of the constellation, a continuous uniform density is just the Riemann integral approximation to the sum that would correspond to the discrete density. Denote as $P(S)$ the variance of a uniform distribution over S. Since the uniform distribution has height $1/V(S)$,

$$P(S) = \frac{1}{V(S)}\int_S \|\boldsymbol{x}\|^2 \, d\boldsymbol{x} . \tag{13.13}$$

This is the *continuous approximation*, first invoked in Chapter 8 to study the properties of transmitter precoding. $P(S)$ is an approximation to the energy per N dimensions, and thus the continuous approximation for the energy per two dimensions is

$$E \approx 2P(S)/N . \tag{13.14}$$

Example 13-3.

When $N = 2$, and the shaping region S is an $2R \times 2R$ square, $V(S) = (2R)^2$, and

$$E \approx P(S) = \frac{1}{4R^2}\int_{-R}^{R}\int_{-R}^{R} (x_1^2 + x_2^2)\, dx_1 dx_2 = \frac{2}{3}R^2 . \tag{13.15}$$

Example 13-4.

When $N = 2$ and the shaping region S is a circle with radius R, then $V(S) = \pi R^2$. If $\mathbf{X} = (X_1, X_2)$ is uniformly distributed over the circle, then by symmetry the marginal distributions of X_1 and X_2 will be the same. The energy is then

$$E \approx P(S) = \mathrm{E}[\|\mathbf{X}\|^2] = \mathrm{E}[X_1^2] + \mathrm{E}[X_2^2] = 2\mathrm{E}[X_1^2] \tag{13.16}$$

where

$$E[X_1^2] = \frac{1}{\pi R^2}\int_{-R}^{R} x_1^2 \int_{-\sqrt{R^2 - x_1^2}}^{\sqrt{R^2 - x_1^2}} dx_2 dx_1 . \tag{13.17}$$

This integral readily evaluates to $R^2/4$, and thus $P(S) = R^2/2$.

The continuous approximation will now be used to study the coding and shaping gain of lattice codes.

Shaping and Coding Gain

The coding gain is a function of the relative spacing of points in Λ, but is independent of S. The transmit power, and hence shaping gain, is a function of both Λ and S, but the continuous approximation discards the dependence on Λ. Based on this simplification, we will now characterize the shaping and coding gains.

The continuous approximation gives a useful approximation for γ, which in turn summarizes the error probability of the constellation based on a lattice code. From (13.10), (13.12) and (13.14),

$$\gamma = \frac{(2^{\nu} - 1)d_{\min}^2(\Lambda)}{2P} \approx \frac{\left(\left(\frac{V(S)}{V(\Lambda)}\right)^{2/N} - 1\right)d_{\min}^2(\Lambda)}{4P(S)/N} = 3 \cdot \gamma_{\Lambda} \cdot \gamma_S , \tag{13.18}$$

where

$$\gamma_{\Lambda} = \frac{d_{\min}^2(\Lambda)}{V^{2/N}(\Lambda)} , \tag{13.19}$$

$$\gamma_S \approx \frac{NV^{2/N}(S)}{12P(S)} \ . \tag{13.20}$$

The approximation for γ_S assumes that the unity term can be ignored, which will be accurate for large constellations. The motivation for the factor of 3 is that $\gamma = 3$ for a two-dimensional square constellation, which we take as a baseline against which to compare other shapes. (It will be shown shortly that the multidimensional cube, which is a generalization of the two-dimensional rectangle, also has $\gamma = 3$.) We will see that both γ_S and γ_Λ are normalized to unity for a square QAM constellation, our reference constellation.

Example 13-5.

For the square constellation of Example 13-3, the shaping gain is

$$\gamma_\Lambda = \frac{2V(S)}{12P(S)} = \frac{2 \cdot 4R^2}{12 \cdot 2R^2/3} = 1 \ . \tag{13.21}$$

Similarly, for a square lattice of points, if the spacing between adjacent points is $d_{\min}$, then the fundamental volume is $V(\Lambda) = d_{\min}^2$ and thus $\gamma_\Lambda = 1$. Thus, for this case $\gamma = 3$, as we knew already.

The factor γ_Λ includes all terms that are a function of Λ, and is defined as the *coding gain*. The factor γ_S includes all terms that are a function of S, and is defined as the *shaping gain*. For lattice codes with a large number of constellation points, the coding gain and the shaping gain are factors that can be manipulated independently of one another, the former by choosing Λ and the latter by choosing S.

The coding gain compares the minimum distance of a lattice Λ against a two-dimensional square lattice Λ_2, our reference lattice. Comparing two lattices is tricky since often both minimum distance and fundamental volume will be different, and the lattices may also have different dimensionality. A fair minimum distance comparison requires that the two lattices have the same density of points, that is, the same number of points per unit volume. If Λ and Λ_2 are the same dimensionality and have the same shaping region S, then this implies the spectral efficiency will be the same (to an accurate approximation).

An additional complication is that the two lattices may have different dimensionality. Thus, we scale the two lattices such that their density of points per two dimensions is the same. This implies that, for the same shaping region, the codes based on the two lattices will have the same spectral efficiency and energy. The energy is the same because, in accordance with the continuous approximation, it is a function the shaping region only.

To compare Λ to the reference lattice Λ_2, first scale Λ_2 by a factor α to get a new square two-dimensional lattice $\alpha \cdot \Lambda_2$ with minimum distance $\alpha \cdot d_{\min}(\Lambda_2)$ and fundamental volume

$$V(\alpha \cdot \Lambda_2) = \alpha^2 d_{\min}^2(\Lambda_2) \ . \tag{13.22}$$

For a square lattice, the fundamental volume is the square of the minimum distance. Second, determine the fundamental volume of Λ per two dimensions, and then set it equal to $V(\alpha \cdot \Lambda_2)$

so that both lattices have the same density of points per two dimensions. According to the continuous approximation, the number of points in Λ per two dimensions is

$$2^{\nu} = \frac{V^{2/N}(S)}{V^{2/N}(\Lambda)}, \tag{13.23}$$

implying that the volume of S per two dimensions is $V^{2/N}(S)$ and the fundamental volume of Λ per two dimensions is $V^{2/N}(\Lambda)$. Finally, set the fundamental volume per two dimensions of $\alpha \cdot \Lambda_2$ and Λ equal,

$$\alpha^2 d^2_{\min}(\Lambda_2) = V^{2/N}(\Lambda), \tag{13.24}$$

which defines the scaling factor α. Now that these fundamental volumes (and hence spectral efficiency for any S) are equal, the coding gain is the ratio of the square of the minimum distance of Λ to that of $\alpha \cdot \Lambda_2$,

$$\gamma_{\Lambda} = \frac{d^2_{\min}(\Lambda)}{\alpha^2 d^2_{\min}(\Lambda_2)} = \frac{d^2_{\min}(\Lambda)}{V^{2/N}(\Lambda)}, \tag{13.25}$$

consistent with (13.19). This establishes an interpretation of γ_{Λ} as the ratio of the square of the minimum distance of Λ to that of a scaled version of Λ_2 (the reference lattice), where the scaling forces the spectral efficiency of the two lattices to be the same for any S.

A slightly different interpretation of γ_{Λ} is to assume that Λ and Λ_2 have the same minimum distances, and compare their fundamental volumes per two dimensions. If we assume that Λ_2 has minimum distance $d_{\min}(\Lambda)$, then the fundamental volume of Λ_2 is $d_{\min}{}^2(\Lambda)$ (since it is a square lattice). Since the fundamental volume per two dimensions of Λ is $V^{2/N}(\Lambda)$, it follows that γ_{Λ} is precisely the ratio of the fundamental volume per two dimensions of Λ_2 to that of Λ. Since the fundamental volume is inversely proportional to the number of points in the constellation per two dimensions (for the same shaping region), the coding gain is the ratio of the number of points in the constellations for equal minimum distances.

A similar interpretation can be applied to the shaping gain. Again, the idea is to compare the energy of the shaping region S against the energy of a two-dimensional square region S_2, where both S and S_2 have the same volume per two dimensions. Let the reference shaping region S_2 be an $2R \times 2R$ two-dimensional square, with each of the two dimensions in the range $x \in [-R, R]$. The "radius" R will be chosen to force the spectral efficiency to be the same as S, which will occur if the two shapes have the same volume per two dimensions. Since S has volume per two dimensions $V^{2/N}(S)$, and S_2 has volume $(2R)^2$, R is chosen to satisfy

$$(2R)^2 = V^{2/N}(S). \tag{13.26}$$

The shaping gain γ_S is then the ratio of the energy per two dimensions of S_2 to that of S. The energy of S per two dimensions is $2/N$ times the energy $P(S)$ per N dimensions. Furthermore, from Example 13-3, $P(S_2) = 2R^2/3$. Thus, the shaping gain is

$$\gamma_S = \frac{2R^2/3}{2P(S)/N} = \frac{NV^{2/N}(S)}{12P(S)} , \tag{13.27}$$

consistent with (13.20). This establishes an interpretation of γ_S as the ratio of the energy per two dimensions of a two-dimensional square shape to the energy per two dimensions for S, where the square shape is scaled so that the volumes per two dimensions (and hence spectral efficiencies) are the same.

It will be shown shortly that the maximum possible shaping gain as $N \to \infty$ is $\gamma_{S,\max} =$ 1.53 dB. As a result, the SNR gap to capacity for lattice codes can be divided into two distinct parts, as shown in Fig. 13-5. The $\gamma = 3$ curve shows the SNR gap to capacity for a square two-dimensional constellation, where $\gamma_\Lambda = \gamma_S = 1$. The maximum shaping gain reduces the SNR gap to capacity, and the remaining SNR gap, labeled the "maximum coding gain" is potentially reduced by the choice of the code. A result of de Buda [1] shows that asymptotically as $N \to \infty$ there exist lattice codes that achieve channel capacity, in the sense that they drive P_e to zero for any bit rate below the Shannon limit. (While there are flaws in the original proof of de Buda, a new proof has been provided by Loeliger.) This answers a long-standing question in information theory; namely, is it possible to achieve the channel capacity limits with codes that have a structure that allows them to be implemented? The answer is yes, although unfortunately it is still not practical to implement soft decoding for lattice codes at high dimensionality.

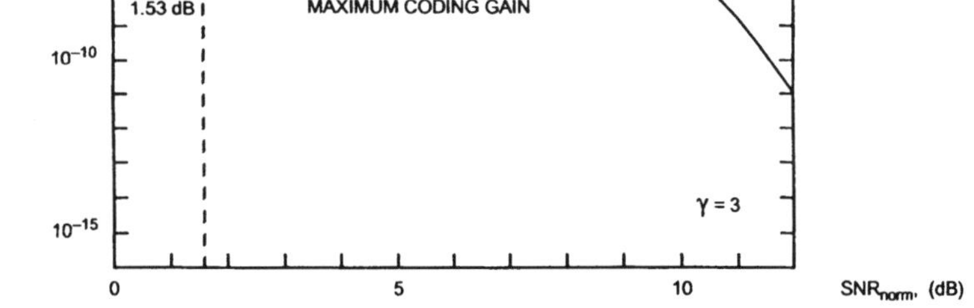

Fig. 13-5. The SNR gap to capacity for lattice codes with large constellations is divided into shaping gain and coding gain. The curve is P_e for a square QAM constellation ($\gamma = 3$). The maximum shaping gain (for S an N-sphere as $N \to \infty$) reduces the SNR gap to capacity by 1.53 dB.

Cartesian Product Constellations

When we use a two-dimensional constellation and transmit a sequence of K symbols drawn from this constellation, we can consider that sequence as a vector in $2K$-dimensional Euclidean space. This sequence is actually drawn from a multidimensional constellation consisting of the Cartesian product of K two-dimensional constellations. Provided that coding and shaping gains are properly defined, we would expect that the coding and shaping gains of this $2K$-dimensional constellation would be equal to the coding and shaping gains of the underlying two-dimensional constellation. This is true, and in fact, we will now prove a more general result: given any lattice code $C = (\Lambda + \boldsymbol{a}) \cap S$, the coding and shaping gains of any K-fold Cartesian product constellation C^K is the same those of C.

Let $C = (\Lambda + \boldsymbol{a}) \cap S$ be a lattice code in L-dimensional Euclidean space. We must define the K-fold Cartesian product C^K. A vector $\boldsymbol{x}$ that is a point in C^K is a vector in KL-dimensional Euclidean space of the form

$$\boldsymbol{x} = (\boldsymbol{x}_1, \boldsymbol{x}_2, \ldots., \boldsymbol{x}_K) , \tag{13.28}$$

where $\{\boldsymbol{x}_i,\ 1 \le i \le K\}$ are each points in C.

First consider the coding gain of C^K. The coding gain is independent of S, so we can work with Λ and Λ^K, where the latter is an LK-dimensional lattice consisting of a K-fold Cartesian product of L-dimensional lattices Λ. (It is simple to see that Λ^K is in fact a lattice.) The distance-squared between $\boldsymbol{x} \in \Lambda^K$ and $\boldsymbol{y} \in \Lambda^K$, $\boldsymbol{x} \ne \boldsymbol{y}$, can be written as

$$\| \boldsymbol{x} - \boldsymbol{y} \|^2 = \sum_{i=1}^{K} \| \boldsymbol{x}_i - \boldsymbol{y}_i \|^2 , \tag{13.29}$$

where the $\boldsymbol{x}_i, \boldsymbol{y}_i \in \Lambda$. Since all the $\boldsymbol{x}_i$ can be chosen independently of one another, and similarly the $\boldsymbol{y}_i$, the minimum distance will occur when all terms but one are zero, and the minimum of the one non-zero term is $d_{\min}(\Lambda)$. Thus,

$$d_{\min}(\Lambda^K) = d_{\min}(\Lambda) , \tag{13.30}$$

so the Cartesian product does not affect the minimum distance. Likewise, we can show that the fundamental volumes per two dimensions are the same,

$$V^{2/LK}(\Lambda^K) = V^{2/L}(\Lambda) , \tag{13.31}$$

establishing that the coding gains are identical for C and C^K. This follows directly from the following exercise, with U equal to the region corresponding to the fundamental volume of Λ.

Exercise 13-1.
Let U be a region of L-dimensional Euclidean space, and let U^K be a region that is the K-fold Cartesian product of U.

(a) Show that $V(U^K) = V^K(U)$.

(b) Let $\mathbf{X}$ be a uniformly distributed random vector on region U^K. Show that its variance is $P(U^K) = KP(U)$.

We can also show that the shaping gain of a Cartesian-product region S^K is

$$\gamma_{S^K} = \frac{LKV^{2/LK}(S^K)}{12P(S^K)} = \frac{LKV^{2/L}(S)}{12KP(S)} = \frac{LV^{2/L}(S)}{12P(S)} = \gamma_S \; . \tag{13.32}$$

In words, the shaping gain of C^K is equal to the shaping gain of C.

This result is fundamental to the understanding of multidimensional constellations (and more generally signal-space coding). Taking a Cartesian product of lattice codes does not affect the coding or the shaping gain. The way to achieve an increase in coding and shaping gains as the dimensionality is increased is to choose the components of the code dependently.

Maximum Shaping Gain

Fig. 13-5 shows the SNR gap to capacity at any P_e. We now know that γ, and hence this gap, is divided between two factors, the coding gain and the shaping gain. The question arises as to how much gain is available from coding and how much is available from shaping. The answer is that the shaping gain is limited to 1.53 dB, which establishes that the remainder must be the available coding gain. We will now derive this result.

The maximum shaping gain is achieved by a spherical multidimensional constellation, as seen in the following argument. For a fixed Λ, a shaping region S that is not spherical and an N-sphere with the same volume will have, to accurate approximation, approximately the same number of lattice points and hence the same spectral efficiency. The continuous-approximation uniform density has the same height in both cases, since the volume is the same. If we move that region of S that is outside the sphere to the inside, the energy will be reduced, since $\| \boldsymbol{x} \|^2$ will be made smaller over that region.

The maximum shaping gain, that of the N-sphere, can be determined as follows. Define $S_N(R)$ to be an N-dimensional sphere of radius R, and let $\boldsymbol{x}$ be an N-dimensional vector with real-valued components. Then

$$S_N(R) = \{\boldsymbol{x} : \|\boldsymbol{x}\| \le R\} = \left\{\boldsymbol{x} : \sum_{i=1}^{N} x_i^2 \le R^2\right\} , \tag{13.33}$$

and the volume is

$$V[S_N(R)] = \int_{S_N(R)} d\boldsymbol{x} \, . \tag{13.34}$$

Changing the variable of integration to $\boldsymbol{r} = \boldsymbol{x}/R$, this volume can be expressed in terms of the volume of a unit sphere,

$$V[S_N(R)] = V[S_N(1)] \cdot R^N \; , \tag{13.35}$$

where $V[S_N(1)]$ will be determined shortly.

Suppose that $\mathbf{X}$ is a random vector uniformly distributed over $S_N(R)$; $\mathbf{X}$ is called a *spherically uniform* random vector. The probability density function of $\mathbf{X}$, $f_{\mathbf{X}}(\boldsymbol{x})$, is $V^{-1}[S_N(R)]$ for $\boldsymbol{x} \in S_N(R)$ and zero elsewhere. The marginal density of one component of $\mathbf{X}$, say X_1, can then be calculated. This marginal density $f_{X1}(x_1)$ is $f_{\mathbf{X}}(\boldsymbol{x})$, a constant, integrated over the region

$$\left\{ \sum_{i=2}^{N} x_i^2 \le R^2 - x_1^2 \right\}. \tag{13.36}$$

This region is itself an $(N-1)$-dimensional sphere $S_{N-1}(\sqrt{R^2 - x_1^2})$, and the integral is the volume of that sphere. Thus, the marginal density is

$$f_{X_1}(x_1) = \frac{V[S_{N-1}(\sqrt{R^2 - x_1^2})]}{V[S_N(R)]} = \frac{V[S_{N-1}(1)]}{V[S_N(1)]} \cdot \frac{1}{R} \cdot \left(1 - \left(\frac{x_1}{R}\right)^2\right)^{(N-1)/2} \tag{13.37}$$

The marginal density allows us to determine the volume of a unit sphere. Since it must integrate to unity,

$$\frac{V[S_N(1)]}{V[S_{N-1}(1)]} = \int_{-1}^{1} (1 - \rho^2)^{(N-1)/2} \, d\rho, \tag{13.38}$$

where the change of variables $\rho = x_1 / R$ has been performed. The volume of a circle ($N = 2$) is πR^2, so $V[S_2(1)] = \pi$. Integral (13.38) is known as a *Beta function*, and can be evaluated in closed form for integer values of N. For even N (the case of greatest interest), the result is,

$$V[S_2(1)] = \pi, \quad \frac{V[S_N(1)]}{V[S_{N-2}(1)]} = \frac{2\pi}{N}, \quad V[S_N(1)] = \frac{\pi^{N/2}}{(N/2)!}, \quad N \text{ even}. \tag{13.39}$$

The energy $P[S_N(R)]$ can also be determined from this marginal density, since the components of a spherically uniform vector are clearly identically distributed,

$$\begin{aligned} P[S_N(R)] &= E[\|\mathbf{X}\|^2] = E[\textstyle\sum_{i=1}^{N} X_i^2] = N \cdot E[X_1^2] \\ &= \frac{NR^2 V(S_{n-1}(1))}{V(S_N(1))} \int_{-1}^{1} \rho^2 (1 - \rho^2)^{(N-1)/2} \, d\rho. \end{aligned} \tag{13.40}$$

Again, the integral can be evaluated for integer N,

$$P[S_N(R)] = R^2 \cdot \frac{N}{N+2}. \tag{13.41}$$

The variance of one component of $\mathbf{X}$ is thus $R^2/(N+2)$.

Finally, the shaping gain of an N-sphere can be determined from the volume and energy; it is

$$\gamma_{S_N(R)} = \frac{\pi(N+2)}{12[(N/2)!]^{2/N}}. \tag{13.42}$$

Using the Stirling limit $k! \to (k/e)^k$ as $k \to \infty$, we find that $\gamma_S \to \pi e/6$ asymptotically as $N \to \infty$. When this maximum shaping gain is achieved without coding gain, then the SNR gap to capacity is closed by $10 \cdot \log_{10}\pi e/6 = 1.53$ dB, as shown in Fig. 13-5. This is the best that shaping the multidimensional constellation can do.

On the other hand, any transmitter that does not use shaping will suffer a 1.53 dB penalty at high SNR. This is illustrated in Fig. 13-6, which compares the capacity with equiprobable QAM symbols to the unconstrained capacity of the AWGN channel. (These curves were discussed in more detail in Chapter 4.) At high SNR, the unshaped QAM alphabets can get no closer than 1.53 dB to the Shannon limit. The only way to close this gap is to discard the uniform distribution in favor of a Gaussian-like distribution that favors constellation points near the origin more than those near the edges.

Marginal Density for Spherical Shaping

For two-dimensional constellations, we saw in Fig. 13-3 that shaping resulted in a non-uniform marginal density. We can now explore the marginal density for higher dimensionality using the continuous approximation. The marginal density $f_{X_1}(x_1)$ is plotted in Fig. 13-7 for some even values of N, assuming spherical shaping. As the dimension N increases, the one-dimensional marginal density gets narrower, as also manifested by a decreasing variance

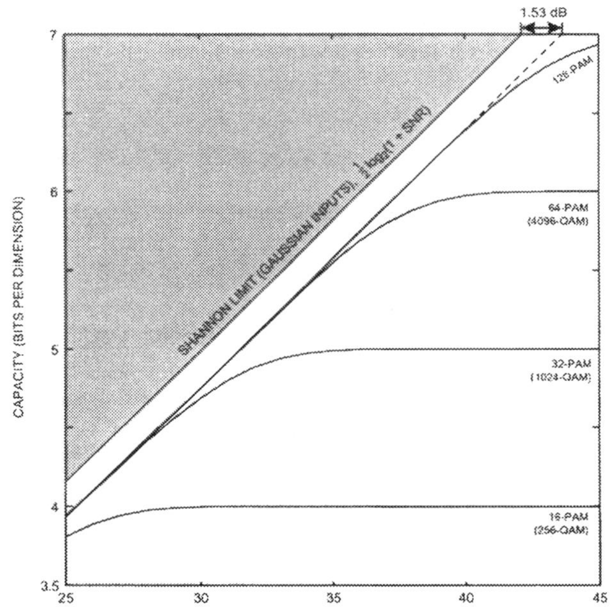

Fig. 13-6. Constraining the input alphabet to be equiprobable (unshaped) PAM or QAM results in an asymptotic penalty of 1.53 dB on the AWGN channel at high SNR. (These curves were shown over a wider range of SNR in Fig. 4-4.)

$R^2/(N+2)$. (This is a natural consequence of spreading a total energy of roughly R^2 over an increasing number of dimensions.) Of more interest is the "bell shaped" appearance, similar to a Gaussian density, that emerges for large N.

Could it be that the density is in fact approaching Gaussian as $N \to \infty$? To find out, define a normalized unit-variance random variable $Y_1 = X_1\sqrt{N+2}/R$. Then Y_1 has a density with the same shape, but its maximum value is $\sqrt{N+2}$ rather than R. Its density is (for some appropriate constant C),

$$f_{Y_1}(y_1) = C \cdot \left(1 - \left(\frac{y_1}{\sqrt{N+2}}\right)^2\right)^{(N-1)/2} . \tag{13.43}$$

As $N \to \infty$, both $(N+2)$ and $(N-1)$ can be replaced by N, and

$$f_{Y_1}(y_1) \to C \cdot \left(1 - \frac{y_1^2}{N}\right)^{N/2} \to C \cdot e^{-y_1^2/2} , \tag{13.44}$$

where the limit $(1 + x/k)^k \to e^x$ as $k \to \infty$ has been used. Thus, X_1 does approach Gaussian. (This is shown in [2] by a less direct conditional-entropy argument.) It can be shown by an identical method (Problem 13-7) that a K-dimensional marginal density of an N-dimensional spherically uniform random vector approaches a joint Gaussian density with independent components for fixed K as $N \to \infty$. In this sense, a spherically shaped lattice code approaches a white Gaussian source for large constellation sizes and high dimensionality.

Approaching channel capacity for the Gaussian channel requires that the transmitted signal be Gaussian. Multidimensional lattice codes with spherical shaping have approximately this property for high dimensionality in accordance with the continuous approximation. This is consistent with the de Buda result [1] that there exist lattice codes that approach capacity as $N \to \infty$.

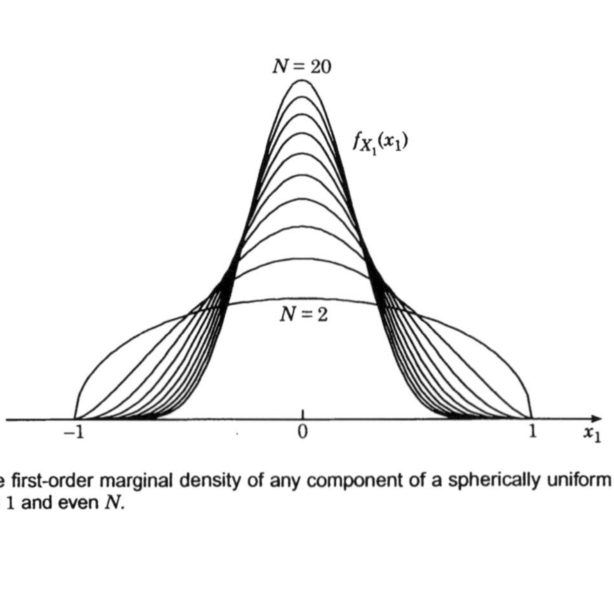

Fig. 13-7. The first-order marginal density of any component of a spherically uniform random vector is plotted for $R = 1$ and even N.

Relation of Spectral Efficiency and Energy

The spectral efficiency of a constellation and its energy are always directly related. For example, for a two-dimensional constellation, if we keep the minimum distance constant but increase the spectral efficiency by increasing the number of points, the constellation gets larger and the energy increases. This basic tradeoff holds for multidimensional constellations as well, and it is important to quantify it.

First, consider the tradeoff between energy and volume. In (13.18), since γ is not affected by any scaling of the signal constellation, and the coding gain γ_Λ is not a function of S, it follows that the shaping gain γ_S is independent of any scaling of S. Since

$$P(S) = \frac{N}{12\gamma_S} V^{2/N}(S) , \tag{13.45}$$

and the first factor is independent of any scaling of the region S, as S is scaled, $P(S)$ is proportional to $V^{2/N}(S)$.

Example 13-6.

For an N-sphere with radius R and fixed dimension N, the volume is proportional to R^N, and the energy is proportional to R^2, which is the volume raised to the power $2/N$.

A fundamental relationship between spectral efficiency and signal energy follows from (13.45). Substituting for the energy per two dimensions,

$$E = 2P(S)/N \tag{13.46}$$

and using (13.12),

$$E = \frac{V^{2/N}(\Lambda)}{6\gamma_S} \cdot 2^\nu . \tag{13.47}$$

Thus, for a fixed lattice Λ (and hence fixed coding gain), the energy per two dimensions is proportional to 2^ν, where ν is the spectral efficiency in bits per two dimensions (bits per complex symbol). Thus, if we add one bit per complex symbol while holding the coding gain constant, the energy per complex symbol is doubled.

13.2. Trellis Codes

Lattice codes are useful for a modest number of dimensions. For large dimensionality, their implementation complexity becomes excessive, because the number of points in the signal constellation grows exponentially and the receiver multidimensional slicer becomes impractical to implement. Significant coding gains can be achieved with lower complexity by using an FSM in the transmitter (as was illustrated in Fig. 13-1), possibly in conjunction with a multidimensional signal constellation. The lower complexity comes from the inherent simplicity of the transmitter FSM and the availability of the Viterbi algorithm for ML detection in the receiver.

The basic advantage of signal-space coding is the same for both approaches in Fig. 13-1. Namely, by going to a higher dimensionality space we can increase the minimum distance in relation to the transmit signal energy. In both cases, the sequence of data symbols is *not* a Cartesian product of two-dimensional symbols. Rather, the symbols are dependent on one another, and, as was seen in Section 13.1, this dependence is the essence of achieving coding and shaping gain. The FSM introduces dependence of the successive symbols by its symbol-to-symbol state memory. The coding gain due to the FSM can augment the coding and shaping gain due to constellation design. A signal-space coder based on an FSM is often called a *trellis coder*, because the FSM is conveniently represented by its trellis (Chapter 5).

The convolutional coder of Chapter 12 is a convenient FSM to use in a signal-space code. In Chapter 12 we thought of the additional bits introduced by the convolutional coder as increasing the bit rate. In signal-space coding, this redundancy is normally mapped into a larger symbol constellation, rather than an increased symbol rate. The bandwidth required for transmission is not increased, nor is total noise admitted by the receive filter. Of course, a penalty is paid in an increase in the number of points per multidimensional symbol, which taken by itself will either reduce the minimum distance or increase the transmitted energy. However, the advantages of working in a multidimensional space more than makes up for this penalty.

A simple form of trellis coding proposed by Ungerboeck [3] uses a two-dimensional constellation, and is illustrated in Fig. 13-8. In Fig. 13-8(a), an uncoded PAM data symbol is generated by a mapper (two-dimensional symbol constellation). The size of the constellation is 2^k, and the information bit rate is k bits per symbol. The trellis coder of Fig. 13-8(b) modifies this configuration by adding a rate k/n channel coder, for example based on the convolutional coder of Chapter 12. The symbol mapper is modified to use a constellation of size 2^n, where $n > k$. Significant coding gains can be achieved in this way.

Example 13-7.

A trellis code designed by Wei [4] is a crucial part of 9600 b/s voiceband data modems compatible with the CCITT V.32 standard. In that standard, the symbol rate is 2400 symbols per second, so a 16 point constellation would suffice without coding to support 9600 b/s. The standard uses a rate-2/3 convolutional code and a 32-point constellation. Most voiceband data modem standards faster than 4800 b/s depend on trellis codes.

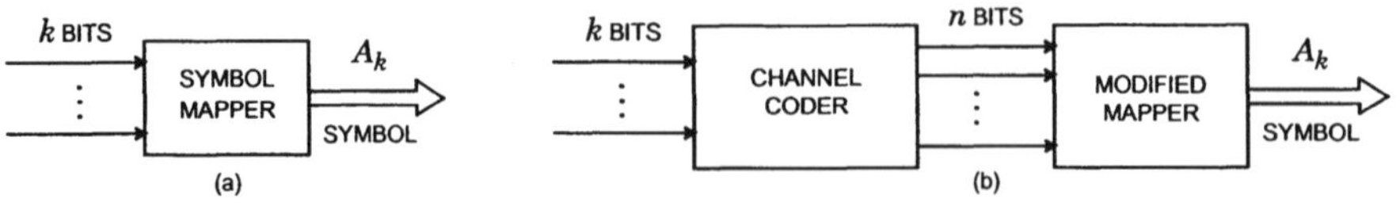

Fig. 13-8. (a) An uncoded system transmitting k bits per two-dimensional symbol with a constellation of size 2^k. (b) A signal-space trellis coder. The rate k/n channel coder introduces redundancy and the mapper accommodates that redundancy by using a constellation of size 2^n.

Assume that for a particular additive white Gaussian noise channel, an acceptable probability of error is achievable *without coding* at some SNR using a constellation of size M. Using coding we can reduce the SNR at the same error probability, or reduce the error probability at the same SNR; the improvement is limited by the Shannon channel capacity. The key observation made by Ungerboeck is that most of this theoretical reduction can be achieved using a constellation of size $2M$ plus a channel coding algorithm. It is not greatly advantageous to use a constellation of size greater than $2M$, and it is not necessary to increase the symbol rate.

The justification for Ungerboeck's observation depends on Fig. 4-4. This figure shows theoretical channel capacity (the maximum information rate for *error free* transmission) under various assumptions. In the figure, the left-most curve is the Shannon bound for discrete-time channels with additive Gaussian white noise. No assumption is made about the input constellation. The rest of the curves constrain the input constellation to be discrete-valued with equally likely symbols, and show the resulting channel capacity as a function of SNR. For example, at low SNR we can get close to the Shannon bound using a four level PAM signal, 4-PAM. As the SNR increases, with 4-PAM the information rate cannot exceed two bits per symbol.

A set of dots are plotted on the curves that show the SNR at which a probability of error of 10^{-5} is achievable for a particular constellation *without coding*. For example, from Fig. 4-4 we see that at about 19 dB SNR we can achieve 10^{-5} with a 4-PAM constellation. Without coding, 4-PAM transmits two bits per symbol. But *with coding*, using an 8-PAM constellation we can theoretically transmit 2 bits per symbol (*error free*) down to about 13 dB SNR. Hence, using a coded 8-AM constellation, we should be able to design a code with a total gain (coding plus shaping gain) of 19 – 13 = 6 dB. Using larger constellations cannot improve the total gain by more than about 1 dB, because the 6 dB gain is already so close to the Shannon bound. Furthermore, since there is no increase in bandwidth with coding, the gain is fully realized. There is no more noise for the coded system than for the uncoded system, unlike some situations considered in Chapter 12, in which the additional noise offset some of the gain.

13.2.1. Simple Trellis Codes

Designing trellis codes with total gains of 3 to 4.5 dB is easy, and the resulting codes are reasonably easy to implement. Simple trellis codes consist of convolutional coders followed by symbol mappers that accommodate the redundancy with a larger alphabet. In this section we will make no attempt to separate the coding gain from the shaping gain. We will also make no attempt to separate the coding gain due to the FSM from the coding gain due to constellation design. We will instead evaluate the overall gain of some simple trellis codes by comparing them to uncoded systems that achieve the same overall bit rate in the same bandwidth. Then in Section 13.3 we develop a model that separates the gains from the various sources.

Example 13-8. ---

Consider the concatenation of a convolutional coder and a 4-PSK mapper shown in Fig. 13-9. The trellis is shown in Fig. 13-10(a). The only difference between this trellis and the one in

Fig. 12-14(b) is that the transitions are labeled (B_k, A_k) rather than $(B_k, [C_k^{(1)}, C_k^{(2)}])$. (Two slightly simpler but equivalent trellis codes are given in Problem 13-9 and Problem 13-10.) The performance advantages apply equally to all three coders.

13.2.2. Total Gain of Trellis Codes

The total gain that can be achieved with a trellis code depends on the number of states in the FSM. Roughly speaking, it is easy to get coding gains of about 3 dB with 4 states, 4.5 dB with 16 states, and close to 6 dB with 128 or more states [5]. In this section we illustrate how to determine the performance of a trellis code by directly comparing it to an uncoded system. In Section 13.3 we will show how shaping, constellation design, and the FSM individually contribute to this overall gain.

Because of the FSM convolutional coder, trellis codes lend themselves to soft decoding using the Viterbi algorithm. This type of decoding is the most common in practice, so we restrict our attention to it. In Appendix 7-B we showed that at high SNR

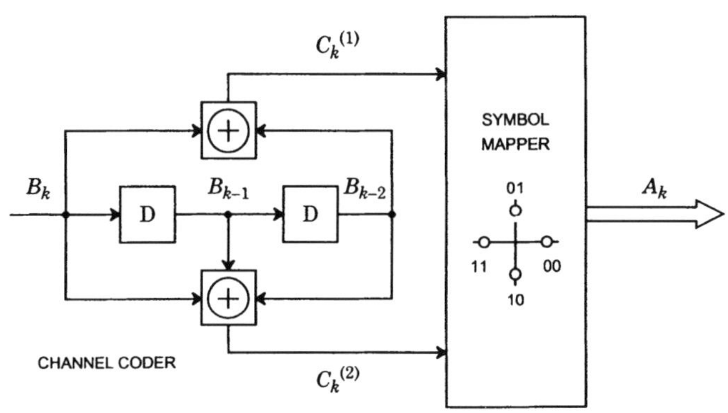

Fig. 13-9. A trellis coder consisting of a convolutional coder followed by a mapper.

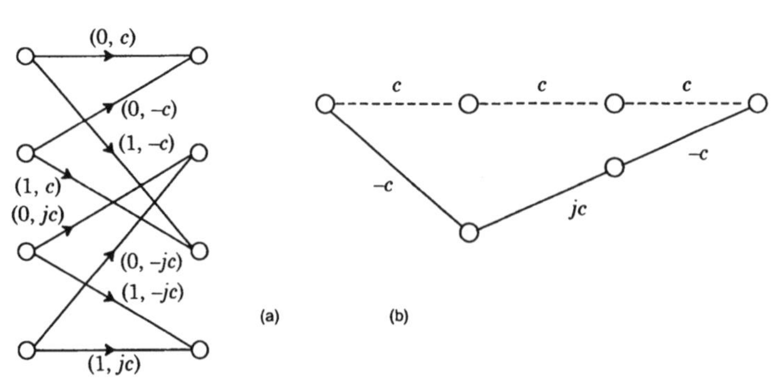

Fig. 13-10. (a) One stage of the trellis for the trellis code in Fig. 13-9, where $c = \sqrt{E}$. (b) The minimum-distance error event, assuming the correct state trajectory is the all-zero state trajectory.

$$\Pr[\text{symbol error}] \approx CQ(d_{E,\min}/2\sigma) \tag{13.48}$$

where C is the error coefficient, lying between P and R defined by (7.157) and (7.150). Hence, as with soft decoded convolutional codes, the performance of trellis codes is dominated by the error events with the minimum Euclidean distance $d_{E,\min}$. However, caution is in order when calculating this distance. In general, trellis codes are not linear even when the underlying convolutional code is linear (see Appendix 12-A and Problem 13-11), so it is usually unsafe to simply find the error event closest to the zero state trajectory (see Problem 13-11). Instead, we may have to be more careful and systematically find the minimum distance error event for each possible correct path through the trellis, as done in Section 7.6.2. (The model of Section 13.3 separates the contribution of the FSM, and in most cases this leads to linearity and simplified methods for calculating minimum distance.)

Example 13-9.

In Fig. 13-10(b) we show the minimum-distance error event assuming the correct state trajectory is zero; it has a distance of $\sqrt{10}c = \sqrt{10E}$. This distance is calculated using complex arithmetic,

$$|c+c|^2 + |c-jc|^2 + |c+c|^2 = 10c^2 . \tag{13.49}$$

For this case it is easy to verify that there is no error event with distance less than $\sqrt{10}\,c$. Observe simply that all branches diverging from the same node have distance $2c$ from each other, and all branches converging on the same node have distance $2c$, so the minimum distance is bounded from below by

$$d_{E,\min} \geq \sqrt{(2c)^2 + (2c)^2} = \sqrt{8}\,c \ . \tag{13.50}$$

We can further determine that after two paths diverge, all possible combinations of subsequent branches have distance $\sqrt{2}\,c$ so

$$d_{E,\min} \geq \sqrt{(2c)^2 + (2c)^2 + 2c^2} = \sqrt{10}\,c \ . \tag{13.51}$$

Hence

$$d_{E,\min} \geq \sqrt{10}\,c \tag{13.52}$$

for all possible state trajectories. It is also easy to verify that for all possible state trajectories there is exactly one error event at distance $\sqrt{10}\,c$.

In Example 13-9 every state trajectory has exactly one error event at distance $d_{E,\min}$, and this error event has exactly one symbol error, so $C = P = R = 1$. Consequently, at high SNR,

$$\Pr[\text{symbol error}] \approx Q\left(\frac{d_{E,\min}}{2\sigma}\right) . \tag{13.53}$$

To compare this performance to an uncoded system, the noise variance σ^2 is the same in both cases because the signal bandwidth is the same. We need to simply find an uncoded system with the same average transmit power that carries the same number of bits.

Example 13-10.

Continuing the previous example, $d_{E,\min} = \sqrt{10E}$ so

$$\Pr[\text{symbol error}] \approx Q\left(\frac{\sqrt{10E}}{2\sigma}\right) = Q\left(\sqrt{5\frac{E}{N_0}}\right). \tag{13.54}$$

Using the pessimistic assumption that Pr[bit error] ≈ Pr[symbol error] (see (7.164)), we get

$$\Pr[\text{bit error}] \approx Q\left(\sqrt{5\frac{E}{N_0}}\right). \tag{13.55}$$

An uncoded 2-PSK system with alphabet $\mathcal{A} = \pm c$ has the same transmit energy as the coded 4-PSK system with alphabet $\mathcal{A} = \{\pm c, \pm jc\}$, $c = \sqrt{E}$, and carries the same number of source bits. It has a probability of error

$$\Pr[\text{bit error}] = Q\left(\sqrt{2\frac{E}{N_0}}\right). \tag{13.56}$$

The coded system is better by approximately

$$10 \cdot \log\left(\frac{5}{2}\right) \approx 4 \text{ dB}. \tag{13.57}$$

We have achieved the same improvement over the uncoded system as was achieved by convolutional coding with soft decoding, but without any increase in the bandwidth!

In more complicated cases (where $P \neq R$) we can find P and R to estimate C in (13.48), as illustrated in Appendix 7-B. It is more common, however, to assume that C is reasonably small and ignore it. A widely used rule of thumb is that at error rates on the order of 10^{-5} or 10^{-6}, if C is not too large, every increase in C by a factor of 2 costs about 0.2 dB of coding gain [6].

Trellis coding is usually preferred over convolutional or block coding for bandlimited channels with additive white Gaussian noise, unless severe nonlinearities or hardware complexity make the increased alphabet size impractical.

13.2.3. More Elaborate Trellis Codes

In the previous examples of trellis codes, one source bit was processed with a rate 1/2 convolutional coder to yield two coded bits. Representing the two coded bits requires an alphabet of size four. The technique is easily extended to make use of alphabets larger than four.

Example 13-11.

A simple extension of the trellis code described in the previous section is illustrated in Fig. 13-11. This trellis code uses an alphabet of size eight, and can be built with any of the rate 1/2 convolutional coders in Fig. 13-9, Problem 13-9, or Problem 13-10. Note that $B_k^{(1)}$ has no effect on the shape of the trellis. The transition from one state to another is controlled entirely by $B_k^{(2)}$. Each transition from one state to another, therefore, occurs for two possible values of $B_k^{(1)}$, zero and one. This can be represented by showing two parallel transitions between every pair of states, as shown in Fig. 13-12.

The general technique, illustrated in Fig. 13-13, can be stated as follows. Given a channel with a bandwidth limitation, determine the symbol rate that can be transmitted. Determine the size

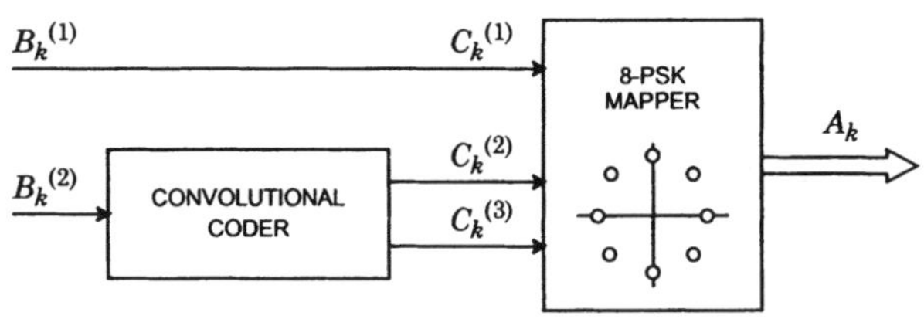

Fig. 13-11. A simple extension of the previous trellis code increases the number of source bits per symbol by using one uncoded bit and a larger alphabet.

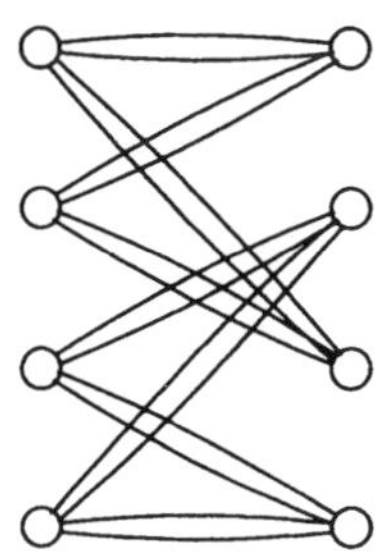

Fig. 13-12. The extra uncoded bit in Fig. 13-11 can be represented in the trellis using parallel branches, as shown. One of two parallel branches is taken, depending on the value of the extra bit.

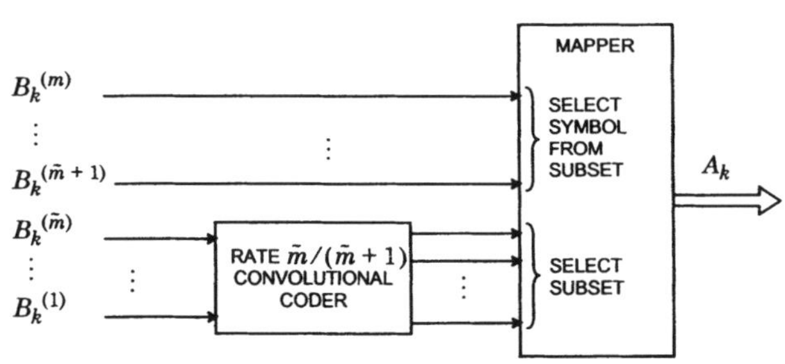

Fig. 13-13. Trellis coders made by cascading a convolutional coder and a mapper can divide the incoming source bits into $\tilde{m}$ bits to code and $m - \tilde{m}$ bits to leave uncoded. The required convolutional coder has rate $\tilde{m}/(\tilde{m}+1)$, and the trellis will have $2^{m-\tilde{m}}$ parallel transitions between every pair of states.

2^m of the alphabet that would be required (without coding) to transmit the source bits at the desired bit rate. Then double the size of the alphabet to 2^{m+1} and introduce a channel coder that produces one extra bit. The coder need not code all incoming bits, as shown in Fig. 13-13. However, leaving some bits uncoded may affect performance.

A trellis with $m - \tilde{m}$ uncoded bits has $2^{m-\tilde{m}}$ parallel transitions between every pair of states. When there are parallel transitions, a very short error event consists of mistaking one of these parallel transitions for the correct one. The coding does not defend at all against this error event. To minimize its probability, we should ensure that the Euclidean distance between the symbols corresponding to parallel transitions is maximized.

Example 13-12. To complete the design of the coder in Fig. 13-11, we need to design the mapping of bits into 8-PSK symbols. Designing the mapping is the same as assigning symbols to transitions in the trellis Fig. 13-12. Divide the 8-PSK constellation into four subsets, as shown in Fig. 13-14. To ensure that parallel transitions in Fig. 13-12 have symbols as far apart as possible, symbols for parallel transitions are selected from the same subset. Hence, in Fig. 13-11, $C_k^{(2)}$ and $C_k^{(3)}$ select the subset, A, B, C, or D, and $C_k^{(1)} = B_k^{(1)}$ selects the point within the subset. Now assign subsets to pairs of transitions to try to maximize the minimum distance. A simple heuristic is to try and keep the distance between diverging or merging branches as large as possible. For example, C is the furthest subset from A, so the two pairs of branches emerging from state zero should be assigned subsets C and A. Using this rule, the mapping is shown in Fig. 13-15.

As usual, the performance of the code will be dominated by the error event with the smallest distance from the correct path. For coders with parallel transitions, we need to check the distance between parallel transitions to see if these are the closest error events.

Example 13-13.

Continuing the previous example, assume $|A_k| = c = \sqrt{E}$ for all symbols in the 8-PSK alphabet. Then using the mapping developed in the previous example, the distance between parallel transitions is $2c$. The next closest error event turns out to be at distance $c\sqrt{6 - \sqrt{2}} = 2.14c > 2c$, so $d_{E,\min} = 2c$ (see Problem 13-13). Hence the probability of an error event is approximately

$$\Pr[\text{error event}] \approx Q\left(\sqrt{2\frac{E}{N_0}}\right) . \tag{13.58}$$

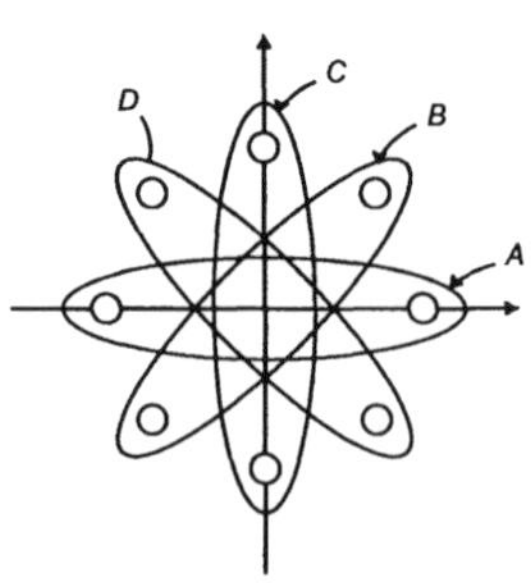

Fig. 13-14. The 8-PSK symbol set is divided into four subsets. To ensure that parallel transitions in Fig. 13-12 have symbols as far apart as possible, symbols for parallel transitions are selected from the same subset.

These most probable error events result in exactly one bit error out of two (the uncoded bit), and there is only one such error event for each correct path, so we can assert

$$\Pr[\text{bit error}] \approx \frac{1}{2}Q\left(\frac{a}{\sigma}\right) . \tag{13.59}$$

To compare this to the uncoded system, note that the uncoded system requires an alphabet of size four to achieve the same source bit rate. Such an alphabet resulting in a signal with identical energy is the 4-PSK alphabet $\{\pm a, \pm ja\}$. So, for the uncoded system, $d_{E,\min} = \sqrt{2}\,a$, and the probability of bit error is

$$\Pr[\text{bit error}] = Q\left(\frac{\sqrt{2}a}{2\sigma}\right) . \tag{13.60}$$

Ignoring the constant coefficient in (13.59), the total gain is

$$20 \cdot \log\sqrt{2} = 3 \text{ dB} . \tag{13.61}$$

Hence the parallel transitions degrade the performance by about 1 dB compared to the system in Fig. 13-9, which does not have parallel transitions.

Mapping by Set Partitioning

In Example 13-13, we developed a mapping between the coded bits $[C_k^{(1)}, C_k^{(2)}, C_k^{(3)}]$ and the transmitted symbols A_k. That mapping has the property that parallel transitions correspond to symbols that are as far apart as possible. A systematic way to design such mappings in general is known as *mapping by set partitioning*, proposed by Ungerboeck [3].

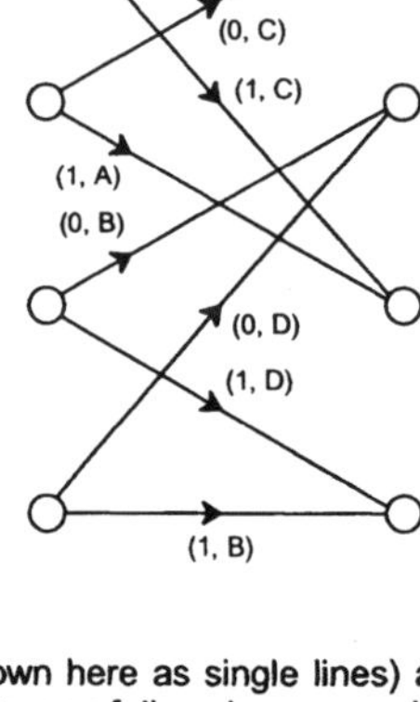

Fig. 13-15. Parallel branches (shown here as single lines) are assigned a subset of the signal set in such a way as to maximize the distance of diverging or merging branches.

From bandwidth and bit-rate considerations we can determine the number of symbols required in the alphabet. If the number is 2^m for an uncoded system, we have proposed using 2^{m+1} for the coded system. However, we still have considerable freedom in choosing how to map the coded bits into symbols. The choice of mapping can drastically affect performance of the code. A good heuristic technique was proposed by Ungerboeck [3].

Example 13-14.

In Fig. 13-14 we divided an 8-PSK constellation into four subsets. Another way to view this partition is illustrated in Fig. 13-16. The constellation is first divided into two subsets that maximize the distance within the subsets, and then subdivided again.

The same principle can be applied to more elaborate constellations. A 16-QAM constellation is partitioned in Fig. 13-17. Consider the trellis coder in Fig. 13-13. It is indicated that the uncoded bits select a signal from the subset, and the coded bits select the subset. Hence there must be $2^{\tilde{m}+1}$ subsets.

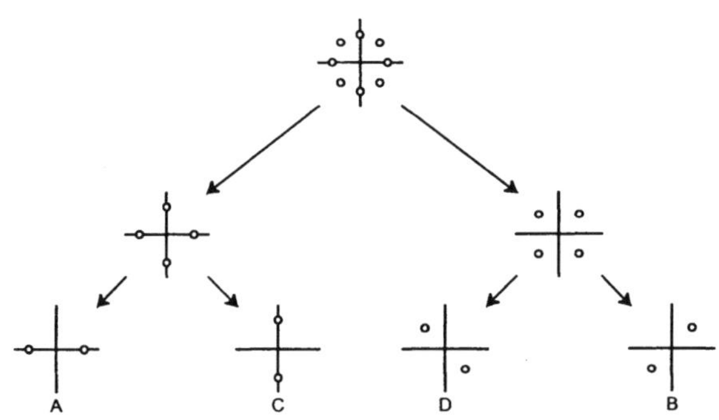

Fig. 13-16. Systematic partitioning of an 8-PSK constellation.

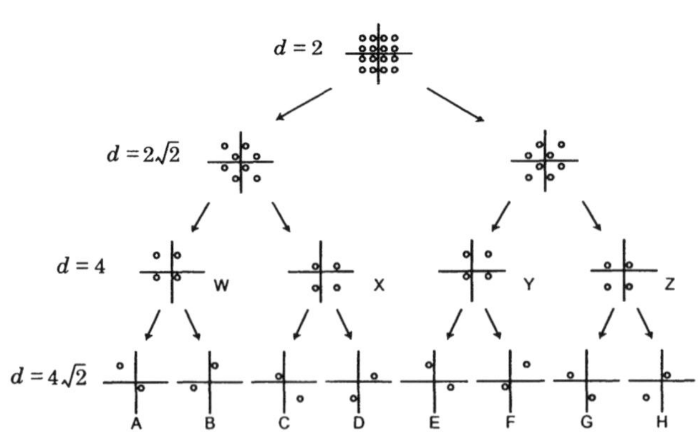

Fig. 13-17. A 16-QAM constellation is partitioned into subsets so that the distance between the symbols within the subset is maximized.

Example 13-15.

If $\tilde{m} = 1$ as in the previous examples, then 4 subsets are required. Thus the subsets in the third row of Fig. 13-17 can be used in the coder shown in Fig. 13-18. The uncoded bits select the symbol within the subset, and since there are two uncoded bits, each subset requires 4 symbols.

To use the subsets in the fourth row in Fig. 13-17, we need a trellis coder with $\tilde{m} = 2$.

Example 13-16.

A trellis coder with one uncoded bit and two coded bits that uses the subsets in the fourth row in Fig. 13-17 is shown in Fig. 13-19. The convolutional coder is from Fig. 12-36(b). It is an 8-state systematic rate $2/3$ convolutional coder of the feedback type. A reasonable mapping between the coded bits and the subsets from the fourth row of Fig. 13-17 is shown in Fig. 13-20. The total gain of this coder is approximately 5.33 dB [7]. We could of course add one more uncoded bit and use a 32-point cross constellation, in which case the total gain reduces to about 3.98 dB. Adding yet one more uncoded bit and using a 64-point QAM constellation reduces the total gain to about 3.77 dB (compared to a 32 point cross constellation used without coding).

It should be emphasized that these total gain coding gains compare the overall performance of the given trellis coder against an appropriate uncoded system. Not all the gain is due to the redundancy introduced by the convolutional coder; some of it is due to the constellations chosen for the comparison. This issue is addressed further in Section 13.3, where absolute measures of coding gain are developed.

The general rules for mapping by set partitioning are:

- First, maximize the distance between parallel transitions.
- Next, maximize the distance between transitions originating or ending in the same state.

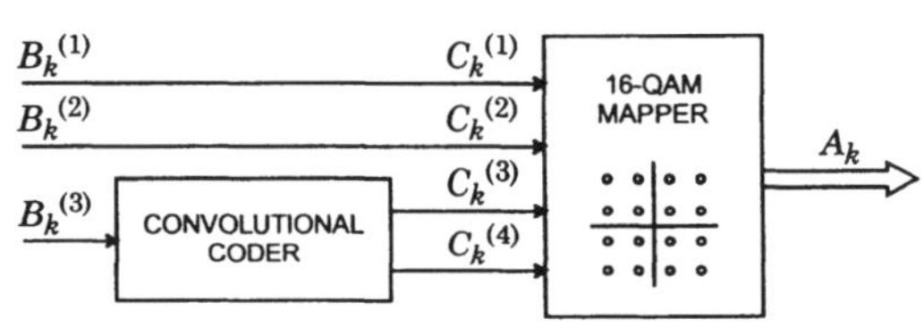

Fig. 13-18. A trellis coder with $\tilde{m} = 1$ and two uncoded bits.

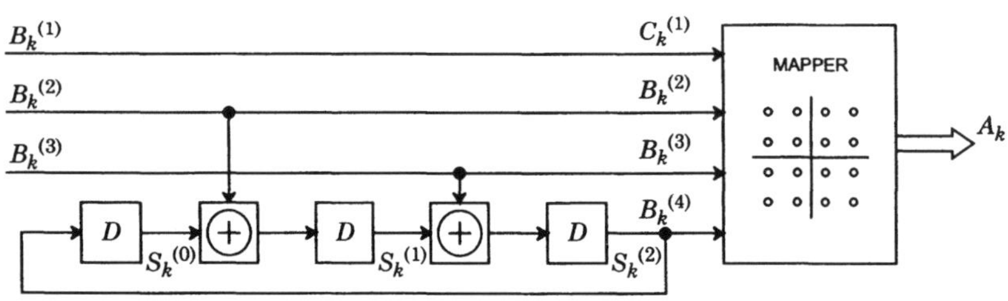

Fig. 13-19. A trellis coder with one uncoded bit and two coded. Also note that the convolutional coder is recursive (it has feedback), which is commonly used.

- Finally, use all symbols with equal frequency.

These rules are heuristic; they do not necessarily lead to a code that is optimal in any sense. In Section 13.3, we develop a more systematic approach to constellation partitioning based on cosets of a lattice.

Catastrophic Codes

Considering only minimum-distance error events has its hazards. In most of our examples, we argued that there was only one minimum distance error event, and consequently it dominates the performance. In this section we give an example at the other extreme, where there are an infinite number of minimum-distance error events. More specifically, R in (7.150) may not be bounded, so C in (13.48) also may not be bounded. Such a code is called *catastrophic*.

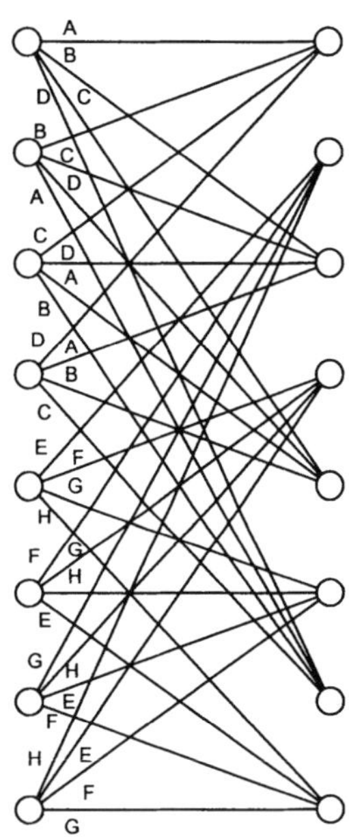

Fig. 13-20. A mapping between the coded bits of Fig. 13-19 and the subsets of the 16-QAM constellation in the fourth row of Fig. 13-17.

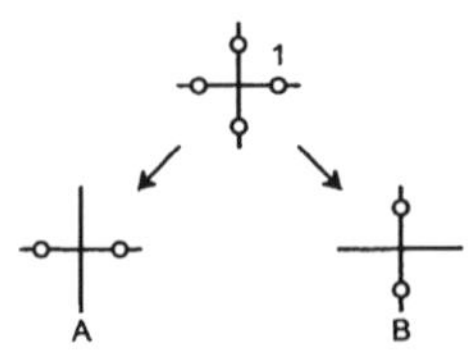

Fig. 13-21. Set partitioning for a four point constellation.

Example 13-17.
A four-point constellation can be partitioned as shown in Fig. 13-21. Consider transmitting 2 bits per symbol, using one bit to select subset A or B, and the other bit to select the point within the subset. This can be represented with a two-state trellis, as shown in Fig. 13-22. A minimum-distance error event is illustrated in Fig. 13-22(b). The minimum distance is 2, which is 3 dB better than the minimum distance of $\sqrt{2}$ in the uncoded system. If we assume that this is the only probable error event, then the total gain is about 3 dB. But we cannot expect this total gain. There are an infinite number of error events with the same minimum distance, some of which are shown in Fig. 13-22(c). In fact, it is easy to show that R in (7.150) is unbounded (see Problem 13-16). We should not expect any gain at all, because we are representing two source bits with an alphabet of size four, so there is no redundancy!

Catastrophic codes have the property that there are error events with finite distance but infinite length. The decoder may get into an error event during a burst of channel noise, and not get out again for a long time, especially if the channel quality improves! This will cause an infinite number of decoding errors. An interesting way of correcting the problem is illustrated in the following example.

Example 13-18.
Suppose that the two-state trellis is forced to return to state zero every fourth symbol, as shown in Fig. 13-23. Note that in the fourth symbol interval the coder does not have a choice of set A or B. Thus only one bit instead of two can be transmitted in the fourth symbol interval. In fact, the resulting code is a simple parity-check block code! Assuming the rate loss is not important, we can compare its performance with that of uncoded 4-PSK at high SNR. Now there are at most three error events with minimum distance starting at any given time, so the probability of a minimum-

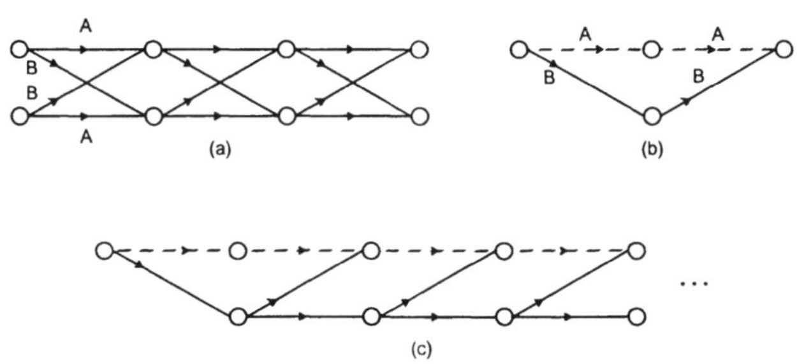

Fig. 13-22. a. A trellis for the code in Example 13-17. b. A minimum-distance error event. c. A set of minimum-distance error events. There are an infinite number of minimum-distance error events, so this code is catastrophic.

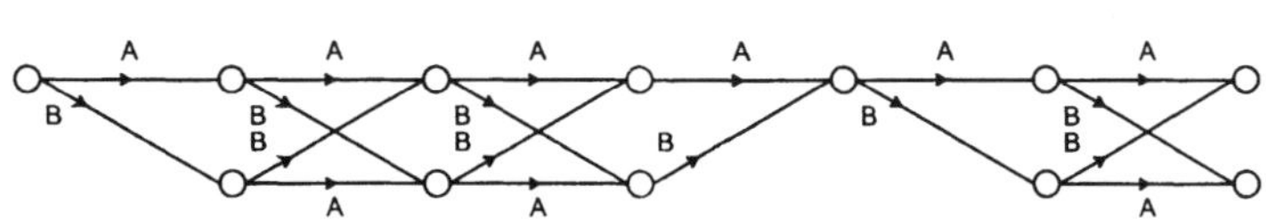

Fig. 13-23. The trellis of Fig. 13-22(a) is modified so that it is forced to return to the zero state every fourth symbol. This converts the code to a block code, and reduces the message bit rate.

distance error event can be approximated as $3Q(d_{E,\min}/2\sigma)$ where $d_{E,\min} = 2$. The factor of 3 is not important at high SNR, so nearly the full 3 dB improvement is realized. The code transmits only an average of $7/4$ source bits per symbol, so there is some room for redundancy in the 4-PSK alphabet.

Any trellis code can be converted into a block code of block size n by forcing the trellis to pass through a particular state every n symbols.

Trellis Codes using Nonlinear Convolutional Codes

In practice, there is little reason to restrict ourselves to linear convolutional coders (see Appendix 12-A). Some desirable trellis codes use nonlinear convolutional coders.

Example 13-19.

A trellis code using a nonlinear convolutional coder is shown in Fig. 13-24. The coder was invented by Wei [4], and has been adopted in CCITT recommendation V.32 for voiceband modems operating at 9600 bits per second. Its main advantage is that the code is invariant under 90 degree phase shifts. This advantage will become clearer when we discuss carrier recovery in Chapter 15. Some of the symmetry is evident from examining the constellation in Fig. 13-24. It is evident that $C_k^{(1)}$ and $C_k^{(2)}$ are the same at all points 90 degrees apart. Furthermore, $C_k^{(3)}$ and $C_k^{(4)}$ differentially encode the quadrant (see Chapter 15). That is, to move -90 degrees, simply add one

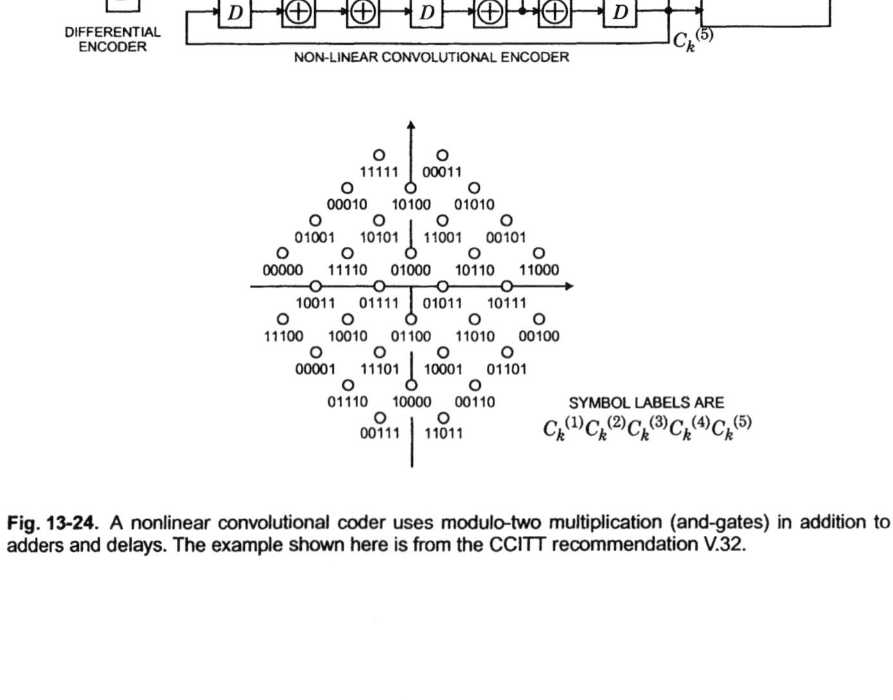

Fig. 13-24. A nonlinear convolutional coder uses modulo-two multiplication (and-gates) in addition to adders and delays. The example shown here is from the CCITT recommendation V.32.

(modulo-four) to these two bits. This differential encoding is deliberately introduced by the differential encoder in the figure. The symmetry of the final bit is more subtle, but can be seen as symmetry in the trellis [4].

Multidimensional Trellis Codes

In Section 13.1, the multidimensional signal constellation was described. It is realized by grouping one- or two-dimensional data symbols and treating them as a multidimensional vector. The general idea of trellis codes was presented in Fig. 13-1 in terms of a multidimensional constellation, but all the examples thus far are two-dimensional.

In the context of a trellis code, the benefit of using a multidimensional constellation can be explained as follows. Recall the observation that doubling the symbol alphabet is sufficient to achieve almost all the available coding gain determined by the Shannon limit. However, doubling the size of the constellation, for the same minimum distance, will increase the signal energy. The coding gain must overcome this immediate disadvantage. In Section 13.1, the continuous approximation predicted that doubling the size of the constellation increases P, the signal energy per two dimensions, by $2^{2/N}$ for an N-dimensional constellation. Thus, as the dimension of the constellation increases, the power penalty decreases. Expressed in dB, this penalty is

$$10 \cdot \log_{10}(2^2/N) \approx 6/N \ . \tag{13.62}$$

Thus, it decreases from 3 dB for a two-dimensional constellation to just 1 dB for a six-dimensional constellation.

Exercise 13-2. ---

Show that the 32-point cross constellation in Fig. 5-14 has 3 dB more energy than the 16-point QAM constellation in Fig. 5-13. Assume the points are all of the form $A_k = [a_1, a_2]$ where $a_i \in \{\pm 5, \pm 3, \pm 1\}$. Thus, as predicted by the continuous approximation, the signal energy is increased by 3 dB for a doubling of the number of points in the constellation.

Exercise 13-3. ---

Consider a four-dimensional symbol

$$A_k = [a_1, a_2, a_3, a_4] \tag{13.63}$$

where $a_i \in \{\pm 3, \pm 1\}$. Show that the average squared energy of this alphabet is 20, assuming all symbols are equally likely. There are $4^4 = 256$ symbols in this alphabet. Now construct a four-dimensional alphabet with 512 symbols by adding symbols of the form:

$$[\pm 5, \pm 1, \pm 1, \pm 1] \tag{13.64}$$

and all its permutations (for a total of 64 possibilities) and symbols of the form

$$[\pm 5, \pm 3, \pm 1, \pm 1] \tag{13.65}$$

and all its permutations (for a total of 192 possibilities). Show that the average squared energy of the 512 point constellation is 27, only 1.3 dB more than the average squared energy of the 256 point constellation. The continuous approximation predicts a 1.5 dB penalty.

While four-dimensional constellations like the one derived in Exercise 13-3 are difficult to draw, they are easy to use. Given a two-dimensional modulation system (a passband system), the first two coefficients a_1 and a_2 of the four-dimensional symbol are transmitted in one symbol interval as the real and imaginary parts of a complex symbol, and the last two coefficients a_3 and a_4 are transmitted in the next symbol interval. It is now easy to construct a trellis coder that makes use of this.

Example 13-20.

Continuing the previous example, the four-dimensional alphabet has 512 symbols, and hence can represent 9 coded bits. The trellis coder in Fig. 13-25 will do the job. Three bits are coded to get four bits, and five bits are used uncoded. Calderbank and Sloane proposed this configuration in 1985 [8], and Forney, *et. al* proposed a similar configuration one year earlier [9]. Using an 8-state convolutional coder, they found a total gain of about 4.7 dB, ignoring the error coefficient C. They compared this performance to a two-dimensional trellis coder with the same source bit rate (4 bits per symbol) which has a total gain of about 4 dB. This suggests that use of a multidimensional trellis code yields an additional total gain of about 0.7 dB in this example. However, the error coefficient C in this case is large enough to nullify much of this advantage.

Ungerboeck tabulates several possible four and eight-dimensional trellis coders and their performance [7]. Many good multidimensional trellis codes are given by Wei [10]. Forney [6] and Ungerboeck [7] also tabulate good multidimensional trellis codes and their properties.

13.3. Coset Codes

In Section 13.2 we compared the performance of trellis coded systems against uncoded systems with the same transmit energy and spectral efficiency. Although such comparisons are useful, they fail to properly account for the sources of improvement in performance. In particular, they mix coding gain with shaping gain, and they mix coding gain due to constellation design with coding gain due to the FSM. Moreover, the methods do not scale well to large constellations and to multidimensional constellations.

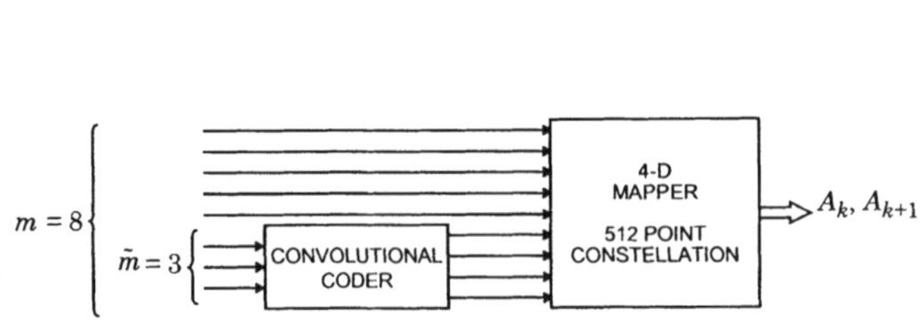

Fig. 13-25. A four-dimensional trellis coder. For every set of 8 bits that come in at the left, one four-dimensional symbol is produced by the mapper. These are actually transmitted, however, as two successive two-dimensional symbols.

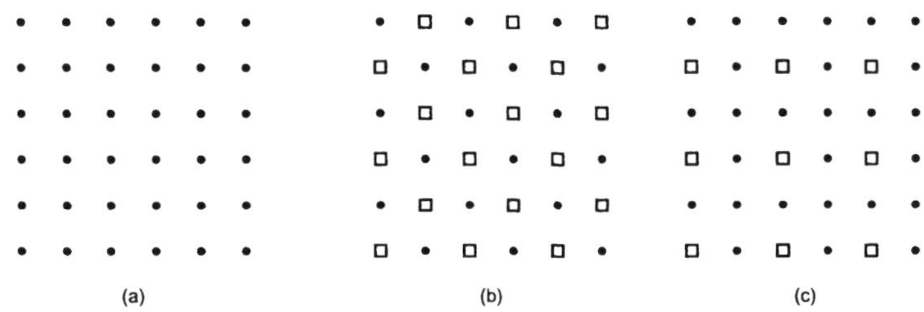

Fig. 13-26. An illustration of portions of a lattice and two sublattices. Assume the lower left is the origin. (a) The integer lattice Z^2. (b) The sublattice RZ^2 (defined by the squares). (c) The sublattice $2Z^2$ (also defined by the squares).

In this section, we introduce a more systematic approach based on the cosets of a lattice, as introduced by Calderbank and Sloane [11]. Trellis codes based on this coset partition are called *coset codes* by Forney [6]. Coset codes allow us to separate the coding gain due to the lattice, the shaping gain, and additional coding gain due to the FSM. Moreover, Ungerboeck's set partitioning generalizes to a simple systematic method that is easy to apply.

13.3.1. Lattice Partitions and Cosets

For a lattice Λ, a *sublattice* Λ' is a subset of the points in the lattice that is itself a lattice. Recall that a lattice is algebraically a group, meaning that it is closed under vector sums and differences. Any lattice must therefore include the zero point, and must be infinite in extent.

Example 13-21.
Let the one-dimensional integer lattice be written $Z = \{\ldots, -1, 0, 1, 2, \ldots\}$. Then the two-dimensional integer lattice in Fig. 13-26(a) is the Cartesian product $\Lambda = Z^2$. A sublattice is defined by the squares in Fig. 13-26(b). Note that this sublattice is equal to a rotated and scaled version of the original lattice. Thus if λ is a vector representing a point in the original lattice (a two-dimensional vector with integer entries), then $R\lambda$ is a vector representing a point in the sublattice, where

$$R = \begin{bmatrix} 1 & 1 \\ 1 & -1 \end{bmatrix}. \tag{13.66}$$

Thus we write the sublattice $\Lambda' = RZ^2$. The sublattice defined by the squares in Fig. 13-26(c) is simply a scaled version of Λ, $\Lambda'' = 2Z^2$. Notice that $2Z^2$ is a sublattice of RZ^2.

Given a sublattice Λ' of Λ, a *coset* of Λ' is the set

$$\{\lambda' + \mathbf{c};\ \text{for all } \lambda' \in \Lambda'\} \tag{13.67}$$

for some $\mathbf{c} \in \Lambda$. Since $\mathbf{c}$ identifies the coset, it is called the *coset representative*. Each coset is written $\Lambda' + \mathbf{c}$.

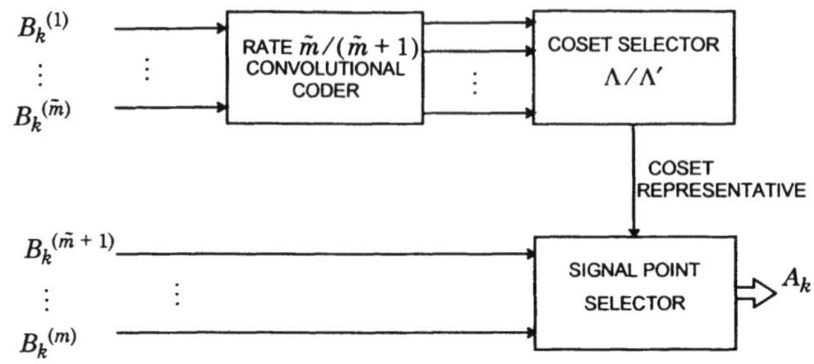

Fig. 13-27. A view of trellis coding that enables deeper understanding of the sources of performance gain.

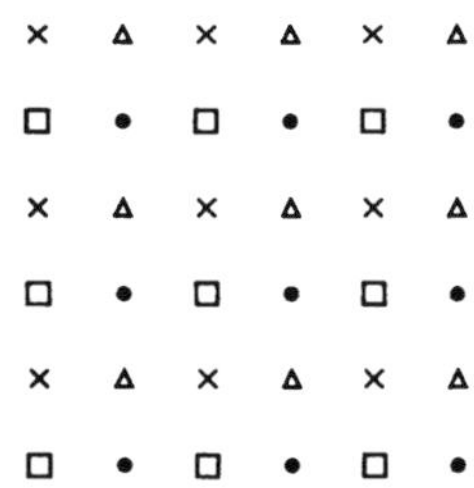

Fig. 13-28. A portion of four cosets of $\Lambda'' = 2Z^2$ identified by four distinct shapes.

Example 13-22.

The sublattice $\Lambda'' = 2Z^2$ in Fig. 13-26(c) has a total of four unique cosets, shown in Fig. 13-28. The coset representatives are $\boldsymbol{c} \in \{(0, 0), (1, 0), (0, 1), (1, 1)\}$. Only the coset identified by squares, the one with $\boldsymbol{c} = (0, 0)$, is itself a lattice, since it is the only coset that includes the zero vector. Similarly, the sublattice $\Lambda' = RZ^2$ in Fig. 13-26(b) has two cosets.

A *partition* of Λ induced by Λ', written Λ/Λ', is the set of all cosets of Λ' in Λ, including Λ' itself. The *order* of a lattice partition, written $|\Lambda/\Lambda'|$, is the number of distinct cosets, and is finite.

13.3.2. Application to Trellis Coding

In contrast to Fig. 13-13, consider the view of trellis coding shown in Fig. 13-27. The output of a convolutional coder (or more generally, an FSM) is used to select a coset from a partition Λ/Λ'. The FSM does not define which point to use within the coset. The output of the coset selector, therefore, is a coset representative.

The point within the coset is selected by the "signal point selector," which therefore applies shaping. It might also translate the lattice by some vector $\boldsymbol{a}$ so that the mean of the points it uses is zero. Thus, given a coset representative $\boldsymbol{c}$, the signal point selector chooses a constellation point in the set $(\Lambda' + \boldsymbol{c} + \boldsymbol{a}) \cap S$, where S is the shaping region and $\boldsymbol{a}$ is a vector not constrained to lie on the lattice Λ.

The model in Fig. 13-27 describes some trellis codes, but not all. The 8-PSK constellation of Fig. 13-11, for example, cannot be described as cosets of a lattice. However, when constellations get large, practical considerations dictate using regular structures, and lattices have compelling advantages. Thus, the model in Fig. 13-27 includes as special cases most practical examples of trellis codes with large constellations.

Example 13-23.

The trellis coders of Example 13-15, Example 13-16 and Example 13-7 can all be described in terms of Fig. 13-27. Multidimensional trellis codes are also generally coset codes, except that the output of the signal point selector will be a vector of $N/2$ complex symbols, rather than just a single complex symbol.

There are potentially three distinct sources of performance gain in Fig. 13-27:

- The lattice Λ may have coding gain, as discussed in Section 13.1.
- The signal point selector will use some shaping region S that may have shaping gain, also as discussed in Section 13.1.
- The convolutional coder will allow only particular sequences of cosets to be sent to the signal point selector, and thus introduces its own coding gain by increasing the minimum distance between sequences.

In Section 13.2, we made no attempt to separate the gain due to these three effects. The first two effects have been thoroughly studied in Section 13.1, so the third is the only addition.

13.3.3. Coding Gain due to Redundancy

The convolutional coder in Fig. 13-27 allows only a subset of all possible sequences of cosets. Thus, there is redundancy in such sequences. This means that the minimum distance between any two allowable sequences will be larger than the minimum distance between pairs of points in the lattice.

Let $d_{\min}(C)$ be the minimum distance between any two sequences of cosets allowed by the convolutional coder C, where the distance between two cosets is taken to be the minimum distance between any point in one coset and any point in the other. This minimum distance will dominate the performance of the overall system. Let $d_{\min}(\Lambda)$ be the minimum distance between any two points in Λ. Then the convolutional coder has increased the minimum distance by a factor of $d_{\min}(C)/d_{\min}(\Lambda)$.

There is a price paid, however, for this increase in minimum distance. Since there is typically one more bit emerging from the convolutional coder than going into it, twice as many points in the lattice will be needed within the shaping region to transmit the coded signal. Thus the size of the shaping region must increase. This *constellation expansion* reduces the gain, since it increases the energy.

To quantify the effect of the constellation expansion, define the *redundancy* $r(C)$ of the convolutional code C to be the number of redundant bits generated by the convolutional coder per N dimensions. In other words, it is the number of bits at the output of the convolutional coder minus the number of bits at its input. Usually $r(C) = 1$, as shown in Fig. 13-27. Define the *normalized redundancy* (per two dimensions) as

$$\rho(C) = \frac{r(C)}{N/2} . \tag{13.68}$$

The transmitted energy is increased by approximately $2^{\rho(C)}$. Note that with the typical $r(C) = 1$, the transmitted energy is increased by $2^{2/N}$. Thus, as explained in Section 13.1, if $N = 2$, the energy is doubled to accommodate the redundancy of the convolutional coder. This is intuitive, because the number of points in the transmitted (two-dimensional) constellation doubles. However, if $N = 4$, then the energy increases by only a factor of $\sqrt{2}$ because the number of points is doubled in a *four-dimensional* constellation, and this increase is divided between two successive two-dimensional symbols.

Combining the positive and negative effects of the convolutional coder, we get the coding gain due to the convolutional coder,

$$\gamma_C = \frac{d^2_{\min}(C)}{d^2_{\min}(\Lambda)2^{\rho(C)}} . \tag{13.69}$$

This is simply the increase in minimum distance due to the convolutional coder divided by the increase in energy (per two dimensions) due to the constellation expansion. We would expect the overall coding gain compared to an uncoded rectangular lattice with a rectangular shaping region to be, from (13.18),

$$\gamma = 3 \cdot \gamma_\Lambda \cdot \gamma_S \cdot \gamma_C , \tag{13.70}$$

where γ_Λ is the coding gain of the lattice defined in (13.19), and γ_S is the shaping gain of the lattice defined in (13.20). Thus, the convolutional coder adds additional coding gain γ_C on top of the lattice coding gain and the shaping gain. This is verified in the following exercise.

Exercise 13-4.

Show that (13.70) reduces to (13.10) where $d^2_{\min} = d^2_{\min}(C)$.
Hint: It might be helpful to use (13.46) and (13.12).

The formulation in Fig. 13-27 leads to another view of Ungerboeck's set partitioning, shown in Fig. 13-29. Given a lattice Λ, form a sublattice Λ' or order two. For example, the sublattice RZ^2 in Fig. 13-26(b) is a sublattice of Z^2 of order two. A coset induced by such a sublattice is therefore selected by one bit. Then form a sublattice Λ'' of Λ' with order two. A coset for this partition is selected by another bit. Continue partitioning the lattice until there are enough partitions to encode all the bits from the convolutional encoder. For example, the set partitioning in Fig. 13-17 is obtained by applying this procedure to $\Lambda = Z^2$.

13.4. Signal-Space Coding and ISI

Trellis coding is advantageous on many media where bandwidth is at a premium. On some, such as radio transmission between fixed antennas, or satellite transmission, ISI is not a significant problem. On others, like wire-pair and coaxial cable, ISI must be equalized or cancelled. The best techniques for countering ISI described in Chapters 8 and 9 are not immediately compatible with trellis coding or other signal-space codes. The reason is simply delayed decisions. In theory, the ML detector using the minimum-distance criterion may have to wait forever before it can make a decision. In practice, the Viterbi algorithm is used with some truncation depth (see Section 5.4.4). However, even a modest truncation depth introduces enough delay in the decision to compromise any decision-directed or decision-feedback technique. Thus, the straightforward combination of equalization for ISI with signal-space coding is not always possible. We now discuss some of these problems and ways around them.

13.4.1. Trellis Coding and Linear Equalization

Linear equalization (Chapter 8) is the simplest way to counter ISI. The LE is fortunately easy to combine with Viterbi decoding of a trellis code, since the input to the Viterbi detector is nominally free of ISI. The price paid for this simple solution, as opposed to ML detection or decision-feedback equalization, is higher noise enhancement.

There is potentially a problem when the adaptation of an LE is decision-directed (Chapter 9). A similar problem occurs with other decision-directed algorithms, such as carrier recovery (Chapter 15). Because of the delay in making decisions, the dynamics of the adaptation algorithm is altered. For example, in the receiver shown in Fig. 5-21, the slicer is replaced with a sequence detector and the decisions are delayed. To ensure stability of the adaptation algorithms, their step sizes must be reduced, so the convergence time and tracking

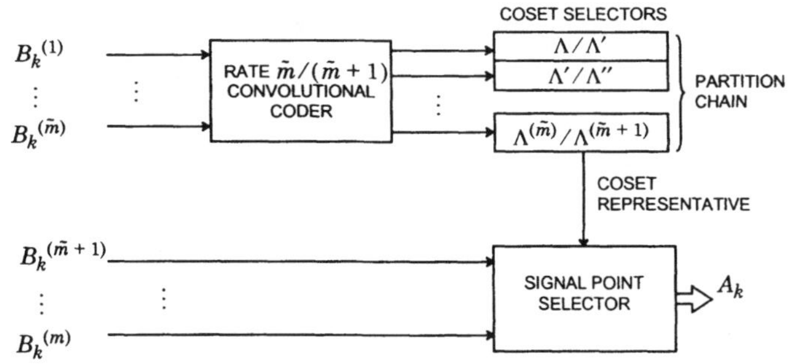

Fig. 13-29. Set partitioning can be viewed as a sequence of second order lattice partitions.

ability suffers. An alternative approach is to make tentative decisions with a conventional slicer, and use those tentative decisions to update the filter taps and carrier phase. Since these tentative decisions do not benefit from the coding gain, a minor degradation is suffered in the performance of the adaptive filters.

13.4.2. Trellis Coding and Transmitter Precoding

In Chapter 8, it was shown that decision-feedback equalization (DFE) results in less noise enhancement than the LE, since postcursor ISI is cancelled rather than equalized. The DFE depends on past decisions to cancel the postcursor ISI, and any delay in the availability of decisions due to sequence detection implies that only postcursor ISI with a compatible delay can be cancelled. For all practical purposes, this renders the DFE useless in the presence of ML trellis decoding, since it is typically the low-delay postcursor ISI that is most consequential. A way to combine the DFE and trellis coding, called "parallel decision-feedback equalization," will be discussed in the next subsection. Here we describe an alternative that is compatible with trellis coding.

Transmitter precoding (Section 8.1.4) was shown to achieve a performance essentially identical to the DFE by doing the postcursor cancellation in the transmitter rather than the receiver. The price paid is the need for knowledge of the channel response in the transmitter, a requirement that is compatible only with channels that are stationary or slowly time varying. This rules out, for example, rapidly fading radio channels.

Transmitter precoding was explained in the context of a one-dimensional signal constellation, although it is compatible with multidimensional constellations as well. For simplicity, we will explain the combination of precoding with trellis coding for one-dimensional data symbols. Recall that if the data symbol a_k is chosen from an alphabet of size M (where M is even) consisting of odd integers less than M in magnitude, then the transmitter is designed in such a way that the *channel output* symbol is an extended data symbol

$$c_k = a_k + 2M \cdot i_k \,, \tag{13.71}$$

where i_k is a sequence of integers chosen to minimize the peak transmitter power. The alphabet of c_k consists of all odd integers. This channel output data symbol is also corrupted by additive noise at the receiver, and in the absence of trellis coding is detected by applying an extended slicer that finds the closest odd integer to the received signal. The original data symbol a_k can be obtained uniquely from c_k by reducing it modulo $2M$.

Transmitter precoding is immediately compatible with trellis codes, if the constellation design and set partitioning is done in a compatible fashion. It requires no modification to the trellis coder, but the trellis decoder does have to be modified slightly. This is illustrated by an example.

Example 13-24.

An eight-point baseband constellation is shown in Fig. 13-30(a). It consists of all odd integers between –7 and 7, and can be partitioned using set partitioning into the four subsets shown. Given a rate one-half convolutional coder with one input bit, and a second input uncoded bit, the two convolutional coder output bits can be used to select the subset (from among four possibilities) and the uncoded bit can be used to select the point (from among two possibilities) within the selected subset. Suppose we are working with the extended constellation for c_k rather than a_k, as shown in Fig. 13-30(b). This extended subset is the lattice on which the constellation is based. Although c_k consists of all odd integers, it can still be partitioned into four subsets, the cosets induced by the specified partition. As before, the two coded bits select the subset, except that now the subsets are extended. The point within the subset, however, is chosen differently. The uncoded bit chooses one of the two points closest to the origin, and then, after i_k is determined, $2M \cdot i_k = 16 \cdot i_k$ is added to that choice. Adding $16 \cdot i_k$ chooses one of the points in the extended partition.

This example illustrates that, to be compatible with trellis coding, the symbol in an extended constellation is chosen in a two-stage process, as illustrated in Fig. 13-31. The trellis coder is unchanged (except to make sure it is compatible with precoding). The output data symbol a_k with an M-ary alphabet is input to the transmitter precoder. The precoder output symbol x_k is determined so as to force the channel output signal component $c_k = a_k + 2M \cdot i_k$ to be a point on the extended constellation, and at the same time minimize the peak power of the symbol x_k.

The trellis decoder must deal with an extended signal constellation. The trellis is unchanged, as determined by the trellis coder, and each pair of successive states in the trellis corresponds to a subset. The only difference is that each subset has an infinite set of points, extended by adding some unknown multiple of *2M*. There are now an infinite number of parallel branches between pairs of states, corresponding to the possible points in that subset. In

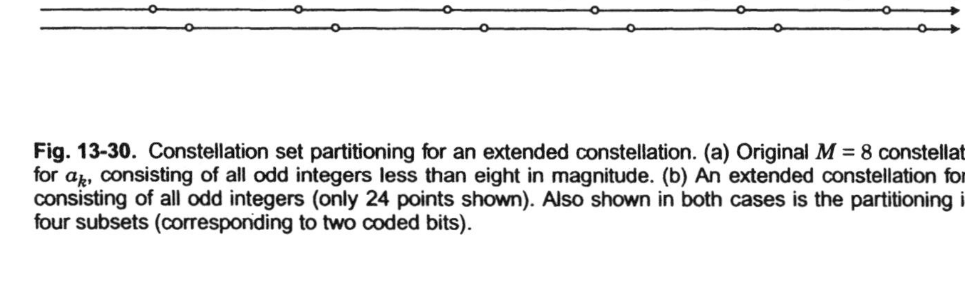

Fig. 13-30. Constellation set partitioning for an extended constellation. (a) Original $M = 8$ constellation for a_k, consisting of all odd integers less than eight in magnitude. (b) An extended constellation for c_k consisting of all odd integers (only 24 points shown). Also shown in both cases is the partitioning into four subsets (corresponding to two coded bits).

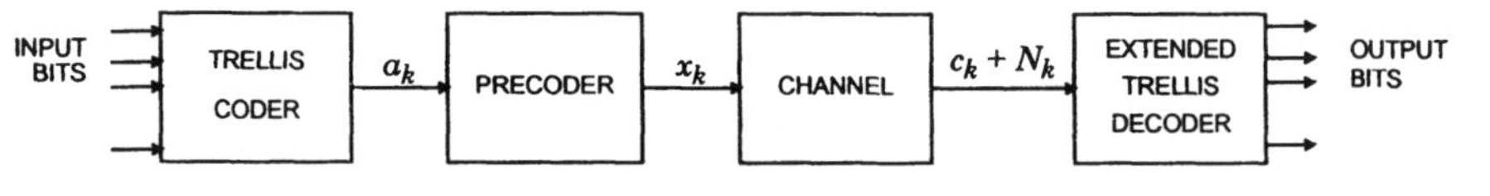

Fig. 13-31. A combination of trellis coding with transmitter precoding for compensation of channel ISI.

principle, the ML detector must consider all parallel branches. In practice, i_k will be bounded, based on worst-case channel assumptions, so that only a finite number of parallel branches need be considered.

This particular example is easily extended to larger constellations, or to two or more dimensions. The trick is to start with a full lattice, rather than a lattice as limited by the shaping region, before performing the set partitioning. The mathematical framework of the sublattice and cosets can be helpful here. The number of sets in the partition is determined by the number of coded bits, and is typically four or perhaps eight. Finally, the transmitter trellis is designed assuming a cubic shaping, but the receiver decoder is designed assuming the full (extended) lattice.

Coding Gains on Channels with ISI

It was established in Section 8.5.7 that if ISI is canceled by the DFE-ZF, then the SNR gap to capacity is the same at high SNR regardless of the nature of the ISI (including no ISI). Of course, the DFE-ZF is not directly compatible with trellis coding. However, in Section 8.1.4 it was shown that transmitter precoding can obtain essentially the same SNR at the slicer (or more generally decoder) input as the DFE-ZF. The conclusion is perhaps a surprising one; namely, the SNR gap to capacity at low probabilities of error can be closed to the same extent on channels with ISI as it can on channels that are free of ISI.

This is not to imply that ISI has no impact on capacity; in fact, as illustrated in Section 8.5.6, ISI does have an impact on the capacity of the channel. Thus, the capacity is affected by the ISI, but what does not change is the SNR gap between uncoded square QAM (at a given P_e) and capacity, and the extent to which that gap can be closed by coding and shaping gain.

Again it should be emphasized that this statement applies only at high SNR. The SNR gap to capacity is constant at all SNR for the MMSE-DFE-U, even at low SNR. It is, however, difficult to exploit this property at low SNR because the combination of noise and residual ISI at the slicer is not white.

Trellis Coding and Shaping Gain

Thus far, we have emphasized the application of trellis coding to obtaining coding gain. Trellis coding in conjunction with transmitter precoding can achieve shaping gain as well.

Recall from Section 8.1.4 that when we apply the continuous approximation to transmitter precoding, the conclusion is that the precoded symbols x_k are uniformly distributed and independent from symbol to symbol. That is, the shaping is cubic; there is no shaping gain. Thus, choosing i_k in the precoder to force the precoded sample x_k to obey $\| x_k \| \le M$ minimizes the energy of each precoded symbol, but doesn't minimize the average power of a *sequence* of symbols.

In fact, if the shaping is spherical, and the energy per complex-symbol is kept constant at E, then the radius of an N-sphere is $R = \sqrt{E(N/2+1)}$. Since each coordinate is limited to R, as we move to spherical shaping the peak value of each coordinate actually increases (in proportion to $\sqrt{N}$), and the marginal distribution approaches a truncated Gaussian. The conclusion is that to obtain shaping gains, we want to increase the allowed peak value of each symbol (which is precisely what we tried to avoid in the design of the transmitter precoder), and we want the distribution of the precoded symbols to be "Gaussian-like."

In light of these observations it is not surprising that shaping gain can be achieved by choosing the precoding i_k appropriately, and in particular *not* choosing them to minimize the peak value of the precoded symbol as we did in Chapter 8. A specific technique is proposed by Eyuboglu and Forney [12]. The basic idea is to choose the i_k to force the transmitted power averaged over time to be approximately equal to the allowed value, rather than the unshaped approach of minimizing the peak transmitted power of each individual symbol.

13.4.3. RSSD of Trellis Codes

On time-varying channels, the transmitter precoding requirement for knowledge of the channel response is problematic. In this case, there is an alternative, *reduced-state sequence detection (RSSD)* [13][14][15] in the receiver. With a trellis coder but no precoding in the transmitter, and channel ISI modeled as an FIR filter, the concatenation of the coder and the channel is an FSM signal-generation model. The received signal is the output of that FSM signal-generation model corrupted by noise. The ML detector for this signal-generation model is the Viterbi algorithm, which can in principle be implemented assuming the channel impulse response is known to the receiver (but not necessarily the transmitter). The problem is the explosion in the number of states, which is the number of states in the trellis coder multiplied by the number of states in the channel model. RSSD reduces the complexity by retaining only a subset of the most important states.

The most attractive of these techniques essentially implements the Viterbi algorithm for the original trellis coder, but performs decision-feedback equalization on each path survivor in the trellis based on the history of that path. This is called *parallel decision-feedback equalization*. If there are n states in the trellis, then n distinct postcursor equalizer filters are used. Each filter uses the decisions from one of the n survivor paths to construct the next decoder input.

13.4.4. Signal-Space Coding and Multicarrier Modulation

In the presence of ISI, multicarrier modulation (MCM) is an interesting alternative. In OPAM (combined orthogonal modulation and PAM), the transmitted signal is represented by

$$S(t) = \sum_{k=-\infty}^{\infty} \sum_{n=0}^{N-1} a_k^{(n)} g_n(t - kT) , \tag{13.72}$$

where the $g_n(t - kT)$ are a set of N orthogonal waveforms (orthogonal for all values of n and k). MCM is a special case where the pulses are time-limited sinusoids at equally spaced frequencies. As described in Chapter 6, in typical applications of MCM the bandwidth of the channel is kept constant, but the dimensionality of the signal set N is increased by increasing the symbol interval and decreasing the spacing between adjacent carriers.

Regardless of the particular choice of orthogonal pulses, if the MCM signal is transmitted through an additive white Gaussian noise channel, and matched filtering is applied at the receiver, then the equivalent channel model for the k-th symbol interval is a received vector of real- or complex-valued symbols $(a_k^{(0)}, a_k^{(1)}, \ldots, a_k^{(N-1)})$ corrupted by additive independent Gaussian random noise samples. This channel is mathematically indistinguishable from an ordinary PAM channel. However, with MCM there are more interesting degrees of freedom. For example, we can take the stream of data symbols corresponding to each dimension, $\{a_k^{(n)}, -\infty < k < \infty\}$, as an independent stream of data symbols for each $n \in \{0, 1, \ldots, N-1\}$, and independently apply a trellis coder/decoder to each one. Alternatively, we can serialize all the data symbols, generating $\{a_k^{(n)}, 0 \le n \le N-1, -\infty < k < \infty\}$, with a single trellis coder. In that case, the trellis coder is operating first across frequencies, and then across time.

As explained in Section 6.4, one of the benefits of MCM is that it offers inherent immunity to ISI, at least as N gets sufficiently large. The effect of ISI is twofold. First, it introduces dispersion within the data stream corresponding to each carrier. Second, it causes crosstalk between adjacent carriers, since they are typically overlapping in frequency and their orthogonality depends on a particular amplitude and phase characteristic. As N increases, each carrier is modulated at a lower symbol rate, and eventually the dispersion of each carrier becomes insignificant. Similarly, adjacent-carrier crosstalk will become insignificant as the distance between carrier frequencies shrinks, and the channel transfer function becomes essentially constant across the bandwidth occupied by each pair of adjacent carriers.

Since MCM offers immunity to ISI, when N is chosen large enough it is also compatible with trellis coding (or other signal space coding). This is an alternative way to achieve significant coding and shaping gains on channels with ISI, at the expense of the delay associated with a long symbol interval.

13.5. Further Reading

The paper that established the importance of trellis coding is by Ungerboeck [3]. It followed a patent by Csajka and Ungerboeck [16]. Useful overviews including tables of trellis codes are [6][5][7][9]. An extensive treatment of trellis codes is given in the book by Biglieri, Divsalar, McLane, and Simon [17]. An alternative method of describing and specifying trellis codes is due to Calderbank and Mazo [18][19]. For an introduction to lattice codes, coset codes, and the combination of coding with ISI, the Dec. 1991 issue of the *IEEE Communications Magazine* is recommended, and the article by Forney and Eyuboglu is particularly helpful [20]. The August 1989 issue of *IEEE Journal on Selected Areas in Communications* includes a number of articles referenced in this chapter.

Some advanced techniques have been incorporated into the V.Fast modem standard [21], including an improved technique for shaping called *shell mapping* [22], and a new method for combining transmitter precoding with trellis coding and shaping known as *flexible precoding* [21].

Problems

Problem 13-1. Describe how you would achieve two-dimensional shaping or coding gain (as in the circular or hexagonal constellations of Fig. 13-2) in a baseband PAM system. Be sure to describe both the transmitter and receiver.

Problem 13-2.

(a) Calculate the continuous approximation for the shaping and coding gains for two-dimensional square lattice constellation with a circular shaping.

(b) Compare the results of (a) against a two-dimensional constellation with square shaping. What is the shaping gain in dB?

Problem 13-3. Calculate the coding gain γ_Λ for the two-dimensional hexagonal constellation of Fig. 13-2. You can use the fact that the area of a hexagon with inscribed circle with radius r is $6r^2 \cdot \tan(\pi/6)$.

Problem 13-4. Show that the coding gain γ_Λ is invariant to the scaling the lattice Λ. Specifically, suppose Λ is scaled by multiplying all the basis vectors by a constant α, and call the new scaled lattice $\alpha \cdot \Lambda$. Show that $\gamma_{\alpha \cdot \Lambda} = \gamma_\Lambda$.

Problem 13-5. Define $C_N(R)$ as an N-cube that is $2R$ on a side; that is, each dimension has range $[-R, R]$. This N-cube is the Cartesian product of N one-dimensional regions $C_1(R) = [-R, R]$. We know that the shaping gain of $C_N(R)$ is unity, the same as the shaping gain of $C_1(R)$. Show this directly by calculating the volume and power of $C_N(R)$.

Problem 13-6. Suppose we transmit a spherically shaped N-dimensional lattice code with spectral efficiency ν bits per complex symbol and energy (variance) E per complex symbol. Suppose hypothetically that you can choose a lattice Λ_N for each N such that the fundamental volume $V(\Lambda_N)$ stays constant.

(a) What is the increase in ν when we go from $N = 2$ to $N = 4$ to $N = 6$? (You can use the continuous approximation.)

(b) What is the asymptotic increase in ν between $N = 2$ and $N \rightarrow \infty$?

Problem 13-7. Let $\mathbf{X}_K$ be a K-dimensional vector consisting of K components of an N-dimensional vector $\mathbf{X}_N$, the latter being a spherically uniform random vector. Show that for fixed K, as $N \rightarrow \infty$, $\mathbf{X}_K$ is a Gaussian vector with identically distributed independent components. *Hint*: Assume the radius is chosen so that each component of $\mathbf{X}_N$ is normalized to unity variance. Then show that $\mathbf{X}_K$ approaches a Gaussian density with unit-variance components.

Problem 13-8. Let $\mathbf{X}$ be a spherically uniform random vector with radius R and dimension N. Show that for any $0 < \varepsilon \le R$,

$$Pr\{\|\mathbf{X}\| \le R - \varepsilon\} \rightarrow 0 \quad \text{as} \quad N \rightarrow \infty \; . \tag{13.73}$$

Thus, as $N \rightarrow \infty$, almost all the volume of a sphere is near its surface, and $\|\mathbf{X}\|^2$ becomes R^2 almost surely. An interpretation is that the multidimensional sphere is making maximum use of the available energy by nearly always transmitting vectors that have this maximum energy.

Problem 13-9. Design a 4-PSK mapper so that the slightly simpler trellis coder in Fig. 13-32 has the same performance as the trellis coder in Fig. 13-9. This convolutional coder (by itself, without the mapper) was shown in Problem 12-9 to be inferior to the one in Fig. 13-9 in that the minimum Hamming distance is 3 instead of 5. However, with a properly designed mapper, the trellis code is not inferior. In fact it is equivalent to the coder in Fig. 13-9. Another equivalent coder is studied in Problem 13-10.

Problem 13-10. Show that the trellis coder in Fig. 13-33 has the same performance as the one in Fig. 13-9. Just as with Problem 13-9, the convolutional coder alone has $d_{H,\min} = 3$, which is inferior to the $d_{H,\min} = 5$ of Fig. 13-9. But the trellis coder is equally good.

Problem 13-11. Show that for the trellis coder in Fig. 13-34 the distance of the minimum distance error event depends on the correct state trajectory. The code is nonlinear even though the convolutional coder used to make it is linear.

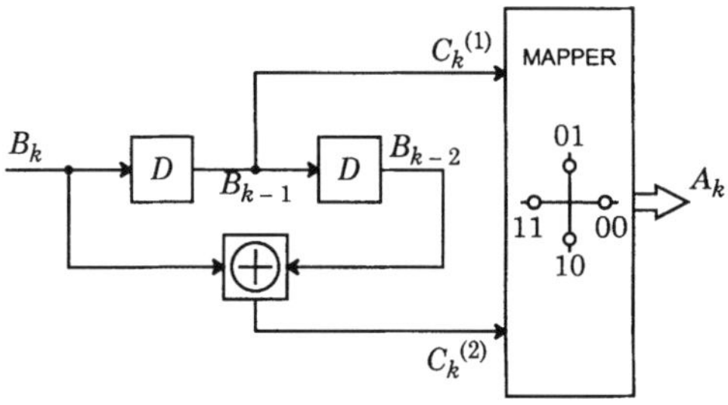

Fig. 13-32. A four-state trellis coder with the same performance as that in Fig. 13-9 can be made with a slightly simpler convolutional coder. The convolutional coder by itself is inferior in Hamming distance to the convolutional coder in Fig. 13-9 (see Problem 12-9).

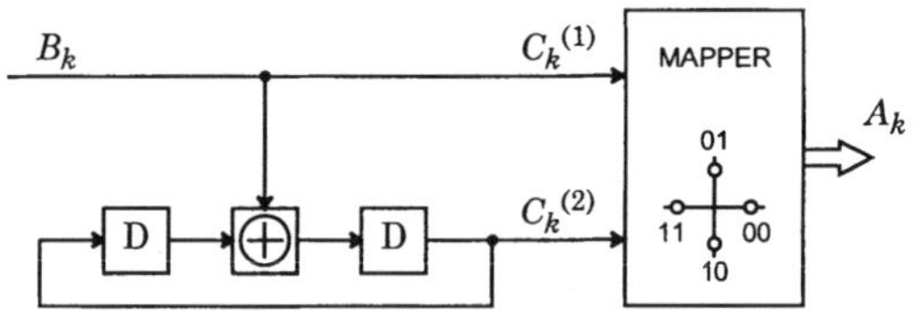

Fig. 13-33. A four state trellis coder made with a feedback-type convolutional coder. This coder is discussed in Problem 13-10.

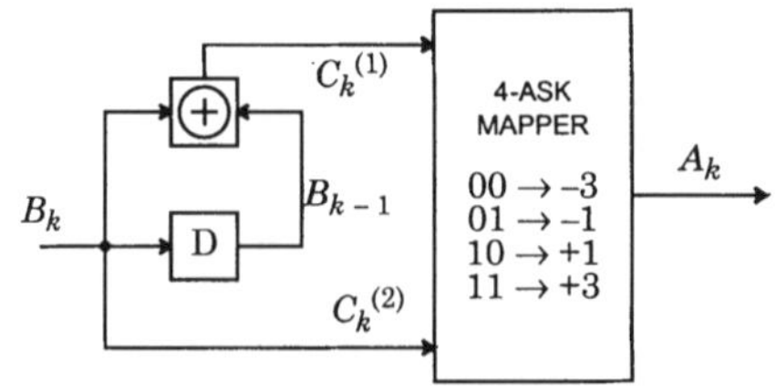

Fig. 13-34. A nonlinear trellis code made using a linear convolutional coder. The minimum-distance error event depends on the correct state trajectory.

Problem 13-12. Suppose you are asked to design a 400 bit per second modem for a noisy bandlimited and power-limited passband channel. You determine that with reasonable excess bandwidth, a symbol rate of 100 symbol per second is possible.

(a) For an uncoded system, what is the required alphabet size? Choose a constellation.

(b) For a two-dimensional trellis coded system, what is the required alphabet size? Choose a constellation.

(c) Suppose that to get an acceptable probability of error the uncoded system requires 4 dB more power than the channel can tolerate. How would you overcome this problem?

Problem 13-13. Find the distance of the second shortest error event for the trellis coder in Fig. 13-11, assuming its mapper uses the 8-PSK constellation in Fig. 13-14, $|A_k| = a$, and the trellis is defined by Fig. 13-15.

Problem 13-14. Consider the $M = 5$ cross constellation in Fig. 5-14. Assume that symbols are of the form $[a_1\ a_2]$ where $a_i \in \{\pm 5, \pm 3, \pm 1\}$.

(a) Do set partitioning and determine the minimum distance between symbols in the set for all levels of partitioning, as in Fig. 13-17.

(b) What is the difference (in dB) in average energy between this 32-cross constellation and a 16-QAM constellation where symbols are of the form $[a_1, a_2]$ where $a_i \in \{\pm 3, \pm 1\}$?

(c) Suppose that you need a total gain of about 3 dB compared to a 16-QAM uncoded system. What is the minimum value of $\tilde{m}$ (the number of coded bits) that you could select for the code? Roughly how many states should the convolutional coder have?

(d) Repeat part (c) assuming that a total gain of about 5 dB is required.

Problem 13-15. Consider the coder in Example 13-15. Assume a convolutional coder equivalent to that in Fig. 13-9.

(a) Assuming that the parallel transitions form the minimum-distance error events, find the total gain of the trellis coder. *Hint:* Compare to an 8-PSK uncoded system.

(b) Assign subsets from the third row of Fig. 13-17 to transitions in the trellis, and find the distance of a length three error event, assuming the correct state trajectory is all zero. Does the assumption in part (a) look reasonable for this coder?

Problem 13-16. Show that R defined by (7.150) is unbounded for the code in Example 13-17.

Problem 13-17. Consider the trellis coder in Fig. 13-35. The mapping is given by

(a) Draw the state transition diagram and trellis with each arc labeled with the pair (B_k, A_k).

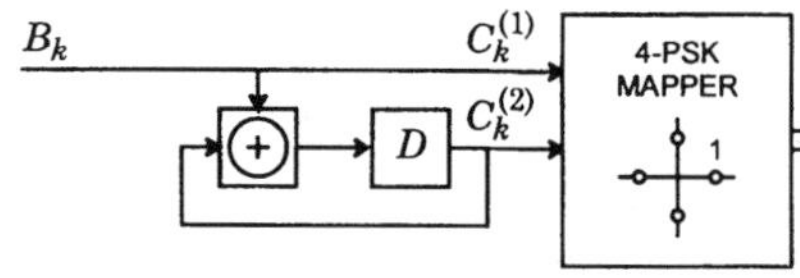

Fig. 13-35. A Trellis coder with a 4-PSK output alphabet.

$C_k^{(1)}$	$C_k^{(2)}$	A_k
0	0	+1
0	1	$+j$
1	0	−1
1	1	$-j$

(b) Find the minimum-distance error events and their distance, and estimate their probability of occurring. Assume the channel adds white Gaussian noise with variance σ^2.

(c) Compare this coded system with an uncoded 2-PSK system.

(d) Consider the related system shown in Fig. 13-36. Use set partitioning to design the mapper.

(a) Estimate the total gain at high SNR (compare to uncoded 4-PSK).

Problem 13-18. Consider the convolutional coder with feedback shown in Fig. 13-37. Use set partitioning to design a mapping so that this trellis code performs as well as the 8-PSK trellis code with trellis shown in Fig. 13-15.

References

1. R. de Buda, "Some Optimal Codes Have Structure," *IEEE Journal on Selected Areas in Communications*, p. 877 (Aug. 1989).
2. G. D. Forney, Jr and L-F Wei, "Multidimensional Constellations, Part I: Introduction, Figures of Merit, and Generalized Cross Constellations," *IEEE Journal on Selected Areas in Communications*, p. 877 (Aug. 1989).

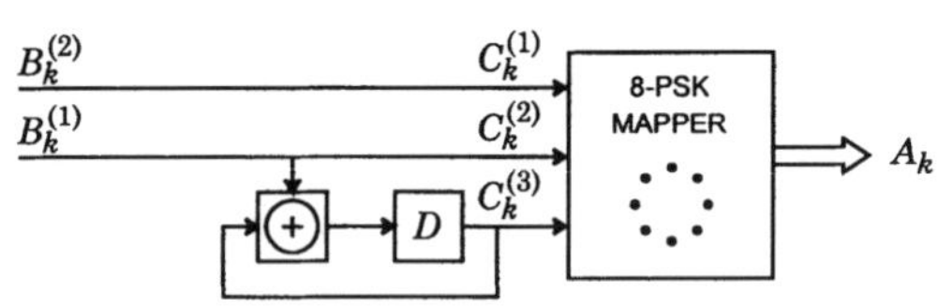

Fig. 13-36. A trellis coder based on the one in Fig. 13-35 but using an 8-PSK output alphabet and parallel state transitions.

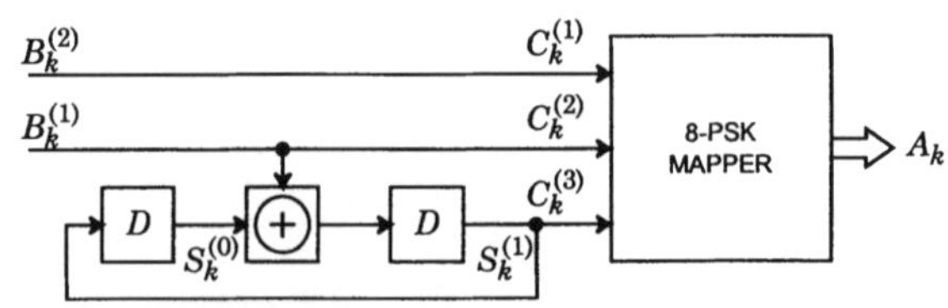

Fig. 13-37. A convolutional coder with feedback, studied in Problem 13-18.

3. G. Ungerboeck, "Channel Coding with Multilevel/Phase Signals," *IEEE Trans. on Information Theory*, Vol. IT-28, No 1, (Jan. 1982).

4. L.-F. Wei, "Rotationally Invariant Convolutional Channel Coding with Expanded Signal Space: Part II: Nonlinear Codes," *IEEE Journal on Selected Areas in Communications*, Vol. SAC-2 (5), (Sep. 1984).

5. G. Ungerboeck, "Trellis-Coded Modulation with Redundant Signal Sets Part I: Introduction," *IEEE Communications Magazine*, Vol. 25 (2), pp. 5-11 (Feb. 1987).

6. G. D. Forney, Jr, "Coset Codes - Part I: Introduction and Geometrical Classification," *IEEE Trans. Information Theory*, Vol. IT-34, p. 1123 (1988).

7. G. Ungerboeck, "Trellis-Coded Modulation with Redundant Signal Sets Part II: State of the Art," *IEEE Communications Magazine*, Vol. 25 (2), pp. 12-21 (Feb. 1987).

8. A. R. Calderbank and N. J. A. Sloane, "Four-Dimensional Modulation With an Eight-State Trellis Code," *AT&T Technical Journal*, Vol. 64 (5), (May-June 1985).

9. G. D. Forney, Jr., R. G. Gallager, G. R. Lang, F. M. Longstaff, and S. U. Qureshi, "Efficient Modulation for Band-Limited Channels," *IEEE Journal on Selected Areas in Communications*, Vol. SAC-2 (5), (Sep. 1984).

10. L.-F. Wei, "Trellis-Coded Modulation with Multi-Dimensional Constellations," *IEEE Trans. on Information Theory*, Vol. IT-33, p. 483 (1987).

11. A. R. Calderbank and N. J. A. Sloane, "New Trellis Codes Based on Lattices and Cosets," *IEEE Trans. Information Theory*, Vol. IT-33, p. 177 (1987).

12. M. V. Eyuboglu and G. D. Forney, Jr, "Trellis Precoding: Combined Coding, Precoding, and Shaping for Intersymbol Interference Channels," *IEEE Trans. Information Theory*, (March 1992).

13. A. Duel-Hallen and C. Heegard, "Delayed Decision-Feedback Equalization," *IEEE Trans. Communications*, Vol. COM-37, p. 13 (Jan. 1988).

14. P. Chevillat and E. Eleftheriou, "Decoding of Trellis-Encoded Signals in the Presence of Intersymbol Interference and Noise," *IEEE Trans. Communications*, Vol. COM-37, p. 669 (July 1989).

15. M. V. Eyuboglu and S. U. Qureshi, "Reduced-State Sequence Estimation for Coded Modulation on Intersymbol Interference Channels," *IEEE Journal Select Areas in Communications*, Vol. SAC-7, p. 989 (Aug. 1989).

16. I. P. Csajka and G. Ungerboeck, "Method and Arrangement for Coding Binary Signals and Modulating a Carrier Signal," U. S. Patent no. 4,077,021, (Feb. 28, 1978).

17. E. Biglieri, D. Divsalar, P. J. McLane, and M. K. Simon, *Introduction to Trellis-Coded Modulation with Applications*, Macmillan, New York (1991).

18. A. R. Calderbank and J. Mazo, "A New Description of Trellis Codes," *IEEE Trans. on Information Theory*, Vol. IT-30 (6), (Nov. 1984).

19. D. Divsalar, M. K. Simon, and J. H. Yuen, "Trellis Coding with Asymmetric Modulations," *IEEE Trans. on Communications*, Vol. COM-35, (2)(Feb. 1987).

20. 20. G. D. Forney, Jr and M. V. Eyuboglu, "Combined Equalization and Coding Using Precoding," *IEEE Communications Magazine*, (Dec. 1991).

21. M. V. Eyuboglu and G. D. Forney, Jr, "Advanced Modulation Techniques for V.Fast," *European Transactions on Telecommunications and Related Technologies*, (to appear).

22. G. Lang and F. Longstaff, "A Leech Lattice Modem," *IEEE Journal on Selected Areas in Communications*, Vol. 7, p. 968 (Aug. 1989).

14

Phase-Locked Loops

In a continuous-time world, establishing a common time base at physically separated locations presents some serious challenges. Typical systems use independent time bases, frequently derived from crystal oscillators, as shown in Fig. 14-1. Although crystal oscillators provide extremely accurate timing references at low cost, "extremely accurate" is not adequate to maintain the integrity of discrete-time data. Timing references often have to be identical, at least in the sense of long term averages. In other words, systems must be *synchronized.* Underlying most synchronization techniques is the phase-locked loop (PLL). In this chapter we derive the basic principles of PLLs. Two practical applications, carrier and timing recovery, are treated in-depth in Chapters 15 and 16.

The basic PLL structure is shown in Fig. 14-1. The *voltage-controlled oscillator* (VCO) attempts to produce a signal $v(t)$ that tracks the phase of the input $y(t)$. A *phase detector* measures the phase error between the input $y(t)$ and the VCO output $v(t)$. The resulting error

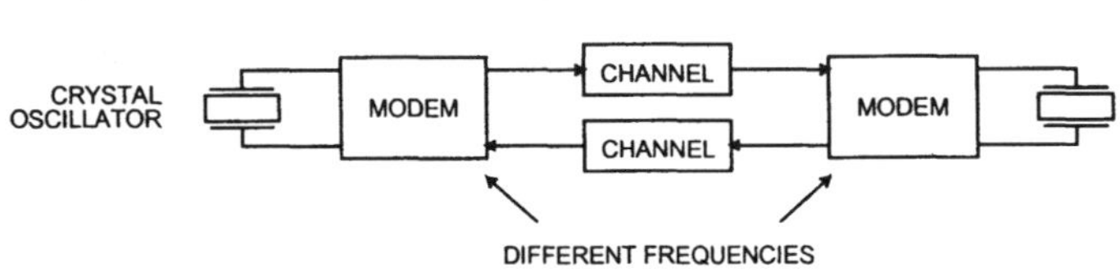

Fig. 14-1. Typical systems use independent time bases, frequently derived from crystal oscillators, at physically separated locations.

signal can be filtered to become a *control signal* that drives the VCO. The basic idea is obvious — if the VCO phase gets ahead of the phase of the input, the control signal should be reduced. If the VCO phase gets behind, the control signal should be increased. As with any feedback system, the parameters must be chosen to ensure stability.

The goal in design of the PLL varies with the application.

Example 14-1.

In timing recovery (Chapter 16) on a single point-to-point link, such as in a voiceband data modem, the objective is to generate a stable single-frequency tone at the output of the VCO. The frequency of this tone should equal the average symbol rate of the input, but transient variations in the symbol rate should be ignored, as should noise or other interference. In fact, any fluctuations in the symbol rate detected by the timing recovery can be assumed to be a consequence of interference such as noise, because there is no mechanism in most channels for introducing significant fluctuations in the symbol rate.

Example 14-2.

In carrier recovery (Chapter 15), by contrast, the objective is to track the phase of the carrier on the input signal as closely as possible, while at the same time minimizing the effect of noise. Unlike timing phase, several important channels can introduce significant fluctuations in the carrier phase and frequency. To properly demodulate the signal, these fluctuations should be replicated on the carrier used by the receiver for demodulation. The output of the VCO is therefore not a single-frequency tone (unless the phase of the carrier on the input does not vary).

Other applications usually fall into one of these two categories as well. The objective is either a single-frequency, or closely-tracked phase, or a compromise between the two.

In practice, many PLLs look very different from that shown in Fig. 14-2. There may be no explicit VCO, or the VCO may be built using digital circuitry arranged as a controllable countdown chain. The phase detector can be very complicated, sometimes actually consisting of an entire receiver, complete with adaptive equalizer, or very simple, consisting of an exclusive-or gate. A PLL may be implemented completely or partly in discrete-time, and completely or partly with digital circuits. Although the relationship between the model in Fig. 14-2 and an actual implementation may be subtle, the basic principles are the same for all implementations.

We will concentrate on the steady-state, in-lock behavior of the PLL, using only linearized analysis, ignoring important issues such as acquisition and non-linear behavior.

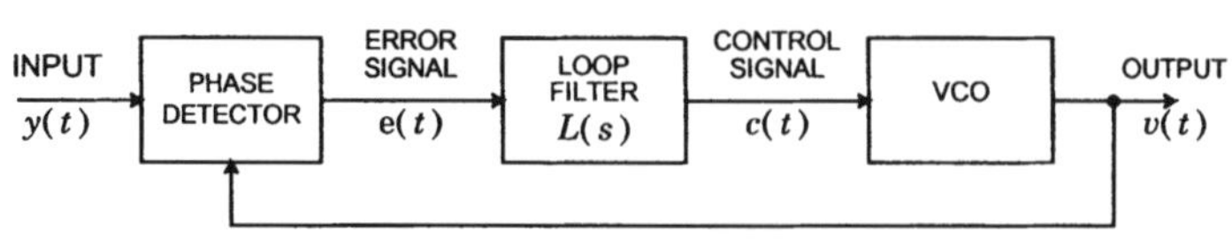

Fig. 14-2. Basic structure of a continuous-time PLL. The VCO produces a signal or clock that tracks the phase of the input signal.

14.1. Ideal Continuous-Time PLL

PLLs are conceptually simple, but they are inherently non-linear systems and their analysis can be difficult. However, with some carefully crafted simplifying assumptions we can develop some powerful analytical tools that simplify the analysis.

14.1.1. Assumptions

First assume a particular form for the input

$$y(t) = A_y \cos(\omega_v t + \theta(t)) , \tag{14.1}$$

where A_y and ω_v are constants. Of course in practice the input is likely to be more complicated, having amplitude variations in addition to phase and frequency, for example, but as long as the design of the phase detector is appropriate for the form of a particular input, our analysis will be valid. The output of the VCO is assumed to have a similar form

$$v(t) = A_v \cos(\omega_v t + \phi(t)) . \tag{14.2}$$

When $\phi(t)$ is a constant the frequency of the VCO output is ω_v, called the ***natural*** or *free-running frequency* of the VCO. It is for convenience that we express the input (14.1) in terms of the natural frequency of the VCO.

14.1.2. The Ideal Phase Detector

Assuming forms (14.1) and (14.2) the output of an ideal phase detector is

$$\varepsilon(t) = W(\theta(t) - \phi(t)) \tag{14.3}$$

where the function $W(\cdot)$, shown in Fig. 14-3, reflects the 2π ambiguity in the phase difference. Because of the shape of $W(\cdot)$, this phase detector is called a *sawtooth phase detector*. We have assumed unity slope for the function $W(\cdot)$, although in practice the phase detector may exhibit some other gain, often written K_p. That gain is easily modeled as part of the loop filter gain, so its explicit inclusion is not necessary. Because of the 2π ambiguity in an ideal phase detector, sudden changes of 2π in $\theta(t)$ or $\phi(t)$ have no effect on the system (they are not detected by the phase detector). Such changes are called *clicks*, and are usually detrimental.

We will see many variations of this basic phase detector. It is also often possible to design *frequency detectors* that do not suffer this 2π phase ambiguity [1].

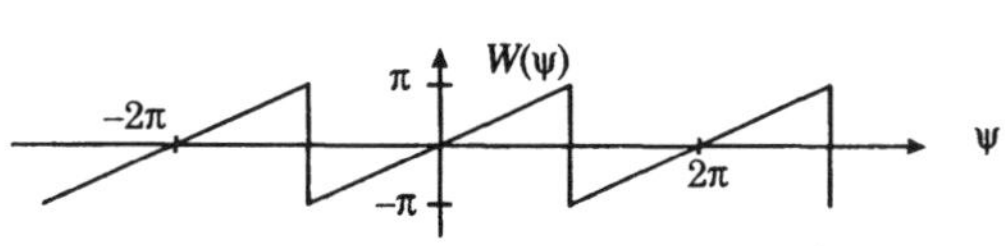

Fig. 14-3. An ideal phase detector can only detect phase errors ψ modulo 2π. This is equivalent to applying this function $W(\cdot)$ to the phase error ψ. Because of the shape of $W(\cdot)$, this phase detector is called a sawtooth phase detector.

14.1.3. The Ideal VCO

The ideal VCO, with properties summarized in Fig. 14-4, produces the output (14.2), which has instantaneous frequency

$$\frac{d}{dt}[\omega_v t + \phi(t)] = \omega_v + \frac{d}{dt}\phi(t). \tag{14.4}$$

Again, a practical VCO may have gain, often written K_v, that can be modeled as part of the gain of the loop filter. Intuitively, we would like to directly control the instantaneous frequency with the control input $c(t)$. The VCO should therefore be designed so that

$$\frac{d}{dt}\phi(t) = c(t)\,. \tag{14.5}$$

Example 14-3.

A constant control signal $c(t) = K$ will produce the constant frequency $\omega_v + K$ at the output.

It is sometimes convenient for analysis to take the Laplace transform of (14.5),

$$s\Phi(s) = C(s) = L(s)E(s)\,, \tag{14.6}$$

where $C(s)$ is the Laplace transform of the control signal and $E(s)$ is the Laplace transform of the error signal $\varepsilon(t)$.

14.1.4. Phase and Average-Frequency Lock

The ideal PLL is *phase locked* if

$$\phi(t) = \theta(t) + \phi \tag{14.7}$$

for some constant ϕ. If $\phi = 0$, the PLL is *perfectly* phase locked. In other words, the VCO output is exactly tracking the phase of the input. It is locked to an average frequency $\omega_v + K$ if

$$\phi(t) = Kt \tag{14.8}$$

for some constant K. The VCO output frequency is presumably exactly the same as the input average frequency. In Example 14-1 it is more important to have average-frequency lock than phase lock, while in Example 14-2 the situation is reversed.

Intuitively, there must be some limitations on the input phase $\theta(t)$ for the PLL to be phase or average-frequency locked because the phase detector output is bounded by $\pm\pi$. To find the limitations, assume a simple form for the phase of the input,

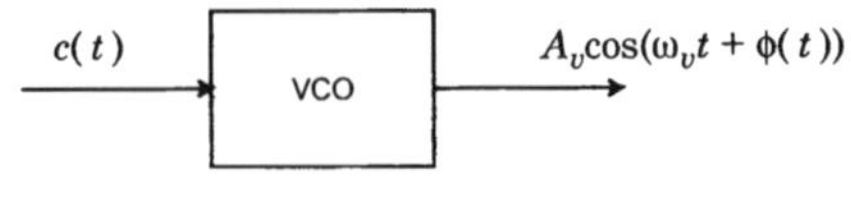

Fig. 14-4. An ideal VCO. The instantaneous frequency of the output is $\omega_v + c(t)$.

$$\theta(t) = \omega_0 t + \theta \ . \tag{14.9}$$

In other words, the input $y(t)$ is a sinusoid with frequency $\omega_v + \omega_0$ and phase θ, a constant. Assume the PLL is phase locked. In order for it to remain phase locked, the *frequency offset* ω_0 must not exceed a limited range called the *lock range* or *hold-in range* of the PLL. It is easy to derive the lock range.

Exercise 14-1.

Assuming that the input phase has the form (14.9), show that the PLL can only maintain phase lock if

$$|\omega_0| \leq \pi |L(0)| , \tag{14.10}$$

where $L(0)$ is the d.c. gain of the loop filter (the Laplace transform evaluated at $s = 0$). Assume an ideal phase detector and VCO.

14.1.5. Analysis of the Linearized Dynamics

Phase and average-frequency lock are static concepts — they assume the PLL is in steady state. If we assume that the phase error is small enough for all t,

$$|\theta(t) - \phi(t)| < \pi \tag{14.11}$$

then the phase detector is operating in its *linear range* (see Fig. 14-3),

$$\varepsilon(t) = \theta(t) - \phi(t) \ . \tag{14.12}$$

and the analysis of the dynamics of the PLL is simple. The transfer function from the phase $\theta(t)$ of the input to the phase $\phi(t)$ of the VCO follows by taking the Laplace transform of Fig. 14-12,

$$E(s) = \Theta(s) - \Phi(s) , \tag{14.13}$$

and from (14.6),

$$E(s) = \frac{s\Phi(s)}{L(s)} . \tag{14.14}$$

Combining these and solving for $\Phi(s)/\Theta(s)$ we get the *phase transfer function*

$$\frac{\Phi(s)}{\Theta(s)} = \frac{L(s)}{L(s) + s} . \tag{14.15}$$

The phase transfer function summarizes many of the important features of the PLL.

Exercise 14-2.

Assume that $L(s) = N(s)/D(s)$ is a rational Laplace transform where the degree of $N(s)$ (the number of zeros) is less than or equal to the degree of $D(s)$ (the number of poles). Show that the number of poles in $\Phi(s)/\Theta(s)$ (called the *order* of the PLL) is one plus the number of poles in the loop filter $L(s)$.

Example 14-4.

A *first-order PLL* is characterized by having the simple loop filter

$$L(s) = K_L . \tag{14.16}$$

In this case, the transfer function can be written

$$\frac{\Phi(s)}{\Theta(s)} = \frac{K_L}{K_L + s} . \tag{14.17}$$

This is a single-pole lowpass filter with its pole at $s = -K_L$. It is stable as long as $K_L > 0$, and its lock range is found from (14.10) to be

$$|\omega_0| \le \pi |K_L| . \tag{14.18}$$

Its lowpass characteristic is a very useful property for many applications. When the application requires average-frequency lock, as in Example 14-1, then a narrowband lowpass PLL will be useful. Even when phase tracking is required, as in Example 14-2, a lowpass PLL can help reject some of the noise, particularly if it is known that the phase variations that we wish to track are relatively slow.

Evaluating transfer function (14.15) at $s = 0$, the PLL has unity gain for d.c. phase errors. In other words, when the input phase is constant, $\theta(t) = K$, then the output phase is the same constant, $\phi(t) = K$. In this case we get perfect phase lock with any loop filter.

Exercise 14-3.

Show that if the input has frequency offset, $\theta(t) = \omega_0 t + K$, $\omega_0 \ne 0$, then the first-order PLL of Example 14-4 cannot achieve perfect phase lock. It can achieve phase lock if ω_0 is within the lock range, but there will always remain a phase error.

The *bandwidth* of a PLL is loosely defined to be the bandwidth of the transfer function $\Phi(s)/\Theta(s)$. Lowering the bandwidth means increasing the attenuation of high frequency components in the input phase or noise, but for the first order PLL, it also reduces the lock range. It is possible to reduce the bandwidth *without* reducing the lock range by using a second-order PLL.

Example 14-5.

Consider a *type I second-order PLL* with loop filter

$$L(s) = K_L \left(\frac{s + K_1}{s + K_2} \right) . \tag{14.19}$$

From Exercise 14-1 the lock range is

$$|\omega_0| \le \pi |K_L K_1 / K_2| . \tag{14.20}$$

From (14.15) the phase transfer function is

$$\frac{\Phi(s)}{\Theta(s)} = \frac{K_L s + K_L K_1}{s^2 + (K_L + K_2)s + K_L K_1} . \tag{14.21}$$

It can be shown that this is stable as long as $K_2 > -K_L$ and $K_L K_1 > 0$ (see Problem 14-3).

Example 14-6.
Suppose that

$$L(s) = \frac{s+1}{s-0.5} . \tag{14.22}$$

The loop filter itself is not stable, but the closed-loop PLL is. The transfer function is

$$\frac{\Phi(s)}{\Theta(s)} = \frac{s+1}{s^2 + 0.5s + 1} \tag{14.23}$$

which has poles at $-0.25 \pm j0.97$, both in the left half plane.

Since the transfer function of a second-order PLL has a zero and two poles, the rolloff at high frequencies is the same (20 dB/decade) as for the first-order PLL. Three possible Bode plots of the phase transfer function are shown in Fig. 14-5. The bandwidth of the PLL is determined primarily by $K_L K_1$. But the lock range (14.20) is determined by both $K_L K_1$ and K_2, as shown in (14.20). In principle, we can make the lock range as large as we like by decreasing K_2 while keeping the bandwidth constant by keeping $K_L K_1$ constant. This is the main advantage of the type I second-order PLL. However, there is potentially one serious problem. Although the gain at d.c. is always unity (evaluate (14.21) at $s = 0$), the closed-loop gain may be greater than unity in some range of frequencies. In this case, some components of the phase of the input will be *amplified*, which is often not desired. This phenomenon is known as *peaking*.

Example 14-7.
In Example 14-6, $K_1 = K_L = 1$, so the bode plot is given by Fig. 14-5(b). The magnitude response is

$$\left|\frac{\Phi(f)}{\Theta(f)}\right|^2 = \left|\frac{j2\pi f + 1}{-4\pi^2 f^2 + j\pi f + 1}\right|^2 . \tag{14.24}$$

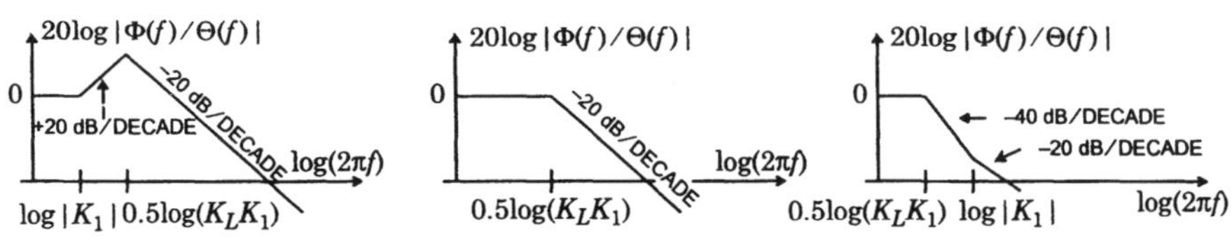

Fig. 14-5. Three possible Bode plots of the phase transfer function (14.21) of the type I second-order PLL, assuming the poles are complex. If $|z|$ is the magnitude of the zero and $|p|$ is the magnitude of the poles, then in (a) $|z|<|p|$, in (b) $|z| = |p|$, and in (c) $|z|>|p|$. Note that the bandwidth of the PLL is independent of K_2.

If we evaluate this at the natural frequency $\omega_n = 2\pi f_n = 1$ (the natural frequency is equal to the magnitude of the poles), we find that

$$\left|\frac{\Phi(1)}{\Theta(1)}\right|^2 = \left|\frac{j+1}{0.5j}\right|^2 = 8 \,. \tag{14.25}$$

If the input phase $\theta(t)$ has a component at $\omega_n = 1$, then the output phase $\phi(t)$ will have that same component eight times larger! The frequency response is sketched in Fig. 14-6.

Peaking is not always a serious impairment, but sometimes it is devastating.

Example 14-8.

In some systems, such as long transmission lines with many repeaters (Chapter 1), many PLLs are cascaded in series. If the PLLs have greater than unity gain for a phase at the input that varies at any particular frequency, the amplification of that phase can become quite severe after just a few PLLs. This issue is explored in Chapter 16 for the specific example of timing recovery PLLs in the repeaters.

The peaking properties of type I second-order PLLs are studied in detail in Problem 14-6. In particular, it is shown that if peaking is disallowed, the lock range of a second-order PLL is actually *smaller* than the lock range of a first-order PLL with comparable bandwidth. Hence, if peaking cannot be tolerated then the primary advantage of second-order loops evaporates.

The following *type II second-order PLL* is a special case of the type I PLL when $K_2 = 0$,

$$L(s) = K_L\left(\frac{s+K_1}{s}\right). \tag{14.26}$$

This is sometimes called a *proportional plus integral* loop filter. From (14.21) the closed-loop phase response is

$$\frac{\Phi(s)}{\Theta(s)} = \frac{K_L K_1 + K_L s}{K_L K_1 + K_L s + s^2} \,. \tag{14.27}$$

As with the previous PLLs, this one has unity gain at d.c. Unlike the previous PLLs, however, it has an integrator in the loop filter. In fact, by convention, the "type" of a PLL is the number of integrators in the loop filter plus one. Its main advantage is that the integrator leads to

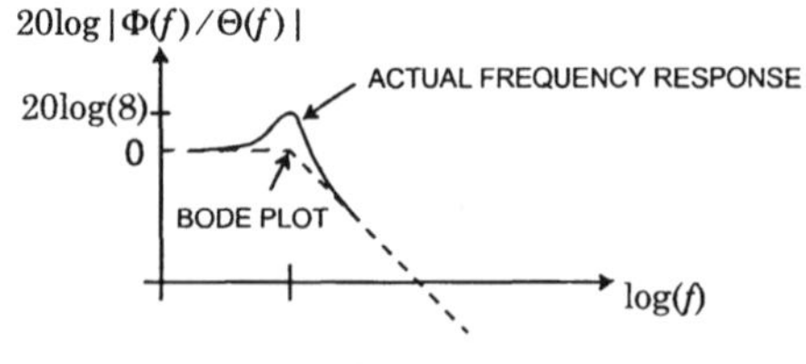

Fig. 14-6. The frequency response of the type I second-order PLL in Example 14-6 exhibits peaking in which the phase at certain frequencies is amplified by the PLL.

perfect phase lock even in the face of frequency offset. A disadvantage is that it always exhibits peaking (see Problem 14-7). Two other second-order PLL loop filters are studied in Problem 14-1 and Problem 14-8.

14.1.6. Steady-State Response

It is often useful to know precisely the steady-state operating point of a PLL given certain inputs. The steady-state phase error is defined to be

$$\varepsilon_{ss} = \lim_{t \to \infty} \varepsilon(t) . \tag{14.28}$$

If the PLL does not achieve perfect phase lock then $\varepsilon_{ss} \neq 0$. If $\varepsilon(t) = 0$ for $t < 0$ then we can (usually) find ε_{ss} using the *final value theorem* for Laplace transforms,

$$\varepsilon_{ss} = \lim_{s \to 0} sE(s) . \tag{14.29}$$

Combining (14.14) and (14.15) we get the Laplace transform of $\varepsilon(t)$ in terms of the input phase,

$$E(s) = \frac{s\Theta(s)}{L(s)+s}, \qquad \varepsilon_{ss} = \lim_{s \to 0} \frac{s^2\Theta(s)}{L(s)+s} . \tag{14.30}$$

Example 14-9.

Suppose the input phase is

$$\theta(t) = \omega_0 t u(t), \qquad \Theta(s) = \omega_0 / s^2 , \tag{14.31}$$

where $u(t)$ is the unit step. In other words, at time $t = 0$ the input suddenly acquires a frequency offset of ω_0. There will of course be transients in the response of the PLL, but after the transients die out, from (14.30) the steady-state phase error will be

$$\varepsilon_{ss} = \lim_{s \to 0} \frac{\omega_0}{L(s)+s} . \tag{14.32}$$

For the first-order PLL, $L(s) = K_L$ and

$$\varepsilon_{ss} = \frac{\omega_0}{K_L} . \tag{14.33}$$

The steady-state error is non-zero, but can be reduced by increasing the loop gain (at the expense of increasing the bandwidth of the loop, see Problem 14-2). For the type I second-order PLL with loop filter given by (14.19),

$$\varepsilon_{ss} = \frac{K_2\omega_0}{K_1 K_L} . \tag{14.34}$$

The steady-state error is again non-zero, but can be reduced this time by making K_2 small. However, this may lead to peaking, or greater than unity gain for inputs at some frequencies (see Problem 14-6). If $K_2 = 0$, we have the type II loop of (14.26) which has zero steady-state error, $\varepsilon_{ss} = 0$. This is the main advantage of type II second-order PLLs. Intuitively, the integrator in the loop filter holds a constant at its output proportional to the frequency offset. As shown in Problem 14-7, there is always peaking in this PLL, but the peaking can be made small by making K_1 small.

At least one integrator in the loop filter is required to get zero steady-state error in the face of frequency offset.

14.1.7. Transients

The previous analysis assumes that the phase error is always small, so that the phase detector operates in its linear range. When the PLL is locked to an input signal, this is a reasonable assumption. But during *capture*, it is not. Consider a PLL that is locked to an input signal with frequency equal to the natural frequency ω_v of the VCO. Then the control signal $c(t)$ is zero. Suppose that suddenly the frequency of the input changes. Because of the loop filter and practical limitations of the VCO, the VCO will not respond instantly to lock onto the new frequency. As a consequence, for a period of time the phase difference between the input and VCO output can get large, and in-fact can easily slip cycles (making sudden jumps of magnitude 2π for the ideal phase detector). During this time, the linearized analysis is not valid.

The lock range is the range of input frequencies over which an in-lock PLL will stay locked. But even within this range the PLL may not be able to capture the input frequency if it starts out unlocked. The *capture range* is defined as the range of input frequencies over which an initially unlocked PLL can lock. The *pull-in* time is the amount of time it takes the PLL to lock. A subset of the capture range into which the PLL can lock without cycle slipping also has a name, the *seize* range. Complete analysis of these effects is complicated by the non-linear nature of the PLL, and is left to references in Section 14.5.

14.2. Discrete-Time PLLs

In digital communications systems, completely analog continuous-time PLLs like those studied in Section 14.1 are rare. Most are hybrid analog/digital or mixed continuous and discrete-time. We will begin with the discrete-time PLL. Mixed systems are considered in Sections 14.3 and 14.4 where alternative phase detector and VCO designs are discussed.

14.2.1. The Basic Model

A discrete-time PLL is shown in Fig. 14-7. Assumptions about the form of the input signal and the output of the VCO are shown in the figure, and are analogous to those for the continuous-time PLL. The phase detector is a discrete-time version of that considered before, and has output

$$\varepsilon_k = W(\theta_k - \phi_k) \tag{14.35}$$

where $W(\cdot)$ is shown in Fig. 14-3.

The discrete-time VCO, although analogous to the continuous-time VCO, is not quite as obvious. Analogous to differential equation (14.5), the phase ϕ_k satisfies the difference equation

$$\phi_{k+1} - \phi_k = c_k \,. \tag{14.36}$$

Using this, the output of the VCO can be written

$$v_{k+1} = A_v \cos(\omega_v (k+1)T + \phi_{k+1}) = A_v \cos(\omega_v kT + \phi_k + \omega_v T + c_k) \,. \tag{14.37}$$

This leads to the structure in Fig. 14-8. Taking the Z transform of (14.36) we get

$$\Phi(z) = \frac{1}{z-1} C(z) = \frac{L(z)}{z-1} E(z) \,, \tag{14.38}$$

where $L(z)$ is the loop filter transfer function and $E(z)$ is the Z transform of the error signal ε_k.

Exercise 14-4.

Assume the input has frequency offset ω_0,

$$\theta_k = \omega_0 kT + \theta \,, \tag{14.39}$$

where θ is some constant and T is the sample interval. Show that the lock range is

$$|\omega_0| \le \frac{\pi}{T} |L(1)| \,. \tag{14.40}$$

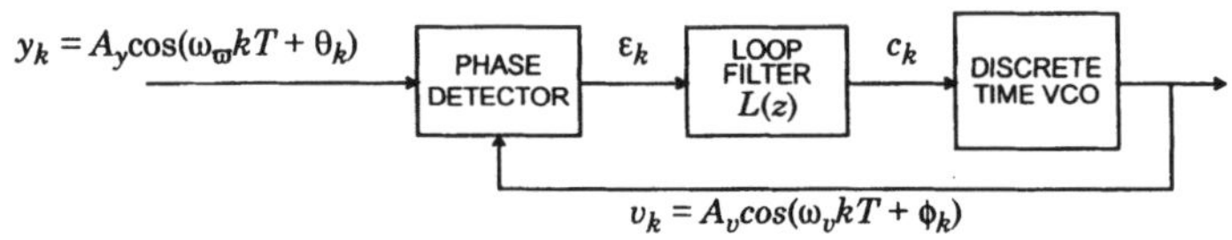

Fig. 14-7. A discrete-time PLL with assumptions about the form of the input and the output shown.

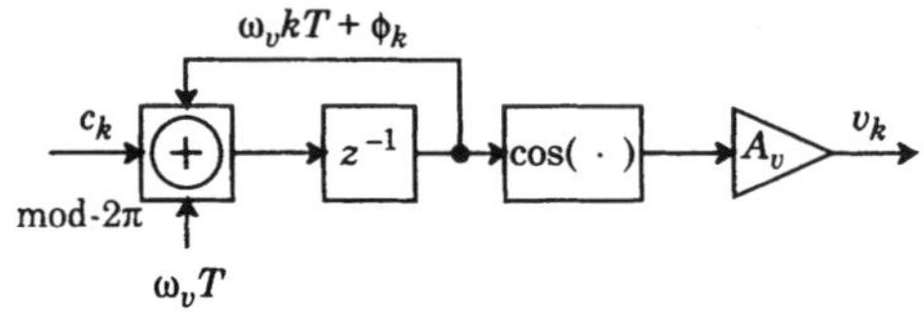

Fig. 14-8. A discrete-time VCO consists of an accumulator and cosine computation, along with some constant additions and multiplications. The modulo-2π adder reflect the fact that the numbers being added are angles (in radians).

14.2.2. Analysis of the Dynamics

As before, to analyze the dynamics we assume that the phase error is small enough that the phase detector is linear. The phase detector output is

$$\varepsilon_k = \theta_k - \phi_k , \tag{14.41}$$

or taking Z transforms

$$E(z) = \Theta(z) - \Phi(z) . \tag{14.42}$$

Combining (14.42) with (14.38) we get the phase transfer function of the PLL,

$$\frac{\Phi(z)}{\Theta(z)} = \frac{L(z)}{L(z) + z - 1} . \tag{14.43}$$

By evaluating this at $z = 1$ we see that just as with the continuous-time PLL, discrete-time PLLs have unity gain to d.c. phase inputs.

Example 14-10.

A first-order discrete-time PLL has loop filter $L(z) = K_L$, so

$$\frac{\Phi(z)}{\Theta(z)} = \frac{K_L}{K_L + z - 1} . \tag{14.44}$$

This has a pole at $z = 1 - K_L$ and hence is stable if and only if $0 < K_L < 2$. Unlike the continuous-time PLL, there is an upper limit on the loop gain imposed by the stability requirement. Also, the phase transfer function is only a lowpass filter if $0 < K_L < 1$ (see Problem 14-11). In fact, if $1 < K_L < 2$ then inputs at all frequencies are amplified, which is probably not what we had hoped for!

Example 14-11.

A general second-order discrete-time PLL has loop filter

$$L(z) = \frac{a_1 z + a_0}{b_1 z + b_0} . \tag{14.45}$$

The transfer function to phase is

$$\frac{\Phi(z)}{\Theta(z)} = \frac{a_1 z + a_0}{b_1 z^2 + (a_1 + b_0 - b_1) z + (a_0 - b_0)} . \tag{14.46}$$

14.2.3. Steady-State Error

Just as with continuous-time PLLs, the steady-state error is

$$\varepsilon_{ss} = \lim_{k \to \infty} \varepsilon_k . \tag{14.47}$$

If $\varepsilon_k = 0$ for $k<0$ we can (usually) use the final value theorem for Z transforms to write

$$\varepsilon_{ss} = \lim_{z \to 1} (z-1)E(z) . \tag{14.48}$$

Combining (14.42) and (14.43) we get an expression for $E(z)$,

$$E(z) = \frac{\Theta(z)(z-1)}{L(z)+z-1} . \tag{14.49}$$

Example 14-12.
Suppose that the input frequency is exactly the natural frequency of the VCO, but a phase offset is introduced at time $k = 0$,

$$\theta_k = \theta u_k , \tag{14.50}$$

where θ is a constant and u_k is the unit step. The Z transform is

$$\Theta(z) = \frac{z\theta}{z-1} . \tag{14.51}$$

Hence

$$\varepsilon_{ss} = \lim_{z \to 1} \frac{(z-1)\theta}{L(z)+z-1} . \tag{14.52}$$

For any $L(z)$ such that $L(1) \neq 0$, $\varepsilon_{ss} = 0$, and the phase error decays to zero.

Example 14-13.
Suppose that the input has frequency offset introduced at time $k = 0$,

$$\theta_k = \omega_0 k u_k . \tag{14.53}$$

The Z transform is

$$\Theta(z) = \frac{z\omega_0}{(z-1)^2} . \tag{14.54}$$

Hence

$$\varepsilon_{ss} = \lim_{z \to 1} \frac{z\omega_0}{L(z)+z-1} , \tag{14.55}$$

which will be zero if and only if $L(z)$ has a pole at $z = 1$. Thus to have perfect phase lock in the face of frequency offset, the discrete-time loop filter must have a pole at $z = 1$, just as the continuous-time loop filter has to have a pole at $s = 0$.

The "type" of a discrete-time PLL is defined to be one plus the number of poles at $z = 1$. From the previous example we see that a type II PLL has $\varepsilon_{ss} = 0$ when there is frequency offset.

14.3. Phase Detectors

So far we have assumed an ideal phase detector with the characteristic shown in Fig. 14-3. As a result of its shape, this phase detector is called a *sawtooth* phase detector. A wide variety of other phase detectors are used, ranging from simple to complicated. Much of the design effort in carrier and timing recovery is in the design of the phase detector (Chapters 15 and 16). In this section we describe some variations on the basic sawtooth phase detector.

14.3.1. Sinusoidal Phase Detectors

Prevalent throughout the history of PLLs is the *sinusoidal phase detector*. In fact it is so prevalent that some books on PLLs never consider any other type. Its dominance is a consequence of its easy implementation in analog circuitry. The structure is shown in Fig. 14-9.

One significant difference between this phase detector and the sawtooth phase detector previously considered is that the VCO will tend to phase lock to the input *in quadrature* (with a 90 degree phase difference). In other words, $\phi(t)$ in (14.2) will have a constant $\pi/2$ term (or equivalently the $\cos(\cdot)$ will be replaced with a $\sin(\cdot)$). Assuming (14.1) and (14.2) are the input signal and the VCO output, the output of the multiplier is

$$\frac{A_v A_y}{2}\left[\cos(\theta(t)-\phi(t))+\cos(2\omega_v t+\theta(t)+\phi(t))\right]. \tag{14.56}$$

Assuming the second term is removed by the LPF in Fig. 14-9 we can write

$$\varepsilon(t)=\frac{A_v A_y}{2}\cos(\theta(t)-\phi(t)). \tag{14.57}$$

At first glance this does not look like a good estimate of the phase error $\theta(t)-\phi(t)$. In fact, $\varepsilon(t)$ is at its *maximum* when $\theta(t)=\phi(t)$ (see Problem 14-14). Suppose that the PLL tries to minimize $\varepsilon(t)$ anyway. It will be minimized (in fact $\varepsilon(t)=0$) when

$$[\theta(t)-\phi(t)] \bmod 2\pi=\frac{\pi}{2}\,. \tag{14.58}$$

Thus, this PLL will minimize $\varepsilon(t)$ by maintaining a constant 90 degree ($\pi/2$ radian) difference between the phase of the input and the phase of the VCO. Such a PLL is said to be phase locked in quadrature. Define

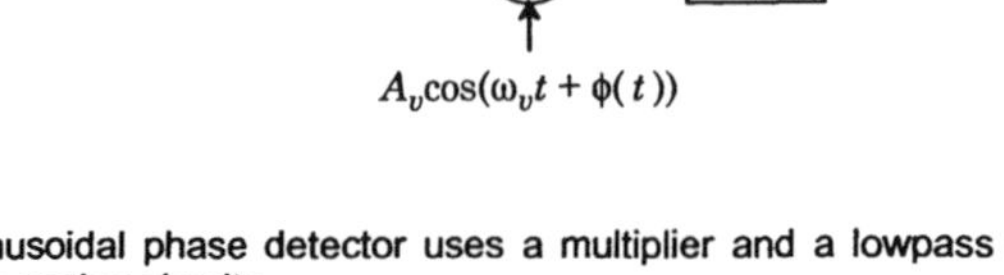

Fig. 14-9. A sinusoidal phase detector uses a multiplier and a lowpass filter. It is relatively easy to implement using analog circuits.

$$\psi(t) = \phi(t) + \frac{\pi}{2} \; . \tag{14.59}$$

Then the PLL will be in quadrature phase lock if $\psi(t) = \theta(t)$. The output of the VCO can be written

$$v(t) = A_v \cos(\omega_v t + \psi(t) - \pi/2) = A_v \sin(\omega_v t + \psi(t)) \; . \tag{14.60}$$

This explicitly shows the 90 degree phase difference. We can now consider $\psi(t)$ to be the phase of the VCO, and write the phase error

$$\varepsilon(t) = \tfrac{1}{2} A_v A_y \sin(\theta(t) - \psi(t)) \; . \tag{14.61}$$

Now $\varepsilon(t)$ is a much more reasonable estimate of the phase error. In fact, the only difference now between this PLL and the continuous-time PLL of Section 14.1 is the function $W(\cdot)$ which is now defined to be

$$W(x) = \tfrac{1}{2} A_v A_y \sin(x) \; , \tag{14.62}$$

where x is the phase error $\theta(t) - \psi(t)$. This is plotted in Fig. 14-10. The polarity of $W(\cdot)$ is the same as that of the sawtooth phase detector for all x. For small phase errors x, the two phase detectors are very similar, and have approximately linear characteristics, since

$$sin(x) \approx x \; , \tag{14.63}$$

so when $\theta(t)$ is close to $\psi(t)$,

$$\varepsilon(t) \approx \tfrac{1}{2} A_v A_y (\theta(t) - \psi(t)) \; . \tag{14.64}$$

This differs from (14.12) only by a constant, so the analysis carried out in Section 3.1 applies without modification if the phase error is assumed to be small.

One major practical difference between the sinusoidal phase detector and the ideal phase detector of Section 14.1 is the dependence of $\varepsilon(t)$ on the input signal level A_y, which in practice may be time-varying. This effect can be analyzed, as can the non-linearity that occurs for larger phase errors, but the analysis is much more difficult, and is beyond the scope of this book (see for example [2].

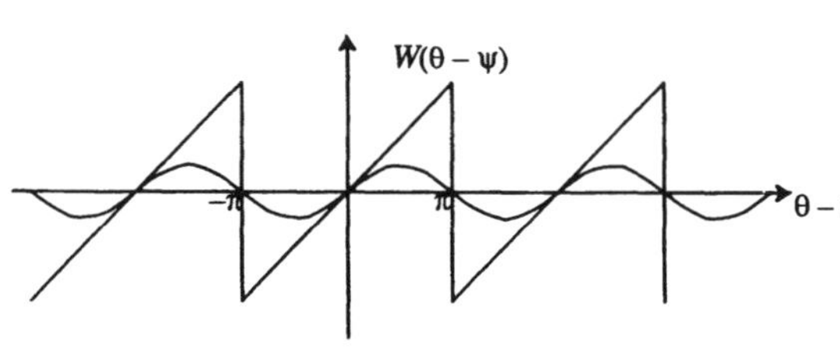

Fig. 14-10. A sinusoidal phase detector operates much like the sawtooth phase detector except for two things: the VCO output is in quadrature with the input, and the $W(\cdot)$ function is sinusoidal, as shown in this figure. Also shown for reference is $W(\cdot)$ for the sawtooth phase detector.

14.3.2. Complex Phase Detectors

For passband PAM systems it is common to do much of the signal processing on the complex-valued baseband equivalent signal, as shown in Chapter 5. A simple extension of the sinusoidal phase detector for complex signals is shown in Fig. 14-11. For small phase errors, the phase detector is approximately linear,

$$\varepsilon(t) \approx A_v A_y[\theta(t) - \phi(t)] \ . \tag{14.65}$$

14.3.3. Sampling Phase Detectors

PLLs in digital communication systems are often a mixture of discrete and continuous-time subsystems. A common scenario is that the PLL determines when the incoming signal is sampled, and the samples are used to estimate the phase error. A simple system of this type is shown in Fig. 14-12. The same forms for the input and VCO output are assumed (although in practical situations a digital VCO is more common, see Section 14.4 below). The sampling instants are at the upward-going zero crossings of the VCO output $v(t)$. Denote these sampling instants $t = \tau_k$, and note that the upward-going zero crossings of $v(t)$ occur when τ_k satisfies

$$\omega_v \tau_k + \phi(\tau_k) = k2\pi - \pi/2 \ . \tag{14.66}$$

Hence we can write

$$\varepsilon_k = y(\tau_k) = A_y \cos(k2\pi - \pi/2 - \phi(\tau_k) + \theta(\tau_k)) = A_y \sin(\theta(\tau_k) - \phi(\tau_k)) \ . \tag{14.67}$$

Thus we *almost* have a discrete-time sinusoidal phase detector; the measured phase error ε_k is *almost* a discrete-time version of (14.61). The reason we say "almost" is that since τ_k is controlled by the VCO, the sampling times are non-uniform! However, we can often assume that $\theta(t)$ and $\phi(t)$ are varying slowly enough that the non-uniform sampling has little effect. With this approximation, the behavior of the PLL will be the same as that of discrete-time PLL with a sinusoidal phase detector.

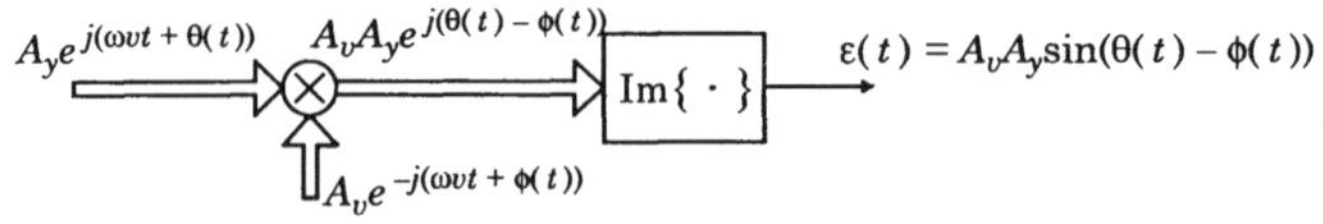

Fig. 14-11. A simple extension of the sinusoidal phase detector for complex signals is shown here.

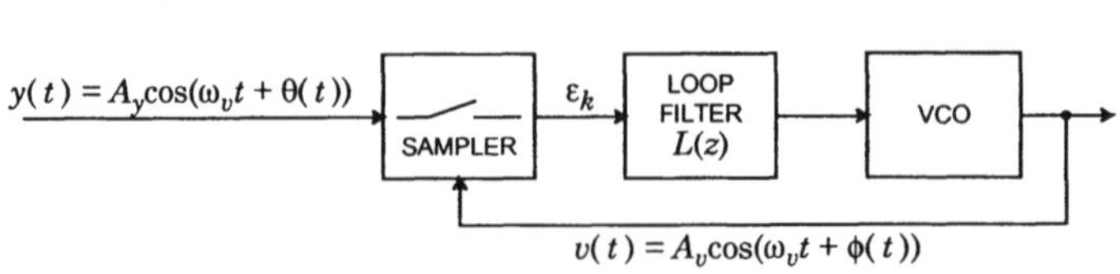

Fig. 14-12. A mixed discrete and continuous-time PLL with a sampling phase detector. The sampling of the input is directed by a continuous-time VCO.

The sampling phase detector in Fig. 14-12 is commonly used for timing recovery (see Chapter 16). The PLL in Fig. 14-12 is sometimes called a *digital PLL*, but this terminology is confusing because many other PLLs have digital elements. For a more detailed analysis, including the effect of the non-uniform sampling, see [3][4][5].

14.3.4. Exclusive-Or Phase Detectors

Another commonly used phase detector is the *exclusive-or phase detector*, illustrated in Fig. 14-13 assuming that the inputs are square waves (digital continuous-time signals) rather than sinusoids (analog signals). In some applications the signals are square waves to begin with (see Section 14.4), but even when they are not, square waves are easily generated from sinusoidal signals using hard limiters. The output $x(t)$ of the exclusive-or gate is high whenever the two input signals are different, as shown in Fig. 14-13(b). The average duty cycle is proportional to the phase error. Thus if $x(t)$ is lowpass filtered to perfectly extract the d.c. component while rejecting the fundamental and harmonics, the result is the function $w(t)$ shown in Fig. 14-13(c). This is not directly useful because it is not linear near zero phase error. Subtracting a constant, the *triangular phase detector* characteristic illustrated in Fig. 14-13(d) is obtained. If the constant K is correct, the VCO phase locks in quadrature, just like the sinusoidal phase detector.

When the input signal is not a square wave, but rather is sinusoidal, a square wave can be synthesized by hard limiting.

14.3.5. Phase Domain PLLs

For many application it is unnecessary to generate the VCO output explicitly. In other words, the sinusoid (or square wave) phase locked to the input is not needed, only its phase is needed. An example of a PLL that works entirely in the phase domain is shown in Fig. 14-14. The phase of the input is measured with respect to a fixed clock, and that phase provides the

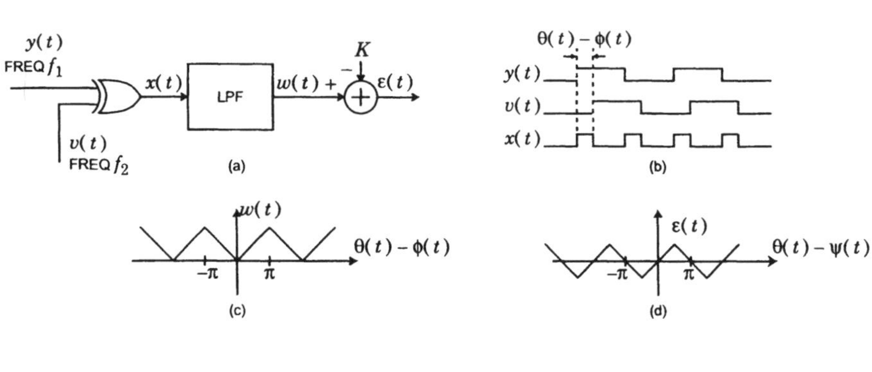

Fig. 14-13. An exclusive-or phase detector (a) assumes that the inputs are square waves (b) (digital continuous-time signals) rather than sinusoids (analog signals). The output of the lowpass filter (c) is proportional to the average amount of time that the two inputs signals differ in each cycle. After subtracting a constant, the result is a triangular phase detector characteristic (d).

input to the loop. We can call the frequency of the fixed reference clock ω_v, and the phase measurement θ_k. The phase measurement itself can be implemented digitally, for example by measuring the time between zero crossings in the reference and the input. The basic operation of this PLL is no different from those considered above.

Exercise 14-5.

Show that the phase transfer function of the PLL in Fig. 14-14 is the same as (14.43).

An application of a phase domain PLL to the demodulation of PSK signals is given in Chapter 15.

14.3.6. Frequency Detectors

The phase detector is sufficient for maintaining phase lock for a PLL when the input frequency is within the lock range. We have also seen that the lock range is dependent on the loop bandwidth and the design of the phase detector. This leads to an undesirable tradeoff between the acquisition properties of a PLL and the in-lock performance. For most designs these problems are manageable. However, there are cases where adequate lock range cannot be achieved while maintaining a sufficiently narrow loop bandwidth.

Example 14-14.

In a digital radio system (Section 14.4) the carrier frequency of the radio is determined by the free-running frequency of a microwave oscillator. The VCO in the receiver is also a microwave oscillator. Even if these oscillators have extremely accurate free-running frequencies, the offset can be large relative to the IF carrier frequency. These carrier frequency offsets are often on the order of 500 kHz or even larger. A PLL design which has a lock range this large would have much too large a closed-loop bandwidth to maintain adequately small phase jitter.

In this instance, something must be done to increase the lock range without affecting the in-lock loop bandwidth. Two techniques are available for this purpose: *frequency sweeping* and the addition of a *frequency detector* to the phase detector.

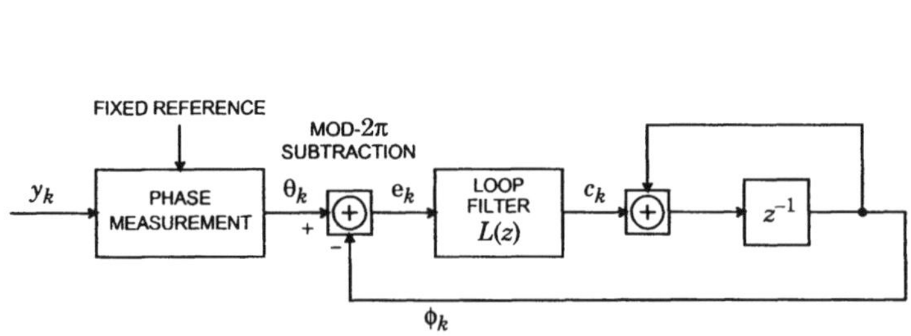

Fig. 14-14. A PLL that works entirely in the phase domain. The phase of the input is measured with respect to a fixed clock, and that phase provides the input to the loop.

A frequency detector is intuitively simple — it compares the frequency of the incoming signal to the local oscillator rather than comparing its phase. This is equivalent to measuring the phase without the 2π ambiguity exhibited by phase detectors. See [1][6] for a description of frequency detectors. A class of phase detectors known as *adaptive phase detectors* can also function as frequency detectors and are described in [7][8][9].

14.4. Variations on a Theme: VCOs

Just as with phase detectors, there are many variations on the design of VCOs. We describe two important examples.

14.4.1. Digital VCOs

Analog VCOs are complicated circuits, difficult to design and sensitive to environmental factors such as temperature. A frequently used alternative is the digital VCO, shown in Fig. 14-15. The VCO produces a square wave that is generated from a local high frequency clock using a variable countdown chain (a frequency divider). By controlling the frequency divider, the frequency of the VCO output can be controlled. The digital logic is such that if $c(t) < -K$, for some threshold K, the frequency of $v(t)$ is decreased by dividing by $N+1$ instead of the nominal N. Similarly, if $c(t) > K$, the divider divides by $N-1$ to increase the output frequency. In a typical operating condition, $c(t)$ is hovering near $\pm K$, and the frequency divider is either alternately dividing by N and $N+1$ or $N-1$ and N.

For digital VCOs, the lock range is not determined by the 2π ambiguity of the phase detector, but rather by N, the nominal divide ratio of the VCO. The reason is obvious: the maximum frequency the VCO can produce occurs when the divider is always dividing by $N-1$, and the minimum frequency occurs when the VCO is dividing by $N+1$. Thus any input frequency outside this range cannot be held (see Problem 14-15).

Although digital VCOs are economical and are commonly used, for some applications they have a serious disadvantage. Namely, since the frequency divider is alternating between two divide ratios at all times, the square wave at its output has considerable jitter. This jitter

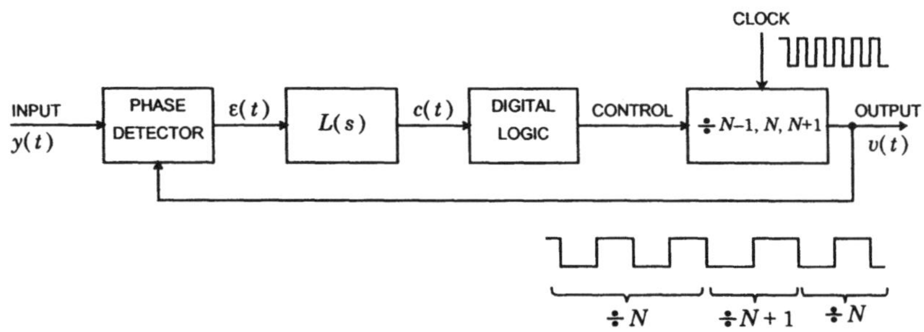

Fig. 14-15. A digital VCO uses a controllable countdown chain (a frequency divider) to generate the output, a square wave.

can be minimized for large N, but large N implies a high frequency oscillator. Depending on the nominal frequency ω_v of the VCO, it may not be practical to implement an oscillator and frequency divider at frequency $N\omega_v$ for N large enough to produce acceptably low jitter. In these situations, it is necessary to either develop algorithms that are insensitive to jitter, or resort to the use of an analog PLL. For a practical example of such a situation, see [10][11][12].

14.4.2. Frequency Synthesizers

It is possible to design PLLs that maintain phase lock when the VCO output frequency is not equal to the input frequency, but is related by a fixed rational multiple. Such a PLL is called a *frequency synthesizer*, shown in Fig. 14-16. When the PLL is phase locked,

$$\frac{f_{\text{in}}}{M} = \frac{f_{\text{out}}}{N}, \qquad f_{\text{out}} = \frac{N}{M} f_{\text{in}}. \tag{14.68}$$

Frequency synthesizers are widely used in multiplexers, where it is often necessary to synchronously generate one frequency from another.

Example 14-15.
Given a clock at 1,544 kHz, what are M and N such that we generate a clock at 2,048 kHz? We could use $N = 2{,}048$ and $M = 1{,}544$, and the inputs to the phase detector would be on the order of 1 kHz. Suppose that an exclusive-or phase detector (Fig. 14-13) is used. Then the output $x(t)$ of the exclusive-or will be periodic with frequency 1 kHz (see Fig. 14-13(b)). The LPF in Fig. 14-13(a) has to have sufficiently narrow bandwidth to remove this fundamental and its harmonics. In addition to complicating the design of the LPF, the narrow bandwidth means that the PLL will respond slowly to changing conditions.

This design is easily improved so that the bandwidth of the LPF can be larger. Observe that $1{,}544 = 193 \times 8$ and $2{,}048 = 256 \times 8$. Consequently we can use $N = 256$ and $M = 193$. This results in inputs to the phase detector on the order of 8 kHz, significantly relaxing the specifications of the LPF.

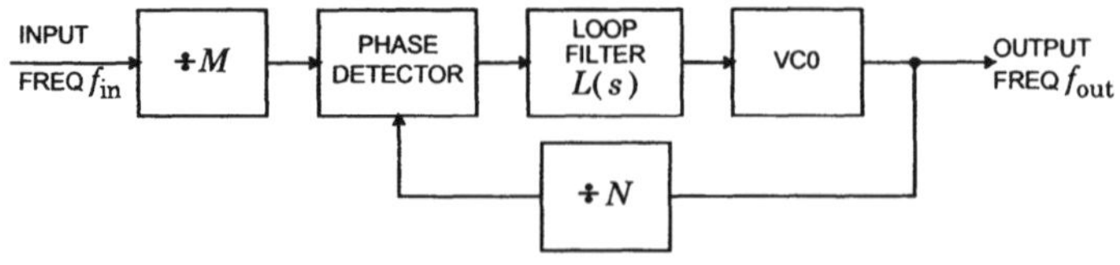

Fig. 14-16. A frequency synthesizer produces an output signal with frequency equal to N/M times the input frequency.

14.5. Further Reading

The understanding of noise performance, acquisition, and non-linear behavior of PLLs is advanced, and we have largely omitted these topics. The available literature, however, is heavily biased towards continuous-time, analog PLLs. In digital communications, many PLLs are either entirely discrete-time or a mixture of discrete and continuous-time. Fortunately, extending the continuous-time results to discrete time is often easy, as illustrated in Section 14.2.

A number of books give comprehensive coverage of analog, continuous-time PLLs. Gardner [3] was the first widely used for design purposes. The first four chapters of Viterbi [13] are devoted to PLL theory. The third chapter gives a particularly good description of an analysis technique called phase-plane analysis. Van Trees [14] uses non-linear estimation theory for the analysis of PLLs. Klapper and Frankle [15] give a detailed treatment biased towards the application of PLLs as FM discriminators. Lindsey [16] gives a thorough theoretical treatment with emphasis on weak signal applications. Blanchard [17] emphasizes design of coherent receivers for analog modulation schemes. Two other important books are by Best [18] and Lindsey and Simon (editors) [19]. A brief overview of advanced PLL topics is given by Gupta [20] with an extensive bibliography. The October 1982 issue of IEEE Transactions on Communications is devoted to PLLs, providing an excellent source of more up-to-date papers. A description of frequency detectors is given by Messerschmitt [1]. Of historical importance is perhaps the first paper on PLLs by Appleton in 1922 [21]. We have not described the adaptive PLL, in which the loop filter is adaptive rather than fixed [22][23]. The filter can adjust itself to changing phase characteristics on the input.

Problems

Problem 14-1. Consider the PLL shown in Fig. 14-2. Suppose $L(s) = K_L/s$.

(a) What is the order of the loop?

(b) Is the loop stable? Why?

Problem 14-2. The lock range of the first-order PLL in Example 14-4 can be increased by increasing the loop gain K_L. Show that as K_L tends to infinity the output of the VCO will achieve perfect phase lock for any input phase $\theta(t)$. What happens to the bandwidth of the PLL? When might this be useful, and when might it not be useful?

Problem 14-3.

(a) Show that $s^2 + as + b$ has all roots in the open left half plane if and only if the real-valued coefficients satisfy $a > 0$ and $b > 0$. This basic result can be used to determine the stability of any second-order continuous-time PLL.

(b) Show that an ideal PLL with loop filter given by (14.19) is stable if and only if $K_2 > -K_L$ and $K_L K_1 > 0$.

Problem 14-4. Suppose the loop filter of a continuous-time second-order PLL with an ideal VCO and phase detector is given by

$$L(s) = \frac{1}{s + \sqrt{2}} . \tag{14.69}$$

(a) Find the poles of the phase transfer function. Is the PLL stable?

(b) Show that the gain $|\Phi(f)/\Theta(f)|$ is never greater than unity, so there is no peaking.

(c) Sketch the magnitude of the frequency response.

Problem 14-5. Consider a type I second-order PLL with the loop filter given by (14.19) and

$$K_L = K_2 = 1 \qquad K_1 = 0.5 . \tag{14.70}$$

(a) Give the closed loop transfer function to phase $\Phi(s)/\Theta(s)$.

(b) Is the PLL stable?

(c) Sketch the Bode plot for phase transfer function.

(d) Show that the PLL does not exhibit peaking.

Problem 14-6. For the type I second-order PLL of (14.19),

(a) Show that there is peaking if and only if

$$K_2^2 \le 2K_L K_1 . \tag{14.71}$$

(b) Find the range of frequencies ω for which the gain is greater than unity, $|\Phi(f)/\Theta(f)|^2 > 1$.

(c) The bandwidth of a type I second-order PLL with complex poles can be estimated as $\sqrt{K_{L2}K_1}$ (see Fig. 14-5) where K_{L2} is K_L for the second-order PLL. By contrast, the bandwidth of a first-order PLL can be estimated as K_{L1}, which is K_L for the first-order loop. Assume these two PLLs have equal bandwidth,

$$\sqrt{K_{L2}K_1} = K_{L1} . \tag{14.72}$$

Show that if there is no peaking, then the lock range of the second-order loop is actually *smaller* than the lock range of the first-order loop.

Problem 14-7.

(a) Use the result stated in Problem 14-6a to show that a stable type II second-order PLL *always* exhibits peaking.

(b) Show that

$$\left|\frac{\Phi(f_n)}{\Theta(f_n)}\right|^2 = \frac{K_1 + K_L}{K_L} , \tag{14.73}$$

where $f_n = \sqrt{K_L K_1}/(2\pi)$ the natural frequency. We see that the magnitude of the peaking at f_n can be made small by choosing a small K_1.

(c) Show that this type II loop will be stable if and only if K_1 and K_L are each greater than zero. This observation combined with (14.73) is another way of showing that this PLL always exhibits peaking, but it also shows that the gain at f_n can be made as close to unity as we like by selecting a small K_1.

Problem 14-8. Consider the type I second-order PLL with a loop filter of the form

$$L(s) = \frac{K_1}{K_2 + s} . \tag{14.74}$$

This is different from (14.19) in that there is no zero for finite s.

(a) Find the transfer function and show that it is a lowpass filter with unity gain at d.c.

(b) Show that this PLL is stable if and only if $K_1 > 0$ *and* $K_2 > 0$.

(c) Find the range of frequencies f over which components $\Theta(f)$ in the input phase will be amplified. Under what conditions is there no amplification (peaking)?

Problem 14-9. Consider the PLL in Fig. 14-2. Suppose that $L(s) = K_L$, a positive constant, and $\theta(t)$ is given below. Find the steady-state response

$$\varepsilon_{ss} = \lim_{t \to \infty} \varepsilon(t) . \tag{14.75}$$

(a) $\theta(t) = \beta u(t)$, where $u(t)$ is the unit step function.

(b) $\theta(t) = \beta t^2 u(t)$.

Compare these results to the result derived in Example 14-9 for $\theta(t) = \omega_0 t u(t)$.

Problem 14-10. Consider the PLL in Fig. 14-2. Show that all loop filters $L(s)$ with one or more poles at $s = 0$ satisfy $\varepsilon_{ss} = 0$ when $\theta(t) = \omega_0 t u(t)$, which corresponds to a steady frequency offset of ω_0. Assume $L(s)$ is chosen so that the PLL is stable.

Problem 14-11. Consider the first-order discrete-time PLL of Example 14-10. Sketch the magnitude of the frequency response $|\Phi(e^{j\theta})/\Theta(e^{j\theta})|$ when:

(a) $K_L = 0.5$.

(b) $K_L = 1$.

(c) $K_L = 1.5$.

(d) Compare the relative merits of the PLLs in (a) through (c).

Problem 14-12. Consider the second-order polynomial with real coefficients

$$D(z) = z^2 + az + b = (z - p)(z - q) . \tag{14.76}$$

Show that the roots p and q lie inside the unit circle (have less than unity magnitude) if and only if $|b| = |pq| < 1$ and $|a| < 1 + b$.

Problem 14-13. Use the result proven in Problem 14-12 to find conditions under which the discrete-time PLL with loop filter given by (14.45) is stable.

Problem 14-14. Consider a continuous-time PLL with a sinusoidal phase detector, Fig. 14-9. Suppose that this PLL is in perfect phase lock, $\theta(t) = \phi(t)$ (not quadrature phase lock). Find an expression for the input phase $\theta(t)$ as a function of the PLL parameters and time.

Problem 14-15. Consider the digital VCO in Fig. 14-15. Suppose that the clock frequency is 1 MHz and $N = 100$. Suppose that the input has the form $\cos(\omega_1 t + \theta)$ where θ is a constant. Find the range of ω_1 such that the PLL can maintain phase lock.

Problem 14-16. Consider the design of a frequency synthesizer (Fig. 14-16) that synthesizes a 2,048 kHz signal given a 1,512 kHz input.

(a) Find the minimum values of M and N.

(b) Consider doing the frequency synthesis with two cascaded frequency synthesizers. Select M_1, N_1, M_2 and N_2 for each of the frequency dividers. What advantages does this design have over the one with a single frequency synthesizer?

(c) What is the maximum number of cascaded frequency synthesizers that can be used effectively?

References

1. D. G. Messerschmitt, "Frequency Detectors for PLL Acquisition in Timing and Carrier Recovery," *IEEE Trans. on Communications,* Vol. COM-27, (9), p. 1288 (Sep. 1979).
2. F. M. Gardner, *Phaselock Techniques*, Wiley, New York (1966).
3. G. S. Gill and S. C. Gupta, "First-Order Discrete Phase-Locked Loop with Applications to Demodulation of Angle-Modulated Carrier," *IEEE Trans. Communication Technology,* Vol. COM-20, pp. 454-462 (June 1972).
4. G. S. Gill and S. C. Gupta, "On Higher Order Discrete Phase Locked Loops," *IEEE Trans. Aerospace Electronic Systems*, Vol AES-8, pp. 615-623 (September 1972).
5. A. Weinberg and B. Liu, "Discrete Time Analyses of Non-Uniform Sampling First- and Second-Order Digital Phase-Locked Lops," *IEEE Trans. Communications Technology*, Vol. COM-22, pp. 123-137 (Feb. 1974).
6. F. M. Gardner, "Properties of Frequency Difference Detectors," *IEEE Trans. Communications*, Vol. COM-33, pp. 131-138 (Feb. 1985).
7. J. C. Haartsen and R. C. Den Dulk, "Novel Circuit Design and Implementation of Adaptive Phase Comparators," *Electronics Letters*, Vol. 23 (11), pp. 551-552 (May 1987).
8. J. Eijselendoorn and R. C. Den Dulk, "Improved Phase-Locked Loop Performance with Adaptive Phase Comparators," *IEEE Trans. Aerospace and Elec. Sys.,* Vol. AES-18, pp. 323-333 (May 1982).
9. J. F. Oberst, "Generalized Phase Comparators for Improved Phase-Lock Loop Acquisition," *IEEE Trans. Communications*, Vol. COM-19, pp. 1142-1148 (Dec. 1971).

10. D. D. Falconer, "Timing Jitter Effects on Digital Subscriber Loop Echo Cancellers: Part I - Analysis of the Effect," *IEEE Trans. on Communications*, Vol. COM-33(8), (Aug. 1985).

11. D. D. Falconer, "Timing Jitter Effects on Digital Subscriber Loop Echo Cancellers: Part II - Considerations for Squaring Loop Timing Recovery," *IEEE Trans. on Communications*, Vol. COM-33 (8), (Aug. 1985).

12. D. G. Messerschmitt, "Asynchronous and Timing-Jitter Insensitive Data Echo Cancellation," *IEEE Trans. on Communications*, Vol. COM-34 (12), p. 1209 (Dec. 1986).

13. A. J. Viterbi, *Principles of Coherent Communication*, McGraw-Hill, New York (1966).

14. H. L. Van Trees, *Detection, Estimation, and Modulation Theory*, Part II, Wiley, New York (1971).

15. J. Klapper and J. T. Frankle, *Phase-Locked and Frequency-Feedback Systems*, Academic Press, New York (1972).

16. W. C. Lindsey, *Synchronization Systems in Communications*, Prentice-Hall, Englewood Cliffs, N.J. (1972).

17. A. Blanchard, *Phase-Locked Loops: Application to Coherent Receiver Design*, John Wiley and Sons, New York (1976).

18. R. E. Best, *Phase-Locked Loops; Theory, Design, and Applications*, McGraw Hill, New York (1984).

19. W. C. Lindsey and M. K. Simon, *Phase-Locked Loops and Their Applications*, IEEE Press, New York (1978).

20. S. C. Gupta, "Phase-Locked Loops," *Proceedings of the IEEE*, Vol. 63 (2), (Feb. 1975).

21. E. V. Appleton, "Automatic Synchronization of Triode Oscillators," *Proc. Cambridge Phil. Soc.*, Vol. 21, p. 231 (1922-1923).

22. R. P. Gooch and M. J. Ready, "An Adaptive Phase Lock Loop for Phase Jitter Tracking," *Proceedings of the 21st Asilomar Conf. on Signals, Systems, and Computers,* (Nov. 2-4, 1987).

23. R. D. Gitlin, "Adaptive Phase-Jitter Tracker," United States Patent, Patent No. 4,320,526, (March 16, 1982).

15

Carrier Recovery

In passband systems, the carrier frequency is generated in the transmitter from a local timing reference such as a crystal oscillator. As we saw in Chapters 5 and 6, *coherent demodulation* of a passband signal requires exactly the same carrier frequency and phase to perform the demodulation. But the receiver usually has an independent timing reference, as illustrated in Fig. 14-1. Deriving the carrier frequency and phase from the data bearing signal is the topic of this chapter.

In previous chapters we have assumed that both the symbol timing and the carrier frequency are known at the receiver. In this chapter we will assume the symbol timing is known, and derive the carrier frequency. Chapter 16 will show that symbol timing can be derived without knowledge of the carrier phase. Hence, when a receiver first starts receiving data, it should first derive timing using the techniques of Chapter 16, then estimate the carrier phase using the techniques in this chapter, and finally adapt the equalizer (Chapter 9).

If the symbol timing is known, it may be that the carrier frequency can be derived from it. If the carrier frequency used at the transmitter is a fixed rational multiple of the symbol rate, then the frequency synthesizer of Fig. 14-16 can derive a high quality carrier, even if there is considerable jitter on the derived timing signal. However, even if the transmitter uses a carrier frequency and symbol rate that are related by a rational multiple, that relationship may be lost by the time they get to the receiver.

Example 15-1.
The telephone channels often introduce frequency offset. If the passband PAM signal

$$x(t) = \sqrt{2}\,\mathrm{Re}\{s(t)e^{j2\pi f_c t}\} \tag{15.1}$$

is subjected to frequency offset of f_0 (and no other impairments) then the received signal will be

$$y(t) = \sqrt{2}\,\mathrm{Re}\{s(t)e^{j2\pi(f_c - f_0)t}\} . \tag{15.2}$$

Frequency offset therefore is indistinguishable from using a different carrier frequency $f_c - f_0$. The symbol rate, however, cannot be changed by the channel because the receiver can only receive exactly as many symbols per unit time as are sent. Consequently, even if the symbol rate and carrier frequency start out as rational multiples of one another at the transmitter, their relationship at the receiver is dependent on the unknown frequency offset.

Example 15-2.
In microwave radio applications where either the transmitter or the receiver is in motion, the carrier frequency is subject to a Doppler shift, but the symbol timing is not. The resulting frequency offset is similar to that found in telephone channels, although it is more likely to be time-varying as the velocity changes.

In addition to frequency offset, it is common for a channel to introduce *phase jitter* which appears as fluctuations in the phase of the carrier. It is desirable to track the phase jitter so that it does not degrade the performance of the system. So even in the absence of frequency offset, it is still desirable to derive the carrier independently from the timing so that phase jitter can be tracked.

We will describe two techniques for tracking the carrier in the receiver. The first is a decision-directed technique, and the second is the *power of N* method.

15.1. Decision-Directed Carrier Recovery

Consider a noiseless passband PAM analytic signal that has been subjected to frequency offset and/or phase jitter

$$e^{j(2\pi f_c + \theta(t))}\sum_{m=-\infty}^{\infty} A_m p(t - mT) , \tag{15.3}$$

where $p(t)$ accounts for the transmit pulse shape, the linear distortion in the channel, and the receive filter, and $\theta(t)$ models the frequency offset and phase jitter. In order to have this analytic signal available at the receiver, we assume either an analytic receiver bandpass filter or a phase splitter, as discussed in Section 5.3.4. If there is frequency offset then $\theta(t)$ will have a linear term $2\pi f_0 t$. Suppose that (15.3) is demodulated with the carrier

$$e^{-j(2\pi f_c + \phi(t))} , \tag{15.4}$$

where $\phi(t)$ is the receiver estimate of the carrier phase. Sample at the symbol rate $t = kT$ to get

$$q_k = e^{j(\theta_k - \phi_k)} \sum_{m = -\infty}^{\infty} A_m p_{k - m} \, , \tag{15.5}$$

where θ_k, ϕ_k, and p_k are samples of $\theta(t)$, $\phi(t)$, and $p(t)$. If the carrier recovery follows a bandpass equalizer, as shown in Fig. 5-21, then the equalized pulse shape should approximately satisfy the Nyquist criterion,

$$p_k = \delta_k \, , \tag{15.6}$$

and consequently

$$q_k = e^{j(\theta_k - \phi_k)} \sum_{m = -\infty}^{\infty} A_k \, . \tag{15.7}$$

Example 15-3.
If the receiver demodulates with a constant phase error $\theta_k - \phi_k = \psi$, and there is no other degradation, then the received constellation will be a tilted version of the transmitted constellation, as shown in Fig. 15-1(a).

If left uncorrected, the tilt will degrade the immunity of the receiver to noise by bringing the received signal points closer to the boundaries of the decision regions.

Example 15-4.
If the receiver demodulates with the wrong frequency $\theta_k - \phi_k = 2\pi f_0 kT$, then the received constellation rotates, as illustrated in Fig. 15-1(b).

If left uncorrected, the rotating constellation will make errors every time a received symbol rotates past the boundary of a decision region. To correct both problems, carrier recovery is needed. Assuming a PLL is used for this function, we will now design a decision-directed phase detector. It is also possible to avoid decision direction in carrier recovery, as illustrated in Section 15.2.

Exercise 15-1.
Given (15.7), show that the phase error in the demodulator is

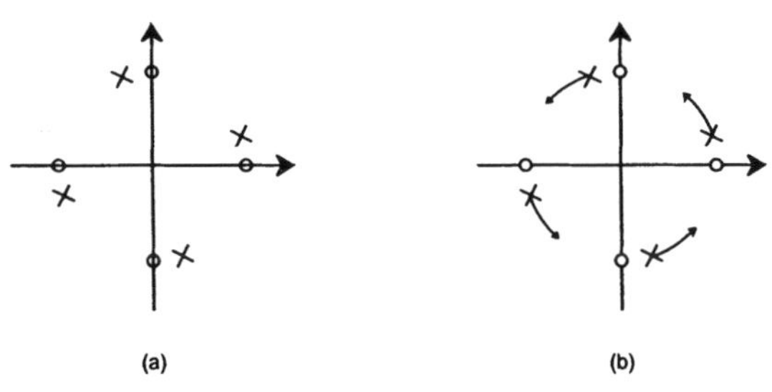

Fig. 15-1. If the receiver demodulates with a constant phase error then the received constellation will be a tilted version of the transmitted constellation, as shown in (a). The O's are the transmit constellation and the X's are the received symbols. If the receiver demodulates with the wrong frequency, the received constellation will rotate, as illustrated in (b).

$$\varepsilon_k = \theta_k - \phi_k = \sin^{-1}\left(\frac{\mathrm{Im}\{q_k A_k^*\}}{|A_k|^2}\right). \tag{15.8}$$

We have the beginnings of a phase detector, but in practical systems the assumptions that $p_k = \delta_k$ and that there is no noise are unrealistic. With any amount of noise and intersymbol interference we can write the received symbols as

$$q_k = c_k e^{j\varepsilon_k} A_k \tag{15.9}$$

where the real-valued $c_k > 0$ accounts for amplitude errors and ε_k accounts for phase errors. Some of the phase error will be due to noise and intersymbol interference, and some will be due to phase jitter and frequency offset.

Exercise 15-2.
Given (15.9), show that

$$\varepsilon_k = \sin^{-1}\left(\frac{\mathrm{Im}\{q_k A_k^*\}}{|q_k| \cdot |A_k|}\right). \tag{15.10}$$

We are now closer to a practical phase detector, but there is still a problem. The symbols A_k are not known in the receiver (or there would be no need to transmit them) except perhaps during a brief training period at the initiation of a connection. Just as we did with adaptive equalizers in Chapter 9, we can use decisions $\hat{A}_k$ instead of the actual symbols A_k. The resulting carrier recovery loop is shown in Fig. 15-2.

This PLL is closely related to those in Chapter 14, of course, but the phase detector is significantly different. One major difference: since the phase detector is decision-directed, errors in the decision will result in errors in phase detection.

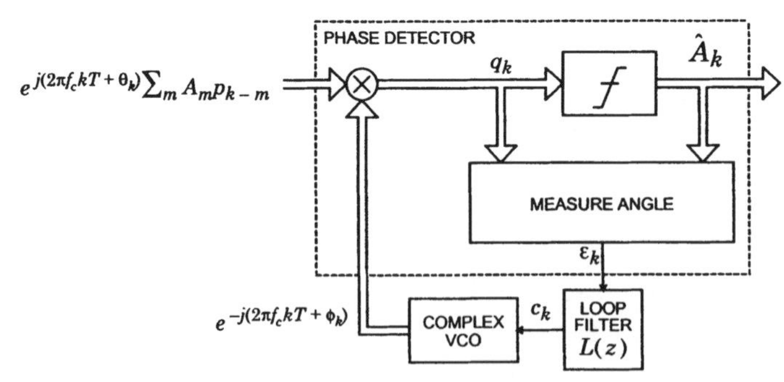

Fig. 15-2. A carrier recovery loop using a phase detector that measures the angular difference between the received sample q_k and the decision $\hat{A}_k$.

Example 15-5.
Consider a 4-PSK signal with constellation and decision regions shown in Fig. 15-3(a). If the received sample has a phase error greater than $\pi/4$ in magnitude, the decision will be incorrect, and the measured phase error will be incorrect. Equivalently, the measured phase error is

$$\varepsilon_k = W(\theta_k - \phi_k) \tag{15.11}$$

where $W(\cdot)$ is shown in Fig. 15-3(b). An immediate consequence is that the derived carrier can have any of four phases, depending on the first few decisions.

Decision-directed carrier recovery generally suffers from a phase ambiguity in the derived carrier, as shown in the previous example. This problem is easily overcome by using a *differential encoder.*

Example 15-6.
A differential encoder and decoder for a 4-PSK signal is shown in Fig. 15-4. Information is carried by the change in phase, rather than by the absolute phase. For example, an increase in the transmitted phase by π indicates the transmission of the pair of bits 10, regardless of the absolute phase.

Exercise 15-3.
Assume the only degradation between the coder and the slicer in Fig. 15-4 is a rotation by a multiple M of $\pi/2$, or in other words a multiplication by $e^{jM\pi/2}$. Such a degradation could be introduced by carrier recovery that uses a phase detector with the characteristic in Fig. 15-3(b). Show that the output bits from the decoder are the same regardless of the multiple M (with the possible exception of the first two bits out of the decoder).

Fig. 15-3. a. The constellation and decision regions for a 4-PSK signal. b. The characteristic of an ideal decision-directed phase detector. It cannot detect angular errors greater than $\pi/4$ in magnitude.

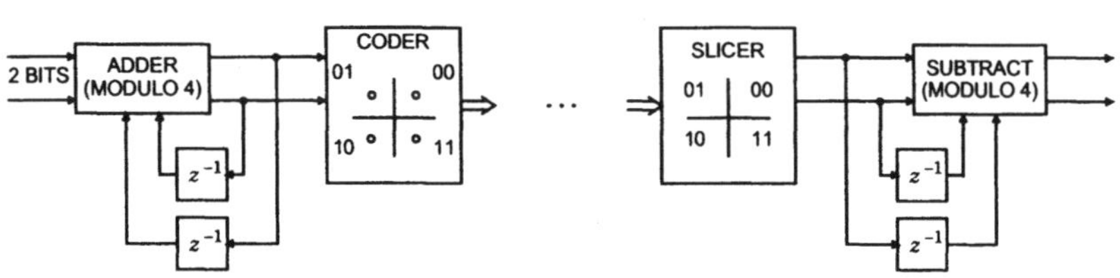

Fig. 15-4. A differential encoder can be used in the transmitter to overcome the $\pi/2$ phase ambiguity in the carrier recovery loop. Information is carried by the change in phase, rather than by the absolute phase. The decoder shown reverses the encoding.

Proper operation of a differential decoder depends on the decisions being correct in the receiver. In fact, if a single decision is incorrect, two symbol intervals will be incorrectly decoded, so there is *error propagation*. The impact of this error propagation is usually minimal.

For constellations larger than 4-PSK, the phase ambiguity of the carrier recovery loop can be more serious (see Problem 15-2). It is common to differentially encode two bits of the M bits needed for a constellation of size 2^M. The coder then uses the two differentially encoded bits to determine the quadrant. Thus phase errors of multiples of $\pi/2$ do not cause decision errors, although smaller phase errors might.

The phase error measurement in (15.10) can be simplified. Observe that for small phase errors, $\varepsilon_k \approx \sin(\varepsilon_k)$, and

$$sin(\varepsilon_k) = \frac{\text{Im}\{q_k A_k{}^*\}}{|q_k| \cdot |A_k|} \quad . \tag{15.12}$$

We can write this as

$$sin(\varepsilon_k) = W'(\theta_k - \phi_k) = \sin(W(\theta_k - \phi_k)) \,, \tag{15.13}$$

where $W'(\cdot)$ is shown in Fig. 15-5 and $W(\cdot)$ is shown in Fig. 15-3(b) for the 4-PSK case. From Fig. 15-5 we see that (15.12) makes a reasonable phase detector. With small angular errors, the characteristic is close to linear. It is common to simplify still further and omit the denominator in (15.12) so as to avoid having to perform the division, using as the estimate of the phase error

$$\varepsilon_k' = \text{Im}\{q_k A_k{}^*\} \,. \tag{15.14}$$

A first-order carrier recovery loop ($L(z) = K_L$) using this angular error estimate is shown in Fig. 15-6. The complex multiplications are shown explicitly so that the total complexity of the algorithm can be understood at a glance. This carrier recovery loop, or a variant of it with a continuous-time VCO, is commonly used, and is quite effective.

Many of the techniques from Chapter 14 can be applied to adapt this basic technique to particular situations. For example, a natural extension of the basic PLL in Fig. 15-6 uses a higher order loop, obtained by inserting a more complicated filter $L(z)$. Careful design of $L(z)$ can lead to carrier recovery loops that perfectly track frequency offset or phase jitter at some frequency.

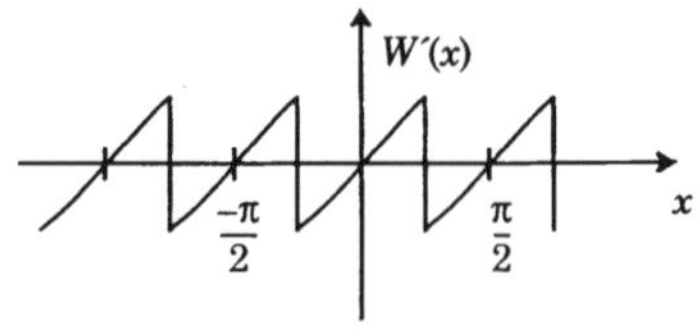

Fig. 15-5. The phase detector characteristic when (15.12)is used to estimate the angular error. Except for a slight curvature, it is very similar to that in Fig. 15-3b.

One practical difficulty arises when an adaptive equalizer is used in conjunction with a decision-directed carrier recovery loop. Baseband adaptive equalizers assume that the input has been demodulated. The solution to this difficulty, given in Section 9.5, is to use a passband equalizer. We can now understand why this is so important. The overall structure of a passband PAM receiver is shown in Fig. 5-21. By placing the forward equalizer before the carrier recovery demodulation, we avoid having the equalizer inside the carrier recovery loop. By contrast, a baseband equalizer would follow the demodulator and precede the slicer. This means that it is inside the carrier recovery loop. Consequently, the loop transfer function of the carrier recovery includes the time-varying equalizer, causing considerable complication. At the very least, the long delay (several symbol intervals) associated with the baseband equalizer would force the loop gain of the carrier recovery to be reduced to ensure stability, impairing its ability to track rapidly varying carrier phase. The passband equalizer shown in Fig. 5-21 mitigates this problem by equalizing prior to demodulation.

Another important practical difficulty arises with decision-directed carrier recovery loops when trellis coding is used (Chapter 13). Unlike a slicer, a trellis decoder does not make immediate decisions. Decisions may be postponed several (say *M*) symbol intervals. This is equivalent to inserting a delay z^{-M} in the carrier recovery loop. The delay can undermine the validity of the PLL (see Problem 15-3). To overcome this practical difficulty, a slicer is added to the receiver and the slicer decisions are used to compute the phase error. The slicer decisions will not be as accurate as the decisions made by the trellis decoder, but at least they are made promptly.

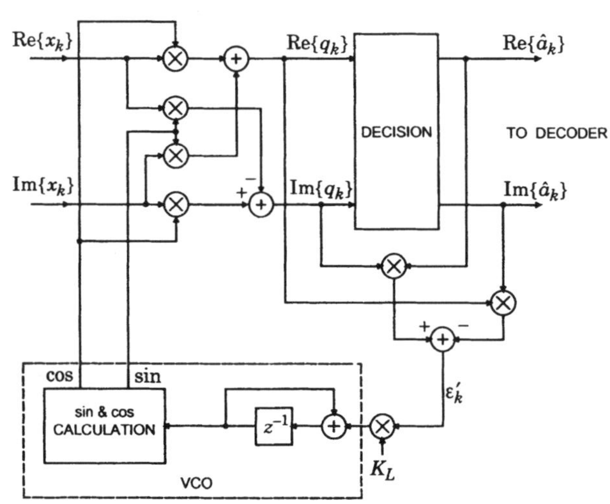

Fig. 15-6. A first-order carrier recovery loop using (15.14), a particularly simple estimate of the angular error.

Fixed Reference Detector

A particularly simple variation on decision-directed carrier recovery is possible for PSK signals. It is a phase-domain PLL like that in Fig. 14-14, but modified as shown in Fig. 15-7. The phase of the input signal is compared against a fixed reference generated by a local oscillator. This phase difference (which is modulo 2π) will reflect the phase modulation of the input plus a drift with time due to the fact that the local oscillator is not synchronized with the remote carrier. The function of the PLL is to remove the drift at the first subtractor. The slicer determines which of N phases was transmitted using a simple threshold test, and the difference between the decision and the input to the slicer is a measurement of the residual drift ε_k (or residual phase error). This phase error is filtered (if desired), and integrated. The integrator comes from the structure in Fig. 14-14. The main advantage of the fixed reference detector in Fig. 15-7 is that it can be implemented using inexpensive and fast digital logic.

15.2. Power of N Carrier Recovery

Although decision-directed carrier recovery is a common technique for coherent demodulation, decision errors will be a problem at low SNR. In this event it is possible to avoid decision direction. We will illustrate this with a popular alternative, *power of N carrier recovery.* Consider again the sampled passband PAM analytic signal

$$x_k = e^{j(2\pi f_c kT + \theta_k)} \sum_{m=-\infty}^{\infty} A_m p_{k-m}$$

$$= e^{j(2\pi f_c kT + \theta_k)} \sum_{m=-\infty}^{\infty} |A_m| e^{j\angle(A_m)} p_{k-m} \,. \tag{15.15}$$

Assume that there is no ISI so that $p_k = \delta_k$ and

$$x_k = e^{j(2\pi f_c kT + \theta_k)} |A_k| e^{j\angle(A_k)} \,. \tag{15.16}$$

If we raise this signal to the Nth power we get

$$x_k^N = e^{jN(2\pi f_c kT + \theta_k)} |A_k|^N e^{jN\angle(A_k)} \,. \tag{15.17}$$

Now suppose that we can find an integer N such that

$$e^{jN\angle(A_k)} = 1 \tag{15.18}$$

for all A_k. For example, this is possible for PSK and AM-PM signals. Then

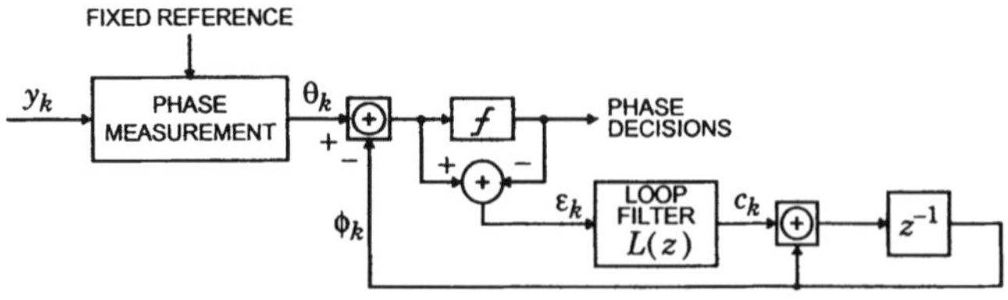

Fig. 15-7. A fixed-reference detector for PSK signals.

$$x_k^N = e^{jN(2\pi f_c kT + \theta_k)} |A_k|^N . \tag{15.19}$$

This signal has a strong spectral line at frequency Nf_c, as is evident from the following breakdown,

$$x_k^N = e^{jN(2\pi f_c kT + \theta_k)} \mathrm{E}[\,|A_k|^N] + e^{jN(2\pi f_c kT + \theta_k)}(\,|A_k|^N - \mathrm{E}[\,|A_k|^N]) . \tag{15.20}$$

The first term is a tone that can be extracted with a bandpass filter or a PLL. The second term may be zero (e.g. for PSK), or may contribute amplitude modulation to the tone.

Except for the possibly time-varying amplitude term $|A_k|^N$, (15.19) has the same form that we assumed in Chapter 14 for the complex phase detector (see Fig. 14-11). It can be fed into a complex phase detector, as shown in Fig. 15-8. The natural frequency ω_v of the VCO should be selected so that $N\omega_c$ is always within the lock range of the PLL (see Problem 15-5).

Example 15-7.

Consider a 4-PSK signal with $|A_k| = 1$, and no ISI or noise, just phase jitter and frequency offset represented by θ_k in (15.15). Let $N = 4$ so from (15.19)

$$x_k^N = e^{j4(2\pi f_c kT + \theta_k)} . \tag{15.21}$$

In the presence of noise or ISI this signal will have phase jitter which can be attenuated by the PLL.

To demodulate the PAM signal, the complex sinusoid with frequency Nf_c must be converted to a complex sinusoid at frequency f_c. This can be done using the frequency synthesizer in Fig. 14-16, which can actually be incorporated into the loop in Fig. 15-8 (see Problem 15-5).

In practical situations, of course, noise and ISI will be present and will contribute to phase jitter in the derived carrier. This jitter can be attenuated with a narrowband filter or a PLL.

Another practical difficulty is that it may not always be possible to find a value of N such that (15.18) is true (see Problem 15-4). In this case, we can raise the signal to the Nth power anyway as in (15.17) and the breakdown similar to (15.20) becomes

$$x_k^N = e^{jN(2\pi f_c kT + \theta_k)} \mathrm{E}[A_k{}^N] + e^{jN(2\pi f_c kT + \theta_k)}(A_k{}^N - \mathrm{E}[A_k{}^N]) . \tag{15.22}$$

The first term will yield the desired tone as long as $E[A_k^N] \neq 0$. It is usually easy to find an N for which this is true (see Problem 15-4).

Not surprisingly, continuous-time versions of power of N timing recovery are also used (see Problem 15-6).

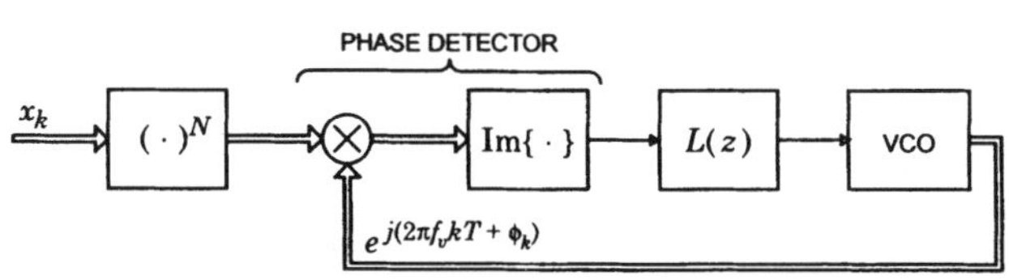

Fig. 15-8. A power of N carrier recovery loop.

15.3. Further Reading

A useful tutorial on carrier and timing recovery is given by Franks [1]. The interaction of carrier recovery and adaptive equalization was resolved in the seminal paper by Falconer [2]. Further information about decision-directed carrier recovery can be obtained from [3][4][5][6][7]. An interesting variation of 4-th power carrier recovery is given in [8]. Simultaneous maximum likelihood carrier and timing recovery is considered in [9]. The use of an adaptive loop filter for carrier recovery is proposed in [10][11].

Problems

Problem 15-1. A common problem that must be avoided in carrier recovery is *false lock*. For example, in Fig. 15-1(b), for four-phase PSK if the constellation rotates precisely $\pi / 2$ radians in every sample time, the samples q_k will fall on the correct signal constellation points and it will be difficult to tell that the carrier frequency is off.

(a) Find the carrier offset frequencies that result in this false lock.

(b) For a carrier frequency of 75 MHz and a baud rate of 15 MHz, what is the minimum false-lock carrier-offset frequency?

(c) Suggest one or more possible ways to deal with this problem.

Problem 15-2. Suppose that a decision-directed phase detector given by (15.10) (with A_k replaced by the decisions A_k) is used in a carrier recovery loop. We can write ε_k as a function of the phase error,

$$\varepsilon_k = W(\theta_k - \phi_k) \ . \tag{15.23}$$

(a) Sketch $W(\cdot)$ for an 8-PSK constellation.

(b) Sketch $W(\cdot)$ for a 16-QAM constellation. Assume radial decision regions, as shown in the following figure:

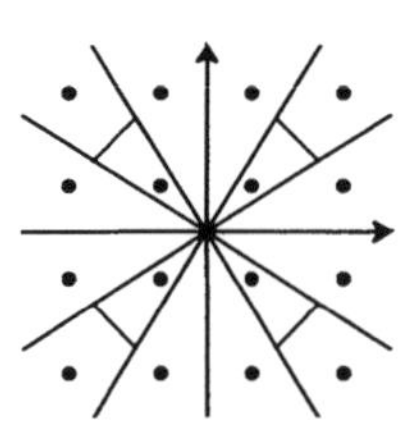

The decision boundary angles are halfway between the angles of the symbols. (These decision regions are far from optimal, but they simplify the problem.) *Hint:* $W(\cdot)$ depends on A_k.

Problem 15-3. Consider the discrete-time carrier recovery loop in Fig. 15-2 with a first-order loop filter $L(z) = K_L$. Suppose that the decisions made by the slicer are delayed by M samples (for example the slicer is replaced with a trellis decoder that has truncation depth M). This extra delay can be modeled as a loop filter

$$L(z) = z^{-M}K_L \tag{15.24}$$

instead of K_L. Suppose that $K_L = 0.1$.

(a) For no delay, $M = 0$, find the pole location of the transfer function to jitter $\Phi(z)/\Theta(z)$ and sketch the frequency response.

(b) Repeat for $M = 1$. Is this PLL useful?

(c) Find the poles when $M = 2$.

(d) Find M such that the loop becomes unstable (may require a computer program).

Problem 15-4.

(a) Show that there is no value of N such that (15.18) is satisfied for a 16-QAM signal.

(b) Find N such that $E[\,|A_k|^N] \neq 0$ for a 16-QAM signal.

Problem 15-5. Consider the design of a power of N carrier recovery loop for a 4-PSK signal with $|A_k| = 1$. Suppose that $f_c = 400$ Hz $\pm\, 2\%$.

(a) For the PLL in Fig. 15-8, find f_v that leads to the smallest required lock range for the PLL. What is the lock range (the range of frequency offset f_0 at the input to the phase detector)?

(b) Suppose you wish to design the PLL with $f_v = f_c = 2400$ Hz. Modify the design in Fig. 15-8 so that this will work.

Problem 15-6. Consider the passband PAM signal

$$R(t) = \mathrm{Re}\{e^{j(2\pi f_c t + \theta(t))}S(t)\}\,, \tag{15.25}$$

where

$$S(t) = \sum_{m=-\infty}^{\infty} A_m p(t - mT)\,. \tag{15.26}$$

(a) Show that

$$R^2(t) = \tfrac{1}{2}|S(t)|^2 + \tfrac{1}{2}\cos(4\pi f_c t + 2\theta(t))\mathrm{Re}\{S^2(t)\}\,. \tag{15.27}$$

(b) Show that for 4-PSK $\mathrm{E}[R^2(t)]$ has no periodicity at the carrier frequency or multiples of the carrier frequency. Hence power of 2 carrier recovery will not be a good choice for 4-PSK.

(c) Find conditions on A_k such that power of 2 carrier recovery will work, assuming that A_k is white.

References

1. L. E. Franks, "Synchronization Subsystems: Analysis and Design," in *Digital Communications: Satellite/Earth Station Engineering*, Prentice-Hall (1981).

2. D. D. Falconer, "Jointly Adaptive Equalization and Carrier Recovery in Two-Dimensional Digital Communication Systems," *Bell System Technical Journal*, Vol. 55 (3), (March 1976).

3. W. C. Lindsey and M. K. Simon, "Data-Aided Carrier Tracking Loop," *IEEE Trans. Communications*, Vol. COM-19 (4), pp. 157-168 (April 1971).

4. W. C. Lindsey and M. K. Simon, "Carrier Synchronization and Detection of Polyphase Signals," *IEEE Trans. Communications*, Vol. COM-20 (2), pp. 441-454 (Feb. 1972).

5. R. Matyas and P. J. McLane, "Decision-Aided Tracking Loops for Channels with Phase Jitter and Intersymbol Interference," *IEEE Trans. Communications*, Vol. COM-22 (8), pp. 1014-1023 (Aug. 1974).

6. M. K. Simon and J. K. Smith, "Offset Quadrature Communications with Decision-Feedback Carrier Synchronization," *IEEE Trans. Communications*, Vol. COM-22, (10), (Oct. 1974).

7. U. Mengali, "Joint Phase and Timing Acquisition in Data Transmission," *IEEE Trans. Communications*, Vol. COM-25 (10), pp. 1174-1185 (Oct. 1977).

8. H. Kurihara, R. Katoh, H. Komizo, and H. Nakamura, "Carrier Recovery Circuit with Low Cycle Skipping Rate for CPSK/TDMA Systems," *Proc. 5-th Int. Conf. Digital Satellite Communication*, (March 1981).

9. M. Oerder, G. Ascheid, R. Hab, and H. Meyr, "An All Digital Implementation of a Receiver for Bandwidth Efficient Communication," in *Signal Processing III: Theories and Applications*, Elsevier Science Publishers, B. V. (North-Holland) (1986).

10. R. D. Gitlin, "Adaptive Phase-Jitter Tracker," *United States Patent*, Patent No. 4,320,526, (March 16, 1982).

11. R. P. Gooch and M. J. Ready, "An Adaptive Phase Lock Loop for Phase Jitter Tracking," *Proceedings of the 21st Asilomar Conf. on Signals, Systems, and Computers*, (Nov. 2-4, 1987).

16

Timing Recovery

The purpose of timing recovery is to recover a clock at the symbol rate or a multiple of the symbol rate from the modulated waveform. This clock is required to convert the continuous-time received signal into a discrete-time sequence of data symbols.

Many digital systems transmit a clock separate from the data stream. This is commonly done in systems which range from the size of an integrated circuit up to perhaps the size of a room. For digital communication systems, however, the transmission of a separate clock would be inefficient, since it requires additional facilities, bandwidth, or power. Hence, it is more economical to implement the additional circuitry which is required to derive the clock from the received modulated waveform itself. This is called *self-timing*, and requires that the timing information be implicit in the received waveform, which is not necessarily the case for all signaling methods.

Example 16-1.

Consider a baseband system using an AMI line code, in which "0" bits are transmitted as a zero voltage, and "1" bits are transmitted alternately as positive or negative-going pulses. If the user transmits a long sequence of zeros, the transmitted waveform will be identically zero, and there will be no timing information.

The need to do timing recovery imposes additional requirements on the modulation technique which are not present where a separate clock is available. The strength of the timing information in a signal is affected by the statistics of the signal, the line code, and the pulse shape.

Example 16-2.

To correct the problem in the previous example, we can use a scrambler to randomize the data bits. In effect, we alter the statistics of the signal to ensure that a timing signal can be extracted at the receiver. The user can still foil the system by transmitting just the right bit sequence so that the output of the scrambler is zero, but for most practical cases, the probability of this occurring is negligible.

Practical timing recovery circuits cannot perfectly duplicate the clock used at the remote transmitter. The most basic requirement is that the average frequency of the derived timing must exactly equal the average frequency of the transmitted signal. Obviously, the receiver must only generate as many bits as were transmitted, over the long term. Although the average frequency of the derived timing must be exact, the timing signal usually has phase jitter, or *timing jitter*. Timing jitter is not a fundamental impairment, but can be reduced to any desired level. We will see that the only barrier is cost.

There are fundamentally two types of timing recovery techniques which we call *deductive* and *inductive*. Deductive timing recovery directly extracts from the incoming signal a *timing tone*, which has an average frequency exactly equal to the symbol rate. The timing tone is used to synchronize the receiver to the incoming digital information, as shown in Fig. 16-1. If the timing tone has unacceptable jitter, that jitter can be reduced using a PLL, as shown in Fig. 16-2. This PLL will usually be of the type considered in Example 14-1 rather than Example 14-1, in that its objective will be produce a single-frequency rather than to track the jitter.

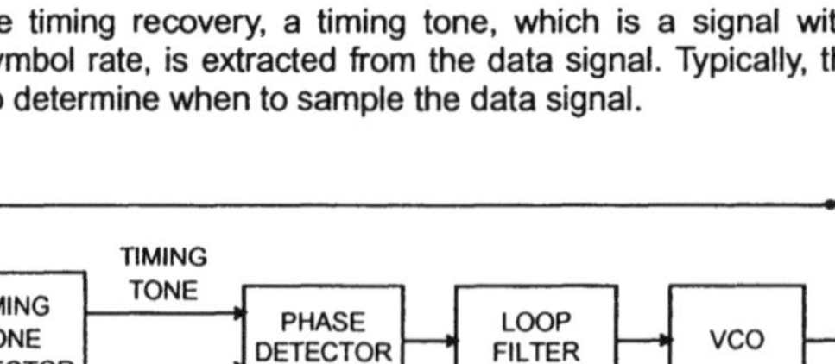

Fig. 16-1. In deductive timing recovery, a timing tone, which is a signal with average frequency is exactly equal to the symbol rate, is extracted from the data signal. Typically, the zero crossings of the timing tone are used to determine when to sample the data signal.

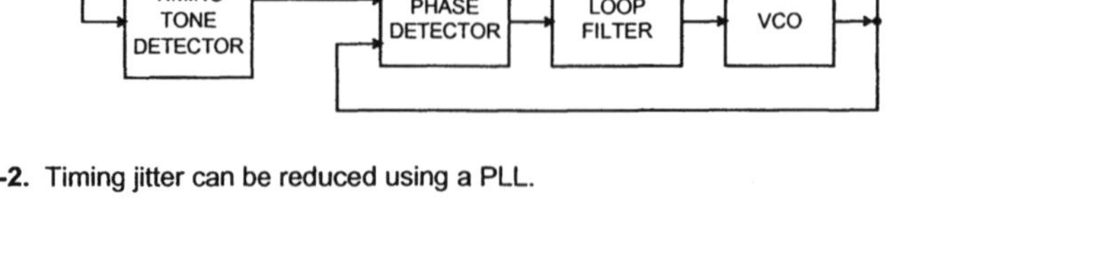

Fig. 16-2. Timing jitter can be reduced using a PLL.

Inductive timing recovery does not directly process the received signal to get a timing tone, but rather uses a feedback loop as shown in Fig. 16-3. Inductive timing recovery uses a PLL not as an added optimization to reduce timing jitter, but rather as an integral part of the method. One obvious advantage of this method is that most of the timing recovery can be done digitally and in discrete-time. A disadvantage is that the sampling rate of the received signal may have to be higher than the symbol rate in order to be able to estimate the timing error (although we will see baud-rate techniques that work). The phase detector in Fig. 16-3 is the *sampling phase detector* of Fig. 14-12.

First we establish the criteria by which we will evaluate timing recovery techniques, and then discuss the most popular deductive timing recovery technique, the *spectral-line method*. After this, we describe an inductive technique, *MMSE timing recovery* and approximations. Finally, we describe a class of inductive techniques, *baud-rate timing recovery*, that allows sampling at the symbol-rate.

16.1. Timing Recovery Performance

Comparing the performance of alternative timing recovery methods analytically is usually very difficult. Some of the criteria for performance are discussed in this section.

16.1.1. Timing Phase

It is necessary to know not only how often to sample the data bearing signal, but also where to sample it. The choice of sampling instant is called the *timing phase*. For some signal schemes and some receivers, the performance of the receiver is critically dependent on the timing phase. The sensitivity to timing phase can be quantified by examining the eye diagram of a signal (see Section 5.1.3). In Fig. 16-4 we show eye diagrams for 25% and 100% excess bandwidth raised cosine binary antipodal PAM signaling. Clearly, the 25% excess bandwidth signal is more sensitive to timing phase because the eye closes more rapidly as the timing phase deviates from the optimum. With zero excess bandwidth, the signal is infinitely sensitive to timing phase, because the horizontal eye opening becomes zero (see Problem 5-5). It should not be inferred, however, that all signals without excess bandwidth have zero eye width. It is

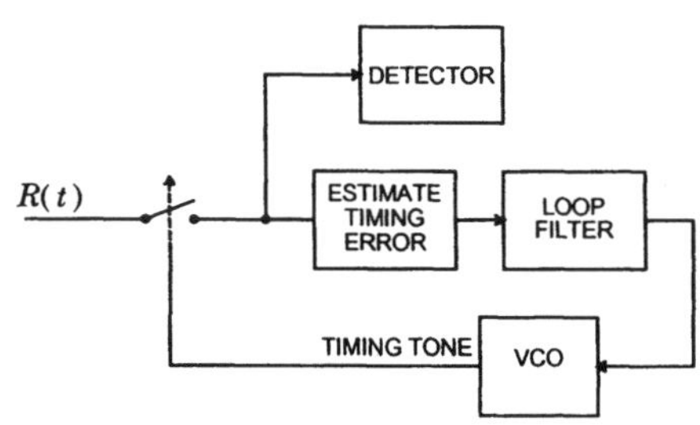

Fig. 16-3. In inductive timing recovery, a current estimate of the timing tone is used to sample the signal. Then the timing error is estimated and the timing tone estimate is updated. This is effectively a PLL.

possible, using partial response line coding, to construct a signal with no excess bandwidth and an open eye. In effect, through coding we can disallow the sequences A_m that lead to large ISI.

Example 16-3.

A zero excess bandwidth modified duobinary pulse $p_{\text{mdb}}(t)$ is formed by subtracting two sinc pulses $p_0(t) = \sin(\pi t/T)/(\pi t/T)$,

$$p_{\text{mdb}}(t) = p_0(t) - p_0(t-2T)\,, \tag{16.1}$$

as shown in Fig. 16-5. Modified duobinary can be used to get a signal with zero excess bandwidth and an open eye. The eye width has been computed by Kabal and Pasupathy [1] and is about 36% of the symbol interval for zero excess bandwidth. To understand intuitively how this can be so, note that the tails of the two sinc pulses tend to cancel each other, so the influence that the MDB pulse can have on distant symbols is considerably less than the influence that either sinc pulse alone would have.

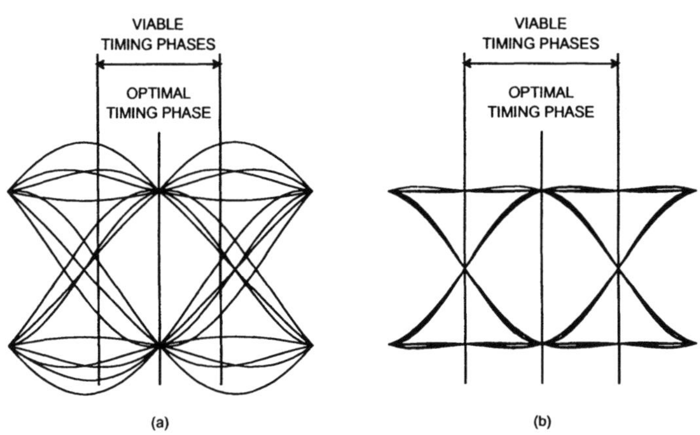

Fig. 16-4. Eye diagrams for (a) 25% and (b) 100% excess bandwidth raised cosine pulses. The vertical lines indicate the range of possible timing phases (sample points) such that positive and negative pulses can be distinguished. In each case, the optimal timing phase is in the center where the eye opening is greatest.

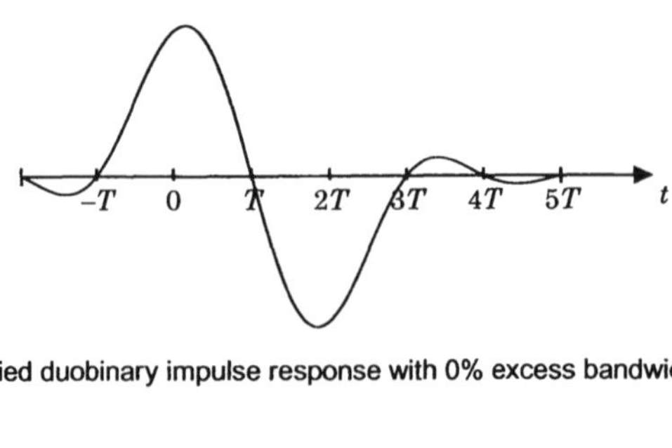

Fig. 16-5. The modified duobinary impulse response with 0% excess bandwidth.

Practical signals with zero excess bandwidth have an important advantage for timing recovery. Zero excess bandwidth signals are bandlimited to half the symbol rate frequency, so no aliasing occurs in sampling at the symbol rate. Even more remarkably, such a signal is *entirely insensitive* to timing phase if an adaptive equalizer (with enough taps) is used (see Problem 16-1). Thus, a fractionally-spaced equalizer (Section 9.4) is not required.

16.1.2. Timing Jitter

The timing tone recovered from a data signal always has timing jitter. This jitter can be reduced to any desired level by the design of timing recovery circuits, or by use of an additional phase-locked loop. Timing jitter introduces two degradations. First, the PAM signal is sampled at a sub-optimal point in the eye, increasing the ISI and thereby reducing noise immunity. Second, the bit stream emerging from the detector will generally have the same timing jitter produced by the timing recovery circuit. Usually, this is not a problem since consumers of the data (such as terminals or computers) are tolerant of limited amounts of jitter. When the data is a digital representation of a continuous-time signal such as speech, however, the jitter will cause the reconstructed speech to have non-uniform sampling, resulting in distortion. Timing jitter can have another serious consequence. For long distance transmission, a signal is often transmitted through a chain of many repeaters (Chapter 1). Each repeater reconstructs the digital signal and retransmits it, including the timing jitter on the input as well as timing jitter introduced in the repeater itself. The accumulation of timing jitter after a number of repeaters must be accounted for in the design of the timing recovery circuits (Section 16.5).

16.2. Spectral-Line Methods

A baseband PAM signal carrying discrete-time digital information

$$R(t) = \sum_{k=-\infty}^{\infty} A_k p(t - kT) \tag{16.2}$$

is not stationary. In fact, it is *cyclostationary*, meaning that its moments vary in time and are periodic with period T, the symbol interval. Consider the new random process

$$Z(t) = f(R(t)) , \tag{16.3}$$

where $f(\cdot)$ is a memoryless nonlinearity. Often we will find that the mean-value of $Z(t)$, $E[Z(t)]$, is non-zero and periodic with period T. We can think of this mean-value as a deterministic component of the random process $Z(t)$, consisting of a fundamental at the baud rate and harmonics. We can exploit this fact for timing recovery by forming $Z(t)$ and passing it through a bandpass filter centered at the baud rate. This is known as *spectral-line timing recovery.*

If the mean-value of $R(t)$ (corresponding to the identity function $f(x) = x$) contains a spectral line at the baud rate, we can simply pass $R(t)$ itself through a bandpass filter to obtain a timing waveform. This is known as the *linear spectral-line method*, and it is discussed in

Section 16.2.1. Most often, however, the mean-value is zero but higher moments of $R(t)$ are periodic. When $f(t)$ is nonlinear, this is called the *nonlinear spectral-line method*, and is discussed in Section 16.2.2

16.2.1. The Linear Spectral Line Method

When the mean-value of the data symbols is non-zero, the baseband PAM waveform may contain a spectral line at the baud rate. To determine whether a spectral line at the symbol frequency exists for a particular signal (16.2), partition the signal into a data independent (deterministic) component and a data dependent, zero mean stochastic component

$$R(t) = \mathrm{E}[A_k]\sum_{m=-\infty}^{\infty} p(t-mT) + \sum_{m=-\infty}^{\infty} (A_m - \mathrm{E}[A_k])p(t-mT). \tag{16.4}$$

The first term is independent of the data A_k, is periodic with period T, and can be thought of as a deterministic timing signal. Its periodicity implies a fundamental at the symbol rate $1/T$ and harmonics. If this fundamental has non-zero amplitude, it can be extracted with a bandpass filter, producing a timing tone. The second (data dependent) term is zero-mean and random, and results in jitter on the recovered timing tone.

Example 16-4.

Consider a binary on-off signal with alphabet $A_k \in \{0, 1\}$ in (16.2). Let the received pulse shape be a 50% duty cycle square pulse,

$$p(t) = \begin{cases} 1, & \text{for } t \in [0, T/2), \\ 0, & \text{otherwise} \end{cases} \tag{16.5}$$

Using this pulse shape, and observing that $E[A_k] = \frac{1}{2}$, the deterministic part of (16.4) is shown in Fig. 16-6. Clearly this signal has a strong spectral component at the symbol frequency.

Even when the mean-value of $R(t)$ is non-zero, it does not contain a spectral line at the baud rate unless the excess bandwidth is at least 100%.

Exercise 16-1.

Define $x(t)$ to be the deterministic term of (16.4),

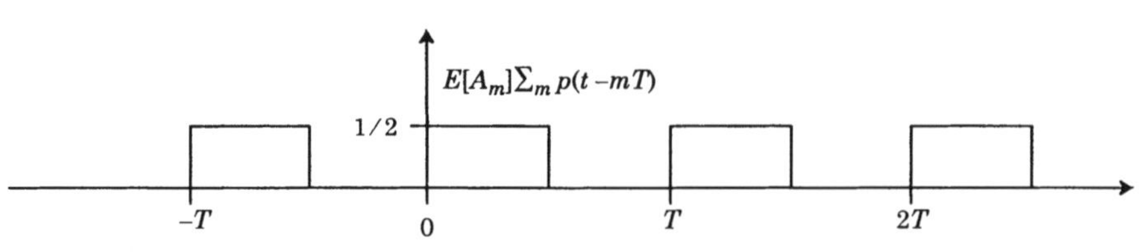

Fig. 16-6. The deterministic part of a binary on-off signal when the pulse shape is a 50% duty cycle square pulse.

$$x(t) = \mathrm{E}[A_k] \sum_{k=-\infty}^{\infty} p(t-kT) \,. \tag{16.6}$$

Show that the Fourier transform $P(f)$ of the pulse must be nonzero at $f = \pm 1/T$ in order for $X(f)$ to have a component at $f = \pm 1/T$. *Hint:* The results of Appendix 16-A may be helpful.

The requirement that the data symbols A_k have a non-zero mean (which implies a power penalty, Section 5.2.2), and that the excess bandwidth be at least 100%, usually rules out the linear spectral line method.

16.2.2. The Nonlinear Spectral Line Method

Often, even when the mean-value of $R(t)$ is zero, the second and higher moments are non-zero and periodic. In 1958, in his classic paper on self-timing, Bennett observed that this can be exploited by passing the received signal though a memoryless nonlinearity [2]. The resulting waveform often has a deterministic mean-value that is periodic in the symbol rate, and a timing tone can be derived from it using a bandpass filter. To illustrate this, we will use the magnitude-squared nonlinearity $f(x) = |x|^2$. Assume a baseband PAM waveform of the form of (16.2), and for generality assume that A_k and $p(t)$ are both complex-valued. The magnitude squared of this process depends on the correlation function of the data symbols, so assume they are white,

$$E[A_m A_n^*] = \sigma_A^2 \delta_{m-n} \,. \tag{16.7}$$

This assumption is reasonable, particularly when a scrambler is used. It is then straightforward to show that

$$E[\,|R(t)|^2] = \sigma_A^2 \sum_{m=-\infty}^{\infty} |p(t-mT)|^2 \,. \tag{16.8}$$

This expected value can again be considered a deterministic component of the process $|R(t)|^2$, and it is obviously periodic with period T, the symbol rate. The fundamental of this signal, if it has non-zero amplitude, will be extracted by the bandpass filter in Fig. 16-7 to yield a timing tone at the symbol frequency. Some of the random component in $|R(t)|^2$ will pass through the bandpass filter and result in timing jitter. In this case, any non-zero excess bandwidth is sufficient to guarantee a non-zero fundamental at the baud rate.

Exercise 16-2.

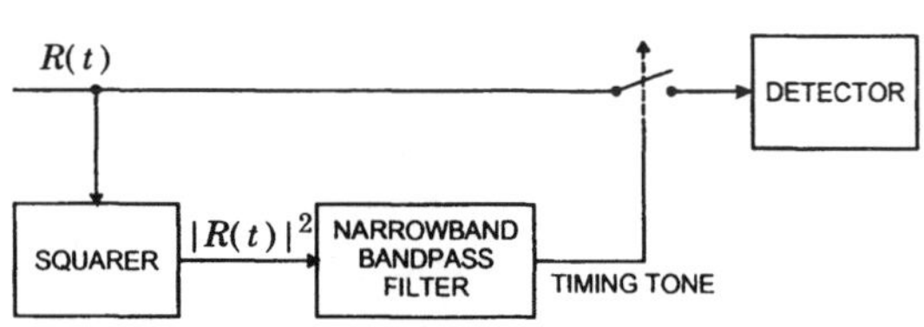

Fig. 16-7. The squarer produces a deterministic periodic component, and the bandpass filter extracts the symbol frequency.

(a) Using the *Poisson's sum formula* (Appendix 16-A), show that

$$\sum_{m=-\infty}^{\infty} |p(t-mT)|^2 = \frac{1}{T}\sum_{n=-\infty}^{\infty} Z_n e^{j2\pi nt/T}, \tag{16.9}$$

where

$$Z_n = \int_{-\infty}^{\infty} P(f)P^*\left(f-\frac{n}{T}\right)\,df\,. \tag{16.10}$$

Note that $|Z_1| > 0$ as long as the bandwidth of $P(f)$ is greater than half the baud rate, $1/(2T)$.

(b) Show that $Z_{-n} = Z_n^*$ for all n.

Usually $p(t)$ has less than 100% excess bandwidth, in which case it is easy to show that the Fourier series coefficients are all zero except for the d.c. and the fundamental, and thus

$$E[|R(t)|^2] = Z_0 + 2\mathrm{Re}\{Z_1 e^{j2\pi t/T}\}\,. \tag{16.11}$$

In this practical case the timing function contains no higher harmonics, although the bandpass filter is still required to reject as much as possible of the random portion of $|R(t)|^2$ and any other noise or interference present.

Example 16-5.

Consider a received real-valued baseband signal formed with a binary antipodal signal constellation $\{\pm a\}$. Since the signal is real-valued, the magnitude-squared signal is the same as the squared signal, and

$$R^2(t) = \sum_{m=-\infty}^{\infty} A_m{}^2 p^2(t-mT) + \Sigma_n\Sigma_{m\neq n} A_n A_m p(t-nT)p(t-mT)\,. \tag{16.12}$$

Since $A_m{}^2 = a^2$, the first term is the deterministic timing tone, and the second term is the random portion that will be reflected in jitter at the bandpass filter output. There is no requirement that the data symbols have a non-zero mean, as in the linear spectral line method.

It is generally desirable to have a larger timing tone, and hence large $|Z_1|$, since then the random components and noise will contribute relatively less to the timing jitter. Observe from (16.10) that the size of $|Z_1|$ can generally be expected to increase as the amount of excess bandwidth increases, since there will be a larger overlap between $P(f)$ and $P(f-n/T)$. We conclude therefore that greater excess bandwidth is generally favorable, and doubly so when coupled with the fact that the eye usually becomes less sensitive to timing jitter (wider eye) with larger excess bandwidth.

The analysis of the timing jitter is a little tricky. It may be tempting to compare the power in the deterministic timing tone to the power in the remaining random component that gets through the narrowband filter. We could find the power spectrum of the random part using the results of Appendix 3-A. Appendix 3-A assumes a random phase epoch in order to get a WSS random process. In timing recovery, we are exploiting precisely the fact that the PAM signal is not WSS to derive the timing tone, and assuming wide-sense stationarity to determine timing jitter leads to significant inaccuracies. Accurate techniques have been developed for assessing the amount of timing jitter compared to the strength of the timing tone [3].

So far we have considered only a magnitude-squared nonlinearity, but there are other possibilities. For some signals, particularly when the excess bandwidth is low, a fourth-power nonlinearity $|R(t)|^4$ is better than the magnitude-squared. In fact, fourth-power timing recovery can even extract timing tones from signals with zero excess bandwidth. An alternative nonlinearity is the magnitude $|R(t)|$, which for a real-valued signal is easily implemented as a full-wave rectifier. Although the analysis is more difficult, simulations show that absolute-value circuits usually out-perform square-law circuits for signals with low excess bandwidth [4]. Like fourth-power circuits, absolute-value circuits can extract a timing tone for signals with zero excess bandwidth. For signals with low excess bandwidth, however, a fourth-power timing circuit may be preferable. Simulations for QPSK [4] suggest that fourth-power circuits out-perform absolute-value circuits for signals with less than about 20% excess bandwidth.

If timing recovery is done in discrete-time, aliasing must be considered in the choice of a nonlinearity. Any nonlinearity will increase the bandwidth of the PAM signal, and in continuous time this is not a consideration since the bandpass filter will reject the higher frequency components. In the presence of sampling, however, the high frequency components due to the nonlinearity can alias back into the bandwidth of the bandpass filter, resulting in additional timing jitter. This is obviously true if the nonlinearity precedes the sampling, but is also true even if the nonlinearity is performed after sampling. This is because preceding a sampling operation with a memoryless nonlinearity is mathematically equivalent to reversing the order of the sampling and the nonlinearity. Therefore, in a discrete-time realization, a magnitude-squared nonlinearity usually has a considerable advantage over either absolute-value or fourth-power nonlinearity. In particular, raising a signal to the fourth-power will quadruple its bandwidth, and full-wave rectifying spreads its bandwidth even more. Each situation must be considered independently to determine whether the aliasing is detrimental.

The advantages of absolute-value and fourth-power circuits over square-law circuits can be at least partly compensated by *prefiltering* of the PAM signal prior to the nonlinearity. This prefilter can reduce the timing jitter substantially, particularly for low excess bandwidth. Prefiltering is based on the observation that often much of the spectrum of the PAM signal does not contribute to the timing tone, so a filter that eliminates the unnecessary part of the spectrum will reduce timing jitter. This is best seen by example.

Example 16-6.

Consider a real-valued baseband PAM signal with 12.5% excess bandwidth and a symbol rate of 1800 baud. We will show that the timing tone of a square law timing recovery circuit is not altered by *prefiltering* the signal using an ideal bandpass filter with passband from 787.5 Hz to 1012.5 Hz. Such prefiltering, however, removes much of the noise and signal components that would only contribute to the jitter. The strength of the timing tone will be determined by the Fourier series coefficient Z_1. This can be found using (16.10), and in particular

$$Z_1 = \frac{1}{T}E[|A_k|^2]\int_{-\infty}^{\infty} P\left(\frac{1}{T}-f\right)P(f)df\,. \tag{16.13}$$

$P(f)$ is shown in Fig. 16-8(a). To get Z_1, the spectrum in Fig. 16-8(a) is multiplied by the spectrum in Fig. 16-8(b) and the product is integrated, per (16.13). Only the cross-hatched regions will contribute to Z_1, which will not change if the received signal is first filtered with the prefilter shown in Fig. 16-8(c). Since the receiver must work with $|R(t)|^2$ and not its expected value, parts of the PAM signal that would contribute to the jitter and not to the timing tone have been removed, as well as undoubtedly some noise components. The bandwidth of the prefilter depends on the excess bandwidth of the pulse. Given 12.5% excess bandwidth, the ideal prefilter bandwidth is 225 Hz, as shown.

Prefiltering is not as helpful for signals with large excess bandwidths (see Problem 16-5).

16.2.3. The Spectral Line Method for Passband Signals

For a baseband PAM system, the previous results apply directly. For a passband PAM system, since a complex-valued baseband signal was considered, they would also apply if demodulation was performed first. However, there are a couple of problems with this approach:

- As described in Chapter 15, demodulation is usually performed last, after timing recovery and equalization.
- Carrier recovery loops are typically decision-directed, and thus depend for their operation on availability a stable timing phase. If the timing recovery depends on carrier recovery, and vice versa, there might be serious problems in initial acquisition.

Fortunately, it is not necessary to demodulate before performing timing recovery. By deriving timing directly from a passband signal, we completely decouple timing recovery from other

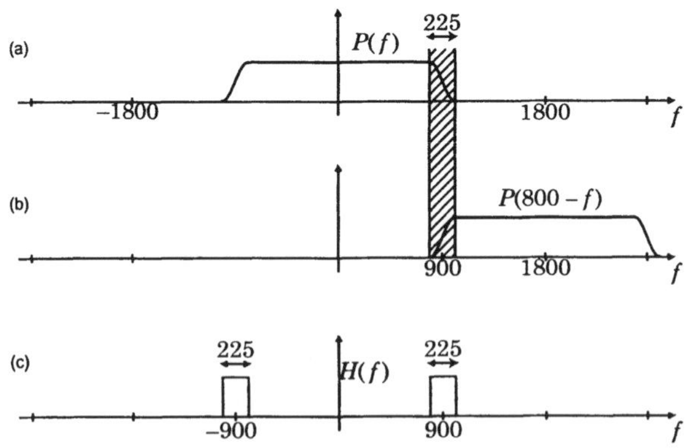

Fig. 16-8. Illustration of the computation of the Fourier series coefficient Z_1 for a baseband PAM signal with a pulse whose Fourier transform is shown in (a). To get Z_1, the function in (a) is multiplied term-wise by the function in (b) and the product is integrated. Only the cross-hatched regions will contribute to the answer. In (c), the prefilter shown applied to PAM signal prior to the square nonlinearity would not affect Z_1 but would reduce jitter.

functions in the receiver. For the spectral-line method, a passband signal can be passed directly through a non-linearity such as a magnitude-square, and a timing tone that can be bandpass filtered will result. This technique is sometimes called *envelope derived timing*, because the squaring and filtering operation is similar to extracting the *envelope* of the signal.

The simplest case results when we have an analytic bandpass filter, or equivalently bandpass filter and phase splitter, at the front end of the receiver (Section 5.3.4). The output of this front end will be an analytic signal

$$Y(t) = R(t)e^{j2\pi f_c t}, \tag{16.14}$$

where $R(t)$ is a complex-valued baseband signal as given by (16.2). Since

$$|Y(t)| = |R(t)| \cdot |e^{j2\pi f_c t}| = |R(t)|, \tag{16.15}$$

we arrive at the conclusion that applying the analytic signal to a magnitude, magnitude-square, or fourth-power nonlinearity is equivalent to applying the demodulated baseband signal.

However, there are some possible difficulties with deriving timing from the analytic signal.

Example 16-7.
Often the front-end of the receiver consists of a bandpass filter, sampler, and discrete-time phase splitter. The sampler would usually be controlled by the derived timing. Thus, the analytic signal depends on the timing recovery output through the sampling phase and frequency, and using this signal to derive timing may lead to undesirable interactions and acquisition problems.

Fortunately, there is a simple alternative to using the analytic signal; namely, to apply just the real part $\mathrm{Re}\{Y(t)\}$ to the nonlinearity. Since $\mathrm{Re}\{Y(t)\}$ is the phase splitter input, and not output, this approach eliminates any problems encountered in Example 16-7. Considering again the squarer nonlinearity,

$$(\mathrm{Re}\{Y(t)\})^2 = \tfrac{1}{2}|Y(t)|^2 + \tfrac{1}{4}Y^2(t) + \tfrac{1}{4}(Y^*(t))^2 . \tag{16.16}$$

Under reasonable assumptions, the expected value of the last two terms in (16.16) is zero.

Exercise 16-3.
Assume the zero-mean complex-valued data symbols A_k are independent of one another and have independent real and imaginary parts with equal variance. Show that $\mathrm{E}[A_m A_n] = 0$ for all m and n. (If this result looks strange, recall that the variance of the symbols is $\mathrm{E}[|A_k|^2]$ and not $\mathrm{E}[A_k^{\,2}]$).

It follows directly from Exercise 16-3 that $\mathrm{E}[Y^2(t)] = 0$ under these same assumptions on the data-symbol statistics. Thus, we get that

$$E[(\mathrm{Re}\{Y(t)\})^2] = \tfrac{1}{2}\mathrm{E}[|Y(t)|^2] , \tag{16.17}$$

and the square of the real part of the analytic signal contains the same timing function (albeit half as large) as the magnitude-squared of the analytic signal. However, we do pay a price:

- The timing function is half as large, making sources of jitter more important.

- In (16.16), the second and third terms are additional inputs to the bandpass filter. These terms have zero mean, and hence do not affect the timing function. However, they do represent an additional source of randomness that contributes to a larger timing jitter.

The same relative merits of squaring, absolute-value, and fourth-power techniques apply to passband timing recovery as to baseband. In particular, absolute-value and fourth-power are usually better than squaring, except when aliasing is a problem in discrete-time implementations. As with baseband signals, it is sometimes advantageous to prefilter the signal before squaring.

16.3. MMSE Timing Recovery and Approximations

Although the spectral-line method is the most popular, it is not always suitable. It can be difficult to use when the timing recovery has to be done in discrete-time.

Example 16-8.

When an echo canceler is used is used to separate the two directions in a full-duplex connection (Chapter 17), the inherently discrete-time echo cancellation must be performed prior to timing recovery. It is often convenient to avoid reconstruction of a continuous-time signal, and perform timing recovery in discrete time. The echo canceler complexity is proportion to the sampling rate, and therefore there is motivation to minimize the sampling rate for the timing recovery function.

In this section we describe *minimum mean-square error (MMSE)* timing recovery, a technique that is not practical in its exact formulation, but which has numerous practical approximations. MMSE timing recovery is also sometimes called *LMS timing recovery*, and is similar to *maximum likelihood* timing recovery [5]. The MMSE family of techniques fit the general framework of Fig. 16-3, so they are *inductive*. The phase detector is essentially the *sampling phase detector* of Fig. 14-12, the main difference being that the input signal has a more complicated form than the simple sinusoid assumed in Chapter 14.

16.3.1. The Stochastic Gradient Algorithm

In Fig. 16-9, the received signal (assumed baseband) is sampled at times $(kT + \tau_k)$. The symbol interval is T, and thus τ_k represents the timing error in the k-th sample. After some possible front-end processing, the notation for the k-th sample is $Q_k(\tau_k)$, rather than just Q_k, to emphasize the dependence on the timing phase. Usually, τ_k is determined by zero crossings of the timing tone (see Problem 16-6). Ideally, τ_k is a constant corresponding to the best sampling phase, but in practice τ_k has timing jitter. Inductive timing recovery is best understood as a technique for iteratively adjusting τ_k.

MMSE timing recovery adjusts τ_k to minimize the expected squared error between the input to the slicer and the correct symbol,

$$E[\,|E_k(\tau_k)|^2] = E[\,|Q_k(\tau_k) - A_k|^2] \ , \tag{16.18}$$

with respect to the timing phase τ_k, where A_k is the correct data symbol. This is the same criterion used for adaptive equalizers in Chapter 9. Just as with adaptive equalizers, MMSE timing recovery can use the stochastic gradient algorithm to try to find the optimal timing phase. If correct data symbols A_k are available at the receiver (for example during a training sequence at system startup), the structure is shown in Fig. 16-9. This criterion directly minimizes the slicer error, and hence should result in close to the minimum error probability. Unfortunately, the input to the slicer $Q_k(\tau_k)$ is a complicated non-linear function of the timing phase τ_k, so unlike the adaptive equalizer case there may not be a well-defined unique minimum MSE timing phase. In addition, finding an analytic closed-form solution may be impossible.

Instead of seeking a closed-form solution, we can try to minimize the expected squared error by adjusting the timing phase in the direction opposite the derivative of the expected value of the squared error,

$$\frac{\partial}{\partial\tau_k}\mathrm{E}[\,|E_k(\tau_k)|^2] = \mathrm{E}\left[\frac{\partial}{\partial\tau_k}|E_k(\tau_k)|^2\right]. \tag{16.19}$$

Exercise 16-4.

Show that for any complex function $E_k(\tau_k)$ of the real variable τ_k,

$$\frac{\partial}{\partial\tau_k}|E_k(\tau_k)|^2 = 2Re\left\{ E_k(\tau_k)^*\frac{\partial E_k(\tau_k)}{\partial\tau_k} \right\}. \tag{16.20}$$

Since A_k does not depend on τ_k, we can write

$$\frac{\partial E_k(\tau_k)}{\partial\tau_k} = \frac{\partial Q_k(\tau_k)}{\partial\tau_k}. \tag{16.21}$$

Hence, to adjust the timing phase in the direction opposite the gradient, we use

$$\tau_{k+1} = \tau_k - \alpha\mathrm{Re}\left\{ E_k(\tau_k)^*\frac{\partial Q_k(\tau_k)}{\partial\tau_k} \right\}. \tag{16.22}$$

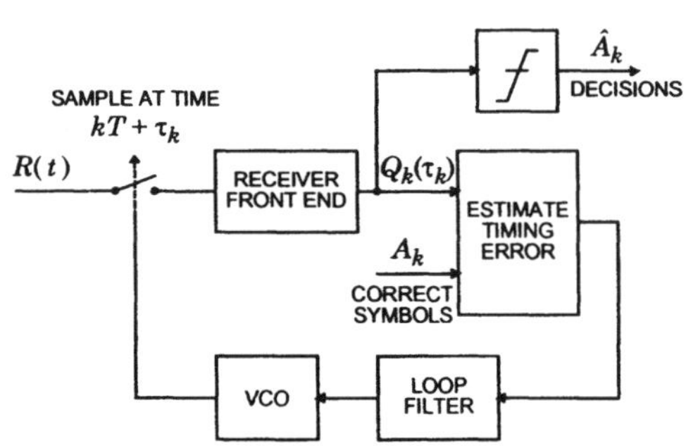

Fig. 16-9. MMSE timing recovery adjusts the timing phase to minimize the squared error between the input to the slicer and the correct symbols.

We have dispensed with the expectation in (16.19), so this is a *stochastic gradient algorithm*. The *step size* α is usually determined empirically to ensure stability, minimize timing jitter, and ensure adequate tracking ability.

The signal $Q_k(\tau_k)$ consists of samples of some continuous-time signal $Q(t)$ taken at $t = kT + \tau_k$, so

$$\frac{\partial Q_k(\tau_k)}{\partial \tau_k} = \left[\frac{\partial Q(t)}{\partial t}\right]_{t = kT + \tau_k}. \tag{16.23}$$

Hence (16.22) is equivalent to

$$\tau_{k+1} = \tau_k - \alpha \mathrm{Re}\left\{ E_k(\tau_k)^* \left[\frac{\partial Q(t)}{\partial t}\right]_{t = kT + \tau_k} \right\}$$

$$= \tau_k - \alpha \mathrm{Re}\left\{ [Q_k(\tau_k) - A_k]^* \left[\frac{\partial Q(t)}{\partial t}\right]_{t = kT + \tau_k} \right\}. \tag{16.24}$$

This update is shown in Fig. 16-10.

As with equalization, there would normally be an acquisition or training period during which the symbols A_k are available, followed by a decision-directed tracking period during which the actual symbols A_k are replaced by decisions $\hat{A}_k$. Furthermore, since $E_k(\tau_k)$ is not a linear function of τ_k, $|E_k(\tau_k)|^2$ is not a quadratic function of τ_k, and the stochastic gradient algorithm is not guaranteed to converge to the optimal timing phase. In practice, some other technique should be used during acquisition to get a reasonable initial estimate of the correct timing phase [6]. This has the disadvantage of requiring two different timing recovery techniques during acquisition and tracking.

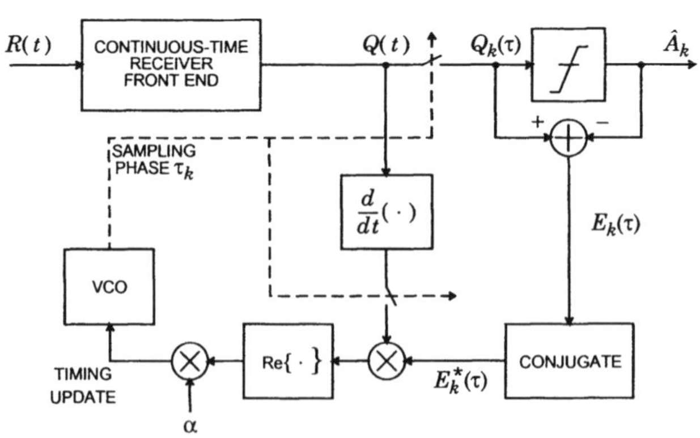

Fig. 16-10. Stochastic gradient timing recovery using a continuous-time version of the input to the slicer.

The method of (16.24) still doesn't accomplish the basic objective of using only *samples* of the received signal, since the receiver front end up to the slicer must be implemented in continuous-time. One approach is to differentiate the continuous-time received signal $R(t)$, sampling this derivative and processing it with a replica of the receiver front end [5], but this approach requires *two* receiver front ends and is often incompatible with echo cancellation. More approximations are needed to eliminate the need for a continuous-time version of the slicer input, as discussed in the next subsection.

The stochastic gradient timing recovery operates on the input to slicer, which has been equalized and demodulated. It therefore relies on the convergence of the equalizer (Chapter 9) and, for passband PAM, the acquisition of carrier (Chapter 15). The equalizer and carrier recovery themselves usually assume that the timing recovery has acquired! This potentially leads to a serious interaction among these three functions. This problem is overcome at least partially if the timing phase is estimated at startup by some other method not requiring equalizer convergence [6]. More recently [7] the interaction with carrier acquisition in passband systems has been eliminated by modifying the error criterion to

$$E_k = |Q_k(\tau_k)|^2 - |\hat{A}_k|^2 \ . \tag{16.25}$$

For passband PAM signals, an incorrect carrier phase will only *rotate* the received signal $Q_k(\tau_k)$ in the complex plane; its magnitude squared is not affected. Hence timing recovery based on this error criterion will not be sensitive to carrier phase.

16.3.2. Other Approximate MMSE Techniques

The need for a continuous-time version of the slicer input, $Q(t)$, can be avoided by one of two techniques. The first, published by Qureshi in 1976, assumes that Nyquist-rate samples of $Q(t)$ are available [6]. Since the derivative is an LTI system, it can be realized as a discrete-time filter. In Appendix 16-B we show that if $Q(t)$ is sampled at the Nyquist rate, then

$$\frac{\partial Q_k(\tau_k)}{\partial \tau_k} = Q_k(\tau_k) * d_k \,, \tag{16.26}$$

where d_k is given by (16.51). It is necessary to assume that τ_k varies slowly for this to be valid (see the appendix). This suggests that if the received signal is sampled at the Nyquist rate, timing recovery can be implemented as shown in Fig. 16-11. The derivative is computed by a discrete-time filter with impulse response d_k, which for practical reasons is approximated by an FIR filter. A particularly simple approximation is given by (16.52), yielding

$$\tau_{k+1} = \tau_k + \alpha Z_k(\tau_k) \,, \tag{16.27}$$

where

$$Z_k(\tau_k) = -\mathrm{Re}\left\{ [Q_k(\tau_k) - \hat{A}_k][Q_{k+1}(\tau_k) - Q_{k-1}(\tau_k)] \right\} \ . \tag{16.28}$$

The assumption of Nyquist sampling often complicates the receiver front end, which may otherwise get away with a lower sampling rate. There are exceptions to this.

Example 16-9.

Using partial response signaling like the modified duobinary response of Example 16-3, it is possible to use zero excess bandwidth. In this case, the symbol rate is the Nyquist rate, and this method imposes no computational penalty.

Example 16-10.

It is common to use fractionally-spaced equalizers (Section 9.4) so that the receiver is relatively insensitive to timing phase. In this case, the equalizer itself may operate on Nyquist-rate samples. Hence all parts of the receiver that precede the equalizer must operate on Nyquist-rate samples. However, there is still a penalty in complexity for the technique in Fig. 16-11. Usually, the output of a fractionally spaced equalizer is immediately decimated down to the symbol rate. This means that the output of the equalizer needs only to be computed at the symbol rate. To use this timing scheme, we would have to compute the output of the equalizer at the Nyquist rate.

A second approach to avoiding a continuous-time front-end to the receiver is explored in [7]. From (16.20), it is sufficient to directly compute the derivative of the *error signal* with respect to timing phase. This derivative can be approximated by taking each sample either slightly ahead or slightly behind the current estimate τ_k of the timing phase. The two phases are alternated, and the difference between the error at even samples and the error at odd samples is an indication of the derivative of the error with respect to timing phase.

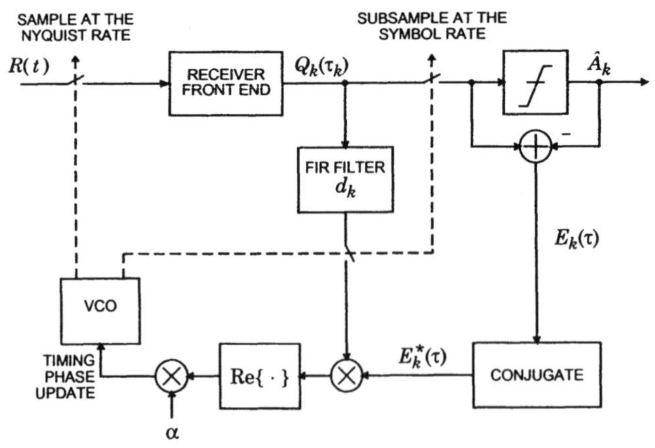

Fig. 16-11. Stochastic gradient timing recovery when the received signal is sampled at the Nyquist rate.

16.3.3. Approximate Maximum-Likelihood Methods

Many *ad hoc* timing recovery techniques have appeared over time, mostly justified first heuristically and then empirically. Several of these have been shown to be approximations to a *maximum likelihood* (ML) timing recovery [5][8], and bear a strong resemblance to MMSE methods.

One such method, called the *sample-derivative method,* is shown in Fig. 16-12 [9] for a baseband PAM system. This technique can be justified heuristically by observing that it will attempt to move the sampling phase until the derivative is zero, which occurs at the peaks of the signal, presumably a good (but not optimal) place to sample. For discrete time systems, approximations to the derivative (Appendix 16-B) are used.

A related technique, shown in Fig. 16-13, is called the *early-late gate method* [8]. In this technique, the received waveform is sampled two extra times, once prior to the sampling instant by $\Delta/2$ and once after the sampling instant by the same amount. The sampling instant is adjusted until the two extra samples are equal.

16.4. Baud-Rate Timing Recovery

Interest in realizing timing recovery using symbol-rate sampling, especially in conjunction with echo cancellation, has led to interest in a class of techniques known as *baud-rate timing recovery.* Interestingly, the timing phase update given by (16.27) works even when the samples

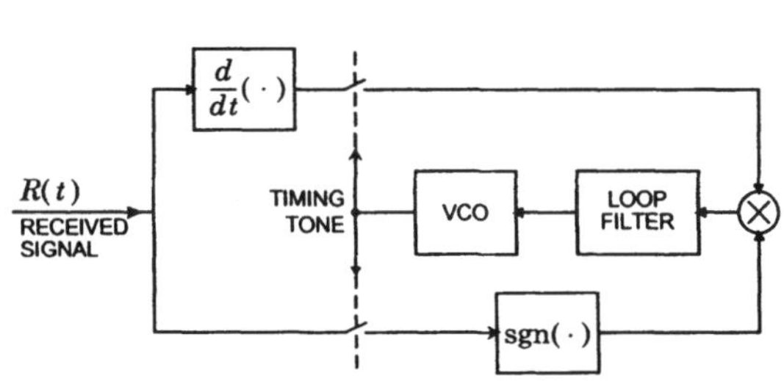

Fig. 16-12. Sample-derivative timing recovery. The box labeled sgn(·) computes the signum of the input, and can be implemented as a hard limiter.

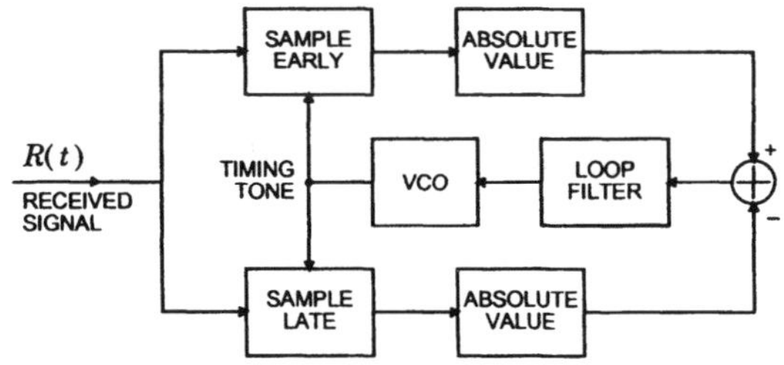

Fig. 16-13. Early-late gate timing recovery.

are taken at the symbol rate and not at the Nyquist rate! First we will justify this claim, and then describe a family of related baud-rate techniques that are similar to but better than (16.27). Any of the baud-rate techniques derived in this section can be used in the receiver configuration shown in Fig. 5-21. An additional frequency synthesizing PLL may be required to generate the multiple sampling frequencies.

Define the *timing function* to be the expected timing update, given the current timing phase τ_k,

$$f(\tau_k) = E[Z_k(\tau_k)] \ . \tag{16.29}$$

From (16.28),

$$f(\tau_k) = -\mathrm{Re}\left\{ \mathrm{E}[Q_k(\tau_k)Q_{k+1}(\tau_k)] - \mathrm{E}[Q_k(\tau_k)Q_{k-1}(\tau_k)] - \mathrm{E}[\hat{A}_k Q_{k+1}(\tau_k)] + \mathrm{E}[\hat{A}_k Q_{k-1}(\tau_k)] \right\}. \tag{16.30}$$

If $Q_k(\tau_k)$ is WSS random process, and τ_k varies insignificantly from one sample to the next, then the first two terms are equal, and

$$f(\tau_k) = \mathrm{Re}\left\{ \mathrm{E}[\hat{A}_k Q_{k+1}(\tau_k)] - \mathrm{E}[\hat{A}_k Q_{k-1}(\tau_k)] \right\} . \tag{16.31}$$

We can show, under benign assumptions for a baseband PAM signal, that $Q_k(\tau_k)$ is WSS.

Exercise 16-5.

Write the input to the slicer as samples of a continuous-time PAM signal, with the kth sample taken at time $t = kT + \tau_k$,

$$Q_k(\tau_k) = \sum_{m=-\infty}^{\infty} A_m p((k-m)T + \tau_k) + N_k \ . \tag{16.32}$$

Assume that A_k and N_k in (16.32) are WSS random processes, and that N_k is zero mean and independent of A_k. Show that $Q_k(\tau_k)$ is WSS. Show that under these assumptions, and also assuming that $Q_k(\tau_k)$ is real-valued, the first two terms in (16.30) are equal and hence cancel.

Exercise 16-6.

Assume that all the decisions are correct, $\hat{A}_k = A_k$, that A_k is white, and that $p(t)$ is real. Show that the timing function is

$$f(\tau_k) = \mathrm{E}[\,|A_k|^2][p(\tau_k + T) - p(\tau_k - T)] \ . \tag{16.33}$$

If $p(t)$ is symmetric about zero, the timing function will be zero at time zero, $f_k(0) = 0$. This is the condition under which the *average* timing phase update in (16.27) is zero, and hence the point to which the timing recovery algorithm will converge. In the case of a symmetric $p(t)$ this is also a good timing phase.

For a given pulse $p(t)$, it is a good idea to check the timing function in (16.33) to ensure that it is monotonic and has a unique zero-crossing, and that this zero-crossing is at a good timing phase. Equation (16.33) is plotted in Fig. 16-14 for raised cosine pulses with various excess bandwidths. Notice that if τ_k is early, the timing function is positive, so the expected timing phase update is positive, which is what we want.

The timing function was derived without assuming Nyquist-rate sampling, so this technique should work with symbol-rate samples. However, it can be improved. In Exercise 16-5 we showed that the first two terms in (16.30) cancel. However, since the timing phase update is Z_k not $E[Z_k]$, these two terms will contribute to the timing jitter. A closely related baud-rate technique that does not have these two terms was given by Mueller and Müller in 1976 [10]. In fact, they gave a general technique that can yield a variety of timing functions, of which (16.33) is only one. Another suitable one is

$$f(\tau_k) = \mathrm{E}[\,|A_k|^2]p(\tau_k + T)\ . \tag{16.34}$$

The general technique uses timing updates of the form

$$Z_k = \mathbf{X}_k'\mathbf{Q}_k\ , \tag{16.35}$$

in (16.27), where $\mathbf{Q}_k$ is a vector of the last m samples and $\mathbf{X}_k$ is a (yet unspecified) vector function of the last m symbols

$$\mathbf{Q}_k = \begin{bmatrix} Q_{k-m+1}(\tau_k) \\ \vdots \\ Q_k(\tau_k) \end{bmatrix}, \qquad \mathbf{X}_k = \begin{bmatrix} x_1(A_{k-m+1}, \ldots, A_k) \\ \vdots \\ x_m(A_{k-m+1}, \ldots, A_k) \end{bmatrix}. \tag{16.36}$$

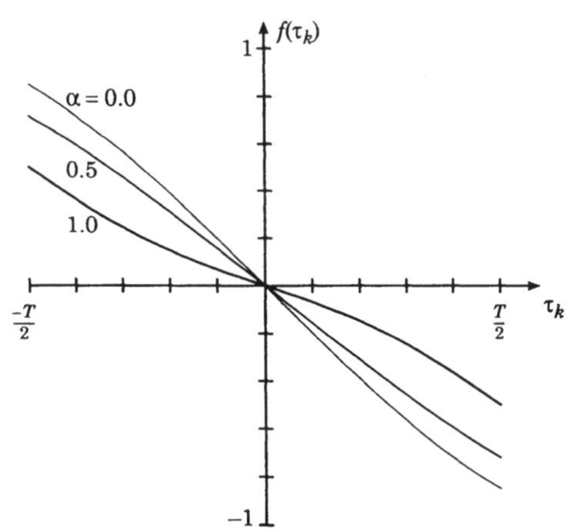

Fig. 16-14. The timing function $f(\tau_k)$ is the expected value (over all possible symbol sequences) of the timing phase update as a function of the timing phase τ_k. It is shown here for the timing phase update in (16.28), for three raised-cosine pulse shapes with rolloff factors α. Notice that in each case it has a unique zero crossing at the optimal timing phase, and that the polarity is correct to ensure that the timing phase will tend to be adjusted towards the zero crossing. (After Mueller and Müller [10]).

To get the timing function in (16.33), simply choose

$$X_k = \begin{bmatrix} -A_k \\ A_{k-1} \end{bmatrix} \tag{16.37}$$

(see Problem 16-7). This technique will have less timing jitter than that given by (16.28). Of course, to make this decision directed, use $\hat{A}_k$ instead of A_k.

16.5. Accumulation of Timing Jitter

In digital transmission systems there are often long chains of regenerative repeaters. We saw in Chapter 1 how beneficial regeneration was in preventing the accumulation of noise and distortion through the transmission medium. However, the regenerative repeaters unfortunately allow an accumulation of timing jitter, which can become a critical problem if it is not properly controlled through the careful design of the timing recovery circuits [11]. This accumulation of jitter will typically result in a limit in the number of repeaters of a given design before the accumulated jitter becomes intolerable.

When considering accumulation of jitter it is important to distinguish between two basic sources of jitter. Consider, for example, the spectral-line method of Fig. 16-7. The timing recovery circuit responds to the timing function, which is the deterministic mean-value component of $|R(t)|^2$. The jitter arises from the random components of this signal. One random component is the random portion due to the PAM signal itself, and this is called the *systematic* or *data-dependent* jitter. The second component of jitter is due to any noise or crosstalk signals present in $R(t)$. This is called the *random* jitter.

In a long chain of repeaters, the systematic jitter greatly dominates the random jitter. This is because, in the absence of transmission errors, the systematic jitter component is the *same* in each and every repeater, and therefore adds coherently. A model for the accumulation of this type of jitter is shown in Fig. 16-15. Any dependency of the jitter introduced in one repeater on the jitter introduced upstream will be slight as long as the total introduced jitter remains modest, and therefore we can assume that the jitter introduced in each repeater is additive. Each repeater has an equivalent transfer function to jitter, $H(f)$.

Example 16-11.

$H(f)$ is typically a low-pass filter response. When timing recovery is implemented using a PLL, the closed-loop response is a lowpass filter, as shown in Chapter 14. For the spectral-line methods, the output of the bandpass filter can have only frequencies in the vicinity of the baud rate, and thus the jitter components are also low frequency.

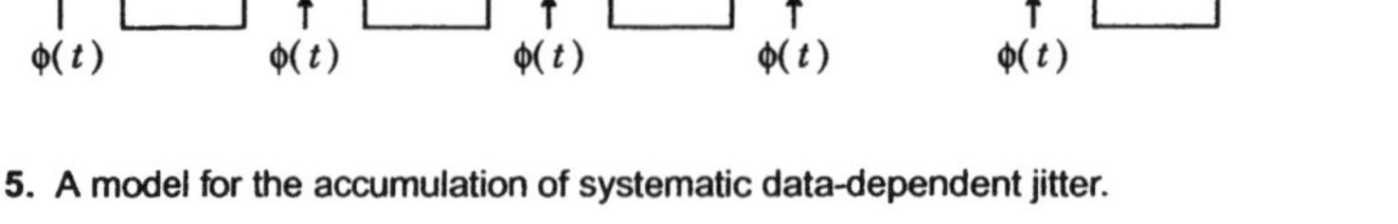

Fig. 16-15. A model for the accumulation of systematic data-dependent jitter.

The response $H(f)$ is applied not only to the jitter introduced in the same repeater, but also the jitter introduced in all repeaters upstream. The overall transfer function to jitter at the output is

$$H_{TOTAL}(f) = \sum_{i=1}^{N} H^i(f) = H(f) \cdot \frac{1 - H^N(f)}{1 - H(f)} , \qquad (16.38)$$

for N repeaters. For a given $H(f)$ and jitter power spectrum, this relation allows us to predict the total jitter as a function of the number of repeaters.

Since the bandwidth of the jitter transfer function is generally narrow relative to the baud rate, the jitter spectrum is not expected to vary substantially within the bandwidth of $H(f)$, and it is a good approximation to assume that it has a white power spectrum Φ_0. The power spectrum of the jitter at the output of the line is then $\Phi_0 \cdot |H_{TOTAL}(f)|^2$, from which we can predict the total mean-square jitter. Our main concern is in how this jitter varies with N, since this accumulation of jitter may limit the number of allowable repeaters for a given timing recovery design.

It is instructive to consider three cases:

- At frequencies where $|H(f)| > 1$, we can readily see that $|H_{TOTAL}(f)| \rightarrow \infty$ as $N \rightarrow \infty$. It is therefore a very bad idea to allow any jitter gain at any frequency for a repeatered line. This rules out any PLL with peaking, for example, many second-order PLLs.
- For frequencies where $|H(f)| << 1$, we can see that $H_{TOTAL} \rightarrow H/(1 - H)$ as $N \rightarrow \infty$. Thus, at frequencies where $h|$ is close to zero there is no jitter accumulation, and the only significant jitter is that introduced in the last repeater.
- The critical case to examine is when $|H(f)| \approx 1$, which will occur within the passband of the single-repeater jitter transfer function. A simple application of L'Hospital's rule establishes that $|H_{TOTAL}(f)| \rightarrow N$ as $|H(f)| \rightarrow 1$. Thus, the jitter at frequencies where the jitter transfer function is precisely unity accumulates coherently as expected — N repeaters results in N times the jitter amplitude.

Example 16-12.

Assume the jitter transfer function of each repeater is an ideal lowpass filter with bandwidth f_1. Then the overall jitter power spectrum is $\Phi_0 \cdot N^2$ within the passband and zero elsewhere, so the jitter power is $2f_1\Phi_0N^2$. Thus, the jitter power increases as the square of the number of repeaters, or the jitter amplitude increases in proportion to N. For any desired N we can limit the accumulated jitter by making the bandwidth of the lowpass filter small.

Example 16-13.

A much more realistic assumption is that the jitter transfer function in each repeater is a single-pole lowpass filter, obtained, for example, by using a first order PLL with loop filter $L(s) = K_L$,

$$H(f) = \frac{K_L}{K_L + j2\pi f} , \qquad (16.39)$$

in which case the accumulated jitter transfer function is

$$H_{TOTAL}(f) = \frac{K_L}{j2\pi f}\left(1 - \frac{K_L^N}{(K_L + j2\pi f)^N}\right). \tag{16.40}$$

The total jitter power in this case can be shown [12][11] after evaluation of a rather complicated integral to be $K_L\Phi_0 N/\pi$. Thus, the jitter amplitude has been reduced from an N dependence for an ideal lowpass filter (see Example 16-12) to a $\sqrt{N}$ dependence. Even though the jitter transfer function is always N at very low frequencies, as N increases the bandwidth over which this dependence is valid narrows, thus slowing the rate of accumulation of jitter (see Problem 16-9). In this case the departure of the lowpass filter from ideality is beneficial!

For digital transmission of continuous-time signals, the limit on accumulated timing jitter will be the distortion suffered due to the irregular spacing of the received samples. While repeatered digital communications systems suffer from this problem of jitter accumulation, it is not a fundamental impairment. To reduce the jitter to any desired level, we can pass the bit stream through a timing recovery circuit with a bandwidth much smaller than the bandwidth of the regenerator timing circuits.

16.6. Further Reading

Tutorial articles on timing recovery are rare, perhaps because the ideas are subtle and the analysis is relatively difficult. One of the few such articles is by Franks [3]. Another basic reference was written by Gitlin and Hayes [13]. The classic article by Bennett, published in 1958 [2], is well worth reading because of its fine craftsmanship and historical value. Another classic article with a lucid discussion of the topic is written by Aaron [14].

The relationship between fractionally-spaced equalizers and timing recovery has been extensively studied [15][16]. Baud-rate timing recovery is described by Mueller and Müller [10] and others [17][18]. The effect of timing jitter on echo cancelers has been examined [19][20][21]. Even the oldest techniques are still being pursued, as for example in [22] where block codes that permit linear timing recovery are studied. Of historical interest is one of the earliest discussions of self-timing by Sunde [23], in which a linear spectral-line method is proposed. Discrete-time timing recovery for voiceband data modems is described in [24]. Other papers with interesting techniques or analysis are [25][9].

Appendix 16-A. The Poisson Sum Formula

In the analysis of the spectral-line method, we need to relate the Fourier series of a periodic signal in summation form

$$x(t) = \sum_{k=-\infty}^{\infty} g(t-kT)\,, \tag{16.41}$$

with the Fourier transform of $g(t)$. In the process, we will get *Poisson's sum formula*, a well known result. Any periodic signal with period T can be written in summation form (16.41).

Because of the periodicity of $x(t)$, we can express it using a Fourier series

$$x(t) = \sum_{m=-\infty}^{\infty} X_m e^{j2\pi mt/T}\,, \tag{16.42}$$

where the Fourier coefficients are

$$X_m = \frac{1}{T}\int_{-T/2}^{T/2} \sum_{k=-\infty}^{\infty} g(t-kT) e^{-j2\pi mt/T} dt\,. \tag{16.43}$$

Exercise 16-7.
Show that (16.43) reduces to

$$X_m = \frac{1}{T}\int_{-\infty}^{\infty} g(\tau) e^{-j2\pi m\tau/T} d\tau\,. \tag{16.44}$$

Except for the $1/T$ factor, this is the Fourier transform of $g(t)$ evaluated at $f = m/T$,

$$X_m = \frac{1}{T} G(m/T). \tag{16.45}$$

Thus, the Fourier coefficients of the periodic signal in summation form are simply samples of the Fourier transform of the function $g(t)$. Putting (16.45) together with (16.42) we get Poisson's sum formula

$$\sum_{k=-\infty}^{\infty} g(t-kT) = \frac{1}{T}\sum_{m=-\infty}^{\infty} G(m/T) e^{j2\pi mt/T}. \tag{16.46}$$

Appendix 16-B. Discrete-Time Derivative

Consider a continuous-time signal $Q(t)$ for which we have available only Nyquist-rate samples

$$Q_k(\tau) = Q(kT+\tau)\,, \tag{16.47}$$

where T is the sampling interval and τ is the sampling phase. If the sampling phase is varying with time, we must assume it is varying slowly enough that we can consider it essentially constant. We wish to compute the derivative of $Q_k(\tau)$ with respect to τ for all k. From the interpolation formula (2.19) we can write

$$Q(t) = \sum_{m=-\infty}^{\infty} Q_m(\tau) \frac{\sin[\pi(t-\tau-mT)/T]}{\pi(t-\tau-mT)/T} \tag{16.48}$$

The derivative of $Q_k(\tau)$ with respect to the timing phase τ is the same as the derivative of $Q(t)$ sampled at the same sampling instants,

$$\begin{aligned}\frac{\partial Q_k(\tau)}{\partial \tau} &= \left[\frac{\partial Q(t)}{\partial t}\right]_{t = kT + \tau_k} \\ &= \sum_{m=-\infty}^{\infty} Q_m(\tau)\left[\frac{\cos[\pi(t-\tau-kT)/T]}{\pi(t-\tau-kT)/T} - \frac{\sin[\pi(t-\tau-kT)/T]}{\pi(t-\tau-kT)^2/T}\right]_{t = kT+\tau} \\ &= \sum_{m \neq k} z_m(\tau) \frac{(-1)^{k-m}}{(k-m)T} \,.\end{aligned} \tag{16.49}$$

The last equality follows from putting the quantity in brackets over a common denominator, and evaluating everywhere except at $m = k$. At that point, use of L'Hospital's rule shows that the value is zero. The sum in (16.49) can be rewritten as a convolution

$$\frac{\partial Q_k(\tau)}{\partial \tau} = Q_k(\tau) * d_k \,, \tag{16.50}$$

where

$$d_k = \begin{cases} 0, & k = 0 \\ \frac{(-1)^k}{kT} & k \neq 0 \end{cases} . \tag{16.51}$$

Hence the derivative of the sampled signal with respect to the timing phase can be obtained by filtering the sampled signal with the filter whose impulse response is given by (16.51).

A filter with impulse response given by (16.51) would be difficult to implement. In practice, the impulse response can be truncated, and the filter can be implemented as an FIR filter. Reducing it to three taps, we get the approximation

$$d_k \approx (\delta_{k+1} - \delta_{k-1})/T, \qquad \frac{\partial Q_k(\tau)}{\partial \tau} \approx (Q_{k+1}(\tau) - Q_{k-1}(\tau))/T. \tag{16.52}$$

Note, however, that the impulse response in (16.51) decays only linearly with increasing $|k|$, so any truncated FIR filter will have substantial error.

Problems

Problem 16-1. Consider a signal $Q(t)$ bandlimited to $|f| < 1/(2T)$ and samples of the signal $Q_k(\tau)$ given by

$$Q_k(\tau) = Q(kT + \tau) ,$$

where the sampling phase τ is arbitrary. Give the impulse response of a filter whose input is $Q_k(\tau)$ and output is $Q_k(\alpha)$, for any arbitrary α different from τ. This suggests that given Nyquist-rate samples of a received signal, an equalizer can compensate for any timing phase. If the equalizer is adaptive, it will automatically compensate for an erroneous timing phase. Most adaptive filters, however, are implemented as FIR filters. Can this filter be exactly implemented or approximated as an FIR filter?

Problem 16-2. Suppose a received real-valued baseband PAM signal has 100% excess bandwidth raised cosine Nyquist pulses given in (5.8) (with $\alpha = 1$). Find the Fourier series coefficients of $E[R^2(t)]$. Would square law timing recovery work well with this signal?

Problem 16-3. Show that a baseband real-valued PAM signal with a 0% excess bandwidth raised cosine Nyquist pulse has no timing tone for the squarer timing recovery circuit of Fig. 16-7. Note that although it has a timing tone for a fourth-power circuit, it has a zero-width eye (shown in Problem 5-5) and hence cannot be used anyway.

Problem 16-4. Consider the WSS random process $R(t + \Theta)$ where $R(t)$ is defined by (16.2) and Θ is a uniformly distributed random variable over the range 0 to T. Assume A_k is white, $R_A(m) = a\delta_m$ (which implies zero mean), and independent of Θ. Show that $R(t + \Theta)$ cannot have any spectral lines not present in $p(t)$. In other words, in order to get a spectral line at the symbol frequency, $p(t)$ would have to be periodic with period T.

Problem 16-5. Consider a 600 baud signal with 100% excess bandwidth. Design a prefilter for square law timing recovery. Will the prefilter improve performance?

Problem 16-6. Consider the inductive timing recovery of Fig. 16-3. Write the output of the VCO

$$v(t) = \cos(2\pi f_v t + \phi(t)) . \tag{16.53}$$

Samples of the input signal are taken at times $kT + \tau_k$, which correspond to the zero crossings of $v(t)$. Consider only the zero crossings from positive to negative, so for small enough $\varepsilon > 0$,

$$\begin{aligned} v(kT + \tau_k) &= 0 \\ v(kT + \tau_k - \varepsilon) &< 0 \\ v(kT + \tau_k + \varepsilon) &> 0 . \end{aligned} \tag{16.54}$$

Find an equation relating τ_k and $\phi(t)$.

Problem 16-7. Suppose that the input to the slicer is given by (16.32). Assume that A_k is real-valued, zero mean, white, and independent of N_k, which also has zero mean. For an implicit timing recovery technique using the update given by (16.27), (16.35), and (16.37),

(a) Show that the timing function (16.29) is given by (16.33).

(b) Suppose that $p(t)$ is the triangular pulse given in the following figure:

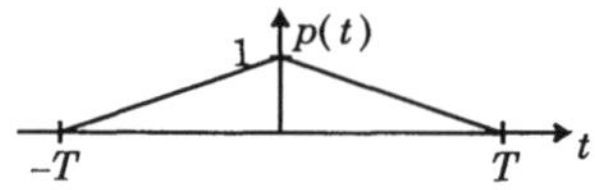

Sketch the timing function.

Problem 16-8. Suppose again that the input to the slicer is given by (16.32), where A_k is real-valued, zero mean, white, and independent of N_k. For an implicit timing recovery technique using the update given by (16.27), (16.35), and

$$\mathbf{X}_k' = A_{k-1} \, . \tag{16.55}$$

(a) Find the timing function (16.29).

(b) Suppose that $p(t)$ is the same triangular pulse used in Problem 16-7. Sketch the timing function.

(c) Will this timing update work with the triangular pulse? Will it work with raised cosine pulses?

(d) Repeat parts (a) through (c) for

$$\mathbf{X}_k' = A_{k+1} \, . \tag{16.56}$$

Problem 16-9. Explain the N dependence of the mean-square jitter in Example 16-13 by examining the bandwidth over which the total jitter transfer function is approximately N. In particular, show that this bandwidth is proportional to $1/N$, and thus, the jitter power is proportional to $N^2 \cdot (1/N) = N$. Thus, the jitter does not accumulate at nearly the same rate as for an ideal lowpass filter of Example 16-12. (*Hint:* Do a Taylor series expansion of $H^N(f)$, retain only first- and second-order terms.)

References

1. P. Kabal and S. Pasupathy, "Partial-Response Signaling," *IEEE Trans. on Communications*, Vol. COM-23 (9), (Sep. 1975).

2. W. R. Bennett, "Statistics of Regenerative Data Transmission," *Bell System Technical Journal*, Vol. 37, pp. 1501-1542 (Nov. 1958).

3. L. E. Franks, "Synchronization Subsystems: Analysis and Design," in *Digital Communications: Satellite/Earth Station Engineering*, Prentice-Hall Inc. (1981).

4. N. A. D'Andrea and U. Mengali, "A Simulations Study of Clock Recovery in QPSK and 9QPRS Systems," *IEEE Trans. on Communications*, Vol. COM-33 (10), (Oct. 1985).

5. H. Kobayashi, "Simultaneous Adaptive Estimation and Decision Algorithm for Carrier Modulated Data Transmission Systems," *IEEE Trans. on Communications Technology*, Vol. COM-19, pp. 268-280 (June 1971).

6. S. U. H. Qureshi, "Timing Recovery for Equalized Partial-Response Systems," *IEEE Trans. on Communications*, (Dec. 1976).

7. H. Sari, L. Desperben, and S. Moridi, "Optimum Timing Recovery for Digital Equalizers," *Globecom '85 Proceedings*, (1985).

8. R. D. Gitlin and J. Salz, "Timing Recovery in PAM Systems," *Bell System Technical Journal*, Vol. 50 (5), p. 1645 (May and June 39 1971).

9. B. R. Saltzberg, "Timing Recovery for Synchronous Binary Data Transmission," *Bell System Technical Journal*, pp. 593-622 (March 1967).

10. K. H. Mueller and M. Muller, "Timing Recovery in Digital Synchronous Data Receivers," *IEEE Trans. on Communications*, Vol. COM-24, pp. 516-531 (May 1976).

11. C. J. Byrne, B. J. Karafin, and D. B. Robinson, "Systematic Jitter in a Chain of Digital Repeaters," *Bell System Technical Journal*, Vol. 42, p. 2679 (Nov. 1963).

12. Bell Laboratories Members of Technical Staff, *Transmission Systems for Communications*, Western Electric Co., Winston-Salem N.C. (1970).

13. R. D. Gitlin and J. F. Hayes, "Timing Recovery and Scramblers in Data Transmission," *Bell System Technical Journal*, Vol. 54 (3), (March 1975).

14. M. R. Aaron, "PCM Transmission in the Exchange Plant," *Bell System Technical Journal*, Vol. 41, pp. 99-141 (Jan. 1962).

15. G. Ungerboeck, "Fractional Tap-Spacing and Consequences for Clock Recovery in Data Modems," *IEEE Trans. on Communications*, (Aug. 1976).

16. R. D. Gitlin and S. B. Weinstein, "Fractionally Spaced Equalization: An Improved Digital Transversal Equalizer," *Bell System Technical Journal*, Vol. 60 (2), (Feb. 1981).

17. O. Agazzi, C.-P. J. Tzeng, D. G. Messerschmitt, and D. A. Hodges, "Timing Recovery in Digital Subscriber Loops," *IEEE Trans. on Communications*, Vol. COM-33 (6), (June 1985).

18. A. Jennings and B. R. Clarke, "Data-Sequence Selective Timing Recovery for PAM Systems," *IEEE Trans. on Communications*, Vol. COM-33 (7), (July 1985).

19. D. D. Falconer, "Timing Jitter Effects on Digital Subscriber Loop Echo Cancellers: Part I - Analysis of the Effect," *IEEE Trans. on Communications*, Vol. COM-33 (8), (Aug. 1985).

20. D. D. Falconer, "Timing Jitter Effects on Digital Subscriber Loop Echo Cancellers: Part II - Considerations for Squaring Loop Timing Recovery," *IEEE Trans. on Communications*, Vol. COM-33 (8), (Aug. 1985).

21. D. G. Messerschmitt, "Asynchronous and Timing-Jitter Insensitive Data Echo Cancellation," *IEEE Trans. on Communications*, Vol. COM-33 (12), p. 1209 (Dec. 1986).

22. C. M. Monti and G. L. Pierobon, "Block Codes for Linear Timing Recovery in Data Transmission Systems," *IEEE Trans. on Communications*, Vol. COM-33 (6), (June 1985).

23. E. D. Sunde, "Self-Timing Regenerative Repeaters," *Bell System Technical Journal*, Vol. 36, pp. 891-937 (July 1957).

24. A. Haoui, H.-H. Lu, and D. Hedberg, "An All-Digital Timing Recovery Scheme for Voiceband Data Modems," *Proceedings of ICASSP*, (1987).

25. D. L. Lyon, "Timing Recovery in Synchronous Equalized Data Communication," *IEEE Trans. on Communications*, Vol. COM-23 (2), (Feb. 1975).

17

Multiple Access Alternatives

Thus far in this book we have discussed the signal processing necessary for digital communications from one point to another over a communications medium. The remainder of the book will begin to address the realization of a *digital communications network* in which many users simultaneously communicate with one another. One of the key issues that must be resolved in moving from a single digital communication system to a network is how we provide access to a single transmission medium for two or more (typically many more) users. In moving from a single transmitter and receiver to the sharing of media by multiple users, we must address two primary issues. First, how do we resolve the *contention* that is inherent in sharing a single resource. This issue is discussed in this chapter. Secondly, how do we *synchronize* all the users of the network as an aid to resolving the contention.

Important practical applications of multiple access techniques include:

- *Full-duplex data transmission on a single medium.* Here we want the two directions of transmission to share a single transmission medium such as a wire pair or voiceband channel. An important application of this is data transmission on the subscriber loop, between telephone central office and customer premises, where only a single wire pair is available for both directions of transmission.

- *Multiple channels sharing a common high-speed transmission link.* By building very high speed transmission links using, for example, coaxial cable or optical fiber, and then sharing that link over many users, economies of scale are realized. The cost of the transmission link when divided by the number of users can be very much lower than the cost of an equivalent lower speed link. The process of sharing many channels on a single high speed link is called *multiplexing.* A network, which provides a communications capability between any pair of users on demand, can be constructed from communications links, multiplexers, and *switches*. The latter provide the rearrangement of the connections through the network necessary to provide this service on demand.
- Many users can share a common transmission medium in such a way that when one user broadcasts to the others, with only the user for which the communication is intended paying attention. By this means, it is possible to provide access between any user and any other user without a switch. This is commonly known as *multiple access*, although we use this term more generally in this part of the book to describe any situation where two or more users share a common transmission medium. Particularly for small numbers of users and within limited geographical areas, multiple access to a single medium can be more economical than using switches. This type of multiple access is often exploited in *local-area networks* within a customer premises, and in *satellite networks* in the context of a wide geographical area.

All these multiple access techniques require that the messages corresponding to different users be separated in some fashion so that they do not interfere with one another. This is usually accomplished by making the messages *orthogonal* to one another in signal space. We can then separate out the different signals using some form of matched filtering or its equivalent, which because of the orthogonality of the signals will respond to only a single signal.

Orthogonality of two or more signals can be accomplished in several ways.

- The messages can be separated in *time*, insuring that the different users transmit at different times.
- The messages can be separated in *frequency*, insuring that the different users use different frequency bands.
- The messages can transmitted at the same time and at the same frequency, but made orthogonal by some other means. Usually this is done by *code division*, in which the users transmit signals which are guaranteed to be orthogonal through the use of specially designed codes.

There are also cases where the signals are not separated by orthogonality. In the particular case of full duplex transmission on a common medium, the signal which is interfering with the received data stream is the transmit data stream generated by the same user. We can use the fact that the interfering signal is known, and use *echo cancellation* to separate the two directions, even though they are not orthogonal (although we do use the fact that they are uncorrelated). In this chapter we cover the first three multiple access techniques, separation by time, frequency, and code division. We also describe the cellular concept, which allows spatial reuse of frequencies for serving large numbers of users.

17.1. Medium Topology for Multiple Access

In any discussion of multiple access techniques, it is appropriate to begin by pointing out the importance of the *topologies* of the medium. This term refers to the geometrical configuration of the medium which is being shared by two or more users. Each of the common media have preferred topological configurations, and each topology suggests appropriate techniques for providing multiple access. Multiple access media are most common over geographically limited areas, such as the local-area and metropolitan-area networks, but can also span large geographical areas when radio and satellite media are used. In this section we discuss briefly the most common topologies, and in the remainder of the chapter describe multiple access techniques in the context of these topologies.

Some representative topologies are shown in Fig. 17-1. In Fig. 17-1(a) the simplest and most common situation is shown, where we have a single unidirectional communications *link* that we wish to share over two or more users. To do so we implement a *multiplexer* and *demultiplexer*, represented by the boxes, which accept information from the users and transmit it over the link. This topology inherently has *contention*, in that two or more users may wish to use the medium simultaneously. The multiplexer, since it has a form of central control, can easily avoid *collisions* (two or more users transmitting simultaneously) since it controls what is transmitted. In Fig. 17-1(b) we show the *bus*, in which two or more users are connected in parallel to a common medium, with the result that every transmission by any user is received by every other user. The most common medium for a bus is coaxial cable or wire-pair, where the users are simply connected in parallel. A set of radio transmitters and receivers with non-directional antennas are also connected, in effect, by a bus, as are a set of transmitters and receivers communicating through a satellite transponder. In the later cases this is known as a *broadcast medium*. The bus has no central control, which makes it more difficult to avoid collisions. The *ring* is shown in Fig. 17-1(c), where the users are connected to their nearest neighbor in a circle. In this case, we cannot allow broadcast, since the information would circulate indefinitely. Therefore, we must have *active nodes* in this topology, which means that

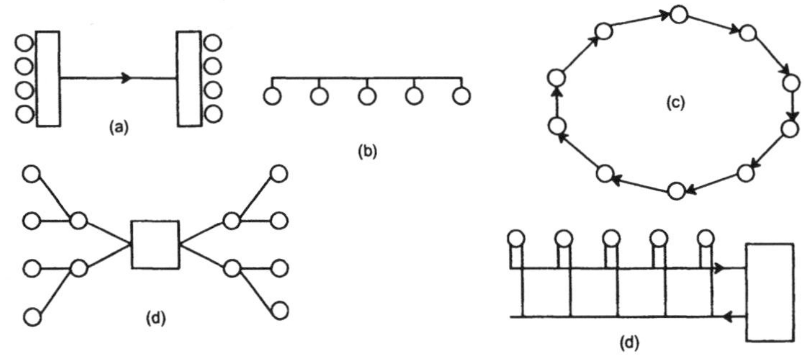

Fig. 17-1. Common topologies for multiple access. A line corresponds to a communication link, one-way if it includes an arrow, a box is a central controller that controls access to the medium, and a circle represents a user. (a) A link, where multiple users share a common one-way communication channel using a multiplexer and demultiplexer. (b) A bus, where every user receives the signal from every other user. (c) A ring, in which each user talks to one neighboring user. (d) A tree, in which users communicate through a central hub. (e) A bus with centralized control.

all communications pass through them rather than just by them. This allows users to remove communications as well as initiate them (typically they remove their own transmissions so that they will not circulate forever). Rings are the topology of choice for local area networks based on fiber optics media, since they require only point-to-point connections and the bus topology is difficult to realize using optical components. The *tree* is shown in Fig. 17-1(d), in which all users are connected to one another through a central controller. Finally, the *bus with central control* is shown in Fig. 17-1(e). This topology is very similar to the tree in that all users communicate through a central controller, and requires two unidirectional busses.

A special application of multiple access is the sharing of a single medium for digital transmission in both directions. This is called *full-duplex* transmission, and is pictured in Fig. 17-2. We have two transmitters, one on the left, and one on the right, and two receivers. The goal is to have the transmitters on each end transmit to the receivers on the opposite end. Unfortunately, most media have the characteristic that each receiver must contend with interference from the local transmitter in addition to the signal from the remote transmitter. In fact, it will often be the case that the local transmitter interference is much larger than the remote transmitter signal. The receiver must find some way to separate the local and remote transmitter signals.

When data signals are transmitted through the network, they encounter echos at points of four-wire to two-wire conversion. In *half-duplex* data transmission (in only one direction), echos present no problem since there is no receiver on the transmitting end to be affected by the echo. In full-duplex transmission, where the data signals are transmitted in both directions simultaneously, echos from the data signal transmitted in one direction interfere with the data signal in the opposite direction as illustrated in Fig. 17-3.

Most digital transmission is half-duplex. For example, the high speed trunk digital transmission systems separate the two directions of transmission on physically different wire, coaxial, or fiber optic media. The two directions therefore do not interfere, except perhaps through crosstalk resulting from inductive or capacitive coupling. However, full-duplex data transmission over a common media has arisen in two important applications. In both these applications, the need for a common media arises because the network often only provides a two-wire connection to each customer premise (this is because of the high cost of copper wire, and the large percentage of the telephone network investment in this facility).

Example 17-1.

The first application shown in Fig. 17-4 is digital transmission on the *subscriber loop*, in which the basic voice service and enhanced data services are provided through the two-wire subscriber loop. Total bit rates for this application that have been proposed are 80 and 144 kb/s in each direction,

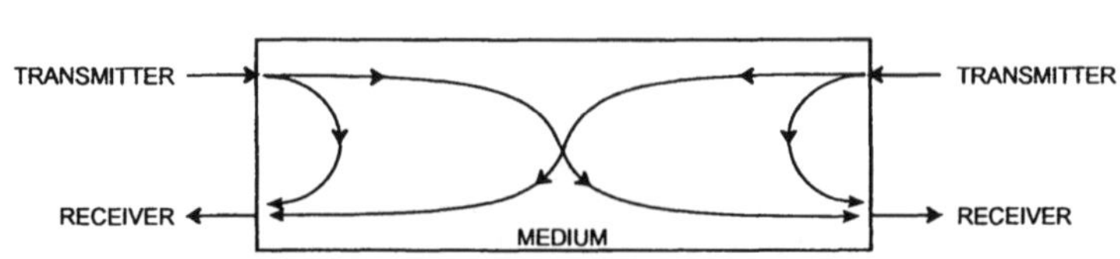

Fig. 17-2. Illustration of full-duplex transmission.

where the latter rate includes provision for two voice/data channels at 64 kb/s each plus a data channel at 16 kb/s, and the first alternative allows only a single 64 kb/s voice/data channel. This digital subscriber loop capability is an important element of the emerging integrated services digital network (ISDN), in which integrated voice and data services will be provided to the customer over a common facility [1]. Voice transmission requires a *codec* (coder-decoder) and antialiasing and reconstruction filters to perform the analog-to-digital and digital-to-analog conversion on the customer premises, together with a *transceiver* (transmitter-receiver) for transmitting the full-duplex data stream over the two-wire subscriber loop. Any data signals to be accommodated are simply connected directly to the transceiver. The central office end of the loop has another full-duplex transceiver, with connections to the digital central office switch for voice or circuit-switched data transmission, and to data networks for packet switched data transport capability.

Example 17-2.

The second application for full-duplex data transmission is in *voiceband data transmission*, illustrated in Fig. 17-3. The basic customer interface to the network is usually the same two-wire subscriber loop. In this case the transmission link is usually more complicated due to the possible presence of four-wire trunk facilities in the middle of the connection. The situation can be even more complicated by the presence of two-wire toll switches, allowing intermediate four-two-four wire conversions internal to the network.

The two applications differ substantially in the types of problems which must be overcome. For the digital subscriber loop, the transmission medium is fairly ideal, consisting of wire pairs with a wide bandwidth capability. The biggest complication is the higher bit rate

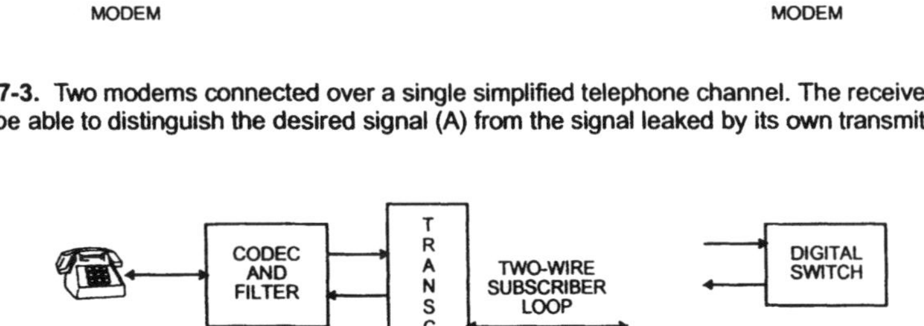

Fig. 17-3. Two modems connected over a single simplified telephone channel. The receiver on the right must be able to distinguish the desired signal (A) from the signal leaked by its own transmitter (B).

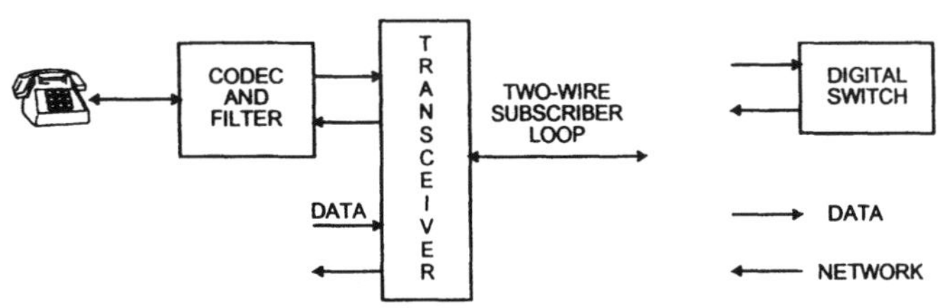

Fig. 17-4. A digital subscriber loop transceiver for full-duplex digital transmission.

and the presence in some countries of bridged taps The voiceband data modem, while requiring a lower speed of transmission, encounters many more impairments. In addition to the severe bandlimiting when carrier facilities are used, there are problems with noise, nonlinearities, and sometimes even frequency offset. Another difference is that the subscriber loop can use baseband transmission, while the voiceband data set always uses passband transmission.

17.2. Multiple Access by Time Division

By far the most common method of separating channels or users on a common digital communications medium is by ensuring that they transmit at different times. This is known as *multiple access by time-division*. This technique has many variations, the most common of which are described in this section. In all these variations, some method is used to avoid *collisions*, or two or more users transmitting simultaneously. Collision avoidance in link access is somewhat easier than in the other topologies, and therefore we discuss link access separately.

17.2.1. Point-to-Point Link Access

It is often desired to divide a high-speed bit stream over a point-to-point communications link into a set of lower-rate bit streams, each with a fixed and predefined bit rate. Where this is desired, it is appropriate to use a technique called *time-division multiplexing (TDM)*. The bit streams to be multiplexed are called *tributary streams*. Where these tributary bit-streams are provided directly to a user, that is they do not themselves consist of tributary streams, then they are called *circuits* or *connections*. We *interleave* these tributary streams to obtain a higher rate bit stream. The purpose of the multiplex, shown functionally in Fig. 17-5, is to take advantage of the economies of scale of a high-speed transmission system.

Example 17-3.

A simple multiplexing function for two tributary streams is shown below:

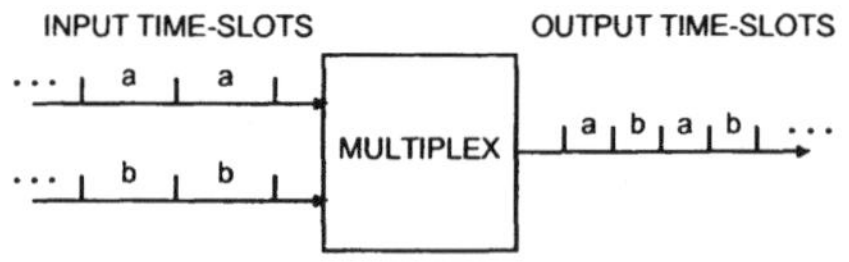

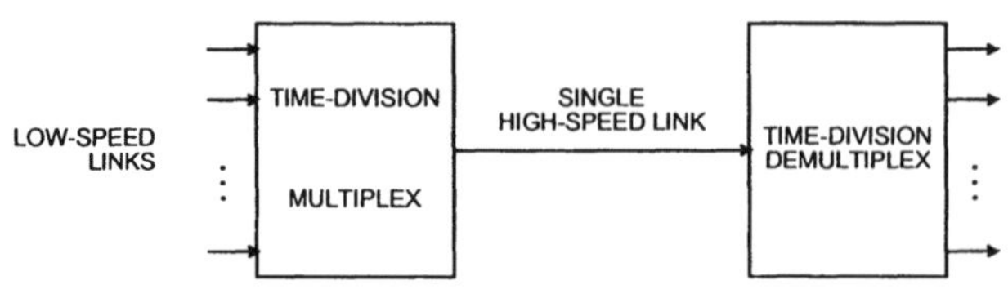

Fig. 17-5. A time-division multiplex, which interleaves a number of lower-speed tributaries on a single higher-speed link.

Each tributary stream is divided continuously into groups of bits, known as *time-slots*, and then these time-slots are interleaved to form the output bit stream. Each slot on the output bit stream occupies half the time of an input slot since the bit rate is twice as great.

In practice any number of tributary streams can be multiplexed, and a defined time-slot can have any number of bits. However, two cases are particularly common: a time-slot equal to one bit, known as *bit-interleaving*, and a time-slot equal to eight bits, which is known as *octet-interleaving*. In multiplexing, an *octet* is a common term for eight bits. Organization around octets is common because voice pulse-code modulation (PCM) systems commonly use eight bits per sample quantization, and because data communications systems typically transfer eight-bit groupings of bits, known in the computer world as *bytes*. In both cases it is necessary to maintain *octet integrity* at the destination, meaning that the bit stream is delimited into the same eight-bit boundaries defined at the origin. This octet integrity is assured by using octet-interleaving, although this is not the only means.

On the high-speed output bit stream, the collection of bits corresponding to precisely one time-slot from each tributary stream is known as a *frame*. In Example 17-3 one frame corresponds to time-slots a and b. At the demultiplex, all we have is a bit-stream originating at the multiplex. In order to realize the demultiplexing function, the boundaries of the time-slots must be known. Furthermore, to ensure that the correspondence between input and output tributary streams is maintained, demultiplexing requires knowledge of the beginning of the frame. For this purpose, the multiplex typically inserts additional bits into the frame known as *framing bits*.

Example 17-4.

In a time-division multiplex, N tributary streams are multiplexed with M-bit time-slots into a single output bit stream. The number of bits in the output frame is $N \cdot M$ plus any added framing bits.

The framing bits follow a deterministic pattern which can be recognized at the demultiplex as distinct from the information bits. Once the demultiplex has located these bits, through a process known as *framing recovery*, it has a reference point that enables it to locate the beginning of the frame.

Since a multiplex cannot store an unbounded number of bits, we must ensure that the minimum bit rate of the output high-speed stream is greater than or equal to the sum of the maximum bit rates of the tributary streams plus the rate required for framing and any other overhead bits.

Example 17-5.

The CCITT 30-channel system (recommendation G.732 [2]) is widely used in Europe and multiplexes 30 tributary streams, each at 64 kb/s, appropriate for a voiceband channel, into a single 2048 kb/s bit stream. Note that $30 \cdot 64 = 1920$, so that 128 kb/s is used for overhead functions such as framing. The organization of the frame is shown in Fig. 17-6. Each frame is divided into 32 eight-bit time slots, 30 of them taken from the tributary streams, and the remaining two used for overhead. Thus, in this case as in the case of most lower-speed multiplexes, octet-interleaving is used. The time for one frame corresponds to an octet on each tributary stream, or 1.25 μsec. There is also defined a *superframe* or *multiframe* of 16 frames, which is used to transmit and frame *on-off hook information* for each of the 30 tributary voiceband channels. This on-off

Table 17-1. Superframe structure of the M12 multiplex. The F bits are the framing bits, M are the superframing bits, and C are stuffing control bits. 48I means 48 information bits, 12 bit-interleaved bits from each of four tributary streams.

M_0	48I	C_1	48I	F_0	48I	C_1	48I	C_1	48I	F_1	48I
M_1	48I	C_2	48I	F_0	48I	C_2	48I	C_2	48I	F_1	48I
M_1	48I	C_3	48I	F_0	48I	C_3	48I	C_3	48I	F_1	48I
M_1	48I	C_4	48I	F_0	48I	C_4	48I	C_4	48I	F_1	48I

hook information, transmitted in frames 0 and 16, is used to communicate between switching machines during call setup and takedown. Time-slot 0 always contains the octet "x0011011" and "x10xxxxx" in alternate frames, indicating the beginning of the frame, and time-slot 16 contains "0000x0xx" in frame 0 of the superframe indicating the beginning of the superframe ("x" indicates bits not assigned, which can be used for other purposes). Time-slot 16 in the remaining frames of the superframe contains the aforementioned signaling information.

Example 17-6.
The CCITT 24-channel system used in North America (CCITT G.733 [2]) has a frame consisting of 193 bits, including 24 eight-bit time-slots for the tributary 64 kb/s channels and one framing/superframing bit. In a superframe of 12 frames, the framing bit contains the pattern "101010,", the framing pattern, interleaved with the pattern "001110," the superframe pattern. The bit rate is $193 \cdot 8 = 1544$ kb/s.

Example 17-7.
The M12 multiplex used in the North American network multiplexes four tributary bit streams at 1544 kb/s (often the G.733 signal of Example 17-6) into a into a 1176-bit superframe shown in Table 17-1 using bit interleaving.

Each line of the table represents one frame as defined by the $F_0F_1\ldots$ pattern, where $F_0 = 0$ and $F_1 = 1$. Similarly, the four-frame superframe is defined by the $M_0M_1M_1M_1\ldots$ superframe pattern.

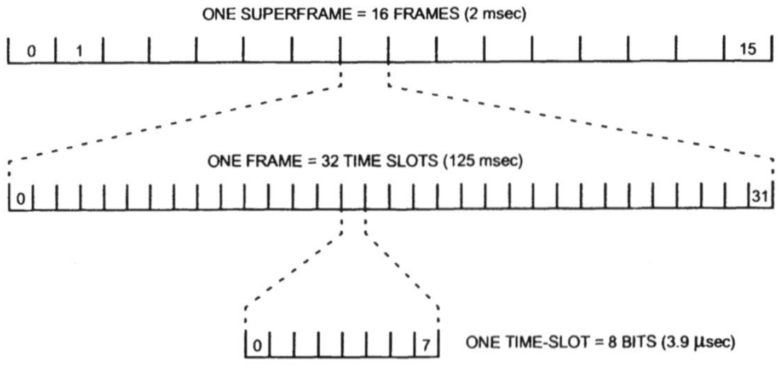

Fig. 17-6. The frame structure for the CCITT G.732 30-channel PCM system.

Digital Circuit Switching

TDM multiplexing provides a basic capability for the network to provide fixed bit-rate bit streams between users. These bit streams, which can be used to provide services such as a voiceband channel or video channel, are called *circuits* or *connections* through the network. This terminology comes from the early days of the telephone network, when voiceband channels were provided by a continuous metallic connection. In order for the network to provide the precise set of circuits requested by the users at any time, it is necessary to provide a set of *digital circuit switches*. The function of a digital circuit switch, and particularly the transmission interfaces, are similar to that of a TDM multiplex, so they will be described briefly here. The switch differs from the multiplex in that it typically has the same number of output bit streams as inputs. Typically each stream is itself composed of time-division multiplexed lower-speed streams, so that it has a defined framing structure with time-slots corresponding to each tributary stream. Further, each of these time-slots typically corresponds to a circuit, or bit stream provided directly to users, since the purpose of the digital circuit switch is to connect different users together. The specific purpose of the switch is to perform an arbitrary permutation of the input circuits (time-slots) appearing on the output.

Example 17-8.
A simple example is shown below:

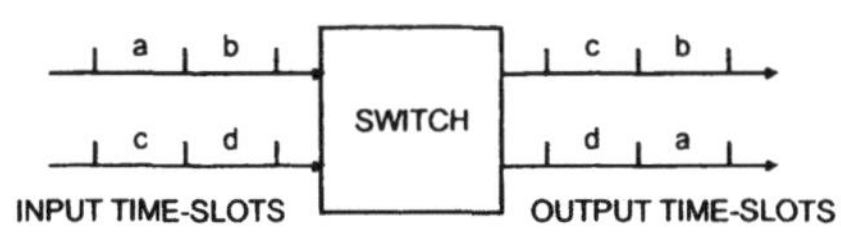

Each of the input time-slots represents a circuit, where there are two circuits corresponding to each TDM input and output in this example. Thus, in total there are four input and four output circuits, and the purpose of the switch is to perform an arbitrary permutation of the input circuits as they appear at the output. (If this switch were used for voiceband channels, each tributary would be 64 kb/s). In the figure, input circuit a replaces b at the output, c replaces a, etc. To perform its function, the switch must be able to transfer the bits corresponding to one time-slot on one of the input TDM streams to a time-slot on a different output TDM stream (for example Ba and Bc in the figure), which is known as *space-division switching*. Also, it must be able to transfer the bits from one time-slots to another (for example Ba and Bd in the figure), which is known as *time-division switching*.

Of course, most practical switches are much larger than this example.

Example 17-9.
The No. 4 ESS is a large toll digital switching machine used in the North American network. It has 5120 input bit streams, each of which is the G.733 signal of Example 17-6 containing 24 tributary voiceband channels, for a total of 122,880 tributary 64 kb/s bit streams (the total input and output bit rate is 7.86 Gb/s, or actually double this because each channel is bidirectional). This switch is not capable of providing all possible permutations of input-output connections, but reasonable traffic demands can be served with high probability.

Circuit switching enables the network to provide a connection between two users for the duration of their need. When the circuit is no longer needed, the switch terminates the connection and uses the associated time-slots on the transmission facility to provide another connection. In this fashion, we avoid provisioning transmission capacity for every possible connection in the network, but rather provide only sufficient capacity for those connections likely to be required at any give time. In Section 17.2.3 we will see an alternative model for providing a connection between two users, called packet switching.

17.2.2. Time-Division Multiple Access

Thus far this section has addressed the use of time-division techniques for access control to a point-to-point link and to a full-duplex channel. Time-division using the circuit switching approach can also be applied to other multiple access topologies, such as the bus or the ring. The appropriate technique is highly dependent on the topology. For example, on the ring topology it is straightforward to define a scheme for time-division multiplexing.

Exercise 17-1.

Describe a method for forming a frame and fixed time-slot structure on a ring topology, and the way in which a circuit could be formed. This approach is called a *slotted ring*.

On the other hand, TDM is difficult to apply to the bus topology. The reason is simply that TDM requires a fixed frame known to all nodes of the network, but in a bus, particularly a geographically large broadcast network such as a satellite network, the propagation delays on the medium will typically be large relative to one bit-time. It is still common to apply TDM techniques in this situation, but they must be modified to account for the significant propagation delays between users. This modification leads to an approach known as *time-division multiple access (TDMA)*. TDMA is applicable to any bus or broadcast topology, where there is a set of transmitters and a set of receivers, all of which hear each transmission. It has been extensively applied to satellite networks in particular [3][4].

TDMA requires a centralized control node, a feature that can be avoided using the random access techniques discussed in Section 17.2.4. The primary functions of the control node are to transmit a periodic *reference burst*, akin to the added framing bits in TDM, that defines a frame and forces a measure of synchronization of all the other nodes. The frame so-defined is divided into time-slots, as in TDM, and each node is assigned a unique time-slot in which to transmit its information. The resulting frame structure is illustrated in Fig. 17-7, including one frame plus the succeeding reference burst. Each node, of which there are N, transmits a *traffic burst* within its assigned time-slot. Thus far, the approach is similar to TDM, but there are significant differences. First, since there are significant delays between nodes, each node receives the reference burst with a different phase, and thus its traffic burst is transmitted with a correspondingly different phase within the time slot. There is therefore a need for *guard times* to take account of this uncertainty. Each time-slot is therefore longer than the period needed for the actual traffic burst, thereby avoiding the overlap of traffic bursts even in the face of these propagation delays. Second, since each traffic burst is transmitted independently with an uncertain phase relative to the reference burst, there is the need for a *preamble* at the beginning of each traffic burst to allow the receiver to acquire timing and carrier phase.

Finally, there must be a centralized control mechanism to assign time-slots and communicate those assignments to the nodes. Time-slots can be *pre-assigned*, implying that changes are infrequent and only as a result of rearrangements, or *demand-assigned*, meaning that frequent reassignments are made to match the ebb and flow of traffic demands.

The reference burst is composed of three parts. A deterministic *carrier and bit-timing* sequence of approximately 30 to 300 symbols enables each node to do accurate carrier recovery and timing recovery for detection of the subsequent information bits. This is followed by a *unique word* with good autocorrelation properties, enabling each node to establish an accurate time-reference within the reference burst. The unique word is entirely analogous to the added framing bits in a TDM frame. Finally, there is the *control and delay channel*, which is a set of information bits used for control of the nodes. It enables the central control to assign time-slots, and can even be used to control the phase of a node's traffic burst within a time-slot, thereby reducing the size of the guard time. The traffic burst preamble has a very similar structure. The control algorithms can get fairly complicated, and the reader is referred to [3] for a more detailed discussion.

Early satellite systems utilized multiple access by FDM, to be described later, but the current trend is to use TDMA.

Example 17-10.

The INTELSAT system uses TDMA for high-volume international traffic. The basic bit rate is 120.832 Mb/s using four-phase PSK so that the baud rate is 60.416 MHz. The basic frame is 2 ms in length, this being 16 times the frame period of both of the standard primary multiplex standards in the world (Example 17-5 and Example 17-6). Each time-slot is assigned to one of these primary bit streams. For a G.732 bit stream (Example 17-5), the traffic burst will contain (2 ms) · (2.048 kb/s) = 4096 information bits. At a bit rate of 120.832 Mb/s, the information portion of the traffic burst consumes 33.9 μsec. Forgetting the reference burst, guard times, and traffic burst preambles, the frame has room for 59 of these G.732 bit streams.

17.2.3. Packetizing

The circuit switching approach described in Section 17.1 allocated a fixed bit rate, corresponding to a reserved time-slot, to each of the multiple users of a given transmission system. This approach is simple to implement, but also cannot provide a time-varying bandwidth or a bandwidth on demand. It is also inflexible in that a fixed maximum number of users can be accommodated on any given transmission system. There are many examples of

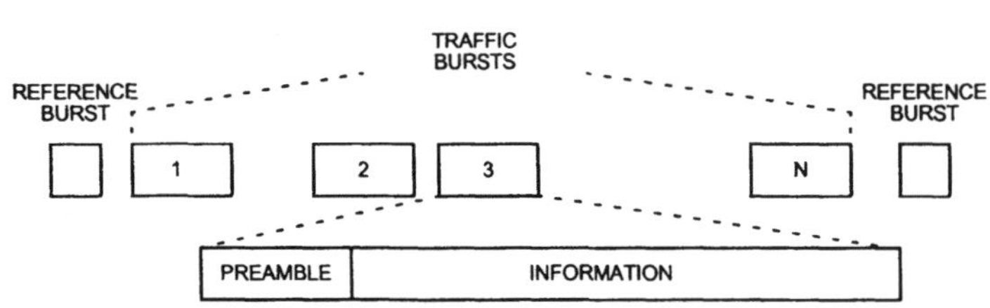

Fig. 17-7. The frame structure of a TDMA system.

services that inherently require a time-varying bit rate, and in this case using circuit switching we must provide a circuit commensurate with the *maximum* bit rate requirements. Circuit switching is therefore somewhat inefficient for these services.

Example 17-11.

In conversational speech, normally only one direction of the conversation is active at any given time. During the resulting silence intervals, a smaller or even zero transmission capacity is required. Circuit switching is therefore no better than 50% efficient.

Example 17-12.

In video transmission, the image can be reconstructed with high accuracy from past images during periods of limited motion. Therefore, for a given fidelity higher bit rates are inherently required during periods of high motion than during periods of reduced motion.

Example 17-13.

In interactive data transmission, transmission capacity from a user typing at a keyboard is sporadic, depending on when keys are pressed. For this case, circuit switching is grossly inefficient.

Example 17-14.

An alarm system connected to an alarm service bureau requires transmission capacity only infrequently to transmit "all is well" messages and even more infrequent "panic" messages.

An additional problem with circuit switching is that mixing different services with different bit rate requirements, even when their bit rates are constant with time, becomes administratively complicated.

Packet switching or *store-and-forward switching* provides these capabilities lacking in circuit switching. It allows us to mix bit streams from different users that vary in bit rate, and dynamically allocates the available bandwidth among these users. In this subsection we will describe the simple technique of packetizing as applied to multiplexing multiple users onto a single communications link, and in the following subsection we discuss multiple access using packetizing.

The basic idea of packetizing a bit stream as a flexible approach for multiplexing a number of such streams together is very simple. The information bits from each user are divided into groups, called *information packets* or just *packets*. A packet is therefore analogous to a time-slot in TDM, although packets are typically much larger (hundreds or thousands of bits as opposed to one or eight). Packets can contain a fixed number of information bits, the same for each packet, or more often the number can be variable from one packet to another (usually with a maximum). The basic idea is then to interleave the packets from different users on the communications link, not unlike in TDM except that by not pre-determining the order of packets from different users or the size of packets we can readily vary dynamically the bit rate assigned to each user.

Example 17-15.

End users sitting at terminals and communicating to central computers typically type a line of characters, followed by a "carriage return," and then expect some response from the computer. Thus, it would be natural to form a packet of user information bits corresponding to this line of characters plus carriage return.

Example 17-16.

Conversational speech consists of active speech intervals interspersed with silence intervals. It would be natural to associate a packet with each active interval. In practice, since active intervals can get quite long, we can associate two or more packets with a single interval of active speech.

In TDM we identified at the output of the link the individual time-slots corresponding to different users by adding framing bits and doing framing recovery at the receiver. We must realize the analogous function in packet switching, although by a somewhat more flexible mechanism. First recognize that since the bit rates provided to each users is varying dynamically, we have a problem if the sum of the incoming bit-rates is instantaneously higher than the total bit-rate of the communications link. We will defer this problem until the next subsection, and for the moment assume that the incoming bit-rates sum to less than the link bandwidth. This implies that there must be *idle-time* on the link. During this idle-time, by mutual convention between transmitter and receiver, we transmit a deterministic sequence, say all-zeros. We must then have some way of identifying at the receiver the beginning and end of a transmitted packet. Since the beginning and end of packets do not occur at predetermined points in time, we say that the packets are *asynchronous*. The process of determining the beginning and end of packets at the receiver is called *synchronization* to the packets, and is analogous to framing recovery in TDM.

For purpose of synchronization, we append to the beginning and end of a packet additional bits, called *synchronization fields*. The combination of the packet, together with the synchronization field and other fields yet to be described is actually the unit of bits transmitted on the link, called a *link frame*. Thus, this frame is analogous to the frame defined in TDM, except that in TDM the frame corresponds to bits from *all users*, whereas in packet switching it corresponds to the bits from a *single user*. The synchronization fields are entirely analogous to added framing bits in TDM, since they provide a deterministic fixed reference in the bit stream for the start and end of the bits corresponding to one user. The term *field* applies in general to any fixed-length collection of bits added to the information packet to form the link frame. We will see examples of other fields shortly.

Example 17-17.

An internationally standardized packet switching format is *high-level data link control (HDLC)* [5][6]. Many specific formats are subsets of HDLC, such as the common international packet switching interface standard *X.25*. HDLC uses one synchronization field at the beginning and one at the end of a link frame. Each field is called a *flag*, and consists of the octet "01111110." Since by convention an idle link contains all-zeros, it is easy to recognize the beginning of a link frame by observing this particular octet. However, we still have a problem with reliably detecting the end of a link frame, since the octet "01111110" may occur in the information packet, thus prematurely ending the link frame. To bypass this problem, we use a form of *variable-rate* coding. We simply insert within the packet any time that five consecutive one-bits appear, an additional zero-bit. Thus, within the packet, before appending the flags to form a link frame, we use the encoding rule

$$11111 \rightarrow 111110$$

and at the demultiplexer we apply the decoding rule

$$111110 \rightarrow 11111$$

to recover the original packet. Note that this *bit-stuffing encoding* increases the number of bits in the packet by the maximum ratio 6/5 or 20%. This is the price we pay for avoiding the replication of the end flag in the information packet. This scheme allows a variable number of bits per packet, since there is no presumption of the length of a packet. In fact, the variable-rate encoding *ensures* that the link frames, if not the packets, will be variable length, since the number of added coding bits will depend on the information bits.

Of course in the presence of bit-errors on the link, we can lose synchronization of the beginning and end of a link frame (Problem 17-3). In this eventuality we must have a *recovery procedure* analogous to framing recovery in TDM.

Once we have synchronized to the beginning and end of a link frame at the receiver, there is still need to identify the user corresponding to each packet. For this purpose, we include the concept of a *connection*, where one user (a human being or computer) may have multiple connections. In TDM each connection corresponded to a time-slot, and the identification of connections was implicit in the location of a time-slot within the frame. In packetizing we usually also identify the concept of a connection, but the flexibility of the packetizing approach results in no fixed correspondence between connections and the sequence of packets. Therefore we must add an *address field* to the link frame. Each connection through a link is assigned a unique address, and the size of the address field places a restriction on the number of simultaneous connections (in contrast to TDM where the number of time-slots in a frame limits the number of connections).

Example 17-18.

For HDLC in Example 17-17, an address field of one octet immediately follows the start flag. Thus, there are 256 possible simultaneous connections. The full HDLC link frame is shown in Fig. 17-8. There are two additional fields, the control and error check fields. Each packet has appended six octets or 48 bits to form the link frame. Obviously, if only very short packets are transmitted, the overhead is substantial. On the other hand, if the packets average thousands of bits, the overhead as a fraction of the entire link bandwidth is insignificant.

Statistical Multiplexing

We have seen a packetizing technique for sharing a link among a set of connections in a much more flexible manner than in TDM. However, we have ignored some important problems that are brought out when we consider the design of a multiplexer that takes a number of lower-speed links, each packetized, and multiplexes them together in a higher-speed link. This device is called a *statistical multiplexer*, and is the packetizing equivalent of a TDM multiplex.

Fig. 17-8. The data link frame defined for HDLC.

The two problems that must be addressed are the probability of two or more packets arriving simultaneously at the statistical multiplexer, or more generally the possibility that since each connection has a variable bit rate, the total incoming bit rate exceeds the bandwidth of the link. If the latter condition persists indefinitely, since the multiplexer has a finite internal memory it is inevitable that some packets must be *lost*. The probability of this occurring can be minimized by using some sort of *flow control*, which is a mechanism for telling the originators of the packets that they must reduce their bit rate due to over-utilization of some link.

Even in the absence of over-utilization in the long-term sense, it is inevitable that the instantaneous bit rate into the multiplex will sometimes exceed the total link bit rate out of the multiplex. If this were not true, the output bit rate would exceed the sum of the maximum bit rates into the multiplex, in which case we would consider using the simpler TDM. This leads to the conclusion that the multiplex must include internal buffer memory for storage of the excess bits. Statistical multiplexing makes more efficient use of the link bandwidth by reducing the output bit rate below the maximum instantaneous sum of incoming bit rates. When the input bit rate exceeds the output rate, it stores the excess bits until the input bit rate is subsequently lower than the output, at which time it transmits the bits in memory. It therefore takes advantage of the *statistics* of the variable-rate inputs to make more efficient use of the output link. The price we pay for this efficiency is the *queueing delay* that is an inevitable consequence of temporarily storing packets in internal buffers before transmission. This delay is statistical in nature, so that we must speak in terms of the average delay, the distribution of the delay, and so forth. The impact of this delay depends on the service that is being offered by packetizing.

Example 17-19.

In interactive data transmission, the queueing delay results in a slower response time from a remote computer. This delay can be disturbing if it gets to be large.

Example 17-20.

Using packetizing for speech transmission results in a random change in the temporal characteristics of the speech unless we take additional measures. One approach is to add to each packet a *time-stamp* which indicates roughly the time elapsed since the end of the last packet (i.e. duration of the silence interval). At the receiver we can restore the rough temporal relationship by adding another buffer that forces a constant rather than variable delay. Based on knowledge of the currently experienced delay variation, this buffer adds a delay to each packet so that the total delay (queueing delay plus buffer delay) adds to a constant. This constant delay will be subjectively better than a variable delay, although it introduces its own conversational and echo problems.

We always have to cope with lost packets in statistical multiplexing. This is because with some non-zero probability, the instantaneous input bit rate will exceed the capacity of the output link for a sufficiently long period of time that the finite buffer capacity of the multiplex will be exceeded, and bits will be lost.

With these concepts in mind, the statistical multiplex is easy to understand, and is illustrated in Fig. 17-9 [7]. Associated with each input link to the multiplex is an input buffer which stores packets as they arrive. The remainder of the link frame, with the exception of the address, can be discarded. The address must be retained and attached to the information packet on the output link. Separate buffers are required because packets may be arriving simultaneously on the input links. A control unit examines the input buffers for packets, and when it finds them places them in the output buffer in the order in which they are to be transmitted. Strictly speaking, the output buffer would not be required, since packets could be removed from the input buffers and transmitted directly. However, the use of an output buffer allows the control unit more freedom in its operation, and also conveniently keeps the packet in memory so that it can be retransmitted if it is lost (should the protocol in use dictate that). Each of the buffers is a *first-in first-out (FIFO) buffer*, meaning that the packets are read out in the order they arrived at the buffer input.

The designer of the multiplex must decide on some strategy for moving packets from input buffers to output. This is called the *queue management discipline*, and affects the distribution of queueing delays experienced by packets passing through the multiplex. The most obvious discipline would be to take the packets out of input buffers in order of arrival, called a *first-in first-out discipline*. The FIFO discipline would make the whole multiplex equivalent to a single FIFO queue, which is simple to analyze. However, this disciple is difficult to implement, since each packet stored in a buffer would have to have an associated time-of-arrival stamp, and the control unit would have to examine this stamp for the oldest packet in every buffer prior to choosing one for transmission. A much more practical discipline is *roll-call polling*, in which the control unit systematically goes through the buffers in order, removing all waiting packets at each buffer. Other disciplines are possible (Problem 17-4). In the evaluation of different strategies, the primary considerations would be implementation and the delay characteristics. Most disciplines are very difficult to analyze, and occasionally we must resort to simulation.

In order to illustrate the essential delay characteristics of a statistical multiplex, we will analyze the discipline that is the most analytically tractable. This is the FIFO discipline, because in this case the multiplex is equivalent to a single FIFO queue, for which we can directly use the analytical results in Chapter 3. For purposes of calculating the delay, two characteristics are important other than the queue management discipline. One is the process modeling the arrival of packets at the multiplex, and the other is the distribution of service times. For a statistical multiplex, the service time is the time required for transmission of the packet, which for a fixed bit rate output link is proportional to the length of the packet. The

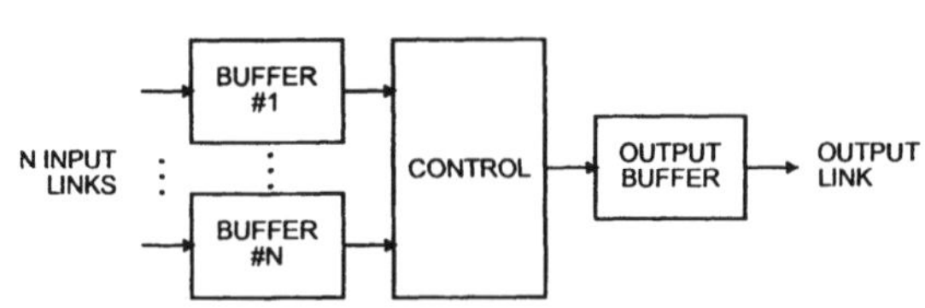

Fig. 17-9. Block diagram of a statistical multiplexer.

analytically simplest case is the M/M/1 queue analyzed in Section 3.4, where the packet arrivals are Poisson, the queue has an unlimited number of waiting positions, and the service time is exponentially distributed. Let the *total* arrivals on all N input links be a Poisson process with arrival rate λ, and let the service time have mean-value $1/\mu$ (the service rate is μ). The quantity

$$\rho = \lambda/\mu \tag{17.1}$$

is the average server utilization, or in this case *link utilization*; that is, the average fraction of the time that the output link is transmitting packets. Obviously we would be happiest if this quantity were as close to unity as possible, but as we will see this leads to large queueing delays. The average queueing delay, or the average time a packet sits in the buffer before beginning transmission, is given by (3.152),

$$D = \frac{1}{\mu} \cdot \frac{\rho}{1-\rho} \ . \tag{17.2}$$

The term $1/\mu$ is the average transmission time, and the queueing delay can be smaller (when $\rho < \frac{1}{2}$) or larger (when $\rho > \frac{1}{2}$). The important point to realize is that the queueing delay gets very large as the output link utilization approaches unity. This places an upper bound on the link utilization that can be achieved. Intuitively, by keeping the utilization low, we enhance the probability that the link will be free when a packet arrives.

Packet Switching

The statistical multiplex can be modified to provide a switching function as well. A *packet switch* has an equal number of input and output links. The function of the switch is to route packets on each incoming link to the appropriate output link. Thus, the switch maintains an internal table which translates from an address and input link number to a corresponding output link and associated address. The internal configuration of the packet switch is similar to Fig. 17-9, except that there are N output buffers and N output links.

In practical implementations packet switching involves many complexities that we have barely touched on here. There are many complex procedures for establishing and taking down connections and for recovery from every conceivable kind of error. These procedures plus the details of the link frame constitute the set of *protocols* used to form a complete communications network based on packet switching. Packet switching has many advantages over circuit switching to exchange for its greater complexity. In addition to the ones mentioned earlier, a particularly important distinction is that packet switching encourages the user to establish many simultaneous virtual connections through a communications network. As such, it is an extremely flexible approach for situations in which users wish to access many computational or peripheral resources simultaneously or in quick succession.

Example 17-21.

In an office environment, many users may wish to access the same printer or storage device. Packet switching encourages this, because it allows that device to establish simultaneous connections to many users.

17.2.4. Packet Switching Multiple Access

Thus far we have discussed packet switching in the context of an important special case of multiple access, the link topology. It is applicable more generally, and in fact most of the practically useful multiple access techniques are based on packet transmission. The techniques are distinguished mainly by the degree of *centralized* vs. *distributed control* and by their *efficiency* and *stability*. Efficiency refers to the degree to which they can successfully use the available bandwidth on the multiple access medium, since they all require some overhead and idle time, and stability refers to the possibility that during high utilization the frequency of lost packets can increase uncontrollably.

Collision Avoidance by Polling

In the typical multiple access situation, as opposed to the statistical multiplex, the nodes transmitting on the medium are distributed, and have significant propagation delays between them. Each node can be considered to contain a buffer, in which packets are queued awaiting transmission. The goal is to allow the nodes to transmit their packets, but to coordinate the transmissions so that they do not collide. In TDMA we accomplished this through the relatively inflexible time-slot assignment. By using packetization approaches, we can allocate the medium bandwidth much more dynamically among the nodes.

We already mentioned in connection with the statistical multiplex a valuable approach directly applicable to this problem — *polling*. Polling comes in two forms, *roll-call polling* which is managed by a central controller, and *hub polling* which is more distributed. In roll-call polling, a central controller sends a message to each node in turn, letting it know that it is allowed to transmit. That node then either transmits the packets waiting in its buffer (or perhaps only one packet maximum, depending on the discipline), or sends back a message indicating that its buffer is empty. Every node is aware of the status of the medium at any given time, and in particular which node (including the central controller) is currently authorized to transmit. There is therefore no possibility of a collision on the medium.

Example 17-22.

Roll-call polling is widely used in dispersed networks of voiceband data modems accessing a centralized computer facility, as for example an airline reservations system. The wide use of polling techniques with voiceband data modems motivates the desire for these modems to acquire timing and carrier quickly, and has led to a lot of research in fast acquisition. As will be shown in a moment, reducing the acquisition time, and therefore the overhead in polling, enhances the performance of the polling technique.

Hub polling, called *token-passing* in the context of local-area networks, eliminates the central controller except for initialization. In this case, a token is possessed by precisely one node in the network at any given time. This token is not a physical object, but rather an authorization to transmit on the medium. The node possessing the token can transmit, and at the end of that transmission must pass the token on to a predetermined next node (the token is passed through a message transmitted on the medium). In this fashion, as the token is passed among all the nodes each has an opportunity to transmit its packets.

Example 17-23.
A popular LAN architecture is the *token-passing ring*. The topology is a ring, and the token is passed around the ring from one station to another. The technique is relatively simple because each node passes the token to its nearest neighbor on the ring by transmitting a generic message that is intercepted and removed by the next node. There is no need to know the details of who the nearest neighbor is, its address, etc.

We can understand polling techniques better by performing a simple calculation of the *average polling cycle time*, or the time required for each node to be given the opportunity to transmit its packets. Assume that the total overhead time for one polling cycle is W — this includes the time for the messages to pass from the controller to nodes and back, or the time to transmit the tokens. Then we get that the polling cycle time is

$$T_c = W + \sum_{i=1}^{N} T_i , \tag{17.3}$$

where T_i is the time it takes node i to transmit all its packets when polled. Taking the expected value of this,

$$E[T_c] = W + N \cdot \mathrm{E}[T_i] , \tag{17.4}$$

assuming that each node has the same utilization. If each node has utilization ρ, that is it transmits a fraction ρ of the time, then considering that each node transmits the packets accumulated during one polling cycle,

$$\mathrm{E}[T_i] = \rho \cdot E[T_c] , \tag{17.5}$$

and combining these two equations we get

$$\mathrm{E}[T_c] = \frac{W}{1 - N\rho} . \tag{17.6}$$

As expected, as the total utilization $N \cdot \rho$ approaches unity, the polling cycle lengthens, and as it approaches zero the polling cycle time approaches the overhead W. The fraction of the polling cycle devoted to overhead is

$$\frac{W}{\mathrm{E}[T_c]} = 1 - N\rho . \tag{17.7}$$

This equation illustrates a very important advantage of polling; namely, as the total utilization approaches unity, the fraction of the time devoted to overhead decreases to zero. Thus, the overhead is only appreciable when the utilization is low, a situation in which we don't care, and is insignificant when the utilization is high, which is precisely what we would hope. Intuitively this is because during high utilization the polling cycle is very long, and the fixed overhead becomes insignificant as a fraction of this cycle.

An undesirable feature of polling is that the overhead time W increases with the number of nodes in the network. Thus, it becomes inefficient for networks with a very large number of nodes, each with low utilization of the medium. In this situation the polling cycle becomes dominated by the overhead. For these types of networks, random access techniques as described in the following subsection are very desirable because they can obtain comparable performance but without the complexity of the centralized control.

Random Access

Thus far in this chapter the focus has been on *avoiding* collisions in multiple access to a common medium. On media that are used with a low utilization, the complexities associated with avoiding collisions can be circumvented by a strategy of allowing collisions to occur, detecting those collisions, and retransmission with a random delay. This enables each node of the network to operate autonomously, with no central control required.

The first and simplest such strategy was invented by N. Abramson of the University of Hawaii in 1970 [8] and is known as *pure ALOHA*. ALOHA is appropriate for broadcast topologies, such as the bus or satellite, where collisions are easy to detect because each node listens to all transmissions including its own. If a node cannot successfully monitor its own transmission packet, it can assume that a collision has occurred. The technique is then very simple: each node simply transmits a packet as desired, regardless of conditions on the medium, and monitors the medium for a collision. When a collision occurs, the node waits for a random delay time, and retransmits. The random delay time makes it less probable that the retransmissions of the colliding nodes will again collide.

We can analyze this system easily if we make a simplifying assumption. Even if the incoming packets to the system have Poisson arrivals, we would not expect that the aggregate of packets on the medium, including retransmitted packets, would be Poisson. However, we assume this to be the case, yielding an approximate analysis that has proven on further examination to be accurate as long as the random retransmission times are long relative to a packet length. Further assume that packets are a *fixed length* $1/\mu$, for simplicity of calculating the probability of a collision. Let the total rate of arrivals of packets to the system be λ_{in} and let the rate of arrivals of packets on the medium including retransmissions be $\lambda_{out} > \lambda_{in}$. Due to the fixed-length packet assumption, if we transmit a packet at time t_0, then a collision occurs if someone else transmits a packet in the interval $[t_0 - 1/\mu, t_0 + 1/\mu]$. Due to the Poisson assumption, the probability of *no* collision is the probability of zero Poisson arrivals for a Poisson process with rate λ_{out} over an interval of time $2/\mu$, or $e^{-2\lambda_{out}/\mu}$. But the probability of no collision is also the ratio of the rate of incoming packets to packets on the medium, $\lambda_{in}/\lambda_{out}$, since the excess are retransmissions due to collisions. Setting these two expressions equal, we get

$$\lambda_{in}/\lambda_{out} = e^{-2\lambda_{out}/\mu}. \tag{17.8}$$

It is convenient to express this in terms of the utilization due to incoming packets and the total utilization of the medium,

$$\rho_{in} = \frac{\lambda_{in}}{\mu}, \qquad \rho_{out} = \frac{\lambda_{out}}{\mu}, \tag{17.9}$$

in which case we get the interesting relation

$$\rho_{in} = \rho_{out} e^{-2\rho_{out}} . \tag{17.10}$$

In this equation, ρ_{in} is the independent variable, the incoming traffic, and ρ_{out} is the dependent variable, the traffic on the medium. This implicit relation is plotted in Fig. 17-10 with the independent variable on the abscissa. Note that the utilization of the medium can exceed unity, as we would expect in the case of a large number of collisions and retransmissions. What is most interesting about the curve is that the incoming traffic can never exceed 0.18, or in other words the random access technique can only work for very low utilizations.

Exercise 17-2.

Show that the maximum input traffic in Fig. 17-10 corresponds to $\rho_{in} = 1/(2e) = 0.18$ and $\rho_{out} = \frac{1}{2}$.

Another interesting feature is the double-valued nature of the curve — for each ρ_{in} in the allowed region, there are *two* operating points possible, one with a small number of collisions and the other with a large number of collisions. This implies a form of *instability*, since the system performance cannot be predicted unambiguously.

Since the original concept of pure ALOHA, a great deal of effort has been expended to increase the throughput of random access techniques and to ensure stability. A simple refinement, known as *slotted ALOHA* (Problem 17-7) results in a doubling of throughput to 0.37. A class of control algorithms known as *collision-resolution algorithms*, first conceived by J. Capetanakis at M.I.T. [9] can ensure stability on a random access channel. The maximum throughput that can be achieved on such a channel is known only to fall in the range of 0.45 to 0.59. Many have speculated that it must be 0.5. See [10] for an excellent review of results on this topic.

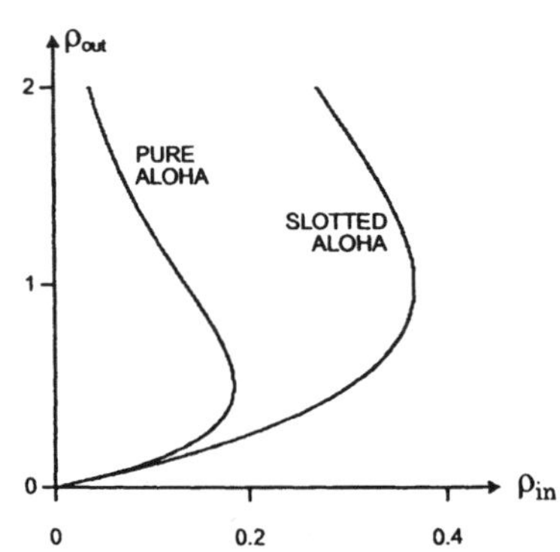

Fig. 17-10. The throughput of the pure ALOHA discussed in the text and the slotted ALOHA derived in Problem 17-7.

CSMA/CD

Random access protocols are widely used in local-area networks, although using a modification of the ALOHA system in which collisions are largely (but not entirely) avoided. The most common approach is for each node to first listen to the medium to determine if a transmission is in progress before proceeding to transmit. This is known as *carrier-sense multiple access (CSMA)*. This greatly reduces the probability of a collision, but does not rule it out because due to propagation delays it is not always possible to detect that another node has just started a transmission. It is therefore common to use *collision-detection (CD)* by listening to one's own transmission, and if a collision is detected to transmit a special *jam signal* that serves to notify other nodes to that effect and then suspend transmission. A substantial improvement in throughput is obtained by reducing the number of collisions and by aborting transmissions in progress when a collision occurs.

Example 17-24.
CSMA/CD schemes differ in the retransmission strategy once a collision occurs [6]. The best-known CSMA/CD system is the popular *Ethernet* local-area network, usually implemented on a coaxial cable at a bit rate of 10 Mb/s. In Ethernet [11] after a collision is detected and transmission is aborted, a node retransmits after a random delay. This retransmission interval is doubled upon each successive collision, up to some maximum.

17.3. Multiple Access by Frequency Division

The separation of many users on a common medium in analog transmission invariably uses frequency division. For example, on the familiar AM radio dial, the stations are selected by tuning a variable frequency filter. Frequency division can also be used for digital transmission, where independent data streams are transmitted in non-overlapping frequency bands. This occurs most commonly when data streams are transmitted over existing analog carrier systems. But it is also used when data streams are transmitted by radio or satellite, and for full duplex data transmission.

Frequency-division multiplexing is very simple, as illustrated in Fig. 17-11, in this case for just two users. The two transmitters sharing the medium have output power spectra in two non-overlapping bands, where they usually use passband PAM modulation to achieve this. To ensure that this is the case, it is common to put bandpass filters at the output of the transmitters, particularly where strict requirements on spectral utilization are in force (as in the case of the FCC mask for terrestrial microwave radio). At the two receivers, similar bandpass

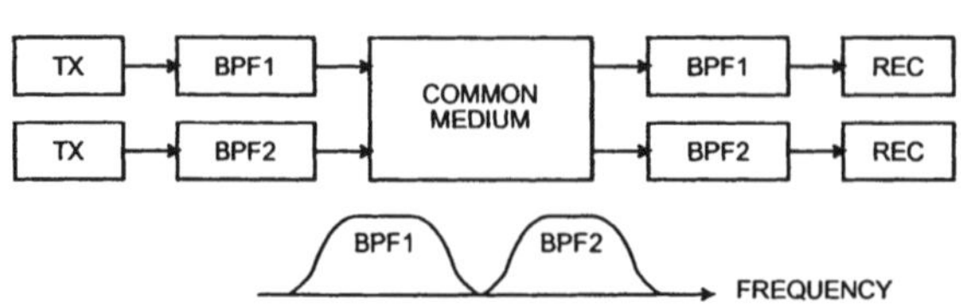

Fig. 17-11. Sharing a medium between two users using frequency-division multiplexing.

filters eliminate all but the desired data signals. The path to a receiver from the undesired transmitter contains two bandpass filters with non-overlapping passbands, and therefore we can make the loss of the crosstalk path as large as we like through the filter design. A consideration in many microwave media (terrestrial radio and satellite) is nonlinearity of the amplifiers in the transmitter or the satellite transponder. These nonlinearities will create out-of-band energy that can interfere with other FDM channels, and it is common to use bandpass filtering *after* the amplifier to reduce these spectral components.

FDM has both advantages and disadvantages over TDM. A major disadvantage on all media is the relatively expensive and complicated bandpass filters required, whereas TDM is realized primarily with much cheaper logic functions. Another disadvantage of FDM is the rather strict linearity requirement of the medium. However, some of the propagation-delay and crosstalk problems of TDM are eliminated by FDM. The technique of choice then depends on the situation.

Example 17-25.

On microwave radio channels, FDM techniques have been used primarily, but TDM is in the process of taking over. The primary advantage of TDM, and particularly TDMA in the case of the satellite channel, is the lowered susceptibility to nonlinearity of amplifiers. Using TDM, the amplifiers can be used nearer saturation, giving a greater available power. The reduction of filtering requirements is an added benefit.

Example 17-26.

In full-duplex data transmission, TDM in the form of TCM is impractical on media with long propagation delays, such as for voiceband channels. Lower-speed voiceband data modems therefore use FDM for full-duplex transmission, as shown in Fig. 17-12. The hybrids turn the two-wire medium into an effective four-wire medium, and the stopband loss of the bandpass filters is added to the hybrid loss resulting in good isolation of the two directions. FSK modulation is used to generate a passband signal, with two different carrier frequencies used for the two directions, at 300 b/s, and passband PAM (PSK and QAM) is used at higher frequencies.

Example 17-27.

For the digital subscriber loop, FDM has the major advantage of eliminating near-end crosstalk, and is not susceptible to propagation delay like TCM. However, it is not used because of the complicated filtering and the larger bandwidth (as compared to echo cancellation) on a medium with an attenuation increasing rapidly with frequency.

Example 17-28.

In direct detection optical fiber systems, FDM in the form of *wavelength-division multiplexing* is sometimes used. A few channels are separated by transmitting them at different wavelengths. In the

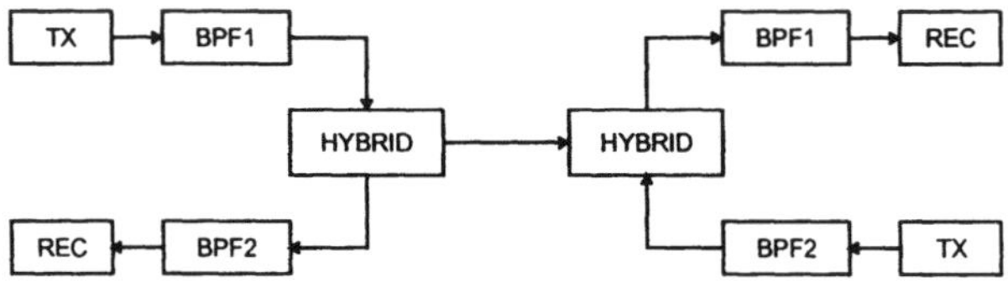

Fig. 17-12. Frequency-division multiplexing of two directions in full-duplex transmission.

future, coherent optical fiber will encourage the use of FDM multiplexing of large numbers (hundreds or thousands) of digital streams. This looks like a very promising use of FDM.

17.4. Multiple Access by Code Division

Separating signals in time or frequency is a relatively simple way to ensure that the signals will be orthogonal. However, it is by no means the only way. In this section we briefly describe the design of orthogonal signals by *code division*, which is closely related to spread-spectrum (Section 6.4).

With *code-division multiple-access (CDMA)*, the objective is to transmit signals from multiple users in the same frequency band at the same time. We can place this alternative in perspective by comparing it to TDMA and FDMA. Suppose we have B Hz of available bandwidth, and want to share this bandwidth over N users. Several options are available:

- *TDMA*. A single stream of PAM pulses with symbol rate $1/T = B$ can be transmitted. (We assume throughout passband PAM with maximum symbol rate consistent with the Nyquist criterion.) We can divide the stream of pulses into N time slots, assigning each timeslot to one of the N users, so that each user has available a net symbol rate of $1/(NT) = B/N$.
- *FDM* or *FDMA*. Divide the frequency band B into N disjoint frequency bands, and assign one to each user. Each user then gets a frequency band B/N Hz wide, and the highest symbol rate that each user can achieve is $1/T = B/N$, the same as with TDMA.
- *CDMA*. In accordance with the generalized Nyquist criterion, design N orthogonal pulses. These pulses each satisfy the Nyquist criterion with respect to symbol interval T, and are mutually orthogonal for all translates by multiples of T sec. A requirement of the generalized Nyquist criterion is that $B \geq N/T$. Each user is assigned one of these orthogonal pulse shapes to use within bandwidth B Hz, and transmits with symbol rate $1/T$ Hz. The receiver for each user consists of a sampled matched filter, which and has no ISI at the output and does not respond to the orthogonal pulses. The maximum symbol rate that each user can achieve is $1/T = B/N$, the same as TDMA and FDMA.

Each of these approaches can achieve, in principle, the same aggregate spectral efficiency, since each one achieves the same symbol rate per user, the same number of users, and the same total bandwidth. Each approach has its advantages and disadvantages. In this section, we concentrate on CDMA.

As shown in Section 6.4, pulses with large bandwidth relative to the symbol rate can be generated using a combination of a chip waveform and a spreading sequence. This is known as *direct-sequence spread spectrum*. Further, maximal-length shift register sequences have the appropriate properties to be used as spreading sequences, and are very easy to generate in the transmitter and receiver. For CDMA, we can assign one period of a maximal-length shift register sequence to each user to use as their spreading sequence. All users use the same chip waveform. (This constrains the spreading factor N to be of the form $2^n - 1$ for some integer n. By assigning different generator polynomials to each user, orthogonality can be achieved. An example of a set of generator polynomials that achieve orthogonality is the *Gold code* [12].

In principle CDMA achieves the same spectral efficiency as TDMA or FDMA, but in practice there are factors that give it much different characteristics. Two considerations are the *near-far* problem, and the *partial correlation* problem. Partial correlation arises where no attempt is made to synchronize the transmitters sharing the channel, or when propagation delays cause misalignment even when transmitters are synchronized. This is shown in Fig. 17-13, where the symbol interval for one transmitter is delayed by τ seconds relative to the other. In order to avoid crosstalk between the two users i and j, two *partial correlations* must be zero,

$$\int_0^{T-\tau} h_i(t)h_j(t+\tau)\,dt = 0\,, \tag{17.11}$$

$$\int_{T-\tau}^{T} h_i(t)h_j(t+\tau-T)\,dt = 0\,. \tag{17.12}$$

Thus, simple orthogonality of the isolated pulses is not sufficient — these two partial correlations must also be zero, or at least small, for any value of τ. This property is not guaranteed by the generalized Nyquist criterion of Chapter 6, which assumes a known relationship between the different symbol intervals. The partial correlations can be reduced by proper choice of the spreading sequences, but cannot be totally eliminated. This implies that in practice, complete orthogonality of the different pulses assigned to different users cannot be maintained. CDMA system capacity is thus typically limited by the interference from other users, rather than by thermal noise.

The near-far problem is analogous to near-end crosstalk on wire-pair media, and results when geographically dispersed users are sharing a common medium such as a radio channel. If all the users transmit at the same power level, then the received power is higher for transmitters closer to the receiving antenna. Thus, transmitters that are far from the receiving antenna are at a disadvantage with respect to interference from other users. This inequity can be redressed by using *power control*. Each transmitter can accept central control of its transmitted power, such that the power arriving at the common receiving antenna is the same for all transmitters. In other words, the nearby transmitters are assigned a lower transmit power level than the transmitters far away.

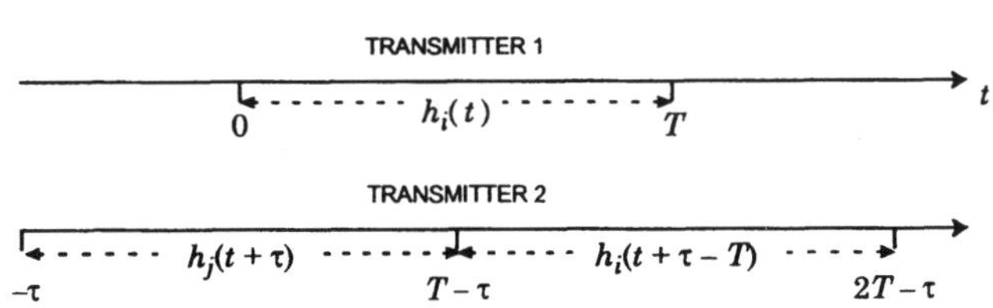

Fig. 17-13. Illustration of the phase relationship between symbol intervals of two users of a CDMA system.

When there are N total users in a CDMA system, then from the perspective of one particular transmitter there are $N-1$ interferers. If power control is used, then each transmission arrives at the receiver with the same power S. The SNR is defined as the ratio of signal power to total interference power, and is

$$SNR = \frac{S}{(N-1)S} = \frac{1}{N-1}. \tag{17.13}$$

By the central limit theorem, the interference, consisting of a superposition of independent transmissions, will be approximately Gaussian. For large N the SNR is very poor, and one might expect unreliable system operation. However, as we saw in Chapter 5, it is not the SNR that matters when a matched filter receiver is used, but rather the ratio of the signal energy per symbol to the interference spectral density. The signal energy per symbol interval is $T \cdot S$ for symbol interval T. If we model the interference signal as white Gaussian noise, then it has total power $(N-1)S$ and bandwidth $2B$ (for positive and negative frequencies), and hence has power spectrum $N_0/2 = (N-1)S/2B$. Thus, the ratio of energy per symbol to noise density is

$$\frac{\sigma_h^2}{N_0/2} = \frac{2BT}{N-1}, \tag{17.14}$$

which reflects the spread-spectrum processing gain $2BT$ (Section 6.4.3). Thus, for a given number of users N, by making $2BT$ large enough we can always make P_e sufficiently small. The minimum value, $2BT = N$ (in accordance with the generalized Nyquist criterion), results in a poor P_e, but we can always make $2BT$ larger than this minimum.

Processing gain is important for CDMA as well as spread spectrum. For N users, where the received power of each user is fixed at S through power control, the $N-1$ interferers represent a signal analogous to the jamming signal discussed in Section 6.4.3. The total power of the interference stays constant as we increase the bandwidth (processing gain), and hence the spectral density of the interference decreases. This decrease in spectral density results in a P_e that decreases as the bandwidth (and processing gain) increases.

Since $2BT \gg N$ in a CDMA system with a small P_e, it appears that CDMA systems have a lower spectral efficiency than TDMA or FDMA systems, which achieve close to $2BT = N$. However, this perspective is oversimplified for many multiple access applications, as illustrated by some examples.

Example 17-29.

CDMA has been proposed for the North American digital cellular telephone network, and a *higher* capacity has been estimated for CDMA than TDMA or FDMA [13][14]. In part, these estimates are based on another factor, voice activity. The transmitter is activated only during periods when the user is talking, and it can be assumed that some fraction of the users are talking at any given time. For a given acceptable interference power (based on the maximum P_e), the number of telephone users can be increased if each user is transmitting only a portion of the time. TDMA and FDMA systems can also take advantage of voice activity, but only by much more complicated mechanisms. Another factor in favor of CDMA is cellular frequency assignment, as discussed in Section 17.5.

Example 17-30.

On a local area network (LAN), each station is typically only transmitting a portion of the time. The CDMA capacity is expressed in terms of the number of active stations at any given time, not the number of total stations. CDMA is advantageous for this type of network because it requires no synchronization of the multiple communications sessions occurring at any given time. It also naturally takes advantage of the low duty cycle of transmission of the stations.

17.5. The Cellular Concept

Thus far in this chapter, we have discussed multiple access by frequency, time, and code division. There is a fourth alternative, which is *space division*. On cables and fibers, this is manifested by parallel communications on physically separate cables or fibers. A more interesting manifestation of space-division multiple access is the *cellular concept* used in mobile radio systems.

In radio systems where a large geographic coverage is desired and large numbers of mobile transceivers must be supported, it is common to divide the region into cells. This is illustrated in Fig. 17-14, where a regular array of base stations (including transmitter and receiver) is deployed, each one dedicated to mobile users in its immediate area. Each mobile transceiver communicates with the nearest base station, resulting in hexagonal regions associated with each base station as shown in Fig. 17-14(a). The motivation is to allow the same carrier frequency to be re-used in different cells, increasing the overall system capacity. Transceivers in two different cells can be assigned the same carrier frequency and time slot, providing that the two cells are far enough apart, since the remote transceiver will suffer a much larger propagation loss. A given mobile transceiver will pass through a succession of

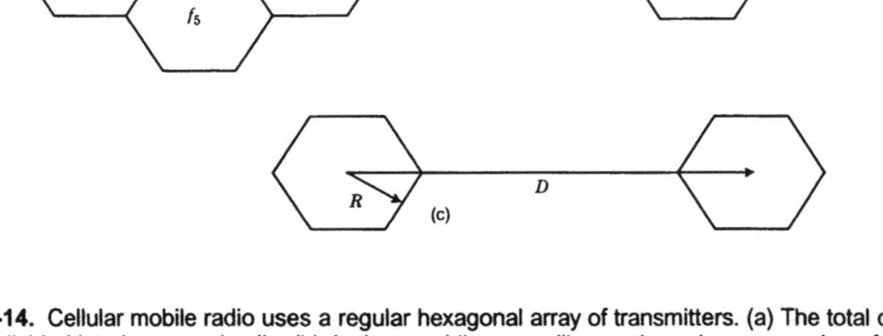

Fig. 17-14. Cellular mobile radio uses a regular hexagonal array of transmitters. (a) The total coverage area is divided into hexagonal cells. (b) A given mobile user will pass through a succession of cells. (c) Each cell has radius R, and is located at a distance D from the nearest cell assigned the same carrier frequency.

cells, as shown in Fig. 17-14(b). As it passes from one cell to the next, it must establish communication with the base station associated with the new cell, possibly requiring a reassignment of carrier frequency and timeslot.

The key parameter of a cellular configuration and frequency assignment is the ratio of D to R, where D is the distance to the nearest cell that uses the same carrier frequency and R is the diameter of one cell (see Fig. 17-14(c)). The larger the ratio D/R, the lower the interference from nearby cells assigned the same carrier frequency.

Example 17-31.
If in a radio system, the propagation obeys the fourth-power law, then given two mobile transmitters with the same transmit power, the desired transmitter at maximum distance R and the interfering transmitter at distance D, then the signal-to-interference ratio is

$$\frac{R^{-4}}{D^{-4}} = (D/R)^4 \ , \tag{17.15}$$

or $40 \cdot \log_{10} D/R$ dB. Thus, the larger D/R, the smaller the interference in relation to the signal. The tradeoff is 12 dB increase in signal-to-interference ratio for each doubling of the D/R ratio.

One goal in designing a modulation scheme for a cellular radio system is to decrease the allowable D/R. All else being equal, this will increase the total system capacity by allowing the same frequency to be assigned to cells closer to one another. Another goal is to design a modulation and multiple access scheme within each cell that is least susceptible to interference from other cells.

Example 17-32.
If FDMA is used, generally the same carrier frequency cannot be assigned to users in adjacent cells. This is because two transmitters using the same carrier frequency can be very close to one another at the boundary between the cells, and cannot be separated at their respective base station sites. However, it is generally permissible to assign the same carrier frequency to cells that are not adjacent, presuming that a D/R ratio on the order of three is permissible. Thus, there is typically a seven-cell frequency reuse pattern, where each carrier frequency is assigned to only one out of each cluster of seven adjacent cells.

Example 17-33.
In the North American IS-54 digital cellular standard for mobile telephony, a combination of FDMA and CMDA is used. Each carrier frequency is modulated with a bit stream that is in turn divided into three (or ultimately six) time slots. Three different users are assigned distinct time slots on that single carrier. The same carrier frequency can be assigned to transceivers in adjacent cells, providing they are assigned to non-overlapping time slots. In this fashion, there will be no interference from adjacent cells.

Example 17-34.
If CDMA multiple access is used, then transceivers *within* cells can be assigned the same carrier frequency and time slot, but are assigned distinct spreading codes. It follows that each carrier frequency can be reused in adjacent cells. The transceivers in adjacent cells manifest themselves as an increase in interference power that can be compensated by increasing the processing gain. That increase in processing gain is reduced by the greater distance to transmitters in adjacent cells. Like

the voice activity factor (Section 17.4), this complete frequency reuse in adjacent cells is a factor that helps compensate for the spectral inefficiency of having to use processing gain to combat interference. A major advantage of CDMA in cellular telephony is the elimination of the need for frequency and timeslot coordination among cells.

Problems

Problem 17-1. For the M12 framing format of Example 17-7, assume that synchronous multiplexing is used so that all the 48I information bits originate on the tributary streams at 1,544 kb/s. (In actuality pulse-stuffing synchronization is used, so this is a hypothetical problem.) For this assumption, determine the following:

(a) The output bit rate.

(b) The time corresponding to one frame and superframe.

Problem 17-2. Rework the parameters of the INTELSAT TDMA system of Example 17-10 assuming that the inputs are G.733 bit streams (Example 17-6). Assume each time slot is assigned to a G.733 bit stream. Given the practical need for guard-times, etc., which primary stream, the G.732 or G.733, is likely to yield the largest number of 64 kb/s voiceband channels over the TDMA system?

Problem 17-3. For the HDLC synchronization method described in Example 17-17, describe qualitatively what will happen or can happen when a single bit-error occurs in the following:

(a) The link frame start flag.

(b) The link frame end flag.

(c) The information packet.

This will identify the situations which will be encountered and suggest recovery procedures that are required.

Problem 17-4. Describe a minimum of two disciplines for the control unit in Fig. 17-9 in addition to FIFO and roll-call polling.

Problem 17-5.

(a) For the FIFO queueing discipline in a statistical multiplex, show that the average number of packets waiting in the multiplex buffers is $\rho/(1-\rho)$.

(b) Show that the probability that the buffer contains M or more packets is ε for $M = \dfrac{\log\varepsilon}{\log\rho}$.

(c) What can you conclude about the size of a finite buffer required to maintain a certain probability of buffer overflow?

Problem 17-6. Assume a statistical multiplex using a FIFO discipline has ten incoming links, each at 1 Mb/s, and one outgoing link at 2 Mb/s. Each incoming link has packets with exponentially distributed lengths, with an average length of 500 bits, and packets arriving on average every 3 msec.

(a) What is the utilization of each of the incoming links?

(b) What is the utilization of the output link?

(c) What is the average queueing delay through the multiplex?

Problem 17-7. *Slotted Aloha* [15]. Show that the following simple modification of pure ALOHA results in a doubling of the throughput. Define time-slots with duration equal to the duration of one packet, and make these time-slots known to each node on the network. Each node then transmits its packets in alignment with the next time-slot after arrival of a packet.

References

1. M. Decina and A. Roveri, "ISDN: Architectures and Protocols," pp. 40 in *Advanced Digital Communications Systems and Signal Processing Techniques*, ed. K. Feher, Prentice-Hall, Englewood Cliffs, N.J. (1987).

2. G. H. Bennett, "Pulse Code Modulation and Digital Transmission," Marconi Instruments, (April 1978).

3. S. J. Campanella and D. Schaefer, *Time-Division Multiple Access Systems (TDMA)*, Prentice Hall, (1983).

4. D. Reudink, "Advanced Concepts and Technologies for Communications Satellites," pp. 573 in *Advanced Digital Communications Systems and Signal Processing Techniques*, ed. K. Feher, Prentice-Hall, Englewood Cliffs, N.J. (1987).

5. D. E. Carlson, "Bit-Oriented Data Link Control Procedures," *IEEE Trans. on Communications*, Vol. COM-28 (4), p. 455 (April 1980).

6. M. Schwartz, *Telecommunication Networks: Protocols, Modeling, and Analysis*, Addison-Wesley, Reading, Mass. (1987).

7. M. R. Karim, "Delay-Throughput and Buffer Utilization Characteristics of Some Statistical Multiplexers," *AT&T Technical Journal*, Vol. 66, (March 1987).

8. N. Abramson, *The ALOHA System*, Prentice-Hall, (1973).

9. J. I. Capetanakis, "The Multiple Access Broadcast Channel," Ph.D. Thesis, Mass. Inst. Tech., (Aug. 1977).

10. J. L. Massey, "Channel Models for Random-Access Systems," *Proceedings NATO Advanced Study Institute*, (July 1986).

11. R.M. Metcalfe and D.R. Boggs, "Ethernet: Distributed Packet Switching for Local Computer Networks," *Communications ACM*, Vol. 19 (7), p. 395 (July 1976).

12. R. L. Pickholtz, D. L. Schilling, and L. B. Milstein, "Theory of Spread-Spectrum Communications – A Tutorial," *IEEE Trans. Communications*, Vol. COM-30 (5), p. 855 (May 1982).

13. K. S Gilhausen, I. M. Jacobs, R. Padovani, A. J. Viterbi, L. A. Weaver Jr, and C. E. Wheatley III, "On the Capacity of a Cellular CDMA System," *IEEE Trans. on Vehicular Technology*, Vol. 40, (May 1991).

14. R. J. Pickholtz, L. B. Milstein, and D. L. Shilling, "Spread Spectrum for Mobile Communications," *IEEE Trans. on Vehicular Technology*, Vol. 40 (May 1991).

15. L. G. Roberts, "ALOHA Packet System With and Without Slots and Capture," *Computer Communications Review*, Vol. 5, p. 28 (April 1975).

Exercise Solutions

Chapter 1

Chapter 2

Exercise 2-1. Addition of two complex-valued signals is illustrated below:

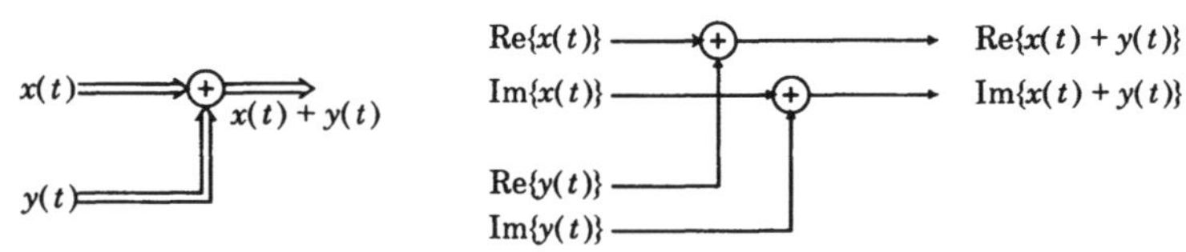

and multiplication below:

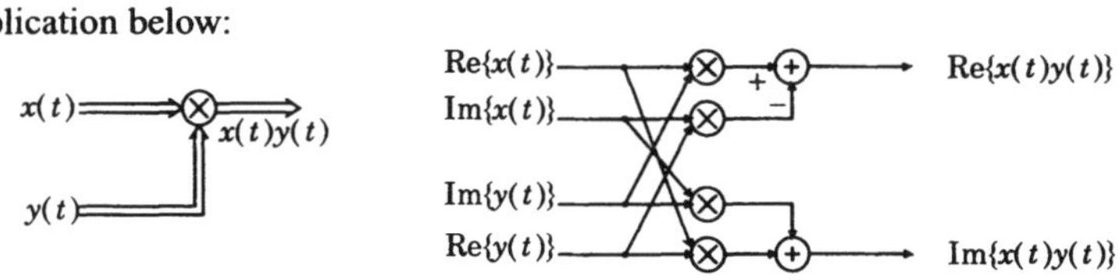

Complex addition is accomplished by two real additions, and complex multiplication by four real multiplications and two real additions.

Exercise 2-2. A complex system with a real-valued input:

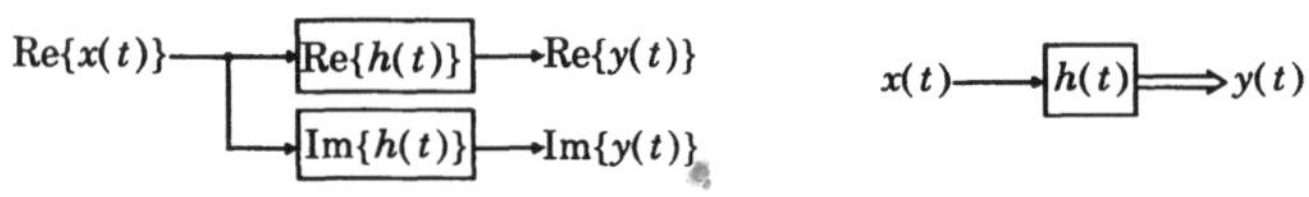

A real system with a complex-valued input:

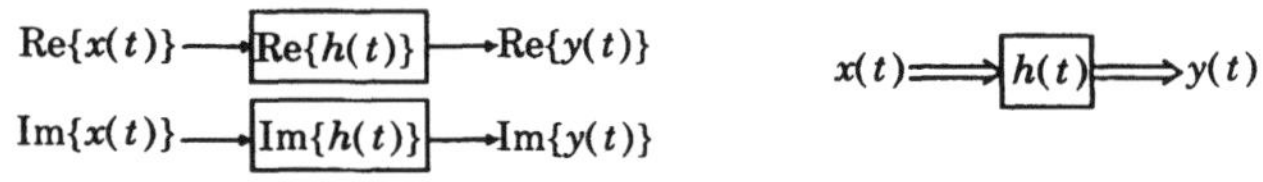

Exercise 2-3. We can treat the convolution $x(t) * h(t)$ just like complex multiplication, since the convolution operation is linear — an integration. To check linearity, for a complex constant A and two input signals $x_1(t)$ and $x_2(t)$,

$$(x_1(t) + A \cdot x_2(t)) * h(t) = x_1(t) * h(t) + A \cdot (x_2(t) * h(t)) \tag{18.1}$$

following the rules of complex arithmetic. This establishes linearity.

Exercise 2-4. $$Y(f) = \int_{-\infty}^{\infty}\int_{-\infty}^{\infty} h(\tau)x(t-\tau)d\tau e^{-j2\pi ft}\, dt\,.$$

Observe that $$e^{-j2\pi ft} = e^{-j2\pi f\tau}e^{-j2\pi f(t-\tau)} \tag{18.2}$$

so that $$Y(f) = \int_{-\infty}^{\infty} h(\tau)e^{-j2\pi f\tau}d\tau \int_{-\infty}^{\infty} x(t-\tau)e^{-j2\pi f(t-\tau)}dt = H(f)X(f)\,, \tag{18.3}$$

after a change of variables.

Exercise 2-5. Take the Fourier transform of both sides of (2.2), getting

$$\begin{aligned} \hat{X}(f) &= \int_{-\infty}^{\infty} \sum_{m=-\infty}^{\infty} x_m \delta(t - mT)e^{-j2\pi ft}dt \\ &= \sum_{m=-\infty}^{\infty} x_m \int_{-\infty}^{\infty} \delta(t - mT)e^{-j2\pi ft}dt \\ &= \sum_{m=-\infty}^{\infty} x_m e^{-j2\pi fmT} = X(e^{j2\pi fT})\,. \end{aligned} \tag{18.4}$$

Exercise 2-6. The impulse response of the system is $g_k = g(kT)$, and hence (2.17) gives the frequency response directly,

$$G(e^{j2\pi fT}) = \frac{1}{T}\sum_{m=-\infty}^{\infty} G(f - m/T)\,. \tag{18.5}$$

Exercise 2-7. Given $X(f) = 0$ for all $|f| > 1/(2T)$, (2.17) implies that

$$X(e^{j2\pi fT}) = \frac{1}{T}X(f) \qquad \text{for all} \quad |f| < 1/(2T)\,. \tag{18.6}$$

To get $x(t)$ from x_k, therefore, we can use, in Fig. 2-1, the following filter:

$$F(f) = \begin{cases} T; & |f| < 1/(2T) \\ 0; & \text{otherwise} \end{cases} \tag{18.7}$$

Exercise 2-8. From (2.24), the modulus-squared of the complex envelope $\tilde{s}(t)$ is

$$|\tilde{s}(t)|^2 = (s^2(t) + \hat{s}^2(t))/2. \tag{18.8}$$

Since the Hilbert transform is a phase-only filter, it doesn't change the energy, so the energy of $s(t)$ and its Hilbert transform $\hat{s}(t)$ are the same. Hence, the complex envelope energy is identical to that of $s(t)$.

Exercise 2-9. From (2.24), the real envelope is $e(t) = \sqrt{2}\,|\tilde{s}(t)| = (s^2(t) + \hat{s}^2(t))^{1/2}$, which is the magnitude of the phase splitter output, before the complex exponential, and is clearly independent of the carrier frequency.

Exercise 2-10. First show that $S < \infty$ implies BIBO. Suppose the input is bounded by $x_k \le L$. Then

$$|y_k| = |\sum_{m=-\infty}^{\infty} h_m x_{k-m}| \le L \sum_{m=-\infty}^{\infty} |h_m| = LS < \infty . \tag{18.9}$$

Then show that if $S = \infty$ there exists a bounded input such that the output is unbounded. Such an input is

$$x_k = \begin{cases} h^*_{-k} / |h_{-k}| ; & k \text{ such that } h_k \ne 0 \\ 0 ; & k \text{ such that } h_k = 0 \end{cases} \tag{18.10}$$

Exercise 2-11.

(a) If the sequence is left-sided (2.37) becomes

$$\sum_{k=-\infty}^{K} |h_k| \cdot |z|^{-k} < \infty . \tag{18.11}$$

for some K. This sum can be rewritten

$$|z|^{-K} \sum_{k=0}^{\infty} |h_K^{-k}| \cdot |z|^k \tag{18.12}$$

and we recognize that the $|z|^{-K}$ cannot affect convergence except at $|z| = 0$ and $|z| = \infty$. All the terms in the sum are positive powers of $|z|$, and hence if they converge for some $|z| = R$ they must converge for all smaller $|z|$ (except possibly $|z| = 0$).

(b) If $K > 0$, the sequence is not anticausal, and there is a K-th order pole at $z = 0$, the ROC does not include $z = 0$. If $K = 0$, then $H(0) = h_K$, the ROC includes $z = 0$. If $K < 0$, there is a K-th order zero at $z = 0$, which is therefore included in the ROC.

Exercise 2-12. This follows directly from the observation that x_{k-l} for a fixed integer l has Z transform $z^{-l}X(z)$. Taking the Z transform of both sides of (2.41), we get the desired results.

Exercise 2-13. This is a straightforward evaluation. For example,

$$\langle y(t), x(t) \rangle = \int_{-\infty}^{\infty} y(t)x^*(t)\, dt = (\int_{-\infty}^{\infty} x(t)y^*(t)\, dt)^* = \langle x(t), y(t) \rangle^* . \tag{18.13}$$

Exercise 2-14. The inequality is obviously true (with equality) if $\mathbf{X} = \mathbf{0}$ or $\mathbf{Y} = \mathbf{0}$, so assume that $\mathbf{X} \ne \mathbf{0}$ and $\mathbf{Y} \ne \mathbf{0}$. Then we have the inequality

$$\begin{aligned} 0 &\le \|\mathbf{X} - \alpha \cdot \mathbf{Y}\|^2 \\ 0 &\le \|\mathbf{X}\|^2 - 2\mathrm{Re}\{\alpha^* \langle \mathbf{X}, \mathbf{Y} \rangle\} + |\alpha|^2 \|\mathbf{Y}\|^2 \end{aligned} \tag{18.14}$$

If we let

$$\alpha = \langle \mathbf{X}, \mathbf{Y} \rangle / \|\mathbf{Y}\|^2, \tag{18.15}$$

then the previous inequality becomes

$$0 \le \|\mathbf{X}\|^2 - |\langle \mathbf{X}, \mathbf{Y} \rangle|^2 / \|\mathbf{Y}\|^2 , \tag{18.16}$$

from which the Schwarz inequality follows immediately.

Chapter 3

Exercise 3-1. This follows from

$$\begin{aligned} \mathrm{E}[e^{s(X+Y)}] &= \mathrm{E}[e^{sX}]\mathrm{E}[e^{sY}] \\ &= \int_{-\infty}^{\infty}\int_{-\infty}^{\infty} e^{sx}f_X(x)e^{sy}f_Y(y)\,dxdy \\ &= \int_{-\infty}^{\infty} e^{sx}f_X(x)dx\int_{-\infty}^{\infty} e^{sy}f_Y(y)dy = \Phi_X(x)\Phi_Y(y)\,. \end{aligned} \tag{18.17}$$

Exercise 3-2. Evaluating the derivatives,

$$\Phi_X(s)\,|_{s=0} = 1 \quad , \tag{18.18}$$

$$\frac{\partial}{\partial s}\Phi_X(s)\,|_{s=0} = \mathrm{E}[Xe^{sX}]\,|_{s=0} = \mathrm{E}[X] \quad , \tag{18.19}$$

$$\frac{\partial^2}{\partial s^2}\Phi_X(s)\,|_{s=0} = \mathrm{E}[X^2e^{sX}]\,|_{s=0} = \mathrm{E}[X^2] \quad . \tag{18.20}$$

Exercise 3-3.

(a) The distribution function can be written in terms of a unit step function $u(x)$ as

$$1 - F_X(x) = \int_{-\infty}^{\infty} u(y-x)f_X(y)dy = \int_x^{\infty} f_X(y)dy\,, \tag{18.21}$$

and since $u(y-x)$ is bounded by $e^{s(y-x)}$ for $s \geq 0$,

$$1 - F_X(x) \leq \int_{-\infty}^{\infty} e^{(y-x)s}f_X(y)dy = e^{-sx}\Phi_X(s). \tag{18.22}$$

(b) Obtained by a similar technique.

(c) Take the derivative of the bound with respect to s and set to zero.

Exercise 3-4. Suppose that y is a discrete value that Y takes on with probability a. Then

$$f_Y(\alpha) = a\delta(\alpha - y). \tag{18.23}$$

Integrate (3.30) over small intervals about y, or over $(y-\varepsilon, y+\varepsilon)$ for small enough ε. Equation (3.32) follows similarly, or it can be easily derived from the definition of conditional probabilities (3.27).

Exercise 3-5. By direct calculation we have

$$\Pr[X > x] = \frac{1}{\sqrt{2\pi\sigma^2}}\int_x^{\infty} e^{-(\alpha-\mu)^2/2\sigma^2}d\alpha = \frac{1}{\sqrt{2\pi\sigma^2}}\int_{(x-\mu)/\sigma}^{\infty} e^{-w^2/2}\sigma dw = Q\left(\frac{x-\mu}{\sigma}\right), \tag{18.24}$$

where we have used the change of variables $w = (\alpha - \mu)/\sigma$.

Exercise 3-6.

(a) The moment generating function can be obtained by evaluating the integral

$$\Phi_X(s) = \frac{1}{\sqrt{2\pi\sigma^2}} \int_{-\infty}^{\infty} e^{sx} e^{-(x-\mu)^2/2\sigma^2} dx \ . \tag{18.25}$$

Combining the two exponents and completing the square in the resulting exponent in the integral, it becomes $-((x-a)^2 - b) / 2\sigma^2$, where $b = 2\mu\sigma^2 s + \sigma^4 s^2$. The $e^{b/2\sigma^2}$ term can be taken outside the integral, and the remaining integrand is just a Gaussian density function and therefore integrates to unity. Thus, the moment generating function is $\Phi_X(s) = e^{b/2\sigma^2}$, and substituting for b we get the claimed result.

(b) The value of s giving the tightest bound can be solved as

$$s = \frac{x-\mu}{\sigma^2}, \tag{18.26}$$

and hence the bound is valid as long as $x \geq \mu$. Substituting this value of s into the bound, we get

$$1 - F_X(x) \leq e^{-(x-\mu)^2/2\sigma^2}, \tag{18.27}$$

which looks remarkably like the Gaussian density. Note than when $x = \mu$, the actual probability is $1/2$ and the Chernoff bound is $e \approx 2.28$, so the bound is rather loose. It becomes much tighter for larger values of x. The relation (3.43) follows by letting $\mu = 0$ and $\sigma = 1$.

Exercise 3-7. Consider a scaled Gaussian, $Y = aX$. If the variance of X is σ^2, then the variance of Y is $a^2\sigma^2$. Hence the moment generating function of Y is

$$\Phi_Y(s) = e^{a^2\sigma^2 s^2/2} . \tag{18.28}$$

The moment generating function is

$$\Phi_Z(s) = e^{(a_1^2 + \ldots + a_N^2)\sigma^2 s^2/2} . \tag{18.29}$$

This is the moment generating function of a zero mean Gaussian random variable with variance (3.46).

Exercise 3-8. We only need to show

$$\mathrm{E}[XY] = \mathrm{E}[X]\mathrm{E}[Y] \Rightarrow f_{X,Y}(x, y) = f_X(x) f_Y(y). \tag{18.30}$$

From (3.48) and the fact that the random variables have zero-mean, $\rho = 0$. Now (3.47) is easily factored into two parts.

Exercise 3-9. Define

$$\mathbf{X} + \mathbf{Y} \leftrightarrow X + Y$$

$$\alpha \cdot \mathbf{X} \leftrightarrow \alpha \cdot X \tag{18.31}$$

$$\mathbf{0} \leftrightarrow 0 \quad \text{(the zero random variable)} \tag{18.32}$$

$$-\mathbf{X} \leftrightarrow -X \tag{18.33}$$

With these definitions, verification of the properties is straightforward, relying on similar linearity properties of random variables.

Exercise 3-10. This is very straightforward. For example,

$$\langle \mathbf{Y}, \mathbf{X} \rangle = \mathrm{E}[YX^*] = \mathrm{E}[XY^*]^* = \langle \mathbf{X}, \mathbf{Y} \rangle^*. \tag{18.34}$$

Exercise 3-11. First show that

$$R_W(\tau) = \mathrm{E}[W(t+\tau)W^*(t)] = h^*(-\tau) * R_{WX}(\tau), \tag{18.35}$$

by substituting for one of the $W(t)$ in terms of $X(t)$. Then show that

$$R_{WX}(\tau) = h(\tau) * R_X(\tau), \tag{18.36}$$

completing the first result. Finally, show that the Fourier transform of $h^*(-\tau)$ is $H^*(f)$, and that the Laplace transform of $h^*(-\tau)$ is $H^*(-s^*)$. Both of these are easily done using the Fourier and Laplace transform definitions.

Exercise 3-12. Calculating first the cross-correlation of the input and output,

$$R_{XW}(m) = \mathrm{E}[X_{k+m}W_k^*] = \mathrm{E}[X_{k+m}\sum_{n=-\infty}^{\infty} X_n^* h_{k-n}^*] = R_x(m) * h_{-m}^* . \tag{18.37}$$

Then calculating the output correlation function,

$$R_W(m) = \mathrm{E}[W_{k+m}W_k^*] = \mathrm{E}[\sum_{n=-\infty}^{\infty} X_n h_{k+m-n} W_k^*] = R_{XW}(m) * h_m . \tag{18.38}$$

Finally, show that the Fourier transform of h_{-m}^* is $H^*(e^{j\theta})$ and the Z transform is $H^*(1/z^*)$.

Exercise 3-13. First we can calculate that

$$R_{WY}(\tau) = R_{XY}(\tau) * h(\tau) \tag{18.39}$$

and then

$$R_{WU}(\tau) = R_{WY}(\tau) * g^*(-\tau) = R_{XY}(\tau) * h(\tau) * g^*(-\tau) . \tag{18.40}$$

Taking the Fourier transform we get the desired result.

Exercise 3-14. Since $A(z)$ is all pass, we must have $|A(e^{j\theta})|^2 = 1$ for all θ. Taking the arithmetic mean, and using Parseval's relationship, implies that $\sum_{k=-\infty}^{\infty} |a_k|^2 = 1$. By itself, this would imply that $|a_0| \le 1$. But since there are at least two nonzero coefficients $\{a_k\}$, we must have $|a_0| < 1$.

Exercise 3-15. This relation can be obtained by exactly the same method as Appendix 3-A, although it is tedious.

Exercise 3-16. We use the fact that the next state Ψ_{k+1} of a Markov chain is independent of the past states $\Psi_{k-1}, \Psi_{k-2}, \ldots$ given the present state Ψ_k to show that *all* future samples of the Markov chain are independent of the past given knowledge of the present.

We wish to show that for any $n > 0$ and any k,

$$p(\psi_{k+n} | \psi_k, \psi_{k-1}, \ldots) = p(\psi_{k+n} | \psi_k). \tag{18.41}$$

This is easily shown by induction. Observe that it is true for $n = 1$, by the definition of Markov chains (3.112). We can assume that it is true for some n and show it is true for $n+1$. A fact about conditional probabilities similar to that in (3.33) tells us that

$$p(\Psi_{k+n+1} \mid \Psi_k, \Psi_{k-1}, \ldots)$$
$$= \sum_{\Psi_{k+1} \in \Omega_\Psi} p(\Psi_{k+n+1} \mid \Psi_{k+1}, \Psi_k, \Psi_{k-1}, \ldots) p(\Psi_{k+1} \mid \Psi_k, \Psi_{k-1}, \ldots). \tag{18.42}$$

Since we assume that (18.41) is true for n,

$$p(\Psi_{k+n+1} \mid \Psi_{k+1}, \Psi_k, \Psi_{k-1}, \ldots) = p(\Psi_{k+n+1} \mid \Psi_{k+1}). \tag{18.43}$$

It is also therefore true that

$$p(\Psi_{k+n+1} \mid \Psi_{k+1}, \Psi_k, \Psi_{k-1}, \ldots) = p(\Psi_{k+n+1} \mid \Psi_{k+1}, \Psi_k). \tag{18.44}$$

Furthermore, from the definition of Markov chains,

$$p(\Psi_{k+1} \mid \Psi_k, \Psi_{k-1}, \ldots) = p(\Psi_{k+1} \mid \Psi_k). \tag{18.45}$$

Substituting (18.44) and (18.45) into (18.42) we get

$$p(\Psi_{k+n+1} \mid \Psi_k, \Psi_{k-1}, \ldots) = \sum_{\Psi_{k+1} \in \Omega_\Psi} p(\Psi_{k+n+1} \mid \Psi_{k+1}, \Psi_k) p(\Psi_{k+1} \mid \Psi_k). \tag{18.46}$$

Using the same fact about conditional probabilities (3.33) we can eliminate the summation to get

$$p(\Psi_{k+n+1} \mid \Psi_k, \Psi_{k-1}, \ldots) = p(\Psi_{k+n+1} \mid \Psi_k), \tag{18.47}$$

which shows that (3.113) is valid for $n + 1$.

Exercise 3-17. Multiplying both sides of (3.117) by z^{-k} and summing from $k = 0$ to $k = \infty$,

$$\sum_{k=0}^{\infty} p_{k+1}(j) z^{-k} = \sum_{i \in \Omega_\Psi} p(j \mid i) \sum_{k=0}^{\infty} p_k(i) z^{-k}.$$

Changing variables and letting $m = k + 1$,

$$\sum_{m=0}^{\infty} p_m(j) z^{-m+1} = \sum_{i \in \Omega_\Psi} p(j \mid i) P_i(z),$$

or

$$z(P_j(z) - p_0(j)) = \sum_{i \in \Omega_\Psi} p(j \mid i) P_i(z).$$

Exercise 3-18. We have

$$f_N = \sum_{k=-\infty}^{\infty} k q_k(N) = \sum_{k=-\infty}^{\infty} k q_k(N) z^{-k} \Big|_{z=1}. \tag{18.48}$$

But the latter summation can be evaluated using a derivative,

$$\frac{\partial}{\partial z} Q_N(z) = \sum_{k=-\infty}^{\infty} (-k) q_k(N) z^{-k-1} = -z^{-1} \sum_{k=-\infty}^{\infty} k q_k(N) z^{-k} . \tag{18.49}$$

The result follows immediately.

Exercise 3-19.

(a) Given the power series $$e^a = \sum_{k=0}^{\infty} \frac{a^k}{k!},$$

and differentiating it once $$e^a = \frac{1}{a} \sum_{k=1}^{\infty} \frac{k a^k}{k!} \tag{18.50}$$

and differentiating it twice $$e^a = \frac{1}{a^2} \left[\sum_{k=1}^{\infty} \frac{k^2 a^k}{k!} - \sum_{k=1}^{\infty} \frac{k a^k}{k!} \right]. \tag{18.51}$$

The moments follow immediately.

(b) The moment generating function is

$$\Phi_N(s) = e^{-a} \sum_{k=-\infty}^{\infty} \frac{(ae^s)^k}{k!} = e^{ae^s - a}, \tag{18.52}$$

and taking the logarithm we get (3.142).

Exercise 3-20. For these initial conditions, we get $q_k(t_0) = 1$, and the Laplace transform becomes

$$\begin{aligned} sQ_j(s) + \lambda Q_j(s) &= \lambda Q_{j-1}(s), \quad j \neq k, \\ sQ_k(s) - q_k(t_0) e^{-st_0} + \lambda Q_k(s) &= \lambda Q_{k-1}(s) . \end{aligned} \tag{18.53}$$

By iteration, we can establish that $Q_j(s) = 0$ for $j < k$ and for $j \geq k$

$$Q_j(s) = \frac{\lambda^{j-k}}{(s+\lambda)^{j-k+1}} e^{-st_0} . \tag{18.54}$$

The result follows immediately by taking the inverse Laplace transform.

Exercise 3-21. The state transition diagram is shown in the following figure:

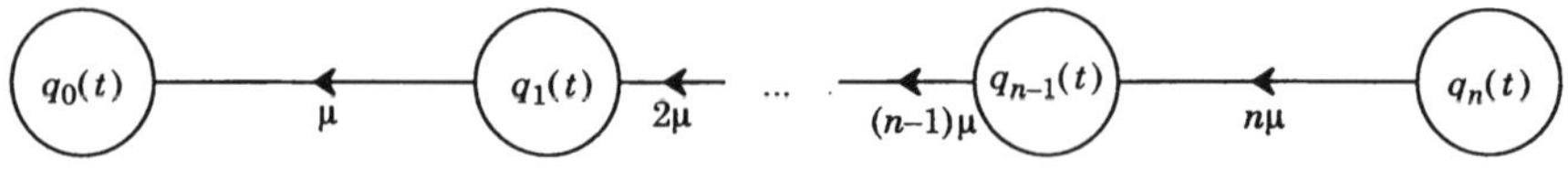

The equations become for this case

$$\begin{aligned} &\frac{d}{dt} q_n(t) + n\mu q_n(t) = 0, \\ &\frac{d}{dt} q_j(t) + j\mu q_j(t) = (j+1)\mu q_{j+1}(t), \quad 0 \leq j < n, \end{aligned} \tag{18.55}$$

with initial condition $$q_j(0) = \begin{cases} 0; & 0 \leq j < n \\ 1; & j = n \end{cases}. \tag{18.56}$$

Taking the Laplace transform, we get

$$sQ_n(s) - q_n(0) + n\mu Q_n(s) = 0 ,$$
$$sQ_j(s) + j\mu Q_j(s) = (j+1)\mu Q_{j+1}(s), \quad 0 \le j < n . \tag{18.57}$$

It follows that

$$q_n(t) = e^{-n\mu t}$$
$$q_j(t) = (j+1)\mu e^{-j\mu t} * q_{j+1}(t) , \tag{18.58}$$

and the reader can verify by induction that (3.144) is valid.

Exercise 3-22. This is a two-state birth and death process with transition diagram shown in the following figure:

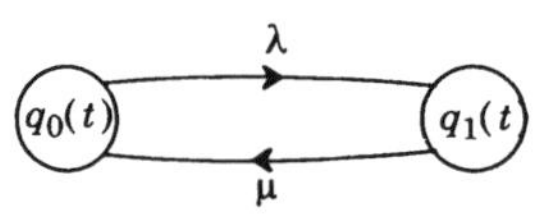

State 0 corresponds to the server idle, and state 1 corresponds to the server busy. When the system is in state $j = 0$, arrivals occur at rate λ, and there are no departures because the server is idle. Similarly, in state $j = 1$, where the single server is busy, arrivals immediately depart from the system, have no effect, and therefore are not reflected on the state transition diagram. Also in this state, departures occur with rate μ due to the completion of service. The differential equations governing the system are

$$\frac{d}{dt} q_0(t) = \mu q_1(t) - \lambda q_0(t) , \tag{18.59}$$

$$\frac{d}{dt} dq_1(t) = \lambda q_0(t) - \mu q_1(t) . \tag{18.60}$$

Since the sum of the two probabilities must be unity,

$$q_0(t) + q_1(t) = 1, \tag{18.61}$$

we can clear one variable to yield a single differential equation for $q_0(t)$,

$$\frac{d}{dt} q_0(t) + (\lambda + \mu) q_0(t) = \mu , \tag{18.62}$$

and the result follows immediately by taking the Laplace transform.

Exercise 3-23. Rewriting (3.149), we get

$$\mu q_{j+1} - \lambda q_j = \mu q_j - \lambda q_{j-1} , \tag{18.63}$$

and since this recursion starts at zero,

$$q_{j+1} = \rho q_j = \rho^{j+1} q_0 . \tag{18.64}$$

Using the fact that the probabilities must sum to unity, we can find that $q_0 = 1 - \rho$.

Exercise 3-24. We get

$$\frac{\partial}{\partial s}\Phi_{X(t)}(s) = \Phi_{X(t)}(s)\lambda(t) * (h(t)e^{sh(t)}) , \tag{18.65}$$

and setting $s = 0$ we get the mean of (3.159). Taking the second derivative, we get

$$\frac{\partial^2}{\partial s^2}\Phi_{X(t)}(s) = \frac{\partial}{\partial s}\Phi_{X(t)}(s)\lambda(t) * (h(t)e^{sh(t)}) + \Phi_{X(t)}(s)\lambda(t) * (h^2(t)e^{sh(t)}) , \tag{18.66}$$

and setting $s = 0$ we get the second moment. Subtracting the square of the mean, we get the variance of (3.160).

Exercise 3-25. The exact moment generating function is

$$\begin{aligned} \log_e\Phi_{X(t)}(s) &= \beta\lambda_0(t) * (e^{sh_0(t)/\sqrt{\beta}} - 1) \\ &\approx \beta\lambda_0(t) * \left(\frac{sh_0(t)}{\sqrt{\beta}} + \frac{0.5s^2h_0^2(t)}{\beta}\right), \end{aligned} \tag{18.67}$$

where all the other terms are of order $1/\sqrt{\beta}$ or smaller and become insignificant as $\beta \to \infty$.

Exercise 3-26. Note that

$$E[G_mG_n] = \begin{cases} E[G^2] ; & m = n \\ E[G]^2 ; & m \neq n \end{cases}, \tag{18.68}$$

and also note that from Campbell's theorem

$$E[\Sigma_m h^2(t - t_m)] = \lambda(t) * h^2(t) = \sigma_X^2(t) , \tag{18.69}$$

since this is a shot noise process with impulse response $h^2(t)$. The rest is straightforward but tedious algebra.

Exercise 3-27. Part (a) is straightforward, so let's concentrate on part (b). Integrating,

$$e^{A(t)}x(t) - e^{A(t_0)}x(t_0) = \int_{t_0}^{t} b(u)e^{A}(u)\,du . \tag{18.70}$$

The solution of (3.194) follows directly from the observation that

$$\Lambda(t) = A(t) - A(t_0) , \tag{18.71}$$

and some algebraic manipulation.

Exercise 3-28. Noting that (3.155) is true for $j = 0$, assume it true for j. Then using (3.196) to determine the distribution for $j + 1$,

$$q_{j+1}(t) = e^{-\Lambda(t)}\int_{t_0}^{t} \lambda(u)\frac{\lambda(u)\Lambda^j(t)}{j!}\,du. \tag{18.72}$$

But noting that $\dot{\Lambda}(t) = \lambda(t)$, this becomes

$$q_{j+1}(t) = e^{-\Lambda(t)} \int_{t_0}^{t} \dot{\Lambda}(u) \frac{\Lambda^j(t)}{j!} du , \tag{18.73}$$

which can be integrated directly because it is a perfect differential to yield (3.155) for $j + 1$.

Chapter 4

Exercise 4-1. Let the K outcomes have probabilities p_i, $1 \le i \le K$. Using the inequality

$$\Sigma_i \, p_i \log_2 \frac{1}{K p_i} \le \Sigma_i \, p_i \left(\frac{1}{K p_i} - 1 \right) = 0 , \tag{18.74}$$

we get the inequality

$$H(X) \le \Sigma_i \, p_i \log_2 K = \log_2 K . \tag{18.75}$$

Exercise 4-2.

(a) From(4.12) and (4.11) we get

$$I(X, Y) = - \sum_{x \in \Omega_X} p_X(x) \log_2 p_X(x) + \sum_{x \in \Omega_X} \sum_{y \in \Omega_Y} p_{X,Y}(x, y) \log_2 p_{X,Y}(x, y) . \tag{18.76}$$

From Bayes rule this becomes

$$I(X, Y) = - \sum_{x \in \Omega_X} p_X(x) \log_2 p_X(x) + \sum_{x \in \Omega_X} \sum_{y \in \Omega_Y} p_{X,Y}(x, y) \log_2 \left[\frac{p_{Y|X}(y|x) p_X(x)}{p_Y(y)} \right] , \tag{18.77}$$

which reduces directly to (4.13) after algebraic manipulation.

(b) We can show this by substituting into (4.14) in terms of the input and transition probabilities, and showing algebraically that the result is equivalent to (4.14).

Exercise 4-3.

(a) The easiest method is to use the formula

$$I(X, Y) = H(Y) - H(Y \mid X) . \tag{18.78}$$

We want to show that the second term is independent of q. When the input is $X = 0$, the two outputs have probability p and $1 - p$, and similarly when input is $X = 1$ the outputs have the same probabilities. Hence the entropy of the output is the same in both cases, and the conditional entropy, the average of these two entropies over the input distribution is independent of the input distribution. The result follows immediately.

(b) We know from Fig. 4-2 that $H(Y) \le 1$ with equality if and only if the two outputs have equal probability. Because of the symmetry of the channel, they will have equal probability when $q = 1/2$, and this is therefore the distribution that achieves channel capacity.

Exercise 4-4. From the inequality $\log(x) \le x - 1$, it follows that:

$$\int_{-\infty}^{\infty} f_Y(y) \log \frac{g(y)}{f_Y(y)} dy \le \int_{-\infty}^{\infty} f_Y(y) \left(\frac{g(y)}{f_Y(y)} - 1 \right) dy$$

$$= \int_{-\infty}^{\infty} g(y) dy - \int_{-\infty}^{\infty} f_Y(y) dy = 1 - 1 = 0. \tag{18.79}$$

This leads directly to (4.22). Equality in (18.79) holds when $g(y) = f_Y(y)$. Substituting a zero-mean Gaussian with variance σ^2 for $g(y)$, we get (4.21). The right-hand equality in (4.21) holds when Y is Gaussian.

Exercise 4-5. From (4.23), using the observation that $f_{Y|X}(y \mid x) = f_N(y - x)$,

$$H(Y|X) = \sum_{x \in \Omega_X} p_X(x) \int_{-\infty}^{\infty} f_N(y - x) \log_2 f_N(y - x) dy$$

$$= \sum_{x \in \Omega_X} p_X(x) \int_{-\infty}^{\infty} f_N(n) \log_2 f_N(n) dn$$

$$= \int_{-\infty}^{\infty} f_N(n) \log_2 f_N(n) dn = H(N) . \tag{18.80}$$

Exercise 4-6.

$$I(X, Y) = H(Y) - H(Y|X)$$

$$= -\int_{\Omega_Y} f_Y(y) \log_2 f_Y(y) dy + \sum_{x \in \Omega_X} p_X(x) \int_{\Omega_Y} f_{Y|X}(y \mid x) \log_2 f_{Y|X}(y \mid x)\, dy$$

$$= -\int_{\Omega_Y} [f_Y(y) \log_2 f_Y(y) - \sum_{x \in \Omega_X} p_X(x) f_{Y|X}(y \mid x) \log_2 f_{Y|X}(y \mid x)] dy. \tag{18.81}$$

Using

$$f_Y(y) = \sum_{x \in \Omega_X} p_X(x) f_{Y|X}(y \mid x), \tag{18.82}$$

we get

$$I(X,Y) = \int_{\Omega_Y} \sum_{x \in \Omega_X} p_X(x) [f_{Y|X}(y \mid x)(\log_2 f_{Y|X}(y \mid x) - \log_2 f_Y(y))] dy$$

$$= \sum_{x \in \Omega_X} \int_{\Omega_Y} f_{Y|X}(y \mid x) \log_2 \frac{f_{Y|X}(y|x)}{f_Y(y)}\, dy, \tag{18.83}$$

from which the result follows.

Exercise 4-7.

(a) Using the inequality $\log(x) \le (x - 1)$,

$$\int f_Y(y) \log_2 \frac{g(y)}{f_Y(y)} dy \le \int (g(y) - f_Y(y))\, dy = 0 , \tag{18.84}$$

for any valid pdf $g(\cdot)$.

(b) In particular, when $g(\cdot)$ is the prescribed Gaussian pdf, we have

$$\log_2 g(\mathbf{y}) = \log_2\left\{ \prod_{n=1}^{N} \frac{1}{\sqrt{2\pi(\sigma_{x,n}^2 + \sigma^2)}} \exp\{-y_n{}^2/2(\sigma_{x,n}^2 + \sigma^2)\} \right\}$$

$$= -\sum_{n=1}^{N} \left\{ \tfrac{1}{2}\log_2(2\pi(\sigma_{x,n}^2 + \sigma^2)) + \log_2 e \cdot \frac{y_n^2}{2(\sigma_{x,n}^2 + \sigma^2)} \right\}. \quad (18.85)$$

With this choice for $g(\cdot)$, (4.31) reduces to

$$H(\mathbf{Y}) \le -\int f_{\mathbf{Y}}(\mathbf{y})\log_2 g(\mathbf{y})d\mathbf{y}$$

$$= \tfrac{1}{2}\sum_{n=1}^{N} \log_2(2\pi(\sigma_{x,n}^2 + \sigma^2))$$

$$+ \tfrac{1}{2}\sum_{n=1}^{N} \frac{\log_2 e}{\sigma_{x,n}^2 + \sigma^2} \int f_{\mathbf{Y}}(\mathbf{y}) y_n^2 d\mathbf{y}$$

$$= \tfrac{1}{2}\sum_{n=1}^{N} \log_2(2\pi e(\sigma_{x,n}^2 + \sigma^2)) . \quad (18.86)$$

(c) Substituting (18.86) and (4.30) into (4.29), we get

$$I(\mathbf{X},\mathbf{Y}) \le \tfrac{1}{2}\sum_{n=1}^{N} \log_2\left(1 + \frac{\sigma_{x,n}^2}{\sigma^2}\right). \quad (18.87)$$

(d) Calculating the difference,

$$\tfrac{1}{2}\sum_{n=1}^{N} \log_2\left(1 + \frac{\sigma_{x,n}^2}{\sigma^2}\right) - \tfrac{1}{2}\sum_{n=1}^{N} \log_2\left(1 + \frac{\sigma_{x,n}^2}{N\sigma^2}\right)$$

$$= \tfrac{1}{2}\sum_{n=1}^{N} \log_2 \frac{1 + \sigma_{x,n}^2/\sigma^2}{1 + \sigma_x^2/(N\sigma^2)}$$

$$\le \tfrac{1}{2}\sum_{n=1}^{N} \frac{N\sigma_{x,n}^2 - \sigma_x^2}{N\sigma^2 + \sigma_x^2} = 0 , \quad (18.88)$$

with equality if and only if $\sigma_{x,n}^2 = \sigma_x{}^2/N$.

Chapter 5

Exercise 5-1.

$$h(t) * h(-t) = \int_0^T h(\tau)h(\tau - t)d\tau . \quad (18.89)$$

Outside the range $[-T, T]$ the two integrands do not overlap, likewise at the endpoints of this interval there is an overlap at only one point, and the integral will be zero.

Chapter 6

Exercise 6-1. Let $x(t) = \sqrt{2}\,\mathrm{Re}\{\tilde{x}(t)e^{j2\pi f_c t}\}$ and $y(t) = \sqrt{2}\,\mathrm{Re}\{\tilde{y}(t)e^{j2\pi f_c t}\}$. By definition, we have $\tilde{X}(f) = \sqrt{2}u(f+f_0)X(f+f_0)$ and $\tilde{Y}(f) = \sqrt{2}u(f+f_0)X(f+f_0)$. Then:

$$\begin{aligned}\int_{-\infty}^{\infty} x(t)y(t)dt &= \int_{-\infty}^{\infty} X(f)Y^*(f)df \qquad = 2\int_{0}^{\infty} X(f)Y^*(f)df \\ &= 2\int_{-\infty}^{\infty} \mathrm{u}(f)X(f)[Y(f)]^*df \;= \int_{-\infty}^{\infty} \sqrt{2}\,\mathrm{u}(f)X(f)[\sqrt{2}\,\mathrm{u}(f)Y(f)]^*df \\ &= \int_{-\infty}^{\infty} \sqrt{2}\,\mathrm{u}(f+f_0)X(f+f_0)[\sqrt{2}\,\mathrm{u}(f+f_0)Y(f+f_0)]^*df \\ &= \int_{-\infty}^{\infty} \tilde{X}(f)\tilde{Y}^*(f)\,df \qquad = \int_{-\infty}^{\infty} \tilde{x}(t)\tilde{y}^*(t)\,dt\,. \end{aligned} \tag{18.90}$$

The first and last equalities follow from Parseval's relationship, while the second equality follows from the Hermitian symmetry of the Fourier transform of real signals. Thus, we see that the inner product of the real signals is precisely equal to that of the corresponding complex envelopes. Hence, orthogonality at baseband implies orthogonality at passband.

Exercise 6-2. The pulses in (6.52) consist of a sinc pulse modulating a cosine. Using the tables in Appendix 2-A, the Fourier transform of the sinc pulses is

$$\sqrt{\frac{E}{T}}\left(\frac{\sin(\pi t/(2T))}{\pi t/(2T)}\right) \leftrightarrow 2\sqrt{ET}\,\mathrm{rect}(f,\, 1/(4T))\,. \tag{18.91}$$

Multiplying the pulse in the time domain by a cosine of frequency $(n - ½)/(2T)$ will scale the Fourier transform by ½ and shift it up and down in frequency by $(n - ½)/(2T)$. The frequency-domain plots in Fig. 6-8 result. These plots make it clear that

$$H_i(f)H_j^*(f) = \sqrt{ET}\,H_n(f)\delta_{i-j}\,. \tag{18.92}$$

Thus, for any $i \ne j$, the sum in (6.56) is zero. So it remains to be shown only that when $i = j$, the sum is constant, or

$$\frac{1}{\sqrt{T}}\sum_{m=-\infty}^{\infty} H_i(f - m/T) = 1\,. \tag{18.93}$$

Equivalently, we require that $\sqrt{T}\,H_n(f)$ satisfy the Nyquist criterion, so that $\sqrt{T}\,h_i(t)$ is zero at all nonzero integer multiples of T. From (6.52),

$$\sqrt{T}\,h_i(kT) = \sqrt{E}\,\delta_k\,, \tag{18.94}$$

verifying that they are Nyquist pulses.

Exercise 6-3. To see this, note first that

$$h_n(t)h_n(t-kT) = \tfrac{1}{2}\,w(t)w(t-kT)\Big(\cos[(n+½)k\pi] + \cos[(2n+1)\pi t/T - (n+½)k\pi]\Big)\,.$$

Since $w(t)$ is bandlimited to $1/(2T)$, $w(t)w(t-kT)$ is bandlimited to $1/T$, and thus when it is multiplied by a cosine with frequency $(2n+1)/(2T) \geq 3/(2T)$, the spectrum does not overlap d.c. This term must therefore integrate to zero, so

$$\int_{-\infty}^{\infty} h_n(t)h_n(t-kT)\,dt = \tfrac{1}{2}\cos[(n+\tfrac{1}{2})k\pi]\int_{-\infty}^{\infty} w(t)w(t-kT)\,dt\,. \tag{18.95}$$

For $k=0$, this integral reduces to the energy of $h_n(t)$. For k even but not zero, (6.57) implies that the integral has value zero. For k odd, the integral is zero because $\cos((n+\tfrac{1}{2})k\pi)$ is zero. Hence, (6.53) holds.

Exercise 6-4. Writing out the convolution in (6.55), we need to show that

$$\frac{2}{T}\int_0^T \sin(\omega_n\tau)\sin(\omega_l(\tau-kT))w(\tau-kT)d\tau \;=\; \delta_k\delta_{l-n}\,, \tag{18.96}$$

for $l, n \in \{0, 1\}$, and for any integer k, where $\omega_n = 2\pi f_n$. When $k \neq 0$, $w(\tau-kT) = 0$ within the range of the integral, so it will suffice to show that

$$\frac{2}{T}\int_0^T \sin(\omega_n\tau)\sin(\omega_l\tau)d\tau = \delta_{l-n}\,, \tag{18.97}$$

or equivalently, that

$$\frac{1}{T}\int_0^T [\cos((\omega_n-\omega_l)\tau)-\cos((\omega_n+\omega_l)\tau)]d\tau = \delta_{l-n}\,. \tag{18.98}$$

Using (6.71), we need to show that

$$\frac{1}{T}\int_0^T [\cos((M_n-M_l)2\pi\tau/T)-\cos((M_n+M_l)2\pi\tau/T)]d\tau = \delta_{l-n}\,. \tag{18.99}$$

When $n \neq l$, both terms under the integral are cosines that are integrated over an integer number of cycles, and hence integrate to zero. When $n = l$, the second term integrates to zero, but the first term integrates to T, thus establishing the result.

Exercise 6-5. The sampled receive filter output is

$$\begin{aligned} g_0(t) * f(t)\big|_{t=0} &= \frac{2}{T}\int_0^T \sin(\omega_0\tau)\sin(\omega_0\tau+\theta)d\tau \\ &= \frac{1}{T}\int_0^T [\cos(\theta)-\cos(2\omega_0\tau+\theta)]d\tau. \end{aligned} \tag{18.100}$$

The second term in the integral integrates to zero for any fixed θ, and the first term is constant, so

$$g_0(t) * f(t)\big|_{t=0} = \cos(\theta). \tag{18.101}$$

Exercise 6-6.

$$\int_{-\infty}^{\infty} g_n(t) g_k(t) dt = \frac{1}{T}\int_0^T e^{j2\pi f_n t} e^{-j2\pi f_k t} dt = \frac{1}{T}\int_0^T e^{j2\pi(n-k)/T} dt . \tag{18.102}$$

When $n = k$, this clearly equals unity. When $n \neq k$, we are integrating over an integer number of cycles of a complex exponential, getting zero.

Exercise 6-7.

$$\int_{-\infty}^{\infty} \tilde{g}_n(t)\, \tilde{g}_k^*(t) dt = \frac{1}{N}\sum_{i=0}^{N-1}\sum_{l=0}^{N-1} e^{j2\pi in/N} e^{-j2\pi lk/N} \int_{-\infty}^{\infty} p(t - iT/N) p^*(t - lT/N) dt$$

$$= \frac{1}{N}\sum_{i=0}^{N-1}\sum_{l=0}^{N-1} e^{j2\pi(in - lk)/N} \int_{-\infty}^{\infty} p(t - iT/N) p(t - lT/N) dt$$

$$= \frac{1}{N}\sum_{i=0}^{N-1}\sum_{l=0}^{N-1} e^{j2\pi(in - lk)/N} \delta_{i,l} == \frac{1}{N}\sum_{i=0}^{N-1} e^{j2\pi(n-k)i/N} \delta_{i,l} . \tag{18.103}$$

We have now reached the discrete-time version of the final equation n in the previous exercise. When $n = k$, this clearly equals unity. When $n \neq k$, we are adding N complex numbers equally spaced on the unit circle, yielding zero.

Exercise 6-8. Evaluate (6.117) at the boundary between two symbols, $t = kT$, and find that for the phase to be the same on either side of the boundary we need

$$2\pi f_c kT + b_k \frac{\pi kT}{2T} + \phi_k = 2\pi f_c kT + b_{k-1} \frac{\pi kT}{2T} + \phi_{k-1} , \tag{18.104}$$

which easily reduces to the desired form.

Exercise 6-9. Letting $x = N(M^{1/N} - 1)$ where $N = WT$, we get

$$M = (1 + x/N)^N \rightarrow e^x \quad \text{as} \quad N \rightarrow \infty . \tag{18.105}$$

Hence $x \rightarrow \log_e M$ as $N \rightarrow \infty$.

Chapter 7

Exercise 7-1. From Bayes' rule we can write

$$p(\psi) = p(\psi_K | \psi_{K-1}, \ldots, \psi_0) p(\psi_{K-1}, \ldots, \psi_0) . \tag{18.106}$$

From the Markov property this becomes

$$p(\psi) = p(\psi_K | \psi_K - 1) p(\psi_{K-1}, \ldots, \psi_0) . \tag{18.107}$$

Repeat with the second factor, and iterate until the desired form results.

Exercise 7-2. This is easily done by considering each of the four possible paths of length two through the trellis and computing the distances of the shortest error events. It is easy to argue that all longer error events have greater distances.

Exercise 7-3. Writing $z = Ae^{j\phi}$, the integral on the right hand side of (7.103) is

$$\frac{1}{2\pi}\int_{-\pi}^{\pi} e^{\mathrm{Re}\{Ae^{j(\theta-\phi)}\}}d\theta = \frac{1}{2\pi}\int_{-\pi}^{\pi} e^{A\cos(\theta-\phi)}d\theta \ . \tag{18.108}$$

Since the integral is over one period of the periodic integrand, the value of the integral is independent of ϕ, and hence the result follows for $\phi = 0$.

Chapter 8

Exercise 8-1. A plot of x_k as a function of a_k, for a given v_k, is shown below:

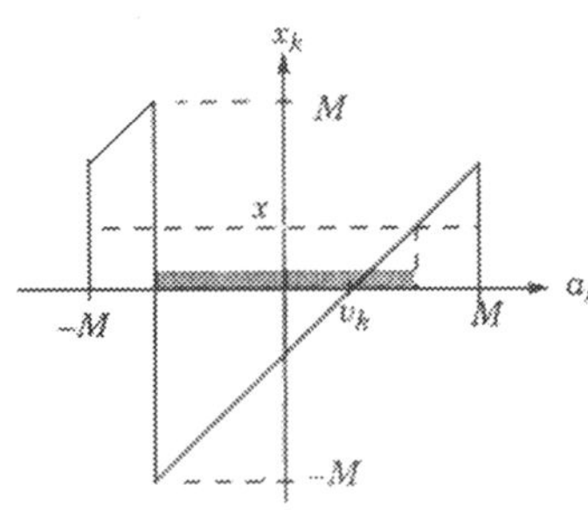

A given threshold x is shown, the probability that $x_k \le x$ corresponds to the range of a_k that is shaded. Assuming that A_k is uniformly distributed, then $\Pr[x_k \le x]$ is easily seen to be proportional to x, regardless of the value of v_k. Hence x_k is uniformly distributed.

Exercise 8-2. By straightforward manipulation,

$$\begin{aligned} S_e &= S_r|C|^2 - S_a(HC + H^*C^*) + S_a \\ &= S_r|C - S_aS_r^{-1}H^*|^2 + S_a - S_a^2S_r^{-1}|H|^2 , \end{aligned} \tag{18.109}$$

and from there it is just some minor algebra.

Exercise 8-3. The Fourier transform of the input pulse is $e^{j2\pi ft_0}H(f)$, and hence at the output of the matched filter the pulse has Fourier transform $e^{j2\pi ft_0}|H(f)|^2$. After sampling we get the result of (8.100).

Exercise 8-4. Assuming less than 100% excess bandwidth, a sampling rate of $2/T$ suffices to represent the matched filter impulse response,

$$H(f) = \sum_{m=-\infty}^{\infty} h_m e^{-j\pi fmT}, \tag{18.110}$$

and the matched filter can be represented in discrete time as a filter with impulse response h^*_{-m}. When we double the sampling rate in $C(z)$, yielding a new filter with coefficients c_k', we have

$$c_k' = \begin{cases} c_{k/2}, & k \text{ odd} \\ 0, & k \text{ even} \end{cases} \tag{18.111}$$

The FSE has impulse response

$$f_k = \sum_{m=-\infty}^{\infty} c_m' h_{m-k}^* = \sum_{r=-\infty}^{\infty} c_{2r}' h_{2r-k}^* = \sum_{r=-\infty}^{\infty} c_r h_{2r-k}^*, \tag{18.112}$$

and the transfer function of this filter is

$$\begin{aligned} \sum_{k=-\infty}^{\infty} f_k e^{-j\pi fkT} &= \sum_{k=-\infty}^{\infty} e^{-j\pi fkT} \sum_{r=-\infty}^{\infty} c_r h_{2r-k}^* \\ &= \sum_{r=-\infty}^{\infty} c_r \sum_{k=-\infty}^{\infty} h_{2r-k}^* e^{-j\pi fkT} \end{aligned} \tag{18.113}$$

and letting $m = 2r - k$ this becomes

$$\begin{aligned} \sum_{r=-\infty}^{\infty} c_r \sum_{k=-\infty}^{\infty} h_m^* e^{-j\pi f(2r-m)T} &= \sum_{r=-\infty}^{\infty} c_r e^{-j2\pi frT} \sum_{m=-\infty}^{\infty} h_m^* e^{-j\pi fmT} \\ &= C(e^{j2\pi fT}) H^*(f) . \end{aligned} \tag{18.114}$$

Exercise 8-5. Consider Fig. 8-18(b), we will show it is equivalent to Fig. 8-18(a). Since the transfer function $C_1(e^{j2\pi fT})$ is periodic in frequency, it can be expanded in a Fourier series,

$$C_1(e^{j2\pi fT}) = \Sigma_m c_m e^{-j2\pi fmT}, \tag{18.115}$$

and hence it has impulse response

$$c_1(t) = \Sigma_m c_m \delta(t - mT) . \tag{18.116}$$

If the matched filter output is called $x(t)$ and the output of the filter $C_1(e^{j2\pi fT})$ is called $y(t)$, then

$$y(t) = \Sigma_m c_m x(t - mT) , \tag{18.117}$$

and therefore

$$y_k = y(kT) = \Sigma_m c_m x((k - m)T) , \tag{18.118}$$

which is evidently Fig. 8-18(a).

Exercise 8-6. We get

$$\begin{aligned} \Pr[w_k \neq 0] &= \tfrac{1}{2}(\Pr[w_k = 2] + \Pr[w_k = -2]) \\ &= \tfrac{1}{2}(\Pr[n_k < -1 - V_k] + \Pr[n_k > 1 - v_k]) \\ &= \tfrac{1}{2}(1 - \Pr[-1 - v_k \le n_k \le 1 - v_k]) \le \tfrac{1}{2} . \end{aligned} \tag{18.119}$$

Chapter 9

Exercise 9-1. Write:

$$|E_k|^2 = E_k{}^*E_k = (A_k{}^* - \boldsymbol{c}^*\bar{\boldsymbol{r}}_k)(A_k - \boldsymbol{r}_k^T\boldsymbol{c})$$

$$= |A_k|^2 - \boldsymbol{c}^*\bar{\boldsymbol{r}}_kA_k - \boldsymbol{r}_k{}^T\boldsymbol{c}A_k{}^* + \boldsymbol{c}^*\bar{\boldsymbol{r}}_k\boldsymbol{r}_k{}^T\boldsymbol{c}\ . \tag{18.120}$$

Taking the expected value, we get the stated result.

Exercise 9-2. R_k is the sum of a signal and a noise term, which are uncorrelated and can thus be considered directly. From Exercise 3-11, the autocorrelation of the signal component is

$$h_k * h^*_{-k} * E_a\delta_k = E_a h_k * h^*_{-k}\ . \tag{18.121}$$

Similarly, neglecting the signal component, the noise component can be determined directly:

$$\mathrm{E}[R_{k+m}R_k{}^*] = N_0\int_{-\infty}^{\infty} f^*(t-(k+m)T)e^{-j2\pi f_c t}f(t-kT)e^{j2\pi f_c t}dt = N_0\rho_f(m)\ . \tag{18.122}$$

Exercise 9-3.

(a) By the definition of Φ,

$$\Phi^* = \mathrm{E}[\bar{\boldsymbol{r}}_k\boldsymbol{r}_k{}^T]^* = \mathrm{E}[(\boldsymbol{r}_k{}^T)^*(\bar{\boldsymbol{r}}_k)^*] = \mathrm{E}[\bar{\boldsymbol{r}}_k\boldsymbol{r}_k{}^T] = \Phi\ . \tag{18.123}$$

(b) The Toeplitz property follows directly from the assumption that the R_k input random process is wide-sense stationary.

(c) The positive semidefinite property follows from

$$\boldsymbol{x}^*\Phi\boldsymbol{x} = \boldsymbol{x}^*\mathrm{E}[\bar{\boldsymbol{r}}_k\boldsymbol{r}_k{}^T]\boldsymbol{x} = \mathrm{E}[|\boldsymbol{r}_k{}^T\boldsymbol{x}|^2] \geq 0. \tag{18.124}$$

Exercise 9-4. The easiest method is to note that if the MSE reduces to a Hermitian form, the matrix must be Φ. Hence, we assume the form of the result,

$$\mathrm{E}[|E_k|^2] = a + (\boldsymbol{c} - \boldsymbol{c}_{\mathrm{opt}})^*\Phi(\boldsymbol{c} - \boldsymbol{c}_{\mathrm{opt}}) \tag{18.125}$$

for unknown constants a and $\boldsymbol{c}_{\mathrm{opt}}$, and multiply out and equate terms to determine the constants.

Exercise 9-5.

(a) From the Hermitian property we get

$$(\Phi_{\mathrm{R}} + j\Phi_{\mathrm{I}})^* = \Phi_{\mathrm{R}} + j\Phi_{\mathrm{I}} \tag{18.126}$$

and the result follows from equating real and imaginary parts.

(b) We have that

$$\boldsymbol{c}^*\Phi\boldsymbol{c} = (\boldsymbol{c}_{\mathrm{R}}{}^T - j\boldsymbol{c}_{\mathrm{I}}{}^T)(\Phi_{\mathrm{R}} + j\Phi_{\mathrm{I}})(\boldsymbol{c}_{\mathrm{R}} + j\boldsymbol{c}_{\mathrm{I}})\ , \tag{18.127}$$

and multiplying out and taking the real part (note we don't need to bother with the imaginary part at all since we know in advance it is zero),

$$\boldsymbol{c}^*\Phi\boldsymbol{c} = \boldsymbol{c}_{\mathrm{R}}{}^T\Phi_{\mathrm{R}}\boldsymbol{c}_{\mathrm{R}} + \boldsymbol{c}_{\mathrm{I}}{}^T\Phi_{\mathrm{R}}\boldsymbol{c}_{\mathrm{I}} - 2\boldsymbol{c}_{\mathrm{R}}{}^T\Phi_{\mathrm{I}}\boldsymbol{c}_{\mathrm{I}} \tag{18.128}$$

or equivalently $$c^*\Phi c = c_R^T\Phi_R c_R + c_I^T\Phi_R c_I + 2c_I^T\Phi_I c_R. \tag{18.129}$$

Taking the gradients, we get the desired results. Similarly,

$$\mathrm{Re}\{c^*\alpha\} = \mathrm{Re}(c_R - jc_I)^T(\alpha_R + \alpha_I) = c_R^T\alpha_R + c_I^T\alpha_I. \tag{18.130}$$

(c) By the given definition of (9.23),

$$\nabla_c c^*\Phi c = 2(\Phi_R c_R - \Phi_I c_I) + j2(\Phi_R c_I + \Phi_I c_R) \tag{18.131}$$

which is the same as $$\nabla_c c^*\Phi c = 2(\Phi_R + j\Phi_I)(c_R + jc_I). \tag{18.132}$$

Similarly, $$\nabla_c \mathrm{Re}\{c^*\alpha\} = \alpha_R + j\alpha_I = \alpha. \tag{18.133}$$

Exercise 9-6.

(a) From the eigenvalue equation

$$\mathbf{A}v = \lambda v \tag{18.134}$$

where λ is an eigenvalue and v is an associated eigenvector, we can take the conjugate and transpose to get

$$v^*\mathbf{A}^* = v^*\mathbf{A} = \lambda^* v^* \tag{18.135}$$

Premultiplying and postmultiplying the two forms by appropriate vectors, we get

$$v^*\mathbf{A}v = \lambda v^* v \tag{18.136}$$

$$v^*\mathbf{A}v = \lambda^* v^* v \tag{18.137}$$

and subtracting these two equations

$$0 = (\lambda - \lambda^*)(v^* v). \tag{18.138}$$

Since the second term is a vector norm and is therefore positive for $v \neq 0$, it follows that $\lambda = \lambda^*$ and λ is real.

(b) Assume that $\lambda_1 \neq \lambda_2$ are eigenvalues of $\mathbf{A}$. Then we get

$$\mathbf{A}v_1 = \lambda_1 v_1 \tag{18.139}$$

$$\mathbf{A}v_2 = \lambda_2 v_2 \tag{18.140}$$

and taking the conjugate-transpose and premultiplying and postmultiplying by the appropriate vector we get

$$v_1^*\mathbf{A}v_2 = \lambda_1 v_1^* v_2 \tag{18.141}$$

$$v_1^*\mathbf{A}v_2 = \lambda_2 v_1^* v_2. \tag{18.142}$$

Subtracting these equations,

$$0 = (\lambda_1 - \lambda_2)(v_1^* v_2) \tag{18.143}$$

which implies that v_1 and v_2 are orthogonal.

(c) From the definition of the modal matrix,

$$\mathbf{V}^*\mathbf{V} = [v_1^*, \dots v_N^*][v_1, \dots v_N] = \mathbf{I} \tag{18.144}$$

from part (b).

(d) This follows from replacing the modal matrix by its constituent eigenvectors and multiplying out.

(e) Let v be an eigenvector and λ be an associated eigenvalue, then we get

$$\lambda = \frac{v^*\mathbf{A}v}{v^*v}, \tag{18.145}$$

which is non-negative by assumption.

Exercise 9-7. This follows directly from multiplying out the diagonalizing transformation of (9.35).

Exercise 9-8. Noting that

$$(\mathbf{I} - \beta\Phi)v_i = v_i - \beta\Phi v_i = v_i - \beta\lambda_i v_i = (1 - \beta\lambda_i)v_i, \tag{18.146}$$

it follows that v_i is an eigenvector and $(1 - \beta\lambda_i)$ is an eigenvalue of $(\mathbf{I} - \beta\Phi)$.

Exercise 9-9. By straightforward manipulations, with $x = c_k - c_{\text{opt}}$, we get:

$$x^*\Phi x = x^*(\textstyle\sum_{i=1}^{n} \lambda_i v_i v_i^*)x = \sum_{i=1}^{n} \lambda_i x^* v_i v_i^* x = \sum_{i=1}^{n} \lambda_i |x^* v_i|^2. \tag{18.147}$$

Exercise 9-10. Substituting into (9.53),

$$q_{k+1} = \Gamma_k q_k + (\Gamma_k - \mathbf{I})c_{\text{opt}} + \beta A_k \bar{r}_k. \tag{18.148}$$

Simplifying the term $(\Gamma_k - \mathbf{I})$, we get

$$q_{k+1} = \Gamma_k q_k + \beta(A_k - r_k^T c_{\text{opt}})\bar{r}_k, \tag{18.149}$$

which is the desired result.

Exercise 9-11. The causal coefficients will be different because the input is data symbols rather than the data symbols filtered by the channel and receive filter impulse responses. The non-causal coefficients will be different because the causal coefficients do not cause noise enhancement, and hence there is more freedom in the choice of non-causal coefficients.

Exercise 9-12. The error is given by

$$E_k = A_k - c^T w_k, \tag{18.150}$$

where

$$w_k = [w_{-(N-1)}, \dots w_0]^T, \tag{18.151}$$

and

$$w_k = \textstyle\sum_{l \in I} h_l A_{k-l}, \tag{18.152}$$

where the index set I consists of all integers except $\{1, 2, \ldots, M\}$. The value of $E[\,|E_k|^2]$ is identical to (9.8) except that Φ and α are

$$\Phi = E[\overline{w}_k w_k^T], \qquad \alpha = E[A_k \overline{w}_k] \,. \tag{18.153}$$

All that remains is to determine the elements of this matrix and vector. We have

$$E[w_{k+i}{}^* w_{k+j}] = E[\sum_{l \in I} h_{l-i}{}^* A_{k-l}{}^* \sum_{m \in I} h_{m-j} A_{k+j-m}] = E_a \sum_{l \in I} h_{l-i}^* h_{l+j-n} \,. \tag{18.154}$$

Similarly,
$$E[A_k \sum_{l \in I} h_{l-i}^* h_{k-l}^*] \; = \; h_i. \tag{18.155}$$

Exercise 9-13. First, let's ignore $f(t)$, and later replace $b(t)$ by $b(t) * f(t)$. Secondly, lets eliminate the real-part and phase splitter by dealing only with the analytic signal. Then the output of the channel is

$$e^{-j2\pi f_1 t} \int_{-\infty}^{\infty} b(\tau) \Sigma_k A_k g(t - \tau - kT) e^{j2\pi f_c (t - \tau)} \, d\tau$$

$$= e^{j2\pi \Delta f t} \Sigma_k A_k \int_{-\infty}^{\infty} b(\tau) e^{-j2\pi f_c \tau} g(t - kT - \tau) \, d\tau \;. \tag{18.156}$$

Exercise 9-14. By direct calculation,

$$\tilde{A} E[\tilde{A}_m \tilde{A}_n{}^*] = e^{j2\pi \Delta f (m-n) T} E[A_m A_n{}^*] \;. \tag{18.157}$$

Hence, the rotated symbols are wide-sense stationary and the power spectrum relation follows directly.

Exercise 9-15. Let $R_k = c + jd$ and write

$$E[\,|c + jd|^4] = E[c^4 + d^4 + 2c^2 d^2] \,, \tag{18.158}$$

and using the fact that
$$E[c^2] = E[d^2] = \tfrac{1}{2}\phi_0 \,, \tag{18.159}$$

$$E[c^4] = E[d^4] = 3 \cdot (\tfrac{1}{2}\phi_0)^2 \,, \tag{18.160}$$

we get (9.107).

Chapter 10

Exercise 10-1. Since $\boldsymbol{x}_k = [x_k^{(1)}, \ldots, x_k^{(n)}]^T$, the (i, j)-th component of the autocorrelation matrix $\mathbf{R}(d)$ is:

$$[\mathbf{R}(d)]_{i,j} = \mathrm{E}[x_{k+d}^{(i)} x_k^{(j)*}]. \quad (18.161)$$

On the diagonal, setting $i = j$, we get:

$$R_{i,i}(d) = [\mathbf{R}(d)]_{i,i} = \mathrm{E}[x_{k+d}^{(i)} x_k^{(i)*}], \quad (18.162)$$

which is the *scalar* autocorrelation function of the i-th random sequence $\{x_k^{(i)}, -\infty < k < \infty\}$. The Z-transform is the power spectrum of the i-th sequence, and also the i-th diagonal element of $\mathrm{S}(z)$, namely:

$$S_{i,i}(z) = [\mathrm{S}(z)]_{i,i} = \Sigma_d R_{i,i}(d) z^{-d}. \quad (18.163)$$

(a) Evaluating $R_{i,i}(d)$ at lag zero ($d = 0$) yields the power of the i-th sequence,

$$R_{i,i}(0) = \mathrm{E}[\,|x_k^{(i)}|^2] = \frac{1}{2\pi}\int_{-\pi}^{\pi} \mathrm{S}_{ii}(e^{j\theta})\,d\theta, \quad (18.164)$$

where the second equality is the inverse-Fourier transform of $\mathrm{S}_{ii}(e^{j\theta})$ evaluated at lag zero.

(b) First, we have

$$\mathrm{E}[\|\boldsymbol{x}_k\|^2] = \mathrm{E}[\Sigma_i |x_k^{(i)}|^2] = \mathrm{E}[\Sigma_i R_{ii}(0)] = \mathrm{tr}\{\mathbf{R}(0)\} = \mathrm{tr}\{\mathrm{E}[\boldsymbol{x}_{k+0}\boldsymbol{x}_k^*]\}. \quad (18.165)$$

That takes care of the first three equalities that were to be proven. The last two follow from (18.164).

(c) By definition, $\mathbf{R}^*(-d) = \mathrm{E}[\boldsymbol{x}_{k+(-d)}\boldsymbol{x}_k^*]^* = \mathrm{E}[(\boldsymbol{x}_k^*)^*(\boldsymbol{x}_{k+(-d)})^*] = \mathrm{E}[\boldsymbol{x}_k\boldsymbol{x}_{k-d}^*]$. Substituting $n = k - d$ yields:

$$\mathbf{R}^*(-d) = \mathrm{E}[\boldsymbol{x}_{n+d}\boldsymbol{x}_n^*] = \mathbf{R}(d). \quad (18.166)$$

(d) Since $\mathrm{S}(z)$ is the Z-transform of $\mathbf{R}(d)$, and since we just showed that $\mathbf{R}(d) = \mathbf{R}^*(-d)$, it follows that:

$$\mathrm{S}(z) = \Sigma_d \mathbf{R}^*(-d) z^{-d} = \Sigma_k \mathbf{R}^*(k) z^k = (\Sigma_k \mathbf{R}(k) z^{*k})^*$$
$$= (\Sigma_k \mathbf{R}(k)(1/z^*)^{-k})^* = \mathrm{S}^*(1/z^*). \quad (18.167)$$

(e) Let $X = \boldsymbol{u}^*\mathrm{S}(e^{j\theta})\boldsymbol{u}$, and let $Y_k = \boldsymbol{u}^*\boldsymbol{x}_k$. Substituting the definition of $\mathrm{S}(e^{j\theta})$ yields:

$$X = \boldsymbol{u}^*(\Sigma_d \mathbf{R}(d) e^{-jd\theta})\boldsymbol{u} = \boldsymbol{u}^*(\Sigma_d \mathrm{E}[\boldsymbol{x}_{k+d}\boldsymbol{x}_k^*] e^{-jd\theta})\boldsymbol{u}$$
$$= \Sigma_d \mathrm{E}[\boldsymbol{u}^*\boldsymbol{x}_{k+d}\boldsymbol{x}_k^*\boldsymbol{u}] e^{-jd\theta}$$
$$= \Sigma_d \mathrm{E}[Y_{k+d} Y_k^*] e^{-jd\theta} = S_{YY}(e^{j\theta}), \quad (18.168)$$

We see that X is equal to the scalar power spectrum of the scalar sequence $Y_k = \boldsymbol{u}^*\boldsymbol{x}_k$. Since a scalar power spectrum cannot be negative, we have $X = \boldsymbol{u}^*\mathrm{S}(e^{j\theta})\boldsymbol{u} \geq 0$.

(f) By contradiction: Suppose $\boldsymbol{u}$ is the eigenvector corresponding to a negative eigenvalue, so that

$$S(e^{j\theta})\boldsymbol{u} = \lambda \boldsymbol{u} \qquad (18.169)$$

for some $\lambda < 0$. This would imply that $\boldsymbol{u}^* S(e^{j\theta})\boldsymbol{u} = \lambda \| \boldsymbol{u} \|^2 < 0$, contradicting property (e).

Chapter 11

Chapter 12

Exercise 12-1. The square of the Euclidean distance is the square of the distance in one component $(2c)^2$ times the number of components that differ, which is the Hamming distance. Since $c = \sqrt{E}$, the result follows immediately.

Exercise 12-2. We need to find the weight of the minimum weight codeword. This is easy to do if we observe that $\boldsymbol{c}\mathbf{H}^T = \mathbf{0}$ is a linear combination of $w_H(\mathbf{c})$ columns of $\mathbf{H}$. Therefore, the minimum number of columns of $\mathbf{H}$ that add to zero is $d_{H,\min}$ For Hamming codes, no two columns add to zero (because they would have to be identical). However, it is easy to find three columns that add to zero for any Hamming code. For example, every Hamming code $\mathbf{H}$ contains the following three columns:

$$\begin{bmatrix} 0 \\ \vdots \\ 0 \\ 0 \\ 1 \end{bmatrix}, \begin{bmatrix} 0 \\ \vdots \\ 0 \\ 1 \\ 0 \end{bmatrix}, \begin{bmatrix} 0 \\ \vdots \\ 0 \\ 1 \\ 1 \end{bmatrix}. \qquad (18.170)$$

Exercise 12-3. Verify that each of the cyclic shifts can be formed by linear combinations of rows of the generator matrix. Alternatively, write down a list of all 16 codewords in the (7, 4) Hamming code, by considering all possible linear combinations of rows of the generator matrix, and you will see that all cyclic shifts of 1000101 are in the list.

Exercise 12-4. Do this by induction. The first $k = m$ bits are source bits. The $(k + 1)^{st}$ bit is a modulo-two sum of a subset of the first k bits, and hence is a parity-check bit. The $(k + 2)^{nd}$ bit is a modulo-two sum of a subset of the second through k^{th} source bits and the $(k + 1)^{st}$ bit, which is itself a modulo-two sum of the source bits, and hence is also a parity-check bit. Etc.

Exercise 12-5. Define the state of the shift register to be the four bits in the shift register at any time. As long as the initial state is not zero, in the process of generating the codeword all $2^m - 1$ nonzero m bit patterns will occur in the shift register. Hence a codeword consists of the right-most bit of all of these m bit patterns, where the exact order is determined by the initial state. For all m, exactly 2^{m-1} of these right-most bits are ones, so the Hamming weight of the code must be 2^{m-1}. This implies that the $(2^m - 1, m)$ code with hard decoder can correct $2^{m-2} - 1$ errors.

Exercise 12-6. Note from Fig. 12-10(b) that

$$c_k^{(1)} = c_k^{(0)} \oplus c_{k-1}^{(0)} \oplus c_{k-1}^{(1)} \qquad (18.171)$$

or

$$c^{(2)}(D) = (1 \oplus D)c^{(0)}(D) \oplus Dc^{(1)}(D) \ . \qquad (18.172)$$

Hence,
$$(1 \oplus D)c^{(0)}(D) \oplus Dc^{(1)}(D) \oplus c^{(2)}(D) = 0 \,, \tag{18.173}$$

which is what we wanted to verify.

Exercise 12-7. (The solution appears in Appendix 12-D, beginning on page 640.)

Exercise 12-8. For the BSC where the input c_n and output r_n are $\{\pm 1\}$, we have:

$$p(r_n \mid c_n) = (1-p)\left(\frac{1-p}{p}\right)^{-d_H(r_n,\, c_n)} . \tag{18.174}$$

As expected, this reduces to $1-p$ when $r_n = c_n$, and it reduces to p when $r_n \neq c_n$. Therefore, according to the first line of (12.108), the intrinsic contribution to the LLR is:

$$\log\frac{f(r_n \mid c_n = 1)}{f(r_n \mid c_n = -1)} = \log\left(\frac{1-p}{p}\right)^{d_H(r_n,\, -1) - d_H(r_n,\, 1)} = \Delta \log\left(\frac{1-p}{p}\right), \tag{18.175}$$

where we have introduced:

$$\Delta = d_H(r_n, -1) - d_H(r_n, 1) \,. \tag{18.176}$$

Looking closer at this expression we see that Δ reduces to 1 when $r_n = 1$, and Δ reduces to -1 when $r_n = -1$. In other words, $\Delta = r_n$. The result follows.

Exercise 12-9. From the distributive law,

$$r_1 \cdot r_2 = r_1 \cdot (r_2 \oplus 0) = r_1 \cdot r_2 \oplus r_1 \cdot 0, \tag{18.177}$$

so $r_1 \cdot 0 = 0$, the additive identity.

Exercise 12-10. All properties are easy to verify. The multiplicative inverses are:

element	inverse
0	none
1	1
2	3
3	2
4	4

Exercise 12-11. There is no multiplicative inverse for 2 under modulo-four multiplication. Trying all possible values, $2 \cdot 0 = 0$, $2 \cdot 1 = 2$, $2 \cdot 2 = 0$, $2 \cdot 3 = 2$, so there is no element in the field that we can multiply by 2 to get 1. Instead of using modulo-four arithmetic, we may instead construct the following multiplication table:

·	0	1	2	3
0	0	0	0	0
1	0	1	2	3
2	0	2	3	1
3	0	3	1	2

Note that in this case there is a 1 in every non-zero row and column, so the multiplicative inverse exists for all elements in the field.

Exercise 12-12. For any $\mathbf{c} \in C$, $\mathbf{c}\mathbf{H}^T = \mathbf{0}$, hence if $\boldsymbol{h}$ is a row of $\mathbf{H}$, then $\boldsymbol{c}\boldsymbol{h}^T = 0$, and the vectors are orthogonal. As a consequence, if $\boldsymbol{h}$ is a row of $\mathbf{H}$ then $\boldsymbol{h}\mathbf{G}^T = \mathbf{0}$. Furthermore, if $\boldsymbol{c}_D$ is any linear combination of the rows of the rows of $\mathbf{H}$ (i.e. a codeword generated by $\mathbf{H}$), then $\boldsymbol{c}_D\mathbf{G}^T = \mathbf{0}$, so $\mathbf{G}$ is a parity check matrix for the code generated by $\mathbf{H}$.

Exercise 12-13. Addition and multiplication over F_D are ordinary polynomial addition and multiplication except that arithmetic on the coefficients is all modulo two. All the properties are easy to verify.

Exercise 12-14. The shift-register output obeys the difference equation $x_k = x_{k-2}$. It is easily shown that with either initial state (0,1) or (1,0) the output alternates between 0 and 1 (period two), and with initial state (1,1) the output is always one (period one).

Exercise 12-15. We can consider that we are observing the first i bits out of n bits in the sequence. Given a particular i-bit sequence which is not all zeros, the remaining $n - i$ bits will assume all of the 2^{n-i} possible values. There are $2^n - 1$ possible n-bit sequences, and therefore the relative frequency of seeing this particular i bit sequence is the ratio of these two numbers. Similarly, if the i-bit sequence is all-zeros, then the remaining $n - i$ bits cannot be all-zeros. Hence for this case there are only $(2^{n-i} - 1)$ possible $n - i$ bit sequences.

Exercise 12-16. The maximal-length binary sequence x_k will have, in a period of r bits, $(r+1)/2$ "ones" and $(r-1)/2$ "zeros." Hence, the average of s_k over one period is

$$\mu_s = \frac{1}{r}\left(\frac{r+1}{2} - \frac{r-1}{2}\right) = \frac{1}{r} . \tag{18.178}$$

Exercise 12-17. In the autocorrelation definition of (12.154), the terms for which $x_k = x_{k+l}$ contribute +1 while the remaining terms contribute –1. Hence the sum is

$$\frac{1}{r}\left(\frac{r-1}{2} - \frac{r+1}{2}\right) = \frac{-1}{r} , \tag{18.179}$$

for $1 \le l \le r-1$. The answer is straightforward when $l = 0$.

Exercise 12-18.

(a) In evaluating the DFT, let $n = k + l$, so that the DFT of s_{k+l} becomes

$$e^{j2\pi ml/r}\sum_{n=l}^{l+r-1} s_n e^{-j2\pi mn/r} . \tag{18.180}$$

The sum can be split into two parts

$$\sum_{n=l}^{l+r-1} = \sum_{n=l}^{r-1} + \sum_{n=r}^{l+r-1} \tag{18.181}$$

where the second summation is

$$\sum_{n=r}^{l+r-1} s_n e^{-j2\pi mn/r} = \sum_{k=0}^{l-1} s_{k+r} e^{-j2\pi m(k+r)/r}$$

$$= \sum_{k=0}^{l-1} s_k e^{-j2\pi mk/r}. \tag{18.182}$$

The substitution $k = n - r$ was used, as well as the periodicity in r of all terms in the summation.

(b) This is straightforward given the results of (a), since

$$\sum_{l=0}^{r-1} \frac{1}{r} \sum_{k=0}^{r-1} s_k s_{k+l} e^{-j2\pi ml/r} = \frac{1}{r} \sum_{k=0}^{r-1} s_k \sum_{l=0}^{r-1} s_{k+l} e^{-j2\pi ml/r}$$
$$= e^{j2\pi mk/r} \sum_{l=0}^{r-1} s_l e^{-j2\pi ml/r}. \tag{18.183}$$

(c) Substituting from (12.157),

$$\sum_{l=0}^{r-1} R_s(l) e^{-j2\pi ml/r} = 1 + \sum_{l=1}^{r-1} \left(\frac{-1}{r}\right) e^{-j2\pi ml/r}$$
$$= 1 + \frac{1}{r} - \frac{1}{r} \sum_{l=0}^{r-1} e^{-j2\pi ml/r} = 1 + \frac{1}{r} \tag{18.184}$$

when $1 \le m \le r - 1$. When $m = 0$ the value can be evaluated directly as $1/r$. We then multiply by r to get $|S_m|^2$.

Chapter 13

Exercise 13-1.

(a) The volume $V(U^K)$ is given by the integral

$$V(U^K) = \int_U \int_U \cdots \int_U dx_1\, dx_2 \ldots dx_K , \tag{18.185}$$

and the integral can be written as the product of K integrals, each equal to $V(U)$.

$$P(U^K) = \mathrm{E}[\|\mathbf{X}\|^2] = \mathrm{E}\left[\sum_{i=1}^{K} \|\mathbf{X}_i\|^2\right] = \sum_{i=1}^{K} \mathrm{E}[\|\mathbf{X}_i\|^2] = K \cdot P(U) . \tag{18.186}$$

Exercise 13-2. The average squared power of the 16-point constellation is 10 and of the 32-point constellation is 20, which is a 3 dB difference.

Exercise 13-3. The 256 point constellation can be thought of as two successive points from the 16 point constellation and hence the average squared power is double that of the 16 point QAM constellation. From the previous exercise, the 16 point constellation has average squared power 10. Now we get the average squared power of the 512 point constellation. Note that the squared power of all symbols of the form (13.64) is $25 + 1 + 1 + 1 = 28$, and the squared power of all symbols of the form (13.65) is $25 + 9 + 1 + 1 = 36$, so the average squared power of the constellation is

$$\frac{1}{512}(256 \times 20 + 64 \times 28 + 192 \times 36) = 27 . \tag{18.187}$$

Exercise 13-4. Substituting (13.69), (13.19), and (13.20) into (13.70) we get

$$\gamma = \frac{NV^{2/N}(S)d_{\min}^2(C)}{4V^{2/N}(\Lambda)P(S)2^{\rho(C)}} \ . \tag{18.188}$$

Recall that the spectral efficiency ν is the number of information bits communicated per two dimensions. Thus the constellation size is

$$2^{\nu + \rho(C)} = \frac{V^{2/N}(S)}{V^{2/N}(\Lambda)} \ , \tag{18.189}$$

analogous to (13.12), so

$$\gamma = \frac{N2^{\nu}d_{\min}^2(C)}{4P(S)} \ . \tag{18.190}$$

From (13.46) we can write this as

$$\gamma = \frac{2^{\nu}d_{\min}^2(C)}{2E} \ . \tag{18.191}$$

Recognizing that for large constellations $2^{\nu} \approx 2^{\nu} - 1$, (13.10) follows.

Chapter 14

Exercise 14-1. Assume the PLL is phase locked,

$$\phi(t) = \theta(t) + \phi = \omega_0 t + \theta + \phi \ . \tag{18.192}$$

Then from (14.5),

$$c(t) = \frac{d}{dt}\phi(t) = \omega_0 \ . \tag{18.193}$$

From (14.7) and (14.3),

$$\varepsilon(t) = W(-\phi) \ . \tag{18.194}$$

This is a constant (d.c.) so

$$c(t) = L(0)\varepsilon(t) = L(0)W(-\phi) \ . \tag{18.195}$$

Comparing (18.195) with (18.190) we see that

$$\omega_0 = L(0)W(-\phi) \ . \tag{18.196}$$

From Fig. 14-3 we see that $|W(-\phi)| \le \pi$ so

$$|\omega_0| \le \pi |L(0)| \ . \tag{18.197}$$

Exercise 14-2.

$$\frac{\Phi(s)}{\Theta(s)} = \frac{N(s)/D(s)}{N(s)/D(s) + s} = \frac{N(s)}{N(s) + sD(s)} \ , \tag{18.198}$$

from which the result follows.

Exercise 14-3. By contradiction. Assume phase lock,

$$\phi(t) = \theta(t) = \omega_0 t + K, \tag{18.199}$$

so
$$c(t) = \frac{d}{dt} d\phi(t) = \omega_0 . \tag{18.200}$$

This is a d.c. signal, so:
$$\varepsilon(t) = W(\phi(t) - \theta(t)) = \frac{1}{L(0)} c(t) = \frac{\omega_0}{L(0)} \neq 0 , \tag{18.201}$$

so
$$\phi(t) \neq \theta(t) . \tag{18.202}$$

Exercise 14-4. Assume phase lock,

$$\phi_k = \theta_k + \phi = \omega_0 kT + \theta + \phi \ . \tag{18.203}$$

Consequently, from (14.36)
$$c_k = \phi_{k+1} - \phi_k = \omega_0 T . \tag{18.204}$$

The phase error is
$$\varepsilon_k = W(\phi_k - \theta_k) = W(-\phi) , \tag{18.205}$$

a d.c. signal, so
$$c_k = L(1)\varepsilon_k = L(1) W(-\phi) . \tag{18.206}$$

Combining (18.206) with (18.204) we get

$$\omega_0 = \frac{1}{T} L(1) W(-\phi) \tag{18.207}$$

so since $| W(\cdot) | \leq \pi$,
$$| \omega_0 | \leq \frac{\pi}{T} | L(1) | \ . \tag{18.208}$$

Exercise 14-5. By inspection,
$$C(z) = L(z)(\Theta(z) - \Phi(z))$$

and
$$\phi_{k+1} = c_k + \phi_k , \tag{18.209}$$

so
$$(z + 1)\Phi(z) = C(z) = L(z)(\Theta(z) - \Phi(z)) , \tag{18.210}$$

which easily reduces to the desired result.

Chapter 15

Exercise 15-1. Multiply both sides of (15.7) by A_k^* and look at the imaginary part of both sides to get

$$\mathrm{Im}\{q_k A_k^*\} = \sin(\varepsilon_k)\,|A_k|^2 . \tag{18.211}$$

The result follows easily.

Exercise 15-2. Multiply both sides of (15.9) by A_k^* and examine the imaginary part to get

$$\varepsilon_k = \sin^{-1}\left(\frac{\mathrm{Im}\{q_k A_k^*\}}{c_k |A_k|^2}\right). \tag{18.212}$$

Then use the fact that $|q_k| = c_k |A_k|$ to get the result.

Exercise 15-3. In the decoder, after the initial conditions are cleared, the output of the z^{-1} boxes is exactly the same as the output of the z^{-1} boxes in the transmitter, regardless of M. Hence the subtractor removes what the adder inserted.

Chapter 16

Exercise 16-1. $x(t)$ is periodic with period T, and so can be written as a Fourier series. Its Fourier series coefficients are

$$X_m = \frac{\mathrm{E}[A_k]}{T} P\left(\frac{m}{T}\right), \tag{18.213}$$

which are scaled samples of the Fourier transform of the pulse $P(f)$. For $X(f)$ to have a component at $f = \pm 1/T$, it is necessary that X_1 and X_{-1} be nonzero, which will only occur if $P(f)$ is nonzero at $f = \pm 1/T$.

Exercise 16-2.

(a) The results of Appendix 16-A apply directly, where

$$g(t) = |p(t)|^2 , \tag{18.214}$$

and using the fact that multiplication in the time domain is equivalent to convolution in the frequency domain, and the fact that the Fourier transform of $p^*(t)$ is $P^*(-f)$, the result follows.

(b) Writing

$$Z_{-n} = \int_{-\infty}^{\infty} P(f)P^*(f + n/T)\, df , \tag{18.215}$$

and changing variables by letting $u = f + n/T$, (18.83) becomes Z_n^*.

Exercise 16-3. For $m \neq n$, by independence

$$\mathrm{E}[A_m A_n] = \mathrm{E}[A_m]\mathrm{E}[A_n] = 0 , \tag{18.216}$$

by the zero-mean assumption. For $m = n$, let $A_m = C + jD$ then

$$A_m^{\,2} = C^2 - D^2 + 2jCD\,, \tag{18.217}$$

which has mean value zero since by assumption C and D have equal variance, and since the real and imaginary parts are independent they are uncorrelated.

Exercise 16-4. Dropping the dependence on k, write $E(\tau)$ in terms of its real and imaginary parts

$$E(\tau) = E_R(\tau) + jE_I(\tau)\,, \tag{18.218}$$

and

$$|E(\tau)|^2 = E_R^{\,2}(\tau) + E_I^{\,2}(\tau)\,, \tag{18.219}$$

so

$$\frac{\partial}{\partial\tau}|E(\tau)|^2 = 2E_R(\tau)\frac{\partial E_R(\tau)}{\partial\tau} + 2E_I(\tau)\frac{\partial E_I(\tau)}{\partial\tau}$$

$$= 2\mathrm{Re}\left[E^*(\tau)\frac{\partial E(\tau)}{\partial\tau}\right]. \tag{18.220}$$

Exercise 16-5.

(a) It easy to show that $\mathrm{E}[Q_k(\tau_k)]$ is independent of k. To show that $R_{QQ}(k, i)$ depends only on $k-i$ we use the assumptions about N_k to write

$$R_{QQ}(k, i) = \mathrm{E}[\sum_{m=-\infty}^{\infty}\sum_{n=-\infty}^{\infty} A_m A_n p((k-m)T + \tau_k)p((i-n)T + \tau_k) + N_k N_i]\,. \tag{18.221}$$

Using two variable changes, $r = k - m$ and $q = i - n$ we get

$$R_{QQ}(k, i) = \sum_{r=-\infty}^{\infty}\sum_{q=-\infty}^{\infty}\mathrm{E}[A_{k-r}A_{i-q}]p(rT + \tau_k)p(qT + \tau_k) + \mathrm{E}[N_k N_i]\,, \tag{18.222}$$

where we have also exchanged expectations with summations. This can be written

$$R_{QQ}(k, i) = \sum_{r=-\infty}^{\infty}\sum_{q=-\infty}^{\infty} R_A(k - i - r + q)p(rT + \tau_k)p(qT + \tau_k) + R_N(k - i)\,, \tag{18.223}$$

which depends only on $k - i$.

(b) With $Q_k(\tau_k)$ real valued, wide sense stationarity implies that

$$\mathrm{E}[Q_k(\tau_k)Q_{k+1}(\tau_k)] = R_Q(1) \tag{18.224}$$

and

$$\mathrm{E}[Q_k(\tau_k)Q_{k-1}(\tau_k)] = R_Q(-1)\,. \tag{18.225}$$

From the symmetry of the autocorrelation function these are equal.

Exercise 16-6. Plugging (16.32) into (16.31) we get that the timing function is

$$f(\tau_k) = \mathrm{Re}\left\{\hat{A}_k\sum_{m=-\infty}^{\infty} A_m p((k - m + 1)T + \tau_k) + \hat{A}_k N_k\right]$$

$$- E[\hat{A}_k \sum_{m=-\infty}^{\infty} A_m p((k-m-1)T + \tau_k) + \hat{A}_k N_k] \Bigg\} \ . \qquad (18.226)$$

Using the assumptions, this simplifies to get the result.

Exercise 16-7. Exchange the integral and summation and change variables inside the integral, replacing t with $t = \tau + kT$, getting

$$X_m = \frac{1}{T} \sum_{k=-\infty}^{\infty} \int_{-T/2}^{T/2} g(t - kT) e^{-j2\pi mt/T} dt$$
$$= \frac{1}{T} \sum_{k=-\infty}^{\infty} \int_{-T/2}^{T/2} g(\tau) e^{-j2\pi m(\tau + kT)/T} d\tau \ . \qquad (18.227)$$

This can be simplified by observing that

$$e^{-j2\pi m(\tau + kT)/T} = e^{-j2\pi m\tau/T} . \qquad (18.228)$$

Observe that the summation is a sum of finite integrals with adjoining limits, which may be replaced with a single infinite integral, getting the desired result.

Chapter 17

Exercise 17-1. We can define a fixed frame structure with a number of time-slots and added bit framing. One node of the ring is defined as the master, and it inserts the framing bits onto the ring. Each of the other nodes can detect this frame, thus defining the same frame and set of time slots. Now suppose a circuit is desired from station A to station B, and that time-slot n is allocated to this circuit. Then station A can identify time-slot n, and insert its information into that time-slot. For every other time-slot and the framing bits, station A simply retransmits whatever bits are incoming. At station B, it knows the position of time-slot n because it also has the framing. It therefore extracts the bits on this time-slot. Every other time-slot it also retransmits incoming bits. Note that a particular time-slot can be *reused*; that is can support two or more circuits as long as those circuits don't overlap one another on the ring topology.

Exercise 17-2. Taking the derivative w.r.t. ρ_{out}, we get

$$\frac{\partial \rho_{in}}{\partial \rho_{out}} = e^{-2\rho_{out}} - 2\rho_{out} e^{-2\rho_{out}} = 0 \ , \qquad (18.229)$$

which leads to $\rho_{out} = 1/2$.

Index

A

B

C

D

E

F

G

H

I

J

K

L

M

N

O

P

Q

R

S

T

U

V

W

Z